Engineering in Rock Masses

Engineering in Rock Masses

Edited by

F G Bell

With specialist contributions

Butterworth-Heinemann Ltd
Linacre House, Jordan Hill, Oxford OX2 8DP

 PART OF REED INTERNATIONAL BOOKS

OXFORD LONDON BOSTON
MUNICH NEW DELHI SINGAPORE SYDNEY
TOKYO TORONTO WELLINGTON

First published 1992

British Library Cataloguing in Publication Data
Engineering in Rock Masses
I. Bell, F. G.
624.1

ISBN 0 7506 1063 8

Library of Congress Cataloguing in Publication Data
Engineering in rock masses/edited by F. G. Bell: with specialist contributions
p. cm.
Includes bibliographical references and index.
ISBN 0 7506 1063 8
1. Rock mechanics. 2. Engineering geology. I. Bell, F. G. (Frederic Gladstone)
TA706.E56 1992
624.1′5132–dc20 92–786
CIP

Composition by Genesis Typesetting, Laser Quay, Rochester, Kent
Printed and bound in Great Britain by Thomson Litho, East Kilbride

Contents

Preface

This book has been produced because there was and is nothing similar available and because its subject matter is important to those who engineer in rock masses. It deals with rocks themselves, their behaviour and investigation, and construction on or in them. It represents a state-of-the-art survey of the subject in both theoretical and practical aspects. Accordingly it should be of value to academics as well as practising civil engineers, mining engineers and engineering geologists.

The editor warmly thanks the contributors for their efforts. He would like to point out, if by any chance the reader is not aware, that they are among the leaders in their subject areas. We are indeed fortunate in being able to have the benefit of their expertise.

F. G. B.
1992

Contributors

Dr A Auld
I W Farmer and Associates, Newcastle-upon-Tyne, UK

Professor F G Bell
University of Natal

D Billaux
Service Géological National, Orléans, France

Dr B H G Brady
ITASCA Consultants Inc, Minnesota, USA

Professor H Brandl
Technische Universitat Wien, Austria

Professor J P Carter
University of Sydney, Australia

P M Cashman
Groundwater Control Consultant

Professor H Duddeck
Universitat Braunschweig, Germany

I Farmer
I W Farmer and Associates, Newcastle-upon-Tyne, UK

B Feuga
Service Géological National, Orléans, France

Professor R Goodman
University of California at Berkeley

Professor T H Hanna
Chartered Civil Engineer

R Holmberg
Nitro Nobel, Sweden

A Houlsby
Consultant Grouting Engineer

Professor F Kulhawy
Cornell University

Dr D McCann
British Geological Society

Professor R Oliveira
Laboratorio Nacional de Engenharia Civil

P G Polsue
Mining and Metallurgical Consultant

Dr L Richards
Golder Associates, Maidenhead, UK

Dr B O Skipp
Soil Mechanics Ltd, Wokingham, UK

R Stacey
Steffen Robertson & Kirsten, Johannesburg

Professor O Stephansson
Royal Institute of Technology, Stockholm

1 Properties and behaviour of rocks and rock masses

Professor F G Bell
University of Natal

1.1 Geological aspects: igneous rocks

Rocks are divided according to their origin into three groups, namely, igneous rocks, metamorphic rocks and sedimentary rocks. Igneous rocks are formed when hot molten rock material, called magma, solidifies. Magmas are developed either within or beneath the Earth's crust, that is, in the uppermost region of the mantle. They comprise hot solutions of several liquid phases, the most conspicuous of which is a complex silicate phase. Hence silicate minerals are quantitatively the most important constituents of igneous rocks (Figure 1.1). Because silica is

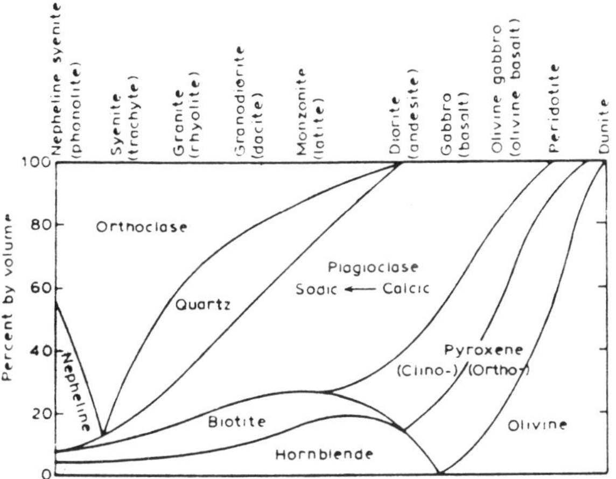

Figure 1.1 *Approximate mineralogical composition of the commoner types of igneous rocks (plutonic types without brackets; volcanic equivalents in brackets)*

the most important constituent in igneous rocks it has been used to distinguish the following groups:

(1) acid igneous rocks, over 65%;
(2) intermediate igneous rocks, 55–65%;
(3) basic igneous rocks, 45–55%;
(4) ultrabasic igneous rocks, less than 45%.

The terms tend to be associated with various groups of igneous rocks, acid with granitic–rhyolitic rocks; intermediate with dioritic–andesitic rocks; basic with gabbroic–basaltic rocks; and ultrabasic with rocks such as peridotite.

However, it would appear that most granitic igneous rocks are developed by either granitization or anatexis. Granitization has been defined as the process by which solid rocks are converted to rocks of granitic character without passing through a magmatic stage. Anatectic processes, which lead to the melting of rocks, are not included within granitization. Rocks formed from remelted material have a mixed or hybrid appearance and have been referred to as migmatites.

The most important rock-forming minerals are often referred to as felsic and mafic, depending on whether they are light or dark in colour, respectively. Felsic minerals include quartz, muscovite mica, feldspars and feldspathoids, whilst olivines, pyroxenes, amphiboles and biotite mica are mafic minerals. Usually, acidic igneous rocks are light in colour whereas basic igneous rocks are dark in colour.

An igneous rock may be composed of an aggregate of crystals, of natural glass, or of crystals and glass in varying proportions.

Igneous rocks may be divided into intrusive and extrusive types according to their mode of occurrence. In the former the magma crystallizes within the Earth's crust, whereas in the latter it solidifies at the surface, having been erupted as lavas and/or pyroclasts from a volcano. The intrusions may be further subdivided by size into major and minor categories. The former are developed in a plutonic, the latter in a hypabyssal environment.

1.1.1 Intrusions

The most important major intrusion is the batholith. Batholiths are very large in size and are generally composed of granitic or granodioritic rock. They are associated with orogenic regions. Some batholiths would appear to have no visible base and have well-defined contacts which dip steeply outwards. Bosses and stocks probably represent upward extensions from deep seated batholiths. Their surface exposures are of limited size, frequently less than 100 km^2.

Dykes and sills are the commonest minor intrusions. The former are discordant, that is, they traverse the host

Figure 1.2 *Composite dyke (basalt and quartz porphyry) running out to sea, south-west of Isle of Arran*

Figure 1.3 *The Whin Sill, Cullernose Point, Northumberland. This is the most notable sill in Britain, outcropping in the north at Bamburgh and the Farne Islands and in the south at Middleton-in-Teesdale. It is composed of quartz dolerite and occurs within Lower Carboniferous strata. It attains a maximum thickness of some 60 m, averaging around 30 m*

rocks at an angle and are steeply dipping (Figure 1.2). As a consequence their surface outcrop is hardly affected by topography and commonly they strike in a straight line. Dykes range in width up to several tens of metres and their length of surface outcrop also varies; dykes have been traced at the surface for distances exceeding 200 km. Dykes often occur along faults, which provide a natural path of escape for the intruded magma. Most dykes are of basaltic composition. However, dykes may be multiple or composite. Multiple dykes are formed by two or more injections of the same material which occur at different times so that the different phases are distinctly discernible. A composite dyke involves two or more injections of magma of different composition.

Sills, like dykes, are comparatively thin, parallel-sided igneous intrusions which frequently occur over relatively extensive areas. Their thickness varies up to several hundred metres. However, unlike dykes, they are injected in an approximately horizontal direction, although their attitude may be subsequently altered by folding. When sills form in a series of sedimentary rocks the magma is intruded along bedding planes (Figure 1.3). Nevertheless, an individual sill may transgress upwards from the horizon to another. Because sills are intruded along bedding planes, they are described as concordant and their outcrop is similar to that of the country rocks. Sills may be fed from dykes and small dykes may arise from sills. Most sills are composed of basic igneous material. Like dykes, they may be multiple or composite in character.

1.1.2 Volcanic activity

Eruptions from volcanoes are spasmodic rather than continuous. Between eruptions activity may still be witnessed in the form of steam and vapours issuing from small vents, namely, fumaroles or solfataras. But in some volcanoes even this form of surface manifestation ceases and such a dormant state may continue for centuries. To all intents and purposes these volcanoes appear extinct. In old age the activity of a volcano becomes limited to emissions of gases from fumaroles and hot water from geysers and hot springs.

Most material emitted by volcanoes is of basaltic composition. Lavas are extravasated from volcanoes at temperatures only slightly above their feezing point. During the course of their flow the temperature falls outwards from within until solidification occurs somewhere between 600 and 900°C, depending upon their chemical composition and gas content. Basic lavas solidify at a higher temperature than do acidic ones.

The rate of flow of a lava is determined by the gradient of the slope down which it moves and by its viscosity which, in turn, is governed by its composition, temperature and volatile content. The higher the silica content of a lava, the greater is its viscosity. Hence basic lavas tend to flow much faster and further than do acidic lavas. Indeed the former have been known to travel at rates of up to 80 km/h.

The surface of a lava solidifies before the main body of the flow beneath. If this surface crust cracks before the lava has completely soldified, then the fluid lava below may ooze up through the crack to form a squeeze-up. Pressure ridges are built on the surface of lava flows where the solidified crustal zone is pushed into a linear fold. Tumuli are upheavals of dome-like shape whose formation may be aided by a localized increase in hydrostatic pressure in the fluid lava beneath the crust. Pipes, vesicle trains or spiracles may be developed in a lava flow depending on the amount of gas given off.

Thin lava flows are interrupted by joints which may run either at right angles or parallel to the direction of flow.

Joints do occur with other orientations but they are much less common. Those joints which are normal to the surface usually display a polygonal arrangement but only rarely do they give rise to columnar jointing. The joints develop as the lava cools. First primary joints form, from which secondary joints arise, and so it continues.

Typical columnar jointing is developed in thick flows of basalt (Figure 1.4). The columns in columnar jointing are interrupted by cross joints which may be either flat or saucer-shaped. The latter may be convex up or down.

Figure 1.4 *Columnar jointing in basalt lavas near Calton Hill, Derbyshire*

These are not to be confused with platey joints which are developed in lava flows as they become more viscous on cooling so that slight shearing occurs along flow planes.

When a magma is erupted it separates at low pressures into lava and a gaseous phase. If the magma is viscous, then separation is accompanied by explosive activity. On the other hand, volatiles escape quietly from very fluid magmas.

Steam may account for 90% or more of the gases emitted during a volcanic eruption. Other gases present include carbon dioxide, carbon monoxide, sulphur dioxide, sulphur trioxide, hydrogen sulphide, hydrogen chloride and hydrogen fluoride.

The amount of and rate at which gas escapes determine the explosiveness of an eruption, an explosive eruption occurring when, because of its high viscosity, magma cannot readily allow the escape of gas. The term pyroclast is collectively applied to material which has been fragmented by explosive volcanic action. Pyroclasts may consists of fragments of lava exploded on eruption, of fragments of pre-existing solidified lava or pyroclasts, or of fragments of country rock.

The size of pyroclasts varies enormously. It is dependent upon the viscosity of the magma, the violence of the explosive activity, the amount of gas coming out of solution during the flight of the pyroclast, and the height to which it is thrown. The largest blocks thrown into the air may weigh over 100 tonnes whereas the smallest consist of very fine ash which may take years to fall back to the Earth's surface. The largest pyroclasts are referred to as volcanic bombs. Lapilli is applied to pyroclastic material which has a diameter varying from about 10 to 50 mm. The finest pyroclastic material is called ash. Rocks which consist of fragments of volcanic ejectamenta set in a fine-grained groundmass are termed aglomerate or volcanic breccia, depending upon whether their fragments are rounded or angular respectively.

After pyroclastic material has fallen back to the surface it eventually becomes indurated. It is then described as tuff. According to the material of which tuff is composed, distinction can be made between ash tuff, pumiceous tuff and tuff breccia. Tuffs are usually well-bedded and the deposits of individual eruptions may be separated by thin bands of fossil soil or old erosion surfaces. Mudflows are frequently interbedded with tuffs, having formed when downpours of rain, associated with eruption, mixed with ash.

When clouds of intensely heated incandescent lava spray fall to the ground, they weld together. Because the particles become intimately fused, they attain a largely pseudo-viscous state, especially in the deeper parts of the deposits. The term ignimbrite has been used to describe the resultant rock. If ignimbrites develop on a steep slope, then they begin to flow. Hence they frequently resemble lava flows.

1.2 Geological aspects: metamorphic rocks

Metamorphic rocks are derived from pre-existing rock types and have undergone mineralogical, textural and structural changes. The latter have been brought about by changes which have taken place in the physical and chemical environments in which the rocks existed. The processes responsible for change give rise to progressive transformation which occurs in the solid state. The changing conditions of temperature and/or pressure are the primary agents causing metamorphic reactions in rocks. Individual minerals are stable over limited temperature-pressure conditions which means that when these limits are exceeded mineralogical adjustment has to be made to establish equilibrium with the new environment. When metamorphism occurs there is usually little alteration in the bulk chemical composition of the rocks involved, that is, with the exception of water, volatile constitutents and organic matter, little material is lost or gained.

1.2.1 Types of metamorphism

Thermal metamorphism occurs around igneous intrusions so that the principal factor controlling these reactions is temperature, shearing stress being of negligible importance. The rate at which chemical reactions take place during thermal metamorphism is exceedingly slow and depends upon the rock type and temperatures involved. It has been estimated that the reaction rate doubles for a rise of 10°C, whilst a rise of 100°C may increase the rate by a thousandfold and 200°C by a millionfold. Equilibrium in metamorphic rocks, therefore, is attained more readily at a higher grade than at a lower grade because reaction proceeds more rapidly.

As remarked above, thermal metamorphism is associated with igneous intrusions and the encircling zone of metamorphic rocks is referred to as the contact aureole (Figure 1.5). The size of the aureole depends upon the temperature and size of the intrusion, the quantity of volatiles which emanated from it and the types of country rocks involved. For example, aureoles developed in argillaceous sediments are more extensive than those found in arenaceous or calcareous rocks. Nevertheless, the capricious nature of thermal metamorphism must be emphasized, for even within one formation of the same rock type the width of the aureole may vary.

Within a contact aureole there is a sequence of mineralogical changes from the country rocks to the intrusion, which have been brought about by the effects of a decreasing thermal gradient whose source was in the hot magma. Indeed aureoles developed in argillaceous sediments may be concentrically zoned with respect to the

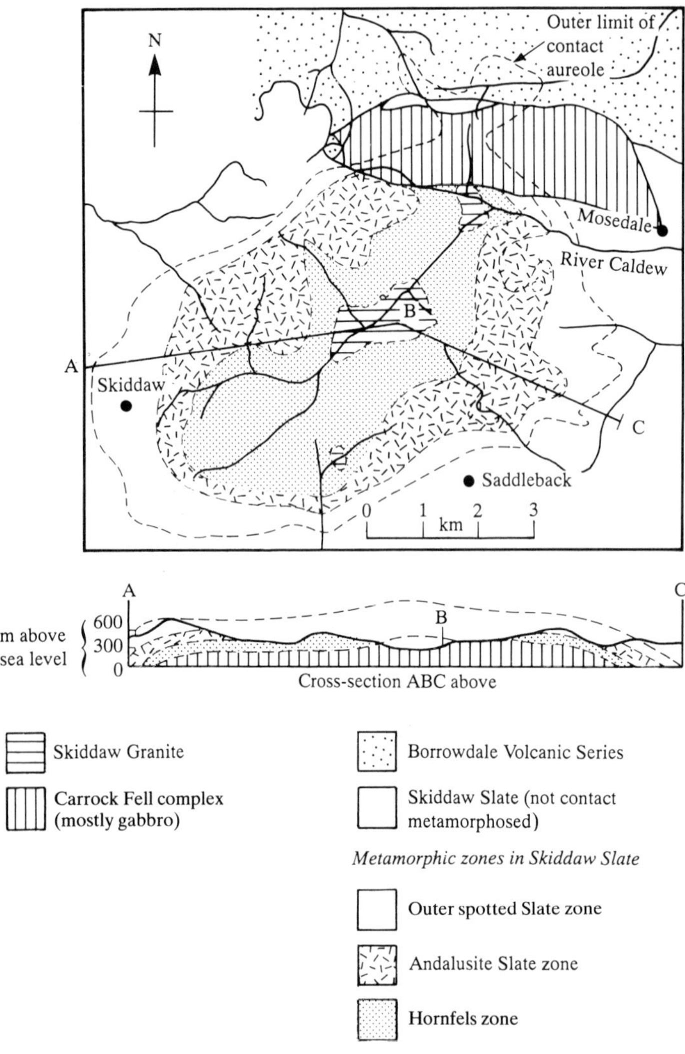

Figure 1.5 *Geological sketch map of the Skiddaw Granite and its contact aureole (after Eastwood* et al. *1968)*

intrusion. A frequently developed sequence varies from spotted slates to schists then hornfels.

Dynamic metamorphism, like contact metamorphism, is usually highly localized, for example, its effects may be found in association with large faults or thrusts. On a larger scale it is associated with folding, however, in such instances it is difficult to distinguish between the effects of dynamic metamorphism and those of low-grade regional metamorphism. What can be said is that at low temperatures recrystallization is at a minimum and the character of a rock is governed by the mechanical processes which have been operative. The processes of dynamic metamorphism include brecciation, cataclasis, granulation, mylonitization, pressure solution, partial melting and slight recrystallization.

Brecciation is a process by which a rock is fractured, the angular fragments produced being of varying size. Crush breccias commonly are associated with faulting and thrusting. The fragments of a crush breccia may themselves be fractured. If, during the process of fragmentation, pieces are rotated, then they are eventually rounded and embedded in worn-down powdered material. The resultant rock is referred to as a crush conglomerate. Mylonites are produced by the pulverization of rocks, which not only involves extreme shearing stress but also considerable confining pressure. In the most extreme cases of dynamic metamorphism the resultant crushed material may be fused to produce a vitrified rock referred to as a pseudotachylite.

Metamorphic rocks outcropping over hundreds or thousands of square kilometres are found in the pre-Cambrian shields and the eroded roots of fold mountains. As a consequence the term regional has been applied to this type of metamorphism. Regional metamorphism involves both the processes of changing temperature and stress. The principal factor is temperature, of which the maximum figure concerned in regional metamorphism is probably around 800°C. Regional metamorphism can be regarded as taking place when the confining pressures are in excess of three kilobars, whilst below that figure, certainly below two kilobars, falls within the field of contact metamorphism. What is more, temperatures and pressures conducive to regional metamorphism have probably been maintained over millions of years.

Regional metamorphism is a progressive process, that is, in any given terrain formed initially of rocks of similar composition, zones of increasing grade may be defined by different mineral assemblages. Slates are the product of low-grade regional metamorphism of argillaceous sediments. As the grade of metamorphism increases slates give way to phyllites which, in turn, are replaced by schists. Gneisses are characteristic of high-grade metamorphism. When sandstones are subjected to regional metamorphism quartzites, schists or granulite may form depending on the original composition of the sandstone and grade of metamorphism. Marbles, of various types, are produced when carbonate rocks are metamorphosed. Schists, gneisses and granulites may be developed from igneous rocks.

1.2.2 Metamorphic textures and structures

Most deformed metamorphic rocks possess some kind of preferred orientation. Preferred orientations are commonly exhibited as megascopic linear or planar structures which allow the rocks to split more easily in one direction than others. One of the most familiar examples is cleavage in slate and phyllites; a similar type of structure in metamorphic rocks of higher grade is schistosity. Cleavage is independent of any original bedding, which it normally intersects at high angles. Where cleavage is developed in a series of beds of different lithologies, its attitude changes as it passes from one bed to another. Cleavage planes do not intersect although they may meet or branch. They are always roughly parallel to each other. Frequently cleavage forms parallel to the axial planes of folds, having developed perpendicular to the direction of maximum principal stress. In other words, recrystallization of minerals of platey habit has occurred in the plane of least stress. Micro-shearing along individual cleavage planes or in narrow zones, together with elongation of parts of the rock mass in the direction of the cleavage, are often present in slates.

Schistosity has been assumed to have been developed in a rock when it was subjected to increased temperatures and stress which involved its reconstitution, which was brought about by localized solution of mineral material and recrystallization. When recrystallization occurs under conditions which include shearing stress, then a directional element is imparted to the newly formed rock. Minerals are accordingly arranged in parallel layers giving the rock its schistose character.

Foliation in a metamorphic rock, which is typically developed in gneiss, is a most conspicuous feature consisting of parallel bands or tabular lenticles formed of contrasting mineral assemblages such as quartz–feldspar and mica–chlorite–amphibole. This parallel orientation agrees with the direction of schistosity, if any is present in nearby rocks. Foliation, therefore, would seem to be related to the same system of stress responsible for the development of schistosity. However, at higher temperatures the influence of stress becomes less and so schistosity tends to disappear in rocks of high-grade metamorphism. By contrast, foliation becomes a more significant feature. It must be pointed out that the term foliation is now frequently used to include cleavage and schistosity. Turner and Weiss (1963) regarded it as penetrative surfaces of discontinuity in deformed rocks which has been formed by metamorphic processes.

1.3 Geological aspects: sedimentary rocks

The sedimentary rocks form an outer skin on the Earth's crust, covering three-quarters of the continental areas and most of the sea floor. They vary in thickness up to ten kilometres. Nevertheless, it has been suggested that they comprise only about 5% of the crust.

Most sedimentary rocks are of secondary origin in that they consist of detrital material derived by the breakdown of pre-existing rocks. Certain sedimentary rocks are the products of chemical or biochemical precipitation whilst others are of organic origin. Thus two groups of sedimentary rock have been distinguished, namely, the clastic or exogenetic and the non-clastic or endogenetic types. The latter group has been further subdivided into precipitate and organic subgroups. However, one factor which all sedimentary rocks have in common is that they were deposited and this gives rise to their most noteworthy characteristic, that is, their bedding or stratification.

1.3.1 Clastic sedimentary rocks

As noted above, most sedimentary rocks are formed from the breakdown products of pre-existing rocks. Accordingly the rate at which denudation takes place acts as a control on the rate of sedimentation which, in turn, affects the character of a sediment. However, the rate of denudation is not only determined by the agents at work, that is, by weathering, or by river, marine, wind or ice action, but also by the nature of the surface. In other words, upland areas are more rapidly worn away than are lowlands. Indeed denudation may be regarded as a cyclic process in that it begins with or is furthered by the elevation of a land surface and as this is gradually worn down the rate of denudation slackens. Each cycle of erosion is accompanied by a cycle of sedimentation. In addition, the harder the rock the more able it is to resist denudation. Geological structure also influences the rate of breakdown. A further point to bear in mind regarding sedimentation is that the amount is affected by the amount of subsidence which occurs in a basin of deposition.

The particles composing most clastic sedimentary rocks have undergone varying amounts of transportation. The amount of transport together with the agent responsible, be it water, wind or ice, play an important role in determining the character of a sediment. For instance, transport over short distances usually means that the sediment is unsorted (the exception being beach sands), as does transportation by ice; with lengthier transport by water or wind not only does material become sorted but it is further reduced in size. The character of a sedimentary rock is also influenced by the environment in which it has been deposited.

The composition of a clastic sedimentary rock depends initially on the composition of the parent material but subsequently it depends upon the stability of that material. As far as the latter is concerned the type of action to which the parent material was subjected and the length of time which it had to suffer this action are important. The least stable minerals tend to be those which are developed in environments very different from those which exist at the surface of the Earth. In fact quartz and, to a much lesser extent, mica are the only common constitutents of igneous rocks which are found in abundance in sediments. Most of the other materials ultimately give rise to clay minerals. The more mature a sedimentary rock is, the more it approaches a stable end product and very mature sediments are likely to have experienced more than one cycle of sedimentation.

The type of climatic regime in which a deposit accumulates and the rate at which this occurs also affect the stability and maturity of the resultant sedimentary product. For example, chemical decay is inhibited in arid regions so that less stable minerals are more likely to survive than they are in humid regions. However, even in humid regions immature sediments may form when basins are rapidly filled with detritus derived from adjacent mountains, the rapid burial affording protection against attack by subaerial agencies.

In order to turn an unconsolidated sediment into a solid rock it must be lithified. Lithification involves two processes, consolidation and cementation. The amount of consolidation which takes place within a sediment depends, first, upon its composition and texture and, second, upon the pressures acting on it, notably that due to the weight of overburden. Consolidation of sediments deposited in water also involves dewatering. The porosity of a sediment is reduced as consolidation takes place and as the particles become more closely packed, they may even be deformed. Pressures developed during consolidation may lead to differential solution of minerals and the authigenic growth of new ones.

Fine-grained sediments possess an initial higher porosity than do coarser types and, therefore, undergo a greater amount of consolidation. For instance, muds and clays may have initial porosities ranging up to 80% compared with 45–50% in sands and silts.

Cementation involves the bonding together of sedimentary particles by the precipitation of material in the pore spaces. This reduces the porosity. The cementing material may be derived by partial intrastratal solution of grains or may be introduced into pore spaces from an extraneous source by circulating groundwaters. Conversely cement may be removed from a sedimentary rock by leaching. The type of cement and, more importantly, the amount, affect the strength of a rock. The type also influences the colour. The matrix of a sedimentary rock refers to the fine material trapped within the pores between the particles. It helps to bind the latter together.

1.3.2 Properties of sedimentary rocks

The texture of a sedimentary rock refers to the size, shape and arrangement of its constitutent particles. The size and shape of sedimentary particles are initially controlled by the fracture pattern of the parent rock. What is more the strength and durability of an individual fragment affects its further comminution, as does the length of transport and the nature of the transporting medium.

Size is a property which is not easy to assess accurately, for the grains and pebbles of which clastic sediments are composed are irregularly shaped, three-dimensional objects. The particle size distribution of gravels and sands normally is derived by sieving, whilst sedimentation techniques are used to assess that of silts. Measurement of individual particles of clay can be done with the aid of the electron microscope.

Several methods have been used in attempts to disaggregate fine-grained rocks like shale but none have proved completely satisfactory (Krumbein and Pettijohn 1938). As a consequence comparatively few size analyses of shales have been carried out. If a rock is strongly indurated, then its disaggregation is impossible without fracturing the grains. In such a case, a thin section of the rock is made and size analysis is carried out with the aid of a petrological microscope and micrometer. The results of particle size analysis are usually represented graphically by drawing cumulative curves (Figure 1.6). The slope of a

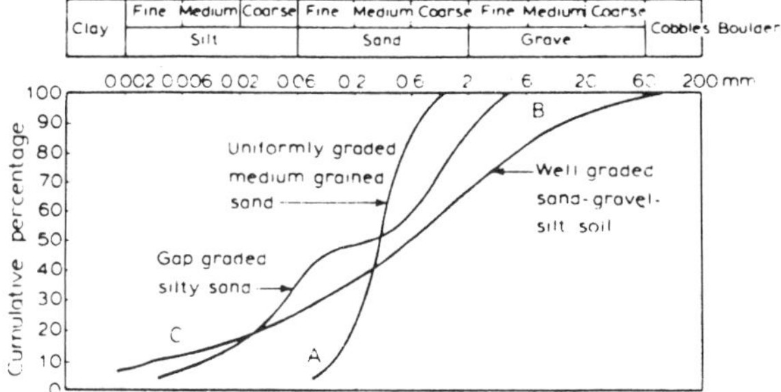

Figure 1.6 *Cumulative curves representing graphically the particle size distribution of three soils, the grading of which is also indicated*

cumulative curve provides an indication of the degree of sorting that a sediment has undergone. If the curve is steep, as curve A in Figure 1.6, then the material is uniformly sorted, whilst curve B represents a well-sorted sediment.

The size of the particles composing a clastic sedimentary rock allows it to be placed in one of three groups which are termed rudaceous or psephitic, arenaceous or psammitic and argillaceous or pelitic (Table 1.1).

Shape is probably the most fundamental property of a particle but unfortunately it is one of the most difficult to quantify. It has frequently been expressed in terms of roundness or sphericity. These two parameters can be estimated visually by comparison with standard images (Figure 1.7).

The degree of grain orientation in a sedimentary rock varies between perfect preferred orientation, in which all the long axes of the grains run in the same direction, and perfect random orientation, where the long axes point in all directions. The latter is only infrequently found as most aggregates posses some degree of grain orientation.

The arrangement of the particles in a sedimentary rock involves the concept of packing which refers to the spatial density of the particles in the aggregate. As such it is

Table 1.1 Classification of size grade of clastic sediment

Boulders		Over 200 mm	
Cobbles		60–200 mm	
	Coarse	20 mm	Rudaceous
Gravel	Medium	6–20 mm	
	Fine	2–6 mm	
	Coarse	0.6–2 mm	
Sand	Medium	0.2–0.6 mm	Arenaceous
	Fine	0.06–0.2 mm	
	Coarse	0.02–0.06 mm	
Silt	Medium	0.006–0.002 mm	
	Fine	0.002–0.006 mm	Argillaceous
Clay		Less than 0.002 mm	

related to the porosity of a rock. The concept of packing has been resolved into two basic aspects termed the unit properties and the aggregate properties respectively. The former include grain-to-grain contacts and the shape of the contact (Figure 1.8). The latter involves the closeness of spread of the particles.

The porosity of a rock can be defined as the percentage pore space within a given volume. In a sedimentary rock the factors affecting its porosity include particle size distribution, sorting, grain shape, fabric, degree of consolidation and cementation, solution effects and, lastly, mineralogical composition, especially the presence of clay minerals. In natural assemblages as the grain sizes decrease so friction, adhesion and bridging become more important because of the higher ratio of the surface area to volume. Therefore, as the grain size decreases, so the porosity increases. Whether the grain size is uniform or non-uniform is of fundamental importance, the highest porosity commonly being attained when all the grains are of the same size. The addition of grains of different size to such an assemblage reduces its porosity and this is, within certain limits, directly proportional to the amount added. Irregularities in grain shape result in a larger possible range of porosity, as irregular forms may, theoretically, be packed either more tightly or more loosely than spheres. After a sediment has been buried and indurated several additional factors help to determine its porosity. The chief amongst these are closer spacing of the grains due to overburden pressure, deformation and granulation of grains, recrystallization and secondary growth of minerals, cementation and, in some cases, dissolution of rock material.

Sedimentary rocks are, as mentioned above, characterized by their stratification and bedding planes are frequently the dominant discontinuity in sedimentary rock masses. As such their spacing and character (are they irregular, waved or straight, tight or open, rough or smooth?) are of particular significant to the engineer. Several classifications of bedding plane separation have been advanced, that given in Table 1.2 being one of the most commonly accepted (see Working Party Report 1970).

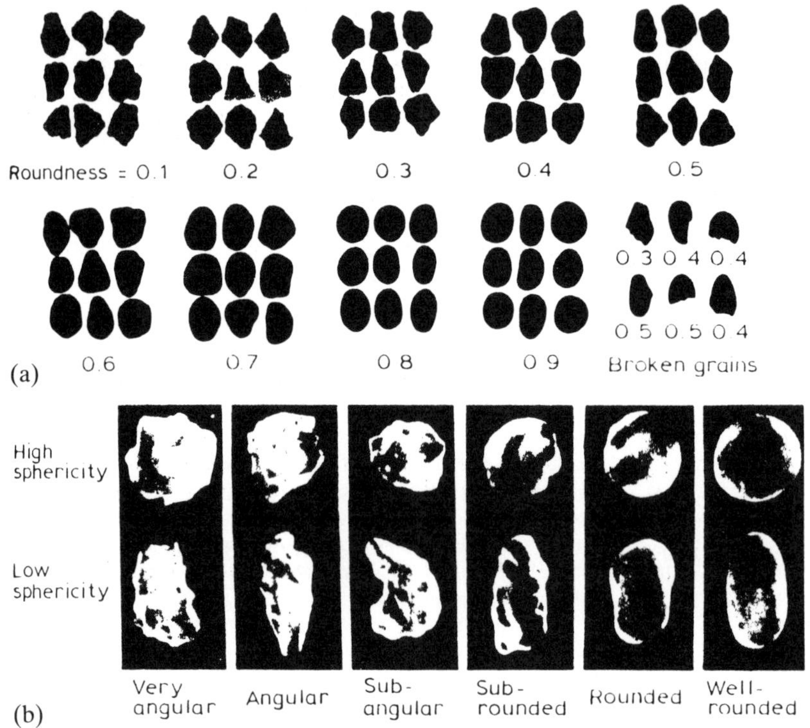

Figure 1.7 (a) *Images for estimating roundness values (after Krumbein 1941);* (b) *models of roundness and sphericity (after Powers 1953)*

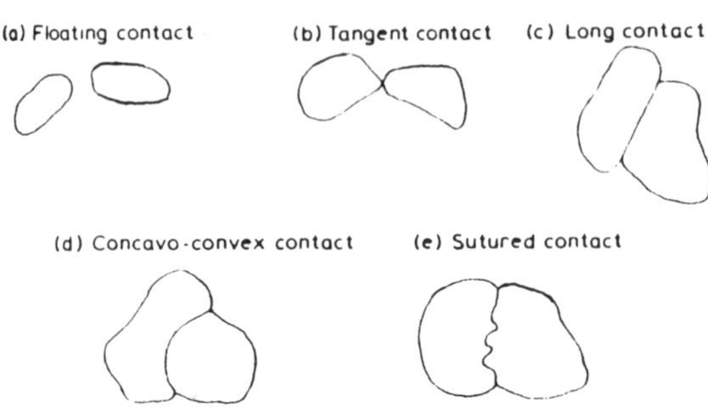

Figure 1.8 *Types of grain contact (after Taylor 1950)*

Table 1.2 Classifications of bedding plane separation

Description	*Bedding plane spacing*	*Soil grading*
Very thickly bedded	Over 2 m	
Thickly bedded	0.6–2 m	Boulders
Medium bedded	0.2–0.6 m	
Thinly bedded	60 mm–0.2 m	Cobbles
Very thinly bedded	20–60 mm	Coarse gravel
Laminated	6–20 mm	Medium gravel
Thinly laminated	Under 6 mm	Fine gravel, sand

An individual bed may be regarded as a thickness of sediment of the same composition which was deposited under the same conditions. Lamination, however, results from minor fluctuations in the velocity of the transporting medium or the supply of material, both of which produce alternating thin layers of slightly differing grain size. Generally lamination is associated with the presence of thin layers of platey minerals, notably micas. These have a marked preferred orientation, usually parallel to the bedding planes, and are responsible for the fissility of the rock. Although lamination is most characteristic of shales, it may also be present in siltstones and sandstones, and occasionally in some limestones.

Cross bedding is a depositional feature which occurs in sediments of fluvial, littoral, marine or aeolian origin and is most notably found in sandstones. In wind-blown sediments it is generally referred to as dune bedding. Cross bedding is confined within an individual sedimentation unit and consists of cross laminae inclined to the true bedding planes. The original dip of the cross laminae is frequently between 20 and 30°. The size of the sedimentation unit in which they occur varies enormously.

1.3.3 Classification of sedimentary rocks

Gravel is an unconsolidated accumulation of rounded fragments, the lower size limit of which is 2 mm. The term rubble has been used to describe those deposits, of similar size, in which the fragments are angular. The composition of a gravel deposit not only reflects the source rocks of the area from which it was derived but is also influenced by the agent(s) responsible for its formation and the climatic regime in which it was, or is, being deposited. When a gravel becomes indurated it forms a conglomerate, when a rubble is indurated it is termed a breccia.

Sands consist of loose mixtures of mineral grains and rock fragments. Generally they tend to be dominated by a few minerals, the chief of which is usually quartz. Sands tend to be close packed and frequently the grains show some degree of orientation. They vary greatly in maturity, the ultimate end product being a uniformly sorted quartz sand with rounded grains. The processes of lithification turn a sand into a sandstone. Several types of sandstone have been distinguished, primarily on a basis of composition (Figure 1.9). For example, greywackes contain more than 15% matrix material and in arkoses more than 25% of the detrital material consists of feldspar.

Silts are mainly composed of fine quartz and occur in a variety of sedimentary environments. Siltstones, their lithified equivalents, may be massive or laminated. They are frequently found as thin ribs interbedded with shales and fine-grained sandstones.

Loess is a wind-blown deposit mainly of silt size and consists largely of quartz particles. It is characterized by lack of stratification and uniform sorting. Fossil root-holes are often present in loess and impart a crude columnar structure.

Deposits of clay are primarily composed of fine quartz and clay minerals. The latter represent the commonest breakdown products of most of the chief rock-forming silicate minerals. Residual clay deposits develop in place, being the products of weathering. The composition of transported clays varies because the materials consist mainly of abrasion products (usually silty particles) and of transported residual clay.

Shale is the commonest sedimentary rock and is characterized by its lamination. Sedimentary rock of

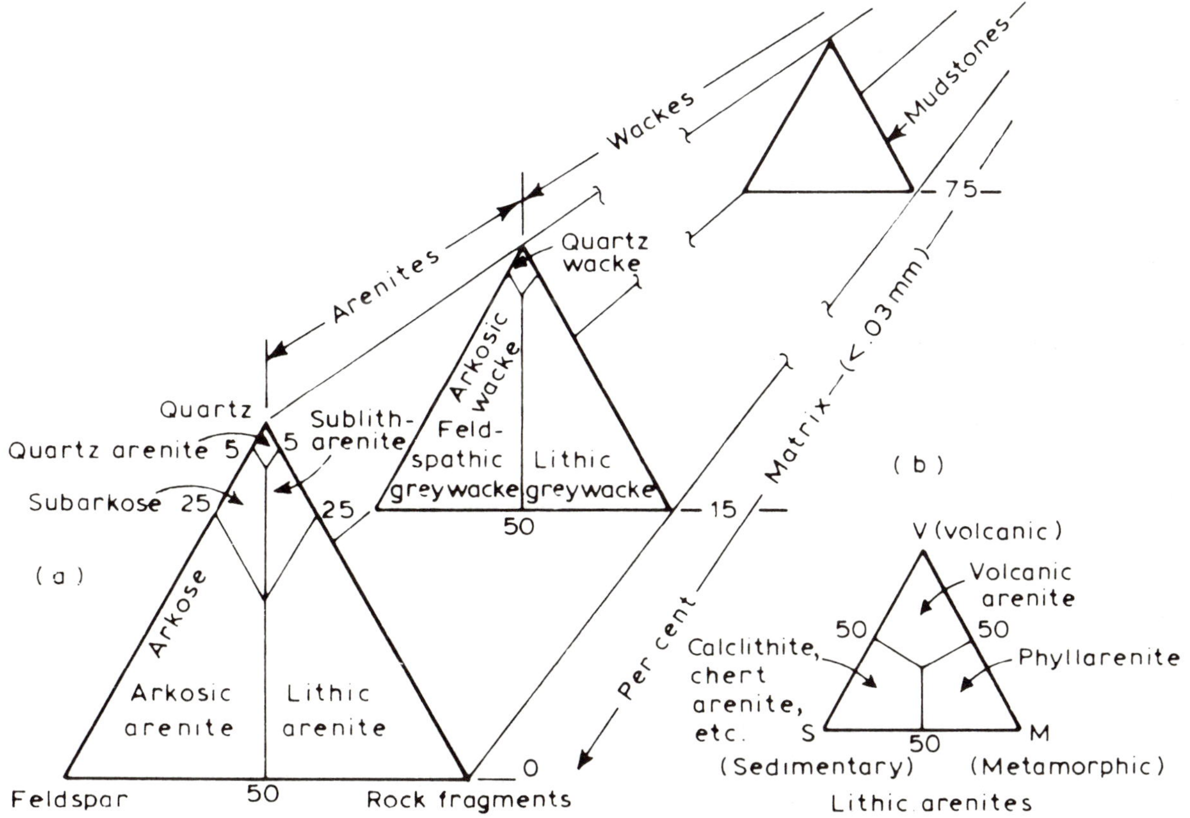

Figure 1.9 *Classification of sandstones (after Pettijohn* et al. *1972)*

similar size range and composition, but which is not laminated, is usually referred to as mudstone. Clay minerals and quartz are the principal constitutents of mudstones and shales. Most shales and mudstones contain a large proportion of silt size material which consists mainly of quartz. In fact in a recent classification of mudrocks, Spears (1980) suggested that the boundary between mudrock and siltstone should be taken as 40% quartz content. His classification was as shown in Table 1.3.

Table 1.3 Mudstone–shale classification

Quartz content	*Fissile*	*Non-fissile*
Over 40% quartz	Flaggy* siltstone	Massive siltstone
30–40% quartz	Very coarse shale	Very coarse mudstone
20–30% quartz	Coarse shale	Coarse mudstone
10–20% quartz	Fine shale	Fine mudstone
Under 10% quartz	Very fine shale	Very fine mudstone

* Parting planes 10–50 mm apart

Shales, in particular, may contain appreciable quantities of carbonate material, indeed calcareous shales frequently grade into shaley limestones.

The term marl has been defined by Pettijohn (1975) as a rock with 35–65% carbonate content and a complementary amount of clay. However, this definition cannot be applied to most rocks referred to as marls in the United Kingdom. Such rocks can be regarded as marly clays or mudstones.

Limestone is the name given to rocks which contain more than 50% carbonate material, over half of which is calcite, or much less frequently, aragonite. If, on the other hand, more than half the carbonate material is composed of dolomite, then the rock is called dolostone. Limestones are polygenetic. Some are of mechanical origin representing accumulations of carbonate detritus, others represent chemical or biochemical precipitates, and yet others are largely of organic origin. Lithification of carbonate sediments is often initiated as cementation at the points of intergranular contact rather than as consolidation. For example, carbonate muds consolidate very little because of this early cementation. The rigidity of the weakest carbonate rock, such as chalk, may be attributed to mechanical interlocking of grains with little or no cement. All the same, cementation may take place more or less at the same time as deposition but cemented and uncemented assemblages may be found within short horizontal distances. Furthermore a recently cemented carbonate layer may overlie uncemented material. Because cementation occurs concurrently with, or soon after, deposition carbonate sediments can support high overburden pressures before consolidation takes place. Hence high values of porosity may be retained to considerable depths of burial. Essentially, however, the porosity is greatly reduced by post-depositional changes which bring about recrystallization.

Limestone is perhaps more prone to post-depositional changes than any other type of rock. For example, after burial, limestones can be modified to such an extent that their original characteristics are obscured or even obliterated. The most profound changes are those which lead to replacement of calcite by dolomite, silica, etc. In addition, carbonate rocks are susceptible to solution (James 1981).

Evaporitic rocks are quantitatively unimportant as sediments. They are formed by precipitation from saline waters, the high salt content being brought about by evaporation from inland seas or lakes. Because of the solubility of most evaporites, they only outcrop at the surface in very arid regions. Gypsum and anhydrite are the exception, they occur at the surface in humid regions and are the commonest evaporites. Nonetheless they are much more readily soluble than carbonate rocks (James and Kirkpatrick 1980).

Organic residues which accumulate as sediments are of two main types, namely, peaty material which, when buried gives rise to coal, and sapropelic residues. Sapropel is silt rich in, or wholly composed of, organic compounds which collect at the bottom of still bodies of water. Such deposits may eventually give rise to cannel or boghead coal. Sapropelic coals usually contain a significant amount of inorganic matter as opposed to humic coals in which the inorganic content is low. A massive deposit of peat is required to produce a thick seam of coal, for example, a seam 1 m thick probably represents 15 m of peat. In order to convert peat to coal the carbon content must be increased, with a concomitant decrease in oxygen and a small reduction in hydrogen. The degree of alteration determines the rank of coal, the rank increasing from peat, through lignite and bituminous coal to anthracite.

1.4 Factors controlling the mechanical behaviour of rocks

The factors which influence the deformation characteristics and failure of rock perhaps can be divided into internal and external categories. The internal factors include the inherent properties of the rock itself, whilst the external factors are those of its environment at a particular point in time. As far as the internal factors are concerned the mineralogical composition and texture are obviously important, but planes of weakness within a rock and the degree of mineral alteration are frequently more important. The temperature–pressure conditions of a rock's environment significantly affect its mechanical behaviour, as does its pore water content. In this respect the length of time which a rock suffers a changing stress and, to a lesser extent changing temperature, and rate at which these are imposed, also affect its deformation characteristics.

1.4.1 Composition and texture

The composition and texture of a rock are governed by its origin. For instance, the olivines, pyroxenes, amphiboles, micas, feldspars and silica minerals are the principal components in igneous rocks. These rocks have solidified from a magma. Solidification involves a varying degree of crystallization: the greater the length of time involved, the greater the development of crystallization. Hence glassy, microcrystalline, fine, medium and coarse-grained types of igneous rocks can be distinguished. In metamorphic rocks either partial or complete recrystallization has been brought about by changing temperature–pressure conditions. Not only are new minerals formed in the solid state but the rocks may develop certain lineation structures. A varying amount of crystallization is found within the sedimentary rocks, from almost complete, as in the case of certain chemical precipitates, to slight, as far as diagenetic crystallization in the pores of, for example, certain sandstones.

Few rocks are composed of only one mineral species, and even when they are, the properties of that species vary slightly from mineral to mineral. Such variations within minerals may be due to cleavage, twinning, inclusions, cracking and alterations, as well as to slight differences in composition. This in turn is reflected in their physical behaviour. As a consequence few rocks can be regarded as homogeneous, isotropic substances. The size and shape relationships of the component minerals are also significant in this respect; generally the smaller the grain size, the stronger the rock. For example, Onodera and Kumara (1980) found that a linear relationship existed between grain size and strength for granite: as the grain size decreased the strength increased.

One of the most important features of texture as far as physical behaviour, particularly strength, is concerned, is the degree of interlocking of the component grains. Fracture is more likely to take place along grain boundaries (intergranular fracture) than through grains (transgranular fracture) and therefore irregular boundaries make fracture more difficult. Willard and McWilliams (1969) noted that when the relative lengths of transgranular (T) and intergranular (I) sections along fractures in Tennessee Marble were summed the $T:I$ ratios varied inversely with failure strength. Subsequently Onodera and Kumara (1980) showed that in granite a linear relationship existed between Young's modulus and the grain boundary surface area per unit volume. The bond between grains in many sedimentary rocks is provided by the cement and/or matrix, rather than by grains interlocking. The amount, and to a lesser extent the type, of cement/matrix is important, not only influencing strength and elasticity, but also density, porosity and permeability.

Rocks are not uniformly coherent materials, but contain defects which occur as visible or microscopic linear or planar discontinuities associated with certain minerals. Defects include microfractures, grain boundaries, mineral cleavages, twinning planes, inclusion trains and elongated shell fragments. As is to be expected defects influence the ultimate strength of a rock and may act as surfaces of weakness which control the direction in which failure occurs. From their investigations on the Barre granite Willard and McWilliams (1969) found that the frequency of defects tended to be inversely proportional to the breaking strength, suggesting that the direction of weakest tensile strength was approximately normal to the direction of most defects.

Grain orientation in a particular direction facilitates breakage along that direction. This applies to all fissile rocks whether they are cleaved, schistose, foliated, laminated or thinly bedded. For example, Griggs (1951) performed tests on the Yule Marble at 1000 MPa confining pressure and ordinary temperatures, to observe the effects of anisotropy. All the specimens tested showed great plastic deformation. When subjected to compression, rock cylinders cut perpendicular to foliation were shown to be stronger than those parallel to the lineation.

Donath (1961) similarly demonstrated that cores cut in Martinsburg Slate at 90° to the cleavage possessed the highest breaking strength, whilst those cores cut at 30° exhibited the lowest. Similar tests were carried out by Brown *et al.* (1977) who showed that the compressive strength of the Delabole Slate is highly directional. Indeed

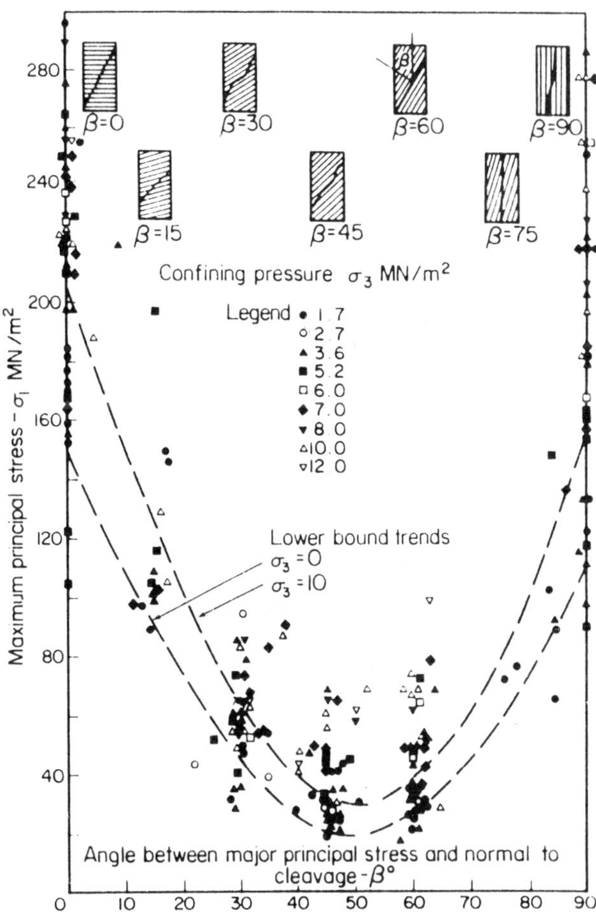

Figure 1.10 *Compressive strength anisotropy in dark grey slate (after Brown* et al. *1977)*

it varies continuously with the angle made by the cleavage planes and the direction of loading (Figure 1.10). They found that even when the cleavage makes high or low angles with the major principal stress direction the mode of failure is mainly influenced by the cleavage. The water content and surface roughness are the principal factors governing shear strength along cleavage planes. For instance, the average friction angle of smooth wet surfaces was determined as 20.5°, which was 9° less than that obtained for the same surfaces when dry. What is more it was found that surface roughness could add up to 40° to the basic friction angles. The degree of surface roughness was shown to vary appreciably according to direction along, and character of, the cleavage plane concerned, which was reflected in the range of shear strength.

1.4.2 Temperature–pressure conditions

Although all rock types show decrease in strength with increasing temperature and an increase in strength as the confining pressure is increased, the combined effect of these factors, as with increasing depth of burial, is notably different for different rock types. Experimental investigation has shown that the effects of temperature changes on sedimentary rocks are of less consequence than those of pressure down to depths of 10 000 m. Griggs (1936) found that the ultimate strength of the Solenhofen limestone was increased by 360% under 10 000 atmospheres (1000 MPa). With increasing temperatures there is a reduction in yield stress and strain hardening decreases. Heating particularly enhances the ductility or the ability to deform permanently, without loss of cohesion, of calcareous and evaporitic rocks. This can be illustrated by the work of Heard (1960) who demonstrated that at lower confining pressures and higher temperatures, the Solenhofen Limestone changed from a brittle to a ductile material. For example, in tension the change over pressure was reduced from 7300 atmospheres (730 MPa) at 25°C to 700 atmospheres (70 MPa) at 700°C; and from 1000 atmospheres (100 MPa) at 25°C to one atmosphere (100 kPa) at 480°C in compression. Granite, peridotite, pyroxenite and basalt become ductile between 300 and 500°C at 5000 atmospheres (500 MPa) confining pressure. They all exhibit a slight decrease in compressive strength, however, above 600°C basalt shows a sudden decrease.

The crushing strengths of the strongest rocks are in excess of 200 MPa but with high confining pressures they become effectively stronger and so more difficult to crush. This is particularly the case with calcareous rocks. At high pressures incipient fractures are closed and indeed the total flow of material without rupture may be indefinitely increased with increasing confining pressure.

Gowd and Rummel (1980) carried out a series of triaxial tests to examine the effect of high confining pressure on the behaviour of porous sandstone. They found that the transition from brittle to ductile deformation is characterized by an abrupt change from dilational behaviour at low pressures to compaction during inelastic axial strain at high pressures. This type of behaviour differs from that of rocks with low porosity. For instance, dilatancy persists well into the ductile field when Carrera marble (porosity about 1%) is subjected to similar conditions (Edmond and Paterson 1972). Gowd and Rummel attributed the compaction which occurs during ductile deformation in porous sandstone at high confining pressure to the collapse of pore space and the rearrangement of quartz grains to give denser packing. At lower pressures, the dilation witnessed in porous sandstones was attributed to fracture along grain boundaries as well as to fracturing of grains, and to the rearrangement of grains. During pre-peak dilation, fracturing is dominant over frictional sliding, which mainly controls post-peak deformation and leads to the formation of microscopic shear planes.

1.4.3 Pore solutions

The presence of moisture in rocks adversely affects their engineering behaviour. For instance, moisture content increases the strain velocity and lowers their fundamental strength. Griggs (1940) demonstrated in experiments with alabaster, subjected to a load of 20 MPa, that a dry specimen soon reached its maximum strain of approximately 0.03%, whereas when a specimen had access to water the strain attained 1.75% in 36 days. Subsequently work done by Price (1960), Bernaix (1969), Parate (1973), Ballivy *et al.* (1976) and Broch (1979) showed that the compressive strength of rock was reduced by saturation with water. More recently Turk and Dearman (1986) have discussed the influence of water content on the engineering properties of weathered rock.

Perhaps the most frequently quoted work in this context was that carried out by Colback and Wiid (1965) who undertook a number of uniaxial and triaxial compression tests at eight different moisture contents, on quartzitic shale and quartzitic sandstone with porosities of 0.28 and 15% respectively. The moisture contents of the rock samples were controlled by keeping them in desiccators over saturated solutions of CaCl at a constant temperature. The tests indicated that the compressive strengths of both rocks under saturated conditions were approximately half what they were under dry conditions. From Figure 1.11 it will be noted that the slopes of the Mohr envelopes are not sensibly different, indicating that the coefficient of internal friction is not significantly affected by changes in moisture content. Colback and Wiid therefore tentatively concluded that the reduction in strength witnessed with increasing water content was primarily due to a lowering of the tensile strength, which is a function of the molecular cohesive strength of the material. Tests on specimens of quartzitic sandstone showed that their uniaxial compressive strength was inversely proportional to the surface tension of the different liquids into which they were placed. As the surface free energy of a solid submerged in a liquid is a function of the surface tension of the liquid, and since the uniaxial compressive strength is directly related to the

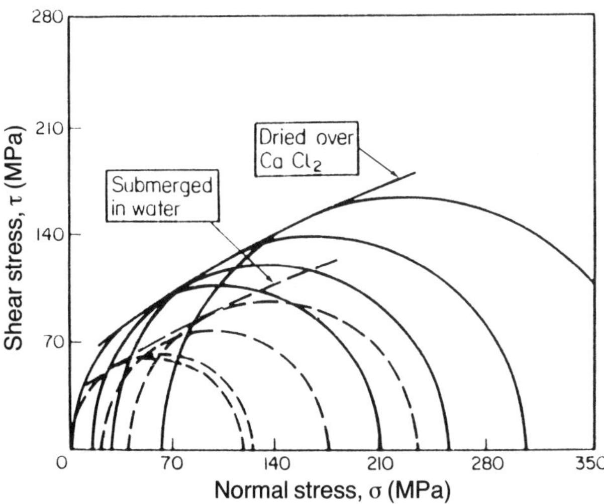

Figure 1.11 *Mohr envelope for quartzitic shale at two moisture contents (after Colback and Wiid 1965) (© Crown copyright reserved. Queen's Printer, Canada, 1965)*

uniaxial tensile strength, and this to the molecular cohesive strength, it was postulated that the influence of the immersion liquid was to reduce the surface free energy of the rock and hence its strength. The authors therefore concluded that the reduction in strength from the dry to the saturated condition of predominantly quartzitic rocks was a constant which was governed by the reduction of the surface free energy of the quartz due to the presence of any given liquid.

Vutukuri (1974) investigated the effects of liquids on the tensile strength of limestone. He reached a similar conclusion to that of Colback and Wiid (1965) regarding the reduction of strength on saturation. In other words he recognized that liquids influence the surface free energy of the rocks and because new surfaces are developed on fracturing, that the strength will depend upon the decrease or increase in surface energy due to the liquid present. For example, he found that as the dielectric constant and surface tension of the liquid increased, the tensile strength of the limestone decreased. In addition, it was suggested that liquids may dissolve material at the apexes of inherent flaws, thereby increasing the state of stress, which aids crack propogation.

As far as rocks consisting of silicate minerals are concerned Atkinson (1984) has suggested that the strength reduction consequent on wetting may be attributable to the replacement of strong silica oxygen bonds by much weaker hydrogen bonds which occurs when silicate lattices are exposed to water. If this takes place at the apex of a microcrack propagating under tension, then it lowers the strength required for failure at the apex by weakening the strength of the crystal lattice which lies in the path of the failure. This phenomenon is referred to as stress corrosion and it gives rise to an increase in the velocity of crack propagation.

More recently it has been recognized that small changes in moisture content can bring about large changes in strength and deformability. For instance, Dobereiner and de Freitas (1986) found that in the case of weak sandstones a 10% change in moisture content may cause a change of approximately 20 MPa in strength and 3000 MPa in deformability. More dramatically Priest and Selvakumar (1982) reported a reduction in the strength of the Bunter Sandstone from 57 MPa to 38 MPa for only an increase of 1% in moisture content above the totally dry state.

The influence of pore water pressure on the behaviour of porous rock in triaxial conditions is illustrated in Figure 1.12. Robinson (1959) carried out a series of triaxial tests

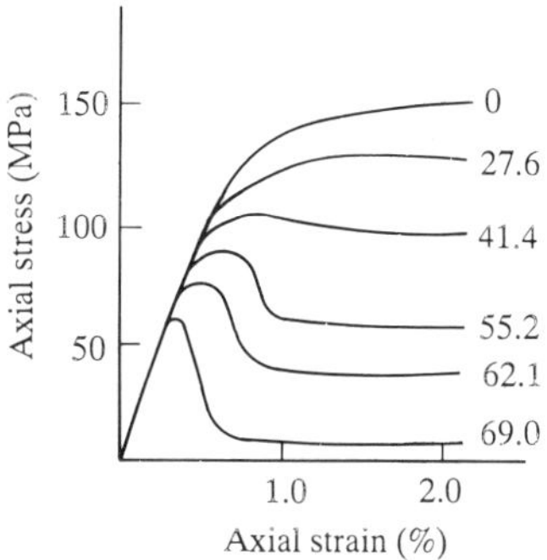

Figure 1.12 *Effect of pore pressure (given in megapascals by the numbers on the curves) on the stress–strain behaviour of a limestone tested at a constant confining pressure of 69 MPa (after Robinson 1959)*

on limestone, with a constant confining pressure of 69 MPa. A number of different pore water pressures were used ranging from 0 to 69 MPa. It was noted that a transition from brittle to ductile behaviour took place as the pore water pressure was increased, the response being controlled by the effective confining pressure. However, the effective stress concept is not necessarily valid for intact rock of low permeability unless the rate of loading is sufficiently slow to allow pore water pressures to equalize (Brace and Martin, 1968).

1.4.4 Time-dependent behaviour

Most strong rocks, like granite, exhibit little time-dependent strain or creep, however, creep in evaporitic rocks, notably salt, may greatly exceed the instantaneous elastic deformation. The time-strain pattern exhibited by a wide range of materials subjected to a constant uniaxial stress can be represented diagramatically as shown in Figure 1.13. The instantaneous elastic strain, which takes place when a load is applied, is represented by OA. There follows a period of primary or transient creep (AB) in which the rate of deformation decreases with time. Primary creep is the elastic effect attributable to

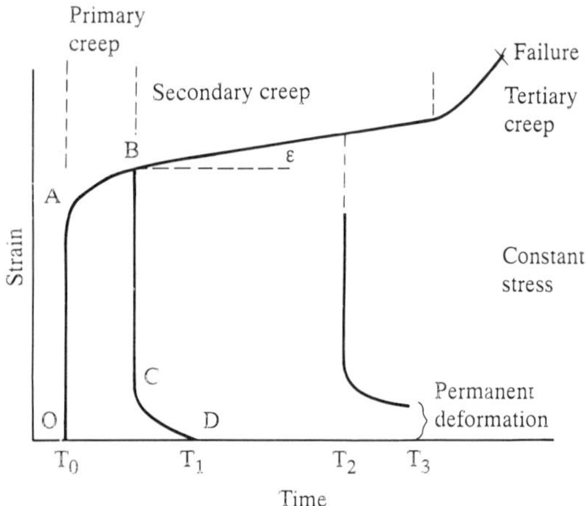

Figure 1.13 *Theoretical strain curve at constant stress*

intragranular atomic and lattice displacements. If the stress is removed the specimen recovers. At first this is instantaneous (BC), but this is followed by a time elastic recovery, illustrated by curve CD. On the other hand, if the loading continues the sample begins to exhibit secondary or pseudo-viscous creep. This type of creep represents a phase of deformation in which the rate of strain is constant and is due principally to movements which occur on grain boundaries. The deformation is permanent and is proportional to the length of time over which the stress is applied. If the loading is further continued, then the specimen suffers tertiary creep in which the strain rate accelerates with time and ultimately leads to failure. Creep deformation is limited at low temperatures and pressures but it may greatly exceed normal plastic flow when the pressures approach the limit of rupture. High temperatures also favour an increase in the rate and extent of creep.

In experiments in which he applied stress to the Solenhofen Limestone Griggs (1936) introduced pauses in the rise of stress. He noted that during these pauses small non-elastic deformation occurred once the differential stress had reached a high enough threshold value (Figure 1.14). It was also observed that for the lowest stress

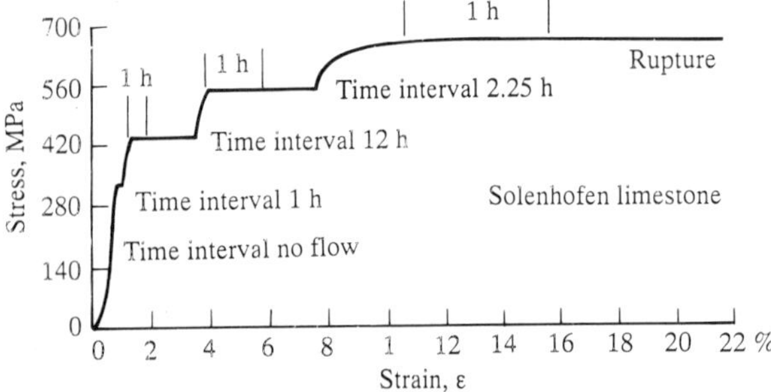

Figure 1.14 *Stress–strain diagram of experiment on Solenhofen Limestone with pauses in increase of stress (after Griggs 1936)*

application the strain did not increase with time as the threshold value was not attained. For each successive pause at a higher stress level the velocity of strain increased. It appeared that flowage in many experiments was masked by the rapid application of increased stress. Griggs also performed experiments which tested the change in ultimate strength with time. He found that time reduced the ultimate strength up to a certain point, beyond which there was no change. Moreover he showed that the amount of plastic deformation before rupture decreased with the duration of time.

Robertson (1960) also carried out experiments on Solenhofen Limestone and showed that if the confining pressure was raised from 100 MPa to 200 MPa, then this caused a hundredfold decrease in the primary creep rate per unit stress. Furthermore he concluded that with increasing confining pressure the number, size and propagation of fractures during creep decreased and that it may facilitate their healing when complete unloading has taken place.

A series of longitudinal strain–time measurements undertaken by Hobbs (1970) on cylindrical specimens of rock subjected to uniaxial compressive stresses ranging from 26.4 to 41.4 MPa for periods ranging from a few minutes to more than a year indicated that after loading the creep rate at first decreased and then became approximately constant. Usually the measured longitudinal time strains were smaller than the instantaneous strains which occurred when the samples were loaded. An instantaneous increase in length occurred when the specimens were unloaded. This was followed by a time-dependent increase which ceased after 15 000 min. This instantaneous increase in length was less than the instantaneous decrease which took place on loading. Hobbs accordingly assumed that both instantaneous strain and primary creep under load were not completely recoverable and that the irrecoverable strain was possibly related to applied stress. During creep the volume of these specimens increased and their volume prior to rupture was larger than the initial unloaded volume.

1.5 Deformation and failure of rocks

1.5.1 Stages of deformation

If the perimeter of a particle is exposed to a force, then internal stresses are developed, and if these are strong enough, they bring about changes in the shape and/or size of the particle. The particle then is said to be strained. Deformation refers to changes in shape accompanied by changes in size.

Four stages of deformation have been recognized: elastic, elastico-viscous, plastic and rupture. The stages are dependent upon the elasticity, viscosity and rigidity of a rock, as well as on stress history, temperature, time, pore water and anisotropy. An elastic deformation is defined as one which disappears when the stress responsible for it ceases. Ideal elasticity would exist if the deformation on loading and its disappearance on unloading were both instantaneous. This is never the case since there is always some retardation, known as hysteresis, in the unloading process. With purely elastic deformation the strain is a linear function of stress, that is, the material obeys Hooke's law. Therefore the relationship between stress and strain is constant and is referred to as Young's modulus (E). Rocks, however, only approximate to the ideal Hookean solid, the stress–strain relationships generally are not linear. Consequently Young's modulus is not a simple constant but is related to the level of applied stress.

The change at the elastic limit from elastic to plastic deformation is referred to as the yield point or yield strength. If the stress on a material exceeds its elastic limit, then it is permanently strained, the latter being brought about by plastic flow. Within the field of plastic flow there is a region where elastic stress is still important and this is referred to as the field of elastico-viscous flow. This term has been used to describe creep or continuous deformation which occurs in rocks when they are subjected to constant stress. Plasticity may be regarded as time-independent, non-elastic, non-recoverable, stress-dependent deformation under uniform sustained load. Solids are classified as brittle or ductile according to the amount of plastic deformation they exhibit. In brittle materials the amount of plastic deformation is zero or very little whilst it is large in ductile substances.

Rupture, or ultimate strength, occurs when the stress exceeds the strength of the material involved. It represents the maximum stress difference a body is able to withstand prior to loss of cohesion by fracturing for constant experimental conditions, fracturing being conceived as the breaking process leading to rupturing. The initiation of rupture is marked by an increasing strain velocity.

Young's modulus is the most important of the elastic constants and can be derived from the slope of the stress–strain curve obtained when a rock specimen is subjected to unconfined compression, it being the ratio of stress to strain. Most crystalline rocks have S-shaped stress–strain curves (Figure 1.15). At low stresses the curve is non-linear and concave upwards, that is, Young's modulus increases as the stress increases. The initial tangent modulus is given by the slope of the stress–strain curve at the origin. Gradually a level of stress is reached where the slope of the curve becomes approximately linear. In this region Young's modulus is defined as the tangent modulus, or secant modulus. At this stress level the secant has a lower value than the tangent modulus because it includes the initial 'plastic' history of the curve.

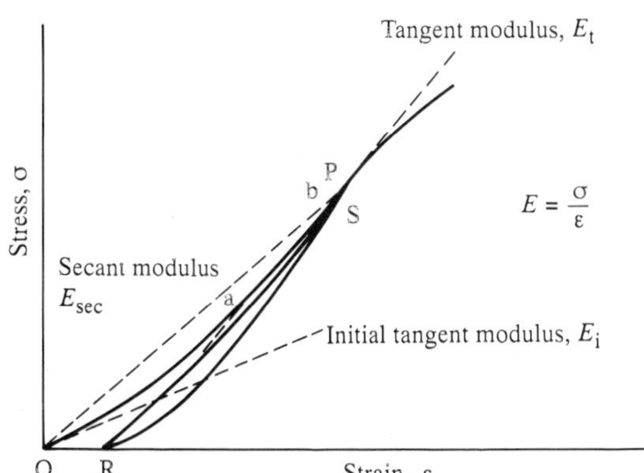

Figure 1.15 *Representative stress–strain curve for rock in uniaxial compression*

Deere and Miller (1966) classified the uniaxial stress–strain curves into six types (Figure 1.16). Types III, IV and V, however, are modifications of the representative S-curve. Type I represents the classical straight-line behaviour of brittle materials which is typical of the more explosive failures of basalts, dolerites, quartzites, and strong dolostones and limestones. Softer limestones, siltstones and tuffs exhibit a more concave downwards curve as illustrated in Type II. These are usually somewhat

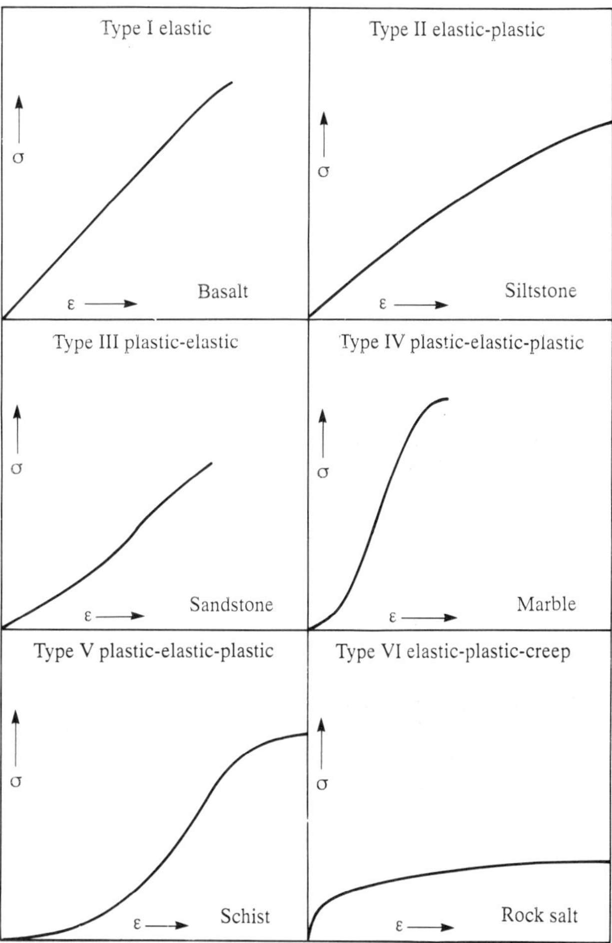

Figure 1.16 *Typical stress–strain curves for rock in uniaxial compression loaded to failure (after Deere and Miller 1966)*

more linear in the earlier and central portions, yielding 'plastically' as failure approaches. Type III is typical of sandstone, granite, some dolostones and dolerites, and schist cut parallel to the schistosity. Metamorphic rocks like marbles and gneiss are represented by Type IV. Schist cored along the schistosity has the long, sweeping S-shaped curve of Type V. Types III, IV and V are characterized by initial 'plastic' crack closing, followed by a steeper linear section. The upper parts of such curves exhibit varying degrees of plastic yield as failure is approached. Type III rocks do not yield significantly, being more explosive with brittle-type fractures (similar to Type I) than Types IV and V. The Type VI curve for rock salt has an initial small elastic straight-line portion followed by plastic deformation and continuous creep.

In addition to their non-elastic behaviour most rocks exhibit hysteresis. Under uniaxial stress the slope of the stress–strain curve during unloading is initially greater than during loading for all stress values (Figure 1.15). As stress is decreased to zero a residual strain, OR, is often exhibited. On reloading the curve RS is produced, which in turn is somewhat steeper than OP. Further cycles of unloading and reloading to the same maximum stress give rise to hysteresis loops, which are shifted slightly to the right. These effects are associated with transient creep. The non-linear elastic behaviour and elastic hysteresis of brittle rocks under uniaxial compression has been explained as due to the presence of flaws or minute cracks in the rocks (Walsh 1965). At low stresses these cracks are open but they close as the stress is increased and the rock becomes elastically stiffer, that is, E increases with stress. Once the cracks are closed the stress–strain curve becomes linear. Nevertheless E is still lower in this portion of the curve than it would be for an uncracked solid and this has been attributed to sliding along crack surfaces. Since these cracks do not immediately slide in the opposite sense as the load is reduced, hysteresis loops are produced.

When hysteresis is large it is difficult to distinguish between elastic and plastic deformation, however, an elastic strain is related only to stress whereas a permanent strain is also related to the period over which the stress is applied. As a consequence, in a strain experiment, with constant load, the elastic deformation is characterized by a gradual decrease of the strain velocity which ultimately leads to a halt in the process. On the other hand a permanent or plastic deformation continues indefinitely with a constant strain velocity. Under high temperature–pressure conditions permanent deformations also may take place by creep.

When a specimen undergoes compression it is shortened and this generally is accompaned by an increase in its cross-sectional area. The ratio of lateral unit deformation to linear unit deformation, within the elastic limit, is known as Poisson's ratio. An idealized value for Poisson's ratio can be obtained by considering an idealized crystal structure, where contraction in one direction automatically leads to extension of the lattice in a perpendicular direction. In such a case, by considering the geometry of the structure, it can be shown that Poisson's ratio is 0.333. The work carried out by Deere and Miller (1966) showed that values of Poisson's ratio for rock must be regarded with some suspicion. These two authors gave the average initial tangent value of Poisson's ratio for all the rocks studied as 0.125 and, at a stress level of 50%, ultimate stress as 0.341.

Rocks subjected to uniaxial compression tend to exhibit a common behaviour in that both Young's modulus and Poisson's ratio increase to more or less constant values as the stress is increased. As the compressive stress approaches the failure limit, Young's modulus falls, eventually reaching zero, while Poisson's ratio increases to a value nearing or exceeding the theoretical maximum of 0.5 for an incompressible solid body. The opposite trend is observed when rocks are placed under uniaxial tension, namely, both Young's modulus and Poisson's ratio are initially high and they fall continuously as stress increases to the failure point. Hawkes *et al.* (1973) found that the initial tangent moduli for the rocks they tested were similar in compression and tension. The value of Young's modulus in compression at half the load failure ($E_{t_{50}}$) is usually greater than the value in tension but there is considerable variation from one rock type to another. At very low stresses Poisson's ratio of a rock in tension can be greater than 0.5, indicating an initial decrease in volume, that is, a decrease in porosity. However, as stress is increased the ratio falls to comparatively low values (0.1).

1.5.2 Theories of brittle failure

Brittle failure is regarded as the sudden loss of cohesion across a plane that is not preceded by an appreciable permanent deformation. It may occur in rock on both microscopic and macroscopic scales.

One of the most popular theories which was proposed to explain shear fractures was advanced by Coulomb (1773). The Coulomb criterion of brittle failure is based upon the idea that shear failure occurs along a surface if the shear stress acting in that plane is high enough to overcome the cohesive strength of the material and the resistance to movement. The latter is equal to the stress normal to the shear surface multiplied by the coefficient of internal friction of the material, whilst the cohesive strength is its inherent shear strength when the stress normal to the shear surface is zero. The relation between the failure criterion, the friction and the cohesion is then expressed by Coulomb's law

$$\tau = c + \sigma_n \tan \phi \tag{1.1}$$

where τ is the shearing stress, c is the apparent cohesion, σ_n is the normal stress and ϕ is the angle of internal friction or shearing resistance. It can be shown that under triaxial conditions (Figure 1.17)

$$\sigma_n = \tfrac{1}{2}(\sigma_1 + \sigma_3) + \tfrac{1}{2}(\sigma_1 - \sigma_3) \cos 2\beta \tag{1.2}$$

and that

$$\tau = \tfrac{1}{2}(\sigma_1 - \sigma_3) \sin 2\beta \tag{1.3}$$

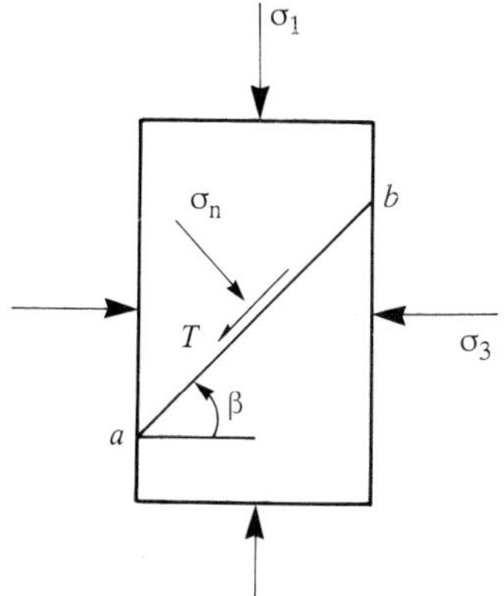

Figure 1.17 *Shear failure on plane* a, b

where σ_1 and σ_3 are the stress at failure and the confining pressure respectively. Substitution of σ_n and τ in the Coulomb equation and rearranging allows the limiting condition of stress, σ_1, on any plane defined by β, to be obtained from

$$\sigma_1 = \frac{2c + \sigma_3 [\sin 2\beta + \tan \phi (1 - \cos 2\beta)]}{\sin 2\beta - \tan \phi (1 + \cos 2\beta)} \quad (1.4)$$

There is a critical plane on which the available shear strength is reached first as σ_1 is increased. For this critical plane $\sin 2\beta = \cos 2\phi$, and $\cos 2\beta = -\sin \phi$, hence the above equation becomes

$$\sigma_1 = \frac{2c \cos \phi + \sigma_3 (1 + \sin \phi)}{1 - \sin \phi} \quad (1.5)$$

In terms of the Coulomb hypothesis, an apparent value of the uniaxial tensile stress, σ_t, can be obtained from

$$\sigma_t = \frac{2c \cos \phi}{1 + \sin \phi} \quad (1.6)$$

However, measured values of tensile strength generally are lower than those predicted by the equation.

The Coulomb criterion has been shown to agree with experimental data for rocks in which the relationship between the principal stresses at rupture is, to all intents, linear. However, experimental evidence indicates that peak strength envelopes generally are non-linear. This may be due to the area of the grains in frictional contact increasing as the normal pressure increases. What is more the criterion implies that a major shear fracture exists at peak strength and this is not always the case (Wawersik and Fairhurst, 1970). It also implies a direction of shear failure which does not always agree with experimental observations.

Coulomb's concept was subsequently modified by Mohr (1882). Mohr's hypothesis states that when a rock is subjected to compressive stress shear fracturing occurs parallel to those two equivalent planes for which shearing stress is as large as possible whilst the normal pressure is as small as possible. This statement assumes that a triaxial state of external stress is applied to a substance and that the maximum external stress is resolved into shear and normal components for any inclined potential shear planes existing in the stressed material.

Griffith (1920) claimed that because of the presence of minute cracks or flaws, particularly in surface layers, the measured tensile strengths of most brittle materials are much less than those which would be inferred from the values of their molecular cohesive forces. Although the mean stress throughout a body may be relatively low, local stresses developed in the vicinity of the flaws were assumed to attain values equal to the theoretical strength. Under tensile stress, the stress magnification around a flaw is concentrated where the radius of curvature is smallest, that is, at its ends. Hence the tensile stresses which develop around a flaw have most influence when the tensile stress zone coincides with the zone of minimum radius of curvature. The concentration of stress at the ends of flaws causes them to be enlarged and presumably with time they develop into fractures. Griffith maintained that if a material is subjected to tensile stress, then the tensile strength, σ_t, is given approximately by

$$\sigma_t = \sqrt{2E\gamma/\pi a} \quad (1.7)$$

where E is Young's modulus, γ is the specific surface energy of the material and a is half the length of a Griffith crack. According to this concept, strength is inversely proportional to the square root of the crack length so that the longest crack in a material determines its strength.

Griffith (1924) extended his theory to the case of applied compressive stresses. Neglecting the influence of friction on the cracks which will close under compression, and assuming that an elliptical crack propagates from points of maximum stress propagation, Griffith derived the following criterion for crack extension in plane compression:

$$\sigma_t = \frac{(\sigma_1 - \sigma_3)^2}{8(\sigma_1 + \sigma_3)} \quad (1.8)$$

if

$$\sigma_1 + 3\sigma_3 > 0$$

or

$$-\sigma_t = \sigma_3 \quad (1.9)$$

if

$$\sigma_1 + 3\sigma_3 < 0$$

The theory predicts that the uniaxial compressive stress at crack extension is eight times the uniaxial tensile strength. This criterion also can be expressed in terms of the shear stress, τ, and the nromal stress, σ_n, acting on the plane containing the major axis of the crack, whereby

$$\tau^2 = 4\sigma_t(\sigma_n + \sigma_t) \quad (1.10)$$

Brace (1964) showed that the fracture in hard rock was usually initiated in grain boundaries which could be regarded as the inherent flaws required by the Griffith theory. He supposed that as stress was increased prior to fracture, grain boundaries at numerous sites in a rock became loosened and that at the instant before fracture the rock was filled with loosened sections along grain boundaries which had various lengths and orientations. Cracks grew in such sections and ultimately gave rise to fractures.

Unfortunately the plane compression theory of Griffith does not provide a very good model for the peak strength of rock under triaxial conditions. Accordingly a number of modifications have been introduced subsequently. For instance, McClintoch and Walsh (1962) argued that in compressive stress conditions the frictional strength of closed cracks must be taken into account and they proposed the modified criterion:

$$\tau = 2\sigma_t + \sigma_n \tan\phi \quad (1.11)$$

Murrell (1963) extended the original Griffith theory and showed that the Griffith criterion of failure corresponds with a parabolic Mohr's envelope defined by the expression

$$\tau^2 = 4\sigma_{tu}^2 - 4\sigma_{tu}\sigma_n \quad (1.12)$$

where σ_{tu} is the universal tensile strength of the material and σ_n is the normal stress. This assumed that the Griffith cracks remained elliptical up to the point of failure, but in some rock materials this is not the case. Murrell also proposed a three-dimensional fracture initiation surface so that in triaxial conditions

$$\sigma_t = \frac{(\sigma_1 - \sigma_3)^2}{12(\sigma_1 + 2\sigma_3)} \quad (1.13)$$

In this case the uniaxial compressive strength is 12 times the uniaxial tensile strength.

Although there is an encouraging agreement between experimental and theoretical results, the Griffith theory does not provide a complete description of the mechanism of rock failure. For instance, Hoek (1968) was able to demonstrate that the original and modified Griffith theories, although adequate for the prediction of fracture initiation in rocks, were unable to describe its propagation and subsequent failure in rocks.

More recently a number of empirical strength criteria have been advanced, because the classical theories do not apply to rock over a wide range of applied compressive stress conditions. These criteria usually take the form of a power law in recognition of the fact that peak σ_1 vs σ_3 and τ vs σ_n envelopes for rocks are non-linear, that is, they are generally concave downwards. In order to ensure that the parameters used in the power laws are dimensionless, these criteria are best written in normalized form with all stress components being divided by the uniaxial compressive strength of the rock (Brady and Brown 1985).

One of the most recent empirical laws is that developed by Hoek and Brown (1980) who proposed that the peak triaxial compressive strengths of a wide range of isotropic rock materials could be described by the expression

$$\sigma_1 = \sigma_3 + (m\sigma_c\sigma_3 + s\sigma_c^2)^{1/2} \quad (1.14)$$

where σ_1 is the major principal stress at failure, σ_3 is the minor principal stress (or in the case of the triaxial test, the confining pressure), σ_c is the uniaxial compressive strength of the intact rock, and m and s are dimensionless constants which are approximately analogous to the angle of friction and cohesive strength of the conventional Mohr–Coulomb failure criterion. The constant m varies with rock type, ranging from about 0.001 for highly disturbed rock masses to about 25 for hard intact rock (Table 1.4). Large values of m (that is, 15 to 25) give steeply inclined Mohr envelopes and high instantaneous friction angles at low effective normal stress levels and are associated with brittle igneous and metamorphic rocks. Lower values of m, around 7, yield lower instantaneous friction angles and tend to be associated with carbonate rocks. For intact rock $s = 1$; for heavily jointed rock masses $s = 0$. Equation (1.14) when normalized becomes

$$\frac{\sigma_1}{\sigma_c} = \frac{\sigma_3}{\sigma_c} + \left(m\frac{\sigma_3}{\sigma_c} + s\right)^{1/2} \quad (1.15)$$

This expression is useful when comparing the shape of Mohr failure envelopes for different rocks.

Rock strength and fracture are influenced by various factors: mineral composition; grain size, shape and packing; amount and type of cement/matrix; degree of grain interlock etc. If these factors are relatively uniform within a given rock type, then a single curve probably will give a good fit to the normalized strength data (e.g. granites; see Hoek 1983). On the other hand if these factors are quite variable, as in limestones or sandstones, then a single curve will give a poorer fit. Nonetheless the empirical criterion formulated by Hoek and Brown (1980) allows preliminary design calculations to be made without testing by using an approximate value of m for a particular rock and by determining a value of uniaxial compressive strength.

According to Hoek (1983) under triaxial conditions a transition from brittle to ductile behaviour usually occurs somewhere between a principal stress ratio (σ_1'/σ_3') of 3 and 5 (Figure 1.18). He suggested a rough rule of thumb, that is, that the confining pressure should not exceed the unconfined compressive strength of the rock for behaviour to be regarded as brittle. However, for those rocks with very low values of m the principal stress ratio may fall beyond the brittle–ductile transition.

Table 1.4 Approximate relationship between rock mass quality and material constants (After Hoek 1983)

Empirical failure criterion $\sigma_1' = \sigma_3' + (m\sigma_c \sigma_3' + s\sigma_c^2)^{1/2}$ σ_1' = major principal stress σ_3' = minor principal stress σ_c = uniaxial compressive strength of intact rock m, s = empirical constants		Carbonate rocks with well developed crystal cleavage, e.g. dolostone, limestone and marble	Lithified argillaceous rocks, e.g. mudstone, siltstone, shale and slate (tested normal to cleavage)	Arenaceous rocks with strong crystals and poorly developed crystal cleavage, e.g. sandstone and quartzite	Fine grained polyminerallic igneous crystalline rocks, e.g. andesite, dolerite, diabase and rhyolite	Coarse grained polyminerallic igneous and metamorphic crystalline rocks, e.g. amphibolite, gabbro, gneiss, granite, norite and quartz diorite
Intact rock samples Laboratory size samples free from pre-existing fractures		$m = 7$ $s = 1$	$m = 10$ $s = 1$	$m = 15$ $s = 1$	$m = 17$ $s = 1$	$m = 25$ $s = 1$
Geomechanics system* rating	100					
Q system (NGI)† rating	500					
Very good quality rock mass Tightly interlocking undisturbed rock with rough unweathered joints spaced at 1–3 m		$m = 3.5$ $s = 0.1$	$m = 5$ $s = 0.1$	$m = 7.5$ $s = 0.1$	$m = 8.5$ $s = 0.1$	$m = 12.5$ $s = 0.1$
Geomechanics system* rating	85					
Q system† rating	100					
Good quality rock mass Fresh to slightly weathered rock, slightly disturbed with joints spaced 1–3 m		$m = 0.7$ $s = 0.004$	$m = 1$ $s = 0.004$	$m = 1.5$ $s = 0.004$	$m = 1.7$ $s = 0.004$	$m = 2.5$ $s = 0.004$
Geomechanics system* rating	65					
Q system† rating	10					
Fair quality rock mass Several sets of moderately weathered joints spaced at 0.3–1 m disturbed		$m = 0.14$ $s = 0.0001$	$m = 0.20$ $s = 0.0001$	$m = 0.30$ $s = 0.0001$	$m = 0.34$ $s = 0.0001$	$m = 0.50$ $s = 0.0001$
Geomechanics system* rating	44					
Q system† rating	1					
Poor quality rock mass Numerous weathered joints at 30–500 mm with some gouge. Clean, compacted rock fill		$m = 0.04$ $s = 0.00001$	$m = 0.05$ $s = 0.00001$	$m = 0.08$ $s = 0.00001$	$m = 0.09$ $s = 0.00001$	$m = 0.13$ $s = 0.00001$
Geomechanics system* rating	23					
Q system† rating	0.1					
Very poor quality rock mass Numerous heavily weathered joints spaced at 50 mm with gouge. Waste rock		$m = 0.007$ $s = 0$	$m = 0.010$ $s = 0$	$m = 0.015$ $s = 0$	$m = 0.017$ $s = 0$	$m = 0.025$ $s = 0$
Geomechanics system* rating	3					
Q system† rating	0.01					

1.6 Strength of discontinuous rock masses and its assessment

Joints in a rock mass reduce its effective shear strength at least in a direction parallel with the discontinuities. Hence the strength of jointed rocks is highly anisotropic. Joints offer no resistance to tension whereas they offer high resistance to compression. Nevertheless they may deform under compression if there are crushable asperities, compressible filling or apertures along the joint or if the wall rock is weathered or altered.

Where discontinuities dip into a rock face, they only impose a direct mechanical instability on the face when they are of the same scale. They do, however, allow the ingress of water into the rock mass and as a result facilitate an increase in pore water pressure. This reduces the effective strength of the rock mass. Conversely, when discontinuities daylight into a rock face they adversely affect stability. In this case the slope of the face is to a greater or lesser extent controlled by the discontinuities.

John (1965) considered that when a jointed rock mass failed by sliding along one joint or a set of joints, the limiting stress ratios could be determined based on the parameters of the joints and the confining pressure. In other words, where a load is applied in a direction parallel or sub-parallel to the joint direction the shear strength depends on the shearing resistance along the joint surfaces. At low normal pressures shearing stresses along a joint with relatively smooth asperities produce a tendency for one block to ride up onto and over the asperities of the other, whereas at high normal pressures shearing takes

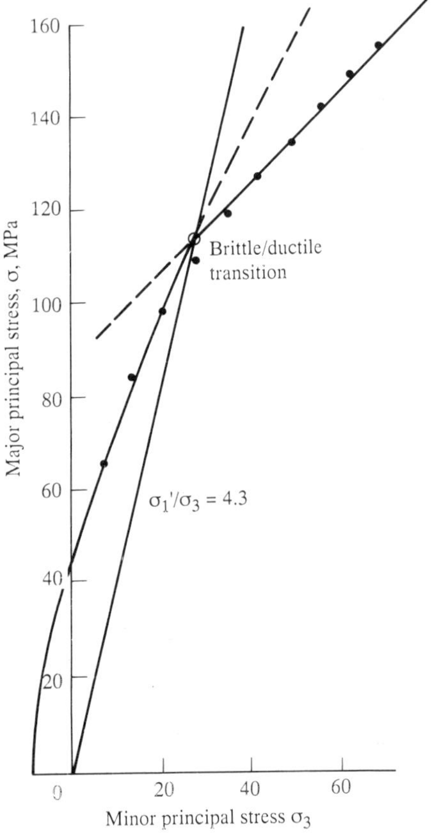

Figure 1.18 *Results of triaxial tests on Indiana Limestone illustrating brittle ductile transition (after Schwartz 1964)*

place through the asperities. When a jointed rock mass undergoes shearing this may be accompanied by dilation, especially at low pressures, and small shear displacements probably occur as shear stress builds up.

It has been suggested (Hoek 1983) that the shear strength, τ, along a surface of failure can be obtained from

$$\tau = (\cot \phi_i' - \cos \phi_i') \frac{m\sigma_c}{8} \tag{1.16}$$

where ϕ_i' is the instantaneous angle of friction at given values of τ and σ' (i.e. the inclination of the tangent of the Mohr failure envelope at the point (σ', τ) shown in Figure 1.19). Figure 1.19 also includes the equations by which ϕ_i', c_i' (instantaneous cohesion) and β (inclination of failure plane) are derived.

Although closely jointed rock is extremely difficult to test, if it is tested under triaxial conditions, then the larger the diameter of the sample the better the results. This is because measurements on small samples give values relating to the intact rock and do not take account of the jointing. Jaeger (1970) described large-scale triaxial tests carried out on closely jointed rock. He suggested that one approach in dealing with such rock was to regard it as randomly jointed on a small scale and to apply soil mechanics theory to it. He found that under confining conditions, adjacent blocks interlocked and the strength of the rock mass was thereby increased. In the sample he tested, movement took place along a large number of planes and some barrelling, and occasional tension gashes, developed. Generally movement on one particular plane tended to become dominant as strain increased. This final surface of shear was usually heavily slickensided.

Under triaxial conditions the peak strengths developed by anisotropic rocks (e.g. those characterized by lamination such as shales, or cleavage such as slates) depend on the orientation of these planes of relative weakness to the principal stress directions. Figure 1.20 shows variations in peak stress in relation to the angle of inclination of the major principal stress to the plane of weakness. Each plane of weakness possesses a limiting value of shear strength in accordance with Coulomb's equation (1.1) and Equations (1.2) and (1.3) allow the normal and shear stresses on the plane to be determined. Substituting for normal stress (σ_n, Equation 1.2) and shear strength (τ, Equation 1.3) in the Coulomb equation and rearranging provides the axial strength, σ_1', of a triaxial specimen from the following equation:

$$\sigma_1' = \sigma'_3 + \frac{2(c_1' + \sigma_3' \tan \phi_i')}{(1 - \tan \phi_i \tan \beta) \sin 2\beta} \tag{1.17}$$

Equation (1.17), however, can only be solved for values of β which are within about 25° of the friction angle, ϕ'. Very high values of σ_1' are obtained from very small values of β whilst values of β which are near 90° yield negative values of σ_1' which are meaningless. Such very high or negative values mean that slip cannot occur along a plane of weakness and that failure will take place through the intact rock. Nonetheless the two-strength model represented in Figure 1.20 gives an oversimplified view of the variation in strength which occurs in anisotropic rocks. After carrying out a series of triaxial tests on slate at a range of confining pressures and cleavage orientations, McLamore and Gray (1967) attempted a fuller explanation by proposing that both cohesion, c_1', and $\tan \phi_i'$ vary according to orientation in relation to the following expressions:

$$c_i' = A - B[\cos 2(\alpha - \alpha_c)]^n \tag{1.18}$$

and

$$\tan \phi_i' = C - D[\cos 2(\alpha - \alpha_\phi)]^m \tag{1.19}$$

where A, B, C, D, m and n are constants, and α_c and α_ϕ are values of α, ($\alpha = \pi/2 - \beta$) at which c_i' and ϕ_i' have minimum values respectively.

Krahn and Morgenstern (1979) carried out a series of direct shear tests on natural and artificially produced discontinuities in limestone in order to demonstrate how the ultimate shearing resistance was influenced by surface

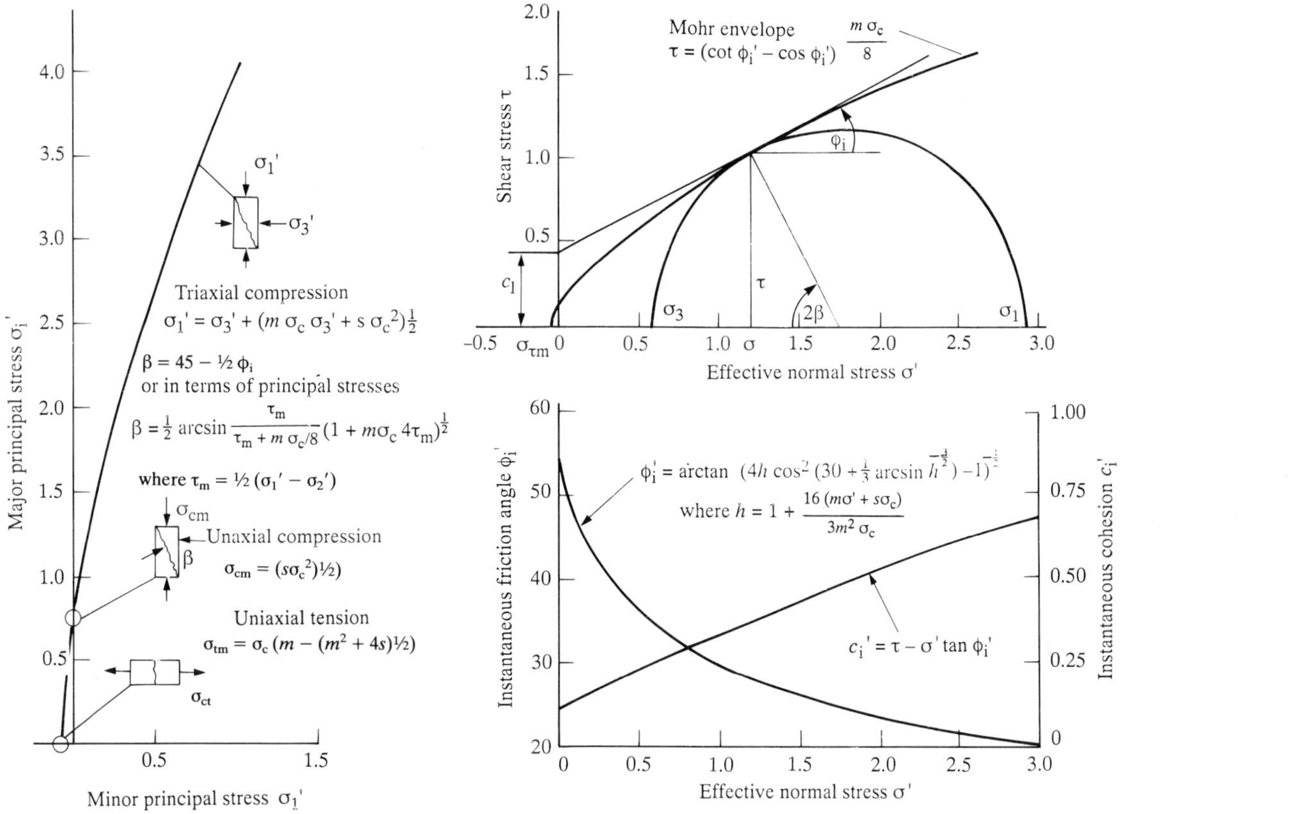

Figure 1.19 *Summary of equations associated with the non-linear failure criterion proposed by Hoek and Brown (1980)*

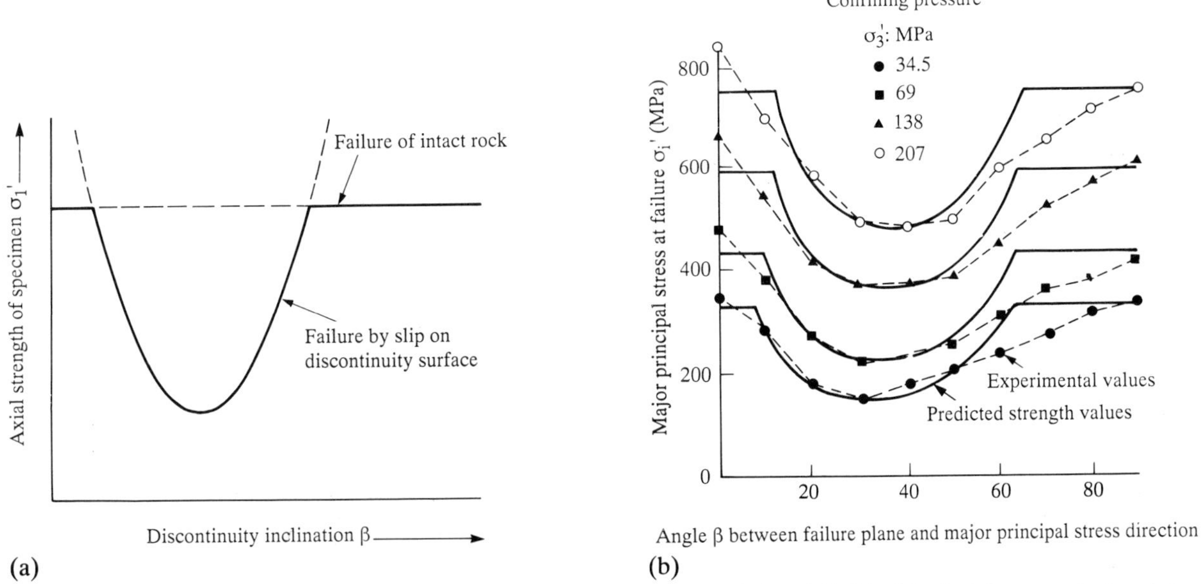

Figure 1.20 (a) *Strength of specimen predicted by means of Equations (1.14) and (1.17);* (b) *triaxial test results for slate with different failure plane inclinations, obtained by McLamore and Gray (1967), compared with strength predictions from Equations (1.14) and (1.17) (after Hoek 1983)*

structure and roughness. They found that except at low normal stresses there was essentially no difference between peak and ultimate or residual resistance for relatively smooth and flat artificial surfaces. All the natural discontinuities, however, showed a significant drop from peak to ultimate strength. The ultimate friction angle (ϕ_r) varied from 14 to 32° for natural discontinuities while the cohesion intercept varied between 55 and 82 kPa. They concluded that the ultimate frictional resistance of jointed hard unweathered rock depends on the initial surface roughness along the joint and the type of surface alteration which occurs during shearing.

Barton (1976) proposed the following empirical expression for deriving the shear strength (τ) along joint surfaces:

$$\tau = \sigma_n \tan (\mathrm{JRC} \log_{10}(\mathrm{JCS}/\sigma_n) + \phi_b) \qquad (1.20)$$

where σ_n is the effective normal stress, JRC is the joint roughness coefficient, JCS is the joint wall compressive strength and ϕ_b is the basic friction angle. According to Barton, the values of the joint roughness coefficient range from 0 to 20, from the smoothest to the roughest surface (Figure 1.21). The joint wall compressive strength is equal

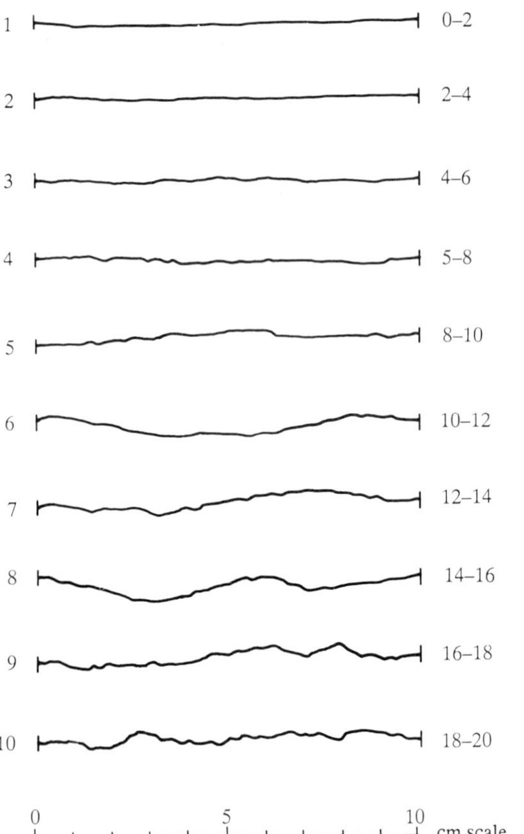

Figure 1.21 *Roughness profiles and corresponding range of JRC values associated with each one (after Barton 1976)*

to the unconfined compressive strength of the rock if the joint is unweathered. This may be reduced by up to 75% when the walls of the joints are weathered. Both these factors are related as smooth-walled joints are less affected by the value of JCS, since failure of asperities plays a less important role. The smoother the walls of the joints, the more significant is the part played by its mineralogy (ϕ_b). The experience gained from rock mechanics indicates that under low effective normal stress levels, such as occur in engineering, the shear strength of joints can vary within relatively wide limits. The maximum effective normal stress acting across joints considered critical for stability lies, according to Barton, in the range 0.1–2.0 MPa.

In practice it is found the JRC is only a constant for a fixed joint length. Generally longer profiles (of the same joint) have lower JRC values. Indeed Barton and Bandis (1980) suggested that mobilization of peak strength along a joint surface seems to be a measure of the distance the joint has to be displaced in order that asperities are brought into contact. This distance increases with increasing joint length. Consequently when testing, longer samples tend to give lower values of peak shear strength. Barton and Choubey (1977) suggested that blocks defined by intersecting disconintuities probably provided the best size of samples for shear testing or joint surface analysis.

Hoek (1983) recommended the use of Equation (1.20) for estimation of shear strength in the field. He went on, however, to point out that this equation was not the only one which could be used for fitting to shear test data obtained in the laboratory. For example, he maintained that the equations for τ and ϕ_i', given in Figure 1.19, provide a reasonably accurate estimation of the shear strength along rough discontinuities in rock masses under a wide range of effective normal stress conditions. Nevertheless Equation (1.20) suggests that there are three components of shear strength, namely, a basic frictional component (ϕ_b), a geometrical component which is governed by surface roughness (JRC) and an asperity failure component which depends upon the ratio JCS/σ_n. From Figure 1.22 it can be seen that the geometrical and asperity failure components together give the net roughness component, $i°$. Accordingly the total frictional resistance can be derived from $(\phi_b + i)°$. The shear strength developed along a rough discontinuity depends upon the scale and amount of stress involved. As the effective normal stress, σ_n, increases, so the term $\log_{10}$ (JCS/σ_n') decreases, as does the net apparent friction angle. The steeper asperities are sheared off and the inclination of the controlling roughness decreases with increasing scale. Increasing scale also means that the asperity failure component decreases since the compressive stength of the rock, JCS, declines with increasing size. Hence the shear force-displacement curves change with increasing scale in that the behaviour along a discontinuity on shearing changes from brittle to plastic as the shear stiffness is reduced (Barton and Bandis 1980).

Tse and Cruden (1979) pointed out that fairly small errors in estimating the joint roughness coefficient could produce serious errors in estimating the peak shear

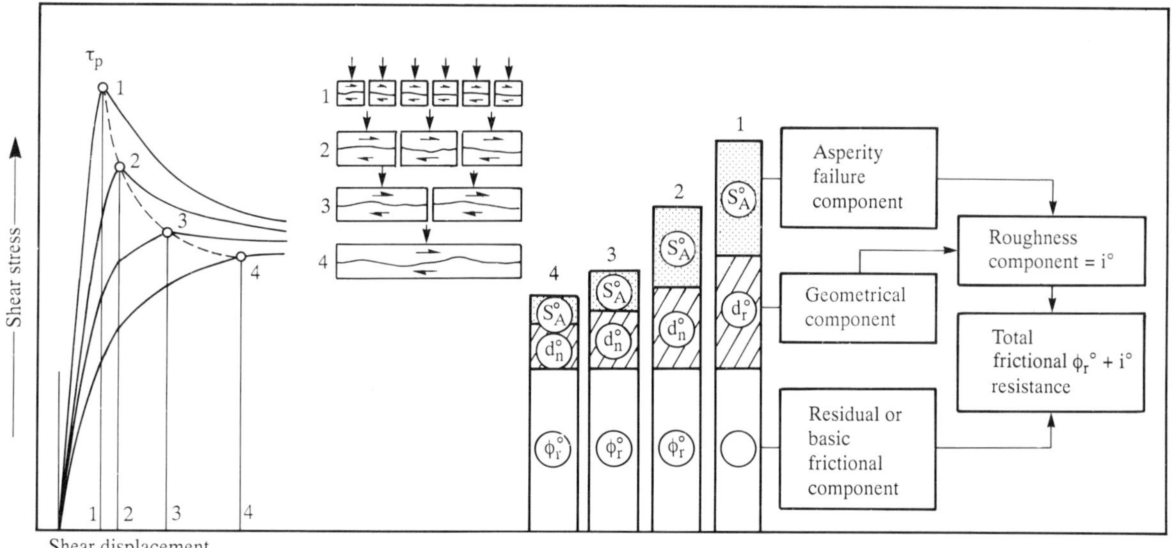

Figure 1.22 *An illustration of the size dependence of shear–stress deformation behaviour for non-planar joints (after Bandis* et al. *1983)*

strength (τ) from Equation (1.20), especially if the ratio JCS/σ_n was large. They therefore recommended a numerical method of checking of the value of JRC, based on a detailed profiling and analysis. Weissbach (1978) developed a profilograph for measuring the roughness of joints.

Previously Barton and Choubey (1977) had suggested that tilt and push tests provided a more reliable means of estimating the joint roughness coefficient than comparison with typical profiles. Barton and Bandis (1980) also supported the use of such tests, particularly in heavily jointed rock masses, when three joint sets are present. In a tilt test, two immediately adjacent blocks are extracted from an exposure and the upper is laid upon the lower in the exact same position as it was in a rock mass. Both are then tilted and the angle (α) at which sliding occurs is recorded (Figure 1.23a). The JRC is extimated from

$$\mathrm{JRC} = \frac{\alpha - \phi_r}{\log_{10}(\mathrm{JCS}/\sigma_{no})} \tag{1.21}$$

where $\sigma_{no} = \gamma H \cos^2 \alpha$ (i.e. normal stress induced by self-weight of block), γ = unit weight, H = thickness of upper block, and ϕ_r = residual friction angle.

In a pull test an external shearing force (T_2) is applied via a bolt grouted into the block in question (Figure 1.23b). The value of JRC is given by

$$\mathrm{JRC} = \frac{\arctan[(T_1 + T_2)/N] - \phi_r}{\log_{10}(\mathrm{JCS}.A/N)} \tag{1.22}$$

where A is the joint area and N is the normal and tangential components of the self-weight of the upper

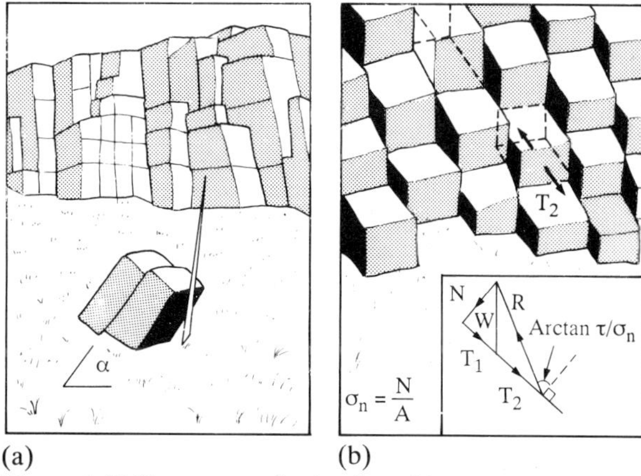

Figure 1.23 *Two extremely simple and inexpensive ways of determining an accurate scale-free value of JRC:* (a) *tilt test;* (b) *pull test (after Barton and Bandis 1980)*

block. In both cases the joint wall compression strength (JCS) and the residual friction angle (ϕ_r) can be estimated by using a Schmidt hammer (Barton and Choubey 1977)

$$\log_{10}\mathrm{JCS} = 0.00088\,\gamma_d R + 1.01 \tag{1.23}$$

where γ_d = dry unit weight, R = Schmidt hammer rebound number, and

$$\phi_r = (\phi_b - 20°) + 20\,(r/R) \tag{1.24}$$

where ϕ_b = basic friction angle, r = Schmidt hammer rebound number on wet joint surface, R = Schmidt

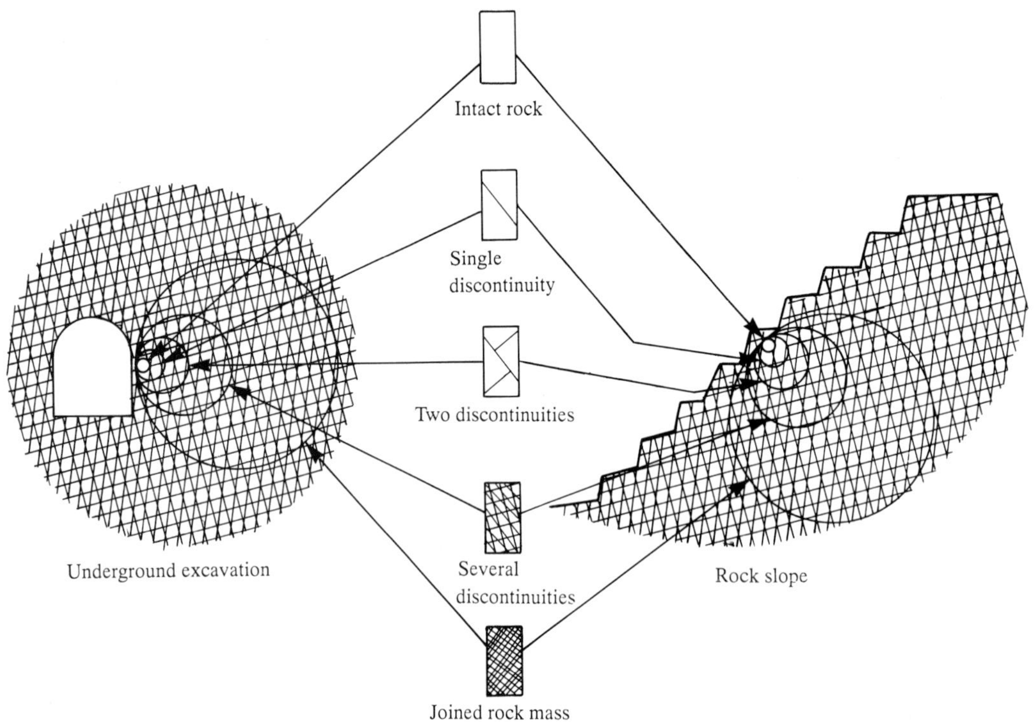

Figure 1.24 *Simplified representation of the influence of scale on the type of rock mass behaviour model which should be used in designing underground excavations of rock slopes (after Hoek 1983)*

hammer rebound number on dry unweathered sawn surface.

Probably the most thorough approach to the assessment of the strength of discontinuous rock masses has been made by Hoek and Brown (1980) and again by Hoek (1983). Their method is summarised in Table 1.4. and involves estimating the values of the empirical constants *m* and *s* from a description of the rock mass. In other words the appropriate box in Table 1.4 is determined from a description of the rock mass or preferably from the Q or Geomechanics systems of rock mass classification (see Chapter 3). However, the size of the structure which is to be constructed influences the significance of discontinuity of spacing in that the larger the structure the greater the number of discontinuities which are going to have an affect on its design (Figure 1.24). Hence the type of model chosen to represent rock mass behaviour also depends on scale. These estimates, together with an estimate of the unconfined compressive strength, can then be used to construct an approximate Mohr failure envelope for the discontinuous rock mass. Hoek (1983) suggested that the values listed in Table 1.4, as far as practical engineering design is concerned, are somewhat conservative and strength estimates derived therefrom can be regarded as lower bound values for design purposes.

References

Atkinson, B. K. (1984) 'Subcritical crack growth in geological materials', *J. Geophys. Res.*, **89,** 4077–4114

Ballivy, G., Ladanyi, B. and Gill, D. E. (1976) 'Effect of water saturation history on the strength of low porosity rocks'. In *Soil Specimen Preparation for Laboratory Testing,* ASTM. STP599, Philadelphia, 4–20

Bandis, S., Lumsden, A. C. and Barton, N. (1983) 'Fundamentals of rock joint deformation', *Int. J. Rock Mech. Min. Sci & Geomech. Abstr.*, **20**, 249–268

Barton, N. (1976) 'The shear strength of rock and rock joints', *Int. J. Rock Mech. Min. Sci & Geomech. Abstr.,* **13**, 255–279

Barton, N. and Bandis, S. (1980) 'Some effects of scale on the shear strength of joints', *Int. J. Rock Mech. Min. Sci & Geomech. Abstr.*, **17**, 69–76.

Barton, N. and Choubey, V. (1977) 'The shear strength of rock joints in theory and practice', *Rock Mechanics,* **10**, 1–54

Bernaix, J. (1969) 'New laboratory methods of studying the mechanical properties of rocks', *Int. J. Rock Mech. Min. Sci.*, **6**, 43–90

Brace, W. F. (1964) 'Brittle fracture of rocks' *Symp. State of Stress in the Earth's Crust,* Judd, W. R. (ed.), Elsevier, Santa Monica, 111–80

Brace, W. F. and Martin, R. J. (1968) 'A test of the law of effective stress for crystalline rocks of low porosity', *Int. J. Rock Mech. Min. Sci.,* **5**, 415–426

Brady, B. H. G. and Brown, E. T. (1985) *Rock Mechanics for Underground Mining,* George Allen and Unwin, London

Broch, E. (1979) 'Changes in rock strength caused by water', *Proc. 4th Int. Cong. on Rock Mechanics,* ISRM, Montreux, **1**, 71–76

Brown, E. T., Richards, L. R. and Barr, M. V. (1977) 'Shear strength characteristics of the Delabole Slates', *Proc. Conf. Rock Engng,* Newcastle University, **1**, 33–51

Colback, P. S. B. and Wiid, B. L. (1965) 'Influence of moisture content on the compressive strength of rock', *Symp. Canadian Dept. Min. Tech. Survey,* Ottawa, 65–83

Coulomb, G. A. (1973) 'Sur une application des règles de maximus et minimus à quelques problèmes de statique rélatifs à l'architecture', *Acad. Roy. des Sci., Mem de Math. et de Phys. par divers Sovans,* **7**, 343–382

Deere, D. U. and Miller, R. P. (1966) 'Engineeirng classification and index properties for intact rock', *Tech. Rep. No. AFWL-TR-65-115,* Air Force Weapons Lab., Kirtland Air Base, New Mexico

Dobereiner, L. and De Freitas, M. H. (1986) 'Geotechnical properties of weak sandstones', *Geotechnique,* **36**, 79–94

Donath, F. A. (1961) 'Experimental study of shear failure in anisotropic rocks', *Bull. Geol. Soc. Am.,* **72**, 985–991

Eastwood, T., Hollingworth, S. E., Rose, W. C. C. and Trotter, F. M. (1968) 'Geology of the country around Cockermouth and Caldbeck', *Mem. Inst. Geol. Sci.,* HMSO, London

Edmond, J. M. and Paterson, M. S. (1972) 'Volume changes during deformation of rocks at high pressure', *Int. J. Rock Mech. Min. Sci.,* **9**, 161–182

Gowd, T. N. and Rummel, F. (1980) 'Effect of confining pressure on the fracture behaviour of a porous rock', *Int. J. Rock Mech. Min. Sci. & Geomech. Abstr.,* **17**, 225–229

Griffith, A. A. (1924) 'The phenomenon of rupture and flaws in solids', *Phil. Trans. Roy. Soc. London,* **A221**, 163–198

Griffith, A. A. (1924) 'The theory of rupture', *Proc. 1 Int. Cong. Appl. Mech.,* Delft, 55–65

Griggs, D. T. (1936) 'Deformation of rocks under high confining pressures', *J. Geol.,* **44**, 541–577

Griggs, D. T. (1940) 'Experimental flow of rocks under conditions favoring recrystallization', *Bull. Geol. Soc. Am.,* **51**, 1001–1022

Griggs, D. T. (1951) 'Deformation of Yale Marble', *Bull. Geol. Soc. Am.,* **62**, 853–862

Hawkes, I., Mellor, M. and Gariepy, S. (1973) 'Deformation of rocks under uniaxial tension', *Int. J. Rock Mech. Min. Sci. & Geomech. Abstr.,* **10**, 493–507

Heard, H. C. (1960) 'Transition from brittle to ductile flow in Solenhofen Limestone as a function of temperature, confining pressure and interstitital fluid pressure' In *Rock Deformation, Geol. Soc. Am. Mem.* **79**, 193–212

Hobbs, D. W. (1970) 'Stress-strain-time behaviour in a number of Coal Measures rocks', *Int. J. Rock Mech. Min. Sci.,* **7**, 149–170

Hoek, E. (1968) 'Brittle fracture of rocks'. In *Rock Mechanics in Engineering Practice,* Stagg, K. G. and Zienkiewicz, D. C. (eds), Wiley, London, 99–124

Hoek, E. (1983) 'Strength of jointed rock masses', Rankine Lecture, *Geotechnique,* **33**, 187–223

Hoek, E. and Brown, E. T. (1980) 'Empirical strength criterion for rock mass', *Proc. ASCE, J. Geot. Engng. Div.,* **106** (GT9), 1013–1035

Jaeger, J. C. (1970) 'Behaviour of closely jointed rock', *Proc. 11th Symp. Rock Mech., Berkeley,* Pergamon Press, New York, 56–68

James, A. N. (1981) 'Solution parameters of carbonate rocks', *Bull. Int. Ass. Engng. Geol.,* No. 24, 19–25

James, A. N. and Kirkpatrick, I. M. (1980) 'Design of foundations of dams containing soluble rocks and soils', *Q. J. Engng Geol.,* **13**, 189–198

John, K. W. (1965) 'Civil engineering approach to evaluate strength and deformability of regularly jointed rock', *Rock Mech.,* **1**, 69–80

Krahn, J. and Morgenstern, N. R. (1979) 'The ultimate frictional resistance of rock discontinuities', *Int. J. Rock Mech. Min. Sci & Geomech. Abstr.,* **16**, 127–133

Krumbein, W. C. (1941) 'Measurement and geological significance of shape and roundness of sedimentary particles', *J. Sed. Pet.,* **11**, 64–72

Krumbein, W. C. and Pettijohn, F. J. *Manual of Sedimentary Petrography,* Appleton Century Crofts, New York

McLamore, R. and Gray, K. E. (1967) The mechanical behaviour of anisotropic sedimentary rocks., *J. Engng for Industry, Trans. Am. Soc. Mech. Engrs.,* Ser B, **89**, 62–73

McClintock, F. A. and Walsh, J. B. (1962) 'Friction on Griffith cracks in rocks under pressure', *Proc. 4th Conf. Appl. Mech.,* 1015–1021

Mohr, O. (1882) *Abhandlungen aus dem Begiete der Technische Mechanik,* Ernst und Sohn, Berlin

Murrell, S. A. F. (1963) 'A criterion for the brittle fracture of rocks and concrete under triaxial stress and the effect of pore pressure on the criterion', *Proc. 5th Symp. Rock Mech.,* University of Minnesota, Pergamon Press, New York, 563–577

Onodera, T. F. and Kumara, A. H. M. (1980) 'Relation between texture and mechanical properties of crystalline rocks', *Bull. Int. Ass. Engng. Geol.,* No. 22, 173–177

Parate, N. S. (1973) 'Influence of water on the strength of limestone', *Trans. Soc. Min. Engrs, AIME,* **254**, 127–131

Pettijohn, F. J. (1975) *Sedimetnary Rocks,* Harper and Row, New York

Pettijohn, F. J., Potter, P. E. and Siever, R. (1972) *Sands and Sandstones,* Springer-Verlag, Berlin

Powers, M. C. (1953) 'A new measurement scale for sedimentary particles', *J. Sed. Pet.,* **23**, 117–119

Price, N. J. (1960) 'The compressive strength of Coal Measures rocks', *Coll. Engr.,* **37**, 283–292

Priest, S. D. and Selvakumar, S. (1986) 'The failure characteristics for selected British rocks', *Report for the Transport and Road Research Laboratory,* Imperial College, University of London

Robertson, E. C. (1960) 'Creep of Solenhofen Limestone under moderate hydrostatic pressure'. In *Rock Deformation, Geol. Soc. Am. Mem.,* **79**, 227–244

Robinson, L. H. (1959) 'The effect of pore and confining pressure on the failure process in sedimentary rock', *Q. Colorado School of Mines,* **53**, 177–199

Schwartz, A. E. (1964) Failure of rock in triaxial shear test, *Proc. 6th Rock Mechanics Symposium,* Rolla, Missouri, 109–135

Spears, D. A. (1980) 'Towards a classification of shales, '*J. Geol. Soc.*, **137**, 125–130

Taylor J. M. (1950) 'Pore space reduction in sandstones', *Bull. Am. Ass. Petrol. Geologists.*, **34**, 701–716

Tse, R. and Cruden, D. M. (1979). 'Estimating joint roughness coefficient', *Int. J. Rock Mech. Min. Sci. & Geomech. Abstr.*, **16**, 303–307

Turk, N. and Dearman, W. R. (1986) 'Influence of water on engineering properties of weathered rock', *Engineering Geology Special Publication No. 3, Groundwater in Engineering Geology,* Sheffield, Cripps, J. C., Bell, F. G. and Culshaw, M. G. (eds), Geological Society, London, 131–138

Turner, F. J. and Weiss, L. E. (1963) *Structural Analysis of Metamorphic Tectonites,* McGraw-Hill, New York

Vutukuri, V. S. (1974) 'The effects of liquid on the tensile strength of limestone', *Int. J. Rock Mech. Min. Sci. & Geomech. Abstr.*, **11**, 27–29

Walsh, J. B. (1965) 'The effects of cracks on the uniaxial elastic compression of rocks', *J. Geophys. Res.*, **70**, 399–411

Wawersik, W. R. and Fairhurst, C. (1970) 'A study of brittle rock fracture in laboratory compression experiments', *Int. J. Rock Mech. Min. Sci.*, **7**, 561–575

Weissbach, G. (1978) 'A new method for the determination of the roughness of joints in the laboratory', *Int. J. Rock Mech. Min. Sci. & Geomech. Abstr.*, **15**, 131–134

Willard, R. J. and McWilliams, J. R. (1969) 'Microstructural techniques in the study of physical properties of rocks', *Int. J. Rock Mech. Min. Sci.*, **6**, 1–12

Working Party Report (1970) 'Logging of cores for engineering purposes', *Q. J. Engng. Geol.*, **3**, 1–24

2 Influence of weathering and discontinuities on the behaviour of rock masses

Professor F G Bell
University of Natal

As far as design and practice in engineering are concerned, rock masses may be grouped into two categories, simply those which are unweathered and those which are weathered. In the case of unweathered rock masses interest centres mainly on the incidence and character of the discontinuities since these adversely affect the engineering performance. This is not so say that discontinuities are of no consequence in weathered rocks, indeed weathering tends to be concentrated along discontinuities.

Weathering

Weathering of rocks is brought about by physical disintegration, chemical decomposition and biological activity. The weathering process is primarily controlled by the presence of discontinuities in that they provide access for the agents of weathering. Hence the earliest effects of weathering are seen along discontinuity surfaces. Weathering then proceeds inwards until the whole of a discontinuity bonded block is affected. The agents of weathering, unlike those of erosion, do not themselves provide for the transportation of debris from a rock surface. Therefore unless this rock waste is otherwise removed it eventually acts as a protective blanket, preventing further weathering taking place. If weathering is to be continuous, fresh rock exposures must be constantly revealed, which means that the weathered debris must be removed by the action of gravity, running water, wind or moving ice.

2.1 Rate of weathering

The type and rate of weathering varies from one climatic regime to another. In humid regions chemical and chemico-biological processes are generally much more significant than those of mechanical disintegration. The degree and rate of weathering in humid regions depends on the temperature, the amount of moisture and organic matter available, and the relief. If the temperature is high, then weathering is extremely active and it has been calculated that an increase of 10°C more than doubles the rate at which chemical reactions occur. The higher the moisture content in the soil mantle, the more readily are silicates and aluminium silicates hydrolysed and substances removed in solution. When organic matter is dissolved by leaching waters carbon dioxide is liberated. Thus the greater the amount of organic matter contributed to the soil, the more active the weathering. In order to allow chemical weathering to proceed the surface layers of rock must remain in place or be removed at a rate which does not inhibit the change of the rock debris from an alkaline to an acid state and impede the removal of soluble material. Such a condition is only found on plains. As the relief increases so mechanical disintegration intensifies and ultimately a point is reached where the rate of slope wash is greater than that of chemical weathering.

The rate at which weathering proceeds depends not only upon the vigour of the weathering agent but also on the durability of the rock mass concerned. This, in turn, is governed by the mineralogical composition, texture and porosity of the rock on the one hand, and the incidence of discontinuities within the rock mass on the other.

The inherent stability of a mineral is influenced by the environment in which it was formed. For example, those minerals which crystallize from a magma at very high temperatures and pressures are relatively unstable when exposed to atmospheric conditions. The least stable minerals occur in ultrabasic and basic igneous rocks such as dunites, peridotites, basalts and gabbros (Figure 2.1). Hence such rocks offer less resistance to weathering than acidic igneous rocks, which are commonly composed of potash feldspar, muscovite and quartz. Muscovite and quartz, in particular, can survive intense weathering and withstand more than one cycle of erosion. They are accordingly found in very mature muds and sands and their lithified equivalents. Most sedimentary rocks are the products of denudation, transportation and deposition and so they frequently contain high proportions of minerals which are stable under atmospheric conditions. Clay

Figure 2.1 *Highly weathered basalt almost completely decomposed along joints, Isle of Arran*

minerals are notable examples; they are the stable end products which form when most of the minerals listed above decompose.

Generally coarse-grained rocks weather more rapidly than do fine-grained types of similar mineralogical composition. The degree of interlocking between minerals is a particularly important textural factor. Obviously the more strongly a rock is bonded together, the greater is its resistance to weathering. The closeness of the interlocking of grains governs the porosity of a rock. This in turn determines the amount of water a rock can hold. Not only are the more porous rocks more susceptible to chemical attack, but they are more prone to frost action than the less porous varieties.

Most studies regarding the rate at which weathering occurs have been made upon stone used for construction purposes and have involved measuring the rate at which the surfaces of stones have been removed (Table 2.1). In addition, a number of tests have been used to simulate and accelerate the rate of weathering of construction stone. The problem with such tests is the difficulty in relating the results obtained to the natural performance of the rocks concerned. Nevertheless, according to Fookes *et al.* (1988) it would appear, firstly, that the rate of weathering declines with time where a residual layer is developed at the surface of the rock and, secondly, that if surface reaction is involved, then the rate of weathering is linear.

2.2 Mechanical weathering

Mechanical or physical weathering is particularly effective in climatic regions which experience significant diurnal changes of temperature. This does not necessarily imply a large range of temperature, as frost and thaw action can proceed where the range is limited.

As far as frost susceptibility is concerned the porosity, pore size and degree of saturation all play an important role. When water turns to ice it increases in volume by up to 9%, thus giving rise to an increase in pressure within the pores. The action is further enhanced by the displacement of pore water away from the developing ice front. Once ice has formed the ice pressures rapidly increase with decreasing temperature, so that at approximately −22°C ice can exert a pressure of 200 MPa (Winkler 1973). Usually coarse-grained rocks withstand freezing better than fine-grained types. Indeed the critical pore size for freeze–thaw durability appears to be about 0.005 mm. In other words rocks with larger mean pore diameters allow outward drainage and escape of fluid from the frontal advance of the ice line and are therefore less frost susceptible. The amount of pore space and the continuity of the pore system also are important. In particular the pore structure governs the degree of saturation and the magnitude of the stresses which can be generated on freezing. Fine-grained rocks which have 5% sorbed water are often very susceptible to frost damage whilst those containing less than 1% are very durable. Alternate freeze–thaw action causes cracks, fissures, joints and some pore spaces to be widened. As the process advances, angular rock debris is gradually broken from the parent body.

It has been alleged that in the case of porous building stones failure will occur if the water in the pores exceeds a certain critical content, this causing expansion on freezing. Critical water contents tend to vary between 75 and 96% of the total volume of the pores. The rapidity with which critical water content is attained depends upon the initial state of saturation of the rock and its ability to absorb further water during thawing. The French have devised a test to assess critical water content in which the test specimens are immersed in water for 22 hours and then submitted to cycles of freezing and thawing until rupture occurs, additional water being absorbed during thawing since this is achieved in water (Honeyborne 1983). The change in length under the influence of freezing from 20°C to −15°C is measured. If the specimens have contracted, then the critical water content has not been reached and frost damage is unlikely. On the other hand, if the specimens expand, then the critical water content has been attained and frost damage is assumed likely. The French use two other tests to assess the potential for frost damage. In the first, specimens are again subjected to cycles of freezing in air and thawing in water. However, the state of the specimens is determined after a given number of cycles by measuring the dynamic value of Young's modulus. The lower the value, the greater is the degradation suffered by the specimens. The second test is an indirect one based on the measurement of porosity and the coefficient of capillarity (i.e. the volume of water absorbed in 48 hours of immersion as a ratio of the volume of voids). These two values provide an indirect indication of the pore dimensions and thereby of the degree of frost susceptibility. The value of these tests has been questioned and they tend not to be used in Britain (Leary 1983).

The mechanical effects of weathering are well displayed in hot deserts, where wide diurnal ranges of temperature

Table 2.1 Records of rates of surface lowering reported for various rock types (From Fookes *et al.* 1988)

Rock type	*Surface studied*	*Period* (years)	*Remarks*	*Average* (mm/year)	*Location*
	Ancient structures	4 000	Fresh	0.009	Aswan
Granite		5 400	Flaking	0.0015	Giza
	Glacial surface	10 000	–	0.00105	Narvik
Slate	Tombstones	90	Engraving clear	–	Edinburgh
Mica schist	Glacial surface	8	–	0.04–0.15	Karkevagge
Marble	Tombstones	90	Crumbling	–	Edinburgh
	Tombstones	500	–	0.051	Yorkshire
	Tombstones	300	–	0.085	Yorkshire
	Tombstones	250	–	0.102	Yorkshire
	Tombstones	240	–	0.106	Yorkshire
	Great Pyramid	1 000	Hard grey	–	Giza
	Great Pyramid	1 000	Soft grey	0.01–0.02	Giza
Limestone	Kammetz fortress	230	–	1.32	Ukraine
	Bare surface	1 000	–	0.009–0.0125	Austrian Alps
	Covered surface	1 000	Under acid soil	0.028	Austrian Alps
	Jetty	16	Intertidal	0.5–1.0	Norfolk Island
	Coastal notch		–	1.0	Point-Perm Australia
	Inscriptions		–	0.5	La Jolla, California
	Inter-tidal notch	155	–	1.0	Puerto Rico
Carb. limestone	Erratic block	12 000	–	0.025–0.042	N. England
Carb. limestone	Glacial striae	13	–	2.2–3.8	N. England
Carb. limestone	Runnels in glacial surface	13	–	11.5	N. England
Limestone	Great Pyramid	1 000	Hard grey	Little	Giza
	Great Pyramid	1 000	Soft grey	0.01–0.02	Giza
Kirkby Stephen limestone	Tombstones	500	–	0.051	Yorkshire
Tailbrig limestone	Tombstones	300	–	0.085	Yorkshire
Penrith limestone	Tombstones	250	–	0.102	Yorkshire
Askrigg limestone	Tombstones	240	–	0.106	Yorkshire
Algal limestone	Tombstones	<1	Microerosion meter	0.11	Aldabra
Limestone (poorly cemented)	Limestone pavement		–	0.003–6.3	Co. Clare
Limestone chert	Glacial surface	70	–	0.02–0.2	Spitzbergen
Limey shale	Great Pyramid	1 000	Rubble	0.2	Giza
Grey shale	Great Pyramid	1 000	–	0.2	Giza
Sandstone	Tombstones	200	Little	–	Edinburgh
Sandstone	Glacial surface	10 000	–	0.34–0.5	Spitzbergen

cause rocks to expand and contract. Because rocks are poor conductors of heat these effects are mainly localized in their outer layers where alternate expansion and contraction create stresses which eventually rupture the rock. In this way flakes of rock break away from the parent material, the process being termed exfoliation. The effects of exfoliation are concentrated at the corners and edges of rocks so that their outcrops gradually become rounded. Furthermore minerals possess different coefficients of expansion and differential expansions within a polymineralic rock fabric generates stresses at grain contacts and can lead to granular disintegration.

Moisture in rock acts as a disruptive agent as well as a means of transporting salts. Capillary movement in rock may be extensive and on rock faces the maximum capillary rise may be indicated by the presence of a white efflorescent rim or a wet margin. Moisture involved in capillary movement does not move out of the capillaries as free flow. Usually sandstones have a more consistent capillary system than carbonate rocks.

The repeated action of wetting and drying together with the expansive force of water can represent a disruptive influence, especially in pores of a rock with a low tensile strength. For instance, when the temperature of water is raised from 0 to 60°C, then it expands by about 1.5% exerting a pressure of up to 52 MPa within the pores and a diurnal temperature difference of 40°C can develop pressures of around 26 MPa. Hence the expansion and contraction of water in narrow capillaries can produce sufficient pressures to disrupt many of the weaker rock types within a year. Moreover water in the pores of granites can cause expansion varying from 0.0004 to 0.009%, in marbles from 0.001 to 0.0025% and in quartzose sandstones from 0.01 to 0.044%. Significant expansion can occur in some dolostones as a result of what has been referred to as moisture cracking. This takes place when the clay mineral content is segregated into thin bands which imbibe water.

Soluble salts which occur in the pores of rock tend to move towards the surface. The more soluble salts, however, such as some chlorides and sulphates, move back and forth in the rock with changes in weather and moisture gradient. The less soluble salts crystallize at or near the surface. Some surface efflorescences may contain potentially harmful salts such as sulphates of Ca, Mg and Na. The development of subflorescences just below the surface can lead to the outer skin of a rock losing its support and exfoliating. Again calcium magnesium sulphates are among the most common salts involved. Resistance to crystallization damage, as with frost damage, is strongly dependent on the internal structure of the rock and decreases as the proportion of fine pores increases. The pressures produced on crystallization in small pores are appreciable, for example, gypsum ($CaSO_4.nH_2O$) exerts a pressure of up to 100 MPa; anhydrite ($CaSO_4$), 120 MPa; kieserite ($MgSO_4.nH_2O$), 100 MPa; and halite (NaCl), 200 MPa; and these are sufficient to cause disruption.

Disruption in rock also may take place due to the considerable contrasts in thermal expansion of salts in the pores. For instance, halite expands by some 0.5% from 0 to 60°C, which may aid rock decay.

As crystallization of salts within the pore space is a principal cause of stone decay, especially in urban atmospheres, the sulphate crystallization soundness test is used to assess performance. This is often preceded by an acid immersion test. The first stage in the latter test involves immersing specimens for 10 days in sulphuric acid with a density of 1.145 Mg/m^3. Stones which are unaffected by the test are regarded as being resistant to attack by acidic rainwater. A more severe test consists of immersing specimens in sulphuric acid with a density of 1.306 Mg/m^3. Experience has shown that this test is of value when it is necessary to determine highly durable stone. If a stone survives the acid immersion test, then it is subjected to the crystallization test.

In the sulphate crystallization soundness test general resistance to weathering is assessed by soaking the stone in a salt solution and then drying in an oven. Repeated crystallization of the salts within the pores during the drying phase simulates the natural weathering process. Test conditions have been standardized (ASTM, 1971) using either magnesium or sodium sulphate; the sodium salt is usually used as it is the cheaper. Either saturated sodium sulphate solution or a 15% solution is used. The saturated solution is very aggressive and is only recommended for use when the natural weathering conditions are particularly severe, or the stone is expected to have a particularly long life.

The crystallization test is carried out on up to six cubic specimens (usually with edges between 40 and 50 mm). The specimens are oven dried at 105°C until they reach a constant weight. After cooling the specimens are completely immersed in a 15% solution of sodium sulphate decahydrate (the specific gravity of the solution at 20°C is 1.055) for 2 hours, the temperature of the solution being kept at 20°C. The specimens then are dried in an oven which has a high relative humidity in the early stages of drying and the temperature of the specimens is raised gradually to 105°C in 10–15 hours. All the specimens remain in the oven for at least 16 hours, after which they are cooled to room temperature, then re-immersed in fresh sodium sulphate solution. The specimens are subjected to 15 cycles of immersion and, after final washing to remove sulphate and drying, are weighed to determine the weight loss.

The results are reported in terms of loss of weight expressed as a percentage of the initial dry weight or as the number of cycles required to produce failure if a specimen(s) is too fractured to be weighed before the fifteenth cycle has been completed. Conclusions relating to durability are obtained by comparing the results of the tests with the performance of stone of known weathering behaviour. This unfortuantely is one of the shortcomings of the test since specific reference stones are seldom available.

2.3 Chemical and biological weathering

Chemical weathering leads to mineral alteration and the solution of rocks. Alteration is principally effected by oxidation, hydration, hydrolysis and carbonation whilst solution is brought about by acidified or alkalized waters. Chemical weathering also aids rock disintegration by weakening the rock fabric and by emphasizing any structural weaknesses, however slight, that it possesses. When decomposition occurs within a rock the altered material frequently occupies a greater volume than that from which it was derived and in the process internal stresses are generated. If this swelling occurs in the outer layers of a rock, then it causes them to peel off from the parent body.

In dry air rocks decay very slowly. The presence of moisture hastens the rate tremendously, first, because water is itself an effective agent of weathering and, second, because it holds in solution substances which react with the component minerals of the rock. The most important of these substances are free oxygen, carbon dioxide, organic acids and nitrogen acids.

Free oxygen is an important agent in the decay of all rocks which contain oxidizable substances, iron and sulphur being especially suspect. The rate of oxidation is quickened by the presence of water; indeed it may enter into the reaction itself, as for example, in the formation of hydrates. However, its role is chiefly that of a catalyst. Carbonic acid is produced when carbon dioxide is dissolved in water and it may possess a pH value of about 5.7. The principal source of carbon dioxide is not the atmosphere but the air contained in the pore spaces in the soil where its proportion may be a hundred or so times greater than it is in the atmosphere. An abnormal concentration of carbon dioxide is released when organic material decays. Furthermore humic acids are formed by the decay of humus in soil waters; they ordinarily have pH values between 4.5 and 5.0 but occasionally they may be under 4.0. The nitrogen acids, HNO_3 and HNO_2, are formed by organic decay or bacterial action in soils. They play only a minor part in weathering. In volcanic regions and in the oxidized zones of sulphide deposits, the sulphur acids, H_2SO_3 and H_2SO_4, become important and in some localities their pH value may be lowered below 1.0.

The simplest reactions which take place on weathering are the solution of soluble minerals and the addition of water to substances to form hydrates. Solution commonly involves ionization, for example, this takes place when salt and gypsum deposits and carbonate rocks are weathered. Hydration and dehydration take place amongst some substances, a common example being gypsum and anhydrite:

$$\underset{\text{(anhydrite)}}{CaSO_4} + 2H_2O = \underset{\text{(gypsum)}}{CaSO_4.2H_2O}$$

The above reaction produces an increase in volume (Table 2.2) and accordingly causes the enclosing rocks to be wedged further apart. These reactions are slow but those involving ferric oxides and hydrates are even slower. Iron oxides and hydrates are conspicuous products of weathering, usually the oxides are a shade of red and the hydrates yellow to dark brown.

Table 2.2 Crystalline solid expansion due to mineral alteration (from Taylor, 1988)

Original material	*Mineral formed*	*Volume increase (%)*
Pyrite, FeS_2	Jarosite	115
	Melanterite	536
	Anhydrous ferrous sulphate	350
Calcite, $CaCO_3$	Gypsum	103
	Bassanite	189
Illite,	Jarosite	10
$KAl_2Si_3O_8(OH)_2$	Alunite	8

Sulphur compounds are readily oxidized by weathering. Because of the hydrolysis of the dissolved metal ion, solutions forming from the oxidation of sulphides are acidic. For instance, when pyrite is initially oxidized, ferrous sulphate and sulphuric acid are formed. Further oxidation leads to the formation of ferric sulphate. Very insoluble ferric oxide or hydrated oxide is formed if highly acidic conditions are produced

$$\underset{\text{(pyrite)}}{FeS_2} + H_2O + 7(O) = FeSO_4 + H_2SO_4$$

$$2FESO_4 + (O) + H_2SO_4 = Fe_2(SO)_4) + H_2O$$

$$2FeS_2 + 15(O) + 4H_2O = Fe_2O_3 + 4H_2SO_4$$

Ferrous sulphate may react with illite to form jarosite which also involved an expansion in volume (Table 2.2).

Perhaps the most familiar examples of rocks prone to chemical attack are limestones and dolostones (Figure 2.2). The solution of carbonates, especially calcium and magnesium carbonate, is principally due to the formation of weak carbonic acid in the soil horizons where CO_2 is dissolved by soil water. Calcium carbonate is readily soluble in water provided that there is an abundant supply of H^+, and the dissociation of carbonic acid, H_2CO_3, represents a major source of H^+. The series of equations expressing the various reactions are as follows:

$$CO_2 + H_2O \rightleftharpoons H_2CO_3 \rightleftharpoons H^+ + HCO_3^-$$

$$HCO_3^- \rightleftharpoons H^+ + CO_3^{2-}$$

$$CaCO_3 + H^+ \rightleftharpoons Ca + HCO_3^-$$

If abundant carbon dioxide is present in groundwater, then dissociation only proceeds as far as the bicarbonate stage. On the other hand if the pH value is increased, due to CO_2 being removed, then the ratio of carbonate (CO_3^{2-}) to bicarbonate (HCO_3^-) ions increases and calcium carbonate may be precipitated.

In water with a temperature of 25°C the solubility of calcium carbonate ranges from 0.01 to 0.05 g/l, depending upon the degree of saturation with carbon dioxide. Dolostone is somewhat less soluble than limestone. When a carbonate rock is subject to dissolution any insoluble material present in it remains behind.

Weathering of the silicate minerals is primarily a process of hydrolysis. Much of the silica which is released by weathering forms silicic acid but where it is liberated in large quantities some of it may form colloidal or amorphous silica. As noted above, mafic silicates usually decay more rapidly than felsic silicates and in the process they release magnesium, iron and lesser amounts of calcium and alkalis. Olivine is particularly unstable, decomposing to form serpentine, which on further weathering forms talc and carbonates. Chlorite is the commonest alteration product of augite (the principal pyroxene) and of hornblende (the principal amphibole).

When subjected to chemical weathering feldspars decompose to form clay minerals, the latter are consequently the most abundant residual products. The process is effected by the hydrolysing action of weakly carbonated waters which leach the bases out of the feldspars and produce clays in colloidal form. The alkalis are removed in solution as carbonates from orthoclase (K_2CO_3) and albite (Na_2CO_3) and as bicarbonate from anorthite ($Ca(HCO_3)_2$). Some silica is hydrolysed to form silicic acid. Although the exact mechanism of the process is not fully understood the equation given below is an approximation towards the truth:

$$\underset{\text{(orthoclase)}}{2KAlSi_3O_6} + 6H_2O + CO_2 = \underset{\text{(kaolinite)}}{Al_2Si_2O_5(OH)_4} + 4H_2SiO_4 + K_2CO_3$$

(a) (b) (c) (d)

Figure 2.2 (a) *Limestone pavement showing the development of fluted clints between solution-opened joint planes (grykes). Note bedding planes also have been opened by dissolution, Malham, North Yorkshire;* (b) *Pinnacles developed in dolostone east of Pretoria, South Africa (Courtesy of Dr Bryan Gregory);* (c) *Gaping Ghyll, Yorkshire, pothole which descends about 150 m in limestone of Carboniferous age;* (d) *Towers developed from massive limestone, south of Guilen, China*

The colloidal clay eventually crystallizes as an aggregate of minute clay minerals. Deposits of kaolin are formed when percolating acidified waters decompose the feldspars contained in granitic rocks.

Clays are hydrated aluminium silicates and when they are subjected to severe chemical weathering in humid tropical regimes they break down to form laterite or bauxite. The process involves the removal of siliceous material and this is again brought about by the action of carbonated waters. Intensive leaching of soluble mineral matter from surface rocks takes place during the wet season. During the subsequent dry season groundwater is drawn to the surface by capillary action and minerals are precipitated there as the water evaporates. The minerals generally consist of hydrated peroxides of iron, and sometimes of aluminium, and very occasionally of manganese. The precipitation of these insoluble hydroxides gives rise to an impermeable lateritic soil. When this point is reached the formation of laterite ceases as no further leaching can occur. As a consequence lateritic deposits are usually less than 7 m thick.

Plants and animals play an important role in the breakdown and decay of rocks, indeed their part in soil formation is of major significance. Tree roots penetrate cracks in rocks and gradually wedge the sides apart whilst the adventitious root system of grasses breaks down small rock fragments to particles of soil size. Burrowing rodents also bring about mechanical disintegration of rocks. The action of bacteria and fungi is largely responsible for the decay of dead organic matter. Other bacteria are responsible, for example, for the reduction of iron or sulphur compounds. It has also been suggested that bacterial action plays an important part in the formation of residual deposits such as laterites and bauxites.

2.4 Slaking and swelling of mudrocks

Physical disintegration of mudrocks is a much more important breakdown process than chemical weathering. The two principal controls on breakdown are slaking and the expansion of mixed-layer clay minerals. Indeed some freshly exposed mudrocks belonging to the Coal Measures can commence breakdown within a few days, weeks or months, depending on the character of the parent rock.

Slaking refers to the breakdown of rocks, especially mudrocks, by alternate wetting and drying. If a fragment of mudrock is allowed to dry out, air is drawn into the outer pores and high suction pressures develop. When the mudrock is next saturated the entrapped air is pressurized as water is drawn into the rock by capillary action (indurated mudrocks with small expandable clay mineral contents require suction pressures in excess of 10 kPa before air entry commences, whereas air entry occurs over the whole suction range in mudrocks with high expandable clay mineral contents). This slaking process causes the internal arrangement of grains to be stressed. Given enough cycles of drying and wetting, breakdown can occur as a result of air breakage, the process ultimately reducing the mudrock involved to tabular-shaped gravel-size particles.

In order to assess the durability of the mudrocks, samples are subjected to slake-durability testing (see Chapter 8). The various categories of the slake-durability index are shown in Table 8.1. However, Taylor (1988) suggested that durable mudrocks can be better distinguished from non-durable types on a basis of compressive strength and three-cycle slake-durability index (i.e. those mudrocks with a compressive strength of over 3.6 MPa and a three-cycle slake-durability index in excess of 60% are regarded as durable).

Intraparticle swelling (i.e. swelling due to the take-up of water not only between particles of clay minerals but also within them, into the weakly bonded layers between molecular units) of clay minerals on saturation can cause mudrocks to break down where the proportion of such minerals is significant, that is, constituting more than 50% of the clay minerals fraction. The expansive clay minerals such as montmorillonite can expand many times their original volume.

It was concluded by Taylor and Spears (1970) that intraparticle swelling of mixed-layer clay during periods of saturation followed by desiccation in the near surface zone was a major control on the breakdown of Coal Measures mudrocks which contained significant amounts of expandable mixed-layer clay. Furthermore the exchangable sodium ion was most abundant in those mudrocks which underwent the greatest breakdown. In fact the exchangeable sodium percentage (ESP = exchangeable Na/cation cation exchange capacity × 100%) has been used as a guide to mudrock behaviour in that it is related to the amount of dispersion in shales.

Sulphur compounds are frequently present in shales, mudstones and marls. An expansion in volume large enough to cause structural damage can occur when sulphide minerals such as pyrite and marcasite suffer oxidation to give anhydrous and hydrous sulphates. According to Fasiska *et al.* (1974) the pyrite structure may be regarded as a stacking of almost close packed hexagonal sheets of sulphide ions with iron ions in the interstices between the sulphide layers. The packing density is related to the radius of the sulphide ion which is 1.85 Å (volume = 26.1 $Å^3$). In the sulphate structure each atom of sulphur is surrounded by four atoms of oxygen in tetrahedral coordination. The packing density is related to the radius of the sulphate ion which is 2.8 Å, giving a volume of 92.4 Å. This represents an increase in volume per packing unit of approximately 350%. Hydration involves a further increase in volume. In fact such a reaction is electrolytic, that is, water is required and the sulphate ion exists in solution. Any cation in the system may cause the precipitation of sulphate crystals. If calcium carbonate is present, gypsum may be formed, which may give rise to an eightfold increase in volume over the original sulphide, exerting pressures of up to about 0.5 MPa. This leads to further disruption and weakening of the rocks involved.

Swelling can also occur as a result of hydration. For example anhydrite is hydrated to form gypsum there is a volume increase of between 30 and 58% which exerts pressures that have been variously estimated between 2 and 69 MPa. It is thought that no great length of time is required to bring about such hydration. When it occurs at shallow depths it causes expansion but the process is gradual and is usually accompanied by the removal of gypsum in solution. At greater depths anhydrite is effectively confined during the process. This results in a gradual build-up of pressure and the stress is finally liberated in an explosive manner. According to Brune (1965) such uplifts in the United States have taken place beneath reservoirs, these bodies of water providing a constant supply for the hydration process, percolation taking place via cracks and fissures. Examples are known where the ground surface has been elevated by about 6 m. The rapid explosive movement causes strata to fold, buckle and shear.

2.5 Engineering classification of weathering

The early stages of weathering are usually represented by discolouration of the rock material which increases from slightly to highly discoloured as the degree of weathering increases. Because weathering brings about changes in engineering properties – in particular it commonly leads to an increase in bulk (and so in porosity) with a corresponding reduction in density and strength – these changes are reflected in the amount of discolouration. In other words the engineering properties of a slightly discoloured rock may differ notably from those of the same rock which is highly discoloured (Dearman, 1986). As weathering proceeds the rock material becomes increasingly decomposed and/or disintegrated until ultimately a soil is formed. Hence various stages in the reduction process of a rock to a soil can be recognized and can be used to form the basis of an engineering classification of weathering. Between five and seven grades of weathering have been recognized in the various schemes which have been proposed, the identification of grades being based upon the presence or absence of discolouration in rock material, the rock to soil ratio and the presence or absence of relict rock fabric in the groups which are predominantly soil. It is almost inevitable that the boundaries between grades are gradational.

However, as was pointed out by Martin and Hencher (1986) weathering processes are rarely sufficiently uniform to give gradual and predictable changes in the engineering properties of a weathered profile. In fact such profiles usually consist of heterogeneous materials at various stages of decomposition and/or disintegration. Nonetheless an indication of the degree of weathering required to describe rock specimens or drillcore is provided by the alteration of individual minerals, the loss of bonding between grains and the development and growth of microfractures. Martin and Hencher suggested that such descriptions should be made in terms of material grades which are uniform and so can be defined within quite precise limits. At a larger scale, for example, when describing rock masses for mapping purposes, then it frequently is necessary to group grades of rock weathering into mass zones which, for engineering purposes, may be regarded as having distinctive characteristics. In addition zonal classifications of weathered rock masses may be of value in terms of general design. Even so, where highly complex ground conditions exist it may be impossible to apply a rigid zonal classification. In such conditions Martin and Hencher recommended that the weathered material should be carefully described and combined with detailed lithological and structural mapping.

Several attempts have been made to devise engineering classification of weathered rock and rock masses. Some schemes have involved quantification of the amount of mineralogical alteration and structural defects with the aid of the petrological microscope. Others have resorted to some combination of simple index tests to provide a quantifiable grade of weathering. Some of the earliest methods of assessing the degree of weathering were based on a description of the character of the rock mass concerned as seen in the field. Such descriptions recognized different grades of weathering and attempted to relate them to engineering permanance. However, grading based on description of the degree of weathering is subjective and accordingly such grading systems now are being coupled with assessments made by index tests to provide better precision and quantification.

2.5.1 Descriptive classifications

Moye (1955) was one of the first to propose a grading system for the degree of weathering as found in granite at the Snowy Mountains scheme in Australia. Similar classifications were advanced by Kiersch and Treacher (1955), Ruxton and Berry (1957), Little (1959), Knill and Jones (1965), Fookes and Horswill (1970), and Fookes *et al.* (1971). These classifications were mainly based on the degree of decomposition exhibited by a rock mass and were primarily directed towards weathering in granitic rocks. Subsequently Dearman (1974) suggested descriptions which could be used to establish the grade of mechanical weathering, and that of solution weathering of relatively pure carbonate rock (Table 2.3). Others, working on different rock types, have proposed modified classifications of weathering grade. For example, Lovegrove and Fookes (1972) made slight variations from the above in their identification of grades of weathering of volcanic tuffs and associated sediments in Fiji. Classifications of weathered chalk and weathered marl have been developed by Ward *et al.* (1968) and Chandler (1969) respectively (Table 2.4).

Nevertheless Dearman (1974) maintained that an ideal profile of weathering, which was irrespective of rock type, would be worthwhile in that each grade would provide an indication of the general engineering properties of the material concerned. Usually the grades will lie one above the other in a weathered profile developed from a single rock type, the highest grade being at the surface but this is not necessarily the case in complex geological conditions (Knill and Jones, 1965).

In a review of recent classifications of weathering advanced by the Geological Society (Anon 1977), the International Association of Engineering Geology (Anon, 1981a), the International Society for Rock Mechanics (Anon, 1981b) and the British Standard (BS 5390: 1981c), Martin and Hencher (1986) were critical of the way in which the term 'grade' was used both as a type rather than a scale of weathering of rock and to classify a zone of heterogeneous rock mass, which they argued led to confusion. Furthermore Martin and Hencher noticed a lack of definition or guidance for the description of rock material grades, in particular they referred to the inadequate definition of the terms 'rock' and 'soil' and pointed out that not all rocks are characterized by the presence of corestones when they weather. They also maintained that boundaries of mass weathering zones were on occasions arbitrarily drawn. Accordingly they proposed the following guidelines to provide a means of describing uniform grades of weathered rock material:

(1) Grade descriptions should apply to uniform materials.
(2) Index tests should be used whenever practical to define boundaries between grades more precisely.
(3) Wherever possible boundaries between grades should be established according to engineering relevance.
(4) Constant grade numbering and nomenclature should be used.
(5) A sixfold division of grades should be used in accordance with common practice.
(6) A single classification should be used whenever possible to cover all types (decomposition, disintegration) and degrees of material weathering.

Table 2.3 Engineering grade classification of weathered rock and their relation to engineering behaviour

Grade	*Degree of decomposition*	*Field recognition (after Little (1969); Fookes et al. (1972); Dearman (1974)*			*Engineering behaviour*		
		Rocks (mainly chemical decomposition)	*Rocks (physical disintegration)**	*Carbonate rocks (solution)*	*After Little (1969)*	*After Hobbs (1975)*	*After Hencher & Martin (1982)**
VI	Residual soil	The rock is discoloured and is completely changed to a soil in which the original fabric of the rock is completely destroyed. There is a large volume change	The rock is changed to a soil by granular disintegration and/ or grain fracture. The structure of the rock is destroyed and the soil is a residuum of minerals unaltered from the original rock		Unsuitable for important foundations. Unstable on slopes when vegetation cover is destroyed and may erode easily unless hard cap is present. Requires selection before use as fill	In completely weathered rock and residual soil it may be possible to obtain fair quality samples depending upon the parent rock type and the consistency of the product. Generally the samples will tend to be less disturbed than when taken in the same rock in the highly weathered state. The bearing capacity and settlement characteristics of rock in these extreme states can be assessed using the usual methods for testing soils	A soil mixture with the original texture of the rock completely destroyed
V	Extremely weathered	The rock is discoloured and is wholly decomposed and friable, but the original fabric is mainly preserved. The properties of the rock mass depend in part on the nature of the parent rock. In granite rocks feldspars are completely kaolinized	The rock is changed to a soil by granular disintegration and/ or fracture. The structure of the rock is preserved	Grades V and VI cannot occur. These grades can be applied to interbedded soluble and insoluble rocks. Void size should be recorded	Cannot be recovered as cores by ordinary rotary drilling methods. Can be excavated by hand or ripping without the use of explosives. Unsuitable for foundations of concrete dams or large structures. May be suitable for foundations of earth dams and for fill. Unstable in high cuttings at steep angles. New joint patterns may have formed. Requires erosion protection		No rebound from N. Schmidt hammer; slakes readily in water; geological pick easily indents when pushed into surface; rock is wholly decomposed but rock texture preserved

They further commented that the end terms 'fresh rock' and 'residual soil' should be adopted as standard and that intermediate grades should be described as slightly, moderately, highly and extremely decomposed or disintegrated according to the dominant type of weathering. Martin and Hencher went on to propose that zonal schemes should be developed and used for the description of weathered rock masses which could be used for mapping purposes and in general engineering design. Any such scheme should take account of the following:

(1) Zones must be recognizable in naturally occurring profiles.
(2) The complete range of expected materials must be accounted for.
(3) Boundaries must be defined so that they separate zones with significantly different engineering properties.

They then presented a general zonal scheme for the description of rock masses whose weathering profiles exhibit a heterogeneous mixture of rock materials (Figure 2.3).

2.5.2 Classifications based on index tests and petrographic techniques

As pointed out above, the mineralogical composition and texture, as well as the engineering properties of rock, alter as it undergoes progressive weathering. Hence simple index tests and petrographic techniques can be used, either separately or in combination, to assess the degree of weathering.

Hamrol in 1961 devised a quantitative classification of weatherability in which he first distinguished two weathering types. Type I weathering excluded cracking of any

Table 2.3 Continued

Grade	*Degree of decomposition*	*Field recognition (after Little (1969); Fookes et al. (1972); Dearman (1974)*			*Engineering behaviour*		
		Rocks (mainly chemical decomposition)	*Rocks (physical disintegration)**	*Carbonate rocks (solution)*	*After Little (1969)*	*After Hobbs (1975)*	*After Hencher & Martin (1982)**
IV	Highly weathered†	The rock is discoloured; discontinuities may be open and have discoloured surfaces (e.g. stained by limonite) and the original fabric of the rock near the discontinuities is altered; alteration penetrates deeply inwards, but corestones are still present. The rock mass is partially friable. Less than 50% rock	More than 50% and less than 100% of the rock is disintegrated by open discontinuities or spheroidal scaling spaced at 60 mm r less and/or by granular disintegration. The structure of the rock is preserved	More than 50% of the rock has been removed by solution. A small residuum may be present in the voids	Similar to Grade V. Sometimes recovered as core by careful rotary drilling. Unlikely to be suitable for foundations of concrete dams. Erratic presence of boulders makes it an unreliable foundation for large structures	In highly weathered rock difficulties will generally be encountered in obtaining undisturbed samples for testing. If samples are obtained the strength and modulus will generally be underestimated, frequently by large margins, even with apparently undisturbed samples. In such rocks *in situ* tests with either the Menard pressure meter or the plate should be carried out to determine the bearing capacity and settlement characteristics. The greatest difficulties in assessing bearing capacity and settlement are likely to be encountered in highly weathered rocks, in which the rock fabric becomes increasingly disintegrated or increasingly more plastic	N. Schmidt hammer rebound value 0 to 25; does not slake readily in water, geological pick cannot be pushed into surface; hand penetrometer strength index >250 kPa. Rock weakened so that large pieces broken by hand; individual grains plucked from surface

kind, whilst Type II weathering consisted entirely of cracking. This represents a division between chemical and physical weathering respectively, but such distinction can be extremely difficult to make. In Type I the void ratio increases as weathering progresses, which means that the saturation moisture content increases and the dry density decreases. These two parameters therefore were used as the basis of an index test, the numerical value (i_I) of which was expressed as the weight of water absorbed by an ovendried rock when it is saturated for a limited period, divided by its dry weight and expressed as a percentage. This is referred to as a quick absorption test. When considering Type II weathering, Hamrol distinguished between unfilled and filled cracks, the index being

$$i_{II} = (x + y + z) \times 100 \qquad (2.1)$$

where x, y and z were the dimensions of the crack along three orthogonal axes. Further indices could be obtained by relating the change in the degree of weathering (j) to a given time (Δt), hence

$$j_I = \Delta i_I/\Delta t \quad \text{and} \quad j_{II} = \Delta i_{II}/\Delta t \qquad (2.2)$$

Table 2.3 Continued

Grade	Degree of decomposition	Field recognition (after Little (1969); Fookes et al. (1972); Dearman (1974))			Engineering behaviour		
		Rocks (mainly chemical decomposition)	Rocks (physical disintegration)*	Carbonate rocks (solution)	After Little (1969)	After Hobbs (1975)	After Hencher & Martin (1982)*
III	Moderately weathered†	The rock is discoloured; discontinuities may be open and have greater discolouration with the alteration penetrating inwards; the intact rock is noticeably weaker, as determined in the field, than the fresh rock. The rock mass is not friable. 50–90% rock	Up to 50% of the rock is disintegrated by open discontinuities or by spheroidal scaling spaced at 60 mm or less and/ or by granular disintegration. The structure of the rock is preserved	Up to 50% of the rock has been removed by solution. A small residuum may be present in the voids. The structure of the rock is preserved	Possessing some strength – large pieces (e.g. NX drill core) cannot be broken by hand. Excavated with difficulty without the use of explosives. Mostly crushes under bulldozer tracks. Suitable for foundations of small concrete structures and rock fill dams. May be suitable for semipervious fill. Stability in cuttings depends on structural features especially joint attitudes	In moderately weathered rock the intact modulus and strength can be very much lower than in the fresh rock and thus the *j*-value will be higher than in the fresh state, unless the joints and fractures have been opened by erosion or softened by the accumulation of weathering products. The intact modulus and strength can be measured in the laboratory and the bearing capacity assessed, in the same way as for fresh rock. Triaxial tests may be more appropriate than uniaxial tests, and it would be advisable to adopt conservative values for the factor of safety	N. Schmidt rebound value 25 to 45; considerably weathered but possessing strength such that pieces 55 mm diameter cannot be broken by hand; rock material not friable
II	Slightly weathered	The rock may be slightly discoloured, particularly adjacent to discontinuities which may be open and have slightly discoloured surfaces; the intact rock is not noticeably weaker than the fresh rock. Some decomposed feldspar in granites. Over 90% rock	100% rock; discontinuities open and spaced at more than 60 mm	100% rock; discontinuity surfaces open. Very slight solution etching of discontinuity surfaces may be present	Requires explosives for excavation. Suitable for concrete dam foundations. Highly permeable through open joints. Often more permeable than the zones above or below. Questionable as concrete aggregate	In faintly and slightly weathered rock it is possible that the *j*-value, owing to the reduction in stiffness of the joints as a result of penetrative weathering alone, will show a fairly sharp decrease compared with that of the same rock in the fresh state. The intact modulus by definition, is unaffected by penetrative weathering. The safe bearing capacity is not therefore affected by faint weathering, and may be only slightly affected by slight weathering	N. Schmidt rebound value >45; more than one blow of geological hammer to break specimen; strength approaches that of fresh rock
I	Fresh rock	The parent rock shows no discoloration, loss of strength or other effects due to weathering	100% rock; discontinuities closed	100% rock; discontinuities closed	Staining indicates water percolation along joints; individual pieces may be loosened by blasting or stress relief and support may be required in tunnels and shafts		No visible signs of weathering; rarely encountered in surface exposures

* Discontinuity spacing should be recorded.
† The ratio of the original rock to altered material should be estimated where possible.

Table 2.4 Weathering classifications of chalk and marl

(a) Grades of weathering in the Chalk at Mundford (after Ward *et al.*, 1968)*

Grade	*Description*
V	Structureless melange. Unweathered and partly weathered angular chalk blocks and fragments set in a matrix of deeply weathered remoulded chalk. Bedding and jointing are absent
IV	Friable to rubbly chalk. Unweathered or partially weathered chalk with bedding and jointing present. Joints and small fractures closely spaced, ranging from 10–60 mm apart
III	Rubbly to blocky chalk. Unweathered medium to hard chalk with joints 60–200 mm apart. Joints open up to 8 mm, sometimes with secondary staining and fragmentary infillings
II	Medium hard chalk with widely spaced, closed joints. Joints more than 200 mm apart. Fractures irregularly when excavated, does not break along joints. Unweathered
I	Hard, brittle chalk with widely spaced, closed joints. Unweathered

* Ward *et al.* emphasized that their classification was specifically developed for the site at Mundford and hence its application elsewhere should be made with caution.

(b) Zones of weathering in Keuper Marl (Mercia Mudstone) (after Chandler, 1969)

Zone		*Description*
V	Fully weathered	Matrix only. Plastic, slightly silty clay. May be fissured
IV	Highly weathered	Matrix with occasional pellets <3 mm diameter. Little trace of original structure, although clay may be fissured. Permeability less than underlying layers
III	Moderately weathered	Matrix with frequent lithorelicts up to 25 mm. As weathering progresses, lithorelics become less angular. Water content of matrix greater than that of lithorelics
II	Slightly weathered	Angular blocks of unweathered marl. First indications of weathering; matrix starting to encroach along joints leading to 'spheroidal' weathering
I	Unweathered	Marl (often fissured). Moisture content varies due to different lithology

Weathering first develops along fissures in Zone II material, the weathered product consisting of a thin veneer of silt. A significant proportion of the marl is weathered in Zone III, the unweathered material occurring as angular fragments set in a weathered matrix which is predominantly silty. The water content of the matrix exceeds that of the lithorelicts. In Zone IV the lithorelicts are mainly of coarse sand size, and the marl has lost much of its silty texture, indeed up to 50% may be composed of clay-sized particles. This indicates that particle aggregation is broken down upon weathering. Little or no trace of the original structure now remains and the material has a lower permeability than has Zone III, 5×10^{-9} to 5×10^{-10} m/s as compared with 1×10^{-8} to 1×10^{-9} m/s respectively. Finally the marl is completely weathered, becoming a plastic, slightly silty, clay. Chandler (1969) showed that highly and fully weathered marls could be distinguished from material from Zones I, II and III by their particle size distribution and plasticity index.

Unfortunately Hamrol gave no scale to his indices so that their meaning in terms of engineering performance is lacking (he did mention that with an $i_I = 10$, a weathered granite would crumble in the fingers, but little else).

Onodera *et al.* (1974) also used the number and width of microcracks (cracks less than 1 mm in width which occur in the rock fabric) as an index of the physical weathering of granite. They found a linear relationship between effective porosity (n_e) and density of microcracks (determined with the aid of a petrological microscope), defined as

$$n_e = 100 \times (\text{total width of cracks/length of measured line}) \tag{2.3}$$

They also found that the mechanical strength of granite decreased rapidly as the density of microcracks increased from about 1.5 to 4%.

Lumb (1962) defined a quantitative index, *Xd*, related to the weight ratio of quartz and feldspar in decomposed granite from Hong Kong, as follows:

$$Xd = (N_q - N_{qo})/(1 - N_{qo}) \tag{2.4}$$

where N_q is the weight ratio of quartz and feldspar in the soil sample, and N_{qo} is the weight ratio of quartz and feldspar in the original rock. For fresh rock $Xd = 0$, whilst for completely decomposed rock $Xd = 1$.

Yet another index or coeffiecient of weathering (K) for granitic rock was developed by Iliev (1967). This coefficient was based upon the ultrasonic velocities of the rock material according to the expression

$$K = (V_u - V_w)/V_u \tag{2.5}$$

where V_u and V_w are the ultrasonic velocities of the fresh and weathered rock respectively. A quantitative index indicating the grade of weathering as determined from the ultrasonic velocity and the corresponding coefficient of weathering is as shown in Table 2.5.

After an extensive testing programme Irfan and Dearman (1978a) concluded that the quick absorption, Schmidt hammer, and point load strength tests prove reliable field tests for the determination of a quantitative weathering index for granite (Table 2.6). This index can be

Zone symbol	Zonal characteristics
6	• Soil derived from *in situ* weathering: 100% soil (grades IV, V or VI) • May or may not have lost rock mass features completely
5	• Soil with corestones: less than 30% rock (grades I, II or III) • Shearing can be effected through matrix • Rock content significant for investigation and construction
4	• Poor quality rock mass: 30% to 50% rock (grades I, II or III) • Corestones affect shear behaviour of mass
3	• Moderate quality rock mass: 50% to 90% rock (grades I, II or III) • Severe weathering along discontinuities • Locked structure
2	• Good quality rock mass: greater than 90% rock (grades I, II or III) • Weathering along discontinuities
1	• Excellent quality rock mass: 100% rock (grades I, II) • No visible signs of rock weathering apart from slight discolouration along joints • Joint surfaces strongly interlocking

Figure 2.3 *Proposed zonal scheme for the classification of heterogeneous weathered rock masses (after Martin and Hencher 1986). The zonal sequence is given in Arabic numerals to avoid confusion with grading systems. The scheme is based on varying proportions of rock and soil. Rock is generally defined as material grades I, II or III, except in zone 1 where all the rock is either grade I or II. Soil comprises grades IV, V and VI*

Table 2.5 Grades of weathering

Grade of weathering	*Ultrasonic velocity* (m/s)	*Coefficient of weathering*
Fresh	Over 5000	0
Slightly weathered	4000–5000	0–0.2
Moderately weathered	3000–4000	0.2–0.4
Strongly weathered	2000–3000	0.4–0.6
Very strongly weathered	Under 2000	0.6–1.0

related to the various grades of weathering recognized by visual determination and given in Table 2.3.

Irfan and Dearman (1978b) also developed a quantitative method of assessing the grade of weathering of granite in terms of its megascopic and microscopic petrography. The megascopic factors included an evaluation of the amount of discolouration, decomposition and disintegration shown by the rock. The microscopic analysis involved assessment of mineral composition and degree of alteration by modal analysis and a microfracture analysis. The latter involved counting the number of clean and stained microcracks and voids under the microscope in a 10 mm traverse across a thin section. The types of microfracture recognized included stained grain boundaries, open grain boundaries, stained microcracks in quartz and feldspar, infilled microcracks in quartz and feldspar, clean transgranular microcracks crossing the grains, filled or partially infilled microcracks, and pores in plagioclase. The data are used to derive the micropetrographic index, I_p, as follows:

$$I_p = \frac{\text{\% sound or primary minerals}}{\text{\% unsound constitutents}} \tag{2.6}$$

The unsound constituents are the secondary minerals together with microcracks and voids. Irfan and Dearman were able to identify five stages and three substages of weathering in granite (Table 2.7).

Olivier (1979) proposed a geodurability classification which was based on the free-swelling coefficient (see Section 8.4) and uniaxial compressive strength (Figure 2.4). This classification was developed primarily to assess the durability of mudrocks and poorly cemented sandstones during tunnelling operations, since the tendency of such rocks to deteriorate after exposure governs the

Table 2.6 Weathering indices for granite (after Irfan and Dearman (1978a)

Type of weathering	*Quick absorption* (1%)	*Bulk density* (Mg/m^3)	*Point load strength* (MPa)	*Unconfined compressive strength* (MPa)
Fresh	Less than 0.2	2.61	Over 10	Over 250
*Partially stained	0.2–1.0	2.56–2.61	6–10	150–250
*Completely stained	1.0–2.0	2.51–2.56	4–6	100–150
Moderately weathered	2.0–10.0	2.05–2.51	0.1–4	2.5–100
Highly/completely weathered	Over 10	Less than 2.05	Less than 0.1	Less than 2.5

* Slightly weathered

Table 2.7 Stages of weathering of rock material in terms of microscopical properties (after Irfan and Dearman, 1978b)

Stage 1: No penetration of brown iron-staining. Microcracks are very short, fine, intragranular and structural. The centres of plagioclases are clouded and slightly sericitized. Altered minerals <6%; microcrack intensity <0.5%; micropetrographic index >12.

Stage 2: Three substages are recognized depending on the amount of discolouration and type and amount of microfracturing.

(1) The rock is iron-stained only along the joint faces. No penetration of iron-staining.
(2) Penetration of iron-staining (brown) inwards from the joint faces along the microcracks. Formation of simple, branched microcracks; tight and partially stained. Slight alteration of the centres of plagioclases. Occasional staining along quartz–quartz and quartz–feldspar grain boundaries. Grain boundaries are sharp.
(3) More inward penetration of brown iron-staining along microcracks and partial staining of plagioclases. Microfracturing of feldspars and quartz by mainly intragranular, but some transgranular microcracks.

Unstained core: Altered minerals 6–9%; microcrack intensity 0.5–1.0%; micropetrographic index 9–12. Stained rims: Altered minerals 9–12%; microcrack intensity 1.0–2.0%; micropetrographic index 6–9.

Stage 3: Complete discolouration of rock by deep brown iron-staining. Partial alteration of plagioclases to sericite and gibbsite (?). Formation of single pores in plagioclases due to leaching. Potash feldspars are unaltered. Slight loss of pleochroism and bleaching of biotite. Grain boundaries are tight but stained brown by iron-oxide. The rock fabric is highly microfractured by complex branched, transgranular microcracks. Altered minerals 12–15%; microcrack intensity 2.0–5.0%; micropetrographic index 4–6.

Stage 4: Nearly complete alteration of plagioclase to sericite and gibbsite and formation of nearly opaque areas in plagioclases. Very slight alteration of potash feldspar. Interconnected pores are formed in plagioclase feldspars due to removal and leaching of alteration products. Some solution of silica forming diffused quartz grain boundaries. Intense microfracturing of the rock fabric by a complex branched and dendritic pattern of microcracks. The whole of the rock is iron-stained. Altered minerals 15–20%; microcrack intensity 5.0–10.0%; micropetrographic index 2–4.

Stage 5: Complete alteration of plagioclases. Potash feldspar is partially altered, but highly microfractured. Biotite is partially and muscovite slightly altered; expansion of biotite. Quartz is reduced in grain size and amount by microfracturing and solution. Almost all the grain boundaries are open. The fabric is intensely microfractured by a dendritic pattern of micro- and macrocracks. Parallel sided, partially filled or clean macrocracks are formed. Highly bleached, highly porous. Rock texture is intact. Altered minerals 20%; microcrack intensity 10%; micropetrographic index 2.

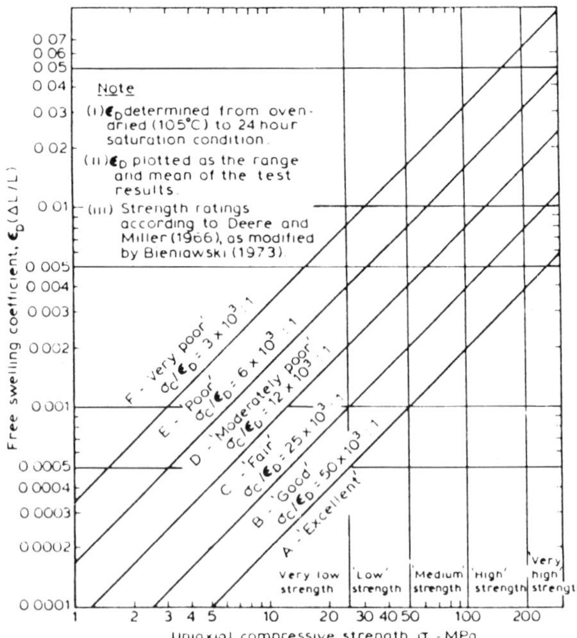

Figure 2.4 *Geodurability classification of the intact rock material (after Olivier 1979)*

stand-up time of tunnels. Olivier suggested that the classification could be used in decisions relating to primary tunnel support, particularly where and when shotcrete should be applied.

More recently Fookes *et al.* (1988) proposed two types of indicators based upon simple engineering tests in order to provide a tentative assessment of in-service durability of rock. The indicators are applicable to either static or dynamic engineering environments and may be used in the assessment of the potential durability at any stage in the life of a structure. The static rock durability indicator (RDI_s) combines four tests as follows:

$$RDI_s = \frac{I_{s(50)} - 0.1\,(\mathrm{SST} + 5\mathrm{WA})}{\mathrm{SG_{ssd}}} \tag{2.7}$$

where $I_{s(50)}$ is the average dry and saturated point load strength; SST is the magnesium sulphate soundness test; *WA* is water absorption; and $\mathrm{SG_{ssd}}$ is the specific gravity surface dried and saturated. The water absorption test result is multipled by 5 to bring the magnitudes of the variables into equivalent terms and to emphasize its importance in assessing the durability of rock. The point load strength is used to give an assessment of the static strength of the material and the magnesium sulphate test provides some indication of the ability of the rock to resist cyclic weathering processes. An estimate of the potential durability of rock based upon the static durability indicator is given in Table 2.8.

When a rock is subject to degradation as a consequence of dynamic loading the dynamic rock durability indicator is used. This includes a modified aggregate impact value (M.AW) which considers petrographic properties as well as the effect of repeated loading in a saturated condition. The indicator is as follows:

$$RDI_d = \frac{0.1(M.AIV + 5(WA))}{SG_{ssd}} \qquad (2.8)$$

The gradings for the dynamic rock durability indicator are given in Table 2.8.

Table 2.8 Potential durability of rock

Potential durability	RDI_s	RDI_d
Excellent	Above 2.5	Less than 0.5
Good	2.5 to −1	0.5 to 2.0
Fair	−1 to −3	2.0 to 4.0
Poor	Less than −3	Above 4.0

Discontinuities

A discontinuity represents a plane of weakness within a rock mass across which the rock material is structurally discontinuous. Although discontinuities are not necessarily planes of separation, most in fact are and they possess little or no tensile strength. Discontinuities vary in size from small fissures on the one hand to huge faults on the other. The most common discontinuities are joints and bedding planes (Figure 2.5). Other important discontinuities are planes of cleavage and schistosity, fissures and faults.

Figure 2.5 *Two sets of prominent joints developed more or less at right angles in Ancaster Freestone, near Ancaster, Lincolnshire*

2.6 Nomenclature of joints

Joints are fractures along which little or no displacement has occurred and they are present within all types of rock. At the surface, joints may open as a consequence of stress release and denudation, espcially weathering.

A group of joints which run parallel to each other are termed a *joint set* whilst two or more joint sets which intersect at a more or less constant angle are referred to as a *joint system*. A conjugate joint system describes two sets of joints which intersect symmetrically about some other structural plane or line. In a complementary joint system two sets of shear joints of the same age, and initiated by the same stress field, intersect at an angle of about 60°. All complementary joint systems are conjugate but the converse is not true.

If joints are planar and parallel or sub-parallel they are described as systematic, conversely, when they are irregular they are termed nonsystematic. If one set of joints is dominant' then the joints are known as primary joints, the other set or sets of joints being termed secondary. Joints commonly ocur in narrow zones where each joint is replaced *en echelon* by another. Cross joints, which are irregular in shape, may cut the rock within these zones. Hodgson (1961) contended that cross joints do not intersect systematic joints or well-developed bedding planes.

The attitude of joints may be referred to that of the enclosing bedding planes when no predominant trend is discernible in the region. Consequently strike joints are parallel to the strike of the bedding planes and dip joints run in the direction of the dip of the bedding. It has been suggested that joints can be classified in the same way as faults, that is, according to the direction in which any movement has taken place (if it can be discerned) along the joint plane. Joints can be grouped as normal dip-slip joints, reverse dip-slip joints and strike-slip joints. In some cases, because of the heavy disruption of strata, joints can only be compared with these groupings when the strata has been restored to horizontal. In such situations these joints are therefore referred to as normal dip-slip equivalent joints, reverse dip-slip equivalent joints and strike-slip equivalent joints. According to Hancock (1968) normal dip-slip and normal dip-slip equivalent joints are rare in all types of structural settings. Reverse dip-slip and reverse dip-slip equivalent joints are locally developed in regions of intense folding and reversed folding. Strike-slip and strike-slip equivalent joints are common in all structural settings.

On a basis of size, joints can be divided into master joints which penetrate several rock horizons and persist for hundreds of metres; major joints which are smaller joints but which are still well-defined structures and minor joints which do not transcend bedding planes. Lastly, minute fractures occasionally occur in finely bedded sediments and such micro-joints may be only a few millimetres in size. Master joints are not usually found in the folds of thin interbedded competent and incompetent rocks. In such instances major and minor joints are generally restricted to the competent rocks.

Joints may be associated with folds and faults, having developed towards the end of an active tectonic phase or when such a phase has subsided. However, joints do not appear to form parallel to other planes of shear failure such as normal and thrust faults. The orientation of joint sets in relation to folds depends upon their size, the type and size of the fold, and the thickness and competence of the rocks involved. At times the orientation of the joint sets can be directly related to the folding and may be defined in terms of the *a*, *b* and *c* axes of the 'tectonic cross' (Figure 2.6). Those joints which cut the fold at right angles to the axis are called *ac* or cross joints. The *bc* or longitudinal joints are perpendicular to the latter joints and diagonal or oblique joints make an angle with both the *ac* and the *bc* joints. Diagonal joints are classified as shear joints whereas *ac* and *bc* joints are regarded as tension joints.

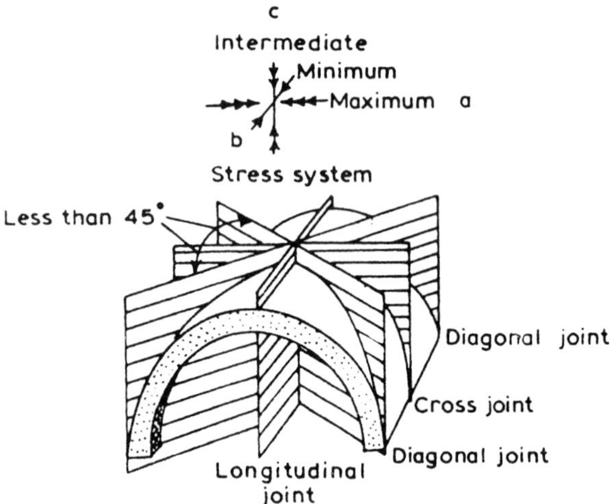

Figure 2.6 *Geometric orientation of longitudinal, cross and diagonal joints relative to fold axis and to principal stress axes (from Willis and Willis 1934)*

2.7 Origins of joints

Joints are formed through failure in tension, in shear, or through some combination of both. Rupture surfaces formed by extension tend to be clean and rough with little detritus. They tend to follow minor lithological variations. If such surfaces are sheared, the resulting load deformation curve displays a peak, rising well above the residual strength which is only reached at large values of displacement. Simple surfaces of shearing are generally smooth and contain considerable detritus. They are unaffected by local lithological changes. Shearing along this type of discontinuity does not yield as great a contrast between peak and residual strength as does a tension joint. The material along shear joints is commonly much more susceptible to alteration than that along tension joints.

Price (1966) contended that the majority of joints are post-compressional structures, formed as a result of the dissipation of residual stress after folding has occurred. Some spatially restricted small joints associated with folds, such as radial tension joints, are probably initiated during folding. The shearing stresses at the time of joint formation probably measure a few hundred megapascals. Such stresses usually can be dissipated by a movement amounting to a millimetre or so along a shear plane. But such movement can only dissipate the residual stresses in the immediate neighbourhood of a joint plane so that a very large number of joints need to form in order to dissipate the stresses throughout a large area. Hobbs (1968) proposed that joint frequency was proportional to the inverse of the bed thickness, the inverse of the square root of Young's modulus of the bed and the square root of the shear modulus of the surrounding beds. Formerly Price (1966) had suggested that joint frequency was related to the lithology of the rock type, the dimensions of the rock unit and the degree of tectonic deformation. On the last point he quoted the work of Harris *et al.* (1960) who found that the highest joint frequencies were associated with areas where the structures exhibited maximum curvature.

Price (1959) suggested that rocks could retain residual strain energy and that the residual stresses associated with the residual strain energy were modified during uplift in such a way as to give rise to tension joints or, in some cases, to both tension and shear joints. During the principal tectonic phase the resistance to the deformation of the rocks grows until it is in equilibrium with the maximum available tectonic stresses. When these conditions are attained there are no further changes in strain and deformation. It was assumed that after tectonic deformation there was usually a phase when the rocks were uplifted and surface material was removed by erosion. As a consequence gravitational loading and lateral stresses were reduced. Moreover when rocks are uplifted they are subjected to tensile stress and Price estimated that the latter would be about 70 MPa if they were elevated some 6000 m. He further argued that since the gravitational load increased by approximately 7 kPa for every 0.3 m increase in depth of overburden, then in an uplift of 6000 m the gravitational load would decrease by about 140 MPa. As a result the tensile stresses which develop when rocks are elevated is equal to approximately half the change in the gravitational load. As uplift continues the tensile strain increases until one of the tensile stresses reaches the tensile strength of the rock concerned and a joint then forms across the bed at its weakest point. In discontinuous strata the initiation of a joint in one bed does not necessarily lead to rupture in the surrounding beds but it does give rise to an increase in stress in the latter beds near the boundaries of the joint. Because of the differences in the elastic properties between the jointed bed and the neighbouring rocks the elastic displacements differ. Shear stresses are therefore produced on planes parallel to the tensile strain. Across the face of the joint the tensile stress is zero.

With the initiation of shear joints a certain amount of residual stress is liberated and as they develop the maximum principal stress gradually decreases, whilst the least principal stress increases. Eventually the vertical load due to gravity may assume the role of principal maximum stress. If the rocks undergo continuous uplift, then they become subjected to tension and tension joints are formed. Shear joints are usually markedly planar fractures which are not affected by local changes in lithology. On the other hand tension joints sometimes present irregular surfaces which may follow the outline of and are deflected by minor changes in lithology such as pebbles in conglomerates.

In horizontal beds which have suffered little tectonic compression two sets of tension joints may be developed, whereas in those rocks which have been subjected to considerable tectonic compression but have remained unfolded, two sets of shear joints may be formed. If uplift follows compression then two sets of tension joints may be developed subsequent to the shear joints.

Joints are also formed in other ways. For example, joints develop within igneous rocks when they initially cool down, and in wet sediments when they dry out. The most familiar of these are the columnar joints in lava flows, sills and some dykes. Cross joints, longitudinal joints, diagonal joints and flat-lying joints are associated with large granitic intrusions. Balk (1938) suggested that most joints in batholiths were caused by the continuing activity of the stresses which were responsible for emplacement and that these features possessed a definite relationship with the shape of the intrusion. Fractures are first developed in the solidified margins of plutonic masses and may be filled with material from the still liquid interior. Balk distinguished four types of joints (Figure 2.7). Cross joints or Q joints lie at right angles to the flow lines. They tend to radiate from the centre of the dome or arch (the term *dome* is used to describe a massif if the flow structures extend over the entire massif, whereas if they are absent from its centre the arrangement is termed an *arch*). Joints which strike parallel to the flow lines and are steeply dipping are known as longitudinal or S joints. Pegmatites or aplites may be injected along both types of joints mentioned. Diagonal joints are orientated at 45° to the direction of the flow lines. Flat-lying joints may be developed during or after emplacement of the intrusion and they may be distinguished as primary and secondary, respectively.

Sheet or mural joints have a similar orientation to flat-lying joints. When they are closely spaced and well developed they impart a pseudo-stratification to the host rock. It has been noted that the frequency of sheet jointing is related to the depth of overburden, in other words, the thinner the rock cover the more pronounced the sheeting. This suggests a connection between removal of overburden by denudation and the development of sheeting. Indeed such joints have often developed suddenly during quarrying operations. It may well be that some granitic intrusions contain considerable residual strain energy and that with the gradual removal of load the associated residual stresses are dissipated by the formation of sheet joints.

Chapman (1958) maintained that other sets of joints developed after the formation of sheet joints and that they were contained between the sheeting surfaces. He showed that the orientation and degree of development of these joints was related to topography. For instance, in Arcadia National Park, Maine, Chapman distinguished six sets of joints in roadside cuttings, yet only a short distance away the number of joint sets was reduced. Observations suggested that the best developed joints formed a pattern where two sets were mutually perpendicular, to which another two sets were diagonally arranged. It seemed that the two perpendicular sets were related to the direction of slope and contour of the topography and Chapman suggested that they may be formed as a result of gravity sliding of rock sheets. The diagonal joints, it was

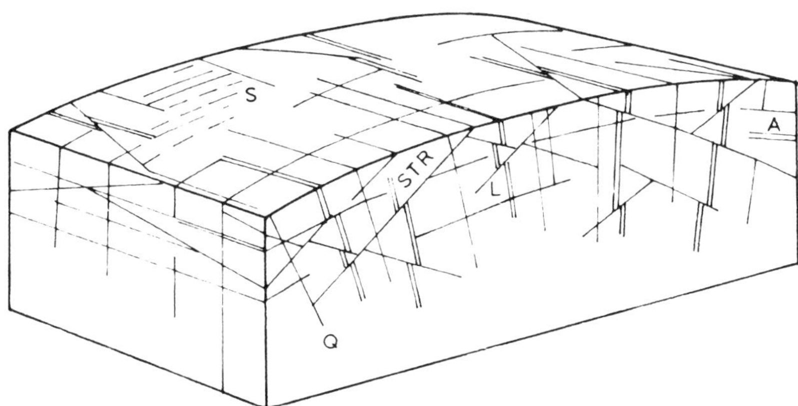

Figure 2.7 *Block diagram showing the types structures in a batholith. Q = cross joints; S = longitudinal joints; L = flat-lying joints; STR = planes of stretching; F = linear flow structures; A = aplite dykes (after Balk 1938)*

tentatively suggested, may represent the opening of incipient joint sets which were already formed.

2.8 Description of jointed rock masses

2.8.1 Incidence of discontinuities

The shear strength of a rock mass and its deformability are very much influenced by the discontinuity pattern, its geometry and how well it is developed. Observation of discontinuity spacing, whether in a field exposure or in a core stick, aids appraisal of rock mass structure. In sedimentary rocks bedding planes are usually the dominant discontinuity and the rock mass can be described as shown in Table 2.9. The same boundaries can be used to describe the spacing of joints (Anon 1977).

Table 2.9 Description of bedding plane and joint spacing

Description of bedding plane spacing	*Description of joint spacing (cf. Barton (1978))*	*Limits of spacing*	*Mass factor (j)*
Very thickly bedded	Extremely wide	Over 2 m	0.8–1.0
Thickly bedded	Very wide	0.6–2 m	0.5–0.8
Medium bedded	Wide	0.2–0.6 m	0.2–0.5
Thinly bedded	Moderately wide	60 mm–0.2 m	0.1–0.2
Very thinly bedded	Moderately narrow	20–60 mm	Less than 0.1
Laminated	Narrow	6–20 mm	
Thinly laminated	Very narrow	Under 6 mm	

The mechanical behaviour of a rock mass is strongly influenced by the number of sets of discontinuities which intersect, since this governs the amount of deformation that the rock mass will undergo. The number of sets also affects the degree of overbreak which occurs on excavation, they therefore may be an important factor in rock slope stability (Figure 2.8).

Systematic sets should be distinguished from non-systematic sets when recording the discontinuities in the field. Barton (1978) suggested that the number of sets of discontinuities at any particular location could be described in the following manner:

(1) massive, occasional random joints;
(2) one discontinuity set;
(3) one discontinuity set plus random;
(4) two discontinuity sets;
(5) two disconinuity sets plus random;
(6) three discontinuity sets;
(7) three discontinuity sets plus random;
(8) four or more discontinuity sets;
(9) crushed rock, earth-like.

2.8.2 Geometry of discontinuities

As joints represent surfaces of weakness, the larger and more closely spaced they are, the more influential they become in reducing the effective strength of the rock mass. The persistence of a joint plane refers to its continuity. This is one of the most difficult properties to quantify since joints frequently continue beyond the rock exposure and consequently in such instances it is impossible to estimate their continuity. Nevertheless Barton (1978) suggested that the modal trace lengths measured for each discontinuity set can be described as in Table 2.10. Simple sketches and block diagrams help to indicate the relative persistence of the various sets of discontinuities.

Table 2.10 Persistence and model trace lengths

Very low persistence	Less than 1 m
Low persistence	1–3 m
Medium persistence	3–10 m
High persistence	10–20 m
Very high persistence	Greater than 20 m

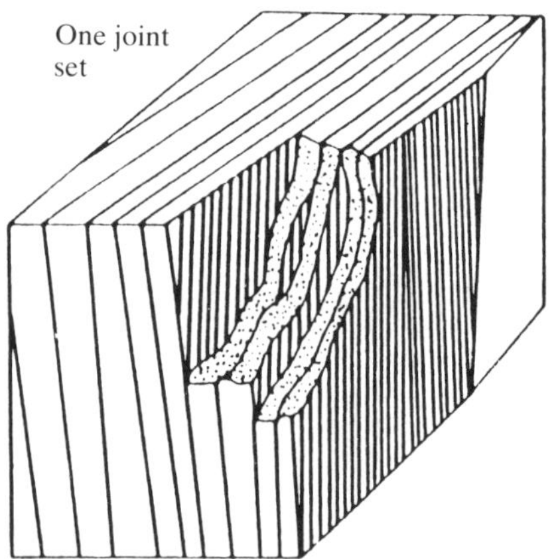

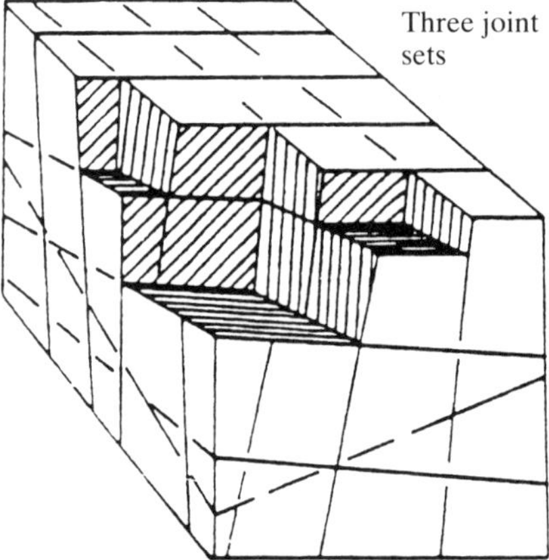

Figure 2.8 *Effect of the number of joint sets on the mechanical behaviour and appearance of a rock mass (after Barton 1978)*

Block size provides an indication of how a rock mass is likely to behave, since block size and interblock shear strength determine the mechanical performance of a rock mass under given conditions of stress. The following descriptive terms have been recommended for the description of rock masses in order to convey an impression of the shape and size of blocks of rock material (Barton, 1978).

(1) massive – few joints or very wide spacing;
(2) blocky – approximately equidimensional (Figure 2.9);
(3) tabular – one dimension considerably shorter than the other two;
(4) columnar – one dimension considerably larger than the other two;
(5) irregular – wide variations of block size and shape;
(6) crushed – heavily jointed to 'sugar cube'.

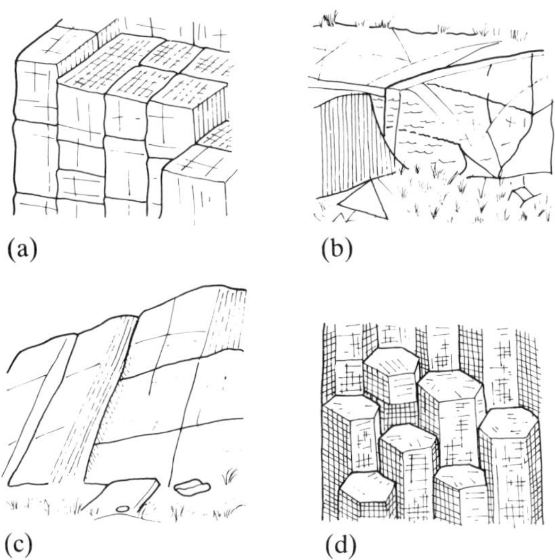

Figure 2.9 *Sketches of rock masses illustrating* (a) *blocky,* (b) *irregular,* (c) *tabular,* (d) *columnar block shapes*

The orientation of the short or long dimensions should be specified in the columnar and tabular blocks respectively. In addition it may be useful to note the ratio of the orthogonal dimensions, for example, 1 vertical: 2 north: 6 east. The block size may be described by using the terms given in Table 2.11 (Anon 1977).

Discontinuities, especially joints, may be open or closed. How open they are (Table 2.12) is of importance in relation to the overall strength and permeability of a rock mass and this often depends largely on the amount of weathering which the rocks have suffered. Furthermore, where the effects of weathering have penetrated deeply into a joint, a wide weak zone may be present. Some joints may be partially or completely filled. The type and amount of filling not only influence the effectiveness with which the opposing joint surfaces are bound together, thereby affecting the strength of the rock mass, but also influence permeability. If the infilling is sufficiently thick – for example, over 100 mm – the walls of the joint will not be in contact and hence the strength of the joint plane will be that of the infill material. Materials such as clay or sand may have been introduced into a joint opening. Mineralization is frequently associated with joints. This may effectively cement a joint, however, in other cases the mineralizing agent may have altered and weakened the rocks along the joint conduit.

Table 2.11 Block size

Term	*Block size*	*Equivalent discontinuity spacings in blocky rock*	*Volumetric joint count* (J_v)* (joints/m^3)
Very large	Over 8 m^3	Extremely wide	Less than 1
Large	0.2–8 m^3	Very wide	1–3
Medium	0.008–0.2 m^3	Wide	3–10
Small	0.0002–0.0008 m^3	Moderately wide	10–30
Very small	Less than 0.0002 m^3	Less than moderately wide	Over 30

* After Barton (1978)

Table 2.12 Description of the aperture of discontinuity surfaces

Anon (1977)			*Barton (1978)*	
Description	*Width of aperture*		*Description*	*Width of aperture*
Tight	Zero		Very tight	Less than 0.1 mm
Extremely narrow	Less than 2 mm	Closed	Tight	0.1–0.25 mm
Very narrow	2–6 mm		Partly open	0.25–0.5 mm
Narrow	6–20 mm		Open	0.5–2.5 mm
Moderately narrow	20–60 mm	Gapped	Moderately wide	2.5–10 mm
Moderately wide	60–200 mm		Wide	Over 10 mm
Wide	Over 200 mm		Very wide	10–100 mm
		Open	Extremely wide	100–1000 mm
			Cavernous	Over 1 m

Infill occupying discontinuities may possess a wide range of physical properties, especially with regard to its shear strength, deformability and permeability. Its short-term and long-term behaviour may differ appreciably. The range of behaviour is influenced by the mineralogy of the infill, its particle size distribution, its water content and permeability, its over-consolidation ratio, any previous shear displacement, width of aperture, and roughness and state of wall rock. The infill may be assessed by using the same method(s) as used to assess the wall rock (see below).

2.8.3 Surfaces of discontinuities

The nature of the opposing joint surfaces also influences rock mass behaviour, as the smoother they are, the more easily can movement take place along them. However, joint surfaces are usually rough and may be slickensided. Hence the nature of a joint surface may be considered in relation to its waviness, roughness and the condition of the walls. Waviness and roughness differ in terms of scale and their effect on the shear strength of the joint. Waviness refers to first-order asperities which appear as undulations of the joint surface and are not likely to shear off during movement. Therefore the effects of waviness do not change with displacements along the joint surface. Waviness modifies the apparent angle of dip but not the frictional properties of the discontinuity. On the other hand, roughness refers to second-order asperities which are sufficiently small to be sheared off during movement. Increased roughness of the discontinuity walls results in an increased effective friction angle along the joint surface. These effects diminish or disappear when infill material is present. The procedure for measuring joint roughness in the field has been given by Barton (1978).

The visual classification of roughness shown in Table 2.13 can be used when quantitative measurements are not made (Anon 1977).

Table 2.13 Visual classification of roughness

Category	*Degree of roughness*
1	Polished
2	Slickensided
3	Smooth
4	Rough
5	Defined ridges
6	Small steps
7	Very rough

This classification only has meaning when the direction of the irregularities on the surface is in the least favourable direction to resist sliding. As a consequence it is necessary to record the trend of the lineation on the joint surface in relation to the direction of shearing. Uniformity of assessment may be obtianed by identifying and photographing each category at the site in question.

An alternative set of descriptive terms has been suggested by Barton (1978) and these should be based upon two scales of observation, namely, small scale (several centimetres) and intermediate scale (several metres). The intermediate scale of roughness is divided into stepped, undulating and planar, and the small scale of roughness, superimposed upon the former, includes rough (or irregular), smooth and slickensided categories. The direction of the slickensides should be noted as shear strength may vary with direction. Barton recognized the following classes (Figure 2.10):

(1) rough (or irregular), stepped;
(2) smooth, stepped;
(3) slickensided, stepped;
(4) rough (irregular), undulating;
(5) smooth, undulating;
(6) slickensided, undulating;
(7) rough (irregular) planar;
(8) smooth, planar;
(9) slickensided, planar;

↑ Increasing shear strength

The compressive strength of the rock comprising the walls of a discontinuity is a very important component of shear strength and deformability, especially if the walls are in direct rock to rock contact, as in the case of unfilled joints. Weathering (and alteration) frequently is concentrated along the walls of discontinuities, thereby reducing their strength. The weathered material can be assessed in terms of its grade (Table 2.3) and manual index tests (Table 3.2). Alternatively, a Schmidt hammer can be used to obtain an idea of the compressive strength of the material concerned (Barton 1978). Samples of wall rock can be tested in the laboratory, not just for strength, but if they are highly weathered, also for swelling and durability.

2.8.4 Flow of water and discontinuities

Seepage of water through rock masses usually takes place via the discontinuities, although in some sedimentary rocks seepage through the pores may also play an important role. The prediction of groundwater levels, probable seepage paths and approximate water pressures frequently provides an indication of stability or construction problems. Barton (1978) suggested that seepage from open or filled discontinuities could be assessed according to the descriptive scheme shown in Table 2.14.

2.9 Discontinuities and rock quality indices

Several attempts have been made to relate the numerical intensity of fractures to the quality of unweathered rock

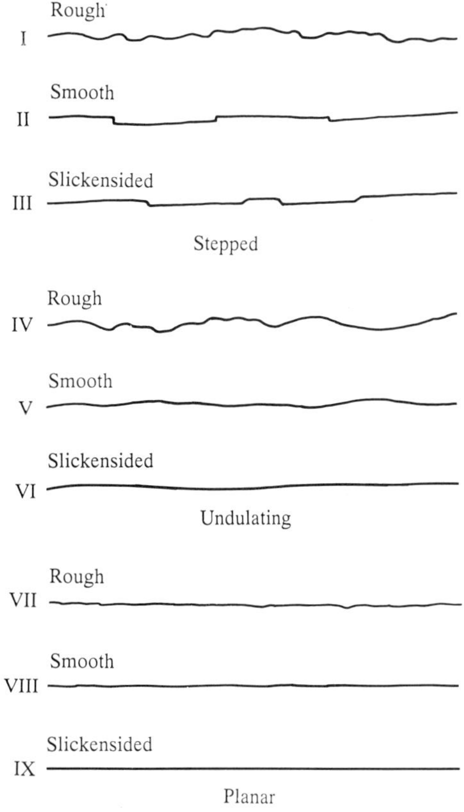

Figure 2.10 *Typical roughness profiles and suggested nomenclature. The length of each profile is in the range 1–10 m. The vertical and horizontal scales are equal (after Barton 1978)*

masses and to quantify their effect on deformability. For example, the concept of rock quality designation (RQD) was introduced by Deere (1964). It is based on the percentage core recovery when drilling rock with NX (57.2 mm) or larger diameter diamond core drills. Assuming that a consistent standard of drilling can be maintained, the percentage of solid core obtained depends on the strength and degree of discontinuities in the rock mass concerned. The RQD is the sum of the core sticks in excess of 100 mm expressed as a percentage of the total length of core drilled. However, the RQD does not take account of the joint opening and condition, a further disadvantage being that with fracture spacings greater than 100 mm the quality is excellent irrespective of the actual spacing (Table 2.15). This particular difficulty can be overcome by using the fracture spacing index as suggested by Franklin *et al.* (1971). This simply refers to the frequency with which fractures occur within a rock mass (Table 2.15).

The concept of fissuration factor was introduced by Hansagi (1974). Like the RQD it refers to the fragmentation of rock cores. However, it is based not only on the total length of the intact fragments and the average length of total core recovery, but also on the number of cylindrical pieces obtained (the lower limit of which is linked with the core diameter and the lengths of core required for strength determination). Furthermore, it is not the total length of all the core fragments obtained which is included in the calculation but a fixed length of the drill hole. Hence the core loss occurring during the drilling process is also taken into account. The fissuration factor (C) is derived from

$$C = 1/2S(pH + Kn) \qquad (2.9)$$

Table 2.14 Seepage from discontinuities

	A. Open discontinuities	*B. Filled discontinuities*
Seepage rating	*Description*	*Description*
(1)	The discontinuity is very tight and dry, water flow along it does not appear possible.	The filling material is heavily consolidated and dry, significant flow appears unlikely due to very low permeability.
(2)	The discontinuity is dry with no evidence of water flow.	The filling materials are damp but no free water is present.
(3)	The discontinuity is dry but shows evidence of water flow, i.e. rust staining, etc.	The filling materials are wet, occasional drops of water.
(4)	The discontinuity is damp but no free water is present.	The filling materials show signs of outwash, continuous flow of water (estimate l/min).
(5)	The discontinuity shows seepage, occasional drops of water but no continuous flow.	The filling materials are washed out locally, considerable water flow along outwash channels (estimate l/min and describe pressure, i.e. low, medium, high).
(6)	The discontinuity shows a continuous flow of water (estimate l/min and describe pressure, i.e. low, medium, high).	The filling materials are washed out completely, very high water pressures are experienced, especially on first exposure (estimate l/min and describe pressure).

Table 2.15 Classification of rock quality in relation to the incidence of discontinuities

Quality classification	*RQD (%)*	*Fracture frequency per metre*	*C factor*	*Mass factor (j)*	*Velocity ratio (V_{cf}/V_{cl})*
Very poor	0–25	Over 15	0.00–0.15		0.0–0.2
Poor	25–50	15–8	0.15–0.30	Less than 0.2	0.2–0.4
Fair	50–75	8–5	0.30–0.45	0.2–0.5	0.4–0.6
Good	75–90	5–1	0.45–0.65	0.5–0.8	0.6–0.8
Excellent	90–100	Less than 1	0.65–1.00	0.8–1.0	0.8–1.0

where S = one investigated until length of drill hole (this is dependent on the diameter of the core and the rock strength), p = the number of cylindrical samples which can be obtained from cores corresponding to length S, H = the height of the cylindrical sample used for compression testing, K = the total length of the core fragments with cylindrical lengths greater than the core diameter, n = the number of these core fragments.

The purpose of determining the C factor is to record the variations in fissuration and strength along the drill hole at each investigation unit at a maximum distance of 1 m. Hansagi maintained that the C factor was more sensitive to changes in the quality of rock than the RQD because the lower limit for the determination of the latter was 100 mm.

An estimate of the numerical value of the deformation modulus of a jointed rock mass can be obtained from various *in situ* tests. The values derived from such tests are always smaller than those determined in the laboratory from intact core specimens and the more heavily the rock mass is jointed, the larger the discrepancy between the two values. Thus, if the ratio between these two values of deformation modulus is obtained from a number of locations on a site, the engineer can evaluate the rock mass quality. In this context the concept of the rock mass factor (j) has been introduced by Hobbs (1975). He defined the rock mass factor as the ratio of the deformability of a rock mass within any readily identifiable lithological and structural component to that of the deformability of the intact rock comprising the component. Consequently it reflects the effect of discontinuities on the expected performance of the intact rock (Table 2.9). The value of j depends upon the method of assessing the deformability of the rock mass, and the value beneath an actual foundation will not necessarily be the same as that determined even from a large-scale field test. According to Hobbs, the greatest difficulties which occur in a jointed rock mass in relation to foundation design are experienced when the fracture spacing falls within a range of about 100–500 mm, in as much as small variations in fracture spacing and condition result in exceptionally large changes in j-value.

The effect of discontinuities in a rock mass can be estimated by comparing the *in-situ* compressional wave velocity with the laboratory sonic velocity of an intact core sample obtained from the rock mass. The difference in these two velocities is caused by the structural discontinuities which exist in the field. The velocity ratio, V_{cf}/V_{cl}, where V_{cf} and V_{cl} are the compressional wave velocities of the rock mass *in situ* and of the intact specimen respectively, was first proposed by Ondodera (1963). For a high-quality massive rock with only a few tight joints, the velocity ratio approaches unity. As the degree of jointing and fracturing becomes more severe, the velocity ratio is reduced (Table 2.14). The sonic velocity is determined for the core sample in the laboratory under an axial stress equal to the computed overburden stress at the depth from which the rock material was taken, and at a mositure content equivalent to that assumed for the *in situ* rock. The field seismic velocity preferably is determined by uphole or crosshole seismic measurements in drill holes or test adits, since by using these measurements it is possible to explore individual homogeneous zones more precisely than by surface refraction surveys.

2.10 Recording discontinuity data

2.10.1 Direct discontinuity surveys

Before a discontinuity survey commences the area in question must be mapped geologically to determine rock types and delineate major structures. It is only after becoming familiar with the geology that the most efficient and accurate way of conducting a discontinuity survey can be devised. A comprehensive review of the procedure to be followed in a discontinuity survey has been provided by Barton (1978).

One of the most widely used methods of collecting discontinuity data is simply by direct measurement on the ground. A direct survey can be carried out subjectively in that only those structures which appear to be important are measured and recorded. In a subjective survey the effort can be concentrated on the apparently significant joint sets. Nevertheless, there is a risk of overlooking sets which might be important. Conversely, in an objective survey all structures intersecting a fixed line or area of the rock face are measured and recorded.

Several methods have been used for carrying out direct discontinuity surveys. Halstead *et al.* (1968) used the fracture set mapping technique by which all discontinuities occurring in 6 by 2 m zones, spaced at 30 m intervals along the face, were recorded. Knill (1971) also suggested using an area sampling method on the rock face concerned. On the other hand, Piteau (1971) and Robertson (1971) maintained that using a series of line scans provides a satisfactory method of joint surveying. The technique involves extending a metric tape across an exposure, levelling the tape and then securing it to the face. Two other scan lines are set out as near as possible at right angles to the first, one more or less vertical, the other horizontal. The distance along a tape at which each discontinuity intersects is noted, as is the direction of the pole to each discontinuity (this provides an indication of the dip direction). The dip of the pole from the vertical is recorded as this is equivalent to the dip of the plane from the horizontal. The strike and dip directions of discontinuities in the field can be measured with a compass and the amount of dip with a clinometer. Measurement of the length of a discontinuity provides information on its continuity. It has been suggested that measurements should be taken over distances of about 30 m, and to ensure that the survey is representative the measurements should be continuous over that distance. The line scanning technique yields more detail on the incidence of discontinuities and their attitude than other methods (Priest and Hudson 1981). A minimum of at least 200 readings per locality is recommended to ensure statistical reliability.

Terzaghi (1965) pointed out that the number of observations of joints of any one set is a function of the angle of intersection between that set and the face under examination. Hence joints intersecting the face at low angles are poorly represented in any one joint survey on that face. She demonstrated, however, that it was possible to correct data for this effect but suggested that the simpler remedy was to choose several exposures with different orientations.

Priest and Hudson (1976) described a line-scanning technique to record discontinuities in a tunnel in order to determine the stability of the rock mass involved, in this case the Lower Chalk. Similar work was done by Young and Fowell (1978) in the Kielder experimental tunnel. In each case a grid of vertical and horizontal lines was laid out over each tunnel face exposed and the intersection of the discontinuities recorded. This enabled a contoured plan of the face to be drawn which showed the spatial density of the discontinuities.

Hudson and Priest (1979) have pointed out that where discontinuities occur in sets, the discontinuity frequency along a scan line is a function of scan-line orientation. They showed that the spacing distributions of discontinuities is a negative exponential distribution with the mean spacing of discontinuities being the reciprocal of the average number of discontinuities per metre (λ). This value can simply be calculated by dividing the number of scan-line intersections by the total scan-line length. According to these two workers the relationship between the RQD and the average number of discontinuities per metre is given by the expression

$$RQD = 100e^{-0.1\lambda}(0.1\lambda + 1) \qquad (2.10)$$

In addition, Hudson and Priest showed how the distributions of block areas, for most locations, can be predicted adequately from discontinuity frequency measurements made along scan-lines, and how to derive cumulative frequency curves for block volumes from scan-line data.

In their joint surveys of sediments of Cretaceous age in south-east England, Fookes and Denness (1969) used the cavity technique, that is, they excavated blocks of material from the faces of exposures, and measurements were taken in the area vacated by the blocks. This meant that only fresh discontinuities were recorded. In addition to recording the dip and strike data for the joints and fissures they also measured the height, and where possible, the length of each joint. A large orientated block was excavated for examination in the laboratory.

2.10.2 Drill holes and discontinuity surveys

The information gathered by any of the above methods can be supplemented with data from orientated cores from drill holes. The value of the data depends in part on the quality of the rock concerned, in that poor quality rock is likely to be lost during drilling. However, it is impossible to assess the persistence, the degree of separation or the nature of the joint surfaces. What is more, infill material, especially if it is soft, is not recovered by the drilling operations.

Core orientation can be achieved by using the Craelius core orientator or by integral sampling (Rocha 1971 and Figures 2.11a and 2.11b respectively).

Drill-hole inspection techniques include the use of drill hole periscopes, drill hole cameras or closed-circuit television. The drill hole periscope affords direct inspection and can be orientated from outside the hole. However, its effective use is limited to about 30 m. The drill hole camera can also be orientated prior to photographing a section of the wall of a drill hole. The television camera provides a direct view of the drill hole and a recording can be made on videotape. These three systems are limited in that they require relatively clear conditions and so may be of little use below the water table, particularly if the water in the drill hole is murky.

The televiewer produces an acoustic picture of the drill-hole wall. One of its advantages is that drill holes need not be flushed prior to its use.

Snow (1968) demonstrated that the discharges from water injection or packer tests in jointed crystalline rocks provide estimates of the spatial frequency of joints, and the mean and variance of the size distributions of the apertures, which are log normal. He also maintained that decreasing permeability in rock masses with depth is more a result of decreasing fracture openings than of decreasing fracture spacing.

(a)

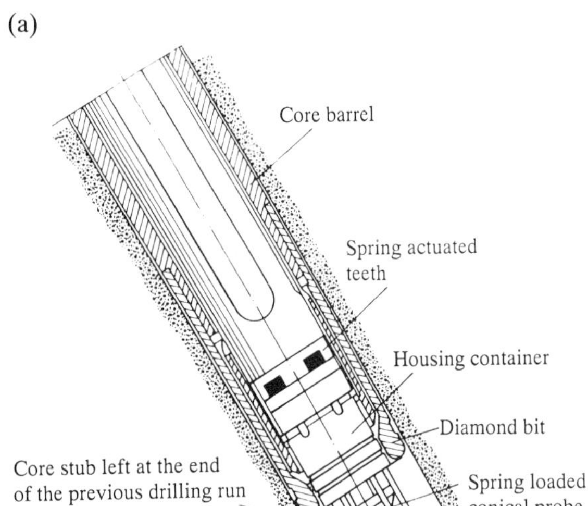

(b)

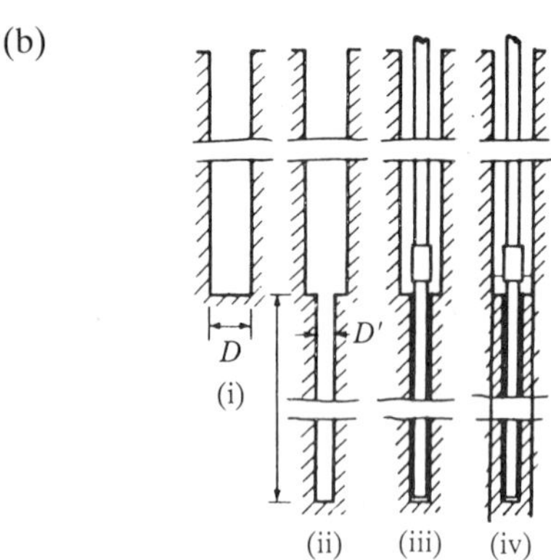

Figure 2.11 (a) *Details and method of operation of the Craelius core orientation. The teeth clamp the instrument into position in the inside of the core barrel until released by pressure on the conical probe. The housing contains a ring against which a ball bearing is indented by pressure from the conical probe, thus marking the position of the bottom of the hole. The probe is released by pressure against the core stub and, when released, locks the probe in position and releases the clamping teeth to allow the instrument to ride up inside of the barrel ahead of the core entering the barrel.* (b) *Stages of the integral sampling method. A drill hole (diameter D) is drilled to a depth where the integral sample is to be obtained, then another hole (diameter D') coaxial with the former and with the same length as the required sample is drilled, into which a reinforcing bar is placed. The bar is then bonded to the rock mass. Drilling of the drill hole is then resumed to obtain the integral sample*

2.10.3 Photographs and discontinuity surveys

Many data relating to discontinuities can be obtained from photographs of exposures. Photographs may be taken looking horizontally at the rock mass from the ground or they may be taken from the air looking vertically, or occasionally obliquely, down at the outcrop. These photographs may or may not have survey control. Uncontrolled photographs are taken using hand-held cameras, stereo-pairs being obtained by taking two photographs of the same face from positions about 5% of the distance of the face apart, along a line parallel to the face. Delineation of major discontinuity patterns and preliminary subdivision of the face into structural zones can be made from these photographs. Unfortunately, data cannot be transferred with accuracy from them onto maps and plans. Conversely discontinuity data can be accurately located on maps and plans by using controlled photographs. Controlled photographs are obtained by aerial photography with complementary gound control or by ground-based phototheodolite surveys. Aerial and ground-based photography are usually done with panchromatic film but the use of colour and infrared techniques is becoming more popular. Aerial photographs, with a suitable scale, have proved useful in the investigation of discontinuities. Photographs taken with a phototheodolite can be used with a stereo-comparator which produces a stereoscopic model. Measurements of the locations or points in the model can be made with an accuracy of approximately 1 in 5000 of the mean object distance. As a consequence, a point on a face photographed from 50 m can be located to an accuracy of 10 mm. In this way the frequency, orientation and continuity of discontinuities can be assessed. Such techniques prove particularly useful when faces which are inaccessible or unsafe have to be investigated.

Moore (1974) outlined a stereo-photogrammetric method by which he recorded the position of major discontinuities in the Oxford Clay as they were exposed during excavation in brick pits. Photographs of the faces were taken with a phototheodolite from three reinforced concrete pillars specially constructed for the purpose. The photographs enabled a series of contours, at 2 m vertical intervals on the face, to be plotted on a plan with a scale of 1:250. Each contour was marked where it was intersected by a joint. Contours on successive faces were referenced to the same eastings line and northings line. A model was constructed from the data in order to illustrate the three-dimensional nature of the joint system.

2.10.4 Recording discontinuity data

The simplest method of recording discontinuity data is by using a histogram on which the frequency is plotted along one axis and the strike direction along the other. Directional information, however, is more effectively represented on a rose diagram. This provides a graphical illustration of the angular relationships between joint sets.

The strikes of the joints and their frequencies are represented by the directions on each rose diagram, the lengths of the vectors being plotted either on a half- or full-circle. Directions are usually plotted for data contained in 5° arcs, while magnitudes are plotted to scale (Figure 2.12).

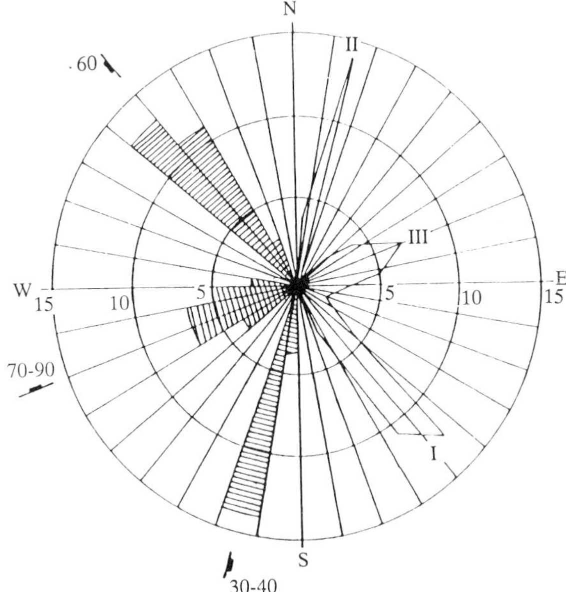

Figure 2.12 *Two methods of representing orientation data on a joint rosette (courtesy of Pergamon Press Inc.)*

Data from a discontinuity survey are now, however, usually plotted on a stereographic projection. The use of spherical projections, commonly the Schmidt or Wulf net, means that traces of the planes on the surface of the 'reference sphere' can be used to define the dips and dip directions of discontinuity planes. In other words the inclination and orientation of a particular plane is represented by a great circle or a pole, normal to the plane, which are traced on an overlay placed over the stereonet. The method whereby great circles or poles are plotted on a stereogram has been explained by Hoek and Bray (1981). When recording field observations of the amount and direction of dip of discontinuities it is convenient to plot the poles rather than the great circles. The poles can then be contoured in order to provide an expression of orientation concentration. This affords a qualitative appraisal of the influence of the discontinuities on the engineering behaviour of the rock mass concerned (Figure 2.13). The orientation of discontinuities can also be illustrated by using a block diagram (Figure 2.14).

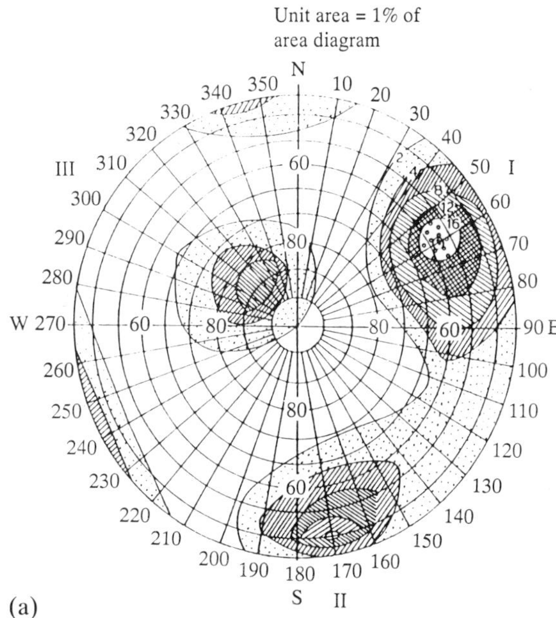

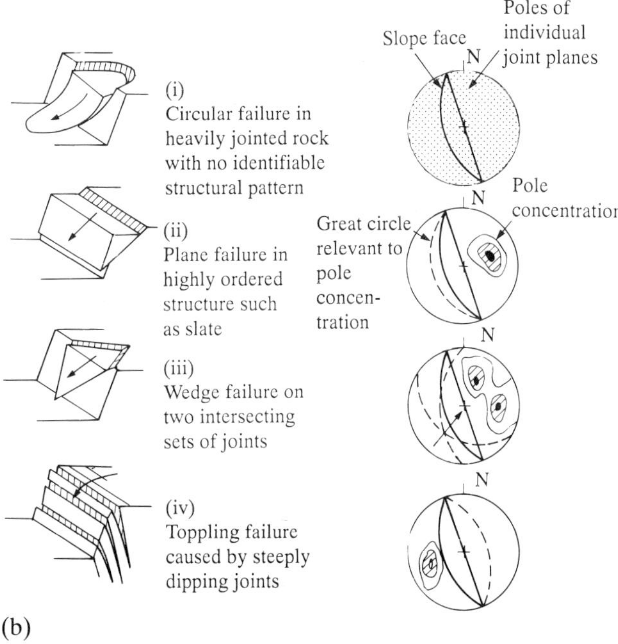

Figure 2.13 (a) *Schmidt contour diagram representing the orientation of three sets of joints plotted on a polar equal-area net. The main sets I and II are approximately normal to each other and the minor set III is nearly horizontal (after Barton 1978).* (b) *Representation of structural data concerning four possible slope failure modes plotted on equal-area nets as poles and great circles (after Hoek and Bray 1978)*

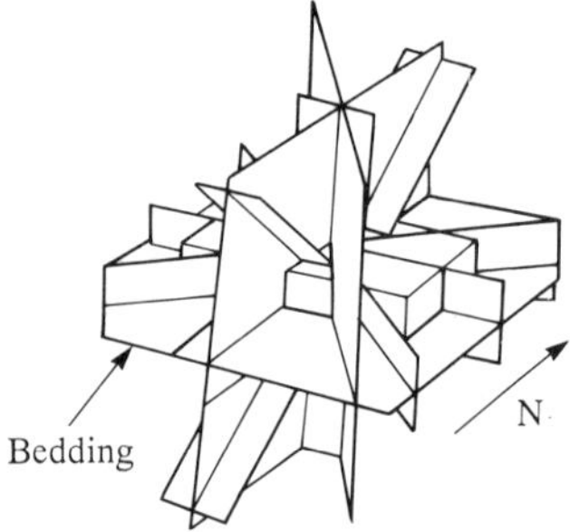

Figure 2.14 *Block diagram of discontinuities*

References

Anon (1971) 'Standard Method of Test for Soundness of Aggregates by use of Sodium Sulphate or Magnesium Sulphate' *American Society for Testing Materials, ASTM* C88–71: 49–53

Anon (1977) 'The description of rock masses for engineering purposes', Geological Society Engineering Group, Working Party Report, *Q. J. Engng. Geol.*, **10,** 355–388

Anon (1981a) 'Rock and soil description and classification for engineering geological mapping. International Association of Engineering Geology Commission on Engineering Geological Mapping, *Bull. In. Ass. Engng. Geol.*, **24**, 235–274

Anon (1981b) Basic geotechnical description of rock masses, International Society for Rock Mechanics, Commission on the Classification of Rocks and Rock Masses, *Int. J. Rock Mech. Min. Sci. & Geomech. Abstr.*, **18**, 85–110

Anon (1981c) *Code of Practice for Site Investigation,* BS 5930, British Standards Institutions, London

Balk, R. (1938) *Structural Behaviour of Igneous Rocks,* Memoir 6, Geological Society of America.

Barton, N. (1978) 'Suggested methods for the quantitative description of discontinuities in rock masses', ISRM Commission on Standardization of Laboratory and Field Tests, *Int. J. Rock Mech. Min. Sci. & Geomech. Abstr.*, **15**, 319–368

Brune, G. (1965) Anhydrite and gypsum problems in engineering geology. *Bull. Ass. Engng. Geologists,* **3**, 26–38

Chandler, R. J. (1969) The effect of weathering on the shear strength properties of the Keuper Marl, *Geotechnique,* **19**, 321–334

Chapman, C. A. (1958) 'Control of jointing by topography', *J. Geol.,* **66**, 552–568

Dearman W. R. (1974) Weathering classification in the characterization of rock for engineering purposes in British practice. *Bull. Int. Ass. Engng. Geol.,* **9**, 33–42

Dearman, W. R. (1986) State of weathering: the search for a national approach. In *Site Investigation Practice: Assessing BS 5930,* Engineering Geology Special Publication No 2, Hawkins, A. B. (ed.), Geological Society, London, 193–198

Deere, D. (1964) 'Technical description of cores for engineering purposes', *Rock Mech. Engg Geol.*, **1**, 18–22

Fasiska, E., Wagenblast, N. and Dougherty, M. T. (1974) 'The oxidation mechanism of sulphide minerals', *Bull. Ass. Engng Geologists,* **11**, 75–82

Fookes, P. G. and Denness, B. (1969) 'Observational studies on fissual patterns in Cretaceous sediments of south-east England', *Geotechnique,* **19**, 453–477

Fookes, P. G. and Horswill, P. (1970) 'Discussion on "The load deformation behaviour of the Middle Chalk at Mundford, Norfolk". In *In Situ Investigations in Soils and Rocks,* British Geotechnical Society, London, 53–57

Fookes, P. G., Dearman, W. R. and Franklin, J. A. (1972) 'Some engineering aspects of weathering with field examples from Dartmoor and elsewhere', *Q. J. Engng Geol.,* **3**, 1–24

Fookes, P. G., Gourley, C. S. and Ohikere, C. (1988) 'Rock weathering in engineering time', *Q. J. Engng Geol.,* **21**, 33–57

Franklin, J. L., Broch, E. and Walton, G. (1971) 'Logging the mechanical character of rock', *Trans Inst. Min. Metall.,* **81**, Mining Section, A1–9

Halstead, P. N., Call, P. D. and Rippere, K. H. (1968) 'Geological structural analysis for open pit slope design, Kimberley pit, Ely, Nevada', Reprint: *Annual AIME Meeting,* New York

Hamrol, A. (1961) 'A quantitative classification of weathering and weatherability of rocks', *Proc. 5th Int. Conf., Soil Mech. Found. Engng,* **2**, 771–773

Hancock, P. L. (1968) 'Joints and faults: the morphological aspects of their origins', *Proc. Geol. Ass.,* **79**, 141–151

Hansagi, I. A. (1974) 'Method of determining the degree of fissuration of rock', *Int. J. Rock Mech. Min. Sci & Geomech. Abstr.,* **11**, 379–388

Harris, J. F., Taylor, G. L. and Walper, J. L. (1960) 'Relation of deformational features in sedimentary rocks and regional and local structure', *Bull Am. Ass. Petrol. Geologists,* **44**, 1853–1878

Hencher, S. R. and Martin, R. P. (1982) 'The description and classification of weathered rocks in Hong Kong for engineering purposes', *Proc. 7th East Asian Geotechnical Conference,* Hong Kong, **1**, 143–149

Hobbs, D. W. (1968) 'The formation of tension joints in sedimentary rocks', *Geol. Mag.,* **104**, 550–556

Hobbs, N. B. (1975) 'Factors affecting the prediction of settlement of structures on rocks with particular reference to the Chalk and Trias' In *Settlement of Structures,* Brit. Geotech. Soc., Pentech Press, London, 579–610

Hodgson, R. A. (1961) 'Classification of structures on joint surfaces', *Am. J. Sci.,* **259**, 493–507

Hoek, E. and Bray, J. W. (1981) *Rock Slope Engineering,* Inst. Min. Metall., London

Honeyborne, D. B. (1983) 'The Building Limestones of France', Building Research Establishment Report, HMSO, Watford

Hudson, J. A. and Priest, S. D. (1979) 'Discontinuities and rock mass geometry', *Int. J. Rock Mech. Min. Sci. & Geotech. Abstr.,* **16**, 339–362

Iliev, I. G. (1966) 'An attempt to estimate the degree of weathering of *intrusive* rocks from their physico-mechanical properties', *Proc. 1st Cong. Int. Soc. Rock. Mech.,* Lisbon, 109–114

Irfan, T. Y. and Dearman, W. R. (1978a) 'Engineering classification and index properties of weathered granite', *Bull. Inst. Ass. Engng Geol.,* **17**, 79–90

Irfan, T. Y. and Dearman, W. R. (1978b) 'The engineering petrography of a weathered granite in Cornwall, England', *Q. J. Engng Geol.*, **11**, 233–244

Kiersch, G. A. and Treacher, R. C. (1955) 'Investigations, areal and engineering geology – Folsam dam project, central California', *Econ Geol.*, **50**, 271–310

Knill, J. L. (1971) *Collecting and Processing of Geological Data for Purposes of Rock Engineering. The Analysis and Design of Rocks Slopes,* University of Alberta, Edmonton

Knill, J. L. and Jones, K. S. (1965) 'The recording and interpretation of geological conditions in the foundations of the Rosieres, Kariba and Latiyan dams', *Geotechnique,* **15**, 94–124

Leary, E. (1983) 'The Building Limestones of the British Isles', Building Research Establishment Report, HMSO, Watford

Little, A. L. (1969) 'The engineering classification of residual tropical soils', *Proc. 7th Int. Conf. Soil Mech. Found. Engng,* Mexico City, **1**, 1–10

Lovegrove, C. W. and Fookes, P. G. (1972) 'The planning and implementation of a site investigation for a highway in tropical conditions in Fiji', *Q. J. Engng Geol., 5*, 43–68

Lamb, P. (1962) The properties of decomposed granite, *Geotechnique,* **12**, 226–243

Martin, R. P. and Hencher, S. R. (1986) 'Principles for description and classification of weathered rock for engineering purposes', In *Site Investigation Practice: Assessing BS 5930,* Engineering Geology Special Publication No. 2, Hawkins, A. B. (ed.), Geological Society, London, 299–308

Moore, J. F. A. (1974) 'Mapping of major joints in the Lower Oxford Clay using terrestrial photogrammetry', *Q. J. Engg Geol.*, **7**, 57–67

Moye, D. G. (1955) 'Engineering geology for the Snowy Mountain scheme', *J. Inst. Engrs.* Aust., **27**, 287–298

Olivier, H. J. (1979) 'A new engineering-geological rock durability classification', *Engineering Geol.* **14**, 255–279

Onodera, T. F. (1963) 'Dynamic investigation of foundation rocks', *Proc. 5th Symp. Rock Mech.*, Minnesota, Pergamon Press, New York

Onodera, T. F., Yoskinaka, R. and Oda, M. (1974) 'Weathering and its relation to mechanical properties of granite', *Proc 3rd Cong. Int. Soc. Rock Mech.*, Denver, **2A**, 71–98

Piteau, D. R. (1971) 'Geological factors significant to the stability of slopes cut in rock', *Symp. Planning Open Pit Mines,* Johannesburg, A. A. Balkema, Rotterdam, 43–53.

Price, N. L. (1959) 'Mechanics of jointing in rocks', *Geol. Mag.*, **96**, 149–160

Price, N. L. (1966) *Fault and Joint Development in Brittle and Semi-Brittle Rock,* Pergamon Press, London

Priest, S. D. and Hudson, J. A. (1976) 'Discontinuity spacings in rock', *Int. J. Rock Mech. Min. Sci. & Geomech. Abstr.*, **13**, 135–148

Priest, S. D. and Hudson, J. A. (1981) 'Estimation of discontinuity spacing and trace length using scanline surveys', *Int. J. Rock Mech. Min. Sci & Geomech. Abstr.*, **18**, 183–197

Robertson, A. M. (1971) 'The interpretation of geological factors for use in slope theory', *Symp. Planning Open Pit Mines,* Johannesburg, Balkema, Rotterdam, 55–71

Rocha, M. (1971) 'Method of integral sampling', *Rock Mech.*, **3**, 1–12

Ruxton, B. P. and Berry, L. 'Weathering of granite and associated erosional features in Hong Kong', *Bull, Geol. Soc. Am.*, **68**, 1263–1292

Snow, D. T. (1968) 'Rock fracture spacing opening and porosities', *Proc. ASCE, Div. Soil Mech. Foundations,* **94** (SMI), 73–91)

Taylor, R. K. 'Coal Measures mudrocks: composition, classification and weathering processes', *Q. J. Engng. Geol.*, **21**, 85–100

Taylor, R. K. and Spears, D. A. (1970) 'The breakdown of British Coal Measures rocks', *Int. J. Rock Mech. Min. Sci.*, **7**, 481–501

Terzaghi, R. D. (1965) 'Sources of error in joint surveys', *Geotechnique,* **15**, 287–304

Ward, W. H., Burland, J. B. and Gallois, R. W. (1968) 'Geotechnical assessment of a site at Mundford, Norfolk, for a large proton accelerator', *Geotechnique,* **18**, 399–431

Willis, B. and Willis, R. (1934) *Geologic Structures,* McGraw-Hill, New York

Winkler, E. M. (1973) *Stone Properties, Durability in Man's Environment.*, Springer-Verlag, New York

Young, R. P. and Fowell, R. J. (1978) 'Assessing rock discontinuities', *Tunnels and Tunnelling,* **10**(5), June, 45–8.

3 Description and classification of rock masses

Professor F G Bell
University of Natal

3.1 Description of rocks and rock masses

Description is the initial step in an engineering assessment of rocks and rock masses. It should therefore be both uniform and consistent in order to gain acceptance.

The complete specification of a rock mass requires descriptive information on the nature and distribution in space of both the materials that constitute the mass (rock, soil, water and air-filled voids) and the discontinuities which divide it (Anon. 1977). The intact rock may be considered as a continuum or polycrystalline solid consisting of an aggregate of minerals or grains whereas a rock mass may be looked upon as a discontinuum of rock material transected by discontinuities. The properties of the intact rock are governed by the physical properties of the materials of which it is composed and the manner in which they are bonded to each other. The parameters which may be used in a description of intact rock therefore include petrological name, mineral composition, colour, texture, minor lithological characteristics, degree of weathering or alteration, density, porosity, strength, hardness, intrinsic or primary permeability, seismic velocity and modulus of elasticity. Swelling and slake durability can be taken into account where appropriate such as in the case of argillaceous rocks.

The behaviour of a rock mass is, to a large extent, determined by the type, spacing, orientation and characteristics of the discontinuities present (see Chapter 2). As a consequence, the parameters which ought to be used in a description of a rock mass include the nature and geometry of the discontinuities as well as its overall strength, deformation modulus, secondary permeability and seismic velocity. It is not necessary, however, to describe all the parameters for either a rock mass or intact rock.

The data collected should be recorded on data sheets for subsequent automatic processing. A data sheet for the description of rock masses and another for discontinuity surveys have been recommended by the Geological Society (Anon. 1977; Figures 3.1, 3.2). Because geological data often have strong spatial inter-relationships they are usually presented in cartographic, graphical or tabulated form which facilitates assessment.

3.2 Properties of rocks and rock masses

3.2.1 Geological properties

Intact rock may be described from a geological or engineering point of view. In the first case the origin and mineral content of a rock are of prime importance, as is its texture and any change which has occurred since its formation. In this respect the name of a rock provides an indication of its origin, mineralogical composition and texture (Figure 3.3). Only a basic petrographical description of the rock is required when describing a rock mass. A useful system of petrographical description has been provided by Dearman (1974) and was further developed by the International Association of Engineering Geology (Anon. 1979a).

The colour of a rock has a composite character attributable to the different minerals of which it is formed, to the size of these minerals, and, in the case of sedimentary rocks, to the type and amount of cement present. Hence the overall colour should be assessed by reference to a colour system or chart since it is difficult to make a quantitative assessment by the eye alone. For example, the Munsell colour system evaluates colour in terms of hue, value and chroma. Hue refers to the basic colour or a mixture of basic colours. The chroma indicates the intensity, strength or degree of departure of a particular hue from a neutral grey of the same value. Value indicates the degree of lightness or darkness of a colour in relation to a neutral grey. Different colour hues and chroma may be obtained from a single type of rock from the same exposure if the rock has been subjected to different degrees of weathering. If a rock is polished this reduces the light value by scattering and enhances the hue by darkening the colour. The value also is reduced by a wet surface. Hence a moist fresh face in an exposure may indicate approximately the hue, chroma and value the rock will have when polished.

The Munsell method of colour notation arranges hue, chroma and value into orderly scales of equal visual intervals so that these properties become dimensions by which colour may be analysed and described under standard conditions of illumination by an observer viewing

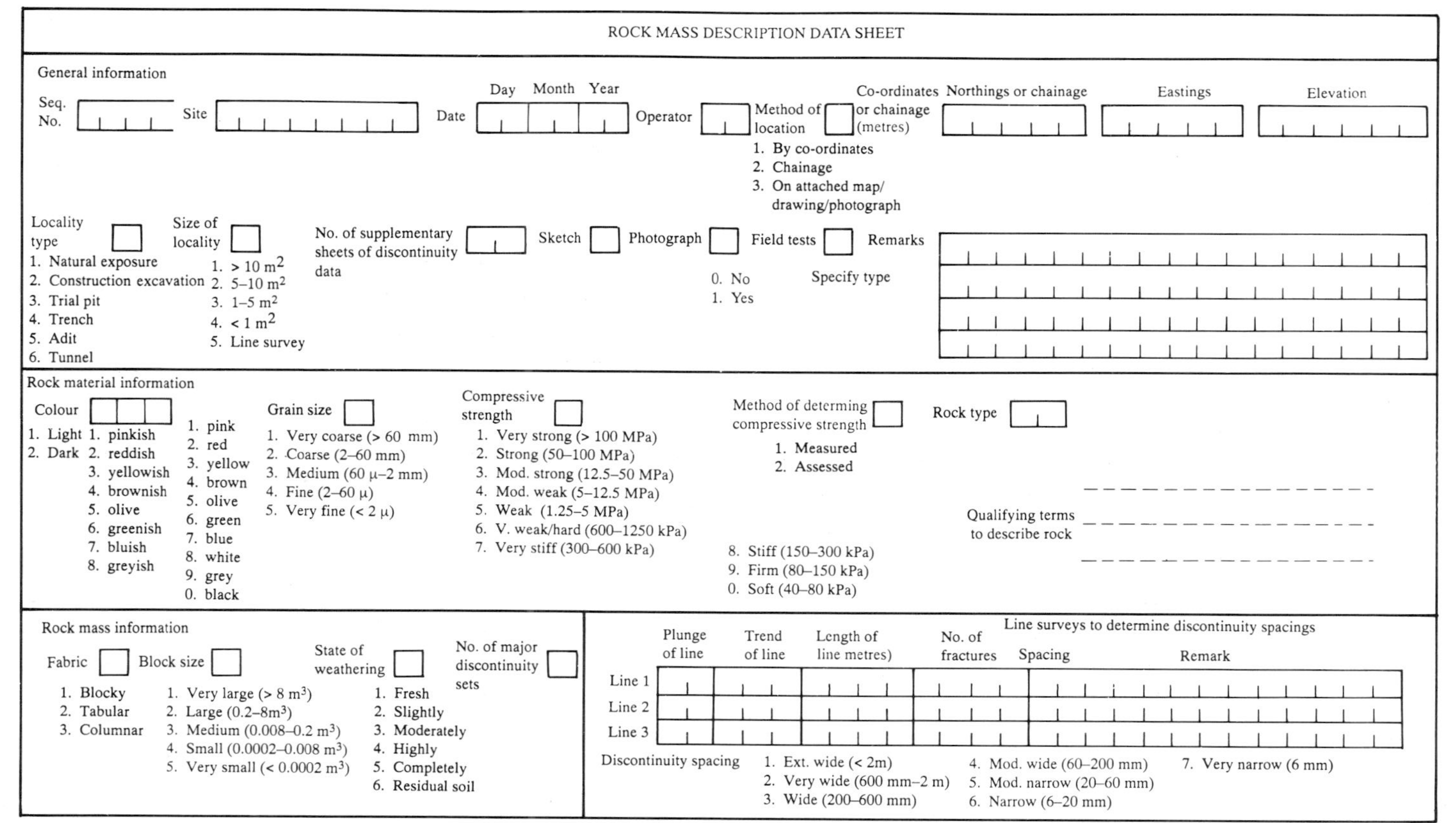

ROCK MASS DESCRIPTION DATA SHEET

General information

Seq. No. | Site | Date (Day, Month, Year) | Operator | Method of location | Co-ordinates or chainage (metres): Northings or chainage, Eastings, Elevation

Method of location:
1. By co-ordinates
2. Chainage
3. On attached map/drawing/photograph

Locality type
1. Natural exposure
2. Construction excavation
3. Trial pit
4. Trench
5. Adit
6. Tunnel

Size of locality
1. > 10 m^2
2. 5–10 m^2
3. 1–5 m^2
4. < 1 m^2
5. Line survey

No. of supplementary sheets of discontinuity data | Sketch | Photograph | Field tests | Remarks

0. No
1. Yes

Specify type

Rock material information

Colour
1. Light
2. Dark

1. pinkish
2. reddish
3. yellowish
4. brownish
5. olive
6. greenish
7. bluish
8. greyish

1. pink
2. red
3. yellow
4. brown
5. olive
6. green
7. blue
8. white
9. grey
0. black

Grain size
1. Very coarse (> 60 mm)
2. Coarse (2–60 mm)
3. Medium (60 μ–2 mm)
4. Fine (2–60 μ)
5. Very fine (< 2 μ)

Compressive strength
1. Very strong (> 100 MPa)
2. Strong (50–100 MPa)
3. Mod. strong (12.5–50 MPa)
4. Mod. weak (5–12.5 MPa)
5. Weak (1.25–5 MPa)
6. V. weak/hard (600–1250 kPa)
7. Very stiff (300–600 kPa)
8. Stiff (150–300 kPa)
9. Firm (80–150 kPa)
0. Soft (40–80 kPa)

Method of determing compressive strength
1. Measured
2. Assessed

Rock type

Qualifying terms to describe rock

Rock mass information

Fabric
1. Blocky
2. Tabular
3. Columnar

Block size
1. Very large (> 8 m^3)
2. Large (0.2–8m^3)
3. Medium (0.008–0.2 m^3)
4. Small (0.0002–0.008 m^3)
5. Very small (< 0.0002 m^3)

State of weathering
1. Fresh
2. Slightly
3. Moderately
4. Highly
5. Completely
6. Residual soil

No. of major discontinuity sets

Line surveys to determine discontinuity spacings

	Plunge of line	Trend of line	Length of line metres)	No. of fractures	Spacing	Remark
Line 1						
Line 2						
Line 3						

Discontinuity spacing
1. Ext. wide (< 2m)
2. Very wide (600 mm–2 m)
3. Wide (200–600 mm)
4. Mod. wide (60–200 mm)
5. Mod. narrow (20–60 mm)
6. Narrow (6–20 mm)
7. Very narrow (6 mm)

Figure 3.1 *Rock mass data description sheet (after Anon. 1977)*

DISCONTINUITY SURVEY DATA SHEET

GENERAL INFORMATION

Seq. no. [ ] Site [ ] Date Day [ ] Month [ ] Year [ ] Operator [ ] Discontinuity data sheet No. [ ] of [ ]

NATURE AND ORIENTATION OF DISCONTINUITY

Chainage or No.	Type	Dip	Dip direction	Persistence	Aperture	Nature of infilling	Consistency of infilling	Surface roughness	Trend of lineation	Waviness wavelength	Waviness amplitude	Water/flow	Remarks

Type
0. Fault zone
1. Fault
2. Joint
3. Cleavage
4. Schistosity
5. Shear
6. Fissure
7. Tension crack
8. Foliation
9. Bedding

Dip, dip direction and trend of lineation (Expressed in degrees)

Persistence (Expressed in metres)

Aperture
1. Wide (> 200 mm)
2. Mod. wide (60–200 mm)
3. Mod. narrow (20–60 mm)
4. Narrow (6–20 mm)
5. Very narrow (2–6 mm)
6. Ext. narrow (< 2mm)
7. Tight

Nature of infilling
1. Clean
2. Surface staining
3. Non-cohesive
4. Inactive clay or clay matrix
5. Swelling clay or clay matrix
6. Cemented
7. Chlorite, talc or gypsum
8. Others-specify

Compressive strength on infilling
1. Very soft (< 40 kPa)
2. Soft (40–80 kPa)
3. Firm (80–150 kPa)
4. Stiff (150–300 kPa)
5. Very stiff (300–600 kPa)
6. Hard/v. weak (600–1250 kPa)
7. Weak (1.25 MPa)
8. Mod. weak (5.12.5 MPa)
9. Mod. strong (12.5.50 MPa)
10. Strong (50–100 MPa)
11. Very strong (100–200 MPa)
12. Ext. strong (> 200 MPa)

Roughness
1. Polished
2. Slickensided
3. Smooth
4. Rough
5. Defined ridges
6. Small steps
7. Very rough

Waviness
Express wavelength and amplitude in metres

Water
1. Dry
2. Seepage
Flow
3. < 10 ml/s
4. 10–100 ml/s
5. 0.1–1 l/s
6. 1–10 l/s
7. 10–100 l/s
8. >100 l/s

Figure 3.2 *Discontinuity survey data sheet (after Anon. 1977)*

GRAIN SIZE (mm)	GENETIC/GROUP		DETRITIAL SEDIMENTARY			LIMESTONE and DOLOMITE (undifferentiated)		PYROCLASTIC	CHEMICAL/ ORGANIC
	Usual structure		BEDDED						
	Composition		Grains of rock, quartz, feldspar and clay minerals			At least 50% of grains are of carbonate		At least 50% of grains are of fine-grained igneous rock	
			Grains are of rock fragments						
60	Very coarse-grained	RUDACEOUS	BOULDERS COBBLES —	Rounded grains: CONGLOMERATE		CARBONATE GRAVEL	CALCI-RUDITE	Rounded grains AGGLOMERATE	SALINE ROCKS Halite Anhydrite
2	Coarse-grained	RUDACEOUS	GRAVEL	Angular grains: BRECCIA				Angular grains VOLCANIC BRECCIA LAPILLI TUFF	Gypsum
			Grains are mainly mineral fragments						
	Medium-grained	ARENACEOUS	SAND	SANDSTONE: Grains are mainly mineral fragments QUARTZ ARENITE: 95% quartz, voids empty or cemented ARKOSE: 75% quartz, up to 25% feldspar: voids empty or cemented GREYWACKE: 75% quartz, 15% fine detrital material: rock and feldspar fragments		CARBONATE SAND	CALC-ARENITE	TUFF (VOLCANIC ASH)	LIMESTONE DOLOMITE CHERT FLINT
0.06	Fine-grained	ARGILLACEOUS or LUTACEOUS	SILT	SILTSTONE: 50% fine-grained particles	MUDSTONE SHALE: fissile mudstone MARLSTONE	CARBONATE SILT	CALCI-SILTITE CHALK	Fine-grained TUFF (VOLCANIC ASH)	
0.002	Very fine-grained	ARGILLACEOUS or LUTACEOUS	CLAY	CLAYSTONE: 50% very fine grained particles	MUDSTONE SHALE: fissile mudstone MARLSTONE	CARBONATE MUD	CALCI-LUTITE	Very fine-grained TUFF (VOLCANIC ASH)	PEAT LIGNITE COAL
	GLASSY AMORPHOUS								

METAMORPHIC		IGNEOUS				GENETIC GROUP	GRAIN SIZE (mm)
FOLIATED		MASSIVE				Usual structure	
Quartz. feldspars, micas, acicular dark minerals		Light coloured minerals are quartz, feldspar, mica		Dark and light minerals	Dark minerals	Composition	
		Acid rocks	Intermediate	Basic rocks	Ultrabasic		
GNEISS (ortho-, para-, alternate layers of granular and flaky minerals	MARBLE	PEGMATITE			PYROXENITE and PERIDOTITE	Very coarse-grained	60
	GRANULITE	GRANITE	DIORITE	GABBRO	SERPENTINITE	Coarse-grained	2
MIGMATITE		MICROGRANITE	MICRODIORITE	DOLERITE		Medium-grained	
SCHIST	QUARTZITE HORNFELS AMPHIBOLITE						
PHYLLITE							0.06
		RHYOLITE	ANDESITE	BASALT		Fine-grained	
SLATE MYLONITE						Very fine-grained	0.002
		OBSIDIAN and PITCHSTONE		TACHYLYTE		GLASSY AMORPHOUS	
		VOLCANIC GLASSES					

Figure 3.3 *Rock type classification (after Anon. 1979)*

in daylight with grey to white surroundings. The chromatic colours in the Munsell system are divided into five principal classes, namely, red, yellow, green, blue and purple. Further division yields five intermediate hues, that is, yellow-red, green-yellow, blue-green, purple-blue and red-purple. If finer subdivisions are necessary the ten hue names mentioned may be combined, as for example, red-yellow-red. Even finer divisions can be obtained by dividing the hues into 10 categories each (again taking red as the example, 1 red is the faintest category and extends up to 10 red, or 10 R, which is the strongest or deepest red). The hue also may be described as reddish, yellowish etc. (Table 3.1). The chroma scale extends from 0 for neutral grey to 10, 12 or 14 depending upon the strength of

Table 3.1 Hue names and abbreviations used in the Munsell Colour Chart

Name	*Abbreviation*	*Name*	*Abbreviation*
red	R	purple	P
reddish orange	rO	reddish purple	rP
orange	O	purplish red	pR
orange yellow	OY	purplish pink	pPk
yellow	Y	pink	Pk
greenish yellow	gY	yellowish pink	yPk
yellow green	YG	brownish pink	brPk
yellowish green	yG	brownish orange	brO
green	G	reddish brown	rBr
bluish green	bG	brown	Br
greenish blue	gB	yellowish brown	yBr
blue	B	olive brown	OlBr
purplish blue	pB	olive	Ol
violet	V	olive green	OlG

the individual colour. The neutral grey scale extends from pure black, symbolized as 0, to pure white symbolized as 10.

The complete Munsell notation for any chromatic colour is written in terms of hue, value and chroma.

The Geological Society of America selected the Munsell book of colours as a basis for its rock colour chart. This represents a semi-quantitative attempt at colour determination. A simple subjective scheme has been suggested by the Geological Society (Anon. 1977) which involves the choice of colour from column 3 below, supplemented if necessary by a term from column 2 and/or column 1.

1	2	3
light	pinkish	pink
dark	reddish	red
	yellowish	yellow
	brownish	brown
	olive	olive
	greenish	green
	bluish	blue
		white
	greyish	grey
		black

The texture of a rock refers to its component grains and their mutual arrangement or fabric. It is dependent upon the relative sizes and shapes of the grains and their positions with respect to one another and the groundmass or matrix, when present. Grain size, in particular, is one of the most important aspects of texture, in that it exerts an influence on the physical properties of a rock. It is now generally accepted that the same descriptive terms for grain size ranges should be applicable to all rock types and should be the same as those used to describe soils (Table 3.2).

Other aspects of texture include the relative grain size and the grain shape. The IAEG (Anon. 1979a) suggested three types of relative grain size, namely, uniform, non-uniform and porphyritic. Grain shape was described in terms of angularity (angular, subangular, subrounded and rounded); form (equi-dimensional flat, elongated, flat and elongated, and irregular) and surface texture (rough and smooth).

Table 3.2 Description of grain size

Term	*Particle size*	*Equivalent soil grade*
Very coarse grained	Over 60 mm	Boulders and cobbles
Coarse grained	2–60 mm	Gravel
Medium grained	0.006–2 mm	Sand
Fine grained	0.002–0.06 mm	Silt
Very fine grained	Less than 0.002	Clay

3.2.2 Rock composition and texture in relation to physical properties

The micro-petrographic description of rocks for engineering purposes includes the determination of all parameters which cannot be obtained from a macroscopic examination of a rock sample, such as mineral content, grain size and texture, and which have a bearing on the mechanical behaviour of the rock or rock mass (Hallbauer *et al.*, 1978). In particular a microscopic examination should include a modal analysis, determination of microfractures and secondary alteration, determination of grain size and, where necessary, fabric analysis. The ISRM recommends that the report of a petrographic examination should be confined to a short statement on the origin, classification and details relevant to the mechanical properties of the rock concerned. Wherever possible this should be combined with a report on the mechanical parameters (Figure 3.4).

Mendes *et al.* (1966) proposed that quantitative micro-petrographic data could be used to formulate rock quality indices which were closely correlated with mechanical characteristics. A model analysis was made of the mineralogical composition of the rock samples concerned, together with an analysis of their texture and microstructure. As a result, the sample could be classified and its type and degree of alteration and the extent of micro-fracturing estimated. The percentage of sound minerals was determined. However, those sound minerals which had an adverse effect upon mechanical behaviour were grouped with the percentage of adverse minerals. Open micro-fissures were distinguished from those which were filled. In the latter type the nature of the cement is important, for example, silica provides a strong bond whilst other materials such as talc may lubricate movement along a micro-fissure. The quality index (K) was defined as

$$K = \frac{\sum_{i=1}^{n} p_i X_i}{\sum_{j=1}^{m} p_j Y_j} \quad (3.1)$$

Project:
Location:
Co-ordinates: Collected by:
Specimen No:
Description of sampling point:
Thin section No: Date:

Geological description
Rock name:
Petrographic classification:
Geological formation:

Photo-micrograph of typical features of thin section

Macroscopic description of sample
Degree of weathering:
Structure (incl. bedding):
Discontinuities:

Qualitative description
Texture:
Fracturing:
Alteration:
Matrix:

Mineral composition (modal analysis)

Major components	Vol. %	Minor components	Vol. %	Accessories	Vol. %

Results of rock property tests	
Point load index:	Porosity ....
......MPa, wet/dry	Density......Mg/m^3
normal/parallel to foliation	Water absorption:
Any other results:	

Significance of results for rock engineering	Grain size and distribution	
	Microns	%

General remarks

Figure 3.4 *ISRM suggested form of petrographic report (after Hallbauer et al.* 1978)

in which n values of X_i are the percentages of sound minerals or minerals having a favourable influence upon mechanical behaviour, and the m values of Y_j are the percentages of altered minerals or sound minerals which have an adverse effect upon mechanical performance together with percentages of micro-fissures and voids. The coefficients p_i and p_j are weights which measure the influence on the mechanical characteristics of the rock sample of one or other mineral or peculiarity. The quality indices of granite and gneiss were correlated with their elasticity modulus (E) values. A good correlation was obtained for the granite material but that of gneiss was not so good.

A number of petrofabric techniques were developed by Willard and McWilliams (1969) in an attempt to gain a better understanding of the mechanical behaviour of rocks in relation to their micro-structure. They first subjected a number of rock samples to non-destructive (pulse velocity) tests and then examined their fabric. Next they took the same rock samples and this time subjected them to destructive (indirect tensile) tests and subsequently observed their fracture characteristics under the microscope. The five techniques they used were diametric mineralogical analysis, defect frequency orientation analysis, grain elongation analysis, macro-grid analysis and transgranular-intergranular analysis. The first three of these techniques record micro-structural features of the rock which are not affected by changes brought about by destructive testing. As a consequence they help to explore the features which play a part in mechanical behaviour and that are not disturbed by cracking due to testing. The last two techniques record micro-structural features in the rock fabric which are brought about as a result of destructive testing. All five techniques have to be performed with reference to a three-dimensional coordinate system. With such a framework micro-structural features can be related to mechanical properties which vary in different directions.

Diametric mineralogical analysis simply consists of a modal analysis of the mineral components along given diameters of a circular thin section of rock, these sections being cut from the rock discs which are tested. In a test carried out on the Salisbury Granite, Willard and McWilliams (1969) showed that the pulse velocity varied with direction and increased with increasing feldspar content. The object of defect frequency orientation (DFO) analysis is to describe and evaluate the frequency of defect occurrence. As far as Willard and McWilliams were concerned defects were either open or closed cracks or sites at which cracks would develop when the rock was subjected to critical or shear stress. Therefore in addition to micro-fractures, defects included grain boundaries, mineral cleavages, twinning planes, inclusion trains and the elongation of shell fragments. Rocks are not uniformly coherent, homogeneous materials and defects occur as visible or microscopic linear or planar discontinuities associated with certain minerals. As is to be expected defects influence the ultimate strength of a rock and may act as surfaces of weakness which control the direction in which failure occurs. Between 500 and 1000 defect orientations were noted in each thin section of rock and each one was regarded as a vector of unit length with a range of 180°. From their investigations on the Barre Granite Willard and McWilliams noted that the frequency of defects tended to be inversely proportional to the breaking strength suggesting that the direction of weakest tensile strength was approximately normal to the direction of most defects. In rocks with a preferred orientation, grain elongation can be used as a method of correlating microstructure with their mechanical properties. Higher failure strengths were found to be associated with line-loading at right angles to surfaces of lineation.

Macro-grid analysis is used to calculate the areal mineral percentages over the fracture surfaces. This is then compared with the volume percentages of minerals in the rock, which are obtained by modal analysis. The comparison reveals any tendency for a fracture surface to include unusually large or small amounts of particular minerals. When a fracture is produced in a granular rock it must either pass through grains or follow boundaries. In the first case it is described as transgranular (T) and in the second as intergranular (I). The relative lengths of the transgranular and intergranular parts along a fracture are summed and the failure expressed as a ratio T/I which can then be compared with the failure strength of the rock. When the T/I ratios were recorded from samples of Tennessee Marble it was found that they varied inversely with failure strength.

Subsequent work on the relationship between cracks in rocks and their elastic properties, by Simmons *et al.* (1975), emphasised the necessity to obtain quantitative petrographic data on crack dimensions, numbers of cracks per unit area or volume, and the distribution and orientation of cracks.

Onodera and Kumara (1980) found a linear relationship between Young's modulus and the grain boundary surface area per unit volume in granite. They also found a linear relationship between strength and grain size, that is, as the grain size of the granite decreased, the strength increased.

3.2.3 Rock masses and weathering

Rock material tends to deteriorate in quality as a result of weathering and or alteration. Weathering refers to those destructive processes, brought about by atmospheric agents at or near the Earth's surface, that produce a mantle of rock waste. Alteration refers to those changes which occur in the chemical or mineralogical composition of a rock brought about by permeating hydrothermal fluids or by pneumatolytic action. Unlike weathering the effects of alteration may extend to considerable depths beneath the surface since the agents responsible may have originated from deeply emplaced igneous intrusions. Although weathering and alteration occur in the rock material, the processes are concentrated along the discontinuities in the rock mass. Qualitative classifications based on the estimation and description of physical disintegration and chemical decomposition of originally sound rock frequently have been used to assess the degree of weathering (see Chapter 2). Such a classification of weathered rock masses as recommended by the IAEG (Anon, 1979a) and ISRM (Anon, 1981) is as shown in Table 3.3.

Table 3.3

Symbol	*Degree of weathering (%)*	*Term*	*Description*
W_0	0	Fresh	No visible sign of material weathering
W_1	Less than 25	Slightly	Discolouration indicates weathering of rock on major discontinuity surfaces
W_2	25–50	Moderately	Less than half the rock material is decomposed and/or disintegrated to a soil. Fresh or discoloured rock is present either as a discontinuous framework or as corestones
W_3	50–75	Highly	More than half the rock is decomposed and/or disintegrated to a soil. Fresh or discoloured rock is present either as a discontinuous framework or as corestones
W_4	Over 75	Completely	Majority of rock material is decomposed and/or disintegrated to soil. The original structure of the rock mass is still intact
W_5	100	Residual soil	All material decomposed. No trace of rock structure preserved

Obviously all grades of weathering may not be present in a given rock mass. Furthermore the classification of weathering grade may have to be modified to suit certain types of rock masses and other classifications have been advanced for the Chalk (see Ward *et al.*, 1968; and Chapter 2) and for Keuper Marl (see Chandler 1969; and Chapter 2).

3.2.4 Physical properties of rock

The IAEG (Anon 1979a) grouped the dry density and porosity of rocks into five classes as shown in Table 3.4.

Table 3.4 Dry density and porosity

Class	*Dry density* (Mg/m^3)	*Description*	*Porosity* (%)	*Description*
1	Less than 1.8	Very low	Over 30	Very high
2	1.8–2.2	Low	30–15	High
3	2.2–2.55	Moderate	15–5	Medium
4	2.55–2.75	High	5–1	Low
5	Over 2.75	Very high	Less than 1	Very low

Determination of the strength and deformability of intact rock is achieved with the aid of some type of laboratory test (see Chapter 8). If the strength of rock is not measured, then it can be estimated as shown in Table 3.5. Obviously such estimates can only be approximate.

Table 3.5 Estimation of the stength of intact rock (after Anon. 1977)

Description	*Approximate unconfined compressive strength* (MPa)	*Field estimation*
Very strong	Over 100	Very hard rock – more than one blow of geological hammer required to break specimen
Strong	50–100	Hard rock – hand-held specimen can be broken with a single blow of hammer
Moderately strong	12.5–50	Soft rock – 5 mm indentations with sharp end of pick
Moderately weak	5.0–12.5	Too hard to cut by hand
Weak	1.25–5.0	Very soft rock – material crumbles under firm blows with the sharp end of a geological hammer

As far as deformability is concerned the five classes shown in Table 3.6 have been proposed by the IAEG (Anon 1979a).

Table 3.6 Deformability

Class	*Deformability* (MPa × 10^3)	*Description*
1	Less than 5	Very high
2	5–15	High
3	15–30	Moderate
4	30–60	Low
5	Over 60	Very low

Hardness can be defined as the mechanical competence of the intact rock. In the strict sense it is a surface property which is measured by using abrasion, indentation or rebound tests (see Chapter 8). These tests tend to reflect the hardnesses of individual grains rather than the integranular bond or coherence of the rock. The point load strength provides a measure of coherence, in other words an indirect measure of tensile strength. Determination of hardness at natural moisture content can give misleading results from those materials, especially argillaceous rocks, that are water sensitive. Indeed most rocks show some hardness reduction when wetted, but mudstones and shales may actually disintegrate when subjected to stress relief combined with moisture content fluctuations. Hence Cottiss *et al.* (1971) recommended that in such instances the durability should be determined to supplement the values of hardness. Durability measures the susceptibility of rocks to weakening and disintegration in water (see Chapter 8, and Anon. 1979b).

The permeability of intact rock (primary permeability) is usually several orders less than the *in situ* permeability

Table 3.7 Estimation of secondary permeability from discontinuity frequency

Rock mass description	*Term*	*Permeability* k (m/s)
Very closely to extremely closely spaced discontinuities	Highly permeable	10^{-2}–1
Closely to moderately widely spaced discontinuities	Moderately permeable	10^{-5}–10^{-2}
Widely to very widely spaced discontinuities	Slightly permeable	10^{-9}–10^{-5}
No discontinuities	Effectively impermeable	Less than 10^{-9}

(secondary permeability), as most water normally flows via discontinuities in rock masses. Although the secondary permeability is affected by the openness of discontinuities on the one hand and the mount of infilling on the other, a rough estimate of the permeability can be obtained from the frequency of discontinuities (Table 3.7). Admittedly such estimates must be treated with caution and cannot be applied to rocks which are susceptible to solution. The IAEG (Anon 1979a) suggested the grades of permeability shown in Table 3.8, which differ slightly from the class limits given in Table 3.7.

Table 3.8 Grades of permeability

Class	*Permeability*	
	k (m/s)	*Description*
1	Greater than 10^{-2}	Very highly
2	10^{-2}–10^{-4}	Highly
3	10^{-4}–10^{-5}	Moderately
4	10^{-5}–10^{-7}	Slightly
5	10^{-7}–10^{-9}	Very slightly
6	Less than 10^{-9}	Practically impermeable

The seismic velocity refers to the velocity of propagation of shock waves through a rock mass. Its value is governed by the mineral composition, density, porosity, elasticity and degree of fracturing within a rock mass. The IAEG (Anon 1979a) recognized the classes of sonic velocity shown in Table 3.9 for rocks.

Table 3.9 Classes of sonic velocity

Class	*Sonic velocity* (m/s)	*Description*
1	Less than 2500	Very low
2	2500–3500	Low
3	3500–4000	Moderate
4	4000–5000	High
5	Over 5000	Very high

Igneous rocks generally possess values of sonic velocity above 5000 m/s, those of metamorphic rocks range upwards from 3500 m/s and those of sedimentary rocks tend to vary between 1500 m/s and 4500 m/s. The latter range does not include unconsolidated deposits. The dynamic value of Young's modulus and Poisson's ratio can be derived from the seismic velocity (Onodera 1963) and both can be correlated with the degree of fracturing (Grainger *et al.* 1973 and Table 2.15).

3.3 Basic geotechnical description of ISRM

The basic geotechnical description (BGD) of rock masses proposed by the ISRM (Anon. 1981) considered the following characteristics:

(1) rock name with a simplified geological description;
(2) the layer thickness and fracture (discontinuity) intercept of the rock mass;
(3) the unconfined compressive strength of the rock material and the angle of friction of the fractures.

It was suggested that, where necessary, the rock mass should be divided into geotechnical units or zones. The division of the rock mass should be made in relation to the project concerned and the BGD should then be applied to each unit. The rock name is given in accordance with Figure 3.3. Although the simplified geological description depends upon the character of the rock masses involved, together with the requirements of the proposed scheme, it usually takes account of the mineralogical composition, texture and colour of the rock on the one hand and the degree of weathering, the nature of the discontinuities and the geological structure of the rock mass on the other. In addition the ISRM recommended that it would be advisable to provide a general geological description for the rock mass as well as one for each geotechnical unit.

The same class limits are used to describe the layer thickness of a geotechnical unit as are used for fracture intercept (Table 3.5; *cf.* Table 2.9). The ISRM defined the fracture intercept as the mean distance between successive fractures as measured along a straight line (Table 2.9). When the fracture spacing changes with direction, the value adopted in the BGD is that corresponding to the direction along the smallest mean intercept. Fractures or discontinuities can be grouped into sets. The average fracture intercept, measured perpendicular to the fractures, is recorded for each set and given as supplementary information.

Table 3.10 Classification of layer thickness and fracture intercept (after Anon 1981)*

Interval (m)	*Layer thickness*		*Fracture intercept*	
	Symbol†	*Description*	*Symbol*	*Description*
Over 2.0	L_1	Very large	F_2	Wide
0.6–2.0	L_2	Large	F_2	Wide
0.2–0.6	L_3	Moderate	F_3	Moderate
0.06–0.2	L_4	Small	F_4	Close
Less than 0.06	L_5	Very small	F_5	Very close

* The IAEG has proposed the same class limits.
† If a unit is not layered, then it is given the symbol L_0.

The unconfined compressive strength of the intact rock within a geotechnical unit of a rock mass represents the mean strength of rock samples taken from the zone. The BGD includes the groupings in Table 3.11; those suggested by the IAEG (Anon 1979a) are given for comparison.

Table 3.11 BGD classification

BGD strength (MPa)	*Symbol*	*Description*	*IAEG strength* (MPa)	*Description*
Over 200	S_1	Very high	Over 230	Extremely strong
60–200	S_2	High	120–230	Very strong
20–60	S_3	Moderate	50–120	Strong
6–20	S_4	Low	15–150	Moderately strong
Under 6	S_5	Very low	1.5–15	Weak

If the rock material is notably anisotropic, the mean strengths obtained in different directions should be recorded, and special note should be made of that direction along which the lowest mean strength occurs.

The angle of friction of fractures as defined by the ISRM refers to the slope of the tangent to the peak strength envelope at a normal stress of 1 MPa. The smallest mean value of the angle of friction is recorded when fracture sets differ in their shear strength. Table 3.12 shows the class limits for the angle of friction of fractures adopted by the BGD.

Table 3.12 Angle of friction of fractures

Interval	*Symbol*	*Description*
Over 45°	A_1	Very high
35–45°	A_2	High
25–35°	A_3	Moderate
15–25°	A_4	Low
Less than 15°	A_5	Very low

The data sheet given in Figure 3.5 is used for the BGD. Each zone is characterized by its rock name, followed by the class symbols corresponding to the parameter values, e.g. Granite L_0, F_3, S_2, A_3. Supplmentary information is incorporated in the BGD when the rock mass concerned exhibits special features or if the requirements of the project so demand.

3.4 Principles of classification

Classifications of rocks devised by geologists usually have a genetic basis. Unfortunately, however, such classifications may provide little information relating to the engineering behaviour of the rocks concerned.

According to Coates (1964) classification is needed in geotechnical engineering in order to assist in making an initial assessment of a problem and to point to areas where additional information must be sought in order to obtain the required answer. Typical engineering problems that require rock classification include the assessment of slope stability, open excavation, subsurface excavation, foundation stability and the selection of rock for construction material.

Franklin (1970) contended that if a different system of rock classification is used for each engineering problem, then much confusion and duplication of effort can arise. He therefore argued that it would be better if one system of classification could be used for a range of problems. However, there has been a trend towards the development of a multiplicity of classifications. Franklin admitted that a classification to some extent should be tailored to suit the application but he suggested that if classification criteria are carefully selected initially, then a change of emphasis from one application to the next, rather than a complete reorganisation, should suffice. Broad terms of reference therefore are necessary in designing such a classification system. Franklin went on to distinguish between basic and supplementary classification tests and observations. The former may be used to establish a universally applicable basis for the engineering classification of rocks. The latter are of less general relevance and may be used for added refinement in particular engineering problems.

Tests which are used for the engineering classification of rocks are termed index tests. If the right index tests are chosen, then rocks having similar index properties, irrespective of their origin, will probably exhibit similar engineering performance. In order for an index test to be useful it must satisfy certain criteria. It should be simple to carry out, inexpensive and rapidly performed. The test results must be reproducible and the index property must be relevant to the engineering requirement. Generally these tests are carried out in large numbers so that a reliable picture of rock variation is obtained.

Classification systems of intact rock may be developed either by selecting individual index properties to represent others that are closely related, or by summing the values of closely related properties to derive a single score (Cottis *et al.* 1971). Obviously if two types of test are closely related, then the values of one may be used to predict the values of the other. Where economy of effort is important there is little to be gained by performing both tests. This has motivated the use of index tests in rock classification. For instance, D'Andrea *et al.* (1965) used nine index observations to predict the unconfined compressive strength of rock, employing multiple linear regression analysis for this purpose. Statistical methods may be used to decide which observations give the best prediction but the final selection of index properties must always take account of which properties are the easiest to evaluate.

Deere and Miller (1966) maintained that information concerning the physical properties of rocks and the nature

Example of application of BGD

Type of work:[1] Concrete Dam

Investigation stage:[2] Preliminary   Exposure:[3] Outcrop

Location: Rocha da Galé Portugal   Observer:[4] Gomes Coelho   Date: June 77

5

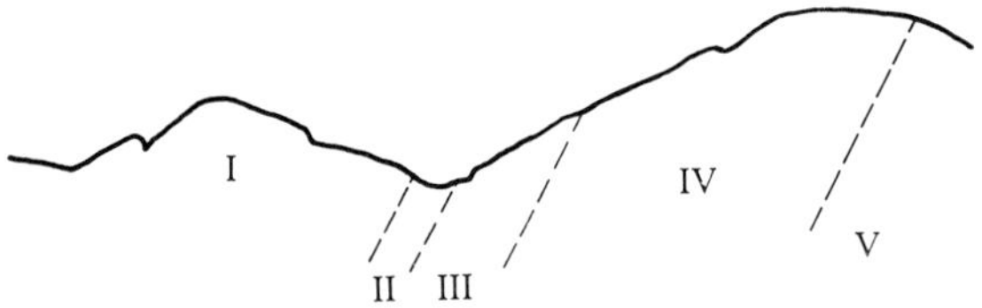

Photograph

Rock name and general geological description:[6]
Isoclinal sequence of metasedimentary and meta-volcanic rocks composed by interbedded siliceous schists and greywackes (I), pyroclastic rocks like tuff and breccia (II), agglomerate with rhyolitic matrix (III), rhyolite (IV), and porphyritic quartz-diorite (V).

Supplementary geological description:[7]
Zone 1: Rock mass formed of grey to red siliceous schist and interbedded greywacke thinly bedded (2–20 cm), and very often thinly laminated (0.6–2 cm). Rock mass is crossed by widely to very widely spaced joints, open, without filling material. Rock is fresh (W_1) and strong.
Zone II: Interbedded zone 30 m thick, composed of pyroclastic tuff and breccia, moderately to highly weathered (W_3–W_4) and moderately weak.
Zone III: Rock mass formed by agglomerate with a matrix of rhyolitic composition; rock is fresh to slightly weathered (W_1–W_2) and strong to very strong.
Zone IV: Rock mass formed by rhyolite massive, fresh to slightly weathered (W_0–W_1) and strong to very strong.
Zone V: Rock mass formed by porphyritic quartz-diorite, with slight foliation and widely spaced bedding (0.6–2 m); rock is fresh to moderately weathered (W_0–W_2) and moderately strong.

Zones	*Occurrence* (%)[8]	*Characterization*[9]	*Zones*	*Occurrence* (%)[8]	*Characterization*[9]
I	20	Siliceous schist L_4; $F_{4.5}$; S_2; A_2	V	45	Quartz-diorite L_2; F_4; S_4; A_2
II	2	Breccia and Tuff L_4; F_4; S_4; A_3	VI		
III	8	Agglomerate L_0; F_4; S_2; A_2	VII		
IV	25	Rhyolite L_0; F_3; S_2; A_2	VIII		

Figure 3.5 *Basic geotechnical description (after Anon, 1981)*

of the discontinuities within rock masses is required in order to make rational predictions about their engineering behaviour under superimposed stresses. They further stated that the properties of intact rock should be investigated initially in an attempt to develop a meaningful system of evaluation of the *in situ* behaviour of rock. Appropriate reduction factors, attributable to the discontinuities, should then be determined for application to the intact rock data.

In fact two types of classification have been developed. First, there are those based upon some selected properties of the intact rock, as mentioned above. Secondly, and more importantly, there are those which take account of the properties of the rock mass, especially the nature of the discontinuities. The specific purpose for which a classification is developed obviously plays an important role in determining whether the emphasis is placed on the physical properties of the intact rock or on the continuity of the rock mass. The object of both types of classification is to provide a reliable basis for assessing rock quality.

Computation of parameters

Zone	Parameters	Samples 1	2	3	4	Average	Std dev	BGD symbols
I	Layer thickness (cm)	4	10	8	6	7		L_4
	Fracture interc. (cm)							$F_{4.5}$
	U. comp. strength (MPa)	66	65	150*	80	70		S_2
	Angle of friction (°)					35		A_2
(II)	Layer thickness (cm)	10	12	16	15	13		L_4
	Fracture interc. (cm)	15				15		F_4
	U. comp. strength (MPa)	15	20	12	15	15		S_4
	Angle of friction (°)					30		A_3
(III)	Layer thickness (cm)	–	–	–	–	–		L_0
	Fracture interc. (cm)	4	12	6	8	7		F_4
	U. comp. strength (MPa)	236	250	150	170	200		S_2
	Angle of friction (°)					40		A_2
(IV)	Layer thickness (cm)	–	–	–	–	–		L_0
	Fracture interc. (cm)	22	25	50	45	36		F_3
	U. comp. strength (MPa)	210	140	180	220	185		S_2
	Angle of friction (°)					40		A_2
(V)	Layer thickness (cm)	80	210	120	160	140		L_2
	Fracture interc. (cm)	4	7	12	18	10		F_4
	U. comp. strength (MPa)	92*	55	60	50	55		S_4
	Angle of friction (°)					40		A_2

Remarks[10]

Layer thickness:	measured on outcrops
Fracture interc.:	measured and estimated
U comp. strength:	lab. test and estimated
Angle of friction:	estimated

Supplementary information

* Normal to layering

(1) Main characteristics of the structure. (2) Preliminary, final, . . . (3) Outcrop, trench, cores . . . (4) Name and qualification. (5) Stereo pair of photographs, with the zones outlined. Other stereo pairs may be added. Ordinary photographs and/or sketches can be resorted to. (6) Rock name, structure (folds, faults). Fracturing (fracture sets, fracture characteristics); weathering. (7) Specific aspects should be considered for each zone. (8) Estimated proportion, by volume, of the occurrence of each zone relative to the observed rock mass. (9) Rock name followed by the interval symbols of the parameter values. (10) Methods followed in the determination of the parameters and difficulties encountered

3.5 Review of classifications

Since 1960 much effort has been devoted to the production of engineering classifications of rocks and rock masses. A review of the development of rock classification for engineering purposes has been provided by Dearman (1974).

Any classification of intact rock for engineering purposes should be relatively simple, being based on significant physical properties so that it has a wide application. For example, Deere and Miller (1966) based their engineering classification of intact rock on the unconfined compressive strength and the modulus ratio, as shown in Tables 3.13 and 3.14 respectively.

The strength categories follow a geometric progression and the dividing line between categories A and B was chosen at 224 MPa since it is about the upper limit of the strength of most rocks. A rock may be classified as CH, BH, DL, etc.

Deere and Miller (1966) found that different rock types, when plotted on Figure 3.6, occupied different positions. For instance, the envelope enclosing sandstones and siltstones indicates that they have a unique position with respect to other rocks. It shows that they are more

Table 3.13 Strength

Class	*Description*	*Unconfined compressive strength* (MPa)
A	Very high strength	Over 224
B	High strength	112–224
C	Medium strength	56–112
D	Low strength	28–56
E	Very low strength	Less than 28

Table 3.14 Modulus ratio

Class	*Description*	*Modulus ratio*
H	High moduius ratio	Over 500
M	Medium modulus ratio	200–500
L	Low modulus ratio	Less than 200

compressible in relation to their strength than most rock types. Granites also occupy a rather special position in the centre of the zone of average modulus ratio. Deere and Miller suggested that specific rock types fall within certain areas on the classification chart because it is sensitive to mineralogy, fabric and direction of anisotropy.

Voight (1968), however, argued that the elastic properties of intact rock could be omitted from practical classifications since the elastic moduli as determined in the laboratory are seldom those which are required for engineering analysis.

Rocks possessing an interlocking fabric and little or no anisotropy generally fall into the medium modulus ratio category. Some limestones and dolostones, however, have a high modulus ratio which Deere (1968) attributed to their mineralogy, as well as to their interlocking fabric, they being composed of calcite and/or dolomite. The sandstone and shale envelopes on Figure 3.6 are open ended in their lower portions because several samples failed at strengths less than 7 MPa. Both sandstone and shale envelopes extend into the zone of low modulus ratio due to the anistotropy attributable to their bedding or lamination. In metamorphic rocks the gneiss envelope overlaps that of quartzite, as well as the two schist envelopes. This transition position indicates an increasing complexity in both mineralogy and fabric in going from quartzite to gneiss to schist.

Deere and Miller (1966) also considered that the Schmidt hammer and Shore scleroscope hardnesses, sonic pulse velocity and unit weight could act as indices of the engineering behaviour of intact rock. In particular they found a correlation between Schmidt hammer hardness and Shore scleroscope hardness on the one hand and unconfined compressive strength and deformation on the other. Moreover they found that the correlation was improved when the unit weight was taken in conjunction with the indices of hardness and sonic velocity. These

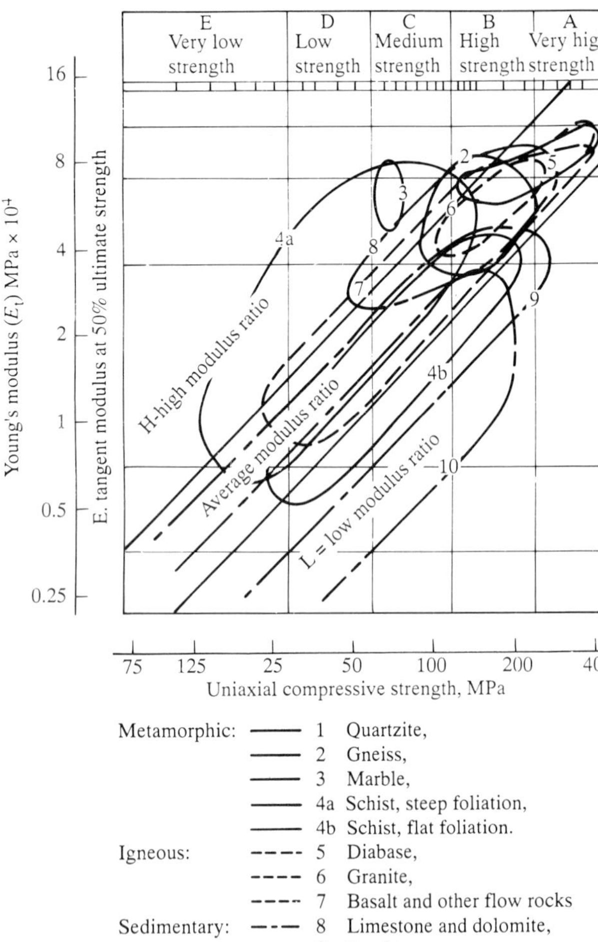

Figure 3.6 *Engineering classification of intact rock based on uniaxial compressive strength and modulus ratio. Fields shown for igneous sedimentary and metamorphic rocks (after Deere and Miller 1966)*

properties were therefore used to plot rock property charts (Figures 8.16 and 8.18). The rock strength chart based on the Shore scleroscope hardness appears to be limited to rocks with unconfined compressive strengths in excess of 35 MPa.

A major advance in the devlopment of the concepts governing engineering classifications of rock was made by Coates (1964). He considered the following five properties to be important:

(1) The principal reason for using the unconfined compressive strength of intact rock is that it indicates whether or not the strength is likely to be a source of trouble in itself. Furthermore there is a rough correlation between compressive strength and the modulus of deformation. Three categories were recognized:

(a) weak, less than 35 MPa;
(b) strong, between 35 and 175 MPa;
(c) very strong, greater than 175 MPa;

(2) The pre-failure deformation characteristics of the intact rock indicate whether creep could be expected in the material at stress levels less than those required to cause failure. In extreme cases it could also indicate the possibility of ground heave. Two categories were recognized:
(a) elastic;
(b) viscous (at a stress of 50% of the unconfined compressive strength the strain rate is greater than two microstrains per hour).

(3) The failure characteristics of the intact rock influence the factor of safety used in design, as well as the precautions to be taken during construction. Two categories were distinguished:
(a) brittle;
(b) plastic (more than 25% of the total strain before failure is permanent).

(4) Gross homogeneity and isotropy of the formation:
(a) massive;
(b) layered (i.e. generally including sedimentary and cleaved and schistose metamorphic rocks, as well as any other rocks with layering which produces parallel lines of weakness).

(5) The continuity of a rock mass, that is, whether it is divided into large or small blocks or is massive:
(a) solid (discontinuity spacing greater than 2 m);
(b) blöcky discontinuity spacing between 75 mm and 2 m);
(c) broken (in fragments that would pass through a 75 mm sieve).

3.6 The rating concept

As far as the classification of rock masses for engineering purposes is concerned most work has been done in relation to tunnelling and the construction of underground chambers. Engineers have been especially concerned with the determination of rock mass quality in relation to the time the rock mass can remain unsupported, and the type and amount of support necessary.

Terzaghi (1946) was one of the first workers to attempt an engineeirng classification of rock masses. In this he recognized the significance of discontinuities, their spacing and their filling materials as well as the influence of weathering (Table 3.15). However, Terzaghi's classification tended to over look the properties of the rock. For instance, both chalk and granite could fall into the same class but their different character is obvious.

Wickham *et al.* (1972), Bieniawski (1973; 1974; 1975a; 1983) and Barton *et al.* (1975) proposed classifications of jointed rock masses which depended on various weighted aspects of both the rock material and the rock mass. Their objective was to obtain rock mass ratings which could be used for design purposes.

Table 3.15 Classification of *in situ* rock for predicting tunnel support requirements (after Terzaghi 1946)

Term	*Description*
Intact	Rock contains neither joints nor hair cracks
Stratified	Rock consists of individual strata with little or no resistance against separation along the boundaries between strata
Moderately jointed	Rock contains joints and hair cracks, but the blocks between joints are locally grown together or so intimately interlocked that vertical walls do not require lateral support
Blocky and seamy	Rock consists of chemically unweathered rock fragments which are entirely separated from each other and imperfectly interlocked. In such rock vertical walls may require support
Crushed	Chemically unweathered rock has the character of crusher run material
Squeezing	Rock slowly advances into the tunnel without perceptible volume change
Swelling	Rock advances into the tunnel chiefly on account of expansion caused by minerals with a high swelling capacity

Wickham *et al.* (1972) introduced the concept of rock structure rating (RSR) which refers to the quality of rock structure in relation to ground support in tunnelling. Although their classification is specifically related to tunnels it did introduce the rating principle which has been adopted subsequently in other classifications. The rating principle allows several parameters to be taken into account and their influence on the rock mass is collectively assessed. The RSR system rates the relative effect on ground support requirements of three parameters in relation to several geological factors and, where applicable, with respect to each other. Parameter A provides a general appraisal of the rock structure; parameter B is related to the joint pattern and the direction of drive and parameter C represents a general evaluation of the effect of groundwater flow on the type of support necessary (Table 3.16). The method allows for three conditions of joint surfaces, namely, tight or cemented, slightly weathered and severely weathered or open; and four types of water inflow. The RSR value of a particular rock mass is given by the numerical sum of the parameters A, B and C, and the values range from 25 to 100, reflecting the quality of the rock mass. For example, Wickham *et al.* concluded that rock masses with RSR values less than 27 would require heavy support, whilst those with ratings over 77 would probably stand unsupported.

The classification of rock masses advanced by Bieniawski (1973) initially incorporated various factors: the RQD; the unconfined compressive strength; the degree of weathering; the spacing, orientation, separation and

Table 3.16 Rock structure rating (after Wickham *et al.* 1972)

(1) Parameter A – Geological structure

Basic rock type	*Massive*	*Slightly faulted or folded*	*Moderately faulted or folded*	*Intensely faulted or folded*
Igneous	30	26	15	10
Sedimentary	24	20	12	8
Metamorphic	27	22	14	9

(Maximum value of parameter A is 30)

(2) Parameter B – Joint pattern and direction of drive

Average joint spacing	*Strike perpendicular to axis*					*Strike parallel to axis*		
	Direction of drive							
	Both	*With dip*		*Against dip*		*Both*		
				Dip of prominent joints				
	<20°	20–50°	50–90°	20–50°	50–90°	<20°	20–50°	50–90°
<0.15 m Clearly jointed	14	17	20	16	18	14	15	12
0.15–0.3 m Moderately jointed	24	26	30	20	24	24	24	20
0.3–0.6 m Moderate to blocky	32	34	38	27	30	32	30	25
0.6–1.2 m Blocky to massive	40	42	44	36	39	40	37	30
>4.0 m Massive	45	48	50	42	45	45	42	36

(Maximum value of parameter B is 50)

(3) Parameter C – Groundwater, joint condition

Anticipated water inflow (l/s per 300 m)	*Sum of parameters A + B*					
	20–45			46–80		
	*Joint condition**					
	1	2	3	1	2	3
None	18	15	10	20	18	14
Slight (<15)	17	12	7	19	15	10
Moderate (15–75)	12	9	6	18	12	8
Heavy (>75)	8	6	5	14	10	6

* 1 = tight or cemented; 2 = slightly weathered; 3 = severely weathered or open
(Maximum value of parameter *C* is 20)

continuity of the discontinuities; the groundwater flow. Unfortunately, no account was taken of the roughness of joint surfaces of the character of the infill material. The unconfined compressive strength of the intact rock has an important bearing on the engineering performance of the rock mass when the discontinuities are widely spaced or the rock mass is weak. It is also important if the joints are not continuous. Because the unconfined compressive strength and the degree of weathering are two interdependent factors, Bieniawski (1974) subsequently revised his views and suggested that both factors should be regarded as one parameter, namely, the strength of the rock material. He chose a somewhat modified version of Deere's (1964) classification of intact strength for his classification (Table 3.16). Bieniawski argued that when assessing importance ratings the strength of rocks less than 25 MPa does not contribute to the overall mobility of the rock mass.

The point load test can be used to determine the intact strength on site as several authors (D'Andrea *et al.*, 1965; Frankline and Broch, 1972; and Bieniawski, 1975b) have found a close correlation between point load strength and the unconfined compressive strength.

Special caution should be exercised in using this classification in the case of shales and other swelling materials. For instance, Bieniawski (1973) suggested that their behaviour under conditions of alternate wetting and drying should be assessed by a slake-durability test (Franklin and Chandra, 1972).

The presence of discontinuities reduces the overall strength of a rock mass and their spacing and orientation govern the degree of such reduction. Hence the spacing and orientation of the discontinuities are of paramount importance as far as the stability of structures in jointed rock masses is concerned. Bieniawski (1973) accepted the classification of discontinuity spacing which was proposed by Deere (1968).

Another revision of Bieniawski's ideas included the continuity and separation of discontinuities which he later (1974) grouped together under the heading 'condition of discontinuities'. This parameter also took account of the surface roughness of discontinuities and the quality of the wall rock. The condition of discontinuities is as important as the discontinuity spacing, for example, tight discontinuities with rough surfaces and no infill have a high strength. By contrast open discontinuities which are continuous mean a plane of weakness and furthermore facilitate unrestricted flow of groundwater. Obviously the condition of discontinuities influences the extent to which the rock material affects the behaviour of a rock mass.

Groundwater has an important effect on the behaviour of a jointed rock mass. However, as the pore water pressures are of greater significance in foundations than groundwater inflow, Pells (1974) suggested that the pore water pressure ratio (r_u), where r_u is defined as the ratio of the pore pressure to the major principal stress, should be included within the classification. This view was accepted by Bieniawski (1974) and the pore water pressure ratio was incorporated into his classification.

Bieniawski (1983) firmly stated that the quality of the output from an engineering rock mass classification is directly related to the quality of the input data. It is essential that the correct engineering geological data are collected and appropriately presented. To this end he devised a special input data form (Figure 3.7). This form represents the minimum information which is required for engineering design.

Bieniawski (1976) grouped each of the chosen rock mass parameters into five classes (Table 3.17). He considered five classes to be sufficient to provide a meaningful discrimination in all the chosen properties. Because these paremeters vary in their relative importance from rock mass to rock mass and can contribute individually or collectively to its engineering performance, Bieniawski used a rating system. In other words a weighted numerical value was given to each class in each parameter. The total rock mass rating is the sum of the weighted values of the individual parameters, the higher the total rating, the better the rock mass condition (Table 3.17c). In describing a rock mass the class rating should be quoted with the class number, for example, class 3, rating 68.

Laubscher (1977) developed a geomechanics classification for jointed rock masses which was based on Bieniawski's classification. This classification also was designed principally for use in relation to subsurface excavation. Laubscher's classification similarly involves the rating concept, the rating extending from 0–100, and is supposed to cover all variations of jointed rock masses from very poor to very good. There are five classes, each one being subdivided into an A and B group, and each group has a 10 point rating (Table 3.18). The classification considers the rock quality designation (RQD), intact rock strength (IRS), joint spacing, joint condition and groundwater. The rating for the spacing of one, two or three joint sets is obtained by reference to Figure 3.8. Joint condition is based on the expression of the joint, its surface properties, the presence of alteration zones and the character of any fill material which is present. The rating of the joint condition is obtained by summing these various factors after they have been adjusted according to Table 3.19a, the total possible rating for joint condition being 30. Groundwater, particularly joint water pressure, was regarded as of paramount importance. The rating should be reduced if an excavation is oriented in an unfavourable direction with respect to geological structures, particularly the weakest joint sets. The magnitude of the adjustment depends on the attitude of the joints with respect to the vertical axis of the block (Table 3.19b). Laubscher's classification also involved adjustment of the class rating according to the influence of weathering, field and induced stresses, changes in stress and the influence of strike and dip orientations (Table 3.19c). For example, weathering affects the RQD, IRS and condition of the joints. As the RQD percentage can be decreased by weathering, Laubscher suggested that a reduction to 95% of the rating value was possible. Similarly the IRS is reduced if weathering occurs along microstructures. This time he suggested a decrease in rating to 96% for the bulk

Name of project
Site of survey
Conducted by
Date

STRUCTURAL REGION	ROCK TYPE AND ORIGIN

DRILL CORE QUALITY R Q D*

Excellent quality	90–100%
Good quality	75–90%
Fair quality	50–75%
Poor quality	25–50%
Very poor quality	<25%

* R Q D Rock Quality Designation

WALL ROCK OF DISCONTINUITIES

Unweathered
Slightly weathered
Moderately weathered
Highly weathered
Completely weathered
Residual soil

GROUND WATER

INFLOW per 10 m of tunnel length litres/minute

or

WATER PRESSURE kPa

or

GENERAL CONDITIONS (completely dry, damp, wet, dripping or flowing under low, medium or high pressure)

STRENGTH OF INTACT ROCK MATERIAL

Designation	Uniaxial compressive strength, MPa	OR	Point-load strength index, MPa
Very high	Over 250		>10
High	100–250		4–10
Medium high	50–100		2–4
Moderate	20–50		1–2
Low	5–25		<1
Very low	1–5		

SPACING OF DISCONTINUITIES

		Set 1	Set 2	Set 3	Set 4
Very wide	Over 2 m				
Wide	0.6–2 m				
Moderate	200–600 mm				
Close	60–200 mm				
Very close	<60 mm				

NOTE These values are obtained from a joint survey and not from borehole logs

STRIKE AND DIP ORIENTATIONS

Set 1 Strike (from to) Dip
(average) (angle) (direction)

Set 2 Strike (from to) Dip

Set 3 Strike (from to) Dip

Set 4 Strike (from to) Dip

NOTE Refer all directions to magnetic north

CONDITION OF DISCONTINUITIES

PERSISTENCE (CONTINUITY)		Set 1	Set 2	Set 3	Set 4
Very low	<1 m				
Low	1–3 m				
Medium	3–10 m				
High	10–20 m				
Very high	20 m				
SEPARATION (APERTURE)					
Very tight joints	<1 mm				
Tight joints	0.1–0.5 mm				
Moderately open joints	0.5–2.5 mm				
Open joints	2.5–10 mm				
Very wide aperture	>10 mm				
ROUGHNESS (state also if surfaces are stepped, undulating or planar)					
Very rough surfaces					
Rough surfaces					
Slightly rough surfaces					
Smooth surfaces					
Slickensided surfaces					
FILLING (GOUGE)					
Type					
Thickness					
Uniaxial compressive strength, MPa					
Seepage					

MAJOR FAULTS OR FOLDS

Describe major faults and folds specifying their locality, nature and orientations

GENERAL REMARKS AND ADDITIONAL DATA

NOTE
(1) For definitions and methods consult ISRM document *Quantitative discontinuities in rock masses*
(2) The data on this form constitute the minimum required for engineering design.
The geologist should, however, supply any further information which he considers relevant.

Figure 3.7 *Input data form for the Geomechanics Classification (after Bieniawski 1983)*

Table 3.17 Engineering classification of jointed rock masses (After Bieniawski, 1989)

(a) Classification parameters and their ratings

	Parameter		Range of values				
1	Strength of intact rock material	Pointed load strength index (MPa)	>10	4–10	2–4	1–2	For this low range, uniaxial compressive test is preferred
		Uniaxial compressive strength (MPa)	>250	100–250	50–100	25–50	5–25 1–5 <1
	Rating		15	12	7	4	2 1 0
2	Drill core quality RQD (%)		90–100	75–90	50–75	25–50	<25
	Rating		20	17	13	8	3
3	Spacing of discontinuities		>2 m	0.6–2 m	200–600 mm	200–600 mm	<60 mm
	Rating		20	15	10	8	5
4	Condition of discontinuities		Very rough surfaces Not continuous No separation Unweathered wall rock	Slightly rough surfaces Separation <1 mm Slightly weathered walls	Slightly rough surface Separation <1 mm Highly weathered wall	Slickensided surfaces or Gouge <5 mm thick or Separation 1–5 mm Continuous	Soft gouge >5 mm thick or Separation >5 mm Continuous
	Rating		30	25	20	10	0
5	Groundwater	Inflow per 10 m tunnel length (L/min)	None	<10	10–25	25–125	>125
			or	or	or	or	or
		Ratio $\dfrac{\text{Joint water pressure}}{\text{Major principal stress}}$	0	<0.1	0.1–0.2	0.2–0.5	>0.5
			or	or	or	or	or
		General conditions	Completely dry	Damp	Wet	Dripping	Flowing
	Rating		15	10	7	4	0

(b) Rating adjustment for discontinuity orientations

Strike and dip orientations of Discontinuities		Very favourable	Favourable	Fair	Unfavourable	Very unfavourable
Ratings	Tunnels and mines	0	−2	−5	−19	−12
	Foundations	0	−2	−7	−15	−25
	Slopes	0	−5	−25	−50	−60

(c) Rock mass classes determined from total ratings

Rating	100 ← 81	80 ← 61	60 ← 41	40 ← 21	<20
Class no.	I	II	III	IV	V
Description	Very good rock	Good rock	Fair rock	Poor rock	Very poor rock

(d) Meaning of rock mass classes

Class no.	I	II	III	IV	V
Average stand-up time	20 yr for 15 m span	1 yr for 10 m span	1 wk for 5 m span	10 h for 2.5 m span	30 min for 1 m span
Cohesion of the rock mass (kPa)	>400	300–400	200–300	100–200	<100
Friction angle of the rock mass (deg)	>45	35–45	25–35	15–25	<15

Table 3.18 Classification of variations in jointed rock masses

Class	1		2		3		4		5	
Rating	100–81		80–61		60–41		40–21		20–0	
Description	Very good		Good		Fair		Poor		Very poor	
Sub classes	A	B	A	B	A	B	A	B	A	B
Item										
(1) RQD, %	100–91	90–76	75–66	65–56	55–46	45–36	35–26	25–16	15–6	5–0
Rating	20	18	15	13	11	9	7	5	3	0
(2) IRS, MPa	141–136	135–126	125–111	110–96	95–81	80–66	65–51	50–36	35–21	20–6
Rating	10	9	8	7	6	5	4	3	2	1
(3) Joint spacing	Refer to Table 3.19									
Rating	30									 0
(4) Condition of joint										
Rating	45 .... Static angle of friction 5									
	30 .... Refer to Table 3.19 0									
(5) Groundwater: Inflow per 10 m length			0		25 l/min		25–125 l/min		125 l/min	
or Joint water pressure / Major principal stress			0		0.0–0.2		0.2–0.5		0.5	
or Description			Completely dry		Moist only		Moderate pressure		Severe problems	
Rating	10		7		4		0			

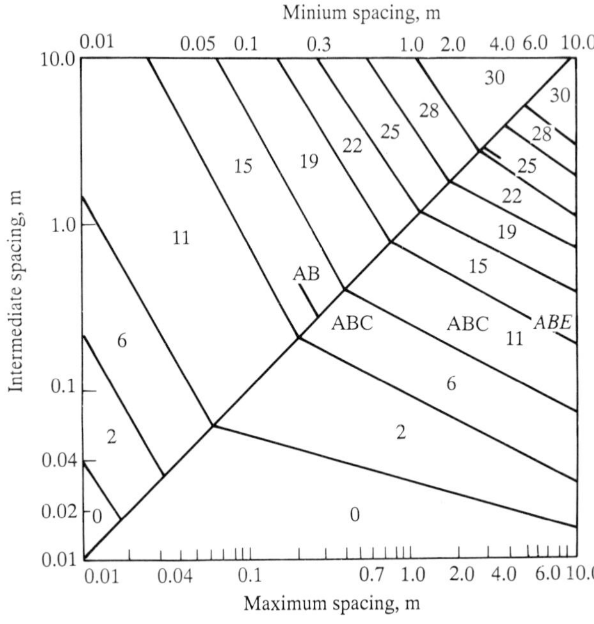

Figure 3.8 *Ratings for multi-joint systems: example. Joint spacing: A = 0.2 m, B = 0.5 m, C = 0.6 m, D = 1.0 m, E = 7.0 m, AB = 15, ABC = 6, ABD = 11, ABE = 15*

of the rock. An adjustment varying down to 82% of the rating was suggested for alteration along joints. Field and induced stresses may be responsible for compressing joints together and consequently the rating can be increased up to 120%. By contrast if the possibility of shear movement is increased, the rating would be decreased to 90% of its value. Similarly if joints open the rating is lowered by 24%.

Barton *et al.* (1975) defined rock mass quality (Q) in terms of six parameters.

(1) The RQD or an equivalent system of joint density.
(2) The number of joint sets (J_n) which is an important indication fo the degree of freedom of a rock mass. The RQD and the number of joint sets provide a crude measure of relative block size (RQD/J_n) with two extreme values (100/0.6 and 10/20) which differ by a factor of 400.
(3) The roughness of the most unfavourable joint set (J_r). The joint roughness and the number of joint sets determine the dilatancy of the rock mass.
(4) The degree of alteration or filling of the most unfavourable joint set (J_a). The roughness and degree of alteration of the joint walls or the filling materials provide an approximation of the shear strength of the rock mass (J_r/J_a). The quotient is weighted in favour of

Table 3.19 Adjustments to class ratings (after Laubscher 1977)

(a) *Assessments of joint conditions (adjustments as combined percentages of total possible rating of 30)*

Parameter	*Description*	*Percentage adjustment*
(a) Joint expression (large-scale)	Wavy unidirectional	99
		90
	Curved	89
		80
	Straight	79
		70
(b) Joint surface (small-scale)	Striated	99
		85
	Smooth	84
		60
	Polished	59
		50
(c) Alteration zone	Softer than wall rock	99
		70
(d) Joint filling	Coarse hard-sheared	99
		90
	Fine hard-sheard	89
		80
	Coarse soft-sheared	79
		70
	Fine soft-sheared	69
		50
	Gouge thickness	49
	< irregularities	35
	Gouge thickness	23
	> irregularities	12
	Flowing material	11
	> irregularities	0

(b) *Orientation adjustments for blocks with exposed bases*

Number of joints defining block	*Number of faces inclined away from vertical and adjustment percentage*				
	70%	*75%*	*70%*	*85%*	*90%*
3	3		2		
4	4	3		2	
5	5	4	3	2	1
6	6		4	3	2,1

(c) *Total possible reductions* (%)

	RQD	*IRS*	*Joint spacing*	*Condition of joints*	*Total*
Weathering	95	96		82	75
Field and induced stresses				120–76	120–76
Changes in stress				120–60	120–60
Strike and dip orientation			70		70

Table 3.20 Descriptions and ratings for the six parameters used to describe rock mass quality Q (from Barton *et al.* 1974)

1. ROCK QUALITY DESIGNATION		*(RQD)*	
A. Very poor		0–25	*Notes*
B. Poor		25–50	(i) Where *RQD* is reported or measured as ≤10 (including 0) a nominal value of 10 is used to evaluate *Q* in Eq. (3.2)
C. Fair		50–75	(ii) *RQD* intervals of 5, i.e. 100, 95, 90, etc. are sufficiently accurate
D. Good		75–90	
E. Excellent		90–100	
2. JOINT SET NUMBER		(J_n)	
A. Massive, no or few joints		0.5–1.0	*Notes*
B. One joint set		2	(i) For intersections use ($3.0 \times J_n$)
C. One joint set plus random		3	(ii) For portals use ($2.0 \times J_n$)
D. Two joint sets		4	
E. Two joint sets plus random		6	
F. Three joint sets		9	
G. Three joint sets plus random		12	
H. Four or more joint sets, random, heavily jointed, 'sugar cube', etc.		15	
J. Crushed rock, earthlike		20	
3. JOUNT ROUGHNESS NUMBER		(J_r)	
(a) *Rock wall contact* and			
(b) *Rock wall contact before 10 cms shear*			
A. Discontinuous joints		4	*Notes*
B. Rough or irregular, undulating		3	(i) Add 1.0 if the mean spacing of the relevant joint set is greater than 3 m
C. Smooth, undulating		2	(ii) $J_r = 0.5$ can be used for planar slickensided joints having lineations, provided the lineations are favourably orientated
D. Slickensided, undulating		1.5	
E. Rough or irregular, planar		1.5	
F. Smooth, planar		1.0	
G. Slickensided planar		0.5	
(c) *No rock wall contact when sheared*			
H. Zone containing clay minerals thick enough to prevent rock wall contact		1.0 (nominal)	
J. Sandy, gravelly or crushed zone thick enough to prevent rock wall contact		1.0 (nominal)	
4. JOINT ALTERATION NUMBER	(J_a)	ϕr (approx.)	
(a) *Rock wall contact*			
A. Tightly healed, hard, non-softening, impermeable filling i.e. quartz or epidote	0.75	(–)	*Note*
B. Unaltered joint walls, surface staining only	1.0	(25–35°)	(i) Values of (ϕ)r are intended as an approximate guide to the mineralogical properties of the alteration products, if present
C. Slightly altered joint walls. Non-softening mineral coatings, sandy particles, clay-free disintegrated rock etc.	2.0	(25–30°)	
D. Slity-, or sandy-clay coatings, small clay-fraction (non-softening)	3.0	(20–25°)	
E. Softening or low friction clay mineral coatings, i.e. kaolinite, mica. Also chlorite, talc, gypsum and graphite etc., and small quantities of swelling clays. (Discontinuous coatings, 1–2 mm or less in thickness)	4.0	(8–16°)	
(b) *Rock wall contact before 10 cm shear*			
F. Sandy particles, clay-free disintegrated rock etc.	4.0	(25–30°)	
G. Strongly over-consolidated, non-softening clay mineral fillings (Continuous, <5 mm in thickness)	6.0	(16–24°)	
H. Medium or low over-consolidation, softening, clay mineral fillings. (Continuous, <5 mm in thickness)	8.0	(12–16°)	
J. Swelling clay fillings, i.e. montmorillonite (Continuous, <5 mm in thickness). Value of *Ja* depends on percent of swelling clay-size particles, and access to water etc.	8.0–12.0	(6–12°)	
(c) *No rock wall contact when sheared*			

Table 3.20 (*Continued*)

K, L, M. Zones or bands of disintegrated or crushed rock and clay (see G, H, J for description of clay condition)	6.0, 8.0 8.0–12.0	(6–24°)	
N. Zones or bands of silty or sandy clay, small clay fraction (non-softening)	5.0.		
O, P, R. Thick, continuous zones or bands of clay (see G, H, J for description of clay condition)	10.0, 13.0, or 13.0–20.0	(6–24°)	
5. JOINT WATER REDUCTION FACTOR	(J_w)	Approx. water pressure (kg/cm^2)	
A. Dry excavations or minor inflow, i.e. <5 *l/min* locally	1.0	<1	*Notes* (i) Factors C to F are crude estimates. Increase J_w if drainage measures are installed (ii) Special problems caused by ice formation are not considered
B. Medium inflow or pressure occasional outwash of joint fillings	0.66	1.0–2.5	
C. Large inflow or high pressure in competent rock with unfilled joints	0.5	2.5–10.0	
D. Large inflow or high pressure, considerable outwash of joint fillings	0.33	2.5–10.0	
E. Exceptionally high inflow or water pressure at blasting, decaying with time	0.2–0.1	>10.0	
F. Exceptionally high inflow or water pressure continuing without noticeable decay	0.1–0.05	>10.0	

6. STRESS REDUCTION FACTOR			(*SRF*)	
(a) *Weakness zones intersecting excavation, which may cause loosening of rock mass when tunnel is excavated*				*Notes* (i) Reduce these values of *SRF* by 25–50% if the relevant shear zones only influence but do not intersect the excavation
A. Multiple occurrences of weakness zones containing clay or chemically disintegrated rock, very loose surrounding rock (any depth)			10.0	
B. Single weakness zones containing clay, or chemically disintegrated rock (depth of excavation ≤50 m)			5.0	
C. Single weakness zones containing clay, or chemically disintegrated rock (depth of excavation ≥50 m)			2.5	
D. Multiple shear zones in competent rock (clay free), loose surrounding rock (any depth)			7.5	
E. Single shear zones in competent rock (clay free; (depth of excavation ≤50 m)			5.0	
F. Single shear zones in competent rock (clay free) (depth of excavation >50 m)			2.5	
G. Loose open joints, heavily jointed or 'sugar cube' etc. (any depth)			5.0	
(b) *Competent rock, rock stress problems*				
	σ_c/σ_1	σ_t/σ_1		
H. Low stress, near surface	>200	>13	2.5	(ii) For strongly anisotropic stress field (if measured) when $5 \leq \sigma_1/\sigma_3 \leq 10$, reduce σ_c and σ_t to $0.8\,\sigma_c$ and $0.8\,\sigma_t$; when $\sigma_1/\sigma_3 > 10$, reduce σ_c and σ_t to $0.6\,\sigma_t$ where: σ_c = unconfined compression strength. σ_t = tensile strength (point load), σ_1 and σ_3 = major and minor principal stresses
J. Medium stress	200–10	13–0.66	1.0	
K. High stress, very tight structure (usually favourable to stability, may be unfavourable to wall stability)	10–5	0.66–0.33	0.5–2.0	
L. Mild rock burst (massive rock)	5–2.5	0.33–0.16	5–10	
M. Heavy rock burst (massive rock)	<2.5	<0.16	10–20	
(c) *Squeezing rock, plastic flow of incompetent rock under influence of high rock pressure*			(*SRF*)	
N. Mild squeezing rock pressure			5–10	
O. Heavy squeezing rock pressure			10–20	
(d) *Squeezing rock, chemical swelling activity depending on presence of water*				
P. Mild swelling rock pressure			5–10	
R. Heavy swelling rock pressure			10–20	

rough, unaltered joints in direct contact, such surfaces being close to peak strength and dilating strongly when sheared.
(5) The degree of water seepage or the joint water reduction factor (J_w).
(6) The stress reduction factor (SRF). Reduction of load due to excavation, stress in competent rock, and squeezing and swelling are taken account of in the stress reduction factor. The active stress is defined as J_w/SRF.

Descriptions and ratings of the six parameters are given in Table 3.20. The rock mass quality (Q) is then derived from

$$Q = (\mathrm{RQD}/J_n) \times (J_r/J_n) \times (J_w/\mathrm{SRF}) \quad (3.2)$$

The numerical value of Q ranges from 0.001 for exceptionally poor quality squeezing ground, up to 1000 for exceptionally good quality rock which is practically unjointed.

In certain situations an assessment of rock mass quality can only be made from core material derived from drilling operations. This raises the question of the reliability of the value of rock mass quality determined solely from core material as compared with that obtained by *in situ* investigation. Unfortunately few attempts have been made to compare borehole predictions of rock mass quality ratings with related values obtained from *in situ* measurements. One of the few comparisons was made by Barton (1976) who determined the rock mass qualities of a massive biotite gneiss in which an underground power house was excavated. He found that the average values of rock mass quality (Q) as obtained from borehole cores were twice those determined *in situ*. By contrast Cameron-Clark and Budavari (1981), after investigations carried out on three tunnels, showed that values derived from core material tended to give a poorer picture of rock mass conditions than those obtained from *in situ* measurements. The opposing conclusions were explained by Cameron-Clark and Budavari in terms of the differing geological conditions examined. Barton had investigated massive rock whilst the rocks they were involved with were predominantly jointed. It was suggested that poorer values of rock mass quality would be obtained from core material obtained from jointed rock than those obtained by *in situ* investigation. Whereas the converse situation applied in massive rocks. When comparisons were made in terms of the geomechanics' classification, Cameron-Clark and Budavari showed that there was an approximate 80% probability of the results derived from core material falling within the same or within one class of those determined *in situ*. When they made comparisons using the Q classification system they found that the correlation between the rock class derived by core material and that derived from *in situ* investigation was somewhat better than that obtained by the geomechanics system.

References

Anon. (1977) 'The description of rock masses for engineering purposes'. Working Party Report, *Q. J. Engg Geol.*, **10**, 355–388

Anon. (1979a) 'Classification of rocks and soils for engineering geological mapping. Part I–Rock and soil materials', *Bull. Int. Ass. Engg Geol.*, **19**, 364–371

Anon. (1979b) 'Suggested methods for determining water content, porosity, density, absorption and related properties and swelling and slake-durability index properties'. ISRM Standardization of Laboratory and Field Tests, *Int. J. Rock Mech. Min. Sci. & Geomech. Abstr.*, **16**, 325–341

Anon. (1981) 'Basic geotechnical description of rock masses'. ISRM Commission on Classification of Rocks and Rock Masses, *Int. J. Rock Mech. Min. Sci. & Geomech Abstr.*, **18**, 85–110

Barton, N. (1976) 'Recent experiences with the Q system in tunnel support design', *Proc. Symp. Exploration for Rock Engineering,* Bieniawski, Z. T. (ed.), A. A. Balkema, Cape Town, **1**, 107–115

Barton, N., Lien, R. and Lunde, J. (1975) 'Engineering classification of rock masses for design of tunnel support', (*Rock. Mech.* **6**, 189–236, 1974) *Norwegian Geot. Inst., Publ.*, **106**

Bieniawski, Z. T. (1973) 'Engineering classification of jointed rock masses'. *Trans S. Afr. Inst. Civil Engrs,* **15**, 335–343

Bieniawski, Z. T. (1974) Geomechanics classification of rock masses and its application in tunnelling, *Proc. 3rd Int. Cong. Rock Mech.*, Denver, **2**, 27–32

Bieniawski, Z. T. (1975a) 'Engineering properties of rock with reference to tunnelling'. (In *Tunnelling in Rock*), S. Afr. Inst. Civil Engrs/S. Afr. Nat. Gr. Rock Mech. CSIR, Pretoria, 105–123

Bieniawski, Z. T. (1975b) 'The point load test in geotechnical practice'. *Engg Geol.*, **9**, 1–11

Bieniawski, Z. T. (ed.) (1976) 'Rock mass classification in rock engineering', *Proc. Symp. Exploration for Rock Engineering,* A. A. Balkema, Cape Town, **1**, 97–106

Bieniawski, Z. T. (1983) 'The Geomechancis classification (RMR system) in design applications to underground excavations', *Int. Symp. Engineering Geology and Underground Construction,* Lisbon (IAEG/SPG/LNEC), **2**, 11.33–11.47

Bieniawski, Z. T. (1989) *Engineering Rock Mass Classifications*, Wiley-Interscience, New York

Cameron-Clark, I. S. and Budavari, S. (1981) 'Correlation of rock mass classification parameters obtained from borecore and *in situ* observations', *Engg Geol.*, **17**, 19–53

Chandler, R. J. (1969) 'The effects of weathering on the shear strength properties of Keuper Marl', *Geotechnique,* **19**, 321–334

Coates, D. F. (1964) 'Classification of rocks for rock mechanics', *Int. J. Rock Mech. Min. Sci.*, **1**, 421–429

Cottiss, G. I., Dowell, R. W. and Franklin, J. A. (1971) 'A rock classification system applied to civil engineering', *Civ. Engg Pub. Works Rev.*, **66**, Part 1–No. 777, 611–714; Part 2–No. 780, 736–43

D'Andrea, D. V., Fischer, R. L. and Fogelson, D. E. (1965) 'Prediction of the compressive strength of rock from other properties'. *Rep. US Bur. Mines,* Invest. No. 6702.

Dearman, W. R. (1974) 'The characterisation of rock for civil engineering practice in Britain'. Centenaire de la Société Géologique de Bélgique, Colloque', *Géologie de L'Ingenieur,* Liège, 1–75

Deere, D. U. (1964) 'Technical description of cores for engineering purposes', *Rock Mech. Engg Geol.,* **1**, 17–22

Deere, D. U. (1968) 'Geologic considerations'. In *Rock Mechanics in Engineering Practice,* Stagg, K. G. and Zienkiewicz, O. C. (eds), Wiley, London, 1–19

Deere, D. U. and Miller, R. P. (1966) 'Engineering classification and index properties for intact rock' *Tech. Rep. No. AFWL-TR-65-116,* Air Force Weapons Lab., Kirtland Air Base, New Mexico

Franklin, J. A. (1970) 'Observations and tests for engineering description and mapping of rocks', *Proc. 2nd Int. Cong. Rock Mech,* Belgrade, **1**, Paper 1–3

Franklin, J. A. and Broch, E. (1972) 'The point load strength test', *Int. J. Rock Mech. Min. Sci.,* **9**, 669–697

Franklin, J. A. and Chandra, A. (1972) 'The slake-durability test', *Int. J. Rock Mech. Min. Sci.,* **9**, 325–341

Grainger, P., McCann, D. M. and Gallois, R. W. (1973) 'Application of seismic refraction techniques for the study of fracturing in the Middle Chalk at Mundford, Norfolk', *Geotechnique,* **23**, 219–232

Hallbauer, D. K., Nieble, C., Berard, J., Rummel, F., Houghton, A. Broch, E. and Szlavin, J. (1978) 'Suggested methods for petrographic description'. ISRM Commission on Standardization of Laboratory and Field Tests. *Int. J. Rock Mech. Min. Sci & Geomech. Abstr.,* **15**, 41–45

Laubscher, D. H. (1977) 'Geomechanics classification of jointed rock masses – mining applications', *Trans. Inst. Min. Metall.,* Section A – Mining Industry, **86**, A1–A8

Mendes, F. M., Aires-Barros, L. and Rodrigues, F. P. (1966) 'The use of modal analysis in the mechanical characterization of rock masses', *Proc. 1st Int. Cong. Rock Mech.,* Lisbon, **1**, 217–223

Onodera, T. F. (1963) 'Dynamic investigation of rocks *in situ', Proc. 5th Symp. Rock Mech,* Minnesota University, Rolla, Pergamon Press, New York, 517–533

Onodera, T. F. and Kumara, A. H. M. (1980) 'Relation between texture and mechanical properties of crystalline rocks', *Bull. Int. Soc. Engg Geol.,* **22**, 173–177

Pells, P. J. H. (1974) 'Discussion: Engineering: classification of jointed rock masses', *Trans. S. Afr. Inst. Civil Engrs,* **16**, 242

Simmons, G., Todd, T. and Baldridge, W. S. (1975) 'Toward a quantitative relationship between elastic properties and cracks in low porosity rock'. *Am. J. Sci.,* **275**, 318–345

Terzaghi, K. (1946) 'Introduction to tunnel geology'. In *Rock Tunneling with Steel Supports.* Proctor, R. & White, T. (eds), Commericial Shearing and Stamping Co., Youngstown, Ohio, 17–99

Voight, B. (1968) 'On the functional classification of rocks for engineering purposes', *Int. Symp. Rock Mechanics,* Madrid, 131–135

Ward, W. H., Burland, J. B. & Gallois, R. W. (1968) 'Geotechnical assessment of a site at Mundford, Norfolk, for a large proton accelerator', *Geotechnique,* **18**, 399–431

Wickham, G. E., Tiedemann, H. R. and Skinner, E. H. (1972) 'Support determination based on geologic predictions', *Proc. 1st N. Am. Tunneling Conf.,* AIME, New York, 43–64

Willard, R. J. and McWilliams, J. R. (1969) 'Microstructural techniques in the study of physical properties of rocks', *Int. J. Rock Mech. Min. Sci.,* **6**, 1–12

4 Groundwater in rock masses

D Billaux and B Feuga
Service Géological National, Orléans, France

4.1 Basic concepts*

4.1.1 The origin of groundwater

Groundwater, in particular that found in the surficial part of the earth's crust (which is of most interest in rock mechanics) comes essentially from infiltrating rainwater. The water cycle can be represented, in a simplified way, by the diagram in Figure 4.1. Movement in the underground part of the water cycle can be extremely slow. The term 'fossil water' is used for groundwater that infiltrated during a past geological age, under climatic conditions differing from those existing today.

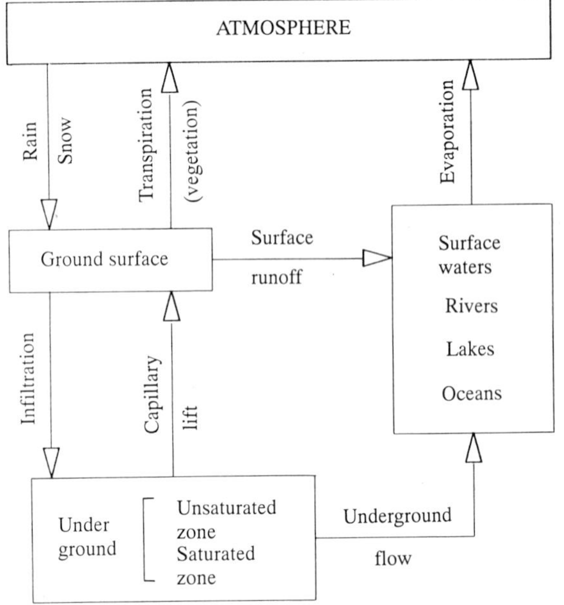

Figure 4.1 *Simplified representation of the water cycle*

A part of the groundwater, most commonly very small, comes, not from the atmosphere, but from deep-seated magmatic phenomena. This is known as 'juvenile water'.

'Connate water' is that which was present in the rocks at the time of their formation and has remained trapped within them. This water was originally either atmospheric or juvenile, depending on the nature of the rocks with which it is associated – sedimentary or igneous.

4.1.2 The forms of water in the subsurface

Water below the Earth's surface is present in a variety of forms. These are defined below in order of increasing mobility:

(1) *Combined water.* This is water that enters into the crystal lattices of minerals, and is thus an essential part of their chemical composition. Such hydrated minerals, which may be present in significant amounts, can lose their combined water under the influence of variations, which may be quite small, in the prevailing thermodynamic conditions. This is the case for certain clay minerals, for example. In contrast, certain anhydrous minerals, such as anhydrite ($CaSO_4$) tend to take up water if and when free water becomes available.

These phenomena, *hydration* and *dehydration*, can have major geotechnical consequences, modifying the mechanical characteristics of the materials – in the case of dehydration, with shrinkage and fracturing, and in the case of hydration, with the development of swelling pressure and the tendency to closure of voids.

(2) *Retained*, or *bound water.* This water is retained in voids by physical bonds, and cannot be mobilized by gravity. In immediate contact with the wall of a fracture or the surface of a grain, it takes the form of a layer of water adsorbed by molecular attraction. This layer is a few tens of molecules (0.1 μm) thick, and pressure in it attains several thousands of megapascals, thus density and viscosity are very much higher than in normal water. With distance from the wall or grain surface, the forces of molecular attraction decrease very rapidly, the water then taking the form of *pellicular water*. This forms a transition layer of the order of 1 μm thick, which has character-

* The definitions of basic terms given in the first part are based largely on Castany and Margat (1977).

istics very similar to those of ordinary water but does not transmit hydrostatic pressure.

(3) Away from this layer, the water is free to move under the effects of differences in hydraulic load, caused, among other things, by gravity. This free water is called *gravitational water*.

(4) *Capillary water* is a particular type of bound water subject to the effect of surface tension. This type of water is found in the unsaturated zone and in the zone of capillarity (the capillary fringe) that overlies the water table.

4.1.3 Groundwater and voids in rock: porous and fractured media

The subsurface can only contain water to the extent that it possesses voids that can be filled by water. These voids can be of different kinds, which, for simplicity, can be grouped into three main categories:

(1) *Pores*. These are very small voids, distributed throughout the body of rock under consideration, none of whose dimensions is dominant over the others. They can be called *three-dimensional voids*. The space density of pores in porous rocks is very high, and at the scale of the problems dealt with by the hydrogeologist or geotechnician, these rocks can be considered as *continuous* and characterized at all points by *total porosity*, porosity being the ratio of pore space to total volume (voids + solid) of the rock sample being considered.

(2) *Two-dimensional voids*. These are characterized by having one dimension (the aperture) much smaller than the other two. They may be of sedimentary origin (e.g. bedding planes), or mechanical origin (e.g. cracks, fissures and joints) and of any size. The presence of joints, certain of which appear during the formation of the rock itself, the others being the result of its tectonic history, is a characteristic common to all rock masses, to the extent that it can be said that rock hydraulics is in large part the hydraulics of *fractures*.

    Two-dimensional voids give a discontinuous character to rock masses (these voids are in fact often called *discontinuities*) and these rock masses are commonly spoken of as discontinuous media.

(3) *One-dimensional voids*. These are essentially karstic conduits, which appear in soluble rocks such as limestone, dolostone and evaporites, and whose dimensions can vary greatly, ranging from a few decimetres to several kilometres in length, and from less than $1\,cm^2$ to several tens of m^2 in section. Mention also should be made here, although from the genetic point of view they belong to joints, of *channels*, which lie in joint planes or at the intersections of such planes, and along which flow in rocks may be concentrated.

The separation of the voids in rock masses into three categories is obviously not as sharp as the above description might lead one to believe. In reality all gradations exist between the three.

The choice of a representative model will depend also on the scale considered. For example, the porosity of a sedimentary, metamorphic, or igneous rock will more often be due to microcracks along grain boundaries or within the grains themselves than to three-dimensional pores. Similarly a highly fractured rock could be considered as continuous, with only the most important fractures being regarded as discontinuities.

In practice, the nature of the problem being dealt with and the volume of the rock mass under consideration will, as much as its structure, determine the mode of representation (continuous or discontinuous) that will be adopted.

4.1.4 Porosity

Whereas the total porosity of a porous medium, defined above, does not distinguish between isolated pores, not connected to each other, and interconnected pores, it is clear that only the latter can accommodate flow. The first type of porosity is known as *closed porosity*, and the second as *open porosity*, the sum of the two obviously being the total porosity.

Part of the pore space, which becomes more important as the size of the pores decreases, is occupied by bound water, which cannot be made to move under the effect of gravity. The amount of water obtained by complete draining of a previously saturated sample corresponds to the *specific yield* (ratio of the volume of gravitational water contained in a saturated porous medium to the total volume of the medium). The complement of specific yield is *specific retention*.

A concept similar to, but distinct from, that of specific yield is that of *effective porosity* (the complement of which is the content of immobile water). This is defined as the ratio of the average rate of displacement of water in movement within the medium to the rate of percolation (see below), and is equivalent to the ratio of the volume of voids actually traversed by the moving water to the total volume of the medium.

4.1.5 Saturated and unsaturated zones. Free surface

The path traversed by groundwater, starting from its point of departure at the surface of the ground, generally begins in an *unsaturated zone*, characterized by the fact that the voids in the rock medium contain not only water, but also air. However, it may happen that no unsaturated zone exists and the ground is saturated up to the surface. The passage of the water from the surface of the ground downward through the unsaturated zone consitutes what is called *percolation*.

Many civil engineering works and shallow mine workings are located in the unsaturated zone, within which

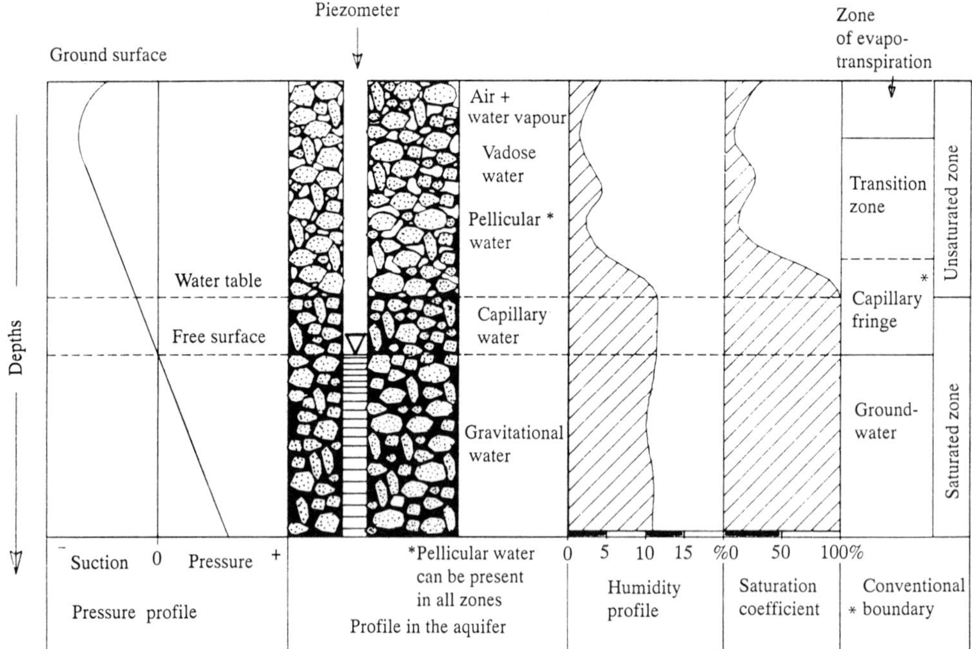

Figure 4.2 *Saturated and unsaturated zones. Diagrammatic representation of basic concepts (after Castany and Margat 1977)*

the forces of surface tension render the water pressure lower than atmospheric pressure (this pressure, negative relative to atmospheric pressure, is called *suction pressure*). Thus, except in the case of re-saturation, no significant forces of hydraulic origin can develop on these works. It should be pointed out that the existence of an unsaturated zone is not limited to the uppermost parts of terrains. Deep underground cavities in weakly permeable media can, as soon as the permeability of terrains in their immediate vicinity is higher than that of the rest of the rock mass, provoke gravity drainage inducing an inverted cone of desaturation above the void.

The phase of percolation ends when the water reaches the saturated zone. The unsaturated/saturated zone interface is known as the *water table*. The upper part of the saturated zone consists of the *capillary (ascent) fringe*, in which the water is maintained by surface tension and is thus still at a pressure lower than the atmospheric pressure. The surface at which the water pressure is equal to the atmospheric pressure is the *free surface*. Figure 4.2 represents diagrammatically the concepts explained above.

Once having reached the saturated zone, the groundwater flows in the subsurface according to the laws of underground hydraulics. The flow takes place in a geological environment which has a structure. The principal elements of the structure are defined, from a hydrogeological point of view, below.

4.1.6 Hydrogeological structures and aquifer systems

An *aquifer* is a geological body or layer of rocks sufficiently permeable to allow the flow of groundwater. Obviously it has boundaries. The lower boundary and the upper boundary are at the interfaces with much less permeable units. The upper boundary may also be quite simply the surface of the ground, in which case the aquifer is said to be unconfined. Lateral boundaries may be impermeable, but also are commonly permeable, corresponding to the limits along which either recharge or emergence take place.

The *sheet of groundwater* within an aquifer comprises all the water in the saturated zone of the aquifer, all parts of which are in hydraulic connection with each other. It may be *unconfined*, that is, be bounded at the top by the *water table*, or *free surface** separating saturated and unsaturated zones, or it may be *confined*, when it possesses no free surface. In this case the water is everywhere at a pressure greater than the atmospheric pressure, and the

* Strictly speaking, 'water table' and 'free surface' are two distinct concepts. The first is the boundary between the saturated and unsaturated zones, and thus is located at the top of the zone of capillarity, while the second is the surface at which water pressure is equal to atmospheric pressure, and thus at the base of the zone of capillarity.

piezometric surface is at a level higher than the top of the aquifer containing the water.

An extension of the aquifer concept is that of the *aquifer system*, which is a finite water-bearing domain all parts of which are in hydraulic continuity, and which is circumscribed by boundaries that form limits to the propagation of any significant influence outside the domain, for a given time constant.

Among the geological bodies or layers the permeability of which is too low for them to be considered as aquifers, hydrogeologists customarily distinguish, in order of decreasing permeability:

(1) *Semi-permeable* layers, which have low permeability, but across which significant transfers of water either coming from or going to contiguous aquifers are possible. These beds may have a considerable capacity, and are therefore called *aquitards*.
(2) *Very weakly permeable* layers, known as *aquicludes*, through which no appreciable flow is possible. Although hydrogeologically unproductive, these layers can, in the same way as the semi-permeable layers, be the site of important heads or hydraulic gradients that can cause serious problems for geotechnicians.
(3) *Truly impermeable* layers, which, in the range of depths that concern the geotechnician, can be considered more a limiting case than a reality: even in a rock in which the matrix is quasi-impermeable, fractures exist that confer to it a permeability which is, on the large scale, measurable.

Speaking of the part played by discontinuities is an opportune moment to mention that the concepts defined above have been developed by hydrogeologists, specialists whose concern is essentially the location of water resources. Their interests thus lie most particularly with permeable rocks, most of which are porous media. The geotechnician, in contrast, often has no choice in the terrain upon which structures are to be sited, and if he does, will most often avoid significantly water-bearing ones. He is in practice therefore very often confronted with poorly permeable rocks, in which any permeability will be due in large part to the discontinuity network, and to which the classical concepts of hydrogeology are not well-adapted.

For this reason, after a final paragraph recalling the basic concepts of subsurface hydrodynamics, this chapter will be divided into two parts, one on continuous media (porous media and those included in the same category), the other on discontinuous media.

4.1.7 Basic concepts of subsurface hydrodynamics

It is customary in subsurface hydraulics to express the potential of a particle of water by its *piezometric head*, ϕ. This piezometric head expressed in metres, is defined by the formula:

$$\phi = z + \frac{p}{\rho g} + \frac{v^2}{2g} \qquad (4.1)$$

in which z = level measured relative to a reference plane (m)
p = pressure (Pa)
ρ = unit mass (kg/m^3)
v = velocity (m/s)
g = gravity acceleration (m/s^2)

The term $z + p/\rho g$ constitutes the *hydrostatic head*, while the term $v^2/2g$ is called the *hydrodynamic head*. In practice, in aquifers where flow is governed by Darcy's law (see below), the hydrodynamic head is negligible compared to the hydrostatic head, which is then equivalent to the piezometric head.

The piezometric head at a point represents the elevation of the *piezometric level*, which is the level of the top of a column of static liquid that balances the hydrostatic head at the point. The piezometric level is represented physically by the level of water in an open vertical tube, called a piezometer, at the point considered. The *piezometric height* is the height of the column in this tube, measured from its lower opening.

Finally, the *piezometric surface* is the location of the piezometric levels. Strictly speaking, it is only possible to define a single piezometric surface for the sheet of water in an aquifer if the piezometric head in the aquifer is constant for all verticals, that is, if flow is horizontal. This situation is the exception rather than the rule, and in practice, in the

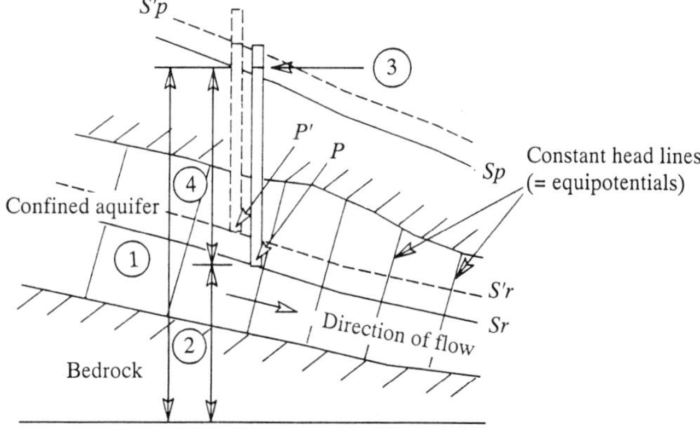

Figure 4.3 *Hydraulic head and piezometric level:*

(1) *hydraulic head at point* P,
(2) *level* z *of point* P,
(3) *piezometric level at point* P,
(4) *piezometric height at point* P,
S_p: *piezometric surface corresponding to the surface* S_r *containing point* P,
S'_p: *piezometric surface corresponding to the surface* S'_r *containing point* P'

general case of the three-dimensional flow, for each surface within the groundwater sheet there is a corresponding specific piezometric surface. The free surface in an aquifer is a particular case of piezometric surface. Figure 4.3 illustrates these different concepts.

4.2 Hydrodynamics of porous and equivalent media

4.2.1 Basic equations in the hydrodynamics of porous media

There are three of these equations:

(1) Darcy's law, published by Darcy in 1856, which defines the relation between the flowrate of fluid flowing in a porous medium and the gradient of the piezometric head.
(2) The equation of continuity, expressing the conservation of mass.
(3) The diffusivity equation, or the fundamental equation of the hydrodynamics of porous media, deduced by combining the preceding two laws.

(a) Darcy's law and permeability

In percolation experiments using tubes filled with various porous materials and under varying head differences (Figure 4.4), Darcy (1856) showed that the flowrate through the tubes was:

(1) proportional to the section of the tube;
(2) proportional to the difference of head between the two ends of the tube;
(3) inversely proportional to the length of the tubes;
(4) dependent upon the type of material, the dependence being expressed by a proportionality factor, always the same for the same material, which Darcy called *permeability*.

These results can be expressed in the form of Darcy's law:

$$Q = KS\frac{\phi_1 - \phi_2}{L} \qquad (4.2)$$

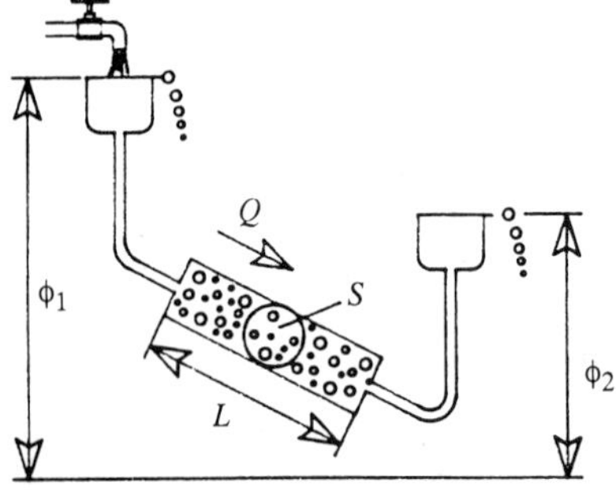

Figure 4.4 *Darcy's experimental set up* (ϕ = *piezometric head)*

where K is the permeability, $\phi_1 - \phi_2$ is the difference in head between the two ends of the tube, and the term $(\phi_1 - \phi_2)/L$ is the *head* or *hydraulic gradient* between these two ends. It can also be noted that the ratio Q/S defines a velocity, V, which is called the *percolation rate*, or the Darcy velocity. It must not be forgotten that this is a fictional velocity. The true average velocity of the particles of water is equal to the ratio between this percolation rate and the kinematic porosity of the medium. Nevertheless the rate of percolation is a very useful concept, making it possible to write Darcy's law in the form

$$\mathbf{V} = -K\,\mathbf{grad}\,\phi \qquad (4.3)$$

The conditions prevailing in a real porous medium are more complex than those of Darcy's experiments. In particular, unmodified natural materials are almost always anisotropic, and their permeability is expressed by a tensor which can be shown to be symmetrical and is represented by the following matrix:

$$\begin{bmatrix} K_{xx} & K_{xy} & K_{xz} \\ K_{xy} & K_{yy} & K_{yz} \\ K_{xz} & K_{yz} & K_{zz} \end{bmatrix}$$

If the coordinate axes correspond to the eigendirections of the tensor (the principal directions of permeability), this matrix takes the form

$$\begin{bmatrix} K_1 & 0 & 0 \\ 0 & K_2 & 0 \\ 0 & 0 & K_3 \end{bmatrix}$$

K_1, K_2 and K_3 representing the three principal permeabilities.

In its more general form Darcy's law can be written

$$\mathbf{V} = -\mathbf{K}\,\mathbf{grad}\,\phi \qquad (4.4)$$

It is valid in both steady and unsteady state conditions.

(b) The equation of continuity

The continuity equation expresses the fact that the amount of water that penetrates into an elementary volume of the porous medium is equal to that which leaves it, corrected by a term of storage or destorage.

It is written

$$\text{div}\,(\rho\mathbf{V}) = -\rho\,(\alpha + n\beta)\frac{\partial p}{\partial t} \qquad (4.5)$$

where t = time
n = open porosity
α = compressibility of the solid framework
β = compressibility of the water.

(c) The diffusivity equation

This is given by the combination of Darcy's law and the equation of continuity, and is written

$$\text{div}\,(\mathbf{K}\,\mathbf{grad}\,\phi) = S_s \frac{\partial \phi}{\partial t} \tag{4.6}$$

$S_s = \rho g\,(\alpha + n\beta)$ (dimension $= L^{-1}$) is called the specific storage; it represents the capacity of the porous medium to store water by compressing it and expanding the solid framework.

If a source term q is introduced (input or offtake flow rate per unit volume), the diffusivity equation is written

$$\text{div}\,(\mathbf{K}\,\mathbf{grad}\,\phi) + q = S_s \frac{\partial \phi}{\partial t} \tag{4.7}$$

This equation is likely to be considerably simplified in practice. Among the possible simplifications are the following:

(1) The case in which the principal directions of permeability are parallel to the coordinate axes, when the equation is written

$$\frac{\partial}{\partial x}\left(K_x \frac{\partial \phi}{\partial x}\right) + \frac{\partial}{\partial y}\left(K_y \frac{\partial \phi}{\partial y}\right) + \frac{\partial}{\partial z}\left(K_z \frac{\partial \phi}{\partial z}\right) + q = S_s \frac{\partial \phi}{\partial t} \tag{4.8}$$

(2) The case of a steady-state regime, for which the second term is nil.

If the term q is also nil, the equation becomes

$$\frac{\partial}{\partial x}\left(K_x \frac{\partial \phi}{\partial x}\right) + \frac{\partial}{\partial y}\left(K_y \frac{\partial \phi}{\partial y}\right) + \frac{\partial}{\partial z}\left(K_z \frac{\partial \phi}{\partial}\right) = 0 \tag{4.9}$$

It can be seen that the solution to this equation, which may be multipled by any given constant factor, is independent of the values K_x, K_y and K_z, depending only on their ratios K_x/K_y and K_x/K_z, that is the anisotropy ratios of the medium.

If the medium is isotropic, the value of permeability does not enter into the solution. This observation is fundamental, for it shows that in a steady-state regime, in the absence of input or offtake imposed externally, the distribution of piezometric head in a porous medium is independent of the value of permeability. This is clearly not the same for flow rates through the medium, which, in accordance with Darcy's law, are proportional to permeability.

(3) The case of shallow, unconfined groundwater, where the compressibility S_s can be ignored, and where the second term is therefore also nil.

4.2.2 Resolution of the diffusivity equation. General considerations

The resolution of the diffusivity equation constitutes the basic problem in the hydraulics of porous media. It governs the distribution in space and time of the piezometric head, knowledge of which enables resolution of virtually any problem that may present itself in practice.

The equation of diffusivity is defined over a given domain at whose boundaries prevail conditions, which may vary in time, knowledge of which is clearly essential for its resolution. These conditions may be to two types:

(1) the Dirichlet conditon (fixed head), for which $\phi(x,y,z,t) = f(x,y,z,t)$, f being a known function;
(2) the Neumann condition (fixed flow), such that

$$\frac{\partial \phi}{\partial_n}(x,y,z,t) = g(x,y,z,t)$$

n being the normal to the boundary and g a known function.

In the cases of unconfined groundwater, the upper boundary of the domain of definition of the head ϕ is constituted by the free surface, whose position is not *a priori* known. This position is sought numerically, in a steady-state regime through the equality of the piezometric head and the level (atmospheric pressure being taken as zero) at the free surface, and in unsteady state through the equation of movement of the free surface, formulated by Boussinesq, which is written:

$$e\frac{\partial h}{\partial t} = K_x\left(\frac{\partial h}{\partial x}\right)^2 + K_y\left(\frac{\partial h}{\partial y}\right)^2 - \left[K_z + K_x\left(\frac{\partial h}{\partial x}\right)^2 + K_y\left(\frac{\partial h}{\partial y}\right)^2\right]\frac{\partial \phi}{\partial z} \tag{4.10}$$

where $h(x,y,t)$ = level of the free surface
e = kinematic porosity
K_x, K_y, K_z = principal permeabilities (assumed to correspond to the x-, y- and z-axes).

A particular type of boundary also should be mentioned which is of great practical importance as it is commonly encountered in the bottoms of deep, unfaced excavations – this is the *surface of seepage*, the interface between the saturated porous medium and the atmosphere. Such a boundary, at which the condition of equality of head and level is true, is not a flow-line since it is crossed by a certain amount of flow. It also should be noted that in the unsteady state the resolution of the diffusivity equation assumes that the distribution of piezometric head within the medium at the initial instant is known. Finally, the resolution of this equation requires that the values of the parameters involved be known, that is, permeability and, in the case of unsteady state, specific storage. The

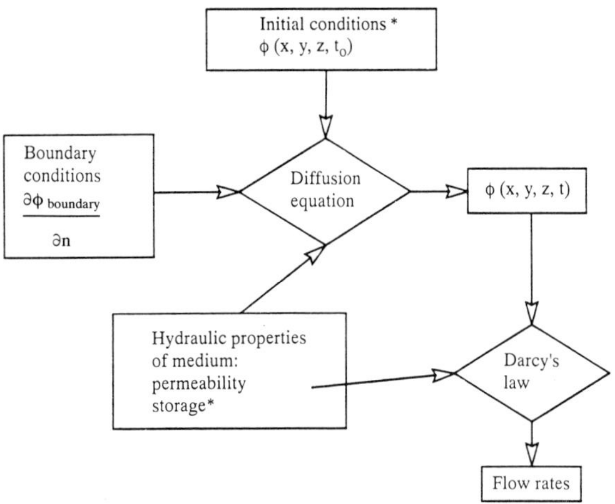

Figure 4.5 *Steps in the solution of a porous medium hydraulic problem. Parameters indicated by an asterisk are used only in transient cases*

concepts presented above are represented diagrammatically in Figure 4.5.

Direct resolution of the equation of diffusivity, even in its most simplified forms, is only possible in a small number of cases; among these are two whose practical importance is such that they will be presented here. They are the solutions of Dupuit (1863) and of Theis (1935). Many other examples of analytical solutions can be found in the reference work of Polubarinova-Kochina (1962). Nevertheless, in most cases there is no analytical solution and recourse must be taken to analogue or numerical (finite differences or finite element) methods whose basic principles will be presented later.

(a) The Dupuit formula

Many similar formulae are known by the name of Dupuit. The best known concerns radial flow towards a pumping well, in the steady state.

We have chosen to give here those of the Dupuit formulae which yield the flow rate, corresponding to a free surface flow, passing through a *dyke* of rectangular cross-section (Figure 4.6), and whose rigorous demonstration is in fact due to the Russian mathematician Charny (in Polubarinova-Kochina, 1962).

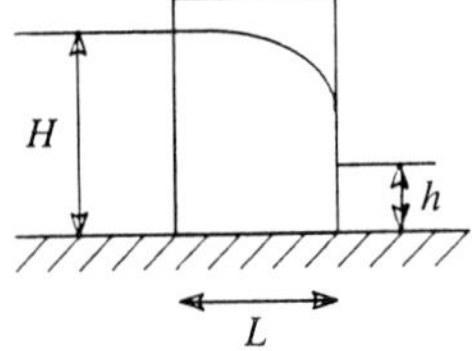

Figure 4.6 *Free surface flow through a rectangular, homogeneous and isotropic bund, on an impermeable bedrock. Notations*

The flow rate, q, through a unit length section of the *dyke* is derived as follows:

$$q = \frac{K(H^2 - h^2)}{2L} \tag{4.11}$$

K being the permeability of the material of which the *dyke* is constituted.

(b) The Theis formula

The formula established by Theis is of great practical importance as it is the basis of the methods used to interpret pumping test data for determining the permeability and specific storage of aquifers. It corresponds to the case of pumping at constant flow rate Q in a well of radius taken as zero, in an aquifer of infinite extent, of thickness b, permeability K and specific storage S_s (Figure 4.7).

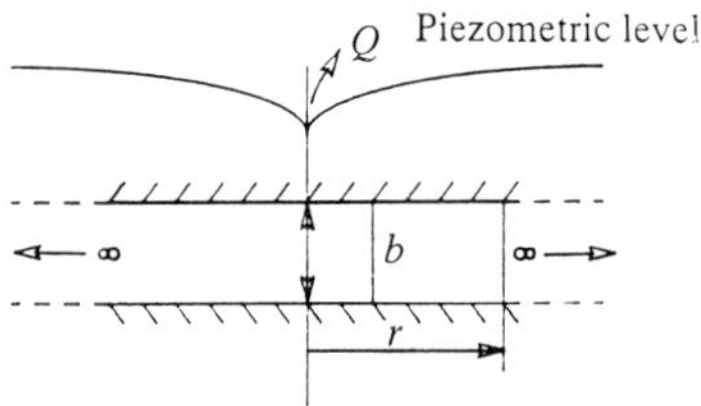

Figure 4.7 *Pumping conditions corresponding to Theis' solution*

The standard way of writing the Theis formula involves two parameters that have not yet been defined:

the transmissivity of the aquifer: $T = Kb$ (dimension L^2T^{-1}), and
its storage coefficient: $S = S_s b$ (no dimension).

The drawdown, s, of the groundwater (lowering of the piezometric level) is given, as a function of time, t, and the distance, r, to the well by the Theis formula:

$$s(r,t) = \frac{q}{4\pi T} W(u) \tag{4.12}$$

where $$u = \frac{r^2 S}{4Tt}$$

and the well function $$W(u) = \int_u^{\infty} \frac{e^{-\tau}}{\tau} d\tau$$

(c) Discharge drained by a circular gallery in a steady state

The formula

$$Q = \frac{2\pi Kh}{\ln \dfrac{h + \sqrt{(h^2 + R^2)}}{R}} \tag{4.13}$$

enables calculation of the flow rate drained per unit length in steady state by a gallery of radius R at a depth h below the piezometric level assumed to be horizontal*, in an aquiferous medium of permeability K (Figure 4.8).

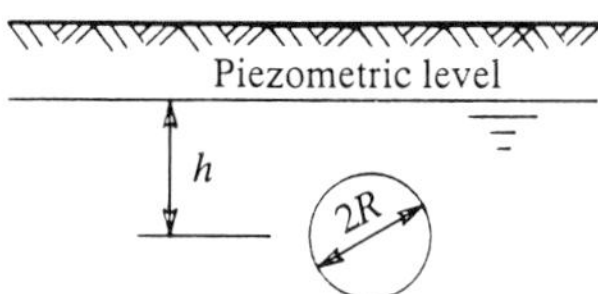

Figure 4.8 *Case of a circular gallery in an aquifer*

If R remains very small relative to h (a deep gallery) this formula is simplified to

$$Q = \frac{2\pi KH}{\ln (2h/R)} \tag{4.14}$$

(d) Numerical resolution of the diffusivity equation by finite differences

The basis of the finite difference method is the discretized form of the diffusivity equation, which is obtained using the formula of Taylor on rectangular (or square, 2D) or parallelepiped-shaped (or cubic, 3D) grids.

In the case of a cubic grid of edge length a, this discretized form of the equation is:

$$Kx\left(x + \frac{a}{2}, y, z\right)[\phi(x + a, y, z) - \phi]$$

$$+ Kx\left(x - \frac{a}{2}, y, z\right)[\phi(x - a, y, z) - \phi]$$

$$+ Ky\left(x, y \frac{a}{2} -, z\right)[\phi(x, y, + a, z) - \phi]$$

$$+ Ky\left(x, y - \frac{a}{2}, z\right)[\phi(x, y - a, z) - \phi]$$

$$+ Kz\left(x, y, z + \frac{a}{2}\right)[\phi(x, y, z + a) - \phi]$$

$$+ Kz\left(x, y, z - \frac{a}{2}\right)[\phi(x, y, z - a) - \phi]$$

$$= a + {}^2 S_s . \frac{\partial \phi}{\partial t} \tag{4.15}$$

where $\phi = \phi(x, y, z)$.

In the case of a transient state regime, time must also be discretized, and is divided into 'time-steps'. The second term can then be written in the form

$$a^2 S_s \frac{\phi(x,y,z,t + \Delta t) - \phi(x,y,z,t)}{\Delta t} \tag{4.16}$$

ΔT being duration of the time-step considered.

The first term also has to be expressed as a function of time. Two basic choices are possible:

(1) The potential values are taken at time t. In this case the equation includes only a single unknown, $\phi(x,y,z,t + \Delta t)$, which can be calculated directly as a function of the values of ϕ at time t. The process obviously begins with the calculation of the potential at each node for the first item-step as a function of the initial potential, which is an input into the problem. Methods of this type are said to be *explicit*. It can be shown mathematically that they are stable only for values of the time-step below a certain threshold value.

(2) The potential values are taken at time $t + \Delta t$. This case is very similar to that of a steady-state regime. For each time-step a system of N linear equations with N unknowns must be solved; the known terms of the equations contain the potentials at the preceding time-step. These methods are termed *implicit*, and are stable whatever the duration of the time-step.

(e) The finite element method

It is not possible, in the framework of this chapter, even in simplifying as much as possible, to present the bases of the finite element method in a rigorous manner. For this the reader should refer to a work such as that of Desai and Abel (1972). It will simply be pointed out that the method is based on the use of a variational principle.

It can be shown that the resolution of the equations for the hydraulics of porous media, which can be written, for the two dimensional case in steady state:

Over A:

$$\text{div } \mathbf{V} = 0 \text{ or } \frac{\partial v_i}{\partial x_i} = 0 \text{ (equation of continuity)} \tag{4.17}$$

$$\mathbf{V} = -\mathbf{K}\,\mathbf{grad}\,\phi \text{ or } v_i = -K_{ij}\frac{\partial \phi}{\partial x_j} \textit{ (Darcy's law)} \tag{4.18}$$

$$\text{on } L_1\ \phi(x_i) = \Phi(x_i) \text{ (boundary potential condition)} \tag{4.19}$$

* In reality, the presence of the gallery should entail a drawdown of the piezometric level; nevertheless in weakly permeable media the flow rate drained by the gallery is small in relation to the recharge of the groundwater from the surface, and the possible drawdown can be ignored.

on L_2 condition) $\qquad v_i(x_i) \cdot n_i(x_i) = V(x_i)$ (boundary flow

(4.20)

where A is the flow domain and L_1 and L_2 are its boundaries, constitutes a problem equivalent to that of the minimization of a functional $F(\phi)$ whose expression, in this case, is as follows:

$$F(\phi) = \int_A \tfrac{1}{2} k_{ij} \frac{\partial \phi}{\partial x_j} \frac{\partial \phi}{\partial_{xi}} \, dA - \int_{L_1} (\phi - \Phi) K_{ij} \frac{\partial \phi}{\partial x_j} n_i \, dL + \int_{L_2} V.\phi \, dL \tag{4.21}$$

The finite element method consists of searching for an approximate discretized form of the function ϕ that minimizes the functional F.

This can be done in one of several different ways. The most commonly used, known as Ritz's method (Zienkiewicz 1977), consists in seeking a solution of the form

$$\phi^N(x_i) = a_n \xi_n(x_i) \qquad (n = 1 \text{ to } N) \tag{4.22}$$

It is important at this stage to make a choice concerning the form of the functions $\xi_n(x_i)$, called *generalized coordinates*, which should be able to take into account the different types of boundary conditions, be adaptable to variations of permeability in the field, etc.

The finite element method consists of dividing the field into small elements, generally triangular or quadrangular, resolving the problem independently for each element and combining the various solutions to obtain an overall solution. To a node is allocated the function $\Phi(x_i)$ equal to 1 at the node and varying linearly over the elements adjacent to this node to attain the value 0 on all the other elements. In these conditions the coefficients a_n are the values ϕ_n of ϕ^N at each node.

Minimization of the function, which consists of seeking ϕ^N such that

$$\frac{\partial F(\phi)}{\partial \phi_n} = 0 \tag{4.23}$$

comes in practice to the same thing as resolving this equation for each element taken separately:

$$\frac{\partial F^e(\phi)}{\partial \phi_n} = 0 \quad \text{(the index (e) represents the element)} \tag{4.24}$$

Writing this equation for each element requires the use of very simple integrations, and aggregation of all the elementary solutions gives an equation of the type

$$\frac{\partial F(\phi)}{\partial \phi_n} = A_{nm}\phi_m - Qn = 0 \tag{4.25}$$

With the matrices A_{nm}, a function of the permeabilities and of the shape of the grid, and Q_n, a function of the boundary flow conditions, depending only on known factors, the resolution of the problem becomes that of the system of N linear equations with N unknowns ϕ_m:

$$A_{nm}\phi_m = Q_n \tag{4.26}$$

which poses no particular problem.

Taking the unsteady state into account in the finite elements method introduces no particular complication.

The finite element method is more complex to program than the finite difference method.

It is still believed by many people that the finite element method provides more flexible discretizations. In fact, formulations of the finite difference method have been proposed long ago (Wilkins, 1964) that allow elements of any shape, with any property.

The finite element method leads in general to programs that run faster than their finite difference counterparts for linear, 'well-behaved' problems. This is generally the case for uncoupled flow simulation.

However, solving highly non-linear and coupled problems, when hydraulic, mechanical and thermal phenoma are all involved, requires a very long calculation time with finite elements, whereas the increase in calculation time is far less dramatic for a finite difference code.

Finally, the use of both methods is not limited to porous media. They are also entirely suitable for the resolution of problems of flow in discontinuities and consequently of those for which both flows within the rock matrix and in the discontinuities are taken simultaneously into account (models incorrectly called 'double porosity models').

4.2.3 The characteristic values and their measurement

Essentially, three types of value enter into the equations for the hydraulics of porous media; knowledge of them is necessary for the resolution of the problems that may be posed in this domain and it is therefore important to know how to measure them. They are:

(1) The rates of influx or of offtake in a system. They are determined, according to circumstances, either by the methods of hydrology, or by the measurement techniques of classical hydraulics. These will not be further discussed.
(2) The piezometric heads. If the variation of heads in the future is often the solution sought for a given problem, knowledge of their present values is no less an essential starting point, either as an initial condition (unsteady state) or as a reference serving for 'calibration' of models in the steady or pseudo-steady state (true steady-state regimes do not exist in nature).
(3) The hydraulic characteristics of the medium – permeability and specific storage.

(a) Specific storage and porosity

The open porosity of a rock is equal to its total porosity minus its closed porosity. Since in most rocks the closed porosity is very small relative to total porosity, it is assumed that open porosity is equal to the latter.

The *total porosity* of a rock depends upon a number of factors, among which is the grain-size distribution and shape of its constituent grains, the presence of cement between the grains developed to different degrees, etc. In a poorly cemented rock composed of slightly elongated grains the total porosity will be higher as the range of grain size is restricted. In such a rock the actual dimensions of the grains will have little influence on the total porosity.

The total porosity can be measured either directly, in the laboratory, by techniques of measuring volume and weighing after saturation and then after drying, or indirectly, *in situ*, with the aid of downhole geophysical techniques (in particular neutron and gamma–gamma logging; the latter also can be used in the laboratory).

Table 4.1 gives some orders of magnitude of total porosities of various rocks (rock matrix alone, excluding discontinuities; from de Marsily 1981).

Table 4.1 Total porosities

	Total porosity %
Unweathered granite and gneiss	0.02–1.8
Quartzite	0.8
Shale, slate, mica schist	0.5–7.5
Limestone, primary dolomite	0.5–12.5
Secondary dolomite	10–30
Chalk	8–37
Sandstone	3.5–38
Tuff	30–40

The *specific storage* depends upon the open porosity (or total porosity if the closed porosity is zero) and the compressibility of the water and of the solid framework. The latter can be measured in a triaxial cell, under isotropic loading and constant total stress:

$$d\sigma' + dP - 0; \quad \alpha = -\frac{1}{V}\frac{dV}{d\sigma'}$$

where σ' = effective stress
P = water pressure
V = volume of sample
α = compressibility of the solid framework.

The compressibility of water, β, is of the order of $5 \times 10^{-10}\,\text{Pa}^{-1}$. That of the solid framework of common rocks is of the same order. The result is that, whatever the nature of the rock and its porosity, the specific storage varies within fairly narrow limits, about $5 \times 10^{-6}\,\text{m}^{-1}$.

Kinematic porosity is a concept that is much more difficult to comprehend in practice. As far as its determination is concerned, it is indistinguishable from effective porosity. These porosities are not often measured in the laboratory, their determination generally being effected by *in situ* measurements using tracers for confined and pumping tests for unconfined aquifers. The kinematic (and effective) porosity is very sensitive to the size of the grains of which the rock is composed (or conversely to the voids which separate them). While very close to total porosity for coarse-grained, non-cemented deposits such as sand and gravel, it is much lower for fine-grained rocks and those with very small pores. This is explained by the fact that the smaller the voids, the greater the proportion of bound water to the total amount of water.

(b) Permeability

Permeability, K, defined on the basis of Darcy's experiments, depends not only on the material but also on the fluid used, or more precisely, on its viscosity. *Intrinsic permeability, k*, which depends only upon the material, is defined by the relation:

$$k = \frac{\nu k}{g} \quad (\text{in m}^2) \qquad (4.27)$$

where ν is the kinematic viscosity of water and g is the acceleration due to gravity.

The influence of viscosity on (Darcy's) permeability can be considerable, for it varies appreciably with variations in temperature. Thus, for water:

$\nu = 1.79 \times 10^{-6}\,\text{m}^2/\text{s}$ at 0°C
$\nu = 1.31 \times 10^{-6}\,\text{m}^2/\text{s}$ at 10°C
$\nu = 1.00 \times 10^{-6}\,\text{m}^2/\text{s}$ at 20°C
$\nu = 0.28 \times 10^{-6}\,\text{m}^2/\text{s}$ at 100°C

The permeability of the matrix of a rock can be measured in the laboratory by a *permeameter*. The measurement can be made either at *constant head*, in the steady state, or under variable head. In the latter case, the dissipation of an overpressure, applied at the initial instant on one of the faces of the previously saturated sample, is recorded through time. This method is used for rocks of low permeability, for which it would take too long for a permanent regime to become established. Note that the particular interest of measuring permeability in the laboratory is that the anisotropy of permeability of a material can be evaluated, which is difficult to obtain by *in situ* tests.

Nevertheless, it is above all *in situ* tests that are used to determine permeabilities. The most commonly used of these is the pumping test, which is carried out, in its classic form, at a constant flow rate. The variation of level is monitored in the pumped well and also, if possible, in observation wells.

If the pumping is carried out in a homogeneous confined aquifer of infinite extension, and intersected through its full thickness by the pumped well, the drawdown observed will obey the Theis equation (4.12). Calibration of experimental curves against the theoretical curve enables evaluation of the transmissivity (and thus of the permeability if the thickness of the aquifer is known) and the coefficient of storage. It should be noted that the interpretation of recovery after pumping has been stopped enables a better determination of the transmissivity (but not of the coefficient of storage) than that of drawdown, since drawdown can be disturbed by phenomena such as head losses that have no effect on recovery.

The Theis formula corresponds to an ideal case which never rigorously occurs. By means of simplifying assumptions or the introduction of a few corrections it can nevertheless be used in the majority of cases, in particular in the case of unconfined aquifers. In this case, however, the coefficient of storage that is determined does not correspond to the specific storage S_s; it corresponds instead to the effective porosity of the aquifer. The water drawn down during pumping comes in fact from dewatering of the zone drained by lowering of the free surface and not from decompression of the water and the solid framework (the latter of course always occurs, but its effect is negligible compared with that of the former).

The development of microcomputers has even seen the appearance of methods of interpretation of pumping tests, so-called 'methods of identification', that can be used for pumping under whatever regime (variable flow rate).

Pumping tests are subject to certain technological constraints, one of which is related to the minimum diameter of the immersed pumps used to carry them out, which is 100 mm for commonly used types. Conventional pumping can thus not in general be carried out in drill holes of less than 100 m diameter. Pumping tests are nevertheless not impossible in such drill holes, but they should be carried out using the air-lift technique. It should be pointed out that it is more difficult to maintain a constant flow rate with an emulsifier than with a pump, and furthermore that with air-lift the level of the water in the pumping well is subject to fluctuations forming 'background noise' that affects the accuracy of the measurements. A further constraint is related to the permeability of the rock, for in poorly permeable ground, pumping, even at a very low flow rate, rapidly causes lowering of the water level such that the pump or the base of the emulsifier is no longer immersed, which obviously prevents continuation of the test. In this case other types of tests must be employed, which will be described in the section on fractured media.

Pumping tests in general apply to a section of ground of definite thickness, and enable an average value of permeability to be determined for that section. But they do not allow variations in permeability through the tested section to be taken into account. More precise information can only be provided by more narrowly located tests, which also will be described in the next section.

Experience has shown that a discrepancy, often very large, exists between permeabilities measured in the laboratory and those calculated from *in situ* tests, the latter being as much as several orders of magnitidue larger than the former. The disparity becomes more marked as the volume affected by the *in situ* measurement is greater, and as the rock matrix is less permeable. This 'scale effect' is due to the fact that an *in situ* test almost never measures the void space (primary) permeability alone, but a permeability that includes a number of discontinuities. This very important point will be developed below.

The primary permeability of a rock is governed by the size and shape of its voids, and by their arrangement. The last factor is difficult to quantify, but is very important, which explains the poor correlation between the permeability and the (open) porosity of a rock.

(c) Piezometric head

The level of water in a vertical drill hole traversing the entire thickness of an aquifer has a precise significance only if the piezometric head in the aquifer is vertically constant (i.e. flow is horizontal). This situation is exceptional in a poorly permeable rock medium, even if there are no abrupt heterogeneities. In such a medium, the flow can in general be represented by a diagram such as that of Figure 4.9. It can be seen from this figure that the

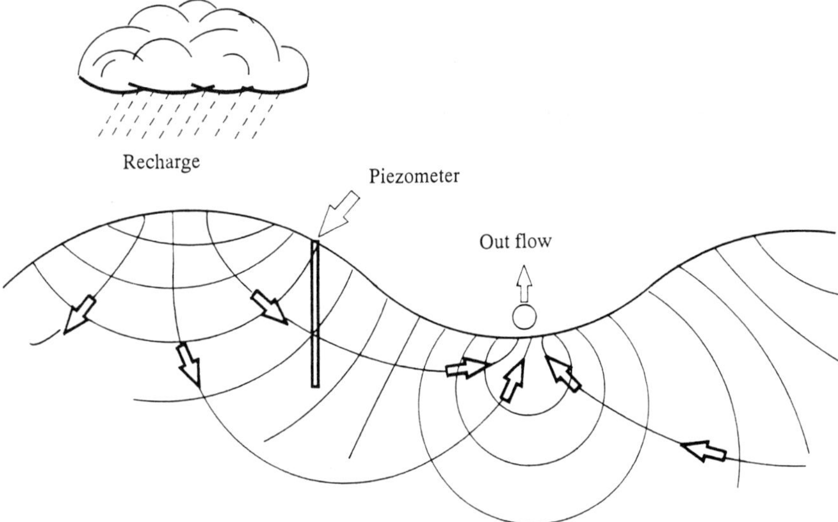

Figure 4.9 *Example of a flow net in a rock mass*

piezometric head varies with depth, either increasing (in a zone of circulation with a vertically ascending component) or decreasing (in a zone with vertically descending circulation).

These variations of head with depth may be even more strongly marked if the medium is characterized by marked inhomogeneities. The sign of the vertical gradient can even in certain cases change with depth.

It is therefore important not to accept uncritically the average value provided by an observation well open throughout its length. More localized measurements often are essential. These can be obtained by equipping the well, either temporarily or permanently, with systems of double packers, separating off short *cavities* (from less than one up to several metres long). These chambers are either equipped with pressure sensors or connected to the well-head by flexible cables enabling head measurements to be made inside the chambers. Double packer systems, which in general give excellent results, nevertheless have one drawback, which is that because of their limited length the packers cannot always provide a guarantee against the risk that the chamber may be connected by discontinuities with the drill hole above or below the packers.

Another possibility is to equip the drill hole, which should be of an appropriate diameter, with a number of piezometric tubes open at the depths at which the measurements are to be made. At these levels the drill hole is filled with sand, while the various chambers are isolated from one another by cementing the drill hole.

Finally, mention should be made of the ingenious device patented by the company Télémac under the name 'Piézofor', which enables continuous piezometric head profiles to be obtained along a drill hole at different times.

4.3 Discontinuous media

The movement of water in jointed rocks is governed by the morphology of the discontinuity field, by the hydraulic properties of the discontinuities and, to a lesser extent, by the hydraulic properties of the pore space of the rock. These are extremely difficult to assess accurately. As there is no way of detecting discontinuities in detail within the rock mass their geometry cannot be determined, and the hydraulic properties of the discontinuities are combined with the effect of the unknown geometry when a perturbation is introduced in the form of a pumping test to measure the response of the medium.

4.3.1 Morphology of the discontinuity field

The shape and size of the individual discontinuities are described by a number of parameters – the orientation of the discontinuity plane, the dimensions of the discontinuity in the plane and its free aperture. The arrangement of discontinuities in space is described by the type of point process their locations follow, and the way they truncate each other.

(a) Orientation of discontinuities

The orientation of discontinuities in a given zone is not entirely random. When discontinuities are generated they appear with preferred directions depending on the stress field, although the inhomogeneity of the rock introduces some scatter. When a rock mass is already cut by several sets of discontinuities, it can accommodate any natural stress field by displacement along pre-existing discontinuities and by the deformability of the pore space, without further fracturing, which is why most of the discontinuities in a rock appear during the earliest tectonic events after its formation. Note that later variations in the stress field, even if they do not create discontinuities, still have a great influence on the hydraulic properties of the rock mass, since they change the morphology of the discontinuities.

In most cases, discontinuity orientations are concentrated in a few preferred directions. This can readily be observed on stereonets. In hydrogeology, two types of projection are commonly used. the Schmidt projection (Figure 4.10a), an 'equal-area projection' preserving true areal relationships, is used for evaluating pole densities. The Wulff projection (Figure 4.10b), which preserves true angles, is used for geometrical investigations.

Figure 4.11 is a Schmidt net diagram of discontinuities. It can be seen that the joints are clustered in sets. Spherical probability distributions can be used to define

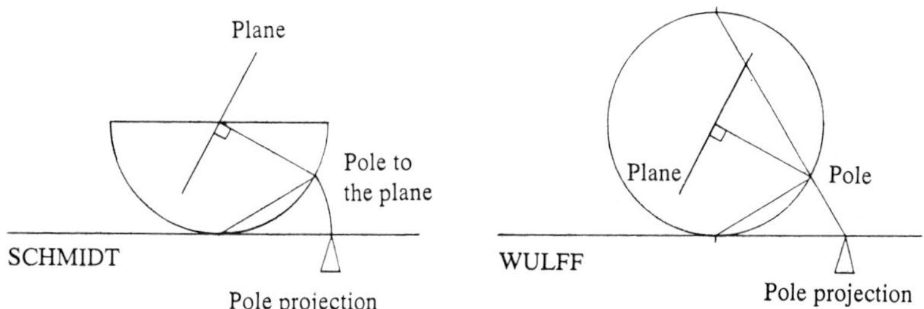

Figure 4.10 *Principle of the Schmidt and Wulff stereonet projections*

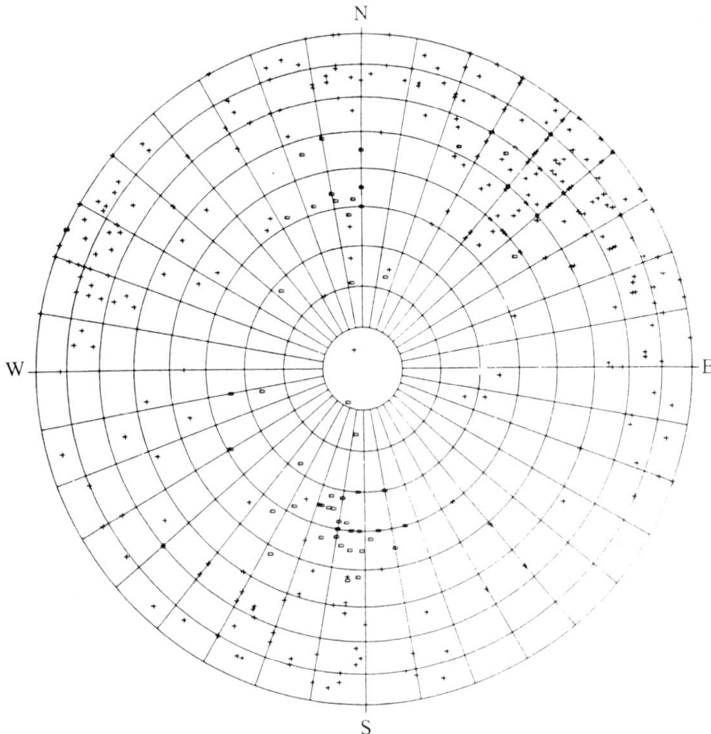

Figure 4.11 *Lower hemisphere Schmidt projection of poles to discontinuities measured in mica schists in Montredon–Labessonié (Tarn, France). Rectangles: schistosity planes, Crosses: joints*

these sets. Mardia (1972) gave a good account of the distribution functions available. Another reasonable method in many cases is to contour the sets on a Schmidt net.

Defining discontinuity sets from orientation data only is, however, too simplistic. Structural geology can give a more rigorous approach. By combining discontinuity morphology and filling, offsets, and the regional geological history, it is often possible to reconstruct the history of fracturing. Then, using the principles of rock mechanics, 'true sets' can be distinguished. That is to say, sets defined only by orientation may be separated into a number of sets formed by separate fracture mechanisms. Information on the successive tectonic episodes also tells us how discontinuities in a given true set are likely to have deformed under stress and to have been filled. This in turn helps in defining those discontinuities which are most important for fluid flow.

(b) Size and shape of discontinuities

In contrast to pores in rock, discontinuities are generally very elonged, and a length/width ratio of 10^4 is common. Most discontinuities are approximately planar along their maximum dimension. Their shape in this plane is not well known, because the only parameter which can be measured is the length of their trace on a surveyed plane. One of two usual assumptions is therefore made when it is necessary to represent discontinuity shape. The discontinuities are assumed to be either circular discs or polygons.

If the circular assumption is used, the distribution of disc diameters can be inferred from the distribution of trace lengths on a plane. This requires a further assumption on the shape of the probability distribution function of the diameters, and this is commonly taken to be lognormal. Figure 4.12 shows an example of trace length distribution, and Table 4.2 shows the parameters of the lognormal distribution of diameters inferred from these data.

The thickness of discontinuities can be assessed on cores or from fracture traces. Such estimates, however, are very unreliable and tend to yield exaggeratedly high values. The aperture can be inferred from indirect measurements, such as water tests on a discontinuity in a drill hole. By assuming that the discontinuity is an infinite parallel-plate void, an 'equivalent hydraulic aperture' can be determined. Table 4.3 gives the results of such studies for two granites. Note that careful direct observation of the Fanay-Augères cores gave a mean of close to 300 µm, an overestimate of an order of magnitude!

Discontinuities are obviously not perfectly smooth

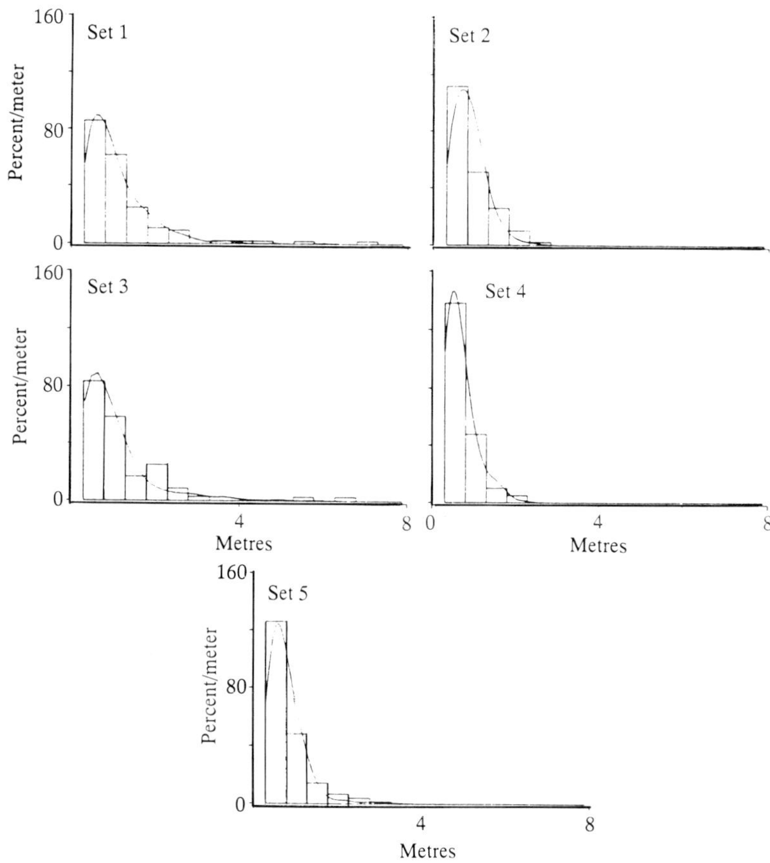

Figure 4.12 *Distribution of trace lengths for discontinuities with both end-points visible, Fanay-Augères mine (Haute-Vienne, France). Histograms represent data. Curves represent distributions obtained by fitting a lognormal diameter distribution (after Billaux et al.* 1989)

Table 4.2 Parameters for lognormal distributions of diameters derived from the distribution of traces shown in Figure 4.12 (Billaux *et al.* 1989)

Drift	*Set*	*Mean* (m)	*Standard deviation* (m)	*Theoretical percentage of traces less than 3 m*	*Average spatial density discontinuities per m³*
	1	0.90	0.65	14.8	0.84
	2	0.87	0.40	18.1	1.12
S1	3	0.60	0.70	26.5	1.46
	4	0.46	0.65	33.2	7.52
	5	0.80	0.40	24.7	5.29
	1	0.55	0.40	24.2	2.44
	2	0.40	0.34	41.6	3.45
S2	3	0.40	0.65	35.5	2.11
	4	0.40	0.35	43.9	11.02
	5	0.50	0.30	38.5	4.06

parallel-plate voids, and Figure 4.13 shows profiles taken on a discontinuity in a fine-grained granite (Maupuy, France).

The roughness of the walls of a discontinuity will have little influence on its hydraulic properties if the aperture is wide, so that in this case, the parallel-plate assumption is justified. However, as depth increases, so stress increases

Table 4.3 Hydraulic apertures in granite; after Carlsson and Olsson (1980) for Forsmark, and after Long and Billaux (1987) for Fanay-Augères

	Forsmark (Sweden)	*Fanay-Augères (France)*
mean (μm)	71	32
standard dev'n (μm)	82	36
sample size	317	3006

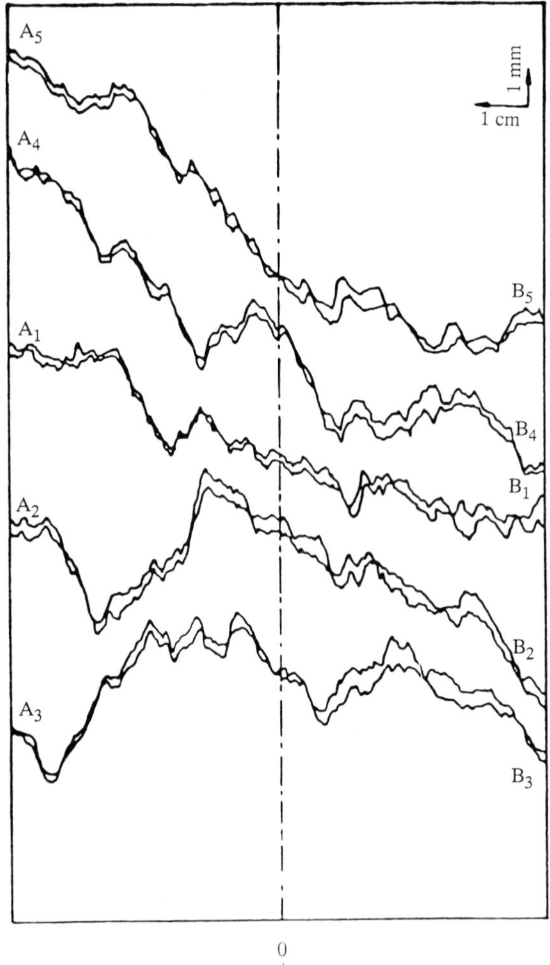

Figure 4.13 *Profiles of a fracture in a fine-grained granite (after Gentier 1987)*

and the discontinuities will tend to close. Flow then may become tortuous as the proportion of the discontinuity still available for flow decreases, and under large compressive stresses the water may be able to migrate only through a few narrow channels, and the main factor governing the hydraulic behaviour then becomes the connectivity of the channels.

(c) The arrangement of discontinuities in space

The main parameter describing the arrangement of discontinuities in space is their density. This is a three-dimensional parameter, but generally it can be measured in two dimensions only, or even only in one. The three-dimensional parameter can be expressed either as the number of discontinuities per unit volume, λ_v, or as the cumulative area of discontinuities per unit volume, $\lambda_{l\,3D}$. In the same way, a two-dimensional density can be expressed either as a number of discontinuities per unit area, λ_A, or a cumulative length of discontinuities per unit area $\lambda_{l\,2D}$. Finally, density along a line is the number of discontinuities per unit length, λ_l. Its reciprocal, the mean spacing between discontinuities, also is used.

Note that only the three-dimensional density, λ_v or $\lambda_{l\,3D}$ is an intrinsic parameter. The two-dimensional density or one-dimensional parameters vary with the direction of the survey plane or line. $\lambda_{l\,3D}$, $\lambda_{l\,2D}$ and λ_l are all expressed in units of the reciprocal of a length. If all the discontinuities were parallel and the scanline or survey plane were perpendicular to them, they would be numerically equal. In reality, a correcting factor must be introduced to take into account the scatter of orientations and the direction of the scanline or survey plane.

If we call $\theta_{S\,3D}$ the three-dimensional angle between the normal to a discontinuity and a scanline, and $\theta_{S\,2D}$ the angle, on a survey plane, between the normal to the trace of a discontinuity and a scanline, then

$$\lambda_{l\,3D} = \lambda_l \left(\frac{1}{\cos\theta_{S\,3D}}\right) \tag{4.28}$$

and

$$\lambda_{l\,2D} = \lambda_l \left(\frac{1}{\cos\theta_{S\,2D}}\right) \tag{4.29}$$

where the overbar means 'average of'.

The parameters λ_V and λ_A are expressed in units of reciprocals of a volume and of an area respectively. If the mean area of the discontinuities is denoted $\bar{A}$ and if the mean length of their trace on a survey plane is denoted $\bar{l}$, then

$$\lambda_v = \frac{\lambda_{l\,3D}}{\bar{A}} \tag{4.30}$$

and

$$\lambda_v = \frac{\lambda_{l\,2D}}{\bar{l}} \tag{4.31}$$

The solution of the above equations gives the classic results

$$\lambda_v = \lambda_l \frac{1}{\bar{A}} \left(\frac{1}{\cos\theta_{S\,3D}}\right) \tag{4.32}$$

and

$$\lambda_A = \lambda_l \frac{1}{\bar{l}} \left(\frac{1}{\cos\theta_{S\,2D}}\right) \tag{4.33}$$

It is possible to define densities for all discontinuities lumped in one sample. They are more commonly determined for each set of discontinuities as defined by the orientation study. The overall density for all the parameters defined above is thus the sum of the densities of the sets.

Density completely describes the arrangement of discontinuities in space only if this arrangement can be considered as perfectly random. In this case, the distribution of the discontinuities is said to be 'Poissonian', that is the location of their centres obey a Poisson point process. In fact, the assumption of Poisson distribution is very often not justified. Two phenomena contribute to 'structure' the arrangement of discontinuities. Firstly, an effect that geostatisticians term 'regionalization' can be expressed simply in the form – 'the difference between the densities of two locations close to each other is likely to be smaller than that between two locations distant from each other'. This causes a curve representing variations in density to be smoother than would be the case for a Poisson process. Secondly, discontinuities tend to cluster together, as can be seen on the example shown in Figure 4.14. This can be detected easily using a scan-line survey, because a Poisson distribution of discontinuities in space yields a negative exponential distribution of their spacing along any scan-line. The mean and the standard deviation of such a distribution are equal. If the discontinuities tend to cluster, this decreases the number of median spaces and increases the number of small and large spaces. The standard deviation of spacings therefore becomes much larger than their mean. The effect of this 'ordering' on the hydraulic properties of the rock mass can be important. The regionalization can be described by the variogram of density $\gamma(h)$. This function represents the variations of the mean semiquadratic difference for the variable under study (here the density λ_v) between two points a distance h apart:

$$\gamma(h) = \frac{1}{2} E\,[\lambda_v(x) - \lambda_v(x+h)] \qquad (4.34)$$

A typical variogram shape is shown in Figure 4.15. The clustering is described by a density of clusters (number of clusters per unit volume), a distribution of the number of discontinuities per cluster, and a distribution in space of the discontinuities in each cluster. Such parameters are

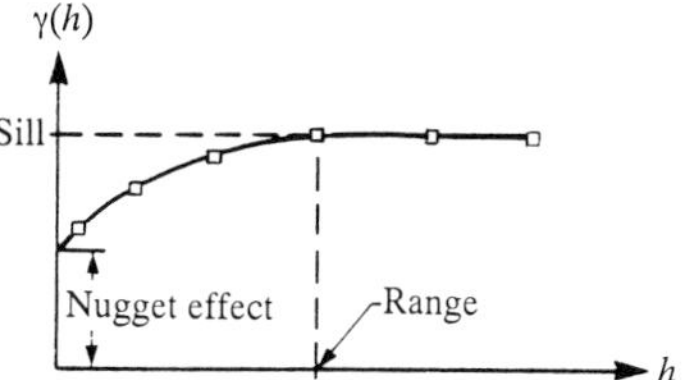

Figure 4.15 *Typical variogram shape: The discontinuity at the origin (nugget effect) is caused by measurement errors, or by variations with a range smaller than the sampling size*

difficult to obtain, and resort generally is taken to a Poisson distribution with or without regionalization.

4.3.2 Flow in a fracture

The confined flow of water in conduits is expressed in terms of three non-dimensional parameters:

(a) the Reynolds number, $R_e = \dfrac{VD_h}{\nu}$ (4.35)

where D_h is the hydraulic diameter, function of the section S and the perimeter, P, of the conduit:

$$D_h = 4\,\frac{S}{P}$$

and ν is the kinematic viscosity of the fluid.

(b) The head loss coefficient, $\lambda = \dfrac{i\,D_h}{V^2/2g}$ (4.36)

where i is the value of the hydraulic gradient.

(c) the relative roughness k/D_h, where k is the height of asperities.

Depending on the value of the Reynolds coefficient, two types of flow are distinguished. For a low Reynolds

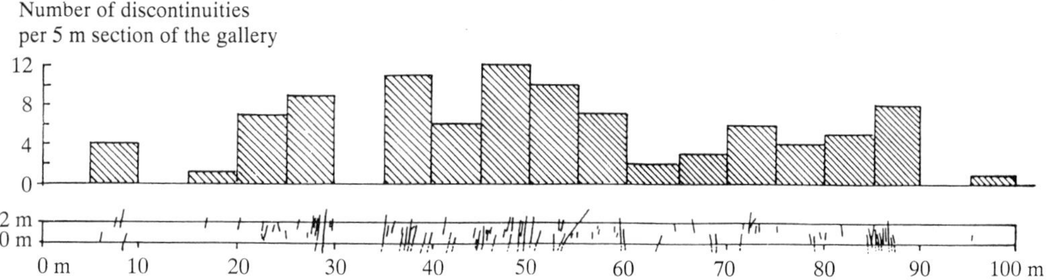

Figure 4.14 *Examples of discontinuity clustering. Discontinuities from one set mapped on the wall of a drift, Fanay-Augères Mine (Haute-Vienne, France)*

number, flow is laminar and the liquid streams can be individualized. For high Reynolds numbers, flow is turbulent, and interactions between the liquid streams level the flow rates.

Louis (1968) has shown that in the range of roughnesses for discontinuities ($k/D_h > 0.033$ in general), flow was 'non-parallel'. In this case, the critical Reynolds number defining the transition between laminar and turbulent flow decreases and can become very low (from 2300 if k/D_h is 0 down to 100 or even 10). The value of R_e is very difficult to determine in fissured rocks, because it can vary enormously from one point to the other along the same fracture. However, it is assumed in most cases that the flow in a discontinuity is laminar. Flow then is governed by Poiseuille's law, which can be expressed as

$$\mathbf{V} = \frac{\tau g e^2}{12 \nu C} \mathbf{i} \tag{4.37}$$

In this equation, τ is the degree of continuity of the discontinuity (ratio of the open surface and the total surface of the discontinuity), g is the acceleration of gravity, e is the mean free aperture of the discontinuity, and C is a coefficient equal to $1 + 8.8\ (k/D_h)^{1.5}$ (Louis 1968).

In the laminar regime, the mean speed of water in a discontinuity is therefore proportional to the square of its free aperture. The flow rate in the discontinuity is thus proportional to the cube of the free aperture. This is the classical 'cubic law'.

Note that in the case of discontinuities with filling, flow in the discontinuities follows Darcy's law, the value of permeability in the discontinuities being the permeability of the filling.

4.3.3 Modelling flow in a rock mass

Because the geometry of fissured rock is very complex, engineers have in the past tried to simplify it by using a hypothetical equivalent porous medium. However, the development of computers has opened the way for the stochastic simulation of discontinuities and the numerical computation of the flow that takes place in the discontinuity networks.

(a) Equivalent porous medium

This type of model is certainly the oldest that is intended to represent the discontinuous rock medium in a form that is amenable to calculations, and is a result of the natural tendency of researchers and engineers to use the existing methods of calculation until their limitations induce them to search for more appropriate ways of modelling.

It is not always possible to define a continuous equivalent hydraulic medium. For such a medium to exist, it is necessary to find a minimum volume of the rock mass, called the 'representative elementary volume', over which the (flow/piezometric head gradient) relation can be expressed, as in a continuous medium, in the form:

$$\mathbf{V} = -\mathbf{K}\,\mathbf{grad}\,\phi \tag{4.38}$$

where $\mathbf{K}$ is a tensor characterizing the 'average' permeability (or hydraulic conductivity) of the medium.

If the equivalent continuous medium exists it will be necessary to calculate its characteristics, essentially the components of its conductivity tensor (the determination of its coefficient of storage poses no particular problem). A first approach to this (Bertrand *et al.*, 1982) assumes that the discontinuities that cut the rock mass are of infinite extent and that the pore space is impermeable.

The permeability of a cubic volume, of edge 1, of continuous medium equivalent to the same volume of discontinuous rock cut by a discontinuity of aperture e is equal, along the plane of the discontinuity, to

$$K_{eq} = \frac{e}{l} k, \tag{4.39}$$

where k is the hydraulic conductivity of the discontinuity; in the direction perpendicular to the plane this is zero.

If the considered volume is cut by N discontinuities of whatever orientation, of thickness e_i and hydraulic conductivity k_i, the matrix representing the permeability tensor of the continuous equivalent medium is equal to

$$\mathbf{k}_{eq} = \frac{1}{l} \sum_{i=1}^{N} e_i \mathrm{k}_i \mathrm{R}_i \tag{4.40}$$

where R_i is a matrix of rotation specific to each discontinuity.

Knowledge of the geometric characteristics (strike, dip, spacing, aperture) and hydraulic characteristics of the discontinuities cutting the mass, from field observations on outcrops and in drill holes or underground galleries and from pumping tests in drill holes enables the coefficients of the matrix to be calculated.

Application of this method to actual cases has led to permeability values greater, sometimes by several orders of magnitude, than values determined by pumping tests on the same sites. The main reason for this is undoubtedly that in reality the discontinuities are not of infinite extent and that a given discontinuity does not in fact cut all the others. The degree of 'connectivity' of the discontinuity network determines in large part the value of the permeability of the equivalent continuous medium.

In order to be more realistic, therefore, the continuous equivalent medium approach should take into account, in determining a permeability tensor, the fact that the discontinuities are finite. This can be effected by coupling this approach with the stochastic approach, described later. Analytical solutions also exist for determining the behaviour of a discontinuous medium during a hydraulic test. These methods make it possible to determine the global characteristics of the medium, taking into account the fact that, even if the medium is treated as continuous, its discontinuous character governs the response in the immediate vicinity of the test well, or modifies it globally.

The first models to take this into account were continuous and homogeneous, termed 'double porosity' models (Barenblatt *et al.* 1960; Warren and Root 1963). They are based on the observation that while the discontinuity system is in general much more permeable than the pore space of the rock, the latter plays a fundamental role in storing the water, owing to its much greater effective porosity. The discontinuity system is therefore assumed to form a continuum. The pore space of the rock does not contribute directly to the flow capacity of the medium, but interacts with the discontinuity system by acting as a source of water. Bourdet and Gringarten (1980) and Gringarten (1984) gave type curves for interpreting pumping tests in the context of this model.

Karasaki (1986) has developed solutions for composite media which take better account of the behaviour in the vicinity of a test well. In fact, if a limited number of discontinuities intersects the well, flow near the well occurs uniquely in these discontinuities and it is thus their characteristics which control flow near the well. This should be taken into account, especially if the hydraulic characteristics of these discontinuities are very different from the overall parameters of the medium. A composite medium is thus composed of two regions, concentric around the well, that have different behaviours. In an aquifer of finite thickness, if the well is intersected throughout the thickness of the aquifer by one or more vertical discontinuities, flow is linear in the central region representing the discontinuities, and radial in the outer region (Figure 4.16). The case where flow in both zones is radial, differing only in the values of its parameters, is also treated. This was completely resolved by Satman (1985). Finally, if a non-vertical discontinuity intersects the well, and is part of a three-dimensional network of well-connected discontinuities, flow can be considered radial near the drill hole and spherical farther away. Figure 4.17 shows an example of a type curve obtained for this case, with fixed geometry and coefficients of storage. Variations of the curve (non-dimensional pressure and time) are given as a function of the permeability contrast between the two regions.

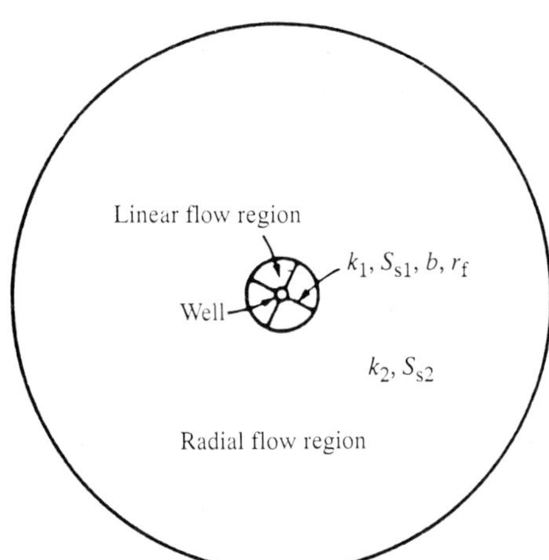

Figure 4.16 *Linear-radial composite model (after Karasaki 1986)*

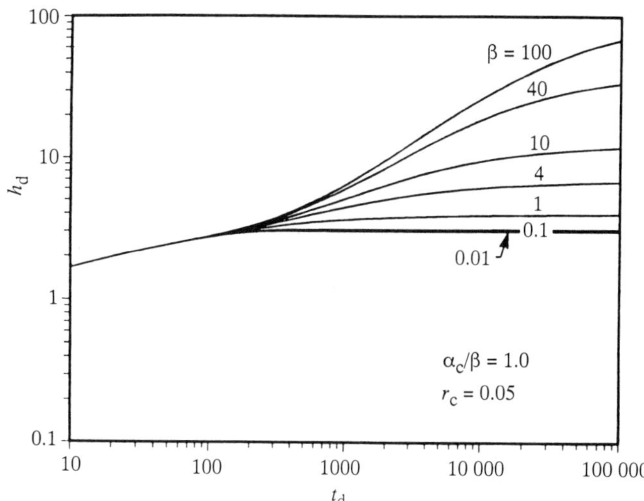

Figure 4.17 *Example of type curve obtained for a radial-spherical model (after Karasaki 1986)* h_d = dimensionless pressure, t_d = dimensionless time, β = permeability contrast, r_c = diameter of well over diameter of inner region, and α_c = dimensionless diffusivity

(b) Random generation of discontinuity fields

Building on the ideas developed initially by Snow (1965), several teams have in the last five years developed tools for generating random discontinuity fields, first in two dimensions (Long 1983; Robinson 1984; Rouleau 1984) and then in three dimensions (Long *et al.* 1985; Billaux *et al.* 1989).

The entry data for these models are the statistical distributions of the various discontinuity parameters – aperture, orientation, size, density – specified in general by sets of discontinuities. The program then generates, in pseudo-random fashion, a field of discontinuities with the specified statistical characteristics, attempting to reproduce the morphology of the discontinuity field as has been described above (Figure 4.18). The flow in the network thus constructed then is calculated by finite elements for standard boundary conditions.

The first use of these models has been in the interpretation of pumping tests and discontinuity surveys to derive the characteristics of the equivalent continuous medium.

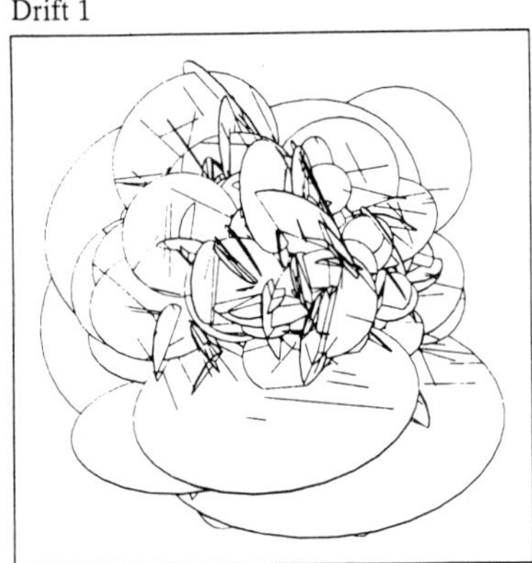

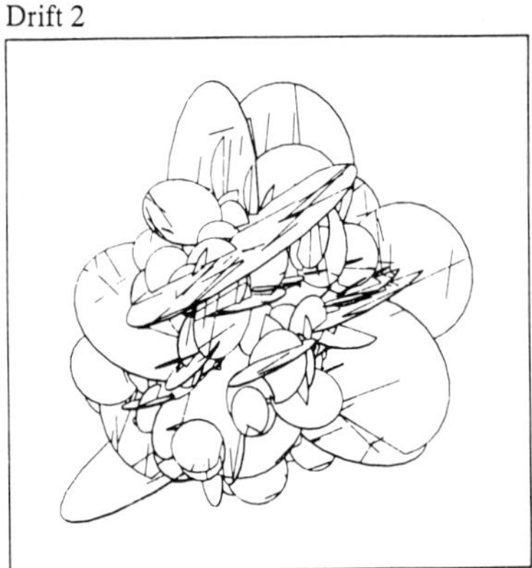

Figure 4.18 *Two discontinuity networks generated to simulate the discontinuities around two drifts, Fanay-Augères Mine (Haute-Vienne, France)*

In effect, on the basis of these discontinuity fields it is possible to calculate, at different scales and in any direction in space the ratio of flow to head gradient, which makes it possible to know whether the equivalent continuous medium exists, to know the size of the representative elementary volume and to determine the components of the permeability tensor. The equivalent continuous medium can then be modelled by one of the techniques mentioned in Section 4.2.2.

The calibration and direct use of these stochastic models on real three-dimensional sites pose no major conceptual problem. However, the cost of simulation quickly increases unless resort is taken to simplifying assumptions such as orthogonal discontinuity sets, the discontinuities in a single set being parallel to one another, for example, or the representation only of large discontinuities, small ones being accounted for by using a continuous medium.

As far as the advantages and disadvantages of these models are concerned, the following points can be made:

(1) *Advantages*. These models are very useful for the study of parameters and for investigating the validity of the concept of the equivalent continuous medium. They are by far the most realistic among the models that enable effective calculation of flows.
(2) *Disadvantages*. Realistic three-dimensional modelling quickly reaches prohibitive costs in calculation time. In general the statistical homogeneity of discontinuities is assumed. The purely statistical description of the medium implies that possible short-circuits cannot be taken into account. The various sets of discontinuities are generated independently, whereas in reality most of the more recent discontinuities either stop at older ones or produce movement on them.

It will be noticed that certain of these inconveniences are inherent in existing models and could be removed by improvement of the procedures without taking recourse to other conceptual models.

4.3.4 *In situ tests and measurements proper to discontinuous media*

All the hydraulic tests that can be used on porous media and presented in Section 4.2.3. can be used in discontinuous media wherever this is technically possible. They then can be interpreted either in terms of 'equivalent continuous medium' or with the aid of models of the 'discontinuous medium' type such as those developed by Bourdet and Gringarten (1980). A certain number of tests, nevertheless, are particularly suitable for discontinuous media, and are presented below.

(a) Injection tests between packers

This type of tests, whose purpose is to determine very local permeabilities, is derived from the Lugeon test. This involves isolating a section of a drill hole between two inflatable packers and measuring the flow rate of water injected into the ground, first at increasing pressures and then at decreasing pressures, maintained constant for time-steps of about ten minutes. The Lugeon test, which was originally developed to test the injectability of dam foundations, has a precise operating procedure. It assumes, among other things, that flow at each pressure step attains the steady state, and its interpretation is based upon the use of the following 'steady-state' type of formula:

$$K = 1.85 \times 10^{-5} \frac{Q/L}{\Delta\phi} \tag{4.41}$$

where K = permeability in m/s
Q/L = flow rate per unit length of the drill hole, in l/min m
$\Delta\phi$ = overpressure in the test chamber, in metres of water.

Technological progress has enabled the Lugeon test procedure to be improved and the use of measurements of permeability between packers to be extended towards very low permeabilities. In particular, testing in the unsteady state, made possible on the one hand by the development of means of *in situ* recording and on the other hand by that of methods of identification that can be used on microcomputers on site gives much more voluminous and more accurate results than the basic Lugeon test. Measurements between packers can currently yield permeabilities of the order of 10^{-11} m/s or even 10^{-12} m/s.

(b) Slug test

The slug test consists of following the variation of water level in a drill hole with time following an abrupt change in level, obtained for example by introducing a certain amount of water more or less instantaneously. The solution to the problem was established by Cooper *et al.* (1973) for assumptions analogous with those for the Theis formula. The calibration of experimental curves against theoretical curves enables transmissivity and storage coefficient to the determined. The slug test is particularly suitable for the measurement of low permeabilities, the scale of investigation being a function of the thickness of the section tested. It can obviously be applied between packers.

(c) Flowmeter

The flowmeter makes it possible, during a pumping test, to locate the productive levels and determine their permeability. It consists in principle in measuring the vertical rate of circulation of water at a given depth by the use of a screw which transmits impulses to the surface according to its rate of rotation. Calibration of the screw enables the rate of rotation of the screw to be converted to the rate of circulation of the water and, if the section of the drill hole is known, the flow rate.

A continuous, point by point profile of flow rate can be established for the drill hole, which, by difference, enables the discharge produced by a given section of the drill hole to be calculated and thus the permeability to be determined.

4.4 Hydromechanical coupling

The presence of water in the rock, whether in motion or not, modifies the stresses to which the rock is subjected (hydraulic–mechanical coupling). Furthermore, the aperture of the discontinuities, and therefore their conductivity, varies as a function of the stresses applied to them (mechanical–hydraulic coupling).

4.4.1 Formulation of the equilibrium of a unit of saturated ground

The two-dimensional case is illustrated, the reasoning being that it is identical to the three-dimensional case. Using the notations given in Figure 4.19, in relation to a reference frame Oxz, the mechanics of continuous media give the following equilibrium equations:

$$\frac{\partial \sigma_x}{\partial_x} + \frac{\partial \tau_{zx}}{\partial_z} = 0 \qquad (4.42)$$

$$\frac{\partial \tau_{zx}}{\partial_x} + \frac{\partial \sigma_z}{\partial_z} = -\gamma \qquad (4.43)$$

where σ_x and σ_z = total normal vertical and horizontal stresses
$\tau_{xz} = \tau_{zx}$ = total shear stress in the plane Oxz and
γ = total unit weight.

These equations are expressed as *total stresses*. If the ground is not saturated, they can be used directly, but for saturated ground, since the laws of behaviour, both for deformation and for rupture, can only be expressed in effective stresses, these equations must be converted in order for the latter to become apparent. We have

$$\left.\begin{aligned} \sigma_x &= \sigma'_x + P \\ \sigma_z &= \sigma'_z + P \\ \tau_{zx} &= \tau'_{zx} \end{aligned}\right\} \text{Terzaghi's law} \qquad (4.44)$$

and

$$\phi = \frac{P}{\gamma_w} + z \qquad (4.45)$$

which represents a definition of the piezometric head ignoring the term for kinetic energy

where σ'_x and σ'_z = effective normal stresses
τ'_{zx} = effective shear stress
P = water pressure
ϕ = piezometric head
γ_w = unit weight of water

therefore: $P = \gamma_w(\phi - z)$

$$\frac{\partial \sigma_x}{\partial x} = \frac{\partial \sigma'_x}{\partial x} + \gamma_w \frac{\partial(\phi - z)}{\partial x} = \frac{\partial \sigma'_x}{\partial x} + \gamma_w \frac{\partial \phi}{\partial x} \qquad (4.46)$$

$$\frac{\partial \sigma_z}{\partial z} = \frac{\partial \sigma'_z}{\partial z} + \gamma_w \frac{\partial \phi}{\partial z} - \gamma_w \qquad (4.47)$$

give

$$\frac{\partial \sigma'_x}{\partial x} + \frac{\partial \tau'_{zx}}{\partial z} = -\gamma_w \frac{\partial \phi}{\partial x} \qquad (4.48)$$

$$\frac{\partial \tau'_{zx}}{\partial x} + \frac{\partial \sigma'_z}{\partial z} = -\gamma + \gamma_w - \gamma_w \frac{\partial \phi}{\partial z} \qquad (4.49)$$

It can be seen that the external volume forces to be taken into account in the calculation of effective stresses are of three types:

(a) the overall specific gravity, directed downwards, $-\gamma \mathbf{k}$ ($\mathbf{k}$ = unit vector as shown on Figure 4.19)

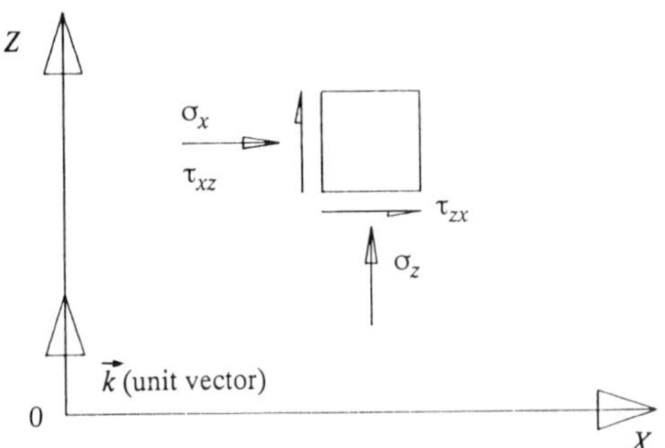

Figure 4.19 *Notations for equilibrium formulation*

(b) the uplift pressure, of intensity γ_w per unit volume, directed upwards: $\gamma_w \mathbf{k}$
(c) the force due to flow, of components:

$$\begin{bmatrix} -\gamma_w \dfrac{\partial \phi}{\partial x} \\ -\gamma_w \dfrac{\partial \phi}{\partial z} \end{bmatrix}$$

If $\mathbf{i}$ is the hydraulic gradient vector, by definition,

$$\mathbf{i} = \begin{bmatrix} -\dfrac{\partial \phi}{\partial x} \\ -\dfrac{\partial \phi}{\partial z} \end{bmatrix}$$

thus the force due to flow is $\gamma_w \mathbf{i}$.

4.4.2 Balance of forces in a mechanical calculation: hydraulic–mechanical coupling

In making a mechanical calculation, the first steps in reasoning are as follows:

(a) In the problem in question (most commonly ground + unconfined water + air), a volume, v, is considered, the conditions at whose boundary ∂v are known. The part of ∂v contained in the ground is termed Σ (Figure 4.20).

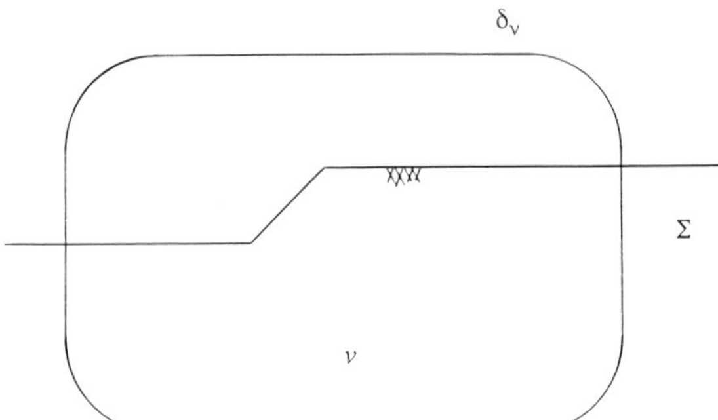

Figure 4.20 *Study domain*

(b) A balance is made of the volume and surface forces acting in equilibrium on v, and on each part of v, if it has been subdivided into a number of volumes (grids in a finite element model, for example).

Therefore:

$$\mathbf{F} = \iiint_v \mathbf{f}_{\text{vol}}\, dv + \iiint_{\partial v} \mathbf{f}_{\text{surf}}\, ds \tag{4.50}$$

is sought.

The volume forces have been shown in the preceding paragraph. In a study using effective stresses, the only surface forces for *a priori* consideration are the normal effective stress σ'_n, and shear stresses τ'_n, on the surface (Σ), plus any load Q imposed on the solid framework through the surface ($\partial v - \Sigma$). This gives, therefore

$$\mathbf{F} = \iiint_v (-\gamma \mathbf{k} + \gamma_w \mathbf{k} + \gamma_w \mathbf{i})\, dv + \iint_\Sigma (\sigma'_n + \tau'_n)\, d_s + \iint_{\partial v - \Sigma} \mathbf{Q}\, ds \tag{4.51}$$

or $\quad \mathbf{F} = \mathbf{F}_v + \mathbf{F}_\Sigma + \mathbf{F}_Q \qquad (4.52)$

The various methods of decomposing the forces correspond to the different possible groupings of terms before integration of the volume forces.

(1) *If the three terms are left separate, this gives:*

$$\mathbf{F}_v = \iiint_v -\gamma \mathbf{k}\, dv + \iiint_v \gamma_w \mathbf{k}\, dv + \iiint_v \gamma_w \mathbf{i}\, dv \tag{4.53}$$

or:

$$\mathbf{F}_v = \mathbf{W} + \mathbf{F}_u + \mathbf{F}_F \tag{4.54}$$

where $\mathbf{W}$ = total weight
$\mathbf{F}_u$ = uplift pressure
$\mathbf{F}_F$ = forces due to flow of water,

and therefore:

$$\mathbf{F} = \mathbf{W} + \mathbf{F}_u + \mathbf{F}_F + \mathbf{F}_\Sigma + \mathbf{F}_Q \qquad \text{(decomposition 1)} \qquad (4.55)$$

The element is thus subjected to its total weight, the uplift pressure, the water flow forces and the surface forces at its boundary.

(2) *If the first two terms are grouped together:*

$$\mathbf{F}_v = \iiint_v -(\gamma - \gamma_w)\,\mathbf{k}\,dv + \iiint_v \gamma_w\,\mathbf{i}\,dv \qquad (4.56)$$

The difference $\gamma - \gamma_w = \gamma'$ is called buoyant unit weight of the soil

$$\mathbf{F}_v = \mathbf{W}' + \mathbf{F}_F \qquad (4.57)$$

where $\mathbf{W}'$ = buoyant weight of the soil

therefore

$$\mathbf{F} = \mathbf{W}' + \mathbf{F}_F + \mathbf{F}_\Sigma + \mathbf{F}_Q \qquad \text{(decomposition 2)} \quad (4.58)$$

In this case, the element is thus subjected to its buoyant weight, the water flow forces and the surface forces at its boundary.

(3) *If the last two terms are grouped together*

$$\mathbf{F}_v = \iiint_v -\gamma\,\mathbf{k}\,dv + \iiint_v \gamma_w(\mathbf{k} + \mathbf{i})\,dv \qquad (4.59)$$

Since $\mathbf{i} = -\mathbf{grad}\,\phi$

and $\mathbf{k} = \mathbf{grad}\,z$

therefore

$$\gamma_w(\mathbf{k} + \mathbf{i}) = \gamma_w\,\mathbf{grad}(z - \phi) = \mathbf{grad}[\gamma_w(z - \phi)] \quad (4.60)$$

however

$$\phi = \frac{P}{\gamma_w} + z$$

therefore $\gamma_w(z - \phi) = -P$

This gives:

$$\mathbf{F}_v = \mathbf{W} - \iiint_v \text{-grad}\,P\,dv \qquad (4.61)$$

Denoting by $\mathbf{n}$ the unit vector normal to ∂v, the volume integral can be transformed into a surface integral on the boundary:

$$\mathbf{F}_v = \mathbf{W} - \iint_{\partial v} P.n\,d\mathbf{S} = \mathbf{W} + \mathbf{F}_p \qquad (4.62)$$

where $\mathbf{F}_p$ = force resulting from the pressures on ∂v.

The volume forces owing to the water are thus converted into surface forces equal to the pressures on the boundary ∂v. This gives

$$\mathbf{F} = \mathbf{W} + \mathbf{F}_P + \mathbf{F}_\Sigma + \mathbf{F}_Q \qquad \text{(decomposition 3)} \quad (4.63)$$

In this case the element is subjected to its total weight, the forces resulting from the pressure of water at its boundary and the surface forces at its boundary.

When the medium is considered to be continuous, or when the rock matrix is assumed to be permeable, decomposition 2 is the one that it is most natural to use, since the flow forces can be deduced directly from the field of hydraulic potentials. It must be pointed out that in this case the pressures owing to the water at the element boundaries *must not* be added. When the rock matrix is assumed to be impermeable, decomposition 3 is used, since in this case the pressures of the water on the rock blocks are deduced from the calculation of flow in the discontinuities.

The influence of water on stresses is one of the major factors in instabilities that may occur in both open-cast and in underground mining. It should therefore be taken into account whenever it is present.

4.4.3 Variations in the conductivity of fractures: mechanical-hydraulic coupling

The conductivity of a discontinuity may vary very considerably according to the effective normal stress that is applied to it. It is in fact on this principle that the measurement of stress by drill hole injection tests (hydraulic fracturing) is based. The sudden increase in conductivity of a discontinuity when its walls move apart makes it possible to identify the instant when the pressure of water balances the total normal stress, and when the effective normal stress therefore becomes zero. This is, of course, an exceptionally extreme case of coupling.

In order to take this coupling into account accurately it must be possible to determine two laws of behaviour, on the one hand the variations in geometry (aperture) of the discontinuities as a function of the stresses applied, and on the other hand the variations in the hydraulic conductivity of the discontinuities as their geometry changes. These can only be measured at the price of costly laboratory tests, and the results cannot be directly applied on the scale of rock masses. This effect is therefore not in general incorporated quantitatively into calculations in rock mechanics. It is nevertheless present in the fact that the overall permeability of the medium decreases with depth, and that those sets of discontinuities perpendicular to the maximum principal stress have every chance of being less conductive than those which are parallel to it.

References

Barenblatt, G. I., Zhelto, I. P. and Kochina, I. N. (1960) 'Basic concepts in the theory of seepage of homogeneous liquids in fissured rocks', *J. Appl. Math. Mech.,* **24**, No. 5 1286–1303

Bertrand, L., Durand, E. and Feuga, B. (1982) 'Détermination en sondages de la perméabilité d'un milieu rocheux fracturé – aspects théoriques et pratiques, *Revue Française de Géotechnique* no. 20, 39–54

Billaux, D., Chilès, J. P., Hestir, K. and Long, J. C. S. (1989) 'Three-dimensional statistical modeling of a fractured rock mass – an example from the Fanay-Augères Mine, *Int. J. Rock Mech. Min. Sci.,* Special Issue on Forced Fluid Flow through Fractured Rock Masses, Vol. 26 No. 3–4, pp. 281–299

Bourdet, D. P. and Gringarten, A. C. (1980) 'Determination of fissured volume and block size in fractured reservoirs by type-curve analysis', *Paper SPE 9293, SPE 55th Annual Technical Conference and Exhibition,* Dallas, Texas, 15 pp.

Carlsson A. and Olsson T. (1980) 'Caractéristiques de fracture et propriétés hydrauliques d'une région au sous-sol cristallin en Suède', *Bull. du BRGM* (2nd series), section III No. 3, 215–233

Castany, G. and Margat, J. (1977) *Dictionnaire français d'hydrogéologie,* Bureau de Recherches Géologiques et Minières, Orléans, France

Cooper, H. H., Bredehoeft, J. D. and Papadopoulos, I. S. (1973) 'On the analysis of slug-test data', *Water Resources Research,* **9**, 1087–1089

Darcy, H. (1856) *Les Fontaines Publiques de la Ville de Dijon,* Dalmont, Paris

Desai, C. S. and Abel, J. F. (1972) *Introduction to the Finite Element Method: a Numerical Method for Engineering Analysis,* Van Nostrand Reinhold, New York, 477 pp.

Dupuit, J. (1863) *Etudes Théoriques et Pratiques sur le Mouvement des Eaux dans les Canaux Découverts et à Travers les Terrains Perméables,* 2 ème édition, Dunot, Paris

Gentier, S. (1987) *Morphologie et Comportement Hydromécanique d'une Fracture Naturelle dans le Granite sous Contrainte Normale. Etude Expérimentale et Théorique,* Bureau de Recherche, Géologiques et Minières, Orleans, France, No. 134, 597 pp.

Gringarten, A. C. (1984) 'Interpretation of tests in fissured and multilayered reservoirs with double-porosity behaviour: theory and practice', *J. Petroleum Technology,* **36**(4), 549–564

Karasaki, K. (1986) 'Well Test Analysis in Fractured Media', Ph.D. thesis, Lawrence Berkeley Laboratory, University of California, 239 pp.

Long, J. C. S. (1983) 'Investigation of Equivalent Porous Medium Permeability in Networks of Discontinuous Fractures', Ph.D. thesis, Lawrence Berkeley Laboratory, University of California 277 pp.

Long, J. C. S. and Billaux, D. (1987) 'From field data to fracture network modelling – an example incorporating spatial structure', *Water Resources Research,* **23**, 1201–1216

Long, J. C. S., Gilmour, P. and Witherspoon, P. A. (1985) 'A model for steady fluid flow in random three-dimensional networks of disc-shaped fractures', *Water Resources Research,* **21**, 1–46

Louis C. (1968) 'Etude des écoulements d'eau dans les roches fissurées et de leur influence sur la stabilité des massifs rocheux', EDF, *Bulletin de la Direction des Etudes et Recherches,* Série A' Nucléaire, Hydraulique, Thermique, No. 3, 5–132

Mardia, K. V. (1972) *Statistics of Directional Data,* Academic Press, London

de Marsily, G. (1981) *Hydrogéologie Quantitative,* Masson, Paris

Polubarinova-Kochina, P. Y. (1962) *Theory of Groundwater Movement,* Translated from Russian by R. J. M. de Wiest, Princeton University Press

Robinson, P. (1984) 'Connectivity, Flow and Transport in Network Models of Fracture Media', Ph.D. thesis, Oxford University

Rouleau, A. (1984) 'Statistical Characterization and Numerical simulation of a fracture system. Application to groundwater flow in the Stripa granite', Ph.D. thesis, University of Waterloo, Waterloo, Ontario

Satman, A. (1985) 'An analytical study of interference in composite reservoirs', *Soc. Petroleum Engineering J.,* **25**(2), 281–290

Snow, D. T. (1965) 'A Parallel Plate Model of Fractured Permeable Media', Ph.D. thesis, University of California, Berkeley, 331 pp.

Theis, C. V. (1935) 'The relation between the lowering of the piezometric surface and the rate and duration of discharge of a well using ground-water storage', *Trans. Am. Geophys. Union,* **2**, 519–524

Warren, J. F. and Root, P. J. (1963) 'The behaviour of naturally fractured reservoirs'. *Soc. Petroleum Engineering J.* **228**, 245–255

Wilkins, M. L. (1964) 'Fundamental methods in hydrodynamics', *Methods in Computational Physics*, **3**, pp. 211–263. Alder *et al.* (Eds) Academic Press, New York

Zienkiewitz, O. (1977) *The Finite eLement Method* (3rd edn), McGraw-Hill, Maidenhead, 787 pp.

5 Block theory in rock engineering

Professor R Goodman
University of California at Berkeley

Engineering an excavation in rock is often accomplished by reference to previous experience, particularly the personal experience of the responsible engineer. Since the range of possible behaviours is vast, this experience needs to be codified and constrained by the properties and structures of the rock mass; accordingly engineering geologists, with an ability to classify and describe the rock mass and to explore its attributes, ought to figure prominently in the design process.

In an effort to place the engineering of rock excavations on a more objective basis, the profession is trying to provide classification systems as keys to empirical design. The engineering classification of rock masses attempts to assign experience to classes so that relevant geological models can be selected in practice in any particular project.

Less satisfactorily, attempts have been made to use formal engineering mechanics to design excavations. The engineer attempts to assign mechanical properties to an idealized rock mass, assumes boundary and initial conditions and prestress and then computes stresses, iterating as needed to hold these within some bounds by means of local and global criteria of failure. In certain well-defined cases, engineering mechanics can add to the engineer's knowledge for design. But the design itself requires judgements and compromises that are not easily amenable to rational solution. Furthermore, it is often difficult to determine the deformability and strength properties of the rock and to characterize the *in situ* stress field with sufficient precision to define a meaningful problem to be solved.

A big stumbling block for both empirical and analytical methods is the fact that typical failure modes in jointed rocks are fully three-dimensional because the orientations of the joint planes govern the displacement directions. Furthermore, these joint directions are particular to each site and cannot easily be codified by the averaging processes of presently available classification systems. Unfortunately, three-dimensional stress/strain computational models embracing joints are not readily available to engineers.

The author prefers to use a rationally based, fully three-dimensional system in support of engineering, one in which the data expected in building a model for the rock are naturally obtainable from the exploration of the rock mass. Such a system is that known as the *keyblock* approach, employing *block theory*. This chapter will explain the basis for block theory and give examples of its application for problems of excavation design. A succession of cases will be examined proceeding from those in which very little is known about the rock mass, to those in which we attach confidence to knowledge of geometrical and physical properties of the joint system.

5.1 Properties of a rock mass and computational possibilities

We assume throughout that the rock is hard and sufficiently strong that it would be excavatable without support were it not for the system of discontinuities. The limestone blocks forming the Roman Pont du Gard, in France, have an average strength of only 7.0 MPa, very weak as rocks go but, in the absence of unfavourably oriented discontinuities, quite enough for this impressive bridge.

In exploring the rock, the geological model must be filled out to determine the configuration and dimensions of the different rock units comprising the volume enclosing the excavation. If the rocks are essentially discontinuous, with three or more sets of extensive joints, and hard enough that new rock cracking is of secondary importance, block theory is appropriate. If the rocks are soft and weak like soils, it may be preferable to elect to use methods other than those discussed in this chapter. For each homogeneous sub-unit, we then attempt to describe and characterize the imperfections and provide geometrical and physical properties for the different sets and types of discontinuities.

(1) At the least, we can hope to characterize the spatial orientations (attitudes) of each prominent set of joints. This can be achieved by mapping natural outcrops, cleaned rock surfaces ('pavements'), test pits, and pilot excavations. Drill holes provide the absolute orientations of discontinuities only if a special effort is made to orient the core samples, either by reference to known structural directions (usually bedding or foliation) or by use of a core-orienting device. The walls of drill holes can be photographed to

determine the absolute attitudes of the discontinuities using drill-hole television or down-hole cameras. The orientations of the joint sets constitute the most important items of information one can hope to learn about the rock to be excavated, and therefore it is usually well worth the extra effort and expenditure necessary to know them. As we shall show, much can be accomplished in excavation layout and design having *only* this knowledge.

(2) Logging of the discontinuities should distinguish between the different geological types of surfaces within the rock mass. These include faults, major shears, shear joints, extension joints, bedding joints, foliation joints, and contacts. The frictional properties of each of these types can be assigned a degree of confidence that depends on the quality of the exposures, the opportunities to perform field or laboratory tests, and the reliability with which one can characterize each joint set's wall-rock type, its surface roughness, its characteristic filling of the aperture, or coating of the wall-rock, and its likely moisture regime. If frictional properties cannot be attached to a joint set confidently, then conservative values must be assumed. The cost penalty of doing this may invite opportunities for special field studies and tests. If frictional properties can be assigned to each joint set, and to the isolated individual discontinuity features of special importance, block theory offers additional conclusions in support of design. Assigning friction values to the joint sets is a tractable problem in most rock masses and can be accomplished satisfactorily if properly motivated and supported.

(3) It is rarely possible to find the absolute location of each relevant discontinuity in a rock mass. The absolute locations of the important *individuals*, like contacts, faults, and shears, should emerge from a successful exploration programme. The various types of *joints*, however, constitute a series of statistical data sets; an individual joint is then merely a statistic. From studies of these data, however, it is possible to characterize the nature of the statistical distributions and their parameters. Most notable among these results are the spacings between individuals of each set. The lengths (extents) of joints of each set cannot be characterized as well, and we know very little yet about the bridge distances between the ends of one joint and its approximately coplanar neighbours. In the absence of data about the spacings and extents of each joint set, conservative assumptions must be adopted. Frequently, we assume the joints to be infinitesimally spaced and infinitely extensive. These assumptions lead to design based upon the largest removable block of each component of the excavation. Thus even without knowledge of the detailed spacings and locations of joints, a meaningful design calculation can be pursued.

If the actual locations of the important joints can be determined, additional design computations can be made. These fall into two classes – *hypothesized* and *real*. In the first case, a statistical simulation is made, based upon the distribution of joint attitudes, spacings, lengths, and bridges. Each realization of the simulation produces a model which can be analysed specifically and after repetitions the suitability of a specific design can be reported probabilistically. In the second case, arising when a specific real excavation is enlarged or reconfigured, a determined optimum support system can be found to assure the stability of every real keyblock at minimum cost.

(4) The case being postulated at this point assumes a data set acquired from exploration that determines the orientation and spacing, and the frictional properties of each joint set. This detail is sufficient for designing a cost-effective and safe support system and for optimal layout. Yet except for the subdivision of the rock mass into homogeneous subdomains, and the naming of the rock types and imperfections, only the system of joints has been characterized. This level of characterization is not sufficient for numerical analysis by finite element or similar methods. Thus, if methods of solid mechanics are to be used to answer specific questions, it will then be necessary to undertake a rock mechanics evaluation of the rock mass to find the deformability properties of each rock unit and, it is hoped, to characterize the *in situ* state of stress. Determining these attributes of the rock mass will be more expensive and more difficult, with less likelihood of success, than the previously discussed geological characterization. We shall show that refinements to block theory are possible to take advantage of information about the state of stress if available, even without knowledge of the deformability properties of the rock.

5.2 Stereographic projection

The ideas of block theory are independent of the specific technique used for practical solutions. Computer programs developed by Gen-hua Shi for block theory are based upon solution of simultaneous inequalities by means of vector analysis. For this paper, results are displayed on the stereographic projection, whether derived from the projection by manipulating overlays on a stereonet, or from the block-theory computer programs.

The net in Figure 5.1 is useful for block theory. Like other stereonets, Figure 5.1 presents a family of planes (great circles) and cones (small circles) having a common axis. The figure is unusual in continuing beyond the equatorial circle (the 'reference circle'). If the stereographic projection is based upon a focus at the bottom of the reference sphere, the upper hemisphere is displayed inside the reference circle, while the lower hemisphere is the region outside the reference circle.

A joint plane is assumed to pass through the centre of the reference sphere and is projected as a great circle, which can be traced from the net. Thus a joint plane is

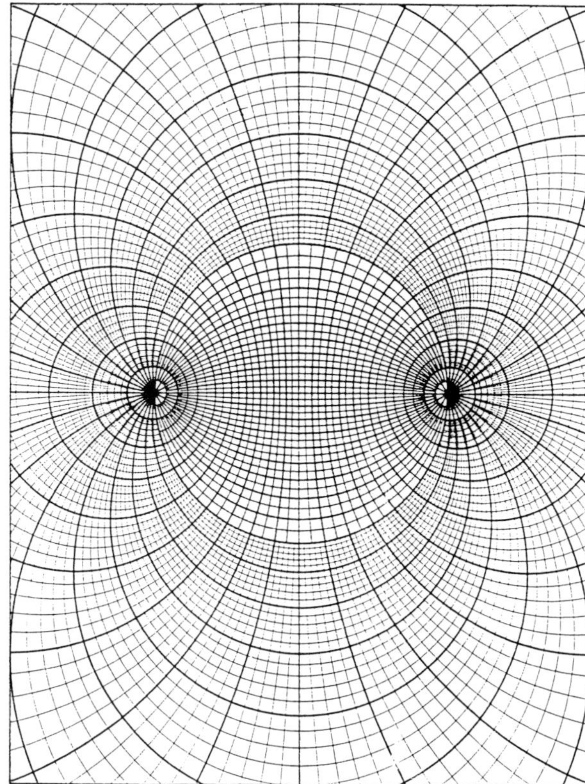

Figure 5.1 *A stereonet*

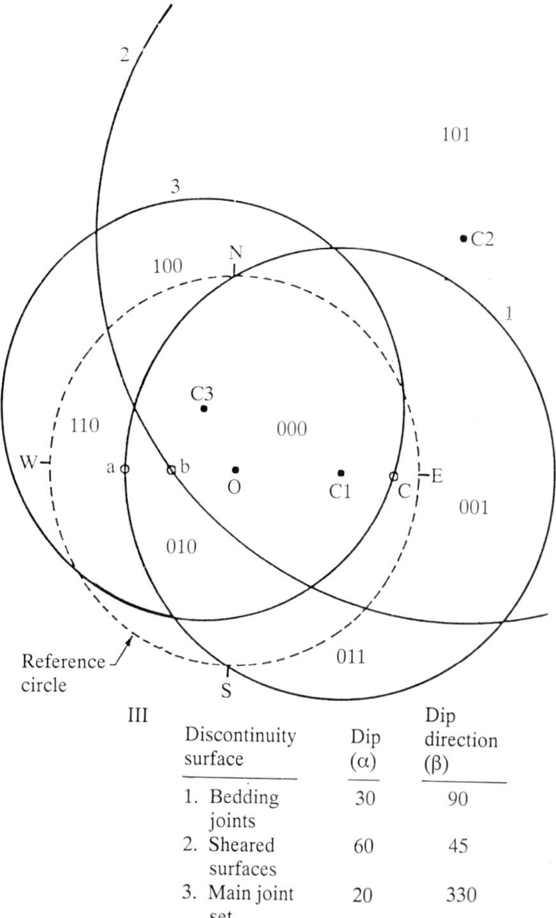

Discontinuity surface	Dip (α)	Dip direction (β)
1. Bedding joints	30	90
2. Sheared surfaces	60	45
3. Main joint set	20	330

Figure 5.2 *Joint plane circles and reference circle*

projected as an entire circle on the tracing. A free surface (an excavation plane or a natural ground surface) can be similarly displayed as an entire great circle. The region inside a joint circle represents the upper half-space of the joint plane and the region outside the joint circle is the lower half-space of the joint plane. Figure 5.2 projects three joint planes and the reference circle. The centres of the three joint circles are labelled C1, C2 and C3. The application of stereographic projection to block theory is discussed by Goodman (1989) and Goodman and Shi (1985).

5.3 Block theory

The three joint plane circles of Figure 5.2 produce eight spherical triangles, each labelled with three ordered digits. A 0 in the first place signifies that the spherical triangle belongs to the upper half-space of joint plane 1; this is true because it is inside the circle corresponding to joint plane 1 (when the focus for the stereographic projection is at the nadir point of the reference sphere). Conversely, a 1 in the first place says that the spherical triangle thus identified is in the lower half-space of joint plane 1, meaning that it is outside the circle for plane 1. The second digit pertains to the second joint, etc.

A block liberated by the intersection of joint planes and free surfaces is actually formed by the intersection of the joint-plane half-spaces. Each joint plane has two half-spaces and the numbers of possible intersections are therefore 8 (for three joints). Each combination corresponds to a particular spherical triangle. The name 'joint pyramid' (JP) has been assigned to a joint-plane spherical triangle and its ordered digits are termed its *JP Code*. Each JP belongs to the infinite set of blocks for which the block-side of each bounding plane is in the half-space of that plane identified by the JP code.

In block theory, the infinite numbers of blocks formed from series of joint sets are all divisible into a small number of JP types. Analysis is necessary only for each JP, not for the infinity of individual blocks. A *keyblock* is a block that will tend to move into the excavation unless support is provided. The identification of the JPs belonging to potential keyblocks is the object of the analysis. Given a particular excavation, the numbers of JPs that can produce keyblocks is limited. When the potentially dangerous JPs are named, the faces of dangerous blocks can be located on the surface of the excavation.

In order for a block to be labelled a keyblock, it must tend to move into the free space created by an excavation. To determine this, a series of tests is applied. First is a test for *removability*. Infinite blocks are not removable; nor are blocks tapered in shape such that there is no allowable direction for the block to move towards the free surface without colliding with a neighbour. Removable blocks are tested for equilibrium under the assumed force field, embracing body forces, pressures applied to the faces, and forces on points of the surface or interior. A block that is in equilibrium without support is a *potential keyblock* while one that requires support is a true *keyblock*.

Mode analysis determines the sliding modes belonging to each JP once the direction of the resultant force on the block has been input. Mode analysis is performed without reference to any specific excavation and serves to reduce the number of interesting JPs. Some JPs will lack a mode and therefore be unable to generate keyblocks in any excavation. Mode analysis is not discussed further here but is presented fully in Goodman and Shi (1985).

When the excavation is described, *equilibrium analysis* then determines which if any JPs produce unstable blocks. Finally, a real or simulated geological map of the excavation surface is inspected to determine if any of the dangerous JPs in fact are yielding real blocks on the surface of the excavation. Timely support can then be installed before the keyblock is isolated by the continuation of the excavation.

In the remainder of this chapter, we will show in more detail how the sophistication of analysis can be tailored to the level of knowledge acquired about the system of discontinuities in the rock mass. The organization by 'cases' conforms to the numbers assigned to levels of knowledge discussed under the rubric 'properties of a rock mass and computational possibilities'.

5.4 Case I: Analysis given only the orientations of the joints

If the joint attitudes are known, a series of JPs is generated as in Figure 5.2. Shi's theorem can now be applied to determine the JPs that yield removable blocks in a given excavation. According to this theorem, which is derived and explored in Goodman and Shi (1985), a block is removable if and only if its JP appears on the stereographic projection and has no intersection with the projection of the block side of the excavation. The latter is termed the *excavation pyramid*, abbreviated EP.

Consider first a simple vertical wall through the rock of Figure 5.2. The stereographic projection of a vertical plane is a straight line through the centre of the reference sphere (since a vertical plane produces a great circle of infinite radius). Suppose the wall runs east–west across the rock and the rock is on its south side. The EP for this excavation is then the ruled region of the stereographic projection shown in Figure 5.3. Drawing such a ruling on

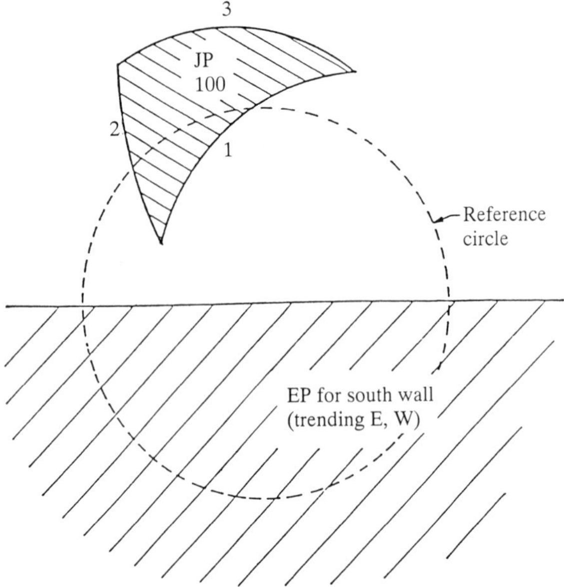

Figure 5.3 *Stereographic projection*

Figure 5.2 shows that only one JP has no region of overlap with the EP – namely JP 100. Therefore, the only blocks that are removable in the south wall are those formed with JP 100. In other words, the potential keyblocks for this excavation are those that are south of the excavation wall, below joint plane 1 and above joint planes 2 and 3. Such a block can be drawn and its maximum size within the limits of the excavation can be established.

In the absence of information about the spacing of the joints, it is prudent to design on the assumption of a *maximum keyblock*, whose dimensions are such that the block just fits inside the excavation limits. The maximum keyblock is adopted as a basis for design because friction tends to become smaller as the size of a block increases; therefore the most unstable of all similar nesting blocks of a given JP will always be the largest of the whole nest. This is an important principle in applied rock mechanics.

In the case of a tunnel excavated through the rock mass, all JPs will yield removable blocks somewhere around the walls, except for those that include the tunnel axis. These removable blocks are not equal, however, since gravity tends to bring down the keyblocks of the roof and high walls.

It is possible to determine the location of the maximum removable block of each JP around the tunnel, given the tunnel shape. To do this one constructs limiting planes that envelope the JP and include the tunnel axis. The intersection of these limit planes with the tunnel

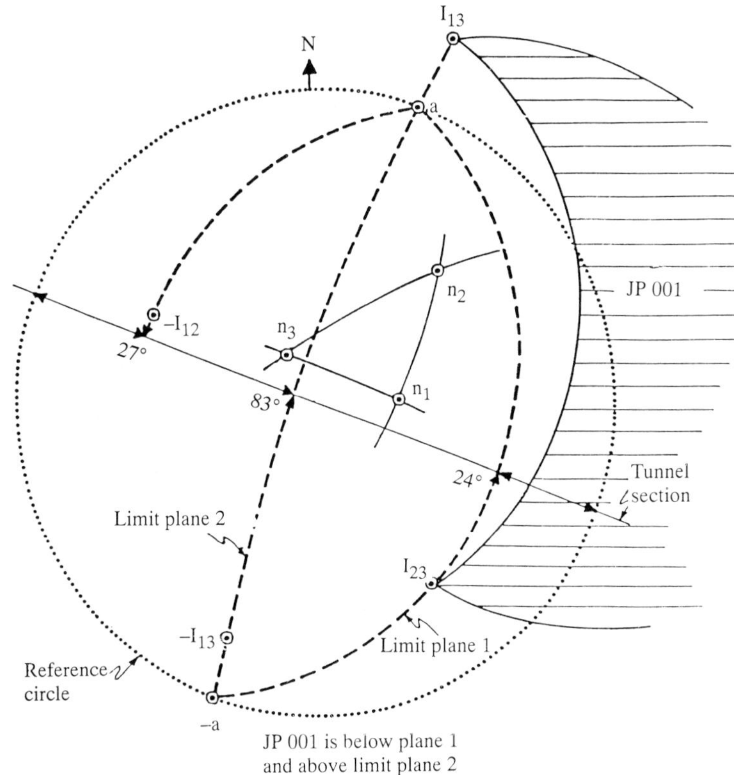

Figure 5.4 *Construction for joint system of Figure 5.2*

cross-section plane determine the orientation of the traces of the limit planes in the tunnel section. The maximum removable block of each JP is found on the tunnel section from these traces and the location of the JP relative to the limit planes on the stereographic projection.

Figure 5.4 presents an example with the joint system of Figure 5.2. JP 001 has been selected (arbitrarily). The axis of a tunnel is at *a*. The limit planes are great circles passing through *a* and each of the corners of the JP. The extreme limit planes are the two shown. These intersect the tunnel section at two points which represent lines plunging 83° above the west and 24° above the east in this section. The JP is above the limit plane common to the tunnel axis (*a*) and the corner I_{13}, and it is below the limit plane common to the tunnel axis (*a*) and corner I_{23} (because it is inside the former great circle and outside the latter). Figure 5.5 shows the region of the maximum removable block corresponding to this JP in the tunnel section.

In the preceding paragraph, we constructed the edges of the maximum removable block for a particular JP in a tunnel of given shape driven in a particular direction. To find all potentially dangerous blocks around this tunnel, it is necessary to repeat the construction for each JP except those two that contain the tunnel axis. Figure 5.6 shows an example of a complete study of all JPs for a system of four

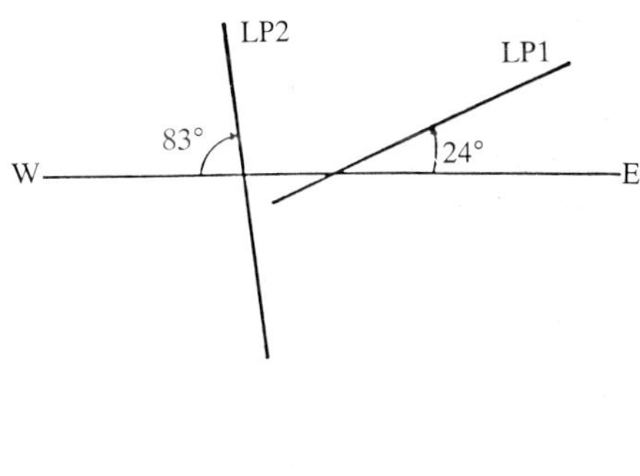

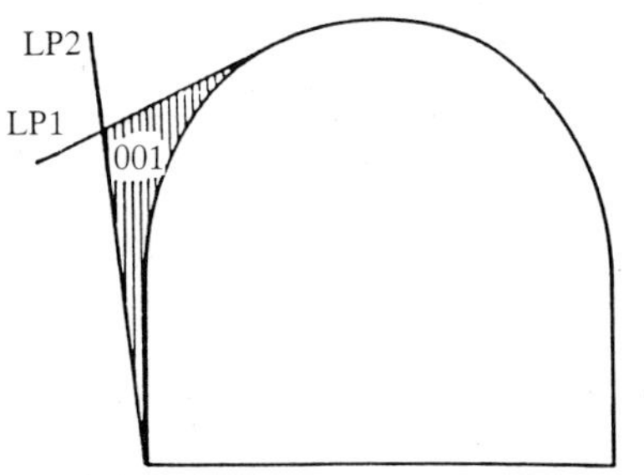

Figure 5.5 *Maximum removable block*

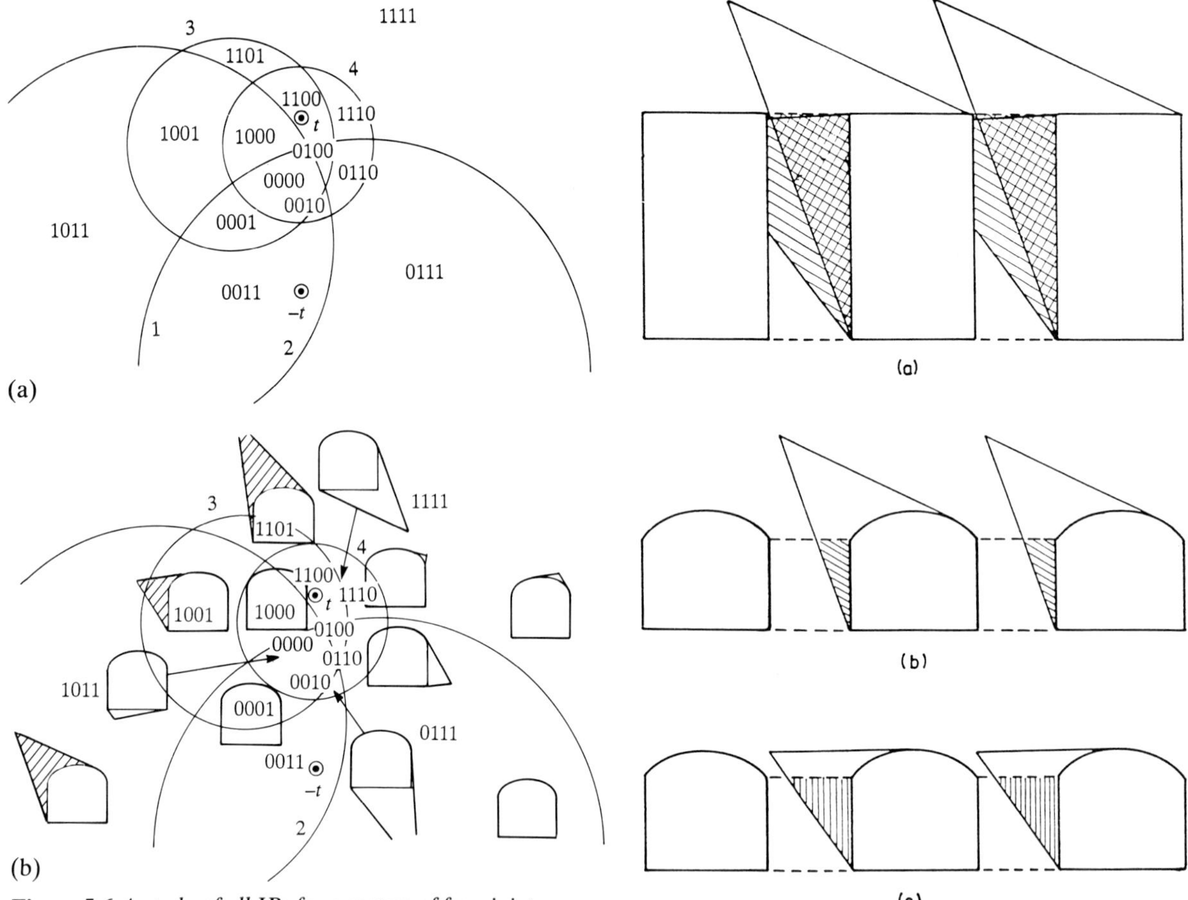

Figure 5.6 *A study of all JPs for a system of four joint planes:* (a) *stereographic projection;* (b) *maximum removable regions*

Figure 5.7 *Spacing between tunnels*

joint planes. Figure 5.6(a) shows the stereographic projection of the joint system and the direction (t) of the tunnel axis. Figure 5.6(b) shows all the maximum removable regions for this tunnel direction by means of a series of figures inset within each JP (or placed close to it if there is insufficient room to do otherwise). Under gravity, the most dangerous removable regions are the three that are shaded, belonging to JPs 1011, 1001, and 1101.

The construction presented in Figure 5.6 produces a rational determination of the minimum spacing between parallel tunnels such that the pillar of rock between them cannot be lost following the removal of any block. In Figure 5.7(a), the spacing is too close for the height and shape of the tunnel selected because the loss of the removable region would destroy the entire wall pillar. In Figures 5.7(b) and 5.7(c), the spacing is seen to be acceptable for blocks of JP 1011 and 1001 respectively. As no other JPs can produce a larger wall deterioration, this tunnel spacing is deemed to be acceptable for the size and shape of the tunnel shown when driven in the direction of t.

5.5 Case II: Further analysis given the joint friction angles

The removability of a block establishes its capacity to be a keyblock but the definition of 'keyness' requires that the block tend to move in the absence of support. To establish whether a JP will produce true keyblocks necessitates an equilibrium analysis. Assuming that the equilibrium is not affected by the deformability of the block, analysis need consider only the forces applied to the block and the reactions on the contacting faces. When these faces are cohesionless, the state of equilibrium can be expressed entirely in terms of the orientation of the resultant reaction forces on each closed face; to be stable, the reaction force on each contacting face must be inclined at an angle from the normal measuring less than the face's angle of friction. Conversely, to be a keyblock, the forces applied to the block, neglecting support forces, must produce reactions on each closed face that incline from the normal by more than the friction angle. Then, the support

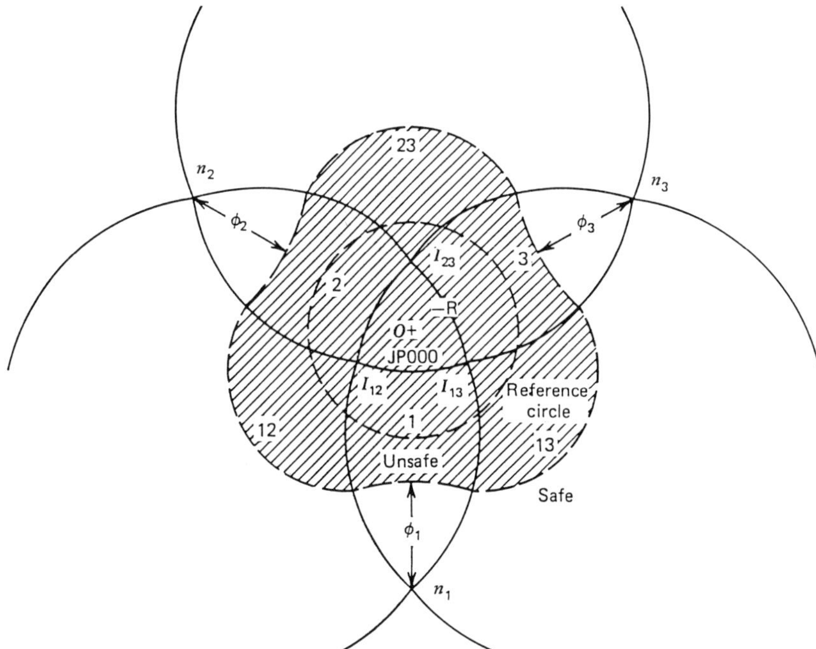

Figure 5.8 *Stereographic projection showing stability analysis*

force required to achieve a particular factor of safety can be calculated, as discussed in Crawford and Bray (1983), Elsworth (1986), Goodman and Boyle (1986), Goodman *et al.* (1982), Karzulovic (1988) and Yow and Goodman (1987).

An example of a stability analysis is presented on the stereographic projection in Figure 5.8. This type of presentation was pioneered by John (1968) and Londe *et al.* (1969) and extended to blocks with more than three joint faces by Shi and Goodman (1989). In this lower focal point projection of a system of three joint sets, the JP to be analysed is the spherical triangle identified by the symbol O, and its corners I_{12}, I_{23}, and I_{13} are the three lines of intersection pointed out of the block, into free space. The analysis is made for each removable JP and then applies to all the infinite blocks that can be produced with this JP and an excavation. While the analysis can be performed independently of any definite excavation, it makes sense only if the JP has already been found to yield blocks that are removable in that excavation.

The outward normals to each joint face are the normal vectors to each face that project on the side not occupied by the JP; they are labelled n_1, n_2, and n_3. The modes of failure by sliding alone include the possibility of sliding on any joint face (mode i for face i) and sliding on any real combination of any two joint faces (mode ij for faces i and j in combination). The modes occupy spherical triangles between block edges and outward normals, as labelled on Figure 5.8. Within each mode triangle, a bounding locus separates stable from unstable regions and in this manner the entire sphere has been divided into a single connected region of stability (the unruled region) and another of instability (the ruled region). The construction of this bounding locus is determined by the applicable friction angles as shown on the figure. In the regions of modes ij, the bounding locus is a great circle, while in regions of modes i it is a small circle about n_i. The direction of the resultant force applied to the block (disregarding support forces) can be projected as a single point on this figure. When that point is in the region of stability, the block is stable, and is only a potential keyblock. When the resultant force applied to the block (not counting support forces) plots in the region of instability, the block is unstable without support and is a true keyblock.

Figure 5.8 was plotted for a JP conveniently located in the centre of the stereographic projection (i.e. JP 000) and applies to the stability of the floor of an excavation. Such a JP usually produces stable blocks under gravity, unless extreme uplift is created by means of water pressures or directly applied structural forces. Its 'cousin' JP 111 will tend to produce weakness in the roof, often by opening of each joint plane. This happens when the applied force resultant plots within the JP itself, and the mode is accordingly labelled with a zero.

Figure 5.9 discusses a more generally important case for a JP (100) arbitrarily located with respect to the centre of the projection. The corners of this JP are still the outward lines of intersection and the outward normals are seen to be those on the non-block side of each of the JP's bounding great circles. The bounding locus is determined by the friction angles, as in Figure 5.8, and the whole sphere of directions is thus subdivided into stable and

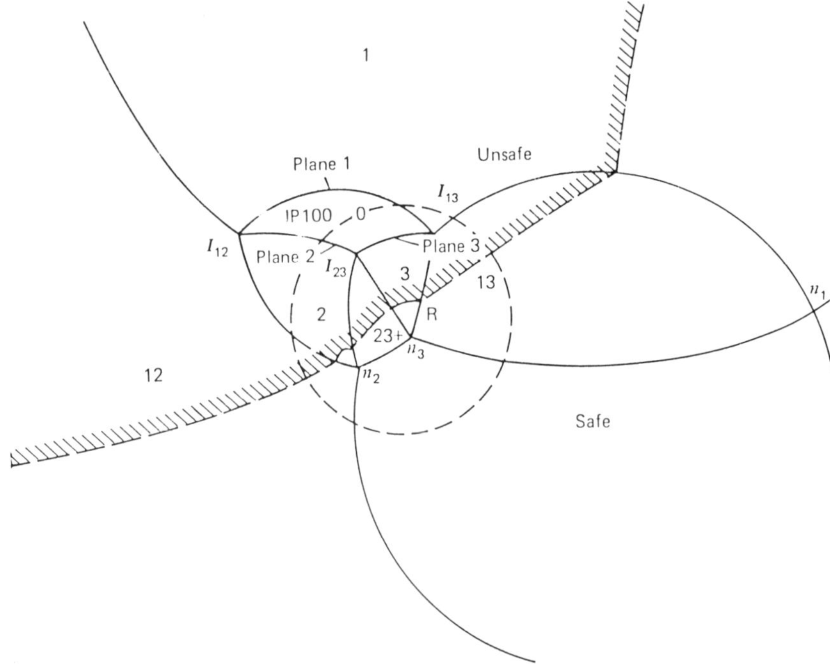

Figure 5.9 *Case of an arbitrarily located JP*

unstable regions. If the resultant force is oriented as shown by the +, the block is safe. Its factor of safety is determined by appropriately reducing the friction angles to shrink the safe zone in order to pass the bounding locus through +. The Factor of Safety concept for block sliding is discussed in Londe *et al.* (1969) and Goodman (1989).

By performing a stability analysis for each removable JP of a tunnel, it is possible to calculate the support requirements for the most dangerous blocks around the tunnel. We now re-examine the removable blocks of a tunnel in the joint system projected in Figure 5.2. As we saw previously (Figure 5.6b), determining the tunnel axis establishes a complete set of maximum removable regions around the tunnel, corresponding to all the JPs that do not contain the tunnel axis. By running a stability analysis for each of these removable regions, it is possible to establish the support requirements for the maximum worst key-block corresponding to the selected tunnel direction.

In Figure 5.10 a series of worst maximum keyblocks is shown for different horizontal tunnel directions through this rock mass. It can be seen that as the tunnel direction passes through azimuth 135° the size and location of the worst keyblock reach a maximum severity. Figure 5.11 is a graph of the data from this analysis showing the maximum required support force for the worst maximum keyblock as a function of tunnel azimuth. We have termed this type of response a 'tunnel support spectrum'. The remarkably directional behaviour of the support requirements for an excavation in jointed rock exhibited in this figure may be understood intuitively by some engineering geologists. However, the quantitative details are in general different from that dictated by the conventional engineering geology wisdom of avoiding a layout that places an excavation wall parallel to a joint set. The problem of choosing an optimum orientation for an excavation in blocky rock is more complex than that and cannot be handled by rules of thumb.

5.6 Case III: Further analysis to define real keyblocks given the locations of joint traces

The locations of actual discontinuity traces across an excavation may be known for an existing work that is to be modified or enlarged. For planning of new excavations, the actual locations of traces cannot be known but they can be assigned statistically as discussed in the introduction. Figures 5.12 and 5.13 show an example of joint trace generation for a tunnel, using a simulation program written by Gen-hua Shi (Shi and Goodman, 1989a). The procedure begins by unrolling the walls of the tunnel to produce a developed base map, as routinely done by engineering geologists for logging the interior surfaces of underground openings. In the unrolling process, the traces of joints cutting diagonally across the tunnel become continuous curved lines, as shown. In this example, the dip/dip-direction data for the four joint sets are as follows: Set 1 – 71/163; Set 2 – 68/243; Set 3 – 45/280; and Set 4 – 13/343. The tunnel has the shape shown and is directed

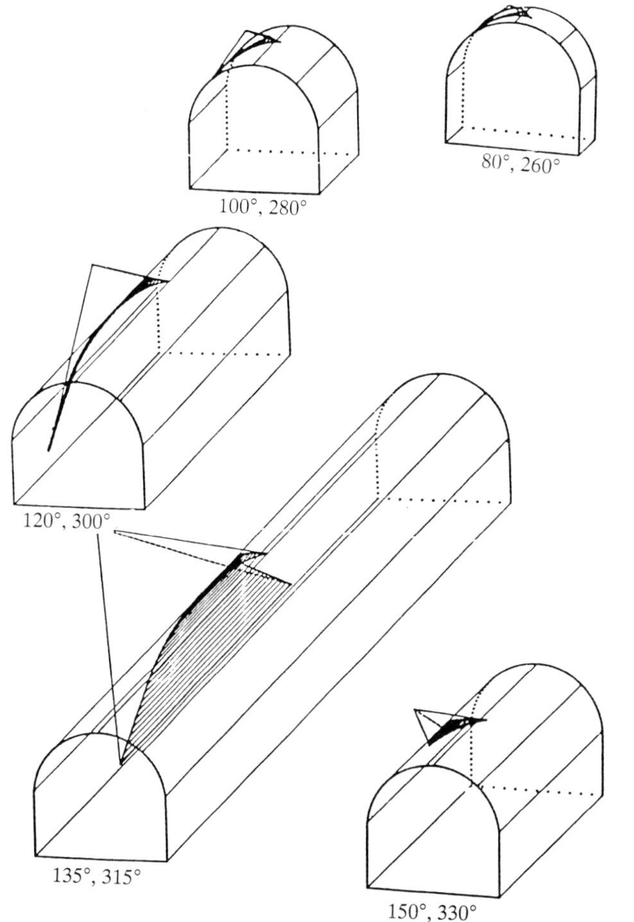

Figure 5.10 *A series of worst maximum keyblocks*

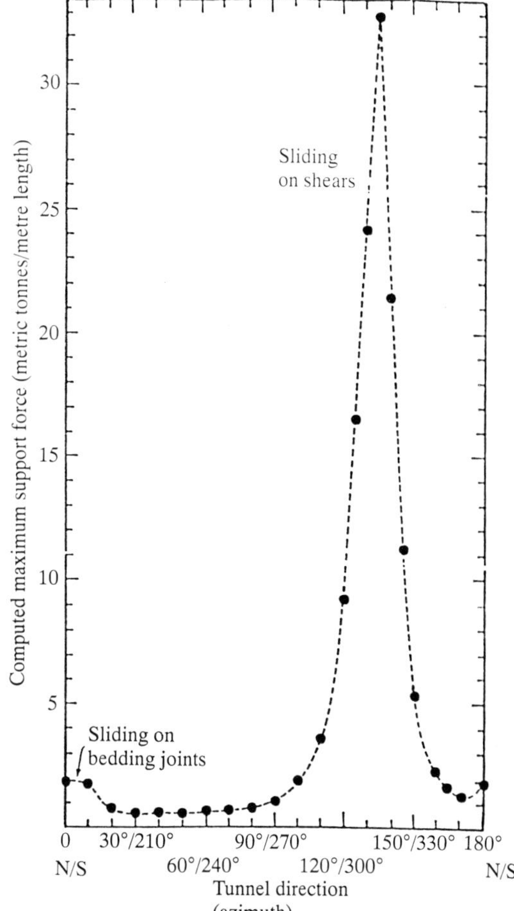

Figure 5.11 *Maximum support force for analysis of Figure 5.10*

horizontally along azimuth 20°. The simulation program assigns an average spacing to each joint set and produces a corresponding regular ruling, which is then perturbed to sample distributions of spacing, length and bridge. Figure 5.13 is one realization produced with a specific set of input values for length, spacing and bridge. Of the many curved polygons belonging to this figure, only some are faces of real removable blocks; block theory allows us to determine which these are. First, we use the construction of Figures 5.4 and 5.5 to determine the intersection of the tunnel cross section with the extreme limit planes that envelope the maximum removable block. For one specific JP, we thus determine the parallel straight lines on Figure 5.13. For this JP, the search for keyblock faces from among the many curved polygons can then be confined entirely to the subregion between the lines thus drawn, for no block belonging to the selected JP can be removable in the tunnel if it extends beyond the limits corresponding to the sides of this subregion.

Shi's unrolling program first restricts the trace map to the subregion for the selected JP, as shown in Figure 5.14. It then removes all joint trace ends that terminate inside a small polygon and all polygons that intersect the bounds of the subregion. This produces a map of real faces shown in Figure 5.15. Tick marks are added to each line of Figure 5.15 to identify the block-side of the trace corresponding to the designated JP code, yielding Figure 5.16. Real faces of removable blocks are those polygons on the ticked side of each bounding trace. The only real faces of removable blocks for the JP of this example, corresponding to this realization of the geometric simulation, are the ruled regions of Figure 5.17. The support needs of the real keyblocks of the tunnel can be determined by repeating this analysis for each JP and conducting equilibrium analyses. In this way the effect of tunnel size and shape on real support needs can be determined.

If the trace map used in this analysis is deterministic, the results give actual support needs, as opposed to maximum

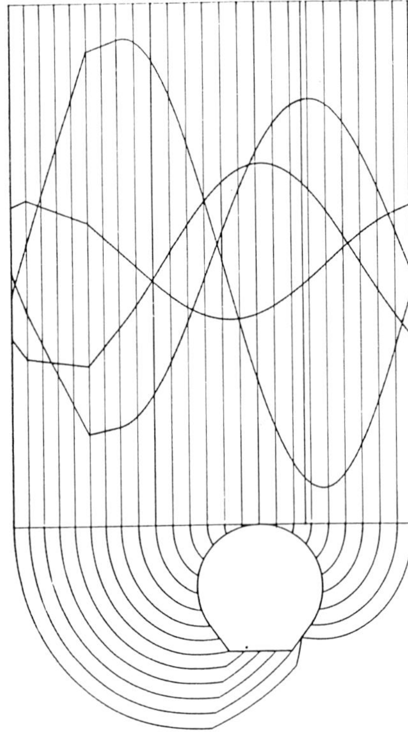

Figure 5.12 *Joint trace generation: developed base map*

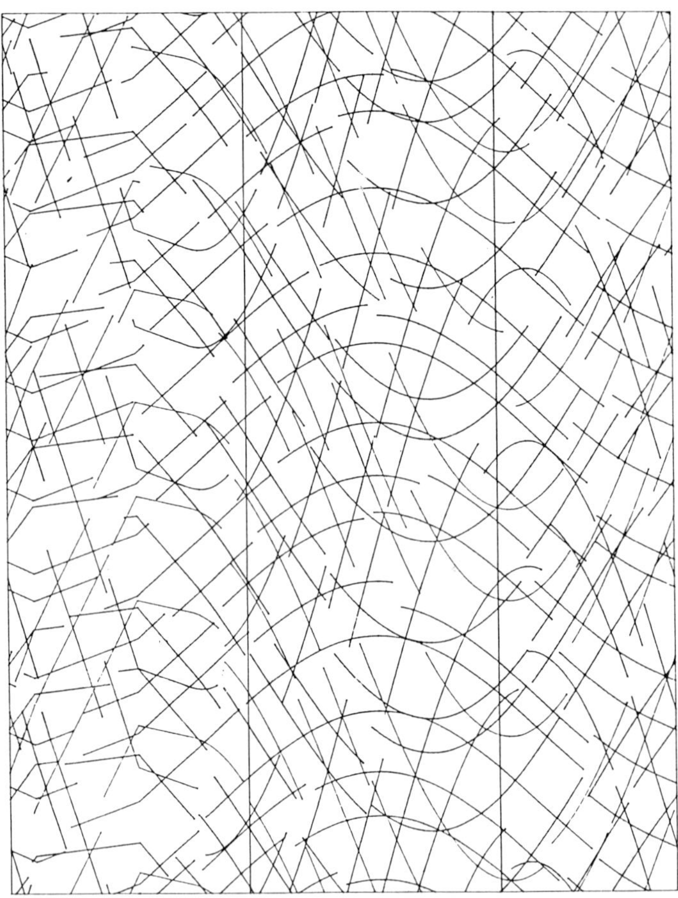

Figure 5.13 *Joint trace generation: a specific realization*

keyblock support needs. If the analysis is conducted on a realization of a geometric simulation, as in the above example, repetitions can yield a distribution of support needs. Since all computation is performed with desk top computers, it is entirely practical and feasible to conduct such a study.

5.7 Case IV: Further analysis for known initial stress, joint dilatancy or rock deformability – The block reaction curve

The equilibrium analysis discussed previously assumes that the block reaches a limiting equilibrium condition dependent only on the external forces on the block and the friction on each of its contacting faces. If the block is acted upon by the resultant traction equilibrating a given stress field in the rock around the excavation, its support requirement may be less than that computed from a limiting equilibrium. As a block tends to move, the frictional properties change on the bounding surfaces, particularly if the joints are non-planar and accordingly possess dilatant tendencies; dilatancy may cause the block to need less support than limiting equilibrium computations show, unless it is highly deformable. A complete solution for equilibrium in these conditions merits a numerical model study as, for example, by discontinuous deformation analysis (Shi and Goodman 1989b). However, short of such studies, an improved computation of block equilibrium can be made embracing initial stress, rock deformability, non-linear friction, and dilatancy, as discussed in Crawford and Bray (1983), Elsworth (1986), Goodman and Boyle (1986), Goodman *et al.* (1982), Karzulovic (1988) and Yow and Goodman (1987). The results provide a curve of required support, as a proportion of the block's weight, as a function of block displacement; such a graph has been named 'the block

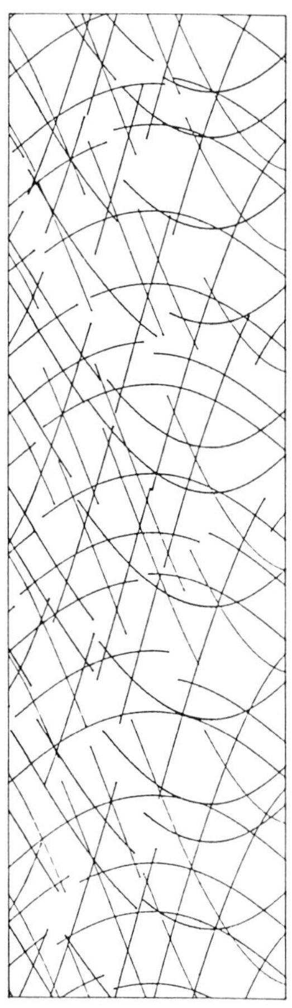

Figure 5.14 *Joint trace generation: restriction to a subregion*

Figure 5.15 *Joint trace generation: a map of real faces*

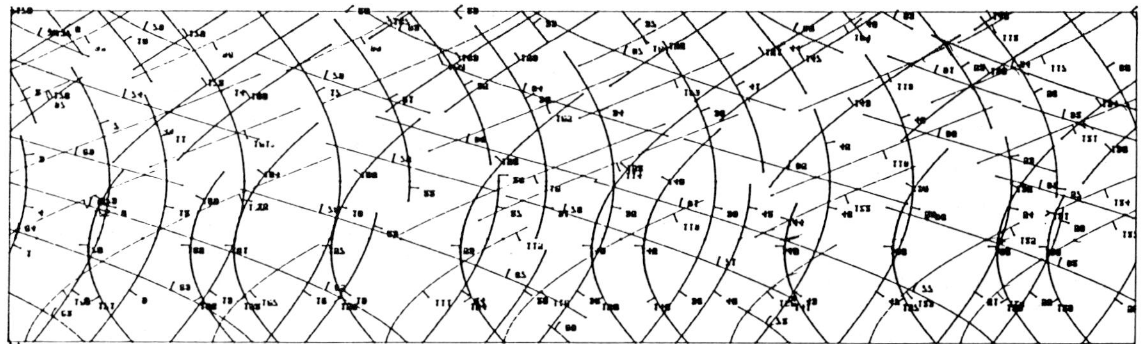

Figure 5.16 *Figure 5.15 with identification marks*

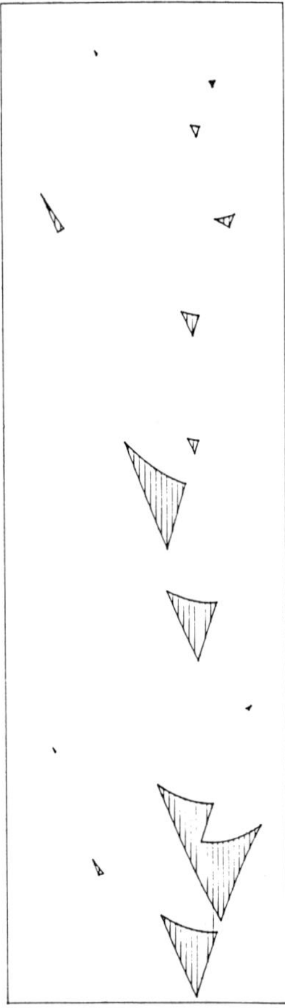

Figure 5.17 *Real faces of removable blocks*

reaction curve', in analogy to the 'ground reaction curve' produced by elastic–plastic solutions for circular tunnels under axisymmetric stress. In this case, the basic mechanism driving the solution is the tendency of a single block to move into an excavation under the action of forces and resistances that vary with displacement.

The concept of a block reaction curve, as described by Karzulovic (1988), is presented in Figure 5.18. In this drawing, k is the acceleration coefficient acting on the block; here k may be assumed to equal unity, corresponding to the acceleration of gravity. According to Bray (unpublished notes on block equilibrium written at Imperial College), A is the support force applied to the block so that the abscissa, A/W, represents the proportion of the block's weight that needs to be supported to maintain equilibrium. In the initial condition, stresses are assumed to flow tangentially around the roof of the opening so that the block receives no direct vertical support from the *in situ* stress. The initial support requirement is assumed to be $A/W = 1$ (condition a). A small downward displacement of the block mobilizes some

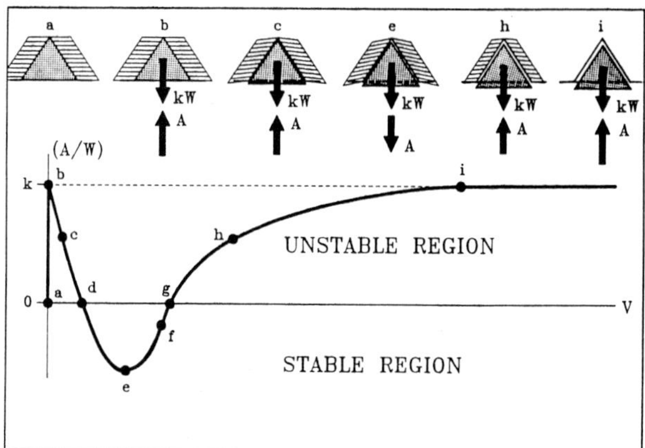

Figure 5.18 *Block reaction curve*

shear stress on its joint faces, effectively reducing the amount of vertical force required from the supports (condition b). However, frictional resistance is reduced as the block moves because it is simultaneously tending to open from its side faces. As the rate of frictional reduction catches up with the strengthening effect of the mobilized shear stresses, the reaction curve turns through a minimum (condition e). Eventually, as the block falls freely, A/W returns to the value of unity.

Yow and Goodman (1987) demonstrated that some blocks tend to yield a block reaction curve like that of Figure 5.18 with the abscissa becoming negative after a certain displacement (d in this case), whereas other blocks reach a minimum entirely in the positive A/W field. The significance of the former is that after a sufficient displacement (between d and g in this example) the block ceases to require any external support and, in fact, an extra weight could be hung safely from an anchor set in the block. However, in the latter case, where the A/W value always remains positive, there is a support minimum but external support must always be provided or the block will fall.

5.8 An example

Figure 5.19 shows the traces mapped by geologists during construction of an underground power-plant chamber and its curving access adit. Analysis of the different JPs for the granitic rock of this excavation indicates that JP 101 should yield removable blocks in the wall of the power-plant chamber and JP 111 should yield removable blocks in the roof of the access adit. Particularly large real keyblocks of these JPs are shown by the shaded polygon on the figure.

A strong rock bolt support system was installed in each block, rotating the resultant force in each case well into the safe zone. The limit equilibrium analysis for the power plant chamber is shown in Figure 5.20. A series of

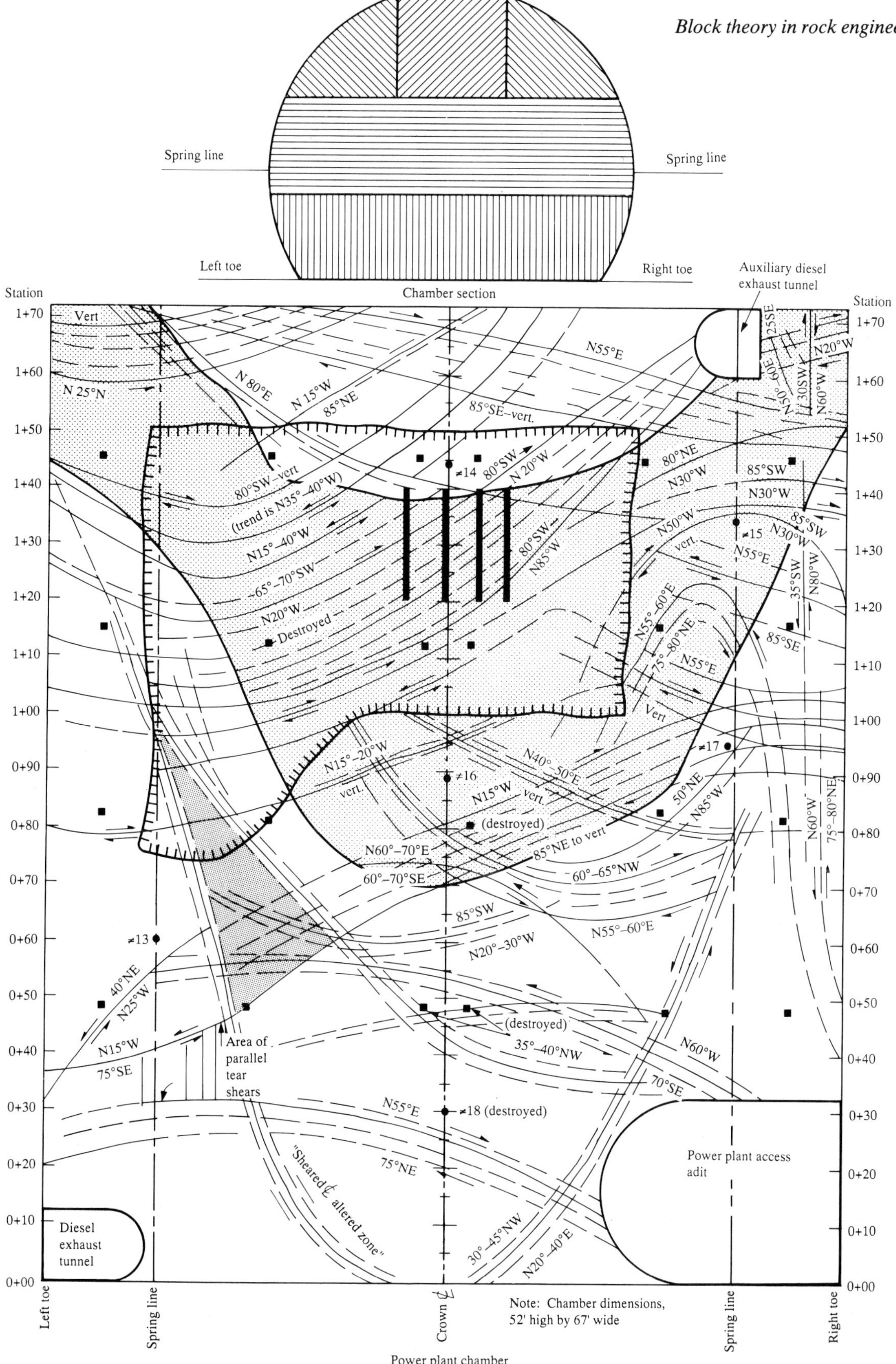

Figure 5.19 *Traces mapped during construction of power-plant chamber and access adit*

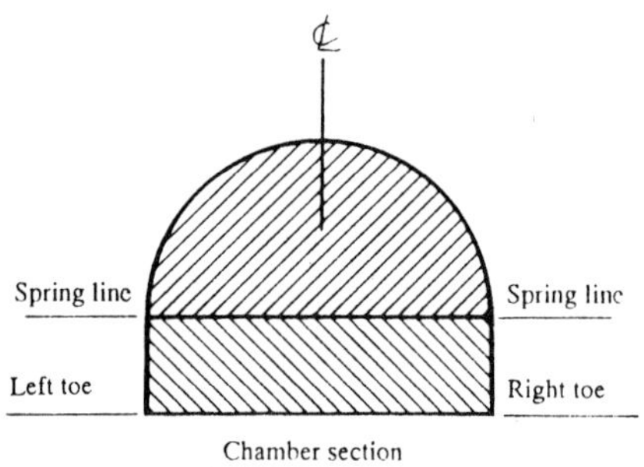

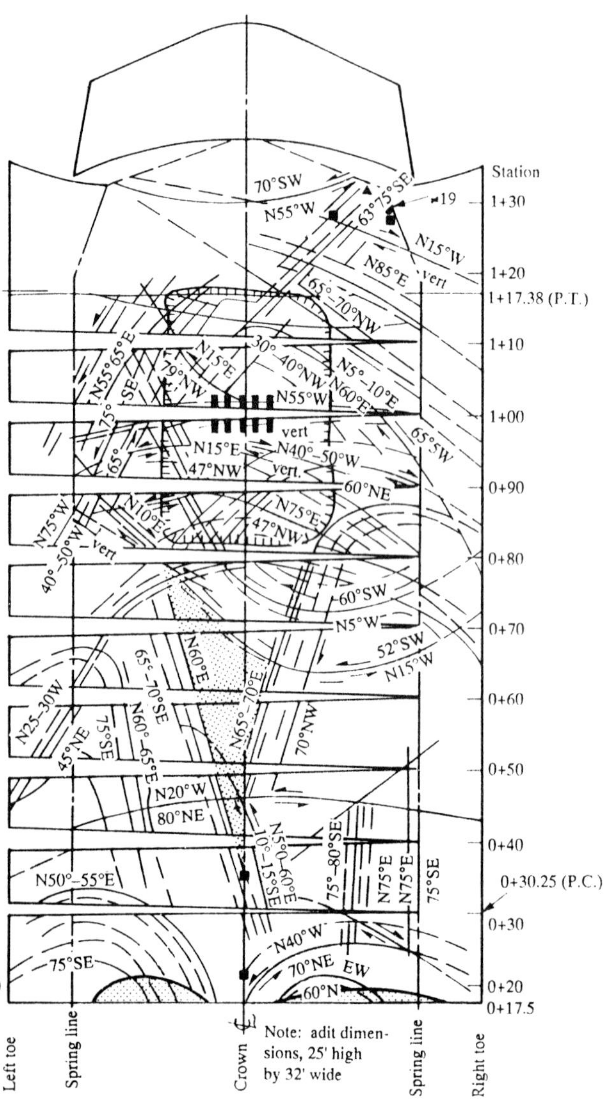

bounding locuses has been drawn for friction angles varying from 0° to 90° in five degree increments. The symbol + indicates the present orientation of the resultant force, including the compressive forces from the bearing plates of the rock bolts. Figure 5.21 shows the limit equilibrium analysis for JP 111 in the access adit, with a similar series of bounding loci for the friction angle. In each case, the present resultant of applied and support forces plots well inside the safe zone; in fact, it lies inside the spherical triangle whose corners are the outward normals to the faces of the JP. This makes the factor of safety infinite in that the block would be safe even if the friction angles were all equal to zero. However, under an increase in acceleration, the block can be brought to failure and the required acceleration to yield the block can easily be calculated.

5.9 Conclusions

In jointed rocks, intersecting joint planes and excavation surfaces create blocks on the boundary of the excavation. Block theory assumes that these blocks can be prevented from moving by providing a design that disallows movement of the most critically located and adversely oriented individuals, which are termed *keyblocks*. Although there are an infinite number of sizes and shapes of such blocks, the number of analyses required is finite and small. Each block belongs to one of a small number of joint pyramids and the analysis of a single joint pyramid satisfies analytical requirements for all blocks derived from it. Failure of keyblocks does not assure collapse but securing all keyblocks does assure general safety.

The level of analysis should be tailored to the degree of completeness of knowledge pertaining to rock-mass properties. There is seldom sufficient information about a rock mass to fully support numerical modelling with methods of continuum mechanics but block theory analysis can be made with limited data of the type one can hope to gain from site studies. When only the orientations of the joint sets is determinable, it is possible to determine the maximum keyblocks of each component of the excavation and to choose a layout that minimizes their impact. Such questions as the orientation, shape and spacing of excavation components can be decided with the minimization of these keyblocks in mind. Then if the friction angle of each joint surface can be assessed, or assumed, a limit equilibrium analysis can be made to design supports that prevent the maximum keyblocks from moving and therefore that render the entire rock mass stable. If the traces of joints can be mapped during construction, or simulated from a statistical description of the joint patterns, real keyblocks can be located and a less conservative support system configured. Finally, if the initial stress state can be characterized, its impact on support forces can be computed by constructing a block reaction curve; this will generally reduce support requirements as a small amount of tangential stress flow tends to

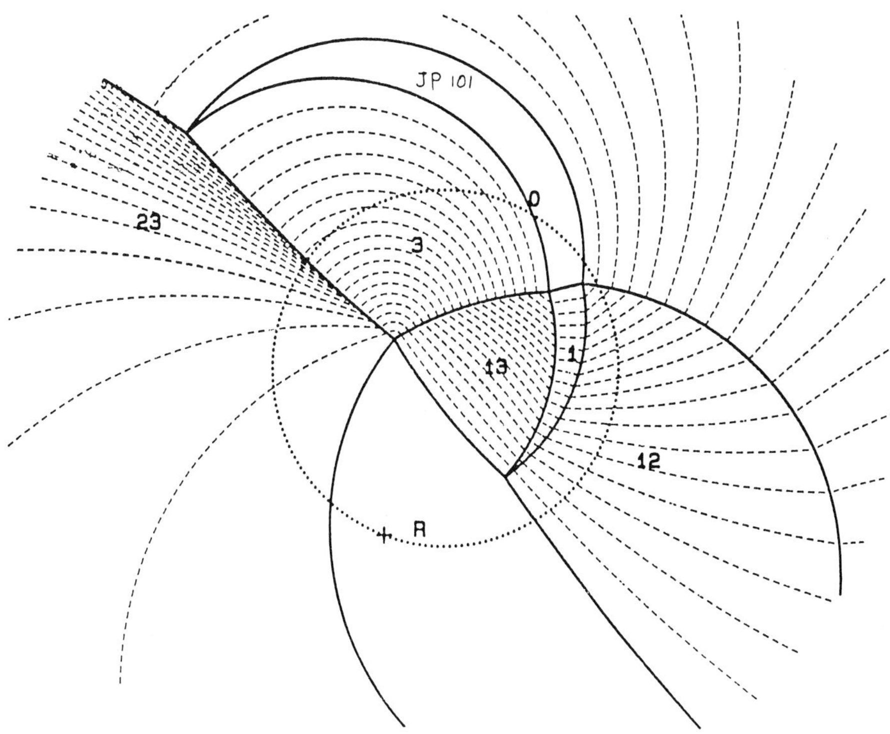

Figure 5.20 *Limit equilibrium analysis for power-plant chamber*

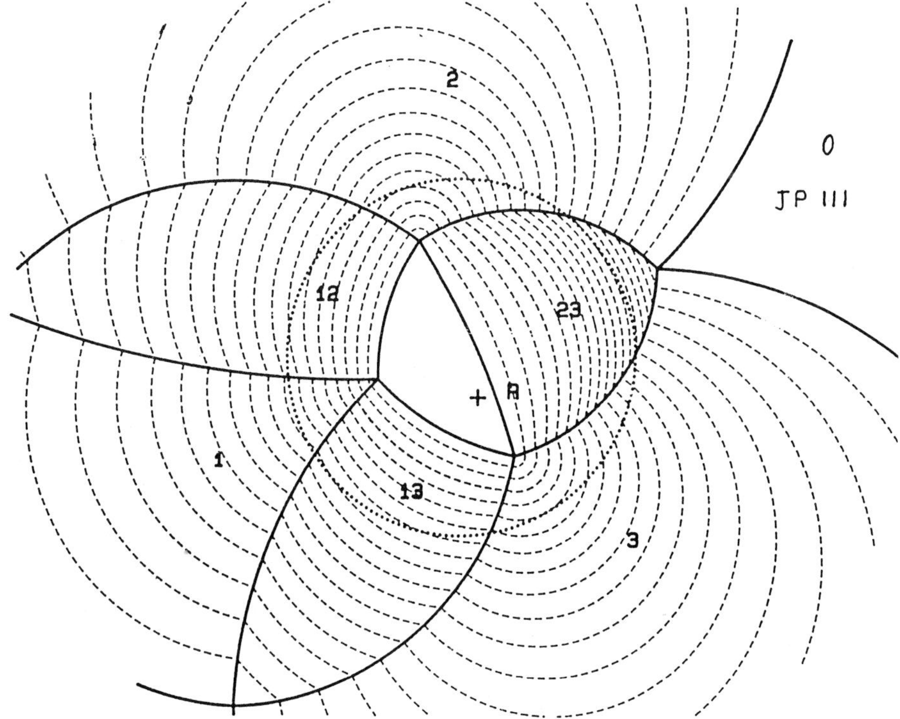

Figure 5.21 *Limit equilibrium analysis for JP 111 in access adit*

stabilize a block, unless it has a very wide JP vertex angle.

Block theory is appropriate for engineering in most hard rocks and its methods can be superimposed on other approaches for soft, jointed rocks that admit failure modes by cracking of previously intact material.

References

Crawford, A. M. and Bray, J. W. (1983) 'Influence of the in-situ stress field and joint stiffness on rock wedge stability in underground openings', *Canadian Geotechnical J.*, **20**, 276–287

Elsworth, D. (1986) 'Wedge stability in the roof of a circular tunnel: plane strain condition', Technical Note in *International J. Rock Mechanics and Mining Sciences & Geomechanical Abstracts,* **23**, 177–181

Goodman, R. E. (1989) *Introduction to Rock Mechanics* (2nd edn), John Wiley, New York

Goodman, R. E., and Boyle, W. (1986) 'Non-linear analysis for calculating the support of a rock block with dilatant joint faces', *Felsbau,* **4**, 203–208

Goodman, R. E. and Chan, L. Y. (1983) 'Prediction of support requirements for hard rock excavations using keyblock theory and joint statistics', *Proc 24th U.S. Symposium on Rock Mechanics,* 557–576

Goodman, R. E. and Shi, G. H. (1985) *Block Theory and Its Application to Rock Engineering,* Prentice Hall, Englewood Cliffs, N.J.

Goodman, R. E., Shi, G. H. and Boyle, W. (1982) 'Calculation of support for hard, jointed rock using the keyblock principle', *Proc. 23rd U.S. Symposium on Rock Mechanics,* 883–898

John, K. W. (1968) 'Graphical stability analyses of slopes in jointed rock', *Proc. Amercian Soc. Civil Engineers, Soil Mechanics and Foundations Division,* **94** (SM2), 497–526

Karzulovic, A. L. (1988) 'The use of keyblock theory in the design of linings and supports for tunnels, Unpublished PhD Thesis, Department of Civil Engineering, University of California., Berkeley

Londe, P., Vigier, G. and Vormeringer, R. (1969) 'Stability of rock slopes – graphical methods', *Proc. American Soc. Civil Engineers, Soil Mechanics and Foundations Division,* **96** (SM4), 1411–1434

Shi, G. H. and Goodman, R. E. (1989a) 'The keyblocks of unrolled joint traces in developed maps of tunnel walls', *Int. J Numerical and Analytical Methods in Geomechanics,* **13**, 131–158

Shi, G. H. and Goodman, R. E. (1989b) 'Generalization of two dimensional discontinuous deformation analysis for forward modelling', *Int. J. Numerical and Analytical Methods in Geomechanics*, **13**, 359–380

Yow, J. L. and Goodman, R. E. (1987) 'A ground reaction curve based on block theory'. *Rock Mechanics and Rock Engineering,* **20**, 167–190

6 Stress analysis for rock masses

Dr B H G Brady
ITASCA Consultants Inc

6.1 Purpose

In the design of an excavation in rock, it is essential that the excavation can perform an intended function under the loads imposed on it during its duty life. This condition is satisfied if displacements of excavation boundary rock are maintained within prescribed limits, under the prevailing operational static and dynamic loads. Thus, assurance of the performance of an excavation is directly related to the capacity to predict rock-mass displacements, and the associated state of stress, around the excavation. It follows that analysis of stress and displacement distribution around excavations is an essential component of excavation design practice. The integration of stress analysis with other components of a design practice appropriate to rock masses is discussed in detail by Brady and Brown (1985).

The purpose of stress analysis is to realize an economic design for an excavation. This involves application of stress analysis to identify, for a proposed design, zones of failure (i.e. the domains in which the displacements are calculated or inferred to be excessive). In the light of results of the analysis, the design may be modified, within the constraints imposed by operational considerations, until the zones of failure are minimized. Alternatively, rock reinforcement or other rock-mass modification techniques are specified, sufficient to achieve control of rock-mass deformation.

A further purpose of stress analysis is to establish a suitable excavation sequence for the construction of an opening. While the final shape of an excavation is the primary concern, it is necessary to establish, by repetitive analysis, the excavation sequence which preserves the integrity and local stability of the adjacent rock mass as the excavation is developed.

In contemporary geomechanics practice, closed-form solutions may be used to estimate the state of stress or excavation-induced displacements around openings of simple geometry, such as those with circular or elliptical cross-section. As will be shown later, such solutions have other valuable applications. However, the convenience and power of computational methods of stress analysis have firmly established them as procedures routinely applicable by a geomechanics engineer. Apart from their utility to an engineer who is not a specialist in deformable-body mechanics, their attraction arises from the capacity to analyze problems with irregular two- and three-dimensional geometry in rock masses which exhibit complex constitutive behaviour, and to take account of fluid pressure in structural features. The capacity to account for dynamic as well as static loads provides the means for analysis of excavation and rock-mass response to earthquakes or explosive loading. For these reasons, the majority of the following discussion is concerned with computer-based, numerical methods of stress analysis.

6.2 *In situ* state of stress

The natural state of stress in a rock mass is a function of the geological history of the medium and the geological setting. Factors such as erosion and isostasy, cooling or recrystallization to generate residual stresses, tectonic setting and the presence of inclusions have a pervasive influence on the virgin state of stress at a site. Depth below ground surface, surface topography and the structural geology of the medium exert a strong local influence on the state of stress. The structure of the rock mass has a particular influence, with major continuous structural features defining domains in which the state of stress may change substantially in passing from one side of a feature to the other. Thus, a heterogeneous state of stress is a natural condition in a discontinuous rock mass, and considerable care may be required to determine the state of stress in a way appropriate to the particular structural conditions and design problem.

The state of stress at a site is determined by direct measurement, using techniques such as stress relief by overcoring biaxial or triaxial strain cells in core- *drilled* holes (Worotnicki and Walton 1976), flatjacks in bored tunnels or raises (Brady and Brown 1985), hydraulic fracturing in *drill* holes (Enever and Chopra 1986), or *drill*-hole slotting (Bock 1986). The result of the measurements, if sufficient are conducted, is specification of the six components of the field stress tensor, expressed relative to a set of global reference axes, as illustrated in Figure 6.1(a). From these can be readily determined the field principal stresses p_1, P_2, p_3, and the orientations of the

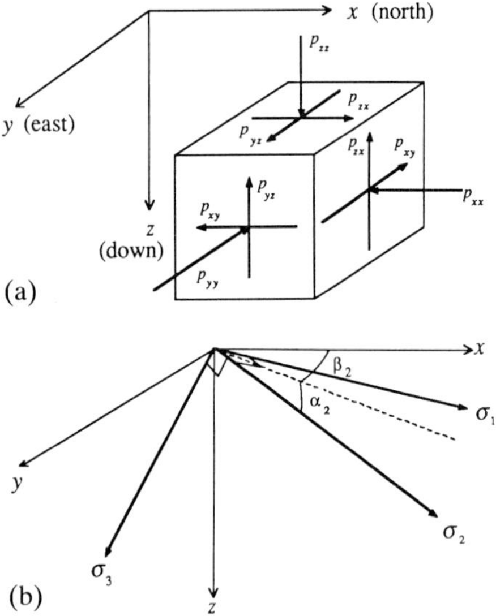

Figure 6.1 *Method of specifying the* in situ *state of stress relative to a set of global reference axes*

principal axes. Each of these is defined conveniently by a dip angle, α_i, and a bearing, β_i, as indicated in Figure 6.1(b).

6.3 Classical stress analysis

Although detailed treatment of the principles of analysis of stress and displacement in deformable bodies is beyond the scope of this text, the fundamental ideas are introduced briefly. This is to provide a basis for evaluation, prior to their use in excavation design, of the correctness of published closed-form solutions and for subsequent discussion of the development of the numerical methods.

For the sake of convenience, the following discussion considers a long, horizontal underground excavation of regular cross-section in an isotropic elastic medium. The analysis can be readily extended, however, to three-dimensional excavation geometry and more general constitutive behaviour, and it applies to surface (half-space) as well as subsurface problems. The plane problem involves far-field principal stresses p_{yy} $(=p)$ and p_{xx} $(=kp)$, which, in general, are variable throughout the problem domain. The problem geometry is illustrated in Figure 6.2. The solutions described subsequently are expressed in terms of the total stresses and excavation-induced displacements operating in the rock mass.

In any solution for the stress and displacement distributions around an excavation in a rock mass subject to a known ambient state of stress, the conditions to be satisfied are the boundary conditions, the differential equations of motion, the constitutive equations for the medium, and the strain compatibility equations. The boundary conditions may be defined in terms of imposed tractions or displacements (or their time derivatives) at excavation surfaces. For example, an excavation surface is usually traction free, so the surface tractions t_x, t_y in Figure 6.2(b) are zero over the complete surface of the excavation. For no induced body forces, the differential equations of equilibrium, the constitutive equations and the strain compatibility equations are, respectively,

$$\frac{\partial \sigma_{ij}}{\partial x_j} = 0 \tag{6.1}$$

$$\sigma_{ij} = \lambda\, \delta_{ij}\, e_{kk} + 2Ge_{ij} \tag{6.2}$$

where λ = Lamé's constant,
δ_{ij} = Kronecker delta,
e_{ij} = strain components,
e_{kk} = volume strain, and
G = shear modulus;

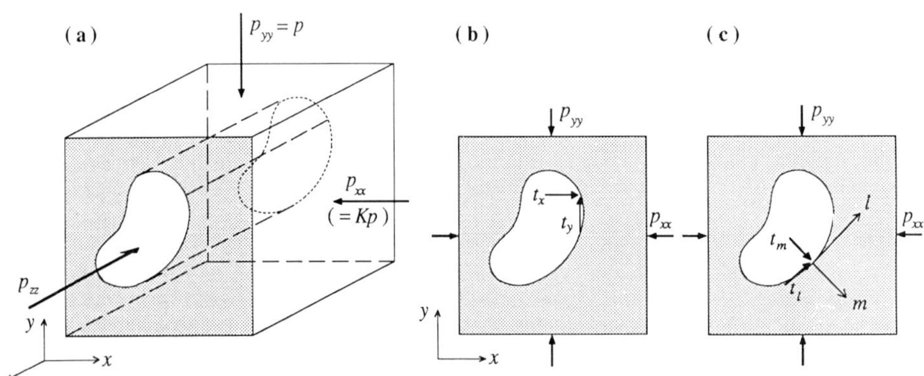

Figure 6.2 *An excavation in a medium subject to initial stresses, for which is required the distribution of total stresses and excavation-induced displacements*

and

$$\frac{\partial^2 e_{11}}{\partial x_2^2} + \frac{\partial^2 e_{22}}{\partial x_1^2} = \frac{2\,\partial^2 e_{12}}{\partial x_1\,\partial x_2} \qquad (6.3)$$

It is a straightforward matter to apply these expressions, particularly Equations (6.1) and (6.3), to demonstrate that a given solution for stresses and displacements around an excavation is correct.

The collection of elastic solutions provided by Poulos and Davis (1974) is a valuable source of applicable formulae for the state of stress and displacement around excavations in rock subject to various types of loading. When applying these solutions in engineering practice, it is frequently the case that rock masses are not elastic, and that some zones may exhibit fracture, yield, or slip on surfaces of low shear strength. However, as noted by Bray (1987), even in these cases, the elastic solutions may be useful in estimating the extent of plastic zones and the influence of excavation design variables on rock-mass response. It is proposed that an elastic analysis, via the published solutions, may be used to assess excavation boundary stresses, to make a first estimate of the extent of zones of fracture or yield, to determine the zone of influence of an opening and to estimate changes in the state of strain energy associated with development of the opening.

6.3.1 Circular excavation

The elastic solution for a circular excavation is important because many important principles related to excavation boundary stresses can be readily illustrated from it. For the problem geometry defined in Figure 6.3, the solutions

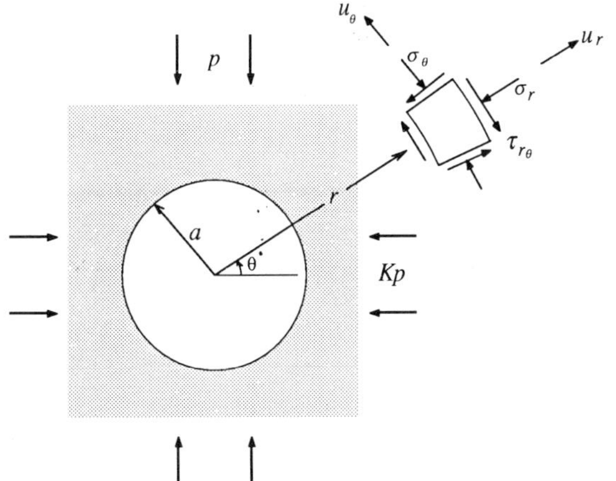

Figure 6.3 *Problem geometry and coordinate system for specifying stress and displacement distribution around a circular excavation in a biaxial stress field*

for stresses and displacements, originally due to Kirsch (1898) are given by

$$\left.\begin{aligned}
\sigma_r &= \frac{p}{2}\left[(1+K)\left(1-\frac{a^2}{r^2}\right) - (1-K)\left(1-4\frac{a^2}{r^2}+3\frac{a^4}{r^4}\right)\cos 2\theta\right]\\
\sigma_\theta &= \frac{p}{2}\left[(1+K)\left(1+\frac{a^2}{r^2}\right) + (1-K)\left(1+3\frac{a^4}{r^4}\right)\cos 2\theta\right]\\
\tau_{r\theta} &= \frac{p}{2}\left[(1-K)\left(1+2\frac{a^2}{r^2}-3\frac{a^4}{r^4}\right)\sin 2\theta\right]
\end{aligned}\right\} \quad (6.4)$$

$$\left.\begin{aligned}
u_r &= -\frac{pa^2}{4Gr}\left[(1+K) - (1-K)\left(4(1-\nu)-\frac{a^2}{r^2}\right)\cos 2\theta\right]\\
u_\theta &= -\frac{pa^2}{4Gr}\left[(1-K)\left(2(1-2\nu)+\frac{a^2}{r^2}\right)\sin 2\theta\right]
\end{aligned}\right\} \quad (6.5)$$

Using cylindrical polar coordinates, it can be readily demonstrated that the expressions for the stresses satisfy an appropriate form of Equation (6.1). Also, boundary stresses are given by the case when $r=a$:

$$\sigma_r = \tau_{r\theta} = 0$$

$$\sigma_\theta = p\,[(1+K) + 2(1-K)\cos 2\theta] \qquad (6.6)$$

6.3.2 Elliptical excavation

Although not of great interest in construction practice, an elliptical excavation can be used to examine some interesting relations between boundary stresses, excavation shape, and boundary curvature. Figure 6.4(a) defines the problem geometry, with the global x-axis parallel to the field principal stress, Kp, and with an axis of the ellipse defining the excavation's local x_1-axis. The width, W, of the ellipse is measured in the direction of the x_1-axis, and the height, H, in the direction of the z_1-axis. The angle β between the x- and x_1-axes defines the attitude of the ellipse in the stress field, and the position of any point in the medium is described by the Cartesian coordinates (x_1, z_1) relative to the local x_1-, z_1-axes.

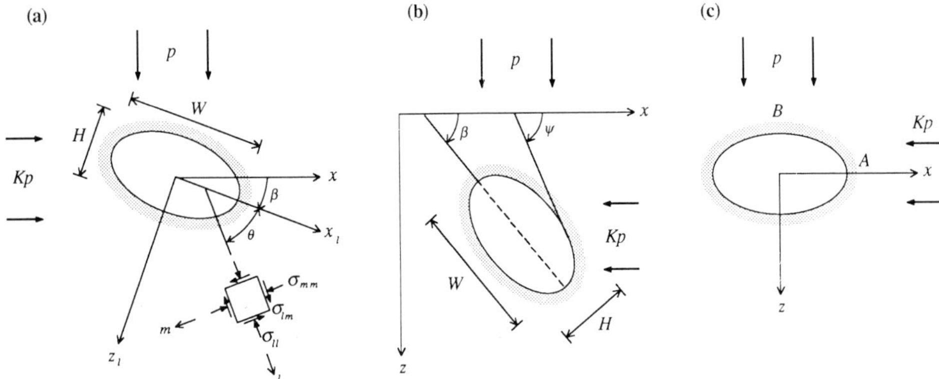

Figure 6.4 *Problem geometry and coordinate axes for specifying the stress distribution around an elliptical excavation in a biaxial stress field*

A convenient and applicable solution to the stress distribution in the medium has been presented by Bray (1977). The simplified form of the solution which follows defines the state of stress at a point in terms of a set of local axes, denoted l and m, centred on the point of interest. Various geometric parameters employed in the solution are given by

$$e_o = \frac{(W+H)}{(W-H)}, \qquad b = \frac{4(x_1^2+z_1^2)}{(W^2-H^2)}$$

$$d = \frac{8(x_1^2-z_1^2)}{W^2-H^2} - 1, \qquad u = b + \frac{e_0}{|e_o|}(b^2-d)^{1/2}$$

$$e = u + \frac{e_o}{|e_o|}(u^2-1)^{1/2}$$

$$\Psi = \arctan\left[\left(\frac{e+1}{e-1}\right)\frac{z_1}{x_1}\right],$$

$$\theta = \arctan\left[\left(\frac{e+1}{e-1}\right)^2\frac{z_1}{x_1}\right],$$

$$C = 1 - ee_o \qquad J = 1 + e^2 - 2e\cos 2\Psi$$

In terms of these parameters, the stress components are

$$\sigma_l = \frac{p(e_o-e)}{J^2}\left[(1+K)(e^2-1)\frac{C}{2e_o} + (1-K)\left[\left[\left(\frac{J}{2}(e-e_o) + Ce\right)\cos 2(\Psi+\beta) - C\cos 2\beta\right]\right]\right]$$

$$\sigma_m = \frac{p}{J}\left[(1+K)(e^2-1) + 2(1-K)e_o(e\cos 2(\Psi+\beta) - \cos 2\beta)\right] - \sigma_l$$

$$\tau_{lm} = \frac{p(e_o-e)}{J^2}\left[(1+K)\frac{Ce}{e_o}\sin 2\Psi + (1-K)\, e(e_o+e)\sin 2\beta + e\sin 2(\Psi-\beta) - \left(\frac{J}{2}(e_o+e) + e^2e_o\right)\sin 2(\Psi+\beta)\right] \tag{6.7}$$

Stresses around the boundary of an elliptical excavation are obtained from Equation (6.7) when the point (x_1, z_1) lies on the boundary countour. In this case, for the problem geometry given in Figure 6.4(b), the tangential component of the boundary stress is given by

$$\sigma = \frac{p}{2q}\left[(1+K)[(1+q^2) + (1-q^2)\cos 2(\Psi-\beta)] - (1-K)[(1+q)^2\cos 2\Psi + (1-q^2)\cos 2\beta]\right] \tag{6.8}$$

where $q = W/H$.

When the axes of the ellipse are oriented parallel to the field principal stress directions, as illustrated in Figure 6.4(c), the boundary stress is

$$\sigma = \frac{P}{2q}\Big[(1+K)\,[(1+q^2) + (1-q^2)\cos 2\Psi] - (1-K)\,[(1+q)^2\cos 2\Psi + (1-q^2)]\Big] \quad (6.9)$$

Two positions of particular interest in Figure 6.4(c) are the sidewall (A) and the crown (B) of the opening. Noting that the radii of curvature at A and B, ρ_A and ρ_B respectively, are given by

$$\rho_A = \frac{b^2}{a}, \qquad \rho_B = \frac{a^2}{b}$$

it can be shown that

$$\sigma_A = p\,(1 - K + 2q) = p\left[1 - K + \left(\frac{2W}{\rho_A}\right)^{1/2}\right]$$

$$\sigma_B = p\left[K - 1 + \frac{2K}{q}\right] = p\left[K - 1 + K\left(\frac{2H}{\rho_B}\right)^{1/2}\right] \quad (6.10)$$

The expressions relating boundary stress and curvature are used later in estimating boundary stresses around excavations whose cross-sections are other than elliptical.

6.4 Zone of influence of an excavation

The concept of zone of influence is important in excavation design in rock because it provides a method for rapid assessment of the degree of mechanical interaction between openings. The notion is best illustrated by an example.

For a circular opening in a hydrostatic stress field, the state of stress in the surrounding medium is obtained from Equations (6.4), with K=1:

$$\sigma_r = p\left(1 - \frac{a^2}{r^2}\right)$$

$$\sigma_\theta = p\left(1 + \frac{a^2}{r^2}\right) \quad (6.11)$$

$$\tau_{r\theta} = 0$$

In the medium, the variation of σ_r and σ_θ shown in Figure 6.5 is axisymmetric. When $r=5a$, Equations (6.11) indicate that $\sigma_\theta = 1.04p$ and $\sigma_r = 0.96p$. Thus, on the contour around the hole defined by $r=5a$, the state of stress is not significantly different from the field stresses, corresponding with them within ±5%. This can be accepted as a useful working definition of the zone of influence of an excavation. In the current example, if a second circular excavation of radius a were developed outside the zone of influence of the first, the state of stress around the boundary of each excavation would be essentially that for a single, isolated excavation.

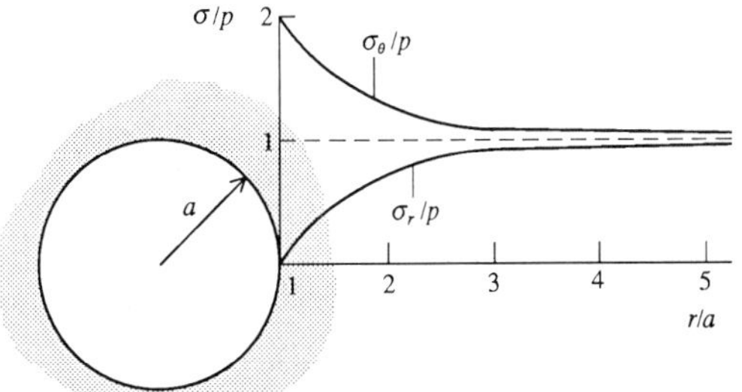

Figure 6.5 *Axisymmetric stress distribution around a circular opening in a hydrostatic stress field*

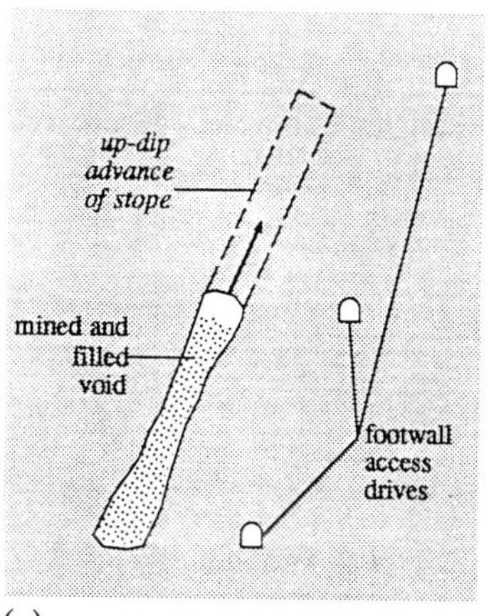

(a)

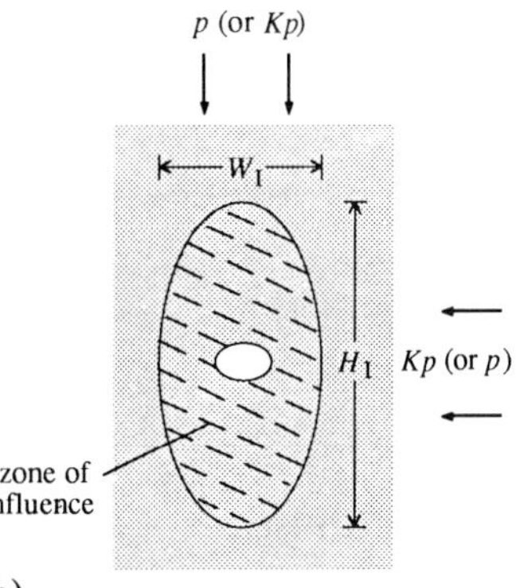

(b)

Figure 6.6 (a) *A design problem involving semi-coupling between a large excavation (a cut-and-fill stope) and smaller access openings;* (b) *definition of the zone of influence of an elliptical opening*

As a general rule, the zone of influence of an excavation is dependent on excavation shape and the field principal stresses. For example, for the small access excavations below the long, narrow slot shown in Figure 6.6(a), the small excavations may be within the zone of influence of the slot, but the slot may be outside the zone of influence of the small excavations. In that case, it would be justifiable to estimate the state of stress around the slot by considering it as a single, isolated excavation. It could be approximated by an ellipse, for example. On the other hand, the boundary stresses around the small access excavations must take account of the local stresses due to the presence of the slot.

The practical value of the concept of zone of influence suggests that its definition should be formalized. It is proposed to be the domain within which

$$|\sigma_1 - p_{max}| > 0.05\, p_{max} \quad \text{or} \quad |\sigma_3 - p_{min}| > 0.05\, p_{max}$$

where p_{max} and p_{min} represent the larger and smaller, respectively, of the field stresses p and Kp. Bray (1987) has used this definition in computing the widths and heights of the elliptical zones of influence of elliptical excavations in various virgin stress fields. A simple case is shown in Figure 6.6(b), where the axes of an elliptical opening of height H are parallel to the field principal stresses. Here, the zone of influence is bounded by an ellipse of overall width W_I and height H_I, noting that, of the two values obtained for each of W_I and H_I from the following expressions, the larger value is taken in each case:

$$W_I = H\,[10\alpha\,|q\,(q + 2) - K\,(3 + 2q)\,|]^{1/2}$$

or

$$W_I = H\,[\alpha\,[10\,(K + q^2) + K\,q^2]]^{1/2}$$

$$H_I = H\,[10\alpha\,|K\,(1 + 2q) - q\,(3q + 2)\,|]^{1/2}$$

or

$$H_I = H\,[\alpha\,[10\,(K + q^2) + 1]]^{1/2}$$

where $\alpha = 1$, if $K < 1$, and

$\alpha = 1/K$, if $K > 1$.

Modification of these expressions is required for extreme values of q and K, as follows. If $K > 5$ and $q > 5$, W_I must be increased by 15%. If $K < 0.2$ and $q < 0.2$, H_I must be increased by 15%.

6.5 Excavation shape and boundary stresses

The purpose of the following disucssion is to demonstrate some useful practical techniques for estimating boundary stresses around an excavation. In particular, it is shown that a useful engineering sense of the prospective performance of excavation boundary rock can be obtained from the simple closed-form solutions.

It was shown previously that boundary stresses for the sidewall (A) and crown (B) of an ellipse are given by

$$\sigma_A = p\left[1 - K + \left(\frac{2W}{\rho_A}\right)^{1/2}\right] \tag{6.12}$$

$$\sigma_B = p\left[K - 1 + K\left(\frac{2H}{\rho_B}\right)^{1/2}\right] \tag{6.13}$$

where ρ_A and ρ_B are the radii of curvature at A and B.

These expressions indicate that, where the radius of curvature is small, the corresponding boundary stress is large. The generalization suggested by this observation is that high boundary curvature [i.e. $(1/\rho)$] leads to high boundary stresses, and that simple inspection of boundary curvature can guide a first-order estimate of relative boundary stresses around an excavation.

The ovaloidal opening shown in Figure 6.7 is oriented with its major axis perpendicular to the major principal field stress. The width/height ratio for the opening is 3,

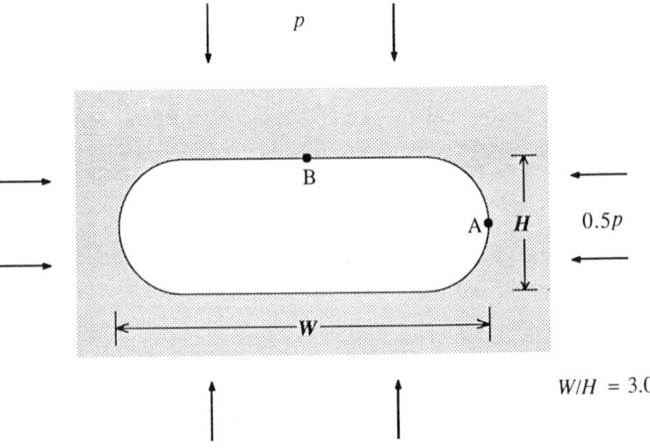

Figure 6.7 *Ovaloidal opening in a medium subject to biaxial stress*

and the radius of curvature for the sidewall is $H/2$, where H is the height of the excavation. If the ratio of the principal field stresses, K, is 0.5, the sidewall boundary stress may be estimated from Equation (6.12):

$$\sigma_A = \left[1 - 0.5 + \left(\frac{2 \times 3H}{H/2}\right)^{1/2}\right]$$

$$= 3.96p$$

In a boundary element analysis of this problem, the sidewall boundary stress was calculated to be $3.60p$, so that the estimated boundary stress is adequate for practical design purposes. Considering the state of stress in the crown of the excavation, the radius of curvature at B is infinite, so that Equation (6.13) indicates little more than that σ_B is tensile. However, if one considers an ellipse inscribed in the ovaloid, with width/height ratio of 6, Equation (6.10) may be used to estimate the boundary stress. The estimated result is $\sigma_B = -0.17p$, while the boundary element analysis predicts a value for σ_B of $-0.15p$. The inference from this result is that the excavation aspect ratio (e.g. W/H ratio), as well as boundary curvature, may be used to establish some useful estimates of the state of stress around the boundary of an opening.

Several other examples are presented by Brady and Brown (1985) in support of the notions relating boundary stress, local boundary curvature, and excavation aspect ratio. They are consistent with the proposal that judicious use can be made of Equation (6.10) in evaluating boundary stresses.

6.6 Rock structure and boundary stresses

The preceding discussion has shown how useful estimates may be obtained for the elastic stress distribution around an excavation in massive, continuous elastic rock. The question that arises is: how valid are the estimates when the boundary of an opening is interesected by one or several planes of weakness? The following discussion indicates that, in some cases, planes of weakness may have little or no effect on the boundary stresses. In cases where the boundary stresses are affected by the rock structure, useful judgements may be formed about the areas subject to modified states of boundary stress.

The assumptions in the analysis presented here are that the structural features have high normal and shear stiffnesses, so that local stress field modification due to joint elastic deformation can be ignored, that the tensile strength of joints is zero and that joints are non-dilatant in shear, with shear strength defined by

$$\tau = \sigma \tan \theta$$

Slip and separation on the planes of weakness are the modes of response by which the stress distribution in a discontinuum is modified from that for the elastic continuum. Clearly, if the boundary stress is notionally tensile near a joint, a crack would open, and the area would be locally destressed. For purposes of illustration of the effects of slip, a circular opening is used to examine local modification of the boundary stresses. It is implied that similar analyses may be performed for arbitrarily shaped openings, using any convenient method of estimating the boundary elastic stresses.

A single flat-lying joint intersecting a circular excavation is illustrated in Figure 6.8(a), with the trace of the feature on the excavation surface defining an angle α above the horizontal diameter. From the small element of the boundary considered in Figure 6.8(b), the normal and shear stress components on the joint are:

$$\sigma_n = \sigma_\theta \cos^2 \alpha$$

$$\tau = \sigma_\theta \sin \alpha \cos \alpha \qquad (6.14)$$

The condition for limiting shear resistance on the plane of weakness is satisfied if

$$\tau = \sigma_n \tan \phi$$

or, introducing Equations (6.14),

$$\sigma_\theta \sin \alpha \cos \alpha = \sigma_\theta \cos^2 \alpha \tan \phi \qquad (6.15)$$

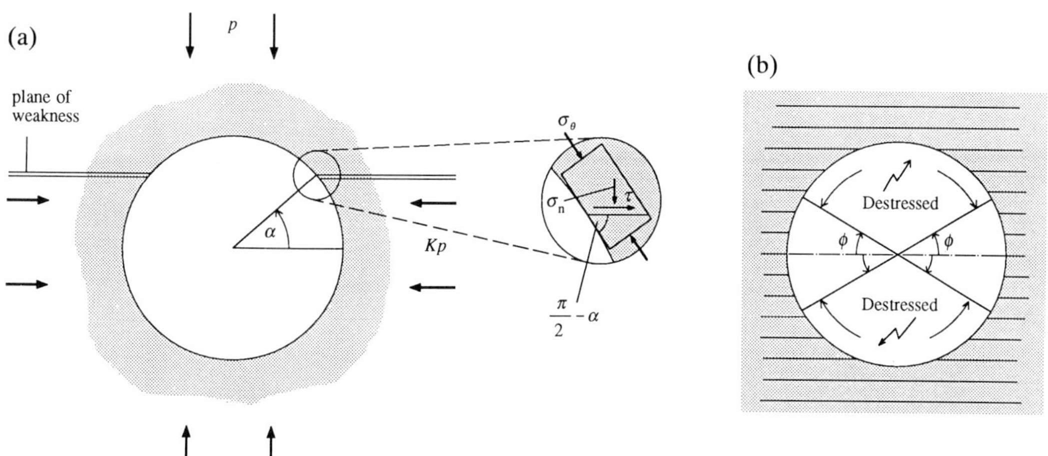

Figure 6.8 *Effect of planes of weakness on the boundary stresses around a circular excavation:* (a) *single plane of weakness;* (b) *multiple parallel planes*

i.e.,

$$\tan \alpha = \tan \phi$$

Thus, if $\alpha<\phi$, the state of stress for the unjointed medium is sustained in the presence of the joint. When $\alpha=\phi$, the condition for slip on the joint is satisfied. Also, the sense of slip, identified from the sense of the shear stress, involves outward displacement of the upper surface of the joint relative to the lower surface. This implies boundary stresses in the crown of the excavation less than those for the continuum.

Equation (6.15) may be recast in the form

$$\sigma_\theta \frac{\sin(\alpha-\phi)}{\cos\phi} = 0$$

For all values of $\alpha>\phi$, this condition can be satisfied only if $\sigma_\theta=0$. This indicates that, for these joints, the regions around the intersections of the joints and the opening are locally destressed.

When a circular excavation is transgressed by a set of parallel features, as shown in Figure 6.8(c), the preceding analysis suggests that slip will occur on all joints for which $|\alpha|>\phi$ and that, for those intersections satisfying the slip criterion, the immediately adjacent boundary will be destressed. The consequences of this are that the crown and floor of the excavation in Figure 6.8(c) can be inferred to be destressed and that the sidewalls are more highly stressed than for the elastic case.

The obvious extension of this analysis is that, for a non-circular excavation intersected by planes of weakness, slip will occur at the boundary when the angle between the boundary and the normal to the joint exceeds the angle of friction for the joint.

A second case of interest involves the effect of a joint adjacent to but not intersecting an excavation. For the problem geometry illustrated in Figure 6.9, it can be shown that the state of stress on the plane of weakness is given by

$$\tau = p\frac{a^2}{r^2}\sin 2\alpha$$

t

$$\sigma_n = p\left(1 - \frac{a^2}{r^2}\cos 2\alpha\right)$$

The variation of the τ/σ_n ratio calculated from these expressions is plotted versus position in Figure 6.9(b). From the plot, it is concluded that if the angle of friction on the joint is greater than about 24° (corresponding to tan $\phi = 0.45$), no slip would occur on the joint, and the elastic stress distribution would be sustained. For an angle of friction of 20°, the predicted range of slip is indicated on the figure. Of course, the real range of slip would be slightly greater than suggested by the elastic analysis. Also, the sense of slip suggests increase in boundary stress near the crown of the excavation. Independent boundary element analysis of this problem by Crotty and Brady (1986), confirms the general conclusions about range and sense of slip and the effect on boundary stresses.

Several other examples of the effect of structural features on the elastic stress distribution around openings are presented by Brady and Brown (1985). In general, it is found that, although the effects of rock structure on the state of stress around an excavation cannot be predicted in detail, some sound engineering jdugements can be made about such effects. Such judgements are frequently valuable in assessing the validity of the results of computational analysis of the state of stress around excavations.

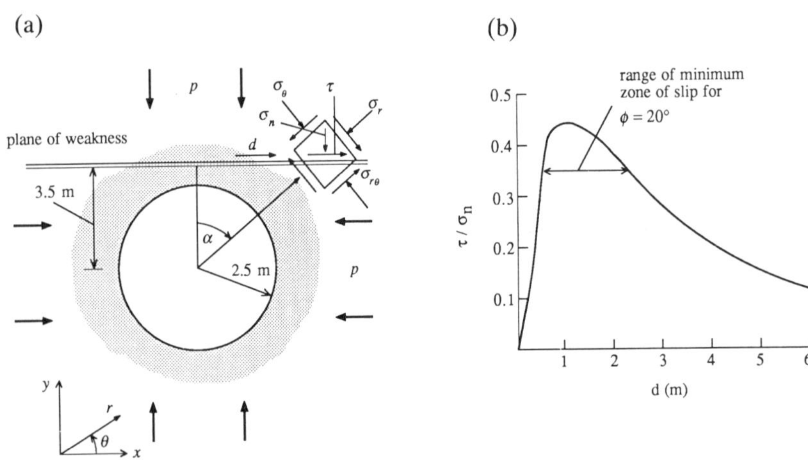

Figure 6.9 *Shear stress/normal stress ratio on a plane of weakness close to, but not intersecting, a circular excavation*

6.7 Computational methods of stress analysis

Analysis of stress and displacement using an appropriate numerical scheme is now regarded as an essential aspect of excavation design in rock. Versatile, efficient, portable and inexpensive stress-analysis packages are now available which will execute on powerful microcomputers, providing a rock mechanics engineer with design tools not conceivable until recently. In spite of this accessible analytical power, the principles considered in the preceding discussion are still highly applicable. Indeed, it would be expected that, prior to computational analysis of a design problem, a design engineer would conduct preliminary scoping studies to examine factors such as the zones of influence of excavations and the likely effects of excavation shapes and rock structure on boundary stresses around openings. In subsequent computational analysis, any gross discrepancy between the results of the preliminary and computational analyses should lead to close scrutiny of the computational model and re-assessment of the conceptual basis for the computational analysis.

The following discussion of computational methods of stress analysis considers two-dimensional problem geometry. The implicit understanding is that the techniques are equally applicable in three dimensions and that suitable computer codes are readily available to perform three-dimensional analysis. However, it is suggested that, although many design exercises may be clearly three-dimensional by virtue of the geometry of the excavation or of the attitude of the rock structure with respect to the excavation boundaries, most such exercises benefit from execution of a prior series of two-dimensional analyses. These are used to define the scope of the problem and to clarify the type of model to be used in the more time-consuming three-dimensional analysis.

Two different types of computational schemes are considered in subsequent sections. Differential methods, represented by finite element, finite difference and distinct element methods, seek solutions for the field equations by dividing the rock mass into elements or zones within which the governing equations are formally satisfied. Integral methods, represented by several versions of the boundary element method, construct solutions to the field equations using fundamental solutions to these equations and by applying some formal solutions from solid mechanics. In these methods, only the surface of an excavation is used in the solution, and the interior of the problem domain (apart, perhaps, from planes of weakness) is not represented explicitly. The methods are therefore highly appropriate to the infinite or semi-infinite body problems posed in rock engineering design.

Differential methods of analysis, particularly finite difference and distinct element methods, are distinguished by a ready capacity to model non-linear constitutive behaviour, such as large-strain plasticity and strain softening. This analytical power is achieved at the expense of either heavy demand on central memory or a heavy demand on execution time. Integral methods, on the other hand, are computationally efficient compared with differential methods (i.e. for the same problem geometry, relatively small demand is made on central memory and execution time), but this is achieved at the expense of requiring relatively simple constitutive behaviour, typically linear behaviour, for the rock mass.

The conditions under which the various methods of analysis are applicable are illustrated with reference to Figure 6.10. The ideas conveyed in the figure are scale, represented by the spacing of joints relative to a linear dimension of the excavation, and continuity of the displacement field induced by development of the excavation. In Figure 6.10(a), the spacing of joints is inferred to be great compared with an excavation dimension – that is, the problem involves excavation design in massive rock. Unless fracture or localized shear failure is induced around the opening, the displacement field is continuous throughout the near field of the excavation. For the rock structure and excavation geometry illustrated in Figure 6.10(b), within large regions of the near-field rock, the displacements are continuous. The planes of weakness define surfaces on which slip or separation may occur, introducing discontinuities in the displacement field. When the rock mass is relatively frequently jointed, as shown in Figure 6.10(c), slip and

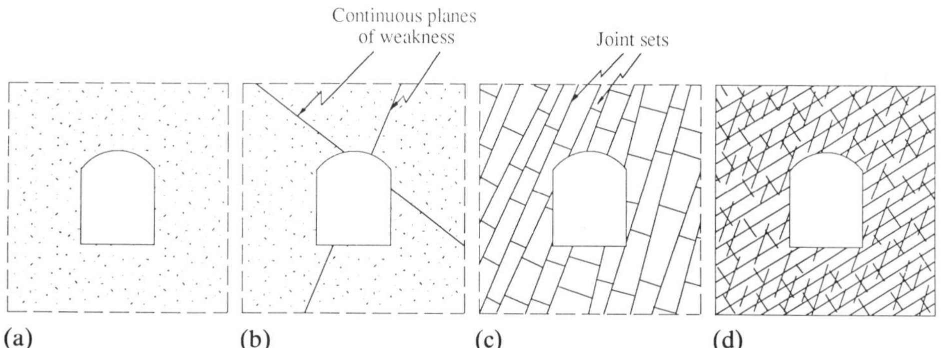

Figure 6.10 *Conceptual models relating rock structure, scale of excavation and joint opening, and rock response to excavation*

separation on joints, and rigid-body translations and rotations, may determine near-field rock displacements. The frequently and randomly fractured rock in Figure 6.10(d) may be subject to pseudo-continuous displacements, because it behaves effectively as a granular, although probably anisotropic, medium.

The different modes of response associated with the various conditions of structure and scale in Figure 6.10 imply that different computational schemes are relevant to the respective design problems. A further distinction may be made on the basis of whether the rock material can be modelled as a linear elastic material or as an elasto-plastic material. A summary of appropriate schemes for each design condition is presented in Table 6.1.

Table 6.1 Appropriate computational methods for various rock-mass and joint-scale conditions

Rock-mass conditions	*Rock material*	
	Elastic	*Elasto-Plastic*
(a) Massive rock	Boundary element	Finite element Finite difference
(b) Sparsely-jointed rock	Boundary element Finite element	Finite difference Finite element
(c) Closely jointed rock	Distinct element	Distinct element
(d) Heavily jointed rock		Finite element Finite difference

In interpreting Table 6.1, it is noted that the methods listed as suitable for analysis of elasto-plastic rock material are clearly suitable for elastic rock material. It is also observed from the table that finite difference methods are more generally applicable than recognized in current rock mechanics practice.

6.8 Boundary element method

The basis of the boundary element method is the construction of a relation between the excavation-induced tractions and displacements on surfaces in a body representing the boundaries of excavations in the medium. Because the principle of superposition is used in the development of the method, it is most appropriate for elastic media. The following discussion of a direct formulation of the boundary element method follows the approaches due to Cruse (1969), Watson (1979), Brady (1979), and Crotty and Wardle (1985), among many others.

The analysis is developed with reference to Figure 6.11. An excavation in the medium, which is subject to stresses p_{ij} at infinity, is shown in cross-section in Figure 6.11(a). Creation of the excavation reduces tractions t_i on surface S to zero and also induces displacements u_i on S. The problem is to determine the unknown boundary stresses and displacements and the distribution of stress and displacement throughout the medium. This is done by treating the problem as the superposition of the two separate problems shown in Figures 6.11(b) and (c). The former represents a continuous, uniformly stressed body, the latter an elastic body containing a contour S subject to excavation-induced tractions t_α and displacements u_α (α=1,2). Because the final surface is traction-free, induced stresses at any point Q on S are given by

$$t_i\,(Q) \;=\; -p_{ij}\,n_j\,(Q) \tag{6.16}$$

A relation between surface tractions and displacements at points Q on S is obtained by considering the fictitious case, shown in Figure 6.11(d), of a unit line load applied at another point P, in directions i (i=1,2) on the surface S. Applying Betti's reciprocal work theorem on the loaded surfaces defined in Figures 6.11(c) and (d) yields

$$c_{ij}(P)\,u_j(P) + \int_S T_{ij}(P,Q)\,u_j(Q)\,\mathrm{d}S = \int_S U_{ij}(P,Q)\,t_j(Q)\,\mathrm{d}S \tag{6.17}$$

In this expression, the term in c_{ij} accounts for the singularity at P and, for doubly subscripted variables, the

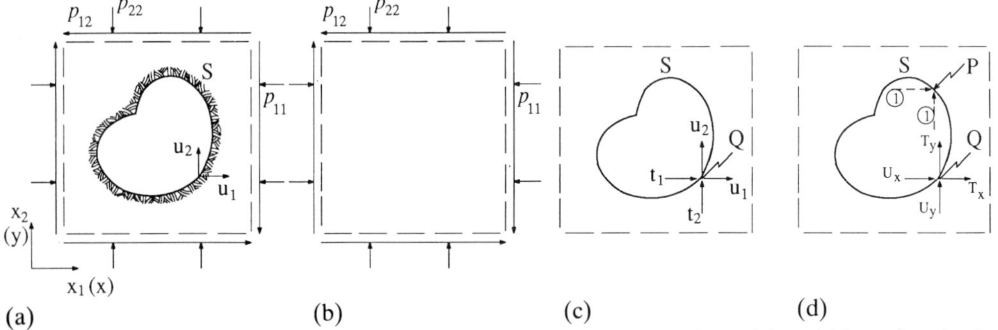

Figure 6.11 (a), (b), (d) *Problem definition for boundary element analysis;* (c) *and* (d) *surface loads for development of boundary integral equation*

first subscript represents the direction of the applied fictitious force at point P.

Equation (6.17) may be recast in a form which allows solution for the unknown surface displacements $u_j(Q)$. This involves dividing the surface S into a set of elements and rewriting Equation (6.17) as the sum of the surface integrals over the individual elements; that is,

$$c_{ij}(P)\, u_j(P) + \sum_{k=1}^{n} \int_{S_k} T_{ij}(P,Q)\, u_j(Q)\, \mathrm{d}S_k = \sum_{k=1}^{n} \int_{S_k} U_{ij}(P,Q)\, t_j(Q)\, \mathrm{d}S_k \tag{6.18}$$

where n is the number of boundary elements.

The solution of Equation (6.18) involves suitable description of the geometry of each element S_k, and assumption of some particular variation of traction, t, and displacement, u, with respect to the element geometry. Figure 6.12 illustrates the division of the surface S into a

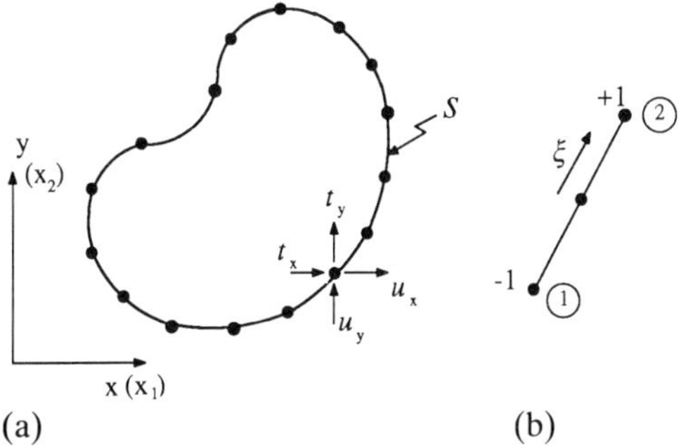

Figure 6.12 *Discretization of a problem surface into elements defined by surface nodes* (a) *and a linear boundary element* (b)

set of linear elements and a typical linear elment with intrinsic coordinate ξ. A linear isoparametric element is defined by linear element geometry and linear interpolation of traction and displacement with respect to ξ. Element geometry is therefore defined by

$$x_i(\xi) = x_i^1\, N^1(\xi) + x_i^2\, N^2(\xi) = x_i^P\, N^P(\xi), \quad -1 \leqslant \xi \leqslant 1, \tag{6.19}$$

where the superscript refers to the nodes 1 and 2 of the element, x_i^P are the nodal coordinates, and $x_i(\xi)$ are the coordinates of a point within S_k.

The linear shape (or interpolation) functions $N^P(\xi)$ are expressed by

$$N^1(\xi) = \frac{1}{2}(1-\xi) \qquad N^2(\xi) = \frac{1}{2}(1+\xi) \tag{6.20}$$

When the element parameters are expressed in terms of the local coordinate ξ, Equation (6.18) may be recast in the form

$$c_{ij}(P)\, u_j(P) + \sum_{k=1}^{n} \int_{S_k} T_{ij}(P,\xi)\, u_j(\xi)\, J_k(\xi)\, \mathrm{d}\xi = \sum_{k=1}^{n} \int_{S_k} U_{ij}(P,\xi)\, t_j(\xi)\, J_k(\xi)\, \mathrm{d}\xi \tag{6.21}$$

where

$$J_k(\xi = \left[\left(\frac{\mathrm{d}x}{\mathrm{d}\xi}\right)^2 + \left(\frac{\mathrm{d}y}{\mathrm{d}\xi}\right)^{1/2}\right]$$

is the Jacobian for the coordinate transformation for element k.

If, in Equation (6.21), the point P is taken as a node on the set of elements defined in Figure 6.12, all the data required to evaluate the surface integrals for each element is provided. The free terms c_{ij} are evaluated by the method of Cruse (1969), which takes account of the need for force equilibrium over any closed surface in a medium under static load. By considering each surface node in turn as a load point, evaluating all the integrals of the kernel (T, U)-shape function products, and combining, for a particular node, the integrals from adjacent elements, a set of $2n$ simultaneous equations is obtained, written as

$$\mathbf{Tu} = \mathbf{Ut} \tag{6.22}$$

In this equation, $\mathbf{u}$ lists the unknown nodal displacements, $\mathbf{t}$ **the known nodal tractions, and the square matrices** $\mathbf{T}$ and $\mathbf{U}$, of order $2n$, are fully populated and diagonal dominant. Multiplying out the right-hand side of Equation (6.22) produces

$$\mathbf{Tu} = \mathbf{v} \tag{6.23}$$

which can be solved directly for the nodal displacements, using an appropriate equation solver.

After solving for the nodal displacements, the directional derivative of the tangential component of boundary displacement can be determined numerically. This strain component can be used, in conjunction with the known boundary tractions, to calculate the tangential component of boundary stress, using the appropriate stress–strain relations.

Displacements at interior points in the medium are obtained by considering the load point P as an interior point. Equation (6.17) becomes

$$u_j(P) = \int_S U_{ij}(P,Q)\, t_j(Q)\, \mathrm{d}S - \int_S T_{ij}(P,Q)\, u_j(Q)\, \mathrm{d}S \tag{6.24}$$

This may be recast in the form of Equation (6.21), indicating that displacements at interior points can be evaluated directly from the known surface tractions and displacements.

The state of stress at interior points is evaluated from the directional derivatives of the displacements:

$$u_{j,i}(P) = \int_S U_{ij,i}(P,Q)\, t_j(Q)\, \mathrm{d}S - \int_S T_{ij,i}(P,Q)\, u_j(Q)\, \mathrm{d}S \qquad (6.25)$$

so that expressions for $U_{ij,i}$, $T_{ij,i}$ must be established from the fundamental solution for a point load. When this is done, the derivatives given by Equation (6.25) can be evaluated directly, following a discretization procedure equivalent to that indicated in Equation (6.21). Stress components then are evaluated from the directional derivatives of the displacements (i.e. the strains) using the relevant stress–strain relations.

6.8.1 Heterogeneous media and discontinuities

The preceding discussion was concerned with a homogeneous, continuous elastic medium. Many engineering problems involve heterogeneous rock masses and sparse discontinuities, as illustrated in Figure 6.10(b). For such cases, boundary element formulations of the types developed by Crouch (1978), Austin *et al.* (1982) and Crotty and Wardle (1985) may be applied. The following discussion is based on that presented by Crotty and Wardle.

When a body is divided into subregions, the conditions to be satisfied on the boundaries between subregions are continuity of displacement and force equilibrium. If the contact between subregions is a discontinuity such as a joint or a fault, the criteria for deformation of the feature must be satisfied, and conditions for slip or separation must be taken into account.

For each homogeneous subregion, s, of a heterogeneous medium, a boundary constraint equation similar to Equation (6.22) may be written in terms of the normal and tangential components of traction and displacement of the nodes; i.e. t_n, t_t, u_n, u_t. Equation (6.22) for each subregion, s, is written

$$\mathbf{T}(s)\,\mathbf{u}(s) = \mathbf{U}(s)\,\mathbf{t}(s) \qquad (6.26)$$

If no slip or separation occurs at an interface between subregions s_1 and s_2, the continuity conditions require

$$\left.\begin{aligned} t_\alpha(s_1) &= -t_\alpha(s_2) \\ u_\alpha(s_1) &= u_\alpha(s_2) \end{aligned}\right\} \alpha = \mathrm{n,t} \qquad (6.27)$$

Taking account of equations of the type of Equation (6.26) for all subregions, a complete set of equations can be constructed for the coupled subregions and the excavations within them. For the excavations, half the set of $(\mathbf{u},\mathbf{t})$ is specified. For the coupled set of subregions and their excavations, Equations (6.27) and (6.22) are recast as

$$\mathbf{A}\,\mathbf{x} = \mathbf{v} \qquad (6.28)$$

in which $\mathbf{x}$ lists the unknown excavation surface variables plus, for each interface, one set of the interface tractions and displacements, $\mathbf{t}(s)$, $\mathbf{u}(s)$, the other set having been eliminated by applying Equation (6.27).

Equation (6.28) can be solved for the unknown surface and interface tractions and displacements. Displacements and stresses within each subregion can then be determined from Equations (6.24) and (6.25), written for the subregion.

When the interface between subregions is a compressible structural feature, with limited tensile and shear strengths, Equation (6.28) must be modified to take account of these properties. The compressibility of the joint between regions r and s is represented by a set of normal and shear springs, of stiffnesses k_n and k_t, joining the interface. Equation (6.27) is rewritten, to express joint compressibility, in the form

$$t_\alpha(rs) = k_\alpha\, \Delta u_\alpha(rs), \qquad \alpha = \mathrm{n,t} \qquad (6.29)$$

where

$$\Delta u_\alpha(rs) = u_\alpha(s) - u_\alpha(r) \qquad (6.30)$$

and k_n,k_t are normal and tangential stiffnesses.

In the analysis of a jointed medium, the parameters of interest for joints are the *total* normal and tangential stresses, $t_\alpha^T(r)$, which are given by

$$t_\alpha(r) = k_\alpha\, \Delta u_\alpha(rs) + t_{\alpha 0}(r) \qquad (6.31)$$

and

$$t_\alpha^T(r) = t_\alpha(r) + t_\alpha^p \qquad (6.32)$$

where $t_{\alpha 0}(r)$ is the induced stress prior to joint deformation, and t_α^p is the initial field stress.

The global boundary constraint equation is constructed by asssembling the stiffness equations ((6.29)–(6.31)) with the subregion boundary element equations, resulting in the expression

$$\mathbf{x} = \mathbf{v} \qquad (6.33)$$

The coefficient matrix $\mathbf{B}$ in this expression contains blocks of $\mathbf{T}(r)$ and $\mathbf{u}(r)$ for each subregion and blocks of elements from the stiffness equations.

Inelastic behaviour of joints arises from their limited tensile and shear strengths. A generalized joint strengh model is described by

$$|t_t^T| \leq |t_t^T|\,(\max) \leq c + t_n^T \tan\phi \qquad (6.34)$$

where c is the cohesion of the joint, ϕ is its angle of friction, and

$$t_n^T \geq t_n^T\,(\min) \qquad (6.35)$$

where t_n^T (min) is the joint tensile strength.

If the tensile failure criterion is satisfied, joint separation occurs, and both t_n^T and t_t^T vanish. Also, joint closure is not unlimited. A limit on the elastic closure is specified as a joint property.

Introduction of the inelastic conditions for joint behaviour results in the need for an iterative solution procedure to satisfy Equations (6.33)–(6.35) simultaneously. This involves, first, direct solution of Equation (6.33) representing subregion and joint elastic behaviour and, then, if required, iterative solution involving Equations (6.34) and (6.35) to determine the equilibrium stresses and displacements at the joint nodes.

6.9 Finite difference and distinct element methods

The summary in Table 6.1 of applicability of various computational methods has already indicated the wide scope for practical application of finite difference techniques. Such techniques are eminently suited for installation on a microcomputer. Because they can model large-strain and rigid-body displacements in rock, they deserve close attention by practising engineers. Also, finite difference and distinct element methods have the same basis in solid mechanics, and it is thus appropriate that they be considered together.

In a finite difference method such as that described by Cundall and Board (1988), attention is focused on the continuum, although several discontinuity surfaces (slip lines) also may be modelled. Alternatively, in the block-jointed medium shown in Figure 6.10(c), the interaction between blocks is the primary concern, and the state of stress in the interior of the blocks is conveniently determined using finite differences. For both methods, the basic problem is the state of stress and displacement in deformable blocks interacting through a deformable interface, as shown in Figure 6.13(b).

A finite difference scheme is developed by dividing a body into a set of convenient, arbitrarily shaped quadrilateral zones, indicated in Figure 6.13(c). For each representative domain, difference equations are developed from the equation of motion and the constitutive equations. Lumping of part of the mass from adjacent zones at a gridpoint or node, as implied in Figure 6.13(b), and a procedure for calculating the out-of-balance force at a gridpoint, provide the technique for constructing and integrating the equations of motion for each gridpoint in the set of zones.

The Gauss divergence theorem is the basis of the method for determining the out-of-balance gridpoint

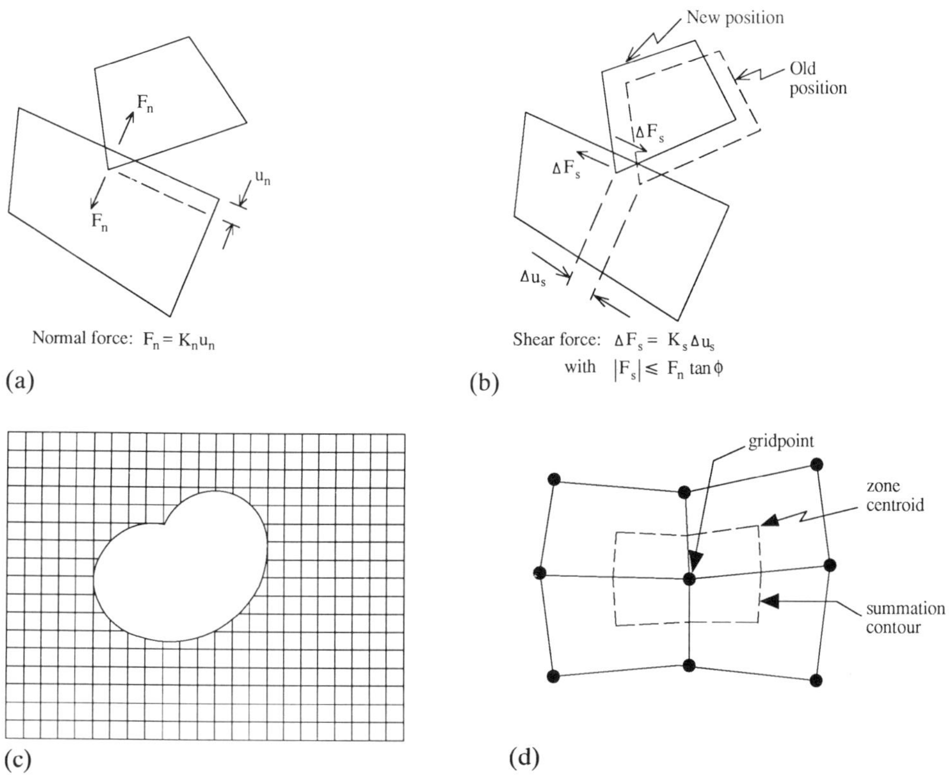

Figure 6.13 (a), (b) *Normal and shear interaction between distinct element blocks;* (c), (d) *zone discretization and local discretization for evaluating contour integrals*

force. In relating stresses and tractions, the theorem takes the form

$$\frac{\partial \sigma_{ij}}{\partial x_i} = \lim_{A \to 0} \frac{1}{A} \int_S \sigma_{ij} n_j \, dS, \qquad i = 1,2 \tag{6.36}$$

where x_i = components of position vector,
σ_{ij} = components of stress tensor,
A = area,
dS = increment of arc length, and
n_j = unit outward normal to dS.

An approximate expression then may be established for the right-hand side of Equation (6.36) involving summation of products of tractions and areas over the sides of a polygon, to yield a resultant force on the gridpoint.

The differential equation of motion is

$$\rho \frac{\partial \dot{u}_i}{\partial t} = \frac{\partial \sigma_{ij}}{\partial x_j} + \rho g_i \tag{6.37}$$

and introducing Equation (6.36) yields

$$\frac{\partial \dot{u}_i}{\partial t} = \frac{1}{m} \int_S \sigma_{ij} n_j \, dS + g_i \tag{6.38}$$

where $m = \rho A$.

Contact forces F_i may arise from interaction between the opposing faces of an extensive discontinuity or from contact between the joint-defined blocks which may be modelled in a distinct element scheme. In those cases, the surface integral in Equation (6.36) is evaluated only within the interior of the solid, and the contact forces are evaluated separately. The procedure for evaluating the contact forces is described by Cundall (1976), and is illustrated in Figure 6.13(a,b). For normal interaction, the contact force, F_n, is evaluated in terms of the normal stiffness, k_n, of the contact, and the notional overlap, u_n, that develops between the blocks, from the expression

$$F_n = k_n u_n \tag{6.39}$$

Due to the common non-linearity of the force–displacement relations for joint deformation in shear, the relevant force–displacement relation is taken to be incrementally linear, and the total shear force is obtained by summing the increments ΔF_s obtained from the expression

$$\Delta F_s = k_s \Delta u_s \tag{6.40}$$

The maximum shear force that can develop at a joint is limited by the shear strength of the surface, which may be described by

$$|F_s| \leq F_n \tan \phi, \qquad F_n > 0 \tag{6.41}$$

When these contact forces have been calculated for interface or block contacts, Equation (6.37) becomes

$$\frac{\partial \dot{u}_i}{\partial t} = \frac{1}{m}\left(F_i + \int_S \sigma_{ij} n_j \, dS\right) + g_i$$

$$= \frac{1}{\mathrm{m}} R_i + g_i \tag{6.42}$$

Equation (6.42) indicates that the acceleration at a gridpoint can be calculated explicitly from an integration of the surface tractions over the contour of the region surrounding the gridpoint.

When the acceleration of a gridpoint has been determined, central difference equations can be used to calculate gridpoint velocities and displacements after a time interval Δt:

$$\dot{u}_i^{(t+\Delta t/2)} = \dot{u}_i^{(t-\Delta t/2)} + \left(\frac{R_i}{m} + g_i\right) \Delta t \tag{6.43}$$

$$x_i^{(t+\Delta t)} = x_i^{(t)} + \dot{u}_i^{(t+\Delta t/2)} \Delta t \tag{6.44}$$

For analysis of a pseudo-static problem, viscous damping terms are introduced in Equations (6.42) and (6.43) to dissipate kinetic energy in the vibrating assembly of lumped masses.

Calculation of changes in the state of stress proceeds by calculation of strain increments and their introduction in the constituive equations. Strain increments are determined from the velocity gradients. Noting that, from the Gauss divergence theorem,

$$\frac{\partial \dot{u}_i}{\partial x_j} = \frac{1}{A} \int_S \dot{u}_i n_j \, dS \tag{6.45}$$

and that the right-hand side of Equation (6.45) can be evaluated as a summation over the contour of a polygon surrounding a gridpoint, strain increments are determined from the expression

$$\Delta e_{ij} = \frac{1}{2}\left(\frac{\partial \dot{u}_i}{\partial x_j} + \frac{\partial \dot{u}_j}{\partial x_i}\right) \Delta t \tag{6.46}$$

Finally, the stress increment in the time interval, Δt, is calculated directly from the existing state of stress, the strain increments and the material constants, k_α, for the medium, through an appropriate constitutive equation:

$$\Delta \sigma_{ij} = f(\Delta e_{ij}, \sigma_{ij}, k_\alpha) \tag{6.47}$$

In typical finite difference codes, the form of the constitutive function, f, may represent isotropic or transversely isotropic elasticity, Mohr–Coulomb plasticity, strain softening, or anisotropic plasticity defined by ubiquitous joints.

Equations (6.35)–(6.47) are solved sequentially through a series of time steps of duration Δt. Thus, the procedure is a time-based integration of the governing equations, to yield the displacement and state of stress at a set of collocation points in the medium.

Some applications of finite difference and distinct element methods to various geo-engineering problems are described by Cundall and Board (1988) and Lemos *et al.* (1985). The methods are shown to be effective and economical in the analysis of displacement and stress in discontinuous and block-jointed media.

6.10 Finite element method

The finite element method is closely related to the finite difference method in that the interior of the problem domain must be discretized completely into separate elements. A representative eight-noded quadrilateral element of a finite element mesh (as might be represented by Figure 6.13(c)) is shown in Figure 6.14. Relative to the

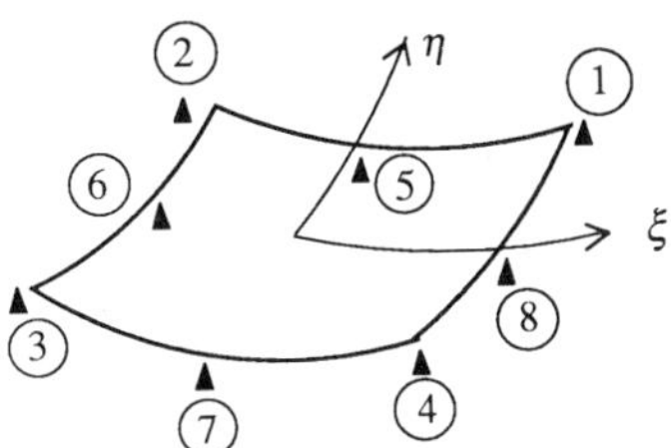

Figure 6.14 *Element geometry for an eight-noded quadratic finite element*

global (x_1, x_2)-axes, the nodal coordinates of node j are x_i^j, and the displacements are u_i^j $(i=1,2)$. The local coordinate axes for the element are ξ, η, with $-1 \leqslant \xi \leqslant 1$, $-1 \leqslant \eta \leqslant 1$.

For each node, an interpolation (or shape) function N^j is defined which takes the value unity at a particular node and zero at the other nodes. For example, for nodes 1 and 5,

$$N^1 = \frac{1}{4}\xi(1+\xi)\,\eta(1+\eta)$$

$$N^5 = \frac{1}{2}(1-\xi^2)\,\eta(1+\eta) \tag{6.48}$$

In the well-known isoparametric formulation described by Zienkiewicz (1977), expressions of the form given in Equation (6.48) are used as weight factors to interpolate, from nodal coordinates and displacements, the coordinates and displacements of a point in the interior of an element – i.e.,

$$\mathbf{X} = \mathbf{N}\mathbf{x}^e \tag{6.49}$$

$$\mathbf{u} = \mathbf{N}\mathbf{u}^e \tag{6.50}$$

where

$$(\mathbf{x})^t = (x_1\ x_2)$$

$$(\mathbf{u})^t = (u_1\ u_2)$$

$$(\mathbf{x}^e)^t = (x_1^j\ x_2^j), \qquad j=1,8$$

$$(\mathbf{u}^e)^t = (u_1^j\ u_2^j), \qquad j=1,8$$

$$\mathbf{N} = \begin{bmatrix} N^j \\ N_j \end{bmatrix}, \qquad j=1,8$$

and superscript t indicates the transpose of the usual column matrix.

Induced strains ε are related to induced displacements by the expression

$$\mathbf{p.141} = \mathbf{Lu} = \mathbf{LNu}^e = \mathbf{Bu}^e \tag{6.51}$$

where

$$\mathbf{L}^t = \begin{bmatrix} \frac{\partial}{\partial x_1} & \mathbf{0} & \frac{\partial}{\partial x_2} \\ 0 & \frac{\partial}{\partial x_2} & \frac{\partial}{\partial x_1} \end{bmatrix}$$

and, for node j,

$$\mathbf{B}^j = \begin{bmatrix} \frac{\partial N^j}{\partial x_1} & 0 \\ 0 & \frac{\partial N^j}{\partial x_2} \\ \frac{\partial N^j}{\partial x_2} & \frac{\partial N^j}{\partial x_1} \end{bmatrix} \tag{6.52}$$

For elastic constitutive behaviour, the total state of stress, σ, within an element is obtained from the induced strains, the elasticity matrix, $\mathbf{D}$, and the initial stresses, σ_o; i.e

$$\begin{aligned} \sigma &= \mathbf{D}\varepsilon + \sigma_o \\ &= \mathbf{DBu}^e + \sigma_o \end{aligned} \tag{6.53}$$

where $(\sigma)^t = (\sigma_{11}\ \ \sigma_{22}\ \ \sigma_{12})$.

The analysis is performed by determining a set of internal nodal forces, $\mathbf{q}^e$, which is statically equivalent to the forces transmitted between elements across their boundaries. For no induced body forces, the principle of virtual work is applied to show that

$$\mathbf{q}^e = \int_{V^e} \mathbf{B}^t \sigma \, dV \tag{6.54}$$

which can be rewritten as

$$\mathbf{q}^e = \mathbf{K}^e \mathbf{u}^e + \mathbf{f}^e \tag{6.55}$$

where

$$\mathbf{K}^e = \int_{V^e} \mathbf{B}^t \mathbf{D} \mathbf{B} \, dV \tag{6.56}$$

and

$$\mathbf{f}^e = \int_{V^e} \mathbf{B}^t \sigma_o \, dV \tag{6.57}$$

Equations (6.55)–(6.57) apply for each element of the discretized domain. The requirement is that, at each node, the sum of the internal forces is an equilibrium with any applied external force, that is, for node α,

$$\mathbf{r}^{\alpha^t} = (r_1^\alpha \;\; r_2^\alpha) = (\Sigma q_1^\alpha \;\; \Sigma q_2^\alpha) \tag{6.58}$$

where the summation is performed for all elements connected at α. The element connectivity exploited in Equation (6.58) is used to construct the stiffness matrix for the complete assembly of elements, by adding the terms of the various element stiffness matrices which relate to a particular nodal displacement component. This procedure and application of Equations (6.56) and (6.57) for each element yield

$$\mathbf{k}\,\mathbf{u}^g = \mathbf{V}^g - \mathbf{f}^g \tag{6.59}$$

Solution of Equation (6.59) yields the vector $\mathbf{u}^g$ of nodal displacements. In practice, some action is required on **K**, taking account of boundary constraints for the mesh, to render it non-singular. After solution for the nodal displacements, Equation (6.53) can be applied to determine the state of stress in each element.

It should be noted that Equation (6.59) has been derived assuming elastic behaviour of the rock mass, whereas it was observed initially that finite elements are of most benefit for elastoplastic or other non-linear rock-mass response. For those conditions, some effort is required to generate a nodal force–displacement relation of the form of Equation (6.59). Because elastoplastic constitutive behaviour is the simplest non-linear material model, it is discussed briefly for purposes of illustration. More extensive treatments are provided by Zienkiewicz (1977) and Owen and Hinton (1980).

Characteristic features of plastic material behaviour are non-uniqueness of the stress–strain curve, hysteresis on load reversal, and energy dissipation during deformation. The incremental theory of plastic deformation due to Hill (1950) was developed to take account of these features of metal plasticity. For frictional materials such as rock, it is necessary to take account of dilatancy of the medium. In all cases, total deformation of a solid is obtained by integration of increments of deformation.

In the incremental theory, the strain during an increment of deformation is taken to be composed of elastic and plastic components:

$$d\varepsilon = d\varepsilon^e + d\varepsilon^p \tag{6.60}$$

The elastic strain increment (and strain rate) is related to a stress increment through the incremental form of Equation (6.53), while the plastic strain increment is derived from a plastic potential $Q(\sigma)$; **i.e.**

$$\mathbf{d}\varepsilon^e = \mathbf{D}^{-1}\, d\sigma \tag{6.61}$$

$$d\varepsilon^p = \lambda \frac{\partial Q}{\partial \sigma} d\sigma \tag{6.62}$$

so that

$$d\varepsilon = \mathbf{D}^{-1}\, d\sigma + \lambda \frac{\partial Q}{\partial \sigma} d\sigma \tag{6.63}$$

where λ is a non-negative constant of proportionality.

Yield occurs in the solid when the state of stress satisfies the failure criterion F; that is, when the failure criterion is mapped as a surface in principal stress space, the state of stress lies on the yield surface. At yield,

$$F(\sigma, K) = 0 \tag{6.64}$$

where K is a hardening parameter denoting change in the yield surface with state of deformation.

Forming the total differential dF from Equation (6.64),

$$dF = \left(\frac{\partial F}{\partial \sigma}\right)^t d\sigma + \left(\frac{\partial F}{\partial K}\right) dk = 0 \tag{6.65}$$

and if a parameter A defined by

$$A = -\frac{\partial F}{\partial K} dK \cdot \frac{1}{\lambda} \tag{6.66}$$

is introduced, Equation (6.65) becomes

$$\left(\frac{\partial F}{\partial \sigma}\right)^t d\sigma - \mathbf{A}\lambda = \mathbf{0} \tag{6.67}$$

Introduction of Equations (6.66) and (6.67) in Equation (6.62), and some elementary manipulation yields

$$\mathbf{d}\sigma = \mathbf{D}^{ep}\, d\varepsilon \tag{6.68}$$

where

$$\mathbf{D}^{ep} = \frac{\mathbf{D} - \mathbf{D}\left(\frac{\partial Q}{\partial \sigma}\right)\left(\frac{\partial F}{\partial \sigma}\right)^t \mathbf{D}}{\mathrm{A} + \left(\frac{\partial F}{\partial \sigma}\right)^t \mathbf{D}\left(\frac{\partial Q}{\partial \sigma}\right)}$$

The form of Equation (6.68) indicates that the elastoplastic matrix can be used in analysis of incremental deformation of a solid.

While the notion of incremental analysis of deformation may appear straightforward, the numerical techniques for the integration through the stress–strain path require substantial effort. The iterative solution of the analogous form of Equation (6.53) poses problems related to convergence, which have been considered at length by Zienkiewicz (1977) and Owen and Hinton (1980).

Because the finite element method was the first widely accepted numerical scheme in stress analysis for rock mechanics design, there are numerous examples of its application in the literature. Some informative examples are provided by Gudehus (1977). However, in the era of microcomputers, it is notable that difficulties with central memory requirements impeded the implmeentation and acceptance of finite element schemes, allowing other methods to gain currency.

References

Austin, M. W., Bray, J. W. and Crawford, A. M. (1982) 'A comparison of two indirect boundary element formulations incorporating planes of weakness', *Int. J. Rock Mech. Min. Sci. & Geomech. Abstr.*, **19**, 338–344

Bock, H. (1986) 'In-situ validation of the borehole slotting stress-meter', *Proc. Int. Symp. Rock Stress and Rock Stress Measurements,* Centek, Stockholm, 261–270

Brady, B. H. G. (1979) 'A boundary element method of stress analysis for non-homogeneous media and complete plane strain' *Proc. 20th US Symp. Rock Mechanics,* University of Texas at Austin, 243–250

Brady, B. H. G. and Brown, E. T. (1985) *Rock Mechanics for Underground Mining,* Allen & Unwin, London, 527 pp

Bray, J. W. (1977) Unpublished note, Imperial College of Science and Technology, London

Bray, J. W. (1987) 'Some applications of elastic theory'. In *Analytical and Computational Methods in Rock Mechanics,* Brown, E. T. (ed.), Allen & Unwin, London, 259 pp

Crotty, J. M. and Brady, B. H. G. (1986) Unpublished note, CSIRO (Australia) Division of Geomechanics

Crotty, J. M. and L. J. Wardle (1985) 'Boundary integral analysis of piecewise homogeneous solids with structural discontinuities;, *Int. J. Rock Mech. Min. Sci. & Geomech. Abstr.*, **22,** 419–427

Crouch, S. L. (1978) 'Solution of plane elasticity problems by the displacement discontinuity method', *Int. J. Num. Meth. Engng.*, **10**, 301–343

Cruse, T. A. (1969) Numerical solutions in three-dimensional elastostatics', *Int. J. Solids Struct.*, **5**, 1259–1274

Cundall, P. A. (1976) 'Explicit finite difference methods in geomechanics'. In *Numerical Methods in Engineering, Proceedings of the EF Conference on Numerical Methods in Geomechanics, Blacksburg, VA,* Desai, C. S. (ed.) Vol. 1, pp. 132–150

Cundall, P. and Board, M. (1988) 'A microcomputer program for modelling large-strain plasticity problems', *Proc. 6th Int. Conf. on Num. Meth. in Geomech. Innsbruck, Austria, 11–15 April*

Enever, J. R. and Chopra, P. N. (1986) 'Experience with hydraulic fracturing stress measurements in granite', *Proc. Int. Symp. on Rock Stress and Rock Stress Measurements,* Centek, Stockholm, 411–420

Gudehus, G. (ed.) (1977) *Finite Elements in Geomechanics,* John Wiley & Sons, London

Hill, R. (1950) *The Mathematical Theory of Plasticity,* Oxford University Press, Oxford

Kirsch, G. (1898) 'Die theorie der elastizitat und die bedürfnisse der festigkeitslehre', *Veit. Ver. Deut. Ing.*, **42**, 797–807

Lemos, J. V., Hart, R. D. and Cundall, P. A. (1985) 'A generalized distinct element program for modeling jointed rock mass (a keynote lecture)', *Proc. Int. Symp. Fundamentals of Rock Joints, Bjorkliden, 15–20 September,* Stephansson, O. (ed.), Centek Publishers, Lulea, Sweden, 335–343

Owen, D. R. J. and Hinton, E. (1980) *Finite Elements in Plasticity: Theory and Practice,* Pineridge Press, Swansea, 594 pp

Poulos, H. G. and Davis, E. T. (1974) *Elastic Solutions for Soil and Rock Mechanics,* Wiley, New York, 411 pp

Watson, J. O. (1979) 'Advanced implementation of the boundary element method for two- and three-dimensional elastostatics'. In *Developments in Boundary Element Methods – 1*, Banerjee, P. K. and Butterfield, R. (eds). Applied Science Publishers, Barking, Essex, pp 31–63

Worotnicki, G. and Walton, R. J. (1976) 'Triaxial "hollow inclusion" gauges for determination of rock stresses in-situ', *Proc. ISRM Symp. on Investigation of Stress in Rock,* Supplement, 1–8, Inst. Engrs. Aust., Sydney

Zienkiewicz, O. C. (1977) *The Finite Element Method* (3rd edn), McGraw-Hill, London, 787 pp

7 Exploration and investigation of rock masses

Professor R Olivieira
Laboratorio Nacional de Engenharia Civil, Lisbon, Portugal

7.1 Introduction

Many civil engineering activities rely on proper exploration and investigation of rock masses. In particular engineering works in which safety is related to the characteristics of rock masses require comprehensive geological study and geotechnical investigation of the ground in order to assess the relevant parameters to be taken into account in design or in the design of their rehabilitation. The quality of the stability analysis in these cases very much depends on the reliability of the methods used during the investigations, on the interpretation of the results and on the selection of the relevant parameters to be used in the analysis.

Decisions concerning the selection of construction rock materials like aggregates, rock fill, dimension stone and the like rely on adequate geological and geotechnical studies of the available rock masses, namely, those aiming at the definition of the quality of the rock and of the volumes of rock masses where a certain homogeneity can be assumed.

The stability of many mining operations, underground excavations and open pits also requires extensive geological and geotechnical investigation, preferably prior to their initiation.

In all these cases, the investigation of the rock masses is thus essential, the relevant properties very much depending on the type, the use, the dimensions and the depth of the structures or works and on the engineering application of the construction materials. Today, the characterization of rock masses in such cases is generally carried out prior to construction or exploration, in order to avoid unpredicted situations, the remedial measures for which can be very costly and time consuming.

However, in spite of the major developments which have occurred in engineering geology and rock mechanics in recent years, situations are still found where large works are started without the necessary investigation of the ground. Problems very frequently occur in such cases and expertise in geological and geotechnical activities at a later stage is required in order to help find appropriate remedial solutions.

On the other hand, situations can still be found where the amount of geological and geotechnical investigations largely exceeds what would be reasonable, this resulting in very high costs, with no further improvement of knowledge of the properties of the rock masses involved. Such cases also constitute very bad examples of professional performance, thus conveying an incorrect image of the value of engineering geological activities.

It is essential that exploration and investigation of rock masses be conducted, at all times, by well-trained and experienced professionals who are able to master on the one hand the structural behaviour of the engineering works and the use of the construction materials, and on the other the most appropriate methods for investigation and testing which can better assess the properties of the rock masses.

7.2 Methodology for rock-mass investigation

A number of new scientific and technical concepts have been developed in recent years, some of them being more important for the evaluation of projects, especially the large ones. Quality assurance, structural safety, cost–benefit analysis and share of contractual risks are examples of the more important ones and, when involved with projects which very much interfere with the ground, they call for high standards of geological and geotechnical performance at all stages, from layout feasibility to design, construction and, more and more, to operation.

At the feasibility stage of any project located in an area where relevant information is available, or where very good outcrops allow a reliable assessment of the rock masses characteristics, fast and low-cost activities generally provide the amount of information required to take the most important decisions. The collection and interpretation of existing geological and geotechnical data oriented for the purpose of the project, photo-interpretation, surface investigation (reconnaissance) of and measurements at outcrops, and sampling and index testing are examples of such activities.

Further engineering geological activities should only be carried out after evaluating the level of the existing data and if the geological complexity or the complexity of the project so requires. This should only be done after the interpretation of all the data preferably in the form of maps and sections predicting the subsurface occurrence and the nature of the rockmasses, accompanied by a comprehensive text where the information is clearly described and the geological and geotechnical* conditions or parameters are anticipated. The degree of confidence of such predictions should always be clearly stated in the report.

In many projects these activities account for only a percentage (although frequently high in industrialized countries or in large urban areas of developing countries) of the required total geological and geotechnical information in order to assess the parameters which control the safety, the effectiveness and the economy of the project. A site investigation programme is then required. An appraisal of the most suitable methods for each case (maximum information/minimum cost), together with a time chart and a cost estimate, will decide the sequence of the investigation works.

The most important purpose of such a programme is the engineering geological zoning of the rock mass, that is, the definition of the spatial limits of quasi-homogeneous volumes of the rock masses in terms of their geotechnical behaviour. This aim should be achieved, as much as possible with the help of the investigation and testing methods which allow for a statistical analysis of the information, this calling for extensive investigation works and/or large amounts of measurements of selected parameters.

The use of sophisticated methods (large volume, costly and time consuming) should be employed only during later stages of the design programme in particular zones of the rock masses, for instance, where high stress levels are likely to be induced by construction or exploration. Large-volume jack testing for deformability, for example, has been performed at many dam sites in zones of the foundation rock mass which, later, have been removed, being considered unsuitable as a foundation for the dam. A correct programme of site investigation would be able, in most cases, to avoid situations like that, by considering such tests only at final design stage, when the zoning is quite firmly decided and these tests are performed on a few locations to more accurately assess the design parameters.

Table 7.1 summarizes the most important basic concepts of the proposed methodology. Some of the considerations above do not apply to cases not requiring large-scale *in situ* tests for correct assessment of rock-mass properties.

* These are general terms which are intended to include neotectonic activity and seismicity of the area, hydrogeological and geomorphological features, cionsiderations on the state of stress, etc.

Table 7.1 Methodology for rock-mass characterization: basic concepts

(1) Engineering geological studies should be conducted by stages starting by simple and fast tasks, progressively using more expensive and time-consuming techniques.
(2) Studies and site investigations should allow for a large number of results in order to provide statistical analysis of measured parameters; this also calls for a good distribution of investigation throughout the rock masses concerned.
(3) Site investigation works should be multi-purpose in order to allow for a large amount of parameter results and still keep the engineering geological study at low cost.
(4) Analysis of the results of relevant parameters should aim at the engineering geological zoning of the rock masses.
(5) Particular localized zones in the rock masses (e.g. large faults, shear gouges, weathered veins, etc.) should be individually investigated and eventually be considered as a separate engineering geological zone.
(6) Detailed design of engineering works usually calls for large volume *in situ* tests, which should be performed in small number and at specific locations of the relevant engineering geological zones of the rock mass.

7.3 Methods for the study of rock masses

7.3.1 General

The most important role of the specialists responsible for the coordination of all the engineering geological activities is the choice of the methods which, for each case, are able to supply the necessary information at the lowest cost and with the minimum delay. This means that a substantial amount of data relating to the geological conditions of the area must be acquired and processed, so that the relevant features and parameters of each project are highlighted.

In view of the general character of this chapter, it is appropriate to consider all possible geological and geotechnical activities within six groupings, as follows:

(1) Desk study of existing data and photo-interpretation.
(2) Surface investigation, geological and engineering geological mapping and sampling of outcrops.
(3) Subsurface investigation, using geophysical and mechanical techniques as well as methods for visual inspection and logging of zones of the rock masses at depth.
(4) Laboratory tests for the characterization of rocks and discontinuities.
(5) *In situ* tests for the assessment of rock-mass properties.
(6) Monitoring the behaviour of rock masses.

A programme of investigation for rock masses should take into consideration the intrinsic properties of the rocks (mineralogical composition, texture and grain size, porosity, specific weight, degree of weathering, etc.) and the properties of the discontinuities (geometrical and physical); and if the project so requires, the mechanical and hydraulic behaviour of the rock masses, resulting from the nature of the rocks; the geological structure and the properties of the discontinuities; the geomorphological and hydrogeological features of the region; and its seismo-tectonic character. Such behaviour is influenced by the deformability, strength, permeability and state of stress of the rock masses, which are the most significant geotechnical properties requiring study.

7.3.2 Desk study

Before starting any investigation programme efforts must be made to collect all the available data of the area concerned, relating to the nature or to the behaviour of rock masses. Generally speaking, it can be said that topographical, geological, seismological, hydrogeological, and geotechnical data are in principle relevant for the study of rock masses and are frequently available in specialized institutions of most countries. In some countries, the existing information, at least in some topics, covers all its territory, but in others it can be very scattered or almost non-existent.

Such information can be presented on maps (typical examples are topographical, geological engineering geological maps), in reports or in literature, and their detail can be very different from place to place. As regards maps, the detail of the information is generally shown by the scale at which they are published. In Portugal, as in most countries, the National Geological Survey produces geological maps at several scales, the most common and general being coverage of the country at the scale 1/50 000.

It is very important to know the available sources of information for each particular problem or place. As they differ from country to country, it is important to be sure that the proper sources are checked before starting any other geological and geotechnical activities. Oliveira (1981) presented a list of Portuguese sources of different data referring to the name and address of the Institution as well as the information which was available. Trautmann and Kulhawy (1983) gave data sources for engineering geological studies in America, as did Chaplow (1975) in the United Kingdom. The general background concerning sources of information for preliminary engineering geological studies is shown in Appendix A of Anon (1981).

Before starting field activities it is mandatory to search for geological information from remote sensing imagery, including aerial photography. In general, this requires that proper photographs be obtained from the respective agencies or institutions and that photo-interpretation or satellite imagery computer-aided interpretation be carried out.

The problem related to the interpretation of images is the problem of scale and of image resolution. In most cases, besides geomorphological features which can be of importance for the characterization of rock masses, only regional tectonic features (like large faults, lineaments, etc.), can be accurately detected. The study of rock masses for engineering purposes based on aerial photography and satellite imagery can be very much improved if appropriate scales and techniques are used. Figure 7.1 shows photographs of the same area, (a) shows the photographs taken at noon (vertical sunlight), and (b) shows the photographs taken at 17.00 h (inclined sun-light). These

(a)

(b)

Figure 7.1 *Aerial photographs of the same area:* (a) *vertical sunlight;* (b) *inclined sunlight*

photographs indicate that the time at which an aerial photograph is taken affects its quality and consequently the amount of data which can be obtained from it.

For the purpose of rock-mass investigation, photo-interpretation can be made from existing aerial photographs, most frequently at scales of 1/15 000 to 1/40 000. Special flights covering restricted areas provide photographs at larger scales (1/2000 to 1/8000) which, of course, have a much better resolution. However, the investigation of rock masses for engineering purposes usually does not justify special flights. An exception could be, for example, a large dam site with excellent rock outcrops or which had been stripped for the purpose of the investigation. In such cases, the study of the fracture and of the weathering conditions of the rock mass would very much benefit from the stereographic interpretation of large-scale aerial photography.

In Portugal, aerial photography of all the country at the scale 1/15000 is available at very low cost and it can be easily obtained from the Forest Department. The city of Lisbon, the Area of Sines (400 km^2) and many highway alignments have been covered in full at the scale 1/5000.

In the past, only black and white panchromatic photography and a stereoscope were used for photo-interpretation. More recently, the use, for example, of colour and infrared photography together with the availability of electronic enhancement techniques has significantly increased the value of this tool.

Satellite imagery is as yet of little interest for general engineering geological investigation of rock masses, mainly owing to poor resolution of the images. Long lineaments, corresponding to regional tectonic features, however, can be very easily depicted (Fig. 7.2). Landsat multispectral scanning (MSS) images are available as colour prints, colour transparencies and digital tapes. Radar and thermographic techniques can also be used but they are inadequate for most rock engineering problems.

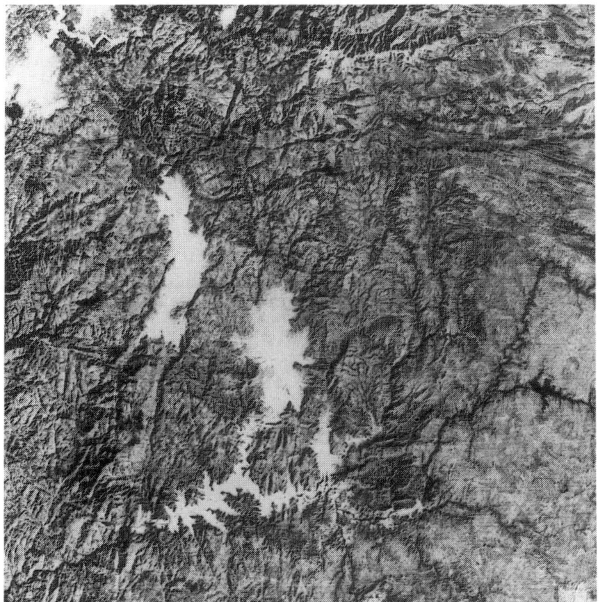

Figure 7.2 *Satellite imagery showing tectonic lineaments*

Only after studying all the available information and doing imagery interpretation should a start be made to the surface and the subsurface investigation of given rock masses, already indicated by the findings of the desk study. This is specially important for sites far away from the office, on account of the high cost of the travel expenses and of the time involved in the trips.

7.3.3 Surface investigation (reconnaissance)

Before starting any surface investigation for the characterization of a rock mass the reconnaissance of the relevant surface features of the ground must be carried out. If there already is information available about the geological and geotechnical properties of the area, the first step should be to appraise the situation *in situ* and to compare it with the existing interpretation. This is specially true for the separation of different rock types and, for the confirmation of faults, shear zones or weathered veins.

This surface investigation can be very effective in areas with very good outcrops and little vegetation. In zones where the rock masses are hidden either by residual soil or by surface deposits or are covered by dense vegetation, the result of the field reconnaissance is generally very poor and it is necessary to perform stripping works before getting into direct contact with the bedrock.

Some of the outcomes of the surface investigation of an area where an engineering project is under consideration are as follows:

(1) information concerning the rock types and boundaries;
(2) the degree of weathering near the ground surface;
(3) the fracture spacing, orientation and persistence;
(4) the bedding and the foliation;
(5) the characteristics of major discontinuities (faults, veins, etc.) and folding;
(6) landslide scars or other disturbances due to the movement of the ground surface.

Mapping is the most effective and comprehensive way of showing the engineering geological information which results from the surface investigation. Most of the information can directly be introduced onto a map, provided an adequate scale base map is available. In many cases the addition of some cross sections or extensive explanations is essential for a good understanding of the map.

However, the mapping of some features (like the joint pattern of a rock mass) can only be done properly after a statistical analysis of the orientation of a large number of joints, replacing the location of all the joints of the rock mass by a few that are well representative of the main joint sets. In these cases the map can accommodate one or more statistical representations of the joints (for example, stereographic pole projection).

At the early stages of an investigation, the engineering geological maps which can be produced are very much based on units which correspond either to stratigraphical boundaries or rock type boundaries, since little engineering information is then available. With the course of the studies true engineering geological maps should be produced. Dearman (1987) distinguished regional engineering geological maps as those with scales larger than 1/10 000 and engineering geological plans as those where the scales are smaller then 1/10 000. Mapping of rock masses for engineering purposes usually calls for direct contact with surfaces excavated below ground surface, where it is important to study the engineering geological conditions in order to assess the stability of the rock mass. Trial excavations (shafts and adits) are the most common access ways to zones of rock masses beneath the surface and they are currently used in many projects. Engineering geological plans are then produced at large scales (1/100, 1/200) and generally comprise very detailed information concerning the fracture and weathering characteristics of the rock mass as well as its geological structure. On account of the shape of excavation surfaces in shafts and adits, there are techniques available for partial or total projection of the observed features in order to show them on a plan (Oliveira 1977). For more information related to mapping and the production of engineering geological maps and plans see Dearman (1987) and Dearman and Matula (1976).

During the field survey for reconnaissance purposes, terrestrial photography is most useful. It can be used as simple photography for recording particular features of a rock mass or for terrestrial photogrammetry if proper ground control is used. In situations where good exposures exist, a good terrestrial photograph can greatly facilitate mapping of the rock mass, provided information concerning the scale is available. In addition good-quality photographs of inaccessible areas can provide a principal means of obtaining relevant information about the rock masses and conditions in those areas.

When it is possible, specially in the case of large linear projects (highways, canals, long tunnels), the observation of the ground surface from a helicopter or a small aircraft flying at low altitude may supply very important information and should always be tried.

During the field trips for surface exploration, sampling of rock outcrops allows proper characterization of the rock types by means of micropetrographic analysis and of laboratory index tests, like density, unit weight and porosity. Sampling is a very important activity and frequently calls for real expertise. Each sample must be representative of a given geological situation and must characterize, as much as possible, the properties of the rock material *in situ*. Careful removal of rock material from outcrops and handling of samples are thus indispensable requirements. If the natural moisture content has to be preserved (which is usual in the case of weak rocks), then the samples should be coated with special paraffin wax immediately after they are obtained.

Once enough samples have been taken from the outcrops, the laboratory investigation can begin and it generally is concerned with determination of the durability and strength of the rocks. Some simple tests are available which can directly or indirectly give information on those properties. In addition portable field equipment is available which can be used during the surface investigation to supply rapid and reliable information about some of the rock properties, especially the strength. The most popular are the Schmidt hammer and the point load equipment. The Schmidt hammer measures the hardness of the rock exposure by the rebound of the spring-loaded plunger. There are charts which allow correlation between Schmidt hardness and uniaxial compressive strength of rocks. Details about the equipment and the test procedure can be found in Anon (1978).

The point-load equipment consists of a loading system, including usually a small hydraulic jack which applies the load through two conical steel loading platens to the rock sample placed between them, until failure occurs. A strength index – point load strength index (I_s)–is calculated, which can be correlated with the uniaxial compressive strength of the rock. Details about the equipment and the test procedure can be found in Anon (1985).

It is clear, after the description of the activities which can be carried out as desk studies and surface investigations that, even before starting any subsurface investigation, much may be known about the engineering properties of rock masses. For some small to medium-size projects (dams, tunnels, slopes, foundations, quarries, etc.) the amount of information obtained can be enough for the design of safe and economic solutions without the need for further investigations.

This is why special attention must be paid to the early stages of engineering geological investigation of rock masses, by having these activities performed under the guidance of capable and well-trained staff.

7.3.4 Subsurface investigation

(a) Introduction

As said before, most large engineering projects in rock masses require good definition of the nature of the ground well below the surface. In underground construction (for example, tunnels and powerhouses) this depth may easily reach hundreds of metres. For many projects (foundations of light structures, excavation slopes, quarries, etc.), the heterogeneity and anisotropy of most rock masses require reliable engineering geological information below the ground surface which cannot, in many cases, be assured by only surface investigation methods.

In these cases, at the design stage of the project, further information has to be collected in order to improve the knowledge of the rock mass conditions gathered by the desk study and the surface investigation of the area. The programme for such subsurface investigation should only be settled after the interpretation of the data resulting from the previous stages of investigation (the first 'model'), and after taking into consideration the aims of the engineering geological study. It should follow the methodology presented in Table 7.1.

The subsurface investigation is based on direct and indirect methods, started or applied at the ground surface, but extending to depths at which information is required. The information they can supply is very different from geological situation to geological situation as well as from engineering problem to engineering problem. The cost of application of the available methods is also very different. These facts call for a careful selection of the methods and techniques able to supply the maximum information about the engineering geological conditions of the rock masses with a minimum cost and time. Appendix I lists the most frequently used methods.

(b) Geophysical methods

The most efficient geophysical methods for application to rock-mass characterization are electrical methods and seismic methods. Gravimetric, magnetic and electromagnetic methods are of little use in most of the engineering problems influencing rock masses, but their application can be very fruitful in some mining problems. Recently, however, developments in ground probing radar have made this an interesting tool for the evaluation of the pattern of discontinuities in crystalline rocks, so that it has particular relevance in investigations for the selection of sites for waste disposal (Olsson *et al.* 1988) and in the detection of cavities in rock masses.

The description of gravimetric, magnetic and electromagnetic methods appears in a large number of textbooks and other publications and is not discussed here. Some are listed in the references at the end of the Chapter (Bell 1980; Griffiths and King 1981; 1987).

(1) *Electric methods* These methods are based on the assessment of the conductivity of the ground as a result of passing a low-frequency electric current through two electrodes (usually steel stakes or ceramic porous pots filled with $CuSO_4$), thus creating a potential field in the zone. The potential difference is measured between two other electrodes generally located in the same line array. The most common electrode configurations for resistivity measurements are the Wenner, the Schlumberger and the pole–dipole arrays shown in Figure 7.3. Several techniques can be used applying the electric resistivity method.

Electrical soundings consist of making a series of apparent resistivity measurements with increasing spacing of current electrodes, which means that for each measurement the current goes deeper into the rock mass. This technique gives good results where there are layers of different rock types more or less parallel to the ground surface, or where a sound rock mass is overlain by residual soil. The interpretation of the results in terms of depth of the separation of different layers and of the contact between the residual soil and the bedrock is possible. However, for long arrays, the heterogeneity of the rock masses, resulting in geometrical differences between the

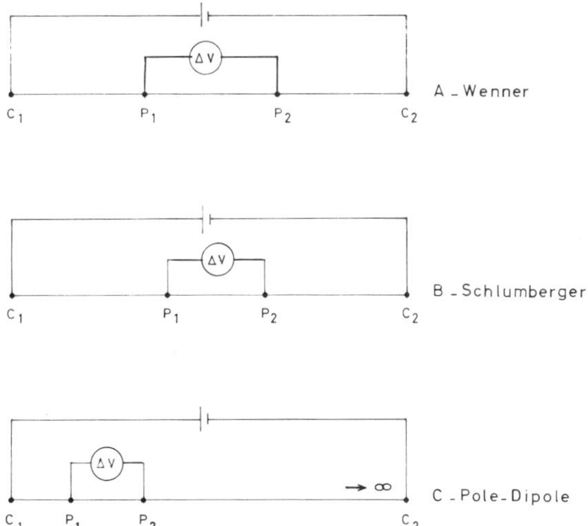

Figure 7.3 *Different electrode configurations for resistivity measurements*

reality and the interpretation model, can create important distortions to the results. This explains why electrical resistivity soundings, although of value in some geological situations, must always be checked by other methods of investigation, preferably direct methods.

Electrical resistivity traverses (profiles) are performed along a fixed line at constant separation of electrodes. This means that the potential field created by the electric current moves along the profile as the electrodes move, but extends to approximately the same depth of investigation. This technique proves very useful for the detection of vertical or near vertical contacts between rocks of different types, as in the case of faults with significant shear zones, veins and separation of rock types. Areas where the same rock type may occur with different degrees of weathering or alteration are also a good example of geological situations which can be detected using electrical resistivity traverses, since the conductivity of the rock mass depends very much on the porosity, mineralogy and water content of the ground, properties which generally change when a given rock is altered. The change is even stronger if many clay minerals develop as a consequence of the alteration process, since clay minerals have low resistivity.

After making a series of electrical resistivity traverses parallel to each other apparent resistivity contour maps can be produced. Figure 7.4 shows a resistivity contour map of a gabbro-diorite rock mass, the higher values corresponding to the sound and compact rock mass and the lower values corresponding to crushed zones which have very deep alteration. The thickness of the residual soil is constant, being about 6 m all over the area. These conclusions have been checked by seismic refraction measurements and by core drilling. Iso-resistivity maps prove very useful for dam foundation studies, for the appropriate location of tunnels, especially their portals, and for the definition of areas to be explored for quarrying.

The pole–dipole technique has been used for apparent resistivity traverses to detect karstic cavities in soluble rock masses (Fialho Rodrigues and Oliveira 1988).

(2) *Seismic methods* These methods are based on the fact that wave velocities depend on the elastic properties of the ground. A source generates volume and surface waves, usually by setting off an explosive charge inside a shallow hole or by a hammer blow, or by dropping a weight onto the ground surface. Volume waves can be compressional (P waves) or shear (S waves) and penetrate the rock mass in a rather spherical way. P waves are faster than S waves (usually 2 to 3 times faster). Seismic data are recorded by seismographs which can be multi-channel or single channel. For most rock mass investigations, 12 or 24 channel seismographs are currently used, each channel being connected to a geophone (receiver). Each seismograph channel records the travel time between the wave source and the receiver located at a pre-established distance. The closer the receivers are, the more accurate is the interpretation of the data in that section of the line. In order to improve the interpretation of the data and to optimize the cost of this method of investigation, it is common practice to generate waves at both ends of the receiver line and also in the middle of the line, without changing the position of the geophones. By doing so for the same seismic line, four time–distance graphs are obtained instead of only one in the case of a single generation of waves.

Waves propagate by direct paths, reflected paths and refracted paths.

The *refraction method*, which uses the first waves to arrive, is the most widely used seismic method for engineering purposes on land. It is applied at the ground surface and the operations involved are fast (including interpretation and calculations). The cost of one seismic profile is approximately the cost of one metre of core drilling.

The results have direct application for some rock engineering problems, since correlations have been tried with some success between wave velocities and elastic properties of the ground, and wave velocities and

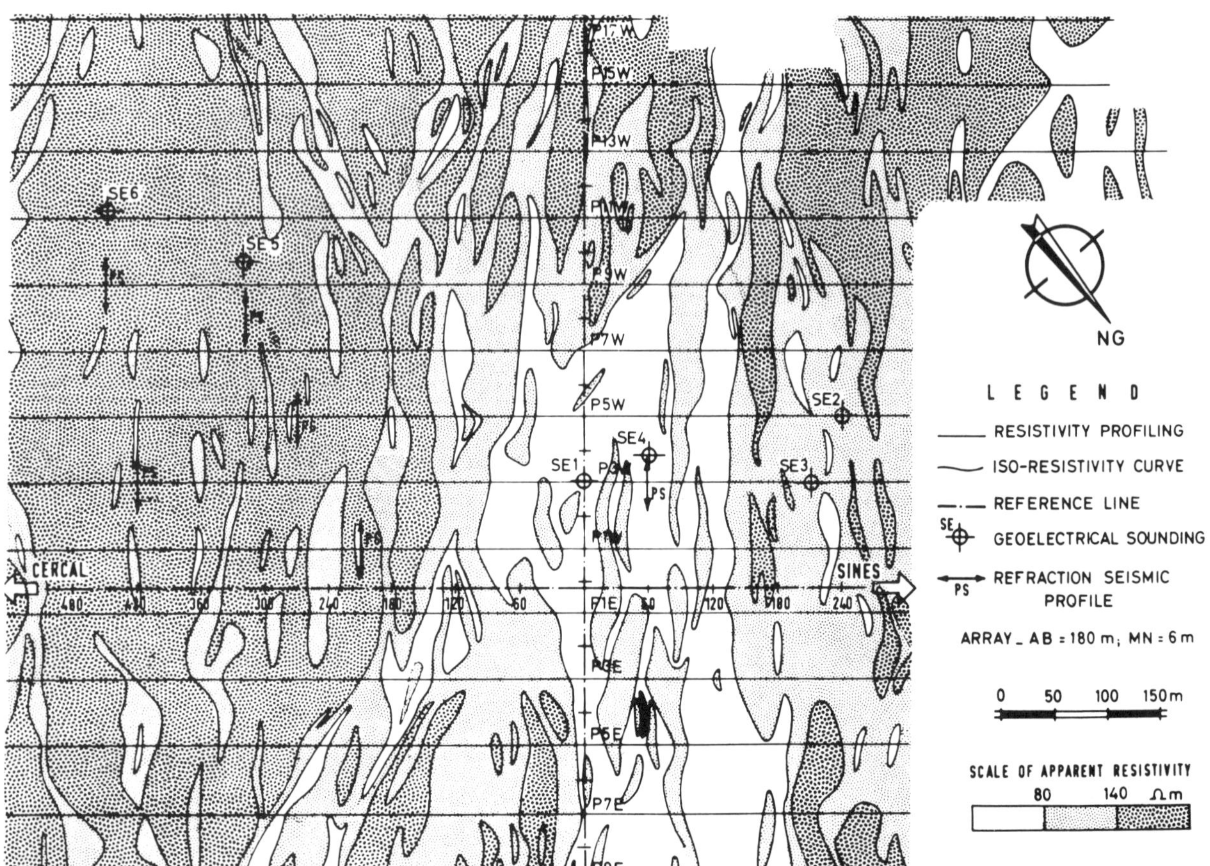

Figure 7.4 *Iso-resistivity map*

rippability and excavatability of rock masses by tractors and rippers (see Figure 21.10). Quality indexes of rock masses based on field and laboratory wave velocities for the same rock type were developed some time ago. Since there is a clear relation between wave velocities and elastic properties of the ground, the seismic method also has been used to measure the effectiveness of the grout take when consolidation grouting is carried out in a fractured rock mass (Fialho Rodrigues *et al.* 1983) (see Chapter 9).

For homogeneous, isotropic and Hookean bodies a correlation exists between wave velocities, and their deformability, which is a very relevant parameter for most engineering problems calling for the stability of rock masses. However, most geological formations behave very differently from that idealized model and show strong anisotropic, heterogeneous and non-elastic character. In these cases the correlations of those same measurements become very difficult and are even impossible for extreme situations. Besides, the levels of stresses applied during dynamic and static tests are very different from each other. This explains why it is not possible to establish universal correlations even for the same lithological type and why the scatter of the results is greater when those features are important in the rock mass.

Nevertheless, as these correlations are of most importance for engineering geological purposes, this subject is under constant research all over the world. Investigators using models, measurements of seismic and mechanical parameters calculated from *in situ* tests at the same depths, are always trying to achieve more reliable and useful correlations (Oliveira 1986).

Seismic refraction techniques applied at ground surface can provide a good assessment of the longitudinal wave velocities but scarce information exists concerning the propagation velocity of shear waves. The use of mechanical sources and of receivers specially designed and properly oriented in the field, for shear-wave generation and detection, can improve the identification of shear-wave arrivals.

For general cases, the volume of a rock mass which is subjected to the propagation of the waves is large when compared with the volume which is loaded in the case of mechanical deformability tests. One first concern when attempting correlations using conventional seismic refraction techniques should be to reduce as much as possible the rock-mass volumes under seismic investigation, by bringing the receivers closer to each other. The *values* of the seismic wave velocities then allow calculation of the dynamic elastic modulus (E_D).

Direct correlations between (E_d) and E are not always very good. Attempts to improve them include measurements of other parameters to allow for 'indirect' correlations.

For example, ultrasonic longitudinal wave velocities have been measured in cores taken from the same drill holes used for geophysical testing at the same depths and a velocity index (I_v) defined by the ratio V_L (field)/V_L (lab) has been determined (Fialho Rodrigues 1986a). Figure 7.5 shows an empirical correlation of I_v versus the ratio between static deformability modulus (E_M) and the dynamic elastic modulus (E_D) of a greywacke rock mass.

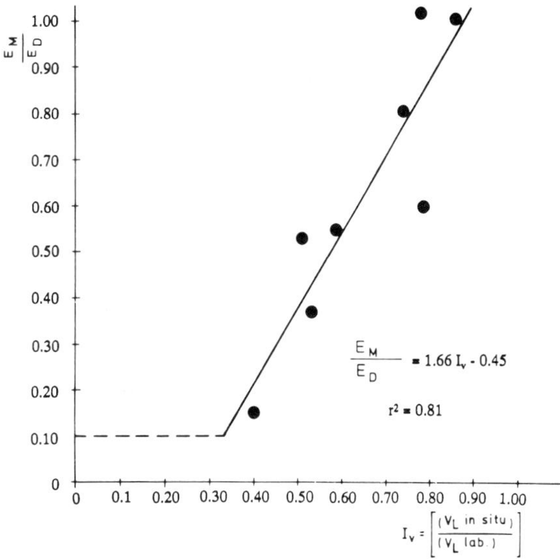

Figure 7.5 *Correlation between velocity index (I_V) and the ratio between deformation static modulus (E_M) and dynamic elasticity modulus (E_D)*

Other seismic parameters have also been used for the purpose of correlation with static deformability modulus. For example, in Figure 7.6 a reliable empirical correlation between the shear-wave frequency and the static deformation modulus of several types of rock masses is shown (Fialho Rodrigues 1979).

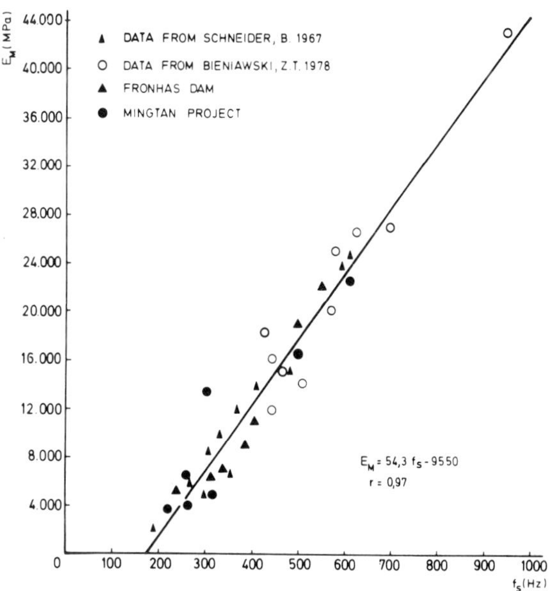

Figure 7.6 *Correlation between shear wave frequency (F_S) and static deformation modulus (E_M)*

But correlations also have been attempted using only geophysical parameters measured *in situ* and those derived by testing samples in the laboratory in order to assess the geotechnical characteristics of rock masses. The work of Onodera in this line is about 30 years old (Onodera 1963).

In a recent study conducted at the Laboratories Nationalo Engenharia Civil (LNEC), the granitic rock mass of the Varosa Dam foundation was characterized initially only by means of seismic methods using both *in situ* and laboratory techniques (Fialho Rodrigues 1986b). Longitudinal wave velocities were measured throughout the rock mass and ultrasonic measurements were performed on cores taken from drill holes sunk in the same area. The 38 granitic cores from 12 drill holes were visually classified according to their degree of weathering (scale W1 to W5) and ultrasonic measurements were made at dry, saturated and natural moisture content. A summary of these results for dry and saturated conditions is shown in Table 7.2.

Table 7.2 Degree of weathering and corresponding ultrasonic velocities for granite from Varosa Dam

Degree of weathering	*Moisture conditions*		V_L (sat) / V_L (dry)
	Dry (average)	*Saturated (average)*	*(average)*
Sound to slightly weathered (W_{1-2})	4020 m/s	5245 m/s	1.31
Moderately weathered (W_3)	2865 m/s	4538 m/s	1.60
Deeply weathered (W_4)	1996 m/s	3644 m/s	1.85

The seismic zoning of the foundation rock mass was based on the seismic *in situ* measurements and on seismic *indices* (*I*) calculated as

$$I = \frac{V_L\ (in\ situ)}{V_L\ (\text{laboratory})} \times 100$$

Table 7.3 shows the values of the velocities and of the indices for the seismic zones considered in Varosa Dam (Portugal).

Table 7.3 Zones distinguished by seismic velocity at Varosa Dam

Seismic zone	*Longitudinal wave velocity (V_L)* (m/s)	*Velocity index (I)*
Zone 1	3000–3500	64–92
Zone 2	3500–4000	96
Zone 3	4000–4500	88–92
Zone 4	4500–5000	85–96
Zone 5	>5000	≃100

Using the same equipment as for the regular seismic refraction technique (source and surface receivers) or introducing some adjustments in the receivers (for example, using hydrophones instead of geophones to pick up waves below the water table) the *fan shooting* technique can be used for the determination of wave velocities. This technique is based on the assumption that the first arrival corresponds to direct waves, that is, that the travel distance is the distance between the source and the receiver. The technique involves access to the rock mass which can be achieved trough adits and drillholes (Figure 7.7).

The *reflection seismic method* has been used extensively for engineering purposes in situations where the rock masses are covered by water, as in the case of marine and fluvial works. The data which can be supplied by reflection seismic investigations, in these cases, are basically the depth of the water, the thickness of deposits above the bedrock and a 'picture' of the geological structure of the rockmass. The main information concerning the rock mass is thus of qualitative nature. Continuous reflection profiles produce a series of seismograms so close to each other that there is a clear relationship between the reflection section and the real geological structural section (Figure 7.8).

The source of energy is different from the sources used in the refraction seismic method, the most common being a 'sparker' which introduces a high-pressure bubble into the water. The receivers are usually hydrophones.

The reflection seismic method also is used in land investigation, but only at great depth (more than 1000 m), for oil and gas prospecting. In these cases it is easy to detect the arrival of the reflected waves. Attempts to apply this method to shallow situations were not successful until recently; the development of powerful multi-channel digital engineering seismographs and the use of microcomputers for data processing has made this technique operational. This subject is still, however, a topic of research.

(c) Drilling, sampling, observation of borehole walls

The most common drilling methods for rock-mass investigations are the core drilling, the rotary-percussive drilling and the rotary drilling (also known as 'open-hole' drilling). The core-drilling method is the only one to supply continuous core samples, and, for that reason, it is the appropriate method for most engineering geological investigations.

Drill holes are usually sunk to investigate the nature of the rock mass at depth. They can reach some hundreds of metres which is more than enough for the type of engineering problems engineering geology has to deal with. They allow *sampling* along the drill hole and supply information about the rock types, the degree of weathering of the rocks, the groundwater conditions, the major discontinuities in the rock mass and the fracture pattern. Only core drilling can 'theoretically' supply direct information about all those conditions. In practice,

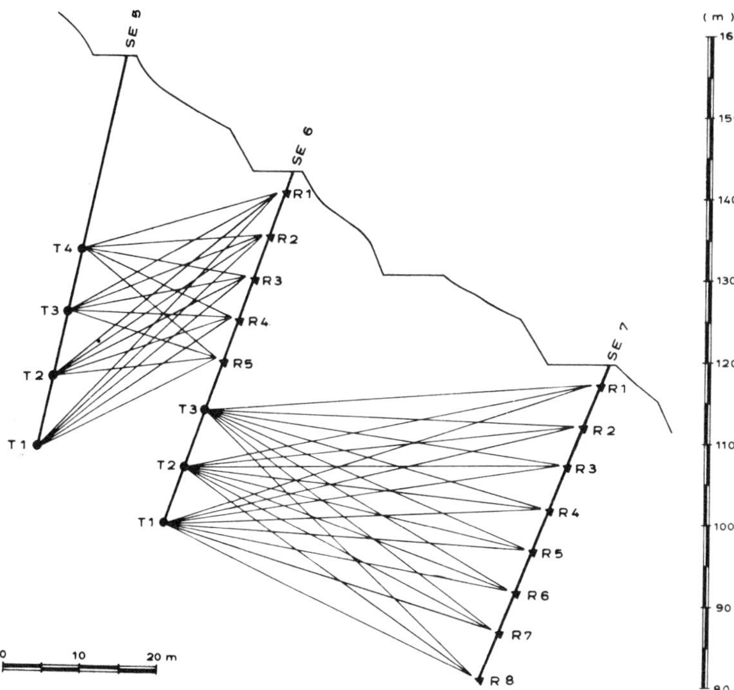

Figure 7.7 *Fan shooting technique*

however, core recovery is not continuous (total). This is the case in many geological situations, especially those characterized by high degrees of rock weathering or by tectonic disturbance. These are generally the zones in rock masses which strongly influence engineering behaviour. The other (destructive) drilling techniques, although faster to perform, supply only cuttings of the rock along the drill hole. Nonetheless cuttings allow indentification of rock type.

The quality of the rock samples obtained by core drilling very much depend on the type and dimensions of core barrels and bits, and also on the skill of the drilling crew and on the quality of the drilling equipment. Core barrels and bits are manufactured to certain standards (European and American), the most commonly used for engineering geological investigations being the double tube core barrels (for sample protection during drilling) with external diameters (hole diameters) of about 86, 76 and 66 mm.

Identification of core from a drill hole forms the basis of a drill hole log which therefore provides an indication of the rock types present and should also record their degree of alteration. Indirect 'parameters' related to the drilling operations are usually recorded as the drill hole proceeds.

The fracture conditions of rock masses frequently represent the most important data for the assessment of rock mass quality for engineering purposes. However, regular drilling techniques do not generally supply information on a large number of fracture parameters like orientation, opening, filling and persistence. Generally the data thereby obtained is restricted to fracture spacing and friction (roughness) of the fracture surfaces. A simple improvement which can be introduced is marking the core tops before drilling in order to extract oriented cores which will permit definition of the relative orientation of fractures.

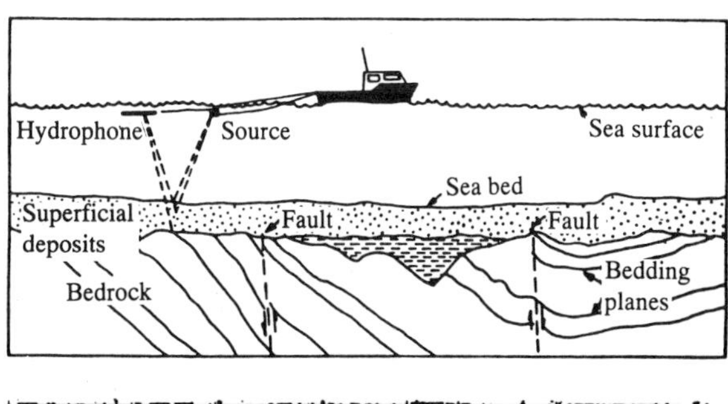

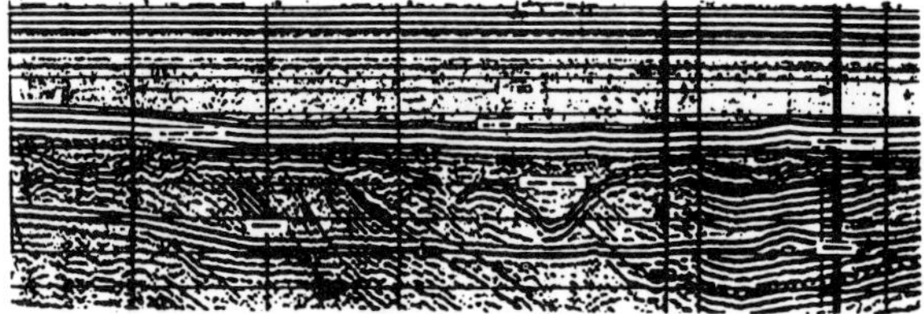

Figure 7.8 *Reflection seismic section*

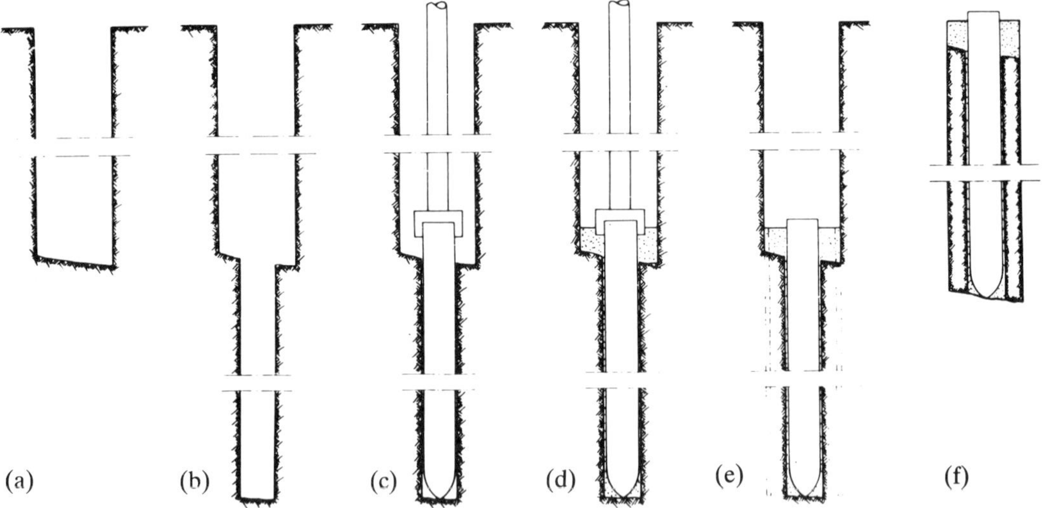

Figure 7.9 *Integral sampling method*

To further improve the quality of the rock samples the *integral sampling technique* was developed at LNEC by Rocha (1971). Besides extracting oriented samples, it is possible to fully recover the rock mass in very disturbed zones and still show the opening of joints and, in many cases, the nature of infilling material. Figure 7.9 shows the main steps of the operations.

The obvious drawbacks of integral sampling are the time spent in the various operations and the cost (average unit cost equals four or five times the unit cost using a regular drilling technique). This explains why the integral sampling technique is always used together with regular core drilling and only in zones or depths where, for a particular problem, the assessment of all those parameters associated with discontinuities are indispensable.

Data supplied by drill holes can be very much inproved by using them for indirect observation of the walls and for down-the-hole logging and for *in situ* testing. The most common techniques for inspection of drill-hole walls are closed-circuit television, borehole cameras and periscopes. Information about the conditions of the drill-hole walls (namely fracture pattern) can also be supplied by impression packers (Barr and Hocking 1976). Although these drill-hole inspection techniques have existed for a long time, their success has always been rather limited. Only closed-circuit television has recently experienced significant improvements in image resolution due to advances in electronics and in image-processing systems.

Drill-holes are also used for the application of continuous logging techniques. The most appropriate for engineering purposes are resistivity, spontaneous potential (SP), sonic (P and S waves), gamma–gamma, neutron and caliper. Figure 7.10 shows a drill-hole log in interbedded sandstones, shales and coal, using caliper, gamma–gamma, SP and resistivity techniques. Application of logging techniques to detect zones of poor-quality rock or voids have been presented by Cooper (1982). Appendix II lists the methods most frequently used.

Other logging techniques have been proposed to use with destructive drilling, called instantaneous logging (Pfister 1985). The idea is to offer, for engineering geological purposes, the full package (destructive drilling and instantaneous continuous logging) instead of core drilling, claiming that it is cheaper, faster and as reliable.

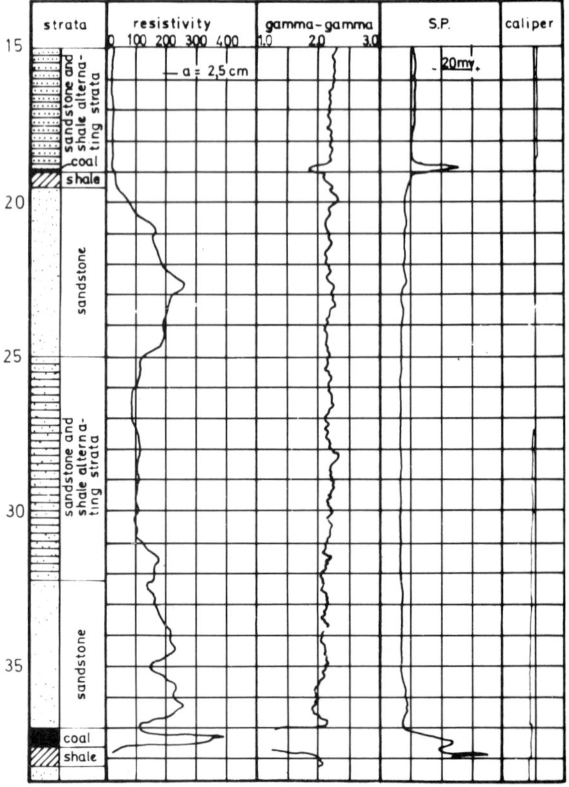

Figure 7.10 *Drill-hole logging*

However, unless for specific engineering geological conditions or specific engineering problems, core drilling is the drilling technique which can supply the most reliable data. Besides, the difference in cost does not justify the replacement (destructive drilling with instantaneous logging costs represent about 80% of core drilling).

The point is not to reduce the cost of investigations by accepting the application of less reliable techniques but, on the contrary, to maximize the multi-purpose concept when planning investigations. This calls, in the case of drill holes, for the proper layout of the site-exploration programme, in terms, for example, of inclination and diameter of drill-holes, in order to make possible their further use for other investigations, like logging and *in situ* testing.

In some problems, however, the joint application of core-drilling techniques and of destructive drilling (with or without instantaneous logging) may be the best solution for adequate investigation of sites. Appendix II lists the geophysical and the drilling parameters most frequently used in the logging of drill-holes.

(d) Trial excavations, trenches, trial pits and adits

Many geological situations related to foundations are characterized by a rock mass overlaid either by a thickness of deeply weathered rock or residual soil. In order to assess the geological conditions of the rock mass below these surface formations, trenching or stripping the ground are frequently performed. Trenches are only a few metres deep and excavation can be carried out manually or mechanically (backhoe excavators are currently used for this purpose) and they are able to remove deeply weathered rock and soils to a depth of 4–5 m.

After excavation of a trench, the bedrock surface and the trench walls can be mapped and photographed and samples can be taken. Geophysical traverses are frequently carried out along a trench and sometimes *in situ* deformation and shear strength tests are done in trenches.

Figure 7.11 shows the trenches which were excavated at the foundation for the Fridão Dam, in Portugal. The layout is typical of such situations, consisting of parallel trenches excavated along contour lines separated by 10–20 m. These trenches were subsequently used for the performance of refraction seismic traverses. In order to excavate, besides the surface formations, the very weathered mica schist and hornfels of the foundation, and thus to penetrate into the more sound zone of the rock mass, these trenches were excavated with a bulldozer D8 type (Caterpillar) and ripper. This has allowed accurate mapping of the structural features of the rock mass (faults, veins and joints).

Trial pits or shafts generally are not used for rock-mass investigation, except in some specific geological situations (namely, weak rocks), specific topographical ground surface arrangements (which require the excavation of trial pits as means of access into the rock mass) or specific engineering problems (for example, the investigation along the alignment of a deep ventilation shaft or of the foundation conditions of buttresses in the case of dams). Shafts or pits are usually vertical and can be excavated with any diameter. As with any other investigation technique, shafts should disturb the bed rock as little as possible, this being easier with small-diameter excavations. However, if testing equipment is required in a shaft it needs to be some metres in diameter.

The site of the nuclear power plant at Ferrel, in Portugal, consists of weak durassic sandstones and siltstones. After the decision about the approximate

Figure 7.11 *Trenches at the Fridão dam site*

location of the power plant was taken, a comprehensive engineering geological programme of the area was set up, including the excavation of four shafts (Oliveira and Fialho Rodrigues 1976). According to the geological conditions, the topography of the area and to the engineering structure of the power plant, the shafts were designed to reach 25 m depth and to allow plate bearing tests to be performed at different depths and geological horizons. The shafts were cylindrical, being 2 m in diameter and they had to be lined with concrete rings in the first few metres, below which the walls were supported with metallic ribs usually one metre apart, to prevent excessive deformation of the ground.

Adits are widely used in the investigation of rock masses. They represent a means of access to zones where observations and tests are carried out. Their orientation can be very flexible, it being very easy to introduce significant modifications in alignment and inclination as excavation proceeds. Branches from the main excavation can be sited according to the geological features which are met as excavation proceeds.

In general, investigation adits should not exceed 1.8–2.0 m in height and 1.2–1.5 m in width, in order to avoid as much destressing of the rock mass as possible. However, localized sections of adits where larger dimensions are required for *in situ* testing can very easily be over-excavated until the necessary dimensions are reached.

For most engineering problems requiring the excavation of adits as part of the engineering geological study of the rock mass, as in the case of large dam foundations (especially concrete dams), their length only seldom reaches 100 m, being often much shorter (30–50 m). However, for specific geological conditions (for example, soluble rock dam foundations) or for specific engineering problems (for example, large underground power stations or any other large underground excavation) the length may reach many hundred metres.

Adit excavations are expensive even when not requiring extensive support, and they usually need blasting. This explains why they are only excavated in specific situations where the accurate assessment of the engineering geological conditions of the rock mass are required for the safe design of engineering structures.

Besides mapping of the excavation surfaces along their alignment, adits serve as access to zones of the rock masses where samples have to be extracted for visual inspection and laboratory testing and where *in situ* static and dynamic tests have to be performed. Information about underground water conditions should also be recorded. Figure 7.12 shows the mapping of an adit excavation in the foundation rock mass of Funcho Dam, in Portugal, using one of the techniques described in Section 3.3, and Figure 7.13 shows the arrangement for the *in situ* static and dynamic tests carried out in the same exploration adit of the Funcho Dam site.

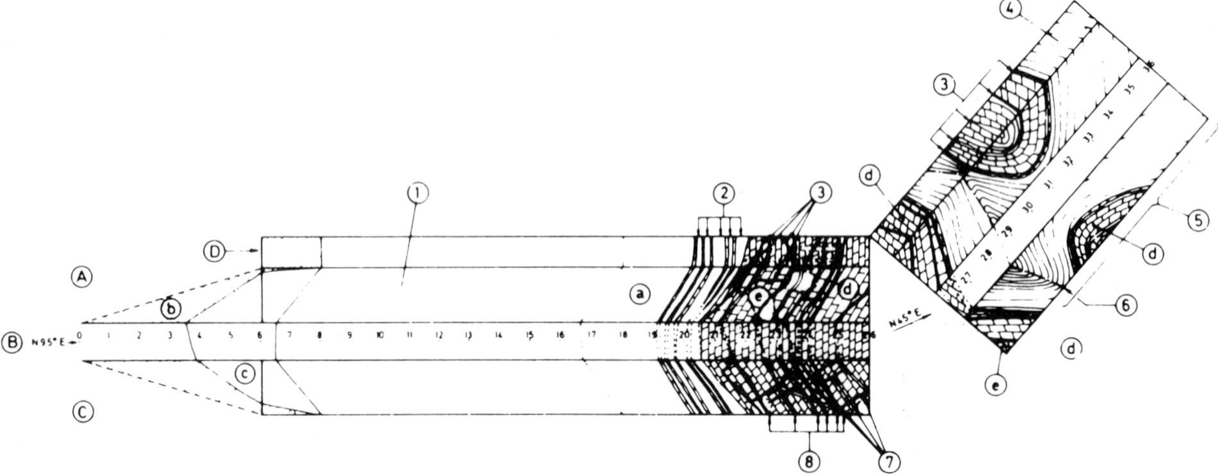

(1) Very soft, dark grey, weathered (W_4), and fractured (F_4) shale.
(2) Sheared shales.
(3) Gouge filling.
(4) Soft shale with interbedded greywacke, disturbed zone.
(5) Very disturbed zone.
(6) Fault through the axial plane of the anticline (thickness 5 to 10 cm).
(7) Shear zones.
(8) Gouge filling and shear zones.
(A) Downstream.
(B) Floor (level 62.3).
(C) Upstream.
(D) Ceiling.
(a) Shale moderately weathered (W_3).
(b) Slope debris.
(c) Black shale very weathered.
(d) Greywacke.
(e) Quartz.

Figure 7.12 *Mapping of an exploration adit*

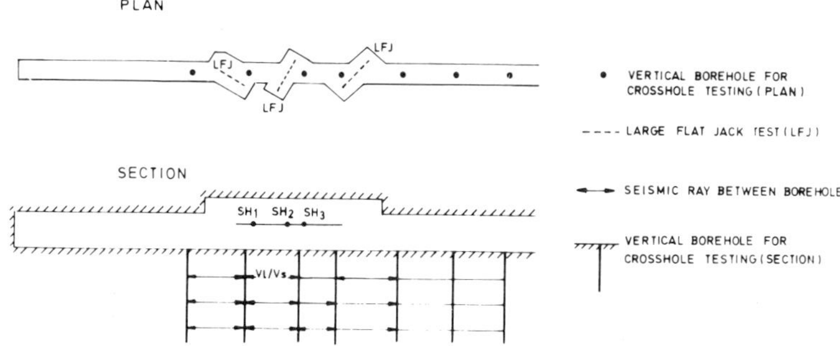

Figure 7.13 In situ *static and dynamic tests inside an exploration adit*

7.3.5 In situ tests

The planning of *in situ* testing of rock masses is a very important activity in any site-investigation programme, and requires good knowledge of the structural behaviour of the works and of the geological formations involved.

Together with the laboratory tests, the *in situ* tests supply quantitative information required to properly assess the relevant geotechnical properties of a rock mass. Although more expensive than laboratory tests, they generally supply more reliable results mainly because they are performed over a larger volume of ground and because the volume tested involves the undisturbed state of stress prevailing at the respective level of testing.

The aim of the tests, in general, is to define the design parameters which can be used in stability analyses of the works and to assess the rock masses. The tests which are usually performed at the design stage are deformation tests, strength tests, permeability tests and, for subsurface works, state of stress tests. Less common at this stage are, for example, grouting tests, which are often carried out only at construction stage, since they require very specialized equipment which site investigation contractors often do not have.

Some authors consider rock anchor tests as *in situ* tests for rock-mass investigation. However, it seems that rock anchor test results, although also dependent on the rock-mass properties, namely, strength and permeability, are very much related to the technology applied to anchor sealing which, if not properly selected, can be the only factor responsible for the results obtained.

A description of the most used methods of *in situ* testing is presented in detail in many publications and many of the methods are already published by the ISRM as standard methods (e.g. Anon, 1986; Anon, 1987a, b). An account of *in situ* testing recently has been provided by Oliveira and Charrua Graça (1987) and so the description and discussion of such tests is not included.

Those *in situ* tests which are mostly used for the geotechnical characterization of rock masses are given in Appendix III. Other tests not included in the list can be used for some of the same purposes but those listed give a good coverage of the needs and are the best known worldwide.

In most cases the correct interpretation of the *in situ* tests requires reliable index tests and mechanical laboratory tests. The prediction of the behaviour of rock masses can be achieved, often, by the performance of durability tests on rock samples.

The most common laboratory tests are given in Appendix IV.

Important developments have to be reported in the interpretation of dynamic test results, namely, using direct cross-hole and fan-shooting seismic methods. The recent application of the tomographic technique to the zoning of seismic wave velocities through a rock mass has given very promising results. In seismic tomography the position of transmitters and receivers are conditioned by the location and orientation of drill holes and adits. Furthermore as seismic wave velocities refract into horizons with higher velocities, hence the trajectories, in complex geological conditions are very difficult to detect. This calls for advanced electronics and for skill in the development of appropriate software.

At LNEC several applications have already been made possible, as a consequence of a number of engineering geological studies of existing dam foundations which need some rehabilitation. In all cases, grouting of the rock masses was considered indispensable, this allowing for a large number of cross-hole measurements between drill holes positioned only a few metres apart, prior to and after the treatment. For each rock-mass foundation, two tomographic sections result from the interpretation, one considering the seismic wave velocity pattern prior to the grouting and the other considering the seismic wave velocity pattern after the treatment. Figure 7.14 shows one tomographic section along the alignment of the Venda Nova Dam, showing the distribution of the wave velocities in the foundation before the treatment.

VENDA NOVA DAM (PORTUGAL)

SEISMIC TOMOGRAPHY

P WAVES VELOCITIES (m/s)

RIGHT BANK

LEFT BANK

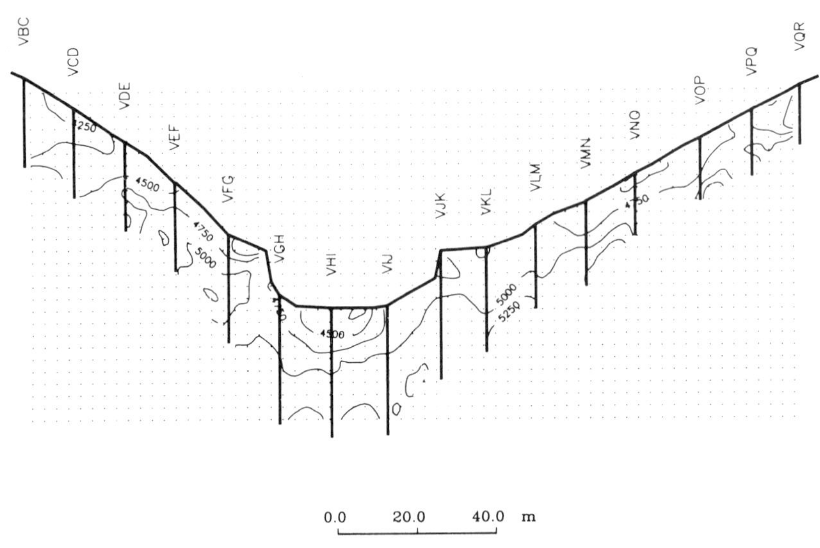

Figure 7.14 *Seismic tomography of the Venda Nova dam foundation*

Appendix I: Site exploration

Geophysical exploration
- Electric resistivity: soundings
  - traverses (isoresistivity maps)
- Refraction seismic
- Reflection seismic
- Electromagnetic
- Gravimetric
- Magnetic

Mechanical exploration
- Trenches
- Trial pits
- Adits
- Drill holes: core drilling
  - rotary drilling
  - rotary percussive drilling
- Sampling: core recovery
  - integral sampling
  - block sample
- Inspection of drill-hole walls: TV
  - cameras
  - impression packers

Appendix II: Logging of boreholes

Geophysical
- Gamma–gamma
- Neutron
- Electric resistivity
- SP (spontaneous potential)
- Sonic (P and S waves)

Drilling parameters
- Fluid pressure
- Rotation torque
- Thrust applied to the bit
- Advance speed
- Rotation speed
- Reflected percussion
- Fall of equipment (voids)
- Caliper

Appendix III: *In situ* tests

Deformability

Static

Large volume
- Plate bearing
- Large flat jack
- Radial jacking
- Pressure chamber

In hole
- Goodman's jack
- Menard's pressuremetre
- LNEC dilatometer

Dynamic (tomography)
- Cross-hole
- Down-hole
- Seismic refraction
- Fan shooting

Strength
- Direct shear (rock and discontinuities)

State of stress

Stress relief
- Doorstopper
- LNEC STT (stress-tensor tube)
- Small flat jack
- USBM deformation gauge
- CSIRO triaxial strain cell
- Hydraulic fracturing
- Dilatometer

Permeability

Large volume
- Pumping test
- Tracer test

In hole
- Lugeon type (LNEC packer)
- TRH (hydraulic register test)
- SHM (hydraulic multitest probe)
- Hydraulic triple probe
- Piezopermeameter

Appendix IV: Laboratory tests

Index tests
- Moisture content
- Water absorption
- Specific weight
- Porosity/porometry
- Permeability
- Point load
- Hardness
- Ultrasonic velocities (dynamic elastic constants)
- Swelling

Mechanical tests
- Uniaxial compressive strength
- Uniaxial tensile strength
- Brazilian test
- Triaxial compressive strength
- Creep
- Shear strength (rock and discontinuities)

Durability tests
- Slake durability
- Soundness (chemical, freeze–thaw, etc.)
- Abrasion (Los Angeles)
- Swelling (strain and pressure)

References

Anon (1978) 'Suggested methods for determining hardness and abrasiveness of rocks', International Society Rock Mechanics Commission on the Standardization of Laboratory and Field Tests, *Int. J. Rock Mech. Min. Sci. & Geomech. Abstr.*, **15** (1), 89–98

Anon (1981) 'Report of the International Association of Engineering Geology Commission on Site Investigations'. *Bull. Int. Assoc. Engng. Geology*, **24**, 185–226

Anon (1985) 'Suggested methods for determining point load strength', International Society Rock Mechanics Commission on Testing Methods, *Int. J. Rock Mech. Min. Sci. & Geomech. Abstr.*, **22** (2), 51–60

Anon (1986) 'Suggested methods for deformability determination using a large flat jack technique', International Society Rock Mechanics Commission on Testing Methods, *Int. J. Rock Mech. Min. Sci. & Geomech. Abstr.*, **21** (2), 131–140

Anon (1987) 'Suggested methods of rock stress determination', International Society Rock Mechanics Commission on Testing Methods, *Int. J. Rock Mech. Min. Sci. & Geomech. Abstr.*, **24** (1), 53–73

Anon (1987) 'Suggested methods for deformability determination using a flexible dilatometer',International Society Rock Mechanics Commission on Testing Methods, Working Group on Flexible Dilatometers, *Int. J. Rock Mech. Min. Sci. & Geomech. Abstr.*, **24** (2), 123–134

Barr, M.V. and Hocking, G. (1976) Borehole structural logging employing a pneumatically inflatable impression packer. In *Exploration for Rock Engineering*, Capetown Z.T. Bieniawski, (ed.), A.A. Balkema, Rotterdam, Vol 1 pp. 29–39

Bell, F.G. (1980) *Engineering Geology and Geotechnics*, Newnes-Butterworths, London

Chaplow, R.C. (1975) 'Engineering and site investigation', *Ground Engineering*, **8**, 34–38

Cooper, S.S. (1982) 'The use of down hole geophysical methods to detect zones of poor quality rock or voids', *Miscellaneous paper GL-82-15*, US Army Engineers Waterways Experiment Station, Vicksburg

Dearman, W.R. (1987) Engineering geological maps and plans', Chapter 28, *Ground Engineer's Reference Book*, F.G. Bell (ed.), Butterworths, London

Dearman, W.R. and Matula, M. (1976) *Engineering Geological Maps. A Guide to their Preparation*, The Unesco Press, Paris

Fialho Rodrigues, L. (1979) 'Métodos de prospecção sísmica em Geologia de Engenharia – A importância da onda de corte', *Thesis LNEC,* Lisboa

Fialho Rodrigues, L. (1980) 'Utilização de métodos sísmicos no estudo da deformabilidade de maciços rochosos', *Geotecnia 48*, Lisboa, November, 41–46

Fialho Rodrigues, L. (1986) 'Caracterização por métodos sísmicos do maciço de fundação da Barragem do Varosa', *Internal Report*, LNEC, Lisboa

Fialho Rodrigues, L., Oliveira, R. and Correira de Sousa, A. (1983) 'Cabril Dam – Control of the grouting effectiveness by geophysical seismic methods' Proceedings 5th Int. Congress ISEM, Melbourne (Australia)

Fialho Rodrigues, L. and Oliveira, M. (1988) 'Prospecção

geofísica no trecho da auto-estrada Torres Novas-Fátima', *Internal Report*, LNEC, Lisboa

Griffiths, D.H. and King, R.F. (1981) *Applied Geophysics for Geologists and Engineers*, (2nd edn), Pergamon Press, Oxford

Griffiths, D.M. and King, R.F. (1987) 'Geophysical exploration', Chapter 27, *Ground Engineer's Reference Book*, F.G. Bell (ed.), Butterworths, London

Oliveira, R. (1977) 'Cartografia geológica de túneis', *Memória do LNEC* 489 Lisboa

Oliveira, R. (1981) 'Introdução à Geologia de Engenharia', *Notas de Aulas*, Universidade Nova de Lisboa, Lisboa

Oliveira, R. (1986) 'Engineering geological investigations of rock masses for civil engineering projects and mining operation', *General report subject 1*, Int. Congress IAEG, Buenos Aires

Oliveira, R. and Charrua Graça, J. 'In situ testing of rocks', Chapter 26, *Ground Engineer's Reference Book*, F.G. Bell (ed.), Butterworths, London

Oliveira, R. and Fialho Rodrigues, L. (1976) 'Colaboração nos estudos geotécnicos e sismológicos para implantação da Central Nuclear do Ferrel', *Internal Report*, LNEC, Lisboa

Olsson, O. *et al.* (1988) 'Fracture characterization in crystalline rock by borehole radar', *Report ID no. 88241*, Swedish Geological Co. Engineering Geology, Uppsala, Sweden

Onodera, F. (1963) 'Dynamic investigation of foundation rock in situ', *Proc. 5th US Sym. Rock Mechanics*, University of Missouri, Rolla, Pergamon Press, New York, 517–533

Pfister, P. (1985) 'Recording drilling parameters in ground engineering', *Ground Engineering*, **18** (3),16–21

Rocha, M. (1971) 'Method of integral sampling of rock masses', *Rock Mechanics*, **3**, 1–12

Trautmann, C.H. and Kulhawy, F. H. (1983) 'Data sources for engineering geological studies', *Bull. Assoc. Engng. Geol,* **20**, 439–454

Further reading

Anon (1975) 'Recommendations on site investigation techniques', *Report of International Society Rock Mechanics Commission*, Lisbon

Anon (1980) 'Basic Geotechnical description of rock masses (BGD)', *Report of International Society Rock Mechanics Commission*, Lisbon

Anon (1981) *Code of Practice for Site Investigation*, BS 5930, British Standard Institution, London

Bergman, S. and Carlsson, A. (1988) *Site Investigation in Rock. Investigation, Prognoses, Reports–Recommendations*, BEFO, Stockholm

Johnson, R. and DeCraff, J. (1988) *Principles of Engineering Geology*, Wiley, New York

8 Laboratory testing of rocks

Professor F G Bell
University of Natal

It must be stated at the onset that there are no fundamental mechanical properties of intact rock, that is, there are no material constants which characterize a particular rock. Obviously there are a number of tests which can be used to determine rock properties and those properties which can be used to describe rocks in terms of engineering classification are referred to as index properties. Index properties frequently show a good correlation one with another. If the correct index tests are chosen, then rock, having similar properties, irrespective of origin, will probably exhibit similar engineering performance. In order for an index property to be useful it must satisfy certain criteria. It should be simple to carry out, inexpensive and rapidly performed. The test results must be reproducible and the index properties must be relevant to the engineering requirement. Generally such tests are carried out in large numbers so that a reliable picture of rock variation is obtained. Hence laboratory tests provide basic information on the physical properties and mechanical reactions of intact rock and help classify the rock, thereby allowing it to be compared with other rock types. Although laboratory tests can provide data which can be used for design purposes in rock masses, such data is more appropriately obtained from field testing.

8.1 Density and porosity

The density of a material is defined as its mass per unit volume. The density of a rock is one of its most fundamental properties. It is principally influenced by the mineral composition on the one hand and the amount of void space on the other; as the proportion of void space increases so the density decreases.

As far as rocks are concerned it is necessary to distinguish four different types of density. Firstly, there is grain density or the mass of the mineral aggregate per volume of solid material. Then, secondly, there is dry density or the mass of the mineral aggregate per volume. Thirdly, bulk density is the mass of mineral aggregate and natural water content per volume. Fourthly and lastly, saturated density is the mass of mineral aggregate and water per volume; in this case the pore spaces are filled with water. The term 'unit weight' can be used instead of density, and is expressed in terms of force (i.e. kN/m^3).

The grain density represents an average density for the assortment of minerals of which a rock is composed and obviously its value depends on the relative proportions of each mineral present. It can be derived in the following manner:

$$\text{grain density} = \frac{\text{dry density}}{(1 - n/100)} \tag{8.1}$$

where n is the porosity.

The grain density is similar to the specific gravity (or relative density) of a rock except that the specific gravity is not expressed in units, it being the ratio of mass of the rock specimen to that of an equal volume of water at a specified temperature.

The specific gravity can be determined by grinding the rock to powder. The powder is oven dried at 105°C and then left to cool. A small quantity of the powder (between 5 and 10 g) is placed in a density bottle of known weight and weighed (British Standards Institution 1975; Brown, 1981). Next, de-aired distilled water is added to the density bottle to cover the powder. It then is placed in a desiccator and vacuum applied to remove any air. After which more de-aired distilled water is added to fill the bottle. The stopper is fitted and bottle, stopper and contents are left at a constant temperature of 20°C for 1 hour, and then weighed. The bottle then is cleaned thoroughly, filled with de-aired distilled water and left for one hour at the same constant temperature. The specific gravity, G_s, of the solid particles is derived from the following expression:

$$G_s = \frac{M_s - M_1}{(M_4 - M_1) - (M_3 - M_2)} \tag{8.2}$$

where M_1 is the mass of the density bottle and stopper, M_2 is the mass of the powder, bottle and stopper, M_3 is the mass of the bottle, stopper, powder and distilled water and M_4 is the mass of the bottle, stopper and distilled water. At least two tests should be carried out and the results averaged.

Alternatively the specific gravity can be determined according to the test outlined in ASTM (1982). In this case the specimens are first dried and weighed. They then are immersed in distilled water at 20°C for 1 hour or until air bubbles no longer form on the specimens within a period of 5 min. After surface drying the specimens are weighed and then immersed for a further 5 min. They then are weighed in water. The specific gravity, G_s, is obtained from

$$G_s = \frac{M_d}{M_{sa} - M_{sw}} \tag{8.3}$$

where M_d is the mass of the dry specimen, M_{sa} is the mass of the saturated specimen in air and M_{sw} is the mass of the saturated specimen in water.

The determination of the dry density, bulk density and saturated density is dependent upon accurately weighing the rock specimen concerned and upon the accurate measurement of its volume. In the case of the dry density the rock specimen is dried in a ventilated oven at 105°C until a constant mass is reached and then allowed to cool in a desiccator. Constant mass is assumed when successive weighings after a number of four-hour intervals differ by less than 0.1%. In order to obtain the saturated density the rock specimen is first saturated by immersing in water in a vacuum for a minimum period of one hour with periodic agitation to remove any trapped air. After saturation the surface of the specimen is dried. This method is not suitable for rocks which slake when immersed in water. The bulk density simply requires the mass of the specimen as it is obtained from the field, that is, with its natural moisture content.

There are two methods by which the volume of specimens can be determined, namely, the caliper method and the buoyancy method. Regular cylindrical or prismatic specimens are required in the caliper method. The volume, V, is calculated from the dimensions obtained by averaging a number of equally distributed caliper or micrometer measurements over the specimen. Regular or irregular shaped specimens may be used in the buoyancy method. The specimen is saturated and then its submerged mass, M_{sw}, is determined. Next the saturated mass, M_{sa}, is found. The bulk volume is given by

$$V = (M_{sa} - M_{sw})/(\text{density of water}) \tag{8.4}$$

The porosity of a rock can be defined as the volume of the pore space divided by the total volume expressed as a percentage. Total or absolute porosity is a measure of the total void volume and is the excess of bulk volume over grain volume per unit of bulk volume. It is usually determined as the excess of grain density over dry density, per unit of grain density, and can be obtained from the following expression:

$$\text{absolute porosity} = \left(1 - \frac{\text{dry density}}{\text{grain density}}\right) \times 100 \tag{8.5}$$

The effective, apparent or net porosity is a measure of the effective void volume of a porous medium and is determined as the excess of bulk volume over grain volume and occluded pore volume. It may be regarded as the pore space from which water can be removed.

The porosity generally is determined experimentally by using the standard saturation method (Brown 1981). Alternatively an air porosimeter can be used (Ramana and Venkatanarayana 1971). Both tests give an effective value of porosity although that obtained by the air porosimeter may be somewhat higher because air can penetrate pores more easily than can water. In the saturation method the specimens are dried and weighed (M_d) and then saturated with water under vacuum. They are then weighed in water (M_{sw}) and again in air (M_{sa}) and the porosity, n, is determined as follows:

$$n = \frac{M_{sa} - M_d}{M_{sa} - M_{sw}} \times 100 \quad (\%) \tag{8.6}$$

The air porosimeter basically consists of two parts (Figure 8.1). On one is a glass sample chamber below

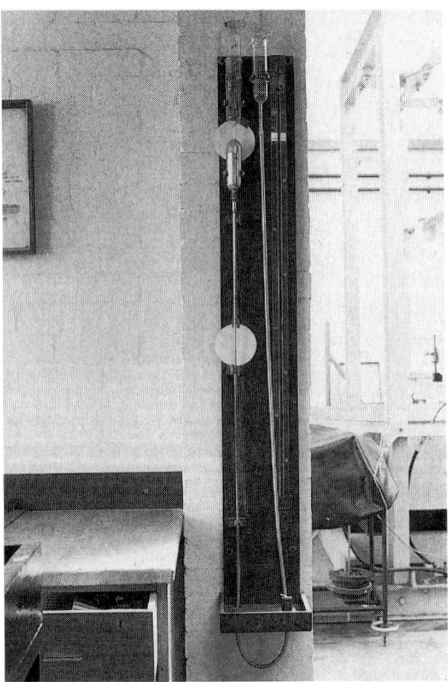

Figure 8.1 *The air porosimeter*

which is a condenser, the two being connected by a small glass tube. The sample chamber possesses a stopcock and a ground glass joint. A stopper, smeared with high vacuum silicon grease, is fitted into the joint when the apparatus is being operated. A graduated thick-walled glass tube leads from the base of the condenser and has sufficiently wide bore to avoid corrections for capillarity effects. The other side of the porosimeter consists of a movable mercury reservoir and the two sides are connected by a flexible polythene tube. When the test is carried out the stopcock on the sample chamber is closed and the reservoir moved to its lowest position. The mercury in the other limb also moves down as the reservoir is open. The height of the mercury in the graduated tube (H_1) is recorded and the reservoir is then moved back to its highest position with the stopcock open. The sample is now placed in the sample chamber, the stopcock is closed, and the reservoir is again moved to its lowest position and the height is read on the graduated tube (H_2). It can be shown that the grain volume, V_g, is equal to $B(H_2 - H_1)$ where B is the apparatus constant. The bulk volume is determined by using a mercury balance and so the porosity can be obtained from

$$\text{porosity } (n) = \left(1 - \frac{V_g}{V_b}\right) \times 100 \quad (\%) \qquad (8.7)$$

The porosity does not provide an indication of the way in which the pore space is distributed within a rock, or whether it consists of many fine pores or a smaller number of coarse pores. Winkler (1973) did classify pores according to size, the largest being megapores (these have a channel diameter greater than 0.062 mm), then came macrocapillaries (between 0.0001 and 0.062 mm), and the smallest were the microcapillaries (the diameter of which is less than 0.0001 mm).

Two tests are used to investigate the distribution of pore sizes within and microporosity of a rock, the suction plate test and the mercury porosimeter test. In the suction plate method a small stone disc is placed on the upper surface of a porous unglazed plate, the lower surface of which is in contact with water in a small reservoir (Figure 8.2). A flexible U-tube containing mercury extends from the base of the reservoir and when the tube is moved it creates a suction effect on the lower surface of the plate. The stone disc is covered with an airtight lid in order to prevent water evaporating from its upper surface. In order to assess microporosity the disc is dried and weighed, M_d, then vacuum saturated and weighed again, M_{sw}. After this it is laid on the plate and subjected to a negative pressure equivalent to 6.4 m of water. The disc is weighed daily until a constant mass, M_c, is reached. The microporosity, μ_n then is determined as follows:

$$\mu_n = \frac{M_c - M_d}{M_{sw} - M_d} \times 100 \ (\%) \qquad (8.8)$$

In the case of the suction plate method microporosity is defined as the volume of water retained (expressed as a percentage of the total available pore space) when a suction equivalent to 6.4 m of head of water is applied to the specimen . In effect this test measures the percentage or pores with an effective diameter of less than 5 μm. Such pores are able to retain water against applied suction and they determine resistance to damage by frost or by the crystallization of soluble salts.

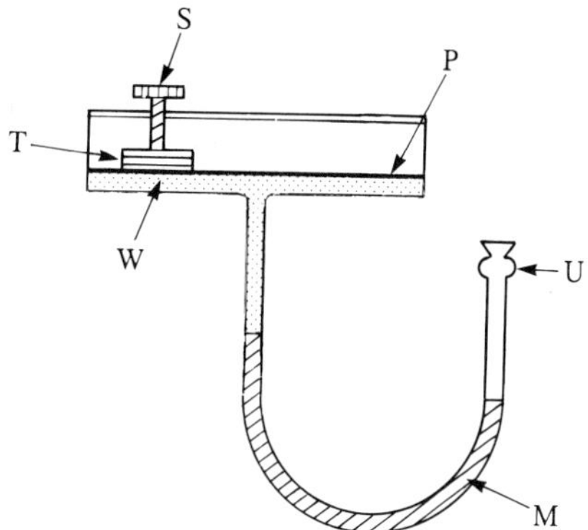

Figure 8.2 *The suction-plate method of determining microporosity. A small slab-like specimen (T) is pressed against the upper surface or a porous plate (P), the underside of which is in contact with a water reservoir (W). A flexible U-tube extends from the base of the reservoir and contains mercury (M). The specimen is held in place by a screw (S). Before being set in place the specimen has been saturated by boiling and then weighed. A series of suctions is applied to the specimen until a head of 6.4 m of water is reached. Each suction draws water out of the specimen, and it is weighed each time. The limiting head mentioned develops sufficient suction to draw water from all pores with diameters exceeding 0.005 mm.*

In the mercury porosimeter test (Figure 8.3) mercury is forced to penetrate the pores of the specimen under applied pressure. Obviously the finer the pores, the higher the pressure which must be used to bring about penetration. In this way it is possible to derive the dimensions and pore-size distribution (Figure 8.4) from a graph showing the distribution of pores sizes. A line is drawn on the curve at the position where 10% of the pore space has been filled with mercury (this is roughly equivalent to the point where 90% of the pore space is filled with water in the water suction method). The pore diameter corresponding to that position is termed d_{10}. The pores below this size limit can be regarded as constituting the microporosity.

8.2 Water sorption and capillarity

Water sorption includes adsorption (i.e. the adhesion of gaseous molecules, gases or ions or molecules in solution, to the surfaces of solid bodies with which they are in contact) on the one hand and absorption on the other (i.e. the take-up, assimilation or incorporation of liquids into solids). The term natural absorption capacity has been used to refer to the quantity of water which readily can be absorbed by a rock and is measured by the water sorption test.

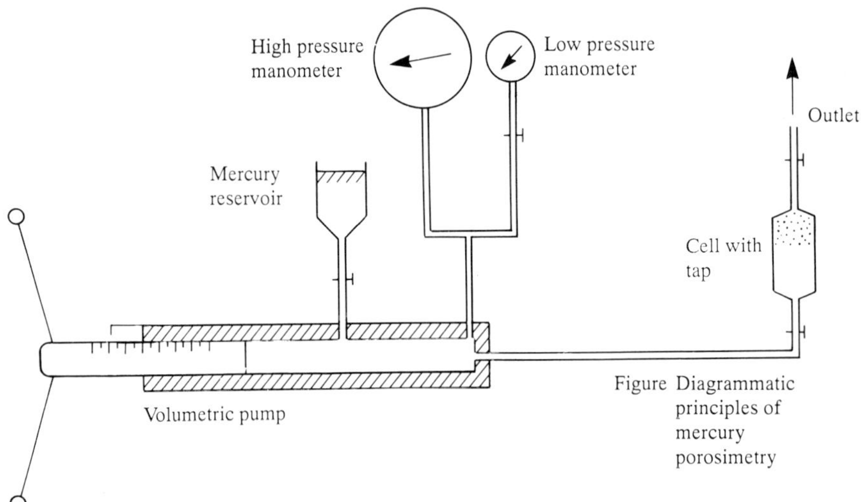

Figure 8.3 *The mercury porosimeter*

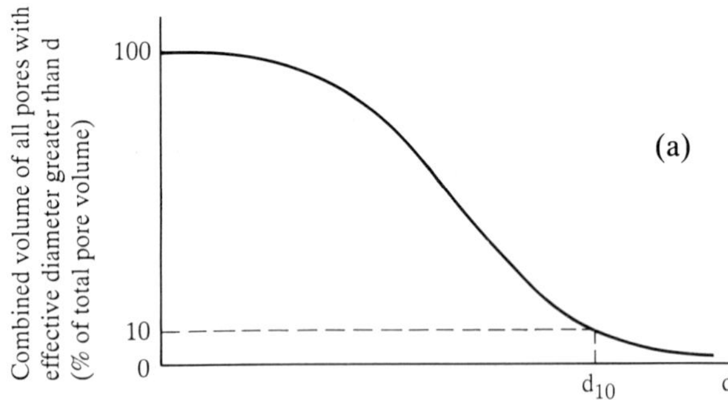

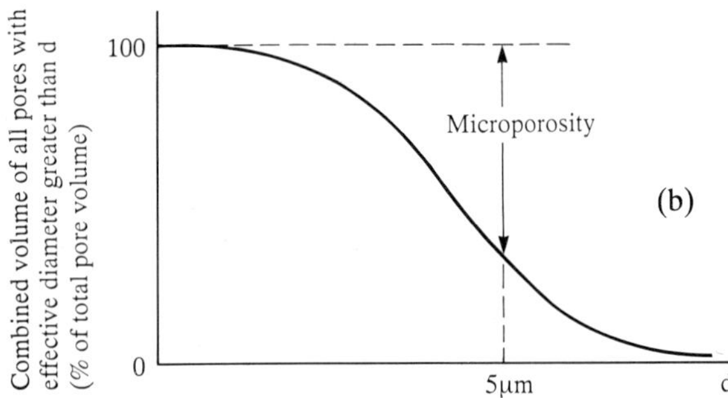

Figure 8.4 *Pore-size distribution.* (a) *In the mercury porosimeter test the specimen is subjected to a range of pressures. The microporosity is obtained by drawing a line at the point on the curve where 10% of the pore space has been occupied mercury.* (b) *The microporosity is defined arbitrarily by the 5 μm pore diameter line in the suction plate method*

The water sorption test outlined in ASTM (1982) recommends that at least three smooth-faced *cubic*, prismatic or cylindrical specimens should be used (with least dimensions not less than 50.8 mm and maximum dimensions not exceeding 76.2 mm and a volume-to-surface area ratio between 0.3 and 0.5). After drying at 105°C for 24 hours and then weighing, the specimens are saturated in distilled water at 20°C for 48 hours and weighed once more. The percentage absorption by weight equals

$$\frac{M_{sw} - M_d}{M_d} \times 100 \tag{8.9}$$

where M_{sw} is the mass of the specimen after immersion and M_d is the mass of the dried specimen. The saturation coefficient is related to the volume of voids accessible to water. Hence it can be regarded as the ratio of effective porosity to total porosity. In other words the saturation coefficient provides a measure of the extent to which the pores of a stone become filled when it is allowed to absorb water for a standard period of time. After the porosity has been measured the stone is dried. It is then immersed in water for 24 hours and subsequently weighed. The mass is denoted by M_s, and the saturation coefficient, S, then is obtained as follows:

$$S = \frac{M_s - M_d}{M_a - M_d} \tag{8.10}$$

where M_a is the mass of the specimen in air.

A rock which has a relatively low coefficient is one in which a relatively high proportion of the pore space remains unfilled with water under normal conditions.

Conversely a high value indicates a rock with a high proportion of fine pores. The saturation coefficient has been used in France, according to Mamillan (1976), to distinguish between stones which are frost susceptible and those which are frost resistant. Those stones which are frost resistant have a saturation coefficient of less than 0.75, whilst susceptible stones possess values exceeding 0.85. Stones with values between 0.75 and 0.80 are considered doubtful.

The coefficient of capillarity measures the rate at which a rock soaks up water. In order to assess the coefficient, prisms of uniform cross-section (70 mm × 70 mm) and 100 mm in height are oven-dried to a constant weight and placed in a shallow tray of water (French Standard NFB 10–502). The base of each test piece is submerged to a depth of 2 mm in water. Throughout the test the level of the water is kept constant. Monitoring of the uptake of water is accomplished by weighing the specimens at regular intervals. The coefficient of capillarity, C, is derived as follows:

$$C = \frac{100M}{S\sqrt{t}} \tag{8.11}$$

where M is the mass of water absorbed in grams, S is the sectional area of the lower face of the test prism in square centimetres, and t is the time in minutes elapsed since the beginning of the capillary rise. The value of the coefficient is equal to the slope of the straight line obtained by plotting $\sqrt{t}$ as abscissa and $100M/S$ as ordinate.

8.3 Permeability

Permeability may be defined as the ability of a rock to allow the passage of fluids into or through it without impairing its structure. In ordinary hydraulic usage a substance is termed permeable when it permits the passage of a measurable quantity of fluid in a finite period of time and impermeable when the rate at which it transmits that fluid is slow enough to be negligible under existing temperature–pressure conditions. The permeability of a particular material is defined by its coefficient of permeability or hydraulic conductivity. In terms of the permeability of a rock mass, its interconnected system of discontinuities is usually far more important than its porosity in transmitting the flow of fluid through it. Indeed the permeability of intact rock (primary permeability) is usually several orders less than its *in situ* permeability (secondary permeability).

Determination of the permeability of many rock types in the laboratory is made by using a falling-head permeameter (Figure 8.5). The sample is placed in the apparatus which is then filled with water to a certain height (h_1) in the standpipe. Then the stopcock is opened and the water infiltrates through the sample, the height of the water in the standpipe falling to h_2. The times at the beginning (t_1) and end (t_2) of the test are recorded. These together with the cross-sectional area (A) and length of sample (l) are then substituted in the following expression, which is derived from Darcy's law, to obtain the coefficient of permeability (k)

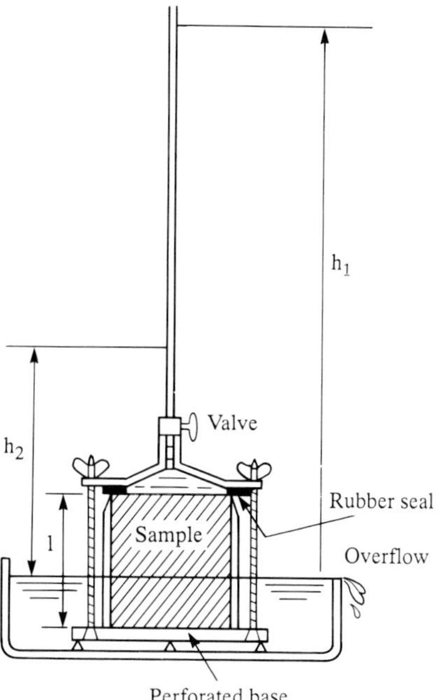

Figure 8.5 *The falling-head permeameter*

$$k = \frac{2.3026a}{A\,(t_2 - t_1)} \times (\log_{10} h_1 - \log_{10} h_2) \tag{8.12}$$

where a is the cross-sectional area of the standpipe.

Bernaix (1969) examined the variations of permeability in rocks under stress by using a radial percolation test. He used a cylindrical specimen 60 mm in diameter and 150 mm in length in which an axial hole, 12 mm diameter and 125 mm in length, was drilled. The specimen is placed in the radial percolation cell, which can contain water under pressure, and the central cavity is in contact with atmospheric pressure; or water can be injected under pressure into the cavity (Figure 8.6). The flow is radial over almost the whole height of the sample and is convergent when the water pressure is applied to the outer face of the specimen, and divergent when the water is under pressure inside the specimen. For radial flow from a cylinder of unfractured rock with interconnected pores the coefficient of permeability (k) is given by

$$k = \frac{q\gamma_w}{2\pi l\,\Delta p} \ln \frac{R_2}{R_1} \tag{8.13}$$

in which R_1 and R_2 are the radii of the inner and outer surfaces respectively, q is the flow discharge, γ_w is the unit

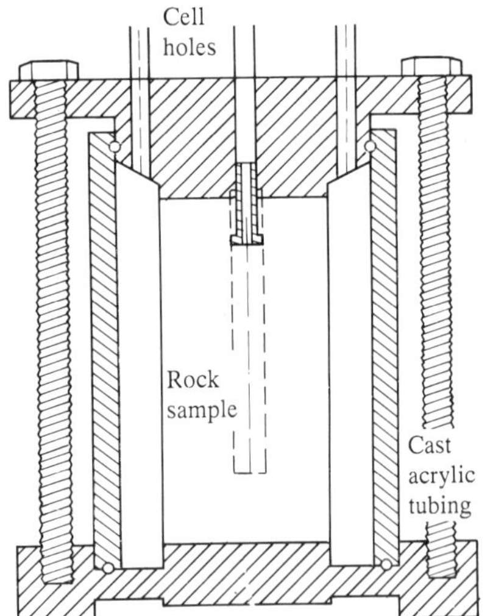

Figure 8.6 *Radial permeameter cell*

Figure 8.7 *The slake-durability apparatus*

mass of water , *l* is the length of cylinder over which flow is occurring and Δp is the difference between external and internal pressures and is positive for convergent flow. Porous rocks remain more or less unaffected by pressure changes. On the other hand, fissured rocks exhibit far greater permeability in divergent flow than in convergent flow. Moreover fissured rocks exhibit a continuous increase in permeability as the pressure attributable to divergent flow is increased. Indeed it would appear that some amount of hydraulic fracturing occurs in divergent flow.

Table 8.1 Description of the degree of slaking

Amount of slaking	*Slake-durability index (%)**	*Liquid limit (%)†*
Very low	0–25	0–20
Low	25–50	20–50
Medium	50–75	50–90
High	75–95	90–140
Very high	over 95	over 140

* After Franklin and Chandra (1972)
† After Morgenstern and Eigenbrod (1974)

8.4 Durability of weak rocks

Durability is a fundamental property of weak rocks. The slake-durability test estimates the resistance to wetting and drying of a rock sample, particularly argillaceous rocks and those which exhibit a certain degree of alteration. In this test the sample, which consists of ten pieces of rock, each weighing about 40 g, is placed in a test drum, oven dried and then weighed. After this, the drum, with sample, is half immersed in a tank of water and attached to a rotor arm which rotates the drum for a period of 10 min at 20 rev/min (Figure 8.7). The cylindrical periphery of the drum is formed of 2 mm sieve mesh so that broken-down material can be lost whilst the test is in progress. After slaking, the drum and the material retained are dried and weighed. The slake-durability index then is obtained by dividing the weight of the sample retained by its original weight and expressing the answer as a percentage. Franklin and Chandra (1972) provided a grading system for the slake-durability test which is given in Table 8.1.

Morgenstern and Eigenbrod (1974) used a water absorption test to assess the amount of slaking undergone by argillaceaous material. This test measures the increase in water content in relation to the number of wetting and drying cycles undergone. They found that the maximum slaking water content increased linearly with increasing liquid limit and that during slaking all materials eventually reached a final water content equal to their liquid limit. Materials with medium to high liquid limits exhibited very substantial volume changes during each wetting stage, which caused large differential strains, resulting in complete destruction of the original structure. Thus materials characterized by high liquid limits are more severely weakened during slaking than materials with low liquid limits. Classes of slaking were therefore defined in terms of the value of liquid limit (Table 8.1). The rate of slaking was grouped into three classes according to the change in liquidity index after immersion in water for two hours, that is, slow–less than 0.75; fast–0.75 to 1.25; and very fast–over 1.25.

Failure of consolidated and poorly cemented rocks occurs during saturation when the swelling pressure (or internal saturation swelling stress, σ_s), developed by capillary suction pressures exceeds their tensile strength.

An estimate of σ_s can be obtained from the modulus of deformation (E):

$$E = \sigma_s/\varepsilon_D \tag{8.14}$$

where ε_D is the free-swelling coefficient. The latter is determined by a sensitive dial gauge recording the amount of swelling of an oven-dried core specimen per unit height, along the vertical axis during saturation in water for 12 hours, ε_D being obtained as follows:

$$\varepsilon_D = \frac{\text{change in length after swelling}}{\text{initial length}} \tag{8.15}$$

When the free expansion of rocks liable to swell is inhibited, stresses of sufficient magnitude to cause damage to engineering structures may develop. A number of tests may be carried out which provide an indication of the swelling behaviour of such a rock. These include the swelling strain index test performed under unconfined conditions and the swelling pressure index test carried out under conditions of zero volume change. Swelling strain measurements can also be undertaken on radially confined specimens under various conditions of axial loading.

The swelling strain represents maximum expansion of an unconfined rock specimen when it is submerged in water. The test specimens may either be oven-dried or contain their natural moisture content. As test specimens are submerged in water it is advisable to prevent those specimens of rock types prone to slaking from collapsing before the ultimate swelling strain has developed. This can be done by wrapping the specimen in muslin. During the test the specimen is supported by a frame between a fixed point and a measuring point. The latter may be a dial gauge or a deformation transducer (Brown 1981). The specimen and frame are placed in a container which is filled with water just above the level of the specimen. Strain is measured in all three perpendicular directions (Figure 8.8). No strain is allowed to develop in the test specimen in the swelling pressure test. This is accomplished by tightly mounting a cylindrical specimen inside a rigid ring which provides radial constraint (Brown 1981). The specimen absorbs water through porous end plates. Any axial strain which develops is monitored and compensated for by increasing the axial load. When equilbirium conditions are established the stress required to prevent expansion is equal to the swelling pressure index. Again the initial moisture content must be recorded prior to testing.

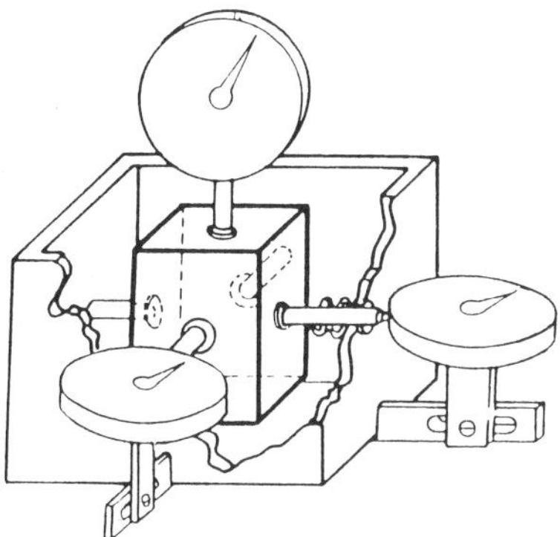

Figure 8.8 *Apparatus for measuring strain in three directions*

The aforementioned test apparatus may be used to determine the swelling strain developed by radially confined specimens under axial loading. In this case axial strain is allowed to take place.

8.5 Compressive and shear strength

8.5.1 Uniaxial compressive strength

The uniaxial compressive strength of a rock is one of the simplest measures of strength to obtain. It may be regarded as the highest stress that a rock specimen can carry when a unidirectional stress is applied, normally in an axial direction, to the ends of a cylindrical specimen. In other words the unconfined compressive strength represents the maximum load supported by the specimen during the test divided by the cross-sectional area of the specimen. Although its application is limited, the unconfined compressive strength allows comparisons to be made between rocks and affords some indication of rock behaviour under more complex stress systems.

The behaviour of rock in uniaxial compression is influenced to some extent by the test conditions. The most important of these is the length–diameter or slenderness ratio of the specimen. Dhir and Sangha (1973) maintained that the most satisfactory slenderness ratio is 2.5. At lower ratios fractures take place in highly restrained specimen ends, while at higher ratios there is an undesirable release of elastic strain energy from the unfractured ends region to the fractured central zone during post-failure stressing. In other words such a ratio provides a reasonably good distribution of stress throughout. However, Obert and Duvall (1967) had previously suggested the use of the following empirical expression to relate the uniaxial strength to the length–diameter ratio:

$$\sigma_c = \frac{\sigma_{act}}{0.788 + 0.222D/L} \tag{8.16}$$

where σ_c is the compressive strength of a specimen of the same material having a 1:1 length-diameter ratio and σ_{act} is the compressive strength of a specimen for which $2>(L/D)>1/3$. Indeed Obert and Duvall reported that as far as the uniaxial compression of cylindrical specimens is concerned, the size of specimen has less effect than the natural variation in the values obtained from testing a given rock type when the specimen length–diameter ratio is kept constant. An approximate relationship between uniaxial conmpressive strength (σ_c) and specimen diameter (for specimens up to 200 mm diameter) is given by

$$\sigma_c = \sigma_{c5o}(50/D) \tag{8.17}$$

where σ_{c5o} is the uniaxial compressive strength of a specimen 50 mm diameter and D is the actual diameter of the specimen in millimetres.

The rate at which loading occurs is another test variable affecting the compressive strength (Hawkes and Mellor 1970). A loading rate of between 0.5 and 1.0 MPa/s is recommended by the ISRM (Brown 1981). The ends of core samples should be lapped so that they are exactly perpendicular to their long axes.

Loading platens should be hardened steel discs with flat surfaces so as to avoid stress concentrations at the contacts between the rock specimen and the platens. One of the two loading platens should incorporate a spherical seat which should be placed on the top end of the specimen; the centre of curvature of the seat should coincide with the centre of the end of the specimen. Both the loading platens and the spherical seat should be aligned accurately in the centre of the testing machine to avoid any eccentricity of loading. If specimens are tested directly from the field, then their natural moisture content should be determined. Otherwise specimens can be tested dry or saturated.

The onset of failure in a rock specimen subjected to compressive loading is first of all marked by the formation of a large number of isolated fractures, both intergranular and intragranular. Such local fracturing characterizes the relief of stress concentration produced by the mechanical inhomogeneities in the rock and most cracks are orientated parallel to the applied stress (Figure 8.9). This is quickly followed, however, by the development of two groups of macroscopic shear failures, at the boundary and in the interior of the specimen, which suggests that most of the major sources of induced lateral tensile stresses are now eliminated. The interior macroscopic shear failures are extended and become interconnected to form a conjugate set of open shear fractures. The central cones, which are now delineated, either abrade during the large shear displacement or produce major fractures in the remnant cores. If one shear failure surface becomes dominant, then cones are not developed and the sample ultimately fails in two parts along a diagonal plane of shear.

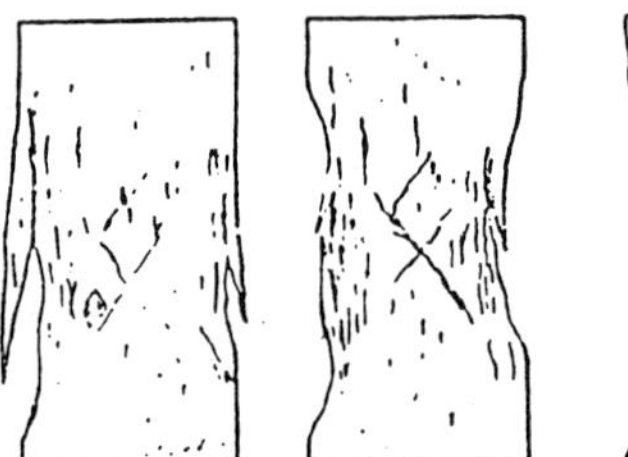

Figure 8.9 *Stages in the development of fracturing with increasing unconfined compressive loading (after Wawersik and Fairhurst 1970)*

The grades of unconfined compressive strength shown in Table 8.2 have been suggested by the Geological Society (Anon 1970), the IAEG (Anon 1979) and the ISRM (Anon 1981).

Where comparative assessment of rock strength has to be made it is sometimes possible to dispense with the conventional uniaxial compressive test in favour of some other form of test. For example, a rock specimen with an irregular shape is used in the Protodyakonov test (Protodyakonov 1963). The weight and volume of each specimen is obtained prior to crushing and the mean crushing strength (P) of the specimen is related to its volume (V) as follows:

$$\log P = \log f + 0.63 \log V \tag{8.18}$$

and

$$f = 0.19\sigma \tag{8.19}$$

where σ is the uniaxial compressive strength. It was suggested that 15–25 specimens should be tested to obtain the mean value.

Hobbs (1964) also carried out a series of tests on irregularly shaped specimens of rock. In this case the

Table 8.2 Grades of unconfined compressive strength

Geological Society		*IAEG*		*ISRM*	
Term	*Strength* (MPa)	*Term*	*Strength* (MPa)	*Term*	*Strength* (MPa)
Very weak	less than 1.25	Weak	Under 15	Very low	Under 6
Weak	1.25–5.00	Moderately strong	15–50	Low	6–20
Moderately weak	5.00–12.50	Strong	50–120	Moderate	20–60
Moderately strong	12.50–50	Very strong	120–230	High	60–200
Strong	50–100	Extremely strong	Over 230	Very high	Over 200
Very strong	100–200				
Extremely strong	Over 200				

maximum height of each specimen was measured prior to testing. The contact area of the specimen ends on the loading platens was measured after the rock had failed. This was done by placing a piece of carbon paper and a piece of graph paper between the specimen and each of the loading platens. The carbon imprint on each of the graph papers is measured and so the average of the two areas is obtained. After an analysis of the results Hobbs found that the relationship between the compressive strength (σ_c) and the average applied stress at fracture (P_{av}) was given by

$$\sigma_c = 0.91 P_{av} - 21.9 \text{ MPa} \tag{8.20}$$

8.5.2 Triaxial compressive strength

Triaxial tests have to be carried out if the complete failure envelope of a rock material is required. A constant hydraulic pressure (the confining pressure) is applied to the cylindrical surface of the rock specimen, whilst applying an axial load to the ends of the sample. The axial load is increased up to the point where the specimen fails. A series of tests, each at higher confining pressure, are carried out on specimens from the same rock. These enable Mohr circles and envelope to be drawn (Figure 8.10). The angle of friction and value of the cohesion are obtained from the Mohr diagram.

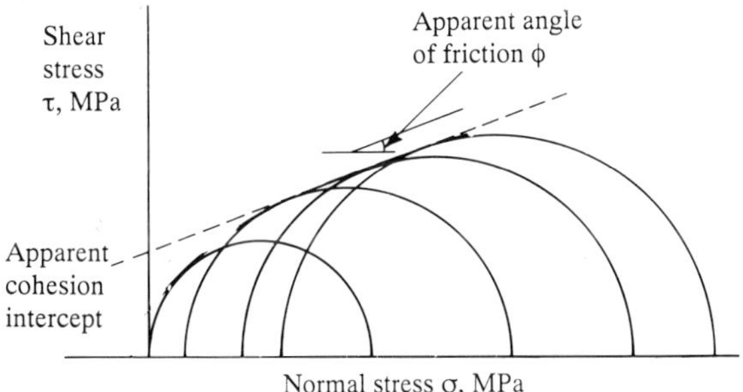

Figure 8.10 *Mohr envelope for rock*

The procedure for determining the strength of rocks in triaxial compression has been outlined by Vogler and Kovari (1978) (see also ASTM 1967). Testing of the rock sample is carried out within a specially constructed high-pressure cell such as the Hoek cell (Hoek and Franklin 1968 and Figure 8.11). In routine triaxial testing of rock it is not usual to measure pore pressures. Indeed after a series of triaxial tests on rocks, Murrell (1963) showed that the pore pressure countered the effect of confining pressure. In other words, if the pore pressure is equal to the confining pressure, then fracture occurs at a constant value of the deviation stress, which is equal to the unconfined compressive strength.

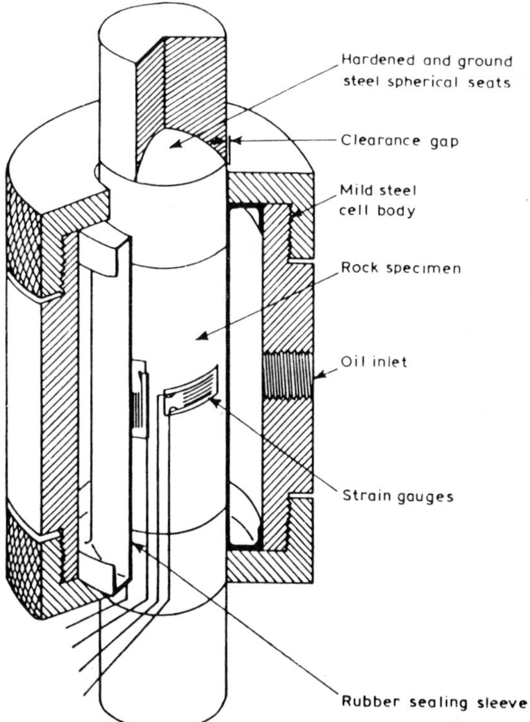

Figure 8.11 *Cutaway view of triaxial cell (after Hoek and Franklin 1968)*

8.5.3 Direct shear tests

A number of tests are used to measure the shear strength of intact rock and include the shear box test, the direct single and double shear tests, the punch shear test, and the torsion test. Shear strength parameters also can be obtained from a triaxial compression test. The most commonly used methods are the shear box, for weaker rocks and along discontinuity surfaces, and triaxial compression (Section 8.5.2) for stronger rocks.

In the shear box test the rock specimen is set in plaster of Paris and if failure is required to occur along a particular direction, as for example along a discontinuity surface, then this direction is aligned with the shear plane of the shear box (Figure 8.12a). A constant normal force is applied to the specimen which is then sheared along the required shear plane. The increments of shear force and resultant displacements are recorded during the test from which a stress–strain curve is produced (Figure 8.12b). A number of tests (preferably more than three), each at a higher normal stress, are undertaken so that a shear strength vs normal stress graph can be drawn. The value of cohesion and angle of friction are derived from this graph (Figure 8.12c). A value of residual strength can be gained by repeating the test over the sheared surface.

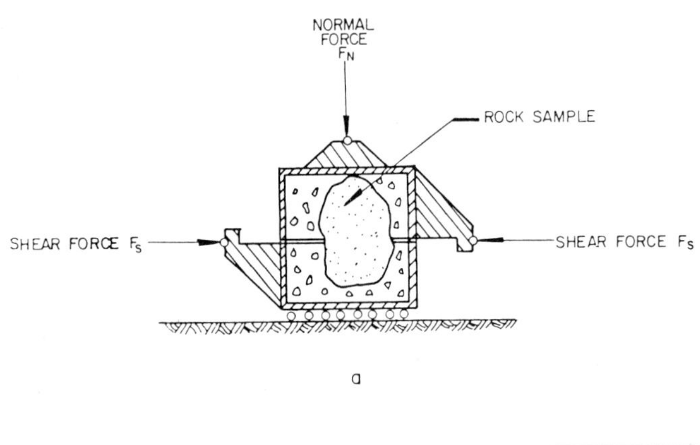

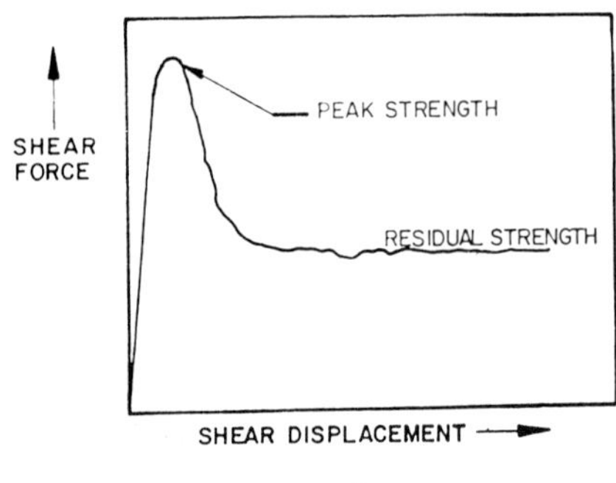

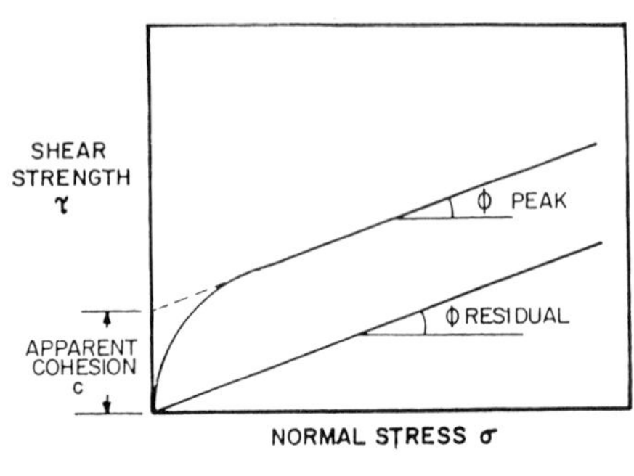

Figure 8.12 *Illustration of a shear test*

The shear strength in the single shear test (Figure 8.13a) is determined from the expression

$$\tau = P/A \tag{8.21}$$

and from

$$\tau = P/2A \tag{8.22}$$

for the double shear test (Figure 8.13b), where A is the cross-sectional area of the specimen and P is the load required to bring about failure.

In the punch test the specimen is placed within a cylindrical guide and a piston is forced through it (Figure 8.13c). The value of the shear strength is then calculated from

$$\tau = P/\pi DH \tag{8.23}$$

where D is the diameter of the punch and H is the thickness of the specimen.

The torsional shear test is illustrated in Figure 8.13(d) the shear strength this time is determined from the expression

$$\tau = 16M/\pi D^3 \tag{8.24}$$

M being the applied torque at failure whilst D is the diameter of the sample.

In all these shear tests the results are determined by the testing arrangement as well as by the rock material. As a consequence Everling (1964) suggested that the shear strength of a rock should only be derived by triaxial testing. He distinguished between the strength of a material in pure shear and the shear stress which is required to cause failure when the normal stress on the plane of fracture is zero. For most rocks there is a considerable difference between these two values.

8.5.4 Tensile strength

Rocks have a much lower tensile strength than compressive strength. Brittle failure theory predicts a ratio of compressive strength/tensile strength of about 8:1 but in practice it is generally between 15:1 and 25:1. The direct tensile strength of rocks has been obtained by attaching metal end caps with epoxy resins to specimens, which are then pulled into tension by wires. In direct tensile tests the slenderness ratio of cylindrical specimens should be 2.5–3.0 and the diameter preferably should not be less than NX core size (54 mm). The ratio of diameter of specimen to the largest grain in the rock should be at least

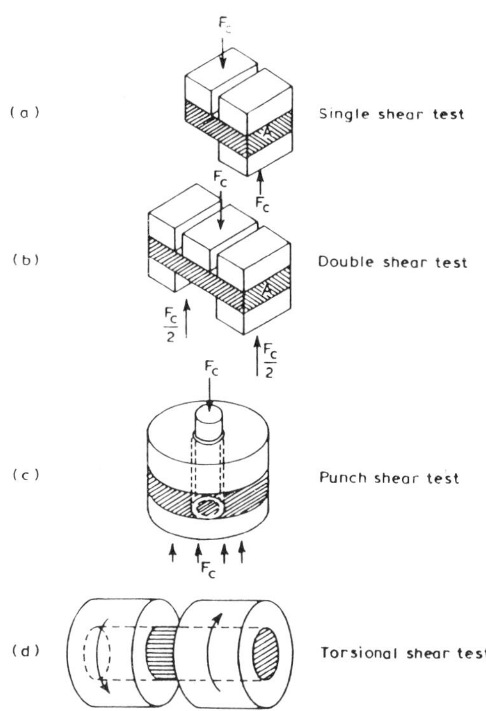

Figure 8.13 *Unconfined shear-strength tests*

10:1 (Bieniawski and Hawkes 1978). Unfortunately the determination of the direct tensile strength has proved difficult since a satisfactory method has not been devised to grip the specimen without introducing bending stresses. Accordingly most tensile tests have been carried out by indirect methods.

In the flexural test a cylindrical specimen of rock is loaded between three points at a rate of 1.4 MPa/min until the sample fails. The flexural tensile strength (T_f) is then given by the expression

$$T_f = 8PL/\pi D^3 \qquad (8.25)$$

where P is the load at failure, L is the length between the supports and D is the diameter of the specimen. The flexural strength gives a higher value than that determined in direct tension.

In the Brazilian test a rock cylinder of length (L) and diameter (D) is loaded (with a load, P) in a diametrical plane along its axis. The sample usually fails by splitting along the line of diametrical loading and the tensile strength (T_b) can be obtained from.

$$T_b = 2P/\pi LD \qquad (8.26)$$

The use of the Brazilian test as an indirect method of assessing the tensile strength of rocks is based on the fact that most rocks in biaxial stress fields fail in tension when one principle stress is compressive. Failure, however, may be brought about by localized crushing along the axis of loading and not by diametral tension.

Disc-shaped specimens are also used in the Brazilian test. Curved jaw loading rigs are sometimes used when discs are tested, in an attempt to improve loading conditions. Uncertainties associated with the premature development of failure are sometimes removed by drilling a hole in the centre of a disc-shaped specimen. This has sometimes been referred to as the 'ring' test.

The International Society for Rock Mechanics (ISRM) recommends that when a disc-shaped specimen is used it is wrapped around its periphery with one layer of masking tape (Bieniawski and Hawkes 1978). In such cases the ISRM also recommends that the specimen should not be less than NX core size (54 mm in diameter) and that the thickness should be approximately equal to the radius of the specimen. A loading rate of 200 N/s was recommended. The tensile strength (T_b) of the specimen is obtained as follows:

$$T_b = 0.636P/DH \qquad \text{(MPa)} \qquad (8.27)$$

where P is the load at failure (N), D is the diameter of the test specimen (mm) and H is the thickness of the test specimen measured at the centre (mm).

Mellor and Hawkes (1971) stated that the Brazilian test is useful for brittle materials but for other materials the test may give wholly erroneous results. Furthermore, Fairhurst (1964) concluded that the uniaxial tensile strengths of materials with low compression/tension ratios is underestimated by Brazilian tests in which radial loading is applied to disc-shaped specimens.

In the point load test the specimen is placed between opposing cone-shaped platens and subjected to compression (Figure 8.14). This generates tensile stresses normal to

Figure 8.14 *The point-load apparatus*

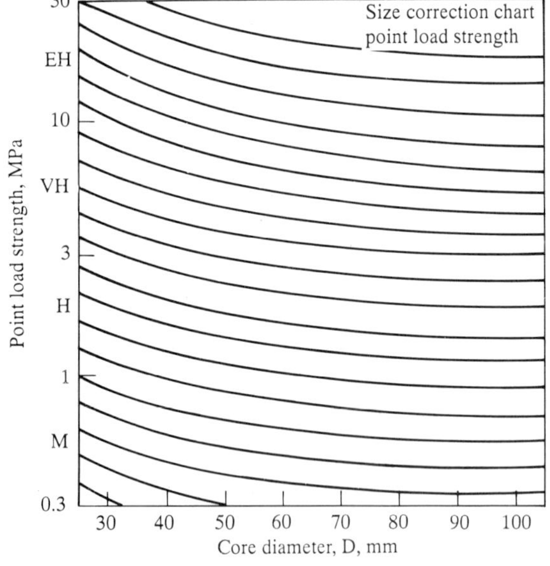

(a)

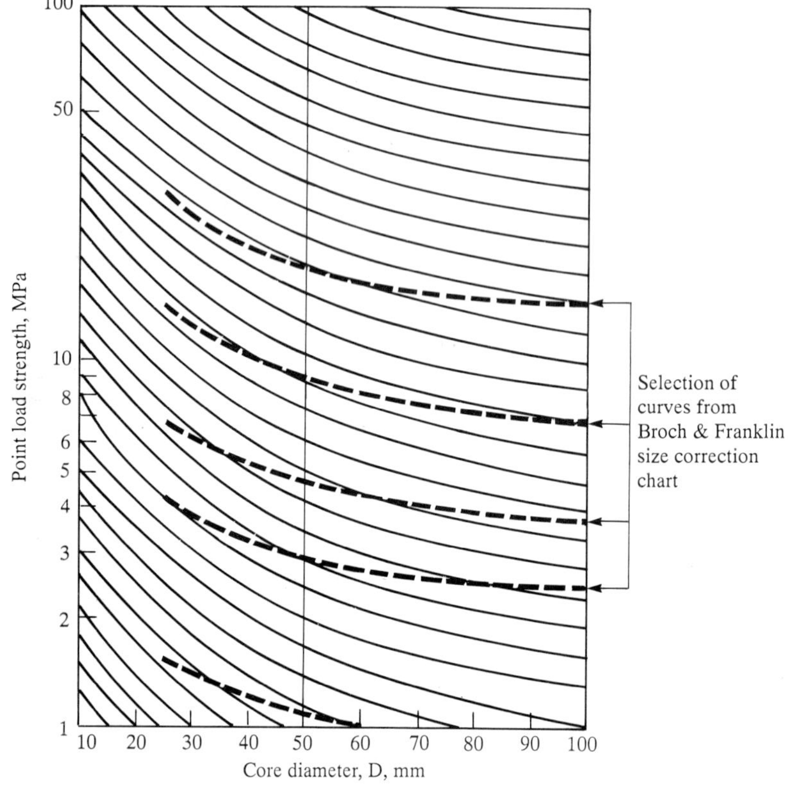

(b)

Figure 8.15 (a) *Size-correction chart for point-load strength testing (after Broch and Franklin 1972);* (b) *modified size-correction chart (after Hassani et al.* 1980)

the axis of loading. The tensile strength (T_p) is then derived by using the empirical expression (Reichmuth 1968)

$$T_p = 0.96P/D^2 \qquad \text{(or more simply } P/D^2\text{)} \qquad (8.28)$$

where P is the load at failure and D is the diameter of the specimen or distance between the cones. The point load test has a number of variations such as the diametral test, the axial test and the irregular lump test. The latter is the least accurate (Bieniawski 1975). The diametral test is more convenient and simpler than the other two. Broch and Franklin (1972) suggested that the distance between the contact point and each end of the specimen should be at least $0.7D$, where D is the diameter of the core in the diametral test. In the axial test core specimens with a length-to-diameter ratio 1:1 were regarded as suitable. They found that the point load strength tended to decrease when progressively larger specimens were tested and that saturated specimens underwent an average reduction in strength of approximately 33%. Rocks which are strongly anisotropic such as slates, schists, laminated sandstones, etc., should be tested both along and parallel to the lineation. Nonetheless axial point load results are sensitive to sample shape, especially in the case of anisotropic rocks. A number of authors have considered this problem of shape. For example, Forster (1983) suggested that the results of axial point load tests could be converted to the strength equivalent to D/L = 1:1 for comparison with diametral strength by using the expression

$$T_{pd} = 0.1723e^{1.5987D/L}\,T_{pa} \qquad (8.29)$$

where T_{pd} is the diametral point load strength that is equivalent to the axial point load strength at D/L = 1:1 and T_{pa} is the recorded axial strength valid between D/L ratios of 0.5 and 1.0. Norbury (1986) commented that the use of such a shape correction factor would permit the adoption of a more flexible shape criterion for the axial point load test. Rock lumps with typical diameter of about 50 mm and with a ratio of longest to shortest diameter of between 1.0 and 1.4 are used. At least 20 lumps should be tested per sample.

The effect of size of specimens is greater in tensile than compression testing because in tension, cracks open and give rise to large strength reductions, whilst in compression the cracks close and so disturbances are appreciably reduced. In the axial and irregular lump tests the effects of size and shape are very pronounced. Accordingly Broch and Franklin (1972) recommended core with a diameter of 50 mm as a reference standard (T_{p50}) to which other sizes should be corrected by reference to a correction chart (Figure 8.15a). Bieniawski (1975), however, considered that NX core (54 mm diameter) would be a better standard and that core specimens with diameter of less than BX size (42 mm) should not be used for point load testing. Hassani *et al* (1980) produced revised size-correction curves (Figure 8.15b) which are steeper than those of Broch and Franklin, providing a higher correction for diameters not equal to 50 mm. The differences are especially marked at diameters greater than 60 mm and at point load strength less than 5 MPa. Norbury (1986) suggested that more reliable results could be obtained with this chart than that of Broch and Franklin.

Brook (1977) discussed the influence of size on the results obtained from testing rocks with the point-load apparatus. He suggested that both the shape effect and size effect of specimens can be accounted for by the expression

$$T_p = \frac{\text{Load at fracture area } 500\,\text{mm}^2}{500\,\text{mm}^2} \qquad (8.30)$$

the load being determined either from a number of standard specimens of minimum cross-sectional area of 500 mm^2, 25 mm cores being the most convenient form, or from a log-log plot of load against area for a variety of shapes and sizes. Turk and Dearman (1985) also used a log-log plot to represent point-load stengths against specimen diameter. They obtained a linear relationship with a negative slope. The slope of the best fit line gave the reciprocal of the Weibull modulus and T_{p50} is found from the intercept with the 50 mm thickness line.

The standard deviation of the results of point load tests averages about 15%, which Bieniawski (1975) maintained was acceptable for practical engineering purposes. The test is limited to rocks with uniaxial compressive strengths above 25 MPa (point load index above 1 MPa). In the case of rocks with apparently lower strengths, uniaxial compression testing is preferred.

Broch and Franklin (1972) suggested that the point-load strength scale shown in Table 8.3 could be used to classify rocks. They also suggested that uniaxial compressive strength (σ_c) could be estimated from the point-load index (T_{p50}) as follows:

$$\sigma_c = 24T_{p50} \qquad (8.31)$$

However, this relationship should be treated with caution since, as Norbury (1986) pointed out, the values of the

Table 8.3 Point-load strength classification

	Point-load strength index (MPa)	*Equivalent uniaxial compressive strength* (MPa)
Extremely high strength	Over 10	Over 160
Very high strength	3–10	50–160
High strength	1–3	15–60
Medium strength	0.3–1	5–16
Low strength	0.1–0.3	1.6–5
Very low strength	0.03–0.1	0.5–1.6
Extremely low strength	Less than 0.03	Less than 0.5

(Compare this classification of strength with that of the Geological Society of London, p. 508.)

multiplier quoted in the literature have ranged between 8 and 45, with most values being between 16 and 24. Hence it is recommended that point lead and unconfined compressive strength testing is carried out to determine a specific site value of the multiplier.

8.6 Hardness

Hardness is one of the most investigated properties of materials, yet it is one of the most complex to understand. It does not lend itself to exact definition in terms of physical units. Indeed the numerical value of hardness is as much a function of the type of test used as a material property. The concept of hardness has usually been associated with the surface of a material. Deere and Miller (1966) considered rock hardness as its resistance to the displacement of surface particles by tangential abrasive force, as well as its resistance to penetrating force, whether static or dynamic. They pointed out that rock hardness depended very much on the same factors as did toughness. The latter is controlled by the efficiency of the bond between the minerals or grains as well as the strength of these two components.

The hardness of a mineral is usually defined as its resistance to scratching and relative hardness has been used as a diagnostic property since the beginning of systematic mineralogy. As long ago as 1822 Mohs proposed a scale of hardness based upon ten minerals and this scale is still widely used today, although more sophisticated techniques are now available. Mohs' scale is as follows:

1. Talc
2. Gypsum
3. Calcite
4. Fluorspar
5. Apatite
6. Orthoclase
7. Quartz
8. Topaz
9. Corundum
10. Diamond

Each mineral in the scale is capable of scratching those of a lower order. The relative hardness of a given mineral can therefore be assessed by using a series of hardness pencils each of which are tipped by one of the minerals in Mohs' scale. The fingernail scratches minerals up to a hardness of about 2.5 and a penknife up to approximately 5.5. Attempts to assess the hardness of rock by summing the hardness values of its principal mineral constituents, according to their relative proportions, has not proved satisfactory.

A wide variety of penetrator and loading devices have been used for assessing the static indentation hardness. For example, the indenters used in the Brinell and Rockwell tests are spherical, and in the Knoop (Winchell 1946) and Vickers tests (Das 1974) they are pyramidal. The loading varies from 70 N in the Vickers test to 20 kN in that of Brinell. The Brinell and Rockwell hardness tests, however, are not generally applicable to rock because of its brittle nature. On the other hand, the Vickers test has been used to determine the microhardness of rock. A pyramidal-shaped diamond is applied to the surface of the material, and the surface area of the impression divided by the applied load provides a measure of the hardness. Because a rock is not a homogeneous material, several hardness tests must be made over the surface of the specimen and the results averaged.

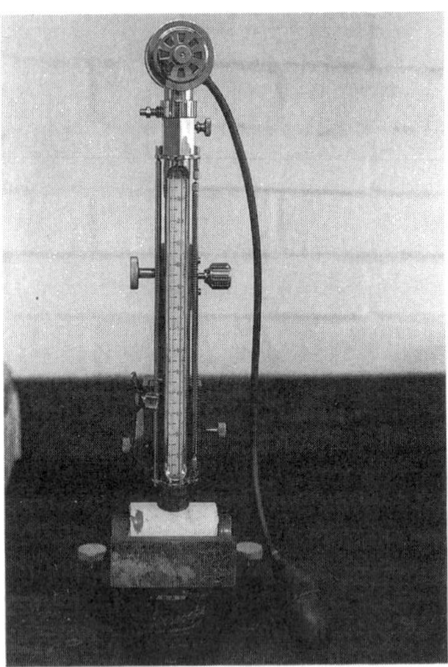

Figure 8.16 *The Shore scleroscope*

The Shore scleroscope is a non-destructive hardness-measuring device which indicates the relative values of hardness by the height of rebound of a small diamond-pointed hammer which is dropped vertically onto a securely clamped test surface from a height of 250 mm (Figure 8.16). The ISRM recommends that the Shore scleroscope should be used to assess the hardness of rock surfaces ground smooth by using No. 1800 grade aluminium oxide abrasive powder. It also recommends that a specimen should have a minimum test surface of 10 cm^2 and a minimum thickness of 10 mm (Atkinson *et al* 1978). At least 20 hardness determinations should be taken and averaged and each point of test should be at least 5 mm from any other.

Rabia and Brook (1979) examined the effects of specimen size on results and concluded that Shore hardness is dependent on the volume of the specimen tested and not simply on length or area. They suggested a minimum volume of 40 cm^3 for each test in order to obtain consistent values and that the mean of at least 50 readings on five specimens was required for a hardness value.

Because they found a very good correlation between Shore hardness and uniaxial compressive strength Deere and Miller (1966) were able to devise the rock strength chart shown in Figure 8.17. They noted, however, that the chart appeared to be limited to rocks with strengths in excess of 35 MPa.

The Schmidt hammer was developed for measuring the strength of concrete and has since been adapted for assessing the hardness of rocks. It is a portable non-destructive device which expends a definite amount of stored energy from a spring and indicates the degree of

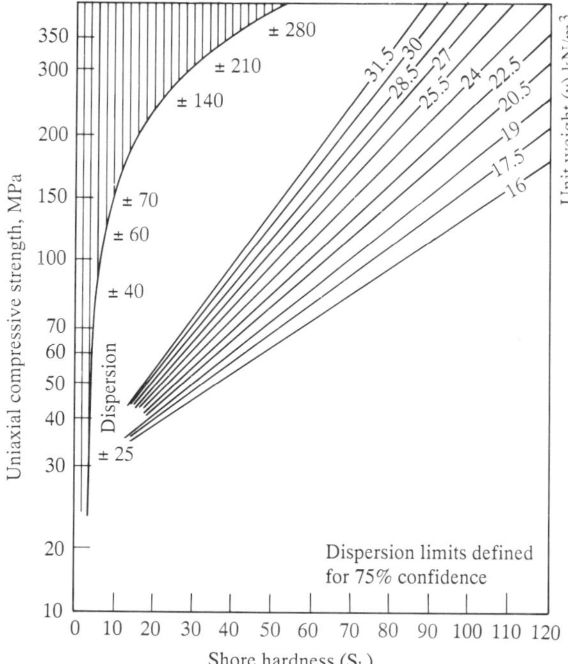

Figure 8.17 *Correlation chart for Shore hardness (H_s) relating unit weight of rock, compressive strength and hardness value (after Deere and Miller 1966)*

rebound of a hammer mass within the instrument, following impact. Tests are made by placing the specimen in a rigid cradle and impacting the hammer at a series of points along its upper surface. The hammer is held vertically at right angles to the axis of the specimen (Hucka 1965). At least 20 readings should be taken from each sample and then averaged to give one value.

The ISRM recommends that specimens used for hardness testing with the Schmidt hammer (Figure 8.18) should have a flat, smooth surface where tested and the

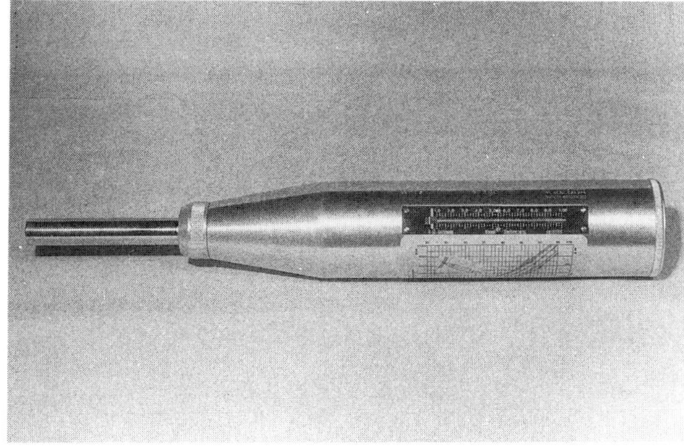

Figure 8.18 *The Schmidt hammer*

rock material beneath this area should be free from cracks (Atkinson *et al.* 1978). Test locations should be separated by at least the diameter of the plunger. The ISRM suggests that a type L hammer with an impact energy of 0.74 N m should be used. Atkinson *et al.* suggested that the lower 50% of the test values should be discarded and the average obtained from the upper 50%. This average is then multiplied by the correction factor to obtain the Schmidt hammer hardness:

$$\text{correction factor} = \frac{\text{specified standard value of the anvil}}{\text{average of 10 readings on the calibration anvil}} \quad (8.32)$$

The Schmidt hammer is not a satisfactory method for the determination of very soft or very hard rocks. However, Schmidt hardness shows a good correlation with compressive strength which allowed Deere and Miller (1966) to design another rock strength chart (Figure 8.19; see also Carter and Sneddon 1977).

Abrasion tests measure the resistance of rocks to wear. Two abrasion tests have been used in the United States to measure hardness–the Dorry and Los Angeles tests. The former test is carried out on a cylindrical rock sample, 25 mm in both length and diameter, which is held against a

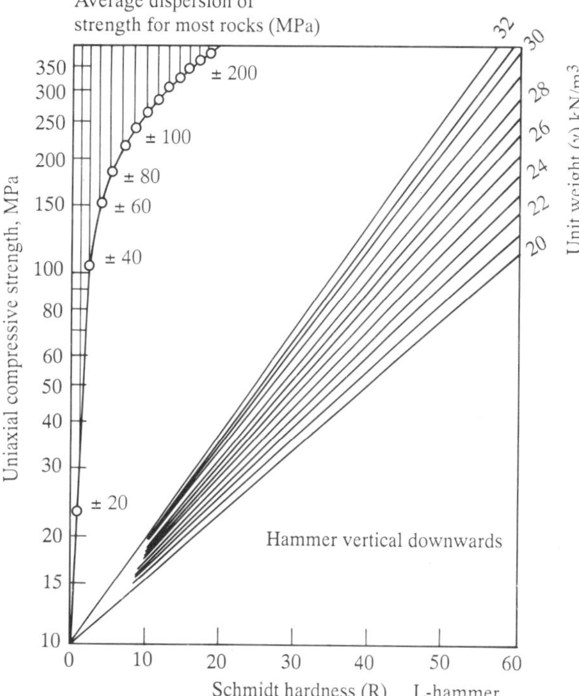

Figure 8.19 *Correlation chart for Schmidt (L) hammer, relating unit weight of rock, compressive strength and rebound number (after Deere and Miller 1966)*

revolving disc and is under a pressure of 2.5 MPa (the total load equals 1.25 kg). Standard crushed quartz, sized between 30 and 40 ASTM mesh screens, is fed onto the revolving disc (ASTM 1951). The loss of weight, obtained by subjecting both ends to a total of 1000 revolutions, gives the hardness index (this is similar to the Dorry aggregate abrasion test (British Standards Institution 1967). The Los Angeles abrasion test subjects a graded sample to wear due to collision between rock pieces and also to impact forces produced by an abrasive charge of steel spheres (ASTM 1969a, b). In the Los Angeles test the rock aggregate and the abrasive charge are placed in the machine and rotated at 30–33 rev/min. If the aggregate particles are smaller than 38 mm they are subjected to 500 revolutions and to 1000 revolutions if they are larger then 19 mm. After the test the sample is shaken through a No. 12 US sieve (approximately 1.7 mm aperture). The amount of wear is the loss in weight expressed as a percentage of the original weight.

Attrition has been defined as the resistance of one surface to the motion of another surface rubbing over it. The apparatus in the Deval test consists of a hollow cylindrical bucket, 340 mm long by 200 mm diameter, which is mounted in a frame which supports it at an angle of 30° to a horizontal axis of rotation (ASTM 1968). The test sample comprises some 50 pieces of rock weighing about 5 kg. After revolving the cylinder 10 000 times at 30 rev/min the quantity of material finer than 0.06 mm is weighed and expressed as a percentage of the original weight, which gives a hardness value. This test is not often used today.

Toughness reflects the ability of a material to absorb energy during plastic deformation. In the impact toughness test a cylindrical sample of rock, 25 mm long by 25 mm diameter, is placed in the apparatus. A weight of 2 kg falls vertically between two guides, upon a spherical ended plunger weighing 1 kg, which rests in contact with the specimen. The height of the first blow is 10 mm and each successive blow is increased in height by that amount. The height of the blow which causes failure represents the toughness of the material. A minimum of six specimens of the same rock should be tested. If the rock material is laminated, cleaved or schistose, then three specimens should be prepared parallel to and three normal to these structural weaknesses. The average toughness in each of these directions is recorded.

8.7 Elastic properties

Young's modulus, E, is the ratio of vertical stress to strain, and Poisson's ratio, ν is the ratio between lateral and axial strain, and these are two of the most important elastic properties. Both Young's modulus and Poisson's ratio can be obtained by monitoring the vertical and lateral strain in rock specimens tested in uniaxial compression. The strains are measured by attaching strain gauges or other displacement transducers to the test specimens and recording their outputs. Dhir and Sangha (1973) investigated the relationship between size, deformation and strength of cylindrical specimens loaded in uniaxial compression. They found that for a slenderness ratio of 2.5 the modulus determined from strain measurement over the central zone is 15% greater than that obtained from overall deflection measurements. A specimen diameter of 50 mm generally represents the transition in the size–strength relationship of a fine-grained material between the predominance of surface and internal flaws. Strain measurements on specimens less than this diameter are high and not representative of material behaviour.

According to Protodyakonov (1963) an approximate value of Young's modulus can be determined indirectly by using the Shore scleroscope hardness value (H_s) in the equation

$$E = 1.07 \times \frac{H_s}{(154 - h)} \times 10^6 \quad (\mathrm{kg/cm^2}) \tag{8.33}$$

Another elastic constant is compressibility, which is the ratio of change in volume of an elastic solid to change in hydrostatic pressure. A fourth elastic constant is rigidity, which refers to the resistance of a body to shear. The modulus of rigidity is the ratio of shear stress to shear strain, in a simple shear. The four elastic constants, Young's modulus. (E), Poisson's ratio (ν), compressibility (K) and rigidity (G) are not independent of each other and if any two are known it is possible to derive the other two from the expressions

$$G = \frac{E}{2(1 + \nu)} \tag{8.34}$$

and

$$K = \frac{E}{3(1 - 2\nu)} \tag{8.35}$$

Young's modulus and Poisson's ratio are more readily determined experimentally.

The values of ultimate compressive (σ_c) and yield strength (σ_{ys}), together with those of Young's modulus, allow the determination of the modulus of resilience (M_r), which refers to the capacity of a material to absorb energy within the elastic range. It is equal to the area under the elastic portion of the stress–strain curve and has been defined as the strain energy absorbed per unit volume when the material is stressed to its elastic limit. Thus the yield strength and elasticity are used to derive the modulus of resilience as follows:

$$M_r = \sigma_{ys}^2/2E_{t50} \tag{8.36}$$

The term resilience should not be confused with the modulus of resilience. Within the elastic limit the resilience is equal to the external work put into the material during deformation. Hence the total resilience of a material is the product of its volume and the modulus of resilience.

The modulus of toughness (M_t) represents the maximum amount of energy a unit volume of rock can absorb without fracture and it can be estimated as follows:

$$M_t = \tfrac{2}{3}\sigma_c\,\varepsilon_f \qquad (8.37)$$

where ε_f is the strain at failure. Toughness consequently reflects the ability of a rock to absorb energy during plastic deformation. In a static test this energy is measured by the area under the stress–strain curve which represents the work required to fracture the test specimen. Rocks with high toughness have high strength and ductility, whilst brittle materials usually have a low toughness since they show only small plastic deformation before fracture.

The constrained modulus of deformation (M_c) can be defined as the rate of change of vertical stress with respect to vertical strain under conditions of zero lateral strain. It is related to Young's modulus and Poisson's ratio by the expression

$$M_c = E\left[\frac{(1-\nu)}{(1+\nu)\,(1-2\nu)}\right] \qquad (8.38)$$

It can also be derived from rigidity (G) and compressibility (K) as follows:

$$M_c = K + \tfrac{4}{3}G \qquad (8.37)$$

A number of techniques have been used to determine the dynamic values of Young's modulus and Poisson's ratio, for example, Obert *et al* (1946) outlined a method of testing cylindrical specimens. These were supported at their centres and vibrated, to obtain the longitudinal frequency (f_1) and the torsional frequency (f_t). With these frequencies and the length (L) of the specimen, the longitudinal and torsional velocities of sound (v_1 and v_t respectively) can be obtained from

$$\nu_1 = 2f_1L \qquad (8.40)$$

$$\nu_t = 2f_tL \qquad (8.41)$$

Young's modulus (E), Poisson's ratio (v) and the modulus of rigidity (G) can be obtained from the longitudinal and torsional velocities and the density (ρ) of the specimen

$$E = \nu_2^1\rho \qquad (8.42)$$

$$G = \nu_t^2\rho \qquad (8.43)$$

$$\nu = \frac{E}{2G} - 1 \qquad (8.44)$$

or

$$\nu = \frac{f_1^2}{2f_t^2} - 1 \qquad (8.45)$$

Hosking (1955) obtained the dynamic values of Young's modulus and Poisson's ratio by determining the velocities of propagation (v_p) in rock of an ultrasonic pulse and the sound at resonance (v_r) where L is the length of the specimen. With v_p and v_r it is possible to find Young's modulus and Poisson's ratio from the following expressions:

$$E = \frac{\nu_r^2\rho}{12g} \qquad (8.46)$$

and

$$\frac{\nu_p}{\nu_r} = \frac{(1-\nu)}{(1+\nu)\,(1-2\nu)} \qquad (8.47)$$

where ρ is the density and g is the acceleration due to gravity.

Deere and Miller (1966) used a similar method to derive the velocity of dilatational waves in rock samples while subjected to axial stress, the rock specimens being subjected to two complete loading and unloading cycles up to 34.5 MPa. They found that generally the denser rocks had higher dilatational velocities and that as axial pressures were increased the propagation velocities also increased, on average by some 10%. Deere and Miller also noticed that the velocities measured during unloading were usually higher than those measured during loading. The relationship between dilatational wave velocity (v_s), density, Young's modulus and Poisson's ratio is given by the expression

$$\nu_s = \rho\,\frac{E(1-\nu)}{(1+\nu)\,(1-2\nu)} \qquad (8.48)$$

This expression assumes that the static and dynamic properties are interchangeable. Indeed the static and dynamic values of Young's modulus obtained by these two authors were similar.

More recently the ISRM (Rummel and van Heerden 1978) have outlined laboratory methods of determining the velocity of propagation of elastic waves through rocks. Three methods were suggested, namely, the high frequency ultrasonic pulse technique, the low-frequency ultrasonic pulse technique and the resonant method. The high-frequency ultrasonic pulse method is used to determine the velocities of compressional and shear waves in rock specimens of effectively infinite extent compared to the wavelength of the pulse used. The condition of infinite extent is satisfied if the average grain size is less than wavelength of the pulse, which in turn is less than the the minimum dimensions of the specimen. The low-frequency ultrasonic pulse method is used to determine the velocity of dilatational and torsional waves in cylindrical or bar-shaped specimens of rock. The length-to-diameter ratio of specimens should be greater than 3 and the ratio of the wavelength of the pulse to the diameter should not be less than 5. By determination of the resonance frequency of both dilatational and torsional vibrations of cylindrical rock specimens (with a length-to-diameter ratio exceeding 3 and a wavelength-to-diameter ratio exceeding 6) the velocity of dilatational and torsional waves can be calculated.

References

Anon. (1970) 'Working party report on the logging of cores for engineering purposes', *Q.J. Engg Geol.*, **3**, 1–24

Anon. (1979) 'Classification of soils and rocks for engineering geological mapping. Part I Rock and soil materials', *Bull. Int. Ass. Engg Geol.*, **19**, 364–71

Anon. (1981) 'Basic geotechnical description of rock masses', International Society for Rock Mechanics Commission on the Classification of Rocks and Rock Masses, *Int. J. Rock Mech. Min. Sci. & Geomech Abstr.*, **18**, 85–110

ASTM, (1951) *Abrasion of Rock by use of the Dorry Machine*, C-241. American Society for Testing and Materials Philadelphia, Pennsylvania

ASTM, (1967) *Standard Method of Test for Triaxial Compressive Strength of Undrained Rock Core Specimens without Pore Pressure Measurement*, ASTM Designation D, 2664-67. American Society for Testing and Materials Philadelphia, Pennsylvania

ASTM, (1968) *Abrasion of Rock by use of the Deval Machine*, D-233. American Society for Testing and Materials Philadelphia, Pennsylvania

ASTM, (1969a) *Resistance to Abrasion of Small Size Coarse Aggregate by use of Los Angeles Machine*, C-131. American Society for Testing and Materials Philadelphia, Pennsylvania

ASTM, (1969b) *Resistance to Abrasion of Large Size Coarse Aggregate by use of Los Angeles Machine*, C-535. American Society for Testing and Materials Philadelphia, Pennsylvania

ASTM, (1982) *Standard Test Methods for Absorption and Bulk Specific Gravity of Natural Building Stone*, C97–117. American Society for Testing and Materials Philadelphia, Pennsylvania

Atkinson, R.H., Bamford, W.H., Broch, E., Deere, D.U., Franklin, J.A., Nieble, C., Rummel, F., Tarkoy, P.S. and Van Duyse, H. (1978) 'Suggested methods for determining hardness and abrasiveness of rocks', ISRM Commission Standardization of Laboratory and Field Tests. *Int. J. Rock Mech. Min. Sci. & Geomech. Abstr.*, **15**, 91–97

Bernaix, J. (1969) 'New laboratory methods of studying the mechanical properties of rocks', *Int. J. Rock Mech. Min. Sci.*, **6**, 43–90

Bieniawski, Z.T. (1975) 'The point load test in geotechnical practice', *Engg Geol.*, **9**, 1–11

Bieniawski, Z.T. and Hawkes, I. (1978) 'Suggested methods for determining tensile strength of rock materials', ISRM Commission on Standardization of Laboratory and Field Tests. *Int. J. Rock Mech. Min. Sci. & Geomech. Abstr.*, **15**, 101–103

British Standards Institution (1967) *Methods for Sampling and Testing of Mineral Aggregates, Sands and Fillers*, BS812, British Standards Institution, London

British Standards Institution (1975) *Methods of Tests for Soils for Civil Engineering Purposes*, BS1377, British Standards Institution, London

Broch, E. and Franklin, J.A. (1972) 'The point load strength test', *Int. J. Rock Mech. Min. Sci.*, **9**, 669–697

Brown, E.T. (ed.) (1981) *Rock Characterisation, Testing and Monitoring*, Pergamon, Oxford

Brook, W.F. (1977) 'A method of overcoming both shape and size effects in point load testing', *Proc. Conf. Rock Engg.*, Newcastle University, **1**, 53–70

Carter, P.G. and Sneddon, M. (1977) 'Comparison of Schmidt hammer, point load and unconfined compression tests in Carboniferous strata', *Proc. Conf. Rock Engg*, Newcastle University, **1**, 197–210

Das, B. (1974) 'Vickers hardness concept in the light of Vickers impression', *Int. J. Rock Mech. Min. Sci. & Geomech. Abstr.*, **11**, 85–89

Deere, D.U. and Miller, R.P. (1966) 'Engineering classification and index properties for intact rock', *Tech. Rep. No. AFWL-TR-65-115*, Air Force Weapons Lab., Kirtland Air Base, New Mexico

Dhir, R.K. and Sangha, C.M. (1973) 'Relationships between size, deformation and strength for cylindrical specimens loaded in uniaxial compression', *Int. J. Rock. Min. Sci. & Geomech. Abstr.*, **10**, 699–712

Everling, G. (1964) 'Comments on the definition of shear strength', *Int. J. Rock Mech. Min. Sci.*, **1**, 145–154

Fairhurst, C. (1964) 'On the validity of the Brazilian test for brittle materials', *Int J. Rock Mech. Min. Sci.*, **1**, 535–546

Forster, I.R. (1983) 'The influence of core sample geometry on the axial point load test', *Int. J. Rock Mech. Min. Sci. & Geomech. Abstr.*, **20**, 291–295

Franklin, J.A. and Chandra, R. (1972) 'The slake durability test', *Int. J. Rock Mech. Min Sci.*, **9**, 325–341

French Standard (1973) *Produits de Carrières–Pierres Calcaires Mesure de l'Absorption d'Eau par Capillarité*, NF10–502, Association Française de Normalisation, Paris

Hassani, F.P., Scoble, M.J. and Whittacker, B.N. (1980) 'Application of the point load index test to strength determination of rock and proposals for a new size correction chart', *Proc. 21st US Symp. Rock Mechanics*, Rolla, 543–553

Hawkes, I. and Mellor, M. (1970) 'Uniaxial testing in rock mechanics laboratories', *Engg Geol.*, **4**, 177–284

Hobbs, D. W. (1964) 'A simple method for assessing the uniaxial compressive strength of rocks', *Int. J. Rock Mech. Min. Sci.*, **1**, 5–15

Hoek, E. and Franklin, J.A. (1968) Simple triaxial cell for field or laboratory testing of rock', *Trans. Inst. Min. Metall.*, **77**, Section A, A22–A26

Hosking, J.R. (1955) 'A comparison of tensile strength, crushing strength and elastic properties of roadmaking rocks', *Quarry Man. J.*, **39**, 200–212

Hucka, V.A. (1965) 'A rapid method for determining the strength of rock *in situ*', *Int. J. Rock Mech. Min. Sci.*, **2**, 127–134

Mamillan, M. (1976) Nouvelles connaissances pour l'utilisation et la protection des pierres de construction. *Annales de l'Instit. Technique du Bâtiment et des Travaux Publics*, Serie Matériaux, No. 48, Supplement No. 335, 18–48

Mellor, M. and Hawkes, I. (1971) 'Measurement of tensile

strength by diametral compression of discs and annuli', *Engg. Geol.*, **5**, 173–225

Morgenstern, N.R. and Eigenbrod, K.D. (1974) 'Classification of argillaceous soils and rocks', *Proc. ASCE. J. Geot. Engg. Div.*, GT10 (**100**), 1137–1156

Murrell, S.A.F. (1963) 'A criterion for the brittle fracture of rocks and concrete under triaxial stress and the effect of pore pressure on the criterion', *Proc. 5th Symp. Rock Mech.*, University of Minnesota, Pergamon Press, New York, 563–577

Norbury, D.R. (1986) 'The point load test'. In *Site Investigation Practice: Assessing BS 5930*, Engineering Geology Special Publication, No. 2, A.B. Hawkins (ed.), Geological Society, London, pp. 326–329

Obert, L and Duvall, W.I. (1967) *Rock Mechanics and the Design of Structures*, Wiley, New York

Obert, L., Windes, S.L. and Duvall, W.I. (1946) 'Standardized tests for determining the physical properties of mine rock', *US Bur. Mines Rep. Invest.*, 3891

Protodyakonov, M.M. (1963) 'Mechanical properties and drillability of rock', *Proc. 5th Symp. Rock Mech.*, University of Minnesota, Pergamon Press, New York, 103–118

Rabia, H. and Brook, N. (1979) 'The Shore hardness of rock', *Int. J. Rock Mech. Min. Sci. & Geomech. Abstr.*, **16**, 335–336

Ramana, Y.V. and Venkatanaryana, B. (1971) 'An air porosimeter for the porosity of rocks', *Int. J. Rock Mech. Min. Sci.*, **8**, 29–53

Reichmuth, D.R. (1968) 'Point load testing of brittle materials to determine tensile strength and relative brittleness', *Proc. 9th US Symp. Rock Mechanics*, 134–159

Rummel, F. and Van Heerden, W.L. (1978) 'Suggested methods for determining sound velocity', ISRM Commission on Standardization of Laboratory and Field Tests, *Int. J. Rock Mech. Min. Sci. & Geomech. Abstr.*, **15**, 55–58

Turk, N and Dearman, W.R. (1985) 'Improvements in the determination of point load strength', *Bull. Int. Ass. Engg. Geol.*, **31**, 137–142

Vogler, U.V. and Kovari, K. (1978) 'Suggested methods for determining the strength of rock materials in triaxial compression', ISRM Commission on Standardization of Laboratory and Field Tests, *Int. J. Rock Mech. Min. Sci. & Geomech. Abstr.*, **15**, 47–51

Winchell, H. (1946) 'Observations on orientation and hardness variations', *Am. Mineral,* **31**, 149–152

Winkler, E.M. (1973) *Stone: Properties, Durability in Man's Environment*, Springer-Verlag, New York.

9 Rock-mass assessment using geophysical methods

Dr D McCann
British Geological Survey

9.1 Geomechanical properties

In construction projects, such as for dams, tunnels, and roads, it is rarely possible to assume that the rock mass is isotropic and homogeneous, since there are usually one or more natural joint patterns resulting from the tectonic stress history with preferential directions and orientations governed by the direction of the regional stresses. Within this overall joint pattern will be zones of highly fractured rock associated with faulting and/or high-intensity fracturing. Further weakening of the integral strength of the rock mass results from the effects of weathering from the ground surface on the fracture and joint surfaces, since the surface roughness controls the shear strength along these surfaces.

As Goodman (1976) stated: 'A rock mass is comprised of the rock, its network of discontinuities and its weathering profile. The behaviour of such a mass reflects all of these components as well as water, and stress regimes, strength, deformability and permeability, which may be unrelated to its material properties'. In addition to the major discontinuities mentioned above the rock mass may also be intersected by dykes and sills and it is necessary to determine the geometry and overall effects of these features, in terms of an engineering characterization of the rock mass.

Bieniawski (1978) reviewed *in situ* methods for determing rock-mass deformability and recommended the following guidelines:

(1) A detailed engineering geological assessment of the rock-mass conditions is essential and this should be expressed in quantitative terms by an engineering classification of the rock masses encountered.
(2) A minimum of two different types of *in situ* tests should be carried out in a sufficient number to determine *in situ* rock deformability in the representative structural regions of the rock mass. The plate-bearing test and the Goodman jack test are particularly recommended.
(3) The stress field should be established in the test areas by either the overcoring technique or a study of the drillhole breakout pattern from a caliper log.
(4) Since *in-situ* tests are performed at only a few localities, seismic surveys should be conducted to determine the continuity of the rock-mass condition throughout the engineering project area.
(5) Diamond drilling of good quality core must be undertaken at all *in situ* test sites to determine rock quality designation (RQD) and obtain samples for the determination of the static elastic moduli and seismic properties of intact rock specimens. The drillholes can also be used for a wide range of geophysical logging methods and for cross-hole geophysical measurements.

Several systems have been developed in an effort to classify the geomechanical properties of a rock mass in such a way that consistent results may be obtained regardless of the observer. The most significant systems to date are those due to Terzaghi (1946), Deere (1964), Wickham *et al.* (1972), Bieniawski (1973) and Barton *et al.* (1974), and are comprehensively summarized in Hoek and Brown (1981). An additional descriptive system has been developed by a Geological Society Working Party (Anon 1977).

Bieniawski (1973) proposed a 'geomechanical classification' incorporating the following parameters:

(1) rock quality designation (RQD);
(2) state of weathering;
(3) uniaxial compressive strength of intact rock;
(4) spacing of joints and bedding;
(5) strike and dip orientation;
(6) separation of joints;
(7) continuity of joints;
(8) groundwater inflow.

The eight parameters listed were incorporated with the original CSIR geomechanics classification (Bieniawski 1973). This classification was later modified (Bieniawski 1974, 1976) by eliminating the state of weathering as a separate parameter and by including the separation and continuity of the joints in a new parameter, the condition of the joints. The five basic classification parameters then become

(1) rock quality designation;
(2) uniaxial compressive strength of the intact rock;
(3) spacing of the joints;
(4) condition of the joints;
(5) groundwater conditions.

These parameters are interlinked, to a considerable extent, since each one contributes to the overall performance of the rock mass. In a brittle rock mass the presence of a fracture network results in

(1) the rock mass being deformable, weaker, and more permeable, depending on the spacing of the fractures, their openness, degree of interconnection etc.;
(2) depending on the inter-relationship between the individual fracture sets, the rock mass being rendered anisotropic in all properties;
(3) all properties of the rock mass becoming highly stress dependent.

The qualification of the nature of the jointing in the rock mass is a key factor in providing an overall assessment of rock-mass conditions. The classification systems mentioned above attempt to do this but require considerable site-investigation data in the form of detailed logs of the nature of jointing, assessment of strength and groundwater condition. This information is time-consuming and expensive to obtain and geophysical methods have been used to give rapid indication of the nature of the rock *en masse* on some civil engineering projects. While it is unlikely that the parameters derived from mechanical testing and observation will be replaced by geophysical parameters, nevertheless an indication of their likely variation in the rock mass can be obtained.

The inter-relationship between fracturing and weathering is extremely important in engineering terms. In a fractured rock mass for instance, the effects of surface weathering can be extended to much greater depths so that the surface of fractures often exhibit decay as a result of weathering. Zones of fractured rock related to joint systems or faults are associated with high permeability conditions in the rock mass and are of considerable significance in crystalline rocks where low permeability conditions prevail. Superimposed on the joint pattern are induced active fractures and faults. Open joints, fractures etc. are normally assumed to be more or less vertical, since it is considered that horizontal fractures will be closed up by the overburden pressure. Horizontal and low-angle oblique fractures are observed in cores and may well provide a path *in situ* to a large vertical fracture network, so their presence cannot be ignored entirely. Where it is important to identify high angle joints, drillholes will be deviated from the vertical and intersect the joint pattern such that both vertical and horizontal fractures or joints will appear in an oblique direction to geophysical logging tools.

9.2 Geophysical properties

From the geological, geomechanical and hydrogeological information obtained from drillholes and by direct observation it is possible to classify a rock mass on the basis of the CSIR classification system (see Chapter 1). However, it is the *in situ* measurements which ultimately provide the most useful information for the design of an underground construction in a particular environment. Direct measurements are made at a few localities only but geophysical surveys can be carried out to assess the continuity of the rock-mass conditions throughout the site of the proposed engineering project. As Hoek and Brown (1981) pointed out, geophysical methods can be used to provide an initial overall assessment of the site, which can assist in the optimization of the site-investigation programme. An excellent review of geophysical methods and their applicability to underground excavation is given by Mossman and Heim (1972). Of the main geophysical methods in use, gravity and magnetic measurements give very little indication of the structural characteristics of the rock mass, but find a limited application in cavity and mineshaft location. Seismic, electrical resistivity, and electromagnetic methods can provide considerable information on the rock mass both in a regional context and the more localized area of a drillhole.

Holes drilled on the engineering site provide access to the environment of interest and geophysical logging methods can be used to give information on the *in situ* conditions which prevail in each drillhole. Most geophysical well-logging methods respond in some way to the presence of a discontinuity in the rock mass. The response of the tool is governed by its resolution, the width and angle of the fracture or joint and the infilling material within the fracture. The effects of a weathered and fractured rock mass on various geophysical properties is discussed in some detail by Sauu and Gartner (1979), by Cratchley (1977), and by McEwen *et al.* (1985); and a general treatment of relevant interpretation procedures is given by Scott Keyes and MacGary (1971). Provided the drillholes are close enough they can be used for cross-hole seismic, electromagnetic and electrical resistivity surveys. Individual drillholes can be used also for surface to drillhole measurements, which can also provide valuable information for the engineering characterization of the rock mass.

The information available on a rock mass can be increased by the use of geophysical techniques, which can provide an indirect assessment of its engineering properties. Of particular importance is the measurement of the seismic properties of the rock mass since the dynamic elastic moduli can be derived from the compressional and shear-wave velocities. While the indirect determination of engineering properties is of particular value, it is the direct assessment of the rock-mass condition, such as the degree of fracturing, where the seismic parameters make their greatest contribution. In this case the seismic and engineering parameters can be viewed as complementary and of equal importance to the overall assessment of rock mass performance.

9.3 Surface geophysical methods

9.3.1 Seismic methods

Seismic surveys provide two types of information on the rock mass at an engineering site as follows:

(1) Seismic refraction and reflection surveys may be carried out to investigate the continuity of geological strata over the site and the location of major discontinuities, such as fault zones.
(2) From measurements of the compressional and shear-wave velocities it is possible to determine the dynamic elastic moduli of the rock mass and estimate its degree of fracturing.

Seismic methods are based on the generation of seismic waves on the ground surface and the measurement of the time taken by the waves to travel from the source through the rock mass to a series of geophones laid out in a straight line from the source. The seismic energy can be generated by a number of sources including a sledge hammer and plate, falling weight, detonator, and explosives, and the resultant ground motion at the surface is detected by geophones.

It will be seen in Figure 9.1 that apart from seismic energy travelling directly through the rock mass to the geophone array two other main paths are possible as follows:

(1) The refracted or head wave, which travels immediately beneath the interface between the two rock types.
(2) The reflected wave from the interface between the two rock types.

Both the seismic refraction and reflection methods are considered in some detail in Telford *et al.* (1976), Dobrin (1976) and Parasnis (1986) and the reader is referred to these books for a comprehensive treatment of the relevant interpretation methods.

The seismic refraction method is widely used in the civil engineering field at the site investigation stage. A typical example of its use in the study of the continuity of geological structure is given in Grainger *et al.* (1973), who described a seismic refraction survey carried out on the Middle Chalk at Mundford, Norfolk, UK. They showed that the compressional wave velocity increased in well-defined steps, which broadly correlated with the rock-mass classification system adopted by Ward *et al.* (1968), which classified the rock mass by observations in drillholes into five engineering grades. The relationship was established in an area of lithologically uniform chalk on seismic lines, which had good drillhole control. It was possible to extend seismic lines into areas of poor drillhole control, both to classify the rock mass in engineering terms from the

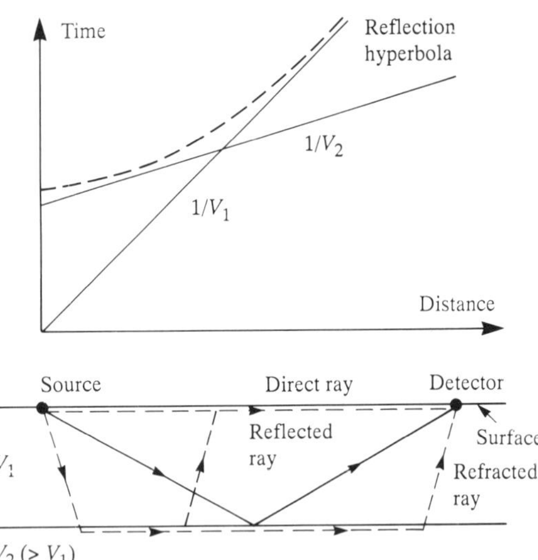

Figure 9.1 *Two-layer case with interface parallel to surface showing refracted, reflected, and direct ray paths for a surface seismic source*

seismic classification and to study the continuity of the seismic section. Grainger *et al.* (1973) did emphasize that the seismic classification was applicable only in the particular lithological section in which it was derived and should not be extrapolated outside the immediate calibrated area.

Cratchley *et al.* (1976) carried out a detailed seismic refraction survey along the line of a tunnel constructed as part of the Foyers pumped storage hydro-electric scheme near Inverness, Scotland. Their results clearly identified a highly weathered fault zone, which intersected the tunnel line on the Loch Mohr heading (Figure 9.2). Measurements in the tunnel itself confirmed the presence of a low-velocity zone where the rock was generally shattered and faulted, which resulted in the use of steel ribbing during the tunnel construction (Figure 9.3).

Mossman and Heim (1972) used seismic methods to provide information on the geological conditions in the Chicago area prior to the excavation of a series of tunnels. The prime concern of the survey was the recognition of any faulting resulting in the vertical displacement of the rocks at depth and the identification of heavily fractured water-bearing zones associated with the faulting. Depth to bedrock, which was usually less than 50 m, was determined from a seismic refraction survey, in which closely spaced survey lines were used to map the irregular erosional surface. However, tunnels also were planned in a deeper dolostone layer at a depth of approximately 180 m; in this case seismic refraction surveys would not give the desired information because of the long length of survey lines required and the large offsets between the seismic source and the geophone array. Hence, a seismic reflection survey was carried out and the top surface of the dolostone layer was mapped over the area of interest to a precision

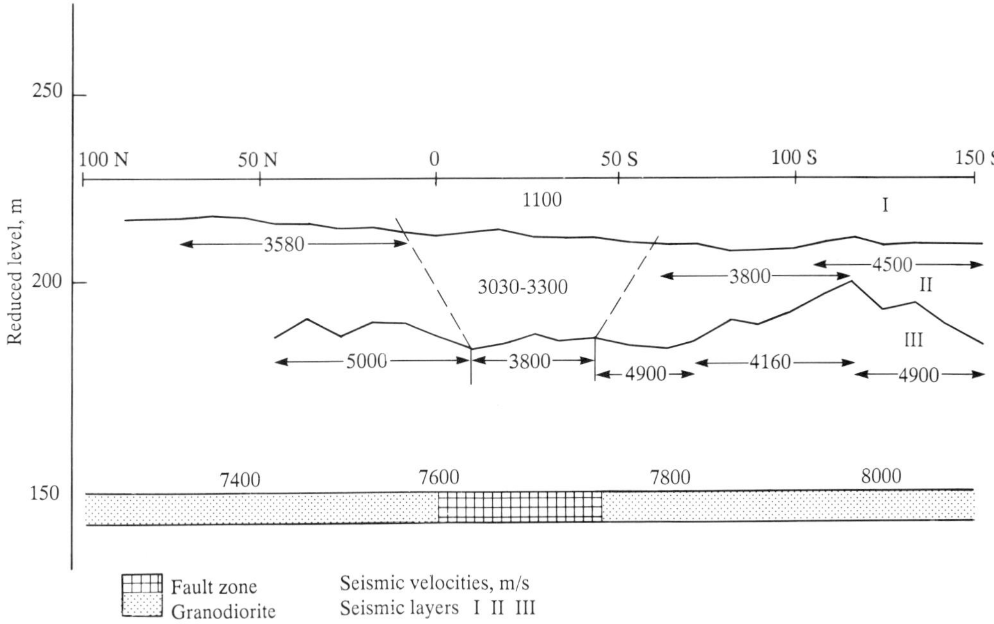

Figure 9.2 *Foyers hydroelectric scheme: seismic refraction interpretation, line 1, over tunnel, Loch Mhor (scale, vetical and horizontal, 1:1000) (after Cratchley* et al. *1976)*

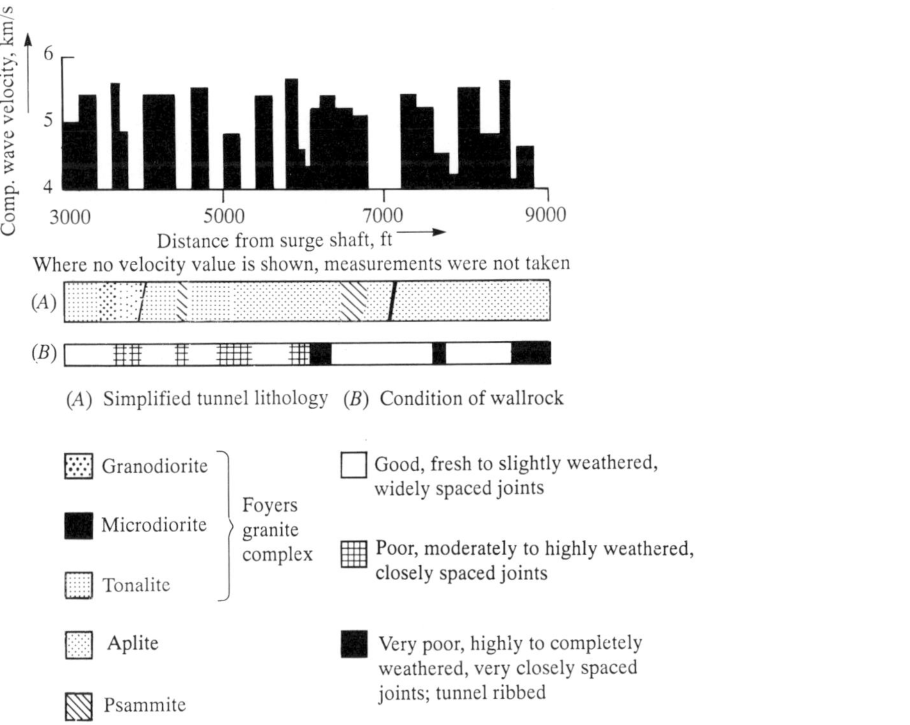

Figure 9.3 *Velocity profiles, Glen Lia – Loch Mhor tunnel (after Cratchley* et al. *1976)*

better than ±6 m. The seismic survey also identified 30 faults, some with vertical displacements as low as ±4 m and most were unknown at the surface. The results of the seismic survey were confirmed for the most part by the subsequent drilling and excavation programme and clearly demonstrated the cost benefits to be derived from a properly organized seismic programme.

In dam site areas, in particular, grouting often is carried out to increase the bearing capacity or reduce the permeability of the foundation rock in the reservoir area, where it is highly fractured. Geophysical surveys may be carried out to locate these zones of fractured rock and these can be treated with a selective grouting programme. Knill (1970) carried out seismic surveys in the rock mass adjacent to 69 concrete dams in the UK to determine the *in situ* compressional wave velocity. A relationship was established between the curtain grout take and velocity using the fracture index as follows:

$$\mathrm{FI} = \frac{V_{\mathrm{FIELD}}}{V_{\mathrm{LAB}}} \tag{9.1}$$

where V_{FIELD} is the compressional wave velocity *in situ*, and V_{LAB} is the compressional wave velocity measured on an intact saturated rock specimen in the laboratory. Knill (1970) showed that seismic measurements can be used to predict the grout take and to assess the depth of foundation deformability and the excavation method used.

Scalabrini *et al.* (1964) also used a sonic method to determine the *in situ* state of the foundation rock at the Frera Dam both before and after grouting of the rock mass took place. The dynamic elastic moduli were measured in the abutment rock at 29 points and the information was used to develop and carry out a comprehensive grouting programme. Following the completion of this programme, the dynamic elastic moduli were determined again at 20 points in the grouted rock mass and it was shown that the measured values were very close to those determined in deep sound rock.

Deere *et al.* (1967) proposed the use of the RQD and the seismic velocity index as an estimate of rock-mass deformability. The seismic velocity index is defined as

$$\mathrm{SVI} = \frac{V^2_{\mathrm{FIELD}}}{V^2_{\mathrm{LAB}}} \tag{9.2}$$

where V_{FIELD} and V_{LAB}, are defined in Equation (9.1).

Coon and Merritt (1970) demonstrated that neither index on its own is sufficient to describe fully the overall condition of the rock mass. Since the above ratio is squared it is equivalent to the ratio of the dynamic elastic modulus (E) but Coon and Merritt showed that when the ratio was used for the prediction of the *in situ* modulus of deformability its resultant value was as much as three times greater than the corresponding laboratory value on an intact rock.

9.3.2 Electrical resistivity

Electrical soundings are suited to horizontal stratified media, since the spatial distribution of the electrical current in the ground and, hence the depth of investigation depends on the configuration and the spacing of the electrodes. When using a standard Wenner or Schlumberger array (Figure 9.4) the depth of investigation increases with the current electrode spacing and this gives rise to a pseudo-electric section which can be related to the geological structure beneath the survey line. It is possible

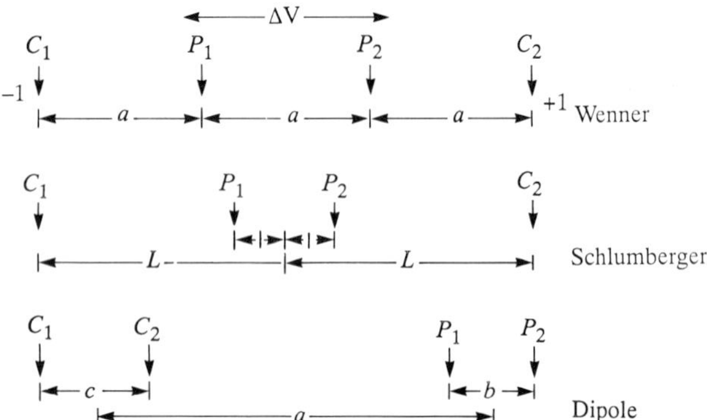

Figure 9.4 *Commonly used electrode configurations (the electrodes are placed in line at the surface of a half-space. A current* I *passes into the ground through* C_1 *and* C_2 *and a potential difference* Δv *is measured between* P_1 *and* P_2*)*

to postulate a geological model from which an electrical section can be computed to fit the experimental data.

An alternative approach is to move the electrodes over the survey area to delineate lateral variations using a constant electrode array. In this way, it is posible to locate discontinuities, such as geological boundaries, dykes infilled sinkholes, or major drainage zones.

Electrical resistivity is a function of the total porosity of the rock mass including fractures, the conductivity of the fluid within the rock mass and the degree of saturation (Mooney 1980). The rock type itself plays a relatively minor role in this respect since it is extremely difficult to resolve layers with resistivities of 10^3 ohm metres from those of 10^4 ohm metres since all rocks with high resistivities tend to look like insulators. This in itself is an extremely useful characteristic since in engineering studies a low-resistivity zone in a rock mass with a relatively constant resistivity may well be associated with extensive fracturing or deep weathering and, thus constitute an area of weakness; this characteristic is exploited in most of the case histories considered below.

Cratchley *et al.* (1976) also reported the use of an electrical resistivity survey at the Foyers scheme. A series of expanding depth probes was carried out along the same lines used on the seismic refraction survey, and again the heavily fractured zone was clearly identified (Figure 9.5).

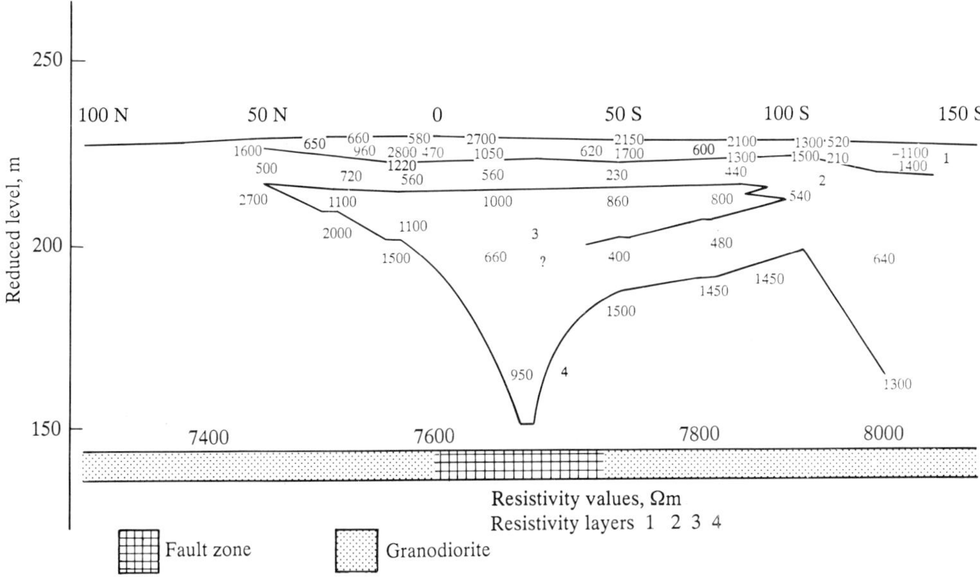

Figure 9.5 *Foyers hydroelectric scheme: resistivity interpretation, line 1, over tunnel, Loch Mhor (scale, vertical and horizontal, 1:1000) (after Cratchley* et al. *1976)*

Scott *et al.* (1968) made seismic refraction and electrical resistivity measurement along the walls of the Straight Creek Tunnel Pilot Base, Colorado, USA, and described the presence of both a low velocity and a high resistivity layer in the disturbed rock mass surrounding the excavation. They showed a consistent correlation between increasing fracture spacing and increasing resistivity of the rock mass at depth. A similar, but less consistent, relationship was observed in the near-surface rock mass but the effects of the blast damage appeared to be more variable in this area. A statistical study of the electrical resistivity measurements indicated good correlation between the amount of steel support used in construction and the magnitude of the resistivity of the rock mass along the walls of the pilot bore. At high resistivity values (>1000 ohm metres) no support was required, while at low resistivity values (<62.5 ohm metres) maximum support was required.

Kennard and Lea (1974) described the use of electrical resistivity traversing to profile the depth of weathering in the granite at the Sheepstor embankment on Dartmoor, Devon, UK. The method was used to detect leakage paths in the rock mass associated with kaolinization in the granite, which may result in associated quartz veins becoming open jointed.

Schwarz (1972) described the successful use of the electrical resistivity method in the Lake Huron Tunnel to determine the thickness of rock above the crown of the tunnel and to identify anomalous geological conditions ahead of the working face. The tunnel was bored in the Antrim Shale, a dense marine shale of Lower Carboniferous or Devonian age, which is characterized by anisotropic electrical properties such that

$$AR = \frac{R_1}{R_2} \tag{9.3}$$

where AR is the anisotropy ratio, R_1, is the resistivity parallel to the bedding and R_2 is the resistivity perpendicular to the bedding. The results showed that the presence of anomalous features in the rock mass, such as fractured rock and major joints, altered the average electrical properties of the shale such that the anisotropy ratio (AR) was reduced.

9.3.3 Magnetic method

In a magnetic survey variations in the Earth's magnetic field are measured. It has the advantage of being one of the easiest and cheapest geophysical techniques to carry out, since large areas can be covered quickly at a low cost and few corrections to the field data are required. The main requirement for achieving an accurate magnetic map of the area of interest is that the diurnal variation in the Earth's magnetic field should be monitored, usually by a recording magnetometer at a base station, so that the actual magnetic readings obtained during the survey can be reduced to a common datum. To be of any value, however, magnetic rocks must be present in the survey area but possible magnetic anomalies on an engineering site can be man-made targets such as brick-lined shafts or the remains of ancient dwellings. In the UK, magnetic surveying has been widely used for the detection of abandoned mine shafts and adits, which are a considerable problem in areas where mining acitivity has ceased and few records are available of the positions of the mines and their associated shafts. Examples of the use of magnetic surveys for the above application are given in Dearman *et al.* (1977), Gallagher *et al.* (1978) and McCann *et al.* (1987). In geological surveying, magnetic dykes within the rock mass can be located and traced in areas where there is superficial cover which masks their presence; a simple example of this application is given in Culshaw *et al.* (1987).

9.3.4 Gravity

Gravity surveying involves the measurement of the Earth's gravitational field, which is a function of:

(1) distance from the centre of the Earth;
(2) centrifugal acceleration due to the Earth's rotation;
(3) the distribution of the mass, which is controlled by the presence of hills and valleys in the vicinity of the observation point and changes in the density of rocks in the Earth.

After making the necessary corrections to the recorded data, the final gravity anomaly map represents variation in the mass distribution in the Earth. Little use has been made of the method in engineering investigations and the method finds its greatest application in the study of deep structures.

With the availabiltiy of more sensitive gravity meters, which are capable of measuring changes in the gravity field of the order of 5 μgal, the method has been used increasingly in the location of cavities, voids and abandoned mineshafts (Neumann 1977; Greenfield 1979). Variations in density of the near surface materials also can give rise to a gravity anomaly and the thickening of the superficial deposits relating to the presence of a buried valley was studied by Cornwell and Carruthers (1986). The delineation of the waste material in landfill schemes is also another practical example of the use of the gravity method (Neumann 1977).

9.3.5 Electromagnetic

The use of electromagnetic methods in geotechnical investigations has increased in recent years. In most cases electromagnetic energy is introduced into the ground by inductive coupling and is produced by passing an alternative current through a coil or loop; the receiver also detects its signal by induction. Most electromagnetic methods involve the use of a portable power source, although use has been made also of powerful long range transmitting stations (Parasnis 1986). Pulse electromagnetic energy also has been used for the detection of geological discontinuities, such as fault zones and open and infilled cavities (Dolphin *et al.* 1974).

The terrain conductivity meter is commonly used in engineering studies. Over a uniform half-space the technique is analogous to carrying out a conventional resistivity survey with a fixed electrode spacing and the meter reads a direct conductivity value which is virtually the inverse of the apparent resistivity valve determined from the fixed array. Penetration between 4 and 6 m can be achieved and McDowell (1981) described the use of the meter for the rapid location of near-surface anomalous ground conditions. Culshaw *et al.* (1987) in their review of the use of geophysical methods in geological mapping described a survey carried out at Bridgend where the presence of fissures in the limestone was detected by a conductivity survey.

In the very low frequency (VLF) method, the electromagnetic waves transmitted by long range radio stations may be used to determine the electrical properties of the Earth. The depth to which radio waves can penetrate into the Earth is quite limited and low frequency transmissions are desirable. Measurement of resistivity based on electromagnetic energy is, thus, mainly applicable to the soil cover and overburden. The value of the method in mineral exploration is that the presence of a conductor gives rise to rapid changes in the measured resistivity. In the engineering field the method has found particular application in the delineation of water filled faults and fault zones in the rock mass (Paterson and Ronka 1971; Phillips and Richards 1975).

Electromagnetic pulse techniques also have been used more in recent years. They have been used both in the reflection mode (Dolphin *et al.* 1974) and in the direct transmission mode (Lytle *et al.* 1979; Laine 1980). As was mentioned above, attenuation of the electromagnetic energy is the biggest problem encountered, but nevertheless the method has found considerable application in the exploration of archaeological sites, which often are located in very dry areas. Considerable penetration of electromagnetic energy is also reported in low porosity rocks, such as limestone, and Laine (1980) described the location of a cavity in limestone by the direct transmission method at a frequency of 10 MHz. The method used is very similar to the cross-hole seismic technique and in Figure 9.6, which is reproduced from Laine (1980), the top and bottom of the cave are clearly visible. Similar experiments were carried out by Lytle *et al.* (1979), who used electromagnetic energy in the frequency range 10 to 70 MHz, and described a number of case histories.

An interesting case history using a reflected electromagnetic pulse is described in Dolphin *et al.* (1974). In this application the technique was applied at an archaeological site and electromagnetic energy in the frequency range 16–50 MHz was used. The application of the electromagnetic pulse reflection system to the detection and identification of geological and man-made anomalies which represent hazards ahead of the working face in

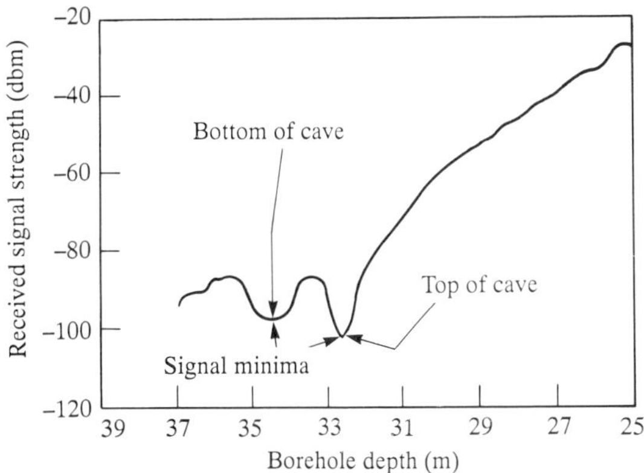

Figure 9.6 *Plot of received signal amplitude vs depth for electromagnetic pulse survey between two boreholes with a cave in the rock mass (after Laine 1980)*

hard-rock rapid-tunnelling operations was described by Moffat and Peters (1972); their experiments with various targets buried in the overburden and also the limestone rock mass demonstrated the feasibility of this approach. Further developments of the method are described in Moffat and Puskar (1976), in which the use of the electromagnetic pulse system discussed in relation to a number of practical field situations.

Ground-probing radar techniques also have been examined by Cook (1972, 1975 and 1977). Using only low-power experimental radar equipment, reflection signals were obtained through 13 m of limestone, through 9–18 m of coal and through 225 m of rock salt. By the late 1970s a commercial ground-probing radar system had been developed and its use in a wide range of field application was discussed by Darracott and Lake (1981). For normal use the transmitting and receiving antennae are mounted within a mobile trolley but for specialized cross-hole or surface to drillhole measurements the two antennae are operated separately. The electromagnetic energy is propagated in pulse form at frequencies in the range 50 MHz to 1 GHz.

The depth of penetration of electromagnetic energy can vary from hundreds of metres in rock salt to less than one metre in clays. A general rule is that for a given material the lower the frequency of transmission, the greater the depth of penetration, while the higher transmission frequencies increase the resolution but reduce the depth of penetration.

Many uses of ground-probing radar have been described in the literature including the determination of permafrost thickness (Annan and Davis 1976), the detection of fractures in rock salt (Unterberger 1978) and the assessment of fracturing in a granite rock mass (Olsson *et al.* 1983). Other civil engineering applications are described by Rubin and Fowler (1978), Benson and Glacum (1979), Leggo (1982), Leggo and Leach (1983) and McCann *et al.* (1988).

9.4 Drillhole geophysical methods during the drilling process

9.4.1 Introduction

The construction of underground structures requires considerable information on the engineering and hydrogeological properties of the rock mass. The site investigation drillholes provide core samples, which can be tested in the laboratory using standard rock- and soil-testing methods (Attewell and Farmer, 1976). However, samples taken from drillholes are often highly disturbed, and even the most carefully taken sample is subject to stress relief.

Laboratory tests on intact core samples from drillholes give an approximate representation of the *in situ* engineering properties of a rock stratum many metres thick and having a lateral extent of hundreds of square metres so that large factors of safety have to be employed in the design of civil-engineering structures. In laboratory tests little or no account can be taken of discontinuities present on anything but a small scale, since bedding-plane partings, jointing and faulting cannot usually be included in such measurements. Hence there has been considerable effort applied to the development of *in situ* testing methods for use in drillholes to optimize the useful information that can be obtained from a drillhole study.

To the geophysicist, a drillhole is an access path to the rock mass, which can be investigated by the use of geophysical methods. The geophysical properties of the rock mass are directly determined properties which can be related to its lithological and geotechnical properties (Cole 1976; Buchan *et al.* 1972; Hamdi and Taylor Smith 1981). Considerable research has been carried out in this field and this has been aimed directly at the concept of determining engineering properties from indirect geophysical measurements made in or between drillholes. The geophysical properties are controlled by the geotechnical properties of the rock mass but the relationships are essentially multidimensional so that the rock fabric and interstitial fluids both play a part in controlling the overall engineering performance. Fracturing, either on a micro-scale in the rock matrix or on a macro-scale in the rock mass, has an additional effect on the engineering performance of the rock mass.

The site-investigation drillholes can be used in two ways for geophysical investigation as follows:

(1) A wide range of geophysical drillhole logging tools can be deployed in each drillhole to provide continuous measurements with depth of a range of geophysical parameters.
(2) Cross-hole seismic or electromagnetic measurements may be carried out between the adjacent drillholes to provide an engineering assessment of the rock mass indirectly from the geophysical properties.

9.4.2 Geophysical well-logging methods

The methods examine the geophysical properties of an annulus of rock around the drillhole, and so act upon a much longer volume of rock *in situ* than that represented by a core sample. The influence of jointing, fracturing, weathering and even lateral lithological variation will all be reflected in the measured parameters. The geophysical logging measurements provide additional data to that obtained from core logging, laboratory testing of cores, and *in situ* observations and measurements. The logging results can be used in conjunction with the geomechanical and hydrogeological information to extend the basic rock-mass classification system used.

(a) Caliper log

The caliper log is used to obtain an accurate profile of the diameter of the drillhole down its length, since correction for any changes in this parameter have to be applied to most other geophysical logs run in the drillhole. The log can be used, to a certain extent, for the identification of lithology and stratigraphic correlation but its most important use arises in the location of zones of fractured rock. The observation of drillhole 'break-outs' also can be noted in the caliper log and this can be used to determine the direction of the *in situ* stress field (Brereton and Evans, in press). The basic principles of caliper logging are considered in some detail by Hilchie (1968).

(b) Sonic log

The sonic log, in its simplest form, is a recording of the time, δt, taken by a compressional wave to travel through one foot of formation plotted against depth. In the oil industry, this is expressed as microseconds per foot, but in civil engineering applications the more familiar units of microseconds per metre are used. The basic sonic log is used mainly to compute the formation porosity via the time-average equation of Wyllie *et al.* (1956), as follows:

$$\frac{1}{V_r} = \frac{\Phi}{V_f} + \frac{(1 - \Phi)}{V_m} \qquad (9.4)$$

where Φ is the effective porosity, V_r is the compressional wave velocity in the rock mass, V_m is the compressional wave velocity in the rock matrix, and V_f is the compressional wave velocity in the fluid, and 100% saturation of the rock mass is assumed.

The P-wave velocity also is affected by the presence of fracture zones on horizontal and sub-horizontal fractures, and the sonic log thus responds to the effective porosity of the rock mass. The presence of vertical fractures does not affect the P-wave velocity to any great extent, and so the sonic log responds only to the porosity of the rock matrix, and too low a value of bulk porosity will be indicated.

With the full wave train sonic log (Figure 9.7) it is often possible to measure the velocities of both the compressional (P) and shear (S) wave velocities (Geyer and Myung 1970). The actual log is usually presented in the form of a variable density plot. It is not possible always to identify the S wave since there is often no distinct break at the onset of the shear-wave pulse or the shear wave is highly attenuated. In the latter case this is extremely useful in picking out zones of highly fractured rock.

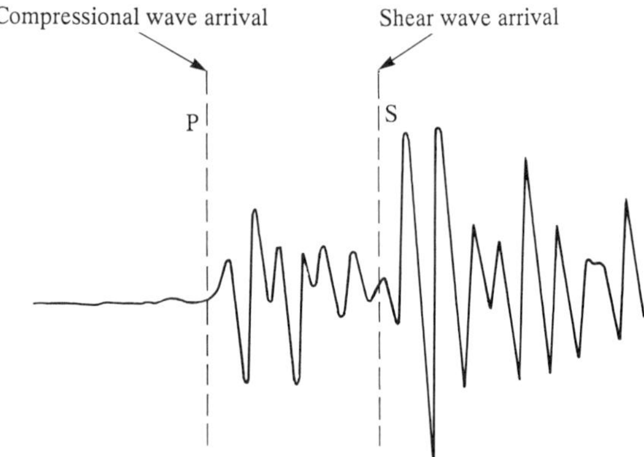

Figure 9.7 *Full wave train sonic log*

Using values of the formation density computed from the gamma–gamma log, it is possible to compute the dynamic elastic properties of the rock mass from the P- and S-wave velocities as follows:

(1) Bulk modulus (K)

$$K = \rho_b (V_p^2 - 4/3 V_s^2) \qquad (9.5)$$

where ρ_b is the bulk density, V_p is the compressional wave velocity, and V_s is the shear wave velocity

(2) Shear modulus (G)

$$\mu = \rho_b V_s^2 \qquad (9.6)$$

(3) Poisson's ratio (δ)

$$\delta = \frac{0.5 (V_p/V_s)^2 - 1}{(V_p/V_s)^2 - 1}$$

(4) Young's modulus (E)

$$E = 2\rho V_s^2 (1 - \delta) \qquad (9.8)$$

The presence of fractures in the rock mass will interfere with the transmission of elastic wave energy along the wall of the drillhole. In highly fractured rock both the velocity of propagation and amplitude of a compressional wave are considerably reduced and similar characteristics have been noted for shear waves.

Laboratory and field values of V_p and V_s, the velocities of the compressional and shear waves respectively, can be used to compute a fracture index following the procedure suggested by Knill (1970) in Equation (9.1). In order to produce an index with a range from 0 to 1, Equation (9.1) has been modified as follows:

$$\text{sonic fracture index (SFI)} = \frac{V_L - V_M}{V_L - V_F} \qquad (9.9)$$

where $V_F < V_M < V_L$, V_F is the compressional wave velocity in the drillhole fluid, V_M is the measured compressional wave velocity at any depth in the drillhole and V_L is the compressional wave velocity measured on intact rock samples in the laboratory.

(c) Radioactive logs

Natural gamma This log is a measurement of the natural radioactivity of a formation and is useful for lithological identification or stratigraphic correlation. The log also can be used for the identification of zones containing readioactive minerals such as potash or uranium-rich ores.

Gamma–gamma This log measures the intensity of gamma radiation from a radioactive source in the sonde, such as cobalt-60 or caesium-127, after it is backscattered and attenuated within the drillhole and the surrounding rock mass. Provided the necessary calibrations are applied to the sonde, the recorded count rate is directly proportional to the formation density. The effects of variation in the drillhole diameter are offset by forcing the sonde against the wall of the drillhole with an excentring arm.

The main use of the gamma/gamma log in the oil industry is the determination of formation porosity, while in engineering studies it is the formation density that is most applicable. The sonde also can be used for lithological correlation and identification of formations within a single geological area. Bulk density can be measured to an accuracy of ± 0.05 tonnes/m^3 but this can be improved by careful calibration of the source, detectors, and instrumentation and considerable care in the preparation of the drillhole.

The gamma–gamma log measures the effective density of the formation close to the drillhole wall and this includes the effects of both the rock matrix and the contained fluids. The log responds to both horizontal and vertical fractures and tends to indicate higher porosities than those derived from a sonic log over a comparable section of drillhole intersected by a high-angle fracture or joint plane. A cross-plot of porosity derived from the two logs can be used to indicate the presence of significant fracture zones within the rock mass.

In a crystalline rock mass, the bulk density, ρ_b, of a rock of porosity Φ is given by

$$\rho_b = \rho_g - \Phi(\rho_g - \rho_f) \qquad (9.10)$$

where ρ_g is the grain density and ρ_f is the fluid density. Hence the formation porosity can be computed from

$$\Phi = \frac{\rho_g - \rho_b}{\rho_g - \rho_f} \qquad (9.11)$$

The gamma–gamma derived density values can be used to produce a porosity Φ_d derived from the density log, and this can be compared with the neutron porosity (Φ_n), The density log is often a better indication of porosity in a fractured rock zone than the neutron log and any large negative values of neutron porosity minus gamma–gamma-derived porosity (Φ_n - Φ_d) can be indicative of potential permeable zones (McEwen and Shedlock, 1985).

Neutron–neutron The neutron–neutron sonde is used to measure formation porosity or moisture content since it responds primarily to the amount of hydrogen present in the formation. A plutonium–beryllium or an americium–beryllium source is used to provide the neutrons. Again the best results are obtained with the neutron source and detector pressed against the drillhole wall to minimize drillhole effects. For accurate determination of moisture content or porosity very careful calibration of the sonde is required, particularly at low porosity values. The neutron sonde responds to the total water content of the formation and, since this would include adsorbed water associated with clay minerals, the porosity measured on a shale, for instance, will actually be greater than the effective porosity of the formation. The log is usually very subdued in low-porosity crystalline rocks but the presence of a fracture zone will artificially increase the effective porosity and a significant reduction in the neutron count will be observed.

Thus, provided there are no significant absorbers of neutron particles present in the rock mass, such as boron, a fracture index based on neutron count (NFI) can be defined as

$$\text{neutron fracture index (NFI)} = \frac{N_A - N_M}{N_A - N_F} \qquad (9.12)$$

Where N_A is the average neutron count for an intact rock mass in the absence of fractures (i.e. $N_A > N_M$)

N_M is the measured neutron count at any depth in the borehole.

N_F is the neutron count in the drillhole fluid.

(d) Electrical methods

Spontaneous potential log The underlying theory of electrical logging devices is considered in detail by Scott Keys and MacGary (1971) and only a brief summary is given below. The spontaneous potential log is a recording versus depth of the natural potential developed between the drillhole fluid and the surrounding rock mass.

The log is used mainly for the detection of permeable beds, the location of their boundaries and their correlation between adjacent drillholes. In a low-porosity rock mass, such as granite, variations in spontaneous potential will be related to zones of fractured rock associated with joint patterns or fissures but the measured values cannot be applied in a quantitative manner to obtain formation permeability.

Single-point resistance log This is the simplest of the electrical logging systems, in which an electrode made of lead is lowered down the drillhole on an insulated cable with an electrode buried at the surface to provide the return path for the current flow. The measured resistance is a function of the formation resistivity so that variations are related directly to changes in lithology. The log has been used as a lithology tool for geological correlations between drillholes. Although the depth of investigations is only between 50 and 100 mm, the log will respond to fractures in the drillhole wall and is used often in conjunction with the caliper log to investigate fracturing in the rock mass.

Normal devices A more realistic value of the apparent resistivity is required to compute the *in situ* porosity of the rock mass and this is achieved by use of the normal logging tool which passes current deeper into the rock mass depending on the electrode separation.

The normal tools are most effective in low- to medium-resistivity media, the short normal log proving particularly useful for geological correlation between drill holes. In low-porosity rocks, such as granite, the resistivity of the rock mass is very high, and the major factor influencing the resistivity log will be joint patterns. The effects of fractures and joints can be observed in a qualitative manner in the rock mass by a comparison of the short and long normal logs.

Focused devices The resistivity sondes described above are all relatively insensitive in high-resistivity formations since most of the current flow is in the drillhole fluid and very little enters the formation. This results in poor resolution of the bed boundaries and the apparent resistivity measured is not representative of the formation resistivity.

This limitation has been overcome in a high-resistivity formation by the introduction of focused resistivity tools. The underlying concept is that if the measuring current is forced to flow radially as a thin sheet of current into the formation being logged, then the influence of the drillhole and the surrounding formation on the resistivity measurement is minimized.

The resistivity tools described above are used for both shallow and deep investigations of the rock mass from the drillhole. In highly resistive crystalline rocks the resistivity logs are normally insensitive to the small changes brought about by variations in lithology. However, where substantial fracturing is present, the rock mass becomes more permeable and the log response will be affected by the fracure orientation, the radial and vertical length of the fracture, and the fluid.

In a very low-porosity rock mass the conductivity network formed by the fracture zone may be largely unidirectional, resulting in a value of apparent resistivity that is significantly different from that measured in an isotropically conductive medium. This effect is observed particularly with deep focused devices (Sauu and Gartner 1979).

It is possible to use both laboratory and field values of electrical resistivity to compute a fracture index following the same procedure developed for the sonic logs. Hence, the fracture index based on the measured focused resistivity values (RFI) is defined as

$$\text{resistivity fracture index (RFI)} = \frac{R_{mat} - R_{for}}{R_{mat} - R_{fl}} \quad (9.13)$$

where R_{mat} is the measured resistivity of the unfractured rock mass, R_{for} is the measured resistivity of the fractured rock mass and R_{fl} is the fluid resistivity.

The above equation is based on the determination of values for R_{mat} and R_{for} in the drillhole, using the same resistivity logging in both cases, so that: RFI $\rightarrow$ 1 when $R_{for} \rightarrow R_{fl}$, RFI $\rightarrow$ 0 when $R_{for} \rightarrow R_{mat}$, R_{fl} values were determined during temperature and conductivity logging (Robertson and Perkins 1979).

(e) Dipmeter

A dipmeter is a 3–, 4– or 6– arm (usually 4–) side-wall micro-resistivity device, which measures fine variations of resistivity in the formation with such a resolution that the relative vertical shift of characteristic patterns of variation on the traces can be used to derive the attitude of a plane intersecting the drillhole (Figure 9.8). Such patterns can be caused by bedding planes, changes in lithology, or fractures and joints in the rock mass. The patterns are arranged across the traces as sinusoids. Computer analysis can identify the patterns by cross-correlation, and perform the trigonometical solution, incorporating the relative shift of patterns, the orientation of the sonde and the drillhole dimensions, to derive the attitude of the planes.

The fracture identification log is a derivative of the dipmeter processing, recognizing that over four sections of the log, stratigraphic dips are reasonably constant. Each trace can be shifted, relative to the datum trace (no. 1) to juxtapose the pattern caused by the dip (Figure 9.9). Opposite traces then can be subtracted, cancelling the effect of dip and leaving anomalies due to planes which are not in the same orientation; fractures are examples of this.

(f) Formation scanning tool

The high-resolution dipmeter log is an effective indication of a high fracture density within a drillhole. In a development of the sonde described by Lloyd *et al.* (1986) two of the four dipmeter pads carry electrical micro-scanners each consisting of 27 buttons. The basic principle

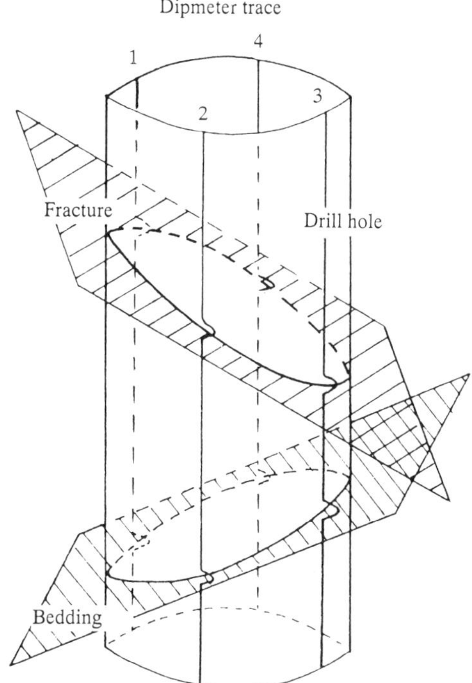

Figure 9.8 *Principles of the dipmeter log for fracture planes in drill hole*

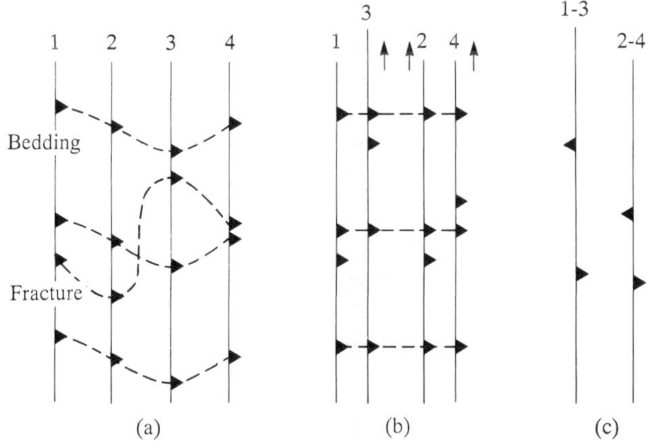

Figure 9.9 *Analysis of dipmeter records to isolate fractures from bedding planes*

is to map the conductivity of the drillhole wall with a dense array of sensors. These give a very high-resolution electrical image of the formation when they have been speed corrected and are displayed on a variable-intensity grey-level scale. The data from the two arrays can be used to produced a very high-resolution (2.5 mm by 2.5 mm) image of two strips down the drillhole wall. The data are produced in digitized form and can be processed by a computer-based interpretational package described by Lanyon (1986), for direct comparison with the records from the drillhole televiewer (Figure 9.10).

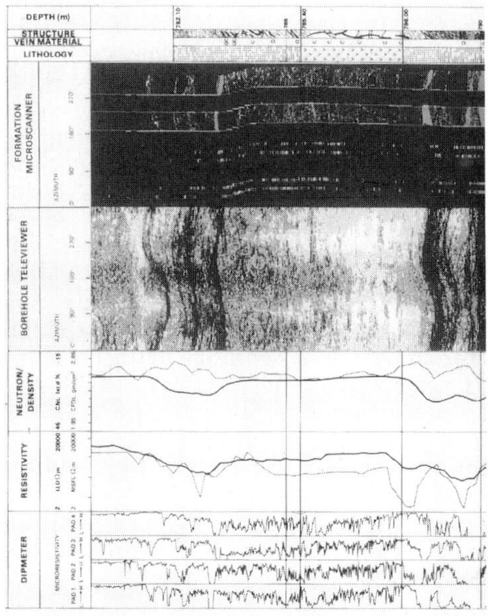

Figure 9.10 *A suite of geophysical logs over a section of cored PreCambrain (Charnian) volcanics in Morely Quarry borehole (central England). (Courtesy of British Geological Survey)*

(g) Televiewer

The televiewer pulses ultrasonic energy from a piezoelectric transducer to the drillhole wall via the fluid in the drillhole, where some of the energy is reflected and detected by the transducer (now acting as a receiver). The transducer is rotated in the drillhole at 3 rev/s and is orientated relative to the Earth's magnetic field by a downhole magnetometer within the sonde. The amplitude of the reflected signal is proportional to the reflected energy, which is a function of the acoustic impedance of the drillhole wall. Presentation of the processed data on an electrostatic plotter gives a flattened picture of the drillhole wall with fractures appearing as sine waves.

When using the televiewer the strength of the acoustic reflection at the drillhole wall is dependent on its surface roughness and mineralogy. Fractures, which are assumed to be approximately planar in the area of intersection with the drillhole, will produce a bilaterally symmetrical image centred about their direction of dip. The dip of the fracture is given by $\tan^{-1}h/d$, where h is the peak-to-peak amplitude of the sinusoid representing the fracture, and d is the diameter of the drillhole (Figure 9.10). Accurate measurements of fracture orientation are not possible for low angles of dip, even at low logging speeds, because of the shape of the tangent curve. Since the accuracy with which the orientation of low-angled fractures can be determined increases with decrease in logging speed, stereoplots of fractures over the same interval of a drillhole are not identical for different logging speeds. When using the electostatic plotter as output, and using the minimum logging speed available of 0.29 m/min, accurate dip angles can be delineated down to values of dip of 20° and dip directions down to values of dip of 15°. If an oscilloscope is used for output, then faster logging speeds can be used with the same accuracy in determining the angle of dip, although the linear resolution is poorer.

A recent development of the acoustic borehole televiewer (BHTV) is the availability of its output in digital form. The data are provided in the form of sampled amplitude and travel time variable density logs (VDLs) plus auxiliary measurements taken every 10 mm; the data are oriented during analysis to magnetic north. Use of the digital output from the televiewer in the development of an interactive computer system to display the relevant amplitude, travel time and filtered travel time VDLs for any section of the drillhole at any scale is described by Lanyon (1986).

9.3.4 Cross-hole geophysical methods

While considerable information on the engineering properties of the rock mass can be obtained from geophysical logging of site-investigation drillholes, the properties of the large volume of rock lying between the drillholes remain largely unknown. The geological structure and the physical properties of the lithological units observed in the drillholes can be extrapolated between the drillholes, but if there is a significant discontinuity such as a zone of fractured rock, its presence may well remain undetected.

Cross-hole geophysical measurements and, in particular, those based on the transmission of seismic energy between drillholes, provide a means by which the engineering properties of the rock mass between the drillholes can, at least, be assessed. Seismic methods are particularly important since the dynamic elastic properties can be obtained from the measurement of the compressional (V_p) and shear (V_s) wave velocities and a knowledge of the formation density.

The use of the cross-hole seismic method in various civil engineering applications is described in McCann *et al.* (1975), Auld (1977), Grainger and McCann (1977), Thill (1978), Paulson and King (1980) and McCann *et al.* (1986). Applications include

(1) continuity of lithological units between drillholes;
(2) detection of old mine workings, shafts and adits;
(3) identification of geological hazards, such as fault zones;
(4) assessment of the degree of fracturing;
(5) determination of elastic moduli.

The simplest form of cross-hole seismic measurements is made between a seismic source in one drillhole and a receiver at the same depth in an adjacent drillhole (Figure 9.11). Several types of seismic source can be used

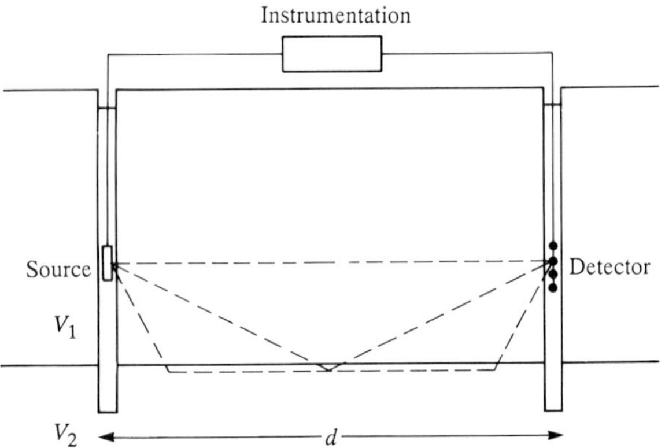

Figure 9.11 *Basic layout of seismic source and hydrophone array for cross-hole seismic measurements*

including a small explosive charge, a detonator, an air-gun, a drillhole hammer, or a sparker probe. The receiver, which detects the seismic signal, can be either a hydrophone suspended freely in the drillhole fluid or a three-component geophone array clamped to the drillhole wall. The latter is important when both compressional and shear wave propagation are being studied, since the clamped geophone will respond to an incident vertically or horizontally polarized shear wave. The choice of source and receiver is a function of the distance between the drill holes, the required resolution, and the properties of the rock mass. The best survey results are achieved with a repetitive seismic source with a high-frequency energy content since this will improve the timing accuracy and allow the use of a signal-averaging seismic recorder.

A number of interesting examples from the civil engineering field are given by McCann *et al.* (1986), who cover most of the applications mentioned above. The study of the rock mass at the hot dry rock (HDR) geothermal project in Cornwall, UK, before and after explosive stimulation and hydraulic fracturing, is a particularly interesting case history, which illustrates the effect of fracturing on the P-wave velocity. The results of the cross-hole seismic measurements in the four 300 m experimental drillholes at the site shown in Figure 9.12 give a good indication of the velocity distribution in the rock mass to a depth of 280 m. The time–distance graph

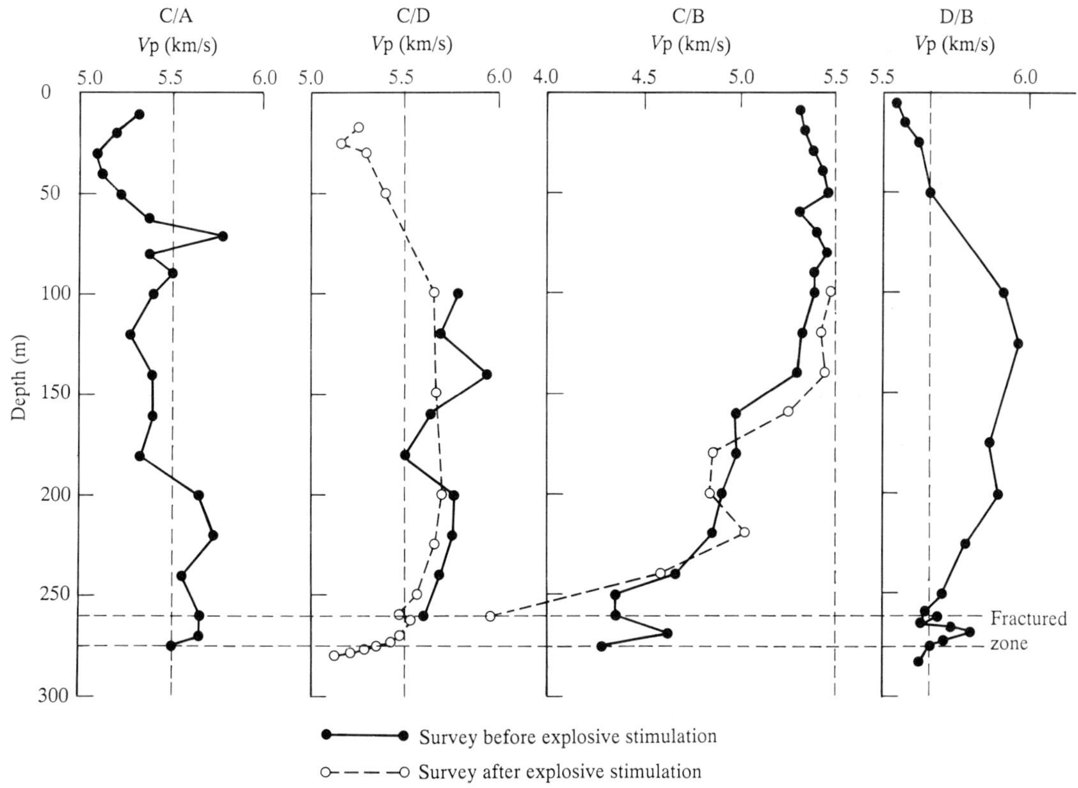

Figure 9.12 *Horizontal cross-hole velocity scans through a granite rock mass before and after fracture stimulation using explosives. The lower velocities after stimulation in Scan C/B suggest the presence of fractures (after McCann* et al. *1986)*

shown in Figure 9.13 gives a mean P-wave velocity for the granite rock mass of 5.49 km/s and 2.49 km/s for the artificially fractured zone below 278 m. This reduction in P-wave velocity is also noted in scans C/D and C/B shown in Figure 9.12.

There are often situations where a two-dimensional anomaly such as a dyke or fault is thought to exist between the two drillholes. These anomalies are often characterized by seismic properties which are significantly different to those of the surrounding rock mass. Thus, the simple velocity profiles referred to above do not categorize the rock mass in an adequate manner and it is necessary to obtain the two-dimensional velocity distribution between the drillholes. The imaging of seismic properties between drillholes generally is referred to as seismic tomography in common with medical applications where computerized tomography (CT) is used to provide real-time images from within the human body using ultrasonic scanning. Considerable development has taken place over the past decade in the development of seismic tomography methods in the geological field and a number of interesting case histories can be quoted (see for instance, La Porte *et al.* (1973); Dines and Lytle (1979); Hatton *et al.* (1986). An interesting example from McCann and Jackson (1988) of a

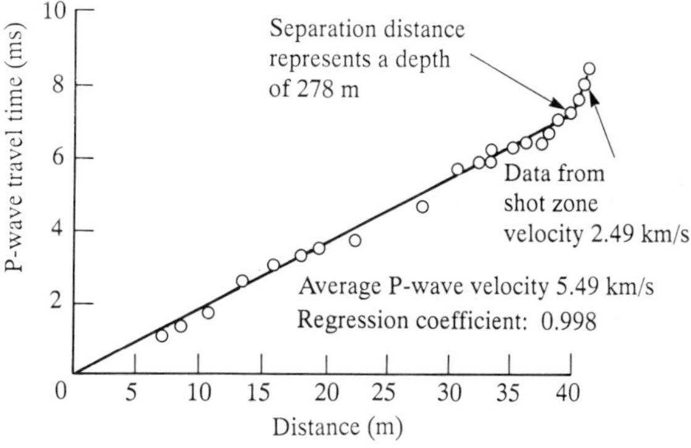

Figure 9.13 *P-wave travel time plotted against distance for the horizontal velocity scans shown in Figure 9.12. The slope of the graph gives the average P-wave velocity in the granite rock mass; the fractured zone is identified by the change in slope (after McCann* et al. *1986)*

dolerite dyke running through Lower Cambrian mudstones and siltstones at Trawsfynydd Power Station is given in Figure 9.14. Tomographic methods are not confined entirely to seismic measurements, since the use

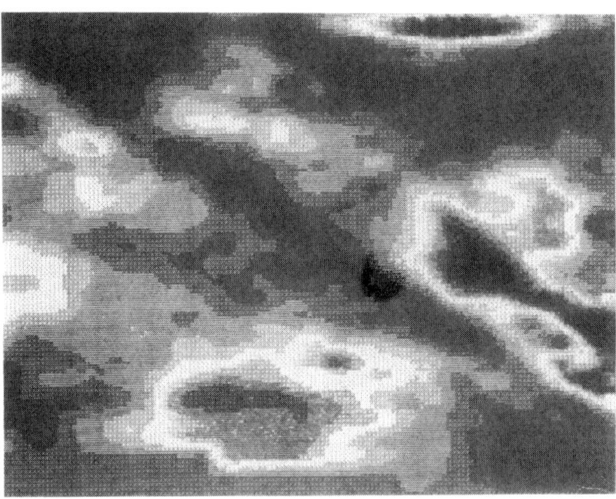

Figure 9.14 *Seismic tomogram for the Trawsfynydd site, central Wales, using curved ray reconstruction (after McCann and Jackson, 1988). Colour scale – dark red (high velocity) through to dark blue (low velocity)*

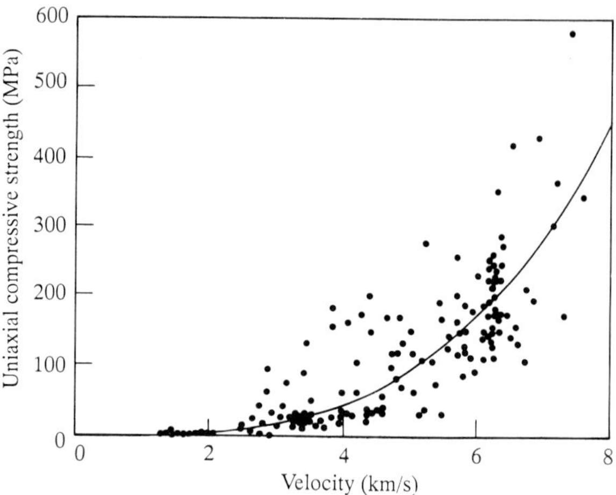

Figure 9.15 *Compressional wave velocity plotted against uniaxial compressive strength for a range of British rocks (measurements made in the laboratory on prepared sample cores) (after McCann* et al. *1990)*

of electromagnetic cross-hole methods also have been described in the literature (Laine 1980).

9.5 Geophysical classification

If the geophysical methods listed above are considered in the light of the rock-mass classification system advocated by Bieniawski (1973, 1974 and 1976) it is clear that seismic and electrical methods are related most closely to the geomechanical and hydrogeological properties of the rock mass. The potential field methods, gravity and magnetic, are useful only for large-scale evaluation of the geological structure or small-scale investigation such as the location of cavities, dykes, or buried valleys.

Electromagnetic methods are becoming increasingly important for the location of water-filled fault zones related to highly fractured rock and in many ways complement electrical measurements on the rock mass. Certainly, the importance of both electrical and electromagnetic methods in the evaluation of water flow in the rock mass should not be neglected.

However, it is seismic methods which make the greatest contribution to the indirect assessment of the geomechanical properties of the rock mass and can be related to the Bieniawski classification. The classification of the intact rock strength proposed by Deere and Miller (1966) based on the uniaxial compressive strength has been used by Bieniawski (1974) in his rock classification system. McCann *et al.* (1990) showed that there is a strong relationship between the uniaxial compressive strength and the compressional wave velocity, and hence this parameter also can be used as a simple index of the rock strength following the CSIR system (Figure 9.15). This can be estimated further by also measuring the shear wave velocity in the rock either by the geophysical logging of the site investigation drillholes or by direct seismic surveys on the ground surface. It is then possible to calculate the dynamic elastic properties, on which a simple rock classification system can be based for the rock mass as a whole.

The dynamic elastic properties, such as Young's modulus or Poisson's ratio, are rarely used in the design of structures in preference to the static values determined in the laboratory except where dynamic loading is involved. The main reason for this is that there is often a wide discrepancy between the modulus values determined by the two methods, and the dynamic values of Young's modulus are usually higher than the static ones, varying on average between 5 and 25% greater for low-porosity, high-density rocks (Ide 1936; Sutherland 1962; Lama and Vutukuri 1978) and as much as 100% for soft rocks. Much of this difference probably can be attributed to the widely different strain levels at which the two sets of measurements are made. Simmons and Brace (1965) showed that the presence of cracks and cavities in the rock fabric, which tend to close under pressure, has a marked effect on the static curves. King (1970) carried out both static and dynamic laboratory tests under pressure and at high stress levels and showed a fairly close agreement between the two values of Young's modulus. Myung and Helander (1972) showed that good agreement was obtained between the dynamic elastic moduli computed from measurements of the compressional and shear-wave velocities, both on samples under simulated formation pressure in the laboratory, and *in situ* drillhole sonic measurements in the field.

Thus there is ample evidence to suggest that the dynamic values of the elastic moduli given in Equations

(9.5–9.8) above are a reasonable indication of the overall elastic porperties of the rock mass (Eissa and Kazi 1988; Van Heerden 1987). These dynamic values are based on the determination of the compressional and shear-wave velocities from the full wave train sonic log and the formation density log. The dynamic elastic moduli values can be used in the following ways:

(1) *In situ* static elastic moduli can be derived from the dynamic values using relationships determined for each rock type.
(2) An assessment can be made of the relevance of the laboratory testing programme on a suite of drillhole core samples by comparing the *in situ* and laboratory dynamic properties.
(3) Using the standard geophysical well-logging methods, the engineering properties can be examined to the full depth of the investigation of the drillhole.

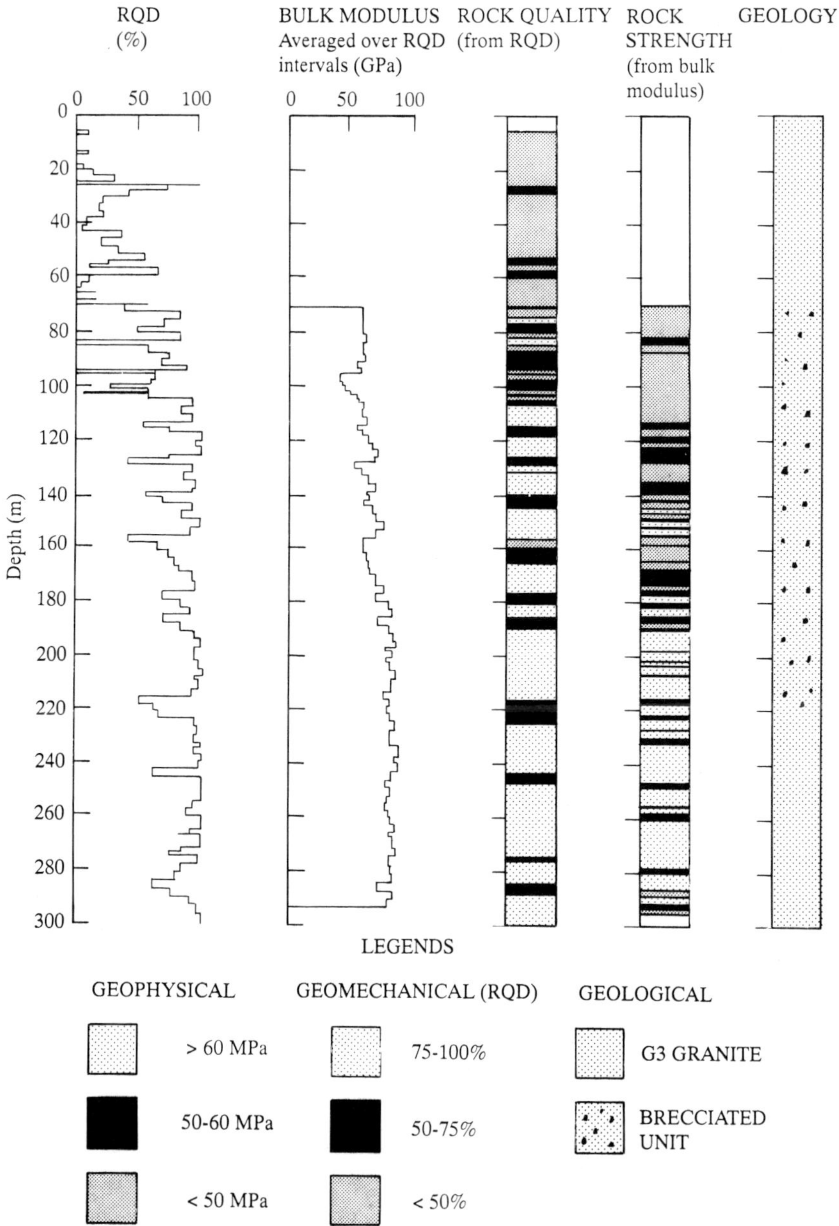

Figure 9.16 *Rock quality and rock strength plotted against depth for borehole A1C at Altnabreac, Caithness*

(4) A comparison of the engineering properties of the various lithological units present in the drillhole can be made based on the dynamic elastic properties.

The geotechnical assessment of the rock mass for engineering design based on the determination of the RQD parameter from drillhole core was considered by McEwen *et al.* (1985) and McCann *et al.* (1990) for a drill hole in the Strath Halladale Granite at Altnabreac in Caithness, Scotland. The granite is a fractured medium-grained rock, predominantly granodiorite in composition and cut by numerous aplite, pegmatite and microgranite veins. The relationship between the numerical value of RQD and the engineering quality of the rock (modified after Deere 1964) is defined as:

RQD Rock Quality
<50% Very poor to poor
50–75% Fair
75–100% Good to very good

The RQD and rock quality variations are shown plotted against depth in Figure 9.16. The main zone of poor-quality rock lies in the top 70 m due to stress relief and weathering processes but fairly poor quality rock exists down to 140 m because of the effect of a fault zone.

The dynamic elastic properties of the rock mass are a reasonable estimation of the overall performance from the engineering design point of view and McEwen *et al.* (1985) introduced the concept of a simplified rock-mass classification system based on the dynamic bulk modulus of the rock mass. This parameter is based on the compressional and shear-wave velocities in conjunction with a formation density and can be determined from geophysical logging measurements in drillholes in the rock mass.

Finally the three fracture indices given in Equations (9.9), (9.12) and (9.13) respectively were considered by McEwen *et al.* (1985). The results for the drillhole are shown in Figure 9.17 and this shows that the main response of the geophysical tools is to open fractures, where the increase in effective porosity has a considerable

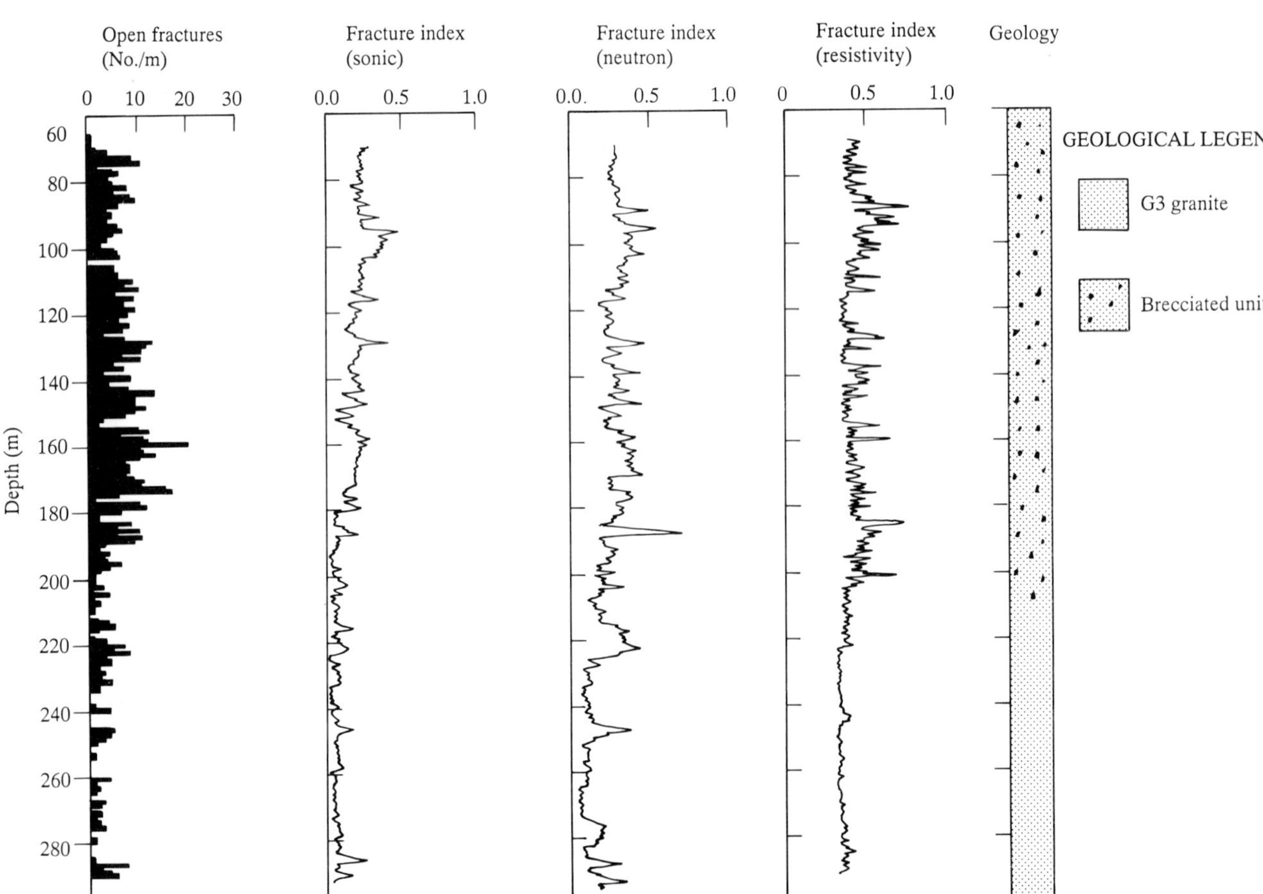

Figure 9.5 *Foyers hydroelectric scheme: resistivity interpretation, line 1, over tunnel, Loch Mhor (scale, vertical and horizontal, 1:1000) (after Cratchley* et al. *1976)*

effect on the gophysical parameters. The changes in the sonic index are quite subdued in comparison with the neutron and resistivity indices, which are more affected by the increase in effective porosity and the properties of the rock mass further from the drillhole wall. These indices are based on average parameters for the solid, intact rock and could be modified by the selection of a more appropriate value for each section of the drillhole.

Unlike the RQD values, the rock strength index is considerably influenced by weathering and alteration effects since there is a considerable reduction in both the compressional and shear-wave velocities in a weathered rock mass. Glenn and Nelson (1979) also observed good correlation between the compressional wave velocity and RQD for moderately to well-fractured rock of one lithological type. As this parameter is used to calculate the dynamic bulk modulus this does confirm that the suggested rock strength index of McEwen *et al.* (1985) is a reasonable approach.

The relationship between the compressional wave velocity and the degree of fracturing for a particular rock type can be used to assess excavation conditions at engineering sites such as roads and pipeline routes. By carrying out a standard seismic survey it is possible to establish whether, and to what extent, blasting is required using charts of rippability versus compressional wave velocity produced by various contractors (Caterpillar 1978). A particularly useful study of the applicability of seismic surveys to the prediction of blasting conditions and, hence, the optimization of the blasting programme is given by Broadbent (1974).

In conclusion, engineering geophysicists have made many attempts to relate empirically the geophysical properties of the rock mass to its geomechanical properties. In this chapter, the progress made in using geophysical properties as part of geomechanical classification systems has been reviewed. It is considered that with the continuing development of geophysical methods their use in the assessment of rock-mass conditions will increase and eventually they will become an integral part of the rock-mass classification process.

Acknowledgement

This chapter is published with the permission of the Director, British Geological Survey (NERC).

References

Annan, A.P. and Davis, J.L. (1976) 'Impulse radar sounding in permafrost'. *Radio Science*, **11**, 383–394

Anon (1977) 'The description of rock masses for engineering purposes', Report by the Geological Society Engineering Group Working Party, *Q. J. Eng. Geol.*, **10**, 355–388

Attewell, P.B. and Farmer, I.W. (1976) *Principles of Engineering Geology*, Chapman and Hall, London

Auld, B. (1977) Cross-hole and down-hole VS by mechanical impulse', *J.Geotech. Eng. Div.*, American Society of Civil Engineers, **GT12**, 1381–1398

Barton, N., Lieu, R. and Lunde, J. (1974) 'Engineering classification of rock masses for the design of tunnel support', *Rock Mechanics*, **6**, 189–236

Benson, R.C. and Glacum, R.A. (1979) 'Radar surveys for geotechnical site assessment', In *Geophysical Methods in Geotechnical Engineering*, American Society of Civil Engineers Speciality Conference, Atlanta, 161–178

Bieniawski, Z.T. (1973) 'Engineering classification of jointed rock masses', *The Civil Engineer in South Africa*, **15**, 335–343

Bieniawski, Z.T. (1974) 'Geomechanics classification of rock masses and its application in tunnelling', *Proc. 3rd Congress of the International Society for Rock Mechanics*, Denver National Academy of Sciences, Washington, D.C., **2**, A, 27–32

Bieniawski, Z.T. (1976) 'Rock mass classification in rock engineering', *Proc. Symp. Exploration for Rock Engineering*, Johannesburg, A.A. Balkema, Rotterdam, **1**, 97–106

Bieniawski, Z.T. (1978) 'Determining rock mass deformability: experience from case histories', *Int. J. Rock Mechanics Mineral Sci. and Geomech. Abstr.*, **15**, 237–247

Brereton, N.R. and Evans, C.J. (in press) 'In-situ crustal stress in the United Kingdom from borehole breakouts'. In *Geological Application of Wireline Logs*, A. Hurst, M.A. Lovell and A.C. Morton (eds), Geological Society Special Publication No. 48

Broadbent, C.D. (1974) 'Predictable blasting with in situ seismic surveys', *Mining Engineering*, **26**(4), 37–41

Buchan, S., McCann, D.M. and Taylor Smith, D. (1972) 'Relation between the accoustic and geotechnical properties of marine sediments, *Q. J. Eng. Geol*, **5**, 256–284

Caterpillar Incorporated (1978) *Handbook of Ripping*, Caterpillar Inc., Peoria, Illinois, USA

Cole, D.I. (1976) 'Velocity porosity relationships in limestones from the Portland Group of southern England, *Geoexploration*, **14**, 37–50

Cook, J.C. (1972) 'Seeing through rock with radar', *Proc. First North American Rapid Excavation and Tunnelling Conference*, Chicago, Volume 1, 89–101

Cook, J.C. (1975) 'Radar transparencies of mine and tunnel rocks', *Geophysics*, **40**, 865–885

Cook, J.C. (1977) 'Borehole radar exploration in a coal seam', *Geophysics*, **42**, 1254–1257

Coon, R.F. and Merritt, A.H. (1970) 'Predicting in situ modulus of deformation using rock quality indexes', *ASTM Special Technical Publication*, 477, ASTM, Philadelphia, 154–173

Cornwell, J.D. and Carruthers, R.M. (1986) 'Geophysical studies of a buried valley system near Ixworth, Suffolk', *Proc. Geol. Assoc.*, **97**, 357–364

Cratchley, C.R. (1977) 'Geophysical measurements in rock mechanic investigations', *Proc. Goguel Colloquium*, Orleans

Cratchley, C.R., McCann, D.M. and Ates, M. (1976) 'Application of geophysical techniques to the location of weak tunnelling ground with an example from the Foyers hydroelectric scheme, Loch Ness', *Trans. Inst. Mining and Metall.*, Section A, **85**, A127–A135

Culshaw, M.G., Jackson, P.D. and McCann, D.M. (1987)

'Geophysical mapping technique in environmental planning, in *Planning and Engineering Geology*, M.G. Culshaw, F.G.Bell, J.C. Cripps and M. O'Hara (eds). Engineering Geology Special Publication No. 4, Geological Society London 171–177

Darracott, B.W. and Lake, M.I. (1981) 'An initial appraisal of ground probing radar for site investigation in Britain', *Ground Eng.*, April, 14–18

Dearman, W.R., Baynes, F.J. and Pearson, R. (1977) 'Geophysical detection of disused mineshafts in the Newcastle upon Tyne area, N E England', *Q. J. Eng. Geol.*, **10**, 257–269

Deere, D.K. (1964) 'Technical description of rock cores for engineering purposes', *Rock Mechanics and Eng. Geol.*, **1**, 17–22

Deere, D.V. and Miller, R.P. (1966) 'Engineering classification and index properties for intact rock', *Technical Report No. AFNL-TR-65-116*, Air Force Weapons Laboratory, New Mexico, USA

Deere, D.V., Hendorn, A.J., Patton, F.D. and Cording, E.J. (1967) 'Design of surface and near-surface construction in rock', *Proc. 8th Symp. Rock Mechanics*, Minnesota, C. Fairhurst (ed.), American Institute of Mining Engineers, New York, 237–302

Dines, K.A. and Lytle, R.J. (1979) 'Computerised geophysical tomography', *Proc. Inst. Elect. Electronic Engineers*, **67**, 1065–1073

Dobrin, M.B. (1976) *Introduction to Geophysical Prospecting*, (3rd edn), McGraw-Hill, New York

Dolphin, L.T., Bollen, R.L. and Detzel, G.N. (1974) 'An underground electromagnetic sounder experiment', *Geophysics*, **39**, 49–55

Eissa, E.A. and Kazi, A. (1988) 'Relation between static and dynamic Young's modulus of rocks', *Int. J. Rock Mechanics, Mineral Sci. Geomech. Abstr.*, **25**(6), 479–482

Gallagher, C.P., Henshaw, A.C., Money, M.S. and Tarling, D.H. (1978) 'The location of abandoned mineshafts in rural and urban environments', *Bull. Int. Assoc. Eng. Geol.*, **18**, 179–185

Geyer, R.L. and Myung, J.I. (1970) 'The 3-D velocity log. A tool for in situ determination of the elastic moduli of rocks', *Proc. 12th Symp. Rock Mechanics*, University of Missouri-Rolla, Missouri, Society of Mining Engineers, American Institution of Mechanical Engineers, 71–105

Glenn, W.E. and Nelson, P.H. (1979) 'Borehole logging techniques applied to base metal ore deposits'. In *Geophysics and Geochemistry in the Search for Metallic Ores*, P.J. Hood (ed.) Geological Survey of Canada. *Economic Geology Report*, **31**, 273–294

Goodman, R.G. (1976) '*Methods of Geological Engineering in Discontinuous Rocks*. St Paul West Publishing Co., 472pp.

Grainger, P. and McCann, D.M. (1977) 'Inter-borehole acoustic measurements in site investigation', *Q. J. Eng. Geol.*, Geological Society, London, **10**, 241–255

Grainger, P., McCann, D.M. and Gallois, R.W. (1973) 'The application of the seismic refraction technique, to the study of the fracturing of the Middle Chalk at Mundford, Norfolk, *Geotechnique*, **28**, 219–232

Greenfield, R.J. (1979) 'Review, of geophysical approaches to the detection of karst', *Bull. Assoc. Eng. Geol.***16**, 393–408

Hamdi, F.A.I. and Taylor Smith, D. (1981) 'Soil consolidation behaviour assessed by seismic velocity measurement', *Geophysical Prospecting*, **29**, 715–729

Hatton, L., Worthington, M.H. and Makin, J. (1986) *Seismic Data Processing Theory and Practice*, Blackwell Scientific Pulications, Oxford

Hilchie, D.W. (1968) 'Caliper logging. Theory and Practice', *The Log Analyst*, **10**(1), January/February, 3–12

Hoek, E. and Brown, E.T. (1981) 'Underground Excavations in Rock', *Inst. Mining and Metall.* London,

Ide, J.H. (1936) 'Comparison of statically and dynamically determined Young's modulus of rocks', *Proc. Nat. Acad. Sci.*, A, **22**, 81–92

Kennard, J. and Lea, J.J. (1974) 'Some features of construction of Fernsworth Dam', *J. Inst. Water Engineers*, **1**, 11–38

King, M.S. (1970) 'Static and dynamic moduli under pressure', In Proc. 11th Symp. Rock Mechanics, W.H. Somerton (ed.) American Institute of Mining Engineers, New York, 329–351

Knill, J.L. (1970) 'The application of seismic methods in the prediction of grout take in rocks', *Proc. Conf. In Situ Investigation in Soils and Rocks*, British Geotechnical Society, London. 93–100

Laine, E.F. (1980) 'Detection of water-filled and air-filled underground cavities', *Publication No. UCRL-53126*, Lawrence Livermore Laboratory, University of California, 15pp.

Lama, R.D. and Vutukuri, V.S. (1978) *Handbook on Mechanical Properties of Rocks*, Trans. Tech. Publications, pp. 196–220

Lanyon, W. (1986) 'Interactive computer analysis of borehole televiewer data', *Proc. 10th European Formation Analysis Symp. of the Society of Well Log Analysts*, Paper S, Aberdeen, 15pp.

La Porte, M., Lavergne, M., Lakishmanan, J. and Willm, C. (1973) 'Measures sismiques par transmission application au Genie civil', *Geophysical Prospecting*, **21**, 146–158

Leggo, P.J. (1982) 'Geological application of ground impulse radar', *Trans. Inst. Mining and Metall.*, Section B, Applied Earth Sciences, **91**, B1–B5

Leggo, P.J. and Leech, C. (1983) 'Subsurface investigation for shallow mine workings and cavities by the ground impulse radar technique', *Ground Eng.*, **16**(1), 20–23

Lloyd, P.M., Dahan, C. and Hutin, R. (1986) 'Formation imaging with micro-electrical scanning arrays: A new generation of stratigraphic high resolution dipmeter tools', *Trans. 10th European Symp. Society of Professional Well Log Analysts*, Aberdeen, Scotland, Paper L

Lytle, R.J., Laine, E.F., Lager, D.L. and Davis, D.T. (1979) 'Cross-borehole electromagnetic probing to locate high-contrast anomalies', *Geophysics*, **44**, 1667–1676

McCann, D.M. and Jackson, P.D. (1988) 'Seismic imaging of the rock mass', *The University of Wales Science and Technology Review*, No. 4, pp. 21–28

McCann, D.M., Grainger, P. and McCann, C. (1975) 'Interborehole acoustic measurements and their use in engineering geology', *Geophysical Prospecting*, **23**, 50–69

McCann, D.M., Baria, R., Jackson, P.D. and Green, A.S.P. (1986) 'Application of cross-hole seismic measurements to site investigation', *Geophysics*, **51**, 914–925

McCann, D.M., Jackson, P.D. and Culshaw, M.G. (1987) 'The use of geophysical surveying methods in the detection of natural cavities and mineshafts', *Q. J. Eng. Geol.*, **20** 50–73

McCann, D.M., Jackson, P.D. and Fenning, P.J. (1988) 'Comparison of the seismic and ground probing radar method in geological surveying', *Proc. Inst. Elect. Engineers*, **135**, Part F, No. 4, 380–390

McCann, D.M., Culshaw, M.G. and Northmore, K. (1990) 'Rock mass assessments from seismic measurements'. In *Field Testing in Engineering Geology*, F.G. Bell, M.G. Culshaw, J.C. Cripps and J.R. Coffey (eds.) Engineering Special Publication. No. 6, Geological Society, London, pp. 257–266

McDowell, P.W. (1981) 'Recent developments in geophysical techniques for the rapid location of near surface anomalous ground conditions', *Ground Engineering*, **14**, 20–23

McEwen, T.J. and Shedlock, S, (1985) 'Fracture delineaiton from borehole geophysics in crystalline rocks', *8th European Formation Evaluation Symp.*, London, Society of Professional Well Log Analysts, Paper R, 1–9

McEwen, T.J., McCann, D.M. and Shedlock, S. (1985) 'Fracture analysis from borehole geophysical techniques in crystalline rocks', *Q. J. Eng. Geol.*, **18**, 413–436

Moffat, D.L. and Peters, L. (1972) 'An electromagnetic pulse hazard detection system', *Proc. 1st North American Rapid Excavation and Tunneling Conference*, Chicago, American Institute of Mining Engineers, New York, **1**, 235–255

Moffat, D.L. and Puskar, R.J. (1976) 'A subsurface electromagnetic pulse radar', *Geophysics*, **41**, 506–518

Mooney, H.M. (1980) Handbook of Engineering Geophysics, Volume 2, Electrical resistivity, Bison Instruments, Minneapolis.

Mossman, R.W. and Heim, G.E. (1972) Seismic exploration applied to underground excavation problems. Proceedings of the 1st North American Conference on Rapid Excavation and Tunnelling, Chicago, American Institute of Mining Engineers, New York, Vol. 1, 169–192

Myung, J.I. and Helander, D.P. (1972) 'Correlation of elastic moduli dynamically measured by in-situ and laboratory techniques', Paper H. *Proc. 13th Annual Logging Symp. Soc. of Professional Well Log Analysts*, Tulsa, Oklahoma, 7–10 May

Neumann, R. (1977) 'Microgravity methods applied to the detection of cavities', *Proc. Symp. Detection of Subsurface Cavities*, Soils and Pavements Laboratory, US Engineering, Waterways Experiment Station, Vicksburg, Mississippi, 142–160

Olsson, O., Sondberg, E. and Nilsson, B. (1983) 'Cross-hole investigations – The use of borehole radar for the detection of fracture zones in crystalline rock', *Stripa Project Internal Report 83–06*, Swedish Nuclear Fuel Supply Co/Division KBS

Parasnis, D.S. (1986) Principles of Applied Geophysics (4th edn), Chapman and Hall, London

Paterson, N.R. and Ronka, V. (1971) 'Five years of surveying with the Very Low Frequency –Electro Magnetic method, *Geoexploration*, **9**, 7–26

Paulson, B.N.P. and King, M.S. (1980) 'Between-hole acoustic surveying and monitoring of a granite rock mass', *Int. J. Rock Mechanics Sci. and Geomech.*, **17**, 371–376

Phillips, W.J. and Richards, W.E. (1975) 'A study of the effectiveness of the VLF method for the location of narrow-mineralised fault zones', *Geoexploration*, **13**, 215–226

Robertson, A.S. and Perkins, M.A. (1979) 'Logging in Three Boreholes at Altnabreac, Caithness', Report, No. WD/ST/79/12, Hydrogeological Unit, Institute of Geological Sciences, (Unpublished)

Rubin, L.A. and Fowler, J.C. (1978) 'Ground probing radar for delineation of rock fractures', *Eng. Geology*, **12**, 163–170

Sauu, J. and Gartner, J. (1979) 'Fracture detection from the logs', *Trans. 6th European Symp. Society of Professional Well Log Analysts*, London, paper L 23p.

Scalabrini, M., Carrugo, A. and Carati, L. (1964) 'Determination in situ of the Frera Dam foundation by the sonic method, its improvement by consolidation, grouting, and verification of the result by again using the sonic method', *Proc. 8th Int. Congress on Large Dams*, Edinburgh, **1**, 585–600

Schwarz, S.D. (1972) 'Geophysical measurements related to tunnelling', *Proc. 1st North American Conference on Rapid Excavation and Tunnelling*, Chicago, American Institute of Mining Engineers, New York, **1**, 195–208

Scott, J.H., Lee, F.T., Carroll, R.D. and Robinson, C.S. (1968) 'The relationship of geophysical measurements to engineering and construction parameters in the Straight Greek Tunnel pilot bore, Colorado', *Int. J. Rock Mechanics and Mining Science*, **5**, 1–30

Scott Keys, W. and MacGary, L.M. (1971) 'Application of borehole geophysics to water resources investigations. US Department of the Interior, *Techniques of Water-Resources Investigations of the United States Geological Survey*, Book 2, Ch.E1

Simmons, G. and Brace, W.F. (1965) 'Comparison of static and dynamic measurements of the compressibility of rocks', *J. Geophys. Res.* **70**(22), 5649–5656

Sutherland, R.B. (1962) 'Sonic dynamic and static properties of rocks', *Proc. 5th Symp. Rock Mechanics*, Minnesota, C. Fairhurst, (ed), Pergamon Press, New York, 473–490

Telford, W.M., Geldart, L.P., Sherriff, R.E. and Keys, D.A. (1976) *Applied Geophysics*, Cambridge University Press, Cambridge

Terzaghi, K. (1946) 'Rock defects and loads on tunnel supports', In *Rock Tunnelling with Steep Supports*, R.V. Proctor and T. White (eds), Commercial Shearing and Stamping Co., Youngstown, 15–19

Thill, R.E. (1978) 'Acoustic cross-borehole apparatus for determining in situ elastic properties and structural integrity of rock masses', *Proc. 19th US Symp. Rock Mechanics*, University of Nevada, **2**, 121–129

Unterberger, R.R. (1978) 'Radar propagation in rock salt', *Geophysical Prospecting*, **26**, 312–328

Van Heerden, W.L. (1987) 'General relations between static and dynamic moduli of rocks', *Int. J. Rock Mechanics Min. Sci. and Geomech. Abstr.* , **24**(6), 381–385

Ward, W.H., Burland, J.B. and Gallois, R.W. (1968) 'Geotechnical assessment of a site at Mundford, Norfolk for a large proton accelerator', *Geotechnique*, **18**, 399–431

Wickham, G.E., Tiedemann, H.R. and Skinner, E.G. (1972) 'Support determination based on geological predictions', *Proc. 1st North American Conference on Rapid Excavation and Tunnelling*, Chicago, American Institute of Mining Engineers, New York, **1**, 43–64

Wyllie, M.R.J., Gregory, A.R. and Gardner, L.W. (1956) 'Elastic wave velocities in heterogeneous and porous media', *Geophysics*, **21**, 41–70

10 Instrumentation and monitoring in rock masses

Professor F G Bell
University of Natal

10.1 Introduction

Instrumentation and monitoring of rock masses not only provides a check on the assumptions made in design, it also can be used when detailed analytical design is not justified but the stability of the rock mass is still questionable. In other words, instrumentation may allow less costly construction, the instrumentation providing a continual check on the stability and safety of a structure during its life-span. This is the observational method as proposed by Peck (1969). In such a situation monitoring can provide advance warning of adverse or unpredicted behaviour, thereby affording time for remedial action. Generally monitoring may be carried out in order to record values of, or variations in, geotechnical parameters such as rock stress, ground movements, groundwater level and pressure, etc. (Hanna 1985).

Rock masses are complex media whose engineering properties are difficult, if not impossible, to predetermine accurately ahead of excavation. The models used to predict the various aspects of rock mass response are based on assumptions, idealizations and simplifications. Hence it is quite freuqently necessary to obtain checks on the accuracy of the predictions made in design calculations.

In order that a monitoring system should fulfil its intended function economically and reliably, it should be easy to install and read, possess adequate sensitivity to provide accurate and reproducible results, and be robust enough to endure the period of installation. Even when instrumentation is as simple, reliable and robust as possible, some proportion of it may fail to work or be destroyed by construction plant or vandals. Hence it is often necessary that additional instrumentation should be available to allow for losses. Furthermore the instrument should be able to determine the required parameter without inducing a change in that parameter due to the presence of the instrument.

10.2 Measurement of groundwater level and pore water pressure

Because of their importance, measurement of groundwater level and pore water pressure represent some of the most frequent types of *in situ* measurements recorded. At least the water table and its seasonal fluctuations should be determined, since such data help to assess the geotechnical information gathered by the investigation and, of course, groundwater conditions, including pore water pressure, influence the type of foundation structure chosen.

A standpipe or a piezometer can be used to assess pore water pressure. Piezometers should be able to respond quickly to changes in groundwater conditions, should cause as little interference to the ground as possible, and should be rugged enough to function over long periods of time.

It was demonstrated by Hvorslev (1951) that as some flow of water from the ground into a piezometer system is required in order for it to record pressure changes, and because the ground around the piezometer offers a resistance to flow, then a time lag exists between the changes of groundwater pressure and their recording. The hydrostatic time lag is proportional to the volume of water that has to flow into the piezometer to produce a given pressure change, and is inversely proportional to the permeability of the ground about the tip of the piezometer.

10.2.1 Standpipes and standpipe (Casagrande) piezometers

The standpipe represents the simplest method of monitoring pore water pressure. It consists of an open-ended plastic tube up to 50 mm in diameter, which is perforated over about 1 m near the base. It is placed in a drillhole. A filter of fine gravel or sand is placed around the tube and the top of the hole is sealed with puddle clay or concrete to keep out surface water. The water level in the standpipe is measured by lowering an electrical 'dipmeter' down the open standpipe. The dipmeter produces a visual or audible signal when it comes into contact with the water level.

The standpipe suffers a number of disadvantages. For example, as no attempt is made to measure pore water pressure at a particular level, it therefore is assumed that a simple groundwater regime exists. However, if flow occurs between adjacent strata with differing permeabilities, then the water level in the standpipe will be meaningless. Another disadvantage is the length of time required for the level of water in the standpipe to stabilize with that in the ground, unless the latter is highly permeable.

Because changing rock types at different depths mean varying permeabilities and pore water pressures, the pore water pressures are measured over a limited zone by sealing off a section of the drillhole. The standpipe piezometer consists of a porous tip surrounded by sand or gravel at the level where the pore water pressure is to be measured and is connected to a plastic tube which extends to the ground surface (Figure 10.1). The filter is sealed above and below with grout, which typically consists of cement and bentonite.

A number of precautions should be taken; in particular, the drillhole should be sealed effectively. During installation of the lower bentonite seal it is recommended that the drillhole be cased to below the level of the bottom of the sand filter, or to about 0.5 m below the proposed level of the tip. Equal quantities of bentonite and cement should be mixed in a grout mixer, and the mix should be as stiff as is compatible with placement by tremie pipe at the bottom

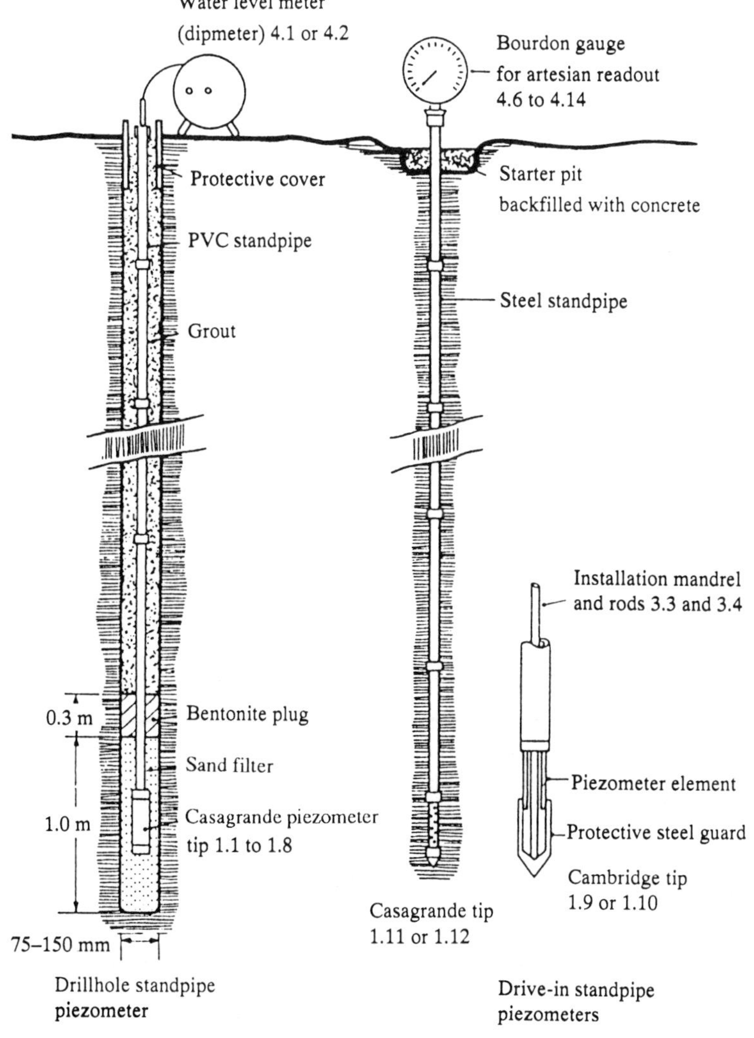

Figure 10.1 *Standpipe piezometers (Soil Instruments Ltd)*

of the hole. Contamination of the filter by the seal must be avoided. This can be achieved by using bentonite pellets (these form a seal with a permeability as low as 10^{-10} m/s). The upper surface of the seal should be tamped. Filter material is dropped from the top of the hole until the level of the porous pot is reached. Once the pot is positioned the casing is raised to the level of the top of the filter, and further sand placed until the correct level is reached. Detailed attention has to be given to sealing the porous tip in the ground, particularly when the pore water pressures are high. The top seal is formed in the same way as the bottom, with the contact between the sand and bentonite being made from bentonite pellets. Backfill above the seal should be gently compacted. The sand filter should be saturated when placed in order to quicken equalization between groundwater levels and the level in the standpipe. The riser pipe should be topped up with water once the bentonite/cement seal has had time to stiffen. If more than one piezometer is installed in the drillhole the likelihood of failure increases. After installation the ability of a piezometer to function should be checked by topping up and it should be observed daily until an apparent equilibrium is attained. Thereafter the intervals between observations can be lengthened to a week or a month but ideally should continue for about a year. Again the response of the system to changes in groundwater pressure is slow.

The most common piezometer tips in use are the ceramic type developed by Casagrande (1949), or the more modern porous plastic equivalents, both typically having an average pore size of 50–60 μm and a low air entry resistance. The main disadvantage of the ceramic tip is its susceptibility to damage. The top of the riser pipe should be protected at ground level by a lockable cap. A bourbon gauge may be attached to the top of the riser pipe when the piezometer water level rises above the ground surface.

10.2.2 The hydraulic piezometer

The pore water pressure in low permeability ground can be measured with a hydraulic piezometer. Such piezometers are reliable and offer long-term performance. Measurements of pore water pressure using the hydraulic piezometer are made at a remote readout station, up to 1 km away, and the measurements are made possible by twin tubes filled with de-aired water connected to the porous tip. The lines are sometimes placed in a conduit to avoid being damaged. The highest level of the tubing should not be more than 5 m above the piezometer level. One of the lines passes to the bottom of the piezometer, the other terminates at the top. After installation, the piezometer is flushed to remove air by pumping water down one line to the bottom of the piezometer tip and allowing it to escape up the other. Once de-airing has taken place, the de-airing line is closed and the pressure line is connected to a manometer or pressure gauge. The system can be de-aired during use and the response lag can be controlled by selection of the tube diameter. Nevertheless the response time of the system generally is low, depending upon the quality of the de-aired water within it and the type of pressure-measuring device employed. In order to avoid the effects of hydraulic fracture, excessive swelling or consolidation, the pressures applied to the two tubes should have an average value equal to the pressure at the tip. Measurement of pore water pressure at the read-out station may be made by mercury manometer,

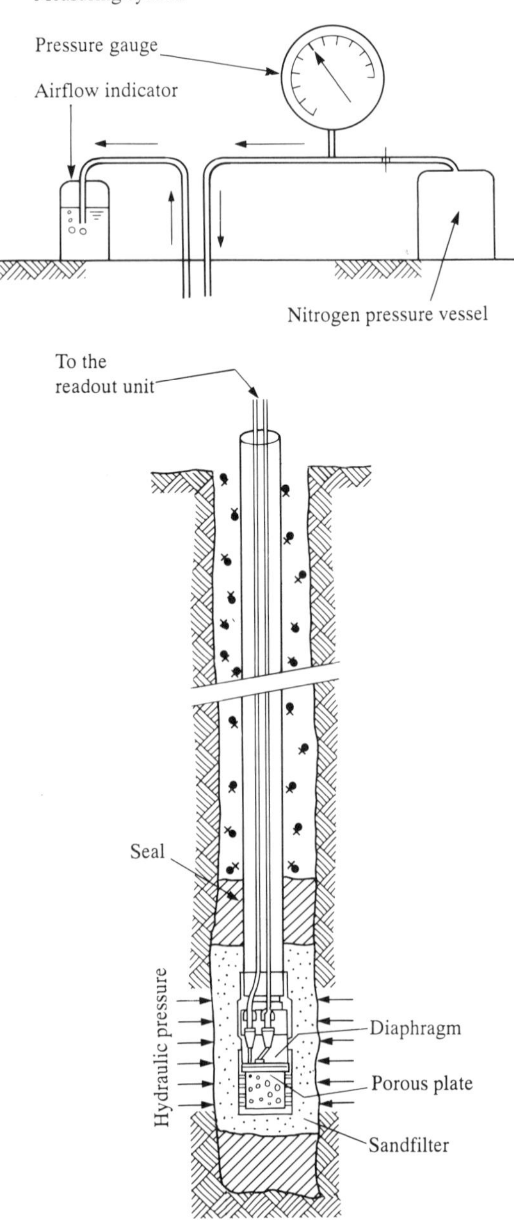

Figure 10.2 *Pneumatic piezometer equipment*

Bourdon gauge or pressure transducer. As the piezometer has two tubes, two independent readings can be made from the same tip, thereby providing a check.

10.2.3 The pneumatic piezometer

Pneumatic piezometers can be used to measure pore water pressures in low-permeability ground (Deardoff *et al.* 1980). The piezometer tip consists of a ceramic porous stone, behind which an air-activated pressure cell is mounted (Figure 10.2). Installation follows the same procedure as that for the hydraulic piezometer. The tip is connected to instruments at the surface by twin tubes. These in turn are connected to a flow indicator, and an air supply and pressure-measuring device. The piezometer is operated by water pressure acting on a diaphragm which is balanced by an externally applied pressure, usually nitrogen. Nitrogen is preferred, particularly in situations where the piezometer is expected to be in use over a period of several years. Compressed air may be employed for short-term usage without seriously affecting the performance of the device; any moisture that does build up in the system can be removed by periodically flushing with nitrogen. When pore water pressures are read, nitrogen is admitted to one line, but is prevented from flowing along the other line by a blocking diaphragm in the tip. When the externally applied pressure equals the groundwater pressure acting on the reverse side of the diaphragm, a valve opens and allows flow along a return line to a detector at the read-out. This read-out may be on a Bourdon gauge or a digital display. The advantages of this type of piezometer are simplicity of operation, accuracy and a fast response time. The latter is due to the fact that the amount of water displaced by the diaphragm is very small ($<0.11\,\text{cm}^3$) and the total fluid volume of the tip is low. These piezometers have an accuracy of about $\pm$ 0.2 m head of water and readings may be taken up to 0.5 km from the piezometer tip.

It is not possible to measure negative pressure and in the long-term some pneumatic piezometers may become unreliable. The fluid chamber inside the piezometer cannot be de-aired once the device is installed.

Sometimes pore pressure measurements are required on the interface of a structure and an illustration of the technique using a disc piezometer is shown in Figure 10.3.

10.2.4 The piezodex system

Various methods have been developed in order to measure the distribution of piezometric head along drillholes in rock masses. The continuous piezometer uses a single continuous membrane as a packer and a probe able to take readings at any position along the drillhole. The modular piezometer works with a chain of packers and with a portable probe, permitting readings in the measurement zones defined by neighbouring packers. The piezodex system also employs a packer chain for defining mutually independent measurement intervals and a portable probe for taking measurements (Figure 10.4). The main difference between the latter two systems lies in the form of the packers and in the measurement principle. With the modular piezometer the packers are activated simply by inflation with gas. Packers in the piezodex system are all interconnected by a central line and are inflated simultaneously with gas. The modular piezometer measures the pressure directly whereas the piezodex system uses a force sensor. In the piezodex packer chain

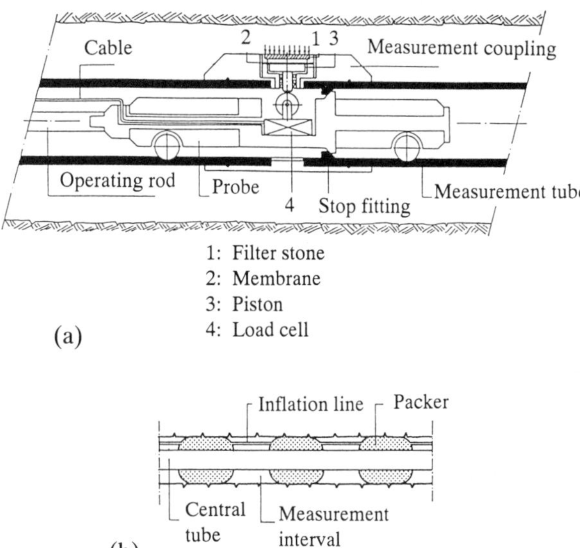

Figure 10.4 *The piezodex system:* (a) *diagram of the probe* (b) *packer chain; (after Kovari and Koeppel 1987)*

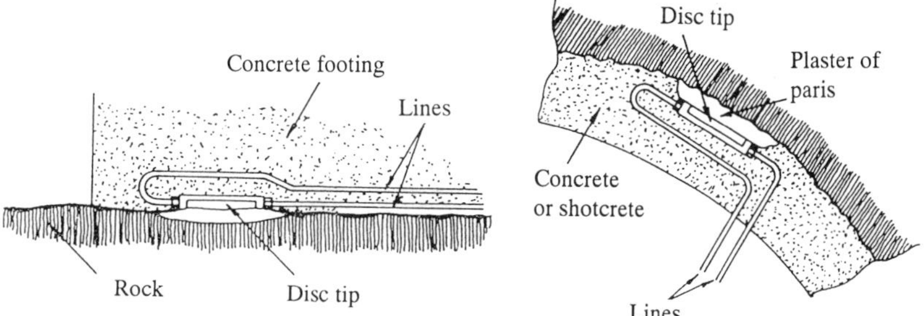

Figure 10.3 *Disc hydraulic piezometers used at an interface (Soil Instruments Ltd)*

nitrogen is used as the inflation gas because of its very low diffusion through the rubber membrane. The inflation pressure is selected in accordance with the anticipated maximum value of liquid pressure along the drillhole. A standard pressure tank is employed for supplying and maintaining the pressure throughout the observation period. Packers of this type can be re-used elsewhere after concluding the measurement programme. The position of the individual packers or the location and length of the measurement intervals are chosen in accordance with the fissure system derived from the rock core. It is possible to arrange the packers at irregular intervals. The number of measuring intervals is limited by the maximum length of the packer chain (approximately 120 m), by the minimum length of a single packer (1.0 m) and by the design minimum length of a measuring interval (0.50 m).

The cells transmit the liquid pressure to the force sensor of the probe during the reading period. The liquid exerts the pressure via a membrane onto the pressure transmitter which takes the form of a piston. When the probe is in measuring position, the piston is raised slightly (0.1–0.2 mm) from its normal position against the membrane. During this phase the water pressure acts directly on the low-friction controlled piston. In this way the pressure is obtained by way of a force measurement. The force is given as the product of the water pressure and the area of the membrane-protected piston surface. By selecting various piston dimensions, it is possible to account for different sensistivities or measuring ranges with the same force sensor. To obtain high accuracy of measurement, the axis of the force sensor must comply with that of the piston. This is achieved by high-precision positioning of the probe using the cone-and-ball principle (Kovari *et al.* 1979). Conical stops are fitted to the central tube and spherical stops to the piezodex probe. A detailed description of the measuring system with measurement examples is provided by Kovari and Koeppel (1987).

10.3 Measurement of stress and strain in rock masses

The state of stress which exists in the Earth's crust is disturbed by engineering operations which produce a new distribution of induced stress within the surrounding rocks. The readjustment of the state of stress in the ground can give rise to stress concentrations which can have a major influence on the stability of surface excavations and underground openings. The original or virgin stresses include gravitational stress attributable to the weight of overburden, tectonic stress due to earth movements and other factors such as weathering, consolidation, dehydration, hydration and pore water pressure. As a consequence the *in situ* state of stress in a rock mass has proved difficult to assess with accuracy.

Stress measurements in rock involve determination of the absolute state of stress on the one hand and measurement of relative stress or change of stress on the other. The measurement of stress, contact pressures and change of stress can be made in two ways, namely, strain may be measured and then converted to stress or stress may be measured directly. Measurement of absolute stress in rock masses which behave more or less in an elastic manner may be achieved by a stress-relief technique, in which rock containing the instrumentation is relieved from the stress produced by the confinement of the surrounding rock. The strain resulting from this stress relief is measured. Change of stress may be determined by measuring absolute stress over a given interval of time. Generally the instruments used in both types of measurement are similar.

The primary objective in determination of rock stress obviously is to obtain precise values. This is easier said than done. Measurements usually are taken in a drillhole, the drilling of which interferes with the state of stress. One of the principal problems associated with stress measurements is attributable to the nature of the rock itself, its behaviour being related to its origin, composition, texture, especially if this involves preferred orientation, geological history and degree of weathering.

Grob *et al.* (1975) reviewed the sources of error involved with the determination of *in situ* stress measurements. Some of these arise in connection with the measuring technique and are closely related to the instrument which is used. The non-elastic, anisotropic and inhomogeneous behaviour of the rocks concerned, even in the case of hard rock, can represent serious difficulties. The extrapolation of local values of stress to large volumes of rock presents further problems such as the effects of discontinuities. Grob *et al.* went on to mention that a large number of measurements with more than one type of instrument and a careful laboratory study of the deformational behaviour of rock cores help to improve the value of the results. Nonetheless the reliable determination of *in situ* stresses is a difficult task, which in many cases cannot be carried out satisfactorily.

In the determination of absolute stress, drillhole deformation meters and drillhole strain-cell devices depend upon assumptions of elasticity, as they measure elastic rebound that occurs on overcoring. Therefore stress-relief strain-measurement techniques are valid only if applied to isotropic homogeneous strong rocks that display near-elastic deformation characteristics. They are of little value in soft sediments, rocks with marked anisotropy or those which undergo time-dependent deformation when loaded. Even rocks which behave in a more or less elastic manner, there is the problem of determining the appropriate deformation modulus. If measurement extends over lengthy periods, then the time-dependent characteristics of the rock must be taken into account in order to interpret the results correctly. Indirect compensation techniques have been used whereby the overcored rock specimen, together with its deformation meter or strain cell, is placed under load in a biaxial pressure jacket. This then is pressurized to restore the meter to its initial reading, the pressure applied being equated to the *in situ* rock pressure. Nonetheless such techniques assume identical stress–strain characteristics on relief as on reloading which is not the case with rocks. It is often more useful, however, to determine the relative rather than the absolute *in situ* stress, in other words, is the stress high or low, increasing or decreasing and if so at what rate?

Instruments which involve bonding electrical strain gauges and rosettes to rock have to be used with great care since the gauges will produce a signal irrespective of the condition of the bond. Hence the condition of the bond should be investigated before taking readings. On the other hand with photoelastic gauges the signal is unreadable if the bond is not good.

The stress-relief technique is illustrated in Figure 10.5. The measuring instrument may be attached to the rock surface forming the wall of the opening or the rock and gauge may be overcored. In the first case the rock to which the gauge is attached is relieved either by cutting slots on all four sides or by drilling a ring of overlapping holes. Instruments used in this way include linear extensometers, strain-gauge rosettes, and photoelastic biaxial gauges. Overcoring requires the measurement of strain at three points around the periphery of a drillhole, three rosette strain gauges being glued around the hole. Zero strain readings are taken and the drillhole then is overcored. Stress-relieved readings are taken after removal of the hollow cylindrical core. The procedure may be repeated at successive depths in the rock wall. Overcoring is a specialist field-measuring technique which tends to be associated with deep rock problems.

10.3.1 Drillhole deformation meters

Three types of drill-hole strain-measuring instruments are used, namely, drillhole deformation meters, drillhole strain cells and drillhole inclusion stress meters. A drillhole deformation meter records changes in the cross-sectional dimensions of a drillhole in rock as a result of stress change. The stresses are calculated with the help of elastic theory.

The USBM single-component drillhole deformation gauge can be used in a drillhole with a 38 mm diameter. Initially a hole is drilled into the rock with a diameter of 150 mm. Then a 38 mm diameter hole is drilled in the centre of the hole to a further depth of 3 m or so. The drillhole deformation gauge is placed, to a depth of from 150 to 250 mm, in the central hole and orientated to take measurements along the vertical diameter of the drillhole. Overcoring takes place after an initial reading is taken and as this proceeds readings of deformation are recorded at regular intervals. The deformation develops gradually as overcoring extends. Stress relief usually is regarded as completed when the drill has passed some 50 mm beyond the gauges. A final reading is taken at this point and then the rock core containing the gauge is removed from the hole. The deformation modulus of the rock is determined subsequently.

As successive measurements are not made in the same plane it is necessary to interpolate between the readings, which limits the usefulness of the instrument in localities where high strain gradients occur. Therefore multiple-

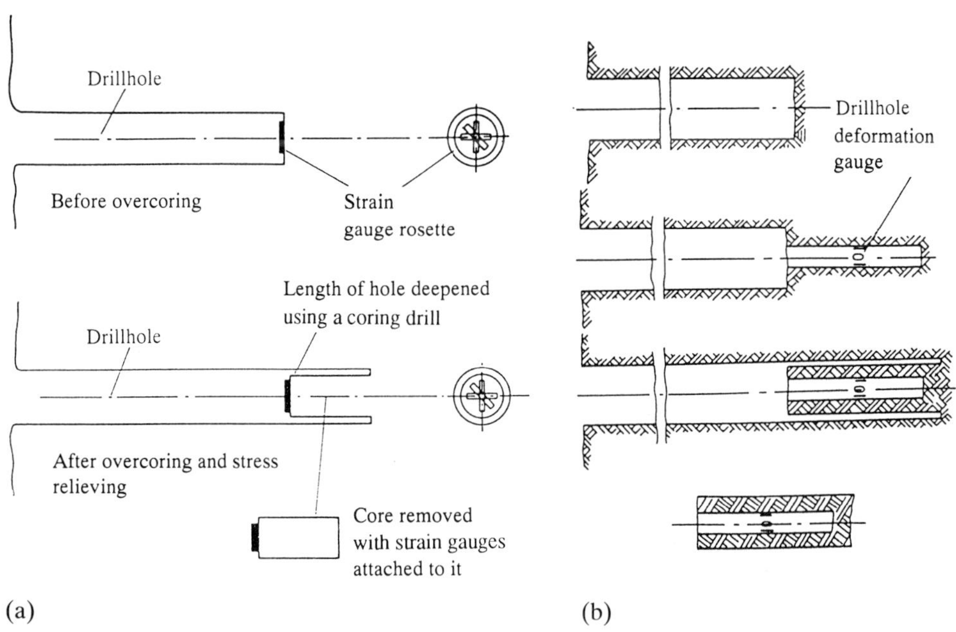

Figure 10.5 *The stress relief technique:* (a) *using drillhole strain cell;* (b) *using drillhole deformation gauge (after Roberts 1977)*

component drillhole deformation meters have been developed which record measurements along three directions 60° apart. These measurements are in a single plane, normal to the axis of the drillhole. The USBM three-component drillhole defomation gauge is one such instrument (Figure 10.6).

10.3.2 Drill hole strain cells

Strain-gauge rosettes are sometimes fixed to the back of a drillhole. However, attempts to overcore electrical resistance strain gauges set at the back of a drillhole sometimes encounter problems, primarily due to the difficulty of insulating the gauges and to contact with flush water from the core bit during drilling.

Leeman (1969) overcame this problem in the CSIR 'doorstopper' drillhole strain cell by encapsulating the electrical connections to the gauges in a silicone rubber moulding (Figure 10.7). The 'doorstopper' can be used in a 60 mm diameter drillhole. It is orientated to measure strains in the vertical, horizontal, and 45° directions.

The CSIR three-component strain cell also was developed by Leeman (1969) and can measure three strains at three points in a drillhole, in the 'roof', the 'sidewall', and an intermediate point. By attaching strain-gauge rosettes to the rock at each of these positions the magnitudes and directions of the principal strains are derived. The stress components then are obtained with the aid of the values of Young's modulus and Poisson's ratio of the rock. The strain cell is inserted in a drillhole which should be inclined upwards at an angle of not less than 5° from the horizontal in order that water can drain out. The hole should be cleaned and dried before attaching the gauge, after which the mouth of the hole is plugged, so as to protect the cells from the flush water involved in overcoring. Overcoring takes place after the gauge is set in place and extends about 50 mm beyond the end of the gauge in the central hole. Next the core, containing the gauge, is broken off and removed. The central hole then is unplugged so that the setting tool can be reconnected and readings of relief-strain taken.

Rocha and Silverio (1969) described another multi-component drillhole strain cell (the LNEC drillhole stress gauge) used to determine the state of stress in a rock mass. The cell consists of a cylinder of epoxy-resin which contains ten strain gauges in the central zone. Four of the gauges province check measurements against the others.

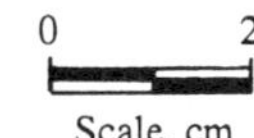

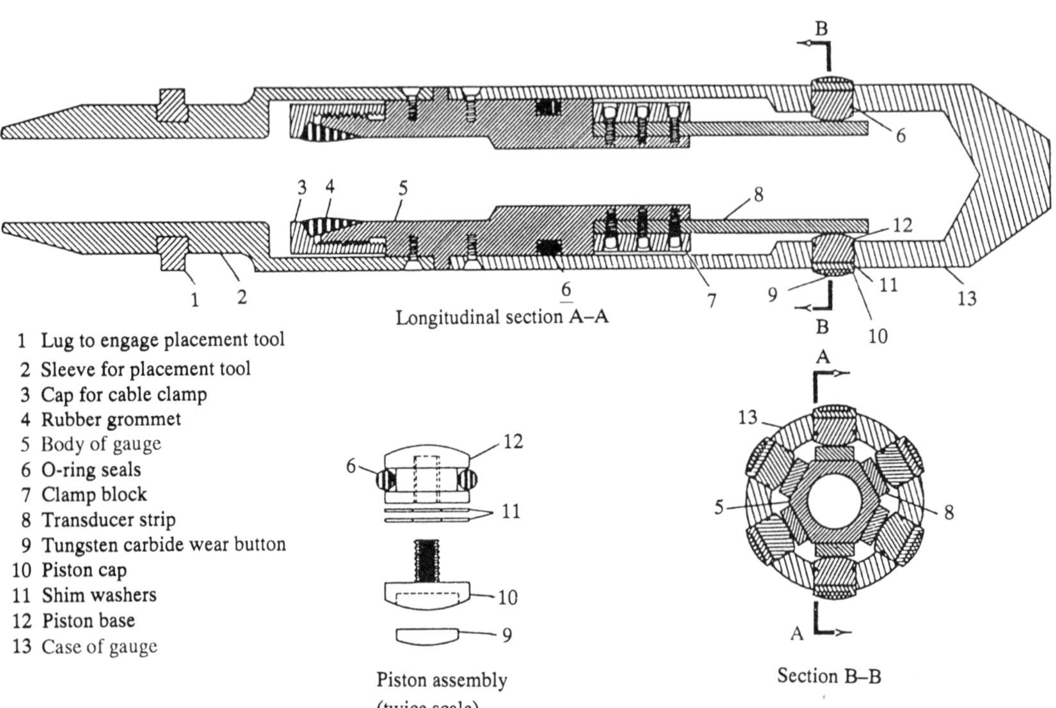

Figure 10.6 *USBM three-component drill-hole deformation gauge (after Roberts 1977)*

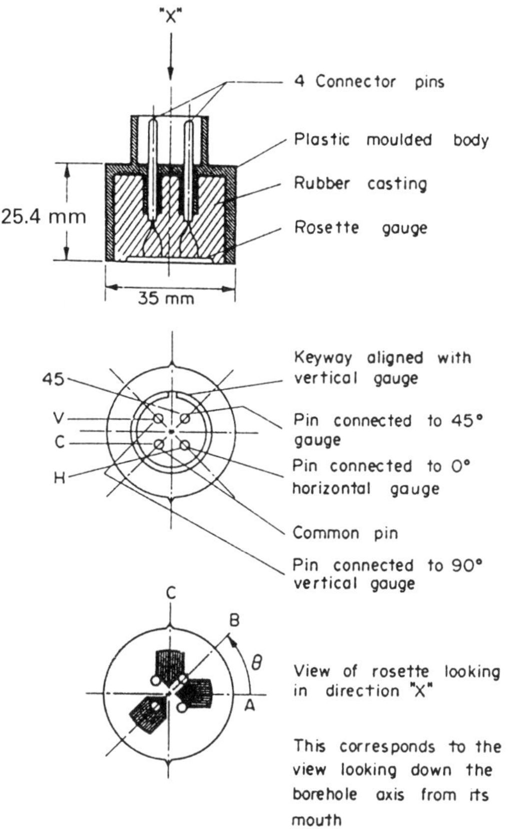

Figure 10.7 *The CSIR 'doorstopper' drillhole strain cell (after Leeman 1969)*

The strain cell is inserted into a vertical or near-vertical drillhole. Epoxy-resin cement is placed in the bottom of the hole and then the strain cell is lowered into the cement. This displaces into the annulus between the strain cell and the drillhole wall. The cell is centred and orientated within the annulus. Overcoring takes place after the cement has set and the relief strains are measured.

Hawkes and Moxon (1966) described a biaxial plastic strain gauge used to determine the *in situ* stress in rock. This also involves using the stress-relief technique. The direct application of electrical strain-gauge rosettes, the Leeman 'doorstopper' and the plastic strain gauge are all entirely dependent on the end of the drillhole being adequately prepared to receive the cell in that there must be no annular socket at the end of the drillhole when the strain cell is inserted. Hence the back of the drillhole is ground by a flat diamond bit to remove any traces of the socket. In addition, it is essential to secure a good cement bond between gauge and rock. Wet holes therefore have to be dried – water is drained, the rock surface cleaned, and acetone sprayed onto the back of the hole to dispel water from the rock surface for enough time to permit the cement to polymerize. Care should be taken to guard against temperature effects. Since differences in the coefficient of expansion between the gauge, its cement, and the rock can give rise to errors, precautions have to be taken. Consequently readings should be made when the gauges attached to the overcored rock are within 2°C of the original temperature at the back of the hole. Furthermore the flush water used during drilling must not be widely different in temperature from that of the rock. The interpretation of the measurements made by these instruments must take account of the effect of stress concentration at the back of the drillhole.

The sliding micrometer is a portable device which can take precise measurements of the strain along a drillhole (Figure 10.8). In fact is is only used to obtain strains, it is not intended to obtain stress values. To do so, a tube of hard PVC is grouted into the drillhole. It contains annular coupling elements at intervals of 1 m, with conical stops fitted on the inside. These stops serve to hold the heads of the probe during the short reading period. The probe heads are spherical in shape. When the two stops touch, the position of the centre of the ball in relation to the cone is uniquely defined. To ensure the probe glides in the tube from one measuring position to the next, the ball (probe) and cone (measuring marker) surfaces are notched and designed to enable the probe to be moved alternately into reading and sliding positions by rotating the probe ± 45°. If neighbouring measuring markers move closer together as the result of deformation in the rock, this is indicated by a change in their spacing (i.e. by the change of two readings). The strain is obtained relative to a length of 1.0 m. To conduct a complete series of readings, the probe is first secured and the reading taken at the mouth of the drillhole. The probe then is moved on, taking readings, until it reaches the bottom of the drillhole. Measurements also are taken in the reverse direction for control purposes. With this measurement system, a change of distance between two neighbouring measuring markers can be determined with an accuracy of ±3 μm which corresponds to a strain of 3×10^{-6}. A detailed description of the measurement system is given by Kovari *et al.* (1979), and Kovari and Amstad (1983).

10.3.3 Drillhole stress meters

A drillhole inclusion stress meter differs from a drill-hole deformation meter in that it may theoretically be calibrated directly in terms of stress, even though its response to stress is a measured strain, or other effect which results from that strain. Stress meters are, in fact, hard rigid or near rigid inclusions, whereas deformation meters and strain cells are either soft inclusions which offer little resistance to drill-hole deformation or simply profile-measuring devices that offer no resistance. This means that if a hard inclusion stress meter is well cemented to the walls of the drillhole changes in rock stress produce a change in the stress meter that only depends to a small extent on changes in the modulus of elasticity of the rock

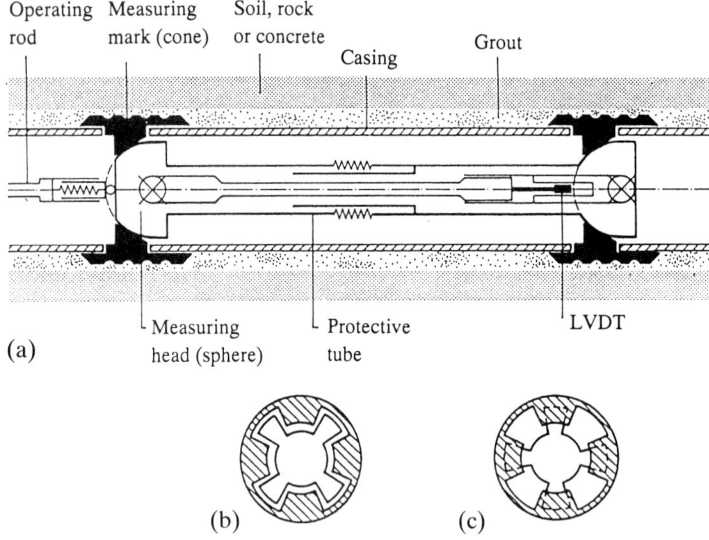

Figure 10.8 *Sliding micrometer:* (a) *schematic diagram of instrument;* (b) *sliding position;* (c) *measuring position (after Kovari et al. 1979)*

so that it is unnecessary to have exact data about the rock modulus. However, the deformability of the rock must be an order of magnitude less than that of the rigid inclusion.

All stress meters can be prestressed after insertion so that they can measure absolute stress by the overcoring technique or monitor relative stresses above and below the initial prestress level.

The calibration characteristics of all prestressed drillhole stress meters are dependent upon the level of prestress as well as the relative elastic properties of stress meter and rock. In some meters it is not easy to control the level of prestress so that it is necessary at the beginning of the test to select a level of prestress to suit the rock. As the *in situ* properties of the rock commonly are unknown the choice of prestress level is largely intuitive. Because of the difficulties of using prestressed rigid insertion stress meters and due to the surface irregularities of the drilled rock surface, they have been used infrequently.

The photoelastic glass-insertion stress meter may take the form of a solid glass cylinder, in which case the observed birefringence is a measure of the shear stress in the glass and therefore of $(\sigma_1 - \sigma_2)$ in the rock in a plane normal to the longitudinal axis of the drillhole (Roberts 1977). When a hole is drilled along the central axis of the glass insertion it forms a biaxial gauge which is calibrated in terms of the major principal stress in the plane of measurement. This stress meter instantaneously measures any increase of stress that occurs around it. In other words it measures relative stress above the initial setting value. The cylinder is placed in the rock wall, and a polarizing light probe is inserted down the central hole when a reading is taken (Figure 10.9). Observations are made with the aid of a small hand-viewer or telescope for distant viewing. This test does not involve overcoring.

10.4 Stress-change measurements

A wide range of instruments has been developed for monitoring the stress changes in rock. These include photoelastic gauges (Roberts 1977), hydraulic pressure cells and rigid inclusion instruments which use electric resistance strain gauges to measure the strains (Worotnicki *et al.* 1980). These are usually drillhole instruments and therefore monitor the stress change in one direction only. Instruments are available which can take measurements in three mutually perpendicular directions, but these measurements cannot give the complete change in the stress tensor unless the three directions correspond with the principal stress directions. Difficulties caused by stiffness incompatibility arise with these instruments. In order to obtain the value required, an experimentally determined calibration curve or a mathematical model must be used. As a consequence the accuracy of the measurements is often much less than that obtainable for displacement measurements.

The vibrating-wire stress meter is widely used (Hawkes and Hooker 1974), the main components of which are illustrated in Figure 10.10. The instrument basically consists of a hollow, hardened steel body. When in use, it is pre-loaded diametrically between the walls of a 38 mm diameter drillhole by means of a sliding wedge and platen assembly. The gauge is installed by using a setting tool. The latter pulls the wedge forward relative to the platen and gauge body which pre-loads the platen and gauge against the drillhole walls. Gauges have been installed in drillholes up to 100 m deep.

The vibrating-wire stress meter is a unidirectional instrument. Changes of stress in the rock in the direction

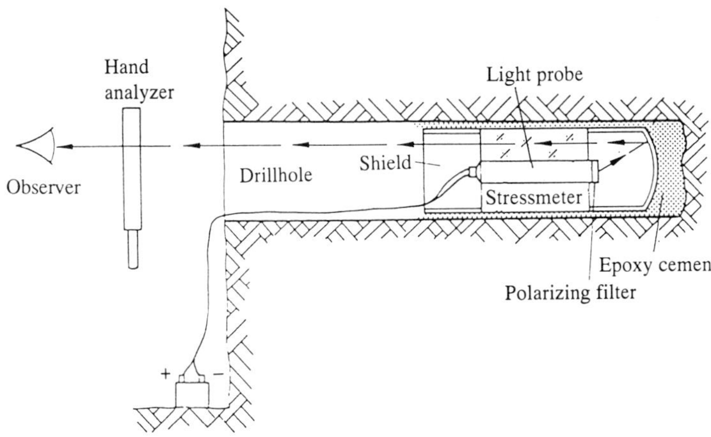

Figure 10.9 *The photoelectric glass insertion stress meter and viewing system (after Roberts 1977)*

of pre-loading induce small changes in the diameter of the gauge cylinder. These changes are measured by means of a change in the frequency of vibration of a high-tensile steel wire which is stretched across the cylinder in the direction of pre-load. The orientation of the stress meter is controlled by means of special location marks on the installation rods. Measurements are taken by plucking the wire by an electromagnetic coil. This also acts as an electronic pick-up to count the number of vibrations the wire makes in a given time. A readout box displays the period of vibration of the stress meter wire, which can be used to determine the change of stress in the rock from calibration tables. If principal stress changes are required in the plane normal to the axis of a drillhole, then three vibrating-wire stress meters, mutually inclined at 45°, may be used. A thermistor (an electrical resistance thermometer) can be built into the gauge to measure the rock temperature so that corrections can be made if necessary. The sensitivity of the gauge varies with the period of the vibration and with the contact angle made by the platen with the drillhole wall. Different platens have been used in 'hard-rock' versions of the instrument: the softer the rock, the larger the platen contact area should be.

A hydraulic fracturing test can be used to measure *in situ* stress (Haimson 1975). The test is carried out by sealing off a section of a drillhole at the depth required by using inflatable rubber packers. Fluid pressure then is applied to the section selected and the pressure is

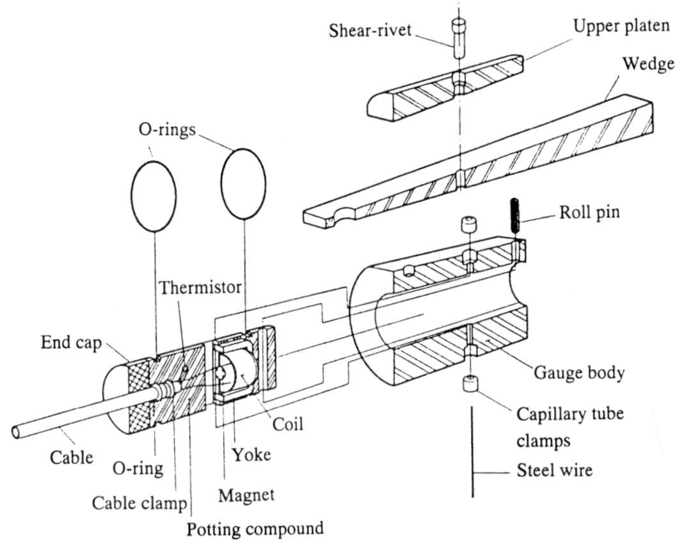

Figure 10.10 *Exploded diagram of a vibrating-wire stress meter (after Hawkes and Hooker 1974)*

increased gradually until the rock fractures. Fracture occurs when the rock wall of the drillhole fails in tension. A shut-in pressure is recorded when the pumps are shut off with the hydraulic circuit closed. The fracture and shut-in pressures are related to the *in situ* state of stress in the rock mass. The direction and inclination of the fracture(s) can be recorded with an impression packer and so the direction and the magnitude of the principal stresses can be determined. Tests can be performed at any depth.

Flat-jacks are hydraulic pressure cells which have been used to measure stress in rock masses. A flat-jack is a constant-dimension measuring device which is placed in a slot cut in a rock wall, the object being to relieve the rock of ambient stress (Figure 10.11). Before cutting the slot, reference points, used to measure deformation or residual stress, are fixed in a suitable arrangement over the rock face and their positions recorded. The slot may be formed by sawing or by drilling parallel holes and shearing the webs of rock left between. The length of the slot is frequently 330–480 mm long, and of similar depth and around 40 mm wide. Wareham and Skipp (1974) enumerated the advantages of the sawcut slot as against the overlapping hole slot, the chief of which being the ease with which the test can be carried out. The flat-jack is inserted in the slot and is grouted into place. The stresses in a biaxial field may be measured by placing two slots at right angles. The slot cutting gives rise to stress relief with the surrounding rock expanding into the slot. The convergence between reference points is measured. The flat-jack then is pressurized until the displacement produced by slot cutting is balanced. The hydraulic pressure then is released and increased in cycles and the mean cancellation pressure determined. The complete operation takes from 2 to 3 weeks. Hence the flat-jack test is a null technique by which rock stresses can be determined directly. The technique depends on the assumption that the same deformation characteristics exist on relief as on reloading, which may not be the case. A disadvantage of the flat-jack test is that the measurements are taken at the edge of the excavation where unknown and irregular stress distributions may exist and which itself may be destressed.

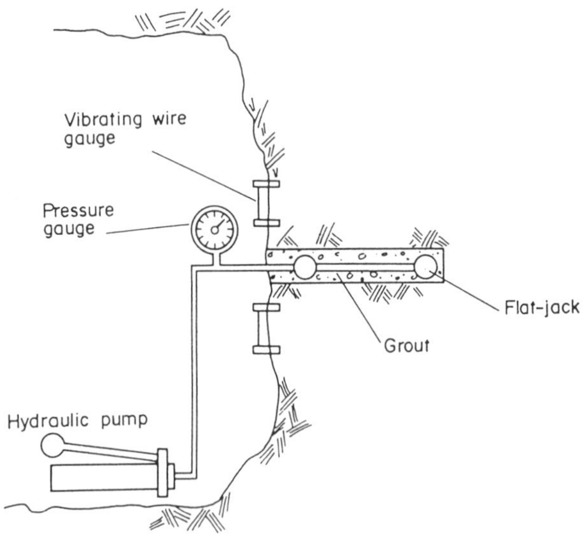

Figure 10.11 *The flat-jack method of measuring rock stress (after Roberts 1977)*

The most widely used hydraulic pressure cell is the Glötzl cell. It is used for measuring contact pressures. The Glötzl cell has a high stiffness at constant temperature and consists of a pressure-sensing pad and a hydrualic bypass valve assembly which functions as a pressure-actuated bypass valve in an individual hydraulic circuit. The cell is maintained in a 'closed' condition by the action of pressure on the sensing pad. In order to measure the magnitude of this pressure, the hydraulic pressure in the cell delivery line is slowly increased, at a constant rate. When the cell delivery pressure equals the pressure on the sensing pad the valve system opens, so by-passing the hydraulic fluid to the cell return line. The pressure at this point is indicated by a manometer on the delivery line.

According to Brady and Brown (1985) the compressibility of the cell should be similar to that of the surrounding rock if the cell is to give a correct measure of the undisturbed normal stress in the rock. If the cell is too stiff for its surrounding, it will register an excessive pressure, whilst one that is not stiff enough will register a pressure that is too low. Cells installed in rock normally are filled with mercury.

Hydraulic pressure cells also may be used to determine absolute *in situ* stress by the stress relief technique if the cell is first calibrated at specific values of internal hydraulic pressure. The calibration may be performed *in situ*, using large flat-jacks to impose a controlled and measured load. The observed change in cell pressure, consequent upon expansion of the rock and cell on relief, then is related to the absolute *in situ* stress level.

Drillhole jacks and pressure cells which are used for the determination of *in situ* deformability of rock also can be used for the determination of *in situ* stress. Dilatometers are used to measure the deformability of rocks in holes formed by diamond drills in order to reduce disturbance and allow a specific zone to be tested. Several tests can be done down the hole. According to Rocha *et al.* (1966) such testing allows an assessment of the distribution of deformability in a rock mass as well as considering the effects of anisotropy. In fact they suggested that the test results could be considered as a quality index. Dilatometer tests can also be carried out beneath the water table and at appreciable depth. Cyclic loading and unloading can be carried out.

Dilatometers may be up to or just over 1.5 m in length and vary from 150 to 300 mm in diameter (Figure 10.12). They can impose loads on the drillhole walls of up to 15 MN/m^2. Diametral strains can be determined either directly along two perpendicular diameters or by measur-

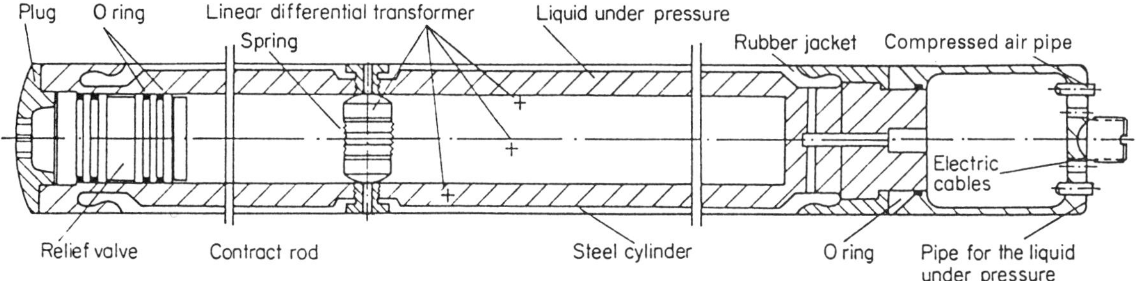

Figure 10.12 *The dilatometer (after Rocha et al. 1966)*

ing the volume of liquid pumped into the instrument. The latter provides a less accurate measurement and tends only to be used when the rock is very deformable.

10.5 Measurement of displacement

Measurements of displacement may be made relative to some datum remote from the point of measurement or to time. Conventional surveying techniques have been used to monitor displacement. For example, precise levelling is used in relation to movements caused by either settlement due to loading or subsidence above mine workings. In addition the absolute movement of the boundaries of excavations, both on the surface and underground, may be determined only be reference to fixed base lines outside the zone of influence of the disturbed ground. The type of datum employed depends on the accuracy required of the measurements. If only low levels of accuracy are required, then a pre-existing datum such as an Ordnance Survey benchmark probably will suffice. In other cases it will be necessary to construct an appropriate datum. This may take the form of a metal point installed on a suitably rigid structure, or a surface monument. Ortlepp and Cook (1964) described the use of reference points established for the measurement of displacement around deep excavations in hard rock. The reference points consisted of stainless steel links attached to rock bolts anchored at a depth of 200 mm in the roof stratum of underground roadways. Measurements made on invar staffs suspended from these links were accurate to ± 1.0 mm. A surface monument may consist of an anchorage plate and a steel bar concreted into a hole in the ground. The steel bar should extend below the level of any seasonal movements in soil. Ideally a steel bar should be driven into bedrock. Any datum should be installed outside the zone of influence of the displacement being monitored.

Electronic distance measurement (EDM) and laser equipment are used for more accurate measurement (Penman and Charles 1972). Some of these instruments are capable of measuring distances to an accuracy of ± 1 part per million over a daylight measuring range of 30 km. They are especially useful in engineering surveys since they enable distances to be measured in a matter of seconds from one instrument setting. The distance to be measured is derived from the time interval which elapses between the transmission and reception of a signal between the instrument and the target.

Photogrammetric techniques, making use of a phototheodolite, can be employed to monitor movements of rock masses, especially where direct access may not be possible or difficult. Time-lapse photographic techniques may be used to record the extent and rate of displacement of rock structures under load in large-scale tests during site investigations, and to observe the development of deformation leading to potential failure of rock masses in open pits and in landslide areas. These techniques have also been used to monitor the extent and rate of ground movements in areas where the underground extraction of minerals or fluids results in subsidence of the ground surface.

10.5.1 Vertical displacements

Settlement can be monitored irrespective of whether the reference point concerned is accessible. Conventional surveying equipment can be used provided an accuracy no better than ± 5 mm is required. If greater accuracy is required, then a precise level, an invar staff, and accurately machined reference points are necessary.

Vertical movements can also be determined by settlement rod or by water-level or mercury-filled gauges. According to Dixon and Clarke (1975) the most convenient method of relating settlement to a central point is by using a pneumatically operated settlement pot, the change in level between the required point and the central datum being indicated by a pressure change in a mercury column. However, this system only provides an indication of relative settlement unless the datum can be related to a bench mark outside the area of influence by precise surveying. A rod settlement gauge can be used to assess vertical movement when the reference point is accessible only from directly above. This may consist of a plate fixed at the required reference point, and coupled to a rod extending through telescopic tubing to ground level (Figure 10.13). Movements of the plate are determined from measurements of the level of the top of the rod made by conventional surveying equipment.

Fluid settlement gauges are needed when the reference point is not accessible in plan. Hydraulic settlement gauges basically consist of a cell containing an overflow pipe (Figure 10.14). Water is passed via a line to the cell until a constant level on a measuring tube away from the cell indicates that the tube is filled, the water level in the tube being that of the overflow in the cell (Penman and Mitchell 1970). A number of precautions must be taken. In particular, all air bubbles in the water line have to be removed by flushing thoroughly with de-aired water. Also tubing should not occur above the cell level and any blockages in drain or air pipes can give inaccurate readings.

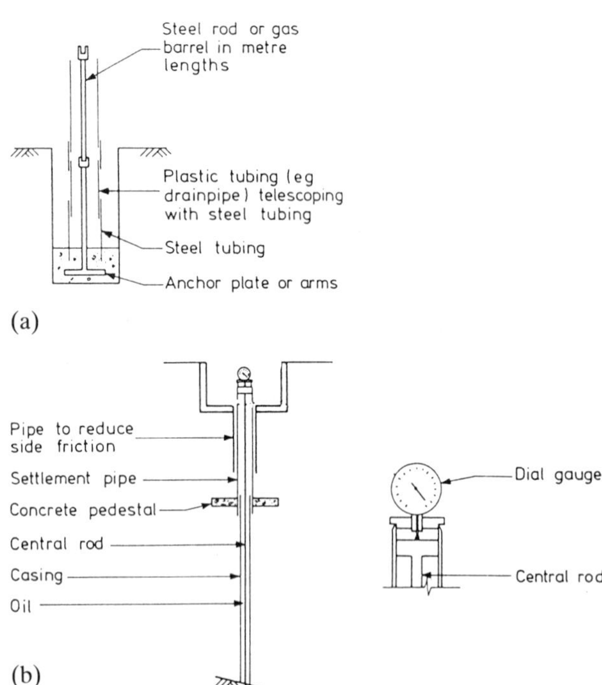

Figure 10.13 *Rod settlement gauges:* (a) *simple rod settlement gauge;* (b) *precise settlement gauge and benchmark*

The mercury settlement gauge is a more sophisticated type of settlement gauge (Figure 10.15). The accuracy of mercury settlement gauges is about ± 2.5 mm, but can often be less than ± 1 mm (Irwin 1967). This compares with about ± 5 mm for hydraulic settlement gauges.

The hydrostatic profile gauge provides a method of measuring a profile of vertical movement (Bergdahl and Broms 1967). This equipment is less accurate than the settlement gauges referred to, having an accuracy of ± 10 mm.

The USBR settlement gauge consists of alternating lengths of large and small-diameter telescopic tubing which are installed in the ground (Figure 10.16). The small-diameter tubes are anchored to the ground by

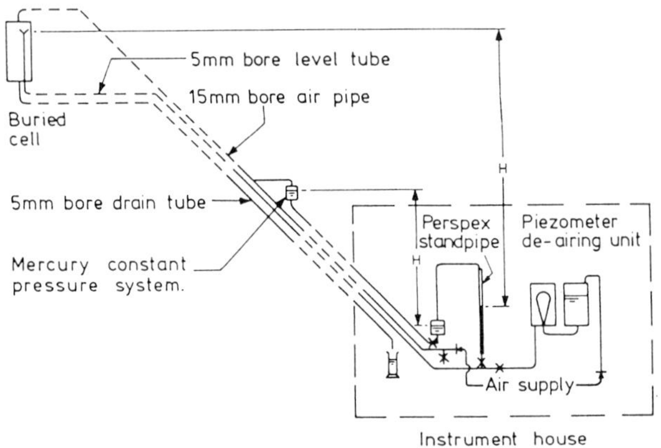

Figure 10.14 *Hydraulic overflow settlement gauge (after Penman and Mitchell 1970)*

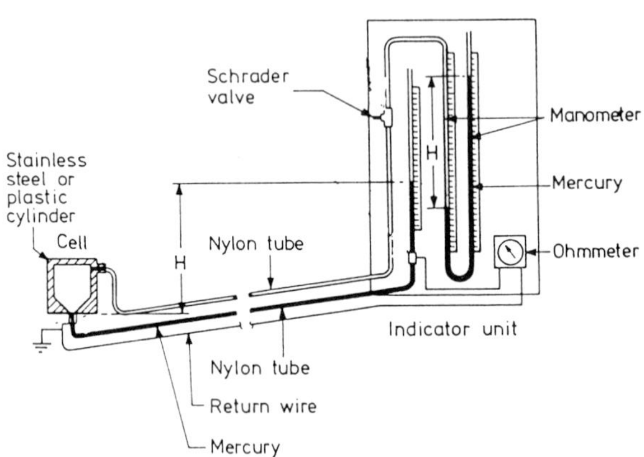

Figure 10.15 *Mercury-filled settlement gauge (after Irwin 1967)*

cross-arms which are fixed to the tubes with bolts. Measurement of the level of the bottom of each tube is made by lowering a probe on a steel tape inside the tubes.

Borehole extensometers can be used to measure the vertical displacement of the ground at different depths. Burland *et al.* (1972) described a precise drillhole extensometer consisting of circular magnets which, when embedded in the ground, act as markers, with reed switch sensors moving in a central access tube to locate the positions of the magnets.

Figure 10.17 shows a multiple-point extensometer which operates on the self-inductance principle. The extensometer consists of a central rod fixed at one end of the drill hole, and carrying a set of inductance displacement sensors passing through coaxial rings fixed by springs to the rock at selected points (Londe 1982). Relative displacement between the sensor and the ring at any point along the axis modifies the resonant frequency of the circuit. The read-out unit is calibrated to give a direct reading of displacement. The sensitivity is 0.01 mm, the

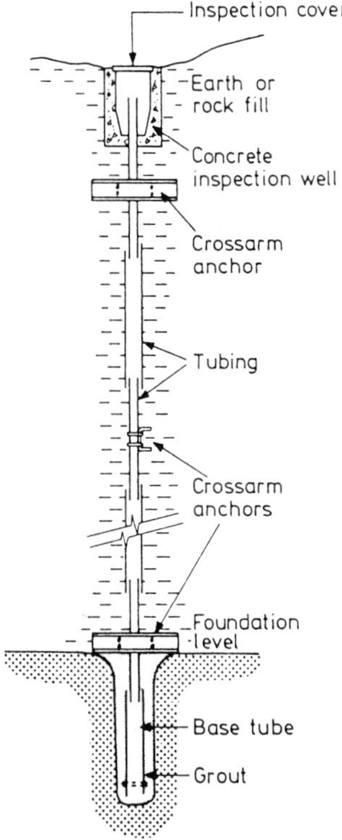

Figure 10.16 *USBR settlement gauge*

precision is ± 0.02 mm and the travel range of each sensor is 125 mm. Sensor spacings can be as close as 1 mm and the drill hole may be at any inclination. The advantage of this instrument over conventional rod and wire extensometers is that it is not necessary to provide a permanent mechanical connection between the sensors and the rock.

The distofor measures displacements at several positions along a drill hole. It has two sensing elements that

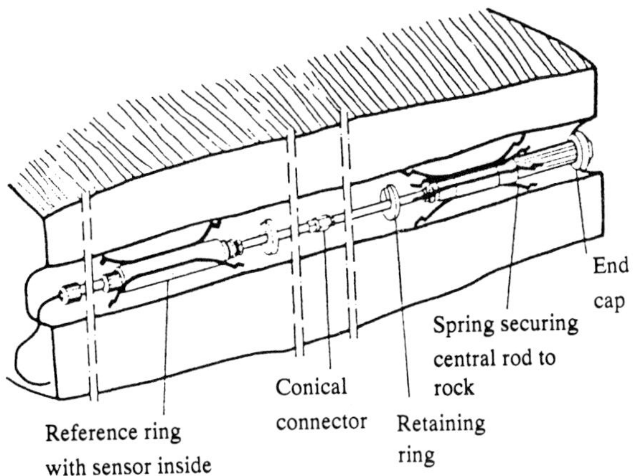

Figure 10.17 *Self-inductance multiplier point drillhole extensometer*

comprise a pair of concentric coils. One coil is mounted on a rod which goes down the drillhole and is anchored at the base. Any movement between the two circuits alters the resonant frequency of the inner one from which the relative displacement can be derived. The other circuit takes the form of a ring of non-ferrous alloy fixed into the ground by springs. The instrument has an accuracy of ± 0.02 mm and is not affected by water. It can operate in temperature varying between 0 and 50°C. A temperature read-out for each sensor enables corrections to be made for thermal expansion of the rod.

10.5.2 Subsidence measurement

Surface displacements are also produced by mining subsidence, the resultant movements of ground taking anything from a few weeks to over a year before they cease. This tends to make measurement procedures more difficult. Traditionally subsidence measurement has been done by precise surveying. However, in order to measure total movement the measurements must be related to a remote datum which is stable and this represents a difficulty since the surface is generally affected over a wide area.

Whittaker and Forrester (1974) contended that surveying techniques need to be supplemented by other methods which provide, in addition to accurate and consistent results, a high degree of sensitivity to detect the onset of initial movements. They therefore developed a rapid technique for measurement of horizontal ground strain over a base length of about 3 m. Steel tape was used, observations being taken between surface stations (vertical steel tubes concreted into the ground, the tops of which were machined to accept special registration plugs designed to carry the tape suspension, tensioning and measuring components). Each station had a protective cover. They also showed that tilt could be measured directly by mounting a sensing head (a BAC electrolevel) on special field stations which could be installed as required. Littlejohn (1975) described the measurement of horizontal strains, differential settlements and changes in slope, monitored at stations at 1.22 m intervals in brick walls subjected to mining subsidence. Measurements were taken each month whilst the site was affected by the extraction of coal and deformation was related to the position of the advancing face.

10.5.3 Convergence measurement

Precision steel tape can be used to measure the distance (up to some 60 m) between adjacent reference points grouted in shallow holes in the ground or on a structure, or between rock bolt expansion shells set into drillholes in rock. The tape may be anchored to one end of the measurement line, and the other end passed over a pulley, to be loaded by deadweight, at the reference point. If the extensometer is anchored to one end of the measurement line, it has to have a spring-loaded connection. The correct tape tension is applied by a tape-tensioning system. An accuracy of 0.1 mm is achievable over short lengths. A calibration bar is used to check the accuracy of the tape periodically.

Rod or tube extensometers may be used where short measurements are to be made. The rod micrometer consists of a micrometer head with extensometer pieces. Special end pieces locate on the reference points. The micrometer is placed between the points, and adjusted until the end pieces make contact with the points. The rod should be rigid and not sag between reference points. The tube extensometer works on a similar principle to that of the rod micrometer except that the distance between the points is measured by a special depth gauge. The accuracy of rod and tube extensometers is in the range 0.1–0.01 mm. The reference points normally are installed in patterns to suit the particular problem. Convergence, or the relative displacement of two points on the boundary of an excavation, is probably the most frequently made underground measurement. A typical arrangement for convergence measurement in a tunnel is given in Figure 10.18.

A convergence meter consists of two telescoping tubes used to span between the floor and roof of a subsurface excavation. Any movements between the roof and floor cause the tubes to move in or out of each other. A magnet is located in each tube and as movement occurs their relative positions are measured by a magneto-restrictive probe. An accuracy of ± 0.03 mm is obtainable.

The distometer is a high-precision convergence-measuring system developed by Kovari *et al.* (1974). It consists of a 1 mm diameter invar wire which is tensioned by a dynamometer (Figure 10.19). The wire is stretched between an anchoring point and the instrument which is fixed to another anchoring point on the rock. The tensioning device moves the end of the wire towards the displacement gauge with a precision thread. The applied tension is read on the dynamometer. The displacement gauge has a readability of 0.01 mm and a range of 100 mm. The overall accuracy of the convergence measurements is ± 0.02 mm for a distance of approximately 10 m.

The radiofor is a drillhole instrument which is used to measure convergence in tunnels or movements in slopes (Hanna 1985). It emits radio signals which are picked up on a read-out box.

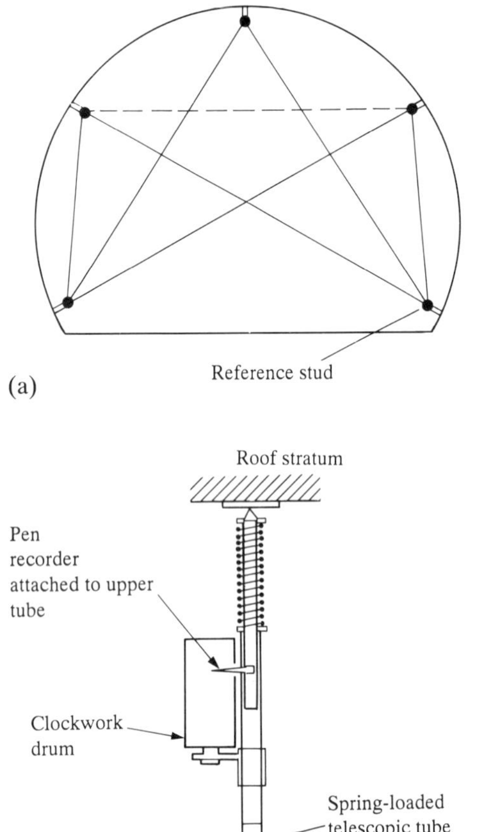

Figure 10.18 (a) *Convergence measurement in a tunnel;* (b) *a convergence recorder*

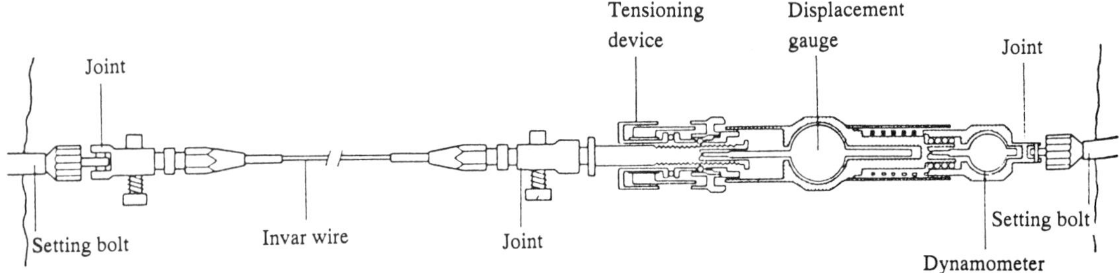

Figure 10.19 *The distometer (after Kovari et al. 1974)*

A deflectometer measures the movement of the ground in relation to the axis of a drillhole (Dunnicliff 1972). The electric chain deflectometer offers precise measurement of movement transverse to the axis of a drillhole. The deflectometers are spaced at a number of centres in a drill hole with movement being measured by inductive transducers. Measurements are made in one or two perpendicular directions.

The extenso-deflectometer can be used to measure movements in three dimensions along drillholes in rock masses (Kovari *et al.* 1979). As the instrument is portable, measurements can be obtained from several drillholes at more than one site in contrast to built-in deflectometers. Nevertheless the extenso-deflectometer possesses high accuracy (to ± 0.01 mm). A casing, made from aluminium or hard PVC, is installed in a drillhole with a minimum diameter of 120 mm. Measuring rings with special stop fittings are fixed at 1.5 m centres along the casing. The measuring rings temporarily accept the three heads of the instrument when readings are made. If a central mark moves relative to the neighbouring ones the displacement is registered as the difference of two readings. During measurement the instrument is moved stepwise from one position to another and thereby set in the measuring rings. Several readings are taken in each position in four orientations by rotating the instrument through 90°.

The instrument consists of an external bracing section and an internal measuring unit (Figure 10.20). The central head, M, of the bracing section is connected to the outside heads A and B by means of two protective sleeves. The connection between these sleeves and M is hinged, with A and B telescoping in a spring-loaded way. The springs at A and B contract the instrument so that the distance between the heads is a few centimetres less than that between the measuring rings. When readings are taken the heads have to be pulled into the measuring rings, causing the springs to stretch. The orientation of the instrument with respect to rotation around its axis is brought about by a spring-supported guide beam at head A. The displacements are picked up by linear displacement transducers.

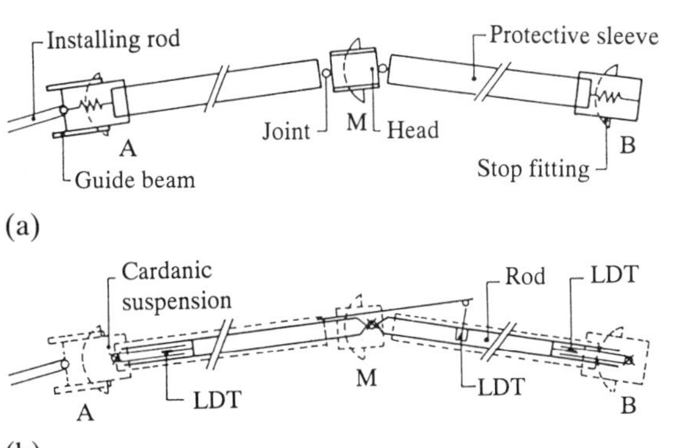

Figure 10.20 *The extenso-deflectometer showing* (a) *external bracing;* (b) *internal measuring system (after Kovari et al. 1979)*

10.5.4 Horizontal movements

A slip indicator may be used when it is necessary to detect the level in the ground at which horizontal movement(s) is occurring, for example, on the shear surface associated with slope failure. One of the simplest slip-plane detectors consists of a flexible tube installed in a drillhole within a temporary casing, and the annulus between is surrounded with sand. Care must be taken to install the tube without bends and as near vertical as possible. The casing is withdrawn as the sand is placed around the tube. A rod on the end of a nylon rope is passed down the tube to the base of the hole. The development of a shear surface bends the tube, so that when the rod is raised it sticks at the bend in the tube and hence locates the position of the slip. Another rod can be lowered down the tube from the ground surface to locate the bend and to check that the only one shear surface has developed. Lines of slip indicators can be used to determine the profile of a slip.

An inclinometer may be installed to detect horizontal movements below ground. It consists of a probe which is fitted with wheels and contains a gravity-operated tilt sensor (Figure 10.21). This produces a signal from which the angle between the axis of the probe and the vertical can be determined. A grooved guide tube is grouted into a drill hole of 100 to 150 mm in diameter. Readings are made by lowering the probe down the guide tube, and making readings about every 0.5 m. The probe transmits signals of tilt and depth to the recorder and a vertical profile is thereby obtained. Sets of readings over a period of time enable both the magnitude and rate of horizontal movement to be determined. For many engineering purposes a range of about ± 20° and a sensitivity of 0.01 to 0.05° is suitable (Green 1974). In other words inclinometers can detect differential movements of 0.17–3.4 mm per 10 m run of hole.

Fixed-position inclinometers monitor differential lateral movements between the drillhole collar and a deep datum. They are most frequently used in slope stability work. Such inclinometers comprise a string of sensors each containing a uniaxial or biaxial tilt sensor which are jointed together by articulated rods. They can provide continuous, automatic or remote detection of movements. Because the sensors are not generally closely spaced, they do not give a detailed profile of ground movements.

The trivec probe can be used in a similar way to an inclinometer but it can measure the distribution of all three displacement components along a drillhole (Köppel *et al.* 1983). In other words the probe measures axial strain and inclinations in two vertical planes normal to each other through the axis of the drillhole. The instrument has been developed from the sliding micrometer, in that it is basically a sliding micrometer with an inclinometer sensor which is rotated by means of two micro-motors to stop fittings 90° apart from one another. By contrast with the borehole inclinometer, which has a smooth casing, the trivec casing is provided with a chain of reference points in the form of ring-shaped measuring marks at intervals of 1 m. The latter serve as coupling elements between two

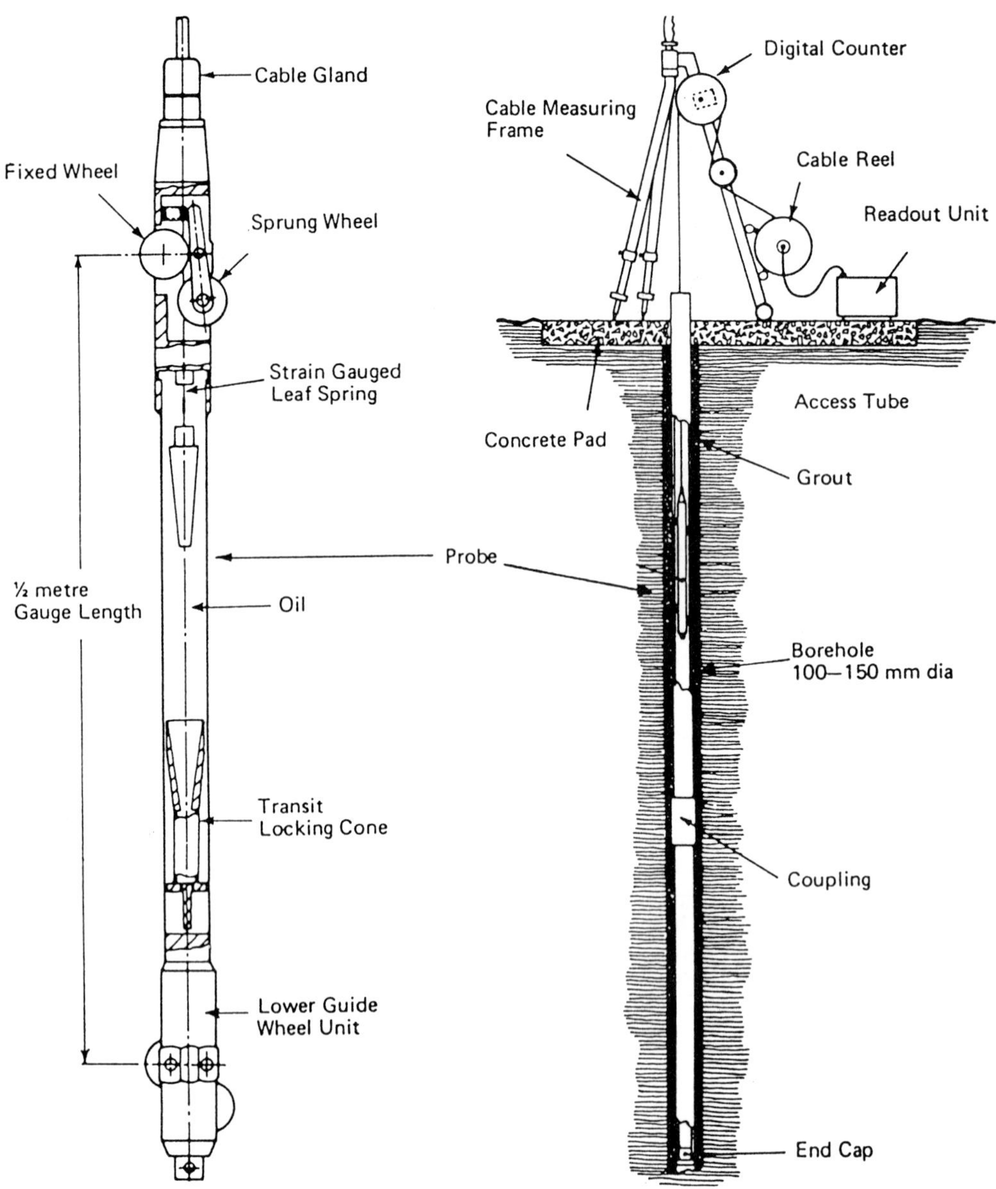

Figure 10.21 *Drillhole inclinometer*

adjacent units of PVC tube and as housing for stop fittings. These stop fittings hold the two heads of the portable probe in position during reading. If the measuring marks displace relative to each other due to the deformation of the rock, then the change in distance (i.e. the strain) and the change in inclinations can be recorded as the difference between two readings.

The rock spy has been used to monitor movements in slopes; in particular, it can be used to monitor movements along discontinuities in rock masses. It consists of an induction sensor between an invar wire and a spring, and measures the relative displacement between two reference points. A continuous record is taken by a strip chart recorder. An alarm can be triggered when a given amount of movement occurs.

Detection and measurement of the location and magnitude of shear movements at depth in a rock mass may be achieved by installing shear strips grouted in drill holes in rock masses. The shear strips are printed circuit resistors which are joined together to make up the required overall length, and coupled by cable to a monitoring read-out system. The latter locates the shear movements.

Monitoring of tension cracks basically involves fixing reference points on either side of the crack and measuring the distance between them accurately. For example, the rod extensometer can be used for such a purpose.

The demec gauge provides reliable measurements of relative movements over short distances. It measures the length between two studs either side of a crack. Readings taken across the crack at regular intervals provide a measure of the relative movement.

Rock movement is accompanied by the generation of noise which is subaudible in the earlier stages of the development of failure but audible at failure. For instance, rock noise is often heard in deep mines and may indicate imminent danger from rock failure. Normally the rate of the micro-seismic occurrences increases rapidly with the development of rock instability. Monitoring of such microseismic or acoustic emissions resulting from mining activity represents an essential part of the monitoring programmes in many deep mines, especially those susceptible to rockburst activity.

The microseismic or acoustic emissions can be picked up by an array of geophones, placed on the faces of the underground excavations or alternatively located in drill holes. The mechanical vibration is converted into its electrical analogue. This electrical signal is then amplified and transmitted to the monitoring station. The data are used to locate the source and determine the magnitude of the events. Noise also is associated with instability of slopes and most movements generating noise originate near or along the plain of failure. Again an array of geophones located in the vicinity of a slope can record the noise and give warning of slope failure.

Acknowledgement

The author gratefully acknowledges the assistance of Professor Kalman Kovari in the preparation of this chapter.

References

Bergdahl, U. and Broms, B. B. (1967) 'The new method of measuring in situ settlements', *Proc. ASCE J. Soil Mechanics and Foundations Division,* **93** (SM5), 51–58

Brady, B. H. G. and Brown, E. T. (1985) *Rock Mechanics for Underground Mining,* Allen and Unwin, London

Burland, J. B., Moore, J. E. A. and Smith, P. D. K. (1972) 'A simple and precise borehole extensometer', *Geotechnique,* **22,** 174–177

Casagrande, A. (1949) 'Soil mechanics in the design and construction of Logan airport', *J. Boston Society of Civil Engineers,* **36**, 176–218

Deardoff, G. B., Lumsden, A. M. and Hefferon, W. H. (1980) 'Pneumatic piezometers: multiple and single installations in vertical and inclined boreholes', *Can. Geotech. J.* **17**, 313–320

Dixon, J. C. and Clarke, K. B. (1975) 'Field investigation techniques'. In *Site Investigation in Areas of Mining Subsidence,* F. G. Bell (ed.), Newnes–Butterworths, London, pp. 40–74

Dunnicliff, C. J. (1971) 'Equipment for field deformation measurements', *Proc. 4th Pan-American Conf. Soil Mechanics and Foundation Engg,* Puerto Rico, **2**, 319–332

Green, G. E. (1974) 'Principles and performance of two inclinometers for measuring horizontal ground movements', *Proc. Symp. Field Instrumentation in Geotech. Engg*, British Geotechnical Society, Butterworths, London, 166–179

Grob, H., Kovari, K. and Amstad, Ch. (1975) 'Sources of error in the determination of in-situ stresses by measurements', *Tectonophysics,* **29**, 29–39

Haimson, B. C. (1975) 'New hydrofracturing measurements in the Sierra Nevada mountains and the relationship between shallow stresses and surface topography', *Proc. 20th US Symp. Rock Mechanics,* Austin, 675–682

Hanna, T. H. (1985) *Field Instrumentation in Geotechnical Engineering,* Trans-Tech. Publications, Clausthal Zellerfeld

Hawkes, I. and Hooker, V. E. (1974) 'The vibrating wire stressmeter', *Proc. 3rd Int. Congress Rock Mecahnics,* Denver, **2**, 439–444

Hawkes, I. and Moxon, S. (1966) 'The measurement of in situ stress using the photoelastic biaxial gauge with the core relief technqiue', *Int. J. Rock Mech. & Min. Sci.,* **2**, 405–419

Horslev, M. J. (1951) 'Time-lag and soil permeability in groundwater observations', *Bulletin No. 36,* US Corps of Engineers, Waterways Experimental Station, Vicksberg

Irwin, M. J. (1967) 'A mercury filled gauge for the measurement of settlement of foundations' *Report No LR62*, Transport and Road Research Laboratory, Crowthorne

Köppel, J., Amstad, Ch. and Kovari, K. (1983) 'The measurement of displacement vectors with the "TRIVEC" Borehole Probe, *Proc. Field Measurements in Rock Mechanics,* **1**, 209–218

Kovari, K. and Amstad, Ch. (1983) 'Fundamentals of deformation measurements', *Proc. 1st Int. Symp. Field Measurements in Geomech.* Zurich, 219–239

Kovari, K. and Koeppel, K. (1987) 'Head distribution monitoring with the sliding piezometer system "Piezodex"', *Proc. 2nd Int. Symp. Field Measurements in Geomech.* Kobi, S. Sakurai (ed), A. A. Balkema, Rotterdam, 255–267

Kovari, K., Amstad, Ch. and Grob, H. (1974). 'Displacement measurements of high accuracy in underground openings', *Proc. 3rd Int. Congress Rock Mechanics,* Denver, 445–450

Kovari, K., Amstad, Ch. and Köppel, J. (1979) 'New developments in instrumentation of underground openings', *Proc. 4th Rapid Excavation and Tunneling Conf.* Atlanta, 817–837

Leeman, E. R. (1969) 'The "doorstopper" and triaxial stress measuring instruments developed by CSIR', *J. South African Inst. Mining and Metall.* **69**, 305–339

Littlejohn, G. S. (1975) 'Observations of brick walls subjected to mining subsidence'. In *Settlement of Structures,* Pentech Press, London, pp. 384–393

Londe, P. (1982) 'Concepts and instruments for improved monitoring', *Proc. ASCE J. Geotech. Eng. Div.* **108** (GT6), 820–834

Ortlepp, W. D. and Cook, N. G. W. (1964) 'The measurement and analysis of the deformation around deep hard rock excavations', *Proc. 4th Int. Conf. Strata Control and Rock Mechanics*

Peck, R. B. (1969) 'Deep excavation and tunnelling in soft ground', *Proc. 7th Int. Conf. Soil Mechanics and Foundation Engg.* Mexico City, State-of-the-art volume, 225–290

Penman, A. D. M. and Charles, J. A. (1972) 'Constructional deformation in a rock fill dam', *Current Paper CP 19/72* Building Research Station. Watford

Penman, A. D. M. and Mitchell, P. B. (1970) 'Initial behaviour of Scammonden Dam', *Trans. 10th Conf. Int. Commission on Large Dams,* Montreal, 723–747

Roberts, A. (1977) *Geotechnology,* Pergamon Press, Oxford

Rocha, M. and Silverio, H. (1969) 'A new method for the complete determination of the state of stress in rock masses', *Geotechnique,* **19**, 116–132

Rocha, M., Da Silveira, A. M., Grossman, N. and De Oliveira, E. (1966) 'Determination of the deformability of rock masses along boreholes', *Proc. 1st Int. Congress Rock Mechanics,* Lisbon, **1**, 697–704

Wareham, B. F. and Skipp, B. O. (1974) 'The use of a flat-jack installed in a sawcut slot in the measurement of in situ stress', *Proc. 3rd Int. Congress Rock Mechanics,* Denver, **2**, 481–488

Whittaker, B. N. and Forrester, D. J. (1974) 'Measurement of ground strain and tilt arising from mining subsidence', *Proc. Symp. Field Instrumentation in Geotech. Engg,* British Geotechnical Society, Butterworths, London, 437–447

Worotnicki, G., Enever, J. R., McKavanagh, B., Spathis, A. and Walton, R. J. (1980) 'Experience with monitoring of crown pillar performance in two Australian mines', *Proc. 3rd Australian–New Zealand Conference on Geomech.,* Wellington, **2**, 161–168

11 Slope stability and rockfall problems in rock masses

Dr L Richards
Goulder Associates Ltd, Maidenhead, UK

11.1 Introduction

This chapter attempts to review the current state of the art with respect to analytical and design methods for the engineering of solutions to slope stability and rockfall problems in rock masses. In view of the fact that the subject of rock slope stability has been fairly comprehensively covered in recent years, attention is devoted mainly to a review of recent developments in the analysis of rockfall problems and the design and execution of measures to contain these.

Engineering assessments of excavations and natural slopes in rock masses generally concentrate on the identification and evaluation of the mechanisms which may lead to failure of significant volumes of rock. However, failures involving relatively small volumes of rock will often pose greater problems, particularly in areas where steep terriain is in close proximity to developed areas or transportation corridors. Spang (1987) reported that rockfalls rather than deep-seated slides are the major causes of interruption to the West German Federal Railway's 28 000 km track network. Martin (1988) stated that rockfalls, small rock slides and ravelling are the most chronic problems on transportation routes in mountainous areas of North America; millions of dollars are spent annually on maintenance and remedial measures to provide protection against such hazards.

Spang (1987) and Spang and Rautenstrauch (1988) have reviewed the definition of the term 'rockfall' and note that there is no general agreement as to the volume which characterizes a rockfall event. However, the phenomenon is generally accepted as having the following characteristics:

(1) The event involves a single block or group of blocks which become detached from the rock face.
(2) Each falling block behaves more or less independently of other blocks.
(3) There is temporary loss of ground contact and high acceleration during the descent.
(4) The blocks attain significant kinetic energy during their descent.

The kinetic energy of a particle with mass m moving with velocity v is defined as

$$K.E = \tfrac{1}{2}mv^2 \qquad (11.1)$$

Spang (1987) suggests that the term 'rockfall' be restricted to events which have a maximum kinetic energy of 500 kN m. This kinetic energy value is equivalent to a 5 tonne block dropping vertically from a height of 10 m. The basis for the above definition is that Spang considered that potential rockfalls with greater kinetic energy would require active stabilization since it would not be practicable to contain them with protective structures such as rock catch fences.

By comparison with Spang's suggested upper limit of 500 kN m for kinetic energy, boulder fences designed by Chan *et al.* (1986a) in Hong Kong were stated to be able to withstand an impact energy of 100 kN m; boulders liable to attain greater energies were required to be stabilized *in situ*.

Rather than adopt arbitrary values of kinetic energy, it is considered preferable to maintain the definition of rockfall as one involving significant velocities and some measure of free flight but without any upper limit on the volume or kinetic energy of individual blocks.

11.2 Slope stability in rock masses

11.2.1 Basic mechanisms of rock slope failure

Rock slopes generally fail along existing geological defects. It is only in very high slopes and/or weak rocks that failure in intact material becomes significant. Most rock-slope problems therefore require consideration of the geometrical relationships between discontinuity planes, the slope and the force vectors involved.

In weathered or highly jointed rock masses, the failure plane may be less controlled by single throughgoing discontinuities. The analytical procedures for such slopes tend to be similar to those for soils.

The most important requirement for rock slopes is to determine the correct failure mechanism. This will generally demand a specific site investigation to evaluate the geology of the slope, the properties of the rocks, the characteristics of the discontinuities and the groundwater conditions.

This chapter provides simple guidelines for the analysis of idealized rock-slope failures. More comprehensive details are provided by Hoek and Bray (1981).

The main failure types in rock slopes are translational, toppling and rotational. Figure 11.1 shows examples of these.

Plane failures are a common type of translational failure and occur by sliding along a single plane which daylights in the slope face (that is, the dip of the failure plane is less than that of the slope). When considered in isolation, a single block may be stable. Forces imposed by unstable adjacent blocks may give rise to *active and passive block failures*. *Wedge failures* are formed by a combination of two planar discontinuities, the slope face and the upper slope surface. The condition for movement to occur is that the line of intersection of the wedge daylights in the slope.

Toppling failures involve the overturning of blocks of rock and are associated with steep slopes and sub-vertical joints dipping back into the slope. In this case, the shape of the block as well as its weight is important.

Rotational failures in rock slopes are usually found only in structureless overburden material, highly weathered rock or very high slopes in closely jointed material. Either circular or non-circular failure surfaces may be generated. *Circular failures* may occur in rock masses which are so intensely fractured in relation to the scale of the slope that they may be considered as randomly jointed and isotropic. In highly weathered materials, *non-circular failures* may occur along a combination of existing joints and failures through weak but previously intact material.

11.2.2 Factor of safety and sensitivity analyses

The most common analytical methods are based on the principles of limiting equilibrium. The stability of indi-

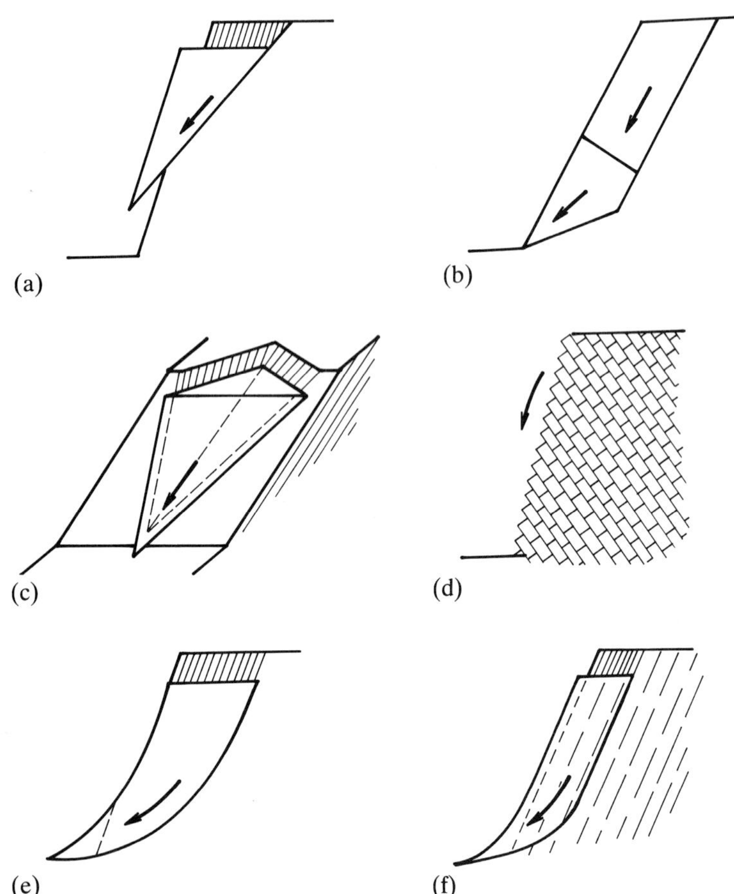

Figure 11.1 *Idealized failure mechanisms in rock slopes:* (a) *plane;* (b) *active and passive blocks;* (c) *wedge;* (d) *toppling;* (e) *circular;* (f) *non-circular*

vidual slopes is expressed as a factor of safety which is the ratio of the forces resisting movement to the forces tending to induce sliding. At limiting equilibrium, the factor of safety is thus equal to unity.

11.2.3 Translational failures

These involve shear movement along single or multiple discontinuities in the rock mass and give rise to planar, multiple block or toppling failures. Stability evaluations are generally carried out by, firstly, assessing the possibility of kinematic mechanisms using discontinuity data presented on stereographic projections. Comprehensive treatment of such mechanisms is contained in Hoek and Bray (1981) and Richards and Atherton (1987) and reference should be made to these publications for fuller information on the subject.

There are three kinds of translational failure: plane failure, active and passive block failures, and wedge failures. In order for plane failure to be kinematically possible, the following conditions must be satisfied:

(1) The failure plane must daylight in the slope face and strike within about ± 20° of the slope crest.
(2) Release surfaces must be present to provide lateral limits for the failure block.

Richards *et al.* (1978) described a method for identifying potential plane failures and for carrying out a simple assessment of stability based on gravity loading and frictional strength alone.

A full analysis of plane failure also requires consideration of cohesion, water pressures, reinforcement loads and acceleration forces. Figure 11.2 summarizes the equations for calculating the factor of safety. Hoek and Bray (1981) give more comprehensive details.

The solution of active and passive block failures is simpler for rock than soil, as the interface between active and passive blocks will be defined by the geological structure and the rock friction angle may be used to determine the inclination of the forces across the interface.

The system can be analysed by assuming each block is a free body and determining the interface force for different factors of safety. The correct factor of safety is that which gives identical interface forces for both blocks. The method is summarized in Figure 11.3. Goodman (1976) described the treatment of this problem by stereographic methods and extended the analysis to the case of two three-dimensional wedges.

Typical wedge failures are bounded by four surfaces: two discontinuities whose line of intersection daylights in the slope, the slope face and the upper slope surface. A tension crack may truncate the upper portion of the wedge.

The general failure mode is by sliding parallel to the line of intersection of the two discontinuities. The geometry of certain wedges is such that sliding occurs on only one plane under gravity loading. These planes will be identified during the kinematic check for planar sliding (see Richards

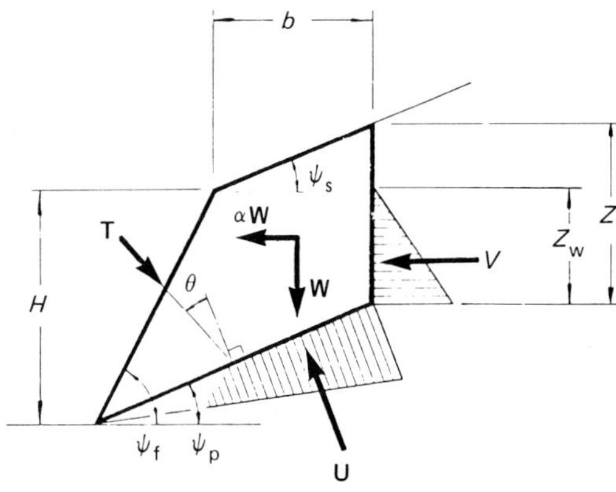

Input data

H = height of slope face
ψ_f = inclination of slope face
ψ_s = inclination of upper slope surface
ψ_p = inclination of failure plane
z_w/z = depth of water in tension crack/depth of tension crack
α = horizontal acceleration (proportion of g)
T = tension in rockbolts or cables
θ = inclination of force T to normal of failure plane
c = cohesive strength of failure surface
ϕ = angle of friction on failure surface
b = distance of tension crack behind slope crest

Factor of safety

$$F = \frac{c.A + [W(\cos\psi_p - \alpha.\sin\psi_p) - U - V.\sin\psi_p + T.\cos\theta]\tan\phi}{W(\sin\psi_p + \alpha.\cos\psi_p) + V.\cos\psi_p - T.\sin\theta}$$

where

$$z = H + b.\tan\psi_s - (b + H.\cot\psi_f)\tan\psi_p$$
$$W = \tfrac{1}{2}\gamma[H^2.\cot\psi_f + 2.b.H + b^2.\tan\psi_s - (H.\cot\psi_f + b)^2\tan\psi_p]$$
$$A = \frac{[h.\cot\psi_f + b)}{\cos\psi_p}$$
$$U = \tfrac{1}{2}.\gamma_w.z_w.A$$
$$V = \tfrac{1}{2}.\gamma_w.z_w^2.$$

Reference: Hoek and Bray (1981).

Figure 11.2 *Analysis for plane failure*

11.2.4 Toppling failures

Toppling of individual blocks is governed by joint spacing and orientation. The condition for stability of an individual block is that the resultant force must be within

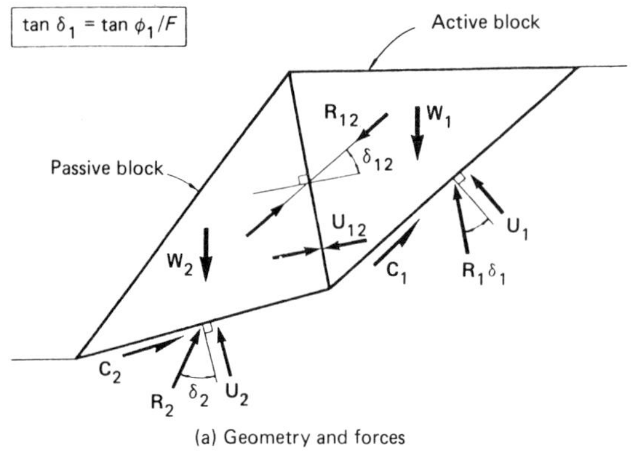

(a) Geometry and forces

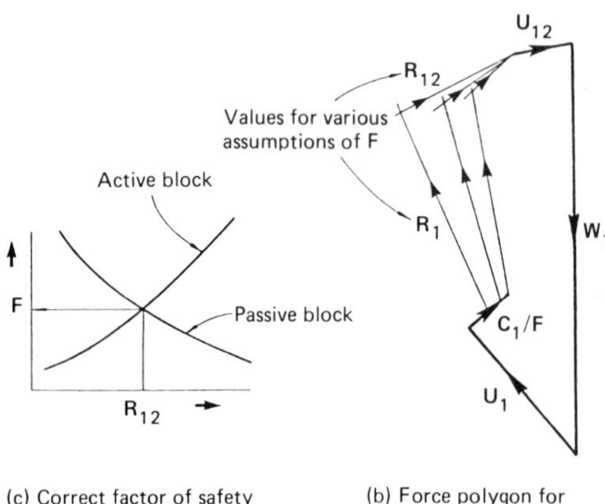

(c) Correct factor of safety

(b) Force polygon for active block

W_1, W_2 = weights of active and passive blocks
C_1, C_2 = total cohesion available on failure surfaces
ϕ_1, ϕ_2 = angles of friction on failure surfaces
ϕ_{12} = angle of friction at interface
U_1, U_2 = water forces on failure surfaces
U_{12} = water force on interface
R_1, R_2 = resultant forces on failure surfaces
R_{12} = resultant force at interface

Procedure

1. Draw force polygon for active block
   (a) Draw vertical force vector W_1
   (b) Draw vector U_1 normal to plane 1
   (c) Draw vector U_{12} normal to plane 12
   (d) Assume initial factor of safety $= F_1$
   (e) Draw C_1/F_1 parallel to plane 1
   (f) From C_1/F_1, draw R_1 at angle δ_1 to normal to plane 1 where $\tan \delta_1 = \tan \phi_1/F_1$
   (g) From U_{12}, draw R_{12} at angle δ_{12} to normal to plane 12 where $\tan \delta_{12} = \tan \phi_{12}/F_1$
   (h) Intersection of R_1 and R_{12} allows magnitude of R_{12} to be determined
   (i) Assume second and third factors of safety (F_2, F_3) and repeat steps (e) to (h)
2. Draw force polygon for passive block using same procedure as above.
3. Plot factors of safety against magnitude R_{12}. Correct factor of safety is when R_{12} for active block is equal to R_{12} for passive block.

Figure 11.3 *Graphical analysis for active and passive block failure*

et al. 1978). Wedges involving a discontinuity susceptible to planar failure should be ignored since they will be less critical than the plane failure. Stability calculations for rock wedges may be carried out by design charts, analytical methods, or stereographic projections (Hoek and Bray 1981). Richards *et al.* gave a simple stereographic technique for identifying and carrying out a preliminary stability assessment on potential wedges. Comprehensive wedge solutions are given in Hoek and Bray (1981). A simplified method is summarized in Figure 11.4.

11.2.4 Toppling failures

Toppling of individual blocks is governed by joint spacing and orientation. The condition for stability of an individual block is that the resultant force must be within the central two-thirds of the base of the block. Hydrostatic forces acting on the near vertical joints will greatly affect the direction of the resultant force.

Analytical techniques are available for a few special cases of toppling failure. The toppling mechanism described by Goodman and Bray (1976) involves the overturning of semicontinuous cantilever beams with interlayer slipping. The failure mechanism for this is described by Richards *et al.* (1978) and a method given for the identification of potential flexural toppling.

In general, methods of toppling analyses are not yet basic design tools and reference must be made to more detailed accounts (see Hoek and Bray 1981; Goodman and Bray 1976) for a better understanding of this topic.

11.2.5 Rotational failures

(a) Circular failures

Design charts for circular failures in homogeneous material are given in Hoek and Bray (1981). These allow the factor of safety to be calculated from a knowledge of the face angle and height of the slope and the friction, cohesion and density of the rock. The charts give factors of safety for the most critical combinations of failure surface and tension crack locations. Five different groundwater conditions are considered.

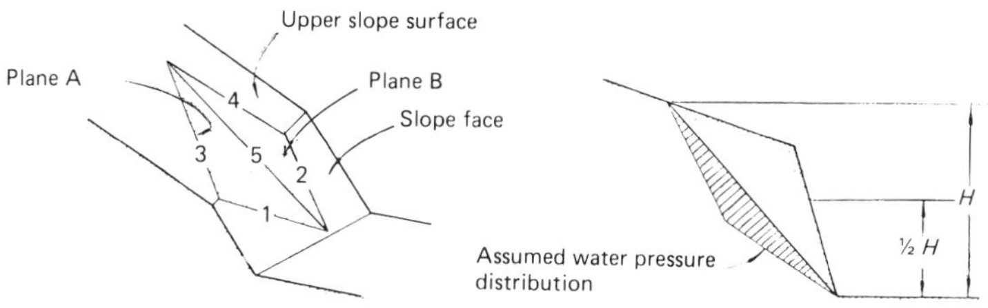

(a) View of wedge showing numbering of planes and intersection lines

(b) View normal to line of intersection 5

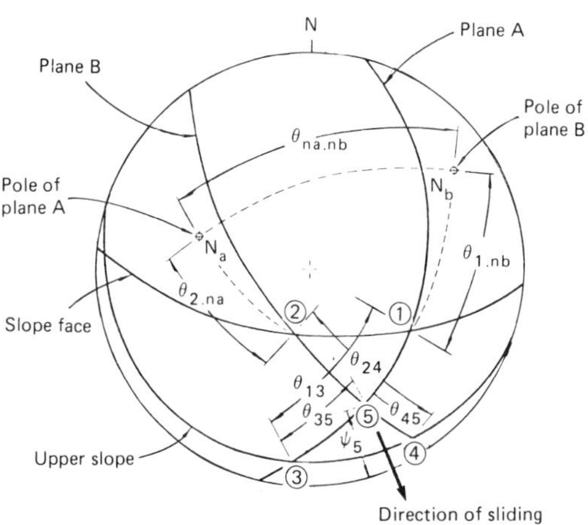

(c) Stereoplot of data required for wedge analysis

Input data

C_A, C_B = cohesive strengths of planes A and B
ϕ_A, ϕ_B = angles of friction on planes A and B
γ = unit weight of rock
γ_w = unit weight of water
H = total height of wedge
ψ_A, ψ_B = dips of planes A and B
ψ_5 = dip of line of intersection 5

Factor of safety

$$F = 3(C_A.X + C_B.Y)/\gamma H + (\mathrm{A} - \gamma_w.X/2\gamma)\tan\theta_A + (\mathrm{B} - \gamma_w.Y/2\gamma)\tan\theta_B$$

where

$$X = \frac{\sin\theta_{24}}{\sin\theta_{45}.\cos\theta_{2.na}}$$

$$Y = \frac{\sin\theta_{13}}{\sin\theta_{35}.\cos\theta_{1.nb}}$$

$$\mathrm{A} = \frac{\cos\psi_A - \cos\psi_b.\cos\theta_{na.nb}}{\sin\psi_5.\sin^2\theta_{na.nb}}$$

$$\mathrm{B} = \frac{\cos\psi_B - \cos\psi_A.\cos\theta_{na.nb}}{\sin\psi_5.\sin^2\theta_{na.nb}}$$

Procedure

1. Plot great circles and poles for planes A and B
2. Plot great circles for slope and upper slope
3. Mark the intersection of the great circles 1–5 as shown and the poles to planes A and B as N_a and N_b respectively
4. Measure required angles (e.g. for angle $\theta_{1.nb}$, rotate overlay so that points 1 and N_b lie on a great circle and measure the angular difference along the great circle)
5. Calculate X, Y, A and B
6. Calculate *F*

Reference: Hoek and Bray (1981)

Figure 11.4 *Simplified method for wedge analysis*

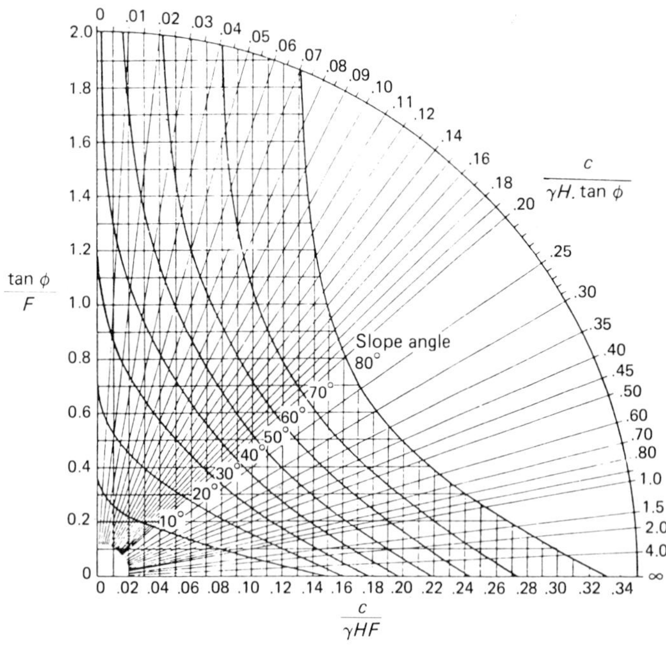

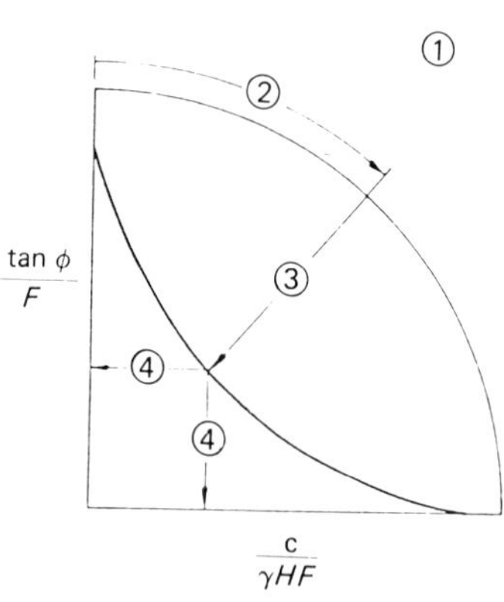

Input data

c, ϕ = shear strength characteristics of failure surface
γ = unit weight of rock
H = height of slope
ψ = face angle of slope

Procedure

1. Define groundwater conditions (only the fully saturated slope is shown here)
2. Calculate dimensionless ratio (c/γ) $H \tan\phi$ and locate this value on the outer circular scale of the appropriate chart
3. Follow the radial line from the value in step 2 to its intersection with the curve for the slope angle
4. Find the corresponding value of $\tan\phi/F$ or (c/γ) HF, whichever is more convenient, and calculate the factor of safety.

Reference: Hoek and Bray (1981)

Figure 11.5 *Design charts for circular failure*

The use of these design charts is illustrated in Figure 11.5 and the chart for a saturated slope is reproduced. Design charts such as these can be readily used for sensitivity studies prior to carrying out detailed evaluation with more comprehensive methods.

(b) Non-circular failures

A convenient method of analysing a general failure surface is Janbu's routine method of slices (Janbu 1954). This gives reasonably accurate factors of safety for shallow failure surfaces but should not be used for deep slip surfaces with low friction angles. The method is described in Figure 11.6.

The Sarma method of analysis (Sarma (1979), Hoek (1987a)) has very significant advantages over these methods in that it allows the analysis of slopes with circular, non-circular or planar sliding surfaces or any combination of these as well as active–passive wedge failures. The inclinication of the slice sides can also be varied to model specific geological features.

The modified Sarma method is ideally suited to the analysis of slopes where the failure mechanism is governed by the rock-mass properties rather than being controlled by one or more dominant discontinuities. In heavily jointed and weathered rock masses slope failures can occur on approximately circular surfaces such as shown in Figure 11.7. The shear strength behaviour of rock masses will generally be highly non-linear and is not amenable to evaluation by conventional testing techniques (as are used for assessing shear strength of specific rock discontinuities). The Hoek–Brown rock-mass failure criterion described by Hoek (1983) provides an empirical basis for ascribing strength properties to jointed rock masses. Inclusion of this criterion in a Sarma type analysis represents the most practicable tool available at present for studying the stability of slopes in heavily jointed and weathered materials.

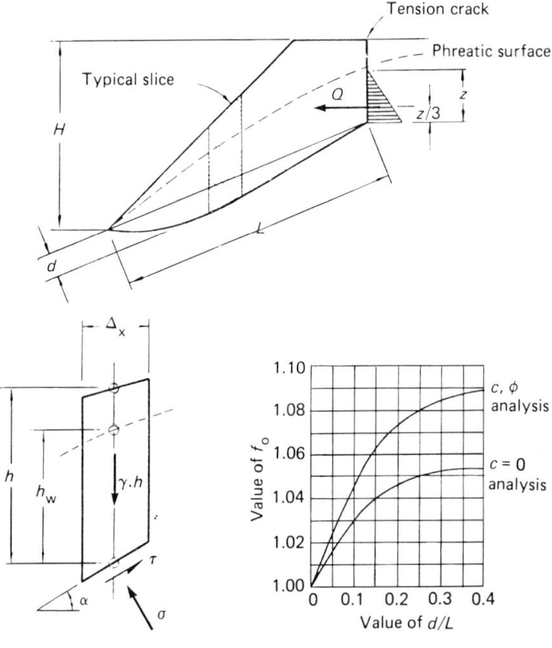

Input data

c, ϕ = shear strength at slice base
z = depth of water in tension crack
L = length of failure surface
d = depth of slide beneath chord line (see sketch)
γ, γ_w = unit weights of rock and water
h = height of slice
h_w = height of water above slice base
Δx = width of slice
α = angle of failure surface at base of slice (positive when sliding is downhill)

Factor of safety

$$F = \frac{f_0 \Sigma X/(1 + Y/F)}{\Sigma Z + Q}$$

where

$$X = \{c + (yh - \gamma_w h_w) \tan \phi\} (1 + \tan^2 \alpha) \Delta x$$
$$Y = \tan \alpha \tan \phi$$
$$Z = \gamma h \, \Delta x \tan \alpha$$
$$Q = \tfrac{1}{2} \gamma_w z^2$$

Reference: Hoek and Bray (1981)

Procedure

1. Define the geometry of the slope
2. Divide the sliding mass into slices and measure α and Δx for each slice.
3. Calculate the vertical stress (γh) and the water pressure ($\gamma_w h_w$) at the base of each slice
4. Assign effective shear strength parameters to the base of each slice
5. Calculate the values of X, Y and Z for each slice and the value of the horizontal water force, Q
6. Assume an initial value of F (say 1.00) and then calculate the new F from the given equation
7. If difference between the assumed and calculated I is greater than 0.001, the calculated I is used as a second estimate of I in a further factor of safety calculation. This iterative process is repeated until the difference between successive values of I is less than 0.001
8. Apply the correction factor, f_0, to the final factor of safety.

Figure 11.6 *Simplified Janbu method for non-circular failures*

Figure 11.7 shows how the Sarma analysis is able to incorporate discontinuity shear strength values along specific structural features, linear Mohr–Coulomb failure criterion in the overburden material and the non-linear Hoek–Brown criterion along the failure surface and inclined slice boundaries in the main body of the rock mass. The Sarma method can, as noted above, also be used to analyse active–passive wedge failures of the type shown in Figure 11.8. The latter figure also shows how the factor of safety will vary considerably depending on whether linear or non-linear failure criteria are used and this emphasizes the need to utilize realistic shear strength parameters in slope analyses.

11.2.6 Stabilization of rock slopes

Slopes which have unacceptable stability conditions may be stabilized by modifications to the slope geometry, artificial reinforcement or reduction of groundwater pressures. Detailed treatment of such measures is beyond the scope of this chapter. Recommended references for information on slope reinforcement are Hoek and Bray (1981), CANMET (1976) and Chapter 18 of this book.

11.2.7 Monitoring of rock slopes

As noted in Section 11.2.1, it is not always possible to fully understand the real failure mechanisms in complex slopes. Monitoring is therefore needed to provide a rational basis for evaluating stability conditions. Comprehensive details of available instruments for monitoring are given in Dunnicliff (1988) and case histories of their application to slope problems in Sharp (1989). Reference should be made to these publications for a fuller insight into the role of monitoring in the assessment of slope performance.

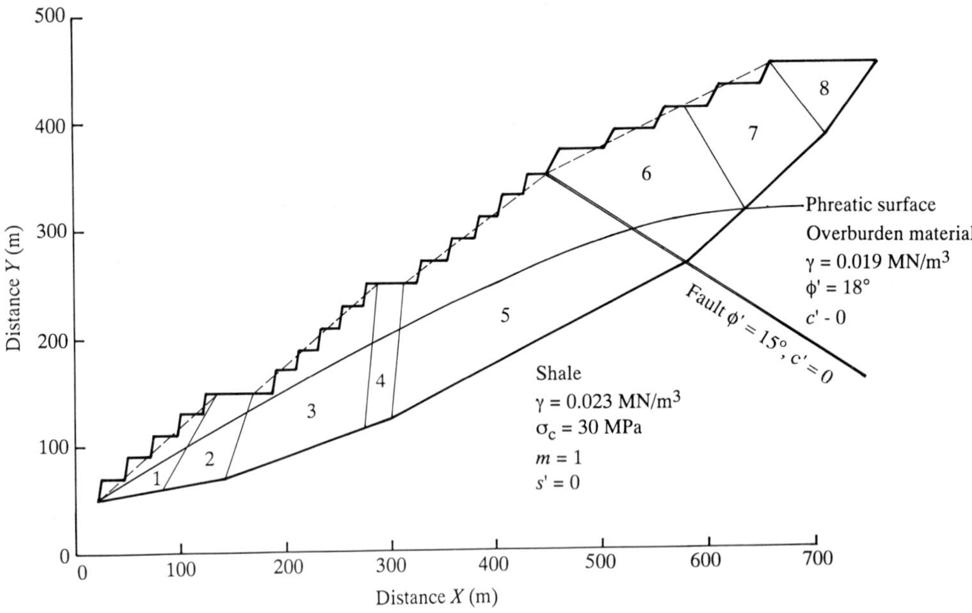

Figure 11.7 *Example of Sarma analysis from Hoek (1983)*

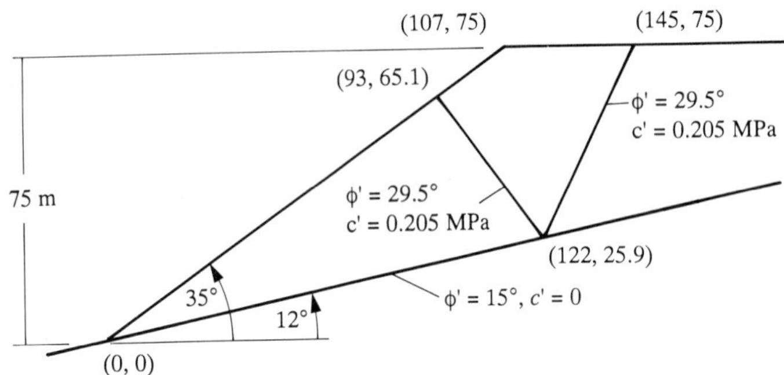

(a)

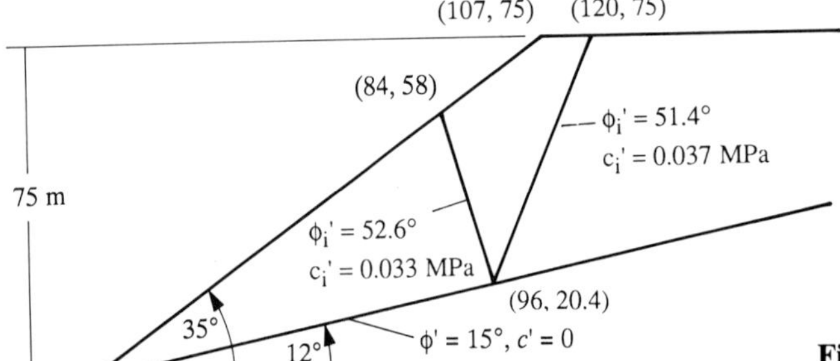

(b)

Figure 11.8 *Active–passive wedge analyses from Hoek (1983):* (a) *Mohr–Coulomb failure criterion: factor of safety = 1.41:* (b) *Hoek–Brown failure criterion: factor of safety = 1.08*

11.3 Rockfalls

11.3.1 Mechanics of rockfall behaviour

The behaviour of falling and impacting rock blocks is governed by Newton's laws of motion and these form a common basis for most of the analytical studies of this phenomenon. These laws are as follows:

(1) Every body continues in its state of rest or of uniform motion in a straight line unless compelled by some external force to do otherwise.
(2) The rate of change of momentum of a body is proportional to the applied force and takes place in the direction in which the force acts.
(3) To every action there is an equal and opposite reaction.

The equations of uniformly accelerated motion are used to calculate the velocities of falling rock blocks:

$$v = u + at \qquad (11.2)$$

$$s = ut + \tfrac{1}{2}at^2 \qquad (11.3)$$

$$v^2 = u^2 + 2as \qquad (11.4)$$

where u and v are the initial and final velocities, a acceleration, t time and s the distance.

Conventional slope failures require identification of relevant failure modes if the appropriate analytical techniques are to be used. Similarly rockfall problems need assessment of the modes of motion. At any given time during its flight, a rock block may have one of the following types of motion:

(1) throw;
(2) free falling;
(3) bouncing;
(4) rolling;
(5) sliding;
(6) rolling/sliding.

The *throw* mode involves the block being given an initial velocity such that it moves out from its starting position in a parabolic trajectory. The initiation of rockfall movement can be triggered by such mechanisms as ice or water pressures, earthquake or blast vibrations, or external factors such as construction activities.

In the *free falling* mode, the block is simply accelerating due to gravitational forces. The effect of air friction is generally neglected in rockfall studies because of its relatively minor significance.

Bouncing is a transitional mode that occurs when the falling block impacts on the slope surface. The horizontal and vertical velocity components at the time of impact are modified to reflected velocity components depending on the local slope angle and the *coefficients of restitution* for the block/slope at the point of impact.

Rolling imparts an angular velocity to the block and this has the effect of reducing the translational velocity as compared with pure sliding for zero friction. Energy losses during rolling are, however, less than those during sliding. The translational velocity of a rolling block is a function of the block shape (see Section S).

The *sliding* velocity of a block is a function of the coefficient of kinetic friction and the inclination of the plane on which sliding takes place.

Rolling/sliding behaviour involves a combination of both modes of motion.

(a) Sliding and rolling modes

Hoek (1987b) provides the following formula for calculating the velocity, v, of a rock which rolls or slides a distance s down a plane inclined at an angle θ:

$$v = (v_0^2 + 2sgK)^{1/2} \qquad (11.5)$$

where v_0 is the initial downslope velocity, g the acceleration due to gravity and K is a constant defined by the slope angle θ and the friction angle ϕ. Values of the cosntant K for various rock shapes and movement modes are as follows:

(1) pure sliding $K = \sin\theta - \cos\theta\tan\phi$
(2) rolling sphere $K = 5/7 \sin\theta$
(3) rolling cylinder $K = 2/3 \sin\theta$

Hoek (1987b) concluded that for most rock shapes encountered in the field, a reasonable general assumption is that $K = \sin\theta$.

Figure 11.9 shows the calculated values of K for the different modes of motion in relation to the inclination of

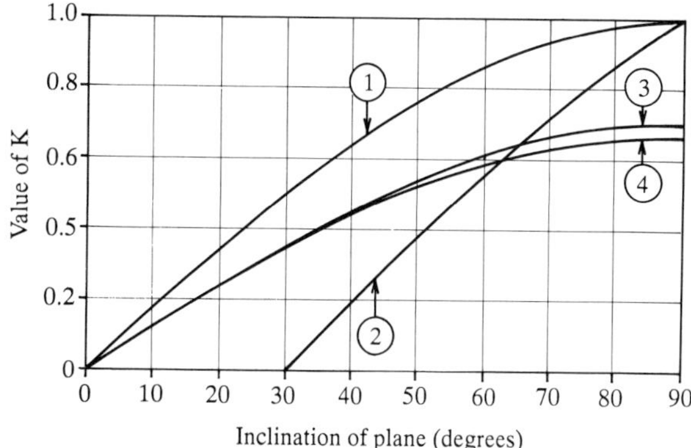

Figure 11.9 *Values of K for different modes of motion*

the sliding plane. It can be seen that for slopes steeper than about 60°, the relevant K values all lie within the range of about 0.60 to 1 and the particular mode of motion does not therefore introduce greatly different velocity characteristics. For slopes steeper than 60°, the sliding mode can produce greater velocities than the rolling mode even with significant values for the friction angle. Figure 11.10 shows the velocities attained by blocks of different shape on a 45° slope. This shows that even on a moderately dipping plane the different sliding/rolling modes can generally attain velocities of about 50–70% of that in the free-falling mode.

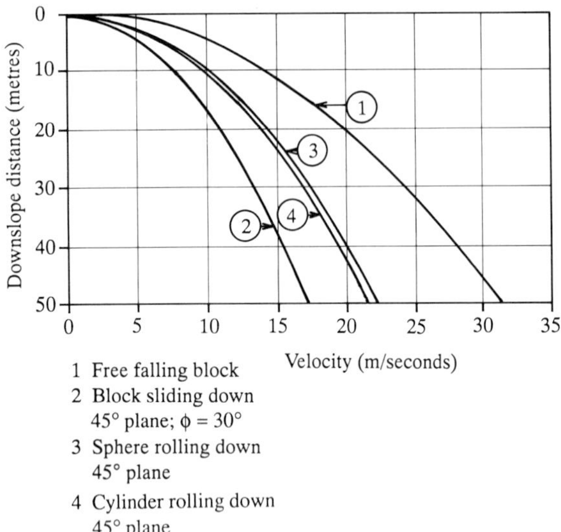

Figure 11.10 *Comparison of velocities for blocks on 45° plane*

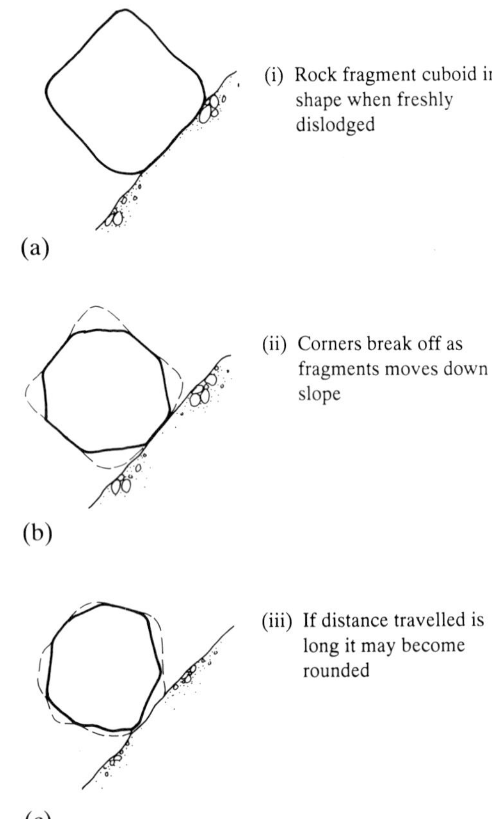

Figure 11.11 *Change in boulder shape with distance travelled:* (a) *rock fragment cuboid in shape when freshly dislodged;* (b) *corners break off as fragment moves down slope;* (c) *if distance travelled is long it may become rounded (Chan et al.* 1986a)

A comprehensive study of boulder protection works in Hong Kong has been reported by Chan, *et al.* (1986a). This was part of a programme to provide remedies against boulders emanating from a large area of 45 000 m^2. Mathematical and modelling studies were carried out to determine boulder velocities during their travel down the slopes. The authors noted that boulders were generally cubical in shape when dislodged from the slope but became progressively more rounded with travel because of attrition at the corners and edges (Figure 11.11). On this basis, the authors considered that it was reasonable to approximate the shape of the boulders to an octagonal prism in their mathematical model. Three different modes of motion were allowed for in their modelling (as shown in Figure 11.12): simple rotation about the corner edges, combined rotation and sliding and bouncing. Figure 11.13 shows an example of a time–displacement curve for an octagonal prism in comparison with that for a cylinder.

Ritchie (1963) carried out detailed field studies of rockfall behaviour and concluded that the 'size and shape of a rock have little bearing on its falling or rolling characteristics. The greatest difference is with large rocks that stay close to the face of a cliff and may from time to time touch on the way down'.

In the light of the above, it is therefore understandable that most analytical studies of rockfall have considered the rock blocks as spheres (e.g. Piteau and Clayton 1977), Spang and Rautenstrauch (1988), and Scavia *et al.* (1988), or made other simplifying assumptions about the block behaviour during sliding and rolling (e.g. Hoek 1987b). Given the fact that none of the analytical models known to the author incorporate breakdown of the rock during its descent, accurate modelling of block geometry would seem to be a relatively low priority at this stage of development in rockfall studies. The most pressing requirement is for a better appreciation of the coefficients

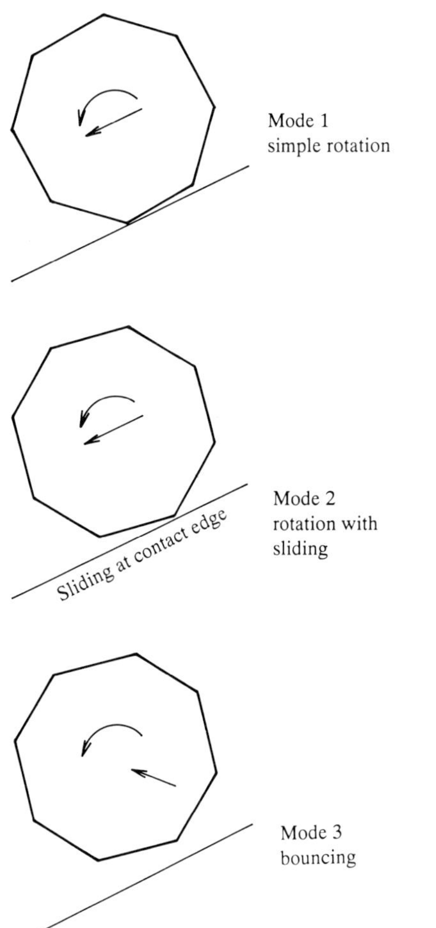

Figure 11.12 *Types of boulder motion (Chan et al. 1986a)*

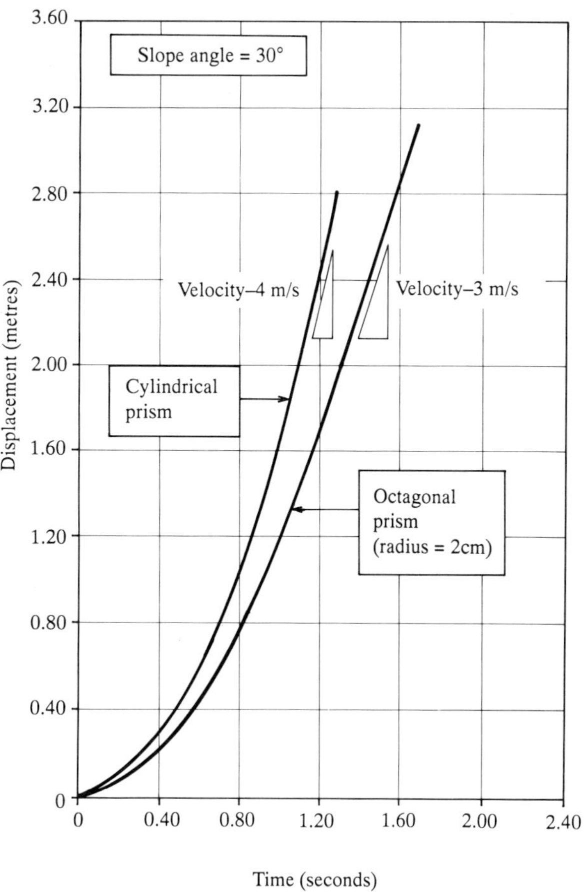

Figure 11.13 *Time–displacement graphs from laboratory experiments (Chan et al. 1986a)*

of restitution which have a far greater effect on rockfall trajectories (see Section 11.3.1 (b)).

(b) Coefficients of restitution

Spang and Rautenstrauch (1988) provided a useful summary of this parameter. The coefficient of restitution, r, can be calculated for impacting bodies from the following formula:

$$r = (v_2' - v_1')/(v_2 - v_1) \tag{11.6}$$

where $1 \geqslant r \geqslant 0$, v_i represents the initial particle velocity and v_i' the final velocity after the collision of two particles 1 and 2. For free-falling tests, the above equation becomes

$$r = (h'/h_o)^{1/2} \tag{11.7}$$

where h_o and h' represent the drop and impact heights respectively.

This coefficient is not a material constant but depends on particle velocities, particle geometries and materials. Empirical measurements of the parameter are therefore necessary.

A number of researchers have used the single coefficient of restitution, r, to predict rockfall behaviour. Experimental evidence shows that a better definition of bouncing rocks is obtained when separate parameters, r_n and r_t, are used for the coefficients of restitution normal and tangential to the slope surface. The latter approach has been used by Piteau and Clayton (1977) according to Pasquero (1987) as well as Kirsten *et al.* (1986) and Hoek (1987b).

Given the great significance of rockfall problems with respect to safety and cost considerations, it is unfortunate that there is so little information available on coefficients of restitution. Results from large-scale *in situ* tests in Lecco, Italy, have been reported by Broili (1976). Camponuovo (1976), reporting on laboratory and model studies for the same site, noted that the coefficient of restitution for sound rock surfaces is very high (>60%). However, rebound tests caried out on 'comparatively sound' rock resulted in the rock spheres breaking when dropped from a height of 1–2 m. According to their own experience, Spang and Rautenstrauch (1988) considered many of the reported values to be relatively high and valid only for sound rock. They noted that for overburden and debris, values of less than 0.1 might be appropriate. Data from other researchers are summarized in Table 11.1.

Table 11.1 shows that there is very little information available which would allow informed estimates to be made of the likely coefficients which would apply in any given rockfall situation. The selected values for this coefficient have great significance in the calculation of maximum trajectories of falling rocks, as can be seen from Figures 11.14 and 11.15. As can be seen from Figure 11.14, assumption of values of r_n greater than 0.6 leads to unrealistically high bouncing modes and long trajectories. Further research on this parameter is therefore crucially important to this topic. In particular, studies are needed which will allow realistic values of r_n and r_t to be determined from other rock and rock mass parameters such as uniaxial compressive strength, tensile strength, modulus of elasticity of the boulder and compressibility and fracture state of the slope surface.

In the absence of satisfactory methods for estimating these parameters, it is currently necessary to resort to back analyses of previous rockfalls at the site. Studies have been carried out by Geoanalysis and Golder Associates for a site at the village of Quincinetto, about 70 km north of Turin. The western side of the village is located close to the foot of a large slope comprising a steep rock wall above a flatter talus slope. The rock wall is nearly vertical, about 50 m in height and 350 m long. The talus slopes are about 100 m in height at an overall angle of about 36°. Various areas of unstable rock have been identified on the wall; these vary from small blocks of about 1 m^3 up to a large mass of about 700 m^3.

A preliminary survey of the talus slope was undertaken to define the locations of terraces, vineyards, trees and rock outcrops. A more detailed investigation of the actual ground conditions was then carried out by means of a heavy penetrometer using a 20 kg weight falling 0.5 m and recording the penetration in centimetres after 20 blows. Penetrometer readings were taken at relatively close distances along four sections oriented down the general dip of the talus slope. A series of computer analyses were then carried out using Hoek's ROCKFALL program to model the flight paths of boulders falling from the rock wall and bouncing/sliding down the talus slope. The model was calibrated by comparing the final positions of the boulders as assessed from computer analyses with the main areas of boulders resting on the talus slope. It was found that the parameters shown in Table 11.2 gave reasonable agreement with the field observations.

Figure 11.16 summarizes part of the results from this study and shows how the maximum reach of the rockfalls is affected by the site topography and the nature of the

Table 11.1 Published values for coefficients of restitution

Reference	*Value for r*	*Value for r_n*	*Value for r_t*	*Remarks*
Habib (1976)	0.75–0.80			Based on experience in Italy
	0.50–0.60			Based on experience in Norway
Descoeudres & Zimmermann (1988)	0.40			Vineyard slopes
	0.85			Rock slopes
Broili (reported in Pasquero 1987)	0.75–0.80			Impact between competent materials (rock–rock)
	0.20–0.35			Impact between competent rock and soil–scree materials
				Type of material on slope surface
Piteau and Clayton (reported in Pasquero 1987)		0.9–0.8	0.75–0.65	Solid rock
		0.8–0.5	0.65–0.45	Detrital material mixed with large rock boulders
		0.5–0.4	0.45–0.35	Compact detrital material mixed with small boulders
		0.4–0.2	0.3–0.2	Grass-covered slopes or meadows
Hoek (1987b) based on unpublished information from Departments of Transportation in USA		0.53	0.99	Clean hard bedrock
		0.40	0.90	Asphalt roadway
		0.35	0.85	Bedrock outcrops with hard surface, large boulders
		0.32	0.82	Talus cover
		0.32	0.80	Talus cover with vegetation
		0.30	0.80	Soft soil, some vegetation

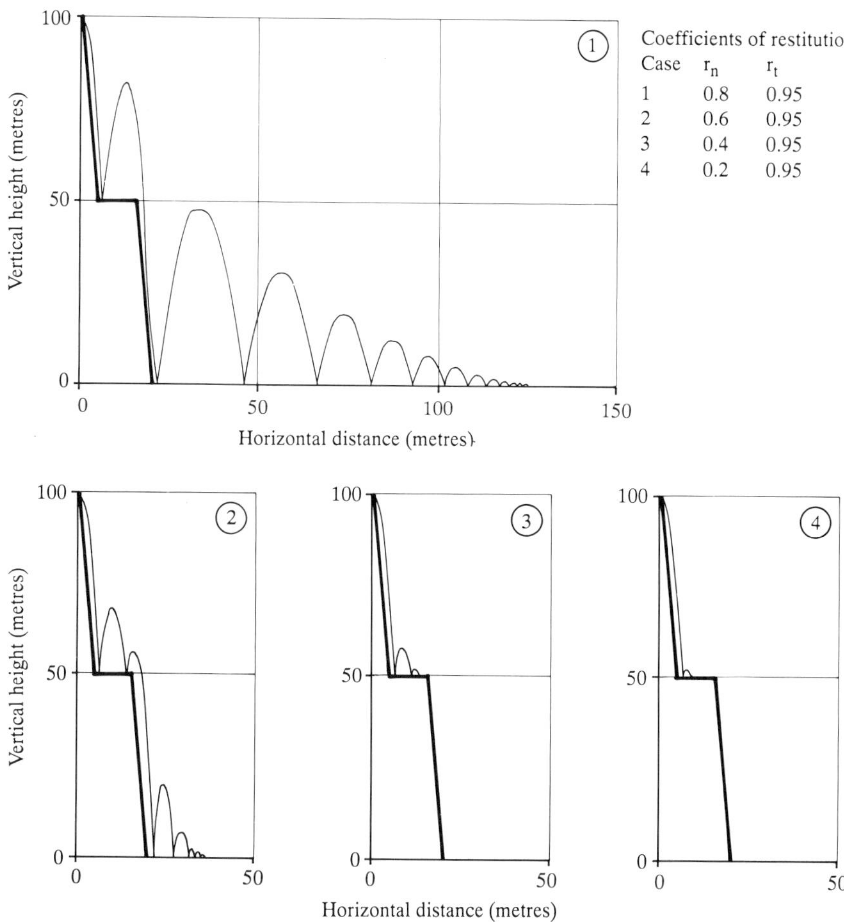

Figure 11.14 *Effect on rockfall trajectory of varying r_n*

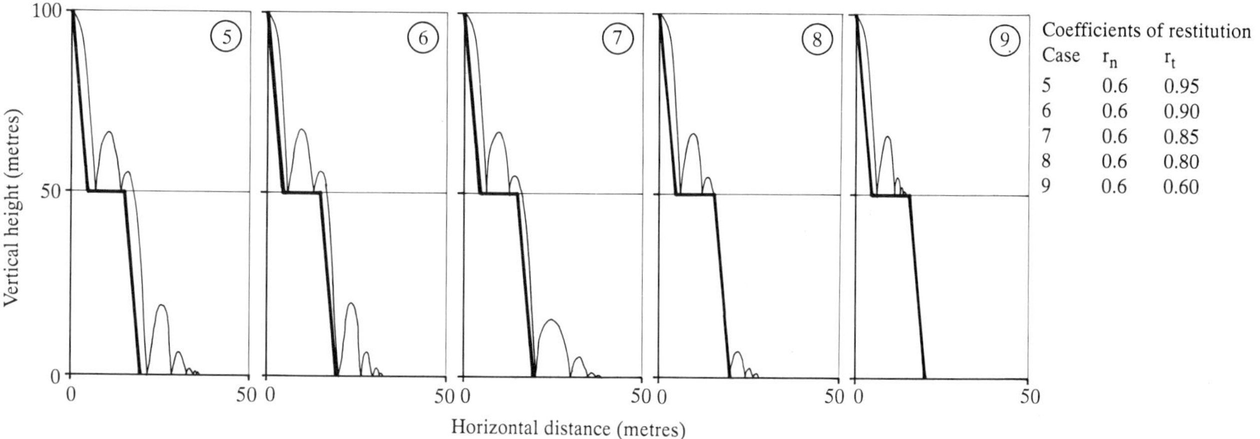

Figure 11.15 *Effect on rockfall trajectory of varying r_t*

Table 11.2 Values of coefficient of restitution

	r_n	r_t
Rock wall	0.8	0.8
Talus slope		
compact rock scree or thin soil cover	0.5	0.7
penetrometer reading <40	0.4	0.7
penetrometer reading >40	0.3	0.7

talus slope. Further information on this study is given in Scavia *et al.* (1988).

11.3.2 Empirical studies of rockfall

The most commonly quoted reference on this aspect is the work done by Ritchie (1963) for the Washington State Highway Commission. He noted that there was a clear need for means of predicting the stability of material on the surface of a rock cut but that 'so far, these factors remain elusive and many engineers approach the problem with apathy, as though walking up to a stone wall and half-heartedly demanding that the wall give up its secrets and come under their slide rule'. Ritchie's work involved the use of slow-motion cameras to record hundreds of full-scale rockfall tests. Most of the study involved hard basaltic rocks and natural or excavated slopes of varying ages. Ritchie observed that the prevalent modes of rockfall motion were as follows:

(1) rolling on slopes at less than 45°;
(2) bouncing on slopes in the range of 45–75°;
(3) falling on slopes steeper than 75°.

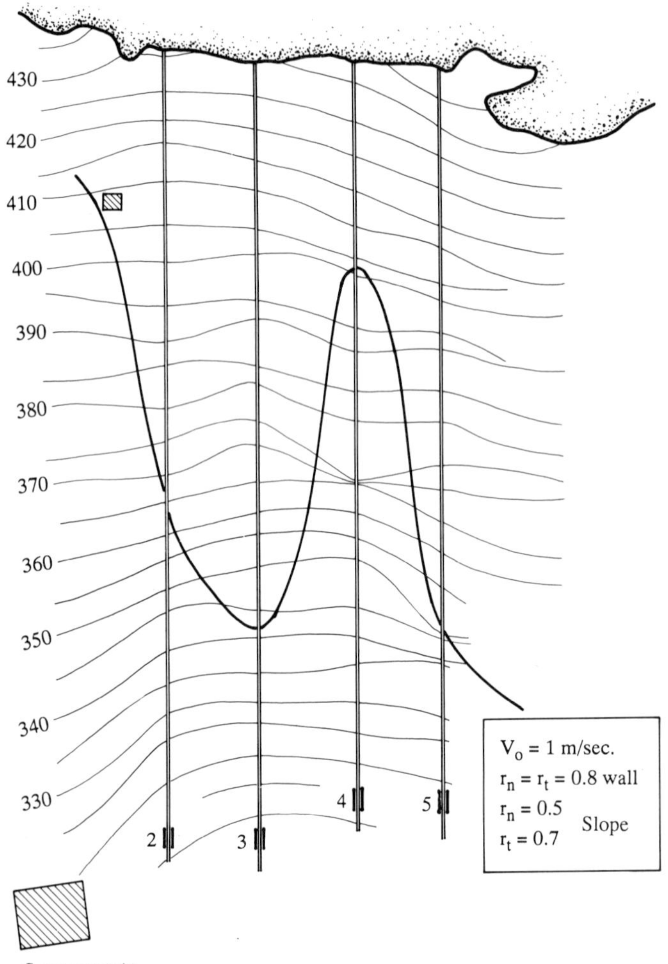

Figure 11.16 *Effect of topographical variation on maximum reach of rockfalls*

The recommendations from this study were published in tabular form and have been reproduced elsewhere on numerous occasions. Fookes and Sweeney (1976) have produced design charts based on Ritchie's tables. However, as noted by Whiteside (1986), the above authors were unduly conservative in their interpretation of Ritchie's work. Figure 11.17 shows the revision to the design chart of Fookes and Sweeney (1976) as suggested by Whiteside (1986).

Ritchie's guidelines have tended to become enshrined in the literature and accepted without reservation by succeeding generations of engineers. Although computational techniques have improved dramatically since the date of Ritchie's work, there has been remarkably little advance on the mechanical aspects of this problem.

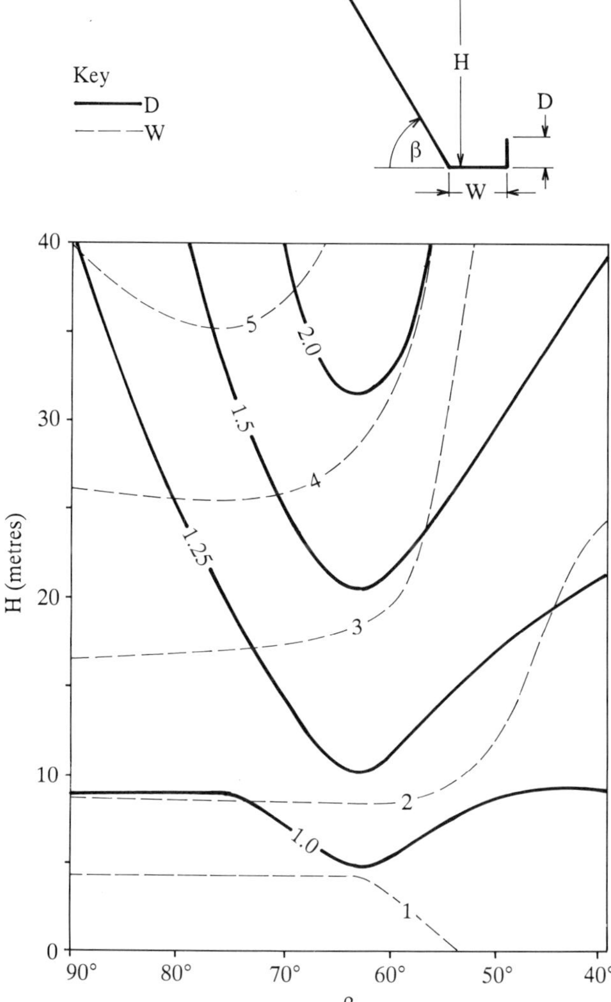

Figure 11.17 *Guidelines for rock traps (revision by Whiteside 1986 of graph from Sweeney 1976 based on Ritchie 1963)*

One of the few detailed studies on rockfall observations in the field has been reported by Mak and Blomfield (1986). The work involved releasing about 1000 boulders on each of thirteen different pre-split slopes ranging from 6 m to 12 m high and at angles of 55–70°. The ground at the toe of each slope was levelled and covered with a layer of compacted rockfill to provide consistent energy-absorbing characteristics. Angular blocks of 100–300 mm were used but the experimental data did not indicate any significant difference in trajectories in relation to the specific block sizes. The authors have presented a summary of their experimental data in a number of graphs. However, Richards (1986) has noted that the data can be more simply summarized as follows: 1 and 1.5 m high barriers at 1.5 m from the slope toe will trap 95 and 100% respectively of all the boulders falling from slopes up to 12 m high in the range of 55–70° (Figure 11.18). Whiteside (1986) has also noted that Mak and Blomfield's results show good correlation with Ritchie's data as given on Figure 11.17.

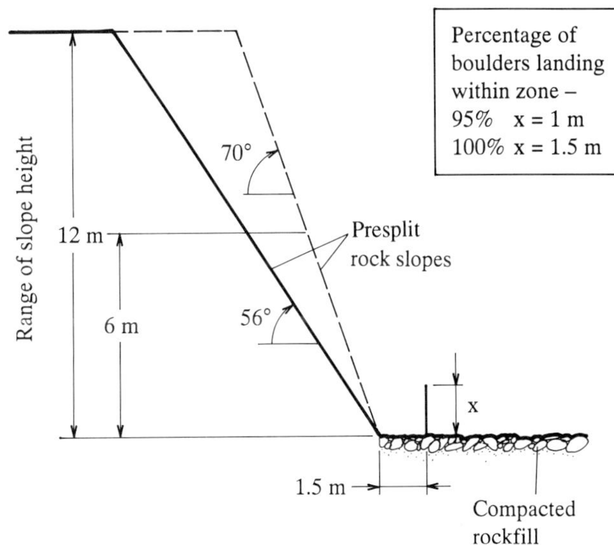

Figure 11.18 *Summary of main findings from Hong Kong rockfall tests (Mak and Blomfield 1986)*

Chan *et al.* (1986a,b) described the results of field experiments in which about 70 boulders from 30 kg to over a tonne were rolled down two 30° slopes, one slope having rock outcrops and thin vegetation, the other being more decomposed with boulder deposits. There did not appear to be a great deal of difference in the measured boulder velocities on these slopes, average velocities being in the range 5–8 m/s. The authors noted that these velocities were less than those predicted by their mathematical model using octagonal prisms and suggested that this was due to the effects of uneven slope surfaces, vegetative cover and existing boulder deposits.

11.3.3 Physical modelling of rockfalls

Probably the most comprehensive scale modelling of rockfall problems to date has been the work carried out by ISMES in Italy in the period prior to 1976 (Fumagalli 1976 and Camponuovo 1976). A detailed model of the mountain was constructed to a scale of 1:160, giving a model height approaching 4.5 m. Camponuovo 1976 described the problems associated with attempting to achieve mechanical similitude between the model and the prototype. The author concluded that, for scaling ratios of about 1:20 to 1:30, it is comparatively easy to get satisfactory results from the model. However, the modelling difficulties increase significantly as the scale model becomes very low.

11.3.4 Computer modelling of rockfalls

Ritchie 1963 in his paper on rockfall experiments, included simple algorithms describing rockfall trajectories. However, the first computer model of rockfall appears to be that mentioned by Piteau and Clayton 1977. This has been referred to in a number of the other references given in this chapter, but a full description of the modelling method is contained only in unpublished company reports. Kirsten *et al.* 1986 described the program as being based on a simple model:spherical boulder, initial horizontal and vertical boulder velocities, boulder obeys conventional laws of reflection with possibility of different normal and tangential coefficents for each bounce. In common with other computer models, the slope geometry is input as straight segments or cells. By varying the input parameters within given frequency distributions, the probability of rocks landing within specific areas could be calculated. Most computer programs for rockfall studies are based on a model along the lines of that shown in Figure 11.19.

Some aspects of the program developed by Hoek (1987b) have been described in the previous section on sliding and rolling modes of motion. The program is written in BASICA for an IBM PC and uses separate parameters for the normal and tangential coefficients of restitution to calculate the rockfall trajectory during bouncing.

The program described by Pasquero (1987) and Scavia *et al.* (1988) has basic algorithms similar to those of Piteau and Clayton (1977) and Hoek (1987b) but also incorporates a Monte Carlo probabilistic routine. The program is written in BASIC 4.0 for an HP 9816 computer with graphic output to an HP 7090A. The input data consists of the following:

(1) coordinates for beginning and end of each cell segment;
(2) coordinates for starting position of rock block;
(3) mean and standard deviation values for both r_n and r_t for each cell;
(4) standard deviation for a roughness angle with respect to the slope angle of each cell;
(5) mean and standard deviations for initial horizontal and vertical velocity components for the block;
(6) number of runs required for each analysis.

The output consists of graphical and numerical output as follows:

(1) minimum and maximum trajectory for rockfall for each set of Monte Carlo simulations;
(2) envelope to maximum height of rockfall trajectories;

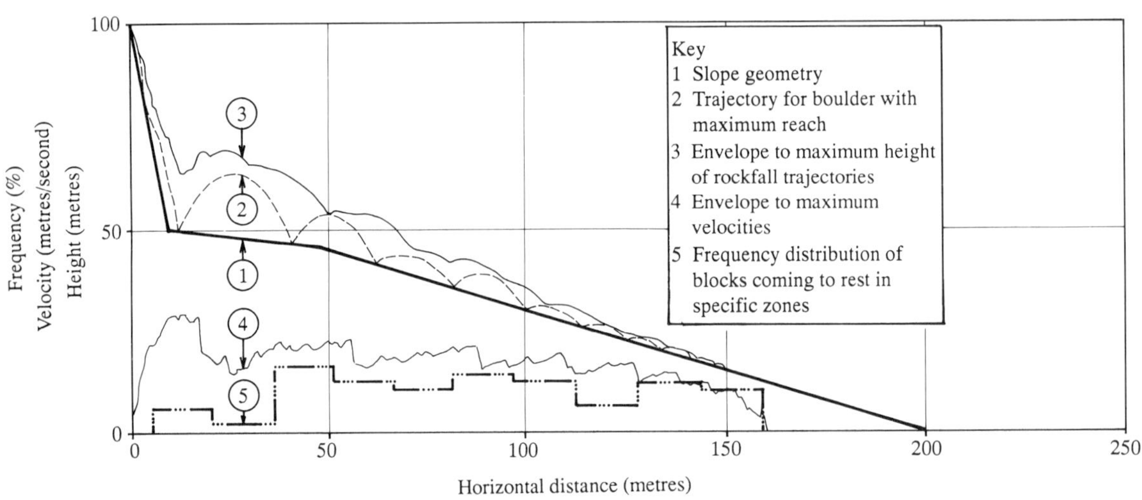

Figure 11.19 *Example of output probabilistic rockfall analysis*

(3) envelope to maximum velocity of rock;
(4) histogram showing frequency distribution of blocks with respect to the locations at which they come to rest.

An example of the graphical output from a simulation of 100 rockfall trajectories is shown in Figure 11.19. This type of approach is ideal for carrying out sensitivity analyses for given slopes. In particular, the use of a 'roughness angle' on the straight-line cell segments is considered to be a convenient and realistic way of accounting for irregular slope profiles.

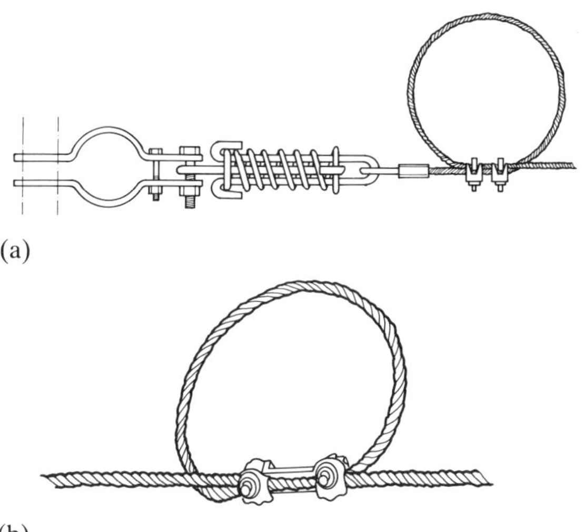

Figure 11.20 *Details of cable brake systems:* (a) *cable spring and brake as manufactured by UMM, Haly;* (b) *cable brake as manufactured by BRUGG, Switzerland*

Spang (1987) and Spang and Rautenstrauch (1988) have described their Fortran 77 Rockfall computer program which models the following rockfall conditions:

(1) blocks modelled as spheres;
(2) rolling as an initial or subsequent mode of motion;
(3) changes in angular momentum during impact;
(4) transitions from one mode of motion to another within a single slope cell;
(5) simultaneous sliding and rolling.

The program uses the normal coefficient of restitution to calculate the normal velocity component after impact. The tangential velocity component is calculated from the tangential translational velocity and the angular velocity of the rock.

Descoeudres and Zimmermann (1988) have developed a numerical rockfall program based on finite elements and incorporating a three-dimensional topographic model. Blocks are modelled as prisms with polygonal sections or as ellipsoids. The blocks are assumed to be rigid and their impacts involve frictional and elastoplastic responses from the topography. The authors state that the model is still under development but that it is proving to be a valuable tool for assisting engineering judgement in the design of protective measures. This type of approach is likely to be of most relevance to complex topographical situations where the usual simplifying assumption of two dimensions is likely to lead to gross errors.

11.3.5 Rockfall protection works

Table 11.3 is reproduced from Martin (1988) and provides a convenient summary of the relevant stabilization, protection and warning methods that are applicable to slopes with rockfall problems. Stabilization methods involve a permanent reduction of the hazard or improvement in the slope stability. Protection methods require the control of rockfalls and are generally not as initially expensive as stabilization but require an ongoing commitment to maintenance. The use of warning methods is generally restricted to railways or other controlled-access systems (Martin 1988).

Table 11.3 Classification of remedial measures for rock slopes (Martin 1988)

Stabilization methods
Excavation
Scaling and trimming
Groundwater control and drainage
Rock reinforcement and rock support
Shotcrete and mortar
Dental treatment
Rockbolts, dowels, rock anchors
Buttresses and bulkheads
Retaining walls and tie-back walls
Anchored beams and strapping
Beam and cable walls
Cable nets, lashings and chains
Protection methods
Relocation
Tunnels and rock sheds
Interception ditches and shaped ditches
Interception berms and shaped berms
Catch walls
Draped and pinned mesh
Catch fences, catch nets
Warning methods
Patrol and signs
Electric fences and wires
Warning lights and sirens
Monitoring systems
Precise surveys
Extensometers, inclinometers, tile meters, load cells etc.
Systems in combination with protection

Detailed information on relevant types of slope treatment is contained in a number of references (e.g.: CANMET, (1976), Chan *et al.* (1986a), Dubin, *et al.* (1986), Fookes and Sweeney (1976), Hoek and Bray (1981), Kirsten *et al.* (1986), Mearns (1976), Peckover and Kerr (1976, 1977), Peckover (1975), Piteau and Peckover (1978), Rochet (1979, 1980), Spang (1987)).

Because of the compreshensive literature already available on rockfall control, it is not proposed to further review such details in this chapter. The following sections deal with a few specific aspects of stabilization and protection.

(a) Rock catch fences

Ritchie (1963) describes the use of a special rock fence which was apparently novel at that time. The fence consisted of chain link suspended 'like a curtain' from a cable mounted on a compressive spring to absorb the impact from rocks. This concept of a yielding fence has subsequently been developed considerably in several European countries.

Yielding catch-fence systems are manufactured and supplied by BRUGG Cable Products in Switzerland and Utensilerie Meccaniche Milanesi (UMM) in Italy. Although the systems differ in specific details, the general concepts are the same with the catch fences being designed as lightweight, flexible and energy-absorbing structures. The common element is the use of a braking device incorporated into the tie-backs and other cables in the catch fences as shown in Figure 11.20. These rope brakes act to limit the loads imposed on the tension and stay-ropes and to absorb the kinetic energy of the rocks impacting on the fence. The brakes consist of a cable loop and a friction brake. The required braking distances and resistances are calculated for specific applications and pre-set to these values. When the tension in the rope reaches the pre-set value, the loop contracts and takes up the energy from the impact.

The principle of the cable brakes is shown in Figure 11.21 where the load versus deformation curve for a rigid beam is compared with that for a wire rope with two braking devices. The failure load of the rigid member may be much greater than that of the wire ropes but the failure will occur at much lower deformation than with the cable system. The energy absorbed by the respective systems is the area under the load versus deformation curves. The energy that can be absorbed by the wire ropes may therefore be much greater than that for the rigid member as can be seen from Figure 11.21. The UMM system also

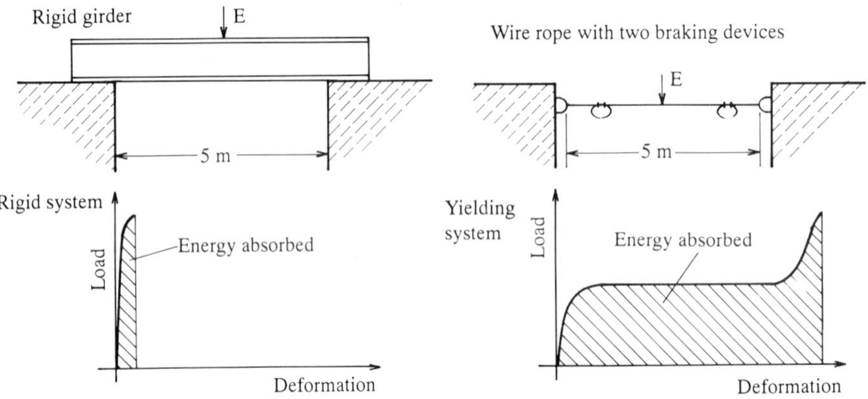

Figure 11.21 *Comparison of load–deformation characteristics for rigid and flexible structures*

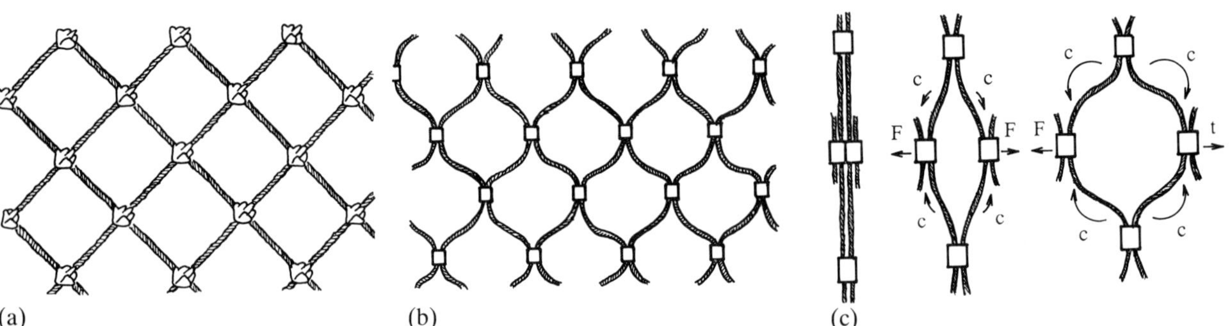

Figure 11.22 *Details of different cable net systems:* (a) *BRUGG cable nets;* (b) *Union cable nets;* (c) *details of Union cable connections*

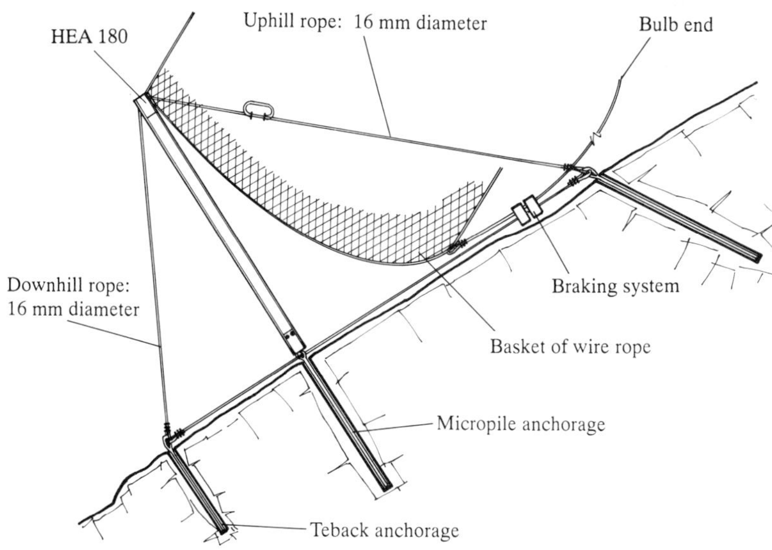

Figure 11.23 *Example of yielding catch-fence design*

incorporates a strong compressive spring to further reduce the shock loading on the cables.

The cable nets provided by the above manufacturers also differ in their construction. The BRUGG nets have a diagonal weave with stamped cruciform connections (Figure 11.22). The impact forces caused by an object striking the net are designed to be distributed over the system's diagonal rope mesh, the border ropes, supports, stay ropes with braking devices and the anchoring structure. By comparison, the UMM nets have parallel and vertical cables which have clamps on alternate pairs of the cables such that the nets have considerable flexibility (Figure 11.22). Both manufacturers have carried out impact loading tests on the mesh sections and demonstrated that these are capable of arresting blocks of over 1 tonne falling from heights in excess of 10 m.

The BRUGG system is extensively used for snow and avalanche protection on scree slopes where explosively anchored tie-backs are often used. These involve inserting a tube into a drilled hole. An explosive charge at the bottom of the tube creates a bulb-shaped cavity into which the splayed end of a looped cable is inserted, the hole then being filled with concrete. Pull-out loads of 70 tonnes for an anchor length of 5 m are reported in BRUGG's literature.

An example of a catch-fence design is shown in Figure 11.23. The components of the fence are relatively light and therefore very appropriate for installation in areas of adverse accessibility. One significant advantage of this type of fence is that maintenance is generally fairly minimal after they are impacted by rocks. In most cases, the fence does not suffer any significant damage and can simply be restored to its original position by resetting the braking devices and loops.

These types of fences have been demonstrated to provide superior rock catching abilities to conventional stiff systems. Although rigid structures can be designed to provide adequate capacity for particular values of kinetic energy, they are often susceptible to failure when

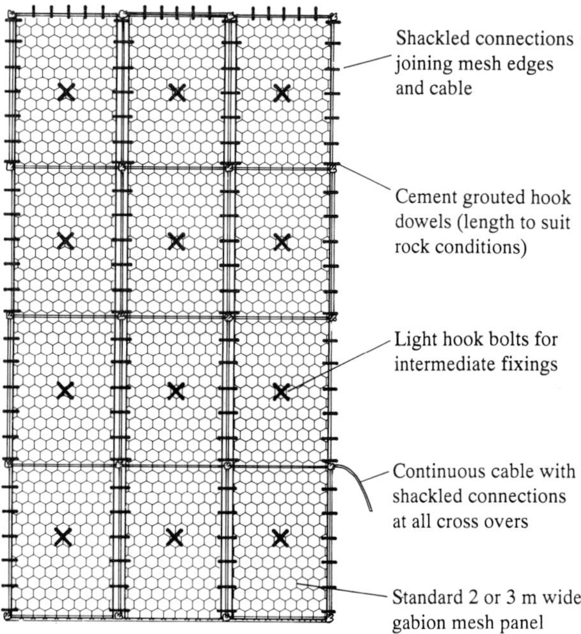

Figure 11.24 *Details of mesh and cable system for loose blocky rock*

impacted by two closely spaced boulders, such that the second one hits the fence when it is already at its elastic limit from the first impact. Despite their advantages, the use of yielding systems appears to be largely restricted to Europe at present with little evidence of their use in other parts of the world where rockfall problems are acute. An example of a yielding system in use is given in Berchten *et al.* (1988). Such catch fences have been designed to absorb up to 500 kN m of kinetic energy (this value coinciding with Spang's suggested definition of the upper limit for defining the term 'rockfall').

(b) Meshing of slopes

Wire meshing of rock slopes is one of the most effective and economical methods of preventing rockfalls from steep slopes. The use of gabion mesh is invariably recommended for this purpose since the double-twist hexagonal weave is much less likely to unravel if individual strands fail. Good durability of the material is assured if galvanized and PVC-coated wire is used. The Maccaferri company provide comprehensive information on their gabion netting together with full details of the recommended methods for laying and fastening it to rock slopes.

The general method of fixing the wire panels together involves lacing it with binding wire and using a double twist every two meshes. Although this is quite satisfactory in most instances, there are occasions where a more robust method is required. A system used on several projects in the UK utilizes horizontal and vertical steel cables shackled to the mesh and fixed to hook dowels as shown in Figure 11.24. This system has the advantage of providing a strong grid of cables at 2–3 m centres to provide greater resistance to large scale block movement. The use of shackles also simplifies the lacing together of the panels and means that it is a simple matter to release loose rock which has become bagged behind the mesh.

(c) Cable lashing and cable nets

This technique is briefly described in Piteau and Peckover (1978) and is relevant to protruding noses of loose unstable rock blocks which are either too loose or inaccessible for rock bolting. Cable nets such as the BRUGG type are commonly used in Europe to restrain large masses of loose blocky rock which are not amenable to bolting or other forms of stabilization.

A particular application of this method is currently being employed on a large nose of loose rock at a site in England. The area of potential instability was first covered with gabion mesh fixed with light cables on a 2 m grid as described in the preceding section. High-capacity horizontal cables are then strung across the nose using anchoring and tensioning methods as shown in Figure 11.25. The method described provides for a positive tension in the cables and does not require rock deformation before it becomes effective.

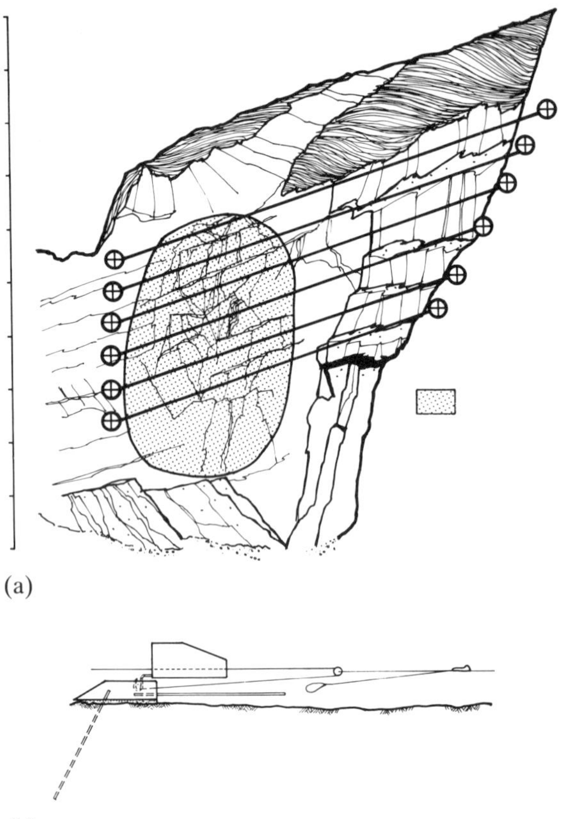

Figure 11.25 *Details of a cable lashing system:* (a) *elevation of face;* (b) *details of cable anchorages and tensioning system*

11.4 Conclusions

This chapter has reviewed the problem of unstable slopes in rock masses. A brief discussion on current design methods for rock-slope stability has been followed by a more comprehensive treatment of rockfalls which are often a more serious problem, especially along transportation corridors. A number of computer programs have been developed which allow reasonably accurate modelling of rockfall trajectories. At present, the detailed applicability of these programs is constrained by inadequate field and experimental data on the dynamic parameters governing the restitution of boulders during bouncing modes.

The analytical methods available, especially where they incorporate probabilistic routines, are a useful aid to engineering judgement in the location and design of rockfall protection measures such as catch fences.

The height and location of catch fences can be selected on the basis of the rockfall trajectories defined by these analyses. Estimation of likely block velocities and therefore block kinetic energies allows the design of catch structures to be undertaken with greater confidence. Unless massive restraining works are provided, catch structures should be designed to allow controlled yield during impact and therefore reduce the likelihood of failure during service. Field observations and trials are essential for critical works.

References

Berchten, J. D., Anatra, S. and Ackermann, T. (1988) 'Protection measures against rockfalls', *Proc. Fifth Int. Symp. Landslides,* Lausanne, Switzerland, 1103–1105

Broili, L. (1976) 'Relations between scree slope morphometry and dynamics of accumulation processes', *Meeting on Rockfall Dynamics and Protective Works Effectiveness,* Istituto Sperimentale Modelli e Strutture, Bergamo

Camponuovo G. F. (1976) 'ISMES' experience on the model of S. Martino', *Meeting on Rockfall Dynamics and Protective Works Effectiveness,* Istituto Sperimentale Modelli e Strutture, Bergamo

CANMET (1976) *Pit Slope Manual: Chapters 1–10,* Publications Distribution Office, CANMET, Ottawa

Chan, Y. C., Chan, C. F. and Au, S. W. C. (1986) 'Design of a boulder fence in Hong Kong', *Proc Conference on Rock Engg in an Urban Environment,* Institution of Mining and Metallurgy, Hong Kong, 87–96

Chan, Y. C., Chan, C. F. and Au, S. W. C. (1986) 'Authors' replies to discussion', *Proc. Conf. Rock Engg in an Urban Environment,* Institution of Mining and Metallurgy, Hong Hong, 495–497

Descoeudres, F. and Zimmermann, T. H. (1988) 'Three-dimensional dynamic calculations of rockfalls', *Proc. Fifth Int. Symp. Landslides,* Lausanne, Switzerland, 337–342

Dubin, B. I., Watkins, A. T. and Chang, D. C. H. (1986) 'Stabilisation of exisitng rock faces in urban areas of Hong Kong', *Proc. Conf. on Rock Engg and Excavation in an Urban Environment,* Institution of Mining and Metallurgy, Hong Kong, 155–171

Dunnicliff, J. (1988) *Geotechnical Instrumentation for Monitoring Field performance,* Wiley-Interscience, New York

Fookes, P. G. and Sweeney, M. (1976) 'Stabilisation and control of local rock falls and degrading rock slopes', *Q. J. Engng Geol.,* **9**, 37–55

Fumagalli E. (1976) Introduction au probleme du Mont St Martino', *Meeting on Rockfall dynamics and protective works effectiveness,* Istituto Sperimentale Modelli e Strutture, Bergamo

Goodman, R. E. (1976) *Methods of Geological Engineering in Discontinuous Rocks,* West Publishing Co., St Paul, Minnesota

Goodman, R. E. and Bray, J. W. (1976) 'Toppling of rock slopes', *Proc. Spec. Conf. Rock Engg for Found and Slopes,* Boulder, Colorado, **2**, 201–234

Habib P. (1976) 'Notes sur le rebondissement des blocs rocheux', *Meeting on Rockfall Dynamics and Protective Works Effectiveness,* Istituto Sperimentale Modelli e Strutture, Bergamo

Hoek E. (1983) 'Strength of jointed rock masses', *Geotechnique,* **33**(3), 187–223

Hoek E. (1987) 'General two-dimensional slope stability analysis'. In *Analytical and Computational Methods in Engineering Rock Mechanics,* E. T. Brown (ed.), Allen and Unwin, London, pp. 95–128

Hoek E. (1987) 'Rockfall – a program in BASIC for the analysis of rockfalls from slopes', *Unpublished notes,* Golder Associates/University of Toronto

Hoek, E. and Bray, J. W. (1981) *Rock Slope Engineering* (3rd edn), Institution of Mining and Metallurgy, London

Janbu, N. (1954) 'Stability analysis of slopes with dimensionless parameters', *D. Sc. Thesis,* Harvard Soil Mechanics Series No. 46, Harvard University, Cambridge, Mass.

Kirsten, H. A. D., Steffen, O. K. H. and Stacey, T. R. (1986) 'Discussion on Rockfall protection measures', *Proc. Conf. Rock Engg and Excavation in an Urban Environment,* Institution of Mining and Metallurgy, Hong Kong, 492–495

Mak, N. and Blomfield, D. (1986) 'Rock trap design for presplit rock slopes', *Proc. Conf. Rock Engg and Excavation in an Urban Environment,* Institution of Mining and Metallurgy, Hong Kong, 263–270

Martin, D. C. (1988) 'Rockfall control: an update: technical note', *Bull. Assoc. Engng Geol,* **25**(1), 137–144

Mearns, R. (1976) 'Solving a rockfall problem in California', *Bull. Assoc. Engng Geol.,* **13**(4), 329–335

Pasquero M. (1987) 'Dinamica della caduta di massi rocciosi nello studio della stabilita dei versanti', *Tesi di Laurea,* Politecnico di Torino, Facolta di Ingegneria

Peckover, F. L. (1975) 'Treatment of rock falls on railway lines', *Am. Railway Assoc.* Bulletin 653, 471–503

Peckover, F. L. and Kerr, J. W. G. (1976) 'Treatment of rock slopes on transportation routes', *Proc. 29th Canad. Geotech. Conf.* Vancouver, 1.16–1.40

Peckover, F. L. and Kerr, J. W. G. (1977) 'Treatment and maintenance of rock slopes on transportation routes', *Can. Geotech. J.,* **14**(4), 487–507

Piteau, D. R. and Clayton, R. (1977) 'Discussion of paper on "Computerised design of rock slopes using interactive graphics for the input and output of geometrical data" by P. A. Cundall, M. D. Voegele *and* C. Fairhurst', *Proc. 16th Symp. Rock Mech.,* Univ. of Minnesota, Minneapolis, 62–63

Piteau, D. R. and Peckover, F. L. (1978) 'Rock slope engineering'. In *Landslides: Analysis and Control* R. L. Schuster and R. J. Krizek (eds.), Transportation Research Board, Special Report 176, National Academy of sciences, Washington, DC, 192–228

Richards, L. R. (1986) 'Rapporteur's comments on "Rockfall protection methods"', *Proc. Conf. Rock Engg and Excavation in an Urban Environment,* Institution of Mining and Metallurgy, Hong Kong, 489

Richards, L. R. and Atherton, D. (1987) 'Stability of slopes in rock'. In *Ground Engineers' Reference Book,* F. G. Bell (ed), Butterworths, London, 12.1–12.16

Richards, L. R., Whittle, R. A. and Ley G. M. M. (1978) *Appraisal of stability conditions in rock slopes,* Chapter 16 'Foundations in Difficult Ground', Newnes–Butterworths

Ritchie, A. M. (1963) 'Evaluation of rockfall and its control', *Highway Research Board Record,* no. 17, Washington, 13–28

Rochet, L. (1979) 'Protection against rock falls by means of metal netting' (in French), *Bull. Liaison Lab Ponts Chaussees,* **101**, 21–28

Rochet, L. (1980) 'Protection against rock-falls. Methodology of specific studies. Application to the study of the La Praz zone of the SNCF railway line between Culoz and Modane' (in French), *Bull. Liaison Lab Ponts Chaussees,* **106**, 57–68

Sarma, S. K. (1979) 'Stability analysis of embankments and slopes', *Proc. ASCE J. Geotech. Engg Div.,* **105**, 1511–1524

Scavia, C., Barla, G. and Vai, L. (1988) 'Analisi di tipo probabilistico', *Proc. Secondo Ciclo di Conferenze di Meccanica e Ingegneria delle Rocce,* Politecnico do Torino

Sharp, J. C. (1989) 'Monitoring of major excavated slopes in stable and unstable rock masses', *Proc. Conf. Geotech. Instrumentation in Civil Engg Projects,* Institution of Civil Engineers, University of Nottingham

Spang, R. M. (1987) 'Protection against rockfall – Stepchild in the design of rock slopes', *Proc. 6th Int. Conf. Rock Mechanics,* Montreal, A. A. Balkema, Rotterdam: 551–557

Spang, R. M. and Rautenstrauch, R. W. (1988) 'Empirical and mathematical approaches to rockfall protection and their practical application', *Proc. 5th Int. Symp. Landslides,* Lausanne, Switzerland, 1237–1243

Whiteside, P. G. D. (1986) Discussion on Rockfall protection measures, *Proc. Conf. Rock Engg and Excavation in an Urban Environment,* Institution of Mining and Metallurgy, Hong Kong, 490–492

12 Settlement and bearing capacity of foundations on rock masses

Professor F Kulhawy, Cornell University
Professor J P Carter, University of Sydney

12.1 Introduction

Significant economy normally can be realized in design if a foundation can be constructed on or in the surface or near-surface rock. However, designers cannot blindly follow the old adage, 'when in doubt, put it on rock', because natural rock masses are among the most variable of all engineering materials. It is mostly for this reason that the design of foundations on or in rock has been largely empirical until recent years. Fortunately, there have been many recent studies that have improved vastly our understanding of the mechanical behavior of rock masses, making possible more rational analytical treatments of rock-foundation systems. In this chapter, the more pertinent of these recent analytical studies are presented. The focus is on the bearing capacity and settlement of foundations on rock. Foundations socketed in rock masses are addressed in Chapter 25. General construction issues are described elsewhere by Kulhawy and Goodman (1987).

12.2 Geological characterization

The first step in the analysis process is geological characterization of the rock mass. Rock masses are inhomogeneous, discontinuous media composed of the rock material and naturally occurring discontinuities such as joints, seams, faults, and bedding planes. The rock materials may have engineering properties that are anisotropic, nonlinear, and stress-dependent. The discontinuities may range from soft and weathered to hard and unweathered, and their spacings and attitudes (strike and dip) may vary considerably. The resulting rock mass exhibits the characteristics of both the material and the discontinuities which, together, tend to make the deformation and strength properties of the rock mass highly directional.

12.2.1 Site exploration

The goal of site exploration is to characterize the rock mass, with particular emphasis on establishing the type of rock material present, the size, frequency, and spatial distribution of the discontinuities, the water table, and perhaps *in situ* measurements of engineering properties. For large or special structures, such as tall buildings, major bridges, high dams, power plants, etc., extensive and specialized exploration programs are normally conducted. Commonly these consist of vertical and inclined borings, detailed geological mapping, down-hole and cross-hole measurements, etc. Extensive laboratory testing is done on the rock, and *in situ* testing may be done to measure the strength and deformation properties of the rock mass. Furthermore, load tests are warranted for these kinds of structures. With the data obtained, the rock mass and its properties can be characterized well.

For more conventional or simpler structures, the exploration and testing is more limited, typically consisting of vertical borings, limited mapping, some simple laboratory core testing, and perhaps a few down-hole measurements. With these more limited data, the characterization is less complete, and increased reliance is placed on prior information and correlations in the literature. Judgment and experience enter more prominently into the analysis process in this case.

Regardless of the level of detail, the exploration program should provide sufficient information to allow the development of a geological model of the site being investigated and establish the engineering properties necessary. These are the necessary input data for analysis and design.

12.2.2 Special geological problems

During the characterization process, special geological features must be identified because they have the potential to influence foundation performance markedly. These features include weathering, chemical effects, solution phenomena, creep, subsidence, and collapse structures, and are described by Kulhawy and Goodman (1987) and Peck (1976).

12.3 Geomechanical models

Analytical predictions in geotechnical engineering require the adoption of a model of the behavior of the real material. For example, to predict the short-term bearing capacity of a foundation on a layer of clay soil, it is customary to assume that the clay behaves in an undrained manner and that the real material can be idealized as being purely cohesive. Predictions of the bearing capacity then are based on the theory for a rigid punch indenting an ideal rigid perfectly plastic half-space. In the application of this theory, the designer must choose an appropriate value for the undrained shear strength of the clay, and the accuracy of the prediction will depend on the accuracy of the strength value and the model for the clay behavior.

The same general principles apply to the prediction of the mechanical behavior of a rock mass, because the success of the analytical prediction depends on the appropriate choice of an ideal material to model the rock behavior, as well as the selection of the basic input parameters for that model. Invariably, the choice of the ideal material depends on the type of prediction to be made and on how much information is available. It would be pointless to adopt a highly sophisticated model for the rock mass if the necessary geological information and test data were not available to support its use.

For predicting the ultimate load capacity of a foundation in rock, a strength model of the rock mass is required. Alternatively, if predictions of the foundation movements caused by the applied loading are required, then a constitutive (or deformation) model must be selected. Usually, the ultimate load capacity and the deformations are considered as separate problems, requiring the adoption of a different material model for each. Constitutive models that combine the strength and deformation issues have been developed, allowing the prediction of the complete load–deformation behavior of the foundation from initial loading to failure. At present, these find major use in the research environment and rarely find application in design practice, although this is likely to change with forthcoming increases in computing power and usage.

In this chapter, the two issues of ultimate load capacity and deformations under working loads will be treated separately. This approach is consistent with current design practice and results in greatly simplified methods of analysis.

12.4 Settlement of foundations on rock

As with foundations in soil, the displacements of foundations in rock normally control the design. With rock masses, however, a model must be established to address their discontinuous nature, taking into account the properties of the rock material and the discontinuities. When dealing with the displacements of foundations in rock, it is usual to idealize the discontinuous rock mass as an elastic continuum.

12.4.1 Geomechanical model for elastic properties

A geomechanical model has been suggested by Kulhawy (1978) to establish equivalent rock-mass properties from the individual elastic properties of the rock material and the discontinuities. In general, the model considers the case of three orthogonal discontinuity sets, as shown in Figure 12.1. The rock material is defined by the Young's

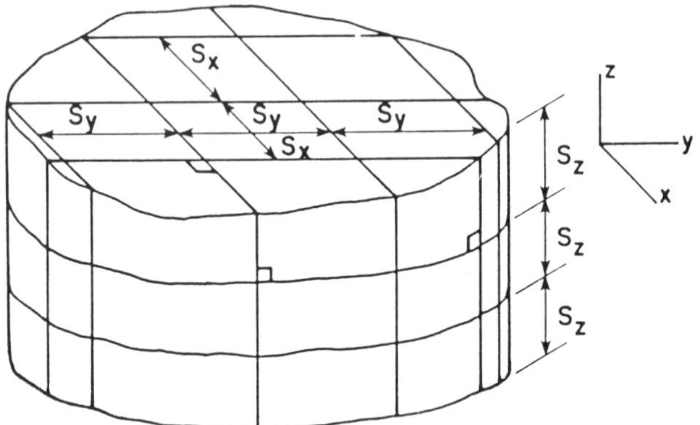

Figure 12.1 *Rock mass model*

modulus, E_r, and Poisson's ratio, ν_r, and subsequently the shear modulus, G_r, while the discontinuities are described by a normal stiffness, K_n, shear stiffness, K_s, and mean discontinuity spacing, S. Anisotropic rock material properties also could be used, as long as the attitude of the property principal planes is coincident with the attitude of the discontinuity planes. For this model, the properties of the equivalent orthotropic elastic mass are given by Duncan and Goodman (1968) as

$$E_i = \left(\frac{1}{E_r} + \frac{1}{S_i K_{ni}}\right)^{-1} \tag{12.1}$$

$$G_{ij} = \left(\frac{1}{G_r} + \frac{1}{S_i K_{si}} + \frac{1}{S_j K_{sj}}\right)^{-1} \tag{12.2}$$

$$\nu_{ij} = \nu_{ik} = \nu_r \frac{E_i}{E_r} \tag{12.3}$$

for $i = x, y, z$ with $j = y, z, x$ and $k = z, x, y$. These equations completely describe the elastic properties of the rock mass.

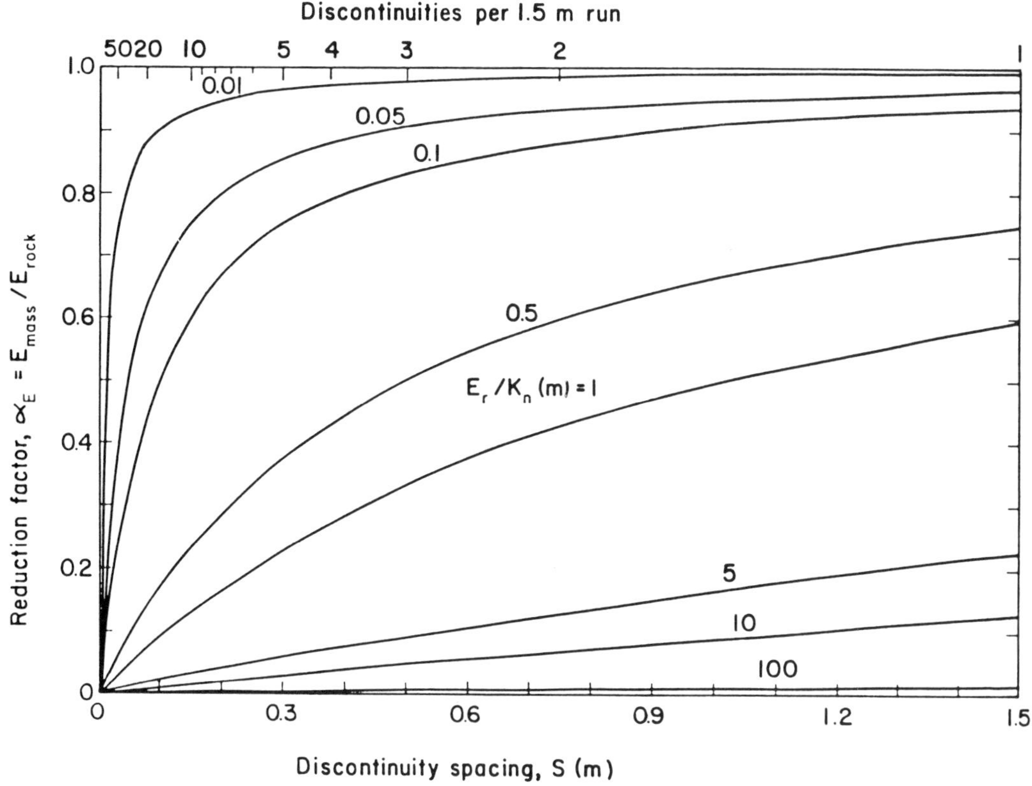

Figure 12.2 *Modulus reduction factor vs. discontinuity spacing (Kulhawy 1978)*

For engineering convenience, it is useful to define a modulus reduction factor, α_E, which represents the ratio of the rock mass to rock material modulus. This factor can be obtained by rewriting Equation (12.1) as

$$\alpha_E = \frac{E_i}{E_r} = \left(1 + \frac{E_r}{S_i\, K_{ni}}\right)^{-1} \tag{12.4}$$

This relationship is plotted in Figure (12.2). This figure shows smaller values of α_E in rock masses with softer discontinuities (larger E_r/K_n values).

Unfortunately, the mean discontinuity spacing is not easy to obtain directly and, in normal practice, rock quality designation (RQD) values are determined instead. Using a physical model, the RQD can be correlated with the number of discontinuities per 1.5 m core run, a common measure in practice. This relationship is shown in Figure 12.3. Other models presented in the literature, using statistical or random number premises, generally have confirmed this approach. Combining Figures 12.2 and 12.3 yields Figure 12.4, which relates α_E to RQD as a function of E_r/K_n.

This model is most effective when good quality geological data are available to define the discontinuity sets adequately. When the data are not sufficient to define the x and y discontinuities adequately, the model must be simplified to deal with the z values only. In the majority of rock masses, S_z is normally less than S_x and S_y and tends to control settlement under compression loading. The

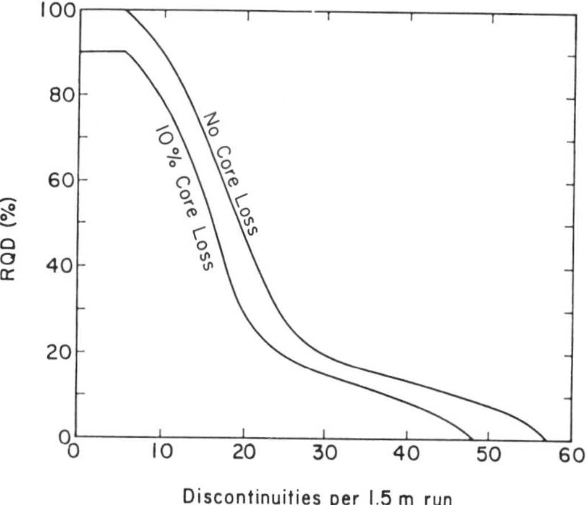

Figure 12.3 *RQD vs. number of discontinuities (Kulhawy 1978)*

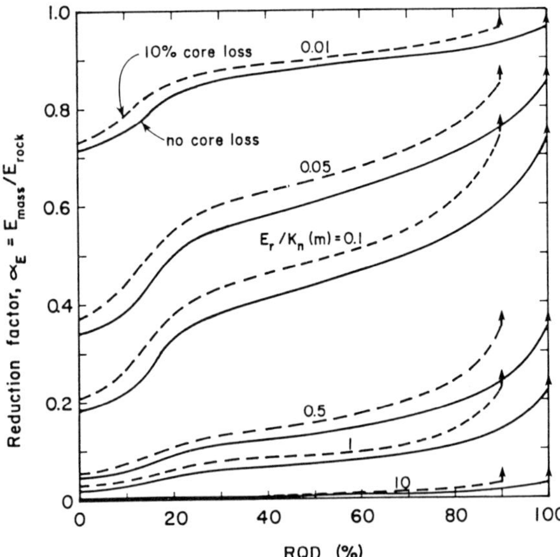

Figure 12.4 *Modulus reduction factor vs. RQD (Kulhawy 1978)*

assumption of an isotropic mass based on S_z and K_{nz} normally gives a conservative upper bound on the settlement.

Good physical property data also are needed. These should be obtained for the anticipated effective stress range to be applied in the field, in tests that impose strains compatible with those expected. If full-scale load tests or field deformation tests are conducted, α_E can be determined easily by dividing the field test E by the value of E from laboratory tests on intact core. When the field and laboratory testing is more limited, the values given in Table 12.1 can provide some general guidelines. The complete details of these data are given by Kulhawy (1975). As a first approximation, the designer can select a mean value of $E_r/K_n = 1$ m, which yields $\alpha_E \approx 0.1$ for an RQD less than about 70%, and progressively higher α_E values varying linearly up to 0.6 at an RQD of 100%.

An alternative to the RQD-based reduction factor has been proposed by Bieniawski (1975). Using the CSIR Geomechanics Classification Rating, which incorporates six rock mass parameters, the modulus reduction factor is given in Figure 12.5. For conventional building and bridge foundations, the field data are likely to be inadequate to provide a CSIR rating. However, for dam and power plant foundations, the field data should be sufficient to provide a CSIR rating.

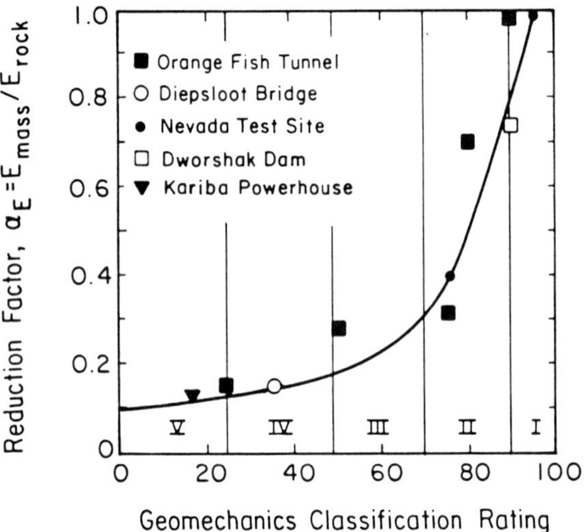

Figure 12.5 *Modulus reduction factor vs. Geomechanics Classification Rating (Bieniawski 1975)*

12.4.2 Isotropic rock mass

Once the rock mass has been characterized, the settlement can be computed. If the data are limited (vertical core or z-direction only), one is almost restricted to the use of a linear, elastic, isotropic, half-space model, in which the settlement (ρ) is given by

$$\rho = \frac{P(1-\nu_m^2)}{\beta_z E_m A^{0.5}} \qquad (12.5)$$

in which P = applied load, ν_m = Poisson's ratio for rock mass (Equation (12.3)), E_m = Young's modulus for rock

Table 12.1 Typical ranges in rock properties (Kulhawy 1978)

Property	*Number of values*	*Maximum*	*Minimum*	*Mean*
Elastic modulus, E_r (GN/m^2)	261	111.6	0.006	34.60
Poisson's ratio, ν_r	138	0.46	0.02	0.20
Normal stiffness, K_n (GN/m^3)	12	67.59	0.24	13.00
Shear stiffness, K_s (GN/m^3)	167	31.60	0.01	2.82
K_n/K_s for same rock	12	83.0	0.84	17.8
E_r/K_n for same rock (m)	9	4.23	0.21	1.22

mass (Equation (12.1)), A = foundation area, and β_z = foundation shape and rigidity factor.

From Table 12.1, it is seen that typical values of ν_r are low; values of ν_m will be lower. For circular to rectangular foundations with length, L, to width, B, ratio up to three, using either flexible or rigid foundation assumptions, the value of β_z is about 1.1, with only a variation up to 5% depending on the assumption (Kulhawy 1978). Using these typical ν_m and β_z values, Equation (12.5) can be simplified as follows:

$$\rho \approx \frac{0.9P}{E_m\,A^{0.5}} = \frac{0.9P}{\alpha_E\,E_r\,A^{0.5}} \qquad (12.6)$$

This isotropic model typically is conservative and therefore should yield upper-bound settlement values.

In these equations and those below, it is assumed that the rock mass does not exhibit significant time-dependent settlements, as in the case of soft weathered shales, salt, etc., and that it does not contain thick soft seams. For these specific cases, consolidation and secondary settlements can be estimated by traditional soil mechanics methods, while the time-independent elastic settlements are computed as above.

12.4.3 Transversely isotropic rock mass

Many rock masses are anisotropic to varying degrees, in which case the isotropic solution given above is really not appropriate. Unfortunately, few anisotropic solutions exist that are applicable to foundation problems. One of the simpler and more useful solutions is given below for the case of a circular rigid load placed on a transversely isotropic mass in which the load axis is coincident with the vertical axis of symmetry of the anisotropic mass. Details of this model are given by Gerrard and Harrison (1970).

To compute the settlement, one of the following equations is used:

$$\text{for } \beta^2 > 0:\ \rho_z = \frac{P(c+G_{zh})d\delta(\delta^2-\beta^2)}{BG_{zh}\,[c+d\,(\delta+\beta)^2]\cdot[c+d\,(\delta-\beta)^2]} \qquad (12.7a)$$

$$\text{for } \beta^2 < 0:\ \rho_z = \frac{P\delta(ad)^{0.5}}{B(ad-c^2)} \qquad (12.7b)$$

$$\text{for } \beta^2 = 0:\ \rho_z = \frac{P(c+G_{zh})d\delta^3}{BG_{zh}\,(c+d\delta^2)^2} \qquad (12.7c)$$

The appropriate equation to use is defined by

$$\beta^2 = \frac{ad-c^2-2cG_{zh}-2G_{zh}(ad)^{0.5}}{4dG_{zh}} \qquad (12.8)$$

in which

$$a = \frac{E_h(1-\nu_{hz}\,\nu_{zh})}{(1+\nu_{hh})\,(1-\nu_{hh}-2\nu_{hz}\,\nu_{zh})} \qquad (12.9a)$$

$$c = \frac{E_h\,\nu_{zh}}{1-\nu_{hh}-2\nu_{hz}\,\nu_{zh}} \qquad (12.9b)$$

$$d = \frac{E_h\,\nu_{zh}\,(1-\nu_{hh}}{\nu_{hz}(1-\nu_{hh}-2\nu_{hz}\,\nu_{zh})} \qquad (12.9c)$$

$$\delta^2 = \frac{ad-c^2-2cG_{zh}+2G_{zh}(ad)^{0.5}}{4dG_{zh}} \qquad (12.9d)$$

These solutions are subject to the following limitations:

$$1-\nu_{hh}-2\nu_{hz}\,\nu_{zh} > 0 \qquad (12.10a)$$

$$1-\nu_{hh} > 0 \qquad (12.10b)$$

$$1+\nu_{hh} > 0 \qquad (12.10c)$$

In these equations, ρ_z = vertical displacement (settlement), P = total load, B = diameter of loaded area, ν_{hh} = Poisson's ratio for the effect of horizontal stress on the complimentary horizontal strain, ν_{hz} = Poisson's ratio for the effect of horizontal stress on the vertical strain, ν_{zh} = Poisson's ratio for the effect of vertical stress on the horizontal strain, E_z = vertical Young's modulus, E_h = horizontal Young's modulus, and G_{zh} = shear modulus between the horizontal and vertical planes. The values of E_h and G_{zh} are often expressed as a function of E_z. This solution may be used for approximating the behavior of rectangular foundations by using an equivalent diameter for the foundation.

12.4.4 Analysis example

To illustrate the use of this approach, an example problem is given below.

(a) Problem statement

A square reinforced concrete footing, 1 m by 1 m in plan, is to be loaded to a vertical stress of 15 MN/m^2 and placed on a horizontally bedded sandstone unit. Rock cores in the sandstone showed a recovery of 100% with an average RQD of 70%. Uniaxial compression tests on the core showed a modulus of 2.5 GN/m^2 in the applied stress range. Surface mapping of the rock showed two orthogonal joint sets normal to the bedding, the first with an average spacing of 1.5 m and the second with an average spacing of 0.5 m. Examination of all three discontinuity faces showed them to be tight and clean. Estimate the settlement of this foundation.

(b) Outline of solution

Given:	$P = 15\,\text{MN}$, $A = 1\,\text{m}^2$ (equivalent $B = 1.128\,\text{m}$), $E_r = 2.5\,\text{GN/m}^2$, RQD = 70%, $S_x = 1.5\,\text{m}$, $S_y = 0.5\,\text{m}$.
Assumed:	$K_n = 5\,\text{GN/m}^3$, $K_s = 10\,\text{GN/m}^3$, $\nu_r = 0.19$ (therefore $G_r = 1.05\,\text{GN/m}^2$) (all assumed from typical values given by Kulhawy 1975); for simplicity, assume that the orthotropic rock mass can be approximated as a cross-anisotropic mass, so that all horizontal properties are the average of the x and y properties.
Spacings:	From Figure 12.3, there are 15 discontinuities per 1.5 m run; therefore $S_z = 0.1\,\text{m}$.
Reduction Factors:	From Figure 12.4 or Equation (12.4), $\alpha_{E_z} = 0.167$; from Equation (12.4), $\alpha_{E_x} = 0.75$ and $\alpha_{E_y} = 0.50$.
Young's Modulus:	Using α values or Equation (12.1), $E_z = 0.417\,\text{GN/m}^2$, $E_x = 1.875\,\text{GN/m}^2$, $E_y = 1.25\,\text{GN/m}^2$; therefore $E_h \approx (E_x + E_y)/2 = 1.563\,\text{GN/m}^2$.
Shear Modulus:	From Equation (12.2), $G_{xz} = 0.495\,\text{GN/m}^2$, $G_{yz} = 0.465\,\text{GN/m}^2$; therefore $G_{zh} \approx (G_{xz} + G_{yz})/2 = 0.480\,\text{GN/m}^2$.
Poisson's Ratio:	From Equation (12.3), $\nu_{xy} = \nu_{xz} = 0.142$, $\nu_{yz} = \nu_{yx} = 0.095$, $\nu_{zx} = \nu_{zy} = 0.032$; therefore $\nu_{hh} \approx (\nu_{xy} + \nu_{yx})/2 = 0.118$, $\nu_{hz} \approx (\nu_{yz} + \nu_{xz})/2 = 0.118$, $\nu_{zh} \approx (\nu_{zx} + \nu_{zy})/2 = 0.032$. Equations (12.10) are satisfied.
Settlement:	From Equations (12.9), $a = 1.593\,\text{GN/m}^2$, $c = 0.0572\,\text{GN/m}^2$, $d = 0.428\,\text{GN/m}^2$, $\delta = 1.313$. From Equation (12.8), $\beta^2 = -0.206$. From Equation (12.7b), $\rho_z = 21\,\text{mm}$.

If there were no orthogonal jointing (i.e. $S_x = S_y = \infty$), the solution would give $\rho_z = 20\,\text{mm}$. If it were roughly assumed that the rock mass was isotropic (only ν_{zh} and E_z are independent), Equation (12.6) would give $\rho_z = 32\,\text{mm}$.

12.4.5 More complex models

Other more general anisotropic models have been given by Gerrard and Harrison (1970), Kulhawy and Ingraffea (1978) and others. General forms of anisotropy also can be addressed readily using finite element models. Although these techniques are available, the major problem is evaluating the rock mass properties reliably. These mass properties are difficult to evaluate under general conditions of anisotropy, so it is wise at the present time to use simpler models for which the properties can be assessed more confidently.

12.5 Axial compression capacity of foundations

Very often, the design of foundations on rock will be governed by displacement considerations. Nevertheless, the ultimate capacity of the foundation always must be evaluated to determine the degree of safety of the proposed design. This evaluation requires consideration of two important issues. First, the foundation element itself must be able to resist the applied loading adequately. Second, the rock mass must be capable of providing resistance to the loading. In both cases, a reasonable margin of safety must be provided. The first case is a structural design problem that is not addressed herein, while the second is described below.

12.5.1 Bearing capacity or tip resistance

In general, compressive loads applied to an embedded foundation are transmitted to the surrounding rock mass through both tip resistance and side resistance. The relative importance of these resistances in determining the capacity depends on the geometry of the foundation and the relative stiffness of the foundation and the rock mass. There also may be some interdependence between the tip and side resistances, especially in jointed rock where both strength and stiffness will depend on the confining stress.

The typical tip resistance or bearing capacity failure modes for rock masses are shown in Figure 12.6. These depend on the discontinuity spacing or rock layering, as described below.

For a thick rigid layer overlying a weaker one, failure may be by flexure. The flexural strength is approximately twice the tensile strength of the rock material, and the tensile strength is of the order of 5–10% of the compressive strength. For a thin layer overlying a weaker one, failure can be by punching which, in effect, is manifested by a tensile failure in the rock material. It is important to realize that, in both of these cases, failure in the underlying layer could occur first by one of the other failure modes.

For loading applied to a rock mass with open joints, where the joint spacing is less than the foundation width (or diameter), failure is likely to occur by uniaxial compression of rock columns. If the rock mass is idealized as a cohesive-frictional material, the ultimate capacity is given by the Mohr–Coulomb failure criterion as

$$q_{ult} = q_u = 2c \tan(45° + \phi/2) \qquad (12.11)$$

in which q_{ult} = ultimate bearing capacity, q_u = uniaxial compressive strength, c = cohesion intercept, and ϕ = friction angle. The values of q_u, c, and ϕ are rock mass (rather than intact rock) properties.

If the rock mass contains closely spaced, closed joints, a general wedge type of failure mode may develop, as shown in Figure 12.6. The ultimate bearing capacity in this case is

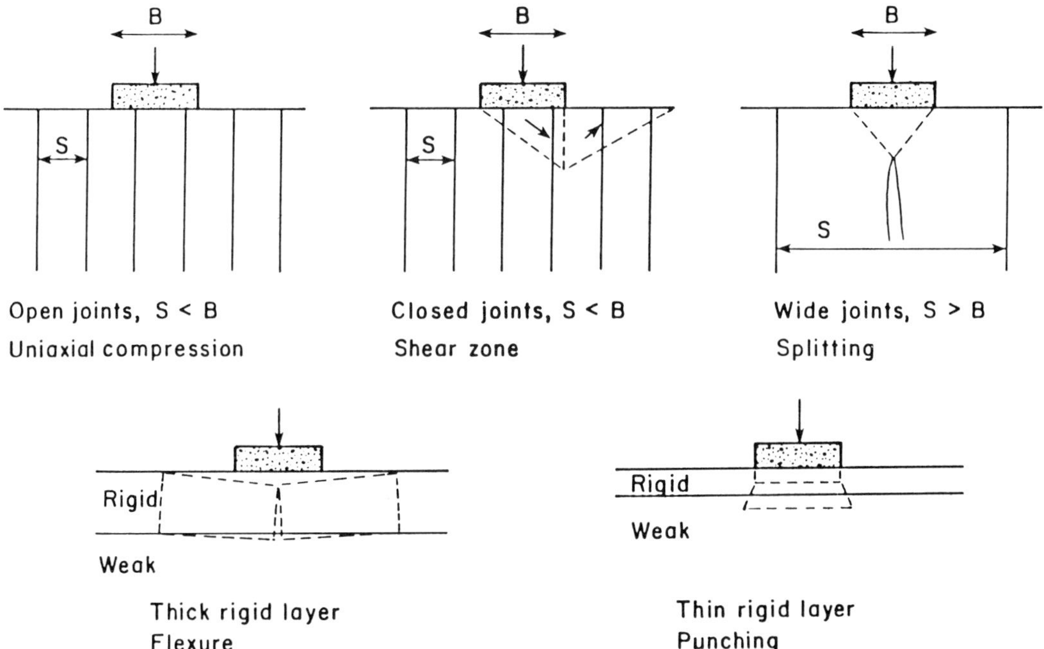

Figure 12.6 *Bearing capacity failure modes (Sowers 1979)*

given by the Bell (1915) solution for plane-strain conditions:

$$q_{\text{ult}} = cN_c + \frac{B}{2}\gamma N_\gamma + \gamma D N_q \qquad (12.12)$$

in which B = foundation width, D = foundation depth, γ = effective unit weight of the rock mass, and N_c, N_γ, and N_q are bearing capacity factors given in Figure 12.7. For a cylindrical or square foundation of diameter or width B and depth D, Equation (12.12) should be modified to allow for the foundation shape:

$$q_{\text{ult}} = \zeta_{cs} cN_c + \zeta_{\gamma s} B\gamma N_\gamma/2 + \zeta_{qs}\gamma D N_q \qquad (12.13)$$

in which $\zeta_{cs} = 1 + N_q/N_c$, $\zeta_{\gamma s} = 0.6$, and $\zeta_{qs} = 1 + \tan\phi$. For foundations at the rock surface, $D = 0$. Also, the second term is normally small compared to the other terms and is often neglected.

For cases in which the joints are spaced more widely than the foundation width, failure occurs by splitting beneath the foundation, which eventually leads to general shear failure. This problem has been evaluated by Kulhawy and Goodman (1980), assuming no stress is transmitted across the vertical discontinuity, to give

$$q_{\text{ult}} \approx JcN_{cr} \qquad (12.14)$$

in which N_{cr} = bearing capacity factor in Figure 12.8 and J = correction factor given in Figure 12.9.

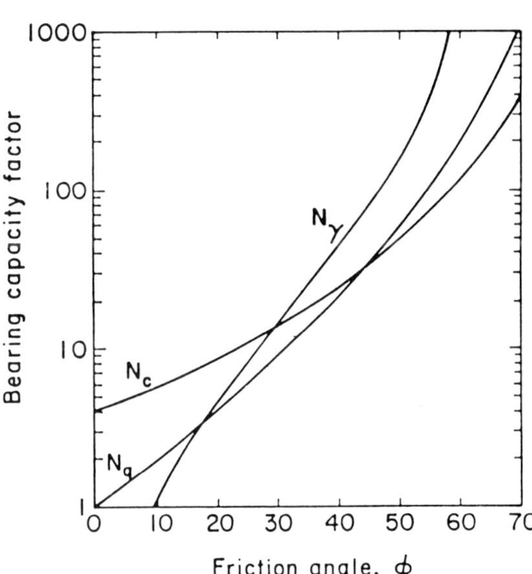

Figure 12.7 *Wedge bearing capacity factors*

In the evaluation of the tip resistance outlined above, it is important to note that the strength parameters of the rock mass, and not of the intact rock, must be used. Values of c and ϕ obtained for the intact rock material are considerably higher than those for the rock mass, and their

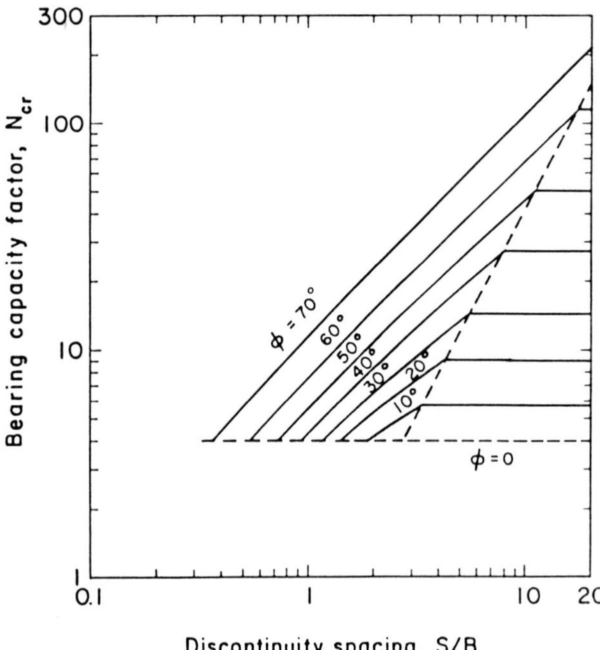

Figure 12.8 *Bearing capacity factor for open joints (Kulhawy and Goodman 1980)*

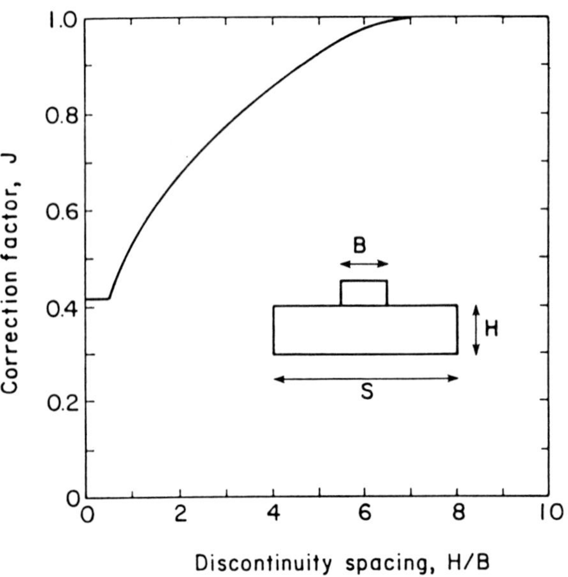

Figure 12.9 *Correction factor for discontinuity spacing (Bishnoi 1968)*

use will result in an overestimate of the actual bearing capacity.

If the actual rock mass properties are not evaluated, Kulhawy and Goodman (1987) suggested that the values of c or q_u for the intact rock may be reduced as shown in Table 12.2.

Table 12.2 Suggested design values of strength parameters (Kulhawy and Goodman 1987)

RQD (%)	Rock mass properties		
	Uniaxial compressive strength	Cohesion	Angle of friction
0–70	$0.33q_u$	$0.1q_u$	30°
70–100	$0.33–0.8q_u$	$0.1q_u$	30°–60°

q_u = uniaxial compressive strength of intact rock core

Alternatively, a strength criterion for jointed rock masses (Hoek and Brown 1980) may be used to determine the bearing capacity. This curved strength envelope can be expressed as

$$\sigma_1 = \sigma_3 + (mq_u\sigma_3 + sq_u^2)^{0.5} \qquad (12.15)$$

in which σ_1 = major principal effective stress, σ_3 = minor principal effective stress, q_u = uniaxial compressive strength of the intact rock, and s and m = empirically determined strength parameters for the rock mass, which are somewhat analogous to c and ϕ of the Mohr–Coulomb failure criterion.

An analysis of the bearing capacity of a rock mass obeying this criterion can be made using the same approximate technique as used in the Bell solution. The details of this approach are described in Figure 12.10. A lower bound to the failure load is calculated by finding a stress field that satisfies both equilibrium and the failure criterion. The rock mass beneath a strip footing may be divided into two zones, with homogeneous stress conditions at failure throughout each, as shown in Figure 12.10. The vertical stress in zone I is assumed to be zero, while the horizontal stress is equal to the uniaxial compressive strength of the rock mass, given by Equation (12.15) as $s^{0.5}$ q_u. For equilibrium, continuity of the horizontal stress across the interface must be maintained, and therefore the bearing capacity of the strip footing may be evaluated from Equation (12.15) (with $\sigma_3 = s^{0.5}q_u$) as

$$q_{ult} = [s^{0.5} + (ms^{0.5} + s)^{0.5}]q_u \qquad (12.16)$$

For a circular foundation, a similar approach may be used, with the interface between the two zones being a cylindrical surface of the same diameter as the foundation. In this axisymmetric case, the radial stress transmitted across the cylindrical surface, at the point of collapse of the foundation, may be greater than $s^{0.5}q_u$, without necessarily violating either radial equilibrium or the failure criterion. However, because of the uncertainty of this value, the radial stress at the interface also is assumed to be $s^{0.5}q_u$ for the case of a circular foundation. Therefore, the predicted (lower-bound) bearing capacity is given by Equation (12.16).

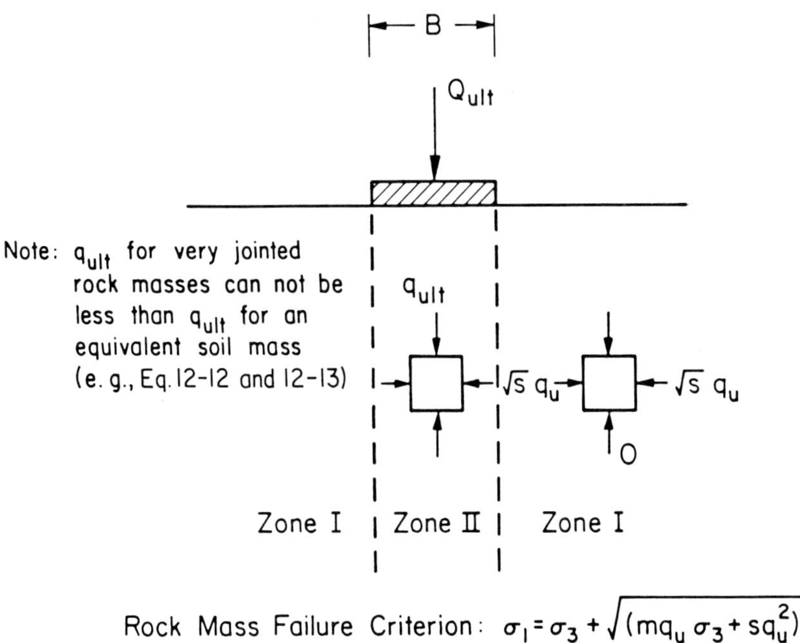

Figure 12.10 *Lower-bound solution for bearing capacity*

Table 12.3 Approximate correlation between rock mass quality and strength constants (based on Hoek 1983)

Rock mass quality	*Approximate joint spacing*	*CSIR rating*	*NGI rating*	*S value*	*m value as function of rock type*				
					A	*B*	*C*	*D*	*E*
Excellent	> 3 m intact	100	500	1	7	10	15	17	25
Very good	1–3 m interlocking	85	100	0.1	3.5	5	7.5	8.5	12.5
Good	1–3 m slightly weathered	65	10	0.004	0.7	1	1.5	1.7	2.5
Fair	0.3–1 m moderately weathered	44	1	10^{-4}	0.14	0.2	0.3	0.34	0.5
Poor	30–500 mm weathered with gouge	23	0.1	10^{-5}	0.04	0.05	0.08	0.09	0.13
Very poor	< 50 mm heavily weathered	3	0.01	0	0.007	0.01	0.015	0.017	0.25

Rock types:
A Carbonate rocks with well-developed crystal cleavage (dolostone, limestone, marble)
B Lithified argillaceous rocks (mudstone, siltstone, shale, slate)
C Arenaceous rocks with strong crystals and poor cleavage (sandstone, quartzite)
D Fine-grained igneous crystalline rocks (andesite, dolerite, diabase, rhyolite)
E Coarse-grained igneous and metamorphic crystalline rocks (amphibolite, gabbro, gneiss, granite, norite, quartzdiorite)

Guidelines for selecting *s* and *m* for jointed rock masses are given in Table 12.3. The categories in this table are determined by the rock type and the conditions of the rock mass, and selecting an appropriate category is easier if either the CSIR (Bieniawski 1974) or NGI (Barton *et al.* 1974) classification data are available. The values in this table should be used only as general guidelines, and they do not replace the need for testing or other means of assessing the strength parameters more reliably.

In another study of the bearing capacity of jointed rock, a plasticity solution was obtained that incorporates the discontinuities in a rock mass (Davis 1980). Although this solution is for a strip foundation, it demonstrates the important effect of jointing. The strength of the intact material was determined by the Mohr–Coulomb criterion, with strength parameters c_r and ϕ_r. Shear failure along the joints also was determined by the Mohr–Coulomb criterion with parameters c_j and ϕ_j. The joints were assumed to be closely spaced (relative to the foundation width, *B*) and parallel, being inclined at an angle ω to the vertical. The results of this analysis for a range of c_j/c_r and ω, with $\phi_r = \phi_j = 35°$, are given in Figure 12.11. The horizontal line for $c_j/c_r = 1$ at the top of the figure gives the bearing capacity without any weakening from discontinuities. For comparison, the uniaxial compressive strength also is shown at the bottom of the figure. It can be seen that, for low and high values of ω, the discontinuities do not cause a very large loss of bearing capacity but, between these limits, the reduction can be large. The vertical drop for ω = 35° when $c_j/c_r = 0$ occurs because no load can be applied to a purely frictional interface if the angle of obliquity is greater than the friction angle. The other kinks in the curves arise from changes in the mode or sequence of plastic regions in the solutions.

The results of Figure 12.11 apply equally well to negative values of ω. The effect of more than one set of discontinuities has not been studied in detail, but there are no theoretical difficulties to such an extension. In general, each additional set will cause further lowering of the bearing capacity, although not necessarily to a great extent. For example, the solution for the two sets ω = 0° and 90°, with $c_j/c_r = 0$, is indicated in Figure 12.11, and gives a value only slightly below that for the single set ω = 90°.

The results in Figure 12.11 are for zero surcharge but, by the following transformation, the effect of surcharge can be included. Consider the case with $\phi_r = \phi_j = \phi = 35°$, discontinuity cohesion = c_j, rock material cohesion = c_r, and surcharge = *q*. For this case, the bearing capacity will be

$$q_{ult} = N_{cs}(c_r + q \tan \phi) + q \qquad (12.17)$$

in which N_{cs} = bearing capacity factor from Figure 12.11, using c_j/c_r modified for the surcharge as follows:

$$(c_j/c_r)_q = (c_j + q \tan \phi)/(c_r + q \tan \phi) \qquad (12.18)$$

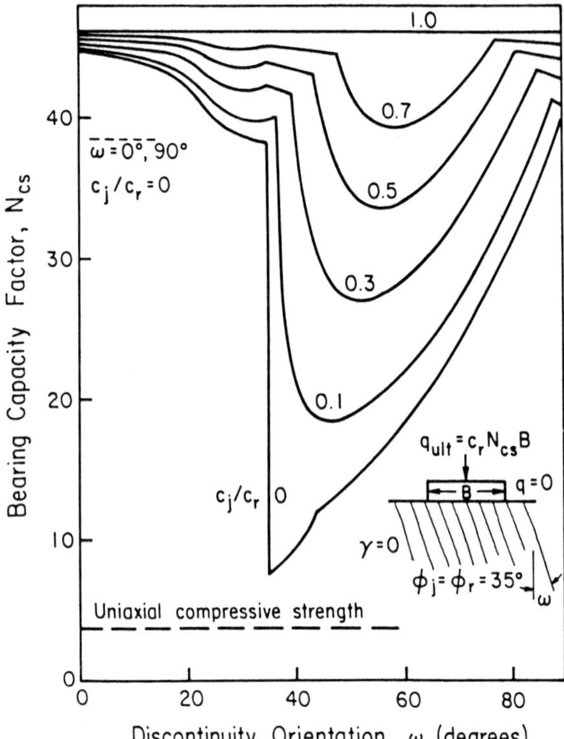

Figure 12.11 *Effect of discontinuity orientation on bearing capacity of strip footing (Davis 1980)*

As mentioned previously, Figure 12.11 applies only for plane strain, but the trends in behavior and the general conclusions implied also should be true for the case of loading over a circular area.

12.5.2 Side resistance of rock sockets

The mechanism of side resistance development in embedded foundations is complex and includes both adhesion and friction effects, as well as dilatancy that may accompany shearing of a roughened interface. Very high normal stresses may develop between cylindrical drilled shafts and the surrounding rock mass, chiefly as a result of dilation at the interface, but also may develop from the effect of Poisson's ratio and the compressive axial stresses in the shaft. In a rock mass that is capable of sustaining these high radial stresses, large shear stresses can develop along the cylindrical interface of a shaft. It is less certain how these stresses can develop in large rectangular excavations. The following discussion focuses on cylindrical drilled shafts socketed in rock.

To predict the side resistance of a rock socket, a constitutive model for interface sliding is required that incorporates the coupling of the shear and normal modes of displacement (e.g. Pease and Kulhawy 1984a,b). With such a model, the development of side resistance from initial loading until full slip of the shaft can be examined. However, these models require, as input data, accurate numerical values for parameters such as the shear and normal stiffness, cohesion intercept and friction angle, and some measure of the dilatancy and perhaps strain-softening. Most of these parameters are not measured routinely in either laboratory or field tests. While the constitutive models may give good insight into the basic mechanism of shear failure, including the progressive development of slip along the shaft, they are not incorporated easily into engineering practice.

The results of most field tests on shafts and direct shear tests in the laboratory are usually summarized in the form of a unit side shear stress. For a load test on a shaft in the field, this stress is obtained by dividing the total force carried in shear along the shaft by the side area of the shaft. Values at peak and residual shear load are quoted commonly but, because of the definition used, they are average values that do not account for the distribution of shear stress along the shaft. For short, rigid shafts, this averaging process will not be important because the stress state will be relatively uniform throughout the loading. The unit side shear resistance corresponding to the peak shaft load is sometimes called, inappropriately, the bond strength of the interface.

Values of the unit side shear resistance have been measured in field and laboratory tests and have been summarized by Amir (1986), Williams and Pells (1981) and Horvath (1978). To date, the most comprehensive summary of the available data is given by Rowe and Armitage (1984), including critical evaluation of the available test data on the basis of the test method, socket roughness, and reliability and type of data recorded. Their suggested correlation between expected unit side shear resistance (τ_{max}) and uniaxial compressive strength (q_u) for most sockets is presented as

$$\frac{\tau_{max}}{p_a} = 1.42\,(q_u/p_a)^{0.5} \qquad (12.19)$$

in which p_a = atmospheric pressure. Particularly rough sockets, defined as those having grooves or undulations of depth greater than 10 mm and width greater than 10 mm, at spacings between 50 mm and 200 mm, were found to be stronger, and the correlation with uniaxial compressive strength was suggested as

$$\frac{\tau_{max}}{p_a} = 1.9\,(q_u/p_a)^{0.5} \qquad (12.20)$$

The data from which Equation (12.19) was deduced have been plotted in Figure 12.12. It can be seen that Equation (12.19) is a reasonable fit to the data; however, it does not represent a lower bound to all data points. A lower bound to most of the observed data is given by

$$\frac{\tau_{max}}{p_a} = b\,(q_u/p_a)^{0.5} \qquad (12.21)$$

in which b = 0.63–0.95. A curve with b = 0.63 also is plotted in Figure 12.12. For design purposes, a conservative estimate of the peak side resistance could be evaluated in most cases using Equation (12.21) with b = 0.63. A conservative approach also is warranted for smooth wall sockets constructed under slurry. However, values of the average stress (τ_{max}) in excess of $0.15q_u$ should be used only when they are demonstrated to be reasonable by a load test, local experience, or adequate *in situ* testing. After selecting a suitable value for τ_{max}, the peak side resistance then is

$$P_{sf} = \tau_{max} A_s \qquad (12.22)$$

in which A_s = area of the socket sidewall.

Relationships as shown in Equations (12.19) and (12.21) are appealing to the designer, because of their succinct form and dependence on a single, easily measured index property (q_u). However, they can be misleading because firstly, they try to correlate a simple rock property (q_u)

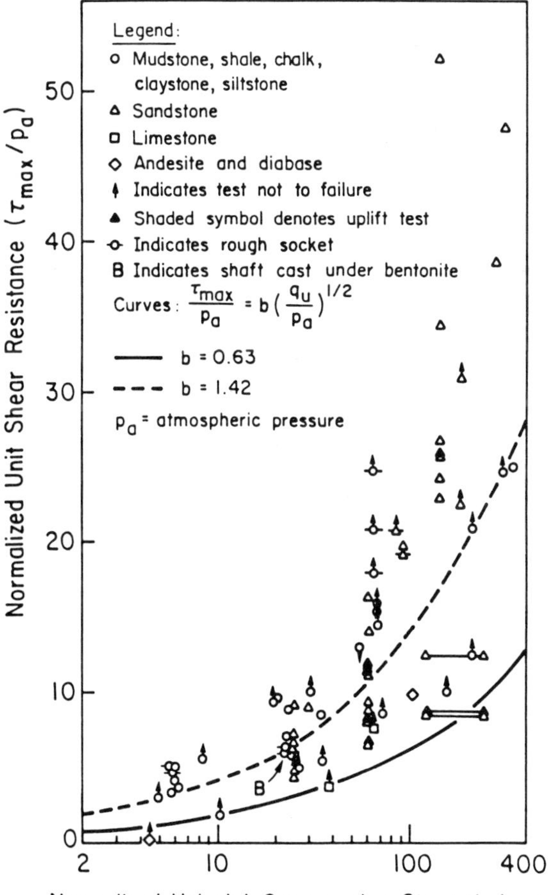

Figure 12.12 *Correlations between bond strength and uniaxial compressive strength (Rowe and Armitage 1984)*

with the mechanical performance of a complicated shaft-rock system, and, secondly, they are probably only applicable to relatively uniform sound rock. In an attempt to extend the correlation to weathered and poor quality (highly jointed) rock masses, an additional empirical factor has been introduced that is based on the parameter α_E, the ratio between the rock-mass modulus and the modulus of the intact rock material. Details of this factor are given by Williams and Pells (1981).

In some cases encountered in practice, the concrete may have a lower compressive strength (f_c') than the intact rock (q_u). For these cases, the average bond strength is governed by the concrete, being approximately equal to $0.05 f_c'$.

12.6 Uplift capacity

Socketed foundations often are used to resist uplift loads, and the uplift capacity is developed from both side and tip resistance. Tip resistance requires the development of a concrete-to-rock tensile bond at the shaft tip. For the ideal case, this tip resistance would equal the tensile strength of the weaker of the rock or concrete, multiplied by the tip area. However, considering the typical construction problems associated with cleaning out the bottom of a socket hole, it is prudent to disregard the tensile resistance developed at the tip.

The side resistance develops from socket shear stresses, as described previously. With compression sockets, there is a positive Poisson's ratio effect, resulting in lateral expansion of the shaft. In tension sockets, the Poisson's ratio effect is negative, resulting in lateral contraction of the shaft. In Chapter 25, it will be shown that the effect of Poisson's ratio of the shaft material is important only in shafts that are relatively flexible, and it is unimportant if the shaft is relatively rigid. In Chapter 25, it also will be demonstrated that a shaft is effectively rigid whenever the ratio, $(E_c/E_r)(B/D)^2$, is greater than about 4. For these cases, the behavior in uplift will be the same as that in compression. For compressible or extensible shafts, it may be prudent to reduce the unit side shear resistance in uplift below that for compression. Based upon examination of elastic solutions, a reduction of up to 30% has been suggested (Poulos and Davis 1980).

12.7 Lateral capacity

The lateral capacity of foundations in rock has received very little attention in the literature. Perhaps the reason is that the lateral design is governed largely by displacement considerations, and therefore the capacity has been assigned lesser importance. Regardless of the reason, reliable evaluation of the lateral capacity still is important, if only to determine the likely margin of safety existing at working load levels. The problem is difficult to solve theoretically, which also may account for the lack of published work in this area.

An approximate theoretical approach for estimating the ultimate lateral capacity is presented here for drilled shaft foundations. It is assumed in the following that the shaft section has sufficient moment and shear capacity to resist the applied loading, and ultimate failure of the shaft occurs when the surrounding rock mass is not able to sustain any further lateral loading, similar to the so-called

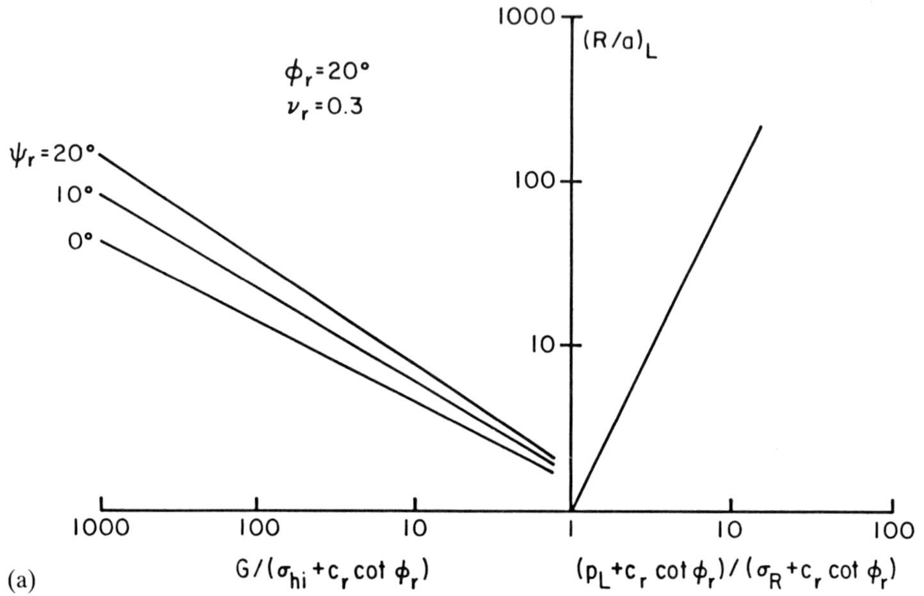

Figure 12.13 *Limit solution for cylindrical cavity (Carter et al. 1986)*

'short pile' failure mode (Broms 1964). This assumption must always be checked; once the limiting state of stress acting on the shaft has been determined, then the calculated maximum bending moment and shear force in the shaft should be compared to the capabilities of the reinforced concrete section. If either of these calculated values exceeds the section properties, then failure will be governed by the strength of the shaft itself.

To determine the ultimate lateral loads acting on a 'short' shaft, the distribution of the limiting reaction (force per unit length acting on the shaft) is required. This may be evaluated as follows.

When a lateral load is applied at the rock-mass surface, the rock mass immediately in front of the shaft will exhibit nearly zero vertical stress, while horizontal stress is applied by the leading face of the shaft. Ultimately, the horizontal stress may reach the uniaxial compressive strength of the rock mass and, with further increases in the lateral load, the horizontal stress may decrease as the rock mass softens during post-peak deformation. Large lateral deformations may be required for the rock mass at depth to exert a maximum reaction stress on the leading face of the shaft. Therefore, it is reasonable to assume that the reaction stress at the rock mass surface, in the limiting case

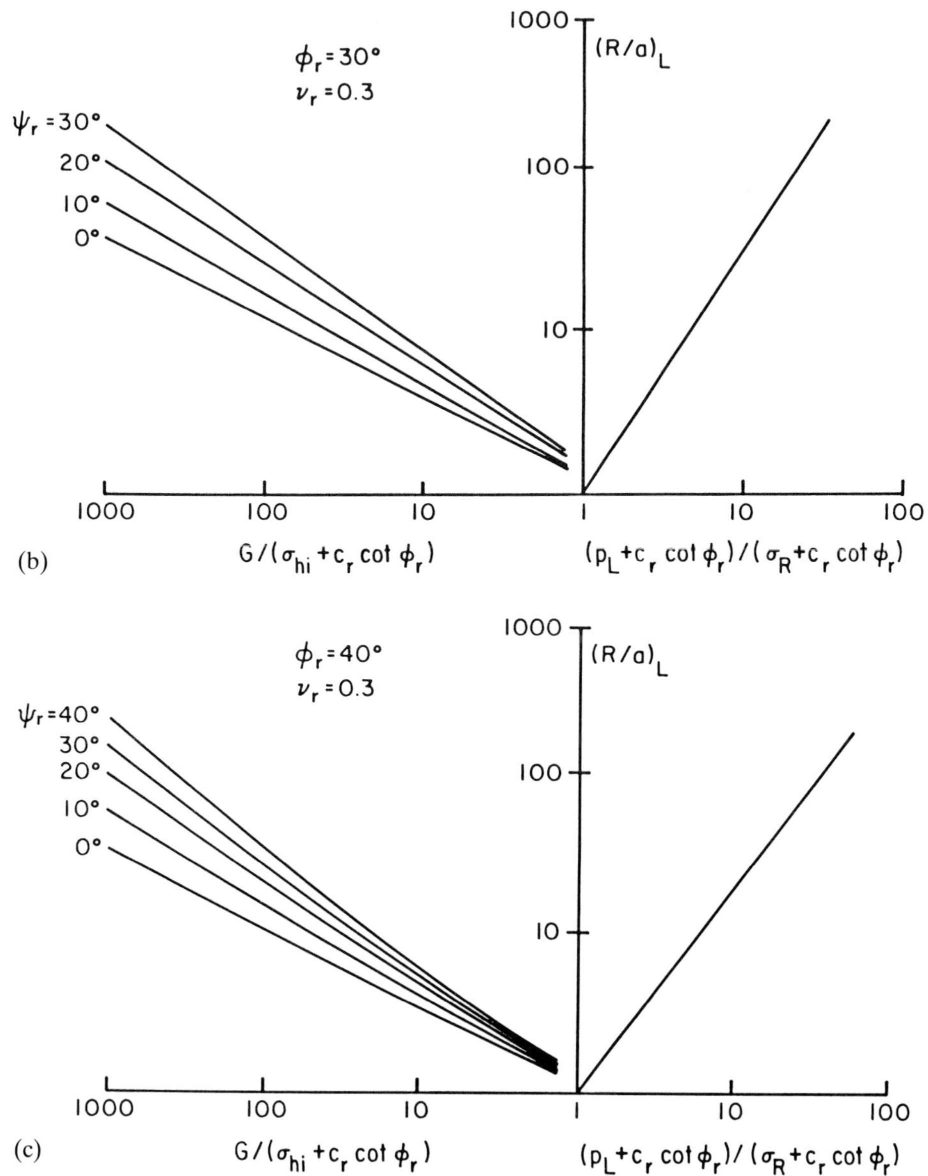

Figure 12.13 *(Continued.)*

of loading of the shaft, is zero or very nearly zero as a result of the post-peak softening. Along the sides of the shaft, some shearing resistance may be mobilized, and this is likely to be of the order of the unit side resistance under axial compression (τ_{max}).

At greater depth, it is reasonable to assume that the stress in front of the shaft may increase from the initial *in situ* horizontal stress level (σ_{hi}), up to the limit stress (p_L) reached during the expansion of a long cylindrical cavity, that is, the plane strain condition will apply. Behind the shaft, the horizontal stresses will decrease, and after tensile rupture of the bond between the concrete and the rock mass, the horizontal stress will reduce to zero. At the side of the shaft, some shearing resistance also may be mobilized. Therefore, at depth, the ultimate force per unit length resisting the lateral loading is likely to be of the order of $B(p_L + \tau_{max})$.

Closed-form solutions have been presented by Carter *et al.* (1986) for the limit stress developed during the expansion of a long cylindrical cavity in an elastoplastic, cohesive-frictional, dilatant material. This limit stress (p_L) may be determined from the following parametric equations in the non-dimensional quantity ρ:

$$\frac{2G_r}{\sigma_{hi} + c_r \cot \phi_r} = \left(\frac{N-1}{N+1}\right)(T\rho^n - Z\rho) \qquad (12.23)$$

and

$$\rho = (p_L + c_r \cot \phi_r)/(\sigma_R + c_r \cot \phi_r) \qquad (12.24)$$

in which

$$T = 2\left(1 + \frac{\chi}{\alpha + \beta}\right) \qquad (12.25)$$

$$Z = 2\left(\frac{\chi}{\alpha + \beta}\right) \qquad (12.26)$$

$$\sigma_R = \left[\left(\frac{2N}{N+1}\right)(\sigma_{hi} + c_r \cot \phi_r)\right] - c_r \cot \phi_r \qquad (12.27)$$

$$\alpha = 1/M \qquad (12.28)$$

$$\beta = 1/N \qquad (12.29)$$

$$n = \frac{1+\alpha}{1-\beta} \qquad (12.30)$$

$$M = \frac{1 + \sin \psi_r}{1 - \sin \psi_r} \qquad (12.31)$$

$$N = \frac{1 + \sin \phi_r}{1 - \sin \phi_r} \qquad (12.32)$$

$$\chi = \frac{(1 - \nu_r)(1 + MN) - \nu_r(M + N)}{MN} \qquad (12.33)$$

and G_r = shear modulus, ν_r = Poisson's ratio, c_r = cohesion intercept, ϕ_r = friction angle, and ψ_r = dilation angle of the rock mass. The rock mass is assumed to obey the Mohr–Coulomb failure criterion, and dilatancy accompanies yielding according to the following flow rule:

$$\frac{\dot{\varepsilon}_3^P}{\dot{\varepsilon}_1^P} = -M \qquad (12.34)$$

in which $\dot{\varepsilon}_1^P$ and $\dot{\varepsilon}_3^P$ = major and minor principal plastic strain increments, respectively.

In most practical cases, the *in situ* horizontal stress (σ_{hi}) will be small compared to the cohesion (c_r) and therefore Equation (12.23) may be simplified slightly by substitution of σ_{hi} = 0. For convenience, solutions for the limit pressure (p_L) have been plotted in Figure 12.13 for selected values of ν_r, ϕ_r, and ψ_r. The central vertical axis on each plot indicates the ratio of the plastic radius at the limit condition (R) to the cavity radius (a). These charts may be used by entering with a value of $G/(\sigma_{hi} + c_r \cot \phi_r)$ and working clockwise around the figure, determining in turn values of R/a and $\rho = (p_L + c \cot \phi)/(\sigma_R + c \cot \phi)$, from which the limit pressure p_L can be calculated.

One final problem remains, and that is determining the depth at which this limit stress is mobilized. In an earlier study by Randolph and Houlsby (1984), it was suggested that, in a purely cohesive material, this depth would be about three shaft diameters ($3B$). In the absence of any other data, this suggestion will be adopted. Therefore, the proposed distribution of ultimate force per unit length resisting the shaft is as shown in Figure 12.14.

The ultimate lateral force that may be applied for the conditions given can be approximated by

$$H_{ult} = (p_L D/6 + \tau_{max} B)D \quad (\text{for } D < 3B) \qquad (12.35a)$$

$$H_{ult} = (p_L/2 + \tau_{max})\, 3B^2 + (p_L + r_{max})(D - 3B)B \quad (\text{for } D > 3B) \qquad (12.35b)$$

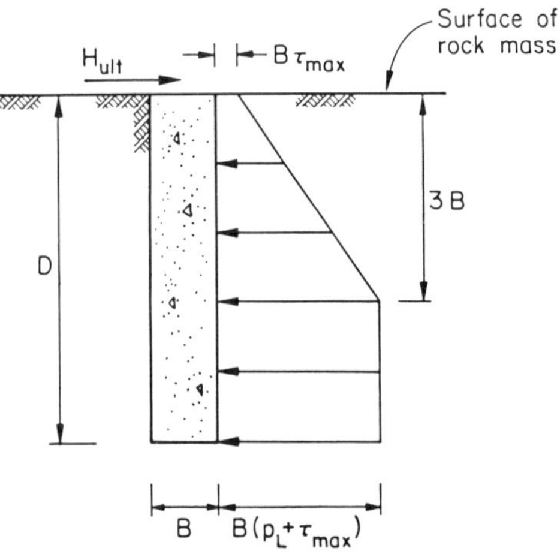

Figure 12.14 *Distribution of ultimate lateral force per unit length*

In most cases, $p_L >> \tau_{max}$. The maximum bending moment in the shaft then is calculated from H_{ult} and the reaction distribution shown in Figure 12.14. If the lateral loading consists of a horizontal force (H) and an applied moment (M), then for purposes of calculating an appropriate bending moment distribution, these may be represented by an equivalent force of the same magnitude, but applied at a height ($e = M/H$) above the rock mass surface.

The theoretical approach suggested above (Equation (12.35)) may be used to calculate the ultimate lateral load for a rock socketed shaft if suitable data are available for the rock mass strength and deformation parameters c_r, ϕ_r, ψ_r, τ_{max}, G_r, and ν_r. This method for predicting the ultimate capacity should be used with caution because it has yet to be tested against field data.

12.8 Acknowledgments

This chapter draws liberally on previous work by the authors and Goodman (Kulhawy and Goodman 1980, 1987; Kulhawy 1978; Carter and Kulhawy 1988). Appreciation is expressed to the Electric Power Research Institute which supported the research in Carter and Kulhawy (1988). L. Mayes typed the text and A. Avcisoy drew the figures.

References

Amir, J. M. (1986) *Piling in Rock,* A. A. Balkema, Rotterdam

Barton, N. R., Lien, R. and Lunde, J. (1974) 'Engineering classification of rock masses for the design of tunnel support', *Rock Mechanics,* **6**, 189–236

Bell, A. L. (1915) 'The lateral pressure and resistance of clay, and the supporting power of clay foundations', *Proc. Inst. Civ. Eng.* **199**, 233–272

Bieniawski, Z. T., (1974) 'Geomechanics classification of rock masses and its application in tunnelling', *Proc. 3rd Int. Congress Int. Soc. Rock Mechanics,* Denver, **2**, 27–32

Bieniawski, Z. T. (1975) 'Case studies: prediction of rock mass behavior by the Geomechanics Classification', *Proc. 2nd Australia–New Zealand Conf. Geomech.*, Brisbane, Australia, 36–41

Bishnoi, B. L. (1968) 'Bearing capacity of a closely jointed rock', *Ph.D. Dissertation,* Georgia Institute of Technology, Atlanta, 120 pp.

Broms, B. B. (1964) 'Lateral resistance of piles in cohesive soils', *J. Soil Mech. Found. Div.*, ASCE, **90**, SM2, 27–63

Carter, J. P. and Kulhawy, F. H. (1988) 'Analysis and design of drilled shaft foundations socketed into rock', *Report EL-5918,* Electric Power Research Institute, Palo Alto, 188 pp.

Carter, J. P., Booker, J. R. and Yeung, S. K. (1986) 'Cavity expansion in cohesive frictional soils', *Geotechnique,* **36**, 349–358

Davis, E. H. (1980) 'A note on some plasticity solutions relevant to the bearing capacity of brittle and fissured materials', *Proc. Int. Conf. Structural Foundations on Rock,* Sydney, **2**, 83–90

Duncan, J. M. and Goodman, R. E. (1968) 'Finite element analyses of slopes in jointed rock', *Contract Report S-68-3*, US Army Engineer Waterways Experiment Station, Vicksburg, 271 pp.

Gerrard, C. M. and Harrison, W. J. (1970) 'Circular loads applied to a cross-anisotropic half-space' and 'Stresses and displacements in a loaded orthorhombic half-space', *Technical Papers 8 and 9,* Division of Applied Geomechanics, Commonwealth Scientific and Industrial Research Organization, Australia (Reproduced as Appendices A and B in Poulos, H. G. and Davis, E. H., *Elastic Solutions for Soil and Rock Mechanics,* Wiley, New York, 1974)

Hoek, E. (1983) 'Strength of jointed rock masses', *Geotechnique,* **33**, 187–223

Hoek, E. and Brown, E. T. (1980) *Underground Excavations in Rock,* Institution of Mining and Metallurgy, London

Horvath, R. G. (1978) 'Field Load Test Data on Concrete-to-Rock Bond Strength for Drilled Pier Foundations', *Publication 87-07,* Department of Civil Engineering, University of Toronto, Toronto, 97 pp.

Kulhawy, F. H. (1975) 'Stress-deformation properties of rock and rock discontinuities', *Engineering Geology,* **9**, 327–350

Kulhawy, F. H. (1978) 'Geomechanical model for rock foundation settlement', *J. Geotech. Engg. Div.,* ASCE, **104**, GT2, 211–227

Kulhawy, F. H. and Goodman, R. E. (1980) 'Design of foundations on discontinuous rock', *Proc. Int. Conf. Structural Foundations on Rock,* Sydney, **1**, 209–220

Kulhawy, F. H. and Goodman, R. E. (1987) 'Foundations in rock', Chapter 55 in *Ground Engineers Reference Book,* F. G. Bell (ed.) Butterworths, London

Kulhawy, F. H. and Ingraffea, A. R. (1978) 'Geomechanical model for settlement of long dams on discontinuous rock masses', *Proc. Int. Soc. Rock Mechanics Symposium on Rock Mechanics Related to Dam Foundations,* Rio de Janeiro, **1**, III.115–III.128

Pease, K. A. and Kulhawy, F. H. (1984a) 'Load transfer mechanisms in rock sockets and anchors', *Report EL-3777,* Electric Power Research Institute, Palo Alto, 102 pp.

Pease, K. A. and Kulhawy, F. H. (1984b) 'Behavior of rock anchors and sockets', *Proc. 25th US Symp. Rock Mechanics,* Evanston, 883–890

Peck, R. B. (1976) 'Rock foundations for structures', *Proc. Rock Engg for Foundations and Slopes,* ASCE, New York, **2**, 1–21

Poulos, H. G. and Davis, E. H. (1980) *Pile Foundation Analysis and Design,* Wiley, New York

Randolph, M. F. and Houlsby, G. T. (1984) 'The limiting pressure on a circular pile loaded laterally in cohesive soil', *Geotechnique,* **34**, 613–623

Rowe, R. K. and Armitage, H. H. (1984) 'The design of piles socketed into weak rock', *Report GEOT-11-84,* University of Western Ontario, London, 366 pp.

Sowers, G. F. (1979) *Introductory Soil Mechanics and Foundations: Geotechnical Engineering* (4th edn), MacMillan, New York

Williams, A. F. and Pells, P. J. N. (1981) 'Side resistance rock sockets in sandstone, mudstone, and shale', *Canad. Geotech. J.*, **18**, 502–513

13 Subsidence in rock masses

Professor F G Bell, University of Natal
R Stacey, Steffen Robertson & Kirsten, Johannesburg

13.1 Introduction

Subsidence of the ground surface can be regarded as ground movement which takes place due to the extraction of mineral resources or the abstraction of fluids. It is an inevitable consequence of mining activities, in the broadest sense, and reflects the movements which occur in the area of exploitation. Unfortunately subsidence can and does have serious effects on surface structures, services and communications, can be responsible for flooding, can lead to the sterilization of land or call for extensive remedial measures or special constructional design in site development. Hence several questions concerning the stability of the surface in potential subsidence areas require answers, these include:

(1) Will subsidence occur and if so, what will be its magnitude?
(2) When will it happen and how will it develop?
(3) What form will the subsidence take?
(4) Is it practical and economic to prevent or reduce its effect?

13.2 Subsidence due to coal mining

13.2.1 Room and pillar workings and subsidence

The room and pillar (bord and pillar, pillar and stall) method, or variations of it, are used in many parts of the world to mine coal, and extensively for working other stratified deposits (Figure 13.1). Pillars support the roof and so have to sustain the redistributed weight of the overburden which means that they and the rock immediately above and below are subjected to added compression. The important factor concerning pillars is that their ultimate behaviour is a function of the pillar height to width ratio, the depth below surface and the size of the extraction area. The height to width ratio of the room also is significant. In addition the greatest stress occurs at the edges of pillars, between pillar and roof or between pillar and floor. Pillars in the centre of the mined-out area are subjected to greater stress than those at the periphery. Individual pillars in dipping seams tend to be less stable than those in horizontal seams since the overburden produces a shear force on the pillar. In the case of coal workings in particular, the mode of failure also involves the character of the roof and floor rocks.

From Figure 13.2 it can be seen that a is the pillar width whilst b is the width of the room. The extraction ratio therefore can be derived from the following expression:

$$r = 2ab + b^2/(a + b)^2 \qquad (13.1)$$

According to Wardell and Wood (1965) determination of the loading on pillars is best approximated by averaging the load on a given pillar due to the weight of overburden. The latter is equal to the weight of the column of strata over an area equal to $(a + b)^2$. It is assumed that the load acts vertically and is uniformly distributed over the cross-sectional area of the pillars. It can be shown that the average loading, P, on a pillar is equal to

$$P = \gamma Z/(1 - r) \qquad (13.2)$$

where γ is the unit weight, Z is the depth and r is the extraction ratio.

Formulae, which have been tested against a significant number of field data, have been developed to describe pillar strength by Salamon and Oravecz (1976) for coal (Equation (13.3)), and Hedley (1978) for hard rock (Equation (13.4)) as follows:

$$P_s = 7.2\, W^{0.46}/H^{0.66}\ \text{MPa} \qquad (13.3)$$

and

$$P_s = 133\, W^{0.5}/H^{0.75}\ \text{MPa} \qquad (13.4)$$

Where P_s is the pillar strength, W is the pillar width in metres and H is the pillar height in metres. The constants in each equation are representative of the strength of the rock, and should be established empirically for different environments.

Figure 13.1 *Pillar and stall workings in salt, Meadowbank Mine, Winsford, Cheshire (courtesy of ICI)*

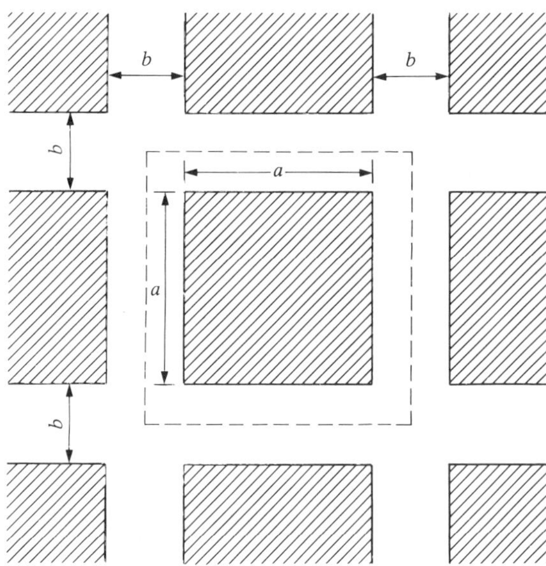

Figure 13.2 *Plan view of pillar and stall workings*

The potential for pillar failure should not be ignored, particular warning signs being strong roofs and floors allied with high extraction rates and moderately to steeply dipping seams. Pillars often experience local failures whilst mining is taking place. If a pillar is highly jointed, then its margin may fail and fall away under relatively low stress. Such action reduces and ultimately removes the constraint from the core, thereby subjecting it to increasing stress. This can lead to pillar failure. Slow deterioration and failure of pillars may take place years after mining operations have ceased. Old workings affect the pattern of groundwater drainage which, in turn, may influence pillar deterioration. Again in the case of coal mines the small pillars of earlier workings may be crushed out once the overburden exceeds 50 or 60 m. Furthermore, old pillars in coal mines at shallow depth have occasionally failed near faults and they may fail if they are subjected to the effects of subsequent longwall mining.

Collapse of one pillar can bring about collapse in others as increased loads are placed on those remaining. A catastrophic multiple pillar collapse of this type occurred at the Coalbrook Colliery in South Africa in 1960. In this case an area of room and pillar workings in excess of 2.5 km^2 collapsed suddenly.

Yielding of a large number of pillars can bring about a shallow broad subsidence over a large surface area. Marino and Gamble (1986) referred to such bowl-shaped depressions as sags. The ground surface in a sag displaces radially inwards towards the area of maximum subsidence. Sag movements depend on the mine layout and geology, as well as the topographic conditions at the surface. They tend to develop rather suddenly, the major initial movements lasting in some instances for about a week, with subsequent displacements occurring over varying periods of time. The initial movements can produce a relatively steep-sided bowl-shaped area. Sag profiles which develop slowly are shallower. In fact the shape of the profile can vary appreciably and because it varies with mine layout and geological conditions, it can be difficult to predict accurately. Normally the greater the maximum subsidence, the greater the likelihood of variation in the profiles. Maximum profile slopes and curvatures frequently increase with increasing subsidence. Marino and Gamble (1986) found that the magnitude of associated surface tensile and compressive strains ranged from slight to severe.

Very often in old workings in coal, pillars were robbed on retreat. Extraction of pillars during the retreat phase simulates the long-wall surface condition although it can never be assumed that all pillars have been removed. As noted above at moderate depths pillars, particularly pillar remnants, are probably crushed and the goaf (i.e. the worked-out area) compacted, but at shallow depths lower crushing pressures may mean that the closure is variable. This causes foundation problems when large or sensitive structrues are to be erected above.

Squeezes or crushes sometimes occur in coal mines, in particular as a result of the pillars being punched into either the roof or floor beds. In such cases the roof or floor may have been weakened or altered by the action of water or weathering. Once again surface subsidence adopts a trough-like or basin form and minor strain and tilt problems occur around the periphery of the basin thereby produced. When pillars are forced into a yielding pavement subsidence may restart. This could take place many years after mining has ceased. Such action becomes more significant with increasing depth or high extraction ratio since in these cases greater loads are imposed upon the pillars.

Even if pillars are relatively stable the surface may be affected by void migration (Figure 13.3). This can take place within a few months of or a very long period of years after mining and normally represents a much more serious problem than that of pillar collapse. Void migration develops when roof rock falls into the worked out areas. When this occurs the material involved in the fall bulks, which means that migration is eventually arrested, unless the fallen rock is dispersed by groundwater flow or moves away in dipping seams. The process can, at shallow depth, continue upwards to the ground surface leading to the sudden appearance of a crownhole. Where superficial deposits occur at the surface they may flow into voids which have reached rockhead, thereby forming features that may vary from a gentle dishing of the surface to inverted cone-like depressions of large diameter.

Figure 13.3 *Void migration in the five-quarter seam (6 cm thick) at Pethburn open-cast site, Co. Durham*

The factors which influence whether or not void migration will take place include the width of the unsupported span, the height of the workings, the nature of the cover rocks particularly their shear strength and the incidence and geometry of discontinuities, the thickness and dip of the seam, the depth of overburden, and the groundwater regime. According to Wardell and Eynon (1968) the maximum width of a self-supporting span (S) can be derived from the expression

$$S = \sqrt{(2t\sigma_t)} \qquad (13.5)$$

where t is the thickness of the seam and σ_t the ultimate tensile strength of the roof rock beam (this should make due allowance for discontinuities).

If a competent rock beam is to span an opening, then its thickness should be equal to 1.75 times the span width in order to allow for arching to develop. Chimney-type collapses can migrate to high levels in massive strata if joints diverge downwards. Walton and Cobb (1984) mentioned that three times the width of the room appeared to be an upper limit to the extent of most void migrations in abandoned coal mines and 1.5 times the span was more common.

Alternatively the maximum height of migration in exceptional cases might extend to 10 times the height of the original roadway, however, it generally is 3–5 times the roadway height. Exceptions, of course, do occur and void migrations at excess of 20 times the seam thickness concerned have been recorded. The self-choking process may not be fulfilled if dipping seams are affected by copious quantities of water which can redistribute the fallen material. Indeed Carter *et al.* (1981) attributed the development of super-voids to such a process. Roof collapse voids that are actually larger than the original stalls could develop in this fashion by engulfing one or two surrounding pillars. Migration of super-voids up to rockhead then produces large-scale subsidence at ground level.

Site investigations in areas of old shallow pillar and stall workings should attempt to determine the extraction ratio, the pattern of the layout and the condition of the workings (Bell, 1986). The sequence and type of roof rocks may provide some clue as to whether void migration has taken place and, if so, its possible extent. The assessment of past and potential future collapse is obviously important.

13.2.2 Subsidence due to longwall working of coal

Longwall mining involves the total extraction of coal (in fact it is not possible to extract 100% of the seam, and 'total extraction' is usually an overall extraction in excess of 80%). Roof support is necessary at the longwall face, and is moved as the coal is extracted, thereby allowing the unsupported roof rock behind the face to collapse into the mined-out void. The classical subsidence which occurs over a completely mined out area in a flat seam is trough-shaped and extends outwards beyond the limits of mining in all directions. In trough subsidence the resulting stratal and surface ground movements are regarded as largely contemporaneous with mining, producing more or less direct effects on any surface development (Brauner 1973). Trough-shaped subsidence profiles develop tilt between adjacent points which have subsided by different amounts and curvature results from adjacent sections which are tilted by differing amounts.

Maximum ground tilts are developed above the limits of the area of extraction and may be cumulative if more than one seam is worked up to a common boundary. Where movements occur, points at the surface subside downwards and are displaced horizontally inwards towards the axis of the excavation (Figure 13.4). Differential horizontal displacements result in a zone of extension on the convex part of the subsidence profile (over the edges of the excavation) whilst a zone of compression develops on the concave section over the excavation itself. Tensile strains usually cause the most significant subsidence damage. Comparatively slight deviations in the subsidence profile are accompanied by appreciable variations in strain. Differential subsidence can cause substantial damage. In fact ground movement is three-dimensional and the vertical and two horizontal components of movement may occur simultaneously.

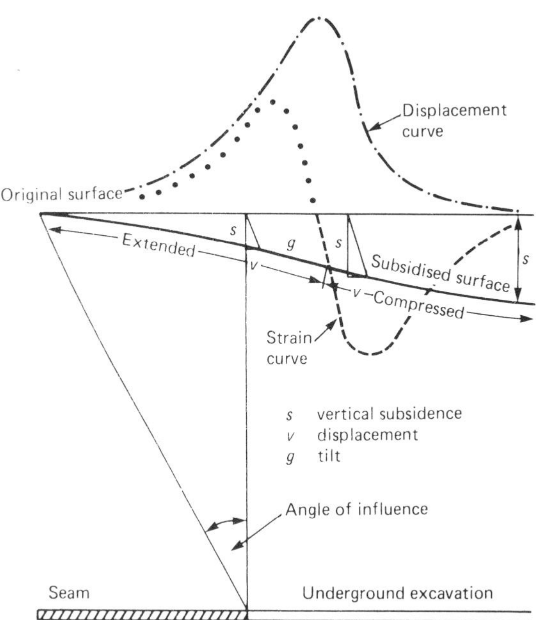

Figure 13.4 *Curve of subsidence showing tensile and compressive strains, vertical subsidence and tilt, together with the angle of influence of draw (not to scale)*

The surface area affected by ground movement is greater than the area worked in the seam. The boundary of the surface area affected is defined by the angle of draw or angle of influence, which varies from 8° to 45° depending on the coalfield (Figure 13.4). Table 13.1 shows frequently quoted angles of draw for different countries. It

Table 13.1 Commonly quoted angles of draw (after Brauner 1973)

Country		*Angle of draw*
Netherlands		35–45°
Germany		30–45°
Northern France		35°
Soviet Union		30°
United Kingdom		25–35°
United States:	Eastern	15–27°
	Central	0–8.5°
	Western	12–16°

would seem that the angle of draw may be influenced by depth, seam thickness and local geology, especially the location of self-supporting strata above the coal seam.

For inclined seams, the surface subsidence trough is displaced towards the down-dip side. Subsidence is at a maximum at the point normal to the centre of the goaf. The angle of draw depends on the dip of the seam; it is least at the up-dip of the goaf and increases towards the down-dip side (Figure 13.5).

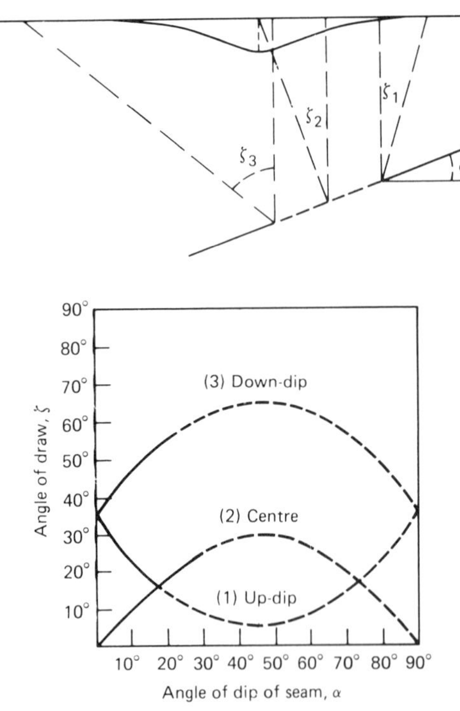

Figure 13.5 *Effect of seam inclination on angle of draw (courtesy of British Coal)*

Table 13.2 Subsidence factor and ratio of maximum horizontal displacement to maximum possible subsidence (after Brauner 1973)

Coal field and method of packing	*Subsidence factor*	*Ratio**
British coal fields		
Solid stowing	0.45	0.16
Caving or strip-packing	0.90	0.16
Ruhr coal field, Germany		
Pneumatic stowing	0.45	0.35–0.45
Other solid stowing	0.50	0.35–0.45
Caving	0.90	0.35–0.45
North and Pas de Calais coal field, France		
Hydraulic stowing	0.25–0.35	0.40
Pneumatic stowing	0.45–0.55	0.40
Caving	0.85–0.90	0.40
Upper Silesia, Poland		
Hydraulic stowing	0.12	–
Caving	0.70	–
USSR		
Donbass district	0.8	0.3
Lvov-Volyn district	0.8–0.9	0.34
Kizelov district	0.4–0.8	0.3
Donets, Kuznetsk, and Karaganda districts	0.75–0.85	0.3
Sub Moscow and Cheliabinski districts	0.85–0.90	0.35
Pechora	0.65–0.90	0.3–0.5
USA		
Eastern (room and pillar)	0.10–0.60	
Central (longwall)	0.50–0.60	
Central (room and pillar)	0.10–0.40	
Western	0.33–0.65	

* Ratio = Maximum horizontal displacement/maximum possible subsidence

All other things being equal, the thicker the coal seam extracted the larger is the surface subsidence. If more than one seam is worked simultaneously beneath the same area, then the subsidence effects are cumulative. The maximum possible subsidence is

$$S_{max} = Ha \tag{13.6}$$

where H is seam thickness and a is a subsidence factor that ranges from 0.1 to 0.9 (Table 13.2). The subsidence factor generally is regarded as being independent of depth. However, it does depend on whether or not the mined out area has been filled or packed. Unfortunately, the character of any infill and the manner by which it is placed is so variable that only qualitative assessments of the 'subsidence factor' can be made.

Usually there is an appreciable difference between the volume of mineral extracted and the amount of subsidence at the surface. For example, Orchard and Allen (1970) showed that where maximum subsidence of 90% of seam thickness occurred, the volume of the subsided ground was only 70% of the coal extracted. This is mainly attributable to bulking.

In a level seam the greatest amount of subsidence occurs over the centre of the working, diminishing to zero at approximatley 0.7 of the depth outside the boundaries of the excavation. But as far as noticeable movement and damage are concerned a distance of 0.5 times the depth is more appropriate.

One of the most important factors influencing the amount of subsidence is the width-to-depth relationship of the panel mined. In Britain it has been shown that maximum subsidence generally begins at a width-to-depth ratio of 1.4:1. This is the critical condition above and below which maximum subsidence is and is not achieved respectively (Figure 13.6). The concept of the area of

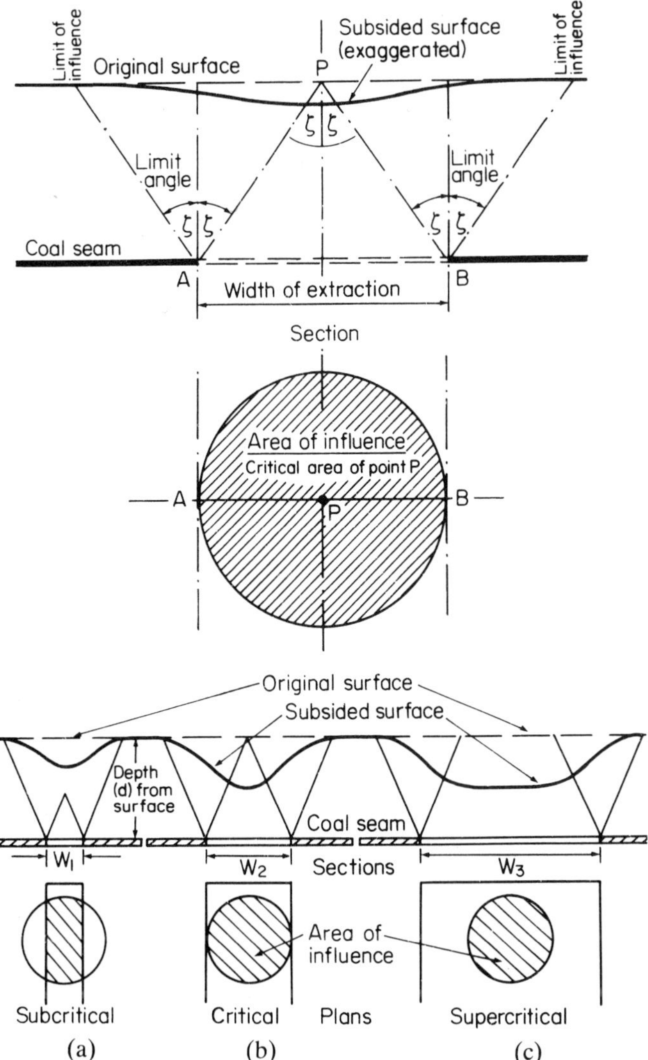

Figure 13.6 *Subcritical, critical and supercritical conditions according to the depth–width ratio. Depth–width ratios* (a) *<1.4;* (b) *1.4;* (c) *>1.4*

influence or critical area of extraction has been explained by Wardell (1954). For a given point P on the surface the area of influence in a level seam is the circle at the base of an imaginary cone with its axis passing upwards through the overlying strata to P at its apex. The diameter of the area equals 1.4 × depth of seam. Any workings outside this area do not affect point P whereas all workings within the area do. However, all the coal from within this critical area must be extracted before point P undergoes maximum subsidence. If only part of the critical area is worked, then point P suffers only partial subsidence. Hence the development of subsidence attributable to a given working is influenced by its width in relation to its depth. Three stages have been distinguished, namely, subcritical (width less than 1.4 × depth), critical (width equal to 1.4 × depth) and supercritical (width greater than 1.4 × depth). It must be pointed out, however, that the width-to-depth ratio for the critical condition does vary from coalfield to coalfield, from country to country. For example, the angle of draw associated with a width-to-depth ratio of 1.4:1 is 35°, therefore reference to Table 13.1 shows how this critical ratio varies.

Wardell (1954) demonstrated that the time taken for subsidence at a surface point to be completed is more or less inversely proportional to the rate of forward advance of the workings. The transition of movement to the surface is almost instantaneous, commencing when the strata at seam horizon begin to relax. Although the precise beginning and end of subsidence are difficult to determine, measurable subsidence occurs when the face is within a distance of $0.75Z$ (Z = depth to coal seam) and reaches approximately 15% of maximum when the face is directly below a given point, P, on the surface. For all practical purposes, it is completed when the face has advanced $0.8Z$ beyond this given point, P. The rate at which subsidence develops is directly related to the rate of face advance.

According to Brauner (1973) in addition to the rate of advance and size of critical area, the duration of surface subsidence depends on geological conditions, depth of extraction, kinds of packings and previous extraction. For example, it lasts longer for thick bedded or stronger overburden and for complete caving. Depth, in particular, influences the rate of subsidence, firstly, because the diameter of the area of influence and therefore the time taken for a working, with a given rate of advance, to traverse it, increases with depth and, secondly, because at greater depths several workings may be necessary before the area of influence is completely worked out. Consequently the time which elapses before subsidence is complete varies according to circumstances.

The slope of the subsidence profile and the applicable critical width of mining are strongly influenced by the properties of the strata overlying the longwall operation. This is particularly the case when the extraction is at relatively shallow depth, and when there are very competent beds of sandstone or dolerite sills in the overburden strata. In such cases it is possible that very little subsidence will occur until a critical span is reached, whereupon the competent bed or sill collapses suddenly, and a large step subsidence profile may be formed on the surface as illustrated in Figure 13.7. The quality of the overburden strata also affects the degree of bulking and subsequent compaction which takes place and therefore has an influence on the value of the subsidence factor.

Residual subsidence takes place at the same time as instantaneous subsidence and may continue after the latter events for periods up to two years. The magnitude of residual subsidence is proportional to the rate of subsidence of the surface and is related to the mechanical properties of the rocks above the seam. For instance, strong rocks produce more residual subsidence than weaker ones. Residual subsidence rarely exceeds 10% of

Figure 13.7 *Abrupt ±0.75 m step in surface profile defining perimeter of subsidence basin (photograph courtesy of Dr J. M. Galvin)*

total subsidence if the face is stopped within the critical width, but falls to 2–3% if the face has passed the critical width.

13.3 Subsidence in metalliferous mining

There are many underground mining methods used for the extraction of metalliferous deposits. These may be divided into three main categories:

(1) Partial extraction methods, in which solid pillars are left *in situ* to provide support in the underground workings;
(2) total extraction methods of the longwall type, usually applicable to narrow tabular deposits;
(3) massive mining methods, in which large ore bodies are mined out completely;

In addition, excavations associated with old mine workings have the potential to cause surface subsidence.

13.3.1 Partial extraction mining

In partial extraction methods, solid pillars are left unmined to provide support to the underground workings. For example, sill pillars and shaft pillars are designed to protect important underground areas such as main accesses, workshops, pumping plant and hoist chambers. Regional or barrier pillars provide regional support for safety of underground operations, and to reduce the potential for, and effect of, seismic events. Pillars may take many forms, depending on the requirements, and may be variable in shape and size. Their geometry will often be determined by the geometry and variation in grade of the ore body. In many cases, since pillars often represent reserves of ore which can be accessed without further underground development, they are extracted or reduced in size towards the end of the life of the mine.

In modern practice, pillars are designed for the purpose required, (Hedley 1978). As for the coal pillars referred to previously, this design takes into account the strength of the pillars themselves, the strength of the foundation rocks in the roof and floor (or hangingwall and footwall), the capacity of the rock to span between the pillars, and the stresses acting on the pillars. For flat-lying deposits the stresses are induced by the overburden and, for steeply dipping deposits, by the *in situ* horizontal stress field. For intermediate dips, the active stresses are a combination of the two, and the pillars may be subjected to significant shear stresses. As a result of the design process, pillar sizes, shapes and spacings are determined. It is therefore possible to design against the occurrence of surface subsidence. However, this procedure was not practised in earlier times. In such situations, and if pillar extraction has taken place, subsidence of the surface is a possibility, and can take several forms:

(1) Collapse of the roof or hangingwall of the workings between pillars, possibly resulting in a localized surface subsidence. This behaviour is most likely to occur in shallow workings, when pillars are relatively competent so that failure of the roof strata occurs preferentially. The subsidence profile may be very severe, with large differential subsidence, tilts and horizontal strains.
(2) Collapse of pillars as a result of their strength being exceeded. This may be restricted to a single pillar, or may involve many pillars. The resulting subsidence may therefore be localized, or may cover substantial areas. With regular pillar systems, failure of a single pillar may lead to overstressing of adjacent pillars and subsequently to a run of pillar failures. Very large areas of subsidence may result. If variable-sized pillars are present, or if larger barrier pillars are provided at regular spacings, the area of subsidence is likely to be restricted. The resulting subsidence profile is likely to be irregular, with large differential settlements, tilts and horizontal strains at the perimeter of the subsidence area. Subsidence of this type may also be influenced by the presence of significant geological structural features such as faults. At such features the subsidence is likely to show a step profile (Stacey and Rauch 1981).
(3) Failure of pillar roof or/and floor foundations (Hagan 1988). This is likely to lead to a similar type of subsidence behaviour, as in (2) above.

In the mining of inclined ore bodies, the steeper the dip, the more localized the subsidence or potential subsidence is likely to be. The stability of pillar supported workings is time dependent – in open or waterlogged workings the pillar, roof and floor rock materials may deteriorate over a period of time, leading ultimately to failure and subsidence.

13.3.2 Total extraction mining of tabular deposits

The term 'total extraction' again implies very extensive extraction over large areas. Complete extraction of the ore body is not usually possible, and is never achieved in practice. This type of mining, using longwalls, or at least tabular stopes with substantial spans, is practised in the gold and platinum mines of southern Africa. Extraction of one, two, and sometimes more reefs has taken place, each typically with a stoping width of 1–2 m. When this mining takes place at considerable depth, substantial subsidence of the surface often occurs. The subsidence profile is smooth, however, and subsidence is not readily noticed. For example, observations of the surface made by Ortlepp and Nicoll (1964) above such mining at a depth of some 1300 m with an effective span of some 1000 m, showed a total subsidence of nearly 0.3 m. Surface tilt was very small, of the order of 0.05%.

When this type of mining is at shallow depth, the situation is different, and detrimental subsidence has commonly occurred (Hill 1981). The effects are dependent on the dip of the reefs, which vary from as little as 10° to vertical, but observation has shown that adverse subsidence is not usually observed when the mining depth exceeds 250 m. Most of the detrimental subsidence that has occurred has been associated with geological structural features such as dyke contacts and faults (Stacey and Rauch 1981). Collapse of workings and subsequent subsidence often occurs after deterioration of artificial support and ingress of water, which washes away waste fill which has provided support in the stopes.

The surface profile resulting from subsidence over shallow 'total extraction' mining is commonly very irregular – tension cracks have often been observed, and large differential subsidence can occur at the fault and dyke contacts, as shown in Figure 13.8.

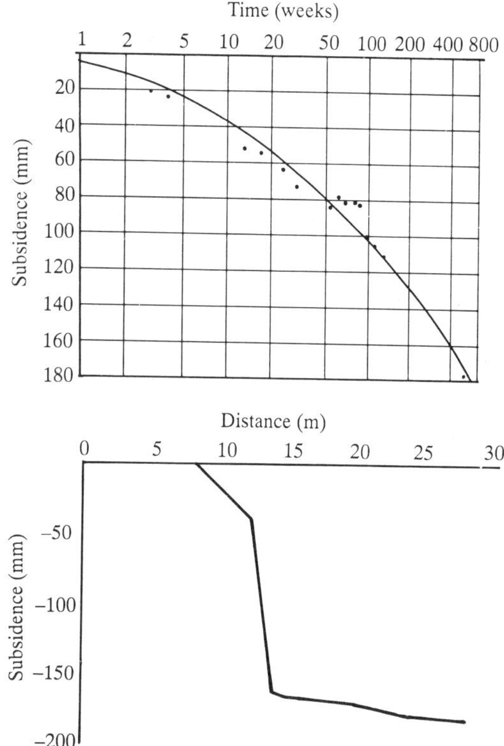

Figure 13.8 *Induced subsidence on a faulted dyke contact (a) development of subsidence with time, (b) differential subsidence across the contact*

13.3.3 Mining of massive deposits

Many metalliferous deposits occur in large disseminated ore bodies, where the size and shape of the ore body is often determined by the market value of the ore at that time and the variation in grade of the ore body. Often the only way in which such ore bodies can be mined economically underground is by means of scale, in which very high-volume production using large equipment can reduce the unit production cost. There are many different mining methods and many variations within each method. Three subdivisions can be considered:

(1) Open stoping: when the rock mass environment is sufficiently competent, large open stopes can be excavated. The benefit of this method is that dilution of the ore is negligible. The sizes of the open stopes are the subject of design, and some form of pillar usually separates adjacent open stopes. The method is therefore a partial extraction approach, but is differentiated on the basis of scale. Open stopes may be very large, with spans of the order of 50–100 m. With correct design, open stopes will remain as stable underground openings, and no subsidence of the surface will occur. However, mining usually aims to maximize extraction, and the long-term stability of the surface over open-stope mining must therefore be in doubt.

(2) Caving: the term implies that material from above is allowed to cave into the opening created by the mining extraction. In block caving, a rectangular 'block' of the massive ore body is undercut and caving of this block is induced. The extraction takes place by drawing the ore from the base of the block without further blasting. A cut-away diagrammatic section of a block caving layout is shown in Figure 13.9.

In sub-level caving the ore is blasted sequentially at deeper levels, and the waste rock from above and the hangingwall and footwall caves onto the ore. This is illustrated diagrammatically in Figure 13.10.

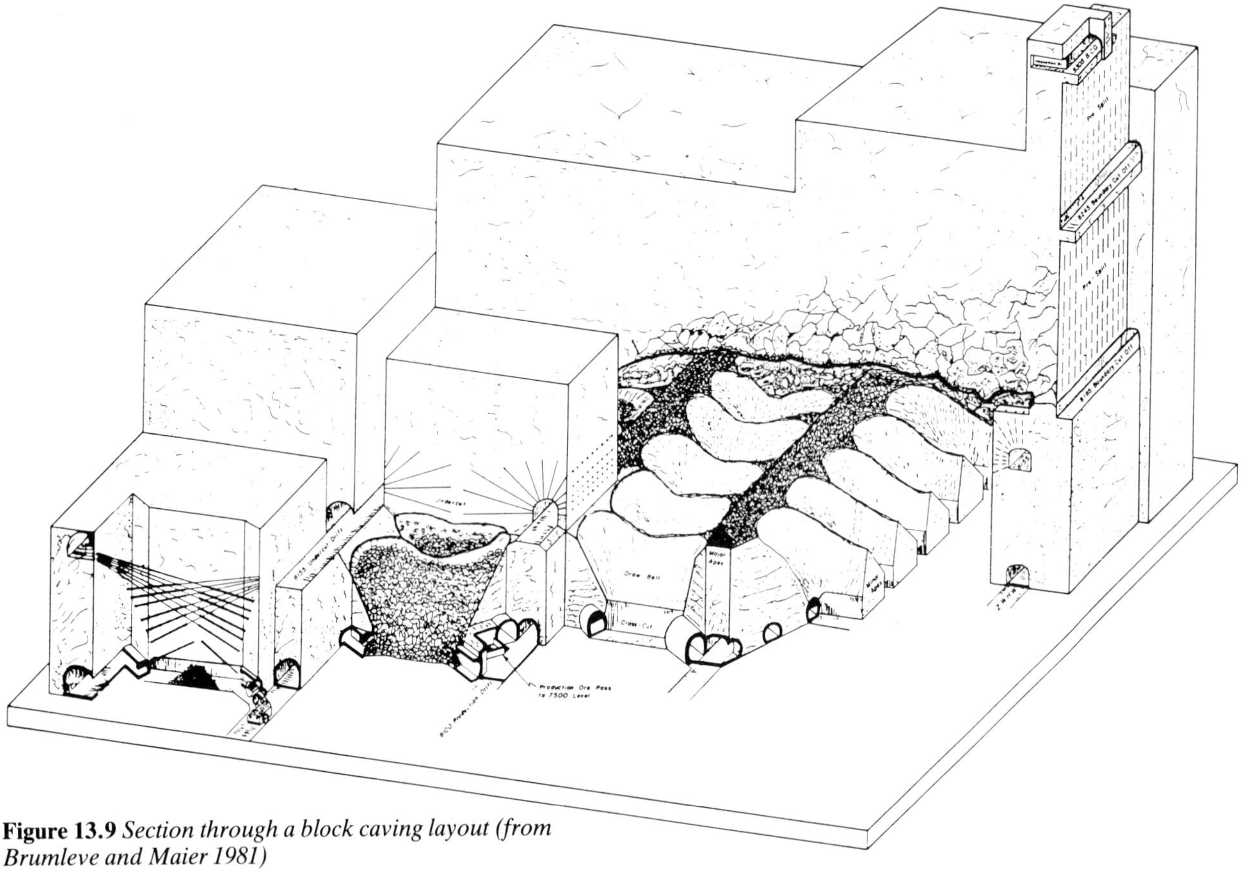

Figure 13.9 *Section through a block caving layout (from Brumleve and Maier 1981)*

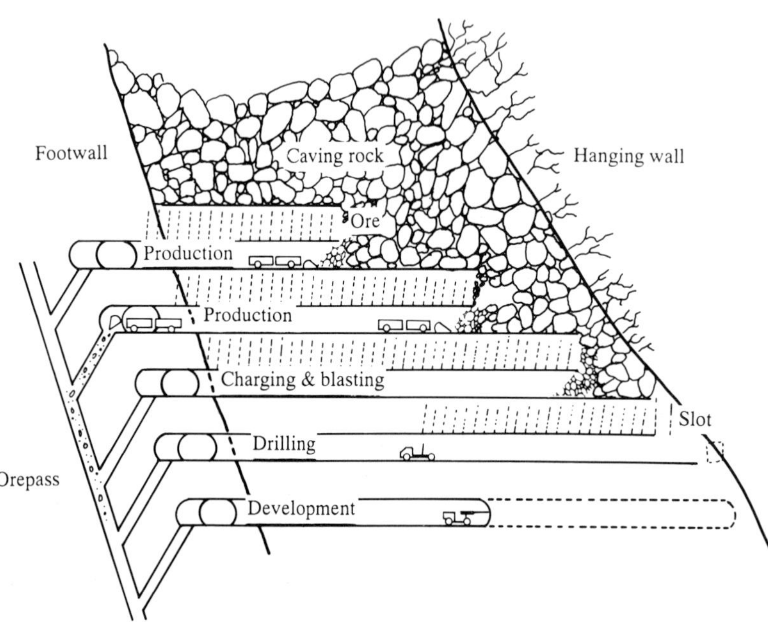

Figure 13.10 *Section of sub-level caving*

Since caving occurs above the ore body in these types of mining, it usually progresses through to the surface to form a subsidence crater. This crater usually has several scarps around its perimeter, and contains material which has subsided *en bloc* overlying caved and rotated material, and loosely consolidated ravelled material. Owing to the large volumes that are extracted during mining, the extent and depth of the subsidence crater can be very large. At San Manuel Mine (Thomas 1971) the crater was approximately 1000 m in length, and its base was some 200 m below the original ground surface.

(3) Fill mining: as the name implies, fill is introduced into the mined-out areas to facilitate continued mining. Depending on the mining method, the purpose of the fill may be to provide confinement and support to solid pillars, to provide support to the hangingwall and footwall, to provide support to the overall workings and be self-supporting so that pillars may be subsequently removed from within the fill, and, as in cut-and-fill stoping, to provide a working platform from which to access the ore above and onto which the ore is blasted.

The implication of the use of fill in mining is that the openings created are substantially backfilled. There remains only a limited potential for collapse of the workings, and therefore catastrophic subsidence of the surface is unlikely. The potential for some subsidence of the surface still exists, however, depending on the area extracted.

13.4 Subsidence due to the abstraction of fluids

The problem of subsidence associated with the abstraction of fluids, notably groundwater, has occurred in many parts of the world (Bell 1988). Frequently such subsidences have taken time to recognize and to do something about, and consequently very costly damage may have occurred during the time elapsed.

13.4.1 Groundwater abstraction and subsidence

In the case of groundwater abstraction, subsidence takes place in areas where abstraction exceeds natural recharge and the water table is lowered, it being attributed to the consolidation of the sedimentary deposits in which the groundwater occurred, consolidation being due to increasing effective stress. The total overburden pressure in partially saturated or saturated deposits is borne by their granular structure and the pore water. When groundwater abstraction leads to a reduction in pore water pressure by draining water from the pores, this means that there is a gradual transfer of stress from the pore water to the granular structure. For instance, if the groundwater level is lowered by 1 m, then this gives rise to a corresponding increase in average effective overburden pressure of approximately 10 kPa. As a result of having to carry this increased load the granular structure may deform in order to adjust to the new stress conditions. In particular, the void ratio of the deposits concerned undergoes a reduction in volume, the surface manifestation of which is subsidence. Surface subsidence does not occur simultaneously with the abstraction of fluid from an underground reservoir, but over a longer period of time than that taken for abstraction. The amount of subsidence attributable to a decline in the level of the water table reflects the compressibility and thickness of the deposits involved, as well as the magnitude of the increase in effective stress and the time over which the increased loading is applied.

Consolidation may be elastic or non-elastic depending on the character of the deposits involved and the range of stresses induced by a decline in the water level. In elastic deformation stress and strain are proportional, and consolidation is independent of time and reversible. Non-elastic consolidation occurs when the granular structure of a deposit is rearranged to give a decrease in volume, that decrease being permanent. Generally recoverable consolidation represents compression in the preconsolidation stress range, whilst irrecoverable consolidation represents compression due to stresses greater than the preconsolidation stress.

Lofgren (1979) recorded that 20 years of precise field monitoring in the San Joaquin and Santa Clara valleys of California have indicated a close correlation between hydraulic stresses induced by groundwater pumping and consolidation of the water-bearing deposits (Figure 13.11). He went on to note that the stress–strain characteristics of the producing aquifer system have been established and that the storage parameters of the system have been determined from such data. What is more this has provided a means by which the response of the groundwater system to future pumping stresses can be predicted. According to Lofgren the storage characteristics of compressible formations change significantly during the first cycle of groundwater abstraction. Such abstraction is responsible for permanent consolidation of any fine-grained interbedded formations, consolidation being brought about by the increase in effective pressure. The water released by consolidation represents a one-time, and somewhat important, source of water to the wells.

Methods which can be used to arrest or control subsidence caused by groundwater abstraction include reduction of pumping draft, artificial recharge of aquifers from the ground surface and repressurizing the aquifer(s) involved via wells, or a combination thereof. The aim is to manage the rate and quantity of water withdrawal so that its level in wells is either stabilized or raised somewhat so that the effective stress is not further increased.

Ground failure is often associated with subsidence due to the abstraction groundwater (and incidentally oil, natural gas or brines). Because the withdrawal of groundwater easily exceeds that of all other fluids some of

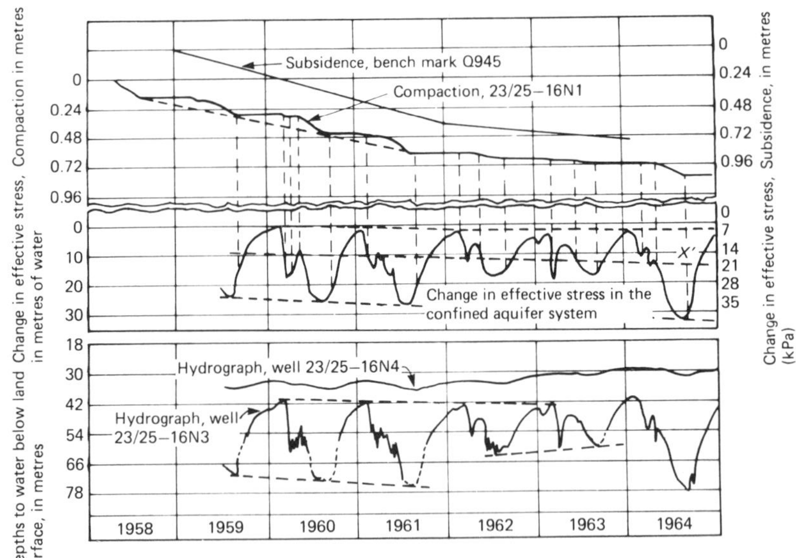

Figure 13.11 *Six-year correlation of land subsidence, compaction, water-level fluctuations and change in effective stresses, near Pixley, California (after Lofgren 1979)*

the most notable failures are found in such subsidence areas. For instance, ground failures associated with subsidence due to groundwater abstraction occur in at least 14 areas in six states in the United States of America.

Fissures may appear suddenly at the surface and the appearance of some may be preceded by the occurrence of minor depressions at the surface. Within a matter of a year or so of their formation most fissures become inactive. In a few instances new fissures have formed in close proximity to older fissures.

Fissure separations tend not to exceed a few centimetres, the maximum reported being 6.4 cm. They, however, commonly are enlarged by erosion to form gullies 1–2 m in width and 2–3 m in depth. The greatest measured depth is 25 m.

Faults suspected of being related to groundwater withdrawal are much less common than fissures. In the United States they frequently have scarps more than 1 km in length and more than 0.2 m high. The largest scarp in the United States is 16.7 km long and the highest is around 1 m; both are in the Houston–Galveston region, which is the region most affected by such faulting in the United States.

In dolomitic environments in the Transvaal, lowering of the groundwater table by mining operations can result in significant localized subsidence, which in some cases can be catastrophic. Two modes of subsidence have been described (Jennings *et al.* 1965). The first is a caving subsidence, in which the compressible material filling a solution 'cavity' consolidates under the increased effective stresses resulting after water-table lowering. The second is a catastrophic subsidence which gives rise to a sinkhole, which forms suddenly when the capping to an infilled solution cavity collapses. A graphic example of this was the disaster involving the disappearance of a three-storey mine crusher plant into a 100 m diameter sinkhole in 1962.

13.4.2 Subsidence due to the abstraction of oil or gas

Subsidence due to the abstraction of oil occurs for the same reason as does subsidence associated with the abstraction of groundwater. In other words pore pressures are lowered in the oil-producing zones due to the removal of not only oil but of gas and water also, and the increased effective load causes consolidation in compressible beds. Such an explanation probably first was advanced to account for the spectacular and costly subsidence at Wilmington oilfield, California. This area is underlain by approximately 1800 m of Miocene to Recent sediments consisting largely of sands, silts and shales. Subsidence was first noticed in 1940 after oil production had been under way for three years. Gilluly and Grant (1949) were among the first to demonstrate that the area of maximum subsidence showed a remarkable coincidence with the productive area of the oilfield. They also indicated that there was a very close agreement between the relative subsidence of the various parts of the oilfield and the pressure decline, thickness of oil sand affected and mechanical properties of the oil sands. By 1966 an elliptical area of over 75 km^2 had subsided more than 8.8 m, the largest amount of subsidence due to the withdrawal of fluids known to have occurred in the world.

By 1947 subsidence was occurring at a rate of 0.3 m per year and had reached 0.7 m annually by 1951 when the maximum rate of withdrawal was attained (i.e. 140 000 barrels a day). Because of the seriousness of the subsidence (and the realization that it was due to declines in fluid pressures) remedial action was taken in 1957 by injecting water into, and thereby repressurizing, the abstraction zones. By 1962 this had brought subsidence to a halt in most of the field. In fact Yerkes and Castle (1970) referred to more than 40 known examples of differential subsidence, horizontal displacement and surface fissuring/faulting, associated with 27 oil and gas fields in California and Texas. The maximum subsidences recorded in these fields are not comparable with that of Wilmington and generally are less than 1 m.

The Groningen gas field has been in production since 1965, the gas reservoir occurring in sandstone belonging to the Permian Rotliegende formation. It is 70 m in thickness in the south increasing to 240 m in the north and occurs at a depth of 3000 m, extending over 900 km^2. It has been estimated that by 2025 the amount of subsidence in several areas will exceed 0.25 m. Although not appreciable, this amount of subsidence still makes it necessary to make provisions for maintaining water management in the polder region of Groningen province. Regular precision levelling surveys have been carried out since abstraction of gas commenced in order to monitor the amount of subsidence occurring at ground surface. Contour maps showing the amount of subsidence have been produced from this data. Another method of subsidence monitoring has involved measuring the consolidation of near surface sediments with the aid of borehole extensometers. Lastly, measurements of reservoir compaction were taken by monitoring the relative displacement of radioactive bullets shot into the formation at regular intervals down a well (Schoonbeck 1976). In fact the initial levelling surveys indicated that subsidence was occurring at an amount considerably less than had been predicted.

13.4.3 Abstraction of brine

Those deposits which readily go into solution, notably salt, can be extracted by solution mining. For instance, salt has been obtained by brine pumping in a number of areas in the United Kingdom, Cheshire being by far the most important. Consequently subsidence due to salt extraction has been an inhibiting factor, especially in parts of Cheshire, as far as major developments were concerned.

Today wild brine operations which were responsible for subsidence, have almost ceased in Cheshire. The production of brine (99%) now is obtained by controlled solution mining from cavities formed in the salt. The most successful wild brine pumping in Cheshire was carried out on the major natural brine runs. Such pumping accelerated the formation of solution channels. Active subsidence normally was concentrated at the head and sides of a brine run where the fresh water first entered the system (Bell serious subsidence occurred at considerable distances, anything up to 8 km, from pumping centres.

Because the exact area from which salt was extracted was not known, the magnitude of subsidence developed could not be related to the volume of salt worked. Consequently, there was no accurate means of predicting the amount of ground movement or strain. However, the rate at which subsidence occurred appeared to be related to the rate at which brine was extracted. Although it is not easy to tell, it seems that subsidence does not necessarily occur immediately after the removal of salt. For example, surface subsidence has continued for six months after the well thought to be responsible ceased to operate. Indeed residual subsidence may continue for one or two years. The maximum strains developed by wild brine pumping ruined buildings. What is worse was the formation of tension scars (small faults) on the convex flanks of linear subsidence hollows. These features are developed in the surface deposits. The vertical displacement along a scar usually is less than a metre.

Wassman (1980) described subsidence which resulted from the solution mining of salt in the area around Hengelo in the Netherlands. Cavities have been produced in the salt and they subsequently have become interconnected. Wassman quoted some 1.6 m of subsidence as having taken place in the centre of one subsidence trough and at Hengelo vertical subsidence amounting to about 50 mm still occurs annually. Some cavities collapsed several years after pumping ceased. In addition, brecciation has taken place in the overlying Red Beds of Bunter age, the claystone or marl losing its strength when wetted. Unfortuantley brine from the cavities in the underlying salt permeates by capillary action into the marl, and eventually the weakened roof material collapses into the cavity. Void migration then takes place and, depending on the bulking factor of the rocks involved, the void may move into the overlying Tertiary clays. Pronounced subsidence occurs when this happens. The subsidence basin in these deposits is demarcated by an angle of influence of 45° which extends upwards from the contact between the Red Beds and Tertiary clays. This explains the relatively small area of influence in relation to the depth of the original cavities. An additional cause of ground movement is attributable to consolidation which occurs in the brecciated marls. This is responsible for the slowly decreasing, but long-lasting, subsidence. Wassman recommended that if the collapse of solution cavities is to be avoided, then abstraction wells should be sealed once pumping ceases.

Deere (1961) described differential subsidence (resulting from the development of concavo-convex subsidence profiles), simultaneous horizontal displacements and surface faulting associated with the extraction of sulphur from numerous wells. The sulphur occurs at a depth of 397–488 m within the cap rock of a salt dome in the Gulf region of Texas. The cap rock is overlain by unconsolidated sands, gravels, clays and clay-shales. Shortly after mining operations commenced, vertical and horizontal movements were noticed. After nine months the maximum vertical subsidence amounted to 0.45 m and the 'diameter' of the subsidence basin was approximately 600 m. During the first 31 months of operation, differential subsidence amounting to 1.75 m occured over an elliptical area exceeding 5 km^2. The subsidence basin was centred directly above the narrow linear producing zone. A normal fault, around 650 m in length, with a downthrow of 0.1 m on the mining side and peripheral to the subsidence basin, developed suddenly during the fifth month of production. By the thirty-first month the length of the fault had increased to 800 m and the displacement to 0.3 m. The fault dipped at about 40° directly inwards towards the top of the producing zone and it formed at a point of maximum surface tension (cf. the tension scars associated with salt extraction referred to above).

13.5 Methods of subsidence prediction

There have been many methods developed for subsidence prediction. These include mathematical solutions based on the theory of elasticity, numerical methods which have been developed in recent years, empirical methods based on the observation of cases of surface subsidence, and combinations of these methods. Comprehensive reviews of methods of subsidence prediction have been prepared by Voight and Pariseau (1970), Brauner (1973) and more recently the subject has been dealt with by Salamon (1988).

13.5.1 Mathematical methods of subsidence prediction for shallow dipping seam mining

Hackett (1959) developed a solution for a two-dimensional horizontal open slit (displacement discontinuity), in an infinite homogeneous, isotropic, elastic medium, for the prediction of the stresses and displacement around a mined seam. Calculated results using this method showed qualitative agreement with field displacements, and several phenomena, previously regarded as anomalous, were shown to be normal elastic movements. Berry (1960) extended this work to consider different closure conditions in the seam. The assumption of isotropy was found to be incompatible with the degree of surface subsidence found, and Berry and Sales (1961) extended the previous work to allow for the case of transversely isotropic material. This showed much better agreement with measured surface subsidence records from British coalfields.

The transversely isotropic approach described above gave reasonable agreement with field observations, but the assumptions required to achieve this may appear to be unreasonable – complete closure was assumed to be about one third of the seam thickness as a result of the bulking of caved material, and the vertical modulus of elasticity was assumed to be twice the horizontal value. According to Salamon (1983), the use of the parameters required to obtain reasonable predictions of subsidence cannot be justified on physical grounds. However, an engineering approach is to use the method with its known weaknesses, since it does give acceptable answers. It can then, in fact, be considered to be an empirical approach.

Salamon (1964a) also developed a means of calculating subsidence assuming elastic behaviour. This approach was based on the suggestion that the stresses and displacements induced by mining of a seam can be predicted from the convergence distribution in the seam, regardless of the complexity of the mining layout. This is a logical generalization of the influence function method, which is dealt with in Section 13.5.2. Salamon introduced the concept of 'face elements' (or displacement discontinuity elements) – the mined-out area is considered to be divided into a network of small areas which are called face elements. By determining the elementary stresses and displacements induced by the closure and ride of a single face element, and, using the principle of superposition, the total displacements and stresses can be obtained by summation of the effects of all face elements in the excavation. Salamon dealt with the exact elastic solutions for the homogeneous isotropic and transversly isotropic cases, and simplified, non-exact, elastic solutions for a multi-membrane model and frictionless laminated model. In the multi-membrane model, the stratified nature of rocks is simulated by a large number of weightless, uniformly tensioned, horizontal membranes connected by a large number of vertical springs. The frictionless laminated model consists of a large number of laminae on top of one another, capable of sliding freely over each other at the interfaces. To use these approaches it is necessary to obtain the convergence profile within the mine workings. For deep mining, this can be done theoretically, using a membrane analogy for the frictionless laminated model, and an electrical analogue for the isotropic, transversely isotropic and multi-membrane models (Salamon 1964b). For shallow applications, measured or estimated convergence contours are used. It has been found that acceptable estimates of subsidence can be obtained using rough estimates of convergence (Jones and Bellamy 1973). Ortlepp and Nicoll (1964) made use of an electrical analogue with a homogeneous isotropic elastic model. Agreement between calculated and measured surface subsidence was not good. For examples reported by Jones and Bellamy (1973) and Salamon (1988), in which the convergence profiles within the mine workings were estimated, agreement between measured and predicted subsidence was very good.

13.5.2 Empirical and semi-empirical methods of subsidence prediction for shallow dipping seam mining

An important and often common feature of subsidence due to longwall mining of coal is its high degree of predictability. Usually movements parallel and perpendicular to the direction of face advance are predicted.

Although adequate for many purposes such treatment does not consider the three-dimensional nature of ground movement.

Empirical methods of subsidence prediction such as those used by British Coal (Anon 1975) have been developed by continuous study and analysis of survey data from British coalfields. They do not take into account the topography, the nature of the strata and geological structure involved and how the rock masses are likely to deform. Consequently Voight and Pariseau (1970) emphasized that such empirical relationships can be applied only under conditions similar to those of the original observations. This subsequently was highlighted by Kapp (1986) who noted that in Australia subsidence prediction curves developed for one area cannot be applied elsewhere. Nevertheless empirical methods are continuously being refined so that they can yield more accurate results. In fact they allow the amount of subsidence due to longwall mining in Britain to be predicted usually within ±10%.

Subsidence prediction by empirical methods is based initially on the width-to-depth ratio of the working face (Figure 13.12a), all other factors being related to this ratio. When a panel is not worked far enough to cause maximum subsidence, in other words when the critical condition is not attained, the subsidence predicted from Figure 13.12a must be adjusted by reference to Figure 13.12b, that is, by considering the distance travelled by the face as a fraction of the depth and applying the appropriate partial subsidence curve.

The observation of shapes of subsidence troughs has revealed that they conform to a general pattern from which standard subsidence profiles can be produced. By itself the value of maximum subsidence over a particular extracted panel is of limited importance, the whole profile of subsidence having to be ascertained in order to study the effects of mining. The shape of a subsidence profile varies with the width-to-depth ratio of the extraction and

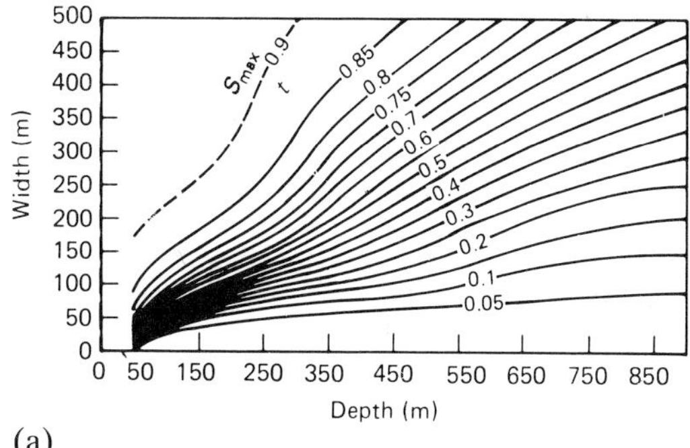

(a)

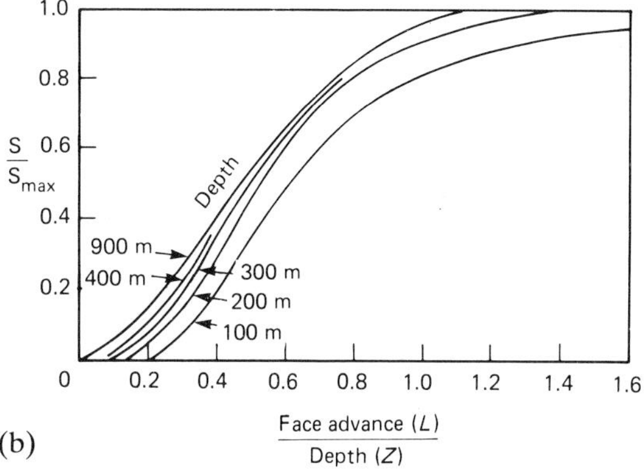

(b)

Figure 13.12 (a) *relationship of subsidence to depth and width; prediction of maximum subsidence,* S_{max}*;* t = *thickness of seam extracted;* (b) *correction graph for limited face advance to be used with* (a): S = *subsidence,* S_{max} = *maximum subsidence (after Anon 1975)*

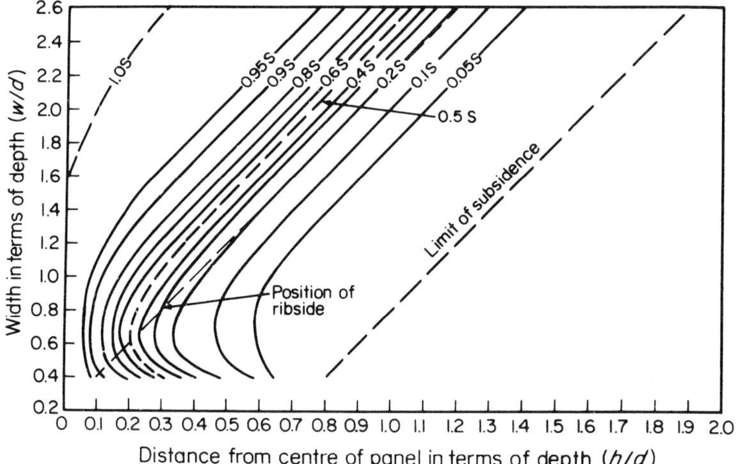

Figure 13.13 *Graph for predicting subsidence profiles:* w = *width;* Z = *horizontal distance (after Anon 1975)*

for any value a curve can be plotted relating partial subsidence at various points to distance from the centre of the panel as a proportion of depth. Figure 13.13 represents a series of profiles from which the subsidence value for any profile can be determined.

The zone of maximum extension coincides with the position of the ribside where the width-to-depth ratio exceeds 1.35:1, lying outside the rib when the ratio is smaller. The maximum compression occurs in the centre of a panel in narrow panels (width less than 0.42 × depth), whilst at higher ratios the profile develops two compression zones (Figure 13.14). Figure 13.15a is a graph for predicting strain profiles and the proportion of extension to compression for a given width-to-depth ratio can be derived from Figure 13.15b, which shows that with narrow panels the intensity of compression greatly exceeds that of extension, whereas that of extension is approximately 25% greater than compression when the critical condition is reached.

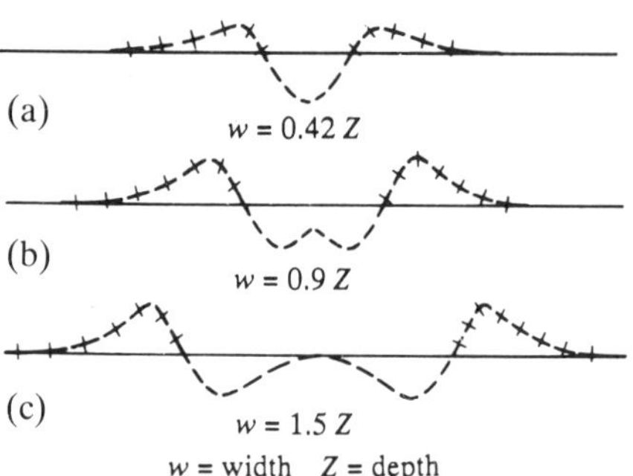

Figure 13.14 *Types of strain profile: Maximum compression occurs at the centre of a panel in the narrow panels* (a), *i.e. when* w = *0.42Z the single compression zone has a greater intensity than the extension. A hump occurs in the compression curve when* w = *0.9Z. Two separate compression zones occur when* w = *1.5Z (after Anon 1975)*

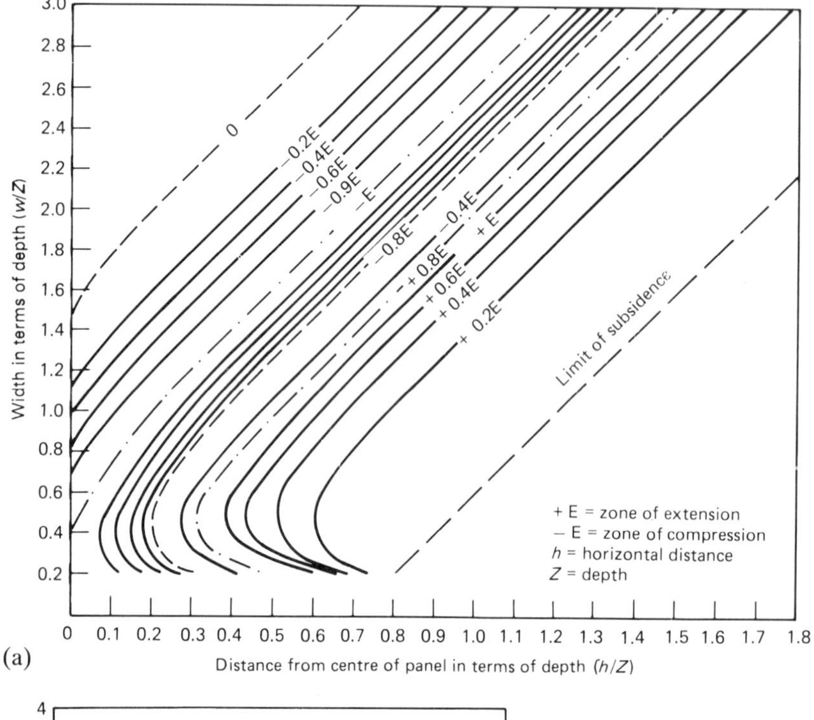

(a)

(b)

Figure 13.15 (a) *Graph for predicting strain profiles;* (b) *graph for predicting maximum slope and strain for various width-depth ratios of a panel:* S = *subsidence;* Z = *depth;* w = *width (after Anon 1975)*

Because all principal factors are related to the width-to-depth ratio this means that to be absolutely reliable this ratio has to remain constant throughout the production life of a working face. For various geological and operational reasons, for example, inclined seams or changes in the width of working face, this is not the case and consequently the above-mentioned prediction methods must be modified.

Over a number of years Burton (1978) developed an empirical three-dimensional model of subsidence prediction which can be represented graphically. In this way a series of average subsidence and strain profile shapes which are related to width-to-depth ratios can be portrayed as a single graph which Burton referred to as a static set. The model consists of two basic patterns, namely, the static loop and the dynamic loop. A static set can be represented by a loop which is related to static values of displacements of lines of points across a subsidence trough. The pattern which is produced indicates the relationship between an increase in vertical and horizontal displacement with an increase in the amount of extraction. The depth of the static loop is equivalent to that of the maximum subsidence and it closes where the surface point concerned is not influenced by subsidence effects. The direction of the slope of the loop shows the general location of tension and compression, the latter occurring towards the centre of the line.

Dynamic loops describe the position of a given point at a certain moment in time. They can be developed either in relation to vertical or horizontal displacement. The vertical pattern of loops becomes smaller on moving from the centre of the trough. Hence a three-dimensional picture of the subsidence basin can be established for any point in time.

Burton (1978) showed that surface points move inwards across the panel and also move backwards and forwards in the line of advance. By fixing one point and then calculating the displacements of a another point in relation to this, a series of differential movements is obtained. Accordingly strains can be calculated along the line between the two points under consideration. In other words strain can be calculated in any direction with respect to face advance. This is important because the value of strain, or change in surface length, is direction related; that is, a zone of tension or compression, which is related to a line normal to the ribside, has no meaning for a line running in another direction. Hence a strain contour map must be related to surface direction (Figure 13.16).

Most methods of subsidence prediction fall into the category of semi-empirical methods since fitting field data to theory often results in good correlations between predicted and actual subsidence. There are two principal methods of semi-empirical prediction, namely the profile function and the influence function methods. The profile function method basically consists of deriving a function which describes a subsidence trough. The equation produced is normally for one half of the subsidence profile and is expressed in terms of maximum subsidence and the location of the points of the profile. In supercritical extraction the central position of the curve is S_{max} and changes to zero subsidence at the edges of the critical area. For subcritical extraction the profile is determined from the critical profile produced from an empirical/mathematical relationship.

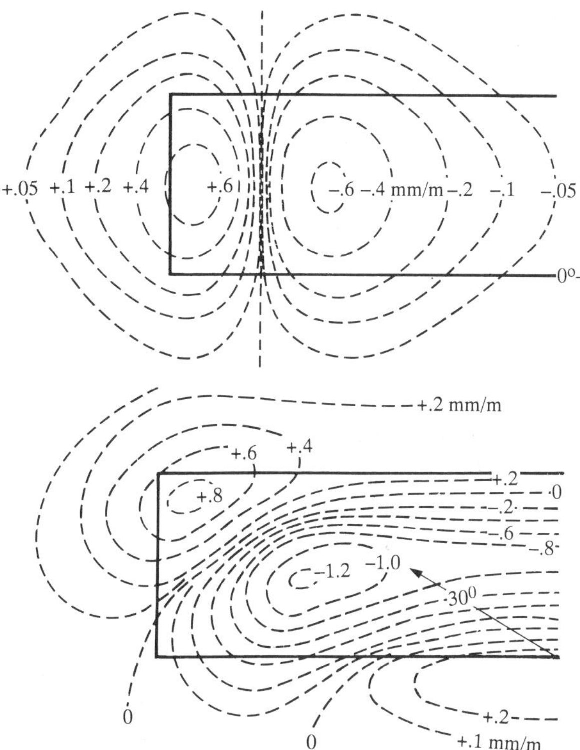

Figure 13.16 *Two different strain contour maps of the same area, produced by the same extracted panel but indicating the difference in arrangement according to the line of direction chosen*

The general relationships of the method are shown in Figure 13.17. The profile is described by a function $s(x)$ normally derived from field measurement or model investigations. Generally the profile is antisymmetrical about the point of inflection, with half full subsidence occurring at this point. The distance, D, from the

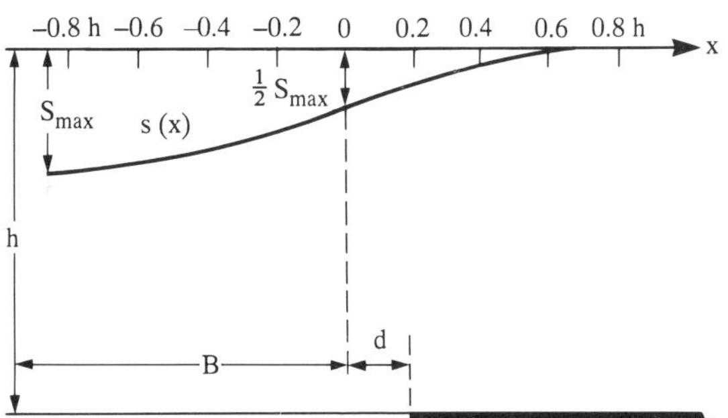

Figure 13.17 *Subsidence profile functions*

inflection point to the ribside must be either determined from empirical relationships or measured. Normally the relationship is given as a proportion of the depth of extraction.

The depth, and therefore critical radius, influence the slope of the profile. Thus the subsidence of a point at position x is a function of the maximum subsidence, S_{max}, the distance, D, and the critical radius, B, that is

$$s = f(x, S_{max}, D, B) \tag{13.7}$$

In subcritical extraction a further parameter, the width-to-depth ratio, may be included. Subcritical extraction also means that the field determination of B is difficult since the slope at the centre of the trough vanishes over a very small distance.

Provided the input data are available, the profile function method can provide an accurate technique for the prediction of subsidence where the mine geometry is relatively simple as in longwall mining. According to Hood *et al.* (1983), however, the influence function approach is better suited to subsidence prediction above irregularly shaped panels.

The influence function approach to prediction of subsidence is based upon the principle of superposition. The subsidence trough is regarded as a combination of many infinitesimal troughs formed by a number of infinitesimal extraction elements (Figure 13.18). The contributions of a single extraction element on a surface point can be derived as the product of its area, dA, and a value p indicating the degree of the influence of dA on the surface point P (Figure 13.19). The influence of p depends on the horizontal distance, r, between the point and element, that is, $p = f(r)$.

The influence value refers to the surface point and hence the extraction element directly beneath the point has the greatest influence. Accordingly the function $p(r)$ reaches a maximum value when $r = 0$. If P is a point at the centre of the critical area, then it is influenced by all the

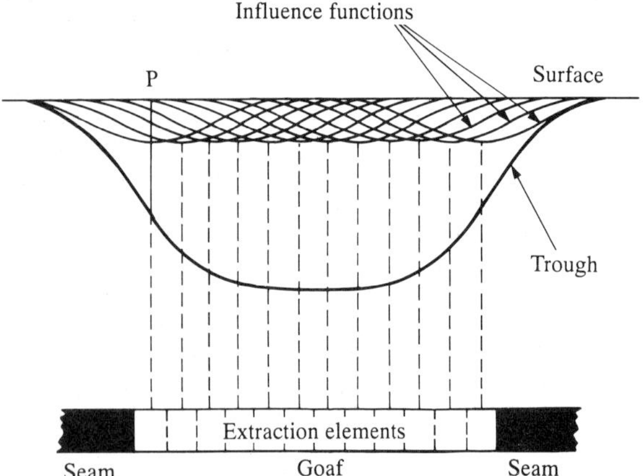

Figure 13.18 *Superposition of infinitesimal influences*

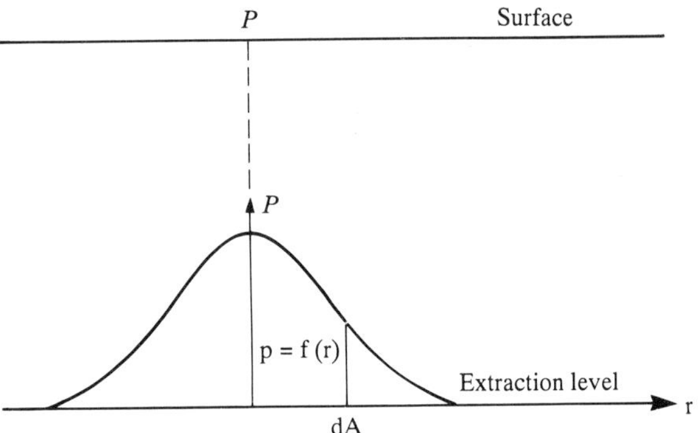

Figure 13.19 *Influence function*

extraction elements and therefore undergoes full subsidence, S_{max}. Thus S_{max} is represented by the complete area under the curve $p(r)$. If P is at the edge (i.e. the ribside) of the extraction, it is influenced by only half the sum of the possible influences and hence only experiences half the full subsidence. This is also the point of inflection of the profile, and so the distance between the point of full subsidence and the inflection point is equal to half the width of the critical area, namely B.

In three dimensions the full subsidence can be referred to as the volume of the solid revolution of $p(r)$ around the z axis, the radius of the solid being B. Hence the full subsidence can be expressed as:

$$S_{max} = 2\pi \int_0^B r\,p(r)dr \tag{13.8a}$$

$$S_{max} = 2\pi \int_0^\infty r\,p(r)dr \tag{13.8b}$$

depending on whether p = 0 at r = B, hence Equation (13.8a), or whether p is asymptotic to zero hence Equation (13.8b).

There are two problems associated with the use of influence functions. First, the subsidence profile obtained is antisymmetric about the inflection points of the sides of the trough and symmetric about the centre of the panel. This often is not the case in an actual subsidence profile. Secondly, the method assumes that the point of inflection is situated immediately above the edge of the panel (the ribside). Special measures have to be taken to relocate the predicted curve when this is not the case and if the distance involved is large, then inaccuracies arise in the subsidence predicted.

The fundamental concept of complementary influence functions is that the separate influence functions describing the response of mined and unmined zones act together to produce the subsidence (Sutherland and Munson 1984). Each influence function is defined by the response of a limit element, that is, an unmined element for the coal left in place and a mined element for the void created by extraction. The amount of subsidence is predicted by approximately summing these elements over the entire seam, it being the sum of the influence of both the mined and unmined response. Hence, complementary influence functions can be used to compute subsidence above mines with complex geometry (i.e. for room and pillar as well as longwall workings). This method, however, tends to overestimate the subsidence directly over the ribside.

Marr (1975) described the zone area or circle method which is another influence function method of subsidence prediction in which surface subsidence is estimated by constructing a number of concentric zones around a surface point, the radius of the outer zone being equal to the radius of the area of influence (Figure 13.20), which can vary with the geological conditions. The subsidence at the surface point is obtained from the summation of the proportions of coal which are extracted in each zone, multiplied by its particular subsidence factor. In practice it has been found that three, five or seven zones are suitable for most estimations. The method permits estimation of subsidence which develops when panels of irregular shape are mined.

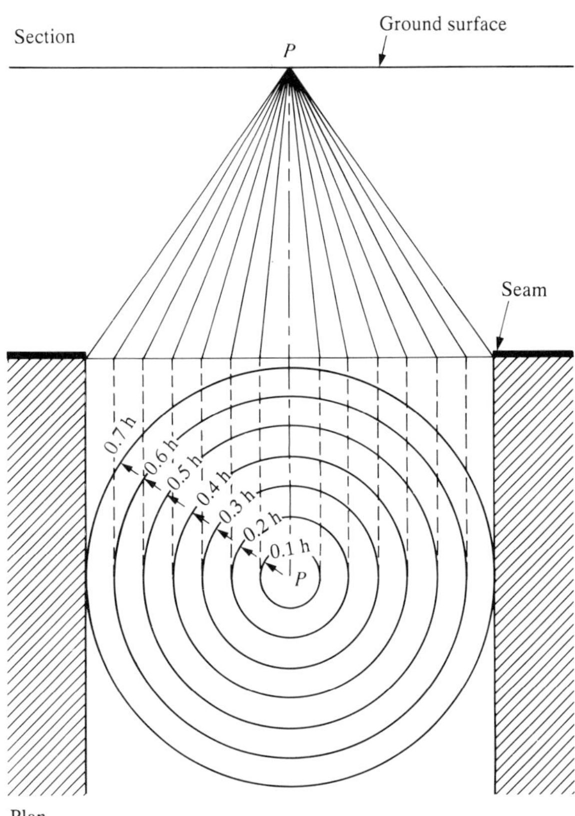

Figure 13.20 *Zone area or circle method of subsidence prediction*

13.5.3 Numerical methods of subsidence prediction for shallow dipping seam mining

Since 1970 there has been an explosion in the availability of computer power, and a commensurate development of software tools for analysis. Major and complex analyses may now be carried out on personal computers, and there is now no longer the same motivation for research into closed-form mathematical solutions. The capability of available numerical techniques exceeds the ability of engineers and scientists to define adequately the input data necessary to permit the modelling of rock mass behaviour with sufficient confidence of accurate foward prediction of subsidence. It can be stated that, if the deformation properties of the rock mass could be defined satisfactorily, the subsidence as a result of mining could be quantified accurately using available numerical analysis techniques.

Finite element and boundary element stress analysis programmes are now readily available for the analysis and prediction of subsidence. These methods have been well documented in the literature, and will not be described here. The finite element method is described in detail by Zienkiewicz (1977), and his book contains the general listing of a computer program which could be applied to subsidence analysis. Boundary element methods appropriate for geotechnical applications are described by Crouch and Starfield (1983). Their book contains several computer program listings, but the listings contained in the publication by Crouch (1976) are probably more appropriate. These numerical techniques have not been widely used for subsidence prediction, probably because considerable experience with, and confidence in, established methods of subsidence prediction (see Section 13.5.2) exists in many affected areas. The potential for application of numerical techniques lies in areas where field data on surface subsidence are limited, and are insufficient to develop empirical relationships.

Mikula and Holt (1983) described the use of two-dimensional finite element analysis for subsidence prediction, and presented several case studies. They allowed for rock material failure, and included both horizontal and vertical joint elements in their model. Laboratory test values of material properties were modified to correlate with *in situ* conditions. For supercritical widths of extraction, the predicted subsidence behaviour was in close agreement with the observed behaviour. However, for subcritical extraction, accurate simulation of subsidence was not achieved. Nevertheless, they concluded positively that, with continuing research, finite element modelling of subsidence can be successful.

The application of the two-dimensional displacement discontinuity method (a type of boundary element method) to subsidence calculation has been described by Crouch (1976), Wardle and Enever (1983), Gurtunca and Schumann (1986), and Gurtunca and Bhattacharyya (1988). The application of this method is most attractive owing to the very limited amount of data preparation required. In these applications the effect of the degree of anisotropy taken into account in the model is emphasized. As for the mathematical models described above, and investigated by Mikula and Holt (1983) using finite elements, it is necessary to include a very significant degree of anisotropy to obtain theoretical subsidence magnitudes and surface profiles which conform with those observed. Crouch (1976) found that, for equal horizontal and vertical elastic moduli, close agreement between observed and predicted subsidence profiles for a British coalfield was obtained with a shear modulus one twentieth of the Young's modulus. This ratio has been adopted by other authors (Wardle 1986; Gurtunca and Bhattacharyya 1988). Gurtunca and Bhattacharyya used one set of measured subsidence data to determine suitable elastic properties for successful prediction of the subsidence profile, and then used these same properties to predict subsidence in other cases. In these cases the measured and predicted maximum subsidence magnitudes differed by less than 10%. Gurtunca and Schumann (1986) used the same method of analysis to model subsidence profiles at several mines in which the overburden strata contained thick, competent, dolerite sills. It was possible to match measured and calculated subsidence profiles closely, but very large ratios of Young's modulus to shear modulus were required to achieve this. The physical admissibility of this may be questioned, but may perhaps be rationalized by the conclusion of Singh (1973), that a jointed rock mass may be represented by an anisotropic elastic material with a low shear modulus.

It may be concluded from the above that, with the correct choice of material properties, the anisotropic elastic boundary element method appears to be well suited to the prediction of subsidence. There remains some doubt, however, whether the calculated stresses within the rock mass model the physical stresses realistically. However, this need not be a restriction on the use of the method to predict subsidence. On the contrary, the method is very easy to use, and results may be obtained rapidly.

Coulthard and Dutton (1988) described the application of two other numerical methods (two-dimensional) for the prediction of subsidence. These are a non-linear finite difference method and a distinct element method, computer programs for which are commercially available (ITASCA). In the finite difference application, the rock mass was modelled as a non-linear medium containing ubiquitous horizontal joints, with slip on the joints being permitted. This represents a horizontally stratified rock mass, and may be considered as a more exact form of Salamon's (1964a) frictionless laminated model. It was found that, although the magnitude of subsidence calculated with the assumed material properties was less than empirically calculated values, the calculated behaviour was broadly consistent with that observed during the transition from subcritical to supercritical spans. The subsidence profile calculated is steeper than that obtained using an anisotropic elastic model, and agrees better with observed profiles. This implies that the modelling of the stresses and deformations within the rock mass is likely to be more realistic. It is expected that correct magnitudes of subsidence could be obtained by appropriate scaling of the rock properties.

The distinct element method models the rock mass as an assemblage of discrete deformable blocks. Blocks may rotate, slide against each other, and separate from each other. Using this model it was found that the transition from sub- to supercritical spans could be modelled realistically. Subsidence magnitudes were not in agreement with empirical values, but this could be corrected by using appropriate material properties. Coulthard and Dutton (1988) noted the sensitivity of the results to the choice of block size and "bedding plane" spacing.

It may be concluded from the above that numerical methods have considerable potential for application to subsidence engineering. Coulthard and Dutton (1988) considered that realistic non-linear representation of the rock mass will be an essential part of any reliable method for subsidence prediction. It appears that most of the numerical methods mentioned above have potential in this regard. It is likely that computer programs will be available in the near future for routine calculation of subsidence in any geological environment, as part of the mine planning process.

Most mining layouts are three-dimensional and cannot be satisfactorily analysed using two-dimensional sections. Two-dimensional analyses therefore have limited application in the analysis of subsidence effects resulting from real mining layouts. Salamon's (1964a,b) approach described earlier allows complex mining layouts and the effects of pillars to be considered. Wardle and Enever (1983) described a three-dimensional displacement discontinuity method for practical analysis of coal-mine layouts. This method can take into account anisotropy and layering, with different rock properties, of the rock mass, non-linear characteristics of the coal seam, and the resistance to closure provided by the caved waste.

Wardle (1986) presented a detailed case study of the application of this method to a real mining situation. The mine was instrumented with stress meters in the coal seam, and subsidence was measured by surface levelling. Rock properties were determined in an extensive programme of laboratory testing and for use in the numerical analysis, all laboratory values were scaled by the same factor, which was based on a back analysis of measured subsidence data. The properties of the coal seam were determined from tests in a large triaxial cell. The rock mass was assumed to be anisotropic, with an artificially high degree of anisotropy, as adopted by Crouch (1976), used to simulate a rock mass with a very low shear stiffness on horizontal planes. The agreement between calculated and measured stresses was excellent, and the predicted subsidence profiles were reasonably close to the measured profiles. Based on the favourable comparison between real and predicted results, Wardle concluded that the displacement discontinuity method can model actual mine performance realistically, and suggested that it can be used for routine design and analysis of mining layouts. In this light it should be noted that the development of Salamon's (1964a,b) approach for the isotropic elastic infinite region case (Plewman *et al.* 1969), and its subsequent refinement, is used routinely for the planning of layouts in deep-level gold mines. It is expected that computer programs with the capabilities described by Wardle and Enever (1983) will be readily available in the near future for routine application to the planning of mine layouts, and for calculation of the resulting surface subsidence behaviour.

13.5.4 Subsidence prediction in partial extraction mining

For partial mining methods, several scenarios may be relevant:

(1) If the mining layout and pillar sizes are correctly designed to prevent adverse surface effects, subsidence will not occur.
(2) If pillar extraction takes place subsequently, the methods of subsidence prediction applicable to total extraction described above can be used for tabular deposits. For non-tabular deposits, the methods described below for massive mining will be applicable. For old workings, if possible, it will be necessary to assess the 'design', that is, the stability of the pillars (Salamon and Oravecz 1976; Hedley 1978) and roof areas in the workings. If this is not possible, or shows that the stability is suspect, estimation of subsidence must be based on observed subsidence in the area or subsidence observed in similar geometrical and geological situations.

Prediction of subsidence as a result of failure requires accurate data regarding the layout of the mine. Such information frequently does not exist in the case of abandoned mines. On the other hand when accurate mine plans are available or in the case of a working mine, the methods outlined by Salamon and Oravecz (1976) or by Goodman *et al.* (1980) may be used to evaluate collapse potential. Basically these methods are the same as that of Wardell and Wood (1965) and involve calculating vertical stress based on the tributary area load concept which assumes that each pillar supports a column of rock with an area bounded by room centres and a height equal to the depth from the surface and then comparing this with the strength of the pillar.

A programme to assess the stability of pillars in limestone mines has been outlined by Agapito (1986). Two types of measurements were made. First, measurements of internal strain in drillholes were obtained for the calculation of stresses. Secondly, measurements of rock mass deformation between roof and floor and within the roof and floor were taken. The objectives of the programme included determination of the stress field, in order to evaluate background mine loading and its possible effects on roof stability and pillar orientation. A second objective considered pillar stress distributions and average loads to help assess mine stability and pillar strengths. In addition, pillar and roof deformation was investigated to help define movement patterns in the roof, pillars and floor, such data being required to determine mine stability and roof strengths. The finite element method of computer modelling was used to evaluate pillar stability. Lastly, an assessment of Young's modulus was required for use in numerical modelling.

When a structure is to be built over an area of old pillared workings the additional load on the pillars can be estimated simply by adding the weight of the appropriate part of the structure to the weight of the column of strata supported by a given pillar. This method is very conservative except when used for large concentrated loads where old workings are located at shallow depth.

Beause of the difficulty of obtaining stall dimensions in abandoned workings, the height of void migration normally is determined from the thickness of mineral extracted and the difference in densities between the roof material *in situ* and when collapsed. Depth of cover should not include superficial deposits or made ground since low bulking factors are characteristic of these materials. Hence the height of void migration can be determined, in general, by using the following expression:

$$H = t[(\rho_1/\rho)/(1 - \rho_1/\rho)] \tag{13.9a}$$

where: t = thickness of the seam
ρ_1 = bulk density of the roof rocks
ρ = bulk density of collapsed roof materials
H = height of migration

Alternatively the simpler expression:

$$H = h/(V - 1) \tag{13.9b}$$

can be used, where h is the height of the working and V is the volume occupied by the collapsed strata divided by the volume it occupied in the roof.

However, if, as Piggott and Eynon (1978) conceived, the void adopts various geometrical forms such as conical, wedge and rectangular collapses, then different expressions must be used to calculate void migration. They showed that for a particular width of mine opening (B), the height of collapse or migration (H) is a function of the original height of the mine opening (h) and the bulking factor (b_f) of the overlying strata. They derived the following simple expressions for obtaining the height of void migration for the three geometrical forms mentioned (see also Figure 13.21):

(1) conical collapse: $H = 3h/b_f$ (13.10)

(2) wedge collapse: $H = 2h/b_f$ (13.11)

(3) rectangular collapse: $H = h/b_f$ (13.12)

It can be seen that in all cases the maximum height of void migration is directly proportional to the thickness of seam mined and inversely proportional to the change in volume

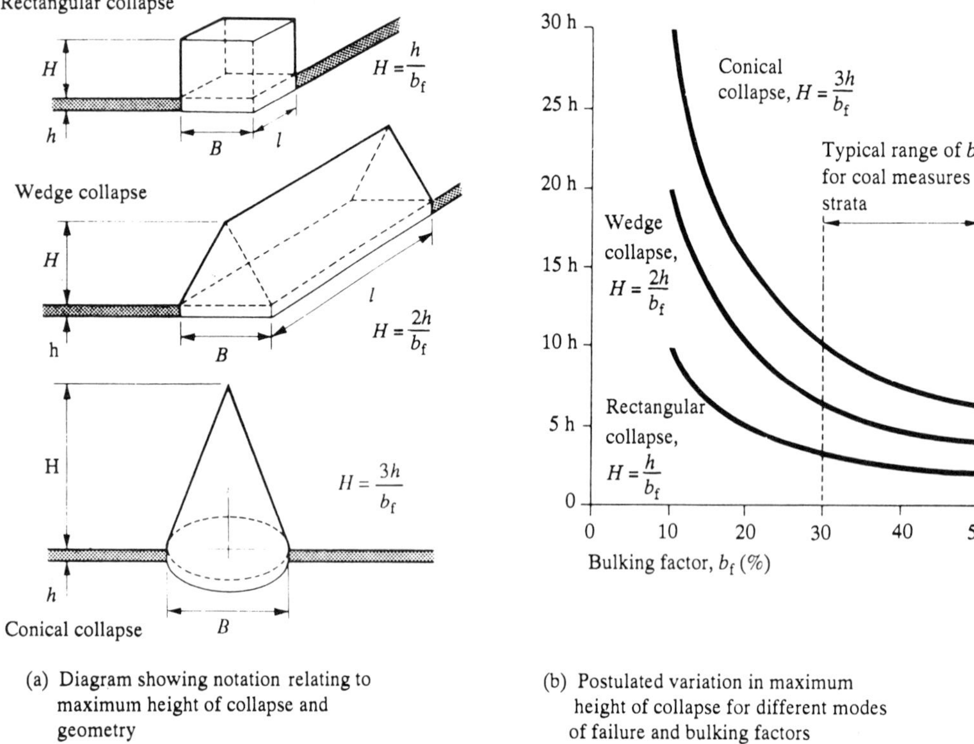

(a) Diagram showing notation relating to maximum height of collapse and geometry

(b) Postulated variation in maximum height of collapse for different modes of failure and bulking factors

Figure 13.21 *Type and amount of void migration (after Piggott and Eynon 1978)*

of the collapsed material. It would appear that the height of collapse is independent of the width of the excavation although it is a limiting factor. In other words the larger the span, the more likely is collapse to occur.

13.5.5 Abandoned mine workings and hazard zoning

When a site which is proposed for development is underlain by shallow old mine workings there are a number of approaches to the problem. The first and most obvious method is to locate any proposed structures on sound ground away from the old workings or over workings proved to be stable. It is not generally sufficient to locate immediately outside the area undermined as the area of influence should be considered. The angle of draw which governs the area of influence tends to vary with the local geology. Of course, such action is not always possible.

After a detailed investigation at a site in Airdrie, Scotland, underlain by shallow abandoned mine workings, Price (1971) was able to establish safe and unsafe zones. In the safe zones the cover rock was regarded as thick enough to preclude subsidence hazards (about 10 m of rock or 15 m of till was regarded as sufficient to ensure that crownholes did not appear at the surface) and normal foundations could be used for the two-storey dwellings which were to be erected. On the boundaries between the safe and unsafe zones, the dwellings were constructed with reinforced foundations, or rafts, as an added precaution against unforseen problems. Development was prohibited in the unsafe zones. In effect, Price produced a thematic mining information plan of the site to facilitate its development. Such maps are now being produced for larger areas. In Britain, thematic geology mapping has been undertaken for a number of areas which are underlain by some form of mine working (Culshaw *et al.* 1988). In relation to mining, three types of map may be produced:

(1) those showing the location of known shafts, adits and other mine entrances;
(2) those showing the known extent of worked and/or unworked mineral at various depths;
(3) those showing the extent of fill and made ground, including infilled opencast workings and expanses of mine or open pit waste.

The scales of these maps vary, depending upon local requirements and the availability of information but, generally they are 1:10 000, 1:25 000 or 1:50 000. Attempts also have been made to zone ground underlain by old mine workings in terms of its suitability for different types of foundation (Gostelow and Browne 1988).

However, it must be borne in mind that thematic maps which attempt to portray the degree of hazard represent generalized interpretations of the data available at the time of compilation. Therefore, they cannot be interpreted too literally and areas outlined as 'undermined' should not automatically be subjected to planning blight (Statham *et al.* 1987). Also, zoning based entirely upon depth of cover above workings cannot be relied on completely, since occasionally subsidences have occurred in zones labelled 'safe'.

13.5.6 Subsidence prediction in the mining of massive deposits

Subsidence prediction in areas of 'total' extraction of tabular deposits can be carried out using appropriate methods described in Sections 13.5.1–13.5.3 above.

When massive mining is being practised, particularly with caving methods, it is almost certain that subsidence will occur, or that the risk of subsidence will be high. Such mining is unlikely to be permitted where it will pose a threat to public safety. It is the stability of mine structures such as shafts, underground accesses and haulages, surface plant and facilities that will usually be the main concern. In such circumstances the prediction of the magnitude of subsidence is unlikely to be required. Rather, a prediction will be required as to whether subsidence will affect the mine structures and adjacent areas or not. This is a common problem in mining, and arises from poor initial planning or from the fact that mining operations often expand beyond those initially planned, as a result of increased ore prices and discovery of additional or extended ore bodies. Although surface plant can be moved relatively easily, waste dumps, tailings dams and shafts and sometimes residential areas may be located too closely to the ore-body limits to be confident of their safety. Owing to the critical importance of such facilities to a mine, and the prohibitive cost of replacing or moving them, it is often very important to determine the potential influence on them of the extended mining. This assessment can be carried out using a theoretical approach or, if applicable field data are available, by extrapolation of the data, or interpretation from similar cases to the problem case.

(a) Theoretical methods

Theoretical methods of analysis of caving and subsidence, based on the study of the failure and flow of solids with a pressure-dependent yield function, have been proposed by Jenike and Leser (1962). Owing to the current availability of suitable numerical methods, however, it is unlikely that these theoretical approaches will be commonly applied in practice.

Hoek (1974) developed a method for the prediction of the extent of caving resulting from mining of an inclined tabular deposit. This approach is based directly on the limit equilibrium analysis of the stability of rock slopes, and allows the failure plane angle and an angle of break to be determined. These define the position on the surface of the furthest influence of mining. Curves from which these values can be obtained are given in Figure 13.22.

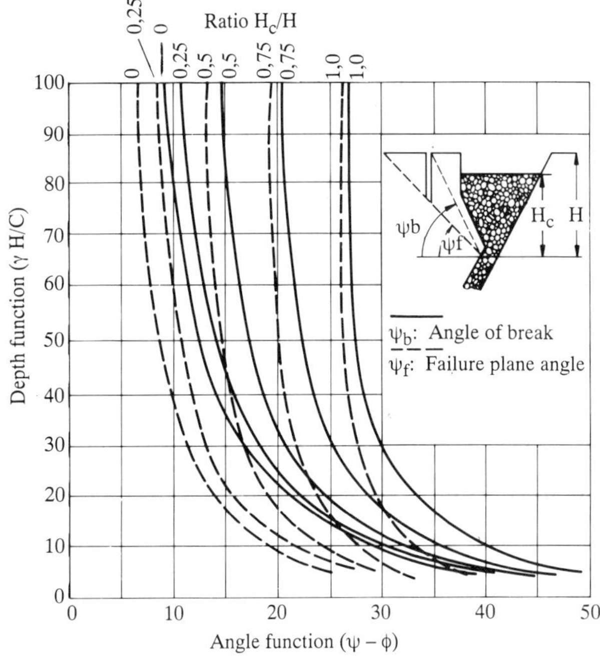

Figure 13.22 *Caving break back angles (modified after Hoek 1974)*

Theoretical methods of subsidence prediction make use of some of the numerical methods described in Section 13.5.3 above. Boundary element and finite element methods are applicable for calculating the deformations around the massive openings. Finite element analysis may be preferable if the geology is substantially non-homogeneous, and non-linear failure behaviour of the mass is to be modelled. However, the openings created in massive mining are rarely sufficiently two-dimensional in nature for a valid two-dimensional analysis to be carried out. It is therefore generally necessary to make use of a three-dimensional stress analysis method. Although both finite and boundary element methods are available, the magnitude of the problem of using the former method for complicated three-dimensional modelling, at present, makes the latter method far more attractive for practical users. A drawback of the boundary element method is its limited capacity to model non-homogeneity. However, from a practical point of view, this is not considered to be a significant drawback, and considerable success has been achieved using a combination of three-dimensional boundary element analyses to model the full geometrical situation, and two-dimensional finite element analyses, based on these results, to model the detailed geological situation (Diering and Laubscher 1987). In such a 'tandem' approach, the three-dimensional results are used as a 'calibration' for the subsequent two-dimensional analyses.

Owing to the limited knowledge available regarding the deformation characteristics of rock masses, and the field stresses, the assumption of linear elasticity for the three-dimensional analyses is considered to be the most suitable choice for practical applications. To predict failure and subsidence, it is necessary to apply a suitable failure criterion to the results. The choice of the criterion should be based on local conditions. The validity of predictions will be enhanced substantially if it is possible to back-analyse a failure, and so establish a credible failure criterion. This approach has proved to be successful in underground situations (Diering and Laubscher 1987). The use of non-linear two-dimensional analyses for more detailed local analysis of failure, in which boundary conditions for the two-dimensional model are derived from the three-dimensional results, is also practical.

(b) Empirical prediction

Methods of empirical prediction dealt with in Sections 13.5.2 will be applicable, or at least sufficiently so for preliminary estimates, in cases in which the horizontal extent of mining is sufficient to result in collapse of the workings.

Besides these established methods, there are no known empirical techniques for prediction of subsidence in massive mining situations. There are, however, some published records of monitored subsidence behaviour, which can be used for comparative purposes (Thomas, 1971; Touseull and Rich 1980; Brumleve and Maier 1981; Panek 1981). These records show that the subsidence behaviour, described by the angle of draw, is very site specific. In many cases the angle is near vertical while in others it approaches values typical of coal mining. At the San Manuel Mine (Panek 1981), angles of draw varied from 33 to 55°. It appears that the area of subsidence on the surface is affected by several factors, including the following:

(1) the ratio between the extracted span and the depth of overburden;
(2) the extent of bulking that takes place during the caving process;
(3) the structural and deformational characteristics of the overburden strata;
(4) the surface topography;
(5) the climate;
(6) major geological structural features. These are known to have a major beneficial effect in some cases, in limiting the area of subsidence, but in others to result in large differential movements (e.g. fault steps).

In summary, whilst empiricial data may provide background comparative information, each case must be evaluated individually. For this purpose the numerical techniques referred to in Section 13.5.3 above are likely to be used to a far greater extent in the future.

13.5.7 Prediction of subsidence due to the abstraction of fluids

Obviously it is necessary to determine the amount of subsidence which is likely to occur as a result of the withdrawal of fluids from the ground, as well as to estimate the rate at which it may occur. Unfortunately a large number of prediction methods have been developed, some of which are relatively simple whilst others are complex. One of the probable reasons for this is that stratal sequences are different in different areas where subsidence has occurred and consequently different models have been devised. Nonetheless a number of steps can be taken in order to evaluate the subsidence likely to occur due to the abstraction of fluids from the ground. These include defining the *in situ* hydraulic conditions; computing the reduction in pore pressure due to removal of a given quantity of fluid; conversion of the reduction in pore pressure to an equivalent increase in effective stress; and estimating the amount of consolidation likely to take place in the formation affected from consolidation data and the increased effective load. In addition to depth of burial, the ratio between maximum subsidence and reservoir consolidation also should take account of the lateral extent of the reservoir in that small reservoirs which are deeply buried do not give rise to noticeable subsidence, even if undergoing considerable consolidation, whereas extremely large reservoirs may develop significant subsidence. The problem therefore is three-dimensional rather than one of simple vertical consolidation. Subsidence prediction in relation to the rate of groundwater pumping therefore should involve relating the consolidation model to a two- or three-dimensional hydrogeological model based on the groundwater flow equation. Variations in the hydraulic head in both time and space in response to groundwater abstraction are obtained from the hydrogeological model. These values then can be used to derive the time-dependent consolidation curve at any point in the system. This provides an indication of the amount of subsidence likely to occur.

Like mining subsidence prediction methods, prediction methods associated with subsidence due to fluid abstraction also can be grouped into empirical, semi-empirical and theoretical categories. Empirical methods involve the extrapolation of available information such as the amount of subsidence, the amount of consolidation or the decline in fluid level being plotted against time in order to determine future trends (Figueroa Vega and Yamamoto 1984).

The semi-empirical approach also depends on the relationship between subsidence and related phenomena. For instance, Castle *et al.* (1969) found that subsidence in six oil fields in the United States varied more or less linearly with net production of liquid but the correlation between the decline in reservoir pressure and subsidence was poor. The most likely explanation of this poor correlation is that pressure decline as measured at individual producing wells probably is not representative of the average decline over a field.

The ratio of subsidence to head decline in coarse-grained permeable beds of consolidating aquifer systems represents the ratio between these two factors over a common interval of time (Poland 1984). This ratio reflects the change in thickness per unit change in effective stress and can be used to predict a lower limit for the amount of subsidence in response to a given increase in virgin stress (stress exceeding the past maximum). If pore pressures in the consolidating aquitards reach equilibrium with those in adjacent aquifers, then consolidation ceases and the subsidence–head decline ratio represents a true measure of the virgin compressibility of the system. Until equilibrium of pore pressure is attained, the ratio of subsidence to head decline is a transient value. Contours of this ratio for a given period of time can be plotted on a map and indicate the amount of head decline required to produce a particular magnitude of subsidence throughout the area concerned.

As remarked above, the abstraction of fluid from the ground reduces the pore pressure which leads to a transfer of load to the granular skeleton and to its subsequent reduction in volume. However, in trying to develop an explanation of the phenomenon one encounters the problems associated with multi-phase systems representing solids, liquids and gases, the properties of which must be inferred from statistical averages or from representative tests. The materials concerned include in their mechanical properties the combined behaviour of their individual components (that is, elasticity and plasticity of solids; viscosity of liquids; compressibility of gases; decay of organic matter; attraction and repulsion of ionic charges etc). Hence the mechanical properties are anisotropic, as well as dependent on stress history and time. Accordingly such material is difficult, if not impossible, to deal with in a theoretical model of subsidence. Resort has to be made to simplifying assumptions in order to develop any type of model of subsidence prediction, in particular, the properties of the ground must be idealized. Other simplifications may include the assumption that strata are horizontal; that flow in aquifers is horizontal whilst it is vertical in aquitards; and that subsidence primarily is due to consolidation of aquitards. Obviously the greater the number of simplifications which are incorporated in a model, the more restricted its use becomes.

Subsidence above water, oil or gas reservoirs frequently is treated as a mechanical problem involving the elastic behaviour of the reservoir and is analysed with the aid of a mathematical model. Unfortunately the amount of data available frequently is limited, which means that, as remarked above, certain assumptions have to be made before the model can be applied. For example, Geertsma (1973) simulated the subsidence above the Groningen gas field by assuming a homogeneous and isotropic semi-infinite porous medium. The reservoir was assigned an idealized shape, that is, a horizontal circular cylinder of limited thickness. The ratio between maximum subsidence and reservoir consolidation was governed by the ratio between depth of burial and the lateral extent of the reservoir. In the same year Gambolati and Freeze (1973) developed a mathematical model to simulate subsidence due to groundwater withdrawal at Venice. First, with the aid of the finite element technique, the regional drawdowns in hydraulic head were determined in a two-dimensional vertical cross-section in radial coordinates, using an idealized ten-layer representation of the geological conditions. Then the values of the hydraulic head calculated for the aquifers were used as time-dependent boundary conditions in a set of one-dimensional vertical consolidation models solved by the finite difference technique and applied to a more refined representation of each aquitard. Subsequently Gambolati *et al.* (1986) developed a method of analysis of subsidence due to the withdrawal of oil and gas from a reservoir overlain by layered anisotropic soils by using the finite element technique. They assumed a disc-shaped reservoir of uniform thickness, which underwent elastic deformation. The model consisted of layered anisotropic soil units which were characterized by five elastic constants. Alternating sands and clays were assumed to occur above the reservoir, their compressibility progressively decreasing with depth. They found that for a given geometry, depth of burial and fluid pressure decline, the subsidence was basically related to the compressibility of the oil-gas bearing strata and of the adjacent overlying/underlying clays. The depth at which a rigid basement occurred beneath a reservoir appeared to have only a limited influence on ground movement.

Pottgens (1986) suggested that a simple yet rapid estimate of subsidence for a disc-shaped reservoir could be obtained from the following expression

$$S = \pm 2(1 - \nu) \times (1 - C)/(1 + C^2) \times m_v \times \sigma \times H \tag{13.13}$$

where ν is Poisson's ratio, $C = Z/R$, where Z is the depth and R is the radius of the disc-shaped reservoir, m_v is the coefficient of volume compressibility, σ is the increase in head due to reservoir pressure reduction and H is the thickness of the reservoir.

References

Agapito, J. F. T. (1986) 'Pillar stability in large underground openings: application from a case study in competent jointed rock', *Colorado School of Mines Quarterly,* **81**, 1–52

Anon (1975) *Subsidence Engineer's Handbook,* National Coal Board, London, 109 pp.

Bell, F. G. (1975) 'Salt and subsidence in Cheshire, England', *Engg Geol.,* **9**, 237–247

Bell, F. G. (1986) 'Location of abandoned workings in coal seams', *Bull. Int. Ass. Engg Geol.*, **33**, 123–132

Bell, F. G. (1988) 'Subsidence associated with the abstraction of fluids'. In *Engineering Geology of Underground Movements*, Engineering Geology Special Publication No. 5, F. G. Bell, M. G. Culshaw, J. C. Cripps and M. A. Lovell (eds), The Geological Society, London, pp. 363–376

Berry, D. S. (1960) 'An elastic treatment of ground movement due to mining – Part I, Isotropic ground, *J. Mech. Phys. Solids*, **8**, 280–292

Berry, D. S. and Sales, T. W. (1961) 'An elastic treatment of ground movement due to mining – Part II, Transversely isotropic ground', *J. Mech. Phys. Solids*, **9**, 52–62

Brauner, G. (1973) 'Subsidence due to underground mining: I – Theory and practices in predicting surface deformation', *US Bureau of Mines Information Circular*, No 8571, 55 pp.

Brumleve, C. B. and Maier, M. M. (1981) 'Applied investigations of rock mass response to panel caving, Henderson Mine, Colorado, USA' In *Design and Operation of Caving and Sublevel Stoping Mines*, Soc. Min. Engrs of AIME, 223–249

Burton, D. (1978) 'A three-dimensional system for the prediction of surface movements due to mining, *Proc. 1st Int. Conf. Large Ground Movements and Structures*, Cardiff, J. D. Geddes (ed.) Pentech Press, London, 209–228

Carter, P., Jarman, D. and Sneddon, M. (1981) Mining subsidences in Bathgate, a town study', *Proc. 2nd Int. Conf. Ground Movements and Structures*, Cardiff, J. D. Geddes (ed.), Pentech Press, London, 101–124

Castle, R. O., Yerkes, R. F and Riley, F. S. (1969) 'A linear relationship between liquid and oil field subsidence', *Proc. 1st Int. Conf. Land Subsidence*, Tokyo, International Association of Hydrological Sciences, Publ. No. 88, **1**, 167–173

Coulthard, M. A. and Dutton, A. J. (1988) 'Numerical modelling of subsidence induced by underground coal mining', *Proc. 29th U.S. Rock Mech. Symp. Key Questions in Rock Mechanics*, 529–536

Crouch, S. L. (1976) 'Analysis of stresses and displacements around underground excavations: an application of the displacement discontinuity method', *Geomechanics Report*, University of Minnesota, 268pp.

Crouch, S. L. and Starfield, A. M. (1983) *Boundary Element Methods in Solid Mechanics*, Allen & Unwin, 322pp.

Culshaw, M. G., Bell, F. G. and Cripps, J. C. (1988) 'Thematic geological mapping as an aid to engineering hazard avoidance in areas of abandoned mine workings', *Mineworkings 88, Proc. 2nd Int. Conf. on Construction in Areas of Abandoned Mineworkings*, Edinburgh, M. C. Forde (ed.), Engineering Technics Press, Edinburgh, 69–76

Deere, D. V. (1963) 'Subsidence due to mining – a case history from the Gulf Coast region of Texas', *Proc. 4th Symp. Rock Mechanics, Bulletin of the Mineral Industries Experiment Station*, The Pennsylvania State University, 59–64

Diering, J. A. C. and Laubscher, D. H. (1987) 'Practical approach to the numerical stress analysis of mass mining operations', *Trans Inst. Min. Metall.*, Section A, **96**, A179–A188

Figueroa Vega, G. and Yamamoto, S. (1984) 'Techniques of prediction of subsidence'. In *Guidebook to Studies of Land Subsidence due to Groundwater Withdrawal*, J. F. Poland (ed.), UNESCO, Paris, 89–117pp.

Gambolati, G. and Freeze, R. A. (1973) 'Mathematical simulation of subsidence of Venice', *Water Resources Research*, **5**, 721–733

Gambolati, G. Gatto, P. and Ricceri, G. (1986) 'Land subsidence due to gas-oil removal in layered anisotropic soils by a finite element model', *Proc. 3rd Int. Conf. Land Subsidence*, Venice, A. I. Johnson, L. Carbognin and L. Ubertini (eds.) International Association of Hydrological Sciences, Publication No. 151, 29–41

Geertsma J. (1973) A basic theory of subsidence due to reservoir compaction: the homogeneous case. *Geologie en Mijubouw*, **28**, 43–62

Gilluly, J. and Grant, U. S. (1949) 'Subsidence in the Long Beach area, California', *Bull. Geol. Soc. America*, **60**, 461–560

Goodman, R. E., Korbay, S. & Buchignani, A. (1980) 'Evaluation of collapse potential over abandoned room and pillar mines', *Bull. Ass. Engg Geol.*, **17**, 27–37

Gostelow, T. P. and Browne, M. A. E. (1988) 'Engineeirng geology of the upper Forth estuary', *Report of the British Geological Survey*, **16**, 8pp.

Gurtunca, R. G. and Bhattacharyya, A. K. (1988) 'Modelling of surface subsidence in the Southern Coalfield of New South Wales', *Proc. 5th Australia–New Zealand Conf. on Geomechanics – 'Prediction versus Performance'*, Sydney, 5pp.

Gurtunca, R. G. and Schumann, E. H. R. (1986) 'Computer simulation of surface subsidence using a displacement discontinuity method', *Proc. Symp. Effect of Underground Mining on Surface*, South African National Group of International Society of Rock Mechanics, 81–87

Hackett, P. (1959) 'An elastic analysis of rock movements caused by mining', *Trans Inst. Min. Engrs*, 118, No. 7, 421–435

Hagan, T. O. (1988) 'Mine design strategies to combat rockbursting at a deep South African gold mine', *Proc. 29th U.S. Rock Mech. Symp., 'Key Questions in Rock Mechanics'*, Minneapolis, 249–260

Hedley, D. F. (1978) 'Design guidelines for multi-seam mining at Elliot Lake', *Canada Centre for Mineral and Energy Technology, CANMET*, Report 78–9

Hill, F. G. (1981) 'The stability of the strata overlying the mined-out areas of the central Witwatersrand', *J. S. Afr. Inst. Min. Metall.*, **81**, 145–191

Hoek, E. (1974) 'Progressive caving induced by mining an inclined orebody', *Trans Inst. Min. Metall., Section A*, **83**, A133–A139

Hood, M., Ewy, R. T. and Riddle, R. L. (1983) 'Empirical methods of subsidence prediciton – a case study from Illinois', *Int. Rock Mech. & Min. Sci. & Geomech. Abstr.*, **20**, 153–170

ITASCA 'FLAC – fast lagrangian analysis of continua, and UDEC – universal distinct element code', *Itasca Consulting Group Inc.*, Minneapolis

Jackens R. C. and Holzer T. L. (1980) Geophysical investigations of ground failure related to ground-water withdrawal, Picacho basin, Arizona. *Ground Water*, **17**, No. 6, 574–585

Jenike, A. W. and Leser, T. (1962) 'Caving and underground subsidence', *Trans Soc. Min. Engrs, AIME*, **67** 73

Jennings, J. E., Brink, A. B. A., Louw, A. and Gowan, G. D. (1965) 'Sinkholes and subsidences in the Transvaal dolomite of South Africa', *Proc 5th Int. Conf. Soil Mech. Found. Eng.*, **6**, 51–54

Jones, C. J. F. P. and Bellamy, J. B. (1973) 'Computer prediction of ground movements due to mining subsidence', *Geotechnique,* **23**, 515–530

Kapp, W. A. (1986) 'Mine subsidence in New South Wales – its effect on surface features and structures', *Proc. Symp. Effect of Underground Mining on Surface,* South African National Group of International Society & Rock Mechanics, Supplementary paper

Lofgren, B. N. (1979) 'Changes in aquifer system properties with groundwater depletion', In *Evaluation and Prediction of Subsidence, Proc. Speciality Conf. Am. Soc. Civil Engineers,* Gainsville, S. K. Saxena (ed.). 26–46

Marino, G. G. and Gamble, W. (1986) 'Mine subsidence damage from room and pillar mining in Illinois', *Int. J. Mining and Geol. Eng,* **4**, 129–150

Marr, J. E. (1975) 'The application of the zone area system to the prediction of mining subsidence', *Mining Engineer,* **135**, 53–612

Mikula, P. A. and Holt, G. E. (1983) 'Prediction of mine subsidence in Eastern Australia by mathematical modelling', *Proc. 5th Int. Cong. Int. Soc. Rock Mech., Melbourne,* Section E, E119–E126

Orchard, R. J. and Allen, W. S. (1970) 'Longwall partial extraction systems', *Mining Engineer,* **127**, 523–535

Ortlepp, W. D. and Nicoll, A. (1964) 'The elastic analysis of observed strata movement by means of an electrical analogue', *J. S. Afr. Inst. Min. Metall.,* **65**, 214–235

Panek, L. A. (1981) 'Ground movements near a caving stope'. In *Design and Operation of Caving and Sublevel Stoping Mines,* Soc. Min. Engrs of AIME, 329–354pp.

Piggott, R. J. and Eynon, P. (1978) 'Ground movements arising form the presence of shallow abandoned mine workings', *Proc. 1st Int. Conf. Large Ground Movements and Structures, Cardiff,* J. D. Geddes (ed.), Pentech Press, London, 749–780

Plewman, R. P., Deist, F. H. and Ortlepp, W. D. (1969) 'The development and application of a digital computer method for the solution of strata control problems', *J. S. Afr. Inst. Min. Metall.,* **70**, 214–235

Poland, J. F. (ed.) (1984) *Guidebook to Studies of Land Subsidence due to Groundwater Withdrawal,* UNESCO, Paris

Pottgens, J. J. E. (1986) 'Ground movements caused by mining activities in the Netherlands', *Proc. 3rd Int. Symp. Land Subsidence,* Venice, International Association of Hydrological Sciences Publication No. 151, 651–664

Price, D. G. (1971) 'Engineering geology in the urban environment. *Q.J. Engng Geol.,* **4**, 191–208

Salamon, M. D. G. (1964a) 'Elastic analysis of displacements and stresses induced by the mining of seam or reef deposits – Part I: fundamental principles and basic solutions as derived from idealized models', *J. S. Afr. Inst. Min. Metall.,* **64**, 128–149

Salamon, M. D. G. (1964b) 'Elastic analysis of displacements and stresses induced by the mining of seam or reef deposits – Part III: an application of the elastic theory: protection of surface installations by underground pillars', *J. S. Afr. Inst. Min Metall.,* **64**, 468–500

Salamon, M. D. G. (1983) 'Linear models for predicting surface subsidence', *Proc. 5th Int. Cong. Int. Soc. Rock Mech., Melbourne,* Section E, E107–E113

Salamon, M. D. G. (1988) 'Developments in rock mechanics: a perspective of 25 years', *Trans Inst. Min. Metall.,* Section A, **97**, A57–A68

Salamon, M. D. G. and Oravecz, K. I. (1976) *Rock mechanics in coal mining,* Chamber of Mines of South Africa, Johannesburg, 119pp.

Schoonbeck, J. B. (1976) 'Land subsidence as a result of natural gas extraction in the province of Groningen', *Soc. Petrol. Engineers J.,* American Institute of Mining Engineers, Paper No. SPE 5751, 1–20

Singh, B. (1973) 'Continuum characterisation of jointed rock masses: Part II – Significance of low shear modulus', *Int. J. Rock Mech Min. Sci. Geomech. Abstr.,* **10**, 337–349

Stacey, T. R. and Rauch, H. P. (1981) 'A case history of subsidence resulting from mining at considerable depth', *Trans S. Afr. Inst. Civ. Engrs,* **23**, 55–58

Statham, I., Golightly, C. and Treharne, G. (1987) 'The thematic mapping of the abandoned mining hazard – a pilot study for the South Wales Coalfield'. In *Planning and Engineering Geology,* Engineering Geology Special Publication No. 4, M. G. Culshaw, F. G. Bell and J. C. Cripps (eds), The Geological Society, London, 225–268pp.

Sutherland, H. J. and Munsen, D. E. (1984) 'Prediction of subsidence using complementary influence functions', *Int. J. Rock Mech. and Min. Sci. & Geomech. Abstr.,* **21**, 195–202

Thomas, L. A. (1971) *Subsidence and Related Caving Phenomena at the San Manuel Mine,* Magma Copper Company, San Manuel Division, 87pp.

Touseull, J. and Rich, C. (1980) 'Documentation and analysis of a massive rock failure at the Bautsch Mine, Galena, III.', *U.S. Bureau of Mines Report of Investigations,* No. 8453, 49pp.

Voight, B. and Pariseau, W. (1970) 'State of predictive art in subsidence engineering', *J. Soil Mech. Found. Div. ASCE,* **96**(SM2), 721–750

Walton, G. and Cobb, A. E. (1984) 'Mining subsidence'. In *Ground Movements and Their Effects on Structures,* P. B. Attewell and R. K. Taylor (eds), Surrey University Press, London, 216–242

Wardell, K. (1954) 'Some observations on the relationships between time and mining subsidence', *Trans. Inst. Mining Engineers,* **113**, 417–483 and 799–814

Wardell, K. and Eynon, P. (1968) 'Structural concept of strata control and mine design', *Trans. Inst. Min. and Metall.,* **77,** Section A, 125–150

Wardell, K. and Wood, J. C. (1965) 'Ground instability problems arising from the presence of shallow old mine workings', *Proc. Midland Soc. Soil Mech. Found. Engg.,* **7**, 7–30

Wardle, L. J. (1986) 'Boundary element methods for stress analysis of tabular excavations', *PhD thesis* Dept of Mining and Metallurgical Engineering, University of Queensland

Wardle, L. J. and Enever, J. R. (1983) 'Application of the displacement discontinuity method to the planning of coal mine layouts', *Proc. 5th Int. Cong. Int. Soc. Rock Mech., Melbourne,* Section E, E61–E69

Wassmann, T. H. (1980) 'Mining subsidence in Twente, east Netherlands', *Geologie en Mijnbouw,* **59**, 225–231

Yerkes, R. F. and Castle, R. O. (1970) 'Surface deformation associated with oil and gas field operation in the United States', *Proc. 1st Int. Symp. Land Subsidence,* Tokyo, International Association of Hydrological Science, UNESCO Publication No. 88, **1**, 55–66

Zienkiewicz, O. C. (1977) *The Finite Element Method in Engineering Science,* McGraw-Hill, New York

14 Seismic movements and rock masses

B O Skipp, Soil Mechanics Ltd, Wokingham, UK

14.1 Introduction

Engineering in rock requires some consideration of dynamic actions. Most parts of the world are subject to earthquakes and in some regions the seismic element in planning and design takes on a major role. Large earthquakes bring with them landslides, rockfalls and ground rupture. Buildings and engineering structures are founded on or in soils or rocks. These materials constitute two poles in the classification of the ground and are traditionally covered by the disciplines of soil mechanics and rock mechanics. Just as there are similarities and differences in those two disciplines so are there similarities and differences when considering the behaviour of the material under ground vibration from natural or man-made earthquakes. In particular, continuum assumptions are more often appropriate for engineering in soils than for engineering in rock.

Man-made or 'technical seismicity' as well as 'natural seismicity' may have to be taken into account. This chapter will cover earthquakes and such man-made vibrations as can be regarded as 'mini-earthquakes'. It will not encompass specifically those aspects of dynamic action which are linked to ground borne or airborne shock waves; the 'hardening' of installations, or the intricacies of dynamic fragmentation, nor will it venture far into the field of acoustic emission, rock bursts or induced seismicity as such.

These restrictions in effect define the period range of ground vibrations under consideration to be from some tens of seconds to some hundredths of a second and the duration range to be from a few hundreds of seconds to a few seconds.

As far as the material is concerned 'rock' will be taken to cover a wide range of consistency – from near 'soil', as in weathered rocks, to the archetypal hard crystalline mass–but rock as material in a rock-fill embankment will not be dealt with.

In order to appreciate the full range of decisions which may have to be made with regard to seismic engineering there are a number of basic definitions to be made. One term 'actions' has already been introduced. It is a word now widely used in earthquake codes and embraces all external processes imposed upon a structure, be they forces, accelerations, velocities or displacements.

These actions can be constant, that is, they will be operative continually during the life of a structure (e.g. gravity forces), variable, that is, they can be expected to occur intermittently but regularly (e.g. live human loading in a concert hall, wind loading), or accidentally (e.g. earthquake in a low seismicity region, gas explosion).

Earthquakes are usually treated as accidental actions in low seismicity countries with consequent implications on the way they are combined in design with other actions. The selection of the earthquake to be used in design brings into play the words 'hazard', 'risk' and 'vulnerability'.

Although 'hazard' and 'risk' are commonly used as if they were synonymous, a narrower set of definitions in use in engineering seismology has been developed by UNDRO (Fournier d' Albe 1981):

(1) *hazard*. The probability of occurrence within a specified period of time of a seismic action of a specified intensity. Here intensity is being used to describe the level of the action–it may be a peak acceleration, a peak velocity or a displacement;
(2) *vulnerability*. The probability that, given a specified level of action a particular consequence, a 'degree of loss', will ensue–for example, that a structure suffers a particular degree of damage;
(3) *elements at risk*. The population, buildings, civil engineering works, economic activities utilities and infrastructure at risk in a given area;
(4) *specific risk*. The expected degree of loss (possibly in monetary terms) due to say an earthquake, as a function of both the hazard and the vulnerability;
(5) *risk*. The expected loss of life, damage, disruption to economic activity as a result of an earthquake and consequently a product of specific loss and elements at risk.

The incorporation of a valuation of loss is especially important in some extractive industries where the choice of an appropriate design earthquake would involve different considerations from those for long-lived civil infrastructure (Skipp 1984). There are also operations such as nuclear power stations where the risk is simply taken as the probability of exceedence of a seismic action multiplied by the 'fragility' where 'fragility' (or frangibility) is vulnerability in terms of the release of ionizing radiation to a particular level at a specified locality without including any value aspect.

14.2 Basic principles of seismic action

14.2.1 Assessment of hazard

Earthquakes present two kinds of hazard: vibratory ground motion and ground rupture. The choice of ground motion is the province of engineering seismology and the effort which needs to be mobilized in a particular instance depends upon many factors. Skipp and Ambraseys (1987) outline work involved in studies at four different levels.

The engineer first looks for some kind of seismic zoning map, often attached to a national earthquake code. A list of such national codes is published at four yearly intervals by the International Association for Earthquake Engineering. Some maps indicate only a historical maximum intensity on such scales as the Medvedev–Sponhauer–Karnik (MSK) or modified Mercalli (MM) scale. Such scales are based on subjective but systematic classification of the effects of an earthquake on man, buildings and the ground itself. Intensity therefore involves the vulnerability of the shaken elements so the probability of a given intensity being exceeded in a given time period does not strictly constitute a hazard. Nevertheless because a very weak correlation can be demonstrated between intensity and some kinematic parameter such as peak particle acceleration and velocity at ground surface, it may sometimes be used to derive a hazard.

Some maps may show zones with intensity in terms of their return period with a stated confidence, others indicate the value of a coefficient to be used in expressions set down in particular earthquake codes. Care must be taken if such information is to be used to characterize a ground motion (Ambraseys 1975; Kriznitzsky and Chang 1988).

There are maps which indicate the probability of exceeding a peak surface acceleration or velocity with stated confidence. They usually refer to a notional firm ground input.

Another kind of map shows the probability of occurrence of the maximum magnitude of earthquake in a given period averaged over an element of the Earth's surface (Burton 1980).

Recently there have been attempts to produce 'seismotectonic' maps which show, in addition to the epicentres of earthquakes, the faults from which the earthquakes emanate, and display relevant geological information such as those faults which can be shown to have moved in recent geological times (Vogt and Godefroy, 1981).

For the most important facilities it is advisable to carry out site-specific hazard evaluations making use of both historical and instrumental records. Seismic hazard evaluation is an important part of engineering seismology.

The choice of the appropriate risk level and hazard level is made by the owner and designer of a facility having regard to the consequences of a failure. Some examples of the levels of risk which have been under consideration for a variety of utilities are shown in Figures 14.1 and 14.2. A figure of 10% for the chance of an action which brings an ordinary building to its ultimate limit state during its 'lifetime' is often in mind.

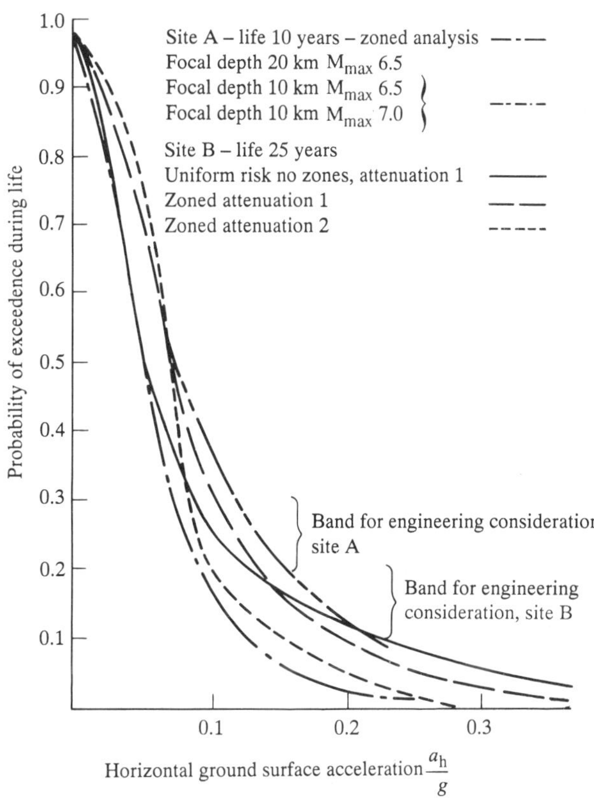

Figure 14.1 *Example of hazard levels under consideration in terms of peak horizontal ground acceleration. – Site A, major new chemical plant complex; Site B, old medium-size chemical plant*

14.2.2 Character of vibratory motion

(a) Quantification of the source

Seismologists were initially interested in the effects of earthquakes and compared them in terms of 'intensities' (see above). Then seismographs became available and some measurements could be made. The instruments, mainly of the pendulum type, were, however, thrown off scale by the shaking in the epicentral regions, where

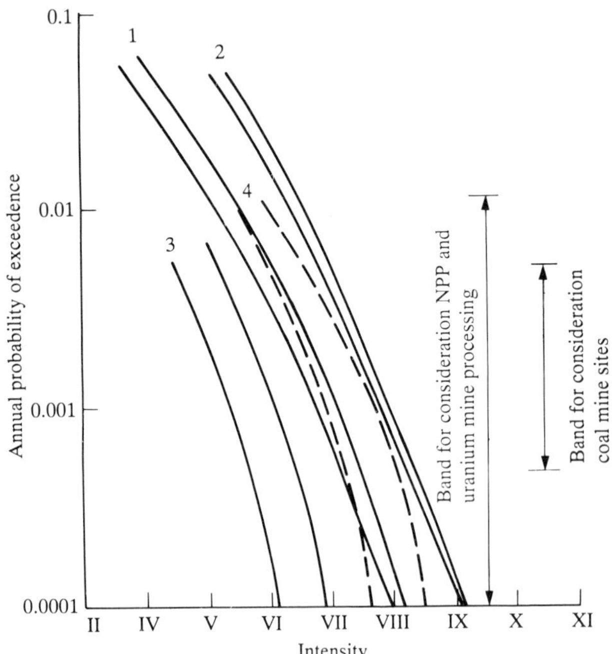

Figure 14.2 *Example of hazard levels as Intensity, Sites C, nuclear related; Site D, open-cast coal mine*

structures were damaged. Seismologists therefore concentrated upon experimental and theoretical work in the 'far field' where vibrations could be treated as travelling elastic waves (Aki and Richards 1980; Hanyga 1985).

With the advent of strong motion seismographs capable of capturing motion having peak accelerations up to 1 g over a frequency range of a few tenths of a hertz to 30 Hz, the last two decades have seen a massive growth in 'strong-motion seismology' with arrays of broad-band instruments capturing the spatial character of the motion (Bullen and Bolt 1985; Loh and Whitman 1985).

The first instrumental characterization of an earthquake which was aimed at describing the earthquake source itself was 'magnitude'. As originally defined it was the logarithm of the ratio of the maximum amplitude derived from the records of a standard Wood–Anderson torsion seismograph at a given distance to the equivalent amplitude of an earthquake of zero magnitude at the same distance. Richter empirically determined the amplitudes of a zero magnitude shock for a range of distances from 25 to 600 km for southern California and this range was later extended to distances nearer than 25 km. 'Richter' magnitude is classed as a 'local' magnitude and in general is fully defined only for a local region (Richter 1958).

A full discussion of the concept of magnitude and its limitations is beyond the scope of this chapter. Several authorities have discussed it extensively (Bath 1981). There are several different kinds of magnitude all based upon the displacement amplitude of different parts of the seismograph record taken at distances from the source which allow the different wave type to develop and be indentified.

Surface wave magnitude, M_s, is based upon the amplitude of the 20 second period surface wave. Body wave magnitude, m_b, is based upon the amplitude of the body wave. After many vicissitudes and competing formulae, standardized procedures were agreed upon by the International Association for Seismology and Physics of the Earth's Interior (IAESPI) in 1967 and it was recognized that M_s and m_b should be reported separately (Willmore 1979).

There are other magnitudes quoted in the literature. Karnik (1969) attempted to homogenize the magnitudes reported in the European catalogues using both surface and body wave depending upon their focal depth.

Another magnitude, 'duration magnitude' is based upon an empirical correlation of duration and local magnitude. It is widely used to describe microearthquakes (Lee *et al*, 1972). There are several dozen expressions to relate empirically the plethora of magnitudes (Bath 1981).

There are shortcomings in magnitude as a descriptor of the size of an earthquake. It depends upon the pathway between the seismic source and the instrument and is tied to a particular band of frequencies, although it can be related to the energy radiated.

A widely used measure of the size is the 'seismic moment', and from that measure another magnitude – moment magnitude, M_w–has been developed:

$$M_w = \tfrac{2}{3} \log M_o - 10.73 \quad \text{(Kanamori 1977)} \qquad (14.1)$$

Seismic moment may be defined (Aki 1967) as

$$M_0 = \int_{\Omega} \mu u \, d\Omega \qquad (14.2)$$

Where μ is the shear modulus of the wall rocks, u is the relative displacement of one side of the generating fault with respect to the other and Ω is the surface area over which rupture takes place when an earthquake is generated. Seismic moment is usually given in dyne centimetres or Joules. Moment can be estimated from the long period radiated seismic spectrum (Bullen and Bolt 1985). Empirical relationships have been established between moment magnitude and other magnitudes (Figure 14.3). Very few moment determinations had been made for earthquakes prior to 1977 although the moment tensor (Backus and Mulcahy 1976) is now being given in the monthly *Bulletin of the National Earthquake Information Services* (NEIS) of the United States Geological Survey.

Seismic moments using data from the most recent seismograph and accelerometer networks have been studied by Ekstrom and Dziewonski (1988) and compared with over 2000 surface wave magnitudes showing that the *Kanamori* relationship* does not fit the full set. Regional

* $M_w = \dfrac{2}{J} \log_{10} M_0 - 10.7$, M_w is the 'moment magnitude'

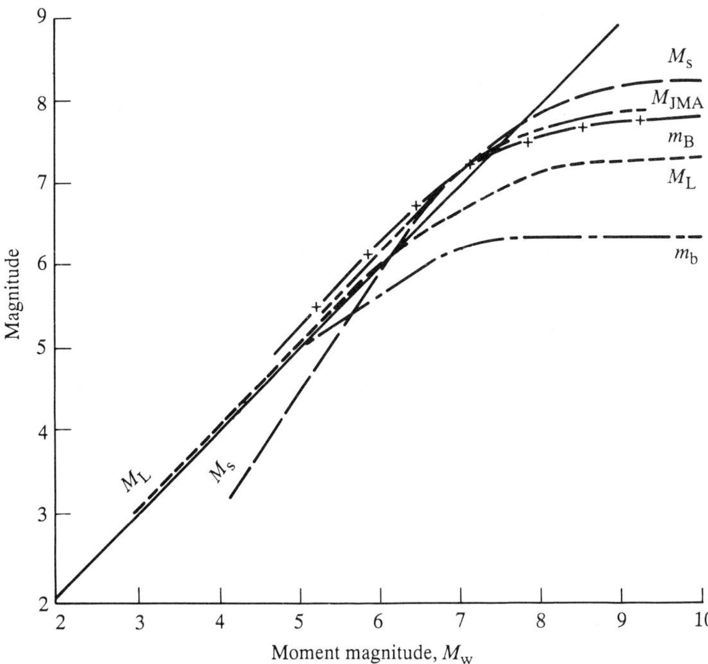

Figure 14.3 *Empirical relationships between various magnitudes and moment magnitude*

variations are such that estimates of the moment from surface wave magnitudes could be wrong by several times. This has implications for seismic hazard and the use of ground motion–magnitude–distance relationships derived from one region in another region.

Homogenization of magnitude scales in the assembled catalogue is a necessary precursor to proper hazard studies.

(b) Strong ground motion

The quantification in magnitude terms so far described depends on the identification of particular wave types, the varieties of which are described in a number of textbooks (e.g. Bullen and Bolt 1985). The closer one approaches to the source of an earthquake the more difficult it becomes to identify these wave types and the motion may appear to be random, or more correctly, non-stationary random, although there is increasing evidence that the direct shear wave predominates (Westaway and Smith 1988). The engineer would, ideally, like to have the time history of surface ground motion in three orthogonal directions, temporally and spatially. This becomes even more desirable when the engineering construction is large compared with the lengths of the seismic waves carrying significant energy. Although it is in principle now possible to undertake sophisticated modelling of the displacements arising from a sudden fault rupture, even to match the spatial form of the wave field, it has only been in the last few years that observations of the spatial and temporal relations of strong ground motion have become available. The deployment of arrays of strong motion instruments has enabled the degree of coherence between vibratory motion at different frequencies to be examined. Coherence is a measure of the order in a transmitted wave; a simple sinusoidal wave on a large deep lake spreading from the small dropped stone is coherent and the motion at one point on the surface can be related deterministically to the motion elsewhere. The relevance of this is apparent later when we consider how far it can be assumed that all parts of a rock mass move at the same time in an earthquake. From the few stong-motion arrays (SMAs) around the world it appears that in the near field of an earthquake the coherence is rapidly lost at frequencies of more than a few hertz.

Many relationships have been developed in the past between peak ground surface acceleration, velocity, distance and magnitude and there are several recent reviews (Ambraseys and Jackson 1984; Skipp and Ambraseys 1987) and they continue to appear (Kriznitzsky *et al.* 1988; Westaway and Smith 1989) (Figure 14.4).

The use of a single kinematic parameter to define a ground motion is clearly a gross oversimplification; frequency content and duration are ignored. A definition of action which contains more information is the ‘elastic response spectrum’. The response spectrum can be

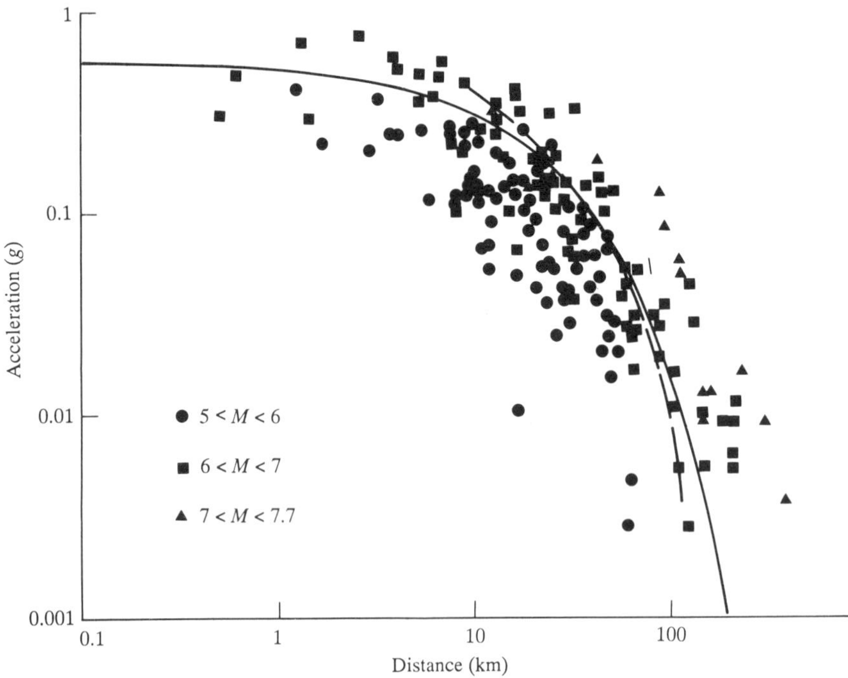

Figure 14.4 *Acceleration versus magnitude and distance (after Joyner et al. 1981)*

thought of as the maximum response of a series of inverted pendulums when their common base is shaken by a particular strong ground motion. A design response spectrum is made up of a number of such spectra using different strong-motion records. The envelope to such a compilation, or some statistical perecentile, is taken as describing the seismic action. A common way of presentation is on tripartite logarithmic paper normalized to 1.0 g (Figure 14.5). For the convenience of earthquake engineers such spectra have been developed from strong-motion data grouped according to the ground conditions at the originating recording seismographs and within a band of magnitudes appropriate to the region. Such spectra are 'anchored' to the zero period peak ground surface acceleration or velocity. Seismic action of this kind is used in conjunction with estimates or calculations of the periods of the structures under excitation. The procedures for the calculation of the complete response of structures to ground motion using response spectra are well described in several texts (Key 1988; Dowrick 1988). The methods can be extended to ductile responding systems but are most appropriate for systems with linear response where natural periods can be calculated.

More information on ground vibration is provided by Fourier amplitude spectra which can be related to response spectra. In recent years there have been attempts to charcterize ground motion and particularly its damage potential in terms of the power spectrum (Chang 1987).

The choice of an appropriate set of strong motion records both to develop a response spectrum and to use directly as time history input is a matter calling for considerable skill and judgement. Data banks are available at various national and academic centres (Imperial College of Science and Technology, London; McMaster University, Toronto, for example).

Ideally the hazard evaluation should point to the kind of earthquake which is most likely to be troublesome. A procedure in which the 'most likely to be felt earthquake' is identified can help. (Burton 1980, 1988). Often two 'design' earthquakes are identified – a small to moderate event which is likely to be experienced at least once during the life of the structure or facility and a less probable but much larger one. Knowledge of the seismotectonics of a region can help inform the judgement as to the suitability of strong motion records from other regions. With some idea then not only of the peak ground accelarations but of the typical magnitude range a rough estimate can be made of likely duration and even 'predominent' frequency of the troublesome earthquakes.

There have been many attempts to define and characterize an 'effective duration' of strong shaking. The *bracketed duration* is defined as the elapsed time at a particular frequency between first and last acceleration excursions greater than a given level (say 0.05 g) and the *cumulative duration* at a particular frequency is the total time for which acceleration exceeds a given value (Bullen and Bolt 1985). Bracketed durations (f>1.0 Hz) within

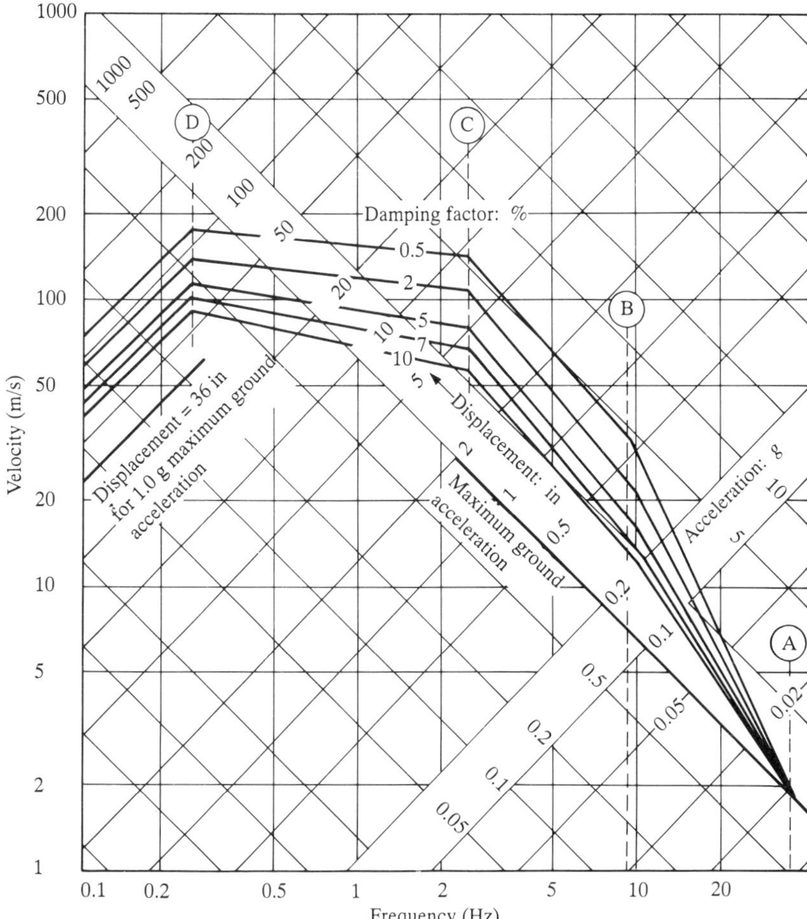

Figure 14.5 *Design response spectra, example nuclear related*

24 km of the fault rupture are unlikely to exceed the following values for accelerations of greater than 0.05 g and 0.1 g respectively:

$$D = 15.5 \tanh (M - 6.5) + 19 \tag{14.3}$$

and

$$D = 7.5 \tanh (M - 6.0) + 7.5 \tag{14.4}$$

Use is then made of peak acceleration (or some kind of 'effective' acceleration which takes account of the energy in the acceleration peaks) with a response spectrum either of a regional character or developed with the specific locality or structure in mind, or a set of suitable time histories, all chosen with reference to the risk and hazard levels calculated or judged to be appropriate.

A statistical approach has been suggested by Vanmarcke and Lai (1980) with duration related to root mean square (RMS) amplitude.

14.2.3 Ground rupture

The popular image of an earthquake contains yawning fissures swallowing houses and people. An earthquake is generated on a rupturing fault and depending on the size of the earthquake, its focal depth, the style and rupture progress on the fault, there can be surface expression, sometimes dramatic. The identification of such ruptures and their differentiation from the effects of slope instability is now an important part of field seismic geological studies.

Empirical relationships between surface fault break and magnitude have been developed (Bonilla *et al*, 1984; Bonilla, 1988). The problem can be treated probabilistically (Campbell *et al*. 1984). Although a way of designing to accommodate the demand of sudden concentrated displacement is conceivable; for many facilities even the remote possibility of such an event would preclude the use of the site (International Atomic Energy Agency 1979).

Engineering geologists are much concerned in the formulation of rational procedures for identifying faults as 'active', 'extinct' or 'unproven' (Skipp 1987; Mallard *et al.*, 1991). The issue of fault rupture is especially relevant to underground works which would otherwise be very tolerant of vibratory ground motion.

14.3 Analysis – an overview

14.3.1 General

The foregoing pages have outlined the hazard and risk considerations and the kinds of seismic action which may be adopted in evaluating the response of rock itself, or of a structure on rock or in rock. But 'rock' as a material has hardly been mentioned except with reference to site-related response spectra.

The transmission of elastic waves through the crust is described by mathematics which assumes the rocks constitute a continuum, albeit with layers of different stiffness, which may be anisotropic and have different attenuation properties. The micromechanics of rocks enters into basic studies of fracture mechanics but even here continuum models have an important role. Except in the modelling of the physics of rupture at the seismic source it is usually possible to accept the assumption that the energy is transmitted through an effectively elastic continuum.

Continuum models predominate in soil mechanics although the constitutive relationships are more complex than that of simple linear elasticity and there are numerical models which now accommodate the micromechanics of the particulate systems. Similarly in rock engineering there are now computer codes which model separate elements of rock and their interactions (Cundall 1976, 1980; Starfield and Cundall 1988).

The degree to which a continuum model is appropriate is largely a function of scale. As long as the domains of different properties within a rock mass are small compared with the wavelengths, continuum assumptions can hold good. This means that first-order estimates of the deformations induced by a travelling elastic wave can be made using the results of classical elasticity. Examples are:

(1) the response of regular shaped holes in an elastic halfspace;
(2) the response of layered strata to body waves of various angles of incidence;
(3) the response of surface features to incident body waves;
(4) the response of surface features to surface waves.

These solutions often assume plane waves and the results are given as an influence factor relating response at positions of interest to the far-field action.

14.3.2 Rock properties

When different layers are involved the body P and S wave velocities and the Poisson's ratio of the 'bulk' rock *in situ* are needed. These properties can be obtained by field geophysical methods (Geological Society 1988). In soil dynamics the *in situ* elastic parameters from seismic veolocity testing are regarded as referring to 'small strain' and are degraded as a function of induced (shear) strain. Similar procedures have been proposed for rocks with very little experimental evidence and some computer codes have such a relationship as a default condition. A recent review of the available data for a 'soil type' hyperbolic relationship between shear modulus and shear strain for rock has cast some doubt upon this procedure (Eldred pers. comm.)

Generally the rate effects in the response of bulk rock to earthquake loading are not significant. This may not be the case on discrete slip surfaces (Hencher 1981, 1985; Crawford and Curran 1981, 1982; Vaughan *et al.* 1985; and Figure 14.6).

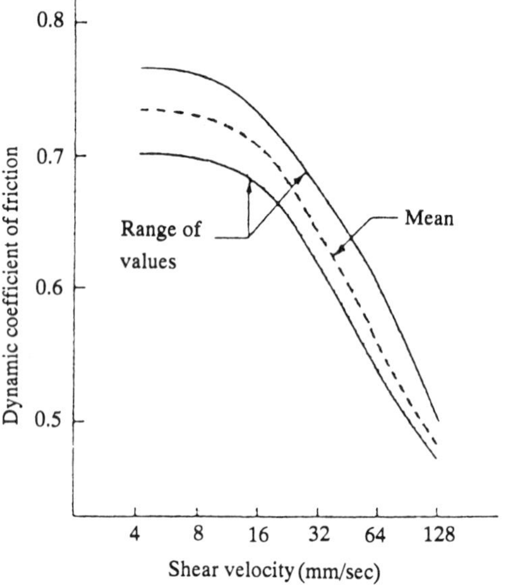

Figure 14.6 *Frictional resistance and velocity of rock surfaces (from Crawford and Curran, 1982)*

Failure of rock under seismic loading usually involves the relative movement of blocks on discontinuities although for purposes of analysis in soft rocks or very disaggregated materials, failure mechanisms similar to those of soil mechanics are adopted. When a 'cohesionless' model is used the angle of shearing resistance of the rock in bulk has to be estimated and values are usually derived from experience with large-size rock fill.

Under static conditions the principle of effective stress is operative and one would expect that a full description of the response of a rock mass with water-filled discontinuities to seismic loading should also consider effective stress, and computer codes are available which do so (Table 14.1). Interaction between pore–water pressure in discontinuities and relative velocity of the blocks is complicated and may also involve heating (Vaughan *et al.* 1985).

As in soils, there is a residual strength to a rock discontinuity which is reached with large relative movement. In rock masses with discontinuities in this condition rock falls may be triggered by relatively small vibrations (Brawner 1980). It has been suggested (Woodward 1988) that the residual angle, φ_j, used in the Goodman (1980) expression for maximum safe slope angle, α ($\alpha = (90 - \delta) + \varphi_j$, where δ is the dip angle of the joint planes) is over-conservative, but could be appropriate for dynamic loading. However, it is rare that the strength of the intact

Table 14.1 Computer programs for dynamic rock engineering

Code name	*General type*	*Origin*	*Dimension*	*Large deformation*	*Seepage flow*	*Jointed rock*
ADINA	General FE	USA, ADINA Eng.	3D	Yes (specl)	Yes	Yes
ANSALT	Geomech. FE	FRG, Control Data GmbH	3D	Yes	No	No
BRAST	FE plane strain	Japan, Osaka University	3D	No	No	No
DAYS 2	FD approximation to equs of motion dyn fract.	Japan, Kyoto University		No	No	No
DEC	FD discrete element	USA, ITASCO	2D, 3D	Yes	Yes	Yes
DSIM	FD wave propagation interaction	FRG, Battelle Institut	2D	No	No	Yes
DY2	FE response analysis	Japan, CRIEPI	3D	No	No	No
ESI-GEOMLIB	FE program library for geomechanics	FRG, ESI ESI GmbH	2D, 3D	Yes	Yes	Yes
FDND	Vertical deformation of rock mass, in slope analysis	USA, MIT	Cylinder	No	No	Yes
FELAN	General FE static & dynamic analysis	Japan, New Japan Eng. Cons. Inc.				
FEST 03	General FE static & dynamic stability analysis	FRG RWTH Aachen	3D	No	Yes	Yes
HONDO III	FE for transient dynamic loads	USA, RE/SPEC Inc	Axisym	Yes	No	No
LAYER	FD shock waves in plane and axisym. horizontal layers	USA, Weidlinger Associates	Axisym	Yes	No	No
NT6PS	General FE for stress wave propagation	China IR&SM Ac. Sinica	Plane strain	No	No	No
NURBM	FD distinct block	US, North Western University	2D, 3D	Yes	Yes	Yes
PAM GEOM	FE non-linear static and dyn. geological media	France, ESI	Plane strain	No	No	Yes
ROSALIE/CESAR	FE library	France, LCPC	2D, 3D	Yes	Yes	Yes
SIGNAS (1982)	General FE	Japan, Toyo Joho System Ltd	Plan strain, stress axisym.	No	Yes	No
SSTIN-2D, 3D	FE soil structure	USA, Univ. Ariz.	2D, 3D	Yes	No	Yes
SWARS-2PM	Sliding wedge for rock slopes seismic forces (also SWARS and SWARS-2MC, Monte Carlo version)	USA, MIT	2 plane	No	Yes	Yes
WONDY	FD wave propagation blasting	USA Sandia National Labs.	1D	Yes	No	No

strong rock need be considered where the loading is from earthquakes although tensile spalling of rocks is a matter of concern when blasting is going on nearby (Labreche 1983) and under some circumstances a tensile failure mechanism would be considered. There are cases where the properties of the intact rock come into the calculations: for example, when a deep cavern in homogenous crystalline rock is analysed using superposition of elastic static and dynamic stresses to define zones of likely departures from elasticity before embarking upon a more elaborate model.

In weaker rocks the design process may call for a determination of the failure properties of the intact rock. The classic Mohr–Coulomb limits have been used but the empirical Hoek–Brown (Hoek and Brown 1980; Hoek 1983) criterion has perhaps found wider favour in general rock mechanics practice including soft rock. The criterion takes into account both intact and irregular jointed rock mass and has been incorporated in non-linear finite element formulations in a simplified form (Pan and Hudson 1988).

Progressive failure has been mapped in terms of a limiting shear strain for a material with a 'drooping' stress–strain curve having a marked difference between peak and residual strength (Scott 1987).

The dominant role of discontinuities in the static stability of rock masses has directed attention to their quantification prior to the application of block mechanics (Goodman and Shi 1985). In its simplest form this has involved detailed mapping and evaluation using hemispherical projection (Priest 1985). Derivation of statistical descriptors of three-dimensional distributions of discontinuities from the one-dimensional (boreholes) and two-dimensional (exposures) is a continuing challenge to statisticians. There is a considerable literature on the issue of joint mapping in three dimensions from one- or two-dimensional exposures (Fossun 1985; Wu 1987).

There is a problem of scale; at one scale a regular set of joints may be evident, at another scale a sequence of kinematically related faults, so before going down a probabilistic route to assessing vulnerability it is prudent to identify such features which may control vulnerability. Rational analysis may then make considerable demands upon the structural geologist (Matheson, 1988).

14.3.3 Numerical procedures — general considerations

Numerical analyses of dynamic rock engineering problems can be divided into the following categories:

(1) *pseudo-static*, where the peak acceleration due to an earthquake is applied at some selected critical position, which my be at the centre of gravity of a potentially sliding element identified by the procedures of static rock stability analysis (Goodman, 1980; Goodman and Shi, 1985; Priest, 1980, 1985; Hoek and Brown, 1980).
(2) *simple dynamic*, where a natural period of the structure is established and an elastic response spectrum is used to estimate the maximum displacements.
(3) *full dynamic*, where solution of the wave equation leads to an estimate of punctual stress, strain or displacement. In the simplest case this may consist of estimating stress and strain directly from particle and propagation velocities in a free field.

The pseudo-static procedures may incorporate estimates of strain or stress magnification deduced from classical continuum expressions and yield criteria which may be for intact rock, the rock mass or critical discontinuities.

There are numerous computer codes available using finite element or finite difference methods and a summary is given in Table 14.1 (see also Christian 1987). Many of these codes are of a general-purpose category with some provision to deal with time histories so as to effect a solution of the wave equation. Others permit the natural frequencies of the system to be established and the point displacements are arrived at by rules of mode combination in a way analogous to the procedures well established in structural dynamics and earthquake engineering (Key, 1988; Dowrick, 1988). In setting up these models great care should be taken to describe the boundary conditions, in particular any radiating boundaries. Modelling of considerable detail with finite elements can be combined in the so-called 'hybrid' methods with lumped half-space modelling of the far field. This permits not only detail but local non-linearities to be dealt with (Wolf, 1985, 1987).

It should also be noted that codes familiar to earthquake engineers dealing with soil structure interaction (e.g. SHAKE, FLUSH) can be used for rock engineering where explicit modelling of discontinuities is not warranted.

In the procedure listed above the dynamic input is based on continuum assumptions and where finite element and finite difference methods are used the detail that can be accommodated depends upon the amount of money available for analysis and the degree to which that detail can be specified.

Just as for soils a particulate approach requires the definition of sets of conditions at particle contacts, for rocks the inherent discontinuities are joints and faults which delineate the distinct elements of the mass. Numerical treatment of the assemblages of elements bounded by such surfaces has been under development since the early 1970s (Cundall, 1976; Dowding *et al.* 1983) and several programs are now in operation allowing inclusion of the constitutive relationships for the joint interfaces, coupling with joint water pressure and compressibility of the distinct elements. An example of the use of such a program has been given by Lemos *et al.* (1987). Programs are also now available which take into account the dilatant behaviour of discontinuous rock masses and will be referred to later in this chapter with respect to caverns.

In all the numerical procedures outlined above there are a number of common questions which have to be posed and if possible answered:

(1) What is at rest stress state?
(2) What is the constitutive relationship for the rock mass, and if needed, the surfaces of the discontinuities?
(3) What failure criterion should be adopted?
(4) Where should the dynamic input be applied and in what manner?

The answers to each of these questions depend strongly on the type of problem facing the analyst and even the specific circumstances of each project. In the following sections which will deal with different categories of engineering works these matters will be addressed.

14.4 Earthquakes and foundations in rock

Foundations in rock present few if any problems to the earthquake engineer if he is dealing with level ground and sound rock. Where there is strong relief there may not only be the effect of the so called 'topographic' magnification but the conjoint circumstance that such relief is also related to geological control and potentially, unstable pinnacles and cliffs (see Ortolan *et al*. 1988; Rodriguez 1988; Zang Chouan and Zhao Chongbin 1988).

Seismic hazard is usually expressed as peak ground acceleration on 'firm' ground or rock and site-dependent response spectra are commonly used (Figure 14.7). The selection of a site-dependent design response spectrum is guided by the ground profile with the properties of the ground often being specified by a shear-wave velocity. *In situ* small strain shear modulus of the rock mass is now usually obtained from crosshole to downhole seismic testing (Geological Society 1988). In Table 14.2 typical values of this shear wave velocity are indicated for a range of rock types.

Large and important structures such as nuclear power stations and dams may warrant a full soil structure interaction analysis. Reference has already been made to some of the more commonly used computer programs and the subject of soil structure interaction has generated an extensive literature and several authoritative treatises (Wolf 1985, 1987).

The numerical procedures used permit the incorporation of variable rock stiffness, but realistic detailed modelling of such items as gouge-filled faults presents formidable difficulties. Fortunately important engineering structures are required to be sited clear of such features although from time to time the need arises to model a situation where a fault is exposed in the excavations for a foundation.

The influence of steeply dipping bedding under a large dam has been considered by Sinitsyn and Samarin, (1985). They described the development of 'domains of unstable motion' using an energy method and allowing for anisotropy. The physical interpretation of the behaviour is that the strata 'bend' (Figure 14.8).

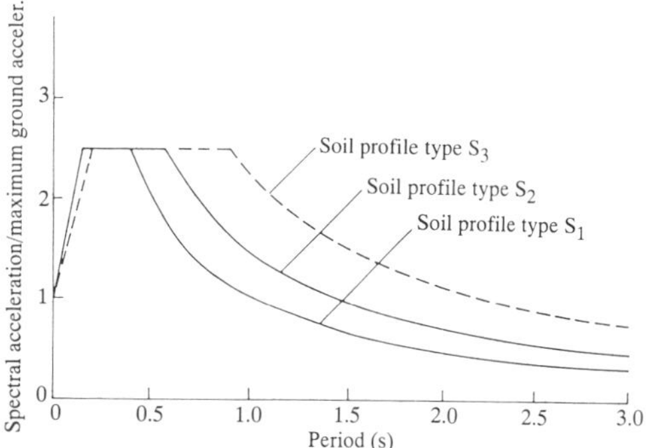

Figure 14.7 *Site-dependent design response spectra (ATC3 1978): S_1 – rock with shear wave velocity greater than 832 m/s (2500 ft/s) or with thin stiff stable soil cover; S_2 – deep (>66 m) cohesionless or stiff clay soils overlying stable deposits; S_3 – 10 m or more of soft to medium stiff clays and sands*

Table 14.2 Shear wave velocities

		(m/s)							
		500	1500	2000	2500	3000	3500	4000	4500
Sedimentary	Argillites	– – ——	——	– – –					
	Sandstones	– – – –	——	——	– –				
	Limestones			– – – ——	——	– – –	– –		
Metamorphic				– – –	– – –	– – – ——	——	– – – –	
Igneous					– – – –	——	——	– – – – –	

Notes: The ranges are for slightly weathered to fresh rock (BS 5930 1981) in the upper 5 km; for lower crust limiting values see Bullen and Bolt (1985), Table A.1

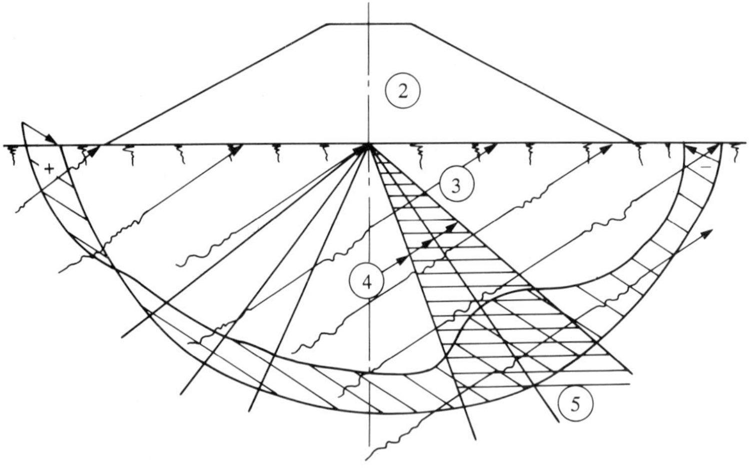

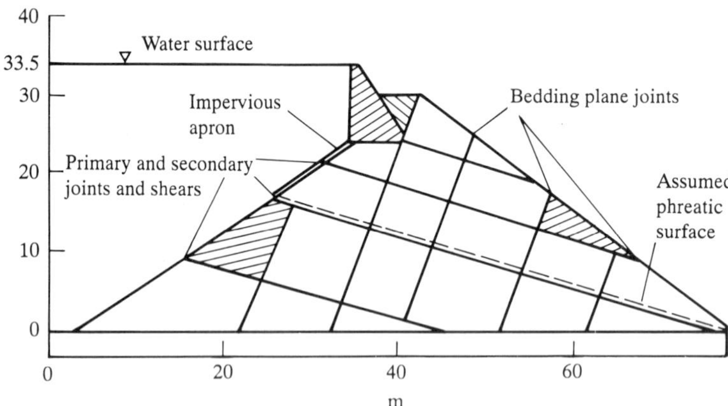

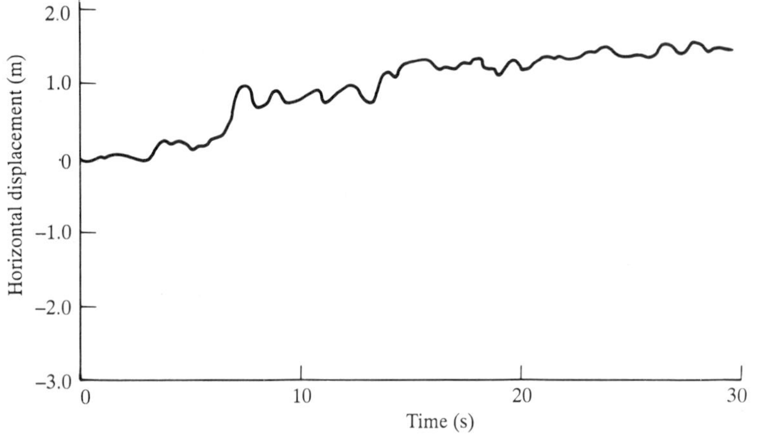

Figure 14.8 *Seismic energy flux normal to a stratified rock dam foundation: 2 = rockfill dam cross section; 3 = multilayer rock foundation; 4 = rays from earthquake focus; 5 = domain of instability (from Sinitsyn and Samarin 1985)*

Where rock is heavily jointed it can be modelled simply by using a reduced shear modulus. Circumstances can, however, be envisaged where a large element bounded by intersecting joint sets or faults can be regarded as 'detached'. Response can be modelled by treating that element as a discrete rigid block with compliant boundaries embedded in a half space. Pseudo-resonant frequencies and damping (material and radiation) are obtained using lumping methods and the displacement responses are calculated using an elastic response spectrum.

14.5 Earthquakes and natural rock slopes

Landslides, avalanches and rockfalls do not need earthquakes to initiate them but it helps. In hilly terrain slope instability can accompany even modest earthquakes and rock falls and slides are widely reported at Intensities above MSK (Medvedev-Sponhauer–Karnik scale) 7 to 8, and an earthquake of magnitude 3.2 is reported as having triggered a slide (Brawner 1980).

14.5.1 A scale classification of landslides

Many landslides involve both rock and soil, the mechanisms of one merging into the mechanisms of the other, and in many classification systems rocks and soils are treated together (Varnes, 1978). Two classification approaches will be considered below; the first on the basis of scale and tractability to engineering intervention, the second on the basis of the style of movement.

(1) *Great landslides*. Volumes of the order 10^9 *cubic metres* with m^3 dimensions of the order of tens of kilometres. Often develop rapidly with high kinetic energy and cannot sensibly be dealt with by engineering intervention or prevention. The environment in which they may occur has to be identified and avoided by major works.
(2) *Major Landslides*. Volumes of the order 10^7 *cubic metres* or less, with m^3 dimensions of the order of kilometres, often develop slowly and may be recognized as old features that are susceptible to reactivation but can be lived with if precautionary measures are taken. The recognition of such old features and planning restraints is usually the most economical course. In some cases they are amenable to back analysis and prognosis although they may involve more than one mechanism.
(3) *Minor landslides*. These usually have dimensions of hundreds of metres and may be numerous and widespread. They may be amenable at some cost to engineered mitigation and are often of a scale and mechanism for which analytical procedures are credible and even worthwhile.

(a) Styles of movement on rock slopes

Rock slope movements can be classified as translational and rotational with the toppling mode, unique to a rock slope being a form of rotational failure (see Chapter 11). 'Rockfall' is often regarded as a separate category with affiliations to 'toppling' (Whalley 1984). Block sliding, either on a plane surface or as a wedge, together with toppling, constitute the mechanisms which have been closely studied. Circular or near circular failures do take place in rock masses and may still be invoked in the design of rock excavations (Gonzales de Vallejo *et al.* 1984).

(b) Great landslides

The significance of great landslides in establishing current topography in regions of active tectonics has been recognized and a common association with plate tectonics, terrain and climate (Palmquist and Bible 1980) is to be expected (Figure 14.9).

The general point has been made by Voight and Pariseau (1977) that earthquakes effects could be cumulative such that quite small ground motions could ultimately trigger a slide. In such active margins as the western coast of South America land elevations may locally be increasing by 10 mm/y and great earthquakes (Ms>8.0), as well as great landslides, take place with significant changes in elevation and tilt.

It has been inferred that earthquakes were instrumental in a number of great landslides in the past. This inference is usually based upon indications of rapid movement. For example the Heart Mountain slide in Wyoming, probably of Eocene age, appears to have involved a bedding plane glide of 50 km on a 2.5 m band of dolomite resulting in scattered plates 1 km thick up to 8 km across jumbled in a way suggesting rapid emplacement (Voight and Pariseau 1977).

The Samar slide in the Phillipines (Figure 14.10) has also been interpreted as earthquakes initiated (Wolfe, 1977), as has the Saidmarreh slide in Iran which affected an area of 166 km^2 and has a volume of 20 km^3 and maximum depth of 300 m (Harrison and Falcon 1937).

The association of old slides with earthquakes relies heavily upon subjective judgement although modern dating methods and the recognition of contemporaneous indicators of palaeoseismicity, ('seismites') are important guides. The application of fuzzy logic (De-yi Feng *et al*, 1988) is improving the confidence of such associations.

Great landslides dissipate vast amounts of potential energy, comparable to that radiated by large earthquakes, and they may generate moderate to small earthquakes themselves. Sokolenko (1977, 1979) has described the main types of great landslides, noting that they are often associated not only with ground shaking but also with ground rupture.

The scale of potential great slides and their geographical setting preclude all but the most general statements on prediction procedure. Old scars and foundered terrains

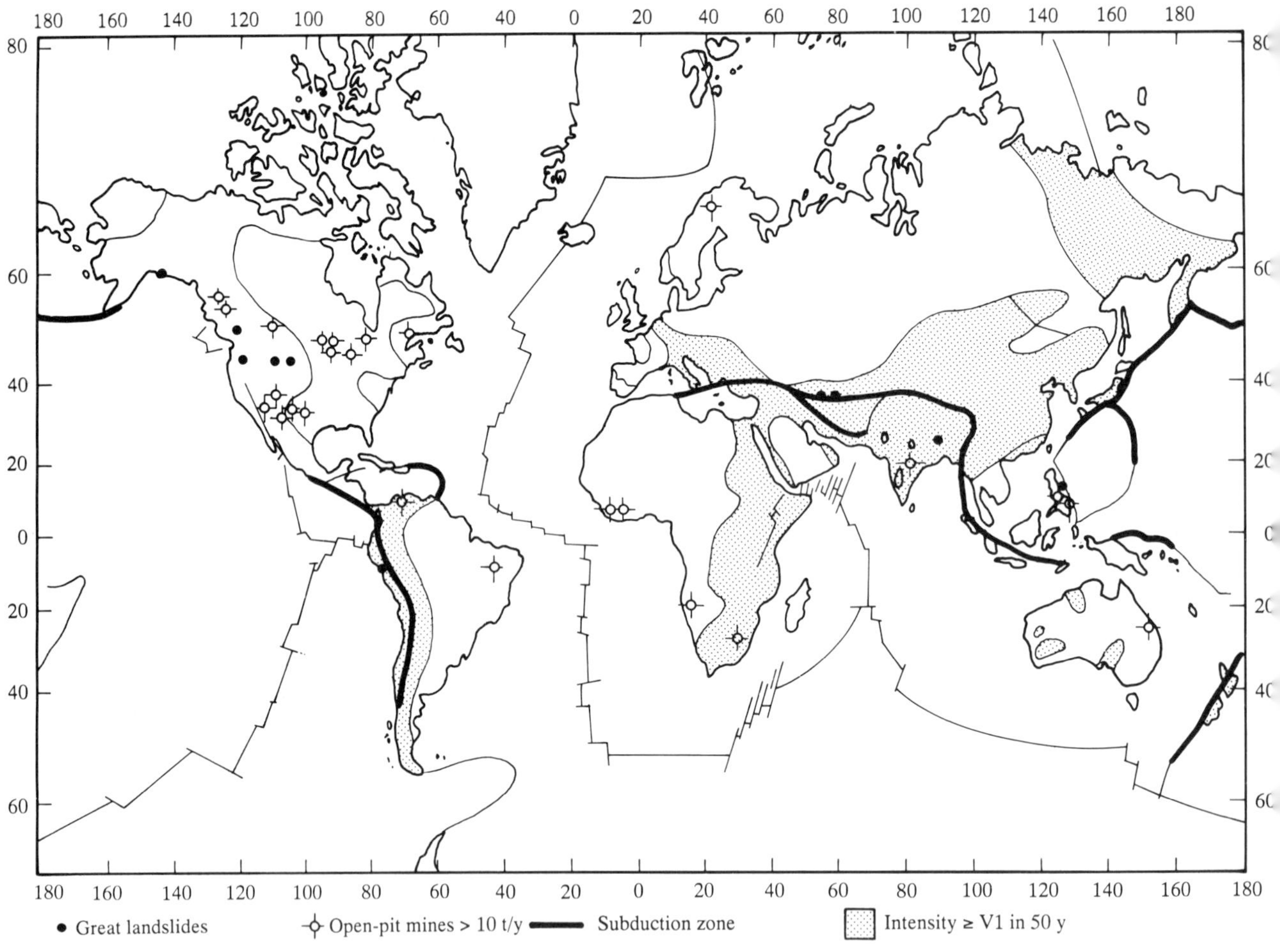

• Great landslides -◇- Open-pit mines > 10 t/y ▬ Subduction zone ▢ Intensity ≥ V1 in 50 y

Figure 14.9 *Great Landslides and global tectonics*

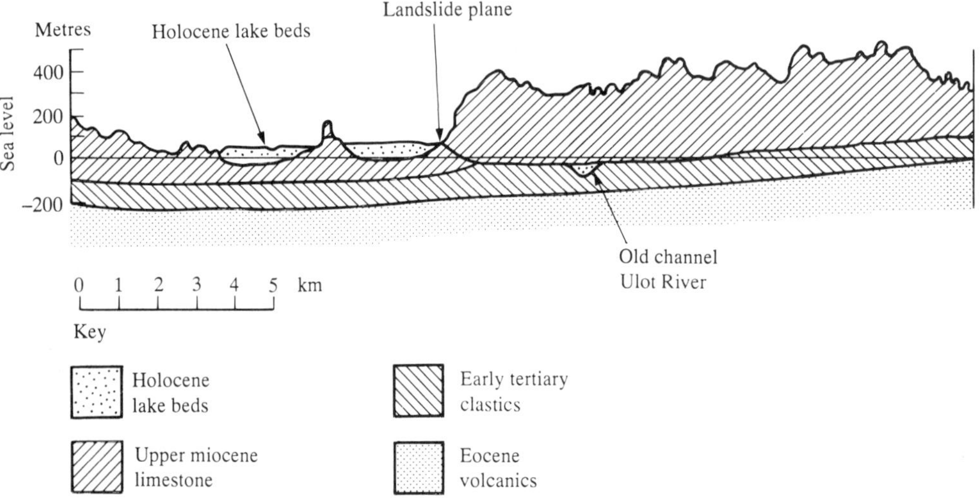

Figure 14.10 *The Samar landslide, Phillippines*

may be apparent on remote sensing and air photography. Sokolenko (1977) has dismissed mathematical modelling and analysis in favour of historical, comparative and statistical study, although recently monitoring and warning systems have been more seriously considered (Voight *et al*, 1987).

(c) Major landslides

In this somewhat ill-defined category rock-slope failures can be more easily isolated. It includes many historic landslips such as the slides at Sardis which were clearly exacerbated by seismic activity (Olson 1977).

Major rock slides may involve individual blocks having volumes of the order 10^6–10^7 m^3 and many currently identified have been active in the past. Major rock avalanches can be devastating; at Nevados Huascarin, Peru on 31 May 1970, the loss was 18 000 lives. Where the mechanism involves block sliding, analysis of current stability requires assessment of residual strength parameters for the potentially sliding surfaces. Individual major landslides can usually be mapped and dated. Some engineering analysis may be worthwhile but full stabilization is only rarely practicable (Anma *et al.* 1987, Ishihara and Hai-Lung 1987; Ito and Watanabe 1985).

(d) Minor landslides

A rock-slope failure may be small in extent but significant in its consequences. Toppling rock falls have been the cause of much loss of life in earthquakes in eastern Anatolia, where villages nestle under steep rock bluffs (Karaesman 1977; Ambraseys 1988). Large earthquakes may cause rock slips and rockfalls too numerous to count. The earthquake of 4 February 1976, in Guatemala, M = 7.5, initiated over 10 000 landslides, mainly rockfalls, on canyon sides and debris slides with individual volumes less than 1500 m^3 (Harp 1978).

The vulnerability of an individual rock slope to a modest scale failure under earthquake loading can be assessed when warranted, given some knowledge of the geology and geomechanics. Although for specific cases this can usefully involve analysis, a more pressing matter in many regions is to identify vulnerable localities over a large area. Statistical study of rock-slope failures and rockfalls has been undertaken by several workers (Youd and House 1978).

Some of the factors which predispose a rock slope to failing and leading to an avalanche under earthquake loading have been set out by Keefer (1984) and Figure 14.11 shows a plot of source slope height against source slope angle for 27 cases studied. Slopes most likely to produce rock avalanches during earthquakes are as follows:

(1) steeper than 25° and higher than 150 m;
(2) undercut by active fluvial erosion or geologically recent glacial erosion;
(3) composed of intensely fractured rocks;
(4) characterized by at least one other geological indicator of instability.

Keefer noted that although these criteria identify slopes that may produce rock avalanches only a small proportion will do so in a given earthquake; however, landslides may be expected up to 40 km away from the epicentre of an earthquake of M_l 6.6 (Wilson and Keefer 1985).

14.5.2 Rock-slope analysis

For a preliminary assessment of the vulnerability of a rock slope to earthquake loading it is useful to note the empirical formula proposed by Hoek (1976) which was based upon a 'pseudo-static' approach; the acceleration due to an earthquake acting horizontally towards the free surface of a slope:

$$F_d = F_s - 2.3a_h/g \qquad (14.5)$$

where: F_d = factor of safety under dynamic conditions
F_s = factor of safety under static conditions
a_h/g = horizontal acceleration as a fraction of gravity.

The factor of safety considered appropriate for a particular slope is largely a matter of engineering judgement, although in recent years the probabilistic implication of the factors of safety used by expert engineers has received considerable attention and the increasing use of probabilistic design has led to compilations of 'acceptable' risks for different kinds of rock slope (McCracken and Jones 1987) subjected to blasting. It is suggested that for unmonitored permanent urban slopes with free access a probability of failure of less than 0.5% could be accepted. This criterion would apply to a public access slope near to a planned excavation and would cover both static and dynamic actions.

More formal attempts to define the amounts of permanent lateral displacement which constitute different scales of disaster in Alaska are as shown in Table 14.3.

A quite modest earthquake may generate 0.2 g peak horizontal surface acceleration in its epicentral area so,

Table 14.3

Failure category	*Annual probability*	*Permanent displacement*	
		(m)	*(ft)*
1 Catastrophic	0.0001	3	10
II Major	0.0005	1.5	5
III Moderate	0.001	0.3	1
IV Minor	0.005	0.15	0.5

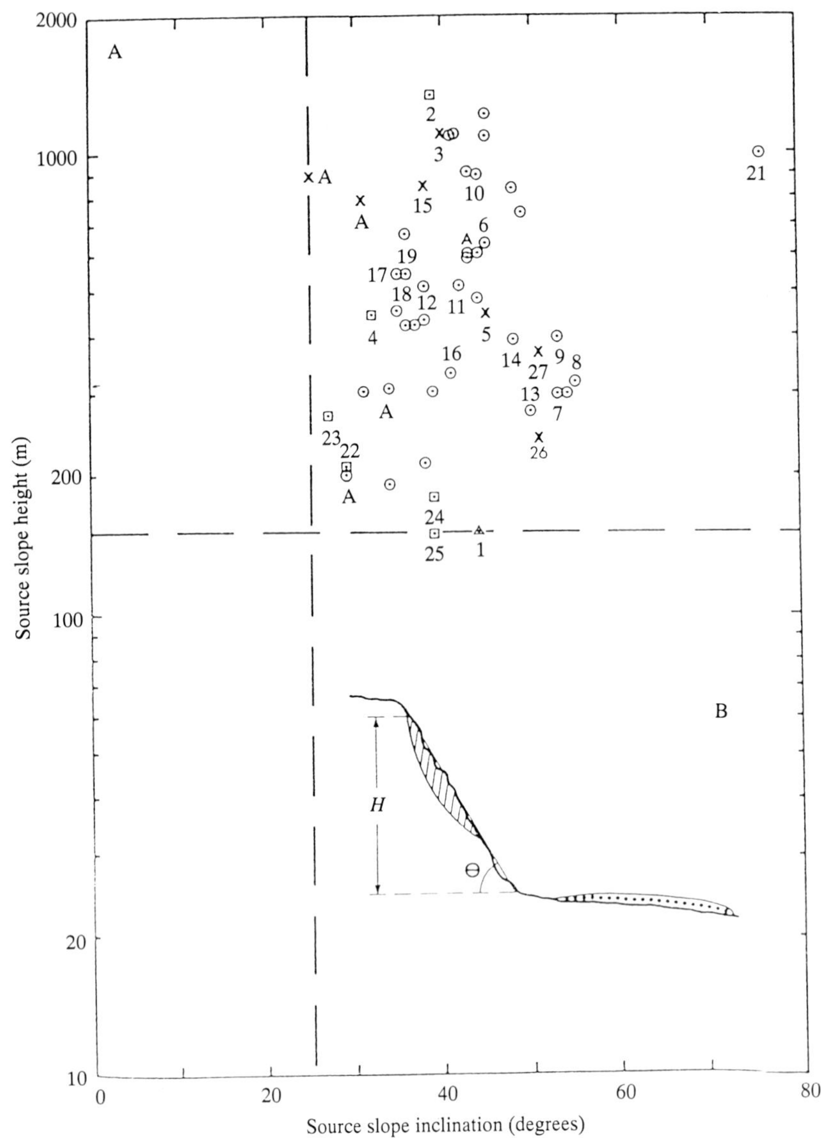

Figure 14.11 *Earthquake-induced rock avalanches, slope angle versus slope height (from Keefer 1984). Circles represent slopes undercut by active glacial erosion: squares by active river erosion: crosses, slopes undercut by Holocene or late Pleistocene glacial or river erosion.*

using Hoek's formula, a safety factor of say 1.5, which might be regarded for many situations as adequate, could apparently be seriously eroded.

(a) Block sliding

The tendency for higher and higher peak accelerations to be reported as strong-motion seismographs of ever-increasing dynamic range undermines the utility of the 'limit state' dynamic design or analysis which cannot live with safety factors less than one. It was realized more than two decades ago that for earthen embankments the criterion for survivability under earthquake loading should be based on deformation.

Newmark (1965) proposed a simple sliding block model where the resolved acceleration imparted to the block by motion of the underlying inclined plane would result in net downward relative motion of the block (Figure 14.12). This sliding-block model, directly applicable to many rock-engineering situations, has become the basis for

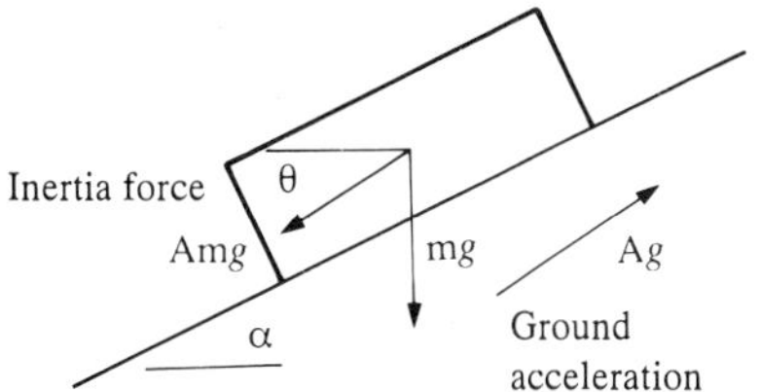

Figure 14.12 *The sliding block*

evaluating ground deformation in earthquakes without specific regard to whether the materials are classed as soils or rocks (Wang Sijing and Zhan Juming 1981).

The model assumes that the sliding resistance, once static friction has been overcome, remains constant. Then the net downward displacement depends upon the number of times the initial resistance is overcome during an earthquake and the duration and mean magnitude of the acceleration which is operative.

The static factor of safety of the block (Figure 14.12) is given by

$$F = \frac{(mg \cos \alpha) \tan \varphi}{mg \sin \alpha} = \frac{\tan \varphi}{\tan \alpha} \qquad (14.6)$$

and the factor of safety during an earthquake acceleration pusle uphill is given by

$$F = \frac{(\cos \alpha - A \sin (\alpha - \theta)) \tan \varphi}{A \cos (\alpha - \theta) + \sin \alpha} \qquad (14.7)$$

and for a down hill pulse by

$$F = \frac{(\cos \alpha + A \sin (\alpha - \theta)) \tan \varphi}{A \cos (\alpha - \theta) - \sin \alpha} \qquad (14.8)$$

The acceleration needed to just start sliding is denoted as A_c and if the factor of safety is taken as unity for both up and down conditions, then

$$A_{c,\,up} = \frac{\cos \alpha \tan \varphi - \sin \alpha}{\cos (\theta - \alpha) - \sin (\theta - \alpha) \tan \varphi} \qquad (14.9)$$

$$A_{c,\,down} = \frac{\cos \alpha \tan \varphi + \sin \alpha}{\cos (\theta - \alpha) + \sin (\theta - \alpha) \tan \varphi} \qquad (14.10)$$

When the earthquake acceleration is greater than the critical value there is relative displacement between the block and the slope;

$$m\ddot{x} = \text{driving force} - \text{resisting force}$$

$$\ddot{x} = \frac{g \cos (\alpha - \theta - \varphi)}{\cos_{\varphi}} (A - A_c) \qquad (14.11)$$

The factor of safety and the displacement are insensitive to Θ and acceleration can be taken as acting parallel to the slope (Sarma 1975)

Relative velocity and displacement are obtained by integration of the above equation of motion. For earthquakes then the total relative displacement is the sum of the displacements taking place during that period of time when the acceleration of each 'pulse' of the time history exceeds the critical acceleration.

Newmark (1965) scaled several strong motion records to obtain an upper bound and a plot of the displacements for a variety of ground motion time histories (not normalized) recorded before 1972 which yielded an upper bound relationship:

$$\log (u) = 2.3 - 3.3 \frac{k_c}{k_m} \qquad (14.12)$$

where u is in centimetres, and k_c (equivalent to A_c but more correctly called a coefficient) and k_m are the critical acceleration and maximum ground motion acceleration respectively. Sarma (1975) using a variety of pulse shapes to simulate earthquake motions and using four strong-motion records and several explosions, arrived at an upper-bound unsymmetrical displacement given by

$$\log \frac{(4u)}{(Ck_m g T^2)} = 1.07 - 3.83\, k_c \qquad (14.13)$$

where u is in centimetres, T is the 'predominant' half period in seconds and C is a factor which depends on the slope and the material properties of the sliding material and for rock slopes the interfacial characteristics of block and slope.

The sliding-block method has been applied by Makdisi and Seed (1977) and Franklin and Chang (1977) to a large number of cases.

The observations of ground displacement used in the above works did not discriminate between rock and soil slopes but in view of the underlying geometry of failure assumed differences would be evident only in the factors affecting post-yield behaviour at the interface.

Lin (1983) has suggested a model of ground motion based on an equivalent stationary process (Vanmarcke and Lai 1980) in which the ground is considered to act as a system with one degree of freedom which is defined by ground damping and ground frequency, together with two parameters associated with the degree of shaking, namely the RMS acceleration and equivalent stationary duration. Ground motion is considered to be random Gaussian and the rate of single direction crossing of the critical acceleration level is derived from random vibration theory. The outcome consists of two curves (Figure 14.13a,b) relating conditional expected permanent displacement and conditional distribution of permanent displacement of rock slopes as a function of the acceleration ratio.

Ambraseys and Menu (1988) have reviewed the observational data obtained in the last decade on level ground, producing a probabilistic estimate of ground displacement versus acceleration ratio k_c/k_m (Figure

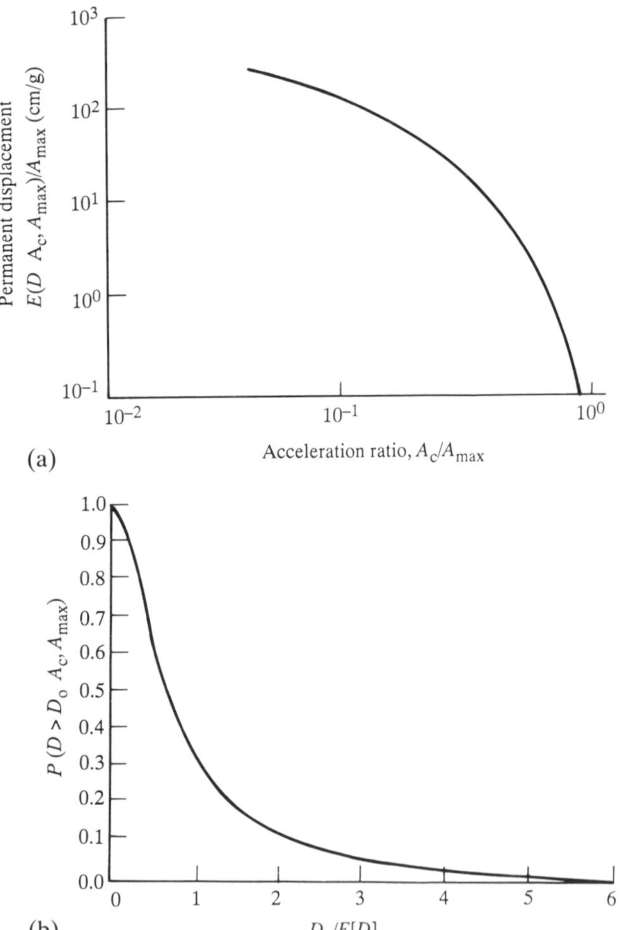

Figure 14.13 (a) *Conditional expected permanent displacement versus acceleration ratio;* (b) *Conditional expected distribution of displacement versus acceleration ratio (from Lin 1983)*

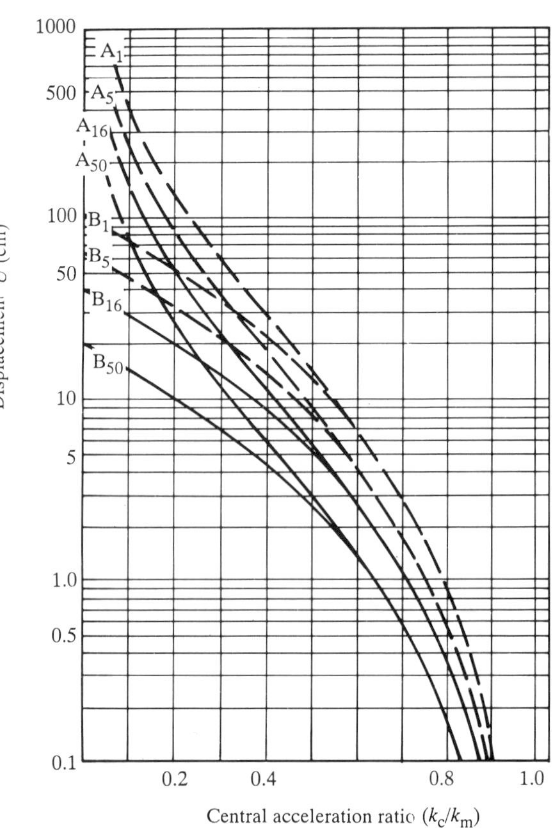

Figure 14.14 *Displacement versus critical acceleration ratio – for probabilities of exceedence 0.1 to 0.5, A, unsymmetrical: B, symmetrical. For* $M_s = 6.9\ (\pm 0.3)$ *(from Ambraseys and Menu 1988)*

14.14) and have suggested that two equations can be used for design purposes providing that k_c is based upon a residual strength:

$$\log(u_1) = 0.9 = \log[(1 - k_c/k_m)^{2.53}(k_c/k_m)^{-1.09}] = 0.3t \quad (14.14)$$

$$\log(u_2) = 1.31 = \log[(1 - k_c/k_m)^{2.96}] = 0.36t \quad (14.15)$$

(b) Limiting equilibrium or deformation criteria

The simple application of a peak acceleration to an identified block or blocks in a slope has been a common analytical procedure and can be extended (simplistically) to three dimensions with allowance for static water pressure (Srivastava 1977). The results may be conservative but require a careful mapping of the discontinuities. The adaptation of block displacement rather than limit equilibrium criteria calls for an appreciation of the mechanisms which could progressively develop and lead to catastrophic failure (Goodman and Shi 1985).

(c) Characterization of discontinuities

The characterization of joints and discontinuities in the rock mass is the essential precondition for any kind of analysis and is covered by a number of texts (Barton 1987), and the methods available for static analysis of stability of slopes has been recently reviewed by Richards and Atherton (1987) and some of the difficulties are pointed out by Hencher (1985). Statistical procedures are being actively studied (Long *et al.* 1987).

(d) Determinism and probabilism

Although in general engineering practice determinism predominates in static rock-slope analysis, probabilistic methods are receiving more attention (Pentz 1982; Priest and Brown 1983; Bolla *et al.* 1987) and fit well with the increasingly favoured probabilistic approach to the definition of seismic action.

(e) A different approach

Although methods based upon the sliding block model are the most widely used and can be extended to more complex (but kinematically admissible) elements, mention should be made of a different energy-based approach suggested by Borg and Schuring (1985) specifically with respect to rock avalanches. Borg and Schuring assumed that the rock mass is effectively cohesionless, the effects of pore water are ignored and the slide type is taken as a combination of shallow rotational and translational slipping (see also the 'clastic models' of Trollope 1968). The phenomenological model underlying the approach is summarized in Figure 14.15a, b, c. Seismic action is expressed in terms of the 'surface energy' which is considered to cause a transitory increase in volume of the near surface resulting in a momentary reduction in the shearing resistance.

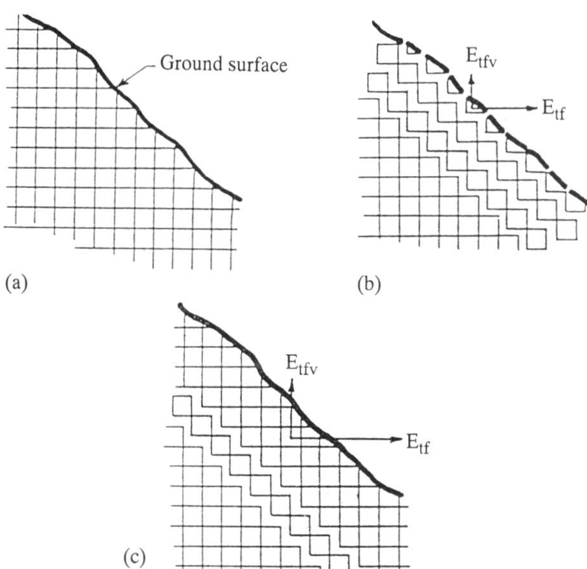

Figure 14.15 *Dilation–energy model for rock slope dynamic failure (from Borg and Schuring 1985).* (a) *Rock mass earthquake* (b) *Rock mass during earthquake, case I: uniform dilation* (c) *Rock mass during earthquake, case II: local dilation*

The reduction in the angle of shearing resistance, φ, due to dilation is taken as

$$\varphi = 0.59\gamma_d - 41 \qquad (14.16)$$

where γ_d is the dry density in lb/ft^3 with a specific gravity of 2.68. The amount of dilation is calculated for assumed geometries according to the axiom that the surface kinetic energy is equal to the change of potential energy of a defined volume of rock resulting in the expression

$$\text{dilatation} = \frac{\text{surface kinetic energy}}{\text{weight of unit column of rock}} \qquad (14.17)$$

Borg and Schuring (1985) developed an empirical relationship between horizontal surface energy in ergs/km^2 and modified Mercalli intensity. The dilation reduces the relative density, which in turn reduces the angle of shearing resistance and so the factor of safety. The procedure has undoubted attraction to a hard-pressed design office and Borg and Schuring reported that it was not inconsistent with three of the cases reported by Keefer (1984) for which adequate data was available.

(f) Rotational failure

Rock slopes, particularly in very shattered, weathered or soft rocks, can fail by rotation. Design charts for static circular failure are given by Hoek and Bray (1981) and for non-circular static failure the non-vertical slice method of Sarma (1979a) is particularly suited to rock slopes. The incorporation of earthquake loading depends upon the fact that the idealization of the sliding block on an inclined plane has a force polygon analogous to that of rigid-body mass sliding defined by a circular arc or a generalized failure surface subject to accelerations. Bromhead (1988) concluded that 'within the general framework of the limit equilibrium method "quasi-static" methods give similar results.'

(g) Toppling

Toppling failure under gravity loading is a principal mode of rock-slope failure. It consists of the overturning of a block and is thus a function of shape as well as mass. The condition for stability of an individual block is that the resultant forces must lie within two-thirds of the base of the block (Richards and Atherton 1987). Analytical procedures are still confined to particular situations; that described by Goodman and Bray (1976) incorporates the overturning of cantilever beams with interlayer slipping. Other toppling styles described by Goodman and Bray are 'block' toppling and 'block–flexural' toppling. Teme and West (1983) described several other secondary toppling styles; slide to toppling, slide-head toppling, slide-base toppling, tension-crack toppling and toppling and slumping.

The stability analyses require the determination of forces exerted on the susceptible element or column by columns higher up a slope and the resisting force by elements lower down. Using limit equilibrium criteria factors of safety for slide-toe toppling and slide-head toppling have been expressed in terms of the ratio of driving to resisting forces and could in principle incorporate earthquake loads as pseudo-static forces.

A precondition for toppling is a degree of columnar separation of an element from the rest of the rock slope and the full dynamic response of such an element, seemingly so well defined and simple, is unexpectedly complicated.

The problem has its analogue in the overturning of rigid blocks and monuments in earthquakes. Clark (1972) has discussed the displacement of boulders during the Borrego Mountain earthquake of 9 April 1968, and has undertaken a theoretical study of rocking and sliding of blocks on the ground. Bolt and Hansen (1977) dealt with the upthrow of objects in earthquakes (which has a bearing on the method of Borg and Schuring 1985, see above). He has demonstrated that in a simple elastic system plausible mechanical properties can be found that entail the separation of the bottom mass from the ground without invoking vertical accelerations greater than gravity. Separation of rigid bodies in a complicated multi-bodied system is probably more likely at modest accelerations. Recognition of such non-linear 'lift off' phenomena has troubled the numerical soil/structure analysts since the limitations of commonly used computer codes were recognized and must cast some doubt upon the adoption of oversimplified models. Tsao and Wong (1989) have examined the steady-state rocking response of rigid blocks to sinusoidal base motion and have shown that there are two classes of response, in-phase and out-of-phase with respect to the excitation, only out-of-phase solutions being stable.

Distinct element modelling (see below) has been used to study toppling failure by Ishida *et al.* (1987) and the analyses have been compared with both theoretical and empirical data.

(h) Finite element modelling

There are some examples of the use of dynamic FEM for slopes and one has been given by Archiprova (1987) where a portion of the slope was treated as an elastic structure and its natural frequency obtained for different boundary conditions. Response was then calculated by coventional normal-mode superposition. Toki *et al.* (1985) described dynamic slope stability analyses with a non-linear finite element method in which joint elements having non-linear properties are introduced between soil elements. Modal analysis requires a calculation of natural frequency which is not very meaningful for a very long slope. Most of the numerical studies of slopes are restricted to short well-defined features.

The role of hydrodynamic forces in rock-slope stability has been studied by Ghaboussi and Hendron (1984) with reference to submerged rock slopes under seismic loading.

(i) Distinct element modelling

Distinct element modelling was introduced by Cundall (1976, 1980) and has been extended to accommodate dynamic loading (Lemos *et al.* 1987). It will be treated at greater length in the section dealing with tunnels and caverns. Scott (1987) described the application of the discrete element program UDEC to the seismic stability of the Juncal ridge, California (Figure 14.16).

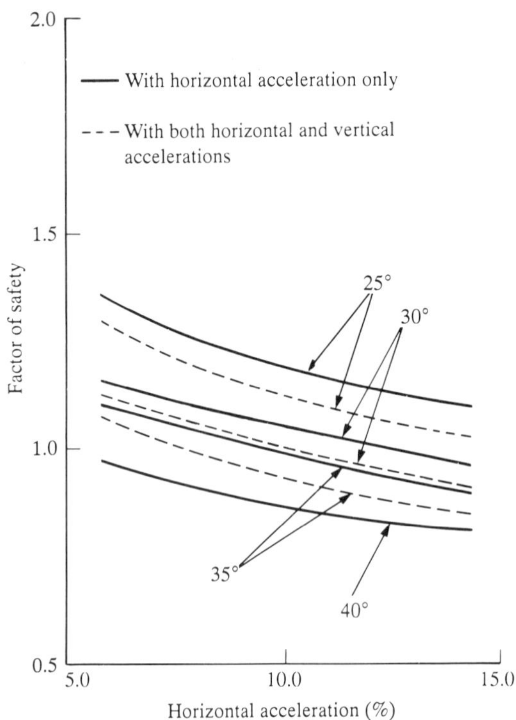

Figure 14.16 *Discrete element model for the Juncal Ridge, California, (from Scott 1987 referring to Lindvall, Richter and Associates 1982)*

(j) Reference point

For all models a point of reference for the free-field seismic action has to be chosen (Wolf 1985). For rock slopes it should be somewhere on the rock surface where the prescribed motion could exist. This precept is often not followed and motion is assigned to some depth where the model is then insensitive to changes in depth. Komada and Hayashi (1979) following observations at the Shiroyama underground power station concluded that it was not necessary to locate the seismic wave input deeper than the bottom of the cavern.

There is also the question of scale and wavelength. If the model assumes a shear wave traversing the basement at say 2000 m/s with a predominent frequency of 5 Hz in the near field of an earthquake, a slope of greater lateral extent than 400 m will not be excited at effectively the same instant depending upon the coherence of the radiation.

Similarly if the excitation is treated as a surface wave in materials of wave velocity say 400 m/s with a frequency of 2 Hz the wavelength of 200 m ensures (given coherence) that different parts of a block longer than 200 m on a slope may not be moving in thc same direction. However, the possible non-coherence of the vibration at such frequencies in epicentral areas has to be remembered, just as for structures the travelling-wave factor should be taken into account for long slopes. Such waves have been examined numerically by Dowding and Gilbert (1988) for an ideal sinusoidal motion with single and multiple block models. They conclude that the length of block that can be coherently excited is about one-quarter of the wavelength. Low coherence can be expected for frequencies above a few hertz.

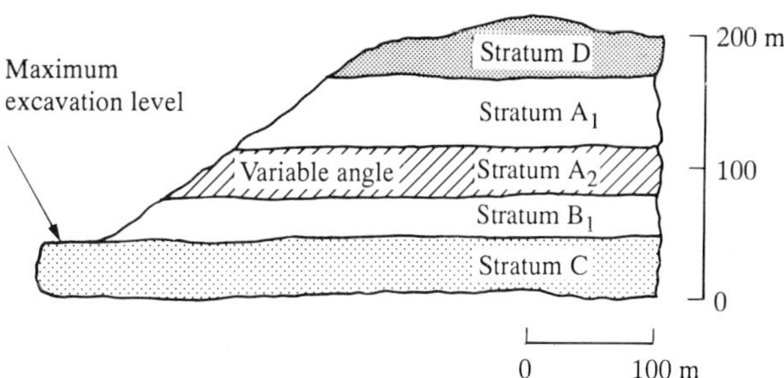

Figure 14.17 *Open pit coal mine (pseudo-static analysis) factors of safety versus assumed friction angle and acceleration. Material: A_1 cemented Sand, A_2 cemented Silt, B_1 Lignite, B_2 Clay, C Marl, D Sandstone (from Gonzales de Vallejo et al. 1984)*

14.6 Earthquakes and open-pit mining

Although the analysis procedures available to the designers of open-pit mines are essentially the same as for rock slopes there are important differences in the way the 'acceptable' risk and hence the design hazard is arrived at.

14.6.1 Items at risk

There are many items within an open-pit complex at risk to earthquakes but only those related to the rock slope will be considered here. There is the working face and especially the haulage roads and the ground upon which costly processing plant may be standing.

14.6.2 Evaluation of vulnerability

Deterministic methods are still widely used (Gonzales de Vallejo *et al.* 1984) with continuum assumptions although probabilistic methods are becoming favoured (Glass 1981); Herget 1982) because it then becomes possible to carry out an economic analysis. The main uncertainty in defining vulnerability lies in the variability of shear strength mobilized on a failure surface. In a static plane-strain analysis the risk of failure on a long slope length B can be described in terms of the ratio B/d where d is the width of the failure zone, the mean factor of safety along the slope and the coefficient of variation of the factor of safety. Typically for $B/d = 5$ and a factor of safety of 1.3 a change in the variance of the shearing strength of from 0.25 to 0.3 increases to probability of failure from 2×10^{-3} to 1×10^{-2}.

Some indication of the ranges of risk involved in the assessment of the economic impact of seismic factors on the design of open-pit mining has been given by Skipp (1984). Gonzales de Vallejo *et al.* (1984) (see Figures 14.2 and 14.17) have described the use of a probabilistic hazard evaluation in conjunction with a deterministic pseudo-static design analysis using a modified Bishop circular slip procedure with vertical and horizontal accelerations applied to each slice. An average seismic action consistent with a return period ten times the operational life of the mine was selected together with an action for the 1000-year return period. A factor of safety of 1.3 was chosen for the average action and 1.0 for the 'peak' action. A simple economic analysis was carried out with the results shown in Table 14.4.

It was concluded that the cost benefit expressed as the simple ratio of the spoil volume cost of the flatter slope to restoration cost was 0.32 for a strip-mine layout although for a non-strip mine operation the cost benefit could be as high as about 2.

Table 14.4

Design condition	*Average slope* (degrees)	*Total spoil* (m^3)	*Cost* (£m)	*Cost difference* (£m)
Static	34	411	380	–
Dynamic	25	467	427	47

14.7 Earthquakes, caverns and tunnels

It has been the general consensus that underground facilities are less likely to be damaged by the vibratory ground motion associated with an earthquake than surface installations (Dowding and Rozen 1978) although some would not agree (Katayama 1985). It is easy to appreciate why this could be so. Horizontally polarized body shear waves which play a dominant role in earthquake strong ground motion emerging at the free ground surface will have double the amplitude that they have in the half-space (Bullen and Bolt 1985). For compressional (P) waves the ratios between the vertical and horizontal components of surface amplitude and the incident amplitudes is a function of the apparent angle of emergence of the rays. Surface Rayleigh waves decline in amplitude with depth.

The motions generated near the surface by emergent body waves, compressional and shear, are complicated by the reflection, refraction and wave-type conversions at stratifications and discontinuities. In many parts of the world, especially where there has been deep tropical weathering of rocks within several hundred metres of the present-day surface this may give rise to an upper layer of reduced shear modulus. Under such circumstances the near surface layer may amplify ground motion.

14.7.1 Analysis–continuum elastic background

The general types of analysis have been outlined above; now those procedures which have found most favour for underground works will be considered. It should, however, be recognized that in this field analysis has not played a key role in the design process, at least not rational analysis. Experience and precedent have ruled, expecting those special projects where strong sustained motion may arise from man-made events such as nuclear explosions (and where sophisticated analyses have been verified by full-scale or near full-scale testing), although advanced numerical modelling has been a feature of design by analysis procedures in the nuclear industry. Nevertheless with the development of new approaches to static cavern design and support and the growing power of computers, incorporation of dynamic loading into the design of more modest civil projects is now being called for.

The response of infinite cylindrical cavities in an elastic continuum to incident plane stress waves has been deduced by numerous workers (Pao and Chao 1973). Dynamic stress concentration factors related to hoop stress for the incident P-wave on a cylindrical cavity has the following complex form:

$$\sigma^{*}_{ee} = (R + iI)e^{-i\omega t} \tag{14.18}$$

The real part R yields the stress at time $t = 0$, the imaginary part I yields the stress at $T/4$, where T is the period of the incident wave. When the frequency $\omega = 0$ the imaginary part vanishes and the real part gives the stress concentration factor for biaxial static loading. Numerical evaluation of the stress concentration factors is undertaken for a range of dimensionless wave number

$$\alpha a = \omega a / c_p \tag{14.19}$$

where α is wave number, a is the radius of the cavity and c_p is the velocity of propagation of the P-wave. In Figure 14.18 the dimensionless hoop stress concentration factor is compared with the Kirsch static solution for Poisson's ratio $v = 0.2$ for two values of the normalized dimensionless wave number; $\alpha a = 0.2$, low frequency, and 3.5, high frequency. Where the wavelength is long compared with the diameter of the cavity the stress field in the vicinity of the cavity is near that of the static field. The dynamic stress concentration factor becomes higher than the static one at wave number between 0.25 and 0.3. With very high wave numbers the stress concentration factor approaches an asymptote – the cavity wall appears to the incident wave to be a plane boundary.

A special circumstance of the dynamic case occurs at $\theta = 45°$ and $135°$ where the stress becomes negative, that is, opposite in sign from the incident stress, where θ defines the radial coordinate.

The maximum dynamic stress concentration factors for steady-state vibration under incident P-waves are shown in Figure 14.18b as a function of dimensionless wave number for a number of values of Poisson's Ratio. It may be noted that there are a number of stationary values corresponding to natural frequencies of the various modes of the cavity and furthermore the maximum dynamic stress concentration factor is always 10–15% higher than for the corresponding static value for any Poisson's ratio.

The plane S-wave has been treated in a similar manner (Figure 14.19).

The transient response of a cylindrical cavity has also been studied by Pao and Chao (1973) who took a step input and examined the 'long-time' and 'early-time' behaviour. The maximum dynamic stress concentration factor occurred 'late', some three transit times after the onset of the wave and was about 10% greater than the static values. Incident plane shear wave steps have also been studied. Steady-state Rayleigh waves impart lower dynamic stress concentration factors but could generate tensile stresses.

In the simplest case, then, for a plane wave impinging parallel to the line of the tunnel (a two-dimensional case) the dynamic amplification of stress at the tunnel wall is negligible if the rise time of the stress pulse is more than twice the transit time across the tunnel diameter. In terms of the wavelengths of a simple periodic time history of particle velocity this reduces to a value eight times the tunnel diameter (Hendron and Fernandez 1983). If such conditions are met the action of an earthquake can be treated as a pseudo-static load.

The implications of this rule of thumb can be explored by assuming a maximum significant strong ground-motion frequency of 10 Hz that is giving the shortest rise time of one-quarter of the period of 0.1 s, that is 0.025 s. Assume that the most significant motion will arise from shear waves. Shear wave velocities in rock may range from 800 to 2000 m/s, giving a range of wavelengths from 20 m to 50 m. Hence unless the spans exceed these values the dynamic amplification factors need not be considered for typical earthquake strong ground motion. This conclusion needs to be qualified, however, for situations very close to a fault source, where there could be significant compressional waves with higher frequencies. Such high-frequency (>8 Hz) spikes are usually of short duration.

The response of long tunnels to other angles of incidence has been deduced by methods which do not account for any dynamic interaction. Kuesel (1969)

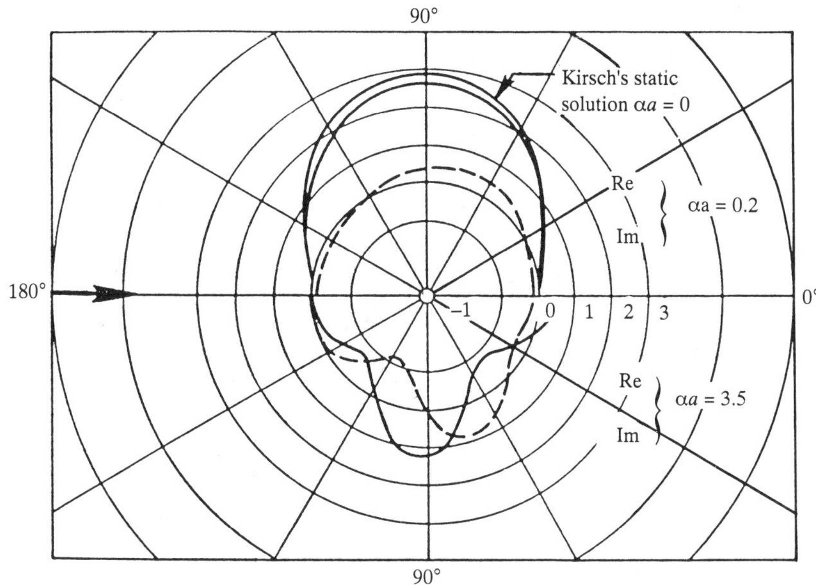

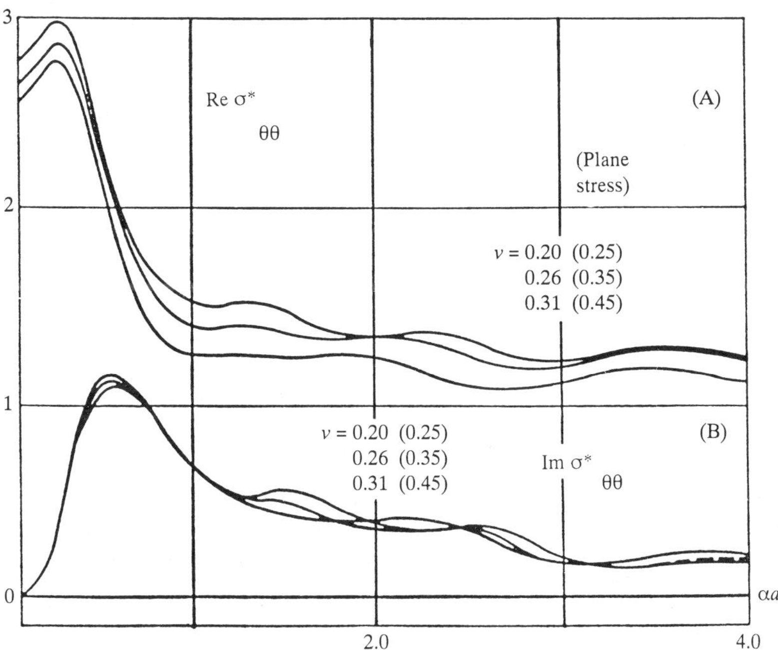

Figure 14.18 *Plane wave normal to cavity – distribution of dimensionless hoop stress* $\sigma_{\theta\theta}{}^*$ *at* r = a, ν = *0.26; solid line* t = *0, dashed lines* t = T/4*;* r = a, *radial distance equal to tunnel radius;* αa = *normalized wave number;* t = *0, time 0 with incident wave at maximum amplitude at* r = *0;* t = T/4, *time one quarter period shift from* t = *0, zero amplitude* r = *0; O = angular component of coordinate (π/2 = 90°)*

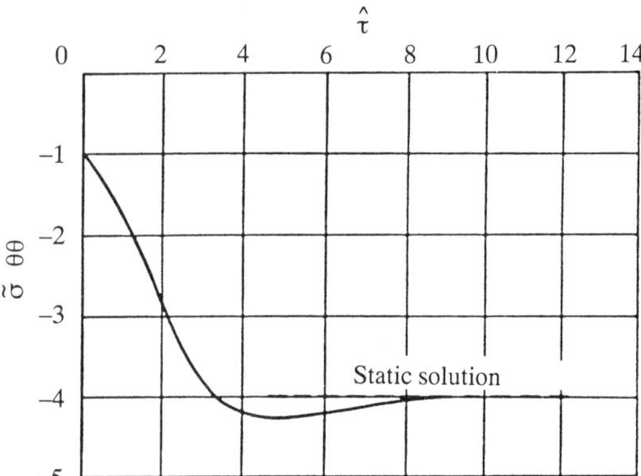

Figure 14.19 *Dynamic stress concentration factor at $\Theta = 3\pi/4$ for incident plane shear wave normal to cavity (from Pao and Chao 1973)*

showed that the maximum longitudinal strain in a tunnel wall arises from a shear wave front making an angle of 45° with the tunnel axis and is given by

$$\varepsilon = V_{max}/2C_s \tag{14.20}$$

where C_s is the shear wave velocity and V_{max} is the maximum particle velocity.

More complex analytical studies of the interaction of elastic wavelets with holes in continua are to be found in the geophysical literature, where such work has a direct bearing upon the interpretation of borehole sonic logging.

(a) Reference point

The location of the free field input, as for slopes and foundations, is a matter of concern. Reference has already been made to the findings of Komada and Hayashi (1979) who located their input at the base of the cavern.

The seismic input is usually represented by a plane wave propagating upward and can be applied as a stress wave at the base of the model which must then be made non-reflecting. The other model boundaries in rock must be far enough away, having regard to the level of material damping, to ensure that free-field conditions are satisfied. Alternatively the free-field condition can be enforced on them.

14.7.2 Simple dynamic stress analysis

Fujita *et al* (1982) examined the seismic stability of a large cavern at large depths compared with the cavern dimensions using a continuum model with seismic input at an assumed rigid base 200 m deep. The stability of the cavern during construction is evaluated using a finite element method assuming a visco-plastic constitutive relationship. Dynamic stresses were calculated, assuming that the strain around the cavern was not perturbed by the presence of the cavern, by establishing the ratio of the particle velocity to the propagation velocity for both P- and S-waves.

The maximum shear strain τ_{max} due to the shear wave is given by U_{max}/V_S and the maximum shear stress is $G\rho$ U_{max} V_M where ρ is mass density and G is the dynamic shear modulus. The maximum stress due to the P-wave is derived in similar fashion as $\sigma_{max} = U_{max}$ V_P.

Using these dynamic stresses the increment due to earthquake loading was expressed as a Mohr–Coulomb diagram and for the particular case studied, a cavern at 80 m depth in homogeneous strong rock was subject to earthquake particle velocity of 200 mm/s. It was shown that the dynamic stresses did not bring the cavern into a yield state.

Fujita *et al* (1982) concluded that even using an earthquake of the same size as would be used in the design of a nuclear power plant at the surface in Japan, the shearing stress increment would be less than about 35% of the static shearing stress. So, allowing for the increase in strength of rock with depth, a deep cavern should be more stable under earthquake than a shallow one.

14.7.3 Equivalent static analysis with non-linearity considered

Armed with some understanding of the likely dynamic effects it is possible to estimate the stability of large caverns even into the stress regime where non-linearities begin to show up. Chesnokov *et al.* (1986) described the analysis of caverns for a pumped-storage project based upon continuum mechanics, isotropic homogenous rock and plane compressional and shear waves with different angles of attack. An approximate solution to the combined static stresses and those imposed by the travelling waves is used to delineate zones where an approximate strength criterion is exceeded. They gave an example of a chamber in granite, 1200 m deep. A range of *in situ* stress conditions was explored with vertical stress up to 1.5 times overburden and horizontal stress twice the vertical stress. The earthquake action was assumed to be that equivalent to a seismic coefficient up to 0.4.

In Soviet earthquake engineering practice this is related to a 'seismic degree' which is in effect the local intensity in MSK units and is selected on historical and seismotectonic grounds, usually with an associated mean return period. The seismic coefficient, $K_c = 0.4$ used in Soviet earthquake engineering codes as a component of the pseudostatic base shear input, is associated with an intensity of 11, and although it carries the dimensions of acceleration it is not strictly ground acceleration at depth.

It was shown that the dynamic stresses had little impact on the development of zones of non-elastic deformation which were actually controlled by static stress ratios. However, another example of a shallower machine hall indicated that the dynamic effects of an earthquake of intensity 11 could extend the non-linear zone surrounding the cavern by 25% as compared with the static case.

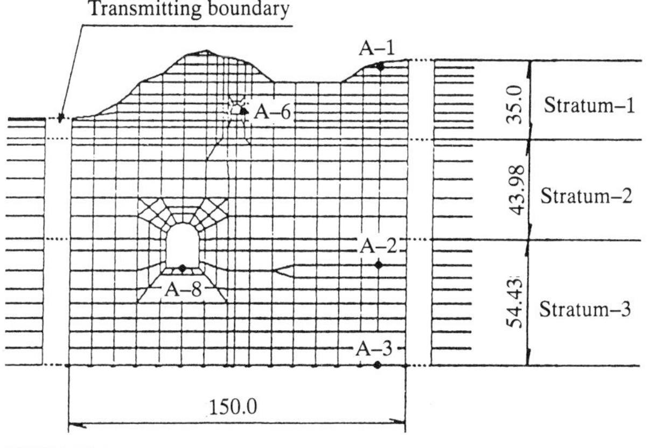

	Vs* (km/s)	Poisson's ratio	Unit weight (t/m^3)
Stratum-1	1.0	0.37	2.65
Stratum-2	2.6	0.31	2.65
Stratum-3	2.8	0.30	2.65

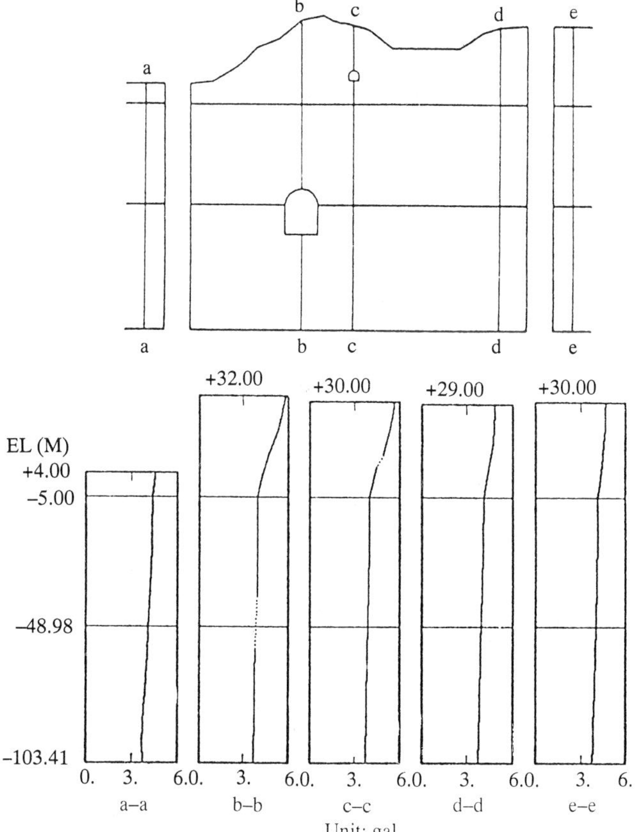

Figure 14.20 *Finite element model of rock storage cavern (from Yamahara et al. 1978)*

14.7.4 Finite-element continuum modelling

Yamahara *et al* (1978) described the design of a rock storage cavern using finite element modelling (Figure 14.20). The natural frequencies, mode shapes and participation factors up to the fifteenth mode were calculated. The acceleration time histories of the El Centro 1940 north–south and Hachinohe (1968) north–south strong motions are used with a layer model to derive a base motion at a depth of 120 m. Shear waves are assumed to propagate upwards and the maximum principal dynamic stress and shear stress increments fround in elements close to the cavern wall were only 20% different from those further away. A further analysis assuming a Rayleigh mode of propagation in a half-space indicated higher stress increments (see also Aoki *et al.* 1986).

14.7.5 Numerical analysis with discontinuities

Only rarely is it possible to be truly content with a continuum model although the relatively simple procedures have, especially where a lining is present, often had to suffice. The continuum studies have as described above enabled, for most earthquake situations, a pseudo-static approach to be justified and local wedges of rock at or near the cavity face can be treated as having a free field acceleration applied at their centre of gravity. If there is a lining which can be regarded as effectively following the continuum, the rock within the 'yield zone' can be accelerated and transmit the transient load to the lining (see below for effect of linings). The free-field acceleration should of course be that which would obtain at the relevant depth.

In the last decade there have been a number of computer programs developed which deal with jointed rocks. These programs are not to be confused with the more familiar finite element modelling which can model the inhomogeneous continuum, albeit at a cost.

Cundall (1980) in his 'distinct element' model assumed initially that the jointed continuum be composed of rigid blocks and that deformation was restricted to the joints between the blocks, which required the formulation of a joint constitutive relation. Later deformable blocks were formulated with each block being discretized as a mesh of triangular or tetrahedral finite difference zones for two- or three-dimensional modelling respectively. An explicit time domain algorithm with a central difference scheme is used and for each time step two sets of calculations are carried out:

(1) Application of the equations of motion allows the new kinematic quantities (acceleration, velocity, displacement) to be determined.
(2) The block and joint constitutive relations enable the new internal stresses and interaction forces between the blocks to be calculated. The computational steps are described by Lemos *et al.* (1987).

(3) The accelerations at each grid point are obtained from the equations of motion:

$$\ddot{u}_i = \int_s \sigma_{ij} n_j \, dS + F_i + \frac{g_i}{m} \qquad (14.21)$$

where S is the Voronoi polygonal (two-dimensional) or polyhedral (three-dimensional) surface surrounding each grid point, σ_{ij} is the stress tensor, n_j are the components normal to S, m is the mass lumped at each gridpoint and g_i are the gravitational accelerations.

The forces F_i include applied external loads and, at the block boundary, the contact forces.

With the integration of the accelerations at each gridpoint the new displacement increments, new strains and new block internal stresses can be obtained using the block constitutive relations and at the boundaries new joint displacements stresses and forces are obtained using the joint constitutive relations. Two joint models are used; an elasto-plastic one with a Mohr–Coulomb yield law and a non-linear continuously yielding one capable of describing a peak residual type of behaviour. High values of viscous damping yield quasi-static solutions and for dynamic analyses both mass proportional and stiffness damping are used. Non-reflecting bondaries are introduced after the manner of Lysmer and Kuhlemeyer (1969).

The distinct element method permits a full stress history of an excavation to be modelled and Lemos *et al.* (1987) described a sequence in which a model of jointed rock is first consolidated under *in situ* stress and gravitational acceleration. Then an excavation is effected, and the model cycled to equilibrium. Thereafter a dynamic load in the form of a stress pulse is applied at one boundary. Displacement versus time plots show the response at two points (Figure 14.21).

A form of distinct element modelling has been combined with finite element modelling of the continuum (Dowding *et al.* 1983). Rigid blocks are defined within about one cavity diameter, at greater distances the rock is modelled as a continuum by finite elements and silent boundaries are introduced to prescribe velocities so as to both initiate motion and absorb waves (Figure 14.22).

The blocks are assumed to interact through an edge-to-edge contact mechanism along a block edge but large relative rotations will lead to corner-to-edge contact and so allow for toppling. The joints are rectilinear and a contact force is generated which is proportional to the penetration of one block into another. Joint normal stresses are taken as linearly related via joint stiffness penetration. Joint shear stresses are defined at each point of contact through a friction law which accounts for deformation history and could include dilatation. The normal force on a joint is limited to a maximum cohesion, normally taken as zero. In this model the block contact relations are not updated so its utility is limited to establishing the behaviour at small relative displacement.

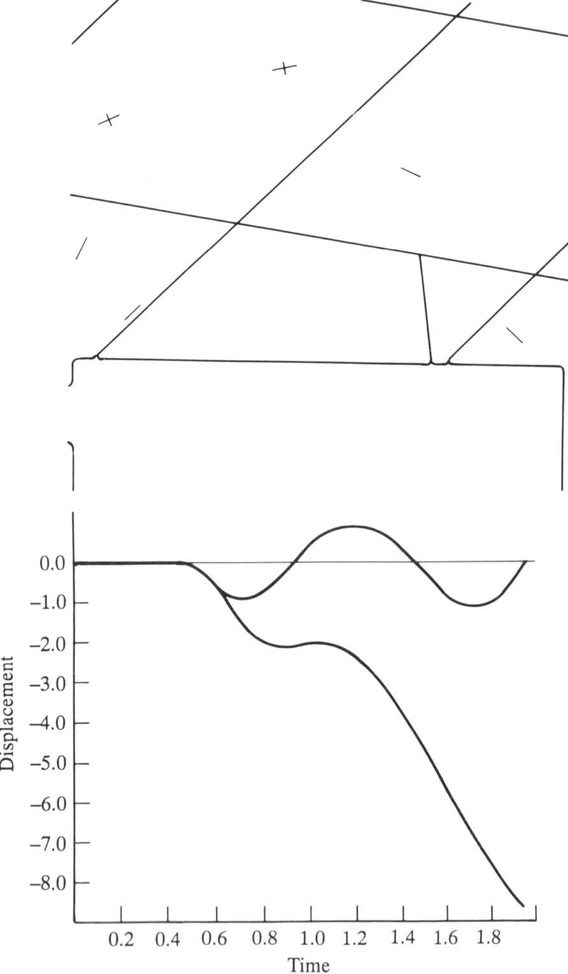

Figure 14.21 *Distinct element model of a cavern (from Lemos et al., 1987)*

Before the dynamic analysis begins it must be established that the static configuration of inter-block forces is stable for the assumed vertical and horizontal stress fields. Once it has been demonstrated that in the static case there are no tensions, an explicit relaxation code is used to calculate the dynamic behaviour of the cavern. The failure modes considered are:

(1) total shear force exceeds peak strength of the joint;
(2) total normal stress decreases to below the small cohesion allowed.

The dynamic action is established by making the non-reflective boundaries follow a prescribed force or stress history on all boundaries and zeroing all non-prescribed stresses from reflected waves after each time step in the calculation. Vertical propagation of the shear wave was modelled by delaying the horizontal stresses to

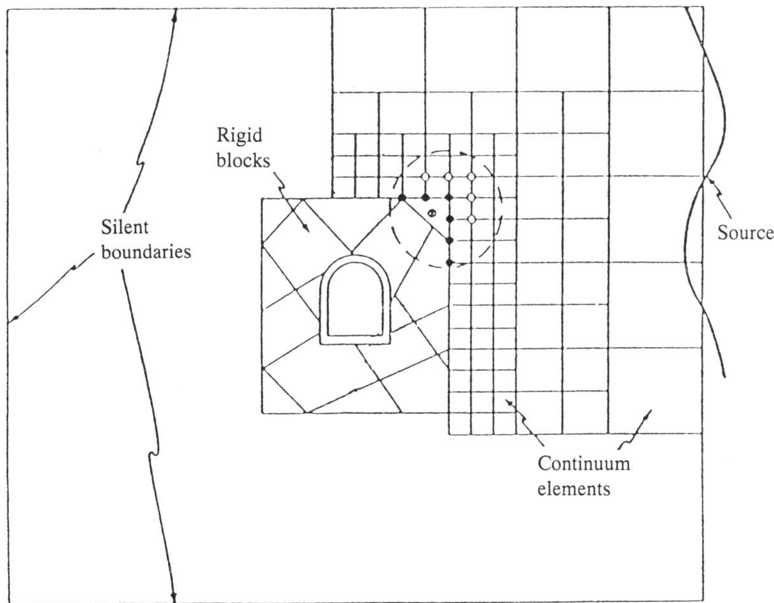

Figure 14.22 *Hybrid block model (from Dowding et al. 1983)*

match the propagation velocity. Dowding *et al* (1983) desribed seven cases of cavern response to vertically propagating shear waves with the exciting wavelength 1, 2, and 8 times the cavern height.

It was found that a shear wavelength twice the cavern height produced the greatest relative displacement, that high particle velocities were needed to cause block displacement at the depths considered, that joint orientation charges of only a few degrees were significant and that a high ratio of stiffness normal to a joint to the stiffness of the joint in shear was conducive to sliding.

The explicit approach outlined above (the North Western University Rigid Block Model–NURBM) has been developed further to allow the dilatational behaviour of rock joints (O'Connor *et al.* 1983) to be modelled with numerical simulation of the results of cyclic direct shear tests on rock joint surfaces. The roughness of the surface is expressed in terms of the angle of microscopic asperities (Plesha 1985) and in the model case studied it was demonstrated that in situations which were not inherently unstable the interlocking roughness was effective in preventing blocks from sliding out at particle velocities of the order of 500 mm/s at 30 Hz.

The growing capacity to model numerically a complex set of discontinuities sets a task for the engineering geologist to map and characterize those discontinuities adequately. The preconditions whereby a set of discontinuities can be subsumed into a continuum of modified bulk properties has already been mentioned. Intuitively one would expect this situation to obtain if some characteristic interval between discontinuities were an order of magnitude smaller than some characteristic wavelength of the earthquake. Stochastic approaches to rock discontinuities are receiving much attention although in most commercial practice a deterministic mapping of potential block failures is carried out.

The constitutive relationship for joints under dynamic conditions is so difficult to assess realistically that numerical modelling must be used primarily in the 'sensitivity mode': that is, the temptation to model in too much detail must be resisted. The simplest credible model should be chosen and the flexible graphic interaction microcomputer programs now available used with a generous variation of parameters. In assessing the behaviour of the joint surfaces attention must be paid to residual properties, possible frequency-related phenomena and the response of pore fluids.

Goodman and Shi (1985) discussed the distinct element method in comparison with the block method, now favoured for static analysis, concluding that block theory, because it is three-dimensional, is of more help in choosing the geometry of an excavation but cannot cope with really large deformations. As the three-dimensional programs for distinct element modelling become available (Hart *et al* 1988), this drawback will disappear and the utility of the method for dealing with dynamic actions should enhance its attractions for earthquake engineers.

14.8 Design to improve resistance to earthquakes

14.8.1 Preamble

After assessing the earthquake hazard it may be considered necessary to incorporate special measures in design.

Some of the approaches adopted by the earthquake engineer with regard to above ground structures may have application but many will not. With an underground facility there is not much room to manoeuvre with respect to providing ductility nor is there much that can be done to modify economically the overall response characteristic of the engineered structure. Serious problems arise where shaking triggers a mechanism which may be progressive so the identification of such potential mechanisms, which does not necessarily require dynamic analysis, and simple precautionary support measures may be sufficient.

14.8.2 Foundations in rock

The provision of capacity for the above ground structure to withstand the ground shaking is outside the scope of this chapter. If on excavation a fault is found which can be regarded as a likely locus of future co-seismic movement, the engineer can either move his sructure, demonstrate that the probability of the fault moving a sufficient amount to be of concern is acceptably low or make provision to accommodate the movement. In principle this may be provided by an interlayer of yielding material. A natural blanket of such material sustained an argument that displacement on a fault under a small test nuclear reactor in California would not damage the unit (Meehan 1984). Such interlayers have been incorporated in the foundations of buildings to decouple them from horizontal ground motion, usually in an experimental manner.

14.8.3 Natural slopes, cuttings

Those natural slopes in rock for which it is judged, perhaps after a regional survey such as is described by Wilson and Keefer (1985), that some kind of human intervention is worthwhile, can be approached using the basic principles of geotechnics; control of disturbing forces, enhancement of resisting forces (Hutchinson 1984). This may involve regrading, at least locally, following some appraisal of vulnerability and measures to control the static pore pressures. Provision of drainage to cope with pore pressures which might be generated during an earthquake is not usually practicable. Local support measures, rock bolts, cable anchorage, may be undertaken where the likely failure mechanism can be identified and one or other of the analytical procedures enables an anchor force needed during an earthquake to be assessed.

A man-made cutting in an earthquake-prone area should be designed with the potential additional forces in mind. Special drainage and rock reinforcement either by local bolting or retaining structures may be incorporated. The design methods which would be used in general practice are unlikely to involve advanced analysis.

14.8.4 Caverns, tunnels

So far the response of caverns and tunnels has been discussed without reference to tunnel support although some of the earliest static stress analyses of tunnels in elastic media did incorporate lining. The dynamic response of a liner is usually calculated by estimating the distortions imposed on the liner by the strains developed in the surrounding rock during the passage of a shear wave. It will be recollected that strong ground motion is mainly a response to shear waves. Allowance is made for the magnification effects associated with the interaction of the wave and the cavity. In the simplest continuum approach the important factor is the relative stiffness of liner to the elastic medium and this is expressed as a 'compressibility ratio' and a 'flexibility' ratio (Peck *et al.* 1972; Hendron and Fernandez 1983). This approach is the same as used for liner analysis in soft ground tunnelling. The compressibility ratio is a measure of the all-round radial pressure needed to cause unit diametral strain without distortion of cross-section and has little effect on behaviour under earthquake loading. The flexibility ratio is a measure of the amount of non-uniform pressure, tension and compression applied at 90° which would cause unit diametral strain leading to the cross-section's becoming oval in shape.

The flexibility ratio for a relatively thin uniform (concrete) liner is

$$F = \{E/[1 + \mu]/6E\}\,\{(\tfrac{1}{12}t^3)/(1 - \mu_1{}^2)\,R^3\} \qquad (14.22)$$

where E, μ, are Young's modulus and Poisson's ratio respectively of the medium, subscripts denote the liner and R and t are the radius of tunnel and thickness of liner respectively. Peck *et al.* (1972) showed that if this value was larger than 20 the liner behaved flexibly, following distortions in the enveloping medium.

Although the analytical procedures are continuum based they constitute a starting point and if, for example, a liner has been designed to cope with static loading using ground-yield concepts (viz. 'New Austrian Tunnelling method'), perhaps in conjunction with rock bolting, the design condition in the ultimate limit state would be a lining which accommodated the travelling distortions by yielding (with adequate continuity of reinforcement), but retaining sufficient integrity to inhibit the development of drop-out mechanisms involving moderate to large discontinuous elements. As discussed above in most situations pseudo-static approaches would have been adopted but distinct-element modelling is now definitely attractive.

Measures taken prior to the installation of a lining which provides a flexible support zone are prudent and rock bolting has received some attention in this respect. Rock bolting and short length anchoring (<20 m) is not amenable to detailed dynamic analysis for earthquake loading. A pseudo-static approach in which the earthquake forces coming on to a defined block of ground are distributed on the anchorages is usually adopted.

Hendron and Fernandez (1983) set down simple precepts for the design and installation of rock bolts with dynamic loading in mind. They discounted distortion of the support system and took account only of the peak stresses in a pseudo-static manner, making allowance for dynamic magnification. thus the total pressure per linear unit of chamber roof, becomes

$$P_t = (1 + a/g)\, n\, B\, \gamma \qquad (14.23)$$

where a/g is the peak acceleration as a fraction of gravity, B is the cavern span,γ is the unit weight of rock and n is a factor such that nB corresponds to an effectively loosened zone of rock to which the earthquake acceleration can be applied.

Of course, the discrete element model codes now give greater scope for more detailed and realistic modelling of caverns and tunnels for earthquake actions but in general, because of the favourable experience of the performance of deep systems in earthquakes, the determining design considerations are usually those related to static loading and stability. Such special facilities as radioactive waste repositories require detailed study and robust design.

References

Aki, K. (1967) 'Scaling law of seismic spectrum', *J. Geophys. Res.* **72**. 1217–1231

Aki, K. and Richards, P.G. (1980) *Quantitative Seismology – Theory and Method*, (2 vols) W.H. Freeman, San Francisco

Ambraseys, N.N. (1975) 'Correlation of intensity with ground motion', *Proc. 14th Assembly, European Seismological Commission*, Trieste, 325–341

Ambraseys, N.N. (1988) 'Engineering seismology', The Mallet-Milne Lecture', *Int. J. Earthquake Engg. and Structural Dynamics*, **17**(1), 1–106

Ambraseys N.N. and Jackson, J. (1984) 'Seismic movements'. In *Ground Movements and their Effect on Structures*, P.B. Attewel and R.K. Taylor (eds), Surrey University Press, Chapman and Hall, New York

Ambraseys, N.N. and Menu, J.M. (1988) 'Earthquake induced ground deformation', *Int. J. Earthquake Engg. and Structural Dynamics*, **16**, 985–1006

Anma, S., Maikuma, K., Yoshimura, M., Fujuto, Y. and Okuma, S. (1987), 'Dynamics of earthquake induced slope failure of Ontake', *Pro. 6th Int. Congress of Rock Mechanics*, Lausanne, **1**, 61–66

Aoki, K. Miyashita, K., Hanamora, T. and Tajuma, T. (1986) 'The first testplant of underground crude oil storage in unlined caverns', *Proc. Int. Symp. Large Rock Caverns*, 3–114

Arkhiprova, E. (1987) 'Seismic stability of rock masses', *Proc. 6th Int. Cong. Rock Mechanics*, **2**, 753–756

Backus, G.E. and Mulcahy, M. (1976) 'Moment tensor and other phenomenological descriptions of seismic sources – continuous displacement', *Geophys. J.* **46**, 341–361

Barton, N. (1987) 'Deformation phenomena in jointed rock', *Publication 168*, Norwegian Geotechnical Institute, Tasen Oslo

Bath, M. (1981) 'Earthquake magnitude – recent research and current trends', *Earth Science Reviews*, **17**, 315–398

Bolla, A., Bonnechere, F. and Arnold, R. (1987) A probabilistic approach to slope stability in fractured rock. *Pro. 6th Int. Congress of Rock Mechanics*, **1**, 301–303

Bolt, B.A. and Abrahamson, N.A. (1982) 'New acceleration relations for peak and expected peak acceleration of strong ground motion', *Bull. Seismological Soc. America*, **72**(6), 2307–2321

Bolt, B.A. and Hansen, R.A. (1977) 'The upthrow of objects in earthquakes', *Bull. Seismological Soc. America*, **67**, 1415–1427

Bonilla, M.G. (1988) 'Minimum earthquake magnitude associated with co-seismic surface faulting', *Bull. Int. Ass. Eng. Geol.* **25** (1), 17–29

Bonilla, M.G., Nash, R.K. and Lienkuentel, J.J. (1984) 'Statistical relationship covering earthquake magnitude surface length and surface fault displacement', *Bull. Seismological Soc. America*, **74**, 2379–2411

Boore, D. and Joyner, W. (1982) 'The empirical prediction of ground motion, *Bull. Seismological Soc. America*, **73**, 543–560

Borg, S.F. and Schuring, J.R. (1985) 'Rational Method for analysing rock slope stability due to an earthquake', *Stevens Institute of Technology*, **5**, 17–27

Brady, B.H.G. and St. John, C.N. (1982) 'The role and credibility of computational methods in engineering rock mechanics', *23rd US Symp. Rock Mechanics*, University of California

Brawner, C.V. (1980) 'Rock slope stability in rail transportation projects', *Transportation Research Record*, part. 749 58–67

Broili, L. (1980) Considerations on rock slopes stability during earthquakes, *Rock Mechanics Supplement*, **10**, 47–61

Bromhead, E.N. (1988) Strategies for the evaluation of seismic slope stability with the limit equilibrium method. *Proc. 6th Int. Congress of Rock Mechanics*, **1**, 561–563

Brunsden, D. and Prior, D.B. (1984) (eds.) *Slope Instability*, Wiley, London

Bullen, K.E. and Bolt, B.A. (1985) *An Introduction to the Theory of Seismology* (4th edn.), Cambridge University Press, Cambridge

Burton, P.W. (1980) 'Variation in seismic risk parameters in Britain, *Global Seismology Report 147* Institute of Geological Sciences, Edinburgh

Burton, P.W. (1988) '"Pathways" to seismic hazard evaluation paper presented to the Sofia 1988 meeting of the European Seismological Commission', *Global Seismology Report WL/88/24*, British Geological Survey, Edinburgh

Campbell, K.W. (1984) 'Probabilistic evaluation of seismic hazard for sites located near active faults', *Proc. 8th World Conf. Earthquake Engg.* **1**, 231–238

Campbell, R.B. and Dodd, J.S. (1966) Estimated rock stresses at Morrow Point underground power plant from earthquakes and underground nuclear blasts', *United States Bureau of Reclamation*, Denver, Rp. **SA -6**, 53pp.

Chang, F. (1987) 'Site characterised by power spectral density', *13th European Association of Earthquake Engineering Seminar*, Istanbul.

Chesnokov, S.L., Sheinkop, Fotieva, N., Bulychev, N. and Sammal, A. (1986) Estimation of stability of large span caverns with respect to seismic effects. *Proceedings of International Symposium on Large Rock Caverns*, Helsinki, 1213–1222

Christian, J.T. (1987) Numerical methods and computing. In *Ground Engineers Reference Book*, Edited by F.G. Bell, 57/1–18 Butterworths, London

Clark, M.N. (1972) Intensity of shaking estimated from displaced stones. *United States Geological Survey, Professional Paper* 787, 175–182

Cotcchia, V. (1987) 'Earthquake prone environments'. In *Slope Stability*, M. G. Anderson and K. S. Richards (eds.), Wiley, Chichester

Crawford, A.M. and Curran, J.H. (1981) 'Rate dependent behaviour of rock joints – black quartz syenite', *Proc. Int. Symp. Weak Rocks*, Tokyo, 231–96

Crawford, A.M. and Curran, J.H. (1982) 'The influence of rate and displacement dependent, shear resistance on the response of rock slopes to seismic loads', *Int. J. Rock Mech. and Mining Sciences and Geomech. Abstr.*, **19**, 1–8

Cundall, P.A. (1976) 'Explicit finite difference methods in geomechanics'. In *Numerical Methods in Geomechanics*, C.S. Desai (ed.), American Society of Civil Engineers, New York pp. 132–150

Cundall, P.A. (1980) 'A generalised distinct element programme for modelling jointed rock', *Report PCAR-1-80*, Contract DAJA 37-79-c-0548, European Research Office, US Army, Peter Cundall Associates

De-yi Feng, Mingh-zhou Lin and Kien-yuan Zhuang (1988) 'Application of fuzzy mathematics in studying natural earthquakes, palaeoseismicity', In *Historical Earthquakes and Seismograms of the World.* W.H.K. Lee, H. Meyer and K. Shimahaza, (eds.) Academic Press, San Diego

Dowding, C.H. and Gilbert, C. (1988) 'Dynamic stability of rock slopes and high frequency travelling waves', *Proc Am. Soc. Civil Engineers J. Geotech. Engg.* **114**, 1069–1088

Dowding, C.H. and Rozen, A. (1978) Damage of rock tunnels from earthquake shaking. *Proc. Am. Soc. Civil Engineers, J. Geotech. Engg. Div.*, **104**, 175–191

Dowding, C.H., Ho, C. and Belytschko, T.B. (1983) 'Earthquake response of caves in jointed rock–effect of frequency and jointing. In *Seismic Design of Embankments and Caverns*, T.R. Howard (ed.), American Society of Civil Engineers, New York, pp. 142–156

Dowrick D. J. (1987) *Earthquake Resistant Design,* (2nd ed.) Wiley, Chichester

Dvorak, A. (1977) 'Landslides caused by blasting', *Bull. Int. Assoc. Engg Geol.* **16**, 166–168

Ekstrom, G. and Dziewonski (1988) 'Evidence of bias in the estimation of earthquake size', *Nature*, **332**, 319–323

Eldredge P.T.E. *et al.* (in press)

Finn, W.D. (1966) 'Static and dynamic stresses in slopes', *First Congress of the International Society for Rock Mechanics*, Lisbon

Fournier d'Albe, E. (1982) 'An approach to earthquake risk management', *Engineering Structures*, **4**, 147–152

Fossun, A.F. (1985) 'Effective elastic properties for a randomly jointed rock', *Int. J. Rock Mech. and Mining Sciences and Geomech. Abstr.* **22**, 467–470

Franklin, A.G. and Chang, F.K. (1977) 'Permanent displacements of earth embankments by Newmark sliding block analysis', *Miscellaneous Paper S-77.17 United States Waterways Experimental Station*. Vicksburg, Mississipi

Fujita, K., Haga, Y. and Ufda, K. (1982) 'A seismic design of large caverns', *Proc. Int. Soc. Rock Mech. Symp. Rock Mechanics and Caverns*, Aachen, 267–274

Geological Society (1988) Engineering Geophysics – Report of a Working Party of the Engineering Group. *Quarterly Journal of Engineering Geology*, **21**(3), 207–272

Ghaboussi, J. and Hendron, A.J. (1984) Seismic hydrodynamic forces on rock slopes, Proceedings American Society of Cave Engineers *Journal of the Geotechnical Engineering Division* 110, 1047–1058

Glass, C.E. (1981) 'Influence of earthquakes on rock slope engineering', *Proc. 3rd Int. Conf. Slope Stability in Surface Mining*, 89–111

Gonzales de Vallejo, L., Eldred, P.J.L., Oteo, C.S. (1984) 'Open pit design in seismic areas', *Institution of Mining and Metallurgy*, Series A, 92

Goodman, R.E. (1980) *Introduction to Rock Mechanics*, Wiley, New York

Goodman, R.E. and Bray, I.W. (1976) 'Toppling of rock slopes', *Proc. Speciality Conf. Rock Engg for Foundations and Slopes Am. Soc. of Civil Engineers*, Boulder, Colorado, **2**, 201–2324

Goodman, R.E. and Shi, Gen-Hua (1985) *Block Theory and its Application to Rock Engineering*, Prentice Hall, Englewood Cliffs, New Jersey

Hanyga, A. (1985) *Seismic Wave Propagation in the Earth*, Elsevier, Amsterdam

Harada, T. and Shinzuka, M. (1986) 'Statistical variables of seismic ground motion and their design implications for buried lifeline services', *Proc. 3rd United States Conf. on Earthquake Engg. Charleston, South Carolina*, 2191–2202

Harp, E.L. (1978) 'Landslides from the February 4 1976 Guatemala earthquake, implications for seismic hazard reduction in the Guatemala City area', *Proc. 3rd Int. Conf. Microzonation for Safer Construction*, National Science Foundation, San Francisco, 353–360

Harrison, J.W. and Falcon, N.L. (1937) 'The Saidmarrah landslide south western Iran', *Geographical J.*, **89**, 42–47

Hart, R., Cundall, P.A. and Lemos, J. (1988) 'Formulation of a three-dimensional distinct element model, Part II–Mechanical calculation for motions and interactions of a system composed of polyhedral blocks', *Int. J. Rock. Mech. and Mining Sciences and Geomech. Abstr.* , **25**, 117–127

Hencher, S.R. (1981) 'Frictional parameters for design of rockslopes to withstand earthquake loading', In *Proc. Conf. Dams and Earthquakes*, Thomas Telford, London, Organized by the Institution of Civil Engineers and the British National Committee on Large Dams

Hencher, S. R. (1985) 'Limitations of stereographic projection for rock slope stability analysis', *Hong Kong Engineer,* **13**(7), 37–41

Hendron, A.J. and Fernandez, G. (1983) 'Dynamic and static design considerations for underground chambers'. In *Seismic Design of Embankments and Caverns*, T.T. Howard (ed.), American Society of Civil Engineers, New York, pp. 157–197

Herget, G. (1982) 'Probabilistic slope design for open pit mines', *Proc. 30th Geomechanical Colloquium*, Australian Society of Geomechanics, in *Rock Mechanics Supplement*, **12**, 163–178

Hoek, E. (1976) 'Rock slopes', In *Rock Engineering for Foundation and Slopes* Proceedings of a Speciality Conference Colorado, Boulder, **2** 157–171, American Society of Civil Engineers, New York

Hoek, E. (1983) Strength of jointed rock masses, **33** 187–223

Hoek, E. and Bray, J.W. (1981) *Rock Slope Engineering* (3rd edn), Institute of Mining and Metallurgy, London

Hoek, E. and Brown, E.T. (1980) *Underground Excavations in Rock* (3rd ed), Institution of Mining and Metallurgy, London

Hutchinson, J.N. (1984) 'Landslides in Britain and their counter measures', *J. Japanese Landslide Society*, **21**, 1–24

International Atomic Energy Agency (1979) 'Earthquakes and Associated topics in relation to nuclear power plant siting', *Safety Series No 80 SG 51*, Vienna

International Society for Rock Mechanics (1988) 'Report of the Commission on Computer Programs', *Int. J. Rock Mech. and Mining Engg. and Geomech. Abstr.* **25**(4)

Ishida, T., Chigira, M. and Hibino, S. (1987) 'Application of the distinct element method for analysis of toppling observed on a fissured rock slope', *Rock Mech. and Rock Engg*, **20**. 277–283

Ishihara, K. and Hsu Hai-Lung (1987) 'Consideration of landslides in natural slopes triggered by earthquake', *Collected Papers, Department of Civil Engineering, University of Tokyo*, **25**, no 87245, 16 pp.

Ito, H. and Watanabe, T. (1985) 'Some consideration of the seismic stability of large slopes by model test and numerical analysis, *Proc. 5th Int. Conf. Num. Meth. Geomech.*, Nagoya, 989–996

Jaeger, C. (1968) 'Discussion to paper by Muller "New considerations of the Vaijont slide – The dynamics of the slide"', FEIN-A **6**, 243–247

Joyner, W.B., Boore, D.M. and Porcella, R.C. (1981) 'Peak horizontal acceleration and velocity from strong motion records including records from the 1979 Imperial Valley California earthquake', *Open File Report,* United States Geological Survey 81-365-1981

Kanamori, H. (1977) 'The energy release in great earthquakes', *J. Geophy. Res.*, **82**, 2981–2987

Karaesman, E. (1977) 'A survey of building damage in the September 6 1975 Lice (Turkey) earthquake', *Proc. 6th World Conf. Earthquake Engg.*, Delhi, **1**, 37–42

Karnik, V. (1969) *Seismicity of the European Area*. Vol 1, Reidel, Dordrecht

Katayama, I. (1985) 'A method to evaluate ultra-long term stability of high-level radioactive waste storage cavern for the development of geological repository and the current status of site selection in Japan. *Proc. 8th Int. Conf. Structural Mechanics and Reactor Technology*, Invited paper to Session C. 6, preprint 1–14

Keefer, D. (1984) 'Rock avalanches caused by earthquakes; source characteristics', *Science*, **1**, 1228–1229

Key, D. (1988) *Earthquake design practice for buildings*, Telford, London

Komada, H. and Hayashi, (1979) 'Earthquake observations around the site of underground power stations 'Research Report of the Central Research Institute of the Electrical Power Industry, No. 379013 (in Japanese), *Proc. Int. Soc. Rock Mech. Symp. Rock Mechanics and Caverns*, Aachen

Kostyuchenko, V.N., (1987) 'Passage of seismic waves through a jointed body', *Translations (Doklady) USSR Academy of Sciences, AGI* Scripta Technica Inc., **285**, 1–6 10–12

Koukis, G., Flores, K. (1988) 'Given foundation conditions and earthquake properties of the Kalemata Calle (Greece)'. In *Engineering Geology of Ancient Works, Monuments and Historical Sites 1*, Balkema, Rotterdam, pp. 137–146

Krinitzsky, E.L. and Chang, F.K. (1988) 'Intensity related earthquake ground motions', *Bull. Int. Ass. Eng. Geol.* **25**(4), 425–435

Kuesel, T. R. (1969) Earthquake design criteria for subways, *ASCE J. Str. Div.*, **95**

Labreche, D.A. (1983) 'Damage mechanisms in tunnels subject to explosive loads'. In *Seismic design of embankments and caverns*, T.R. Howard(ed.) American Society of Civil Engineers, New York, pp. 128–141

Ladanyi, R. and Archimbault, G. (1980) 'Direct and indirect determination of the shear strength of rock mass, Paper No 80–25, *Las Vegas Annual meeting, American Institution of Mining Engineers*, New York, 16 pp.

Lee, W.H.K., Bennet, R.E. and Meagher, K.L. (1972) 'A method of estimating magnitude of local earthquakes from signal duration', *United States Geological Survey. Open File Report*, 30pp.

Lemos, L., Skempton, A.W. and Vaughan, P.R. (1985) 'Earthquake loading of shear surfaces in slopes', *Proc. 11th Int. Conf. Soil Mech. and Foundation Engg.*, San Francisco, 7B, **4**, 1955–1958

Lemos, J., Hart, R., Lorig, L. (1987) 'Dynamic analysis of discontinuities using the distinct element method', *Proc. Int. Congress in Rock Mech.*, Montreal, 1079–1884

Lin, J-S, (1983) 'Evaluation of earthquake induced permanent displacement of rock slopes', *Proc. 24th United States Symp. Rock Mechanics*, Texas A & M University, AEG 187–191

Lin, J.S., Whitman, R.V. (1988) 'Earthquake induced displacement of sliding blocks', *J. Geotech. Engg*, American Society of Civil Engineers, **112**(1), 345–359

Loh, CH and Whitman, (1985) 'Analysis of the spatial variation of seismic waves and ground movement for the SMART 1 array data', *Int. J. Earthquake Engg and Structural Dynamics* **13**, 561–568

Long, J.C.S., Bilaux, D., Hestir, K. and Chiles, J-P. (1987) 'Some geostatistical tools for incorporating spatial structure in fracture network modelling', *Proc. Int. Congress of Rock Mechanics*, Montreal, **1**, 171–176

Lysmer, J. and Kuhlemeyer, R. (1969) 'Finite element dynamic model for infinite media', *Proc. Am. Soc. Civil Engineers, J. Engg. Mech.*, **95** (EMA), 859–877

McCracken, A. and Jones, G.A. (1987) 'The use of probabilistic stability analysis and cautious blasting in an urban environment', *Proc. Conf. Rock Engg. in an Urban Environment*, Inst. Mining and Metall. and the Hong Kong Inst. Engineers, Hong Kong, 24–27

Makdisi, F.I. and Seed, H.B. (1977) A simplified procedure, for estimating induced deformations in dams and embankments', *Report of the Earthquake Engineering Research Institute EERI 77/19*, University of California

Mallard, D.J., Higginbottom, I.E., Muir-Wood, R. and Skipp, B.O. (1991) 'Recent developments in the methodology of seismic hazard assessment' in *Civil Engineering in the Nuclear Industry,* R. Dexter-Smith (ed.) Telford, London

Matheson, G.D. (1988) 'The collection and use of discontinuity data in rock slope design', *Q. J. Engg. Geol.*, **22**, 19–30

Meehan, R.L. (1984) *The Atom and the Fault*, MIT University Press, Cambridge, Massachussets

Melikyan, A.A. (1964) 'Seismic pressure on underground

constructions according to the data of experiment', (in Russian), *SPSE* – N 1–2

Mizuno, E., Chen, W.F., (1884) 'Plasticity models for seismic analysis of slopes', *Int. J. Soil Dynamics and Earthquake Engg.* 3,1,2–7

Muir-Wood, R. and Mallard, D.J. (in press)

Newmark, N. (1965) 'The effects of earthquakes on dams and embankments', The Fifth Rankine Lecture, *Geotechnique* **15**

O'Connor, K., Zubelewicz, A., Dowding, C., Belytschko, T.B. and Plesha, M. (1983) 'Cavern response to earthquake shaking with and without dilation', *Proc. 27th US Symp. Rock Mechanics, Key to Energy Production*, North Western University, Evanston, American Society of Mechanical Engineers, 891–895

Okonato, S., Mizukoshi, T. (1958) 'Vibrations in the base of a power station during an earthquake', (in German), *GEBW* -A v24, 2, 1958

O'Leary, P.M., Datta, S.K. (1985) 'Dynamics of buried pipelines', *Int. J. Earthquake Engg and Structural Dynamics*, **4**(3), 151–159

Olson, G.W., (1977) 'Landslides at Sardis in western Turkey', *Rev. Eng. Geol.*, Geological Society of America, **3**, 253

Ortolan, Z., Stanic, B., Tusñr, Cuijhnovic, Z. and Cukvak, N., (1988) 'Condition des foundations, degradation et mode de reaction des bâtiments du monsteri Savin a Acrug Novi, Yugoslavia, *Engineering Geology of Ancient Works, Monuments and Historical Sites*, P.G. Marinus and G.K. Koukis (eds.), Balkema, Rotterdam, Vol. 1, pp. 469–477

Palmquist, R.E. and Bible, G. (1980) 'Conceptual modelling of landslide distribution in space and time', *Bull. Int. Ass. Engg. Geol.*, **21**, 178–186

Pao, Y.H. and Chao, C.M. (1973) *Diffraction of Elastic Waves and Dynamic Stress Concentrations*, Crane Ruscak, New York; Adam Hilger London

Peck, R.B., Hendron A.J. and Mohraz, B. (1972) 'State of the art in soft ground tunnelling', *Proc. 1st North American Rapid Excavation and Tunnelling Conf.,* ASCE–AIME, Chicago, 259–286

Pentz, D.L. (1982) 'Slope stability analysis of variable rock slopes'. In *Stability in Surface Mining*, Brawner, W. (ed.), American Society of Mining Engineers, New York

Plesha, M.E. (1985) 'Constitutive modelling of rock joints with dilation', *Proc. 26th Symp. Rock Mechanics*, **1**, 383–394

Priest, S.D. (1980) 'The use of inclined hemisphere projection methods for the determination of kinematic feasibility, slip direction and volume of rock blocks', *Int. J. Rock Mech. and Mining Sciences and Geomech. Abstr.*, **17**, 1–23

Priest, S.D. (1985) *Hemispherical Projection in Rock Mechanics*, Allen and Unwin

Priest, S.D. and Brown, E.T. (1983) 'Probabilistic stability analysis of variable slopes', *Trans. Inst. Mining and Metallurgy*, Section A1–12

Ramirez, J. (1975) 'Histoire de los terremotos en Colonbia', *Institute Geographicos 'Augustin Coderzi'*, Bogota

Rice, J.R. (1984) 'Shear instability in relation to the constitutive description of fault slip', *Proc. First Int. Conf. Rock Mechanics*, Johannesburg

Richards, L.R. and Atherton, D. (1987) Stability of rock slopes. In *Ground Engineer's Reference Book*, F.G. Bell (ed.), 12/1–12/16, Butterworth-Heinemann, Oxford

Richter, C.F. (1958) *Elementary Seismology* Freeman, San Francisco

Riznichenko, J.V. (1965) 'The relationship of the flow of rock masses to seismicity', (in Russian) DAN – A, **V161**, 96–98

Rodriguez Ortiz, J.M. (1988) 'Stability problem of historic sites atop degrading cliffs', In *Engineering Geology of Ancient Works, Monuments and Historical Sites*, P.G. Marinos and G.K. Koukis (ed.), Balkema, Rotterdam, Vol. 1, pp. 3–9

Sacramento District, United States Army Corps of Engineers (1952), Underground Test Program, *Technical Report No 4*, Granite, Vols I and II

Sacremento District, United States Army Corps of Engineers (1953) Underground Test Program, *Technical Report No. 5*, Sandstone, Vol I and II

Sarma, S.K. (1975) 'Seismic stability of earth dams and embankments', *Geotechnique*, **25**, 743–761

Sarma, S.K. (1979a) 'Stability analysis of embankments and slopes *Proc. J. Geotec. Engg. Div.*, American Society of Civil Engineers, **105**, 1511–1524

Sarma, S.K. (1979b) 'Response and stability of earth dams during strong earthquakes', *Miscellaneous Paper GL 79–13, United States Waterways Experiment Station*, Mississippi

Sarma, S.K. (1987) 'A note on the stability analysis of slopes *Geotechnique*, **77**(1), 107–111

Scott, R.F. (1987) 'Failure', Twenty Seventh Rankine Lecture, *Geotechnique*, 421–466

Simonett, D.S. (1987) 'Landslide distribution and earthquakes in the Bavani and Torriolli mountains, New Guinea', In *Land-forms from Australia, New Guinea*, J.M. Jennings and J.A. Mappuit (eds.), Australian National University Press, Canberra, pp. 64–84

Sinitsyn, A.P. and Samarin, V.V. (1985) 'Numerical investigation of stresses in rock structure with deformation beyond the elastic limit due to seismic waves', *Proc. Fifth Int. Conf. Num. Meth. Geomech.*, Nagoya, 1329–1335

Skipp, B.O. (1984) 'Seismic hazard and risk analysis for open pit mining', *Trans. Inst. Mining and Metall.*, Section A, **93**, 180–192

Skipp, B.O. (1987) 'Geological faults and earthquake codes', *Proc. Seminar of the European Ass. Earthquake Engg*, Istanbul, Turkish National Committee on Earthquake Engineering, preprint

Skipp. B.O. and Ambraseys, N.N. (1987) 'Engineering seismology'. In *Ground Engineers Reference Book*, F.G. Bell (ed.), 18/1–18/25, Butterworths, London

Sokolenko, V.P. (1977) 'Landslides and collapses in seismic zones and their prediction', *Bull. Int. Ass. Engg. Geol.*, **15**, 4–8

Sokolenko, V.P. (1979) 'Mapping the after effects of disastrous earthquakes and estimation of hazard for engineering construction', *Bull. Int. Ass. Engg. Geol.*, **19**, 138–142

Srivastava, I.S. (1977) 'A method of design of rock slopes subjected to strong ground motion', *Proc. 6th World Conf. Earthquake Engg.*, Delhi, 2375–2376

Starfield, A. M. and Cundall, P.A. (1988) 'Towards a methodology for rock mechanics modelling, *Int. J. Rock Mech. and Mining Sciences and Geomech. Abstr.*, **25**, 99–112

Teme, S.C. and West, T.R. (1983) 'Some secondary toppling failure mechanisms in discontinuous rock slopes', North

Western University, *Proc. 24th US Symp. Rock Mech.*, AEG, 193

Toki, K., Miura, F. and Oguni, Y. (1985) 'Dynamic slope stability analysis with non-linear finite element methods', *Earthquake Engg. and Structural Dynamics*, **13**, 151–172

Trollope, D.H. (1968) 'The mechanics of discontinuous or clastic mechanism in rock slopes. In *Rock Mechanics and Engineering Practice*, OC. Zienkowicz and K. Stagg (eds.), Wiley, London, 275–320

Tsao, W.K. and Wong, C.M. (1989)'Steady state response of rigid blocks', *Earthquake Engineering and Structural Dynamics,* **18**, **1**, 107–120

Vanmarcke, E. H. and Lai, P. S. (1980) 'Strong motion duration and RMS amplitude of earthquake records', Bull. Seismological Soc. America, **1**, 1293–1307

Varnes, D.J., (1978) 'Slope movement types and processes'. In *Landslides Analysis and Control*, Transportation Research Board, National Academy of Sciences, R.L. Schaster and R.J. Krizek (eds.), Washington Special Report 176, 11–33

Vaughan, P.R., Lemos, L., Tikke, T. (1985) 'Strength loss on sheer surfaces during rapid loading', 11th Int. Conf. of Soil Mechanics and Foundation Engineering, San Francisco, Session 7b, Slope stability – natural slopes

Vogt, J., Godefroy, P. (1981) 'Carte Sismotectonique de la France', *Memoire BRGM, 111*

Voight, B. (1978) 'Lower Gros Ventre Slide, Wyoming, USA', In *Rockslides and Avalanches*, B. Voight (ed.), Elsevier, Amsterdam

Voight, B. and Pariseau, W.G. (1978) 'Introduction to Rock slides and avalanches'. In *Rock Slides and Avalanches*, B. Voight (ed.), Elsevier, Amsterdam, pp. 1–67

Voight, B., Calvache, M.L. and Herrera, O.O. (1987) 'High altitude monitoring of rock mass near the summit of volcano Nevado du Ruiz, Colombia', *Int. Conf. Rock Mech.*, **1**, 275–279

Wang, Sijing and Zhan Juming (1981) 'On the dynamic stability of block sliding on rock slopes', *Proc. Int. Conf. Recent Advances in Geotech. Earthquake Engg*, 413–434

Westaway, R. and Smith, R.B. (1989) 'Strong ground motion in normal faulting earthquakes, *Geophys. J.* **96**, 529–559

Whalley, W.B. (1984) Rockfalls. In *Slope Instability*, D.Brunsden and D.B. Prior (eds.), Wiley, London, pp. 217–256

Willmore, P.L. (1978) *Manual of Seismological Observatory Practice*, World Data Centre for A Solid Earth Geophysics, Denver

Wilson, J.A. (1981) 'Physical modelling to assess the dynamic behaviour of rock slopes', *International Conference on Recent Advances in Geotechnical Earthquake Engineering*, St. Luis, 447–452

Wilson, R.C. and Keefer, D.K. (1985) 'Predicting area limits of earthquake induced landsliding', *Evaluation of earthquake hazard in the Los Angeles region*–An earth science project, United States Geological Survey, 317–345

Wolf, J.P. (1985) *Dynamic Soil Structure Interaction* (Vol 1), Prentice Hall, Englewood Cliffs, New Jersey

Wolf, J.P. (1987) *Dynamic Soil Structure Interaction* (Vol 2), Prentice Hall, Englewood Cliffs, New Jersey

Wolfe, J.A. (1977) 'Late Holocene low angle landslide, Samar Island, Phillipines', *Geological Society of America, Reviews in Engineering Geology*, **3**, 149–153

Woodward, R.C. (1988) 'Investigation of toppling slope failures in welded flow tuff at Glaman Creek Dam, NSW, Australia', *Q. J. Geol. Soc.*, **21**, 288–298

Wu, F. (1987) 'A 3D model of a jointed rock mass and its deformation problem', *Int. J. Rock Mechanics and Mining Engg*, **6**, 169–176

Yamaguchi, Y., Tsujita, M. and Wakita, K. (1987) 'Seismic behaviour of rock tunnel', *Int. Congress Rock Mechanics*, 1325–1328

Yamahara, H., Hisatomi, Y. and Mori, J. (1978) 'A study of earthquake safety of rock cavern', *Rockstore 77, Proceedings of First International Symposium*, H. Bergin (ed.), Pergamon, Oxford, *2*, 377–382

Youd, T. L. and House, S.H. (1978) Historic ground failures in northern California triggered by earthquakes', *Professional Paper 993, United States Geological Survey*, pp. 177

Zang Chuhan and Zhao Chongbin (1988) 'Effects of canyon topography and geological conditions on strong ground motion', *Int. J. Earthquake Engg. and Structural Dynamics*, **1**, 81–88

15 Control of groundwater in rock masses by pumping systems

P M Cashman, Groundwater Control Consultant
P G Polsue, Mining and Metallurgical Consultant

15.1 Introduction

The economics of an excavation, whether for mining or for civil engineering, are likely to be influenced greatly by the quantity of groundwater encountered within the overburden of the area to be excavated. For civil engineering construction operations there is usually a requirement that excavations such as tunnels and shafts be lined and waterproofed. Generally such measures are less usual in mining because of the cost implications involved with the significantly greater depths of mining shafts, lengths of tunnel drives and numbers of levels. However, lining is often undertaken at the top of mine shafts where the excavation passes through overburden rather than rock. Also, in underground mining, lining is sometimes used to progress through wet areas, e.g. where a drive in rock passes through a fault zone which has water filled fissures. Hence it is prudent to emphasize the necessity for an awareness of the problems that can arise during mining operations due to groundwater and the precautionary and control measures that are appropriate to achieve workable conditions. Due to a number of criteria the problems of groundwater control are generally of greater significance in mining than in civil engineering but the principles to be observed to achieve control are the same. It is a fact that stability problems and difficulties with excavations are almost entirely attributable to the water contained in the voids – the pore water – not to the soil particles.

The major source of replenishment of groundwater is precipitation and the amount available for recharge varies with the prevailing climatic regime. In very arid areas of the world it is unlikely that near surface deposits will contain groundwater except immediately following the occasionally occurring freak rainstorms. Initially rainfall infiltrates into the ground and thence percolates, under the influence of gravity, towards the water table. The rate of percolation (i.e. the rate of recharge of the groundwater) depends primarily upon the permeability of the overburden deposits.

In rock masses percolation and flow is primarily via discontinuities. The likelihood of encountering underground water during excavation is influenced greatly by the type of rock(s) encountered and their geological history. In residual deposits percolation and flow occur almost entirely in the voids between soil particles. Accordingly the geological character of the ground within which mining or civil engineering excavations are to be made will affect the probability, or otherwise, of the presence of groundwater. Hence, an understanding of the relevant geology facilitates realistic assessments of the likely stability problems and the associated cost implications of controlling the groundwater.

15.2 The overburden and open-cast mining

Often the rock mass to be mined is not exposed at ground surface but is overlain by water bearing residual deposits. If, in the overlying deposits, the groundwater is not controlled adequately, its presence could cause mining operations to be expensive, difficult or even hazardous. In particular water-bearing unconsolidated ground generally proves difficult to excavate and so in order to achieve economic and safe excavation, appropriate measures are necessary to control the flow of both surface water and groundwater in the vicinity of an excavation. Also, measures may be needed to control the pore-water pressures. The necessary control may be achieved either by pumping techniques or by exclusion techniques or a combination of both (Cashman and Hawes 1970). The techniques available for groundwater control are listed in Table 15.1. In order to optimize profitability it is essential that conditions within the operational area be 'workable' and to achieve this it is necessary that both the surface water and the groundwater be controlled. Obviously once groundwater has been collected it should be disposed of expeditiously.

15.2.1 Surface water run-off

The basic rule of good practice to be adhered to by the project management is to collect and control the surface water run-off as soon as, or better still even before, it enters the area of work.

Table 15.1 Methods of groundwater control (after Somerville 1988: reproduced with kind permission of CIRIA)

Method	*Soils suitable for treatment*	*Uses*	*Advantages*	*Disadvantages*
Group 1: Surface water control				
1. Ditches 2. Training walls 3. Embankments	All soils if used in conjunction with polythene sheeting	Open excavations	Simple methods of diverting surface water	May be an obstruction to construction traffic
Group 2: Temporary groundwater control				
Internal pumping				
4. Sump pumping	Clean gravels and coarse sands	Open, shallow excavations	Simple pumping equipment	Fines easily removed from ground. Encourages instability of formation
5. Gravity drainage	Impermeable soils	Open excavation especially on sloping sites	Simple pumping equipment	
Groundwater lowering				
6. Wellpoint systems with suction pumps (including the machine-laid horizontal system)	Sandy gravels down to fine sands (with proper control can also be used in silty sands)	Open excavation including progressive trench excavations. Horizontal drain system particularly pertinent for pipe trench excavations outside urban areas	Quick and easy to install in suitable soils. Economical for short pumping periods of a few weeks	Difficult to install in open gravels or ground containing cobbles and boulders. Pumping must be continuous and noise of pump may be a problem in a built-up area. Suction lift is limited to about 4.0–5.5 m, depending on soils. If greater lowering is needed, multi-stage installation is necessary
7. Eductor system using high-pressure water to create vacuum as well as to lift the water	Silty sands and sandy silts	Deep excavations in space so confined that multi-stage wellpointing cannot be used. More appropriate to low-permeability soils	No limitation on amount of drawdown. Raking holes are possible	Initial installation is fairly costly. Risk of flooding excavation if high-pressure water main is ruptured
8. Shallow bored wells with suction pumps	Sandy gravels to silty fine sands and water-bearing rocks but particularly suitable for high permeability soils	More appropriate for installations to be pumped for several months or for use in silty soils where correct filtering is important	Generally costs less to run than a comparable wellpoint installation, so if pumping is required for several months costs should be compared. Correct filtering can be controlled better than with wellpoints to prevent removal of fines from silty soils	Initial installation is fairly costly. Pumping must be continuous and noise of pump may be a problem in a built-up area. Suction is limited to about 4.0–5.5 m, depending on soils. If greater lowering needed, multi-stage installation is necessary
9. Deep-bored filter wells, i.e. those with submersible pumps (line-shaft pumps with motor mounted at well head used in some countries)	Gravels to silty fine sand and water-bearing rocks	Deep excavations in, through or above water-bearing formations	No limitation on amount of drawdown as there is for suction pumping. A well can be constructed to draw water from several layers throughout its depth. Vacuum can be applied to assist drainage of fine soils. Wells can be sited clear of working area. No noise problem if mains electricity supply is available	High installation cost

Table 15.1 (*Continued*)

Method	*Soils suitable for treatment*	*Uses*	*Advantages*	*Disadvantages*
10. Electro-osmosis	Silts, silty clays and some peats	Open excavations in appropriate soils or to speed dissipation of pore water during construction	In appropriate soils can be used when no other water-lowering method is applicable	Installationa nd running costs are usually high
11. Drainage galleries	Any water-bearing strata underlain by low-permeability strata suitable for tunnelling	Removal of large quantities of water for dam abutment, cut-offs, etc.	Very large quantities of water can be drained into gallery and disposed of by conventional large-scale pumps	Very expensive, galleries may need to be concreted and grouted later
12. Collector well	Clean sands and gravel	Dewatering deep confined aquifers	Minimizes number of pumping points	Suitable only for large excavations
Group 3: Exclusion methods (not covered in detail in this Chapter)				
Temporary methods				
13. Ammonium/ brine refrigeration	All types of saturated soils and rock, particularly mixed strata	Formation of ice in the voids stops water flow	Imparts temporary mechanical strength to soils. Treatment effective from working surface outwards. Better for large applications of long duration	Treatment takes time to develop. Initial installation costs are high and refrigeration plant is expensive. Requires strict site control. Ground may heave
14. Liquid nitrogen refrigeration	As for (13)	As for (13)	As for (13), and better for small applications of short duration or where quick freezing is required	Liquid nitrogen is expensive. Requires strict site control. Ground may heave
15. Compressed air	All types of saturated soils and rock	Confined chambers such as tunnels, shafts and caissons	Gives stability to sides of chamber by limiting ingress of water. Reduces pumping to a minimum	High set-up costs; possible health hazards
16. Slurry trench cut-off with bentonite or native clay	Silts, sands, gravels and cobbles	Practically unrestricted. Extensive curtain walls round open excavation	A rapidly installed, cheaper form of diaphragm wall. Can be keyed into impermeable strata such as clays or soft shales	Must be adequately supported. Cost increases greatly with depth. Costly to attempt to key into hard or irregular bedrock surfaces. Upper limit in soils of permeability 5×10^{-3} m/s.
17. Impervious soil barrier	Silts, sands, gravels and cobbles	As for (16)	Relatively cheap. Local materials may be used	Must be placed some distance from excavation. Restricted depth of installation
18. Sheet piling (can be permanent)	All types of soil (except boulder beds and where natural or unnatural obstructions exist – particularly timber baulks)	Practically unrestricted	Well-understood method using readily available plant. Rapid installation. Steel can be incorporated in permanent works or recovered	Difficult to drive and maintain seal in boulders. Vibration and noise of driving may not be acceptable. Capital investment in piles can be high if re-usage is restricted. Seal may not be perfect

Table 15.1 (*Continued*)

Method	*Soils suitable for treatment*	*Uses*	*Advantages*	*Disadvantages*
Permanent methods – Diaphragms				
19. Diaphragm walls (structural concrete)	All soil types including those containing boulders (rotary percussion drilling suitable for penetrating rocks and boulders by reverse circulation using bentonite slurry)	Deep basements. Underground car parks. Underground pumping stations. Shafts. Dry docks, etc.	Can be designed to form part of a permanent foundation. Particularly efficient for circular excavations. Can be keyed into rock. Minimum vibration and noise. Treatment is permanent. Can be used in restricted space. Can be put down very close to existing foundation	High cost may make it uneconomical unless it can be incorporated into permanent structure. There is an upper limit to the density of steel reinforcement that can be accepted
20. Secant (interlocking) and contiguous bored piles	All soil types, but penetration through boulders may be difficult and costly	As for (19). Underpasses in stiff clay soils	Can be used on small and confined sites. Can be put down very close to existing foundations. Minimum noise and vibration Treatment is permanent	Ensuring complete contact of all piles over their full length may be difficult in practice. Joints may be sealed by grouting externally. Efficiency of reinforcing steel not as high as for (19)
Permanent methods – Grouted cut-offs				
21. Thin, grouted membrane	Silts and sands	As for (16)	As for (16)	The driving and extracting of the sheet pile element used to form the membrane limits the depth achievable and the type of soil. Also as for (16).
22. Jet grouting	All types of soil and weak rocks	Practically unrestricted	As for (16)	May be expensive
23. Cementitious grouts	Fissured and jointed rocks	Filling fissures to stop water flow (filler added for major voids)	Equipment is simple and can be used in confined spaces. Treatment is permanent	Treatment needs to be extensive to be effective
24. Clay/ cementitious grouts	Sands and gravels	Filling voids to exlcude water. To form relatively impermeable barriers (vertical or horizontal). Suitable for conditions where long-term flexibility is desirable, e.g. cores of dams	Equipment is simple and can be used in confined spaces. Treatment is permanent. Grout is introduced by means of a sleeved grout pipe which limits its spread. Can be sealed to an irregular or hard stratum	A comparatively thick barrier is needed to ensure continuity. At least 4 m of natural cover needed (or equivalent)
25. Silicates, Joosten, Guttman and other processes	Medium and coarse sands and gravels	As for (24), but non-flexible	Comparatively high mechanical strength. High degree of control of grout spread. Simple means of injection by lances. Indefinite life. Favoured for underpinning works below water level	Comparatively high cost of chemicals. Requires at least 2 m of natural cover or equivalent. Treatment can be incomplete in silty material or in presence of silt or clay lenses

Table 15.1 (*Continued*)

Method	*Soils suitable for treatment*	*Uses*	*Advantages*	*Disadvantages*
26. Resin grouts	Silty fine sands	As for (24), but only some flexibility	Can be used in conjunction with clay/cementitious grouts for treating finer strata	High cost so usually economical only on larger civil engineering works. Requires strict site control
Permanent methods – soil strengthening				
27. Electrochemical consolidation	Soft clays	Improved shear strength of soft clay without causing settlement	See 'Uses'	Installation and running costs are usually high

Depending upon climatic conditions it may be prudent to make major or minor contingency plans. For instance, in some parts of the world it is known that monsoon rains will occur and the consequential high-volume surface run-off may be very significant. Hence, good planning dictates that appropriate measures to control the surface run-off and prevent slope erosion must be incorporated into the mining excavation plan at design stage. A judgement has to be made whether to cater for the once in 10 year occurrence, or the 50/100 year occurrence.

Surface run-off must be controlled, as far as is practicable, so as not to be an encumbrance to the excavation, the spoil banks or the overburden. Soil erosion can cause instability of the side slopes and lead to blockage of collector drains as well as adding to the groundwater control problems. There is now an expanding range of geotextiles available which should be considered for slope protection against erosion damage.

The gradients of excavated floor area of an open-cast working should be formed so that the run-off water is directed to collector drains and the risk of ponding assiduously avoided. Run-off from slopes should be collected by the surface drainage system, suitably sited at each berm and/or toe of slope, and thence be directed to pumping sumps for ultimate disposal. Also, the drainage system should incorporate appropriate measures to collect and direct run-off from land areas surrounding and adjacent to an excavation so as to prevent surface water run-off from encroaching into the project area. Adoption of this philosophy requires the installation of adequate interceptor drains sited at original ground level upstream of the mine excavation (Figure 15.1). Any such interceptor or collector drain should be lined to obviate the risk of instability.

15.2.2 Slope stability of open excavations

When planning open-cast workings, consideration must be given to maintaining stable side slopes throughout the construction period and, if the excavation is not to be backfilled, also after completion of mining.

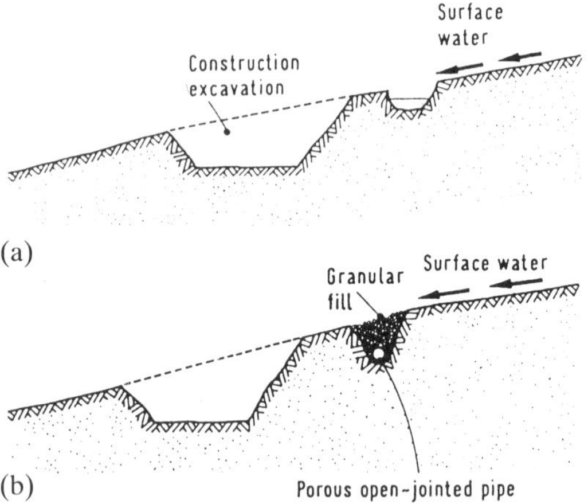

Figure 15.1 *Typical drains:* (a) *open ditches and* (b) *agricultural drains (reproduced with kind permission of CIRIA)*

It is unlikely that there will be slope stability problems due to groundwater where the host rock and ore-body are well-consolidated rock formations. Often, if these geological conditions prevail, the side slopes and working face may be steep or even near vertical.

In other rock formations that initially appear to be well consolidated such as weakly cemented sandstones, mudstones and coal depositis, slopes will be stable but with the passage of time, due to the effects of weathering and continuous seepage of groundwater, there will be deterioration that could result in progressive instability. Such instability is almost certain to occur in residual deposits at an early stage if appropriate groundwater control measures are not implemented so as to effect adequate dissipation and control of pore water pressures to prevent transportation of fines (i.e. no loss of ground).

Generally, to establish stable and workable conditions for an open-cast excavation, in deposits composed of granular and unconsolidated particles, it is desirable to drain the ground from water abstraction points positioned outside or immediately adjacent to the area to be excavated rather than from within the area. Under certain overburden conditions these measures may be quite simple types of installations.

Open-cut slopes should be such that instability does not occur due to groundwater seepage (Somerville 1988) either because the slopes are too steep and/or the hydraulic head is such that seepage flows emerging at the face causing continuous transportation of fines (Figure 15.2a). Often stable slopes can be ensured by providing a flatter batter, a drainage system at the toe of the slope and a suitable filter layer or geotextile fabric laid on the face of the slopes (Figure 15.2b).

15.3 Some aspects of soil structure of alluvials

In practice soil conditions are never homogeneous nor isotropic. Soil structures both in granular and cohesive alluvials vary significantly due to a great variety of factors. A study of the mode of deposition (i.e. their geology/lithology) is an important starting point to an engineering appreciation of measures that it would be prudent to implement.

Alluvial deposits are composed of particles that have been released by weathering and later transported, sorted and collected by natural action of flowing water. Placer deposits which occur in alluvium represent valuable souces of minerals that are usually of high specific gravity and resistant to abrasion. Examples are gold, platinum, diamonds, tin (cassiterite) and ilmenite.

Alluvial deposits have been laid down under varying climatic conditions and hence exhibit lithological variations. These variations have an influence on the soil macro-structure and consequential non-homogeneity of the deposits. In simplistic terms detrital deposits tend to be laid down in more or less horizontal layers and so the horizontal radial permeability tends to be some ten or more times greater than the vertical permeability, and often the effect of this is not adequately recognized. However, sometimes this is not the case: for example, fissures in till deposits enhance vertical permeability and so tend to negate this generalization.

Appreciation of the significance of 'soil structure' is vital to realistic assessment of the groundwater control

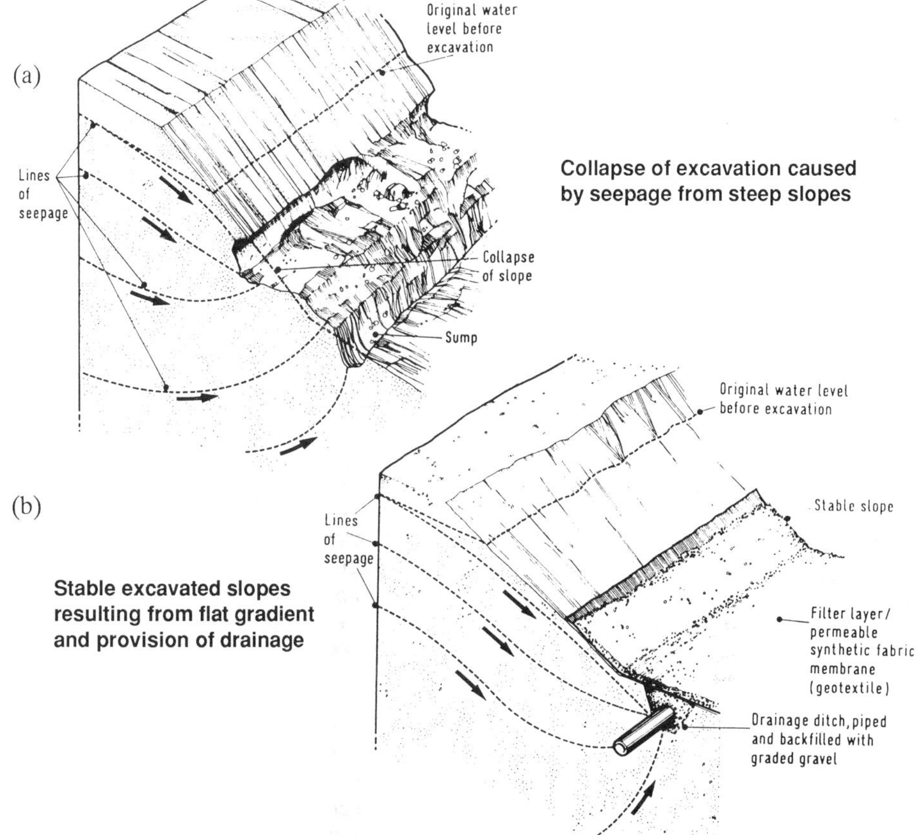

Figure 15.2 *Sketches of* (a) *unstable and* (b) *stable slopes (reproduced with kind permission of CIRIA)*

measures likely to be needed for stabilizing excavations in alluvial deposits.

In order to obtain a reliable view of soil structure sometimes it may be necessary to use continuous sampling but normally a few widely spaced boreholes provide adequate information (Anon 1981). Also, borehole pumping tests should be carried out to determine the permeabilities, storage coefficients and transmissivities of the various permeable strata to provide data for pumping systems analyses and subsequent cost assessments and to determine any hydrogeological boundaries (Anon 1975).

15.3.1 Standing-water levels/pore-water pressures

It is essential to coherent planning of groundwater control measures that factual data relevant to the site be obtained on a continuing basis. Initial and all subsequent soils investigations should be structured to obtain data about strata succession, soil classifications and soil structure as well as about standing-water levels and piezometric pressures in confined and unconfined aquifers, and seasonal variations thereof.

15.3.2 Local trouble spots

Superficial deposits are invariably heterogeneous and so the possibility of exposing local trouble spots in the sides of excavations is ever present . When observed, these should be dealt with immediately, otherwise they will get progressively worse. Often these local trouble spots are due to a layer or lens of lesser permeability within a more permeable stratum above which, despite the installed groundwater control system, there is some residual water flow which transports fines (Figure 15.3). In some

Figure 15.3 *Thin layer of silt within sandy gravel stratum indicated by arrow (reproduced with kind permission of the CEGB Barnwood)*

Figure 15.4 *Example of 'spring sapping' or 'seepage erosion'*

circumstances, if this condition is allowed to persist, it will result in spring sapping or seepage erosion (Figure 15.4).

The prevention principles to be applied to this and all similar groundwater control problems are to stop the removal of fines by the application of a suitable filter blanket (or geotextile membrane) and to avoid the build-up of pore-water pressures (i.e. do *not* stop the flow of water). These principles have been stated earlier in the section dealing with slope stability: provided that fines are not removed and pore-water pressures are dissipated, the slopes will be stable and there only remains the pumping and disposal costs of the seepage water collected.

15.4 Rock types and groundwater

The presence of groundwater should always be allowed for when planning mining underground and civil engineering-rock tunnelling. The form of the rock type will be relevant to a realistic assessment of the magnitude of the groundwater control measures likely to be required since the composition and mode of interlocking of the constituent particles will affect permeability greatly.

In igneous and metamorphic rocks the component mineral clastic usually forms a crystalline interlocking texture, whilst in clastic sedimentary deposits (which constitute the bulk of sedimentary rocks) the particles are bound together by natural cement. Heavily cemented sedimentary deposits, like igneous and metamorphic rocks, possess very low primary permeability. But all rocks contain joints and may be faulted. The incidence and character of the discontinuities govern the flow of groundwater in rock masses (see Chapter 3). A point to bear in mind is that some sedimentary deposits, notably the evaporites, undergo dissolution when they come in contact with water which is unsaturated with the mineral in question. Fortunately such evaporative deposits occur within sequences in which marls (mudstones) predominate and which possess low permeabilities. Nevertheless because of the high solubility of evaporite deposits there is a great necessity to ensure *total* exclusion of groundwater from associated mines and not just to control the flow.

15.5 Engineering precautions

Many excavations are made at shallow depths and in dry rock formations located above the water table but in such cases it is often necessary to undertake precautions against inflows of surface water. For instance, when designing subways for pedestrians or traffic it is advisable to ensure that the tunnels are concrete lined and also that precautions are taken to direct inflows of rain water at the tunnel entrance to a sump pump. A system of gutters covered with steel mesh or underground drains with small entrances to the road surface at intervals within the tunnel are usual methods of ensuring that the water flows to a sump pump chamber at the lowest point of the tunnel. Likewise for underground mining developments it is usual to install the main pump sump at the bottom of a vertical shaft and for the water to be pumped therefrom to the surface with provision for collection from the various upper levels of the mine.

Modern mining developments of major deposits endeavour to contain costs by arranging for large tonnages to be handled by the use of LHD (load–haul–dump) trackless vehicles entering and leaving the workings via inclined shafts–often constructed in the form of descending spirals. However, a second vertical shaft for emergency use and ventilation purposes must be included in the developments and the bottom of this shaft is usually an appropriate location for the pump sumps.

15.5.1 Cofferdam shafts

Shaft sinking for mining operations often involves extensive precautions against inflows of water. In some cases the greatest risk of water entering the workings is from the drift deposits overlying the rock in the area of the shaft. In virtually all modern mines the shaft is concrete lined and keyed into the bedrock in order to form a watertight seal against groundwater.

In some shafts a system of water rings is provided for collecting water running down shaft sides (Figure 15.5). The rings are built into the shaft linings at intervals and pipes are arranged to conduct water to the next lowest ring and ultimately to the sump located at the bottom of the shaft.

Cofferdams can be used for sinking shafts through saturated or cohesive sediments of limited depth. A cofferdam is an enclosure with the top open to the air and the sides comprising an enclosing wall which often is constructed by driving down interlocking steel sheet piles (figure 15.6). This arrangement keeps water out of the

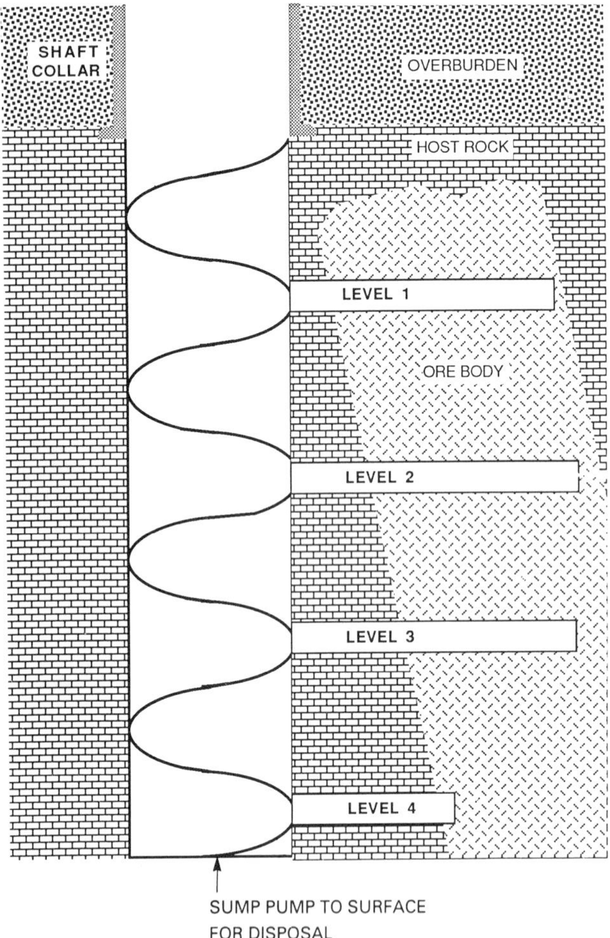

Figure 15.5 *Schematic of a typical water ring layout for collecting water in a mine shaft*

shaft so as to allow excavation to proceed expeditiously. Alternatively cast *in situ* diaphragm walling or secant piles can be used in lieu of steel sheet piles. Normally cofferdams are used for relatively shallow depths to toe into bedrock or other impervious strata so that limited sump pumping can keep the shaft area sufficiently dry for working and installing the permanent arrangements. Cast *in situ* diaphragm wall shafts have been constructed to depths of 125 m. in North America to reach rock head.

A recent example of countering the problems of sub-surface water during shaft sinking is the method employed at the Point of Ayr colliery in North Wales (Anon 1985). Production capability had exceeded the winding capacity of the original shaft and it was decided to construct an incline in which a conveyor belt would bring coal to the surface. The incline ran to the bottom of the pit. It had to penetrate 15 m of water bearing alluvium overlying 15 m of glacial clay before reaching bedrock.

After detailed investigation it was decided that the best method of constructing the portal and drifting through the saturated overburden would be by the installation of two concentric cofferdams. The walls of the first and larger were toed into the glacial clay stratum in order to form a seal. Coincidental with dewatering of the granular material within the first cofferdam, excavation proceeded to the upper surface of the glacial clay stratum and then an inner cofferdam was installed and keyed into the underlying bedrock.

15.5.2 Diaphragm wall shafts

The cast *in situ* diaphragm wall technique has been used in civil engineering for well over three decades for such construction projects as underpasses and below ground circular pumphouses as well as for working shafts for tunnelling. Also the technique is very efficient for mine shaft sinking through overburden to bedrock. This construction technique involves trench excavating from the surface by mechanical methods and supporting the soils by a suitable slurry, usually a conventional bentonite slurry or a water based biopolymer slurry having pseudo-plastic fluid viscosity properties. A steel reinforced cage is lowered through the slurry and when this is in position, concrete is introduced into the trench using a tremie tube and simultaneously the tremied concrete displaces the support slurry.

A plastic diaphragm cut-off wall using a suitable mix of concrete and bentonite slurry can be similarly constructed except that the slurry in this instance, because of its lesser strength than structural concrete, requires a berm support to be retained at the inside face of the excavation.

Diaphragm walls can be constructed to considerable depths. Recently diaphragm walls were installed at Sizewell 'B' Nuclear Power Station, Suffolk, to approximately 150 m depth.

15.5.3 Heave

If there is a confined permeable stratum under piezometric pressure at some depth beneath the mine floor or formation level, overlain by a relatively impermeable layer, then as excavation from ground level deepens there may come a stage where the downward weight of overburden remaining above the interface is no longer greater than the upward pressure of the pore water in the aquifer and so the ground may heave or be subjected to a blowout (Figure 15.7a and b). In order to prevent such hydraulic uplift some form of piezometric pressure relief system should be installed previously into the confined aquifer. This may take the form of sand-filled relief holes or bleeder wells or of a system of water abstraction points that are pumped to reduce piezometric pressures in the underlying confined aquifer.

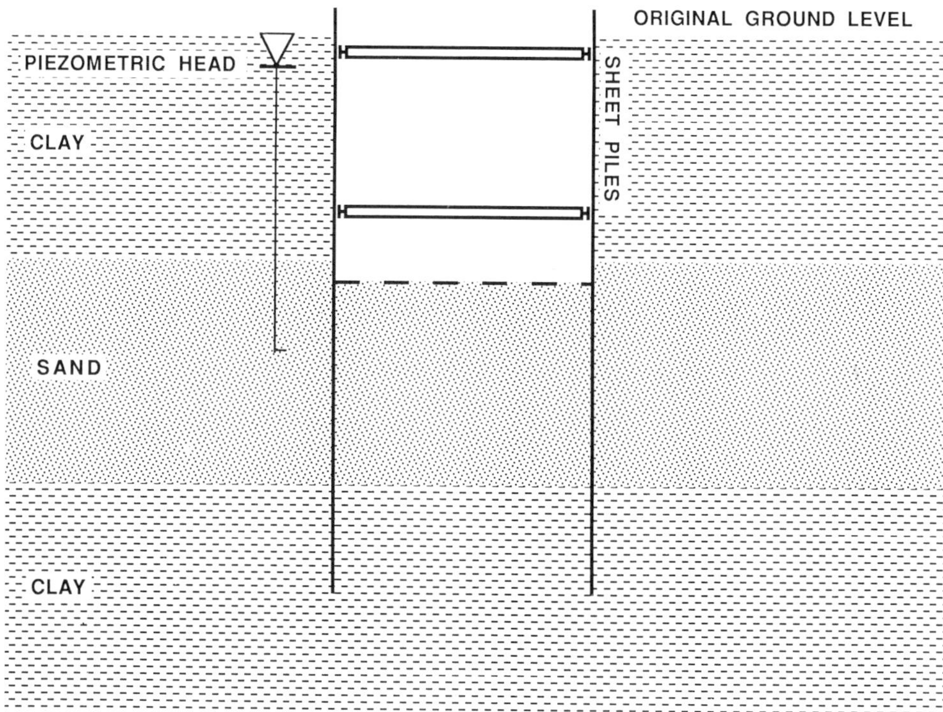

Figure 15.6 *Cofferdam: interlocking sheet piles*

15.6 Initial investigation

Water is likely to be encountered in most underground mining operations. Even when mining in igneous rock masses unexpected and copious quantities of water can be encountered especially in association with fissure zones. Indeed the amount of water which has to be removed from some mines has an important bearing on their economic operation. There are cases on record where mineral mines have been closed because the cost of pumping water from them exceeded the value of ore obtained. Review of the information available concerning the type of rocks, mineral formations and strata can indicate the probability of water being encountered.

The initial investigation for any below ground construction should provide a realistic indication of potential problems of dealing with groundwater throughout the proposed mining period. When undertaking such initial evaluations it is advisable to review the history and costs of pumping in adjacent mines over a period of years in order to asesss seasonal variations of water inflows and the effect of exceptionally heavy rainfalls. The estimates for pumping costs should allow for the worst conditions anticipated. The depths and distances over which water has to be pumped affect equipment selection and the captial and operating cost estimates.

In the case of ore bodies, major and medium-sized mining developments are normally preceded by a phase of underground exploration into the deposit via small shafts and adits. This phase allows for checks on water inflows and for estimates of pumping costs to be assessed and modified.

15.6.1 Evaluation of likely pumping costs

The pumping costs of a mine normally form a significant proportion of total operating costs and must be incorporated into the technical and economic feasibility studies that precede commitment of funds for the capital costs of a new mine or significant expansions to existing operations. The feasibility and desirabiltiy of installing stand-by pumps and also an alternative power source should be evaluated during the preliminary studies. Stand-by pumps involve increases in capital costs and decisions have to be made in accordance with the circumstances prevailing in individual mines. Principal considerations are the anticipated rate of inflow of water and whether the time taken for the lowest workings to flood will allow for emergency pumps to be mobilized. The cost of having permanent stand-by pumps is generally justified in deep mines that have water problems.

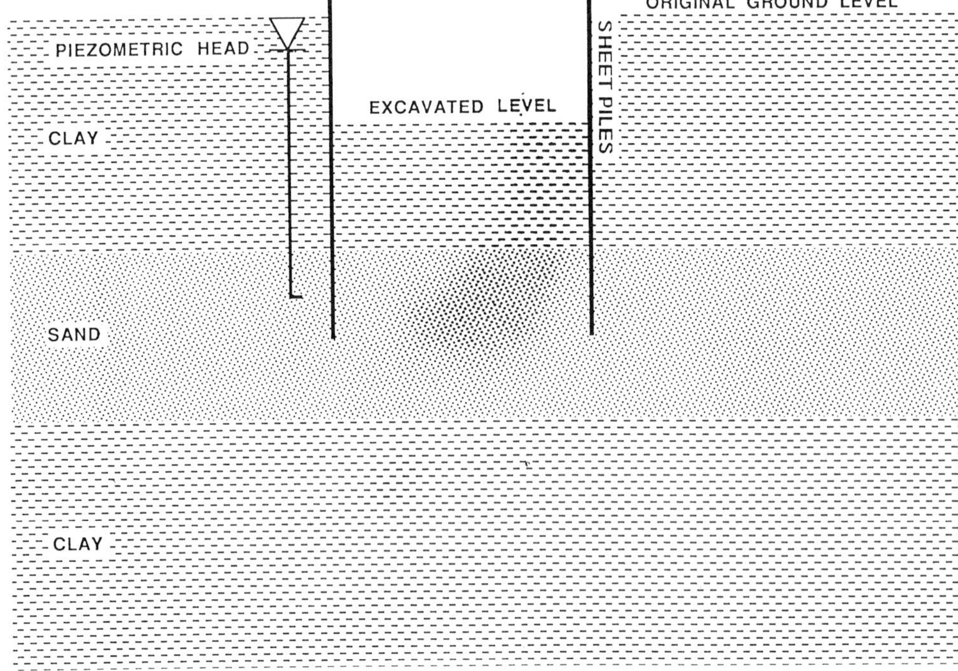

(a)

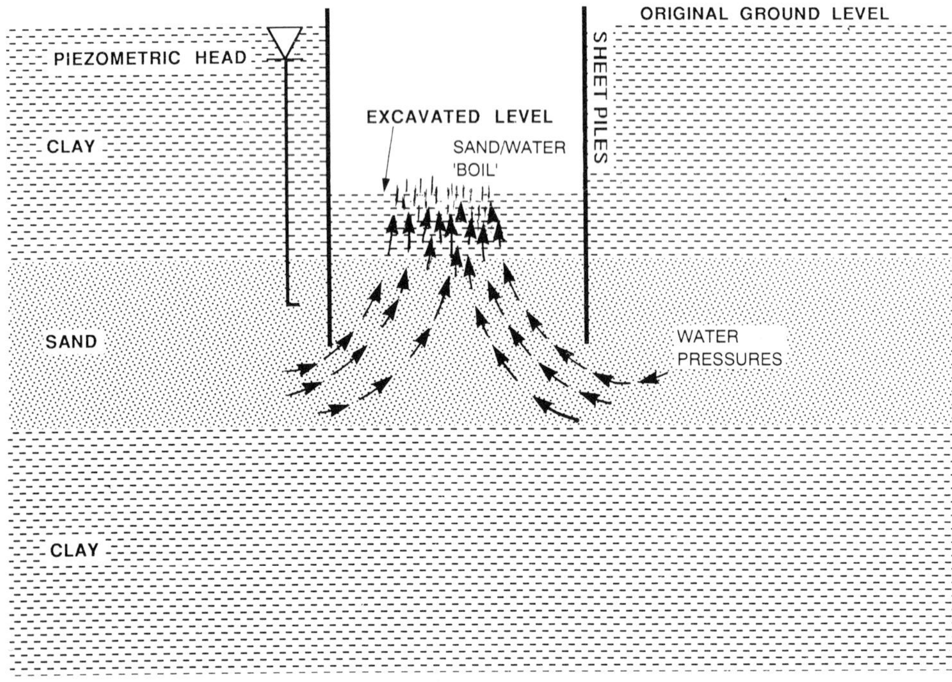

(b)

Figure 15.7 *Confined aquifer condition:* (a) *stable;* (b) *heave situation*

15.6.2 Mine pumping installations

The removal of water from underground workings is usually from a pump station located at the lowest level of the mine. In deep mines it is often economically desirable to install an arrangement of stage pumping as mining progresses downwards. It is good engineering practice to arrange for some of these pump installations to be suitable for permanent stage pumping installations. This system has the advantage of flexibility and enables capital costs of the system to be phased during the development period. Also when metal and/or mineral prices fall it may become financially advisable to abandon mining and pumping from the deepest levels in order to contain operating costs.

Pumps are divided into two main categories, namely, displacement and centrifugal pumps. There are several variations of each type but centrifugal pumps have tended to be preferred in new deep mines.

The selection of the type of pump for a particular application is often based on experience, preferences and known back-up servies in a particular territory. Consultations with experienced pump suppliers' technical personnel to assist in selection of appropriate systems and layout is strongly recommended. Calculations can be checked and assisted by various pumping and pipework data publications.

The down-the-hole pump can deal with the water at or about the level of its pump intake and is therefore an energy-efficient unit for removal of water. Consequently this is the preferred pumping system, whether it is an electro-submersible or a down-the-hole shaft-driven unit with prime mover at ground level. This latter type is often preferred in countries where sophisticated repair facilities are at a premium. However, where practical, the basic energy principle should be to avoid letting water fall to a lower level and then having to expend extra energy costs to pump from that lower level. Where groundwater control is required it should be pre-planned in order to maximize production.

The simplest and most reliable method of monitoring that the system continues to operate efficiently is the regular measurement of the rate of pumping of the total system. Flume measurement is likely to be the most appropriate. In addition, where practicable, the rate of pumping of selected individual wells should be measured regularly so as to monitor performance.

It is good practice to provide an adequate number of observation wells and/or piezometers installed in each aquifer and by means of these, to monitor regularly what is being achieved throughout the operation.

Initially, based on the pumping test results, a groundwater control system and its pumping capacity has to be designed and installed, but bearing in mind the relatively long-term pumping requirements, a computer and/or analogue study may be considered desirable. Today computers can and should be used to collate, analyse and control mine drainage and pumping systems and to ensure that the most suitable sizes of pumps and pipes are selected for specific operations. Software systems have been developed to establish the effects of variations in water flows, the number of stages and pump speeds, and thus assist selection of optimum combinations of pump and pipe work sizes.

Automatic monitoring systems of pump performance and water levels also are available and the data can be fed along a single telephone line to a control centre. Such systems incorporate automatic alarms for pump failures or abnormal flows of water and thus enable the pump stations to be worked unattended for long periods. The systems are obviously advantageous, if not mandatory, for large mines where there are several shafts, for extensive workings and where the current working operations may be some distance from the site of the pump sumps originally installed.

The mining plan proposed should have built into it as much flexibility as possible so that the groundwater control installations can be modified as more operating experience becomes available.

The likely effect of the groundwater control system should be studied to determine:

(1) whether the effluent is acceptable to the river or other relevant authority;
(2) the risks of corrosion to the components of the system;
(3) the risks of incrustation of components of the system;
(4) the likely routine maintenance costs;
(5) which alternative power supply source and standby units would it be prudent to provide.

15.7 Underground water in rocks

The presence of underground water in rocks is manifest in various degrees. In some mines that water merely drips from the roof but in others there can be working places where the water appears to pour from the apparently solid walls of rock.

In recent years there have been efforts to reduce the effects of water from rock faces by waterproofing with impervious coatings. Various types of spray-on plastic coatings have been employed and it is necessary to ensure that the material is safe in the event of fire in the mine. Poisonous fumes resulted when a recent fire in a South African gold mine set light to material coating the wall rock, and a large number of deaths resulted.

15.7.1 Plugs in underground workings

The use of concrete plugs or dams in underground workings is commonplace in deep mines where severe inflows of water may occur. When excavating horizontally or at an inclination in areas of rock where the presence of water is anticipated it is a sensible precaution to have a concrete plug in the drive behind the working area, the access to the working face being via a waterproof door built into the plug, similar to an air lock as used in civil engineering construction for soft-ground tunnelling. Such precautions are necessary in the event of an inrush of water where a sudden and often overwhelming flow may occur by striking unsuspecting old workings or faults containing large volumes of water in levels above the working drive.

The installation of a waterproof plug is justified when working highly priced minerals in wet areas where the rocks are known to contain fissures or where there are old workings that may be waterlogged and are possibly shown inaccurately on mine plans. Without the precaution of a concrete plug it is possible for a sudden inundation to flood a large area of a mine.

A well-known example of the use of concrete to stem inflows of water involved the plugging of a hole in the sea bed at the workings of the old Levant tin mine near Lands End in Cornwall. Sea water broke through the roof of undersea workings during the 1920s and flooded the mine. Increases in the price of tin during the 1960s led to studies being carried out by the new owners, Geevor Tin Mines, of methods which would allow the flooded workings to be reopened. Divers located the hole in the sea bed and directed pumping of cement grout from a barge located above the hole (Batchelor and Wardle 1969). Simultaneously miners drove a tunnel in solid rock to reach the area from underground. The combined efforts enabled a cement mat to be successfully positioned. Subsequently the flooded area of the mine was dewatered by pumping which was commenced at the lowest point in the unflooded area of the mine accessible from the mine shaft and progressively lowered as the water level declined (Figure 15.8).

If the presence of water is known to exist in adjacent old workings or in adjacent faults, then it is prudent to leave a mineral zone to protect against entry of water. Such arrangements are known as water barriers or barrier pillars. It is advisable to allow for possible inaccuracies in old plans and allow a generous margin of safety.

15.7.2 Water blast

An unexpected and disturbing occurrence in mining opertions is a 'water blast'. This involves the explosion of water under pressure into mine workings and is caused by trapped pressurized air expanding as the water level is lowered or by the sudden escape of air pent up in rise working under pressure from a head of water in a connected area. Water blasts are generally unexpected but usually can be avoided after careful study of existing plans of mine workings so as to anticipate areas where air may be trapped and by drilling holes, thereby preventing build-up of air pressures that could ultimately cause water blast.

15.7.3 Groundwater chemistry

The chemical constituents of the groundwater should be determined at an early stage of the investigation and feasibility studies, to assess whether the discharge waters need to be processed prior to release into the adjacent river system and also to evaluate the likely degree of corrosion and/or incrustation of the proposed groundwater control systems.

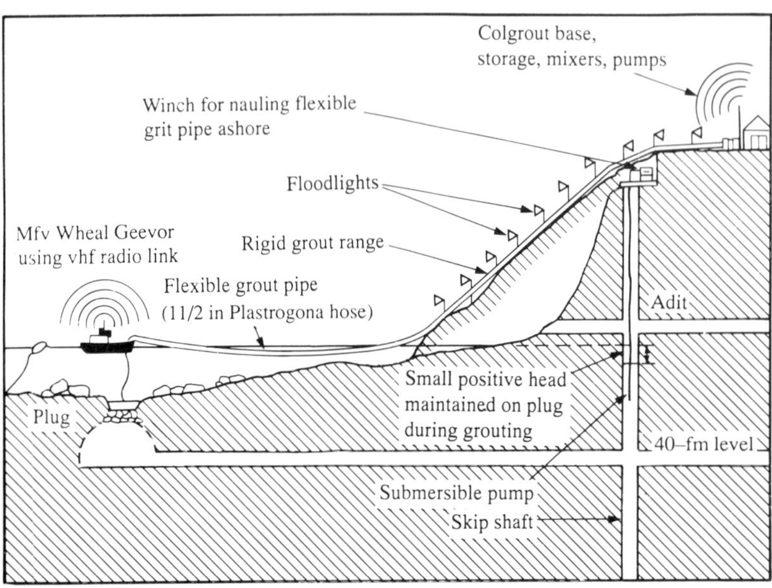

Figure 15.8 *Section of part of Levant mine – the sealing of the breach (reproduced with kind permission of Geevor plc)*

The degree of treatment of the effluent and discharge waters will be influenced by the relevant water authority or government requirements. In addition, the possible long-term effects of the chemical constituents of the surface and groundwater to be disposed of must be acceptable to the overall mine plan. Hence, it is advisable to study the likely analysis of effluent before commencement of mining operations and to reach preliminary understandings with local authorities.

The presence of small traces of metals such as lead, zinc, arsenic and cadmium is generally regarded as objectionable and since lead is the most commonly encountered it is usually regarded as the most objectionable impurity. Its maximum acceptable limit is approximately 0.05 mg/l. Less objectionable impurities such as zinc should have a maximum limit of approximately 0.50 mg/l in effluent. However, legislative 'limits' vary for different countries.

15.7.4 Some potential uses of discharge waters

The many problems that can arise due to the presence of groundwater have been described above together with systems for control. All of these seem to indicate that groundwater is a nuisance. However, often some benefits can be derived. It may be used to supplement water supply to the site infrastructure (depending upon its chemical composition) or for wet drilling, mill processing and pipeline transportation of tailings.

15.8 System design guidelines

Evaluation of the installation and operating costs of a groundwater control system will form an esential part of the feasibility study. It is necessary to assess, for instance, the number, depth and size of deep wells to be installed together with pump performance and energy consumption. The cumulative pumping rates will dictate the requisite sizing of pipework.

15.8.1 Groundwater flow regimen

Seepage of groundwater towards a pumping well may be idealized as a case of axially symmetrical three-dimensional flow of a homogeneous fluid through a porous medium. All analytical models are based primarily on this idealized premise. However, in practice such idealized conditions do not exist. The variance therefrom is greater when applied to groundwater flow in rock masses, where the flow is predominantly through joints, than flow in porous residual deposits, where flow is through pores between individual soil particles.

It follows, therefore, that all computations of groundwater flows, whether analytically or analogue based should be scrutinized and probably modified using expereinced engineering judgement.

Whether considering an exclusion technique such as grouting or a pumping technique such as a deep-well system, a key parameter to determining which groundwater control technique is appropriate for use at a particular site is the permeability of the soils/rock formations.

In practice reliable determination of permeability is difficult. A pumping test with observation wells and flow-rate measurements is the most reliable method (Kruseman and De Ridder, 1979). Since it is the most costly, it is far too often discounted, but its superior reliability very often ultimately justifies the apparently high initial cost more than adequately.

Other assessment techniques (Wenzel 1942; Powers 1981; Hazen 1893; Hvorslev 1951; Louden, 1952) well used for alluvial soils are based upon:

(1) rising and falling head testing in site investigation boreholes;
(2) evaluation based upon particle-size analysis of individual undisturbed samples or disturbed samples where it is assumed that no fines have been lost;
(3) determination of permeability of individual soil samples in a soils laboratory test rig.

The reliability of values assessed by each of the foregoing methods is questionable on several counts. Hence when contemplating any excavation into a water-bearing rock mass one or more deep-well pumping tests should be mandatory.

15.8.2 Steady-state analyses

The theory of steady-state flow through permeable soils originates with the researches of the French hydraulic engineer Henri Darcy (1846). The equations of Dupuit (1863) incorporating some modifications by Forchheimer (1930), deal with flow through porous media for confined and unconfined conditions (Figure 15.9) and enable one to compute total rate of pumping. These were based on Darcy's work.

Whilst it is known that the Dupuit–Forchheimer simplifying assumptions impose limitations on the conditions for which the formulae are valid, Hantush and Jacob (1955) demonstrated that for many practical purposes Dupuit gives an acceptable assessment of quantity to be pumped. Boulton (1951) and many others expressed similar views.

Forchheimer (1930) also propounded the concept of an hypothetical equivalent single well for a system of wells enclosing an area of length a and width b:

$$r_s = \sqrt{\left(\frac{a \times b}{\pi}\right)}$$

where r_s = radius of equivalent single well.

Next Weber (1928) determined that the radius of influence (R_o) of a pumping system is dependent upon permeability (k) and the amount of drawdown ($H - h_o$).

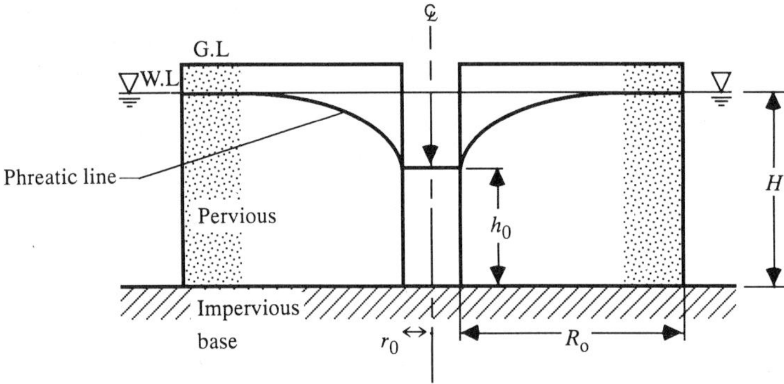

Dupuit's equation for the water table condition is:

$$Q = \frac{\pi k\,(H^2 - h_0^2)}{\log_e R_0/r_0} \qquad (1)$$

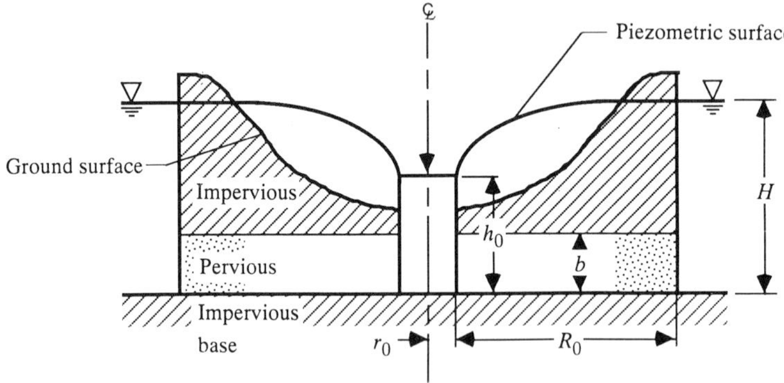

Dupuit's equation for the confined aquifer condition is:

$$Q = \frac{2\pi k b\,(H - h_0)}{\log_e R_0/r_0} \qquad (2)$$

Where:

Q total rate of pumping (m^3/s)
k coefficient of permeability (m/s)
b thickness of the confined pervious stratum (m)
H elevation of the original piezometric surface above the impermeable base (m)
h_0 elevation of the operating level of the pumping well above the base (m)
R_0 radius of the Dupuit island (m)
r_0 radius of the pumping well (m)

Figure 15.9 *Confined and unconfined aquifer conditions – Dupuit Equations*

Based upon analysis of much field data Weber proposed the empirical relationship

$$R_0 = C(H - h_0)\sqrt{k}$$

For radial flow, a value of $c = 3000$ is commonly acceptable; while for line flow a C value of 1500 to 2000 is often used. However, these are empirical and generally accepted average values which have been demonstrated to be reasonable for most cases, but where there is an adjacent source of recharge, such as a nearby water course, they may need to be modified.

The Dupuit formulae for computing total flow (Q) to the well system are shown on (Figure 15.9). The potential yield from an individual well will be dependent upon bore diameter (r_o), permeability (k) and wetted depth (h_w). There are published data sheets available for evaluation of well yield where the above three factors have been determined and so the number of wells required and individual pump outputs may be determined.

The Dupuit formulae are not appropriate for determining rate of flows to a line drain such as at the toe of a slope or on a berm. Based on model studies, Chapman (1956) proposed for partial penetration by a single row of wellpoints (or of a continuous slot) into an unconfined aquifer fed from a line source:

$$Q = \left[0.73 + 0.27\frac{(H - h_o)}{H}\right]\frac{kx}{2R_o}\left[H^2 - h_o^2\right]$$

where x = length of trench (m) and all other symbols are as given previously (see Figure 15.9).

All of the foregoing steady-state mathematical models are easy to use and assume isotropic homogeneous soil permeability/rock fissure pattern. Additional allowance should be made for non-homogeneity plus an allowance for additional pumping capacity required to draw the water down to steady-state levels. Fifty per cent additional pumping capacity is a reasonable rule of thumb.

When required to evaluate a large-scale and/or long-term groundwater control project, use of the mathematical models outlined above will give speedy indication of the order of the costs but the design proposals should be re-examined using the more refined analyses outlined by Mansur and Kaufman (1962). These take account of the cumulative effects of non-symmetrical well arrays and, if applicable, partial penetrations, and they lend themselves to computer studies.

15.8.3 Non-steady state analyses

Theis (1935) developed the non-steady state well formula whereby the drawdown at any given time after commencement of pumping can be predicted. The aquifer coefficients can be determined by means of a single observation well (i.e. multiple observation wells are not necessary though can be useful to determine if there is directional groundwater flow).

The radius of influence R_o can be determined from field observations by a plot of drawndown (s) against log distances of observation wells from the pumped well (i.e. multiple observation wells are needed) instead of having to use empirical determination of the radius of influence.

The Theis formula is based on assumptions similar to the Dupuit–Forchheimer formulae, namely, a homogeneous isotropic water-bearing formation of uniforms thickness and full penetration of the aquifer.

The Theis formula was refined by Jacob (1940). However, the non-steady-state analyses are heavily oriented to determining likely yields from single wells for domestic and industrial water supplies.

15.9 Summary

Excavation in water-bearing ground has always been a difficult operation and even with modern methods it is possible still to get into serious trouble, unless before work is started, sound engineering judgement has been exercised concerning the problems likely to be encoutered during the project and particularly the properties of the water-bearing soils and rock formations. Hence, in order to optimize mining or construction progress it is essential that the groundwater control systems be installed and operated so as to be capable of ensuring stable and workable conditions. This fundamental requirement dictates that there has to be adequate data concerning soils, rock formations and groundwater (including seasonal variations) available at the pre-planning stage.

Thus it should be possible to determine which of the groundwater and surface water control techniques are appropriate and thence to proceed to evaluate the installation and operating costs of the control systems proposed. These costs form an important part within the overall feasibility and viability reports.

The project management team should be experienced, have pragmatic philosophies and ensure that there are available on-site adequate alternative standby units and facilities in case of individual pump failure, power failure, etc.

Acknowledgements

The authors are grateful that

1. permission to reproduce parts of CIRIA Report No 113, namely Table 15.1 and Figures 15.1 and 15.2, has been given by the Director General of CIRIA;
2. permission to reproduce Figure 15.3 has been given by the CEGB Barnwood – Dungeness 'A';
3. permission to reproduce Figure 15.8 has been given by the Chairman of Geevor plc.

References

Anon (1975) *Groundwater and Wells,* Johnson Division, UOP Inc., Saint Paul, Minnesota, Chapter 6

Anon (1981) *Code of Practice for Site Investigation*, BS 5930, British Standards Institution, London p. 131 (line 6)

Batchelor D.H. and Wardle, D.H. (1969) 'Sealing of the undersea breach in Levant tin mines, Cornwall', *Trans. Inst. Mining and Metall.*, **78**(A752), A65–A89

Beard, G.L. (1986) 'Point of Ayr Colliery Surface Drift Development' *The Mining Engineer*, November, pp. 225–232

Boulton, N.S. (1951) 'The flow pattern near a gravity well in a uniform water bearing medium', *J. Inst. Civil Engineers*, **36**, 534–550

Cashman, P.M. and Hawes, E.T. (1970) 'Control of groundwater by water lowering', *Proc. Conf. Ground Engg.*, Institution of Civil Engineers, London, 23–32

Chapman, T.G. (1956) 'Groundwater flow to trenches and wellpoints', *J. Inst. Engs. Australia*, **191**, 275–280

Darcy, H. (1846) *Les fontaines publics de la ville de Dijon*, Dalmont, Paris.

Dupuit, J. (1863) *Etudes théoretiques et practiques sur les mouvements des eaux*, Dunod, Paris

Forchheimer, P. (1930) *Hydraulik*, Teubner, Leipzig and Berlin

Hantush, M.S. and Jacob, C.E. (1955) 'Non-steady state radial flow in an infinte leaky aquifer', *Trans. Am. Geophys. Union*, **36**, 95–112

Hazen, A. (1893) 'Some physical properties of sands and gravels', *24th Annual Report*, Massachusetts State Board of Health pp. 541–556

Hvorslev, M.J. (1951) 'Time lag and soil permeability in groundwater observations', *Bulletin 36*, U.S. Waterways Experiment Station, U.S. Army Corps. of Engineers, Vicksburg, Mississippi

Jacob, C.E. (1940) 'On the flow of water in an elastic artesian aquifer', *Trans. Am. Geophy. Union*, **21**, 574–586

Kruseman, G.P. and De Ridder, N.A. (1979) 'Analysis and evaluation of pumping test data', *Bulletin No. 11*, International Institute for Land Reclamation and Improvement, Netherlands, 401 pp.

Louden, A.G. (1952) 'The computation of permeability from simple soil tests', *Geotechnique*, **3**, 165–183

Mansur, C.I. and Kaufman, R.I. (1962) 'Dewatering', In *Foundation Engineering*, G.A. Leonards (ed.), McGraw-Hill, New York, pp. 241–350

Powers, J.P. (1981) *Construction Dewatering: A Guide to Theory and Practice*, Wiley – Interscience, New York

Somerville, S.H. (1988) 'Control of Groundwater for temporary works', *Report 113*, Construction Industry Research and Information Association, London

Theis, C.V. (1935) 'The relation between the lowering of the piezometric surface and the rate and duration of discharge of a well using groundwater storage', *Trans. Am. Geophys. Union Reports and Papers, Hydrology*, **16**, 519–524

Weber, H. (1928) *Die Reichweitte von Grundwasserabsenkungen Mittels Rohrbrunnen*, Springer, Berlin

Wenzel, L.K. (1942) 'Methods for determining permeability of water bearing materials with special reference to discharge well methods, *Water Supply Paper No. 887, United States Geological Survey*, Washington D.C., 192 pp.

16 Ground freezing

Professor F G Bell
Univerisity of Natal

16.1 Introduction

Ground freezing involves the artificial lowering of ground temperatures so that pore water is converted into ice, thereby reducing the permeability and increasing the strength of the ground so treated. Basically it is an exercise in heat transfer and in the absence of moving groundwater, heat transfer in the ground is by conduction. Thus the thermal conductivity of the ground governs the rate at which freezing proceeds. Fortunately the values of thermal conductivity for all types of frozen soil fall within quite narrow limits. This is one of the reasons why artificial ground freezing is a versatile technique, being able to deal effectively with a great variety of types of ground.

Artificial freezing is generally employed as an exclusion technique to stop the flow of groundwater into excavations. In this context, frozen walls can be used in conjunction with bored pile walls or slurry trenches. The frozen ground permits the excavation to be free of bracing and earth supports. The barrier to groundwater provided by an ice wall means that there is no need for drawdown of the water table. Consequently the lateral flow of groundwater must be given full consideration during the planning stage of a freezing project since such flow can give rise to the failure of a frozen wall.

Although the process of ground freezing has been used chiefly in connection with sinking of mine shafts, it has several other useful applications. These have included stabilizing tunnels; sinking caissons delayed by poor ground conditions; underpinning buildings undergoing rotational settlement; containing slides; and backfreezing piles placed in permafrost. Maishman (1975) mentioned an instance where freezing was instrumental in stabilizing a long abandoned flooded mine shaft, which was in a dangerous condition, so that it could be safely plugged at rock-head. Other uses have included mat freezings to provide temporary roads over marshland, and freezing the bottom of a lake to aid the mining of ore reserves below. Paradoxically, artificial ground freezing also is used to keep the ground below heated buildings in permafrost regions permanently frozen, thereby avoiding foundation problems.

According to Jumikis (1979) the following factors influence artificial freezing of the ground:

(1) The type of ground to be frozen, its porosity, water content and amount of ice present, if any. Although complete saturation of the ground is desirable, freezing may be carried out safely when the degree of saturation is as low as 10%.
(2) The position and fluctuations of the water table.
(3) The velocity and temperature of groundwater flow.
(4) The physical and thermal properties of ground, water and ice. Thermal conductivity varies vertically with each change in water content and lithology. It also changes horizontally from the frozen to the unfrozen ground.

The mechanical properties of frozen ground are much more dependent on time and temperature than on the geology of the strata involved (Jessberger 1981). In fact artificial ground freezing is less sensitive to geological prediction than other methods of ground treatment and may be used in any moist rock formation, irrespective of its structure or permeability (Shuster 1972). Rock profiles, too heterogeneous to grout predictably, also are amenable to freezing. However, in heterogeneous ground the frozen zone tends to be irregular in shape. Nonetheless the mechanical properties and thermal performance of frozen ground are influenced by the thermal coefficients of expansion of the various components of the ground. More specifically the engineering performance of ice depends on the amount of impurities and dissolved gases present in the groundwater; the formation of air bubbles at the ice–water interface; the kind and amount of solute; the rate of crystal growth; the charge separation during the process of solidification; and thermal deformation which is a function of temperature. A crystal of ice is made up of parallel platelets weakly bound together. Hence, ice cleaves perfectly along the contacts between adjacent platelets. In this way it forms an ideally flowing solid, which grows in the direction of heat flow.

Frozen ground can undergo creep and suffer a reduction in strength as time proceeds, therefore the design of a frozen wall must take account of the dimensions of the workings at which the continuity of the frozen ground is not impaired, with a given overburden pressure and within the proposed time over which the wall will be in existence. Laboratory testing of frozen samples may be required to obtain the creep stength and deformation under stress conditions expected to prevail during the lifetime of the frozen wall.

Artificial freezing of ground is usually carried out in two stages, which are referred to as the *active* and *passive* stages of freezing. Active freezing involves freezing the ground to form the ice wall whilst passive freezing is that required to maintain the established thickness of the ice wall against thawing. Initially columns of frozen ground form around the freeze probes. As the columns develop they coalesce to form the frozen wall. The refrigeration plant has to operate at a much higher capacity during the active than the passive stage of freezing. The choice of refrigeration plant and coolant, according to Shuster (1981), errs significantly on the conservative side. In other words a refrigeration system used to form and maintain a wall 1.5 m thick normally will also form a wall of similar temperature 2–3 m thick. Shuster went on to argue that this, together with the fact that a frozen ground structure changes continuously, means that sophisticated material testing, characterization and structural analysis are rarely justified. He maintained that it usually is easier, less expensive and more certain to lower the temperature and increase the size of an artificial frozen ground structure than to develop a refined analysis. Thus the successful execution of a ground freezing project depends on the experience of the individuals concerned in its design and construction and, in particular, on the field engineering quality assurance. Indeed Shuster concluded that a ground freezing system and associated frozen ground structure are indeterminate from the point of view of accurate engineering computations.

16.2 Freezing methods

16.2.1 Conventional use of supercooled brine

The primary plant with pumped loop secondary coolant is the system most frequently used (Figure 16.1). An ammonia or freon refrigeration plant provides the primary source of refrigeration. The cold source consists of a primary circuit which, by compressors and condensers, bring the ammonia or freon to its liquid state. The refrigeration fluid is turned to gas in an evaporator by extraction of heat from the brine flowing in a secondary circuit, which includes the freeze probes. The advantages of ammonia are that leaks can be detected easily by smell and it is considerably cheaper than freon, which is odourless. The system is worked by either electricity or diesel motor and the condensers may be air or water cooled. These plants have a wide range of capacities. However, the refrigeration capacity frequently is limited by the condenser since a larger refrigeration compressor and motor are worthless if the condenser is insufficient. Two stages are required to produce temperatures below −25°C.

Several different types of coolant have been used with this system including glycol–water mixtures, diesel oil and propane, as well as brines. However, the most common type of refrigerant is calcium chloride brine. Other brines have been made from chlorides of sodium, magnesium or lithium. The crystallization point of the chosen brine should be at least 5°C lower than the ultimate temperatures at which ground freezing will proceed. Calcium chloride brines are dense (specific gravity 1.24–1.28) relatively viscous and corrosive, with high specific heat. Any coolant has a temperature beyond which it cannot be chilled. This is about −40°C in the case of brine. The

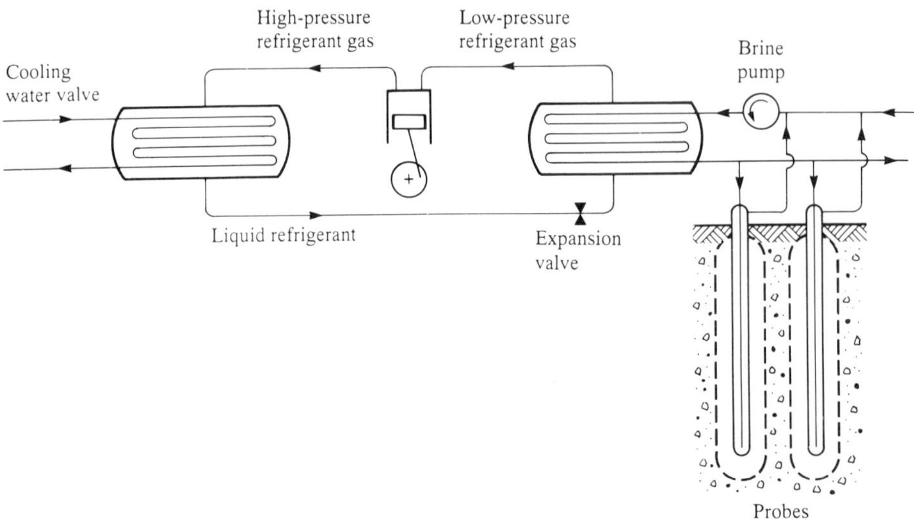

Figure 16.1 *Layout for circulating brine cooling system*

lowest temperatures are achieved by using lithium chloride brine. Typical brine temperatures for commercial refrigeration plants are around −20°C to −40°C. A temperature of −5°C can normally be attained throughout the zone to be frozen in reasonable freezing times and freeze-probe spacings.

Heat is removed from the ground by way of probes placed in drill holes, or alternatively drilled or thrust into the required positions. In fact the installation of the freeze probes represents the most costly, time consuming and risky part of a ground freezing project.

Rotary drilling with mud flush represents one method by which freeze probes can be installed rapidly and efficiently. However, oil or salt-water-based muds should be avoided as they are likely to affect ground freezing adversely or reduce its strength considerably. Loss of drilling mud below the water table suggests the possible existence of problems due to groundwater flow. Consequently when this occurs it may be necessary to undertake remedial grouting to reduce permeability or to fill any voids which are present.

A coolant distribution system normally is assembled in a series parallel configuration. The amount of freeze pipe in a single series in the ground generally should not exceed an area of around 50 m^2 of heat transfer. Each probe consists of an external pipe, closed at the lower end, and contains an open-ended inner tube of slightly shorter length. Cooling fluid is normally introduced through this inner tube. The supply and return temperatures of the coolant in the freeze probes usually are monitored by thermocouples to determine the amount of energy removed. The difference between these two temperatures (commonly referred to as the 'split') varies more or less in proportion to the number of probes in the series (e.g. the more probes, the larger the split and vice versa). In the absence of unusual sources of heat, a balanced heat flow is obtained by balancing the split temperatures.

Each group of probes in series is connected in parallel between an insulated coolant supply manifold and a coolant return manifold. The configuration of the manifold around the perimeter of the area to be frozen should be arranged so that multiple sections are of approximately the same length to try to ensure an equal distribution of coolant. Valves have to be incorporated in the main supply and return manifold system at selected positions to allow any part of the system to be dismantled for repair or correction without coolant being lost or the system having to be closed down.

Heat is transferred between the coolant and freeze probes by convection. Large quantities of coolant have to be circulated to bring about freezing. Unfortunately the circulating pumps required to distribute the coolant contribute heat to the system in direct opposition to the primary refrigeration plant. The refrigeration capacity of each freeze probe can be increased only by increasing the capacity of the entire system. However, the ability to increase the capacity of individual probes is desirable in order to control localized conditions, especially unexpected water flows.

A temperature gradient exists along the length of the freeze probe which is greater when the content is circulated slowly. The coolant emanating from the bottom of the inner pipe is warmed by the abstraction of heat from the ground as it moves upwards. Consequently the growth of ice is more rapid at the base of the freeze probe. This, to some extent, can be overcome by circulating the coolant more quickly, which gives rise to turbulent flow in the freeze piping, which, in turn, is more favourable to efficient heat exchange than laminar flow.

As the brine rises in the annular space between the freeze probe members, heat is extracted from the environment and a warmer brine is discharged from the head of each probe into a collection main, whence it is piped back to the refrigeration plant. Hence it is pumped through a chiller and is delivered, recooled, via a distribution main to the inner tubes of the probes. The coolant is therefore confined in a closed, recirculatory flow path.

All piping above the surface in a freezing system must be insulated to minimize heat loss. The insulation, preferably a closed-cell foamed plastic, should be protected against weather and damage and should be painted with white or aluminium paint. Insulation also is necessary in large shallow excavations.

16.2.2 Use of expendable refrigerants

Expendable refrigerants are very attractive for individual projects which last only a few days or for projects where the cost of delay is high. Frequently such systems are used in emergency conditions where time is critical in that the formation of a frozen wall is required quickly. They are advantageous where only small volumes of ground require freezing. They have also been used where groundwater flow is heavy.

Expendable refrigerant systems normally use liquid nitrogen or to a lesser extent, carbon dioxide in an open system. The refrigerant is lost to the atmosphere after it has absorbed energy and vapourized. Such systems can freeze the ground in a matter of hours. However, they are relatively expensive since the cost of refrigerant is high. However, the use of expendable refrigerant to freeze ground proves difficult to control. For example, venting of liquid nitrogen in vertical or horizontal freeze probes produces an irregular frozen zone as well as wasting refrigerant. The irregular nature of the frozen zone is due to the variation in the heat-transfer coefficient, it being governed by the quality and velocity of the liquid–vapour mixture. A supply and exhaust manifold with appropriate valves at each end end of a series of freeze probes allow a reversal of flow and so help smooth irregularities in freezing.

Liquid nitrogen (LN_2) provides the fastest and thermally most efficient means of freezing the ground. The use of carbon dioxide is thermally less efficient than liquid nitrogen and normally is more difficult to control. For instance, liquid carbon dioxide tends to plug orifices, values and piping. Moreover, carbon dioxide (dry ice), even in pellatized form, is bulky and difficult to handle. On the other hand if dry ice is used with a mixing tank and a circulating secondary coolant it probably provides one of the most efficient ways of using expendable carbon dioxide.

Freeze pipe systems may consist of either a concentrically arranged downpipe and riser or a freezing lance, consisting of a perforated pipe, or three concentric pipes for localized icing (Figure 16.2). The pipes are composed

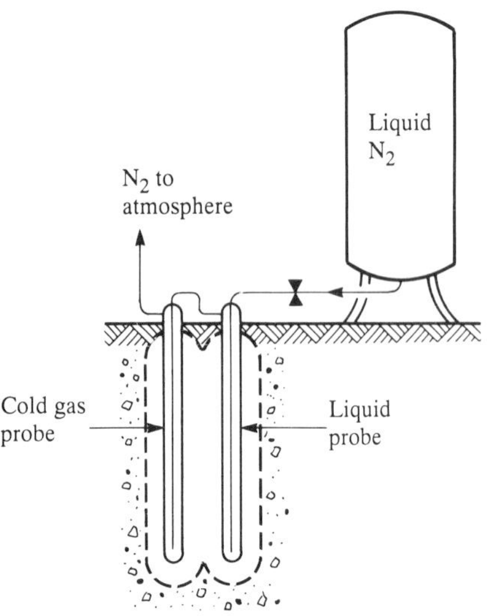

Figure 16.2 *Freezing system using liquid nitrogen*

of metals, such as V2A steel, copper or brass, which retain their toughness at sub-zero temperatures, and materials which are semi-flexible such as non-rigid polyamide. They are protected by high-grade insulation. A tanker carrying liquid nitrogen, from which it is injected into the freeze pipes, may be used on small or emergency projects otherwise one or more double-walled storage tanks are needed on site (Gallavresi 1985). Liquid nitrogen is introduced directly into the circuit, by utilizing the pressures developed in the storage tanks, through a distribution pipe carrying the liquid nitrogen to the freeze probes. These are grouped and connected in series. There is a valve downstream from the terminal freezing pipe through which the nitrogen gas is vented to the atmosphere.

The pressure and consumption of liquid nitrogen are regulated at the storage tank in accordance with the desired rate of heat extraction to form the frozen wall. Each freeze pipe and assembly incorporates shut-off values, enabling the liquid nitrogen to be delivered to specific areas. Automatic delivery apparatus regulates supply, and functions according to the exhaust gas temperature. This tends to stabilize the amount of refrigerant used. Once ground freezing has been achieved, liquid nitrogen is introduced whenever the thermometers indicate temperatures in the ground higher than those acceptable.

According to Veranneman and Rebhan (1979) 800 kg/ LN_2 per cubic metre of ground represents a useful mean value for calculating the quantity needed to freeze a certain volume, under normal conditions, that is, with a low water content and no groundwater movement. This value can vary between 600 and 1900 kg/m^3 of ground. The same quantity of liquid nitrogen is required to maintain an ice wall in the frozen state for two to three weeks as that needed for its initial formation.

The most important technical advantages of using liquid nitrogen for freezing include:

(1) Freezing with liquid nitrogen is independent of expensive refrigeration units, and site equipment and installations are relatively simple to erect. Accordingly material turn around is fast and site upkeep and energy running costs are low.
(2) The time required to form a frozen wall of given thickness is comparatively short.
(3) The higher strength of the ice wall formed by freezing with liquid nitrogen (which is directly proportional to the negative temperature reached) means that the wall can be thinner.
(4) Little heaving, or subsequent settlement, results from this method of freezing.
(5) Because of the very low operating temperature the method is less affected by groundwater flow.
(6) Rupture or leakage of freeze pipes cannot cause local thawing of the frozen wall. Indeed liquid nitrogen can be pumping into the ground to deal with fractured pipes.
(7) Liquid nitrogen proves an ideal refrigerant in densely populated areas subject to stringent emission and noise-control requirements.
(8) Unlike brine freezing there is no risk of groundwater pollution when using liquid nitrogen.

The main disadvantage in using liquid nitrogen is that it is difficult to design and control the system to give a regularly shaped frozen wall.

It is possible to combine brine and liquid nitrogen freezing (Gallavresi 1981). In this way the freezing capacity can be raised to meet peak demands, or liquid nitrogen can be used as a stand-by in the event of failure. Another possibility involves freezing the ground with liquid nitrogen and then keeping it frozen with cooled brine. Liquid nitrogen can be very quickly applied again if unforeseen difficulties arise. Depending upon the circumstances, a decision will have to be made as to whether it is better to inject both the liquid nitrogen and brine into one system of pipes, or into two separate systems.

The liquid nitrogen method can speedily produce almost unlimited quantities of energy to meet individual demands. Hence freezing with liquid nitrogen increases the speed of ice formation. It also gives rise to less heaving and greater strength than freezing with brine. Table 16.1

Table 16.1 Characteristics of LN_2-freezing and brine freezing (after Stoss and Valk 1979)

	Quality	*LN_2*	*Brine*
Site installations	Electric power	Not required	Required
	Water for cooling	Not required	Required
	Refrigeration plant	Not required	Required
	Storage tank	Required	Required
	Circulation pumps	Not required	Required
	Pipe system for distribution of coolant	Supply only	Supply and return
	Low-temperature material for surface pipes, valves, etc.	Required	Not required
	Low-temperature material for freeze pipes	Not required	Not required
Execution of freezing	Physical condition of coolant	Liquid/vapour	Liquid
	Minimum temperature achievable (theoret.)	−196°C	−34°C $MgCl_2$ −55°C $CaCl_2$
	Re-use of coolant	Impracticable	Standard
	Control of system	Difficult	Easy
	Shape of freeze wall	Often irregular	Regular
	Temperature profile in freeze wall	Great differences	Small differences
	Frost penetration	Fast	Slow
	Impact on freeze wall in case of damage to freeze pipe	None	Thawing effect
	Noise	None	Little

outlines the characteristics of freezing ground with liquid nitrogen or brine.

Liquid nitrogen evaporates at a temperature of −195.8°C at atmospheric pressure. As it evaporates it draws 162.4 kJ/l from its surroundings, and about the same amount of heat is needed to raise the cold gas to ambient temperature. Nitrogen is non-inflammable and non-poisonous.

16.3 Hydrogeology and ground freezing

Hydrogeology is the most important technical factor in an evaluation of the feasibility of ground freezing. For example, according to Braun *et al.* (1979) the late and/or untimely detection of flowing groundwater has resulted in most of the delays which have occurred during artificial ground freezing projects. The reason for this is that the natural enemy of a frozen ground structure is heat, especially that contained in moving groundwater. Hence, according to Shuster (1981), stresses and strains and related ground movements are all secondary to the economical, efficient insulation from, or collection and disposal of, heat.

As the most important technical factor affecting the feasibility of a ground freezing project is the amount, quality and velocity of flow of the groundwater at the site, then three criteria must be established. First, there must be sufficient water present in the ground to bond it together so that it possesses adequate strength when frozen. Secondly, the groundwater must not be contaminated since this can lower the point at which freezing takes place significantly and thereby reduce the strength of the frozen ground. Thirdly, the velocity of groundwater flow must be such that it does not introduce enough energy into the zone which is to be frozen so that the required amount of heat cannot be removed by the refrigeration system.

The best situation in terms of moisture content for a ground freezing project is when the ground is saturated, although freezing may take place effectively when the degree of saturation is as low as 10%. Even when it is less than 10% certain measures can be taken to allow freezing to proceed.

Where dissolved salts are present in the groundwater, chemical analysis should be made and the temperature at which freezing occurs should be ascertained. Sea water presents no special problems in this respect since its freezing point is only about 3°C below that of fresh water. However, when a volume of saline water in the ground has been confined by a perimeter array of freeze probes, further inward advance of the frozen wall causes a progressive increase in the salinity of the unfrozen fluid. In narrow excavations the salinity of the water in the unfrozen core is commonly several times that of the original groundwater.

Artificial ground freezing has been used in formations containing saline water derived from evaporitic deposits. However, the temperature of the coolant must be much lower than when groundwater is fresh. Maishman (1975) noted that the ground had to be lowered to −21°C in order to freeze naturally occurring brines associated with salt domes in Louisiana. In this situation the strength of the frozen soil is lower at any given temperature than it would be otherwise.

Movement of groundwater normally leads to an uneven development of an ice wall in the direction of the flow. If the velocity of groundwater flow is too large, the frozen columns in the first stage of freezing will not merge, thereby leaving windows in the wall. Indeed a frozen wall may be 99% impervious but small windows may not close no matter what refrigeration power is used. Obviously it becomes impossible to freeze soil if the amount of heat conveyed to the freeze zone by flowing groundwater exceeds that being removed by the freezing system.

It is generally accepted that the normal freezing process is not likely to be successful where the flow of water exceeds 1.5 m/day. In such situations special provisions have to be taken either to lower the flow rate or to cope with the excess heat. Sanger and Sayles (1979) suggested the following expression for determining the critical velocity of groundwater flow (v_c):

$$v_c = \frac{k}{4S \ln (S/2d_0)} \frac{t_s}{t_0} \text{ m/day} \tag{16.1}$$

where k is the thermal conductivity of the frozen soil, S is the probe spacing, d_0 is the diameter of the freeze probe, t_s is the difference between the temperature at the surface of the freeze probe and the freezing point of water and t_0 is the difference between the original temperature of the ground and the freezing point of water. Braun *et al.* (1979) recommended that if groundwater flow is greater than 1.5 m/day but less than 3 m/day, then the spacing between freeze probes should be reduced or a second row of probes should be placed on the upstream side. An alternative is to reduce the temperature of the coolant. When the flow exceeds 3 m/day, either the groundwater gradient or the permeability of the formation must be reduced. This is brought about by grouting before or during the installation of the freeze probes, or by intercepting the flow with a sheet pile wall or with wells. Generally the latter method provides the most reliable solution. Liquid nitrogen has been used successfully where groundwater flows of more than 50 m/day have been encountered.

The position of the water table plays a significant role in ground freezing because of the importance of water content as far as the mechanical properties of the frozen ground are concerned. A fluctuating water table may occur near rivers or lakes which have significant seasonal variations in level, or along coasts where the rise and fall of the tides are appreciable.

The ground beneath the base of excavation is extremely important. Ideally the base of a frozen wall should be located in an impervious layer which forms a seal preventing the incursion of groundwater into the excavation. In such a situation there is no need for any significant pumping to control groundwater. Sanger (1968) suggested that freeze probes should penetrate such a layer to a depth of 3 m. Where a frozen wall is not founded in an impervious horizon the excavation has an open base. Consequently extreme caution must be exercised to minimize water movement beneath the frozen wall and to avoid heaving of the floor. Ground freezing is usually much more risky when an open base condition is likely to be met with beneath the water table. Braun *et al.* (1979) mentioned that continuous pumping from an open sump should not be used within a circular frozen wall because of the relative lack of control. Furthermore dewatering from within an ice wall, could draw in a significant quantity of water from beneath the toe and this could cause thermal erosion. Deep wells have been used around shafts to provide a reliable means of disposing of seepage water. The drain should be located at more than three times the thickness of the frozen wall from its outer circumference.

Surface water must be collected and drained away from a frozen wall, otherwise severe thermal erosion may be caused if water pours over the wall and floods the excavation or ponds against the wall. Once the wall is eroded by water it generally is difficult and costly to repair the area affected.

As an ice wall grows this leads to an increase in pore pressure in the area enclosed. This is due to the expansion which is consequent upon the change of water to ice. Holes can be drilled within the central area to dispose of the water and thereby relieve the pressure (Cleasby *et al.* 1975; Wild and Forrest, 1981). Pressure relief holes may have to be cased in order to prevent their collapse.

16.4 Design of a frozen wall

The first step in the design and construction of an ice wall is to find out whether the site is suitable for the freezing process. This entails carrying out a site investigation. Then, if the site proves satisifactory, a structurally stable configuration is chosen for the ice wall. Lastly, the arrangement of the freeze pipes is determined along with the refrigeration requirements (Collins and Deacon 1972).

The exploration holes sunk during the site investigation should extend well below the proposed base level of the excavation. A drillhole is normally sunk at the centre of a shaft area. Undisturbed samples should be taken and the ground temperature, groundwater level, flow and piezometric pressure should be measured.

When the depth of artificial ground freezing has been finalized, the internal dimension(s) of the area enclosed by the freeze probes is usually determined by surface construction requirements. The frozen wall then has to be designed to ensure safety in the various water-bearing formations. Most ground freezing contractors have evolved their own method of calculating the thickness of a frozen wall. When the thickness of the frozen wall has been decided, the number of freeze holes and their spacing, the refrigeration capacity, the diameter of the inner tubes, brine mains and ring mains, and the capactiy of pumps required, is ascertained.

As mentioned above, one of the first items requiring consideration as far as the design of a frozen wall is concerned is its thickness. Wall thickness takes account of the creep strength of the frozen ground and the amount of deformation likely to take place during the lifetime of the wall. If investigation indicates that the amount of deformation probably will be excessive, then the design must accommodate this deformation within acceptable limits. This can be achieved by changing the geometry of the frozen wall especially by increasing its thickness or by lowering the temperature of the wall, hence increasing its strength. More than one row of freeze pipes may be necessary for efficient freezing if a very thick ice wall is required.

The spacing of the freeze probes, as indicated by the ratio of their spacing to their diameter, is critical for the successful and timely completion of a frozen wall. Generally speaking this ratio should not exceed 15 (Braun *et al.* 1979). Subsequently Shuster (1981) suggested that this ratio should be 13. Not only can wider spacing than this produce excessive delays in the development of the frozen wall but it can suffer progressive deterioration after being exposed on excavation. In fact field observations suggest that the radius at which the ground temperature is unaffected by freezing is 3–5 times the radius of the frozen column.

Sanger (1968) contended that a spacing between freeze probes of 0.9–1.5 m has proved satisifactory. Smaller spacings have been used in ground of low conductivity and high water content, or where there was a significant flow of water which meant that closure of the ice wall was critical. Trial designs may be required in such situations.

The centreline of each freeze probe should not deviate from the vertical by more than 150 mm throughout the entire depth of a hole in a safe and timely ground freezing operation. Consequently each hole drilled should be checked for deviation, especially if it is deeper than 10 m or if there is a flow of water in the ground. Drill collars are now being used in open hole drilling and the use of stabilizers fitted to the drill collar string has provided a more effective method of obtaining a vertical hole (Wild and Forrest, 1981). If it is found that two adjacent holes are too far apart, then the offending holes should be redrilled or another should be sunk nearby to minimize the risk of a window forming in the ice wall.

Shuster (1972) demonstrated the influence of the size and spacing of freeze probes on the time required to freeze a wall by various methods of freezing (Figure 16.3). In fact the size and spacing of the freeze probes are the most important factors governing the cost of an artificial freezing project. Obviously the amount of drilling necessary, the capacity of the refrigeration plant and the time taken for freezing depend on these factors. The period of time during which active freezing is in progress, that is, the time it takes the frozen zone around a freeze probe to form and expand until it connects with those around adjacent probes, is exponentially proportional to the relative spacing of the freeze probes.

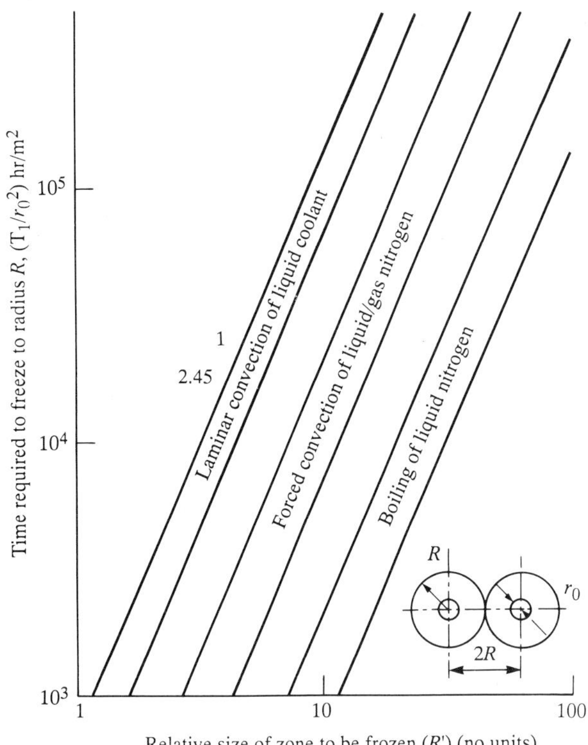

Figure 16.3 *Generalized relationship between freeze-pipe size and required freezing time*

General notes:

1. *Indicated bands represent normal range of observed field and laboratory results. However, results with forced convection of LN_2 may vary more widely than indicated due to variables in control of freezing process.*
2. *$R' = R/r_0$*

Other factors which affect the time taken to form a frozen ground structure include the type of ground and its thermal properties, as well as those of the coolant, the moisture content and the velocity of groundwater flow. The freezing time is directly proportional to the amount of heat which has to be extracted from the ground and inversely proportional to the temperature selected for freezing. Furthermore in any given type of ground the relative temperature of the refrigerant and spacing of the freeze probes are influenced significantly by the efficiency with which heat is transferred between the refrigerant and the freeze probe.

Halving the distance between the freezing columns reduces the freezing time by a factor of four. Doubling the numerical value of the negative coolant temperature (for example, from −20° to −40°C) shortens the freezing time by the same ratio. This can be achieved by a double-stage refrigeration unit. In order to minimize the possibility of inadequate refrigeration capacity, it is desirable to ensure that the refrigeration system has an actual heat-rejection capacity, at a coolant temperature of −23°C, in excess of about 165 kcal/h/m^2 of refrigeration pipe surface, under the ambient conditions prevalent at the site (Braun *et al.* 1979).

16.4.1 A note on thermal analysis

Although mechanical design is one of the chief problems in artificial ground freezing, Frivik (1981) pointed out that it should be preceded by a thermal analysis in order to obtain the temperature distribution, elapsed freezing time and amount of energy required to carry out the work. Unfortunately, however, any thermal analysis for a ground freezing system is fundamentally crude. Indeed complex three-dimensional transient heat transfer in a heterogeneous porous medium including phase change, essentially is indeterminate. Moreover the specific refrigeration system and procedures which the contractor employs frequently determine the performance of the system and the *in situ* thermal regime that will exist. Frivik considered that these factors mean that refined analysis is a pointless exercise and that determination of refrigeration requirements prior to construction remains largely an art. Nonetheless he added that thermal design in artificial ground freezing should fulfill four requirements, namely:

(1) It should provide the information needed to optimize the size and spacing of the freeze probes in order to minimize drilling costs.
(2) It should give the time required for freezing the volume specified and the energy requirement.
(3) It should present the temperature distribution within the frozen wall as input for subsequent mechanical design.
(4) It must be performed within reasonable cost limits.

Two important thermal properties as far as the design of a frozen wall is concerned are the thermal conductivity and the heat capacity of the ground. The thermal conductivity of wet porous ground is difficult to assess since many parameters, not the least groundwater flow, influence the value. Nevertheless thermal conductivity basically depends upon the volumes of materials involved, that is, of the mineral particles, water–ice and ice and their particular thermal conductivities, as well as the fabric or arrangement of the component materials.

The heat capacity of the ground (Figure 16.4) outside the principal freezing interval can be derived from the capacities and weight fractions of its constituents so long as a phase change does not take place. However, as the apparent heat capacity is very much influenced by the amount of unfrozen water in the ground as well as being temperature dependent, then the amount of unfrozen water in relation to the temperature should be known within the freezing interval. The latent heat of fusion can be included in the apparent specific heat capacity, and the latent heat of fusion for ice, although it is temperature dependent, for practical purposes can be regarded as 334 kJ/kg.

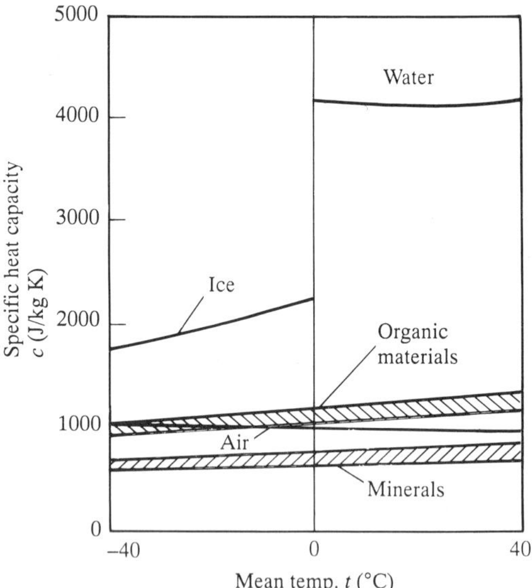

Figure 16.4 *Specific heat capacities for minerals and other constituents (after Frivik 1981)*

The thermal regime also is influenced by the refrigeration process (i.e. whether brines or cryogenic liquids are used since these influence the rate and amount of heat transfer), by the cooling, freezing and subsequent thawing of the ground and by the flow of groundwater. The specific refrigeration capacity (i.e. the ratio of the capacity, W, to the total area of the outer surface of the freeze probe, m^2) for the primary refrigeration system with pumped loop secondary coolant is shown is Figure 16.5 and relates the

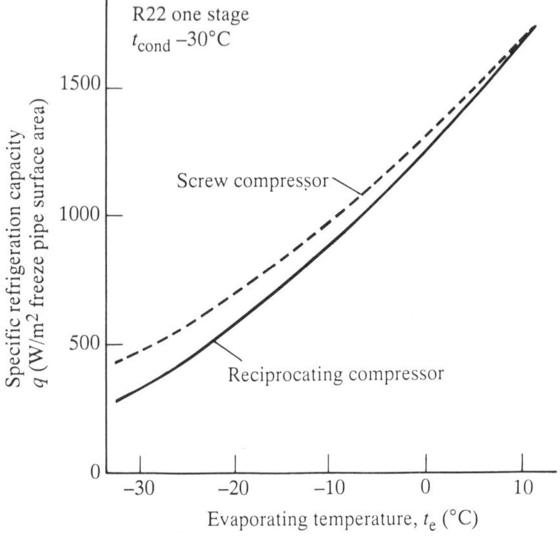

Figure 16.5 *Mechanical refrigeration with pumped-loop secondary coolant-specific refrigeration capacity for plant with screw/reciprocating compressor*

size of the system to the volume of ground required to be frozen. As can be seen from Figure 16.5, the specific refrigeration capacity is influenced greatly by evaporating temperature and consequently by ground temperature and length of time freezing. Because of changes in the properties of brine and refrigeration, changing velocities of flow in pipes and heat fluxes during freezing, the overall heat-transfer coefficients are temperature dependent. A low coefficient of heat transfer in the freeze pipes gives rise to lower evaporating temperatures. A pressure drop in the freeze pipes means an increase in the pump energy requirement. Both are related to the velocity of flow and physical properties of the brine (Figure 16.6). Heat transfer between the outer and inner pipes of the freeze probe can be estimated from Figure 16.6. As long as the refrigeration plant is run at full capacity, there is a fixed relationship between capacity and the temperature of the freeze pipe.

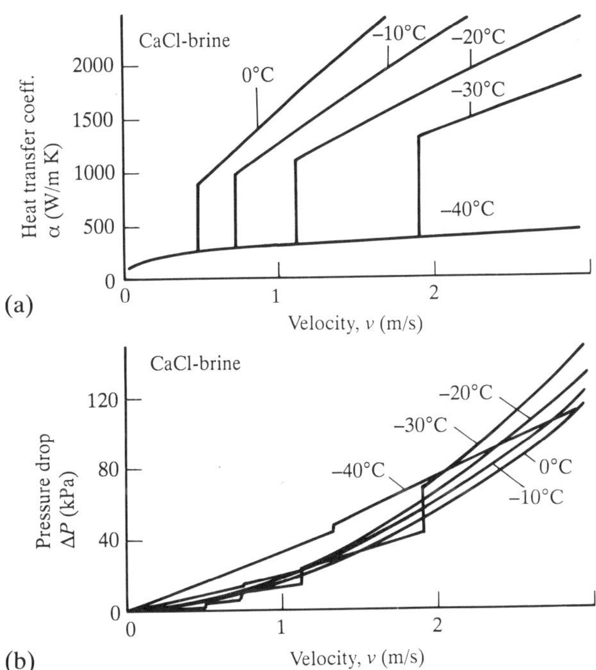

Figure 16.6 *Heat transfer coefficients* (a) *and pressure drops,* (b) *vs. coolant velocities, and temperatures for calcium chloride brine at eutectic concentrations (after Frivik 1981)*

Little data is available relating to the boiling heat transfer of cryogenic liquids in freeze pipes. The different thermal regimes are illustrated in Figure 16.7(a) and depend on the difference in temperature between the liquid and the pipe wall. As can be seen in Figure 16.6(a) film boiling occurs when a cryogenic liquid (i.e. LN_2) is introduced into a freeze pipe. It may take 3–4 hours before boiling moves into the nucleate phase. Frivik

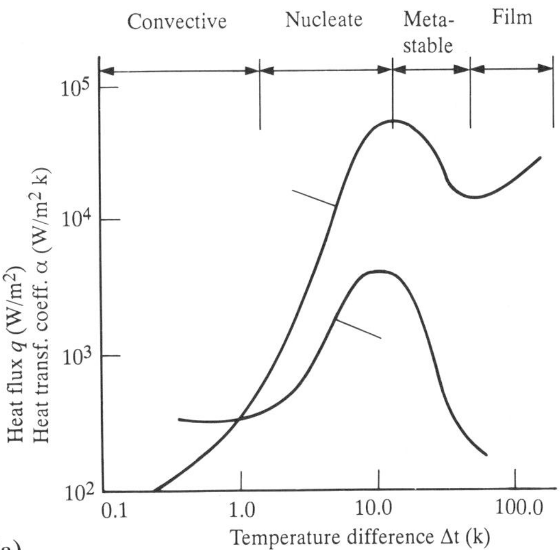

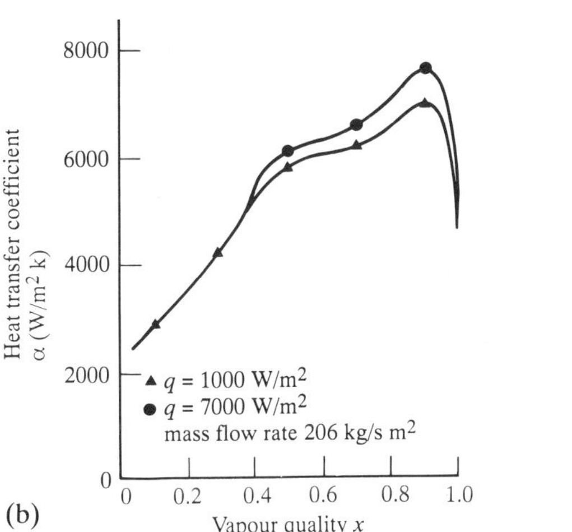

Figure 16.7 *Boiling processes and heat transfer coefficients for nitrogen* (a) *boiling regimes and heat transfer vs. temperature difference* (b) *heat transfer coefficient vs. vapour quality*

(1981) suggested that the heat flux decreases from about 7000 W/m^2 to 3000 W/m^2 within four days of freezing. The corresponding heat transfer coefficients and temperature difference range between 800 and 1500 W/m^{2}K, and 2 and 5 K respectively. Frivik therefore felt that the temperature difference between saturated liquid and freeze pipe could be assumed to vary between 5 and 10 K. Klein (1976) argued that any evaluation of heat transfer should take account of the main flow rate and the quality of the cryogenic vapour; the latter refers to the ratio between vapour mass flow and the total refrigerant (i.e. LN_2) flow (Figure 16.7(b)). The vapour quality reaches unity somewhere in the freeze pipe and beyond that point the remainder of the pipe has a low coefficient of heat transfer producing an uneven distribution of temperature. This problem can be overcome by connecting pipes in series so that the first is cooled by evaporation and the next by the sensible heat in gas.

Khakimov (1966) advanced an empirical method of design for determining the time taken to form a frozen wall around a row or circle of probes. All other things being equal, the time taken to freeze the ground varies inversely with the temperature of the coolant and the conductivity of the frozen ground, and is roughly in a direct ratio to the square of the freezing radius (the square of half the distance between the columns up to the time they merge). These relationships make it possible to select the coolant temperature in advance, as well as the distance between the holes to bring about freezing in a specified time. Khakimov also suggested how changes in the moisture content of the ground could be taken into account during freezing operations and examined the influence of the diameter of the freeze probes on the freezing time. The method provides a very conservative estimate of the length of the freezing time and it cannot predict how a frozen wall will really develop.

Khakimov (1966) maintained that it is possible to obtain a uniform removal of heat from the ground throughout the freezing period by lowering the temperature of the coolant in relation to the increase of the thermal resistance of the frozen ground. A single-stage ammonia refrigeration unit is sufficient for this purpose if the distance between freeze holes does not exceed 1.7 m. A two-stage unit suits any practically used distance between the holes. According to Khakimov freezing by multistage refrigeration units should be widely used since it reduces the freezing period, water migration and unproductive heat losses.

A general procedure for designing a retaining structure composed of frozen ground was proposed by Sanger and Sayles (1979) and was based upon the Khakimov method. They suggested that in order to determine the development of a frozen wall around vertical freeze probes, thermal calculations can be made in three stages. The first stage concerns the growth of frozen columns around each freeze probe. In the second stage these columns have merged to form a continuous wall which thickens as time proceeds. The third stage involves the formation of two walls from two rows of frozen columns; these then merge to produce a single wall which continues growing until the required thickness is reached.

16.4.2 Design of frozen walls for shafts

In a review of strength and stability design problems of frozen walls for deep shafts, Auld (1985) noted that a multiplicity of expressions have been used to determine the thickness of a frozen wall. These depend upon structural considerations (i.e. on the length of the wall), on whether support is provided to the inside of the wall (i.e. whether it is lined with a temporary or permanent lining during construction) and on the condition of the material comprising the frozen wall (i.e. is it in an elastic, partially plastic–partially elastic, fully plastic or viscous state).

One of the frequently used design methods was proposed by Domke (1915), for example, this method was used in the design of the frozen wall for Wistow No 2 Shaft in the Selby Coalfield, Yorkshire (Wild and Forrest 1981). Domke considered the case of a long thick unlined cylinder for which he suggested the following expression to derive the thickness ($t = R - r_0$) of a frozen wall around a mine shaft:

$$t = r_0 \left[0.29 \left(\frac{p_a}{\sigma_c} \right) + 2.3 \left(\frac{p_a}{\sigma_c} \right)^2 \right] \qquad (16.2)$$

where σ_c is the unconfined compressive strength and p_a is the horizontal load of earth pressure on the wall (Figure 16.8). This expression was based on the shear-stress

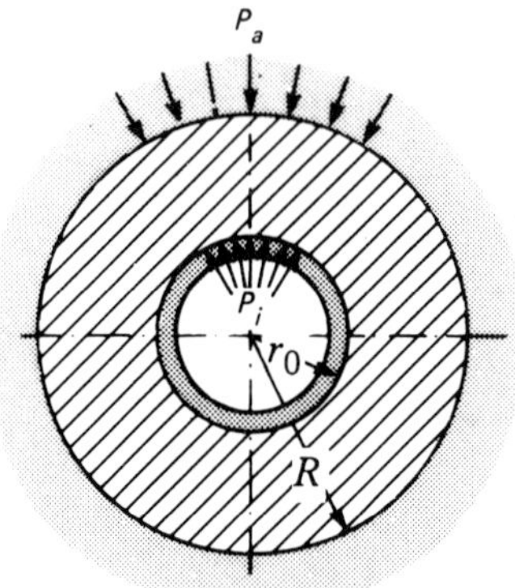

Figure 16.8 *Geometry of and load on an annular frozen wall*

criterion and on the assumption that the frozen wall is in a partly elastic, partly plastic condition. The boundary, s, between these two states in the wall is given by

$$s = \sqrt{(r_0 R)} \qquad (16.3)$$

Klein (1981) modified Equation (16.2) by including internal friction, thereby giving

$$t = r_0 \left[(0.29 + 1.42 \sin \phi) \left(\frac{p_a}{\sigma_c} \right) + (2.3 - 4.6 \sin \phi) \left(\frac{p_a}{\sigma_c} \right)^2 \right] \qquad (16.4)$$

where ϕ is the angle of internal friction.

Vyalov (1962) assumed that the frozen wall was plastic and was provided with temporary support. In such a situation design can be undertaken in accordance with the Mohr–Coulomb failure criterion. The angle of internal friction, ϕ, and the cohesion, c, represent the strength parameters. In addition, account is taken of the horizontal loading imposed upon the frozen wall by earth and water pressure, p_a. Accordingly Vyalov proposed that the thickness of the frozen wall, t, could be derived from

$$t = r_0 \left[\left(p_a \frac{N_\phi - 1}{2c\sqrt{(N\phi)}} + 1 \right)^{\frac{1}{N_\phi - 1}} - 1 \right] \tag{16.5}$$

where $N\phi = (1 + \sin\phi)(1 - \sin\phi)$.

Klein (1980) modified the Domke expression to take account of support afforded by an inner lining. This was done by replacing p_a by $p_a - p_i$ (the intended pressure). Then assuming that the frozen wall is completely plastic, the following design equations are available.

Tresco criterion:

$$t = r_0 \left[\exp\left(\frac{p_a - p_i}{\sigma_c} \right) - 1 \right] \tag{16.6}$$

Mises criterion:

$$t = r_0 \left[\exp\left(\frac{\sqrt{3}}{2} \frac{(p_a - p_i)}{\sigma_c} \right) - 1 \right] \tag{16.7}$$

Mohr–Coulomb criterion:

$$t = r_0 \left\{ \left[\frac{p_a + \sigma_c \left(\dfrac{1 - \sin\phi}{2\sin\phi} \right)}{p_i + \sigma_c \left(\dfrac{1 - \sin\phi}{2\sin\phi} \right)} \right]^{\left(\frac{1}{2\sin\phi} - \frac{1}{2} \right)} - 1 \right\} \tag{16.8}$$

Klein produced a series of curves for long thick frozen walls using various design expressions and showed that the thickness derived in relation to depth by these expressions varied considerably.

In the case of short thick frozen walls, their thickness can be obtained by using one of the appropriate expressions given in Figure 16.9. Where the unconfined

SHORT THICK CYLINDER (UNLINED)

1. Full and equal restraint at both ends $t = \sqrt{3}/2p_a \frac{h}{K}$
   Q = $\Sigma\gamma H$ = overburden weight
   where
   H = overburden depth

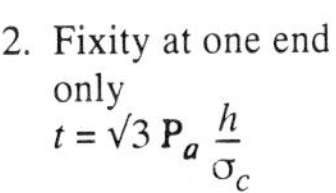

2. Fixity at one end only
   $t = \sqrt{3}\,P_a \frac{h}{\sigma_c}$

3. Partial fixity at both ends
   $t = 1.3P_a \frac{h}{\sigma_c}$

4. $t = \frac{\gamma H h}{\sigma_c}$
   where
   γH = overburden pressure

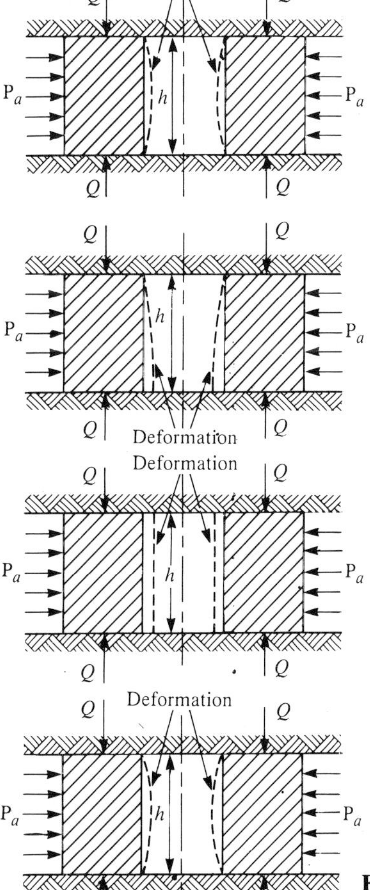

Figure 16.9 *Freeze wall strength formulae based on short, thick cylinder theory (after Auld 1985)*

compressive strength is used in any of the above mentioned equations (including those in Figure 16.9) it may have to be reduced by a suitable factor of safety.

Auld (1985) quoted eight expressions which can be used to determine wall thickness when deformation is taken into account. The same conditions are considered as are in the strength equations, that is, structural geometry, internal support and state of the frozen wall. Auld emphasized that it is essential to determine accurately the behavioural characteristics of the material *in situ* before any equation can be used.

The long- and short-wall approaches are essentially different as the height of excavation is a dominant factor in the latter case. The expression chosen to determine the thickness of the frozen wall depends on the method of sinking and lining the shaft. The long thick cylinder expressions are employed when the shaft is sunk with a temporary lining and the permanent lining is then constructed upwards from a foundation. When the permanent lining is emplaced as construction of the shaft proceeds, then the short thick cylinder expressions are more appropriate. Short cylinder methods of design result in thinner frozen walls.

16.4.3 Computer methods

Various computer programs have been developed, especially for use in the design of larger freezing projects so that the basic errors in the conventional design methods can be avoided. For example, computer programs are available to calculate the thickness of a frozen wall and the temperature distribution within it. Different boundary conditions related to the positions of the freeze probes and the influence of the type of coolant are taken into consideration, as are the effects of groundwater flow and excavation procedure. In particular, the finite element method has been used to determine the geometry of frozen ground structures and to consider the complex relationship between frozen and unfrozen ground (Jessberger 1985). Braun *et al.* (1979) found that for irregular geometries, arches and various curved shapes, a two- or three-dimensional, non-linear elastic, finite element analysis is the most useful analytic method for developing designs, although the results are not highly accurate.

16.5 Placing concrete against frozen ground

In order to capitalize on the self-supporting nature of excavations in frozen ground and the dry working conditions, concrete is normally placed directly against the frozen boundary. Numerous tests have confirmed that this is a perfectly legitimate practice. At the junction of the two masses, heat conduction causes the frozen ground to rise well above freezing point for a considerable period. In other words, generally under normal ground freezing applications, if concrete is placed at 15–18°C the adjacent frozen ground is likely to thaw to a depth approximating to 50–100% of the thickness of the concrete section.

The ground and concrete must not be allowed to refreeze before the concrete hydrates sufficiently to achieve its initial set. Secondly, the reduced curing rate of the concrete when frozen must not prevent it from developing adequate strength before it has to carry working loads. Neither of these two points normally presents a problem if ordinary concrete is placed in sections over 250 mm thick. In the case of thinner sections the heat of hydration and/or rate of set must be increased. This can be achieved by using a lower water–cement ratio, a richer mix design, an early or regulated-set cement, accelerating additives or high concentrations of calcium chloride (9–15%).

Insulation of some type should be used over the frozen ground before concrete is placed. The concrete mix should be tested after 48 h at 4–10°C to find out if it will attain a minimum strength of 4 Pa in this initial period prior to any danger of freeze-back. No special measures need to be taken if the mix performs satisfactorily and the section exceeds 250 mm in thickness.

16.6 Monitoring frozen ground

The refrigeration system and developed frozen zone are monitored 24 hours a day during the formation of a frozen wall. The refrigeration load is greatest during this period and freezing is continued at full capacity until the frozen wall reaches its design thickness. After the frozen wall is completed the refrigeration plant only need run at a reduced capacity or intermittantly in order to maintain the thickness of the frozen wall. Nonetheless it is necessary to monitor temperatures at critical locations in order to control the freezing process. For example, temperatures are monitored at locations where deviations have produced an increased spacing between freeze pipes since these points represent weak spots in the wall.

Heinrich *et al.* (1979) provided a review of some of the techniques used in monitoring ground freezing. This included a survey of instruments used to determine drillhole deviation, for temperature measurement and for assessment of the degree of freezing.

Drillhole inclinometers are used to record deviations from vertical in freeze holes. They also can be used to measure displacements due to freezing or thawing by comparing the results of repeated measurements. In addition stationary gauges and movable probes can be used for this purpose.

Temperature measurement provides the simplest method of monitoring the rate of frost advance in ground freezing (Cleasby *et al.* 1975). Sensors can either monitor the coolant at the refrigeration plant and in the freeze pipes or record ground temperatures in observation drill holes. Sensors may be stationary or movable. Gradient probes, which consist of several closely spaced temperature sensors with an insulating material of extremely low thermal conductivity in between, are used for the determination of the temperature gradient over short distances.

Ultrasonic measurements have been used to monitor the advance of the frozen front during freezing operations. The technique is based on the fact that ultrasonic waves travel in water at a velocity of 1500 m/s, while in ice their velocity is 4000 m/s. Thus the higher the velocity the more pore water has been transformed to ice. The freezing process may be controlled by measuring the ultrasonic velocity of the ground by placing an ultrasonic transmitter in one drillhole and a receiver in another. As the ice content increases the compressive strength of frozen ground increases, the ultrasonic velocity also provides a measure of the compressive strength. Ultrasonic measurements have also been able to locate sources of disturbance such as windows and to monitor the effectiveness of remedial treatment.

The pressure drop across delivery and return brine pipes can be measured by means of orifice plates. Measurement of brine flow in a pipe can be achieved by locating flow meters at intervals in the pipe. Flow also can be measured by ultrasonic equipment. This detects the velocity of solids and impurities in the flow. Alarm systems are housed in the plant room to indicate any breakage or fault in the system.

References

Auld, F.A. (1985) 'Freeze wall strength and stability design problems in deep shaft sinking. Is current theory realistic?' *Ground Freezing*, S. Kinosita, and M. Fukuda (eds), Balkema, Rotterdam, pp. 343–349

Braun, B., Shuster, J.A. and Burnham, E. (1979) 'Ground freezing for support of open excavations'. *Ground Freezing, Engg Geol.*, **13**, 429–454

Cleasby, S.P., Pearse, G.E., Grieves, M. and Thorburn, G. (1975) 'Shaft sinking at Boulby Mine, Cleveland Potash Limited', *Trans. Inst. Min Metall.*, Section A, **84**, A7–A28

Collins, S.P. and Deacon, W.G. (1972) 'Shaft sinking by ground freezing, Ely-Ouse, Essex Scheme'. *Proc. Inst. Civ. Engrs Supp. Paper 7506S*, 129–155

Domke, O. (1915) Uber die Beanspruchung der Frostmauer beim Schachtabteufen nach Gefrierverfahren, *Gluckauf* **51**, (47), 1129–1135

Frivik, P.E. (1981) State-of-the-art report. Ground freezing: thermal properties, modelling of processes and thermal design', *Ground Freezing, Engg Geol.*, **18**, 115–133

Gallavresi, F. (1981) 'Ground freezing – the application of the mixed method (brine-liquid nitrogen)', *Ground Freezing, Engng Geol*, **18**, 361–376

Gallavresi, F. (1985) 'Soil improvement by means of ground freezing'. In *Recent Developments in Ground Improvement Techniques*, A.A. Balasubramanian, S. Chandra, D.T. Bergado, J.S. Younger and F. Prinzl (eds), Balkema, Rotterdam, pp. 459–468

Heinrich, D., Müller, G. and Voort, H. (1979) 'Ground freezing monitoring techniques', *Ground Freezing, Engg Geol.*, **13**, 455–472

Jessberger, H.L. (1981) 'Ground Freezing. Mechanical properties, processes and design. State-of-the-art report', *Ground Freezing, Engg Geol.*, **18**, 5–30

Jessberger, H.L. (1985) 'The application of ground freezing to soil improvement in engineering practice'. In *Recent Developments in Ground Improvement Techniques*, A.A. Balasubramaniam, S. Chandra, D.T. Bergado, J.S. Younger, and F. Prinzl (eds), Balkema, Rotterdam, pp. 469–482

Jumikis, A.R. (1979) 'Cryogenic texture and strength aspects of artificially frozen ground', *Ground Freezing, Engg Geol*, **13**, 125–136

Khakimov, K.R. (1966) *Artificial Freezing of Soils: Theory and Practice*, Israel Program for Scientific Translations, Jerusalem

Klein, G. (1976) Heat transfer for evaporating nitrogen streaming in a horizontal tube', *Proc. 6th Int. Cryogenic Conf.*, Grenoble, IPC Science and Technology Press.

Klein, J. (1980) 'Die Festigkeitsberechnung von Frostwanden im Gefrierschachtbau', *Gluckauf-Forschungshefte*, **41**, (H5), 3–8

Klein, J. (1981) Zur Berechnung der erforderlichen Dicke der Frostwand im Gefrierschacht', Vortrag am 17.11.81 im Rahmen des 'Bergmannischen Kolloquiums' am Institut für Bergbaukunde I der RWTH Aachen

Maishman D. (1975) 'Ground freezing', In *Methods of Treatment of Unstable Ground*, F.G. Bell (ed), Butterworths, London pp. 159–171

Sanger, F.J. (1968) 'Ground freezing in construction', *Proc. ASCE J. Soil Mech. Found Engg Div.*, **94**(SMI), 131–157

Sanger, F.J. and Sayles, F.H. (1979) 'Thermal and rheological computations for artificially frozen ground construction', *Ground Freezing, Engng Geol*, **13**, 311–338

Shuster, J.A. (1972) 'Controlled freezing for temporary ground support', *Proc. 1st North American Rapid Excavation and Tunneling Conf.*, ACSE, Chicago, 863–879

Shuster, J.A. (1981) 'Engineering quality assurance for construction ground freezing', *Ground Freezing, Engng. Geol.*, **18**, 333–350

Stoss, K. and Valk, J. (1979) 'Uses and limits of ground freezing with liquid nitrogen', *Ground Freezing, Engg Geol.*, **13**, 485–494

Veranneman, G. and Rebhan, D. (1979) 'Ground consolidation with liquid nitrogen (LN_2)', *Ground Freezing, Engg Geol.*, **13**, 473–484

Vyalov, S.S. (1962) *The Strength and Creep of Frozen Soils and Calculation for Ice-Soil Retaining Structures*, US Army Cold Regions Res. Engg. Lab., Hanover, NH.

Wild, W.M. and Forrest, W. 'The application of the freezing process to ten shafts and two drifts at the Selby Project', *The Mining Engineer*, **140**, 895–905

17 Grouting in rock masses

A C Houlsby
Consultant Grouting Engineer

17.1 Nature and purposes of grouting

Grouting is the process of injecting a fluid, which subsequently sets or gels, into cracks and voids. Grout materials include: cement with water, clay with water, chemicals which react after a pre-determined time, fly-ash, pozzolans and combinations of these. Sand and other bulking agents are sometimes added.

In rock foundations, the main uses of grouting are for the construction of curtains to restrict seepage and for the strengthening of foundations at dams, tunnels, shafts and heavy structures. Cement is usually the most suitable material for this. Drilled holes provide access to the openings to be grouted.

There are other uses of grouting outside the scope of this book. Houlsby (1990) indicated some.

17.2 Site investigation

Grouting design requires knowledge of site geology and permeability. The amount of effort put into investigating these generally relates to the size and importance of the job and to any previously obtained data which is available.

17.2.1 Geological investigation

Determination of jointing patterns, strikes and dips enables grout holes to be aligned to give optimum interception. Useful tools for examining jointing *in situ* include borehole periscopes and closed circuit television. These instruments permit direct observation of the width of open cracks and the condition of crack walls, and assessment of joint patterns and continuity. Information on these helps when designing high-quality grouting programmes. Each site has individual requirements, but in general terms, aspects of the geology can influence the design in the following ways.

(a) Spacing of open joints

If the open, groutable joints are widely spaced, the grouting is usually easier than if closely spaced, where troubles such as frequent surface leaks, collapsing holes, and patchy penetrations can happen. These make for more expensive grouting, perhaps requiring special surface treatment or unusual methods.

(b) Width of openings

The easiest joints to grout have widths in the range between about 6 mm and 0.5 mm. Joints wider than 6 mm permit the grout to travel too easily and precautions have to be taken to inhibit excessive penetration, such as recourse to multiple applications of limited volume at 24 hour intervals. At the other extreme, very fine cracks can be difficult to grout, as discussed at Section 17.12.

(c) Dip

Vertical grout holes are the easiest to drill and can give good interception of joints having dips between 0° and about 60°. Steeper jointing usually requires the use of inclined holes which should not be more than 45° off the vertical unless drilled into a steep face.

(d) Rock strength and soundness

Grouting of strong, massive, well-anchored rock is usually easier than working in weak, broken, loose materials where holes repeatedly collapse and move under grouting pressure.

(e) Stress in rock

If the rock contains tectonic stress, the grouting will need to take account of this; early recognition of the condition is necessary. Section 17.13 gives appropriate grouting methods.

(f) Uniformity of foundation

Geological uniformity permits a regular layout of grout holes, whereas irregular jointing, presence of dykes, shears, non-conformities, and so on may require placement of holes at various inclinations and spacings, and weak features may require especially intensive localised grouting.

(g) Piping

If material in joints can be removed by seepage, either in solution or by erosion, the grouting needs to be more intensive than would otherwise be the case, with the purpose of virtually eliminating seepage through the areas prone to pipe.

17.2.2 Permeability testing

A good knowledge of site permeabilities is essential if one of the purposes of the grouting is seepage control. Permeability exploration is done in holes drilled at representative locations — often those used for obtaining rock cores.

Amongst permeability tests in use for various purposes the Lugeon test has found wide acceptance in grouting work. It was introduced by Lugeon (1933), and is a 'pump-in' test where the volume of water taken in a section of test hole is measured during given time intervals. One lugeon unit is defined as one litre per metre of test length per minute at 10 bars pressure. In the best version of the test, a single packer is used in the 'Downstage with packer' mode (as explained in Section 17.6) to isolate the test stage below it. A length of 5–6 m is convenient for the stage length. The pressure was originally specified by Lugeon as 10 bars (1 MPa approx) but this is too great for stages near the surface and in weak rock, so lesser pressures are used in these circumstances and an adjustment is then made in accordance with the ratio of pressures as follows:

$$\text{Lugeon value} = \text{litres/metre/minute} \times \frac{10\ \text{(bars)}}{\text{actual pressure (bars)}}$$

To obtain a sense of proportion for the lugeon unit:

- 1 lugeon = 1.3×10^{-5} cm/s;
- 1 lugeon is the degree of permeability of a foundation which requires almost no grouting at all;
- 10 lugeons warrants grouting for most types of dams.
- 100 lugeons is the type of permeability met in heavily jointed sites with relatively open joints, or in sparsely cracked foundations where joints are very wide open.

The test has its maximum accuracy in low-permeability situations. The realism diminishes in a mode resembling logarithmic: in the 1–5 lugeon range, each variation of 1 unit is meaningful; between 5 and about 10, meaningful increments are about 2 lugeons; between 10 and 15 they are 5 lugeons, in the 15–50 range ± 10 lugeons is quite accurate enough for meaningful use, and between 50 and 100 lugeons ± 30 units is adequate.

Lugeons should never be given in fractions; accuracy is not good enough to justify this, and yet in spite of its relative roughness, the test is ideally suited to grouting work. The nature of things in grouting is such that finesse can often be illusory.

Although the lugeon scale has no upper limit, after 100 lugeons is reached it is meaningless to distinguish betweeen different values: they all have the same significance in terms of a grouting design; that is, that grouting is very necessary. Therefore recommended practice is to report all values above 100 as '>100 lugeons', without further distinction.

There are two different phases of lugeon testing in grouting work:

(1) *Exploratory testing*, where permeability exploration of a site is carried out before preparation of a design;
(2) *Grout hole testing*, where each stage of every grout hole is water tested before grouting.

The first type of testing needs to be more thorough than the second. Accordingly the exploratiory testing should be carried out by a method which enables determination of the flow regime of the water passing along the cracks, because selection of a correct lugeon value requires this knowledge. A satisfactory method of doing this uses five changes of pressure, immediately following each other, as follows:

(1) a *low* pressure for the first 10 minutes, followed by
(2) a *moderate* pressure for the next 10 minutes, then
(3) a *peak* pressure for the next 10 minutes, then
(4) the *moderate* pressure again for 10 minutes, then
(5) the *low* pressure for the final 10 minutes.

In summary form, this pressure range is described as 'a–b–c–b–a'.

The lugeon values are calculated for each of these five pressures, and are then inspected to determine the flow regime. Figure 17.1 shows how this is done. For example, if the lugeon values are the same for all five pressures, then the flow can be regarded as mainly laminar, and any of the values can be reported as the test result, or their average used. A different flow regime is indicated if the lugeon values at the higher pressures are less than at the lower ones; this indicates turbulent flow, and instead of averaging the five values, the one for the peak pressure should be reported as the test result. The other test possibilities are dilation, wash-out and void filling, and Figure 17.1 shows how these can be recognized. The method is further explained in Houlsby (1976). The peak pressure is related to the depth and rock strength and need not exceed Lugeon's specified pressure of 10 bars. It is measured at the surface, and in sound rock is determined by the same relationship as used for conservative grout pressure: one p.s.i. per foot depth (to the bottom of the

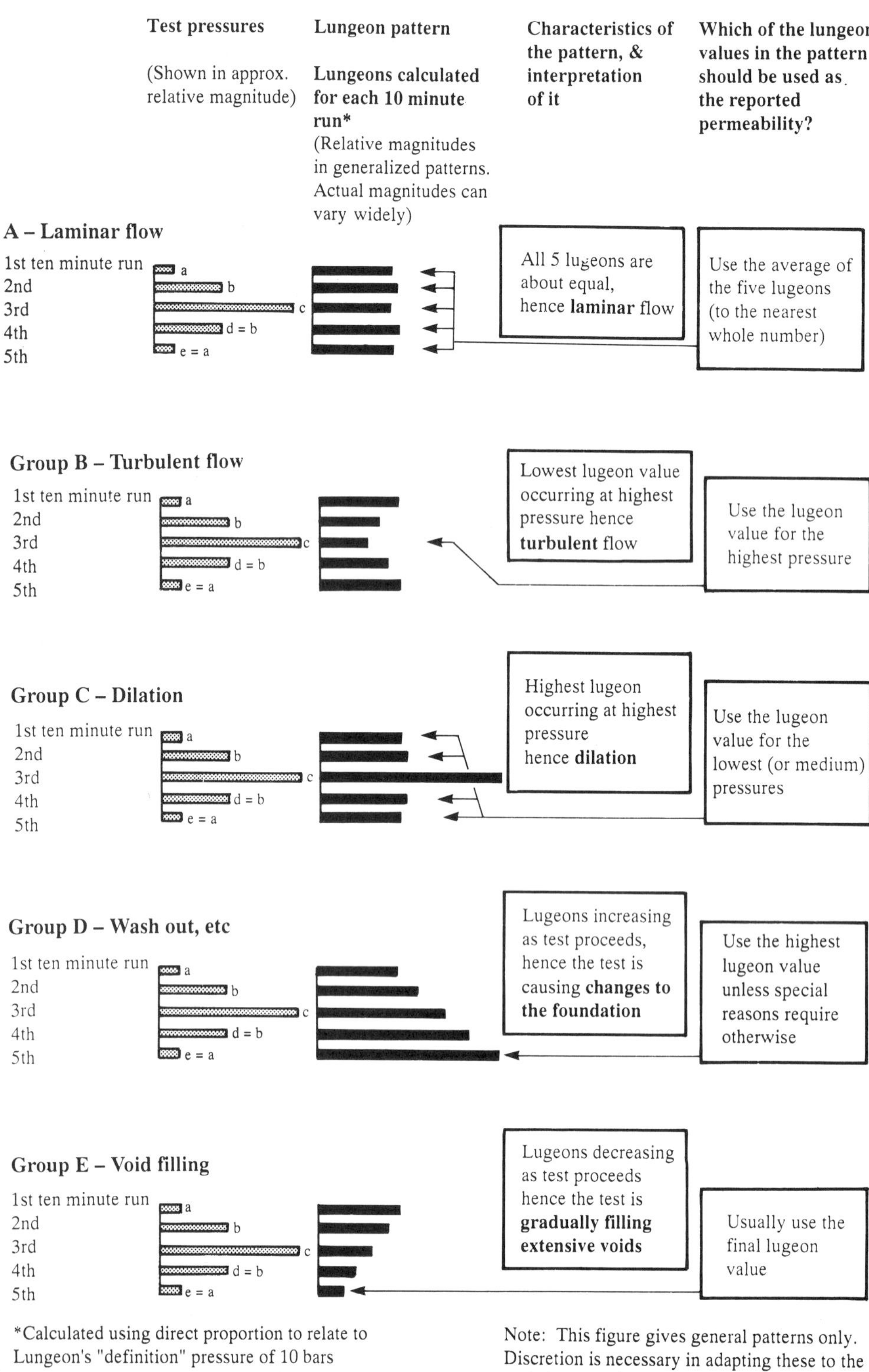

Figure 17.1 *Permeability testing for exploratory purposes. The use of five pressure sequences for determining the type of flow and hence the lugeon value to report for the test*

test stage). This is equivalent to about 0.23 bars/m. Further discussion on pressures in connection with grouting is given in Section 17.9. Correction for head loss in packers and rods is optional; the resulting improved accuracy is rarely relevant in grouting decisions.

There is no need to plot graphs depicting the relationship between the water taken and the pressures. Although such plots have been used frequently, they are superfluous if the above method is used, and it can be argued that they may be misleading (Houlsby 1990).

The other form of lugeon testing — that for grout hole water testing — needs only use of one pressure. By running the test for 15 minutes and recording the water take every 5 minutes, a comparison of the three results gives a good enough idea of the flow regime and hence which value should be adopted. The author uses 1 bar pressure for all tests of this type unless the rock is too weak for it. The use of this single (generally) safe pressure simplifies site calculations because lugeons can be read off a table, and it is quite high enough to show up any problems which might arise during the subsequent grouting, such as surface leaks or connections. There is no need to use the same pressures for water testing as for grouting, and on some sites, such usage could move rock which would not have moved at the lower 1 bar pressure.

17.3 When is grouting necessary?

In terms of seepage control, there are occasions when grouting can be omitted quite safely, and there are borderline cases where the need for it is uncertain, but likewise there are many sites where it is most definitely needed. Then, if grouting is considered necessary, there arises the further question of just how much grouting to do.

In terms of dams, Figure 17.2 provides a guide on this. It shows a flow chart commencing with the query 'how valuable is water lost by leakage?' If water is precious or very expensive, grouting could be warranted if the ungrouted foundation has a permeability generally in excess of 1, 2 or 3 lugeons, as indicated on the chart. Examples of extra-tight grouting such as this include pumped storages, and storages in arid climates where water may be in short supply.

If water lost by seepage is of negligible value, which is the usual case, the flow chart next asks 'does piping of foundation material need to be prevented?' A positive answer leads to a fairly tight curtain such as 3 lugeons; a negative one, the usual case, leads on to considerations concerning the type of dam. Embankment dams, including

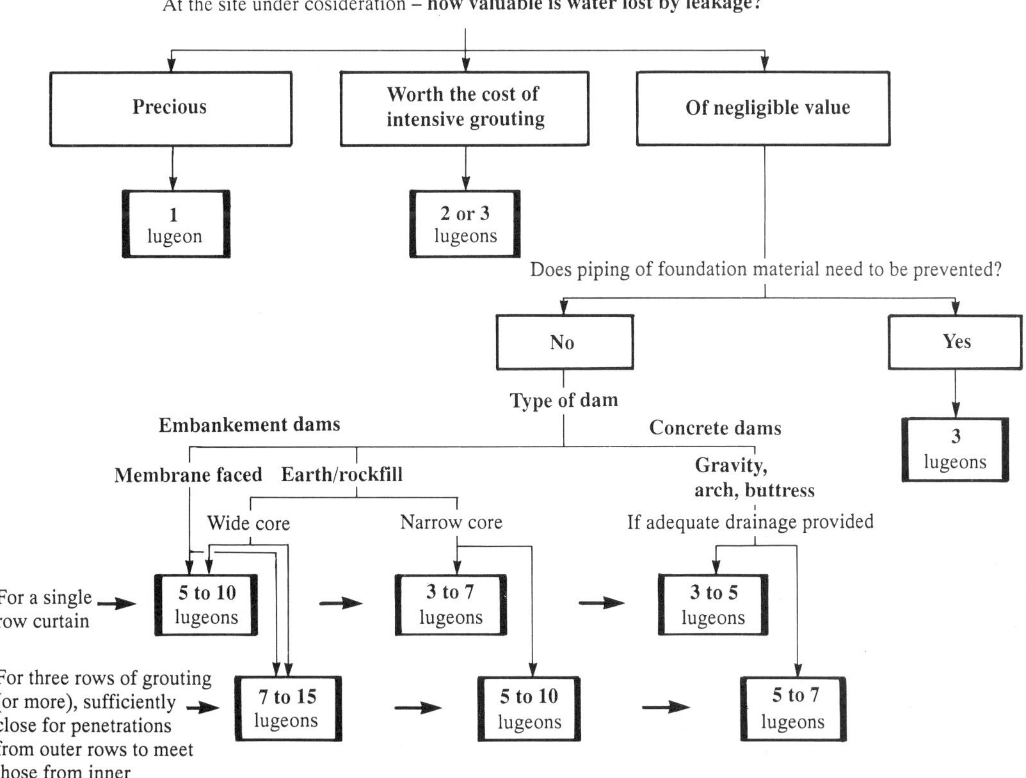

Figure 17.2 *Criteria for deciding whether a grout curtain is necessary, and if so, to what standard of tightness the grouting should be taken*

membrane-faced ones, require degrees of tightness depending on the core thickness (if zoned), and on the number of rows in the grout curtain. The chart shows these different intensities of grouting. For concrete dams there are likewise differing requirements as shown. Note particularly that Figure 17.2 is only a guide. Its derivation is explained in Houlsby (1990). After deciding the degree of tightness required, any need for grouting is determined by comparing this standard with the natural permeability of the site. Grouting is likely to be needed where the natural permeability is greater than the standard required. In some cases this can mean that only parts of the site need grouting.

In fairly tight sites, where the need for grouting may be borderline, a nominal curtain has sometimes been used having only primary holes. The primary holes are then essentially probes intended to find any weak areas not picked up in exploratory drilling. Any weak areas found are then treated through additional holes. This method can be an economical means of augmenting what may have been a meagre exploratory programme. This concept has been described as 'proving, rather than improving, the site'.

Different factors from these have to be considered if the grouting is for strengthening purposes. The need for this type of grouting is usually decided by considering whether sufficient poorly filled jointing exists to benefit from grouting for the purpose of preventing undue settlement, or dilation, as the result of structural loads.

Tunnels and conduits passing under dams may require special grouting treatment to provide a cut-off; blasting during tunnelling can leave loose rock surrounding the tunnel, needing grouting for consolidation.

17.4 Types of grout

The various types of grouts can be grouped as follows:

- Cement grouts;
- Clay grouts;
- Chemical grouts.

Mixtures of these are sometimes used.

The subject of this chapter is grouting in rock masses; for this, cement grout is the most suitable usually and accordingly is the type of grout dealt with here. It is generally the cheapest grout and requires less expertise, but is not generally suitable for grouting of alluvium, sand, silt or clay; chemical or clay grouts are more successful with these. On most sites satisfactory grouting can be obtained with ordinary Portland cement, ASTM Type I. Section 17.12 comments on the use of finer cements that this.

The use of additives to cement grout finds varied acceptance. Some grouting practitioners, including the author, have found them unnecessary in general.

Additives can be grouped as follows:

- for bulk filling;
- reactive materials;
- for producing particular effects.

Bulk fillers include sand, clay (usually bentonite), lightweight and heavyweight aggregates, and these have the purpose of saving cement, and hence cost, where cracks or voids are sufficiently wide to permit the resulting stiffer grout to penetrate. Sometimes penetration is purposely inhibited by use of fillers: in extreme cases of runaway takes, additives such as straw, sawdust, wood shavings, corn stalks, etc. can be useful, or a mixture of sodium silicate + bentonite + cement (which quickly gels), or the drillers' trick of bentonite + oil + a filler. The light and heavyweight aggregates produce light or heavy grout for special purposes.

Reactive additives include materials of the pozzolan group, such as fly-ash, silica fume, etc.. Care is required with fly-ash because its effects on cement grout can vary widely — from providing improvement to providing the opposite. Reliable and proven sources are essential. Silica fume can assist fluidity greatly. It is about 100 times finer than cement, and its spherically shaped particles are like minute ball bearings in the grout. It is more reactive than fly-ash.

Additives to produce special effects include those intended to inhibit bleed, or reduce the water content, or improve penetration. Bleed inhibiting additives have the purpose of controlling the settlement of the cement and the rise of bleed water out of the grout. There are differences of opinion about whether these additives are really necessary. The use of mixers producing high-quality grout (see Section 17.5) largely avoids the need for such additives. The issue is discussed in Houlsby (1990) where it is pointed out that they are largely a palliative to compensate for poor mixing. The main additive used for this purpose is bentonite. When mixed with cement, sodium bentonite changes to calcium bentonite which then flocculates. The flocs are gelatinous and large, and prevent sedimentation of the cement particles. Drawbacks against its use include difficulties in getting it thoroughly mixed with the cement and the possibility that the bentonite will separate from the cement underground, leaving some cracks filled with bentonite only. This has been known to happen.

17.5 Cement grouting equipment

Figure 17.3 shows the main items of grouting equipment and their layout. The mixer mixes the grout in batches and then sends each batch to an agitator where it is stored until drawn off by a pump and sent along a circulation line to the grout hole. Here fittings on a standpipe control the amount of grout going down the hole. Surplus grout, not

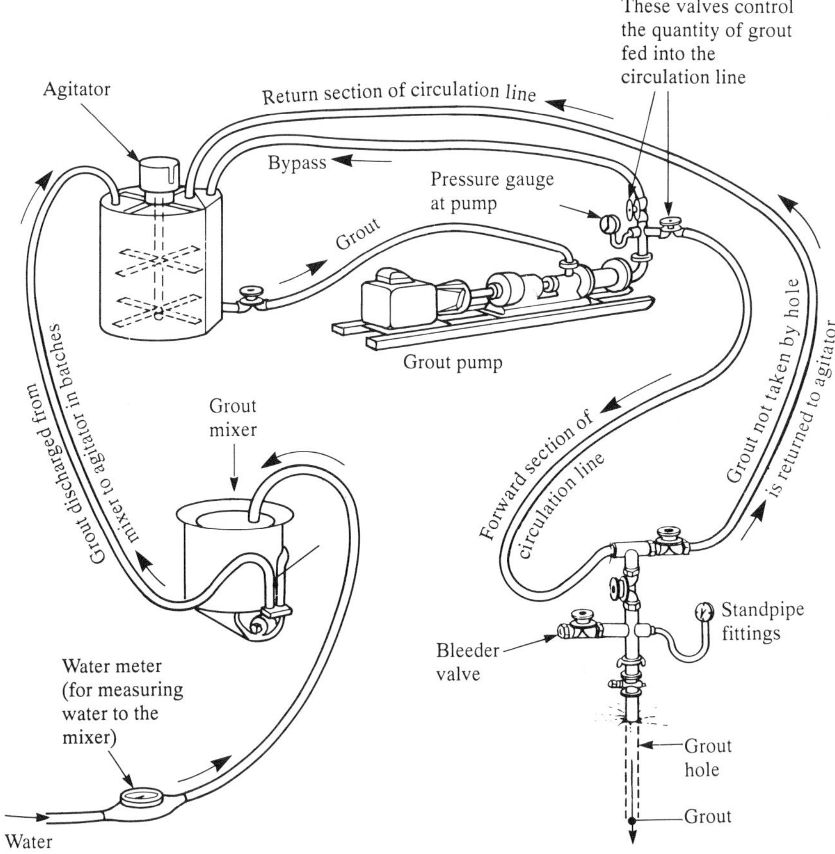

Figure 17.3 *Layout of equipment for cement grouting*

used down the grout hole, circulates back to the agitator for later use.

17.5.1 Grout mixers

The quality of the grouting depends greatly on the type of mixer used. For comparative purposes, the author has grouped mixers into:

- those able to produce high-quality grout;
- those of lesser efficiency;
- those for lower-quality grout.

In this grouping, high-quality grout is regarded as having the following properties, not necessarily found in lesser-quality grouts:

(1) Cement grains are separate from each other: there are no flocs or clumps of grains.
(2) Each cement grain is thoroughly wetted and surrounded by a film of water. As well as improving penetration in fine cracks, this chemically activitates each particle, giving the full hydration necessary for strength and durability.
(3) The grout is uniform throughout.
(4) It has less settlement of the cement when stationary than lower-quality grout and has less bleed.

The mixers able to produce these grouts make use of the violent turbulence and high shearing action of a high speed rotor. They are called 'high-speed', 'semi-colloidal' or 'high-shear' mixers. A rotor revolving at 1500 revolutions or more per minute in a small chamber mixes the grout ingredients rapidly and passes them to a vortex chamber. Figure 17.4 shows a typical machine. The vortex acts as a

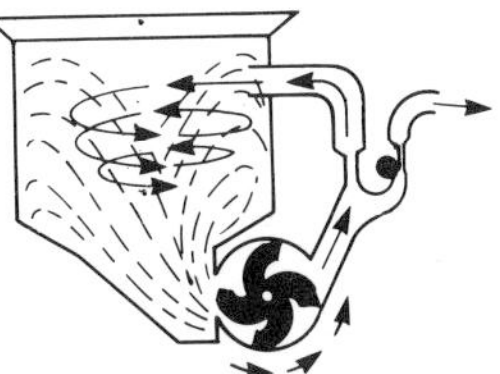

Figure 17.4 *Mixer able to produce high-quality grout. Rotor speed >1500 rev/min. Vortex formed in drum by tangential discharge from rotor*

centrifugal separator and helps to ensure that the grout is finally uniform. Mixing time is only 15 s in the best machines and during this time the ingredients pass through the rotor several times. Manufacturers of mixers of this general type include Colcrete, Cemix (Craelius), Chemgrout and Hany.

Mixers of lesser efficiency cost less to buy, but can be more expensive to use. Examples are the combined type and the roller mixer. The combined type have a small grout pump of the helical rotor type combined with a mixing rotor mounted in the main bowl of the mixer. In some models, the same vertical shaft drives both the rotors and the pump. Speed is about 700–1000 rev/min and mixing time is 3 minutes or more. Roller mixers are of several types; one has a pair of rubber rollers revolving towards each other in the bowl.

Mixers for lower-quality grout can be of the paddle or propeller types or sometimes are improvised on the job. Paddle mixers have single or double paddles sweeping around in a drum at about 100–700 rev/min. Mixing time is about 5 minutes. Propeller mixers use a large propeller on a shaft to provide a strong mixing action in the vicinity of the blades. The portable type enables the propeller to be moved around in a drum; the fixed type does not — it depends on the turbulence at the blades reaching out to the remainder of the drum. Other types of mixers include the jet mixer, as used on drilling rigs for mixing mud, and the pump mixer, where mixing is done in the grout pump.

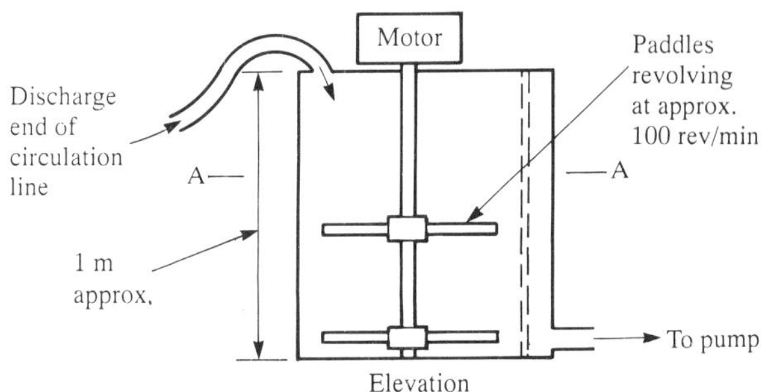

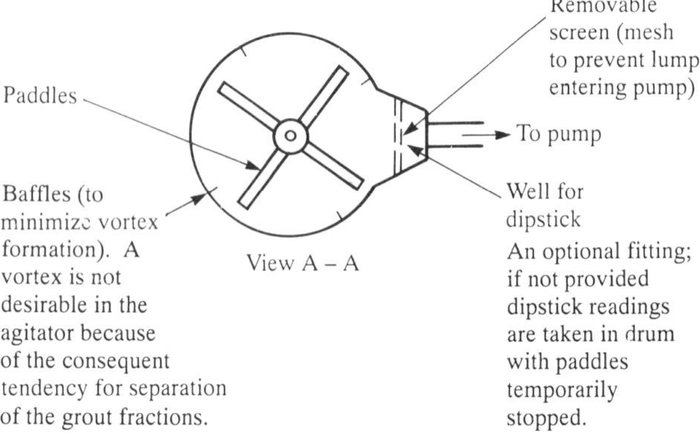

Figure 17.5 *Typical agitator for storing and stirring grout awaiting pumping*

17.5.2 Agitators

After each batch of grout is mixed it is discharged into an agitator for holding until needed. Agitators stir the grout slowly using paddles in much the same fashion as a paddle mixer. Figure 17.5 shows a commonly used type. Paddle speed is about 100 rev/min. Baffles projecting into the drum break up any tendency for a swirl or vortex to develop; these are undesirable at this stage of the grout handling. Dipstick readings can be taken in the agitator with the paddles temporarily stopped, and when done from time to time, give a measure of the amount of grout taken by the hole. Alternatively, computer-controlled equipment can be used for measurement (Houlsby 1990).

17.5.3 Grout pumps

Pumps for grout service have to be able to withstand the very abrasive nature of cement grout and must be free from low-velocity passages where cement can settle and set. Pumps used include helical rotor, piston, ram, diaphragm and hand pumps. Helical rotor pumps are of the progressive cavity type where a steel rotor of helical configuration rotates slowly inside a softer stator having a double internal helix shape. Figure 17.6 indicates the arrangement. Valves are not needed. This type of pump has come to be the most popular grouting pump in a

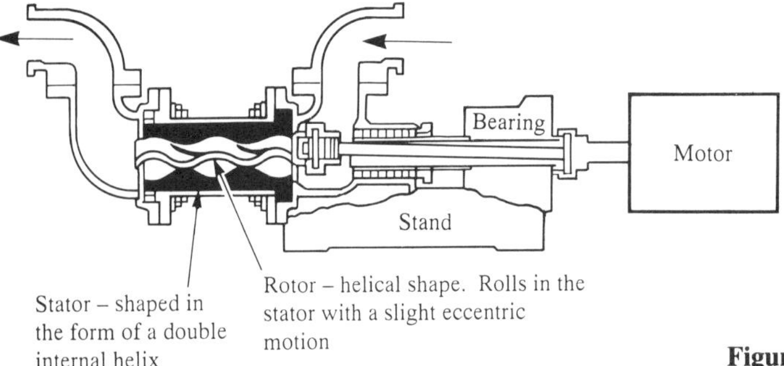

Figure 17.6 *Grout pump of progressive cavity type*

number of countries. Manufacturers include Mono (Moyno in USA).

Piston pumps are usually duplex, with pistons operating in close-fitting cylinders. Apart from the need to have pistons and cylinders made of material able to withstand the wear, the weak point of these pumps is their need for valves. Cleanliness and care are needed to keep the valves operating over long periods of time.

Ram pumps also reciprocate and have valves. However, their pistons move in loose-fitting cylinders and then into a chamber so that the displacement caused by the volume of the piston provides the pumping action.

Diaphragm pumps usually have pairs of diaphragms flexing under the stimulus of pistons or cranks. As they flex in and out, they suck in grout and then expel it to provide the pumping action. Valves are necessary.

Where jobs are too small to warrant a powered pump, a hand pump can be useful. These have a long handle operated by a person, providing the reciprocating motion for a small piston or diaphragm pump.

17.5.4 Fittings on the grout hole

Various names are given to the fittings mounted on the grout hole to control the grout injection, measure the pressure and enable thin grout and water to be bled out of the hole periodically, such as: standpipe fittings, Christmas tree, and header. Figure 17.7 shows a practical layout for them.

17.5.5 Circulation line

The circulation line carries grout from the pump to the grout hole. On many jobs, a return line then carries surplus grout back to the agitator as shown on Figure 17.3. This provides an arrangement where the grout is kept moving briskly along in the line and the pump can be set at a fairly steady pressure because control of injection pressures is done at the hole. However, the return part of the line is sometimes omitted: this requires control of

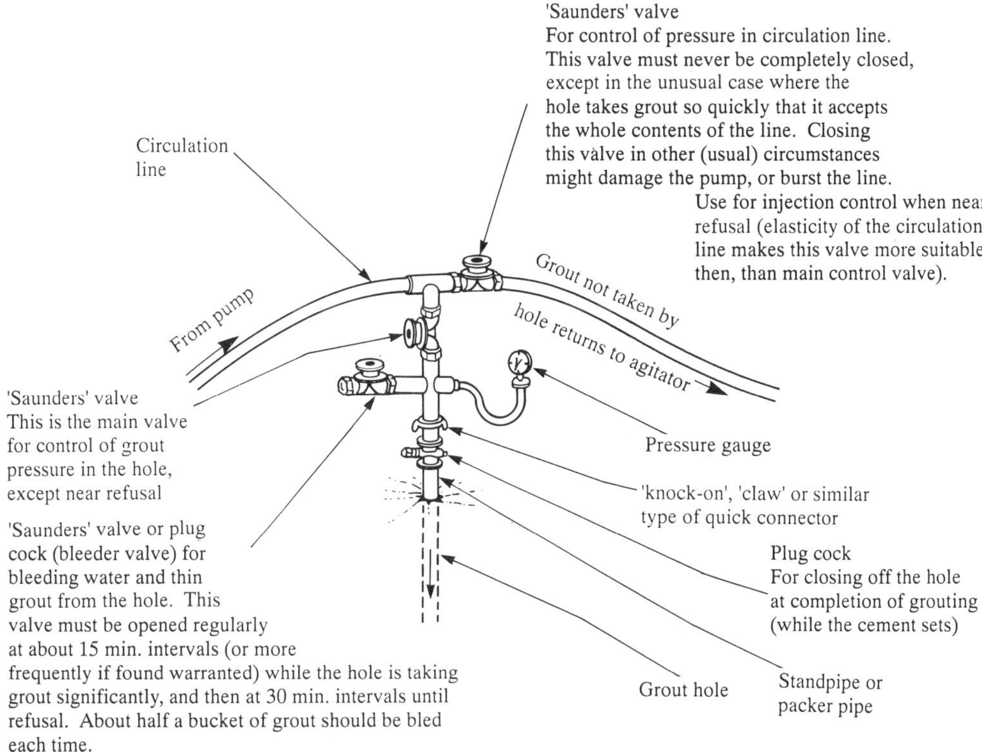

Figure 17.7 *Fittings on the grout hole for injection control, pressure measurement and bleeding*

injection to be at the pump, perhaps using automatic pressure-limiting equipment. Piston pumps operated by variable-speed drives are then needed to be able to handle the stalling conditions encountered when injection of an application is reaching refusal, and when holding pressure. Blockages in the line are possible in this situation and there can be waste when changing mixes. There are very good practical reasons for not omitting the return, apart from when grouting fast holes.

Circulation lines are made up of hoses or pipes. To minimize blockages the grout velocity should be kept high; internal diameter of the lines should not be larger than 25 mm.

17.5.6 Valves, pressure gauges

Ordinary valves and pressure gauges are unsuitable for grout. Some means is necessary to keep grout out of working parts. This is usually done with a diaphragm. In the case of pressure gauges, an alternative is to use a pipe bent into 'U' shape and filled with oil; the oil prevents grout getting to the gauge.

17.5.7 Packers

These are devices for isolating part of a grout hole. The two main types are the inflated and the mechanical. Inflated packers have a soft rubber tube which can be expanded against the wall of the hole by inflating it with compressed air, compressed gas or water. This type has most application in irregular holes, but is liable to be torn when lowering or raising in the deflated condition. Mechanical packers are much tougher but will only seat properly in circular holes. They have stubby rubber sleeves which are expanded against the hole wall by longitudinal compression using screw devices at the surface. Single packers are used almost exclusively in grouting work: double packers find very little use because of uncertainty about the seating of the lower packer.

17.6 Stage grouting and methods

Grout holes longer than about 10 m should be grouted in stages unless the grouting is intentionally of low standard. Stages provide greater control over the area being grouted, and help to minimize segregation and bleed in grout holes. They also facilitate assessment of grouting while in progress as noted later.

There are several methods for doing stage grouting. The main ones are illustrated on Figure 17.8. Sometimes combinations of them are used.

For high-quality grouting, the 'downstage without packer' method is recommended. Although the drill has to be set up afresh for each new stage, this is not a very expensive item if modern track drills are in use: setting these up is very quickly and easily done. Omission of packers means that, as the hole gets deeper, the surface stages are subjected to the greater pressures used for the deeper stages; this finds any weaknesses left in the upper stages and remedies them. A proving process such as this is of value because the upper stages, particularly of a curtain, are the most important.

The method called 'downstage with packer' is similar in sequences but, it is hoped, separates lower from upper stages by means of a packer. This means that upper stages are not checked, but the method finds use where excessive surface leaks occur or where protection is required for nearby structures against the grout pressure. This is the most expensive of the stage grouting methods. It involves some uncertainty about the effectiveness of the packer.

'Upstage' grouting is the simplest and cheapest in sites where the holes do not collapse and are suitable for the seating of packers. Holes are drilled to their full depth in one drill setting, and then grouted upwards using a single packer. Drawbacks include difficulties in water testing each stage before grouting, handling stressed rock and surface leaks, and the lack of checking of upper stages.

Packers can be troublesome in some sites and are better avoided if possible. Problems come from such unknown happenings as grout bypassing them through cracks and entering the hole above; packers can then be inadvertently grouted in. Another problem is that water and thin grout in the hole cannot escape up to the bleeder valve on the standpipe fittings but become trapped in the hole at the top of the stage. Cracks in that region only receive this watery grout instead of the intended much thicker grout. The 'downstage without packer' method avoids these problems.

'Circuit' grouting is a special method for foundations where holes collapse so badly that they will not stop open long enough for a downstage method. A pipe or tube is pushed down the grout hole immediately the drill rods are pulled, and forms part of the circulation line. The method is slow and prone to problems, such as blockages to the circulation if the hole collapses extensively, as it is very likely to do.

With all methods, additional stages are desirable to allow for individual grouting of large cracks if drilling water is lost, if the hole collapses or if groundwater comes out of it under pressure.

17.7 Closure grouting

This is a commonly used method for checking that grouting is proceeding properly and for deciding when to stop grouting. It finds most application in curtain construction, but is also useful in blanket or consolidation grouting.

The method grouts widely spaced holes initially in an area. Then the next holes are placed half-way between those, and after that centres are halved again and so on

The steps when working downstage without packer are:-

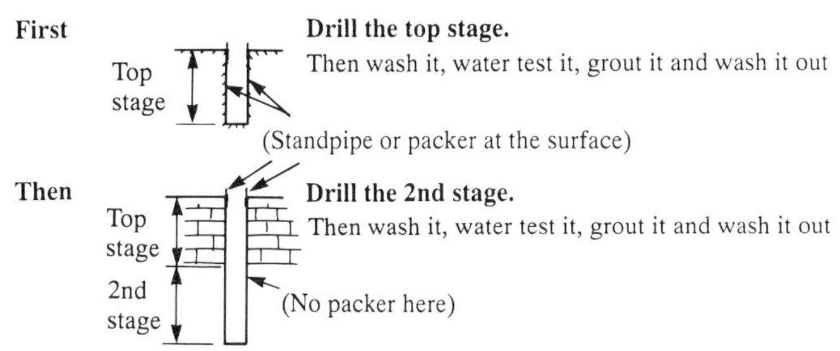

Then

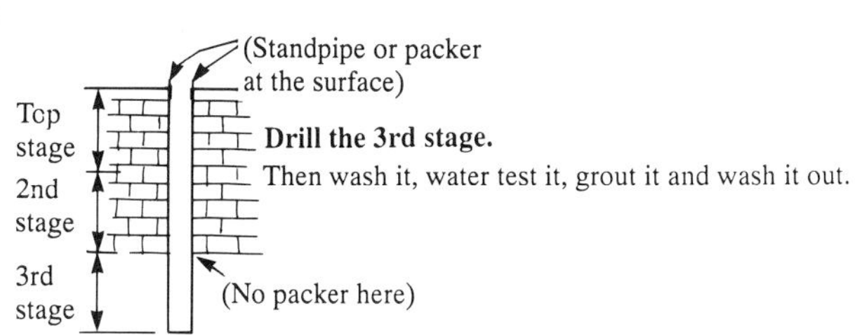

And so on – for the remainder of the hole

When grouting of the bottom stage is finished, do not wash it out (leave the hole full of grout)

The steps when upstage grouting are:-

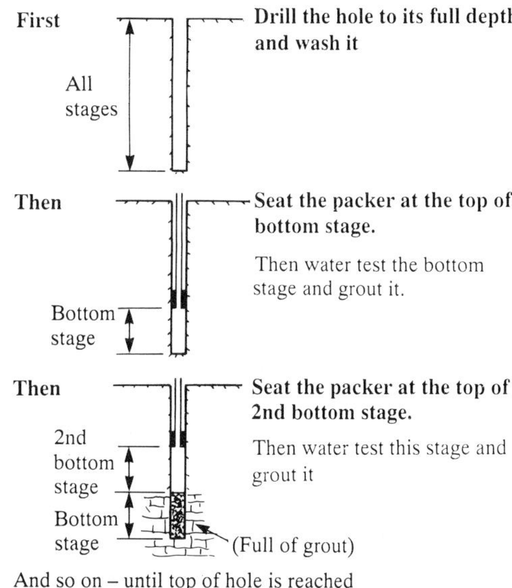

The steps when working downstage with packer are:-

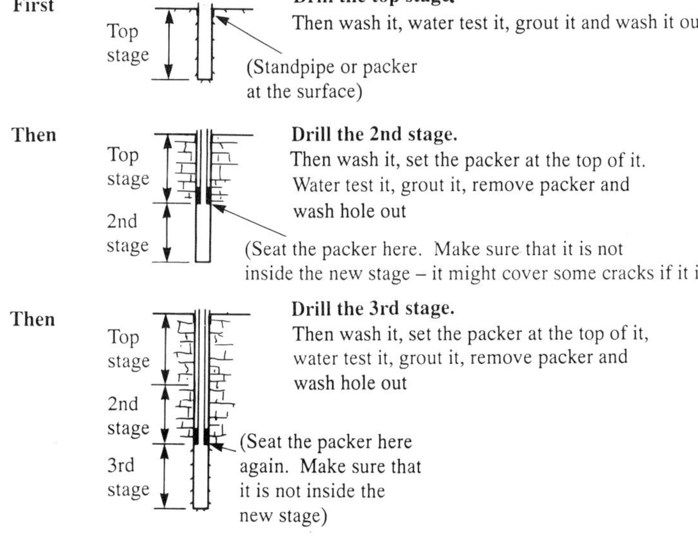

And so on – for the remainder of the hole

When the bottom stage is finished, the hole is then filled by one of a number of methods as instructed in each case

The steps when circuit grouting downstage are:-

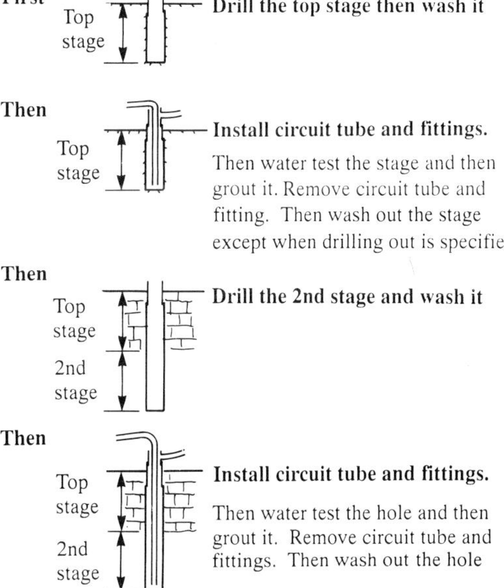

And so on – for the remainder of the hole

When the bottom stage is finished, don't wash it out. Pull the circuit tube and later top up the hole as necessary

Figure 17.8 *Methods for grouting of stages in a grout hole*

until further closure becomes unwarranted. The first holes are known as primary and they should be sufficiently far apart that connections do not occur between them and nor should penetrations from them overlap. In curtain grouting, typical primary spacing is 12 m, or 6 m in fairly tight foundations.

Secondary holes follow and are midway between primaries. Then tertiary, quaternary, quinary, etc. continue the closure pattern until the specified standard of grouting is reached as described later. Examination of water and cement takes while closure is proceeding gives an idea as to whether the grouting is progressing satisfactorily or not: if penetrations are inadequate changes may be necessary to the water:cement ratio, or to the pressure or other aspects of the techniques.

17.8 Grout curtain design

A typical design for a single row grout curtain is shown in Figure 17.9. Holes are inclined to give optimum intersection of open jointing which is assumed to be vertical and horizontal in this example. In the valley floor, the change from the inclination in one abutment to the inclination of the other is not done gradually (which would give poor intersection of cracks in the transition area) but is located at a crossover which should be placed at the weakest geological feature in the floor if possible. The curtain is divided into stages, with primary holes going the full depth, secondaries not quite so deep, and subsequent sequences shallower. This provides a flexible arrangement whereby secondaries can be deepened into the bottom stage at any locations where the primaries show takes which are sufficiently great to warrant more work in their area. Likewise holes of later sequences can be deepened into areas below their design depth where the planned holes are inadequate to bring the permeability down to the desired standard and more grouting is evidently needed. The final curtain might look rather different from the design layout. Flexibility in layout is very advisable; unforeseen weaknesses can then be treated where and as they are found.

At dams, the position of grout curtains depends on the type of dam. Figure 17.10 shows the usual positions.

17.9 Grout pressures

In order to get maximum grout penetration, pressures should be as high as possible. There are two schools of thought about pressures: one prefers moderate pressures which do not disrupt the foundation, while the other uses deliberately high pressures to displace rock and widen cracks.

Deeper stages can usually accept greater pressures than surface stages. The pressures in surface stages are largely governed by the rock strength and jointing: the pressures must not move surface rock nor must they produce excessive surface leaks. Both schools of thought use about the same pressures in these stages.

At greater depths, however, the moderate-pressure philosophy takes the view that the pressure should be conditioned to the strength of the rock on the premise that the weight of the overlying rock, plus some beam and cantilever strength, acts to hold the foundation against dislodgement by grout pressures. A well-known rule of thumb for these pressures is one lb/in^2 per foot depth (0.23 bars/m), and this is usually satisfactory in average and weak rock, although in very sound rock double this amount can sometimes be used. These pressure relationships are shown in Figure 17.11.

Also shown in Figure 17.11 is the rule of thumb sometimes used in countries where metric has been traditional, 1.00 bars/m. It is the pressure relationship used in displacement grouting philosophy. Pressures at depth

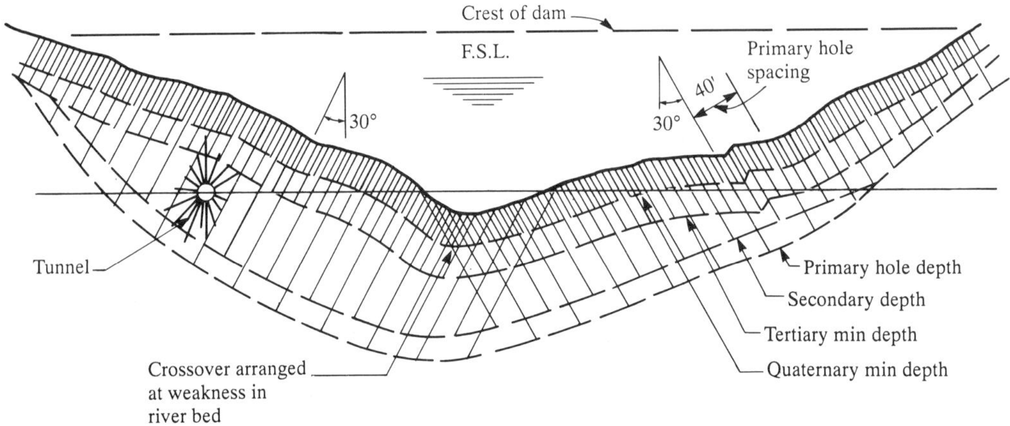

Figure 17.9 *Example of simple grout curtain*

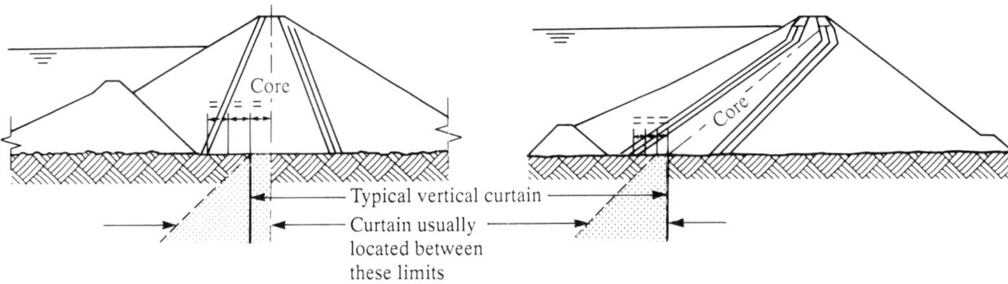

Earth core dams

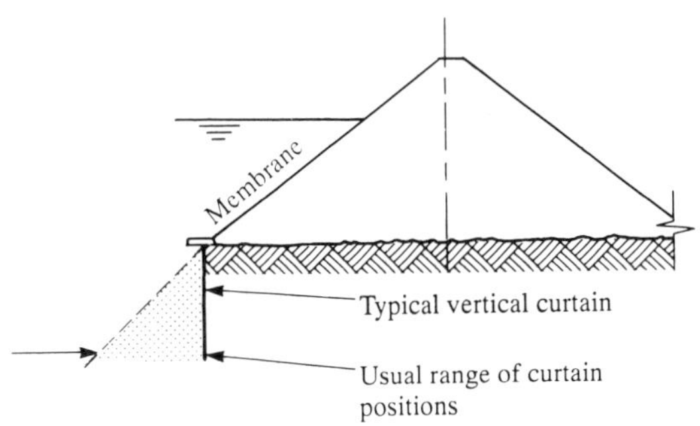

Sloping membrane dams

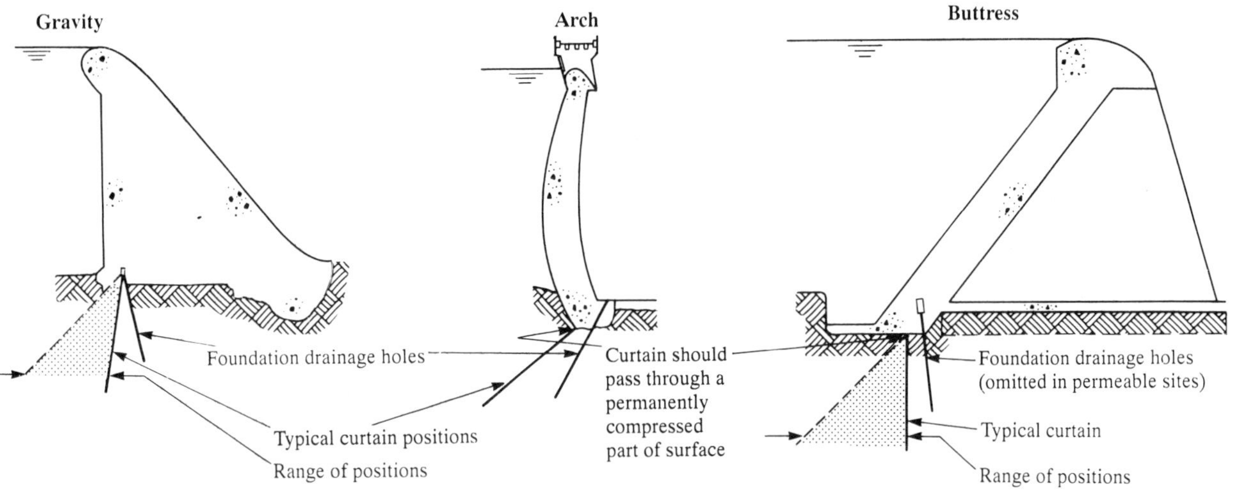

Concrete dams

Figure 17.10 *The usual positions of grout curtains at the main types of dams*

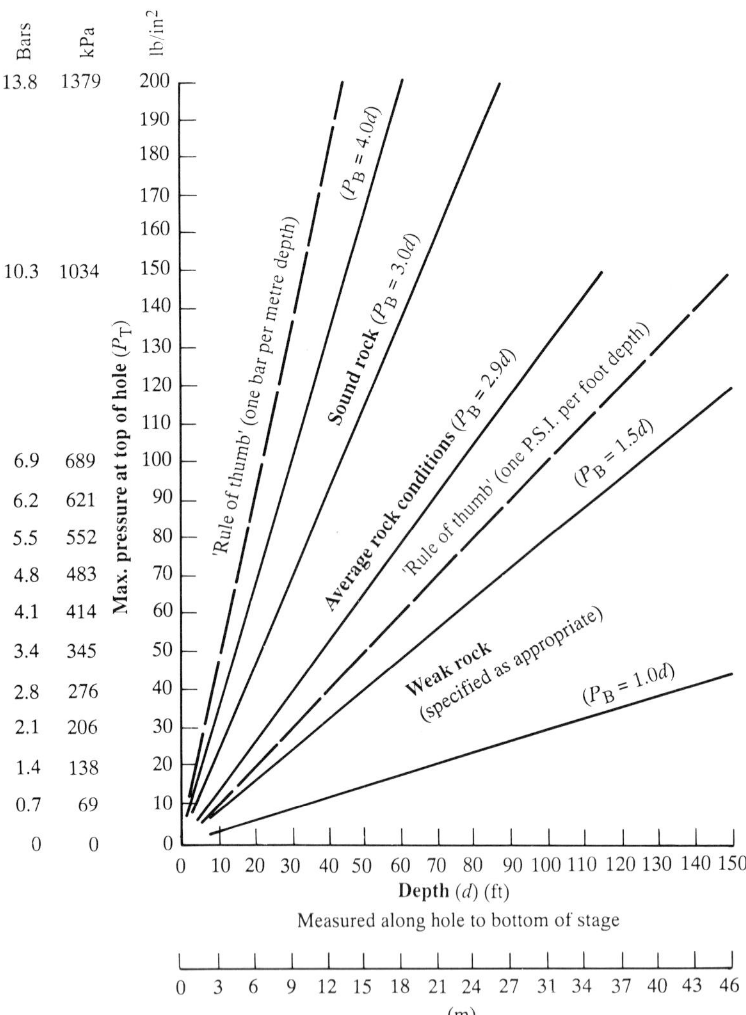

Figure 17.11 *Maximum grout pressures shown against depth to bottom of stage for various rock strength conditions*

are more than four times as great as those of the moderate rule of thumb. Displacement grouting has the purpose of opening cracks and helping with penetration. When pressure is removed after grouting, cracks are expected to tend to close and 'prestress' the grout films. The greater penetration means that fewer grout holes are needed than when moderate pressures are used, but there is always the doubt that the grout intrudes all parts of the ruptured area. It has been said that this type of grouting is like 'making the patient worse in trying to cure him'.

Sabarly (1968) suggested that the choice of grout pressures depends not so much on technical issues of safe injection pressures, as on which system of units has been traditionally in use: those of the metric tradition have used the metric rule of thumb while the 'Anglo-saxon' (his description) uses the lb/in^2 per foot depth rule.

In applying pressure in a grout application, good practice is to start with a much lower pressure for a few minutes while a check is made for any leaks, connections or any other problem likely to become worse if pressure is built up. If there are no troubles, the full pressure should be reached gradually, all the time watching for anything warranting reduction.

17.10 The nature of grout penetration in cracks

The principle is to get the thickest possible grout into the cracks as quickly as possible — so as to reach the limit of penetration while the grout is still able to move freely and transmit the pressure effectively. The overall time available for this ranges from half an hour in the case of trouble-free injection in fine cracks, to three hours or so for fast, high takes in wide cracks. Except when delays such as leaks, connections or rock movements occur, applications can generally be expected to reach refusal within about three hours from start.

Some cracks are easier to grout than others. The easiest are near-vertical, the hardest are those which are near horizontal. Figure 17.12 gives examples.

The mode of grout travel along a near-horizontal crack where there are no upward cracks through which bleed water can escape is indicated in Figure 17.13. At the start of grouting (Step (1)) the grout travels from the grout hole freely along the open crack under the action of pressure from the grout hole. The pressure falls off as the distance from the hole increases until at the tip of penetration it is relatively slight. As the grout continues its penetration along the crack (Step (2)), the pressure profile follows in much the same way until the limit of penetration is reached for the particular pressure and water:cement ratio used (Step (3)). Up to this stage grout further back in the crack has been in motion and additional grout has been entering the crack from the hole. There has been little opportunity for bleeding to start. However, the cessation of movement now permits separation of water from cement. Because there is little room for this to occur in the upward direction, it takes place in a lateral fashion, leaving ridges of grout separated by meanders of bleedwater. There is a visual resemblance to delta formation at the mouths of large rivers.

If there are frequent cracks venting upwards, the bleedwater tends to escape through them and this permits fresh grout to enter from the hole. However, if there are not, as in the example given in Figure 17.13, the bleedwater can become trapped amongst the grout in an

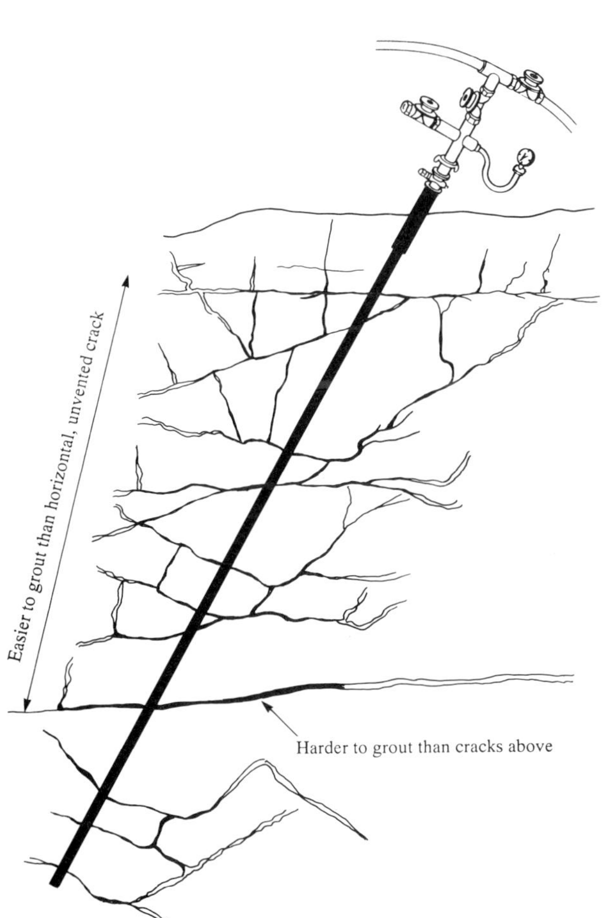

Figure 17.12 *Cracks of differing grouting ease*

1 At start of the grouting

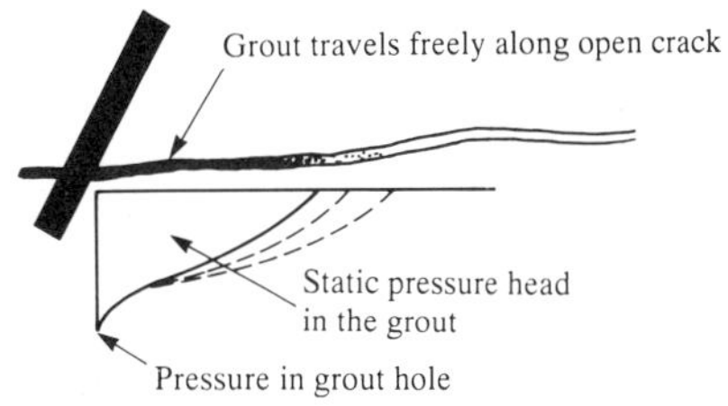

2 Grout reaches limit of penetration

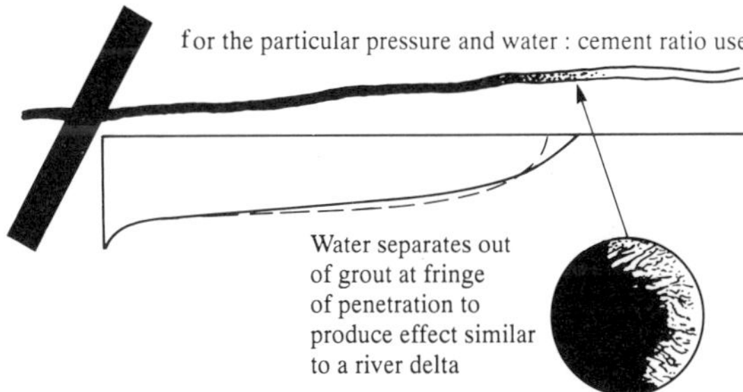

3 Grout stiffens:
Radius of pressure transmission contracts to vicinity of hole and bleed pockets develop

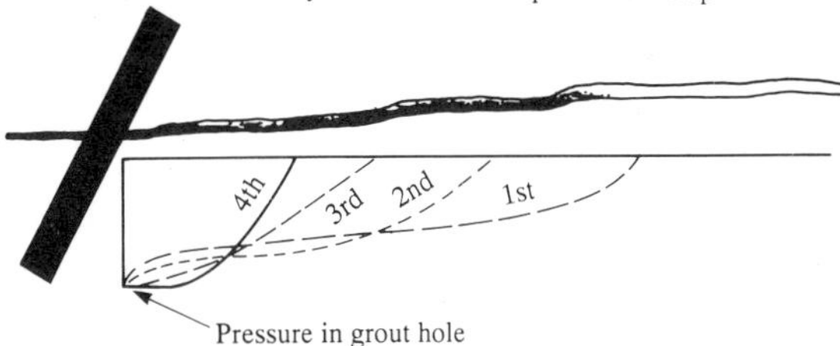

Figure 17.13 *Mode of grout travel in a near-horizontal crack if there are no upward cracks able to take bleed water off*

imperfect filling once it becomes too stiff to further transmit pressure from the hole.

The rheology of grout flows in various crack conditions is further discussed in Houlsby (1985a) and test results are reported there.

17.11 Water:cement ratio and its effect on durability

Water is necessary to give grout mobility but becomes a nuisance after its transporting role is finished. The water needed for mobility exceeds the amount required for the chemical action of cement hydration. An axiom of grouting is that the minimum amount of water should be used consistent with achieving adequate penetration.

If there is too much water the structure of the grout when set can be so weak that it lacks durability and can be chemically attacked by groundwater and sometimes biologically attacked by calcium-feeding bacteria. Therefore care is required in deciding just how much excess water to use in each grout application.

There are various opinions on whether grouts should be thick or thin. It seems that the very divergent attitudes of earlier years are converging to common ground where water:cement ratios thinner than about 5:1 (by loose volume) are avoided because of doubts about durability, and 3:1 is regarded by many people as the thinnest desirable. Many sites can be effectively grouted with 2:1 or thicker.

These water:cement ratios are by volume in accordance with usual rock grouting practice, where the cement volume is taken as the loose volume as it comes in bags. For instance, the volume of cement in a 50 kg bag can be rounded off at 35 litres for batching; likewise a 40 kg bag can be batched as 27 litres; and a 94 lb bag is taken as 1 ft^3 of cement.

The choice of water:cement ratios is a matter of experience. Examples are given in Houlsby (1982c, 1990), and the matter is discussed at some length in Houlsby (1985a). Recommended practice is to start with a mix which is a little thinner than the majority of cracks will accept, and then to thicken up during the course of the application, and to carry on to refusal.

At the start of a grout application, the hole often contains water from the water test, or groundwater. Additional water and watery grout can separate from the grout during the application, particularly if a thin mix is in use. These can sometimes be driven off through cracks by the grout, but at other times must be bled off through the bleeder valve on the standpipe fittings (unless a packer is in use, which unfortunately inhibits this bleeding process). An example of unsatisfactory grouting caused by omission of bleeding at the standpipe is quoted in Houlsby (1985a), and cases of non-durable grouting because of use of grouts thinner than 3:1 are also quoted.

17.12 Grouting of fine cracks

As mentioned earlier, grouting of cracks finer than about 0.5 mm can present problems. In routine production work, it is usually possible to go a little finer, to 0.4 mm, with ordinary cement, provided high-quality mixing is used. If a foundation contains significant jointing finer than this the question arises as to whether grouting is the best form of treatment for the site: grouting materials become appreciably more expensive when ordinary cement cannot be used. Other types of cement used in fine cracks, with varying degrees of success, include finely ground separated and microfine cement.

17.13 Grouting of stressed rock

If surface rock moves during grouting, use of special grouting methods becomes necessary and precautions to prevent continuation of movement are required. These measures can inhibit the production of good grouting and therefore movements must be minimized. They may take the form of individual slab dislodgement or of widespread heaving of sheets of rock. The individual slabs might move because of insufficient embedment amongst neighbouring slabs; sheet movements usually arise from tectonic stress fields in the rock which are marginally in equilibrium before grouting but become unbalanced by the grout pressure and then relieve themselves by dilating.

Precautions for controlling movements include a close watch for sudden increase in grout take and sudden loss of pressure in the hole. These may be accompanied by cracking of rock slabs, and by surface leaks.

The general principle when grouting stressed rock is to control the amount of energy conveyed into the cracks via the grout. This energy can be loosely regarded as the product of the pressure and of the rate of take. Control is not merely a matter of restraining the pressure (Wong and Farmer 1973). By limiting the supplied energy to an amount which does not cause stress relief or movement, effective grouting can be produced. The limiting amount is found by experience. The basis for it is that the rate of take is restricted to that volume needed to fill initially open voids plus that subsequently needed to compensate for bleeding and leakage losses. The injection is at low pressure. Intermittent grouting using a very thick mix may have to be used. Sometimes stressed slabs creep for hours after grouting. Several applications at 24 hour intervals may be necessary to deal with this and with bleeding losses. A flow meter is useful for continuously monitoring the take when handling movements.

Bolting of surface slabs sometimes helps, but can be worthless in heavily stressed sites such as deep canyons in massive rock.

17.14 Assessment of grouting

Grouting is not a process where a recipe or prescribed procedure can be followed. Every hole has its own particular variations on the general theme and in the case of large foundations, grouting characteristics can vary across it. Successful grouting requires adaptability and experience.

Tailoring the work to the site conditions involves assessment of each application of grout during injection, and of neighbouring holes to decide whether further closure is needed.

If thorough grouting is required there is no substitute for hole-by-hole assessment. Averaging of several holes is sometimes resorted to in lesser-quality work, but can leave poorly grouted windows in the work.

17.14.1 Assessment during injection

The purpose is to make sure that the hole is taking the maximum amount of cement possible. This may become secondary though to dealing with leaks, connections or rock movements if these occur.

When grouting a hole, steps in the injection and assessment processes are as follows:

(1) Decide the starting mix and the maximum pressure.
(2) For the first few minutes of grouting use a low pressure and check for any problems such as leaks, connections and rock movements. If these occur, they may require appropriate special treatment. If they do not occur, slowly increase the pressure to the maximum and continue to check for trouble.
(3) After 15 minutes (or thereabouts) of grouting, measure the grout taken and from the reading decide whether the hole will take a thicker mix or whether it is better to continue with the starting mix. This decision should be based on experience guided by comparing the rate of take with the water test figure, and by observing whether the hole pressure is sensitive or not to changes in the control valve opening (this indicates whether the hole is tight or open).
(4) Every 15 minutes or so review the rate of take for the purpose of deciding whether the mix should be thickened.
(5) Periodically bleed water and watery grout out through the bleeder valve.
(6) Most holes show reduced take as the injection proceeds. When this happens, grouting should continue until refusal is reached. When it does not happen, such as the case of a runaway hole, it is not good practice to continue supplying the hole indefinitely: the grout is probably travelling far and wide rather than doing useful work in the vicinity of the hole where it is required. In this event, recommended practice is to stop grouting after a fairly substantial amount of cement has been injected, and then to let this set before resuming 24 hours later. Further limited applications are given like this until eventually refusal is reached. The amount of cement stipulated for each such injection varies with the site and is usually between 100 and 200 bags.
(7) When refusal is reached, it is advisable to hold pressure for a further 15 minutes to ensure that thixotropic stiffening of the grout has developed sufficiently to enable the grout to withstand dislodgment by pressurized groundwater or by other grouting activities nearby.
(8) A 'post mortem' review of the application should then be carried out to find any deficiencies which require remedying by placement of new holes nearby.

A fuller treatment of assessment methods is in Houlsby (1990).

17.14.2 Assessment related to closure

This is the process of reviewing the grouting of several holes of a closure sequence. The water and cement takes in the latest holes are compared with those in earlier holes and with the desired standard of grouting to decide if enough grouting has been done or whether further closure is needed. If leaks, connections or movements have required special treatment which may not have given good grouting of other cracks, extra holes may be necessary.

This reviewing is applied to each stage in turn. Inevitably the resulting grouting finishes with uneven spacing and depths of final holes.

17.15 Some words of caution and of encouragement to the inexperienced

A casual reading of this chapter might give the impression that grouting is easy — just a matter of following the rules. In fact it can be difficult and complex, calling for much experience and skill.

Grouting is an engineered process, but involves a significant amount of art; perhaps more so than most engineering. The best way to acquire the requisite degree of competence is to spend much time on the job, at the grout holes, at the pumps, personally operating the control valves, doing the reports, and all the other odds and ends. It is even better if this essential field experience can be followed by site investigation, grouting design and monitoring of the efficacy of completed work.

Acknowledgement

This chapter is adapted from material in *Construction and Design of Cement Grouting* by A. C. Houlsby (© 1990 John Wiley & Sons, Inc.), and is included by permission of John Wiley & Sons, Inc.

References

Fergusson, F. F. and Lancaster-Jones, P. F. F. (1964) 'Testing the efficiency of grouting operations at dam sites', *Proc. 8th Int. Congress on Large Dams*, Edinburgh, **1**, 121–139

Houlsby, A. C. (1976) 'Routine interpretation of the lugeon water test', *Q. J. Engg. Geol.*, **9**, 303–313

Houlsby, A. C. (1977) 'Engineering of grout curtains to standards', *Proc. Am. Soc. Civil Engineers, J. Geotech. Engg Division*, **103**, (GT9), 953–970

Houlsby, A. C. (1982a) 'Cement grouting for dams', *Proc. Conf. 'Grouting in Geotech. Engg'*, Speciality Conference, New Orleans, American Society of Civil Engineers 1–34

Houlsby, A. C. (1982b) 'Optimum water:cement ratios for rock grouting'. *Proc. Conf. 'Grouting in Geotech Engg'*, Speciality Conference, New Orleans American Society of Civil Engineers, 317–331

Houlsby, A. C. (1982c) 'A digest of typical cement grouting takes', *Proc. Conf. 'Grouting in Geotech. Engg'*, Speciality Conference, New Orleans, American Society of Civil Engineers, 1000–1014

Houlsby, A. C. (1985a) 'Cement grouting: water minimizing practices', *Issues in Dam Grouting*, American Society of Civil Engineers, 34–75

Houlsby, A. C. (1985b) 'Design and construction of cement grouted curtains', *Proc. 15th Int. Congress on Large Dams*, Lausanne, International Commission on Large Dams, **3**, 999–1015

Houlsby, A. C. (1990) *Construction and Design of Cement Grouting*, Wiley-Interscience, New York

Lugeon, M. (1933) *Barrages et Géologie*. Dunod, Paris

Pratt, H. K., McMordie, R. C., and Dundas, R. M. (1972) 'Foundations and abutments — Benett and Mica dams', *Proc. Am. Soc. Civil Engineers, J. Soil Mechanics and Foundation Division*, **98**(SM10), 1053–1072

Sabarly, F. (1968) 'Grouting and drainage of dam foundations in rock of low permeability', *Geotechnique*, **18**, 229–249

Wong, H. Y. and Farmer, I. W. (1973) 'Hydrofracture mechanisms in rock during pressure grouting', *Rock Mechanics*', **5**, 21–41

Bibliography

Baker, W. H. (1982) 'Grouting in Geotechnical Engineering', *Proc. Speciality Conference*, New Orleans, American Society of Civil Engineers, New York

Karol, R. H. (1983) *Chemical Grouting*, Marcel Dekker, New York

18 Reinforcement and support of rock masses

I W Farmer
I W Farmer & Partners, Newcastle-upon-Tyne, UK

18.1 Introduction

The purpose of reinforcement is to preserve the natural competency of the rock mass in which it is used and, to some extent, to add the required increment of strength to prevent or limit failure. The requirements for reinforcement are governed by the geological conditions, the most important of which are the state of stress, the properties of the rock such as strength, hardness and chemical stability, as well as the character, frequency and orientation of the discontinuities present. Reinforcement strengthens a rock mass by increasing the shear resistance along discontinuites, by enhancing the interlock between individual blocks and by preventing the detachment of loose blocks. The formation of a reinforced zone at the excavated surface of a rock mass should maintain its integrity and allow for the redistribution of stresses within the excavation. It also should possess sufficient stiffness to minimize the dilation of discontinuities within the rock mass surrounding an excavation. In fact as discontinuities tend to dilate after excavation, reinforcement should be installed as soon as possible thereafter. Otherwise delay can reduce the stability of the rock mass, thereby allowing blocks to move and become loosened and eventually to fall from the face of the excavation. Rock masses can be reinforced and supported by using dowels, bolts, mesh, shotcrete, cables or anchors, frequently in some combination with one another.

Reinforcement of rock masses is used in both civil and mining engineering practice but generally more exacting standards are demanded in the former than in the latter. This is because excavations in civil engineering are normally permanent and often used by or for the public. Moreover such excavations are frequently larger and more often constructed in poor ground conditions. On the other hand violent failure is rarely experienced in civil engineering practice.

The nature of the geotechnical data which needs to be gathered for the design of a system of rock mass reinforcement is outlined in Table 18.1. In order to keep a check on the ground conditions and to see if they are as assumed in the design, such data should continue to be collected throughout the period of construction.

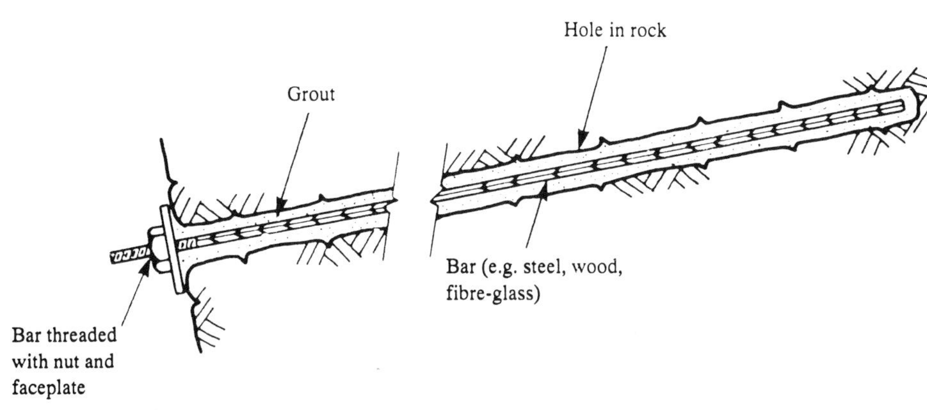

Figure 18.1 *Grouted dowel (after Douglas and Arthur 1983)*

Table 18.1 Geotechnical data (after Douglas and Arthur 1983)

Geotechnical data	*Source*
Preliminary information	Published and unpublished geological maps, reports, memoirs, etc. Records, logs of excavations, boreholes etc. in close proximity to proposed project Published and unpublished reports on projects in similar ground
Geological description	Surface mapping, core logging, etc.
Rock material strength and deformation characteristics	Laboratory testing (uniaxial and triaxial)
Rock mass index properties	Assessment of rock quality (e.g. rock quality designation (RQD)) Point load testing, etc.
Discontinuity data: description orientation spacing continuity	Surface mapping Structural logging of drillholes Core orientation Mapping of surface outcrops and completed excavations
Discontinuity shear strength characteristics	*In situ* shear testing Shear box tests
Rock mass deformation moduli	Geophysical methods Drillhole modulus gauges Plate bearing tests Monitored excavations
In situ stress measurement	Drillhole stress meters Deformation gauges Strain cells
Unit weight	Laboratory testing
Groundwater conditions	Piezometric measurements Packer tests Visual inspection of excavation
Seismic data	Published seismic records Vibrograph measurements

18.2 Dowels

A dowel is an unstressed bar which is grouted, generally along its whole length, into a drill hole in a rock mass (Figure 18.1). Grouted untensioned dowels have been used extensively and effectively for support and stabilization of rock surfaces. Dowels may be 1–2 m in length and 15–30 mm in diameter. They may be made from plain or deformed steel, fibreglass or hardwood. Composite dowels of various materials also are available.

The fact that they are unstressed means that grouted rock dowels behave very differently from rock bolts. In other words because dowels are not tensioned, they act passively rather than actively in stabilizing the rock mass, that is, the rock must move before the dowel can exert any restraining force (Lang 1971). A dowel is especially useful where the main component of the load applied is normal to the direction of the installed bolt or where, because of creep or plastic flow, it is difficult to maintain constant tension between a fastening device in the rock mass and a bearing plate at its surface during the expected life of the excavation.

Because dowels rely exclusively on self-tensioning developed as a result of rock movement the grouting must be fully effective. Dowels can be grouted into place by inserting cement or resin grout into the drillhole and pushing the bar into the grout by rotating the bar into cement cartridges or resin capsules in the drillhole, or by injecting grout into the hole after the bar has been installed.

It is usually only necessary to test dowels when they are employed as the chief means of reinforcement. In such cases a special dowel, which incorporates strain gauges, is installed in a drillhole in order to assess dowel performance. The strain gauges provide a measure of the distribution of loading along the bar.

One of the simplest methods of rock mass reinforcement is to use dowels as shear keys to hold together medium and thinly bedded rocks dipping parallel to a slope. Holes, wherever possible, are drilled normal to the bedding and the dowels are grouted into place, hopefully with any potential shear force occurring at mid-depth. As dowels are unstressed and weak in bending, they therefore should only be used where the discontinuities are narrow.

A faceplate and nut commonly are used in conjunction with a dowel when it acts as part of the permanent support system. This arrangement ensures an adequate surface anchorage and helps prevent the dowel debonding from the drillhole at the rock face. Mesh reinforcement can be used with faceplates and dowels to provide further support for the rock face (Douglas and Arthur 1983).

A number of proprietary dowels have been developed, primarily for providing rapid reinforcement in mining operations. For example, in the perforated sleeve dowel, cement mortar is placed in a perforated sleeve which then is inserted in the drillhole. By inserting the dowel bar into the perforated sleeve, the cement grout is forced via the perforations into the annular space of the drillhole. Obviously the relationship between the diameters of the drillhole, sleeve and dowel bar are critical (Figure 18.2). The friction anchor or pre-set dowel (Figure 18.3) consists of a split tube that is forced into a drillhole which has a diameter which is slightly less (35 mm) than that of the tube (38 mm). The spring section of the compressed tube applies a radial force so that frictional resistance to sliding is developed over the entire length of the bar within the rock mass (Hoek and Brown 1982). Although quick and easy to install, the dowel cannot easily be protected against corrosion. Nevertheless, according to Brady and

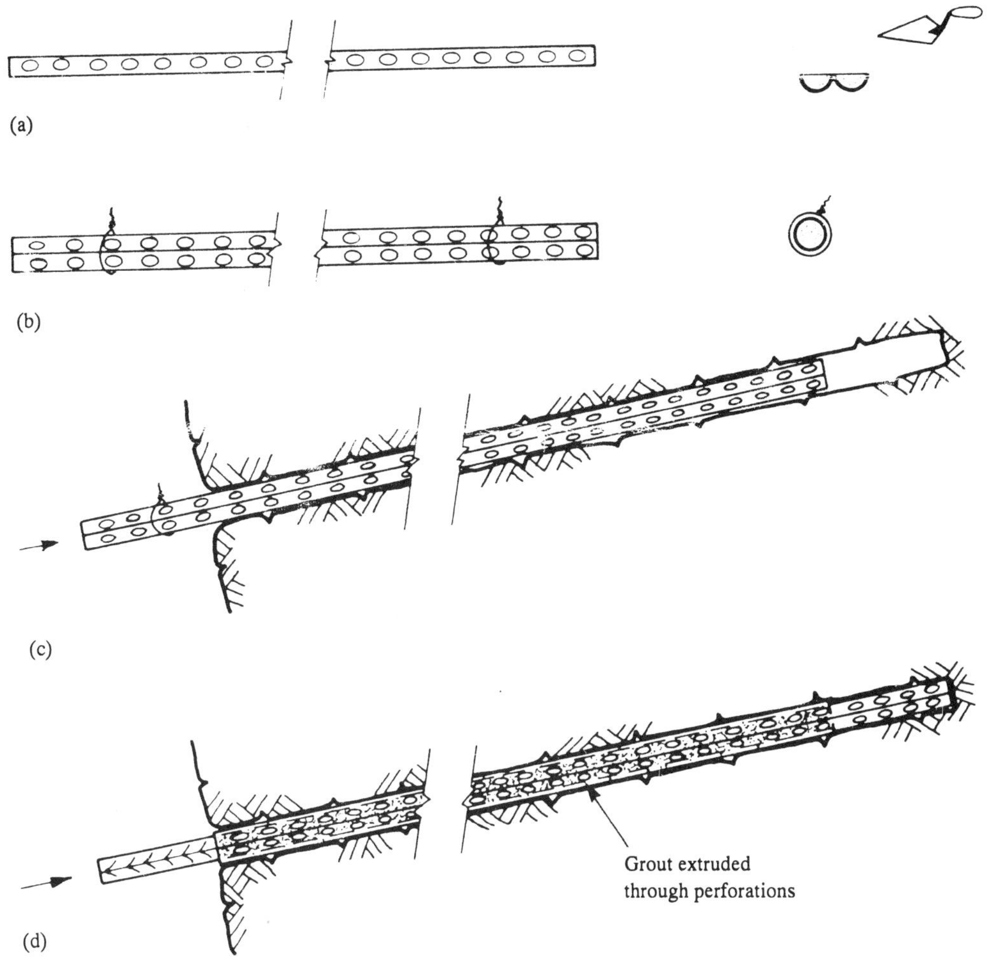

(a)

(b)

(c)

Grout extruded through perforations

(d)

Figure 18.2 *Perforated sleeve dowel:* (a) *sleeves filled with grout;* (b) *sleeves wired together;* (c) *sleeve inserted into hole in rock;* (d) *bar inserted into sleeve (after Douglas and Arthur 1983)*

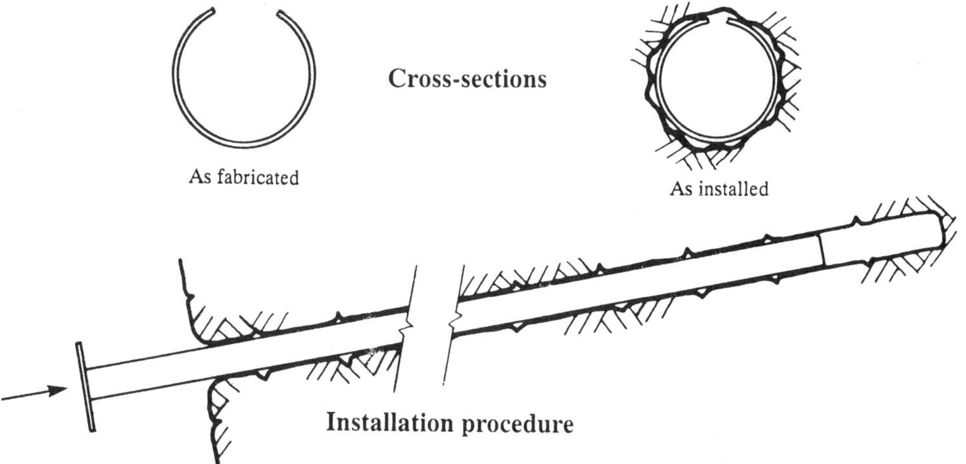

Figure 18.3 *Friction rock stabilizer (after Douglas and Arthur 1983)*

Brown (1985), the frictional resistance is increased as the outer surface of the tube rusts.

The Swellex dowel is essentially a thin-walled tube which is folded into a collapsed shape with a diameter between 25 and 28 mm (Figure 18.4). When placed within the drill hole (33–39 mm in diameter) it is expanded by the injection of water under high pressure (20–30 MPa). This not only produces a radial force against the rock which generates frictional resistance but also causes a contraction in the length of the tube which, in turn, pulls the faceplate tight against the rock face. A small amount of tension also is induced in the dowel. The diameter of the drill hole is not as critical as for other types of dowels. Again protection cannot readily be afforded against longer-term corrosion.

18.3 Rockbolts

18.3.1 Introduction

Rockbolts are the simplest method of rock support or reinforcement. It is surprising that such a large literature including books (Hobst and Zajie 1983) and conference proceedings (Stephansson 1984) should have grown up around something so inherently easy to understand. The conventional way to describe the action of bolts is best approached through the typical theoretical stress distributions around two shapes of openings illustrated in Figure 18.5. This shows the principal stress trajectories and also the major and minor principal stresses around a rectangular opening in a hydrostatic stress field and a circular

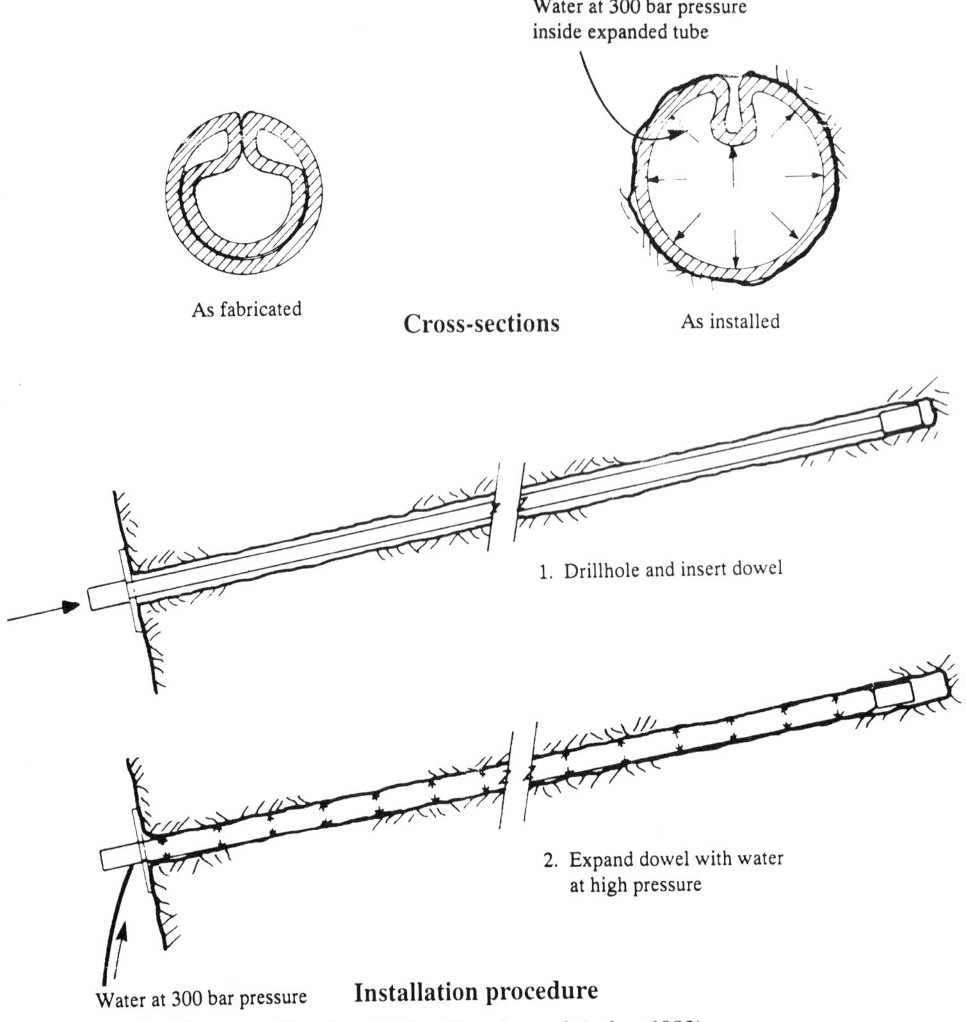

Figure 18.4 *Swellex expanding dowel (after Douglas and Arthur 1983)*

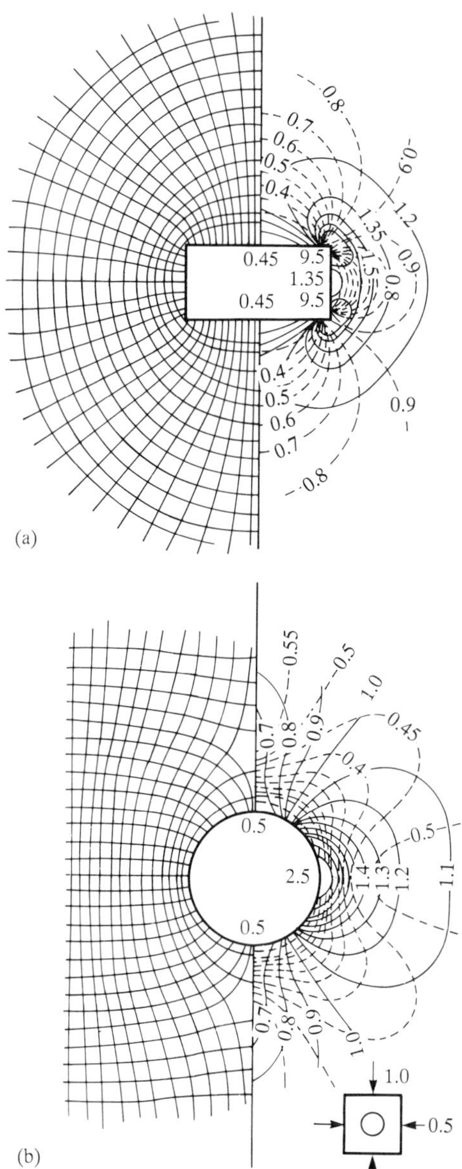

Figure 18.5 *Principal stress trajectories (LHS) and contours (RHS) of the ratios of major (solid line) and minor (dotted line) principal stress to vertical applied stress around* (a) *a rectangular opening with a 2:1 width–height ratio in a uniform stress field* (b) *a circular opening in a stress field with a vertical–horizontal stress ratio of 2 (from Farmer 1985, after Hoek and Brown 1980).*

opening in a non-uniform stress field in a homogeneous, continuous elastic material. In both cases the surface of the opening is subjected to compressive tangential stress and zero radial stress. Further away from the surface both the radial and tangential stresses approach the primitive stress levels in the rock mass disturbed by excavation.

In the case of the rectangular excavation, the tangential compression stresses are high at the corners and in the sidewalls, and low in the roof and floor. This condition will be exacerbated if the height–width ratio is reduced further. This will have two effects — there will be crush at the corners and possibly squeeze in the sidewalls, and the reduced radial compression will allow sag of the roof and uplift of the floor. The most important of these is the reduced roof compression — particularly if combined with bedded and jointed strata — which will create conditions for release of blocks from the roof.

In the case of the circular excavation, a similar set of circumstances can be created by a high vertical and low horizontal geostatic stress, however unlikely this may be in practice. However, in a low-stress environment, typical of tunnelling, it is easy to see that radial compressive stresses will be insufficient to maintain stability where well-defined or occasionally weathered release surfaces are present in the rock.

Rockbolts are the cheapest and most obvious way of maintaining stability in such circumstances. Provided that the rocks are suitable for an anchorage location, are not subject to swelling or slaking and there are no high pore pressures or water flows, then they have two main functions acting either singly or as a pattern. These are to maintain the stability of sagging roofs, particularly in weaker stratified rocks, and to restrain blocks in well-jointed or blocky rocks where release surfaces daylight into the exposed roof. The former application is principally for roof support in room and pillar mining in stratified rocks. This is the most common use of rockbolts and up to 100 million bolts (Peng and Tang 1984) are used annually in the US coal mining industry alone. The latter application is principally in civil engineering works, such as tunnel and cavern construction and occasionally slopes, where quite large-capacity anchors are often used.

18.3.2 Rockbolt design

(a) Stratified deposits

The most simple assumption for design purposes is to consider a sagging roof plate or beam of thickness, L, span, B, and length, X, supported by rows of bolts with separation, a, between rows and spacing, S. The bolt tension force, P, to support the roof will then be given by

$$P = \frac{\gamma BXL}{\left(\frac{x}{a} + 1\right)\left(\frac{B}{S} + 1\right)} \qquad (18.1)$$

where γ is the unit weight of roof rock.

This equation, suggested by Obert and Duvall (1967), is valid if the roof above the excavation is completely suspended by bolts. For an assumed bolt load it can also be used to estimate spacing, and the number of rows. It represents the upper limit of bolt force since it ignores the important supporting effect of the abutments. It also ignores the interaction of a series of roof beds.

A more accurate approximation can be obtained by considering the effects of friction between beds and also by considering the roof span as a series of thin beams, fixed at each side of the opening. Panek (1962a, b, 1964) in a series of seminal papers considered this condition both experimentally by centrifugal testing and analytically, and developed the nomograph illustrated in Figure 18.6, which has been used extensively in mine design. It is explained in detail by Panek and McCormack (1973) in the *SME Mining Engineering Handbook*. The basic variable is a reinforcement factor, RF, which is used to evaluate the inter-bed friction effect due to bolting. The roof is considered a series of beds of equal thickness, of the same material, and without bond between them. The bolts are assumed normal to the beds and tensioned to give normal compressive loading across the beds. Then

$$RF = \left(1 + \frac{\Delta\sigma_f}{\sigma_{fs}}\right)^{-1} \tag{18.2}$$

where $\Delta\sigma_f/\sigma_{fs}$ is the *decrease* in bending stress from frictional resistance induced by bolting, expressed as a ratio of the maximum bending stress in the unbolted strata, and is given by the empirical equation

$$\frac{\Delta\sigma_f}{\sigma_{fs}} = 3/8\mu(aB)^{-0.5}\left[\frac{B}{S}P\left(\frac{L}{t}-1\right)\frac{1}{\gamma}\right]^{0.33} \tag{18.3}$$

where μ is the interbed coefficient of friction and t is the average roof-layer thickness. P is the assumed bolt tension and L is assumed equal to bolt length or support thickness. For typical thin bedded mine roof strata RF should be greater than 2 and bolt spacing should be less than 1.5 m (5 ft).

(b) Non-stratified deposits

Lang (1971), Rabcewicz (1969) and Alexander and Hosking (1971) were among the first to develop a general empirical approach to rock bolting in non-stratified deposits. The optimum ratio of bolt length to bolt spacing can be determined by simple analysis of stresses beneath an elastic half space. By using the Boussinesq distribution beneath a point force, it can be shown (Figure 18.7a) that

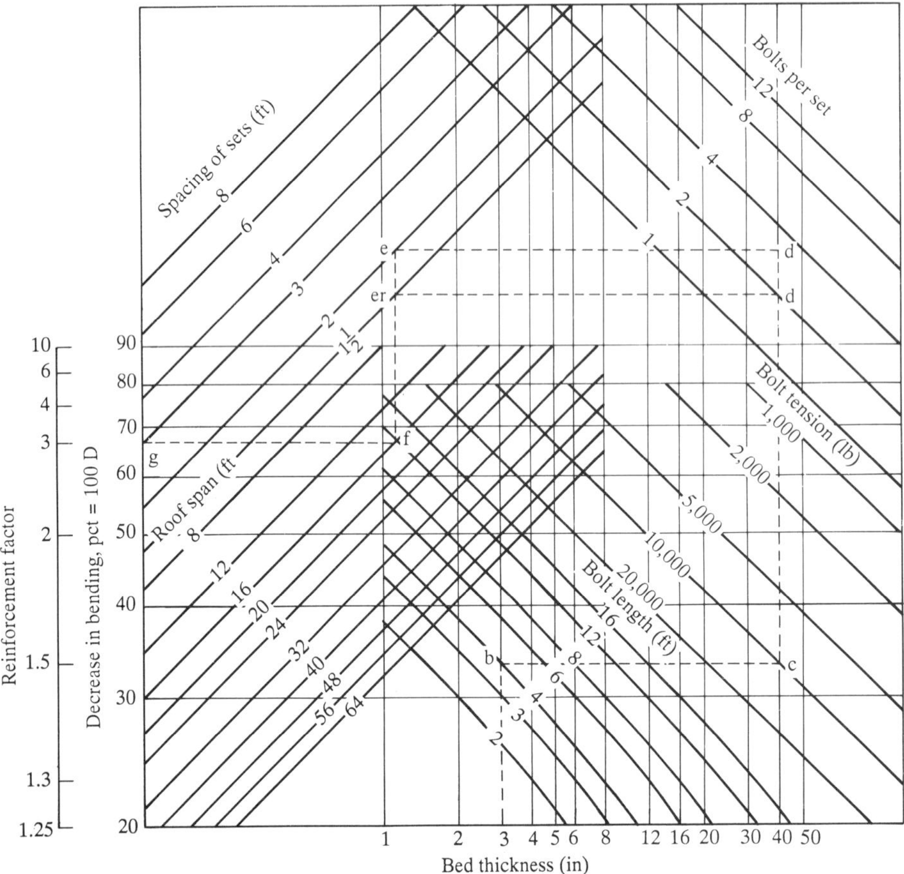

Figure 18.6 *Nomograph by Panek (1962a) to determine friction effect for bolting in mine roofs. Note 1 in = 25.4 mm, 1 ft = 0.3048 m, 1000 lbf = 4.448 kN.*

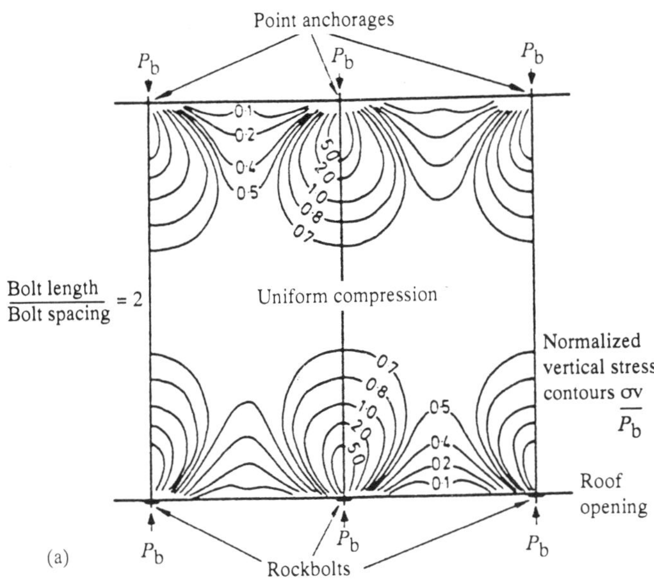

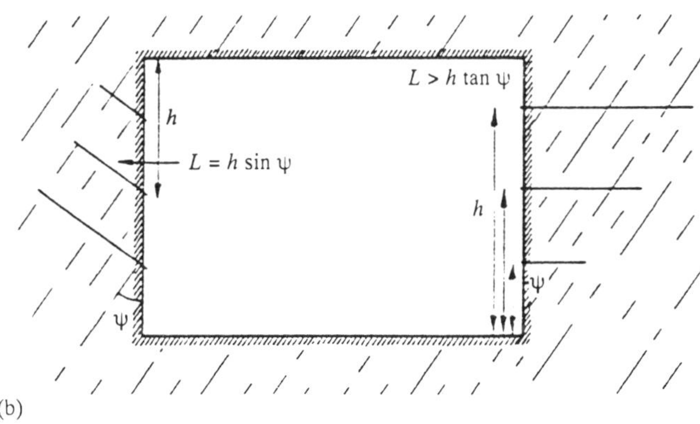

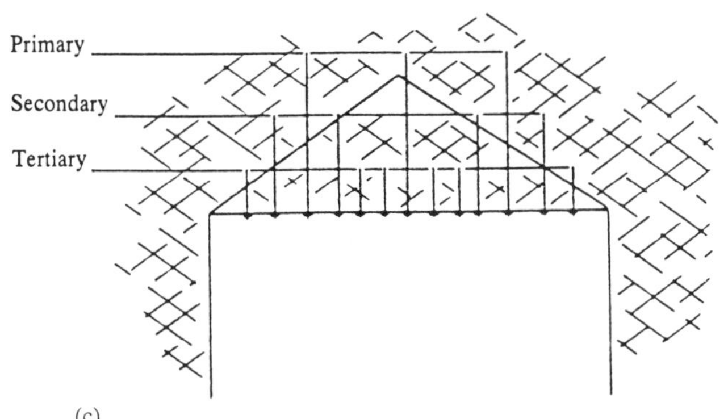

Figure 18.7 (a) *Two-dimensional representation of stress distribution in an elastic roof beam induced by point loads with a spacing equal to half the beam thickness;* (b) *determination of bolt lengths in excavation sidewalls;* (c) *primary, secondary and tertiary bolting patterns in a large span excavation.*

a bolt length to bolt spacing ratio of 2 is required to create a zone of uniform (radial) compression. Ideally bolt forces can be calculated to give a uniform compressive stress of about 30 kPa per metre of thickness of compression zone to support the body forces exerted by the compressed layer. This is the basis of most empirical methods which are summarized in Table 18.2 and Figure 18.7 (b, c).

(c) Single block

An interesting but more academic exercise is to consider the stability of individual blocks, where discontinuity sets combine to form blocks which are kinematically capable of moving into an excavation, whether by falling or sliding from the roof or side (Figure 18.8). An outline of a simple stability analysis for these conditions using stereographic projections and taken from Farmer and Shelton (1980) is summarized in Figure 18.9. To distinguish a falling block from a sliding block from the roof, poles are plotted to each discontinuity set. For a falling block, poles must surround the centre of the net; thus, *abd* and *acd* represent falling blocks in Figure 18.9(a) and *abc* and *bcd* represent sliding blocks. For a falling block, weight and reinforcement may be calculated from Figure 18.10. To analyse a sliding block, planes are plotted to represent the sliding wedge and its lines of intersection. The plot may be rotated so that the two points of intersection closest to the centre lie on a single great circle. If the line of maximum dip of the plane lies between these two points, the block tends to slide along that plane only, as in Figure 18.9(b). If not, then it tends to slide along the line of intersection nearest to it, as in Figure 18.9(c). The friction cone is plotted to represent the angle of friction of discontinuities. If the sliding direction is plotted within the cone, the wedge may be unstable, if outside it may be stable.

In the case of the sidewall, to distinguish between a block that slides along a single plane and a block that slides along a line of intersection (wedge failure), planes and intersections are plotted to define the rock block and the excavation surface. If two lines of intersection dip out of

Table 18.2 Rockbolt parameter design rules for rock masses with <2 and <3 discontinuity sets with clean tight interfaces (after Farmer and Shelton 1980)

Excavation span	*Number of discontinuity sets*	*Bolt design*	*Comments*
<15 m	<2 inclined at 0–45° to horizontal	$L = 0.3B$ $S = 0.5L$ (depending on thickness and strength of strata). Install bolts perpendicular to lamination where possible with wire mesh to prevent flaking	The purpose of bolting is to create a load-carrying beam over span. Fully bonded bolts create greater discontinuity shear stiffness. Tensioned bolts should be used in weak rock; sub-horizontal tensioned bolts where vertical discontinuities occur
	<2 inclined at 45–90° to horizontal	For side bolts: $L > h \sin \psi$ (installed perpendicular to discontinuity) $L > h \tan \psi$ (installed horizontally). See Figure 18.7b for h, ψ; L, bolt length; S, bolt spacing; B, excavation span	Roofbolting as above. Side bolts designed to prevent sliding along planar discontinuities. Spacing should be such that anchorage capacity is greater than sliding or toppling weight. Bolts should be tensioned sufficiently to prevent sliding
	>3 with clean tight interfaces	$L = 2S$ $S = 3\text{–}4 \times$ block dimension Install bolts perpendicular to excavation with wire mesh to prevent flaking	Bolts should be installed quickly after excavation to prevent loosening and retain tangential stresses. Prestress should be applied to create zone of radial confinement. Sidewall bolting where toe of wedge daylights in sidewall
>15 m	<2	$L_1 = 0.3B_1$ Primary bolting $S_1 = 0.5L_1$ $L_2 = 0.3S_1$ Secondary bolting $S_2 = 0.5L_2$ Install wire mesh to prevent spalling	Primary bolting conforms to smaller excavation design. Secondary (and tertiary) bolting supplements primary design (Figure 18.7c)
	>3 with clean tight interfaces	$L_1 = 0.3B_1$ Primary bolting $S_1 = 0.5L_1$ $S_2 = 3\text{–}4 \times$ block size. Secondary bolting $L_2 = 2S_2$	Primary bolting should have sufficient capacity to restrain major blocks. Decisions on block size for secondary bolting should be left to the section engineer

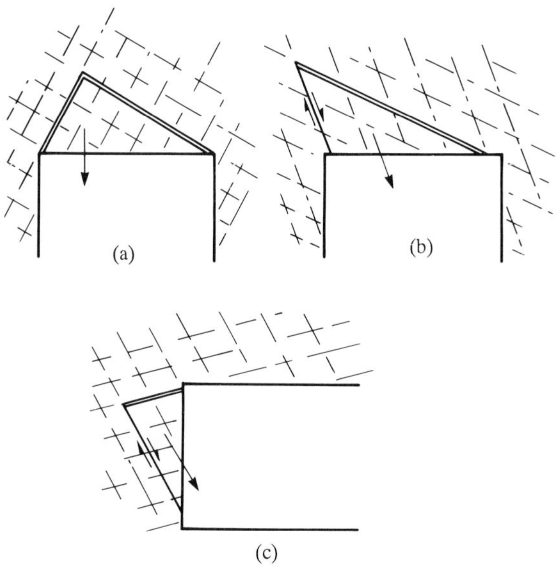

Figure 18.8 *Two-dimensional representation of rock blocks kinematically free to fall or slide into an excavation:* (a) *falling from the roof;* (b) *sliding from the roof;* (c) *sliding from the sidewall.*

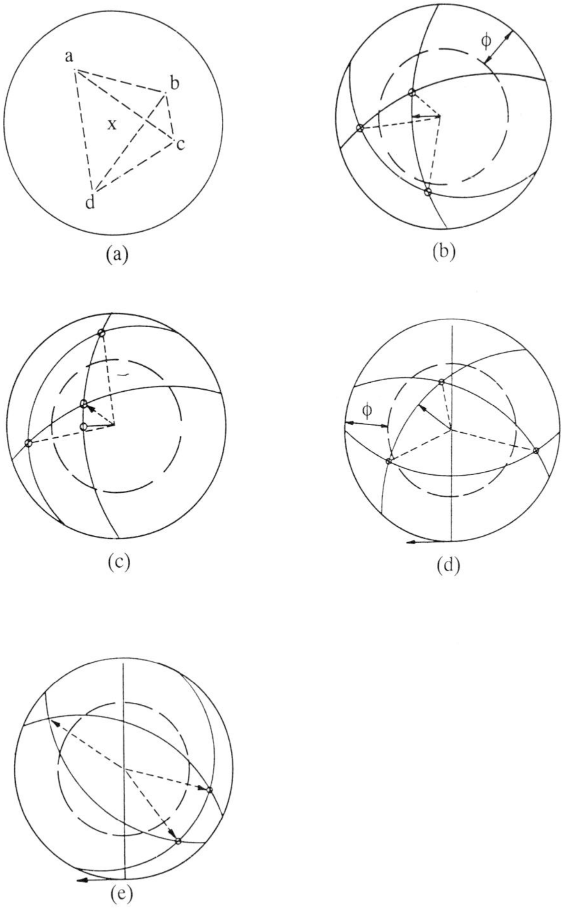

Figure 18.9 *Stability analysis for rock blocks falling or sliding into an excavation roof and sliding from an excavation sidewall using stereographic projections (from Farmer and Shelton 1980).*

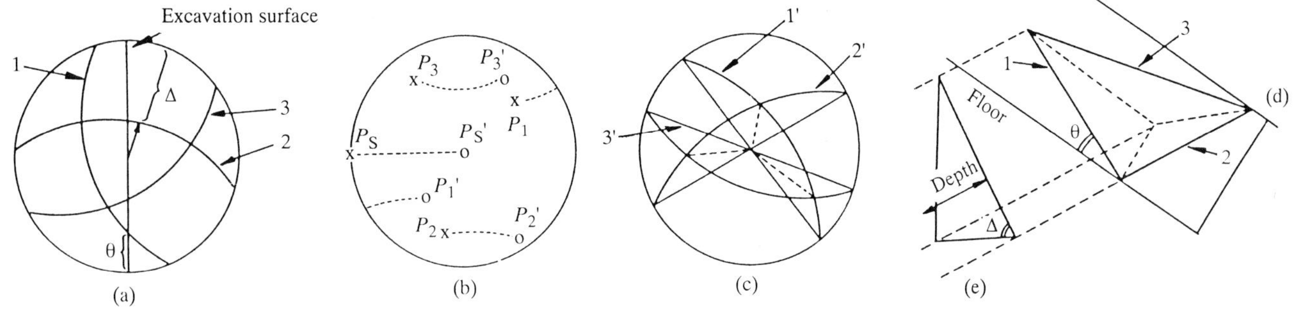

Figure 18.10 *Calculation of volume and weight of a rock wedge (from Farmer and Shelton 1980).*

the face and the line of maximum dip lies between them, as in Figure 18.9(d), then planar sliding may occur. If two lines of intersection do not dip out of the face or do not enclose the direction of maximum dip, sliding may occur along the line of intersection as in Figure 18.9(e). For a first approximation of stability, a friction cone is marked for the angle of friction of discontinuities φ (dashed line). If the direction of sliding is contained in the cone, as in Figure 18.9(d), it may be unstable; if it plots outside the cone, as in Figure 18.9(e), it may be stable.

The size of a block can be determined in either of two ways: as a function of the orientation of the discontinuities and the dimensions of the openings or by consideration of the geotechnical data to establish the likely maximum continuity of the members of each joint set. A method of calculating the volume, and hence the weight, of a rock block, is summarized in Figure 18.10. To simplify the procedure, a stereoplot should be drawn so that the excavation surface lies in a horizontal plane. For a vertical sidewall, all planes must be rotated by 90°; for sidewalls with angles of intermediate values, rotation depends upon the angle to the horizontal plane. For a horizontal roof, no rotation is required. For a sidewall, the plot is rotated so that P_s lies on an east–west axis (Figure 18.10a). Poles to each of the planes are plotted and each pole is moved by 90° in angular units along the respective small circles in the same directional sense (Figure 18.10b). The strike lines of each plane are then replotted to form a closed figure (Figure 18.10c) and these are plotted onto a scaled plan of the excavation (Figure 18.10d). This figure represents the area of the block in surface projection. The depth of block is determined (Figure 18.10e) by simple construction and the volume and weight are calculated.

If the block has the potential to fall or slide from the excavation roof, for simplicity the restraining load through the rockbolts should be calculated from the full weight of the block of rock. If a block is sliding, the disturbing forces, because of the weight of the block, may be reduced by an amount equivalent to the shear resistance of the sliding surfaces. Dividing the block weight by the exposed face area of the block determines the support pressure that must be applied to the block for stability. If this support pressure were applied throughout the excavation, all blocks of the largest size and of smaller sizes would be retained.

This method is similar to the 'ubiquitous joint method' developed by Goodman (1967). The length of the rockbolt is governed by the size of the largest block, so the rockbolt is anchored in stable ground to ensure that the full required bolt load is mobilized. The spacing of rockbolts, as in Figure 18.7(c), depends on the discontinuity spacing and the size of the block, which can either be allowed to fall or be retained by wire mesh or shotcrete.

Goodman and Shi (1985) have extended this approach in their key block theory which considers the interaction of interlocked blocks and isolates the blocks which if supported will prevent loosening of other blocks. It has a tendency to be an academic exercise, but is important in understanding the behaviour of roofs in blocky rocks.

18.3.3 Bolt types

Although design is important, the type and method of installation of bolts can also have a significant effect on performance. Classification of rockbolts into types is difficult. Conventionally there are two methods — either as grouted or mechanically anchored bolts or as point anchored or fully grouted bolts. A list of available bolt types, from Peng and Tang (1984), is given in Table 18.3. A point anchored bolt is usually tensioned, a fully grouted bolt is usually untensioned. A mechanical anchor can be installed easily, but is unreliable over a period of time; a resin bolt requires precision in installation — whether point or fully grouted.

Conventional rockbolts are made from 16 mm (⅝ in), 19 mm (¾ in), 25 mm (1 in) or 32 mm (1/¼ in) steel rebar with an approximate yield force respectively of 6, 8.5, 15, 23.5 tonnes. Normally the installed bolt tension is 50% of this load. Steel bearing plates at the hole collar are usually 150 mm (6 in) square and 6–9.5 mm (¼–⅜ in) thick and are flat or bell shaped with a centre hole. The main function is to distribute stress to the rock at the collar through a bolt threaded on to the top of the bolt and tensioned through a drill chuck. Angle or spherical washers are used to create a uniform bearing surface. To prevent falls of rock between bolts — an important factor in weak rock — mesh or bench bars are placed behind the anchor bearing plates. For long-term installations shotcreting is essential.

18.3.4 Bolt and anchorage performance

(a) Bolts

Bolts are usually considered temporary supports. At bolt forces close to working load, they are, like all rock stress systems, prone to strength deterioration with time. At diametral deformations greater than 1–1.5% they usually cease to function, although performance can be improved with shotcreting. The reduction in capacity with time is not well documented and relies to a great extent on ground conditions.

Certainly in the case of mechanical bolts, installation is always accompanied either by reduction in tension with time or excessive roof deformation during loading with untensioned bolts. These phenomena were investigated by De la Cruz (1964) and Parson and Osen (1969) among others and were attributed principally to slippage of serrations on the anchor shell, rock deformation and rock breakage at the anchorage, and collar and ground movement following excavation. In addition dynamic vibration due to blasting is a major cause of tension loss. This means that constant monitoring and retensioning of bolts is needed if long-term installation is required.

It has been claimed that resin or cement anchors give improved performance, both long and short-term, and there is some evidence for this. Figure 18.11 compares short-term performance of resin grouted and mechanically anchored bolts, from well-known experiments by Franklin

Table 18.3 Types of roof bolt (after Peng and Tang 1984)

Types of bolt	*Types of anchor*	*Suitable strata type*	*Comments*
Point anchor	Slot-and-wedge	Hard rock	Used in the early stages
	Expansion shell:		Most commonly used in USA
	Standard anchor	Medium-strength rock	
	Bail anchor	Soft rock	
	Explosive set	Lower-strength rock	Limited use
	Resin grout:	All strata especially for weak rock	Increased usage recently
	Pure point anchor		Resin length less than 0.6 m (24 in)
	Combination system		Resin length greater than 0.6 m (24 in)
	Combination anchor (expansion shell and no mix resin)	Most strata	Good anchorage with 'no mix resin'
Full-length anchor	Cement: Perfo, Injecto, Berg jet, Cartridge	Most strata	Disadvantage: (1) shrinkage of cement (2) longer setting time
	Resin: Injection, Cartridge	All strata	Increased use recently especially for weak strata
Yieldable	Expansion shell	Medium-strength rock	An expansion-shell bolt with yielding device
Pumpable	Resin	Weak strata	Complex installation
Helical	Expansion shell	Most strata	In experimental stage
Split set	Full-length fraction	Weak strata	Cheap, but needs special installation equipment
Roof truss	Expansion shell	Adverse roof	Recommend use at intersections and/or heavy pressure areas
Cable sling	Cement anchor and full-length friction	Weak strata	Substitute for timber, steel or truss support
Lateral force system	Full-length Two-piece steel	Soft strata	Applying full-length lateral force (compression) to the strata
Swellex bolt	Full-length holding	Water-bearing strata	Uses high-pressure water to swell steel tube

and Woodfield (1971). It can be seen that the reliance on bond rather than friction means that the force take-up is much quicker, and by extrapolation the possibility of slippage is much less. There remain dangers associated with faulty installation, excessive annulus thickness and poor bonding in wet holes which in practice can make resin grouting less attractive.

Deterioration of underground structures is much better documented (Farmer 1985) than individual anchor performance. Some classic experiments were described by Ward (1978) and Ward *et al.* (1976) in the Kielder Aqueduct experimental tunnel, where 50 m lengths of drilled and blasted and machine excavated 3.3 m diameter tunnel were driven at a depth of 90 m in Coal Measures mudstones. Deformation data in Figure 18.12 shows typical long-term deformation of rockbolt support systems (in this case 1.8 m long, resin bonded, untensioned at 0.9 m centres) in weaker rocks. It should be noted that in less disturbed ground, deformation is reduced, and also that shotcrete with bolting has a stabilizing effect. There are many adequately supported tunnels using rockbolts and shotcrete which have performed well over a long period of time in various rock types. There are no tunnels supported solely by rockbolts which have performed *over the long-term* better than they would have performed without support.

(b) Anchors

Large-capacity anchors are usually treated differently from bolts. Cement or resin is usually used as the anchorage material and a tension pile design approach is used. A simple approach to design recommended by Coates (1981) is to assume that load transfer from the grouted anchorage to the rock is evenly distributed as shear stress along the cylindrical grout/rock interface. Failure will then occur if this shear stress ($F_R/\pi dL$) exceeds the shear strength, c, of the weaker substrate

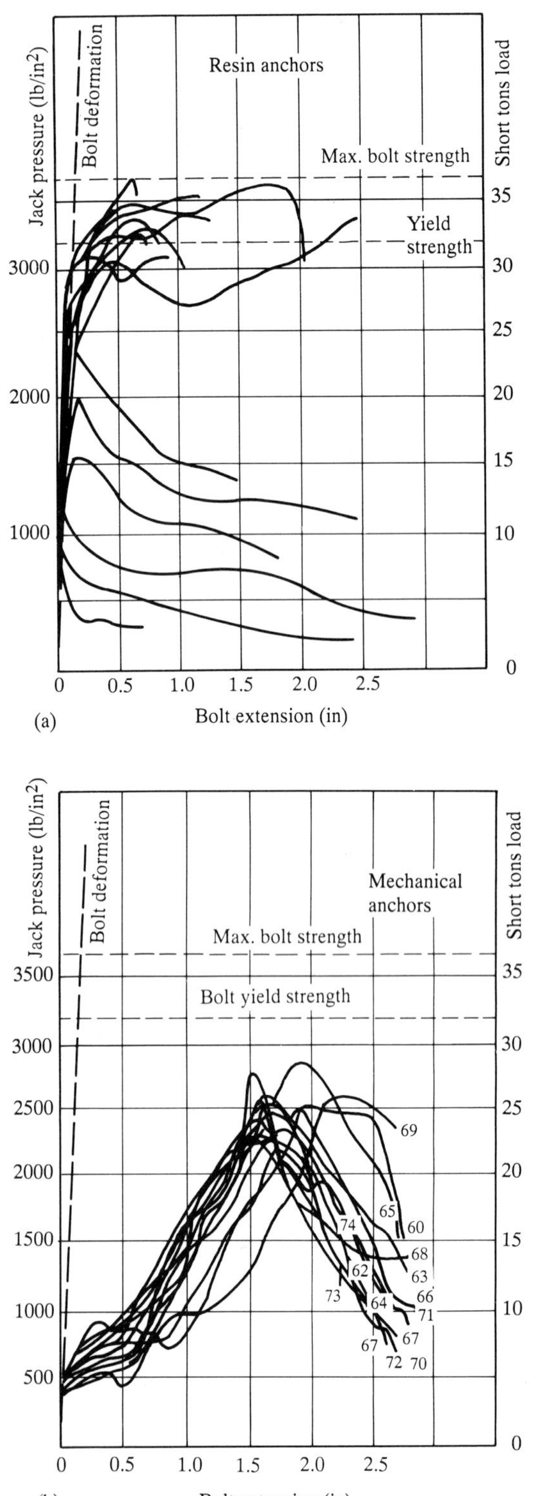

Figure 18.11 *Comparison of pullout test results on resin and mechanical anchors (after Franklin and Woodfield 1971).*

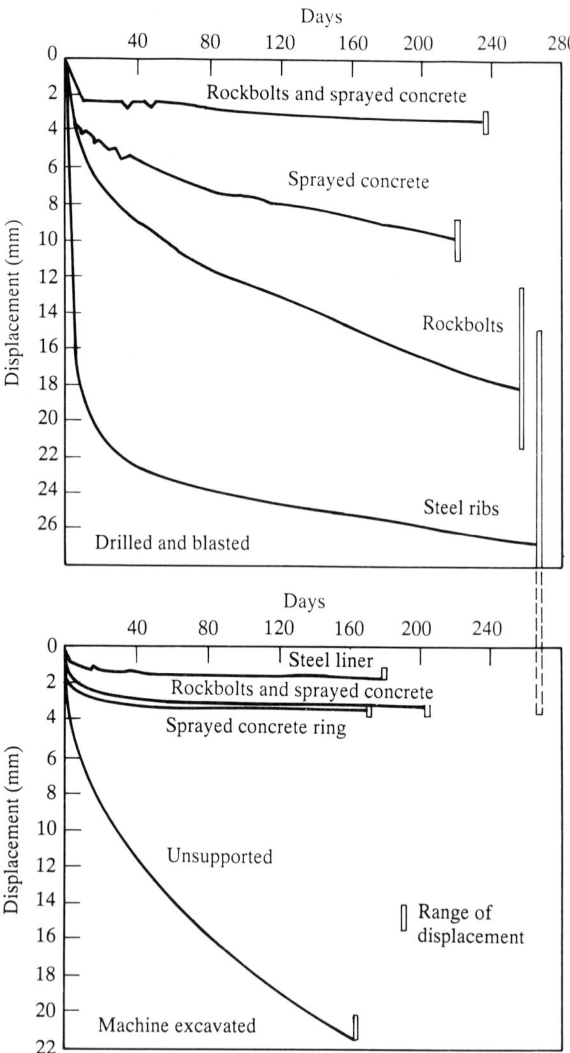

Figure 18.12 *Typical vertical displacements of rockbolts and other support systems in a 3.3 m diameter tunnel in mudstones at a depth of 90 m (after Ward 1978 and Ward et al. 1976).*

(where F_R is frictional resistance, d is diameter and L is length). In weak rocks where $c < 3$ MPa, the weaker substrate will be the rock, whilst in stronger rock it will be the grout ($c = 3$–4 MPa for 0.4 water–cement ratio grout at 7 days).

A safe assumption that c is approximately one quarter of the magnitude of the unconfined compressive strength of the rock produces a simple equation for anchor resistance in which any end-bearing is ignored:

$$F_R = \pi d L \frac{(\sigma_{cf})}{4} \tag{18.4}$$

where σ_{cf} is the unconfined compressive strength of the weaker material.

The inherent assumption in the foregoing analyses that load transfer from the anchorage to the surrounding ground is evenly distributed as shear stress along the anchorage length can be challenged in a situation where full skin-friction mobilization is not obtained. For instance, work on reinforced concrete bars (Hawkes and Evans 1953) has shown that shear stress may decay exponentially along a reinforcement bar according to the relationship

$$\frac{\tau_x}{\tau_0} = \exp\left(-\frac{Ax}{d}\right) \tag{18.5}$$

where τ_0 is the shear stress at the top of the anchor (Figure 18.13), τ_x is the shear stress at length x from that end, and A is the ratio between the shear stress in the receiving substrate and the axial stress in the anchor rod.

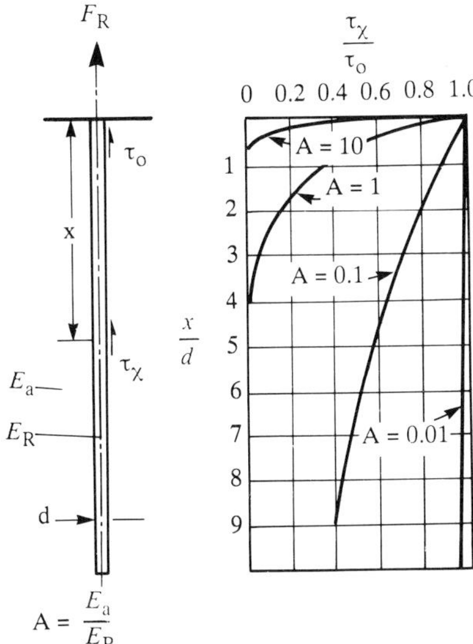

Figure 18.13 *Stress distrubution along an anchored rod (after Farmer 1975).*

If the strains in the anchor are small and the stress-strain relationship linear, A is by definition directly proportional to and arguably similar to the modulus ratio E_a/E_R, where the subscripts a and R refer to the anchoring material and the rod. Further analysis by Farmer (1975) has shown that for a rod of radius r anchored in a hole of radius a in a rigid material by a shearable grout, at small strains

$$\frac{\tau_x}{\sigma_0} = \tfrac{1}{2} r\alpha \exp(-\alpha x) \tag{8.6}$$

where

$$\alpha^2 = \frac{E_a/E_R}{r(a-r)} \quad \text{with} \quad (a-r) < r$$

and σ_0 is the normal stress at the free end of the anchor.

Figure 18.13 shows that the implications of this analysis are quite important. If A is high, the major part of the anchor load is transferred as a peak shear stress at the top of the anchor. If A is low, the shear stress at the top of the anchor is comparatively low and is uniformly distributed over the anchor length.

A rock anchor is, of course, a three-phase system and the anchored rod analysis that has been used refers to a two-phase system. If, however, the stress distribution along the rod in a three-phase grouted rod system is reasonably uniform, then there is some justification for considering the grouted anchor as a two-phase system. Table 18.4 gives typical A-values for conventional grouting materials and substrates. An interesting anomaly is the high E_a/E_R ratio for a resin-hard rock interface, which indicates high stress concentration at the top of an anchor in this material was confirmed experimentally by Farmer (1975). This indicates a problem in using fully grouted bolts in this type of rock.

Table 18.4 Range of modulus ratios for typical anchor materials (after Farmer 1975)

Anchorage material	*Rod material*	E_a (MPa)	E_R (MPa)	$A = E_a/E_R$
Cement grout	Steel	20 000	200 000	0.1
Polyester resin	Steel	3 000	180 000	0.02
Hard rock	Cement grout	90 000	20 000	4.5
Soft rock	Cement grout	5 000	20 000	0.25
Hard rock	Polyester resin	90 000	3 000	30
Soft rock	Polyester resin	5 000	3 000	1.7
Dense rock	Cement grout	800	20 000	0.04

18.4 Sprayed mortar and concrete

18.4.1 Introduction

The technique of spraying a sand–cement mixture was developed in the early 1900s (under the trade name of gunite). This same process using coarse aggregate (now called shotcrete) was not developed until the mid-1950s when it was used for tunnel support in Austria. However, the American Concrete Institute now defines shotcrete as a stiff mortar or concrete conveyed from a gun through a hose and pneumatically projected onto a surface on which it hardens.

Both sprayed mortar and concrete must be sufficiently stiff at impact to support themselves without sagging from an overhead surface of sloughing from a vertical surface. From some standpoints, for example, strength, rebound problems, control problems and equipment maintenance problems, mortar is probably superior to concrete. On the other hand gunned concrete undergoes less shrinkage, is more economical of material and is generally better than mortar for thicker sections. However, the application technique affects the quality of shotcrete more than all the other factors combined and the mix should be tested in the equipment to show that it is satisfactory. In other words shotcrete bonds effectively with any kind of rock surface when applied skillfully and can mould itself around differently shaped and inclined surfaces.

Shotcrete fulfills a number of functions. It can be forced into open discontinuities, thereby binding the opposing rock surfaces together, so providing reinforcement. Its adhesion to rock surfaces and its own shear strength provide resistance against the fall of loose blocks. In addition a thick layer of shotcrete (150 to 250 mm) provides structural support either in the form of a closed ring or as an arch.

Shotcrete used for ground support generally should possess quick-setting properties in order to allow a rapid build-up of the shotcrete section. It should develop very early strength to provide immediate support and should seal off water leakage, where required. By inhibiting water seepage from joints, this also prevents piping and erosion of joint-filling materials. The aforementioned properties are commonly attained through the use of powerful accelerators in the mix. Because shotcrete sets quickly and develops a high early strength, only a relatively thin layer need be used to provide support to the surface on which it is applied. Nevertheless, where required, layers of different thickness can be applied. What is more, early application of shotcrete onto a newly excavated surface helps prevent it loosening.

Shotcrete has been used to help stabilize slopes and as primary support in tunnels. Support in a tunnel generally takes the form of a structural ring or arch. However, the ability of a thin layer of shotcrete to inhibit water softening has not always proved successful. Moreover shotcrete, if it does not reach sufficient strength quickly enough, may not be satisfactory when used in highly overstressed ground where large-scale failures occur rapidly. On the other hand shotcrete is particularly appropriate for tunnels excavated in hard rock since it provides support for loosened blocks and can smooth the irregular surface produced by blasting. This provides a better distribution of stress around the periphery of the tunnel.

The application of shotcrete may be used in conjunction with steel mesh and rock bolts and in abnormal situations with steel ribs.

18.4.2 Material and Mixing

The aggregate–cement ratio of shotcrete is usually about 4:1, by weight, with some 385 kg to 450 kg of cement (usually ordinary Portland cement) and 1920 kg to 1950 kg sand and aggregates used per cubic metre. Sulphate-resisting cement is advisable where the sulphate content of the surrounding ground and/or groundwater is excessive. Shotcrete is characterized by a high degree of compaction as a result of the pneumatic application, hence it requires a relatively low water–cement ratio. Indeed the water–cement ratio generally is around 0.35:1, increasing with decreasing aggregate size to 0.55:1, the upper limit being applicable in the case of sand–cement mixtures.

The setting time of shotcrete can be controlled reasonably well by using controlled setting time cement with heated mix water and/or addition of soda ash. For example, regulated set cement (reg-set) contains calcium fluoroaluminate which significantly alters the behaviour of the cement during the first few hours after mixing. At normal temperatures and without additive, reg-set cement sets in about 10 minutes. Set occurs immediately if it is mixed with water at about 38°C. Because regulated-set cement begins to hydrate more or less immediately after it is mixed with moist aggregate, shotcrete should be used as soon as possible after it is mixed. However, the cost of regulated-set cement is higher than that of ordinary Portland cement. The initial rate of gain of strength is very rapid up to a compressive strength of about 7–21 MPa, which can be achieved in a few hours.

Shotcrete with high early strength is required to ensure rapid development of the composite concrete–rock structure to provide stability and to enable relatively thick applications to be built up quickly without sagging and sloughing, and with less fallout, it is also required in applications against rock from which water is seeping. As noted, high early strength is achieved by the addition of rapid setting agents. An accelerator may be unnecessary when only a thin layer of shotcrete is needed on a clean dry rock surface.

Accelerators are available in powder or liquid form. The quantity of powder accelerator required generally lies in the range 3–6% by weight of cement, depending on the type of agent used and acceleration. The addition of accelerator in powder form has to be controlled by an accurate dispenser and thorough mixing is necessary to ensure its uniform distribution within the mix. The actual amount of fast setting agent should be determined at the site according to the materials, mix and temperature prevailing. Obviously the proposed accelerator and the cement must be compatible as incompatibility can result in drastic loss in quality of the shotcrete. Compatibility tests indicate the additive concentration which provides the shortest possible set time. A check for possible alkali–aggregate reactions also is important. In addition, the effect of various percentage addition on the ultimate strength of the shotcrete should be determined.

Liquid accelerators are more expensive than powder accelerators. Nevertheless they are superior to powder accelerators in that not only do they give smaller reduction in ultimate strength but they also give a higher initial strength, up to 50% reduction in rebound and general all-round improvement in the overall quality of the work. The use of liquid accelerators enables a prescribed amount of the accelerator to be added to the mix water at the nozzle (the volume ratio to water ranging from 1:20, very low, to 1:1). In fact liquid accelerators added to the nozzle now tend to be used rather than powder accelerators added into the shotcreting machine. Obviously the possible premature reaction of powder accelerators in the machine due to moisture in the sand or between the machine and the nozzle is avoided.

Set accelerators normally are used in wet conditions

where only quick sealing is necessary. Calcium sulphate, sodium carbonate, sodium and potassium hydroxides, potassium silicate and fluorosilicate provide the most rapid set but have a detrimental effect on ultimate strength. There is no improvement in early strength.

Hardening accelerators usually are used only when a rapid gain of strength is required, for example, in spraying layers not more than 30–40 mm in thickness. Calcium chloride or admixtures of calcium chloride and sodium sulphate are the most common accelerators in this category. Depending on the proportion of additive used, the ultimate strength may be enhanced but the shrinkage may be increased. Compounds of set and hardening accelerators normally are used to achieve high early strength when spraying layers thicker than 40 mm in one application.

Initial and final sets of 30 s and 120 s respectively can be achieved. Additives giving faster setting times should be introduced at the nozzle.

The inclusion of steel fibres in shotcrete affords considerable tensile and flexural strength, the fibres acting as crack arrestors. The flexural and tensile strength is improved because the fibres tend to lie flat in the matrix. About 1–2% by volume (3–6% by weight) of steel fibre is added. However, if the fibres are not dispersed correctly in the mix they collect into a ball which must be removed prior to shooting. The use of steel fibres has led to thinner sections being used, the elimination of conventional mesh reinforcement and the ability to follow more closely the profile of the exposed rock. The resultant surface offers more impact and abrasion resistance.

The fibre content forms a loose matting when projected into the freshly formed layer of shotcrete, which assists the build-up process and helps to reduce rebound. Such matting encourages build-up even at low application pressure and consequently low delivery velocity. Lowering the pressure (which would be deleterious to normal shotcrete) dramatically reduces rebound, partly because the particle velocity is lower and partly because the nozzle wetting action is more effective. Moreover, fibre shotcrete can be sprayed wetter than normal due to the effectiveness of the matting action. However, the top surface of steel fibre shotcrete is extremely uneven and is very abrasive unless covered with a fibre-free coat to smooth it out.

Shotcrete may be applied as a dry or wet mix. In the dry mix method the material is pre-mixed and fed into the gun, where an accelerator is added. The material then is carried in suspension by compressed air through a hose to a nozzle. Water is injected at the nozzle into the material in several fine streams and is mixed as it passes through the final 250–300 mm of the nozzle. Mixing continues as the shotcrete and water pass between the nozzle and the point of contact. The quality of the dry mix depends on a well-graded aggregate curve and the mix, prior to adding further water at the nozzle, containing between 2 and 5% moisture (less than 2% aggravates dust problems and more than 5% tends to plug hoses). The cement–water ratio is the most important factor in the operation and is controlled by the nozzle man.

Rebound also is influenced by grading and moisture content. In addition the angle of spraying a dry mix should be normal to the rock face, the nozzle being positioned 0.9–1.2 m from the surface. The air and water pressures should be constant at 550 kPa and 485 kPa respectively. Lastly the accelerator feed should be constant and the feeder capable of varying the rate immediately, according to the demand of the rock surface (more is required when the rock surface is wet).

In the wet mix method the material is mixed with water before it is pumped to the nozzle. As accelerators are soluble in water and may have initial setting times as low as 30 s, they can only be added via the air line at the nozzle. Greater productivity can be achieved by the wet than the dry mix method. There is no problem of dust with the wet mix method and the quality of the work is no longer dependent on the water–cement ratio produced by the nozzle man.

The gradation of a mix greatly affects its pumpability flow through the hoses, hydration at the nozzle and adherence to the area sprayed. However, aggregate grading seems to have little effect on compressive strength, although flexural strength, shear strength and bond qualities appear to increase in mixes containing maximum aggregate of at least 12.5 mm. Normally gravel is used rather than crushed rock since it gives less rebound and wear on equipment. The only acceptable method of introducing fibre is by batching.

18.4.3 Properties

The strength and other properties of shotcrete are more or less the same as those of conventional mortar or concrete with the same mix proportions and void content. In fact the strengths are high since mixes have a high cement content and a low water–cement ratio. According to Anderson and Poad (1974) there is no clear evidence to suggest that either the wet or the dry shotcreting process is superior in terms of finished product. Compressive strengths exhibit significant differences which are dependent upon the amount of fast setting agent (accelerator), regulated set Portland cement and cement content. The compressive strength at 28 days is inversely related to the concentration of accelerator and directly related to cement content. However, the effect of cement content is smaller when fast setting agent is added or the mix is made using regulated-set Portland cement. Shotcrete made with regulated-set cement produces higher early strengths than shotcrete made with fast setting agents. Brekke (1972) suggested that the required minimum final strength should be around 17.2 MPa with high accelerator content, and about 20.1 MPa with minimum or no accelerator. A moderate strength requirement can be compensated by an increase in the average thickness of shotcrete. Normally shotcrete possesses a strength exceeding 35 MPa but under certain conditions it may be greater than 70 MPa after 28 days. When accelerators are added to shotcrete the strength attained after 28 days is some 30–45% lower than that of an ordinary mix. For example, Ward and Hills (1977) found that the 28 day compressive strength of shotcrete containing 15% sodium silicate concentration developed 85% of the strength of shotcrete without it. On the other hand it was noticed by Reading (1974) that the addition of 2% calcium chloride gave a moderate acceleration for all types of shotcrete and that the strengths at later stages were improved compared with the ordinary mix. Unfortunately, however, the strengths during the first few hours were much less than those obtained by commercial accelerators. Hence calcium chloride does not fully meet the requirements for good underground support material. Nevertheless Blank (1974) maintained that by careful testing of cements and additives, a high early strength can be achieved (i.e. 3.5 MPa in 3 hours, 7 MPa in 5 hours) using a 2–3% admixture. At the same time the loss of ultimate strength can be minimized to approximately 25% at 28 days and 15% at 90 days. In fact the loss in ultimate strength decreases with time.

The flexural strength for wet process shotcrete is about two thirds that of dry process shotcrete. For instance, Singh and Bortz (1974) showed that the flexural strength for dry-mix shotcrete varied from 10.5 MPa to 21 MPa after 28 days, whereas for wet process shotcrete it ranged between 7 MPa and 14 MPa. The addition of a fast setting agent gave lower strengths. They quoted values of shear strength ranging from 3.5 MPa to 8.5 MPa. The compressive strength, tensile strength and density given by Ward and Hills (1977) for 75 mm diameter cores, cured in water, for replicate coarse and fine aggregate large test panels (1 m^2) are provided in Table 18.5. They found no significant difference in the transverse and longitudinal compressive strength. This is in accord with the findings of Anderson and Poad (1974) who recorded that there is no significant difference in shotcrete fabricated in the vertical and horizontal position. Anderson and Poad also found no difference in strength due to thickness when comparing panels of 50 and 150 mm thickness.

The highest values of Young's modulus obtained by Singh and Bortz (1974) for shotcrete with no additives varied between 35 GPa and 40 GPa after 28 days. With fast-setting additives these values were reduced to 20 GPa and 23 GPa.

Shotcrete creeps under load. It also contracts thermally during curing by varying amounts, depending on the type of accelerator used and the rate and thickness of application. Drying shrinkage normally is not large, although it tends to be a little higher than that of concrete, especially for mortar mixes. This ordinarily does not create problems in terms of ground support. The durability of shotcrete generally has proved good. It is relatively impermeable to the flow of gas such as methane.

According to Brekke (1972) the quality control of shotcrete should always be made on shotcrete as applied, not on casted cylindrical specimens or even on specimens from shotcrete panels. This is necessary because the quality of the shotcrete in place is not only dependent on a correct mix but also on proper application. Hence the most reliable method of achieving quality control is by testing cores removed from the layer of shotcrete. Cored specimens have the advantage that they show lamination between layers of shotcrete, trapped rebound and the quality of the shotcrete rock bond. However, quality testing by coring is possible only after the placed shotcrete has reached a certain strength. Testing during the first hour after application can be carried out by a penetration device (Sällstrom, 1970). Penetration tests also have been

Table 18.5 Strength and density of cores, coarse and fine aggregate sprayed concrete – laboratory trials (after Ward and Hills 1977)

		Mean 35-day equivalent cube strength compression (MPa)					*Mean 35-day tensile strength in direction of spraying only*				*Density* (kg/m^3)		
	Trial number	*In direction of spraying*	*Normal to direction of spraying*	*All specimens* *n*		*Coefficient of variation*	*In sprayed concrete* *n*		*At interface* *n*		*Oven dry* *n*		*Saturated surface dry*
Coarse aggregate mix	32*	15.4	17.3	16	16.3	20%	9	0.9	4	1.1	4	2005	2188
	33*	22.3	23.4	16	22.9	21%	11	1.5	2	1.3	4	2018	2203
	34	22.5	21.6	17	22.0	15%	9	1.0	3	1.4	7	1986	2185
Fine aggregate mix	36	33.1	29.7	16	31.4	23%	12	0.9	3	1.4	6	2028	2227
	37	29.7	32.3	16	31.0	28%	10	1.1	4	1.0	7	2027	2222
	38	27.6	34.0	16	30.8	28%	12	1.0	3	1.2	7	2017	2217

n = number of specimens tested
* test cores not capped

used to assess the effectiveness of shotcrete. For example, the Schmidt hammer has been used for testing shotcrete in place, as has the Windsor probe. This is a stud driven into shotcrete by an explosive charge, the depth of penetration being correlated with strength. It is not suitable for strengths less than 10.5 MPa. Pull-out tests have been used to determine the strength of placed shotcrete. For the range 1.4–9.7 MPa, Sällstrom (1970) found good correlation between unconfined compressive strength and the ratio of pull-out force to depth of penetration of a 3.5 mm diameter probe, and up to 1.4 MPa the correlation was very good.

18.4.4 Application

In the wet-mix process, pre-mixed wet shotcrete is conveyed by pump or pneumatic placer along a pipe to a nozzle where compressed air is injected. Conversely in the dry-mix process, which is more widely used, a dry mix of shotcrete is pneumatically conveyed at high velocity to the nozzle, where water is injected. If the material being used has a fairly constant moisture content, then the precise water–cement ratio can be preset on the equipment so that no alteration has to be made by the nozzleman. In fact if adjustments are made, then the machine is automatically cut off.

The appearance of the Dial-a-Mix concrete mobile occurred in America in the early 1970s, which could be used for wet or dry concrete production, and has enabled a single unit to produce up to 40 tonnes of material an hour. Similarly in Britain the Steelcrete mobile plant with its associated cement silo wagon represents a versatile unit, capable of feeding spraying machines or of placing wet concrete. A hopper arrangement has been developed for the Steelcrete mobile which, with the aid of a conveyor feed system, can supply six spraying machines. In the case of tunnelling operations boom-mounted nozzles and conveyor feeds, all operated from a central control cab, can be linked with tunnel-cutting equipment so that a lining can be emplaced rapidly whilst the rock is newly exposed.

A reasonable amount of scaling prior to the application of shotcrete appears to enhance results where extruding rocks have been produced by blasting. The rock face may be cleaned with air and water sprayed from the nozzle prior to the application of shotcrete. However, Brekke (1972) suggested that when shotcrete is applied directly onto a rock face cleaning with water jet may be unnecessary or even detrimental since the jet may give rise to overbreak. Indeed shotcrete itself produces a cleaning action through its impact on the rock surface.

A sufficient volume of clean compressed air must be available at the spraying equipment so that fluctuations do not occur in the delivery of material and the mix should be proportioned consistently and uniformly mixed. Non–uniform distribution of the mix from the nozzle gives rise to layering in shotcrete. Such non–uniform distribution also can come about due to the cyclic nature of the discharge of a dry mix from each rotor chamber into the conveying pipe, which results in an uneven conveyance of material through the pipe and nozzle. As the water is injected at a constant rate close to the discharge, the spray and hence the deposit tend to consist of alternating quantities of mix with varying water content. It would appear that the pulsing severity is reduced the longer the conveyor pipe. Transverse segregation of materials on discharge from the nozzle may give rise to non-uniform distribution. This may be due to a combination of expansion of the conveying air and inadequate mixing of the water and finer particles, so that these are carried towards the edge of the spray cone. Eccentric discharge can produce poor distribution. The irregular flow of solids in the short-term (that is, in hundredths of a second) can give rise to an uneven distribution. The mixing time in the nozzle is incredibly short (approximately 0.008 s). Increasing the mixing time by injecting the water into the materials stream 5 m back from the nozzle tip leads to a remarkable improvement. In addition if the mixing time is increased by up to 25 times, then a substantially uniform deposit is produced with very little evidence of layering.

Some of the most important factors in the effective use of shotcrete as support relate to spraying, notably the technique of nozzle operation and the direction of application. The nozzle should be held at right angles to and about 1 m from the surface for best results and not further away than 1.5 m. Closer work may be necessary when applying shotcrete to a wet rock surface or around steel reinforcement. The pressure in the delivery line therefore should not be excessive. Spraying ideally involves a systematic clockwise rotation of the nozzle starting at the lowest point and working upwards. If the nozzleman is too far away from the point of impact, then it becomes difficult for him visually to follow the application. The experience of the machine operator is just as important as that of the nozzleman. He must ensure that blockages and partial clogging do not occur, the correct operation and maintenance of the spraying machine obviously being all important. The use of automatic remote-controlled equipment for spraying ensures that nozzle distance and angle of impact are correct. It is particularly valuable when shotcreting is required immediately after blasting.

When coatings exceeding 30 mm in thickness are required, the shotcrete, depending upon the additives used, should be applied in two or more layers to prevent sloughing or sagging of the freshly placed material. However, with the use of suitable accelerators layers up to 150 mm may be sprayed into place in one application (Kidd 1974). The use of accelerators makes it possible to apply shotcrete in wet conditions. When applying more than one layer, the preceding layer should be allowed to reach a strength of at least 3.5 MPa before the next is applied. As noted the setting time of shotcrete has to be short, otherwise it will be incapable of providing the necessary rapid support to the rock face.

The most important factor in the use of shotcrete in tunnelling operations is not thickness but adhesion. It is very important for thickness to be controlled, and if the layer is made thicker than necessary it is wasteful and produces inferior workmanship. Sem (1979) suggested that a layer 30 mm thick usually is adequate, successive layers being formed if required.

Due to the high velocity of shotcrete at the moment of impact a considerable proportion bounces off the surface on which it is being applied. Such rebound is the chief enemy of sound shotcrete. Many factors influence the amount of rebound, such as nozzle distance and orientation; grading of the mix; effects of admixtures; and the nature of the surface being sprayed. When spraying on a hard surface only the finer particles coated with cement adhere initially (Ward and Hills 1977). For example, samples of rebound may contain more than 60% coarse aggregate and only a minor amount of cement. As this layer increases in thickness a sufficient depth is reached at which progressively larger particles can be embedded and so the proportion of material rebounding diminishes. Rebound may be of the order of 10–35% for horizontal spraying, zero when spraying more than 30° below the horizontal and between 20 and 50% on overhanging surfaces. Singh and Bortz (1974) found that rebound from the wet mix process was about half that from the dry mix method. This probably is due to the better mixing and wetting of cement and aggregate which is achieved by the former method. They also found that both vertical and horizontal faces exhibited nearly equal rebound with the wet process. For the dry process horizontal panels displayed about 25% greater rebound than the vertical panels, the variation being attributable to gravity forces and the inhomogeneity of moisture in the mix.

Rebound tends to increase with increasing particle size, shape and velocity and varies inversely with the water–cement ratio. At increasing water contents the rebound loss is smaller. Within the practical range of water–cement ratio from 0.45 to 0.65 the relationship is almost linear, varying from 25 to 30% at a water–cement ratio of 0.45 to about 10% at a water–cement ratio of 0.65.

Trapping rebound produces weak sandy lenses which give shotcrete a laminar structure. Hence it is important to allow it to escape. On the other hand the peening effect causes compaction of shotcrete and penetration into any open cracks and irregularities in the surface.

If fibre is used in shotcrete, it tends to have a variable distribution unless some care is taken with spraying. The first 10–15 mm generally contains only a low fraction of its intended fibre content, whilst the next 15–20 mm has a good two-dimensional array. From then on to the outer surface the shotcrete is less compacted, the voids increasing and the array tends to become more three-dimensional.

Particle velocity also influences the degree of compaction and therefore the density and strength of the placed shotcrete. Particle velocities as high as 150 m/s have been recorded, mean particle velocities, when discharging from the nozzle, varying from 15 to 35 m/s, with occasional particles travelling as slow as 10 m/s and as fast as 56 m/s. Particle velocity increases with particles of smaller size so that grains of 3 mm diameter are generally up to 20% faster than 20 mm pebbles.

It is often necessary to provide drainage through the layer of in-place shotcrete to prevent the build-up of pore water pressure behind it. This can be done by inserting small-diameter flexible pipes of sufficient length, at appropriate positions into heavily dipping discontinuities so that the pipes project through the subsequently sprayed shotcrete. Alternatively drainage channels with an open outlet running down the surface may be incorporated into the work by fixing perforated pipes or slender troughing to the surface before spraying the shotcrete. If heavy flows of water are anticipated, they should be staunched by the injection of cement or chemicals before blasting.

Shotcrete has been reinforced with steel mesh or ordinary steel reinforcement. However, Heuer (1974) maintained that the benefits of incorporating mesh or other reinforcement with shotcrete were controversial. Fox (1973) went further, remarking that shotcrete which is applied with wire mesh reinforcing cannot be trusted. He argued that the peening action of the coarse aggregate against the mesh frequently had ruined the density of the product. Accordingly he recommended the use of reinforcing steel as reinforcement which, because of its weight, does not vibrate so that a good bond is obtained. A diametrically opposite view was recommended by Kidd (1974). He contended that the use of heavy bars should be avoided where possible as they tended to split the jet of shotcrete and inhibit its placement on the leeside of the bar. When steel mesh is used, it should be installed after one pass and then shotcreted. This gives a better adhesion of the shotcrete behind the net. As far as Kidd was concerned if the openings in steel mesh are sufficiently large, then they prevent bridging of the shotcrete, causing voids between the mesh and the rock surface. He suggested that mesh openings should exceed 150 mm^2. Furthermore the reinforcement should be fixed to the surface with steel pins or bolts sufficiently close together (usually 300 mm is adequate for steel mesh) to prevent excessive vibration or sagging of the reinforced work as the shotcrete is sprayed into place. The nozzle should be held close to the work to avoid initial build-up on the outer face of the reinforcement and ensure that the shotcrete is forced behind each bar, so eliminating voids.

Heuer (1974) gave examples where reinforcement may or may not be suitable for use with shotcrete in tunnel construction. For instance, if the excavated perimeter of a tunnel breaks to a fairly smooth arch shape during blasting, the layer of shotcrete may act predominantly in compression as an arch structure and in such a case the value of reinforcing is questionable. On the other hand if the perimeter breaks to a very irregular shape with large flat surfaces or prominent re-entrant angles and surfaces, a thin layer of shotcrete cannot function as a compression arch. In this situation reinforcement of the shotcrete may improve its load-carrying capacity. However, Heuer stated that it is simpler and cheaper to apply a thicker layer of shotcrete or to use rock bolts to stabilize flat or re-entrant surfaces.

Acknowledgment

Grateful thanks to F. G. Bell for his contribution to this chapter.

References

Alexander, L. and Hosking, A. D. (1971) 'Principles of rockbolting — formation of a support medium', *Symp. Rockbolting, Aust IMM*, paper 1

Anderson, G. L. and Poad, M. E. (1974) 'Early aged strength properties of shotcrete', *Proc. Engg Foundation Conf. on Use of Shotcrete for Underground Structural Support*', South Berwick, Maine, American Society Civil Engineers/American Concrete Institute, ACI Publication **SP-45**, 277–296

Blank, J. A. (1974) 'Shotcrete durability and strength — a practical viewpoint', '*Proc. Engg Foundation Conf. on Use of Shotcrete for Underground Structural Support*', South Berwick, Maine, American Society Civil Engineers/American Concrete Institute, ACI Publication **SP-45**, 320–329

Brady, B. H. and Brown, E. T. (1985) *Rock Mechanics for Underground Mining*, Allen and Unwin, London

Brekke, T. L. (1972) 'Shotcrete in hard rock tunnelling', *Bull. Ass. Engg Geol.*, **9**, 241–264

Coates, D. F. (1981) *Rock Mechanics Principles*, CANMET, Ottawa

De la Cruz, R. V. (1964) 'Mechanism of Bolt Anchorage', *MS Thesis*, Pennsylvania State University

Douglas, T. H. and Arthur, L. J. (1983) 'A guide to the use of rock reinforcement in underground excavations',*CIRIA Report No. 101*, Property Services Agency, DoE, London.

Farmer, I. W. (1975) 'Stress distribution along a resin grouted anchor', *Int. J. Rock. Mech. & Min. Sci. & Geomech. Abstr.* **12**, 347–351

Farmer, I. W. (1985) *Coal Mine Structures*, Chapman & Hall, London

Farmer, I. W. and Shelton, P. D. (1980) 'Factors that affect the stability of underground rockbolt reinforcement systems', *Trans. Inst. Min. Metall.* **89**, A68–83

Fox, K. H. (1974) 'Use of shotcrete from the standpoint of the contractor', *Proc. Engg Foundation Conf. on Use of Shotcrete for Underground Structural Support*, South Berwick, Maine American Society Civil Engineers/American Concrete Institute, ACI Publication **SP-45,** 22–28

Franklin, J. A. and Woodfield, P. F. (1971) 'Comparison of a polyester resin and mechanical rockbolt anchor', *Trans. Inst. Min. Metall.*, **80**, A91–A100

Goodman, R. E. (1967) 'Analysis of Structures in Jointed Rock', *Report No. 3 for Omaha District*, US Army Corps. of Engineers

Goodman, R. E. and Shi, G.-H. (1985) *Block Theory and its Application to Rock Engineering*, Prentice Hall, Englewood Cliffs, N.J.

Hawkes, J. M. and Evans, R. H. (1953) 'Bond stresses in reinforced columns and beams', *Structural Engineer*, **29**, 323–327

Heuer, R. E. (1974) 'Selection/design of shotcrete for temporary support', *Proc. Engg Foundation Conf. Use of Shotcrete for Underground Structural Support*, South Berwick, Maine, American Society Civil Engineers/American Concrete Institute, ACI Publication **SP-45**, 160–174

Hobst, L. and Zajie, J. (1983) *Anchoring in Rock and Soil*, Elsevier, Amsterdam

Hoek, E. and Brown, E. T. (1980) *Underground Excavations in Rock*, Institution of Mining & Metallurgy, London

Kidd, B. C. (1974) 'Use of shotcrete in tunnel construction'. In *Tunnelling in Rock*, Z. T. Bieniawski (ed.), South African Institution Civil Engineers/South African Institution Mining and Metallurgy SANGORM, Pretoria, 389–416

Lang, T. A. (1971) 'Rock reinforcement', *Bull. Ass. Engg Geol.* **9**, 213–219

Obert, L. and Duvall, W. I. (1967) *Rock Mechanics and the Design of Structures in Rock*, Wiley, New York

Panek, L. A. (1962a) *The Effects of Suspension in Bolting Bedded Mine Roof*, US Bureau of Mines, Rept. Invest. 6138, Washington D.C.

Panek, L. A. (1962b) *The Combined Effects of Friction and Suspension in Bolting Bedded Mine Roof*, US Bureau of Mines, Rept. Invest. 6139, Washington D.C.

Panek, L. A. (1964) Design for bolting stratified roof, *Trans. Soc. Mining Engrs.* **229**, 113–119

Panek, L. A. and McCormack, J.A. (1973) 'Roof/rock bolting'. In *SME Mining Engineering Handbook*, A. B. Cummins and I. A. Given (eds.), 13-125–13-134

Parson, E. W. and Osen, L. (1969) *Load Loss from Rockbolt Anchor Creep*, U.S. Bureau of Mines. Rept. Invest. 7220, Washington D.C.

Peng, S. S. and Tang, D. H. Y. (1984) 'Roof bolting in underground mining — a state-of-the-art review', *Int. J. Min. Eng.*, **2**, 1–42

Rabciewicz, L. (1969) 'Stability of tunnels under rock load', *Water Power*, **21**, 266–273

Reading, T. J. (1974) 'Corps of Engineers studies of shotcrete', *Proc. Engg Foundation Conf. on Use of Shotcrete for Underground Structural Support*, South Berwick, Maine, American Society Civil Engineers/American Concrete Institute, ACI Publication **SP-45**, 263–276

Sallström, S. (1970) 'Improving initial compressive strength of shotcrete by accelerating agents', *Proc. Conf. Large Permanent Underground Openings*, Oslo, 227–232

Sem, B. (1979) 'How to achieve effective roof reinforcement with sprayed concrete', *Tunnels and Tunnelling*, **12**,(6) 61–62

Singh, M. M. and Bortz, S. A. (1974) 'Use of special cement in shotcrete'. In *Use of Shotcrete for Underground Structural Support*, American Society Civil Engineers, New York, pp. 200–231

Stephansson, O. (1984) *Rock Bolting*, Balkema, Rotterdam

Ward, W. H. (1978) 'Ground supports for tunnels in weak rocks', *Geotechnique*, **28**, 133–171

Ward, W. H. and Hills, D. C. (1977) 'Sprayed concrete: support requirements and the dry mix process', *CP108*, Building Research Establishment, Watford

Ward, W. H., Coates, D. J. and Tedd, P. (1976) 'Performance of tunnel supports in the Four Fathom Mudstone', *Tunnelling '76*, Inst Min & Metal, London, 329–340

19 Rock anchors

Professor T H Hanna
Chartered Civil Engineer, Sheffield, UK

Anchor construction in rocks started with the pioneering work of A. Coyne for the strengthening of a concrete dam in Algeria in the early 1930s. Since that time many changes have taken place, not only in the principles of design but also in the range of rock conditions in which anchoring is feasible. In general an anchor is a suitable foundation element where uplift and/or pull-out forces have to be resisted. Good examples are to be found with transmission tower foundations, rock-slope strengthening, deep excavations, concrete dams and associated structures, dry docks, underground openings and reaction for *in situ* load tests.

A rock anchor is a tensile resisting foundation element which may be used in isolation or as part of a cluster to resist pull-out loading. The tensile resisting element normally comprises a tendon attached to the walls of a drill hole over a fixed length at its distal end and attached to the structure at the other end. The tendon normally is formed from bar, wire or strand steel in a particular configuration to give efficient bond and ensure that the steel in the tendon is stressed uniformly. The fixing medium is normally a cement grout. The magnitude of the load which may be resisted is dependent on the sectional area of the tendon and the average bond stress which can be mobilized in the ground. This review considers the basic types of anchor in use, the principles of design, corrosion protection, anchor testing and anchor-test evaluation. Consideration is also given to problems of anchor-hole alignment, failure of anchors, and the behaviour of anchor-supported structures. Reviews of ground anchors are given by Hobst and Zajic (1977) and Hanna (1982). Codification of anchoring principles will be found in the recently produced British Code, Anon (1989), which also reviews other relevant standards.

19.1 General approach to anchoring

The success of any anchor depends on the ability of the bond developed between the rock and the anchor hole grout to be maintained throughout the life of the anchor with very small amounts of creep occurring. To achieve this general objective it is essential that the factors which affect the behaviour of an anchor are understood, particularly bond mechanics, rock-mass behaviour, corrosion tendon behaviour, grout, testing and construction methods. In all anchor work it is essential to evaluate the forces to be resisted. These include loads due to earth and rock pressures, water pressures, pressures due to dilation of the rock mass, surcharge effects and dynamic effects such as nearby blasting works. In certain ground types there is the ever-present problem of ground movements due to physicochemical reactions causing volume change. Examples are clay mineral hydration in shale rocks; the presence of soluble minerals which may cause heaves or settlements due to hydration or dehydration; the expansion of rock masses caused by mineral alteration; and movements due to the presence of industrial materials.

In order to assess the likely problems that may have to be dealt with it is essential to carefully examine the ground as an engineering material and how the ground will respond to anchor construction. The starting point, therefore, is a thorough site investigation with the anchor project in mind. General principles of site investigation will be found in most standard texts and useful guidance is given in Anon (1981). The feature of an anchoring scheme which differs from all other forms of foundation engineering is that each anchor is subjected to a quality control via a field load test. Consequently the geotechnical data on the ground is of extreme importance if the load test data generated are to be assessed reliably. The specific information required is

(1) The soil and rock succession and the position of the groundwater table(s).
(2) The quantification of the different strata encountered and their engineering behaviour, particularly strength, quality (RQD), fractures, joints, imperfections, shear zones and orientation of the discontinuities relative to the directions of the proposed anchor holes. In addition, as the drilling operations are likely to alter the characteristics of the rock near to the sides of the anchor hole, an assessment needs to be made of the susceptibility of the ground to alteration (softening) during and after drilling of the anchor hole.
(3) Most anchor tendons are fixed to the rock by a cement

grout and every effort must be made to ensure that there is no loss of grout into joints and openings in the rock. The tightness of the rock, particularly in shear zones is of great importance.

(4) Rock as a natural material varies both in plan and in elevation as a consequence of its mode of formation and subsequent alteration. This variability should be established by the site investigation because ground anchors apply high load values locally in different directions. Consequently the orientation of the anchor hole relative to the dip and strike of the rock strata may be critical.

Because of the range of ground types which routinely have to be anchored in, high grout pressures are sometimes used in an attempt to generate high loads. The efficacy of such techniques is dependent on the rock mass to accept these high radial pressures. The radial stress/strain characteristics of the ground become important, particularly in weak rocks. It is these quality and strength characteristics of the walls of the anchor hole which will control the behaviour of the anchor. The shape of the anchor hole, therefore, is controlled by the potential of the rock to swell on stress release and hydration, to erode under the action of drilling fluids, to collapse due to the inherent joint pattern and to yield due to the applied radial pressures. Photography, television camera surveys and drill-hole calipers enable the quality of the drill-hole walls to be assessed. The quality of the rock mass can be assessed via uniaxial testing, point load testing, durability testing and deformability testing. Useful guidance will be found in Anon (1989). From a practical consideration it is prudent to follow a general check list (Table 19.1) which covers all of the relevant factors that might be of importance on an anchoring scheme. It is important also to be aware of the dangers of using a general site investigation directly for an anchoring project where all of the relevant site data may not have been procured.

Table 19.1 Summary of data required during a site investigation

(1) Soil and rock succession across the site and in particular in the vicinity of the fixed anchor lengths.
(2) Position of the groundwater tables and their variation with time and with construction activity.
(3) Strength of the various strata encountered and their variability.
(4) Nature of fissures and joints in clay and rocks and their orientation relative to the direction of anchoring. Assessment of the influence of construction on the strength and permeability of the joints.
(5) Plasticity of clays and clay shales, softening and loosening potential of rocks during anchor-hole drilling.
(6) Chemical composition of the soil, rock and groundwater; organic content; temperature of groundwater; an assessment of changes expected in the future.
(7) Bulk and local permeability of the ground particularly in the fixed anchor zone.
(8) Geological history of the ground, particularly an assessment of the *in situ* stress state; seismic activity.
(9) Need for casing during anchor-hole drilling; an assessment of the problems of sealing a hole with a packer unit during waterproof testing and grouting operations.
(10) Use of adjacent sites, properties and their sensitivites to anchor construction works and associated ground movements.
(11) The influence of vibrations and other construction works that may affect the performance of the anchors.
(12) The proposed use of the site at present and in the future.

19.2 The components of an anchor system

The load-carrying capacity of an anchor is controlled by the geometry of the anchor, the transfer of stress from the fixed anchor to the adjacent ground and the methods of anchor-hole formation and tendon fixing (grouting) in the ground. The most common type of anchor is a straight-shafted hole in which the centrally located tendon unit is fixed by grout placed by tremie methods or cartridge. Such anchor holes may have to be temporarily lined, depending on the hole stability. Resistance to withdrawal relies on side shear at the rock–grout interface on a cylinder over the fixed anchor length. When the rock is weak and fissured the bearing capacity of the anchor is increased by the permeation of the grout through the pores and joints in the natural rock by a low but controlled grout pressure (up to 1 MPa). The effect is to increase the diameter of the fixed anchor and hence enhance the load capacity of the anchor. The capacity of rock anchors may also be enhanced by the use of high pressure grouting (greater than 2 MPa) by use of the tube à manchette or an equivalent method of control or by the use of under-reams to increase the diameter of the fixed anchor in weak rocks.

To achieve the necessary bond between tendon and grout and between grout and rock it is essential that the materials used are suitable. The grout used should be formed from cement which has been established over the years as suitable. Those acceptable are Ordinary Portland Cement Anon (1958), Rapid Hardening Portland Cement Anon (1958), Portland Blast Furnace Cement Anon (1973), Low Heat Portland Cement Anon (1979), Sulphate Resistant Portland Cement Anon (1980), Low Heat Portland Blast Furnace Cement Anon (1974). High Alumina Cement Anon (1947) may be used for temporary anchors or for test anchors. Useful guidance on sulphate and chloride limits, fillers and admixtures is given in Anon (1989).

Generally the water–cement ratio of the grout for rock anchors is less than 0.45 to ensure good bond. Generally the unconfined compressive strength at 28 days should be at least 40 MPa. It is usual to establish the following properties for any project:

(1) water cement ratio;
(2) the cement type;

(3) the nature, if any, of admixture or filler used;
(4) the flow characteristics of the grout;
(5) the strengths of cubes at 3, 7, 14 and 28 days;
(6) the setting time, bleed, free shrinkage and expansion.

Manufacturers of resinous grouts normally provide capsules for use primarily with rock bolts. Tests to confirm the adequacy of the grout may be required and good guidance is provided by Hanna (1982) or Anon (1989).

Irrespective of the grout used it is essential that it can be delivered and placed around the tendon in the anchor hole to provide the strength necessary to mobilize the applied load.

The tendon usually comprises steel bar, strand or wire either singly or in groups. Because load loss or load change with time can be important in the acceptance of the performance of an anchor under test, it is necessary to understand the relaxation characteristics of the tendon steel. Appropriate values are included in Anon (1989) for typical steels.

The ground anchor is in a potentially hostile environment, the ground, and protection against corrosion is essential. The object of protection is to ensure that during the life of the anchor metal loss through corrosion is kept very small. In general protection to the complete anchor is provided. Also, centralizers are normally located on all tendons to provide a minimum grout cover of 10 mm to the tendon or at least 5 mm cover to an encapsulated tendon. Spacers must also be provided in the fixed anchor zone to give separation between the individual parts of the tendon and allow grout penetration to give appropriate bond. Both the centralizers and the spacers should be made of materials which do not react with the tendon material.

The anchor head comprises a unit in which the anchor tendon is fixed and a bearing plate to transfer the applied load to the ground or the structure without causing large bearing plate displacements. The design requirements for the stressing head are the same as those for prestressed concrete. It is normal to allow for an angular deviation of the tendon of up to 5° for its axial position and without causing any overstressing of the head. The dimensions of the bearing plate must be sufficient to prevent any significant deformation of the rock on which they bear. Normal principles of structural design should be applied. General guidance will be found in Anon (1989)

19.3 The tendon system

As explained earlier all tendons for permanent use must be protected. The protection system provided must be durable and have a life expectancy equivalent to that of the tendon yet permit free movement of tendon-free length. In addition it must be applied once only to the tendon, not fail during anchor stressing and overloading, accomodate the handling and installation stresses on site and allow inspection after assembly. Over the years a range of protective systems have evolved, the primary objective being to envelop the tendon within an impervious sheath and elminate any moist or gaseous environment around the tendon steel. Because of the mode of load mobilization different details of protection are required for the free and fixed lengths and for the anchor head. The most common protection is a plastic sheath installed under factory conditions. Strand tendons are usually greased before application of the sheaths. For bars continuous loose-fitting sheaths are used and are normally filled with a grease or resin to exclude moisture. Over the fixed anchor length the tendon is normally placed within a corrugated encapsulation with a grout or resin to transfer the tendon bond stress through the corrugated encapsulation and the drill-hole grout to the drill-hole wall rock. With tendon bundles formed from bar it is necessary to form joints. The joint must be detailed such that there is continuity of the protective system along the length of the tendon. At the anchor head there are two parts of the tendon which require separate protection in addition to the bearing plate. These are the sections of the tendon above and below the bearing plate. At the inner head protection of the short exposed length of tendon is necessary yet free movement of the tendon must be provided. At the outer head protection of the exposed tendon, friction grips and locking nuts is determined by whether the anchor has the capacity for re-stressing. In general the outer head is placed within a sealing cap or box and the void filled with an inert grease or other suitable compound. Usually the bearing plate is painted with a bitumastic layer. General examples of the 'double' corrosion protection of tendon of a Dywidag anchor are shown in Figure 19.1, of anchor coupling devices in Figure 19.2, and of the protection of anchor heads in Figure 19.3. There are many methods in use. For comprehensive details Anon (1989) should be consulted.

19.4 The mechanics of load mobilization

The transfer of load from the tendon via the surrounding grout annulus to the drill-hole wall is complex and the following failure modes are usually checked:

(1) failure of the grout/tendon bond;
(2) failure of the grout/rock bond;
(3) failure within the rock mass;
(4) failure of the tendon steel;
(5) crushing or bursting of the grout column;
(6) failure of a group of anchors.

The mechanics of bond between the grout and the tendon steel have been studied extensively and an international review was conducted by Littlejohn and Bruce (1976). Based upon this survey and other supplementary data the ultimate bond stress over the tendon bond length should not be greater than 1.0 MPa for clean plain wire or plain bar, 1.5. MPa for clean crimped wire, 2.0 MPa for clean

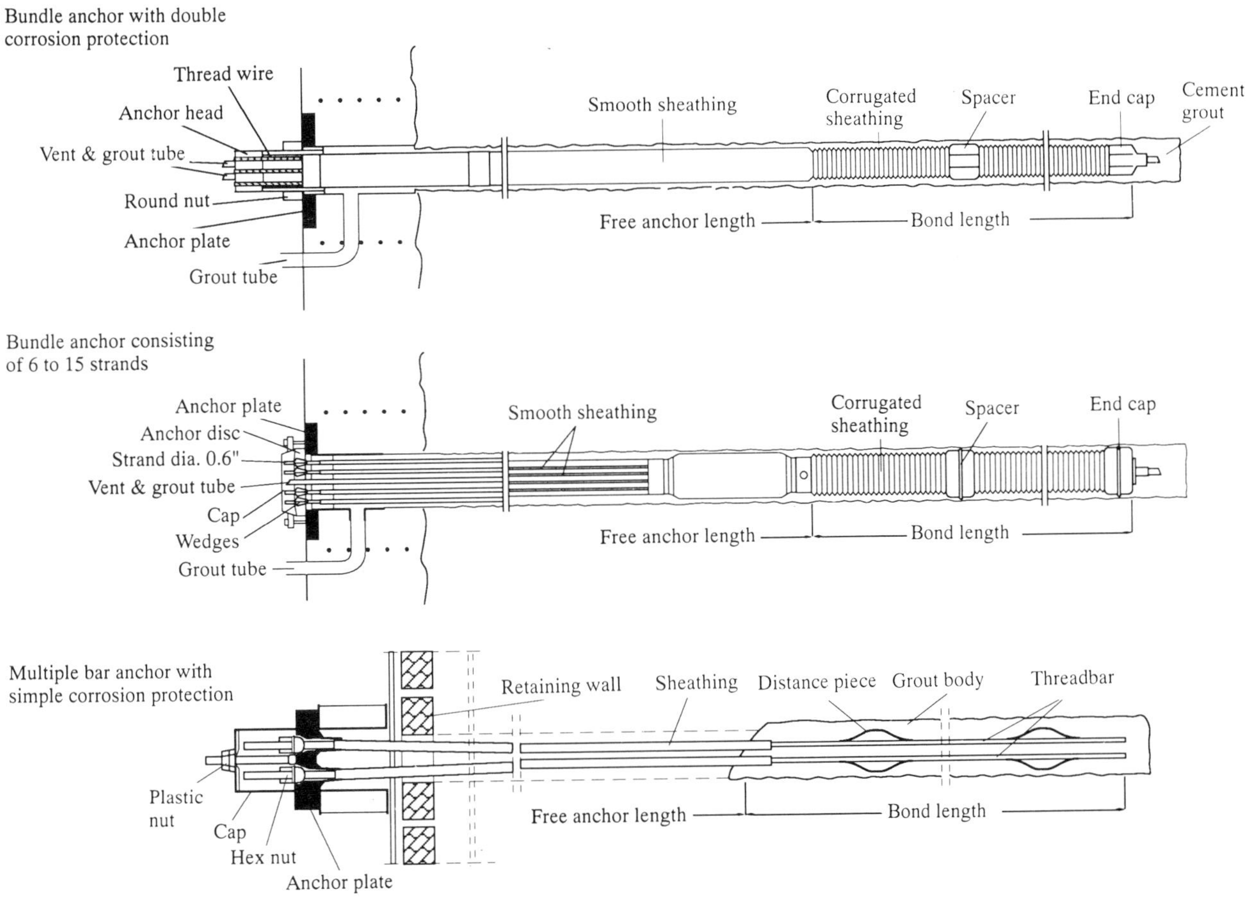

Figure 19.1 *Double protection corrosion system–Dywidag bar anchors*

strand or deformed bar and 3.0 MPa for locally noded strands. These values are based on a minimum grout compressive stength of 30 MPa. In general anchor manufacturers provide sufficient bond area in their anchors to provide an appropriate factor of safety.

The transfer of load from the grout column to the anchor hole walls is complex dependent on the stiffness of the grout, the stiffness of the adjacent rock and the radial yield of the anchor hole under load. The general problem was examined by Coates and Yu (1970) who clearly demonstrated that the bond stress distribution was non-linear, varying as a function of anchor tendon to rock modulus ratio. It is well known that drilling may alter the properties of the rock, especially where the rock is weak and susceptible to softening. Littlejohn (1979) has given an overview of fixed anchor design rules and values of bond used in design have been assembled by Littlejohn and Bruce (1976). Because of the non-uniform distribution of bond along a fixed anchor length there is a theoretical length of about 8 m of fixed anchor. Lengths greater than

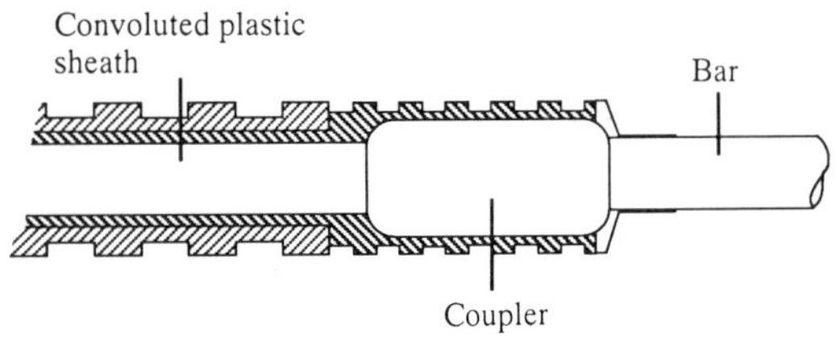

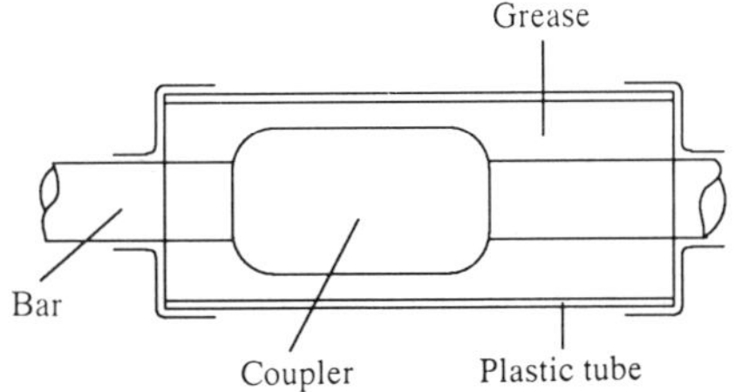

Figure 19.2 *Detail of coupler protection for end of fixed anchors (top) and free anchor length (bottom)*

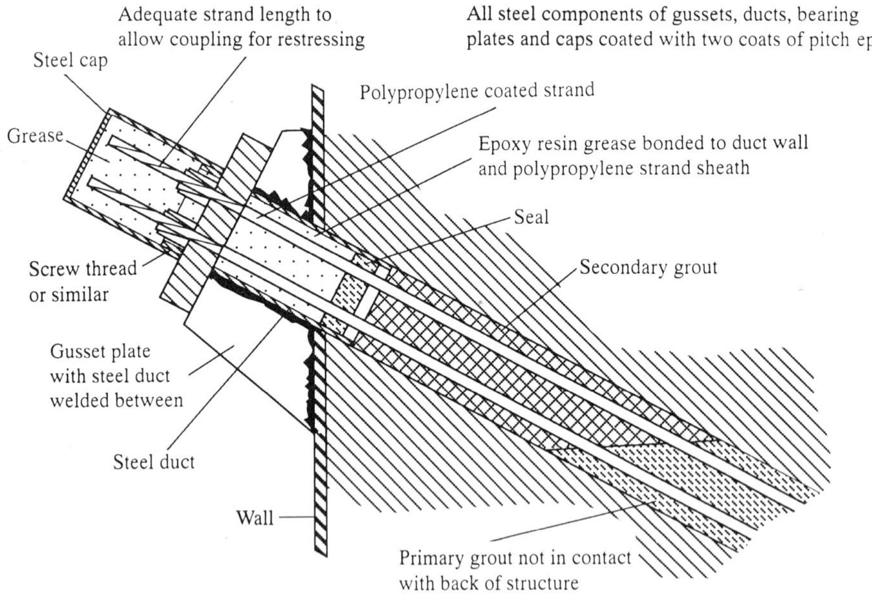

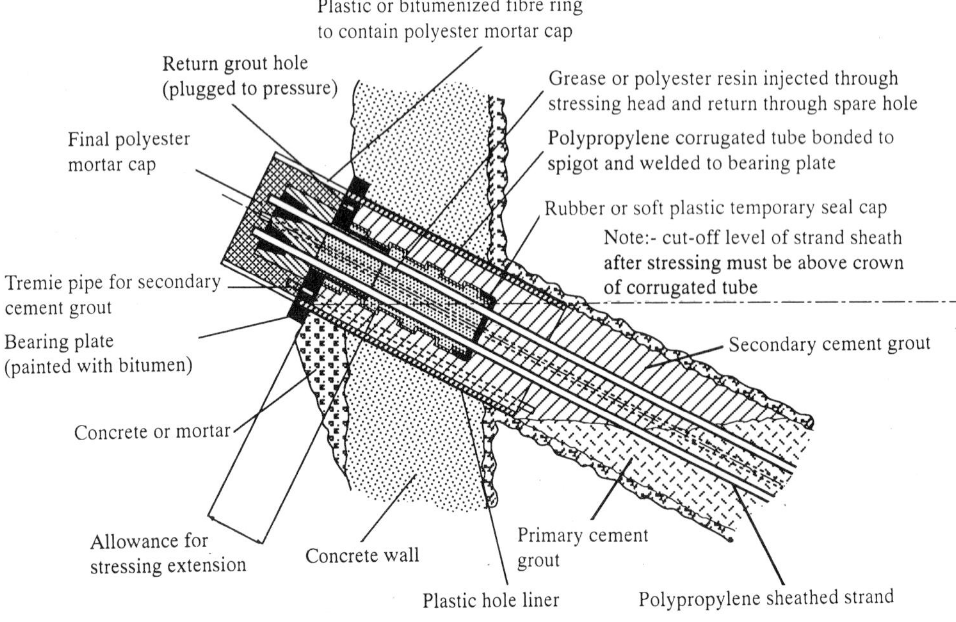

Figure 19.3 *Anchor head details for re-stressable multi-stand tendon (top) and non-restressable tendon (bottom)*

this value do not achieve any increase in load capacity. Attempts have been made to overcome this limitation in anchor-bond efficiency by providing either encapsulation with multi-unit load transfer or a multi-encapsulation system in the fixed anchor zone. With this system the applied load is not concentrated at the front end of the fixed length but distributed over the fixed length (Figure 19.4). This system allows efficient fixed lengths in weak rocks in excess of 20 m to be used.

To prevent an anchor failure within the rock mass the fixed anchor should be located well away from the rock face. Most authorities assume that a cone of rock will be pulled out by the anchor, and a number of empirical formulae have been proposed by Hobst (1985) for the determination of the anchor embedment depth. The structure of the rock mass, particularly bedding planes, laminations and joint patterns, is a controlling factor. An overview is given by Hanna (1982).

In soft rocks it is possible that the triaxial stress system generated in the grout column loads the rock adjacent to the drill-hole wall to a level where yield of the rock occurs, allowing the normal stress at the grout/rock interface to decrease and the anchor tendon to yield. Consequently, for an anchor to perform efficiently the radial stress level must be kept low relative to the stress necessary to cause yield and/or crushing of the drill-hole walls. With the development of the self-boring pressuremeter for soft rocks it is now possible to determine the radial yield pressure of the rock, but anchor theory has not been developed to the stage where this information can be used. The theory of Coates and Yu (1970) is applicable provided the materials behave elastically, that is, at low loads where bond is developed between the grout and the rock. At higher applied loads the top end of the anchor moves relative to the distal end and by a significant amount, causing a stretching between the grout and tendon and the grout and rock, leading to cracking of the grout and partial de-bonding of the grout-rock interface. As a consequence the 'effective' fixed anchor length decreases. Research work in progress on this and related topics reveals that the progressive distortion of the grout column cause the zone the grout wedging, that is the area over which the greatest load can be mobilized, to migrate towards the distal end of the anchor (Casanovas, 1989). One of the most important factors is the effective radial modulus of the rock and it may be shown that the anchor load capacity decreases with increase in the required load. As a consequence an unstable situation may airse whereby the factor of safety of the anchor is reduced, in weak rocks particularly, with the increased risk of a high proportion of anchor tests failing to reach the test specification. Further study of the behaviour of anchors in weak rocks is justified.

To check against the failure of a group of anchors, a failure mode (realistic) is postulated and the resistance of the block ground is factored by the uplift force to give the overall factor of safety. Discontinuities, joint patterns and planes of weakness in the rock must be carefully evaluated if realistic assessments of the factor of safety are to be achieved. In principle, therefore, the methods advocated in the French Code (1972) are followed. It is necessary, however, to be aware of the changes which may be caused to the rock mass by adjacent excavation works and by high-pressure grouting of anchors.

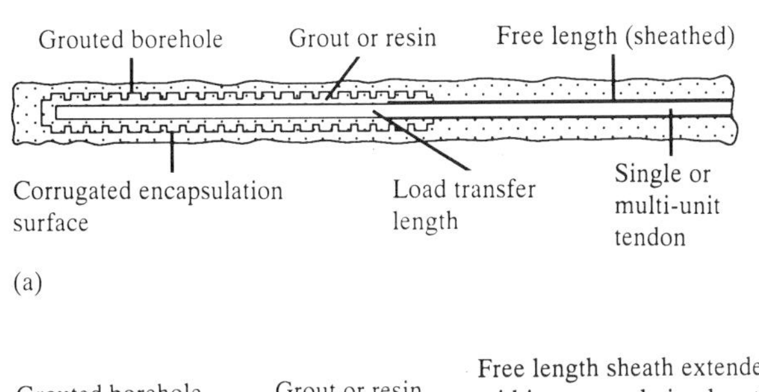

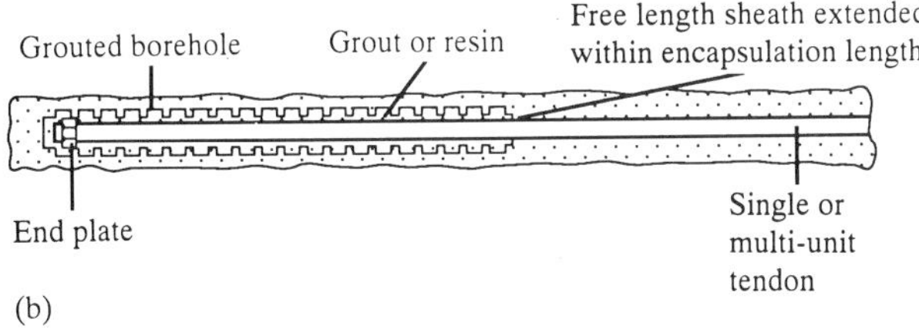

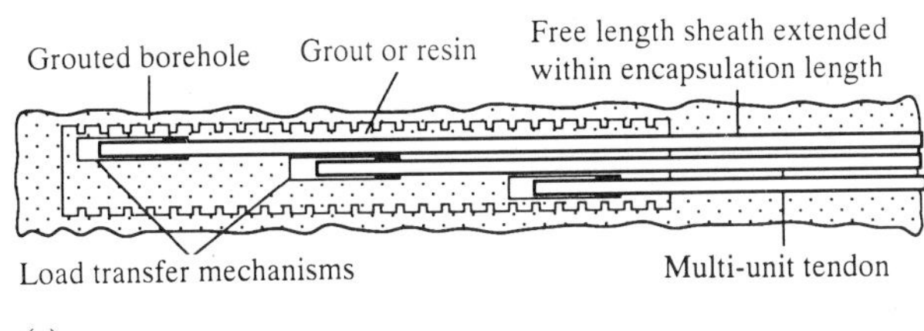

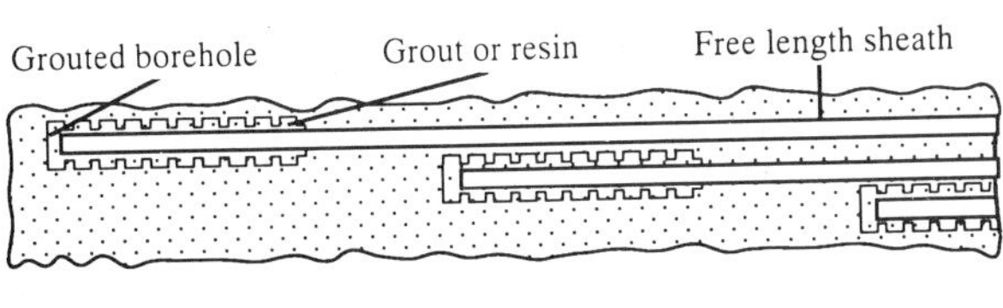

Figure 19.4 *Load transfer mechanisms:* (a) *encapsulation with full-length tendon load transfer;* (b) *encapsulation with distal end load transfer;* (c) *encapsulation with multi-unit load transfer;* (d) *multiple encapsulation (after Barley 1988)*

19.5 Anchors in soft rocks

It is recongnized that the transfer mechanism of load from the grout column to the adjacent rock is not uniform but from a practical design consideration it is useful to adopt the simple relationship $Q_{\text{ult}} = \pi d \ell \tau_{\text{ult}}$, where Q_{ult} is the ultimate anchor capacity, d the anchor diameter, ℓ the fixed anchor length and τ_{ult} the ultimate rock–grout bond stress. The limitations of this treatment include:

(1) failure occurs by shear (i.e. sliding) at the grout/rock interface;
(2) there is a uniform bond stress over the total fixed anchor surface area;
(3) there is no local debonding at the rock/grout interface.

Bearing in mind these general restrictions it is necessary to develop a reliable, yet simple, method of relating τ_{ult} to the properties of the rock mass. Following the work on the bearing capacity of piles in weak rock, the ultimate bond stress may be related to the SPT *N*-value by means of a stress factor *F*. Thus $\tau_{ult} = FN$ kN/m^2. There are many limitations to the use of the SPT test in weak rocks and reference should be made to Stroud (1974). A large number of anchor test results for weak rocks have been assembled by Barley (1988). These data complement and extend the survey of Littlejohn and Bruce (1976). For anchors in chalk the working bond stress for straight-

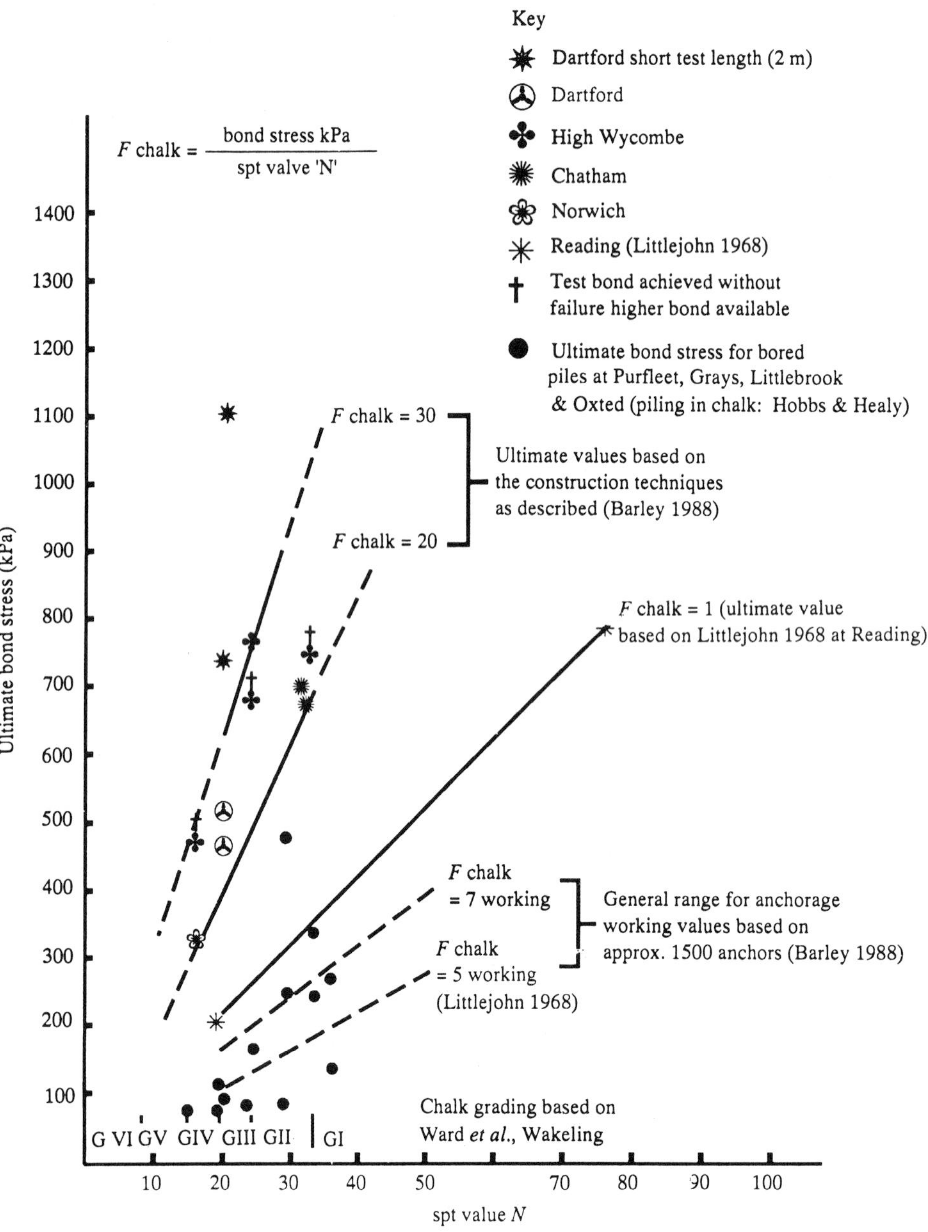

Figure 19.5 *Average bond stress values for chalk (after Barley 1988)*

shafted fixed anchors is between $4N$ and $7N$ KPa. Relatively high factors of safety, of three or more, were applied in order to eliminate or control creep losses. The test data on which this conclusion is based are reproduced as Figure 19.5. Except for very high-capacity anchors where very low levels of creep loss have to be maintained there appears to be no need to use under-reaming methods.

An even more difficult ground in which to form anchors reliably is the group comprising siltstones, mudstones, shales and marls. For many years all anchors in these strata were constructed either by under-reaming methods or pressure grouting. With advance in knowledge through the results of numerous load tests the accepted method of construction was augering with water flush. To achieve greater depths rotary percussion drills are normally used. Following a similar procedure of tests data analysis to that of anchors in chalk, the stress factor F for these rocks was arrived at. The degree of weathering of the rock has a major influence of the F-value as well as local variations in the rock strength. The test data from some mudstone rocks are given in Figure 19.6. For weak rock values of τ working = 1.0–1.5N KPa are recommended. The work shows that this rock is extremely difficult to anchor in reliably. Consequently large factors of safety (>3.0) to cater for the extreme variability of this ground material are required. Where there is doubt about the capacity of straight-shafted anchors in mudstones, under-reaming methods should be considered. With all of these rock types a very careful evaluation of the rock cores should be carried out to assess the degree of weathering, the nature of the rock and its susceptibility to 'damage' during anchor construction work.

Much other data has been accumulated for sandstones and other rocks. Reference to Barley (1988), Anon (1989) and Hanna (1982) will provide useful guidance on both ultimate and working bond values.

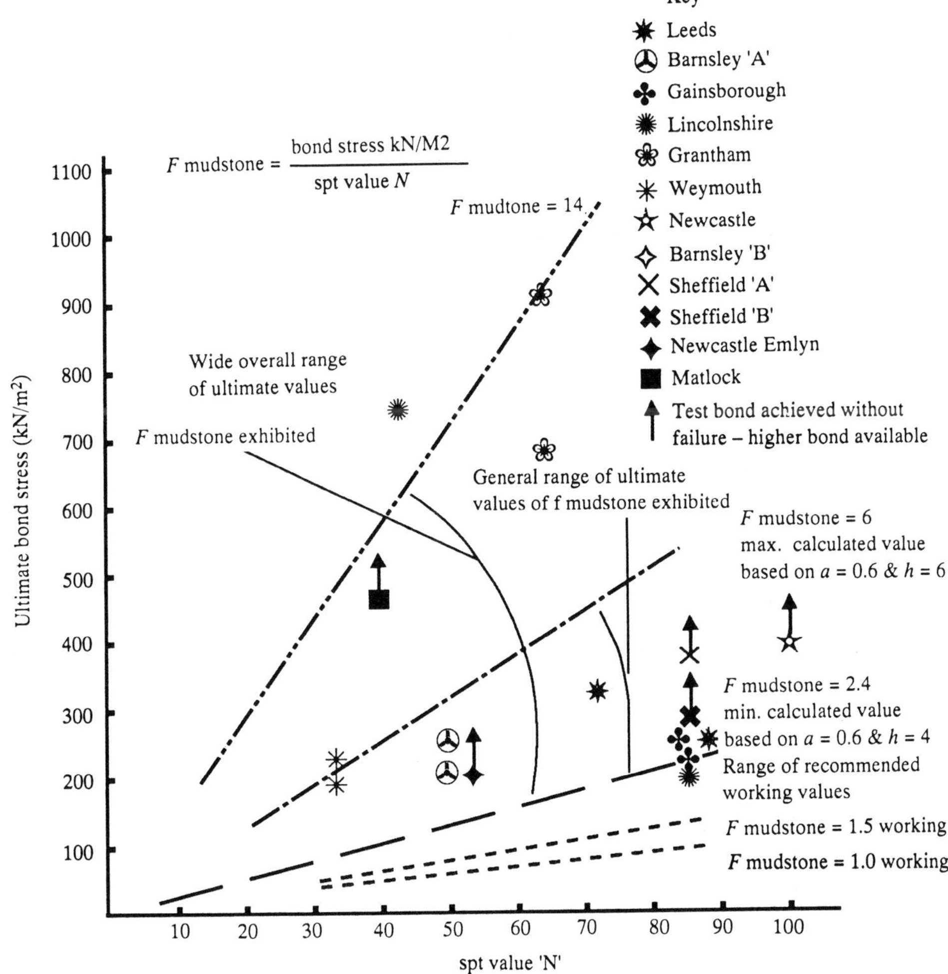

Figure 19.6 *Average bond stress values for mudstone (Barley 1988)*

19.6 Anchor construction

Anchor construction comprises four main operations: hole drilling and preparation to accept the tendon; tendon fabrication and homing into the hole; grouting; and anchor stressing.

In rocks drilling is normally by percussive and rotary–percussive techniques. Throughout the operation care must be exercised to prevent damage to the adjacent ground or nearby anchors. When drilling through over-burden deposits casing may be required to support the hole. The flushing medium normally is water. When deciding upon the diameter of the hole allowances should be made for swelling in strata such as marl and shale, to give the necessary clearance around the tendon bundle. The alignment of the hole is normally within ± 2° of the theorectical except for very closely spaced anchors where interference of the fixed anchors must be prevented by staggering the inclination of adjacent anchors. On critical sites it may be prudent or necessary to check the deviation of the anchor hole. Special inclinometer devices may be used to survey the as-constructed drill hole (Raison, 1987; Harris, 1987). Current practice is to overdrill about 1 m to collect drilling waste which cannot be removed from the hole by the flushing medium. Good practice is to avoid delays, which may cause deterioration of the rock. After satisfactorily drilling the anchor hole it may be advisable to check the potential for grout loss through large fissures in the drill-hole sides. Over the years water testing (the Lugeon test) was used and very often anchor holes were penalized for failing such a water acceptance test. Littlejohn (1975) established the basis for rock pre-grouting. In general pre-grouting is not required if the water loss in the fixed anchor length does not exceed 5 litres/min at an excess head of 0.1 MPa measured over a ten-minute period. Where pressure grouting of the fixed anchor length is not carried out, it is usual to prefill the hole with grout and the grout level observed until it becomes static. When the level continues to fall with time, the hole is normally redrilled and retested. This operation is normally a part of the anchor-grouting stage.

Normally tendons are assembled under factory conditions and delivered to site for installation. Immediately before installation they should be carefully inspected for damage. Particular attention should be given to the corrosion protection system. The tendon is then lowered slowly into the hole either manually or mechanically for heavy anchors, care being taken not to damage the tendon against the sharp top edge of the casing.

The grouting operation is to fix the anchor in the ground. A check list is normally followed by the anchoring contractor to ensure quality control. Records are usually taken of the quantity of grout injected, the pressure applied and cubes recovered for subsequent testing. A useful record sheet for anchor installation is given in Appendix C of Anon (1989).

The purpose of anchor stressing is twofold : to apply the design load to the anchor and to record the loading behaviour of the anchor for comparison with the behaviour of the other anchors and the control anchors. Detailed records of anchor stressing are recorded including displacements of the anchor head, losses in load. From these records load–displacement records can be plotted and assessed against control test results to check that the necessary quality control is being achieved.

19.7 Anchor testing

Perhaps the most important feature of an anchor is that each unit constructed is subjected to a load test. The level of sophistication of test depends upon the purpose of the test and the information to be obtained. In the UK most anchor work is executed according to the general recommendations of Anon (1989). Three types of testing are covered: proving tests; on-site suitability tests; and on-site acceptance tests.

The proving tests are designed to check the quality and the adequacy of the design in advance of the installation of the working anchors. Consequently an overall check on the suitability of materials, components, methods of construction and workmanship is being performed. Examples of tests performed in the laboratory/factory are trial grout mixes, component tests such as characteristic strength of tendons, relaxation and creep of steels, damage to corrosion protection systems, grout-tendon bond evaluation and assessments of the adequacy of the assembly of the tendon. A good example of such tests is reported by Raison (1987(a,b)). Full-scale tests are normally performed to demonstrate (1) the integrity of the corrosion protection system, (2) the magnitudes of the grout/tendon and grout/ground bond values, (3) the development of cracks within the fixed anchor length caused by testing. The results of such trials are usually in report form including details of test certificates, drawings, trial fabrications, loading results and ground data. Following the satisfactory performance of the anchor design full-scale anchor tests are performed to provide data on ultimate load capacity, load transfer, creep and the performance of the corrosion protection system. For this latter purpose it may be necessary to exhume one or more anchors after installation and stressing.

The load tests are designed to secure data on load transfer within the fixed anchor, load loss with time and the apparent free length of the tendon. The test detail parallels the procedures developed for pile testing many years earlier but with significant differences. Load–displacement data are plotted in increments over the range 5–80% fpu (fpu is the characteristic strength of the tendon) with unload cycles. The purpose of these plots is to establish the 'linearity' of the load diagram and the residual displacement on unload. Load–time data are also secured with load periods of up to 10 days. The purpose of these tests is to quantify acceptance criteria for load loss or displacement at constant load. Because relatively small changes in quantity are being observed a very high degree of accuracy is demanded. The apparent free length of the tendon may be determined from the load-displacement data by use of the manufacturer's value of the modulus of the tendon steel as follows: apparent free length $= AE_s\,\Delta_e/T$, where A is the sectional area of the tendon, E_s is the modulus of the tendon, Δ_e is the elastic displacement of the tendon and T is the applied load. The limits of the apparent free length are usually set between 90% of the free length and free length plus 50% of tendon bond length. Normally the observed tendon behaviour will fall within these limits (Figure 19.7). When it does not,

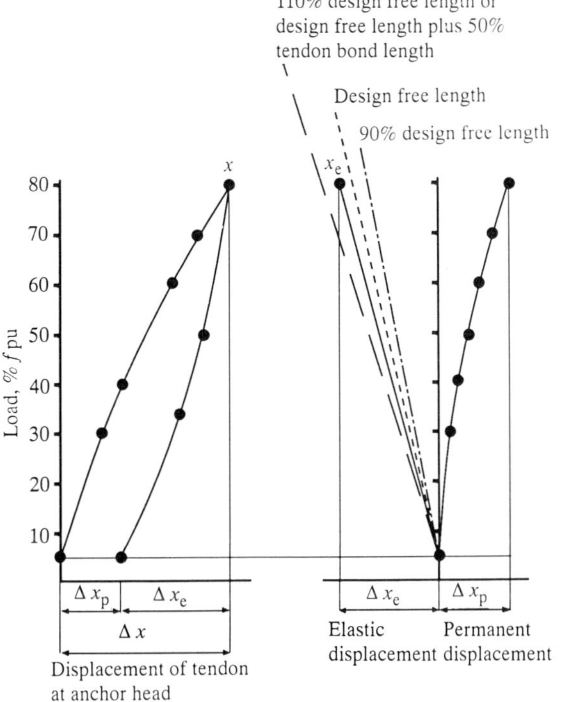

Figure 19.7 *Acceptance criteria for anchor displacement at the anchor head (BS 8081:1989)*

further tests are normally performed to assess the linearity of the load–displacement data under further cycles of loading. The rate of load loss with time is recorded accurately with appropriate correction made for temperature effects, structural movements and stress relaxation of the tendon. For an anchor to meet the acceptance of the standard the load losses are as shown in Table 19.2. An alternative to load loss is rate of displacement with time. The displacement equivalent to the amount of tendon shortening, Δ_e, caused by a load loss of 1% of the specified load test is now used as the acceptance criterion (Table 19.2).

The on-site suitability tests are normally performed at the start of an anchoring project, being constructed as per contract anchors and in the same ground. A minimum of three anchors are tested with additional anchors for categories such as variation in inclination, different rock strata, and different loads. Normally three load cycles are applied (Table 19.3) the load level for temporary anchors being 1.25 the working load and 1.5 times for permanent anchors. Both residual load-time data and displacement–time data may be procured following the procedures for the proving tests. The purpose of these tests is to confirm the designs for the various load levels and strata to be anchored in, in particular that the design criteria can be achieved on production anchors.

Quality control of all production anchors is met by the requirements of on-site acceptance testing. Normally two cycles of load are applied (Table 19.3).

Details of all of these tests will be found in Anon (1989). There are many other methods of load testing anchors and a general review has been provided by Hanna (1982). In all test work the engineer is looking for 'linearity' in the

Table 19.2 Acceptable criteria for (a) residual load–time behaviour (b) displacement–time behaviour at residual load (after Anon 1989)

(a)

Period of observation		*Permissible loss of load (% initial residual load)*
min		%
5		1
15		2
50		3
150		4
500*		5
1 500	(approx. 1 day)	6
5 000	(approx. 3 days)	7
15 000	(approx. 10 days)	8

* 500 min reading is not observed in routine practices

(b)

Period of observation		*Permissible displacement (% of elastic extension Δ_e of tendon at initial residual load)*
min		%
5		1
15		2
50		3
150		4
500*		5
1 500	(approx. 1 day)	6
5 000	(approx. 3 days)	7
15 000	(approx. 10 days)	8

* 500 min reading is not observed in routine practices

Table 19.3 Recommended load increments and time intervals of observation for (a) on-site suitability tests; (b) on-site acceptance tests (after Anon 1989)

(a)

Temporary anchorages Load increment (% T_w)		*Permanent anchorages Load increment (% T_w)*		*Minimum period of observation (min)*
*1st load cycle**	*2nd and 3rd load cycles*	*1st load cycle**	*2nd and 3rd load cycles*	
10	10	10	10	1
50	50	50	50	1
100	100	100	100	1
125	125	150	150	15
100	100	100	100	1
50	50	50	50	1
10	10	10	10	1

* For this load cycle, there is no pause other than that necessary for the recording of displacement data.

(b)

Temporary anchorages Load increment (% T_w)		*Permanent anchorages Load increment (% T_w)*		*Minimum period of observation (min)*
*1st load cycle**	*2nd and 3rd load cycles*	*1st load cycle**	*2nd and 3rd load cycles*	
10	10	10	10	1
50	50	50	50	1
100	100	100	100	1
125	125	150	150	15
100	100	100	100	1
50	50	50	50	1
10	10	10	10	1

* For this load cycle, there is no pause other than that necessary for the recording of displacement data.

load–displacement diagram. Where there is a departure from linearity, which is usually an indication that load has reached the distal end of the fixed anchor length and plastic deformations are taking place, there is a need to carefully assess the loading curve. Much useful data are contained in the relevant sections of several standards.

19.8 Anchor performance

Unless extra tendon steel is incorporated in an anchor it is not possible to load that anchor to failure. The various tests referred to earlier guaranteed factors of safety on the working load of between 1.25 and 1.5, although with proving tests special units may be loaded to rupture. Much additional data may be secured from load tests not taken to failure by applying one of the extrapolation techniques which has been successfully used for pile load-test interpretation. It is now accepted that the load–settlement relationship of a foundation element in the ground to a first approximation is of a hyperbolic form. Chin (1970) recognized this fact and pioneered a technique for determining the ultimate capacity of a pile where the pile test has not been taken to failure. The same technique has been shown to be valid for anchors in soils (Hanna, 1987). Essentially a plot of Δ/T against Δ, where Δ is the anchor head displacement and T the applied load, provides a straight line, the inverse slope of which is the ultimate capacity of the anchor. For the method to be reliable the values of Δ must be accurate. This phenomenon has implications in other aspects of anchor testing where often a load change in an anchor may result from the structure deforming rather than from any inherent deficiency in the anchor. Load loss in an anchor is usually a combination of several factors including stress level, temperature and deflection of the structure. The significance of structural deflection increases with decrease in the anchor length. Data for a rock anchor are reproduced in Figure 19.8.

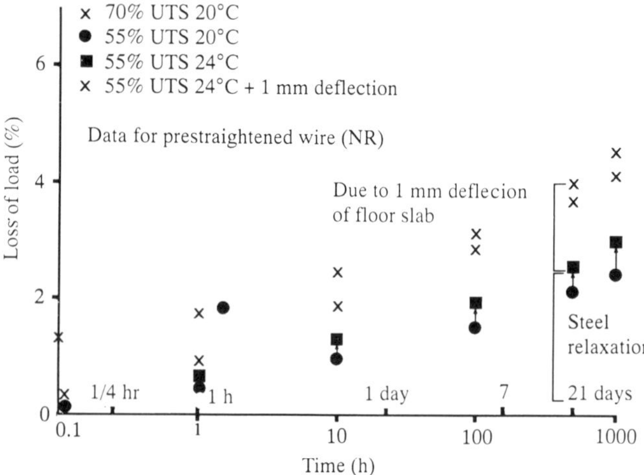

Figure 19.8 *Load loss with time for a range of conditions*

These data show the absolute need to record not only the movements of the anchor head accurately but also the movements of the structure against which the anchor bears. On large rafts tied down to resist hydrostatic uplift forces the load testing of high-capacity adjacent anchors can affect the performance of nearby stressed anchors because of small structural movements, even of a 1 m thick reinforced concrete slab. Figure 19.9 shows load changes induced in a stressed rock anchor from testing two anchors located 3 m from the stressed anchor. In special situations very sophisticated tests may be performed to assess load loss with time and ground strain around the anchor. By instrumenting the anchor and the adjacent ground mass it is possible to determine either load loss or displacement

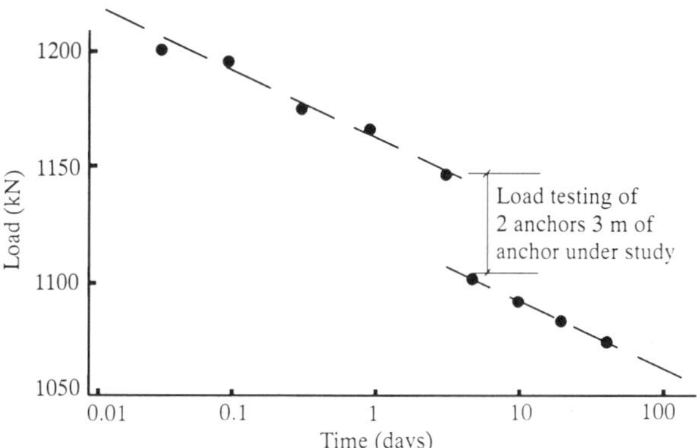

Figure 19.9 *Load loss with time — effect of nearby anchoring works*

with time. Data for a 70-day test on a 255-tonne working load anchor in a very strong shale are given in Figure 19.10. There was a gradual creep of the rock mass but there is also a scatter in the readings. This scatter is not due to measurement error but is the direct result of the time being out of phase with the daily 9.0 a.m. readings taken. This example demonstrates how any local changes in the ground mass will be reflected in the performance of the anchor, (Hanna, 1987b).

Corrosion performance is always a worry to anchor users yet the statistical data show that very few failures are due to this cause. Littlejohn (1987) has reviewed all of the known corrosion failures, 35 in total. He concludes that provided standards are adhered to and quality assurance applied on site, then satifactory performance of anchors can be assured.

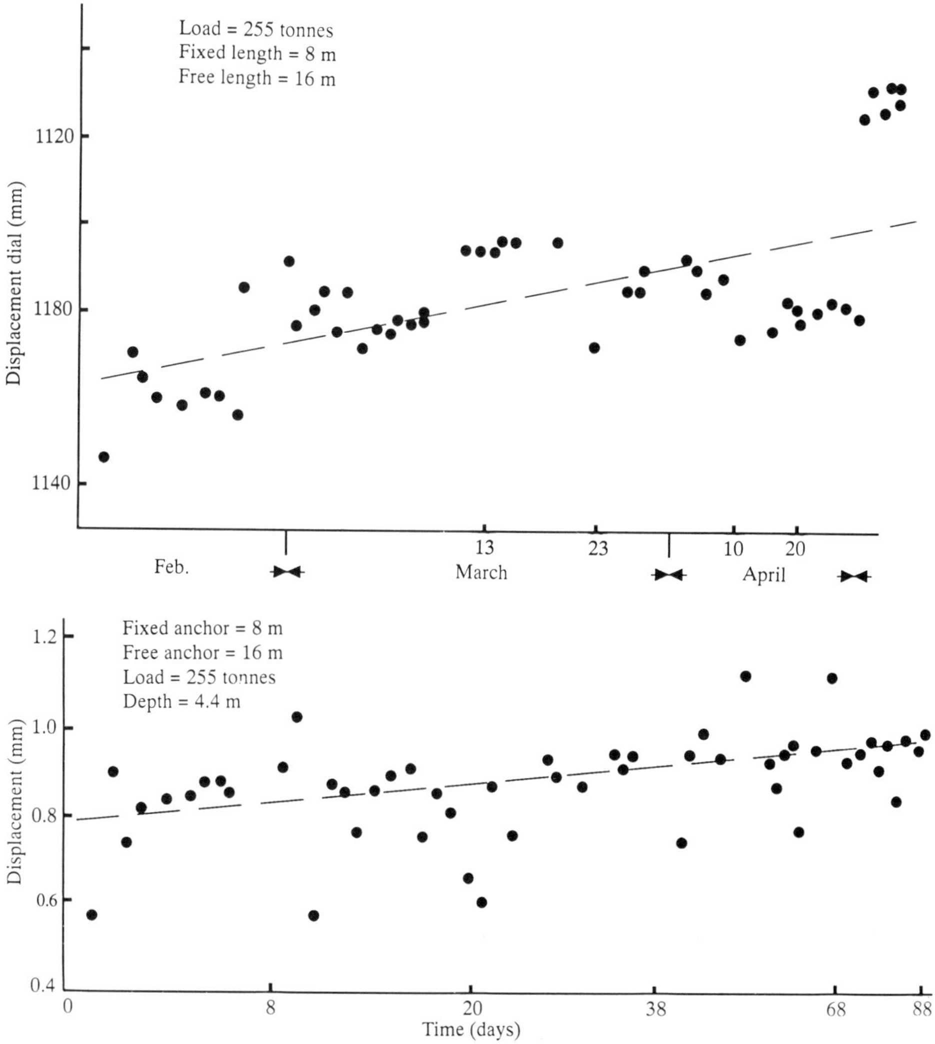

Figure 19.10 *Ground displacement against time for loaded anchor at top end of fixed anchor (top) and 11.6 m above fixed anchor (bottom)*

19.9 Anchor Monitoring

The purpose of anchor monitoring should be clearly defined. Either individual anchors may be instrumented to record load change with time or parts of a complete structure may be monitored including the anchors. Priniciples of field monitoring are explained by Hanna (1986) and Dunnicliff (1989). In general anchor monitoring is useful when there is uncertainty about the behaviour of the anchors within a structure and where malfunctioning of the anchor would not be detected before the structure became distressed, other than by monitoring; when the anchors are in ground strata which may creep; when the anchors are in ground where the ground may swell and/or dilate and cause the anchor loads to increase with time; when the anchored structure is of sufficient scientific interest to warrant detailed field evaluation, the results of which may be used to calibrate predictive models; and when the failure of one or more anchors could put the complete structure at risk.

To monitor an individual anchor requires the installation of a load cell at the anchor head, which may require to be corrosion protected. In addition the absolute movement of the anchor head may have to be measured. Access for reading will normally be required. Details are given by Hanna (1982). The monitoring of an anchored structure is much more ambitious because the performance of all or part of the complete structure may require study. For example, an anchor-supported deep excavation will require movement measurement in three dimensions, water pressures in the ground, as well as the monitoring of the wall structure and the anchors. For the wall structure settlement, tilt and lateral displacement may be required. Contact stress between the structure and the ground may also be required. All of these quantities can be measured provided that the correct questions are posed and answered at the appropriate time (Table 19.4). Where

Table 19.4 General checklist for field monitoring of structures

(1) Has the problem been defined in engineering terms? What is the quality of the site investigation? Have adjacent structures and services been surveyed? Is the ground understood? Are the ground watertable(s) defined?

(2) What is the overall purpose of the proposed field monitoring? To enable the future behaviour of the structure to be predicted? To check that the contractor has fulfilled the requirements of the contract? As a research activity to provide data for future design? To check the adequacy of the design, the construction technique, the performance of the structure during and after construction? To quantify a particular detail where an uncertainty exists?

(3) What quantities have to be measured and why? Anchor and wall loads, pore water pressures, pore air pressures, total earth pressure, settlement (total and differential), heave, horizontal movement, distribution of movements below the surface, temperature, rainfall, acceleration, vibrations.

(4) How much is already known about the expected behaviour? What is the maximum value of the parameter to be measured (this defines the range of the instrument required)? What is the minimum value of the parameter which is of engineering interest (this defines the sensitivity of the instrument required)? How does the construction environment and the long-term use of an instrument influence these decisions? How often have readings to be taken and how long will it take? Can they be taken from a central station or is access to the instrument station required? How will the readings be evaluated and processed?

(5) What is the line of responsibility for purchase, installation and maintenance of the instruments; taking readings, processing them and analysing data; taking decisions concerning future instrumentation needs including reading?

(6) What criteria are to be used for instrument selection? Have the following factors been considered: maximum simplicity and durability; vandal-proof; accuracy; reliability; calibration and its checking; total cost; previous performance record of the instrument; comparison of all instruments available? Are all parts of the instrumentation system compatible? Will a true or a false reading be provided? Does the instrument installation interfere with construction work? Are suitable skills available? Is the instrument durable?

(7) Provide a checklist of factors which may influence the measured quantity. How can these influences be quantified and/or minimized? Has a policy been implemented to record in detail all construction activity, distress to the structure; environmental factors such as temperature, pressure, wind, rainfall? Is a strict supervisory policy implemented on site?

(8) How are readings taken? Is a duplicate measuring system required? Are redundant instruments needed? If so, where are they to be located? Are the readings reliable — can the instruments be recalibrated?

(9) By how much may the measured quantity differ from the estimated value before action has to be taken by the site engineer? Have these differences been arrived at in a logical manner and is the arrangement of the support instruments such that decisions can be taken with confidence?

(10) Has consideration been given to where measurements are required? How have these decisions been arrived at? What are the most critical parts of a project and what are the most critical stages during construction? Is it possible to gain a feel for the project early only in the construction? Is the instrument layout rigid or can it be altered as construction progresses and more specific needs are identified? Does the layout of the instruments provide for cross-checking? Is it best to instrument in detail at a few critical locations with a few instruments at several other locations as cross-checks? What is the purpose of each instrument? Why is it needed? What will it provide? What significance does its malfunctioning have on the overall monitoring programme?

(11) Specifiy and plan installation and recording procedures. Record all details of instrument installation. Have methods available to deal with emergencies should problems arise during installation. Provide a detailed list of all instruments, materials and tools.

(12) Planning considerations. Make arrangements for taking readings, processing and analysis. Make arrangements for maintenance of instruments and their replacement, if necessary. Make arrangements for procurement of instruments, their installation and reading.

such field work is comtemplated reports on examples of instrumentation schemes should be consulted — see, for example, Haws *et al.* (1973), Bundred (1973), Henauer and Otto (1976), Josseaume and Stenne (1979).

19.10 Anchor maintenance

The assessment of the adequacy of an anchor scheme relies heavily on the records made at the time of the anchor installation. In particular data should be available on the exact positioning of anchors to ensure that future construction activities do not cause distress to them. In some designs provision is made for extra anchor heads to give the capacity for increase in anchor loading in the future. The efficiency of all anchors relies on lack of corrosion in the future. When there is uncertainty regarding corrosion, a number of anchor heads should be exposed to allow examination of the tendon steel in the vicinity of the inner head. If regular monitoring is carried out, the data plotted and evaluated, any change in behaviour of the anchors can be detected and appropriate remedial action taken. It is important, however, to be aware of the causes of a change in anchor behaviour before initiating action; for example, changes in the groundwater table can cause significant movements of an anchored slope.

19.11 Uncertainty in anchor use

The introduction of standards and codes of practice in several countries has resulted in a very high level of quality control in anchor use. Some designers continue to use anchors in more and more difficult ground conditions. On production runs allowances must be made for repeatability and construction error. There are areas where knowledge is still limited and good-quality field case records are needed. These include the following:

(1) Corrosion is an emotive term. Littlejohn (1987) has shown that there are few anchor failures from corrosion. Means of monitoring the rate of corrosion at present rely on load recording, yet such methods are not sufficiently sensitive to pinpoint the start of corrosion. Research on non-destructive test methods of corrosion measurement are warranted.
(2) The mechanics of load mobilization in weak rocks needs more study. Can a different grout of very dilatant properties be developed to counteract the effects of drill-hole wall radial yield? Further work must be directed to a classification of the rock.
(3) Dynamic and seismic loading on anchor performance is sparse. The effects of nearby blasting are still not fully understood.
(4) What is the maximum load which an anchor may safely carry without causing distress to the grout column or to the adjacent drill-hole walls? At present damage to both components occurs in weak rocks where large loads are specified. More efficient distribution of the load within the fixed anchor length should be a feature of all anchors in the future.

References

Anon (1947) *Specification for High Alumina Cement*, BS915, British Standards Institution, London

Anon (1958) *Specification for Ordinary Portland Cement*, BS12, British Standards Institution, London

Anon (1972) *Recommandations concenant la conception, la calcul, l'execution et le contrôle des tirants d'ancrage*, Recommendations TA72, Bureau Securitas, Paris

Anon (1973) *Specification for Portland Blast Furnace Cement*, BS146, British Standards Institution, London

Anon (1974) *Low Heat Portland Blast Furnace Cement*, BS4246, British Standards Institution, London

Anon (1979) *Specification for Low Heat Portland Cement*, BS1370, British Standards Institution, London

Anon (1980) *Specification for Sulphate Resisting Portland Cement*, BS4027, British Standards Institution, London

Anon (1981) *Code of Practice for Site Investigations*, BS5930, British Standards Institution, London

Anon (1989) *Ground Anchorages*, BS8081, British Standards Institution, London

Barley, A. (1988) 'Ten thousand anchorages in rock', *Ground Engg* September–November

Bundred, J. (1973) 'In-situ measurements of earth pressure and anchor forces for a diaphragm retaining wall', *Proc. Symp. on Field Instrumentation in Geotech. Engg*, Butterworths, London, 52–69

Casonovas, J. S. (1989) 'Bond strength and bearing capacity of injected anchors — a new approach', *Proc. 12th Int. Conf. on Soil Mechanics and Foundation Engg*, Rio de Janeiro, **2** 1005–1008

Chin, F. K. (1970) 'Estimation of the ultimate load of piles from tests not carried to failure', *Proc. Second South East Asian Conf. on Soil Engg*, Bangkok, 81–90

Coates, D. F. and Yu, Y. S. (1920) 'Three dimensional stress distribution around a cylindrical hole and anchor', *Proc. Second Congress International Society for Rock Mechanics*, Belgrade, **2** 175–182

Dunnicliffe, C. J. (1988) *Geotechnical Instrumentation for Monitoring Field Performance*, Wiley New York

Hanna, T. H. (1982) *Foundations in Tension : Ground Anchors*, Trans Tech Publications, Clausthal

Hanna, T. H. (1986) *Field Instrumentation in Geotechnical Engineering*, Trans Tech Publications, Clausthal

Hanna, T.H. (1987a) 'Ground anchorages : ultimate load estimation by the Chin method', *Proc. Inst. Civil Engineers*, **82**(1), 601–605

Hanna, T. H. (1987b) 'Ground anchorages : anchor testing — measuring absolute movement', *Proc. Inst. Civil Engineers*, **82**(1), 639–644

Harris, J. S. (1987) 'Ground anchorages : drillhole accuracy determining device — the Fotobor', *Proc. Inst. Civil Engineers*, **82**(1), 635–638

Haws, E. T., Lippard, D. C., Tabb, R. and Burland, J. B. (1973) 'Foundation instrumentation for the National Westminster Bank Tower', *Proc. Symp. on Field Instrumentation in Geotech. Engg*, Butterworth, London, 181–193

Henauer, R. and Otto, L. (1976) 'Retaining walls and supervision systems for a 16 m deep excavation', *Proc. Sixth European Conf. on Soil Mechanics and Foundation Engg*, Vienna, **1.1**, 149–156

Hobst, L. (1969) 'Stabilization of slopes by prestressing', *Inzenyrske Stavby*, **17**(19), 353–359

Hobst, L. and Zajic, J. (1977) *Anchoring in Rock*, Elsevier, Amsterdam

Josseaume, H. and Stenne, R. (1979) 'Experimental study of a trench wall anchored by four rows of tie backs', *Revue Française Géotechnique*, **8**, 51–64

Littlejohn, G. S. (1975) 'Accpetable water flows for rock anchor testing', *Ground Engg*, **8**(2), 46–48

Littlejohn, G. S. (1979) 'Design estimation of the ultimate load holding capacity of ground anchors', *Symp. on Prestressed Anchors*, The Concrete Society of South Africa, Johannesburg

Littlejohn, G. S. (1987) 'Ground anchorages : corrosion performance', *Pro. Inst., Civil Engineers*, **82**(1) 645–662

Littlejohn, G. S. and Bruce, D. A. (1976) *Rock Anchors : State of The Art*, Foundation Publications,

Raison, C. A. 'Ground anchorages : component testing at the British Library, Euston', *Proc. Inst. Civil Engineers*, **82**(1), 615–625

Raison, C. A. (1987b) 'Ground anchorages : drillhole alignment determination at the British Library, Euston', *Proc. Inst. Civil Engineers,* **82**(1), 627–634

Stroud, M. A. (1974) 'The standard penetration test in insensitive clays and soft rocks', *Proc. European Seminar on Penetration Testing*, Stockholm

20 Drilling and blasting of rock masses

R Holmberg
Nitro Nobel, Sweden

20.1 Drilling of rock masses

Drilling and blasting are essential and integrated components in the rock excavation technique. The drill hole provides space for placing explosives inside the rock mass. The explosive is located so that its power breaks the rock according to the specified requirements. Blasthole drilling is a matter or providing holes in the rock mass at a predetermined geometrical pattern.

20.1.1 Rock drilling methods

Rock drilling bores a circular hole in rock. The rock is a hard and compact material, whose strength must be overcome by the drilling tool. There are several ways to accomplish this.

Rotary crushing breaks into the rock by high point-load impact, accomplished by a toothed drilling bit, pushed downward with high force. The drill rod is at the same time rotated and the drill cuttings removed by blowing with compressed air. The drill rig for rotary drilling is a large and heavy machine. Rotary drilling is mainly used in open-pit mines, where large-diameter holes are used for blasting.

Rotary cutting creates the hole by shear forces, breaking the tensile strength of the rock. Drill bits are fitted with cutter inserts of hard metal. The torque for rotating drill rods provides the drilling energy. The rotary cutting technique is feasible only in soft rock with low tensile strength, not containing abrasive quartz minerals. Rock salt and limestone are easily drilled with this technique.

Percussion drilling breaks the rock by hard, hammering impacts. The impact is generated by the rock drill, inside which a piston bounces back and forth, each time striking the end of a steel rod. The impact creates a shock wave, which propagates at high speed through the rod. This rod may consist of one piece only, or a number of rods joined to a drill string. The drill bit, with hard metal inserts, is attached to the front of the drill string. Here energy of the shock wave is released, to crush the rock into small fragments, that is, drill cuttings. Cuttings are removed from the hole by air or water flushing. The drill string rotates the bit, so that each blow hits a fresh section of rock. A feeding device holds the rock drill and forces the drill bit into the rock.

Percussion drilling can be used in any type of rock, hard or soft. It is the most common drilling technique for rock excavation in underground mining and civil construction. Two variants of the technique have found practical use, distinguished as top-hammer drilling and down-the-hole drilling, DTH for short.

The most common rock drills work on the top-hammer principle (Figure 20.1). Here the drill hammer stays outside the hole, while the drill string is extended to drill deeper holes. Some impact energy is lost in the joints between the extension rods, the penetration rate therefore drops gradually in longer holes.

The down-the-hole hammer follows the drill bit down the hole, as a component of the drill string. A piston strikes directly on the bit, hence no energy is lost in the transmission (Figure 20.1). The drill tubes convey compressed air to the impact mechanism and transmit rotation torque and feed pressure. The exhaust air blows the bottom of the hole clean and carries the cuttings up the hole. The penetration rate for the DTH hammer is close to independent of hole depth. The air pressure is increased from the normal 6 bars, so that the impact mechanism produces more energy, and the penetration rate increases. DTH drills are designed to operate at pressures of 20 bars and higher. Holes drilled with DTH hammers are generally straight, with a minimum of hole deviation. The DTH hammer works better in fractured rock than a top hammer, as the smooth tube string does not get struck if rock fragments fall into the hole.

20.1.2 Rock drillability

Rock drillability depends on several factors, including the hardness of minerals, grain size and how tight minerals are cemented together. For example, quartz is a common mineral in rock formations. Because quartz is very hard, a

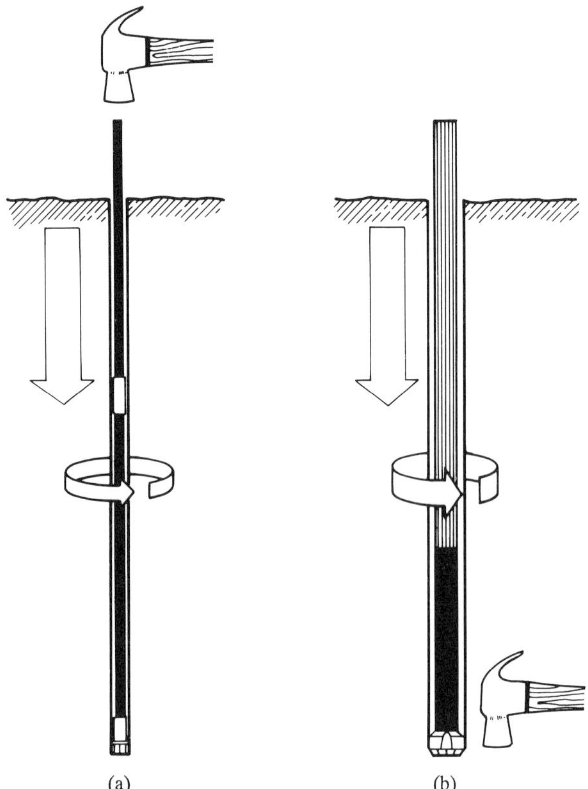

Figure 20.1 *Percussion drilling;* (a) *top-hammer;* (b) *down-the-hole*

high quartz content, as in a quartzite, is both slow to drill and abrasive, with heavy wear on drill bits. Dull inserts on the worn-out bit cause the penetration rate to drop. The bit must be replaced and sharpened frequently to maintain performance. In a limestone formation the drilling is fast, and bit wear minimal. The drill rig performance is higher, and drilling can carry on for long periods without need for bit change.

Rock drillability is used to predict drilling performance and costs at rock excavation projects. A true drillability rating must be checked out in a laboratory or by field testing, and is normally a rather complicated procedure.

If tolerances are accepted, the compressive strength of the rock can give a short-cut estimate of the penetration rate. Figure 20.2 illustrates how the COP 1238 hydraulic rock drill performs in rock of differing hardness. The drill rate decreases with increasing rock hardness in a sloping ski-jump curvature, as illustrated. The diagram is valid for the COP 1238 rock drill only, but would be analogous for other types, accepting that this is an estimate.

20.1.3 Rock drilling technique

Escalating labour costs together with the universal demand for a better, more acceptable working environment, has forced designers and manufacturers of rock drilling products into their present position of output and automation.

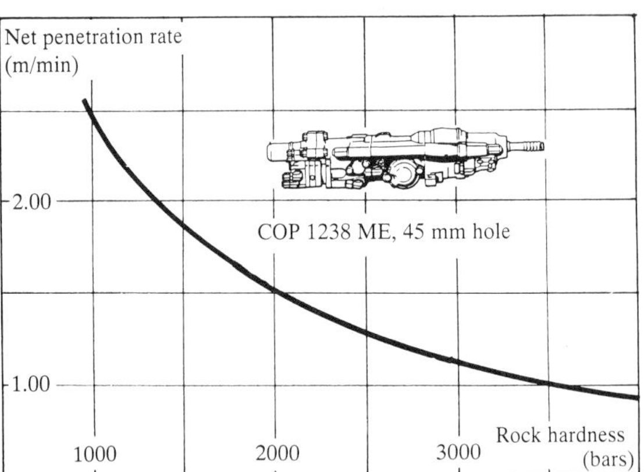

Figure 20.2 *Penetration rate of COP 1238 ME rock drill related to compressive strength of the rock*

(a) Drilling equipment

The light rock drill handled by the driller still survives, but production of drill holes for rock-mass excavation has been taken over by large and specialized machines. Mechanized drill rigs handle heavy rock drills, many times more productive than those of the past, and drill the pattern with more precision and accuracy. The operator is comfortably seated in an air-conditioned, sound-insulated cabin, and directs the drilling from a control panel.

Less effort is required for increased production of a higher-quality product. Figure 20.3 illustrates the influence of mechanization on tunnel drilling productivity over the last 50 years.

The drill rig is a large and complex unit, in which the rock drill is a small, but vital component. The remainder of the drill rig contains a superstructure with power pack, traction gears, positioning facilities, controls, etc.

The range of rock drills available to anyone engaged in rock excavation is extensive and diversified. There are rock drills for all applications, to meet the demands for variations in output and type of drilling. Manufacturers' catalogues are recommended reading for in-depth studies of this subject.

Figure 20.4 shows the Atlas Copco COP 1440 hydraulic rock drill, a high-performance hydraulic drifter. This rock drill will penetrate through hard granite at a rate of 3–4 m/min. It is powered by an electric motor rated 70 kW.

Tools for drilling rock have developed dramatically over the last decades, to keep up with the more hard-hitting rock drills. There has also been a diversification and specialization. Each class of rock drill has a matching set of

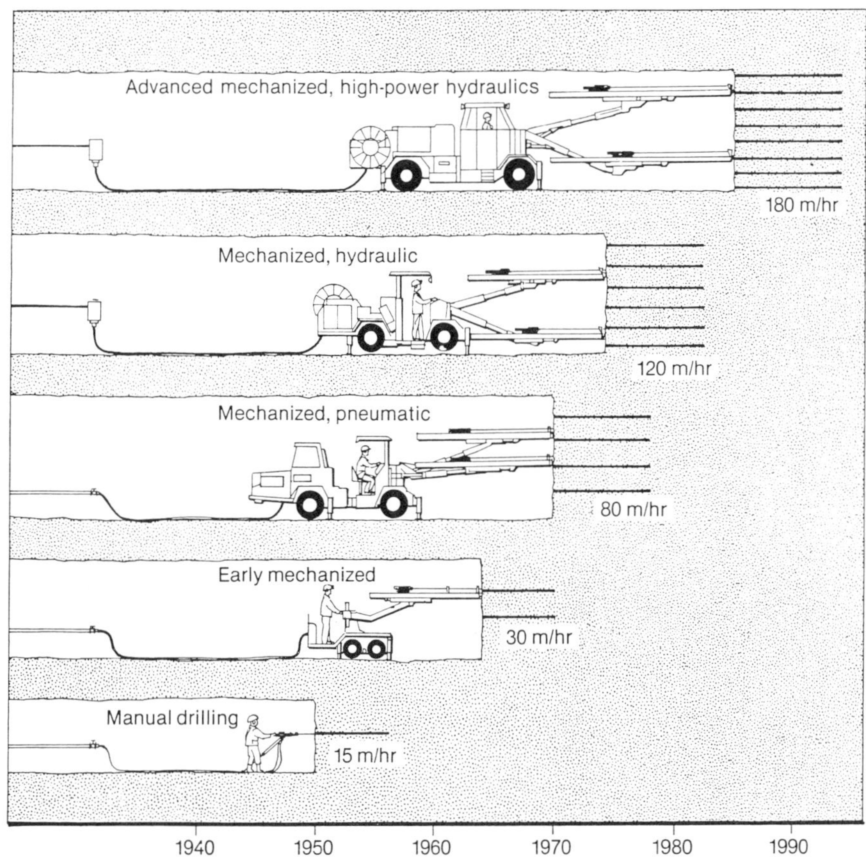

Figure 20.3 *Development of tunnel-drilling productivity*

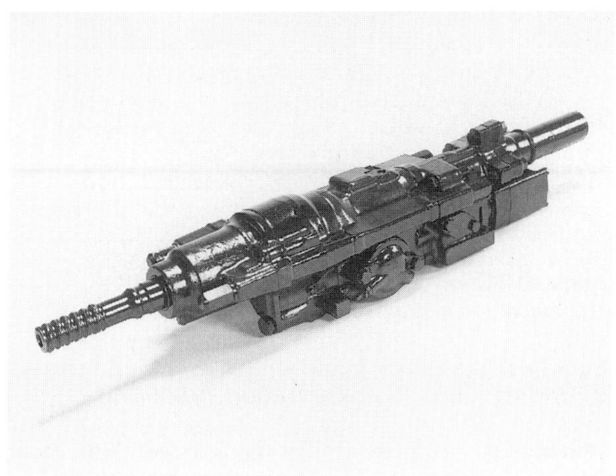

Figure 20.4 *Hydraulic rock drill type COP 1440*

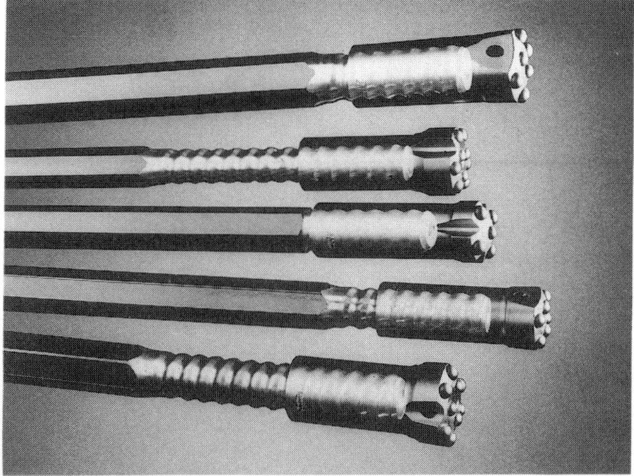

Figure 20.5 *Drifter rods with button bits for tunnel drilling*

drilling tools, which gives optimum performance and service life.

Drifter rock drills are matched with drifter rods. The drifter rod has a bumped-up thread at the drill end, to extend service life and absorb energy. The rod has a hexagonal cross-section, to retain stiffness and drill the long hole straight. The front end thread is slimmed down, to accept a small-diameter bit. The aim is to transmit maximum drilling energy to the bit, without sacrificing drill steel life.

The drill bit has button type inserts with a rounded head (Figure 20.5). These have proved to drill faster and last longer than the usual edged inserts. Furthermore, sharpening intervals have been extended.

Extension drill steel with various dimensions and threads has been developed for bench drilling, to match different rock drills and blasthole diameters. Drill steel variants are recognized by the thread type and rod diameter, for instance, T38 means a 'T' thread and 38 mm rod.

Speedrods are often used in connection with drill rigs featuring mechanical rod-handling. The Speedrod has a male thread at one end and female thread at the other. No coupling is needed, which favours use of a mechanical rod-handling device.

(b) Surface rock excavation with bench blasting

Bench blasting is a technique used for surface rock excavation, as in open pit mines, quarries and on civil engineering construction projects. Blastholes are drilled downward from the surface to a depth normally not exceeding 20 m. The hole diameter varies from 33 mm for the small shot to 350 mm in the large open-pit mines.

The drill rig used for bench drilling is equipped with tracks for mobility, and recognized as a crawler rig. The range of crawler rigs features a choice of rock drills. A simple rig comes with a pneumatic rock drill, where the operator handles the extension rods manually. The sophisticated crawler rig comes with air-conditioned operator's cabin and fully automatic rod handling.

In civil engineering construction projects the object is often to remove a rock mass of irregular shape: for instance, a rock cutting in a mountain side to prepare a base for a road. Bench height and hole depth varies, from zero up to maybe 20 m. Here the crawler rig shown in Figure 20.6 proves useful. The oscillating tract undercarriage enables it to climb steep terrain. The folding boom arm permits holes to be positioned on a rough surface. Mobility and versatility in drill positioning are important features.

Conditions in a quarry are different from those met with in construction projects. The quarry operator wants a steady production of blastholes to feed his crusher, and has an unlimited rock mass at his disposal to penetrate. A quarry is designed with a flat bottom, and/or benches, on which heavy machines can operate. The bench height is fixed, all blastholes are of the same depth and placed in a regular pattern. Conditions are staged for mass production.

Quarry crawlers are available which give a range hole diameter, with increasing diameter generally indicating a heavier rig with a larger production capacity (Figure 20.7). Mobility is required, so that the track system can be less elaborate, as the rig travels on road quality surfaces. The drilling ability can be limited to vertical or inclined holes from a fixed position. The benefits of more sophisticated mechanization options, such as mechanical rod handling, can be fully exploited on the quarry rig.

Figure 20.6 *Crawler rig for civil engineering construction projects*

Figure 20.7 *Crawler rig with operator's cabin and mechanical rod handling for production drilling in quarries*

(c) Underground rock excavation: tunnelling

Underground rock excavation is a larger and more complex operation than that of surface excavation. The drill-blast procedure aims to create openings inside the rock mass, with set geometrical dimensions, without disturbing the strength in the surrounding rock. The first step is to enter a solid block of rock and is the most difficult one. The rock mass is constricted in that there is only one free surface to blast against. Once the opening is there, the rock can expand and excavation continue. The opening is created by a cut, drilled with holes parallel to each other. A small hole is formed which is enlarged step by step until the full tunnel profile is reached. The main objective for tunnel drilling is to produce blastholes, parallel to each other, directed straight ahead into the tunnel face.

Drilling used to be done manually, the driller holding a pusher leg mounted rock drill. Today, the mechanized drill jumbo has taken over this laborious operation at most sites (Figure 20.8). The tunnelling drill jumbo is a complex tool, capable of drilling holes in all directions. As parallel drilling is of importance for the performance of the cut, the drill jumbo has an automatic positioning system. As it moves from one hole to the next, the direction of the drill feed is maintained. This saves time and relieves the rig operator from tricky adjustments.

Figure 20.8 *Two-boom hydraulic drill jumbo, set up for drilling*

Accurate parallel holding has enabled tunnellers to increase the effective blasting length. Standard feeds and drill steels are available for drilling 5.1 m deep holes, breaking the round to 4.8 m.

The jumbo is equipped with a drifter rock drill with high output, as performance is of importance in tunnelling projects. A normal penetration rate in hard rock is between 1.5 and 2.0 m/min; higher rates are not unusual.

The jumbo may mount two, three or several drilling units, each one adding to its output. For tunnelling projects, the contractor has the option to decide for himself (e.g. a jumbo with two booms needing three hours to drill the round, or three booms to drill out in two hours). With known parameters, time can be predicted accurately.

The hydraulic booms, which carry the rock drills, determine the drill jumbo's reach. A large jumbo may have the ability to cover a tunnel width of more than 100 m^2 face area, enough to drill a road tunnel from one set-up (Figure 20.9).

20.2 Blasting of rock masses

20.2.1 Bench Blasting

In many surface operations the geological conditions are such that ripping or digging can be used for excavation of large volumes. When the strength or homogeneity of the

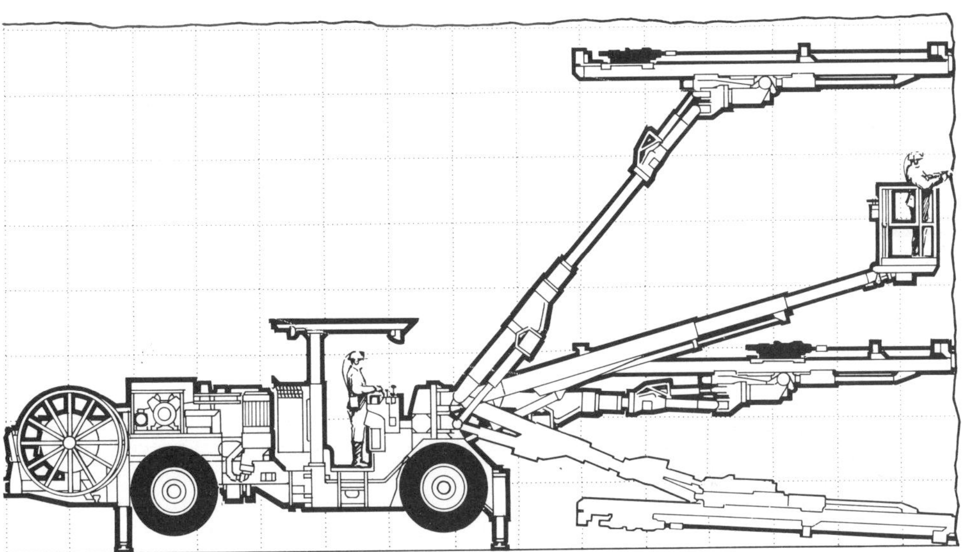

Figure 20.9 *Boomer H 145 operating range*

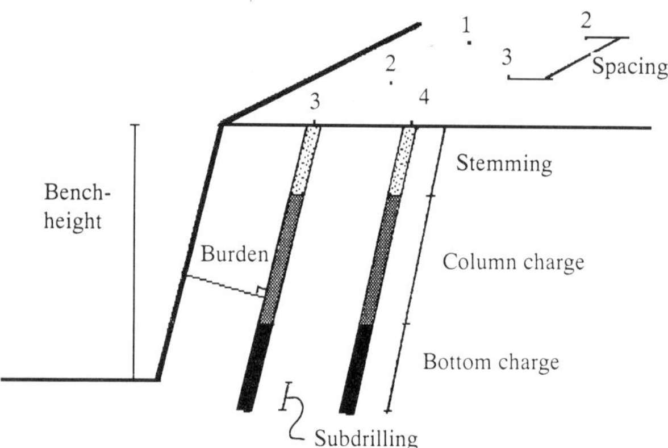

Figure 20.10 *Nomenclature used in bench blasting*

rock mass is such that mechanical excavation is not possible in an economic or technical way, then bench blasting is the main method.

In bench blasting (Figure 20.10) vertical or slightly inclined holes are drilled from the top of the bench in a specific drill pattern based upon bench height, hole diameter, hole deviation, hole inclination, rock type, structural properties, explosive properties, fragmentation demands, muckpile profile and environmental restrictions. In each drill hole a precalculated weight of explosive is loaded normally divided into a bottom charge and a column charge. The bottom charge always has a higher-energy density than the column charge as it is much tougher to break and fragment the toe. To achieve maximum breakage and the best results the explosive column should be bottom initiated. One or two detonators are placed in the drill hole for initiation of the charge at the correct time. All the holes in the drill pattern must be initiated with a proper delay timing between each hole for the optimum blasting result.

In order to provide the blasting engineer with a reliable procedure for calculation of the blasting pattern semi-empirical formulae have been established.

Bench height, K is the vertical distance between the upper and lower level of the bench. In shallow quarries or in road cuttings, for example, where only one bench is excavated, the bench height varies with the nature of the surface terrain. The thickness of the formation may sometimes restrict the bench height. Otherwise, in open-pit mines with thick deposits the bench height is determined in advance. This bench height is normally determined by quarry safety regulations.

It is not unusual to see quarries, where a hole diameter of 75 mm is used, for bench heights around 30 m. This is possible if the explosive is well distributed in the rock mass and a small number of rows are blasted in the round which promote a well-controlled throw creating a low, well-distributed muckpile which is easy and not dangerous to load and haul with available diggers, front-end loaders and trucks.

When large hole diameters of 150 mm and above are used the bench height normally does not exceed 15 m. Blasting with heavy-loaded spread-out holes results in a tremendous heave, leaving a muckpile in which the rock-mass movement is vertical rather than horizontal. The height of the muckpile usually is some 20% higher than the previous bench height and it can be difficult to muck it, even with the giant equipment used today.

The inherent stability of the rock mass can be a controlling parameter when bench height is chosen. Rock slides from the bench face above the working area must be avoided.

A rule of thumb for percussion drilling with hole diameters ranging from 25 to 100 mm is that the bench height normally is equal to 3–8 times the burden. Figure 20.11 shows bench height versus hole diameter for the majority of bench blasting operations.

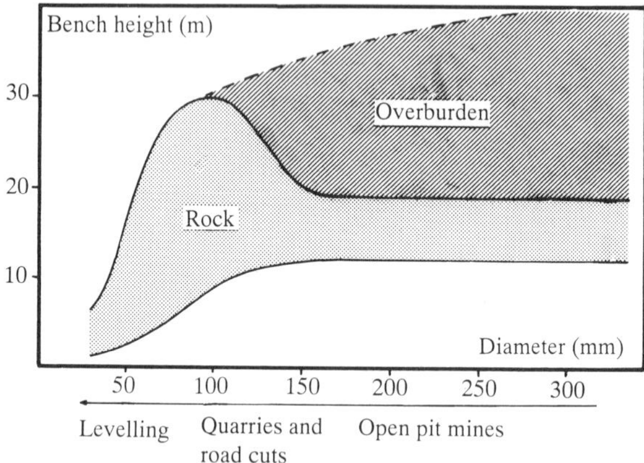

Figure 20.11 *Bench height versus hole diameter for bench blasting*

The hole diameter for the stock blasts is normally chosen with regard to the production rate. Some restrictions in choice of hole diameter size may occur if ground vibration or fly-rock problems exist. Occasionally the fragmentation demand or the geology (caving of hole walls) may control the hole diameter.

The drilling cost is reduced significantly when a larger hole diameter is introduced . However, it is important to realize that the effects of joints and fractures play a more dominant role upon fragmentation when large diameter holes are used. This implies that the powder factor increases if the fragmentation degree is to be kept constant after the change over to larger holes. The powder factor is defined as the weight of explosive used per cubic metre of rock. Figure 20.12 shows powder factor versus hole diameter for some hard rock quarries and open pit mines.

Environmental aspects often control the choice of hole diameter in the blast planning stage. Smaller hole diameters increase the blasting costs but it is easier to

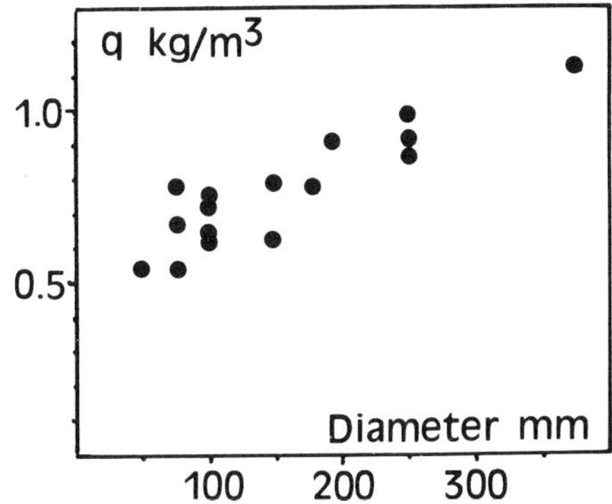

Figure 20.12 *Powder factor, (q), versus hole diameter. Examples from hard rock mining*

control ground vibrations and fly-rocks. In densely populated areas it is necessary to have safe blasting operations.

The burden is equal to the perpendicular distance between the drill hole and the free face. When burden is defined in a drill pattern, it is the perpendicular distance between the drill hole and the adjacent row. A very simple rule of thumb states that the burden is about equal to 40 times and drill-hole diameter for normal ratios between bench height and burden. The calculated maximum burden for a given combination of rock type, hole diameter, hole inclination and explosive is reduced by faulty drilling.

The distance between the blastholes in a row is defined as the spacing. In normal bench blasting the ratio between the spacing and the burden is equal to 1.25. If the hole spacings in the row are increased and the product spacing times burden is kept constant, the fragmentation in short delay multiple-row rounds usually is improved. This method is named wide-space blasting. Model-scale and full-scale tests have shown that the spacing could be increased to up to eight times the burden with improved fragmentation. However, if faulty drilling is large it plays a more dominant role and influences the fragmentation and the toe breakage to a higher extent when wide-space blasting is used.

For an acceptable toe breakage where stumps are to be avoided above the planned floor level it is necessary to drill somewhat deeper than the bench height. Poor fragmentation due to insufficient drilling may lead to inefficient shovel operations and costly levelling blasting. This extra drilling is defined as the subdrilling and usually this depth is equivalent to 30–40% of the burden.

In open-cast coal mining operations where there is a pronounced low-stength boundary between the coal seam and the overburden strata subdrilling is not needed. In fact it is normal to backfill the drilled holes some metres above the coal seam depending upon the strength of the overburden strata. This prevents blast damage and poor recovery of the coal.

The breakage of the burden and the fragmentation is affected by the rock mass properties and the direction of blasting relative to the dip and strike. Test blasting is worthwhile to achieve information about the blastability of the rock. It is not possible to determine blastability from small-scale laboratory tests on intact rock specimens as they do not represent the rock-mass properties.

The rock constant used in the calculations below is a measure of the powder factor used in a test blast with a predetermined bench geometry. As a first input to the charge calculations one can use a rock constant (c) equal to 0.3, 0.4 and 0.5 for rock that is easy, normal and difficult to blast respectively.

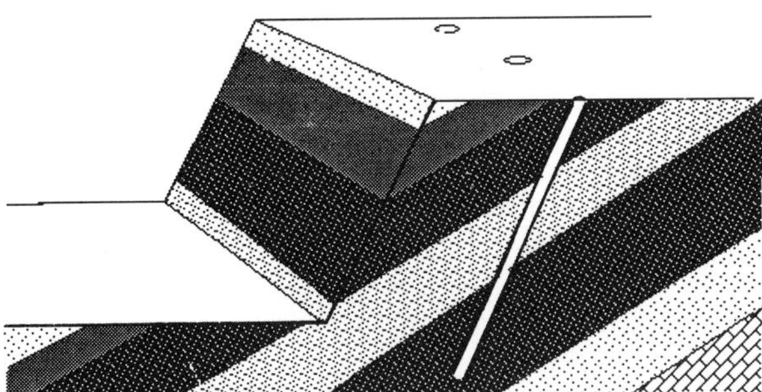

Figure 20.13 *Blasting with the dip direction*

The following guidelines can be used when the blasting direction is considered;

(1) *Blasting with the dip direction (Figure 20.13):*
- good utilization of the explosive energy;
- few toe-problems;
- good rock displacement and mucking conditions;
- back-break problems can occur.

(2) *Blasting against the dip direction*:
- reduced back-break but risk for overhang and toppling;
- potential toe problems;
- reduced rock displacement;
- risk for flyrocks from bench face.

(3) *Blasting parallel to the strike direction*:
- If possible this should be avoided.

The economy of the blasting operation is very much affected by the achieved fragmentation distribution. Sometimes coarse fragmentation is needed for dam construction or wave rip-rap, but usually the fragmentation should be such that the overall cost for blasting, mucking, hauling and crushing is minimized. This optimization is never simple to carry out. Fragmentation distribution curves must be established, loading and hauling capacities must be monitored, energy consumption and maintenance work must be determined and a model must be implemented in which it is possible to simulate what happens if some parameter is changed in the operation.

The best way to obtain the fragmentation distribution

curve is to sieve some of the material. This is, however, extremely expensive and therefore various techniques have been established to judge the fragmentation. One way is to use photographic analysis where photographs are taken of the muck pile or continuously during the loading operation. Through this it is possible to obtain information of the stone sizes. One often mentioned quality factor of the fragmented material is the k_{50}-factor which tells the size of the sieve through which 50% of the blasted material passes.

For safety aspects it is important to drill the front row parallel to the bench face to maintain a constant burden along the face. This is easier to control if the holes are inclined. Inclined holes are favourable to use as they normally give less back-break, fewer toe problems and greater displacement. A greater burden can also be blasted compared with when vertical holes are used.

In the charge calculations the fixation factor is used for adjustment of the burden due to the use of inclined holes. Inclined holes demand an experienced and skilled drilling crew to align the holes properly. Many drill rigs have difficulties handling inclined holes: for example, when rotary drilling with a high thrust is used.

To provide for the use of various explosives, it is necessary to have a basis of comparison. Several methods have been developed to characterize the strength of an explosive. In the charge calculations described in this chapter, the Swedish weight–strength relationship is used. The weight–strength relative to ANFO for an explosive is given by

$$s = Q_{ex}/5.04 + Q_{gas}/4284$$

where

Q_{ex} = heat of explosion for 1 kg of actual explosive (MJ/kg)

Q_{gas} = released gas volume at STP (standard temperature and pressure) for 1 kg explosive (l/kg).

Table 20.1 Some weight-strength values of explosives

Explosive	Q_{ex} (MJ/kg)	Q_{gas} (l/kg)	s
PETN	6.12	780	1.39
Dynamex M	4.65	877	1.13
Prillit (ANFO)	3.92	973	1.00
Nabit	4.42	904	1.09
Gurit	–	–	0.83
Emulite 150	4.06	835	1.00

Simplifed formulae for calculation of the drilling and charging pattern are shown below. The calculated values achieved by the formulae are to be used as guidelines for the blaster as site-specific parameters may influence the blast extensively. The guidelines should be used as a qualified estimate for design. The blasting results should be observed and if necessary corrections made for achieving the optimum blasting method.

The formula for calculation of the burden is valid only for bench heights greater than 2.3 times the burden. For lower bench height–burden ratios the calculations become somewhat more difficult to handle as an iterative process is involved when solving the equations. A diagram of burden versus bench height for various hole diameters is given in Figure 20.14(a) for guidance.

Bench blasting formulae and an example of a calculation for a 12 m bench height are given below.

Bench height (m)	K	12.0
Hole diameter (m)	d	0.076
Hole inclination (m/m)	n	0.33
Fixation factor	$f = 3/(3+n)$	0.90
Type of rock		Limestone
Rock constant (kg/m^3)	c	0.4
Type of explosive, bottom		Dynamex M
Weight strength rel. ANFO	s_b	1.13
Density (kg/m^3)		1450
Degree of packing (kg/m^3)	P_b	1250
Maximum burden (m)	$B_{max} = d\sqrt{\left(\frac{P_b s_b 0.6}{f(c+0.05)}\right)}$	3.48
Subdrilling (m)	$U = 0.3B_{max}$	1.04
Hole depth (m)	$H = K\sqrt{(1+n^2)} + U$	13.68
Collaring deviation (m)	F_c	0.2
Alignment deviation (m/m)	F_a	0.03
Drilling error (m)	$F = HF_a + F_c$	0.61
Practical burden (m)	$B_p = B_{max} - F$	2.87
Spacing (m)	$S = 1.25B_p$	3.59
Height of bottom charge (m)	$h_b = 1.3B_{max}$	4.52
Weight of bottom charge (kg)	$Q_b = \pi d^2 P_b h_b/4$	25.6
Stemming (m)	$h_0 = \sqrt{B_p}$	1.69
Type of explosive, column		ANFO
Degree of packing (kg/m^3)	P_c	900
Height of column charge (m)	$h_c = H - h_0 - h_b$	7.47
Weight of column charge (kg)	$Q_c = \pi d^2 P_b h_b/4$	30.5
Total charge per hole (kg)	$Q = Q_b + Q_c$	56.1
Powder factor (kg/m^3)	$q = Q/((H-U)B_p S)$	0.43
Specific drilling (m/m^3)	$g = H/((H-U)B_p S)$	0.105

Figure 20.14 gives an example of some graphs plotted from computer calculations where the following input data were given:

Explosive: Bottom Dynamex M, $P_b = 1200\,kg/m^3$
Column ANFO, $P_c = 900\,kg/m^3$
Vertical holes, $F_a = 0.02\,m/m$, $F_c = 0.1\,m$
Stemming, $h_0 = B_p$ if B_p less than 1 m;
otherwise $h_0 = \sqrt{B_p}$
Spacing–burden ratio = 1.3

The way in which the holes in a round are delayed is of great influence to the blasting result. It is important that the rows and also the individual holes of the same row are delayed correctly in order to control fly-rock, minimize ground vibrations and achieve the best possible fragmentation. A rule of thumb is to keep the interval time between rows to 10–20 ms/m burden and the holes in the same row can be delayed between 5 and 10 ms/m spacing.

20.2.2 Blasting methods for wall control

All types of cautious blasting have one common objective, namely, to better distribute the explosive energy, generated in the rock mass by the explosive detonation, so as to reduce stressing, fracturing and back break of the remaining rock. Most methods for this have been developed in the field, mainly by trial and error methods.

Smooth blasting (Figure 20.15) is a method where the row adjacent to the planned contour is fired after the rest of the round with a light charge per hole, with a small spacing and usually a spacing to burden ratio of 0.8.

In pre-splitting, the contour charges are initiated before the rest of the charges, not often in a separate round. The crack running from hole to hole then has to be accommodated by elastic deformation of the rock on both sides, because no rock is broken loose. Therefore pre-splitting needs a closer spacing of contour holes, about 50–70% of that for smooth blasting, and pre-splitting thus

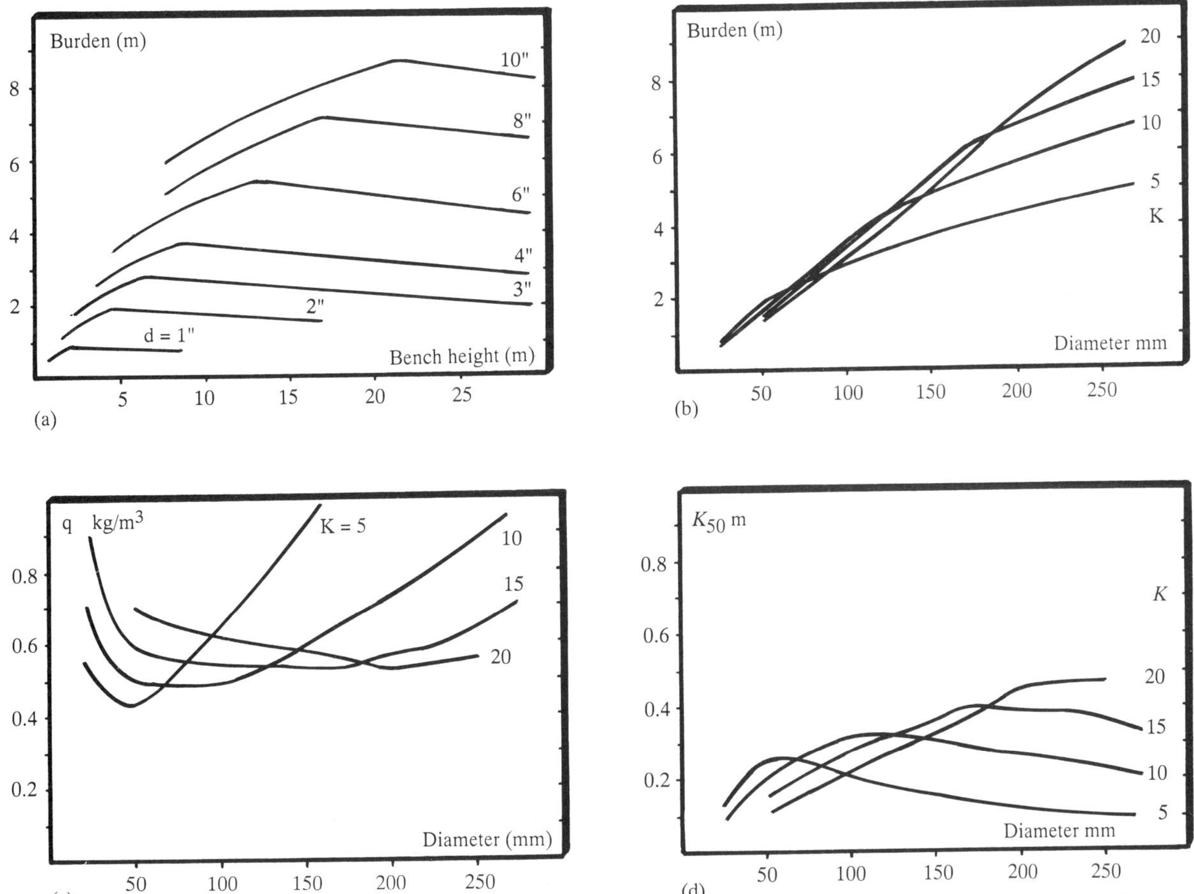

Figure 20.14 (a) *Burden versus bench height for various hole diameters;* (b) *burden versus hole diameter for various bench heights;* (c) *powder factor versus hole diameter;* (d) *fragmentation* (k_{50}) versus hole diameter

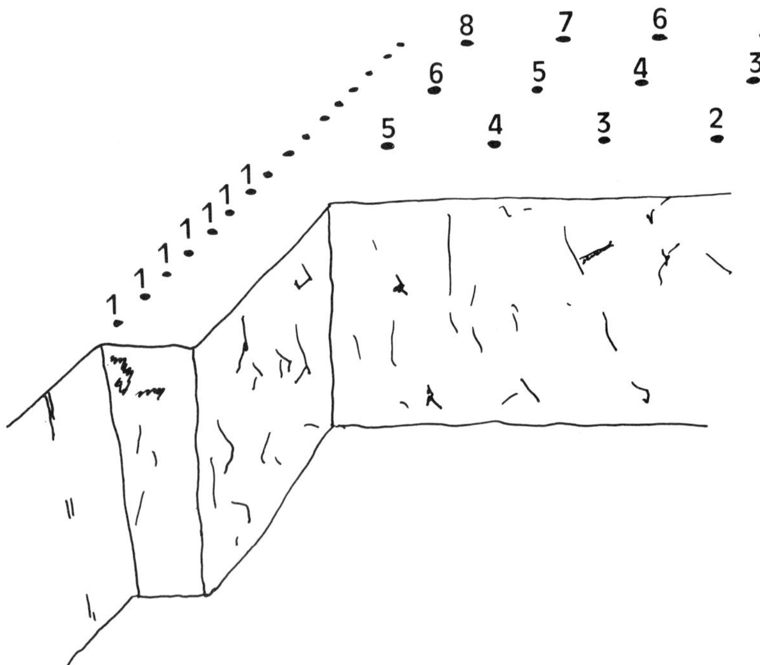

Figure 20.15 *Example of smooth blasting together with a main round for a road cut*

is more expensive. Because of this, and also because it is often difficult to fit in an extra blasting operation in the shift cycle, pre-splitting is very seldom used underground. Smooth blasting is the main method, particularly in tunnelling.

Empirically it has been shown that the minimum required linear charge concentration (ANFO-equivalent) for smooth blasting and pre-splitting is a function of the hole diameter;

$$\ell = 90d^2$$

where ℓ is charge concentration in kilograms per metre of drill hole and d is the hole diameter in metres (Table 20.2). This means that a decoupled charge with the same properties as ANFO should have a charge diameter about one third of the blasthole diameter. Normal spacing for pre-splitting is 8–10 times the hole diameter and for smooth blasting 12–16 times the hole diameter.

For best results in smooth blasting, the charges in the contour should be initiated simultaneously so that they can cooperate fully. However, because the extension of the crack between the holes involves relatively slow dynamic motion of the considerable mass of material within the burden, the latitude for time scatter in the times of detonation is larger than might be expected.

Successful wall control by pre-splitting and smooth blasting require extremely good precision drilling and a fairly good rock quality. However, it is worth noting that even if the rock mass contains such structural features as bedding planes, joints, fractures or if the rock mass contains some poorly consolidated material, a cautious blasting method will always result in less overbreak and less disturbances of the rock mass.

Table 20.2 Approximate values of drill-hole diameter, charge concentration, spacing and explosive for pre-splitting

Drill-hole diameter d (m)	*Charge conc.* $l = 90\ d^2$ (kg/m)	*Spacing* $S = 10d$ (m)	*Explosive per* m^2 *pre-split wall* $9d$ (kg/m^2)
0.025 (1 in)	0.06	0.25	0.23
0.038	0.13	0.4	0.34
0.045	0.18	0.45	0.41
0.051 (2 in)	0.23	0.5	0.46
0.064	0.37	0.65	0.58
0.076 (3 in)	0.52	0.75	0.68
0.102 (4 in)	0.94	1.0	0.92
0.127 (5 in)	1.5	1.3	1.14
0.152 (6 in)	2.1	1.5	1.37

Line drilling is a method where very closely spaced holes are drilled along the desired final wall. It is not unusual to have spacings around four times the hole diameter. The holes are left unloaded and act as crack arresters and gas-pressure relievers for the adjacent blasthole row. The distance from the blasthole row should be of the same order as for a pre-split line, that is, about 0.5 times the spacing of the closest blasthole row. This method is more costly than smooth blasting or pre-splitting due to the high drilling cost.

In the region close to the charge permanent damage occurs at a given critical level of particle velocity. Whether the damage affects the stand-up time of the rock contour or not depends on the character of the damage, the rock structure, the groundwater flow and last, but not least, on the orientation of the damaged weakness planes in relation to the contour and the static load.

The rock damage can be described by the peak particle veolocity, which is proportional to strain as a measure of the damage potential of the wave motion. The surrounding rock mass, of course, contains a number of potential weak planes, each of which is able to withstand a different level of peak particle veolcity.

It is not unusual for blasters to fail to consider the effects of the charges in the rows adjacent to the often well-planned smooth blasted contour row. Charging the adjacent rows with a heavy charge results in cracks spreading further into the remaining rock than from the smooth blasted row. It is better to optimize the charge calculations such that the damage zone from the contour holes is limited. This can easily be done by use of Figure 20.16(a), where the damage zone is given for different charge concentrations.

Table 20.3 contains recommended burdens for some Swedish explosives for use in smooth blasting in tunnelling operations. For a hole diameter of 48 mm with 17 mm GURIT pipe charges, a burden of 0.8 m is normal. From Figure 20.16(a) it can be seen that this charge results in a damage zone of about 0.3 m. Choosing a fully charged hole of ANFO (ℓ = 1.6 kg/m) in the next row with a damage zone of 1.5 m is of no value because this results in a damage zone that extends 0.4 m further into the rock (1.5–0.8–0.3 = 0.4 m) than to use a charge concentration that results in a damage zone equal to that caused by the GURIT plus the burden, that is, 1.2 m (such a charge should have ℓ = 0.8–1.2 kg/m).

It is apparent from this example that a reduction of the damage zone can be obtained by a reduction of the charge concentration per metre of drill hole. This obviously results in increased costs for drill and blast operations, but these are balanced, for example, in tunnelling by the advantage of a safer roof and decreased costs for grouting and maintenance.

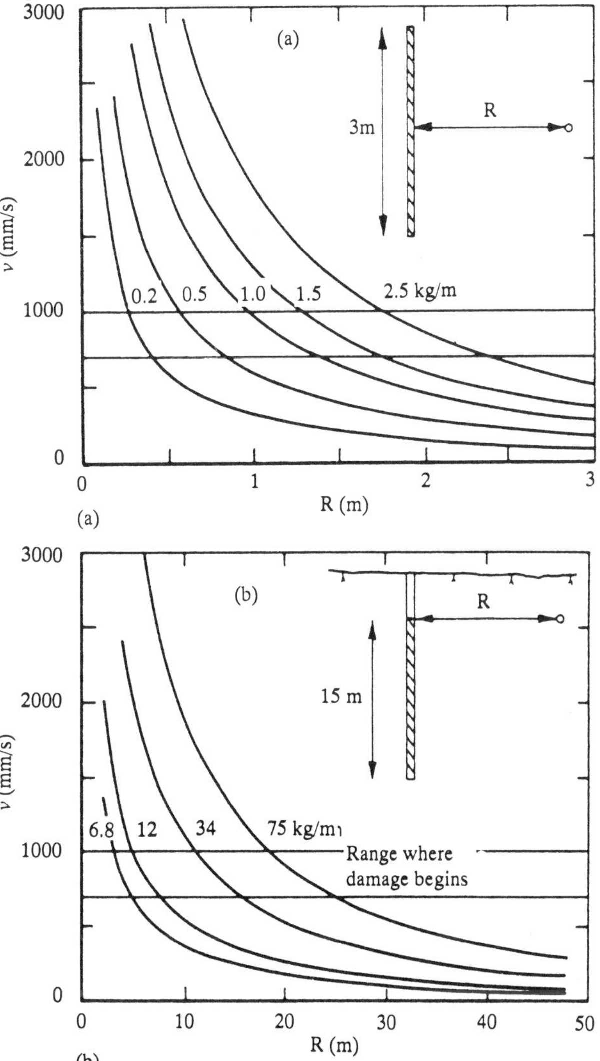

Figure 20.16 *Estimated peak vibration velocity as a function of distance for different linear charge densities, in kilograms per metre of drill-hole length and distance ranges typical of* (a) *small-diameter hole tunnel blasting and* (b) *large-diameter hole open-pit blasting*

20.3 Explosives

An explosive is a material that can undergo an exothermal chemical reaction resulting in a rapid expansion of the reaction products into a volume larger than the original.

The reaction rate of an explosive or the velocity of detonation is mainly dependent upon chemical composition, particle size, density, charge diameter and confinement. For detonating explosives at pressures in the range of 10 000–200 000 bars, temperatures are around 4000°C and the velocity of detonation may range from 2000 to 7000 m/s.

The energy released per kilogram is typically in the order of 4 MJ, which is equivalent to the energy needed for heating 10 litres of water from 4°C to boiling point, 100°C. This is only about one tenth of the energy released when 1 litre of fuel oil is burnt. An explosive must contain all the oxygen needed for forming the detonation products but when fuel oil is burnt all oxygen is taken from the surrounding atmosphere. However, it is the very rapid transformation of explosion energy into mechanical work that makes explosive outstanding for blasting rock.

Table 20.3 Some common explosives used for tunnelling in Sweden

Explosive	*Dynamex M*		*ANFO*	*NABIT*	*GURIT*
Hole diameter (mm)	45	45	45	45	45
Charge diameter (mm)	40	32	45	22	17
Explosive density (kg/dm³)	1.45	1.45	0.95	1.1	1.2
Weight strength	1.13	1.13	1.0	1.09	0.83
Charge concentration					
real (kg/m)	1.6	1.12	1.5	0.40	0.24
(kg ANFO/m)	1.82	1.28	1.5	0.44	0.20
Borehole pressure					
(bars)	32 000	13 000	21 000	3300	900
(MPa)	3 200	1 300	2 100	330	90

Commercial explosives are usually close to oxygen-balanced, implying that the main detonation products formed are steam (H_2O), carbon dioxide (CO_2) and nitrogen (N_2). The amount of gaseous products released for an ANFO type of explosive is normally around 40 moles/kg or equal to 900 litres/kg at standard temperature and pressure. The technical grade of ammonium nitrate used in ANFO should be about 94.5% and the fuel oil should be about 5.5% when the explosive is oxygen-balanced. The decomposition would then be approximately

$$3NH_4NO_3 + \tfrac{1}{12} C_{12}H_{24} = 7H_2O + CO_2 + 3N_2$$

$$(94.5\%AN + 5.5\%FO)$$

As fuel oil and ammonium nitrate prills around the world have different properties the oxygen balance may be shifted somewhat. This also affects the amount of toxic fumes formed at detonation. An oxygen-deficit results in increasing amounts of carbon monoxide (CO) and an ANFO mixture with a low fuel content (oxygen excess) will form higher amounts of noxious gases (NOx) (Figure 20.17). In underground works it is important to have oxygen-balanced explosives in order to minimize the amount of toxic gases.

In rock blasting it is accepted among the blasters that the high-velocity explosives give a better fragmentation result in hard homogenous rock (high P-wave velocity) than low-velocity explosives. On the other hand in soft rock formations the blaster usually chooses a low-velocity explosive with a high gas volume to achieve a better heave.

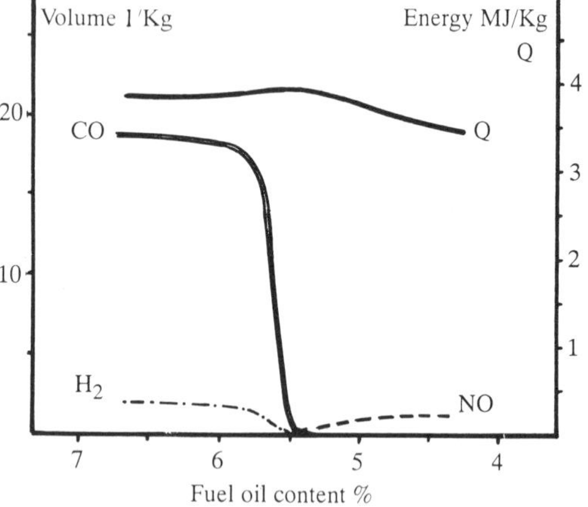

Figure 20.17 *Relation of toxic fumes and energy-relative oxygen balance for ANFO*

20.3.1 Blasting performance

A number of methods exist today to evaluate the blasting performance of an explosive. When selecting his explosive the blaster should look for testing methods where blasting has taken place under conditions similar to the rock breakage operation he is concerned with (i.e. charge sizes, confinement and energy transfer). It is by no means always best to choose the explosive with the highest relative bulk strength.

In the blasting operation the fragmentation mainly depends upon how much of the available chemical energy has been transferred to expansion energy before the rock breaks up and the high-temperature gases escape through the fractures. This depends upon the components in the explosive and the strength of the rock.

In rock blasting where the explosive is charged in a drill hole the gaseous reaction products escape to the atmosphere after they have expanded about 10 times their original volume. This corresponds to a pressure in the neighbourhood of 1000 bars and the expansion work utilized for direct fragmentation is of the order of 60–70% of the total available chemical energy. For softer rock the useful expansion may be continuing up to 20 times the original volume.

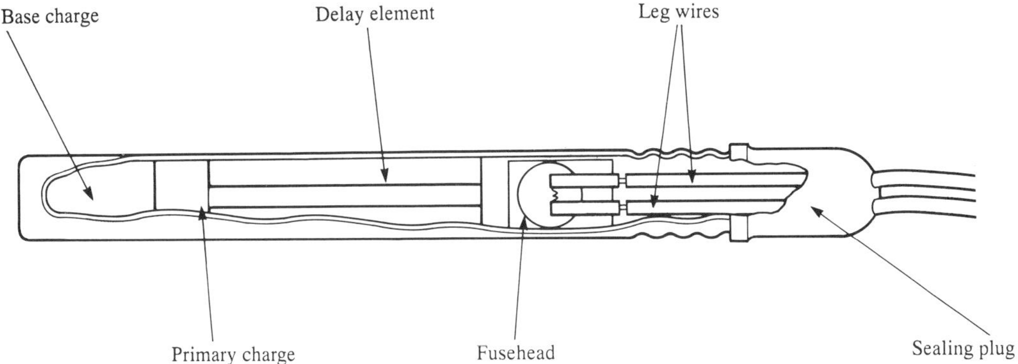

Figure 20.18 *Electrical detonator with pyrotechnic delay element*

20.3.2 Today's detonators

The majority of delay detonators manufactured today use a pyrotechnic delay element, that is, an element which burns inside the cap (Figure 20.18). The timing is determined by the length of the element and the chemical composition of the pyrotechnic element. A drawback with this system is that precision diminishes with increased delay times.

For safety and handling reasons there has been a change over from electrical detonators to the non-electric system NONEL. This can be seen in underground mines and construction worksites where more and more electrical equipment is introduced and the electrical hazard is increased. The non-electric system is also advantageous in surface blasting operations where thunderstorms may be a safety risk giving rise to unintentional ignition of electrical detonators.

A large amount of detonating cord is still used for initiation of blastholes but blasters are more and more realizing the environmental problems which occur when the surface cord is detonated or when the detonating cord causes stemming to be released with subsequent air shock waves. The cord will often side initiate the explosive string in the blasthole and the underdriven detonation can cause a slow energy release during the crucial rock breakage phase.

The NONEL system is based on a plastic tube, the inside of which is coated with a reactive substance that maintains the propagation of a shock wave at a rate of 2000 m/s. This shock wave has sufficient energy to initiate the primary explosive or delay element in a detonator. Since the reaction is contained in the tube, this has no blasting effect and acts merely as a signal conductor.

In surface operations it is convenient to use the NONEL UNIDET system shown in figure 20.19. The timing

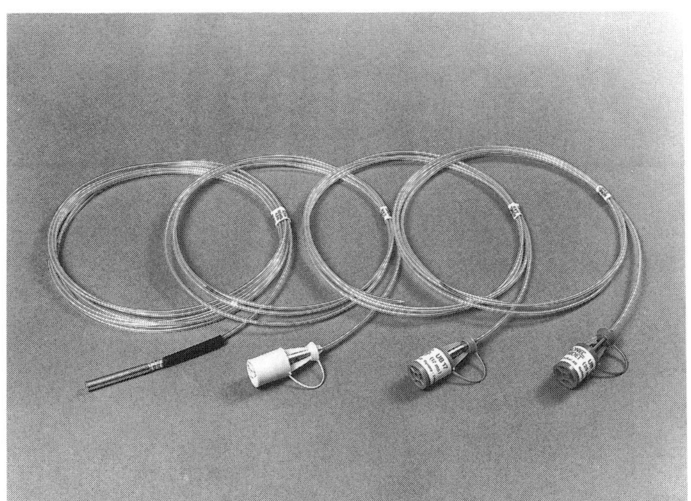

Figure 20.19 *The NONEL UNIDET initiation system*

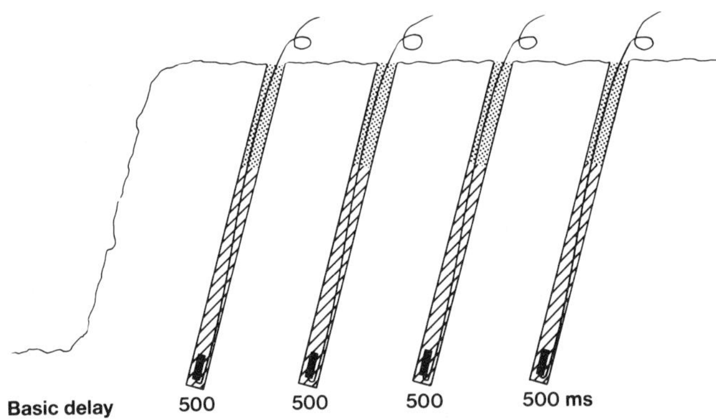

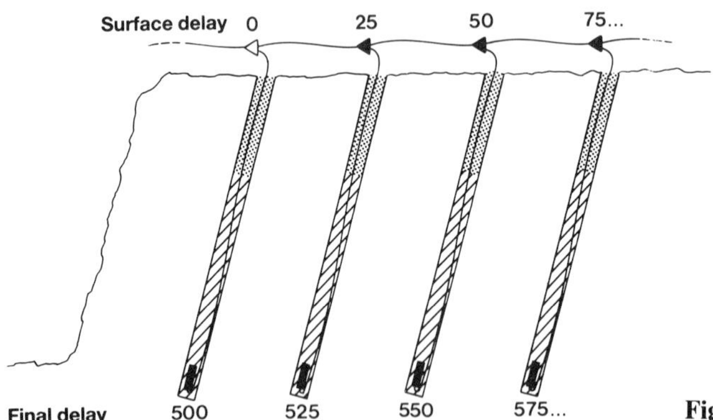

Figure 20.20 *Principle of function for NONEL UNIDET*

sequence does not have to be defined beforehand. In the first instance, each drill hole receives a detonator with a basic delay of 500 ms. When the charging operation is finished, the round must be connected up. Figure 20.20 shows how to start with a UB 0 connector which has no delay. The next step is to connect in a UB 25 which relays the activation after 25 ms, partly to the detonator and partly to the next connector. The operation is then repeated so as to build up a sequence of delays with 25 ms interval. When connection is complete the detonators are conditioned and their final delay is now the total of basic and surface delays. The basic 500 ms delay allows for the activation to run so far ahead of the blasting process that any risk of cut-off is eliminated. The surface connectors can be chosen with delay times of 0, 17, 25, 42, 67, 109 or 176 ms.

Symbols

B_{max} = maximum burden (m)
B_p = practical burden (m)
S = spacing (m)
U = sub-drilling (m)
H = hole depth (m)
K = bench height (m)
Q_b = bottom charge (kg)
Q_c = column charge (kg)
Q = total charge (kg)
P_b = degree of packing, bottom charge (kg/m^3)
P_c = degree of packing, column charge (kg/m^3)
F_a = alignment deviation (m/m)
F_c = collaring deviation (m)
F = drilling error (m)
d = drill-hole diameter (m)
c = rock constant (kg/m^3)
l_b = charge concentration, bottom charge (kg/m)
l_c = charge concentration, column charge (kg/m)
q = powder factor (kg/m^3)
g = specific drilling (m/m^3)
s = weight strength of explosive
f = fixation factor
n = hole inclination (m/m)
h_b = height of bottom charge (m)
h_c = height of column charge (m)

Bibliography

Aflces Copco, Product Catalogue, Stockholm, Sweden

Bjarnholt, G. and Holmberg, R. (1976) 'Explosive Expansion Works in Underwater Detonations', *Proc. 6th Symp. on Detonation*, ACR–221, pp. 450–550, Office of Naval Research-Department of the Navy, Arlington, Virginia, USA

Holmberg, R. and Persson, P. A. (1978) *The Swedish Approach to Contour Blasting* – 4th Conference on Explosives and Blasting Technique, Society of Explosives Engineers, New Orleans, USA

Holmberg, R. and Persson, P. A. (1979) *Design of Tunnel Perimeter Blasthole Patterns to prevent Rock Damage – Proc. Tunnelling '79,* M. J. Jones, ed., Inst. of Mining and Metallurgy, London

Johansson, C. H., and Persson, P. A. (1970) *Detonics of High Explosives,* Academic Press, London

Langefors, U. and Kihlstrom, B., (1963) *The Modern Technique of Rock Blasting,* Wiley, New York

Nitro Nobel, Nonel Manual, Gythorp, Sweden

Persson, P. A. and Holmberg, R. (1983) 'Rock Dynamics', General Report, 5th International Congress on Rock Mechanics, ISRM, Melbourne Australia

Wijk, G. (1982) *The Stamp Test for Rock Drillability Classification*, SveDeFo Report DS 1982: 1, Stockholm, Sweden

21 Open excavation in rock masses

Professor F G Bell
University of Natal

21.1 Introduction

Open excavation refers to the removal of material, within certain specified limits, for mineral exploitation and construction purposes. For this to be accomplished economically and without hazard, the character of the rocks involved and their geological setting must be investigated. Indeed, the method of excavation and the rate of progress are very much influenced by the geology of the site. Another factor which is important is the position of the water table in relation to the base level of the excavation, as are any possible effects of excavation on the surrounding ground and buildings.

The cross-section of an open excavation is influenced by the dimensions of the base, its depth and the profile of its slopes. Slopes should be as steep as possible, consistent with safety, in order to minimize the volume of material to be removed. Allowance must be made for any drainage works as far as the dimenisons of excavations are concerned.

Other factors apart, the maximum height which can be safely developed in a rock slope is roughly proportional to its shearing strength, that is, the stronger the rock, the steeper the slopes which may be cut into it. For instance, excavations in fresh massive plutonic igneous rocks such as granite can be left more or less vertical after the removal of loose fragments.

In stratified rocks which are horizontally bedded excavation usually is straightforward and slopes can be determined with some degree of certainty. However, slopes may have to be modified in accordance with how the dip and strike directions are related to the excavation when it occurs in inclined strata. The most stable excavation in dipping strata is one in which the face is orientated normal to the strike, since in such situations there is a low tendency for the rocks to slide along their bedding planes. Conversely, if the strike is parallel to the face, then the strata dip into one slope. This is most critical where the rock dips at angles between 30° and 70°. If the dip exceeds 70° and there is no alternative to working against the dip, then the face should be developed parallel to the bedding planes for safety reasons.

Sedimentary sequences in which thin layers of shale, marl or clay are present may have to be treated with caution, especially if the bedding planes dip at a critical angle. Indeed, wherever weaker strata underlying stronger rocks are exposed upon excavation, then undermining of the latter is likely to occur as the former are readily removed by agents of denudation. Ultimately this action will produce a rockfall or slide.

Faults which traverse the area in which excavation is to be made may cause serious trouble. This is principally because of the greater freedom afforded rock masses to move along fault planes. In particular, if a fault intersects a prominent joint or bedding plane in such a way that it produces a wedge which daylights into the excavation, then this is likely to slide.

Open joints in rocks facilitate weathering and generally aid slope failure. Fissure zones usually represent zones of weakness along which rock may have been altered to appreciable depth by weathering.

21.2 Groundwater and excavation

Groundwater frequently provides one of the most difficult problems during excavation and its removal can prove costly. Not only does groundwater make working conditions difficult, but flow of water into an excavation can lead to erosion and failure of the sides. Collapsed material has to be removed and the damage has to be made good. Subsurface water is normally under pressure, which increases with increasing depth below the water table. Under high pressure gradients weakly cemented rock can disintegrate. Hence data relating to the groundwater conditions should be obtained prior to the commencement of operations.

Groundwater flow in rock masses in controlled primarily by the discontinuities present rather than by permeation

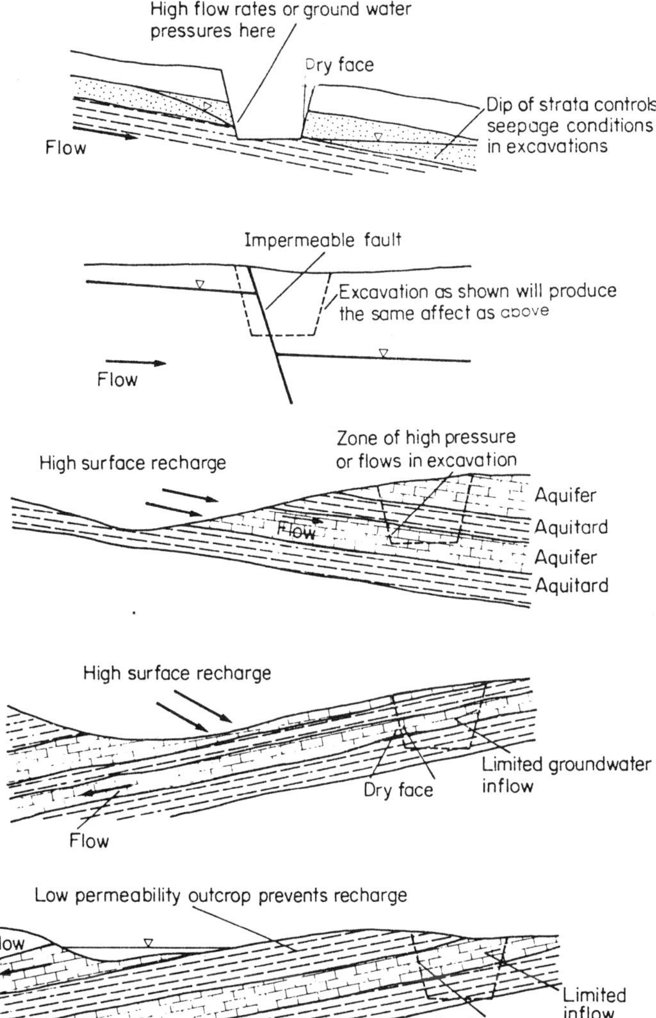

Figure 21.1 *Some effects of geological structure on groundwater conditions*

through the intact rock. Since discontinuities generally fall into 'sets' having a certain orientation, the permeability of the rock mass varies with direction.

Since measurements of permeability represent spot values in the field, an adequate description of the geology of the rock mass and its discontinuity pattern is essential before an assessment of groundwater conditions can be made or means of controlling groundwater designed. Geological features such as faults or dykes may have a significant effect on groundwater flow patterns. For example, faults containing clay gouge impede flow and may cause high groundwater pressures behind them. Alternatively shear zones may form zones of high permeability, concentrating groundwater flow and effectively draining the rock on either side. The geological structure and topography will indicate the potential for recharge of the rocks which are to be exposed in an excavation (Figure 21.1)

The permeability of rock masses should be determined from detailed studies of drillhole core, information on water or air loss during drilling and field permeability tests. In the falling head test, an uncased section of the drillhole is subjected to a pressure head greater than static groundwater pressure. This head then is allowed to dissipate to the static value and measurements of the loss of head with time are taken. A relationship between flow and pressure is determined, from which the permeability is calculated. In the constant head test, an uncased section of the drillhole is subjected to a steady excess pressure head (either by maintaining a constant excess water level in the hole above the test section or by pumping water at a steady pressure directly into the test section once it has been isolated by a mechanical or pneumatic packer unit). The resultant flow out of the test section is measured and the permeability is calculated from the relationship between flow and pressure. Analytical methods to determine the groundwater pressure distribution within a slope formed by excavation from permeability and piezometric measurements include graphical flow net sketching, electrical resistance analogues, and computer analyses.

Groundwater pressures may adversely affect the stability of rock masses by reducing their shear strength along discontinuities and by producing hydraulic driving forces in steeply dipping discontinuities which are not subject to shear failure (e.g. tension cracks). Hence the groundwater pressure distribution in a slope must be known before the effect of groundwater on stability can be evaluated. This means that measurements have to be taken with piezometers at representative locations. Piezometric heads within rock masses often vary considerably from one location to another and a sufficient number of piezometer tips must be installed to determine the overall conditions. Tips should be located with reference to the local geology, and especially in relation to intersecting sets of discontinuities, and must be properly sealed in the drillhole above and below the tip so that measurements of piezometric head represent only conditions encountered at the tip. For this reason, open holes rarely give reliable results. There must be a sufficient number of piezometer installations to allow for malfunction of individual instruments, as well as to give a reasonable statistical assessment of piezometric conditions.

Artesian conditions can cause serious trouble in excavations and therefore if such conditions are suspected it is essential that both the position of the water table and the piezometric pressures should be determined before work commences. Otherwise excavations which extend close to strata under artesian pressure may be severely damaged due to blow-outs taking place in their floors. Such action may also cause slopes to fail and could lead to the abandonment of the site in question. Sites on which such problems are likely to be encountered should be dealt with, prior to and during excavation, by employing either dewatering or exclusion techniques.

The most commonly used method of groundwater control is drainage which, by reducing the detrimental effects of groundwater, can increase the stability of excavated slopes. The maximum benefit is obtained by reducing groundwater pressures within the slope prior to movements taking place during excavation and while the rock mass still possesses its undisturbed strength. The most probable failure zone must be determined in order to delimit the zone of the rock mass which needs to be drained.

Drainage involves the removal of water from an area which may be critical with respect to stability. In some cases, it may be more expedient to prevent the water from reaching the critical zone by creating a cut-off which retains the water behind it and allows natural drainage to occur within the slope. Cut-offs are advantageous where internal erosion of soft material may be caused by increased water flows induced by drainage measures. In rock masses cut-offs can be formed by the construction of a grout curtain, or in soft rocks by excavating a trench and backfilling with concrete. Alternatively a row of pumped vertical wells can be sunk to form a well curtain which intercepts flow towards the excavation and provides 'dry' conditions on its downstream side.

Drainage of the slope itself includes such measures as drainage trenches constructed down and/or along the slope face; horizontal or near horizontal drain holes drilled into the slope face; vertical pumped wells drilled behind the slope crest or on the slope face; and drainage galleries excavated in the rock mass behind the slope, with or without supplementary holes drilled from the gallery (Figure 21.2). The method chosen to drain an excavated slope depends on slope geometry as well as permeability of the slope material.

Effectiveness of drainage installations can be enhanced by controlling surface water before it infiltrates the slope. This is usually accomplished by constructing trenches at the top and toe of the trench and, where necessary, surface drainage over the slope (Anon 1987). If open cracks occur on a slope, laying impermeable membranes may prove advantageous.

Control of water is particularly important when individual benches are drained by horizontal holes in order to prevent water from percolating into the bench below. Each row of drain outlets should issue into a properly graded interceptor or trench, lined if necessary with impermeable material to prevent erosion.

Some of the worse conditions are met in excavations which have to be taken below the water table. In such cases the water level must be lowered by dewatering. The method adopted for dewatering an excavation depends upon the permeability of the ground and its variation within the stratal sequence, the depth of base level below the water table and piezometric conditions in underlying horizons. Pumping from a sump within the excavation can generally be achieved when the excavation is made in rock where the rate of inflow does not lead to instability of the sides or base. Ditches are dug in the floor of the excavation which lead the water to a sump located at a lower level. Sump pumping, however, cannot dewater confined aquifers and may not be able to cope with flow from aquifers with large storage capacities or which transmit copious quantities of groundwater. In such situations internal drainage measures, as mentioned above, may provide an answer. Alternatively groundwater may be removed by abstraction via vertical wells to form interconnected cones of depression, thereby lowering the water table (Figure 21.2). A bored filter well consists of a perforated tube surrounded by an annulus of filter media and the operational depth may, in theory, be unlimited. Generally wells are placed in a 600 mm diameter hole. Having drilled the hole to a sufficient depth within the aquifer (in the case of a thin aquifer the well bore is often taken some 1–2 m into the impervious stratum beneath the aquifer) the perforated screen (well-screen) is lowered into place and the appropriate filter medium then is placed. An electric submersible pump is lowered into each well on its own riser pipe and connected to a common discharge main. Deep wells are particularly suited to multi-layer aquifers, as well as to the control of groundwater under artesian and subartesian conditions.

21.3 Methods of excavation: drilling and blasting

The method of excavation is very much determined by the geology of the site however, consideration must also be given to the surroundings. For instance, drilling and blasting, although generally the most effective and economical method of excavating hard rock, are not desirable in built-up areas since damage to property or inconvenience may be caused. Neither is it wise to blast where landslides or rockfalls might result.

21.3.1 Drilling

If drilling is not carried out properly, blasts are unable to provide muckpiles having the characteristics required for subsequent operations. Optimum drilling is therefore a prerequisite of optimum blasting.

According to McGregor (1967) the properties of a rock mass which influence drillability include strength, hardness, toughness, abrasiveness, grain size and discontinuities. The strength of a rock has an appreciable influence on the drilling force required. The hardness of a rock depends not only upon the hardness of the individual minerals involved but also upon their texture. The harder the rock, the stronger the bit which is required for drilling since higher pressures need to be exerted. Toughness is related to hardness and represents the work required to bring about fracture.

With respect to drilling, abrasiveness may be regarded as the ability of a rock to wear away drill bits. This property is closely related to hardness and in addition is influenced by particle shape and texture. The size of the fragments produced during drilling operations also influences abrasiveness. For example, large fragments may cause scratching but comparatively little wear of a bit whereas the production of dust in tougher but less-abrasive rock causes polishing. This may lead to the development of high skin hardness on tungsten carbide bits which in turn may cause them to spall. Even diamonds lose their cutting ability upon polishing. Rock fabric determines the characteristics of the chippings produced. Generally, coarse grained rocks can be drilled more quickly than can fine grained varieties or those in which the grain size is variable. Fish (1960) maintained that a realistic assessment of rock drillability should also include the projected penetration rate at a given thrust and also the economic life of the bit.

The ease of drilling in rocks in which there are many discontinuities is influenced by their orientation in relation to the drillhole. Generally the rate of drilling is less difficult and therefore quicker if the hole runs at a high angle to the discontinuities. Drilling over an open discontinuity means that part of the energy controlling drill penetration is lost. Where a drillhole crosses discontinuities at a low angle this may cause the bit to stick. It may also lead to excessive wear and to the hole going off line. If the ground is badly broken, then the drillhole may require casing. Where discontinuities are filled with clay this may penetrate the flush holes of the bit, causing it to bind or deviate from alignment.

In addition to the properties of the rock mass, the penetration rate is influenced by the power of the rock drill, the shape of the cutting edge of the bit, the air pressure at the rock drill and the diameter of the drillhole. In this context, the relative proportions of transmitted and reflected energy are important and are governed by the bit-rock contact and the reaction of the rock to impact. There is a certain critical level of stress for any particular rock below which penetration cannot be brought about by impact. The type of flush used and the experience of the drilling crew also influence the rate of penetration.

Penetration is effected by the drill bit causing the rock to be pulverized and chipped. The later action is the more effective as far as drilling is concerned and if chipping action is to be successful, then the pulverized rock material must be removed as this has a cushioning effect on drilling. Hence an important function of the flushing medium is to clear the cutting edge of the bit of rock fragments.

A bit with a wide wedge angle is only able to achieve a small depth of penetration. Unfortunately, however, although a bit with a sharp angle is more effective at chipping, it suffers rapid wear (Furby 1964). Therefore the shape of the bit represents a compromise between a high initial rate of penetration and the rate of wear.

Percussion-type churn drills will penetrate the hardest rocks but they are slow and inefficient. Consequently they have been largely superseded by the down-the-hole and rotary drills, since the penetration rate of both is much faster. In softer rocks, rotary drills give faster penetration than the percussive types, but in hard to medium rocks the rotary drill is not economical because of the severity of bit wear. The best compromise according to Antill and Ryan (1972) is the rotary percussion drill.

Blastholes of up to 50 mm can be sunk in soft to medium hard rocks (for example, some sandstones, shale, coal, gypsum and rock salt) with rotary drills. Larger holes up to 100 mm diameter, can be made in soft rocks by reaming the original hole.

Rotary percussion drills are designed for rapid drilling in rock. They provide a continuous rotary cutting action upon which axial percussive blows are superimposed. In this way the rock is subjected to rapid high-speed impacts whilst the bit rotates, which causes fracture at a lower value of rotary thrust and torque. Consequently faster rates of penetration are achieved at lower thrust than with rotary drilling (the latter can still produce a faster rate of penetration but with much higher thrust). The technique is most effective in brittle materials since it relies upon chipping the rock. The impact mechanism is coupled to the bit in a down-the-hole hammer drill and accompanies the

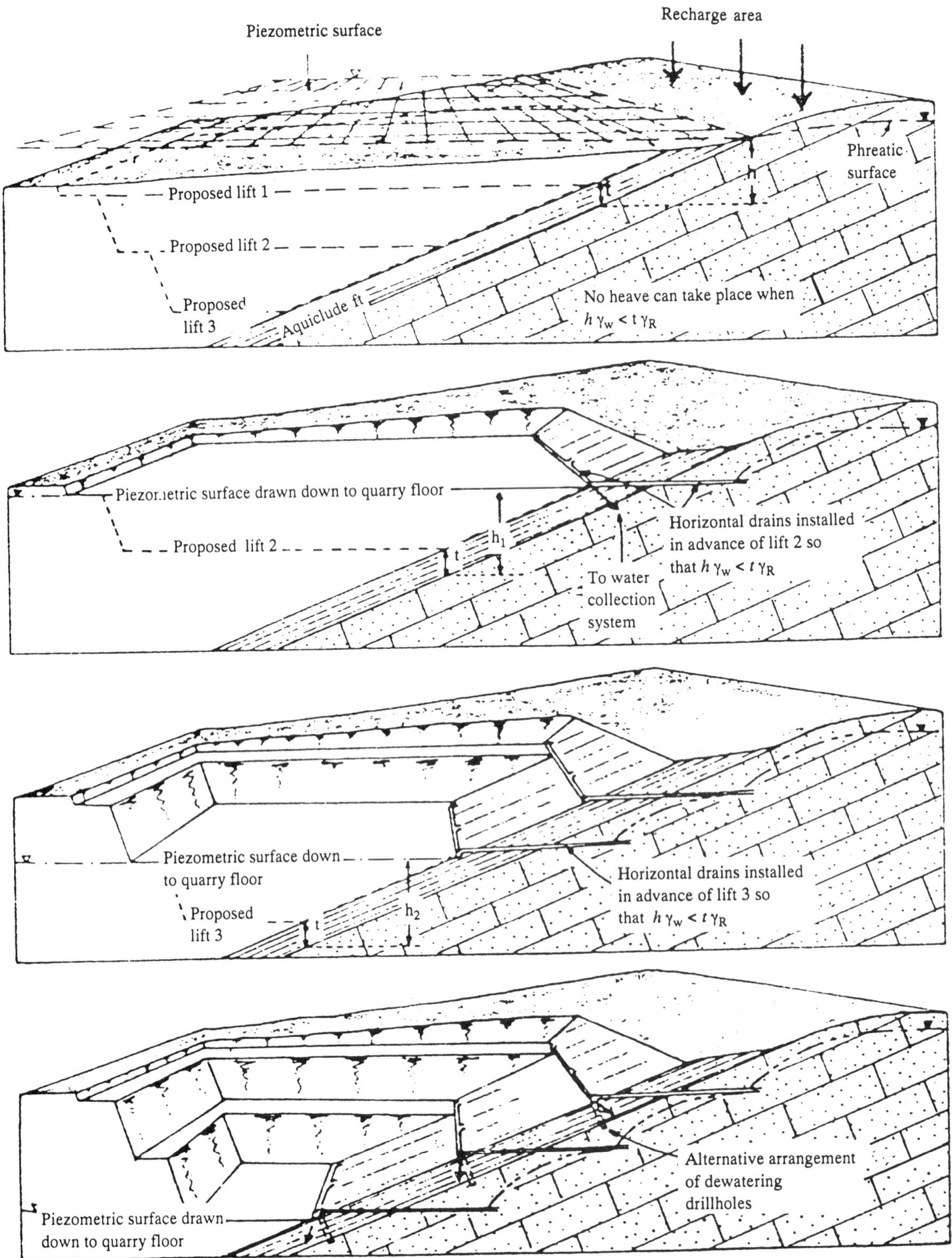

Figure 21.2 *Dewatering in open-cast mines and quarries. Progressive dewatering (a) of confined aquifers below steeply dipping footwalls (γ_w = unit weight of water; γ_R = unit weight of rock); (b) using vertical wells; (c) using horizontal drains (after Anon 1987)*

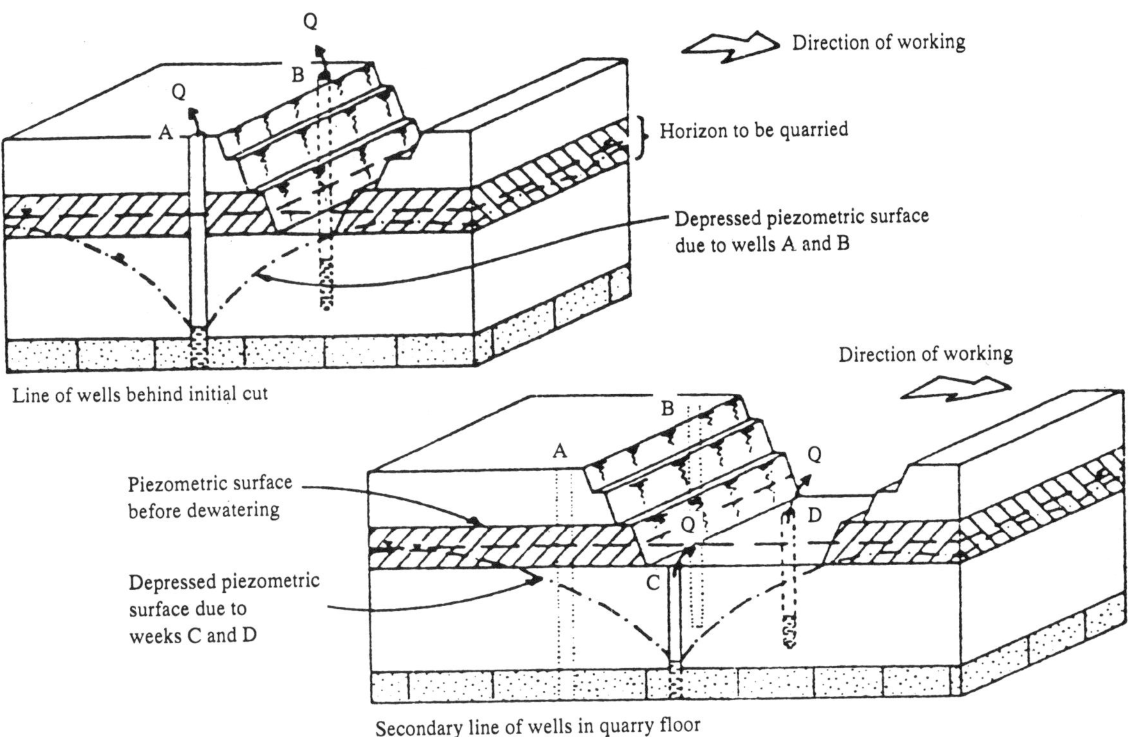

(b) Progressive dewatering using vertical wells

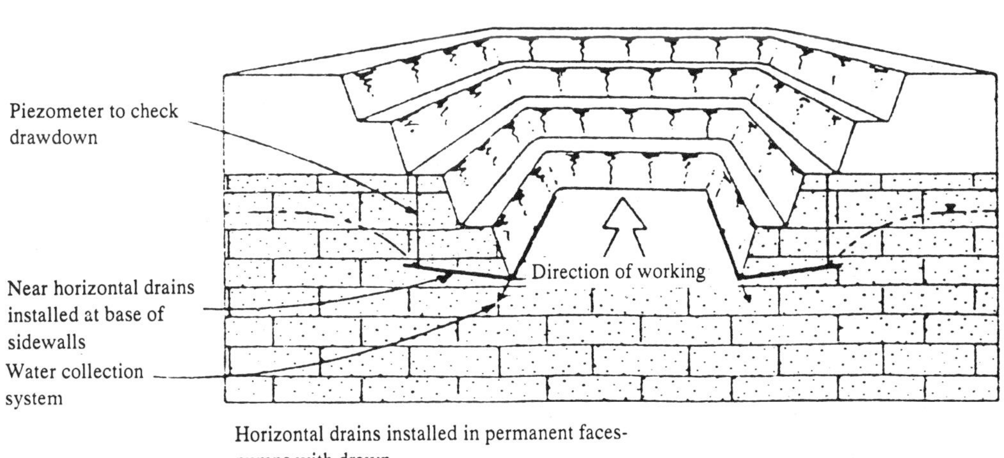

(c) Progressive dewatering using horizontal drains

Figure 21.2 *Continued*

bit down the hole. Generally, compressed air is used with a down-the-hole hammer, the air pressure being around 1.7 MPa, but in some cases pressures of up to 2.8 MPa have been used. High operating pressures, however, lead to heavy abrasive wear and impact damage to the bit.

Where the depth of excavation does not exceed 3.5 m, the use of jack-hammers to drill holes achieves good results. The holes are normally 38 mm in diameter. Mounted drilling machines are required for deeper excavations. Holes 50–76 mm diameter are suitable for depths between 3.5 and 9 m.

Holes with a diameter of 100 mm cater for all normal requirements. As the diameter of blastholes is increased, so the charge is increased. Ground vibrations may present a problem where charges are fired. However, vibrations generally can be reduced to an acceptable level by, for example, using short-delay blasting. Also as the diameter of a blasthole increases, the burdens and spacings are increased, so that the discontinuities in the rock mass become more significant in terms of fragmentation.

Depending on the pattern of holes and firing technique. subgrade drilling, that is drilling blastholes some 300 to 900 mm below excavation grade, is usually necessary to ensure that the area being blasted is broken down to grade (Anon 1972). Instantaneous blasts usually require greater subgrade drilling than short-delay blasts. It is only when there is a notable parting, such as a prominent bedding plane which is coincident with the grade of the excavation, that subgrade drilling is unnecessary.

The dip of a drill hole can be measured with an inclinometer. If the deviation is excessive, then this will entail redrilling. Generally speaking, the possibility of deviation tends to increase with distance from the top of the hole and its incidence tends to be greater in angled than in vertical holes.

21.3.2 Explosives for blasting operations

The essential characteristic of an explosive material is that, on initiation, it reacts suddenly to form large volumes of gases at high temperatures, the almost instantaneous release of these gases giving rise to very high pressures (Anon 1972). This reaction proceeds very rapidly and is self-sustaining in that it continues throughout the explosive when it is set off at any point within it.

Initiating explosives are used to detonate high explosives and, as such, they are extremely sensitive and relatively easy to explode. Small quantities of initiating explosives are contained in copper or aluminium tubes, thereby forming detonators, when ignited they produce an intense local shock which starts the reaction in less-sensitive high explosives. Initiating explosives do not necessarily produce large volumes of gases and alone are not suitable for blasting operations.

High explosives detonate at velocities from about 1500 to 7500 m/s, depending upon the explosive composition, and large volumes of gases are formed at exceptionally high pressures. The reaction is started and sustained by a shock wave initiated by a detonator. The rate at which the detonation wave travels through a column of explosive is termed the velocity of detonation. The shock wave developed by the explosive is transmitted through and produces fractures in the rock mass concerned. The movement of the high pressure gases through the rock mass completes its breakage.

The performance of a high explosive depends on the velocity of detonation and on the volume and temperature of the gases which form. Its power is governed by the amount of energy released when it is detonated.

Although ammonium nitrate fuel oil (ANFO) mixtures have proved popular in quarrying and open-cast mining, they suffer several disadvantages. For example, ANFO has a low bulk strength and a low density. It also has a low velocity of detonation and is not water resistant. These limitations are to some extent compensated for by its low cost per unit weight and high order of safety. Where ANFO mixtures have to be packed in order to provide water resistance, the addition of metallized powder gives extra density as well as energy. Sandy (1989) described the use of ANFO at Rössing Mine, Namibia. In the case of dry holes bulk explosive trucks were used to mix ammonium nitrate with fuel oil and load the explosive down the hole. Wet holes were charged with slurry which was almost twice as expensive as ANFO. Blastholes were dewatered by pump and plastic dry-liner sleeves were inserted. These, however, were not always successful as occasionally the sleeve would tear causing water to dissolve the AN prill, giving coarse fragmentation on blasting. At times the detonating cord downlines snapped under the strain of the slumping explosive column.

Digging conditions at the tow of a blast can be improved by increasing the bulk strength of ANFO through the addition of aluminium fines to the toe charge of the explosive column. The use of aluminized ANFO (ALANFO) in the main charge allows the drill pattern to be enlarged. Sandy (1989) found that the optimum composition of ALANFO explosive was 4% by weight of aluminium in the main charge and 8% in the toe charge for dry holes and 9% aluminized slurry for wet holes.

Heavy ANFO has a higher bulk strength and therefore blastholes can be positioned further apart. In addition its water-resistant properties mean that it can be used instead of more expensive slurry for charging wet holes. Heavy ANFO has two main components, an oxidizing agent (porous ammonium nitrate prill) and a dense high-energy fuel (HEF). The latter is an emulsion made from a hot solution of ammonium nitrate and calcium nitrate, which then is mixed with diesel and patented emulsifier. Prill and HEF are loaded separately into a bulk explosives truck and are mixed during discharge into the blasthole. A range of products of different density can be produced by varying the amount of HEF mixed with ammonium nitrate. The proportion of HEF also affects the water resistance of the product (Table 21.1). As the sensitivity of all these products is extremely low, they require a heavy boost to initiate detonation.

The density of an explosive is dependent upon its composition. If it has a high density, then the energy of the

Table 21.1 The relative properties of various HEF products (after Sandy 1989)

Product	*Proportion of HEF (%)*	*Density* (tonne/m^3)	*Weight-strength index*	*Bulk-strength index*	*Water resistance*
HEF 202	20	1.01	1.02	1.23	None
HEF 203	30	1.15	1.03	1.41	Poor
HEF 204	40	1.25	1.03	1.52	Slight
HEF 205	50	1.33	1.04	1.64	Good
HEF 275	57	1.36	1.00	1.62	Waterproof

explosion is concentrated. On the other hand, a low-density explosive can be used when excessive fragmentation has to be avoided, since a low explosive distributes the energy more evenly along the blasthole. Slurry explosives have a similar range of density to conventional explosives. However, as they are designed to fill a blasthole, they afford a higher effective concentration of energy than normal cartridge explosives used under the same conditions. On the other hand, ammonium nitrate explosives possess a lower density than many of the conventional explosives. Nonetheless, as they also can be used to fill a blasthole completely, they therefore can produce a similar concentration of energy to cartridge explosives when employed in the same situations.

If water is present in a drillhole but the time involved in loading and firing is limited, then an explosive which possesses a good water resistance proves satisfactory. However, a very good to excellent water resistance is required of an explosive if the exposure of water is prolonged. Usually the best water resistance is provided by blasting gelatines. The higher-density dynamites possess fair to good resistance, whereas low-density dynamites offer little or none.

An extensive range of explosive types and grades is available to satisfy all blasting requirements (Anon 1972). The gelatine group of explosives is unrivalled as far as use in the hardest rocks such as dolerite, basalt and granite are concerned. They can be used in wet conditions because of their high resistance to water. In addition, because they have a relatively high density they can be used where a powerful concentration of explosive energy is required.

Open-cast gelignite is a powerful high-density explosive with good water resistance. It was developed for blasting operations in open-cast coal mining.

The powder explosives have low densities and are of relatively high weight strength. They are especially suitable for use in soft to medium strength rocks which are moderately dry.

The ammon dynamites have lower densities than the gelatines of the same strength grade. Because they have good spreading action they are suitable for blasting conditions which are not too severe. They have satisfactory water-resistant properties.

Belex explosives are characterized by high weight strength but with decreasing density. Hence they represent explosives with varying bulk strengths from which a particular type may be chosen for certain given conditions. The lower density of these explosives can provide advantages as far as fragmentation is concerned. This is because the explosive charge is spread over a greater length of the blasthole and, when compared with an equal weight of gelatine explosive, a larger area of the rock mass comes into direct contact with the explosive charge. Belex explosives possess good water resistance.

Trimonite powders are medium-density explosives which must be used in dry conditions since they are not moisture resistant. However, if they are packed in sealed containers they may be used in wet conditions.

Blasting agents can be used in open-cast mining and quarrying where large-diameter drilling is employed for overburden blasting. Nobelite and Anobel are common types. Because blasting agents are insensitive, they are very safe to handle. They are not sensitive to detonators or detonating fuse and are therefore initiated by primer cartridges of a conventional high explosive.

Like ANFO, slurry explosives can be pre-packed or mixed on site. However, they involve a higher initial explosive cost than does ANFO. Water is an essential ingredient in slurry explosives, since it makes a solution which turns into a gel and holds solids in suspension. The gel provides the water resistance.

Special gelatine (80% strength) and open-cast gelignite are recommended for most types of excavation (Anon 1972). They are medium-strength nitroglycerine gelatines. Because they possess good water resistance, they offer reliable performance in wet conditions. In addition they have a high velocity of detonation and accordingly produce good fragmentation. Anon (1972) recorded that where exceptionally good fragmentation is required the high-strength nitroglycerine gelatine, special gelatine 90%, should be used. This explosive has excellent water resistance.

Charges are decked by separating zones of the explosive column by using inert materials, that is, stemming. The peak blasthole pressure within each charge deck is not reduced but the rate of decay of pressure can be increased appreciably (Hagan 1979). Because of availability, drill-hole chippings are most frequently used as stemming. In dry blastholes, however, angular crushed rock (around 15–25 mm in size for holes with diameters between 225 and 380 mm) is better than chippings since it possesses a higher effective (air) void ratio, thereby increasing the rate of decay of pressure within the adjacent charge deck. But even crushed rock is not as good as air decks in this respect. These are constructed by locating closed rigid empty cylinders between the charges in the blasthole.

Plain detonators and safety fuse, electric detonators (instantaneous and short delay), Cordtex detonating fuse and detonating relays may be used for initiation. Because remote control is easy, electric shotfiring is used for most operations.

Short-delay firing is recommended for most excavation work. This is achieved by using either short-delay electric detonators or Cordtex and detonating relays. The former have a nominal delay interval between consecutive numbers of 22 ms in the early numbers with a slight increase in the later numbers. As far as detonating-relays are concerned, a range of five relays with delay times of 10–45 ms is available. The introduction of short delays yields the highest efficiency from the explosive, giving maximum rock fragmentation and minimum ground vibration.

21.3.3 Blasting operations

As Hagan (1986) pointed out, blasting can affect drilling. When a blast causes considerable overbreak, for example, the mean inclination of the newly created face is often so small that the toe burden for front-row vertical blastholes in the next blast is excessive. These front-row charges then fail to perform properly, and blasting results are sub-optimal. In any blast, the front-row blastholes are the most important. This is because they are most prone to errors in burden distance (especially at bench floor level) and, to a lesser extent, blasthole spacing. Therefore, every possible step should be taken to ensure that overbreak does not cause collaring errors, blocked blastholes or the redrilling of caved blastholes.

If a rock is to be blasted efficiently it must be capable of transmitting the explosive energy some distance from the blastholes. When this does not happen, then the rock immediately surrounding the hole is pulverized and the area between holes is not fractured. The early movement of the rock face is most important for an efficient blast, that is, one which achieves good fragmentation per drill hole with minimum quantity of explosive. Rock breakage in blasting, apart from the character of the rock itself, especially the fracture index, depends largely on the relation between the burden and the hole spacing and also on the time of ignition between the holes.

Obviously efficient blasting should produce rock fragments sufficiently reduced in size so that they can be easily loaded without resort to secondary breakage. Accordingly blastholes should be drilled accurately to the requisite pattern and proper depth to ensure satisifactory fragmentation. The greater the capacity of the buckets of the loading machines, the larger the fragment size that is acceptable. Good fragmentation also reduces wear and tear on loading machinery.

The quantity of a particular explosive required to blast a certain volume of rock is difficult to estimate since it depends upon the strength, toughness and incidence of discontinuities within the rock. Consequently no rigid sequence for progressive resistance to blasting offered by different rock types can be formulated. Nevertheless it can be said that an explosive with the greatest energy and concentration is required for removing very hard rock; in medium hard rocks a high-velocity detonation produces a shattering effect; a medium to high explosive can be used in medium to hard laminated rocks; and the greatest efficiency is obtained with fairly bulky explosive in soft to medium rocks. *Table 21.2* gives an idea of the amount of charge in relation to burdens for primary blasting.

Table 21.2 Typical charges and burdens for primary blasting by shot-hole methods (adapted from Sinclair 1969)

Minimum finishing diameter of hole (mm)	*Cartridge diameter* (mm)	*Depth of hole* (m)	*Burden* (m)	*Spacing* (m)	*Explosive charge* (kg)	*Rock yield* (tonnes)	*Blasting ratio*	*Tonnes of rock per metre of drillhole*
25	22	1.5	0.9	0.9	0.3	3	10.0	2
35	32	3.0	1.5	1.5	1.8	19	10.5	6.3
57	50	6.1	2.4	2.4	9.5	97	10.2	15.9
75	64	9.1	2.7	2.7	18.0	180	10.0	19.8
75	64	12.2	2.7	2.7	25.0	245	9.8	20.1
75	64	18.3	2.7	2.7	36.3	365	10.1	20.0
100	83	12.2	3.7	3.7	43.0	430	10.0	35.3
100	83	30.5	3.7	3.7	104.3	1120	10.7	36.7
170	150	18.3	6.7	6.7	216.5	2235	10.3	122.1
170	150	30.5	6.7	6.7	363.0	3660	10.1	120.0
230	200	21.3	7.6	7.6	329.0	3350	10.2	157.3
230	200	30.5	7.6	7.6	476.3	4670	9.8	153.1

Roberts (1981) suggested the following expression for obtaining the thickness of the burden (B), in metres:

$$B = 0.024d + 0.85 \quad (21.1)$$

where d is the diameter of the blasthole. Vutukuri and Bhandari (1961) suggested that the spacing (S), in metres, between blastholes could be derived from the expression

$$S = 0.9\mathrm{B} + 0.91 \quad (21.2)$$

However, careful trials provide the only means of determining the burden and blasting pattern in any rock. As a rule the spacing of blastholes will vary between 0.75 and 1.25 times the burden.

Calculation of the burden and charge must also take account of the toe section since this is confined at the back and base. In other words, there needs to be a higher concentration of charge at the base than in the column, whilst at the top of the hole no charging is required. Hence the concentration of explosive in the bottom of the hole generally is approximately 2.5 times greater than in the column. The basal charge may be regarded as extending from the bottom of he hole to a point above the floor level equal to the burden. The column charge extends from there to a point below the crest equal to the burden.

Changes in blasthole diameter usually necessitate changes in burden distance, blasthole spacing and stemming length. The blasthole diameter is governed by the properties of the rock mass being blasted, the degree of fragmentation required, and the bench height. If the blasthole diameter is too small, the costs of drilling, priming and initiation are too high, and charging, stemming and connecting-up operations are too time consuming. On the other hand if the blasthole diameter is too large, then excessively large blasthole patterns give inadequate fragmentation, especially in rock masses which contain widely spaced open discontinuities. An increase in blasthole diameter normally has to be accompanied by an increase in energy factor for the degree of fragmentation to remain unchanged, the increase being greatest for blocky and least for highly fissured rock masses.

Blasting of two or more dissimilar rock masses frequently occurs in open excavation. Where it is impractical to select a different blasthole diameter for each rock mass a compromise value of blasthole diameter has to be chosen. Inevitably, this will be sub-optimum for at least one of the rock types. According to Hagan (1986) in blocky areas, blasthole diameter may be so unsuitable that it requires the use of pocket charges (Figure 21.3).

With vertical front-row blastholes, the burden distance generally increases from the top to the bottom of the face, particularly when the face is high and/or shallow dipping. In such instances front-row blastholes often are collared as close to the crest as safety permits, in order to provide toe burden resembling the design toe burden. However, the top of a charge of normal length then is unburdened, which frequently means that explosive gases burst from the upper face, with resultant high levels of airblast and/or flyrock. This, in turn, reduces blasthole pressure at bench floor level and makes it more difficult to break and move the toe.

Vertical hole blasting generally is preferred when blasting to a free face but in special circumstances horizontal holes may be used. For example, horizontal blastholes (or angled holes drilled from the bottom of the face) are frequently used when the ground surface makes access for drilling equipment difficult. In such cases horizontal holes can be used to prepare the surface so that access is provided for drilling vertical holes in the main excavation. Horizontal drilling sometimes is used when excavating steeply dipping rocks but generally it is undesirable since it can lead to dangerous overhangs and less stable faces.

Inclined blastholes are more difficult to drill, but are very effective in elminating excessive front-row toe burdens. Indeed the success of a multi-row blasting depends largely on the ability of front-row charges to heave their burden forwards. If front-row charges fail to displace their burden, progressive relief is not achieved and the blast never recovers, irrespective of the number of rows of blastholes.

Inclined blastholes also yield good fragmentation, displacement and muckpile looseness. Fragmentation usually improves as a result of the reduction in premature venting of explosive gases to atmosphere from alongside the tops of front-row charges, and to faster movement of the front-row toe, and hence improved progressive relief of burden. Inclined blastholes also reduce disruption of, and drilling difficulties in, the bench beneath, and produce smoother and more stable faces. They reduce the tendency for stumps being left at the base of the excavation after blasting. Stumps generally occur in hard rock when the bottom charge generates insufficient energy at the toe of the hole. For example, if vertical holes are drilled to grade only, then a large proportion of the blast is dissipated as compression pulses that are absorbed by the bedrock. Hence the remaining tension energy component may be insufficient to break out to a level base, so stumps are left at the toe. Accordingly blastholes should be taken below grade level.

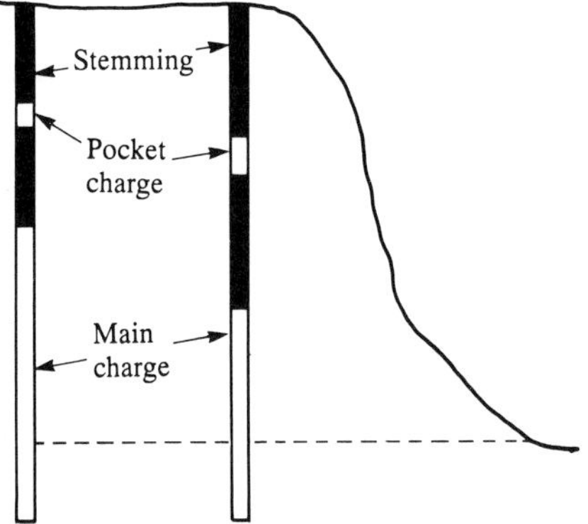

Figure 21.3 *Use of pocket charges to assist in breaking collar rock (after Hagan 1986)*

If improved fragmentation is not required, then inclined blastholes permit slightly larger blasthole patterns and lower energy factors to be used. Moreover inclined blastholes cause less surface overbreak (through cratering behind the collars of blastholes) and therefore allow the use of longer inter-row delays, which give greater progressive relief of burden and hence better fragmentation and muckpile looseness.

In order to obtain good fragmentation and thereby ease loading operations, the explosive consumption in excavation is somewhat greater than in quarrying. When firing is confined to a single row of blastholes in soft laminated strata, the charging ratios may be as low as 0.15–0.25 kg/m^3. In harder sedimentary strata the charging ratios generally are around 0.45 kg/m^3 while they may be about 0.6 kg/m^3 in jointed igneous rocks. Even higher charging ratios have been necessary in order to obtain satisfactory results in some metamorphic rocks such as mica schists, which absorb much of the energy of blasts. Generally 1 kg of high explosive will bring down about 8–12 tonnes of rock.

The strain produced by explosion on the equatorial plane of a clyindrical charge increases as the length:diameter ratio of the charge increases in the approximate range 0–20. Strain remains constant when length:diameter ratio exceeds about 20. When it decreases below about 20 the optimum burden distance for the charge decreases.

Staggered blasthole patterns are more effective than square or rectangular patterns. Hagan (1986) maintained that in hard rocks blastholes should be drilled on equilateral triangular grids, since these provide the optimum two-dimensional distribution of energy within the rock mass, whilst allowing a high degree of flexibility in initiation sequence (and consequently direction of firing). In other words, for a given cost of drilling and blasting, equilateral triangular patterns produce the best fragmentation. However, when blasting weak rocks, good fragmentation is readily achieved by using square or rectangular patterns.

Initiation should commence at that point in a blast which gives best possible progressive relief for the maximum number of blastholes. If there is no free end, initiation should commence near, but not at, one end of the blast block (Figure 21.4a). This produces less overbreak and lower ground vibrations. If there is a free end, initiation should commence at the end (Figure 21.4b). If this end is choked by muck broken by a previous blast, initiation should commence near that end of the blast which is remote from the buffer (Figure 21.4c).

The interaction between adjoining blastholes when short-delay firing is used is an important factor in the design of the blast pattern. Hence the delay interval between blastholes should allow sufficient time between shots for the fracture process to be completed. In other words, the time factor must provide for the travel of compressive and tensile waves through the burden, for the adjustment of the ambient stress field and for the movement of rock fragments. A cratering type of blast is produced if the delay time is too short. Roberts (1981)

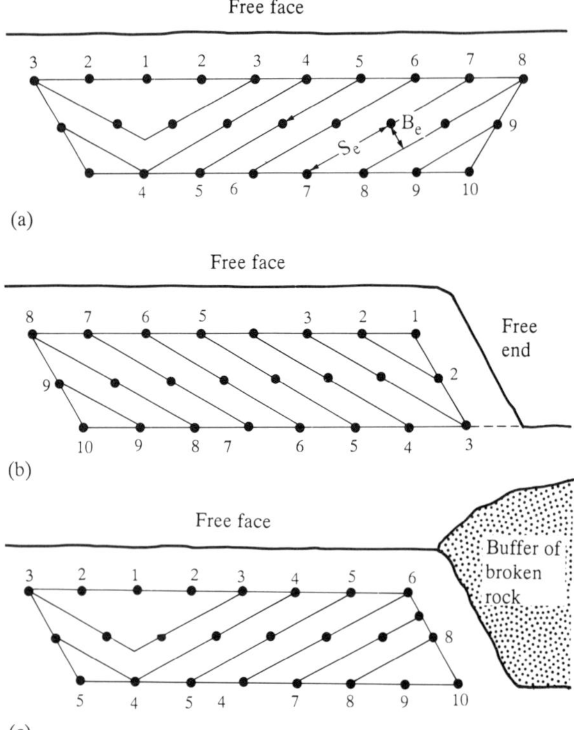

Figure 21.4 *Staggered blasting to (a) a free face; (b) a free end; (c) a buffered end (after Hagan 1986)*

therefore suggested that the minimum delay should not be less than 10 to 20 ms or 1.5 to 30 ms/m of burden. The maximum delay time is that required for optimum movement of rock frangments and is around 30 to 50 ms or 4.5 to 7.5 ms/m of burden.

One of the fundamental requirements of blasting is that a free face should exist in front of the blastholes when they are fired. When an excavation has to be made into solid rock from the surface, then the rock is broken into so as to provide free faces at which subsequent blasting can take place. Because there is no free face the blast must occur upwards from the ground surface. For efficient operation, the width of the cut should be at least twice the depth and the area broken by each charge should be in the form of a cone with its apex below the bottom of the cut. Hence there are a large number of holes drilled to overlap below grade level. The spacing of blastholes is often about half the depth to grade and nearly double the quantity of explosive is needed as is used in blasting a free face in the same rock.

The blasting in a sinking cut can be done by instantaneous shots or by short-delay firing. The former method can be used where excessive ground vibration, air-blast, fly-rock and overbreak are not likley to constitute problems. Either instantaneous electric detonators or Cordtex detonating fuse can be used for initiation. A comparatively large number of holes are fired simultaneously and the spacing of blastholes, subgrade drilling and width of burden should ensure that breakage occurs to the required depth. Moreover the blastholes must be well stemmed. There may be inadequate fragmentation if hard bands of rock occur in the upper part of the excavation. Line drilling is required around the perimeter of the excavation in order to prevent excessive overbreak if such a blast is fired over the entire excavation area.

Figure 21.5 illustrates a typical pattern of blastholes for a short-delay sinking cut. With such an arrangement the zero delay shots fracture the central area producing an upward movement of rock. This then provides sufficient relief of burden to permit succeeding shots to blast towards the central area.

6 4 3 2 2 2 3 4 6
4 3 2 1 1 1 2 3 4
3 2 1 0 0 0 1 2 3
3 2 1 0 0 0 1 2 3
4 3 2 1 1 1 2 3 4
6 4 3 2 2 2 3 4 6

Figure 21.5 *Typical pattern of blastholes for a short-delay sinking cut (after Anon 1972)*

In large excavations, additional blasting is necessary to expand the preliminary excavation area. These succeeding blasts can be regarded as firing to a free face and can be fired either before or after the rock from a sinking cut has been removed. In the former case, although broken rock remains in the area initially blasted, the burden has been sufficiently relieved to allow satisfactory performance of succeeding blasts. The sinking cut need not necessarily be located at the centre of the excavation and especially where buidings are in close proximity, the initial cut should be located as far away from them as possible.

Two of the principal methods of drilling and blasting used in large excavations are benching and well-hole blasting. As the name suggests, in benching the face is worked in a series of levels or benches, the maximum height of each bench generally not exceeding 10 m. Individual benches can be worked simultaneously, the rows of blastholes being drilled parallel to the free face. When fired, a row of holes forms a new free face for the succeeding row. Single or multiple rows of holes may be used.

Single face with well-hole blasting is commonly used for working large excavations, indeed it is the standard method for working quarries. Here a line of holes is drilled parallel to the face and they usually are inclined, often between 10° and 15°, for safety reasons. Sometimes multiple rows of holes may be used. The height of the face brought down is greater than that in benching and tends to vary between 15 and 35 m.

Under certain conditions heading blasting may be used. Small tunnels are driven into the base of the face and cross headings are driven from them. Chambers are excavated at intervals and filled with explosive. The headings are then backfilled. This is time consuming and there is little control over fragmentation.

The drilling pattern adopted for excavating trenches in rock depends upon their width. In wide trenches the ground is first opened up with a pattern of angled holes drilled to successively greater depths, the object being to develop a free face for benching. Conversely in narrow trenches the rock may be blasted free with holes drilled along the edges of the excavation. The simultaneous detonation of charges produces a linear crater.

When the explosive in a blasthole is fired it is transformed into a gas, the pressure of which may sometimes exceed 100 000 atmospheres* (Langefors and Kihlström 1962). The tremendous energy liberated shatters a zone around the blasthole and exposes the rock beyond to enormous tensile stresses. This takes place under the influence of shock waves which radiate from the explosion at between 3000 and 5000 m/s. A zone of intense deformation occurs about the blasthole, its thickness frequently approximating to the diameter of the hole, and radial cracking extends appreciably further (Persson *et al.*, 1970).

Under the influence of the pressure of gases from the explosive the primary radial cracks expand and the free face yields and is moved forwards. When this occurs the pressure is unloaded and tension increases in the primary cracks which incline obliquely forward. If the burden is not too great several of these cracks expand to the exposed surface and the rock is completely loosened. In this way the burden is torn from the rock mass. The lateral pressure in the shock wave is initially positive but falls rapidly to negative values implying a change from compression to tension. Hence the area beyond the hole is exposed to vast tangential stresses. The shock wave itself is not responsible for breakage of rock but it does provide the basic conditions for the process. Shock wave energy when using high explosive may only account for 5 to 15% of the total energy liberated.

The discontinuities within a rock mass act as free surfaces which reflect shock waves generated by an explosion. They also provide paths of escape along which energy is dissipated. The geometry of the discontinuity pattern is very important since the greatest loss of energy occurs where most discontinuities intersect. At such

* 1 atmosphere = 100 kPa

locations the explosive energy opens existing breaks in the rock but generates few new ones. Secondary blasting is therefore required to break large masses of rock which have only been loosened.

21.3.4 Controlled blasting

In many excavations it is important to keep overbreak to a minimum. Apart from the cost of removing extra material which then has to be replaced, damage to rock forming the walls or floor may lower the bearing capacity and necessitate further excavation. Also smooth faces allow excavation closer to the payline and are more stable.

The properties of the rock mass have a very great influence on the amount of overbreak resulting from a blast. The most important of these properties are the *in situ* dynamic tensile breaking strain and, more particularly, the nature, frequency, orientation and continuity of structural features, notably discontinuities.

The type of explosive and its density also affect the amount of overbreak. A reduction in the density of an explosive produces a significant reduction in the velocity of detonation. For instance, when the density of ANFO is reduced, typically by the addition of polystyrene beads, the peak blasthole pressure amounts to a small fraction of its initial value. Accordingly, when multi-row blasting is undertaken and overbreak has to be minimized, such low-density mixtures can be used in back-row blastholes.

Overbreak also can be reduced by decoupling and/or decking back-row charges (Hagan 1979). In both cases the charge weight in the back-row blastholes is lowered so as to reduce the strain wave energy, density and volume of explosion gases. However, less overbreak can only be guaranteed when the effective overburden on the back row decreases in proportion to the lower energy yield per blasthole. Charges are decoupled where the charge diameter is less than that of the blasthole. In particular, decoupling provides a means by which the amount of overbreak can be reduced in densely fissured rocks. Furthermore the use of highly decoupled charges in the upper sections of back-row blastholes brings about a large reduction in the amount of crest damage.

Starfield (1966) indicated that the replacement of end-initiated by central-initiated charges produces an increase of around 37% in peak strain in the rock mass. Consequently minimization of overbreak, and of ground vibrations, are achieved by individual charges being initiated at one end (preferably the bottom) rather than at the centre. Both overbreak and ground vibrations are increased by multiple-primed charges.

Line drilling is most commonly used to improve the peripheral shaping of excavations. It consists of accurately drilling alternate small-diameter holes between the pattern blastholes forming the edge of the excavation. The quantity of explosives placed in each line hole is significantly smaller and, indeed, if these holes are closely spaced, from 150 to 250 mm, then explosive may be placed only in every second or third hole. The closeness of the holes depends upon the type of rock being excavated and on the payline. These holes are timed to fire ahead, with or after the nearest normally charged holes of the blasting pattern. The time of firing is largely dependent on the character of the rock involved. Line drilling is not always successful in preventing overbreak although it helps to reduce it. Generally, line holes in sedimentary and some metamorphic rocks are not as effective as in igneous rocks.

Fissile rocks such as slates, phyllites and schists tend to split along the planes of cleavage or schistosity if these run at a low angle to the required face. Paine *et al.*, (1961) suggested that the raggedness of faces in sedimentary and metamorphic rocks may occur more readily when the blastholes are fired from 1 to 2 m away from the line holes. Compressive strain pulses moving out from the blastholes are reflected from joint, cleavage, schistosity and bedding planes and the reflected tensile strain pulses cause rupturing of the rock beyond the holes. On the other hand, homogeneous igneous rocks sometimes split quite readily from hole to hole when line drilled, because the major free surfaces are the holes and not a series of natural shear planes.

Cushion blasting is a type of line drilling where large holes, about 170 mm diameter, are located between the line holes to act as guides to the crack direction. Frequently one large hole to three small holes are used with either all three small holes or only the large central one being charged. The holes are loaded with light charges at intervals separated by stemming. The trimming holes are fired after the main blast. Cushion blasting often gives better results than line drilling.

Pre-splitting can be defined as the establishment of a free surface or a shear plane in rock by the controlled usage of explosives in approximately aligned and spaced drillholes. A line of trimming holes around the perimeter of the excavation is charged and fired to produce a shear plane. This acts as a limiting plane for the blast proper and is carried out prior to the blasting of the main round inside the proposed break lines. The spacing of the trimming holes is governed by the type of rock and the diameter of the holes. The diameter of the trimming holes is smaller than that of the main blastholes in order to avoid the extra radial cracking of the walls that would be produced by large charges. Continuously loaded, moderately coupled charges are recommended.

Pre-splitting also requires accurate drilling. Air decking is sometimes used, that is, the charges are separated by spacers instead of stemming. The air space provides effective decoupling and damps the shock wave transmitted to the rock mass. Nevertheless this is time consuming and can prove troublesome. It is not recommended when ANFO mixtures are used. In weak rock alternate advance pre-split holes may be left uncharged. The uncharged holes form relief holes which guide the shear fracture along the required plane of separation. Maximum effectiveness is achieved when the pre-split holes are fired simultaneously. A trial should be made to determine whether the site conditions require modifications in the hole spacing or charging prior to extensive pre-splitting taking place.

In most rocks a shear plane can be induced to the bottom of the trimming holes, that is, to base level, but in very tight unfissured rocks difficulty may be experienced in breaking out the main blast to base level. In such instances the spacing between the outer blastholes and the shear plane may need to be reduced by 50–75%. Once pre-split the rock can be blasted with a normal pattern of holes. If a rock mass is heavily fractured the trimming and primary blastholes can be fired together, delays causing the latter to fire immediately after the former.

21.3.5 Blasting and vibrations

When excavation in rock has to be carried out in urban areas the proximity of existing structures may determine whether or not blasting is used, since noise, air concussion and ground movement may cause inconvenience or the latter may even damage structures. Experience has shown that the great majority of complaints and even law suits against blasting operations are due to irritation and that subjective response to vibrations normally leads a person to react strongly long before there is any likelihood of damage occurring to his property (Figure 21.6). The duration of the operation and the frequency of occurrence of the blasts are almost as important as the level of the physical effects.

The three most commonly derived quantities relating to vibration are amplitude, particle velocity and acceleration. Of the three, particle velocity appears to be the one most closely related to damage in the frequency range of typical blasting vibrations.

Edwards and Northwood (1960) defined three categories of damage attributable to vibrations:

(1) *Threshold damage*: widening of old cracks and the formation of new ones in plaster, dislodgement of loose objects.
(2) *Minor damage*: damage does not affect the strength of the structure, it includes broken windows, loosened or fallen plaster, hairline cracks in masonry.
(3) *Major damage*: damage seriously weakens the structure, it includes large cracks, shifting of the foundations and bearing walls, distortion of the superstructure caused by settlement and walls out of plumb.

They proposed that a vibration level with a peak particle velocity of 50 mm/s could be regarded as safe as far as structural damage was concerned, 50–100 mm/s would require caution, and above 100 mm/s would present a high probability of damage occurring (see also Figure 21.7). The US Bureau of Mines subsequently lent support to the idea that 50 mm/s provides a reasonable safety from the possibility of damage (Duvall and Fogelson 1962; Nicholls *et al.*, 1971). However, Oriard (1972) maintained that no single value of velocity, amplitude or acceleration could be used indiscriminately as a criterion for limiting blasting vibrations. In particular, low vibration levels may disturb

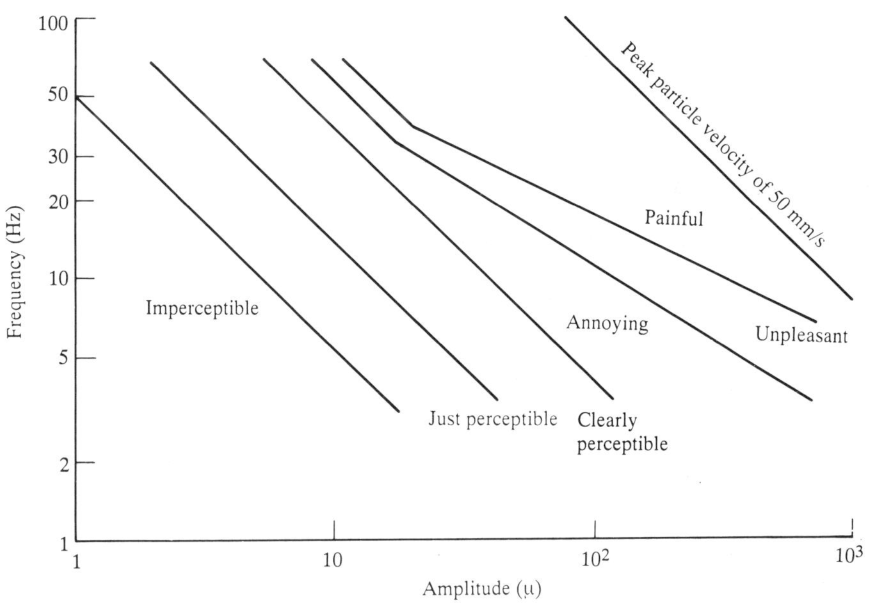

Figure 21.6 *Human sensitivity to vibration. If the peak particle velocity of 50 mm/s is taken as the threshold of damage, then the figure indicates that man reacts strongly long before there is any reason to apprehend damage to his property*

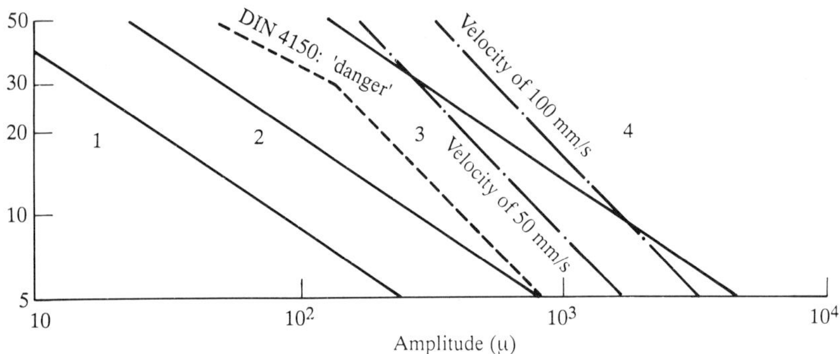

Figure 21.7 *Possible damage to buildings for frequencies between 5 and 50 Hz, the range most frequently encountered in buildings: 1 = no damage; 2 = possibility of cracks; 3 = possible damage to load-bearing structural units 4 = damage to load-bearing units*

sensitive machinery and in this case it is impossible to specify a limit of ground velocity; each instance should be separately assessed (Skipp and Taylor 1970).

Moreover, blasting vibrations at 50 mm/s particle velocity would, in terms of human response, be regarded as highly unpleasant or intolerable (Nicholls *et al.*, 1971). Indeed, Crandell (1949) had previously concluded that the average person would consider a vibration to be 'severe' at about one-fifth the level which might damage structures. Subsequently Roberts (1971) indicated that the threshold of subjective perception had been variously placed from as low as 0.5 mm/s to 10 mm/s. People will react more unfavourably to large-amplitude vibrations of long duration than to low-amplitude short-duration vibrations of the same intensity. It is likely that this sensitivity is increased in the low-frequency range 3–10 Hz.

When dealing with high levels of shock and vibration, the time history of the motion and the characteristic response of the structure concerned to the type of motion imposed become increasingly important. For instance, a structure with a slow response, such as a tall chimney, when subjected to vibrations with large amplitude, low frequency and long duration, would come closer to the resonant response of the structure and therefore this would be more dangerous than vibrations with small amplitude, high frequency and short duration, even though both may have the same acceleration or velocity. Because of the dependence of response on frequency, conservative limits should be accepted when applying single values of velocity or acceleration as criteria for different types of structures subjected to different kinds of motion.

Vibrations associated with blasting generally fall within the frequency range 5–60 Hz. The types of vibration depend on the size of the explosive charge, the volume of the ground set into vibration, the attenuation characteristics of the ground and the distance from the blast. A small explosive charge generates a low vibration with relatively high frequency and relatively low amplitude. By contrast, a large explosive charge produces a vibration with relatively low frequency and relatively high amplitude. The shock waves are attenuated with distance from the blast, the higher frequencies being maintained more readily in dense rock masses. In other words, these pulses are rapidly attenuated in unconsolidated deposits which are characterized by lower frequencies.

Vibrographs can be placed in locations considered susceptible to blast damage in order to monitor ground velocity. A record of the blasting effects compared with the size of the charge and the distance from the point of detonation is normally sufficient to reduce the possibility of damage to a minimum (Figure 21.8). A pre-blast survey informs the owners of buidings as to the condition of their buildings before the commencement of blasting and offers some protection to the contractor against unwarranted claims for damage.

The effects of vibration due to blasting operations can be reduced by:

(1) Time dispersion, which includes the use of delay intervals in ignition. Delay intervals mean that shock waves generated by individual blasts are mutually interfering, thereby reducing vibration. For instance, Duvall (1964) noted that a delay interval of around 9 ms reduced ground vibrations and Oriard (1972) suggested that even shorter delay intervals were important as far as the control of vibration and potential damage was concerned.
(2) Spatial dispersion, which involves the pattern and orientation of the blastholes.
(3) The way in which the charge is distributed in the blasthole, the diameter of the hole and the depth of lift.
(4) The confinement of the charge, which involves the type of decking and the powder factor as well as the width of burden and the spacing of the blastholes.

If the use of conventional explosives in urban areas is prohibited, then alternative methods of rock breakage must be used. These include hand-held pneumatic breakers or lorry-mounted diesel-powered breakers. Unfortunately compressed-air or diesel breakers are very noisy and cause vibrations. Consequently in some

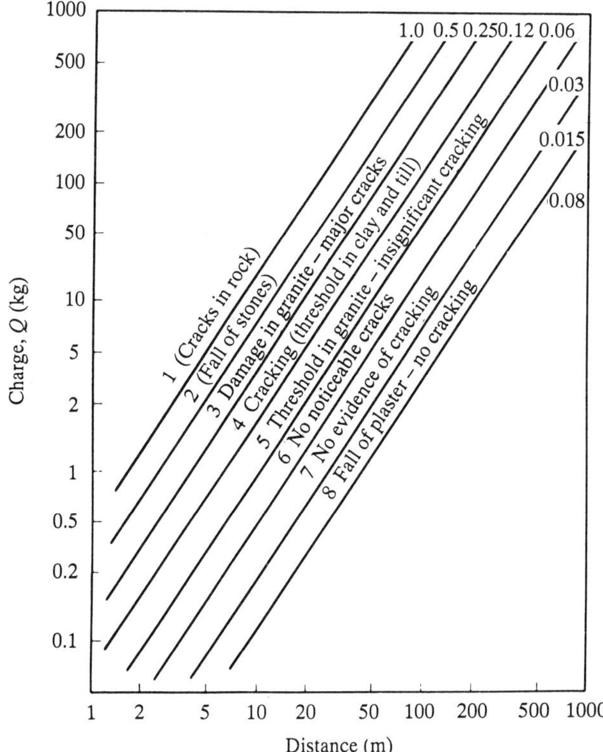

Figure 21.8 *Charge as a function of distance for various charge levels. Numbers 3–8 inclusive describe damage expected in normal houses (after Langefors and Kihlstrom 1962)*

situations the use of these tools is also excluded. In such instances the employment of hydraulic bursters has been found to be the most suitable method for producing large excavations.

A burster is essentially a multiple hydraulic jack with several circular rams operating from one side. A series of holes is drilled parallel to the rock face and one or more bursters inserted. Pressure is then exerted by the rams causing the rock to split along the line of holes. With holes about 150 mm diameter, going to a depth of 2–3 m, the burden and spacing of holes should be about 1 m. Obviously use is made of discontinuities, especially bedding planes, when this method is employed. The holes for the hydraulic burster can be drilled without much noise or vibration by rotary diamond core drills or by oxygen lance if no noise is permitted. Splitting can also be accomplished by freezing water in drillholes with the aid of liquid nitrogen.

21.4 Methods of excavation: ripping

Ripping is an inexpensive method of breaking discontinuous ground or soft rock masses, the fragmented material being removed by bulldozer. The geological factors which influence rippability in rock masses include rock type and fabric (e.g. coarse-grained rocks are easier to rip than fine-grained rocks), intact strength and degree of weathering, rock hardness and abrasiveness, and the nature, incidence and geometry of discontinuities (Table 21.3). The latter reduce the overall strength of the rock mass, their spacing governing the amount of such reduction. Obviously the continuity of discontinuities within a rock mass has a significant influence on its engineering character. The greater the amount of gouge or soft fill along discontinuities, the easier it is to penetrate the rock mass and so to rip it. In other words strong massive rocks and hard abrasive rocks do not lend themselves to ripping. On the other hand, if sedimentary rocks such as sandstone and limestone are well-bedded and jointed or if strong and weak rocks are thinly interbedded, then they can be excavated by ripping rather than by blasting (Figure 21.9). Indeed, some of the weaker

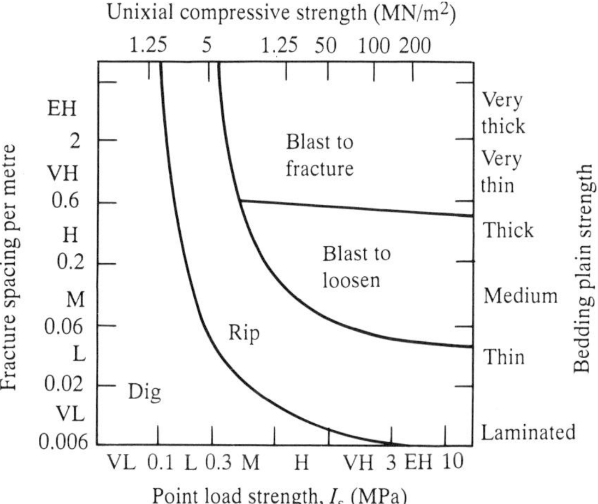

Figure 21.9 *Rock quality classification in relation to excavation (after Franklin et al., 1971)*

sedimentary rocks (less than 15 MPa compressive strength or 1 MPa tensile strength) such as mudstones are not as easily removed by blasting as their low strength would suggest. The reason for this is that they are pulverized in the immediate vicinity of the blasthole. What is more, blasted mudstones may lift along bedding planes to fall back when the gas pressure has been dissipated. Such rocks, especially if well jointed, are more suited to ripping.

Field examination of all available exposures in the vicinity of the site to be excavated along with studies of available geological maps and aerial photographs contribute data for a geological interpretation and the engineering application of the material at the site in question.

According to Atkinson (1970) the most common method for determining whether a rock mass is capable of

Table 21.3 Excavation characteristics

(a) In relation to rock hardness and strength

Rock hardness description	*Identification criteria*	*Unconfined compression strength* (MPa)	*Seismic wave velocity* (m/s)	*Excavation characteristics*
Very soft rock	Material crumbles under firm blows with sharp end of geological pick; can be peeled with a knife; too hard to cut a triaxial sample by hand. SPT will refuse. Pieces up to 3 cm thick can be broken by finger pressure.	1.7–3.0	450–1200	Easy ripping
Soft rock	Can just be scraped with a knife; indentations 1 mm to 3 mm show in the specimen with firm blows of the pick point; has dull sound under hammer.	3.0–10.0	1200–1500	Hard ripping
Hard rock	Cannot be scraped with a knife; hand specimen can be broken with pick with a single firm blow; rock rings under hammer.	10.0–20.0	1500–1850	Very hard ripping
Very hard rock	Hand specimen breaks with pick after more than one blow; rock rings under hammer.	20.0–70.0	1850–2150	Extremely hard ripping or blasting
Extremely hard rock	Specimen requires many blows with geological pick to break through intact material; rock rings under hammer.	>70.0	>2150	Blasting

(b) In relation to joint spacing

Joint spacing description	*Spacing of joints* (mm)	*Rock mass grading*	*Excavation characteristics*
Very close	<50	Crushed/shattered	Easy ripping
Close	50–300	Fractured	Hard ripping
Moderately close	300–1000	Block/seamy	Very hard ripping
Wide	1000–3000	Massive	Extremely hard ripping and blasting
Very wide	>3000	Solid/sound	Blasting

being ripped is seismic refraction. The seismic velocity of the rock concerned is compared with a chart of ripper performance based on ripping operations in a wide variety of rocks (Figure 21.10). In fact the limit of ripper operations can be regarded as a seismic velocity (V_p) of 2 km/s. However, the ground can be loosened by blasting if the strength of the rock mass is high or if ripper production has to be increased. Such a pre-blasting process uses light charges to open the discontinuities. Hydraulic fracturing has also been used to break surface rock prior to ripping.

Kirsten (1982), however, argued that seismic velocity could only provide a provisional indication of the way in which a rock mass could be excavated, pointing out that in terms of overall assessment seismic velocity cannot be determined to an accuracy better than about 20%. Furthermore seismic velocity may vary by as much as 1 km/s in apparently identical materials.

As far as the method of excavation is concerned, whether it be blasting, ripping or digging, its assessment should be made in such a way as to avoid any misinterpretation. Hence any classification related to excavation should be based on the fundamental properties of the rock mass concerned. A basic requirement of any classification system is that it should be applicable both to drill-hole core and field investigation alike. Two classifications systems have been developed in South Africa for the excavation of rock masses, one based on the geomechanics system of Bieniawski (1973) and the other on the Q system of Barton *et al.*, (1974).

Weaver (1975) proposed the use of a modified form of the geomechanics classification of rock masses as a rating system for the assessment of rippability. His rippability rating chart is given in Table 21.4.

Subsequently Kirsten (1982) adapted the Q system of classification for defining an excavatability index (N) as follows:

$$N = M_s \frac{(\mathrm{RQD})}{(J_n)} \times J_s \times \frac{(J_r)}{(J_a)} \qquad (21.3)$$

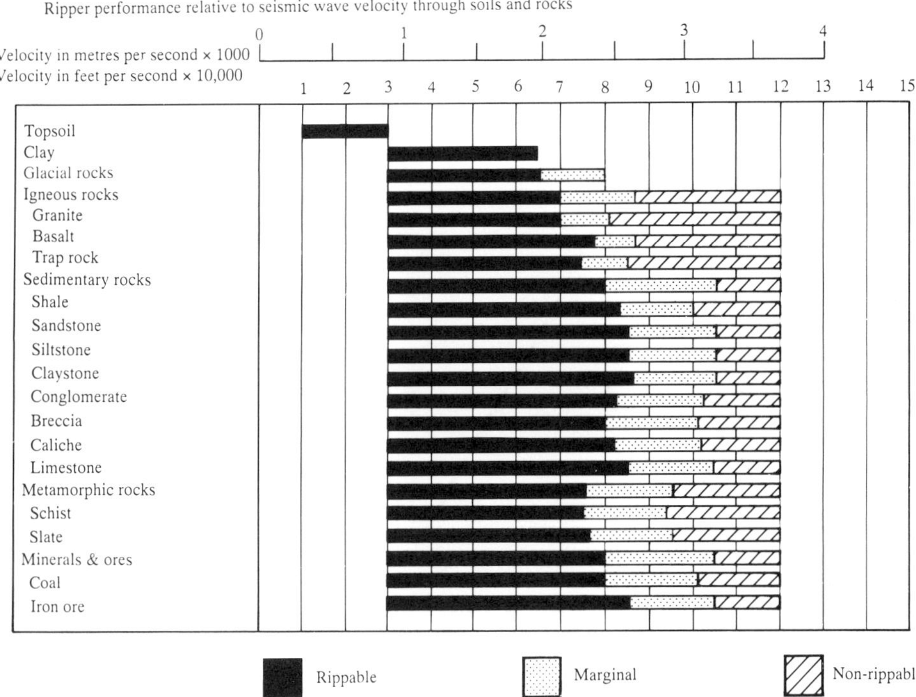

Figure 21.10 *Seismic velocities in realtion to rippability*

The mass strength number, M_s, indicates the amount of effort needed to excavate dry, homogeneous material which contains no discontinuities. The term (RQD/J_n) brings together the rock quality designation (RQD) which affords an equivalent estimate of joint density and the number of joint sets (J_n) which provides an indication of the degree of freedom of rock mass. Hence together they offer a crude measure of block size and the smaller the size of the block the less is the effort required to excavate dry, homogeneous, continuous rock. The term J_s represents the reducing effect which the block shape and orientation relative to the excavating force has on the effort to excavate the perfect material. The roughness of the most unfavourable joint set (J_r) and the degree of alteration or filling of the most unfavourable joint set are combined in the last term and as such provide an approximation of the shear strength of the rock mass. In this way it represents the reducing effect which the deformability and weakness of the joints have on the effort to excavate the perfect rock.

Once the type of rock mass has been identified, the ratings for the individual terms constituting the excavatability index are determined. Each parameter is divided into a number of categories and given a rating. It is admissible to interpolate intermediate values.

As far as the mass strength number, M_s, is concerned, five categories are recognized and their ratings are given in Table 21.5(a). As far as the block size number is concerned only joints which affect the excavation process should be taken into account. The RQD can be determined in the conventional way from core stick or from the joint count, J_c, that is, the number of joints per cubic metre, the two being related by the following expression:

$$RQD = 115 - 3.3J_c \tag{21.4}$$

Table 21.5(b) gives values of J_c in relation to RQD. The joint set number, J_n, is given in Table 21.5(c) in terms of the same categories identified by Barton *et al.* (1974).

The spacing and orientation of discontinuities affect the effort required to penetrate the ground as well as that needed to dislodge individual blocks. It is easier to rip the ground in the direction in which rocks dip than in the opposite direction. The possibility of ripper penetration along a discontinuity is directly related to its inclination. As more than one set of discontinuities usually are present in a rock mass the overall kinematic possibility of penetration may be regarded as the average discontinuity inclination weighted by the number of discontinuities per unit length. The value of J_s for various ranges of angle of dip and dip direction of discontinuities are given in Table 21.5(d). The angle of dip and direction in this table refer to the closer spaced joint set.

The joint roughness number is given in Table 21.5(e) for

Table 21.4 Rippability rating chart (after Weaver 1975)

Rock class	I	II	III	IV	V
Description	Very good rock	Good rock	Fair rock	Poor rock	Very poor rock
Seismic velocity (m/s)	>2150	2150–1850	1850–1500	1500–1200	1200–450
Rating	26	24	20	12	5
Rock hardness	Extremely hard rock	Very hard rock	Hard rock	Soft rock	Very soft rock
Rating	10	5	2	1	0
Rock weathering	Unweathered	Slightly weathered	Weathered	Highly weathered	Completely weathered
Rating	9	7	5	3	1
Joint spacing (mm)	>3000	3000–1000	1000–300	300–50	<50
Rating	30	25	20	10	5
Joint continuity	Non-continuous	Slightly continuous	Continuous – no gouge	Continuous – some gouge	Continuous – with gouge
Rating	5	5	3	0	0
Joint gouge	No separation	Slight separation	Separation < 1 mm	Gouge < 5 mm	Gouge > 5 mm
Rating	5	5	4	3	1
**Strike and dip orientation*	Very unfavourable	Unfavourable	Slightly unfavourable	Favourable	Very favourable
Rating	15	13	10	5	3
Total rating	100–90	90–70†	70–50	50–25	<25
Rippability assessment	Blasting	Extremely hard ripping and blasting	Very hard ripping	Hard ripping	Easy ripping
Tractor selection	–	DD9G/D9G	D9/D8	D8/D7	D7
Horsepower	–	770/385	385/270	270/180	180
Kilowatts	–	575/290	290/200	200/135	135

* Original strike and dip orientation now revised for rippability assessment.
† Ratings in excess of 75 should be regarded as unrippable without pre-blasting.

various joint conditions, and the joint alteration number is set out in Table 21.5(f). No rigorous test can be proposed for the determination of these parameters on their own. As a last resort Kirsten (1982) suggested that the joint strength number could be determined from

$$J_r/J_a = \arctan(\tau_p/\sigma_n) \qquad (21.5)$$

which represents the total friction angle as defined by Bandis *et al.* (1981). If several joint sets of different strengths are present in the rock mass the values of (J_r/J_a) should be determined for each set. The equivalent joint strength for the mass may be determined as the average of the different values of J_r/J_a weighted according to the number of joints in each set per cubic metre.

The class limits for the various categories of excavatability are given in Table 21.6. As the limits differ by an order of magnitude any minor inaccuracies in the determination of magnitudes of any of the basic parameters generally

Table 21.5 Parameters for determination of excavatability index (after Kirsten, 1982)

(a) Mass strength number for rocks (M_s)

Hardness	*Identification in profile*	*Unconfined compressive stength* (MPa)	*Mass strength number* (M_s)
Very soft rock	Material crumbles under firm (moderate) blows with sharp end of geological pick and can be peeled off with a knife. It is too hard to cut a triaxial sample by hand	1.7	0.87
		1.7–3.3	1.86
Soft rock	Can just be scraped and peeled with a knife; indentations 1–3 mm show in the specimen with firm (moderate) blows of the pick point	3.3–6.6	3.95
		6.6–13.2	8.39
Hard rock	Cannot be scraped or peeled with a knife; hand-held specimen can be broken with hammer end of a geological pick with a single firm (moderate) blow	13.2–26.4	17.70
Very hard rock	Hand-held specimen breaks with hammer end of pick under more than one blow	26.4–53.0	35.0
		53.0–106.0	70.0
Extremely hard rock (very, very hard rock)	Specimen requires many blows with geological pick to break through intact material	106.0–212.0	140.0
		212.0	280.0

(b) Joint count number (J_c)

Number of joints per cubic metre (J_c)	*Rock quality designation* (RQD)	*Number of joints per cubic metre* (J_c)	*Rock quality designation* (RQD)
33	5	18	55
32	10	17	60
30	15	15	65
29	20	14	70
27	25	12	75
26	30	11	80
24	35	9	85
23	40	8	90
21	45	6	95
20	50	5	100

(c) Joint set number (J_n)

Number of joint sets	*Joint set number* (J_n)
Intact, no or few joint/fissures	1.00
One joint/fissure set	1.22
One joint/fissure set plus random	1.50
Two joint/fissure sets	1.83
Two joint/fissure sets plus random	2.24
Three joint/fissure sets	2.73
Three joint/fissure sets plus random	3.34
Four joint/fissure sets	4.09
Multiple joint/fissure sets	5.00

Note: For intact granular materials take $J_n = 5.00$

(d) Relative ground structure number (J_s)

Dip direction of closer spaced joint set (degrees)*	*Dip angle† of closer spaced joint set (degrees)*	*Ratio of joint spacing, r* 1:1	1:2	1:4	1:8
180/0	90	1.00	1.00	1.00	1.00
0	85	0.72	0.67	0.62	0.56
0	80	0.63	0.57	0.50	0.45
0	70	0.52	0.45	0.41	0.38
0	60	0.49	0.44	0.41	0.37
0	50	0.49	0.46	0.43	0.40
0	40	0.53	0.49	0.46	0.44
0	30	0.63	0.59	0.55	0.53
0	20	0.84	0.77	0.71	0.68
0	10	1.22	1.10	0.99	0.93
0	5	1.33	1.20	1.09	1.03
0/180	0	1.00	1.00	1.00	1.00
180	5	0.72	0.81	0.86	0.90
180	10	0.63	0.70	0.76	0.81
180	20	0.52	0.57	0.63	0.67
180	30	0.49	0.53	0.57	0.59
180	40	0.49	0.52	0.54	0.56
180	50	0.53	0.56	0.58	0.60
180	60	0.63	0.67	0.71	0.73
180	70	0.84	0.91	0.97	1.01
180	80	1.22	1.32	1.40	1.46
180	85	1.33	1.39	1.45	1.50
180/0	90	1.00	1.00	1.00	1.00

* Dip-direction of closer spaced joint set relative to direction of rip
† Apparent dip angle of closer spaced joint set in vertical plane containing direction of ripping

For intact material take $J_s = 1.0$
For values of *r* less than 0.125 take J_s as for $r = 0.125$

(e) Joint roughness number (J_r)

Joint separation	*Condition of joint*	*Joint roughness number* (J_r)
Joints tight or closing during excavation	Discontinuous joint	4.0
	Rough or irregular, undulating	3.0
	Smooth undulating	2.0
	Slickensided undulating	1.5
	Rough or irregular, planar	1.5
	Smooth planar	1.0
	Slickensided planar	0.5
Joints open and remain open during excavation	Joints either open or containing relatively soft gouge of sufficient thickness to prevent joint wall contact upon excavation	1.0

(f) Joint alteration number (J_a)

Description of gouge	*Joint alteration number* (J_a) *for joint separation* (mm)		
	<1.0†	1.0–5.0†	>5.0‡
Tightly healed, hard, non-softening impermeable filling	0.75	–	–
Unaltered joint walls, surface staining only	1.0	–	–
Slightly altered, non-softening, non-cohesive rock mineral or crushed rock filling	2.0	4.0	6.0
Non-softening, slightly clayey non-cohesive filling	3.0	6.0	10.0
Non-softening strongly over-consolidated clay mineral filling, with or without crushed rock	3.0§	6.0§	10.0§
Softening or low-friction clay mineral coatings and small quantities of swelling clays	4.0	8.0	13.0
Softening moderately over-consolidated clay mineral filling, with or without crushed rock	4.0§	8.0§	13.0§
Shattered or micro-shattered (swelling) clay gouge, with or without crushed rock	5.0	10.0	18.0

* Joint walls effectively in contact
† Joint walls come into contact after approximately 100 mm shear
‡ Joint walls do not come into contact at all upon shear
§ Values added to Barton's data

should not result in a change of the class of excavation, unless the index is near a class boundary. Kirsten (1982) did not propose that the machines quoted in Table 21.6 should be specified with respect to the different classes of excavation in any conditions of contract but rather that the excavation classes be specified in terms of class intervals for the excavatability index.

As far as ripping is concerned the run direction should be normal to any vertical joint planes, down-dip to any inclined strata, that is, normal to the strike, and on sloping ground it should be downhill. Atkinson (1970) suggested that long ripping runs of 70–90 m usually gave the best

Table 21.6 Excavation classification system for rock masses

Material type	*Class*	*Excavation class boundaries*	*Description of excavatability*	*Bulldozer characteristics*					*Backhoe characteristics*			
				Type	*Operating mass** (kg)	*Flywheel power* (kW)	*Drawbar pull†* (kN) *Stalling speed*	*Drawbar pull†* (kN) 1.6 km/h	*Type*	*Operating mass* (kg)	*Flywheel power*	*Man-draw bar pull* (kN)
Rock	1	1.0–9.99	Easy ripping	D7G	20230	149	376	220	Cat 235	38297	145	263
	2	10.0–99.9	Hard ripping	D8K	31980	224	500	323	Cat 245	59330	242	472
	3	100.0–999	Very hard ripping	D9H	42780	306	667	445	RH 40	83200	360	–
	4	1000.0–9999	Extremely hard ripping/blasting	D10	77870	522	1230	778	–	–	–	–
	5	Larger than 10 000	Blasting	–	–	–	–	–	–	–	–	–

Quoted in *Caterpillar Performance Handbook*†, *or equivalent*
All machines referred to are track mounted

results. He further suggested that where possible the ripping depth should be adjusted so that a forward speed of 3 km/h could be maintained, since this is generally found to be the most productive, reduces track wear significantly and avoids impact shocks. Adequate breakage in rock depends on the spacing between ripper runs which in turn is governed by the fracture pattern in the rock. The output of a ripper also depends upon the capacity of the bulldozer. Output generally falls within the range of 40–230 m^3/h.

21.5 Diggability

Some of the softer rocks such as shales can be excavated by digging machines. The diggability of ground is of major importance in the selection of excavating equipment and depends primarily upon its intact strength, bulk density, bulking factor and natural water content. At present there is no generally accepted quantitative measure of diggability, assessment usually being made according to the experience of the operators. However, a fairly reliable indication can be obtained from similar excavations in the area concerned, the behaviour of the ground excavated in trial pits or from tests on core samples.

Attempts have been made to evaluate the performance of excavating equipment in terms of seismic velocity. It would appear that most earthmoving equipment operates most effectively when the seismic velocity of the ground is less than 1 km/s and will not function above 1.8 km/s, but in areas of complex geology seismic evaluation may prove difficult if not impossible.

When material is excavated it increases in bulk, this being brought about by the decrease which occurs in density per unit volume. The amount of bulking which takes place when a given rock or soil is worked can be ascertained by filling large boxes of known volume with the excavated material and averaging the results of several tests. This can then be compared with the *in situ* density to give the bulking factor:

$$\text{bulking factor} = \frac{\text{intact density per unit volume}}{\text{disturbed density per unit volume}}$$

References

Anon (1972) *Blasting Practice*, Nobel's Explosives Co. Ltd (ICI), Stevenson, Ayrshire

Anon (1987) *Hydrogeology and Stability of Excavated Slopes in Quarries*, Department of the Environment, HMSO, London

Antill, J. M. and Ryan, P. W. S. (1967) *Civil Engineering Construction*, Angus and Robertson, Sydney

Atkinson, T. (1970) 'Ground preparation by ripping in open pit mining', *Min Mag.*, **122**, 458–469

Bandis, S., Lumsden, A. C. and Barton, N. (1981) 'Experimental studies of scale effects on the shear behaviour of rock joints', *Int. J. Rock Mech. Min. Sci. & Geomech. Abstr.*, **18**, 1–20

Barton, N., Lien, E. and Lunde, J. (1974) 'Classification of rock masses for the design of tunnel support', *Rock Mechanics*, **6**, 189–236

Bieniawski, Z. T. (1973) 'Engineering classification of jointed rock masses', *Trans. SA Inst.Civil Engrs.*, **15**, 335–344

Crandell, F. J. (1949) 'Ground vibrations due to blasting and its effect on stressmeters', *J. Boston Soc. Civ. Engrs.*, **36**, 222–225

Duvall, W. I. (1964) 'Design requirements for instrumentation to record vibrations produced by blasting', *Rep. Invest. No 6487*, US Bureau of Mines, Washington, D.C.

Duvall, W. I. and Fogelson, D. E. (1962) 'Review of the criteria for estimating damage to residences from blasting vibrations', *Rep. Invest. No. 5968*, US Burea of Mines, Washington, D.C.

Edwards, A. T. and Northfield, R. D. 'Experimental studies of the effects of blasting on structures', *The Engineer*, London **210**, 539–546

Fish, B. G. *'Studies in Rotary Drilling'*, *Rept. No. 2161* National Coal Board, Mining Research Establishment, Bretby

Franklin, J. A., Broch, E. and Walton, G. (1971) 'Logging the mechanical character of rock', *Trans. Inst. Min. Metall.*, **80**, Section A – Mining Industry, A1–9

Furby, J. (1981) 'Tests for rock drillability', *Mine and Quarry Engg.*, **30**, 292–298

Hagan, T. N. (1979) 'Designing primary blasts for increased slope stability', *Proc. 4th Int. Cong. Rock Mech. (ISRM)*, Montreux, **1**, 657–664

Hagan, T. N. (1986) 'The influence of some controllable blast parameters upon muckpile characteristics and open pit mining costs', *Proc. Conf. Large Open Pit Mining*, Australia Inst. Min. Metall/Inst. Engrs., 123–132

Kirsten, H. A. D. (1982) 'A classification system for excavation in natural materials', *The Civil Engr. in South Africa*, **24**, 293–306

Langefors, U. and Kihlstrom, B. (1962) *The Modern Technique of Rock Blasting*, Wiley, New York

McGregor, K. (1967) *The Drilling of Rock*, C.R. Books Ltd (A. McClaren and Co.), London

Nicholls, H. R., Johnson, C. F. and Duvall, W. I., (1971) 'Blasting vibrations and their effects on structures', *US Bur. Mines, Bull*, 656, Washington, D.C.

Oriard, L. L. (1972) 'Blasting operations in the urban environment', *Bull. Ass. Engg. Geologists*, **9**, 27–46

Paine, R., Holmes, D. and Clarke, H. (1961) 'Controlling overbreak by pre-splitting', *Int. Symp. Min. Res. Univ. Missouri*, Pergamon, **1**, 179–209

Persson, R., Lundborg, N. and Johansson, C. H. (1970) 'The basis mechanism in rock blasting', *Proc. 2nd Int. Cong. Soc. Rock Mech.*, I.S.R.M. Belgrade, Paper 5–3, 19–33

Richards, R. L., Leg, G. M. M. and Whittle, R. A. (1978) 'Appraisal of stability in rock slopes'. In *Foundation Engineering in Difficult Ground*, F. G. Bell (ed.), Butterworth-Heinemann, Oxford, pp. 449–512

Roberts, A. (1971) 'Ground vibrations due to quarry blasting and other sources – an environmental factor in rock mechanics',

Proc. 12th Symp. on Rock Mechanics, Rolla, University of Missouri, AIME, New York, 427–456

Roberts, A. (1981) *Applied Geotechnology*, Pergamon, Oxford

Sandy, D. A. (1989) Drill, blast, load and haul practices at Rösing Mine', Namibia. *Trans. Inst. Min. Metall.*, **98**, Section A, Mining Industry, A98–A104

Sinclair, J. (1969) *Quarrying, Opencast and Alluvial Mining*, Elsevier, London

Skipp, B. O. and Taylor, J. W. (1970) 'Blasting vibrations – ground structure and response'. In *Dynamic Waves in Civil Engineering*, D. A. Howell I. P. Haigh and C. Taylor (eds.), Wiley Interscience, New York

Starfield, A. M. (1966) 'Strain wave theory in rock blasting', *Proc. 8th Symp. Rock Mech.*, University of Missouri, Rolla, Suppl. Paper No. 4

Vutukuri, V. S. and Bhandari, S. (1961) *Some aspects of design of open pits*, Colorado School Mines Q., **56**, 51–61

Weaver, J. M. (1975) Geological factors significant in the assessment of rippability', *The Civil Engr. in South Africa*, **17**, 313–316

22 Tunnelling in rock masses

Professor H Duddek
Universitat Braunschweig, Germany

22.1 General approach for tunnelling projects

For tunnelling in rock, the ground actively participates in providing the stability of the opening. Therefore, tunnelling requires the interdependent participation of at least the following disciplines: geology, geotechnical engineering, excavation technology, structural engineering, surveying and monitoring. Although the experts working in these subjects may be responsible only for their specific knowledge, the decision on the main features should be the outcome of the cooperative integration of all the disciplines. Thus, it is ensured that the project is seen in all its details as a unity and not as a consecutive addition of the separate work of each of the experts. Failures in tunnelling or too conservative an approach are very often caused by lack of mutual understanding of the special results, for example, by insufficient interpretation of the geotechnical features and/or the monitored *in situ* deformations, which contribute to the selection of the structural support measures.

A general outline of the tunnelling process (Duddeck 1987) is given in Figure 22.2.

(1) The site investigation must determine the geology and confirm the line, orientation, depth etc. of the tunnel.
(2) Ground probing and applied rock mechanics determine the ground characteristics, such as primary stresses, rock strength, fissures, faults, water conditions etc.
(3) Experience and preliminary estimations or calculations lead to the cross-section required and to the choice of the excavation method or the tunnel driving machine, as well as to the methods of dewatering the ground, and to the selection of the supporting structural elements.
(4) Then the tunnelling engineer has to derive, even to invent, a structural model from which he draws, by applying equilibrium and compatibility conditions, those criteria which decide whether the design is safe or not. There may be different models for each excavation step or for the preliminary or the final tunnel lining, for different ground behaviour (discontinuous rock or homogeneous ground). Modelling of the geometric features may vary greatly, depending on the desired intensity of the analysis.
(5) A safety concept drawn from failure hypotheses may be based, for example, on strains, stresses, deformations, or failure modes.
(6) The bypass in Figure 22.2 may indicate that for many underground structures as in mining or in self-supporting hard rock, no design models at all are applied, past experience alone may be sufficient.
(7) Risk assessment by the contractor as well as by the owner is necessary at the time of contract negotiations. Risks involve possible structural failures of the tunnel support and lining, functional failures after completion of work, or financial risks. The contractual aspect also cover risk sharing and risk responsibilities.
(8) *In situ* monitoring can be applied only after tunnelling has begun. If the displacements no longer increase with time, it generally may be assumed that the structure is designed safely. Yet monitoring provides only one part of the answer. It does not tell how close the structure may be to sudden collapse or non-linear failure modes. The results of field measurements and experiences during excavation may compel the engineer to change the design model by adjusting it to the real behaviour.

For the design of structures in the ground, which employ the participating strength of the ground, an iterative step-by-step approach (loops in Figure 22.2) is characteristic. The designer may begin by applying estimated and simple behavioural models. The model should gain more closeness to reality and more refinement (if consistent with the purpose) by adjustments based on the actual experiences during the tunnelling excavation as by the initial section in same ground conditions or by driving a pilot tunnel. The interpretations of *in situ* measurements (and some back analyses) also may assist in making the adjustments. All the elements as given by Figure 22.2 should be considered in unison. The scattering of parameters or the inaccuracy in one part of the model will

Figure 22.1 *Tunnelling for the two-lane high speed railway line in central Germany 1984, Rollenbergtunnel (Photo courtesy of Hildegard Steinmetz)*

affect the accuracy of the model as a whole. Therefore, the same simplicity or refinement should be provided consistently through all the elements of the design model. For example, it is inconsistent to apply very refined mathematical tools and at the same time rough guesses of important ground characteristics.

22.2 Site investigations and ground probings

A sufficiently intense site exploration, from which geological and hydrogeological maps and ground profiles are produced, is most important for selecting the appropriate tunnel design and excavation method (Duddeck 1988). A well-documented geological report should provide as much information as can be obtained about the physical features along the tunnel axis and in the adjacent ground. The amount of information should be much greater than that directly required for a structural analysis.

The appropriate amount of ground investigation on site and in the laboratory may vary considerably from project to project. The types of ground exploration and probing depend on the special features of the tunnelling project, its purpose, excavation method and so on. Therefore, they should be chosen by the expert team, especially in consultation with the design engineer. The intensity of the ground exploration will depend on the complexity of the ground, the purpose of the tunnel, the cost of drilling, etc. (e.g. for shallow or deep cover).

The geological investigation should include the following basic geotechnical descriptions.

(1) *Zoning*. The ground should be divided into geotechnical units for which the design characteristics may be considered uniform. However, relevant characteristics may display considerable variations within a geotechnical unit.

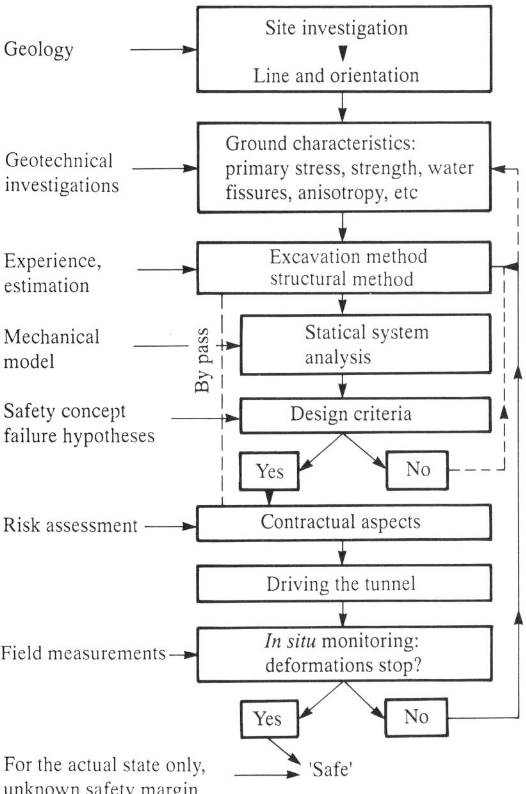

Figure 22.2 *Tunnelling procedure*

For the geological description of each zone the following aspects should be considered:

(a) name of the geological formation in accordance with a genetic classification;
(b)geological structure and fracturing of the rock mass with strike and dip orientations;
(c) colour, texture and mineral composition;
(d) degree of weathering.

(2) *Parameters of the rock mass* (e.g. in five classes of intervals as suggested by Bieniawski (1984), see Chapter 3)
(a) thickness of the layers;
(b) fracture intercept;
(c) rock classification;
(d) core recovery;
(e) uniaxial compressive strength of the rock, derived from laboratory tests;
(f) angle of friction of the fractures (derived from laboratory direct shear tests);
(g) strength of the ground in on-site situations;
(h) deformation properties (modulus);
(i) effect of water on the rock quality;
(j) seismic velocity.

(3) *Primary stress field of the ground.* For larger tunnel projects, tests evaluating the primary stresses in the rock mass may be required. For usual tunnel projects one should at least estimate the stress ratio σ_h/σ_v at tunnel level, where σ_h is the lateral ground pressure and σ_v the major principal stress (usually in the vertical direction), for which the weight of the overlying rock generally may be taken. Tectonic stresses should be considered.

(4) *Water conditions.* Two types of information are required:
(a) permeability, given by coefficient *k* (m/s) (from field tests) and Lugeon unit (from tests in drill holes);
(b) water pressure at tunnel level (hydraulic head) and at piezometric levels in drill holes.

(5) *Deformability of the rock mass*: *In situ* tests are required to derive the two different deformation moduli, which can be determined either from static methods (dilatometer tests in drill holes, plate loading tests in adits, or radial jacking tests in chambers) or from dynamic methods (wave velocity by seismic refraction or by geophysical logging in drill holes). Engineering judgement should be exercised in choosing the value of the modulus most appropriate for the design; for instance, by the relevant tangent of the pressure–deformation curve at the primary stress level in the static method.

(6) *Properties needed for employing tunnel boring machines*:
(a) abrasiveness and hardness;
(b) mineral composition especially quartz content;
(c) homogeneity.

(7) *Swelling potential of the rock*: The presence of sulphates, hydroxides or clay minerals should be investigated by mineralogical testing. A special oedometer test may be used to determine the swell test curve of a specimen subjected first to a load–unload–reload cycle in a dry state, and then unloaded with water.

(8) *Groundwater conditions*:
(a) water levels, piezometric levels, variations over time, pore-pressure measurements in confined aquifers;
(b) water chemistry, expecially with regard to aggressiveness;
(c) water temperatures;
(d) expected amount of water inflow.

(9) *Evaluation of parameters* by ground probing and laboratory tests. The properties of the ground that are relevant for the tunnel design should be evaluated as carefully as possible. *In situ* tests, which comprise larger ground masses, generally are more significant than laboratory tests on small specimens, which often are the better preserved parts of the core. The natural scattering of ground properties requires an appropriate number of parallel tests.

Results of laboratory tests require an adjustment to the site situation including the consideration of the size of specimen, the effects of groundwater, the inhomogeneity of the ground on site, the effects of scattering. The conclusions drawn from tests should consider whether the specimens were taken from disturbed or undisturbed ground.

In many cases, the first part of the tunnelling operation may be interpreted as a large-scale test, from which experiences may be drawn upon not only for the subsequent excavations but also for predicting ground behaviour. In certain cases, long horizontal drill holes may facilitate ground probing ahead of the face, or a pilot tunnel may serve as a test tunnel providing drainage at the same time. The on-site investigations provide valuable results for checking the correlation of large-scale *in situ* tests with laboratory tests.

Special tests that correspond directly to the proposed tunnelling method may be required, such as freezing. The evaluation of the parameters should indicate the expected scattering. From probabilistic considerations of normally distributed quantities it can be deduced that a mean value or a value corresponding to a moderately conservative fractile of a Gaussian distribution is more appropriate than the worst-case value.

A set of all the parameters describing the ground behaviour of one tunnel section with regard to tunnelling should be seen as a comprehensive unit and should be well balanced in relation to each of the parameters. For example, a small value of ground deformation modulus indicates a tendency to plastic behaviour, which corresponds to a ratio of lateral to vertical primary stress that is closer to 1.0. Hence, for alternative investigations some complete, balanced sets of parameters should be chosen instead of considering each parameter alone, unrelated to the others.

The available methods for ground probing and laboratory tests, their applicability and accuracy are given in Chapters 7 and 8.

(10) *Interpretation of test results and documentation*. The field and laboratory tests should be given in well-documented reports, in the form of actual results. Based on these reports, an interpretation of the tests is necessary, that is relevant to the actual tunnelling process and the requirements of the design models for the structural analysis. At the time the tests are planned, the team of experts referred to above should decide which ground properties and ground characteristics are necessary for the general geotechnical description of the ground and for the projected design model. Thus, a closer relationship may be achieved between ground investigations and tunnelling design, and between the amount of refinement of tests and tunnelling risks.

The documents should offer a rational interpretation in which design values are derived from test results. It is useful, especially for tendering, to condense the relevant data for the description of the ground and for the design of the tunnel on a band along the tunnel axis beneath a graphical representation of the tunnel profile (see Figure 22.14). Such condensed tables may be prepared first for tendering and the preliminary design, and then improved through experiences gained from incoming monitoring results. However, it should be clearly stated, especially in the contract papers, that much relevant information is lost or oversimplified in such tables, and that therefore the geotechnical reports and other complete documents should be considered the primary documents.

22.3 Excavation and support methods for rock tunnelling

In applying the advances of rock mechanics and tunnel engineering, the excavation and support methods should be selected so that the following principles (Müller-Salzburg 1978) are observed:

(1) The natural strength of the rock should be preserved as much as possible, even at the tunnel surface, for example, by controlled smooth blasting.
(2) The ground (not the lining) is the principal contributor to the stability of the tunnel opening.
(3) Hence, improving the rock quality (e.g. by rock bolts or by grouting) is more effective than strengthening the lining.
(4) By controlled decompression of rock stresses before the lining is acting, the required lining may be thinner, thus, more flexible and less endangered by bending moments.
(5) By closing the invert, a ring stiffness of the lining is provided that is much more effective.
(6) The excavation and the support measures should be controlled and adjusted to rock conditions by continuous monitoring of deformations and stresses.
(7) The lining and the rock should act permanently (no wood allowed in between) as a composite structure with full bond.

In order to select an appropriate excavation method an optimum may be sought, which complies to these principles in the specific ground as well as to economical considerations. In recent years tunnelling has progressed very much by applying these principles.

A full-face tunnel boring machine (TBM) may be suitable, when a competent rock alone provides the stability of the tunnel excavation for a long enough section without structural support, and when the ground is fairly homogenous along the entire tunnel length to be driven by the boring machine. Ideally, a TBM which cuts the ground mechanically preserves the rock strength. However, when there are tunnel sections where the ground converges and where support or ground improvement is necessary immediately at the tunnel face, tunnel boring machines may fail.

If the ground is so weak that even the tunnel face needs permanent support during excavation, then the tunnel may be driven only by employing a shield and by providing immediate support in errecting prefabricated segmented ring linings.

For most tunnelling projects in rock the ground reacts to tunnelling somewhere between these two extreme conditions. The face may have a certain self-supporting stability, yet support measures are necessary, say half or one tunnel diameter close to the face. Here the classical drill and blast excavation method may be the most efficient one, now of course improved in many aspects, especially so that the principles given above are met (smooth blasting, applying rock bolts and shotcrete etc.).

There are many other varieties of excavation methods, such as the following:

(1) excavation within a partially supporting shield;
(2) mechanical rock excavation;
(3) freezing or grouting of the ground.

An example of a present-day method of excavation, which proved to be successful in medium hard rock and even in difficult ground conditions, has been provided by tunnelling for the new rapid railway line in central Germany took place through sandstone and limestone, clayey layers of both and under groundwater in flow. Sixty-one tunnels of 121 km total length have been excavated between 1982 and 1988 along the 327 km long

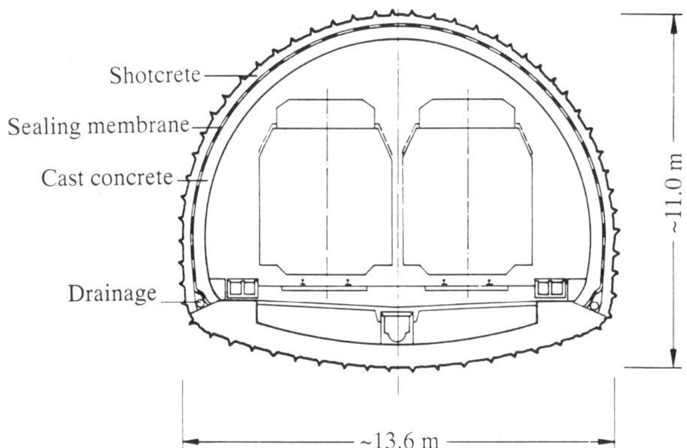

Figure 22.3 *Tunnel cross-section for double tracks of the German railway line, left: flat bottom. right: deep invert*

Figure 22.4 *Drilling holes for top heading advance, German high-speed railway line 1984*

line from Hannover to Würzburg. Figure 22.1 shows an example in very difficult ground conditions, where a smaller drainage pilot tunnel has been driven ahead of the excavation of the top heading, the bench, and the invert. Figure 22.3 shows a representative cross section. The tunnels are designed for 250 km/h speed, requiring an air cushion, which results in an open cross area of 145 m^2. Figure 22.4 explains a typical tunnel face in good rock. Figure 22.5 shows the sequence of excavation and support phases. In fairly good quality rock, firstly, a top heading of 6.0 m height was driven by the drill and blast method (Figure 22.4), each round excavating 1.0–2.0 m, advancing at 4–8 m/day. Secondly, often some 150–200 m behind the top heading, the bench was excavated either with a ramp in the centre or successively on each side. The advance rate was 6.0–15.0 m/day. Thirdly, the invert was excavated and the final invert concrete structure was cast on half the cross-section, providing traffic on the other half. Finally the inner lining was cast, often some months later, in shorter lengths, after finishing the excavation and the outer lining. The inner lining was placed by 11 m long sections of cast concrete. The advance rate for the inner lining reached up to six sections (60 m) per week. Figure 22.6 shows the work sequence of the support measures (for the top heading), and Figure 22.7 illustrates the consecutive steps in erecting the inner lining above the invert concrete.

In general, the method of excavation and the choice of preliminary support are related to each other. For the tendering as well as for the actual tunnelling procedure on site and the final price settlements, a classification system is useful which provides well-defined classes of excavation and support measures (e.g. 7–9 classes). Internationally there are many different classification systems (Müller-Salzburg 1978; Maidl 1988; and Hoek and Brown 1980). They should be adjusted to each tunnelling project

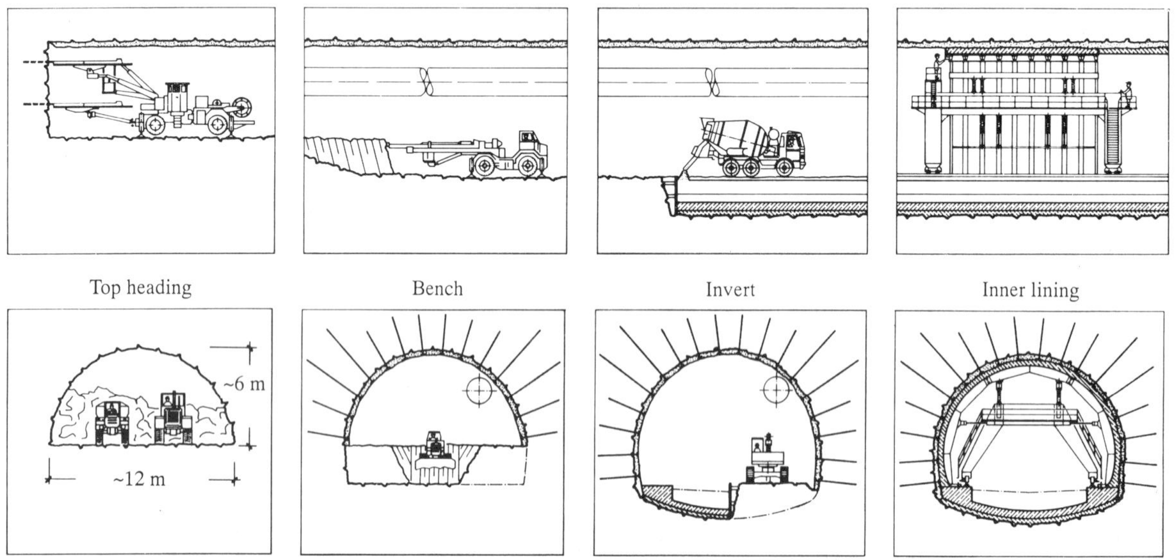

Figure 22.5 *Drill and blast excavation for large tunnel cross-sections (after Leichnitz, 1987)*

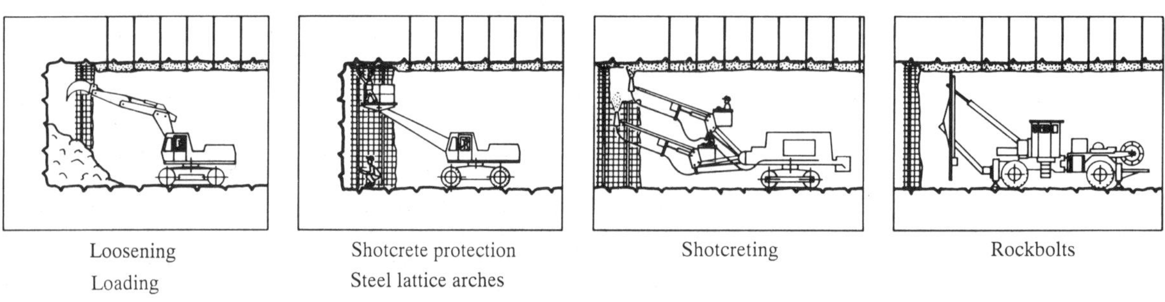

Figure 22.6 *Work sequences for outer lining (after Leichnitz, 1987)*

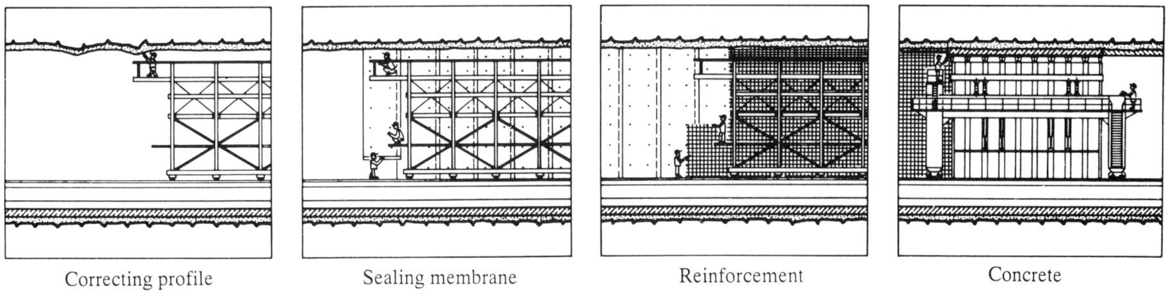

Figure 22.7 *Work sequences for inner lining (after Leichnitz, 1987)*

individually and may, for excavations as in Figure 22.5–22.7, include:

(1) length of excavation advance;
(2) amount, length, and kind of rock bolts and/or anchors;
(3) distance and kind of steel arches;
(4) thickness and strength of shotcrete;
(5) allowed length and time interval for unsupported section;
(6) amount of steel wire mesh, one or two mesh layers;
(7) forepoling;
(8) amount of grouting;
(9) invert concrete arch also for top heading;
(10) subdivision of excavation face, e.g. by driving side galleries first;
(11) special support features for the open tunnel face.

In addition to these support measures the corresponding rock qualities of each class may be defined verbally.

Tunnel classification systems based on ground conditions only, such as those based on RQD criteria, or that of Bieniawski (1984) or Barton, *et al.* (1974), stop somewhat short of most European systems, which include support elements. In tenders, unit price bids may be asked for those alternative support elements which very likely may be employed. The unit price — say, for an additional 5 cm of shotcrete thickness — should include the costs of work delays, time considerations and all other secondary cost effects related to these units.

To some extent, the different classification systems are influenced by the different contractural concepts. In Europe, the design of the tunnel itself usually is included in the tendering papers. The choice of excavation and of support is the result of the contractor's design. The actual work at the tunnel face is decided on in joint meetings of an expert team that includes the owner, the owner's geotechnical and structural engineering consultants and the contractor. The unit price scheme for alternative items in the tender ensures that even unexpected ground conditions and unpredicted heavy structural support measures are covered by the tender documents.

Excavation in difficult rock may require special ground treatment ahead of excavation. The ground can be stabilized by applying the following methods either individually or in combination:

(1) Dewatering either from the tunnel face or from the ground surface or even by driving a pilot tunnel.
(2) Grouting from inside or outside the tunnel.
(3) Freezing the ground around the tunnel (e.g. by insertion of freeze pipes as shown in Figure 22.8). The ground must have a certain water content.
(4) Compressed air, usually applied only for tunnelling under open water or where groundwater lowering is endangering structures as in cities.

22.4 Structural design of tunnels

22.4.1 Ground-lining interaction and design approaches

The engineering approach for designing a tunnel structure differs very much from structures above the ground. Whereas for example for a bridge, loads are mostly well defined and strength is provided by manufactured materials, for tunnels stress release in the ground and the strength of the rock mass are the main features for design considerations. The response of the ground to excavation of an opening can vary widely. Based on the type of ground in which tunnelling takes place, three principal types of tunnelling may be defined:

(1) In soft ground, immediate support must be provided by a stiff lining (e.g. in the case of shield-driven tunnels with segments for ring support and pressurized slurry for face support). In such a case, the ground usually participates only by providing resistance to outward deformations of the lining.
(2) In medium-hard rock or in more cohesive soil, the ground may be strong enough to allow a certain open section at the tunnel face. Here, a certain amount of stress release may permanently be valid before the supporting elements and the lining begin acting

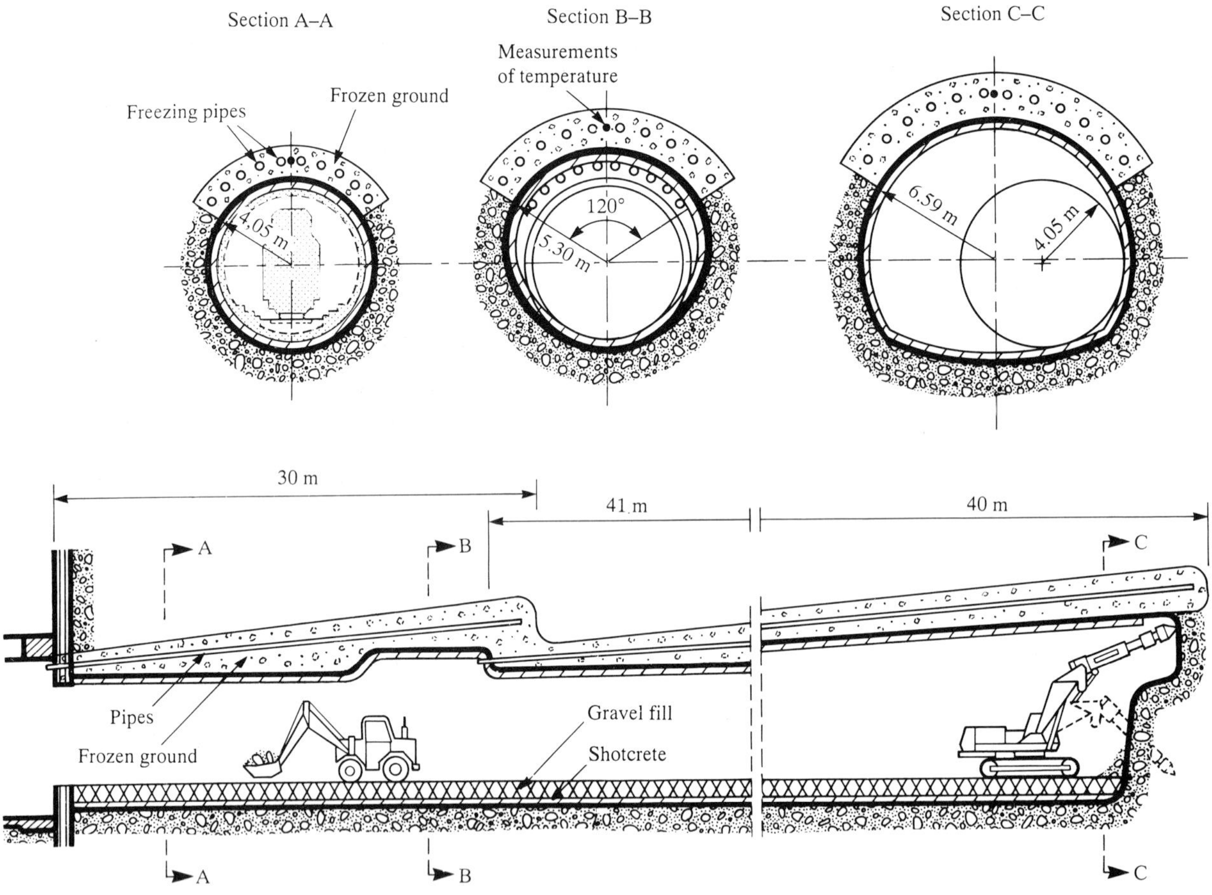

Figure 22.8 *Freezing the ground for the S-Bahntunnel in Stuttgart (Ph. Holzmann AG)*

effectively. In this situation, only a fraction of the primary ground pressure is acting on the lining.

(3) When tunnelling in hard rock, the ground alone may preserve the stability of the opening, so that only a thin lining, if any, is necessary for surface protection. The design model must take into account the rock around the tunnel in order to predict and verify stability, safety considerations, and deformations.

Particularly in ground conditions that change along the tunnel axis, the ground may be strengthened by grouting, anchoring, draining, freezing, etc. Under these circumstances, case (1) may be improved, at least temporarily, to case (2).

The characteristic stress release at the tunnel face is shown in Figure 22.9 The relative crown displacement, w, is plotted along the tunnel axis, where $w/w_0 = 1.0$ represents the case of an unsupported tunnel. In medium-stiff ground nearly 80% of the deformation has already taken place before the lining (shown here as shotcrete) is stiff enough to participate. For a simplified

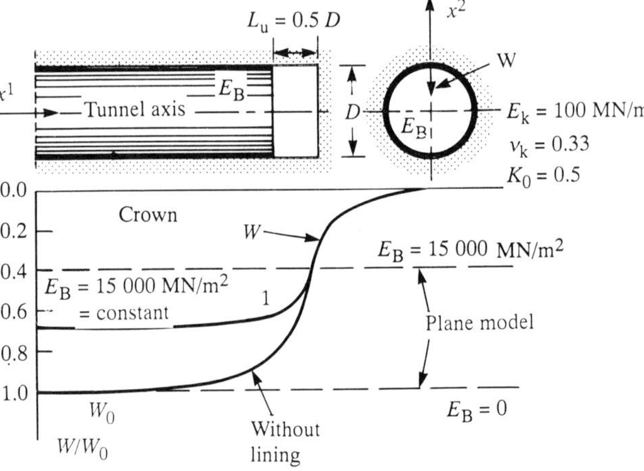

Figure 22.9 *Crown displacement, w,* along the axis, ahead and beyond the tunnel face (after Erdmann, 1983). E_B = modulus of elasticity for concrete E_K = that of ground

plane model with no stress release, where the full primary stresses are assumed to act on a lined opening, the displacement may be only 0.4 of that occurring in the unsupported case. If a lining is providing support at a distance half the tunnel diameter from the face, then the deformation may follow curve (1) in Figure 22.9. The difference between w_0 and w for curve (1) is responsible for the stresses in the lining. If the shotcrete is developing its strength with time, then the deflection curve (1) may be much closer to w_0.

For tunnels with very pronounced stress release due to inward deformations, for example, for deep tunnels in rock, a simple approach to design considerations is given by the convergence–confinement approach (Anon. 1980; Duddeck 1980; Kerisel 1981; Lombardi 1980), which is based solely on the interaction of the radial inward displacement and the support reaction to deformation by resisting ring forces and the corresponding outward pressure, s (Figure 22.10). The primary stresses, p_0, in the ground are released with progressive inward displacement, u. The acting pressure may even increase when rock joints are opening with larger displacements. In self supporting rock, the ground characteristic (the upper curve in Figure 22.10) meets the axis. Because the primary stresses are released completely, a supporting lining is not necessary.

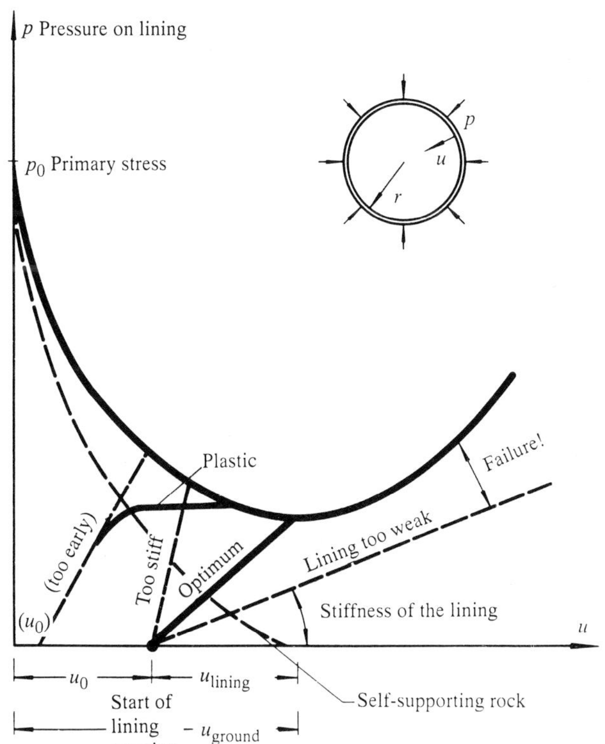

Figure 22.10 *Characteristics for the interaction of ground and support for circular tunnel cross-sections (Fenner–Pacher curves)*

Before the supporting members are installed, it is unavoidable — even desirable — that decompression associated with the predeformation u_0, will occur. The stiffness of the lining determines where both curves (characteristic lines) will intersect. At this point, equilibrium as well as compatibility conditions are fulfilled. If the ground characteristic is known, for example by *in situ* monitoring, the predeformation u_0 and the stiffness of the lining (including its development over time and as tunnelling advances) are very decisive for the actual stresses in the lining.

Both curves in Figure 22.10 may vary considerably. Some possible variations of the support characteristic are shown in Figure 22.10. If the support system is placed too early or if it is too stiff, the possible optimum is missed. If the support is placed too late or if it is too weak, equilibrium will not be achieved. Plastic behaviour helps to meet the ground curve at smaller acting pressure.

For rock tunnelling, the stress release within the rock mass at the advancing face is most important. Therefore, a design model for the assessment of the excavation procedure and the required support must consider these features. However, the characteristic curves, such as those in Figure 22.10, are too simplified an approach for most of the actual underground structures. The convergence–confinement curves are derived by assuming circular openings and constant stiffness relations along the circumference. Only constant radial ground pressure and hence only ring forces are considered, and no bending moments at all.

The results of an analysis of a tunnelling procedure depend very much on the assumed model and the values of the significant parameters. The main purposes of the structural analysis are to provide the design engineer with:

(1) a better understanding of the ground-structure interaction induced by the tunnelling process;
(2) knowledge of what kinds of principal risks are involved and where they are located;
(3) a tool for interpreting the site observations and the *in situ* measurements.

The available mathematical methods of analyses are much more refined than are the properties that constitute the structural model. Hence, in most cases it is more appropriate to investigate alternative possible properties of the model, or even different models, than to aim for a more refined model. For most cases, it is preferable that the structural model employed and the parameters chosen for the analyses are lower-limit cases. This may prove that even for unfavourable assumptions, the tunnelling process and the final tunnel are sufficiently safe. In general, the structural design model does not try to represent exactly the actual conditions in the tunnel, although it covers these conditions.

In situ monitoring is important and should be an integral part of the design procedure, especially in cases where stability of the tunnel depends on the ground properties. Deformations and displacements can be measured generally with much more accuracy than stresses. The geometry of the deformation and its development over time are most significant for the interpretation of actual events. However, *in situ* monitoring only evaluates the very local and actual situation in the tunnel. Therefore, in general the conditions taken into account by the design calculations do not coincide with the conditions that are monitored. Only by relating measurement results and possible failure modes by extrapolation may the engineer arrive at considerations of safety margins.

In many cases, exploratory tunnelling may be rewarding because of the information it yields on the actual response of the ground to the proposed methods for drainage, excavation, TBM driving, support, etc. In important cases a pilot tunnel may be driven. Such a tunnel may even be enlarged to the full final tunnel cross-section in the most representative ground along the tunnel axis. For larger projects, it may be useful to excavate a trial tunnel prior to commencing the actual work. More intensive *in situ* monitoring of the exploratory tunnel sections should check the design approach by numerical analysis.

An underground structure may lose its serviceability or its structural safety by the following cases of failure:

(1) The structure loses its watertightness.
(2) The deformation is intolerably large.
(3) The tunnel is insufficiently durable for its projected life and use.
(4) The material strength of the rock or the structural elements is exhausted locally, necessitating repair.
(5) The support technique fails or causes damage.
(6) Exhaustion of the material strength of the system causes structural failure, although the corresponding deformation develops in a restrained manner over time.
(7) The tunnel collapses suddenly because of instability.

The structural design model should yield criteria related to failure cases, against which the tunnel should be designed safely. These criteria may be:

(1) Deformation and strains.
(2) Stresses and utilization of plasticity.
(3) Cross-sectional lining failure.
(4) Failure of the strength of the rock mass.
(5) Limit analysis failure modes.

In principle, the safety margins may be chosen differently for each of the failure cases listed above. However, in reality the evaluation of an actual safety margin is most complex and very much affected by the scatter of the properties of the ground involved and those of the structure and, furthermore, by the interacting probabilistic characteristics of these properties. Therefore, the results of any calculation should be subject to critical reflection on their relevance to the actual conditions. National codes for concrete or steel structures may not always be appropriate for the design of tunnels and the supporting elements. Computational safety evaluations should always be complemented by overall safety considerations and risk assessments employing ciritical engineering judgement, which may include the following aspects:

(1) The ground characteristics should be considered in the light of their possible deviations from average values.
(2) The design model itself and the values of parameters should be discussed by the design team, which consists of all the experts involved.
(3) Several and more simple calculation runs with parametric variations may uncover the scatter of the results. In general, this approach is much more informative than a single over-refined investigation.
(4) The *in situ* measurements should be used for successive adjustments of the design model.
(5) Long-term measurement of deformation and extrapolation may reveal to a large extent the final stability of the structure, although sudden collapse may not be announced in advance.

22.4.2 Design models for tunnels

The excavation of a tunnel changes the primary stress field into a three-dimensional pattern at the tunnelling face. Farther from the face, the stress field will eventually return to an essentially two-dimensional system. Therefore, the tunnel design may consider only two-dimensional stress–strain fields as first approximations.

Because the design of a tunnel should take into account the interaction between ground and lining, the lining must be placed with closest possible bond with the ground. The deformation resulting from the tunnelling process (see Figure 22.9) reduces the primary ground pressure and creates stresses in the lining corresponding to that fractional part of the primary stresses in the ground which act on the sustaining lining. The stresses depend on the stiffness relationship of the ground to the lining, as well as on the shape of the tunnel cross-section. The latter should be selected such that an arching action in the ground and the lining may develop. Figure 22.11 presents four different structural models for a plane-strain design analysis. The cross-sections need not be circular.

In soft ground, immediate support is provided by a relatively stiff lining. For tunnels at shallow depth (as for underground railways in cities), it is agreed that a two-dimensional cross-section may be considered, neglecting the three-dimensional stress release at the face of the tunnel during excavation. In cases (1) and (2) in Figure 22.11, the ground pressures acting on the cross-section are assumed to be equal to the primary stresses in the undisturbed ground. Hence, it is assumed that in the final state (some years after the construction of the tunnel), the ground eventually will return to nearly the same condition as that before tunnelling commenced. Changes in groundwater levels, traffic vibrations etc. may provoke this readjustment.

In case (1), for shallow tunnels and soft ground, the full overburden is taken as load. Hence, no tension bedding is allowed at the crown of the tunnel. The ground reaction is simplified by radial and tangential springs, arriving at a bedded beam model.

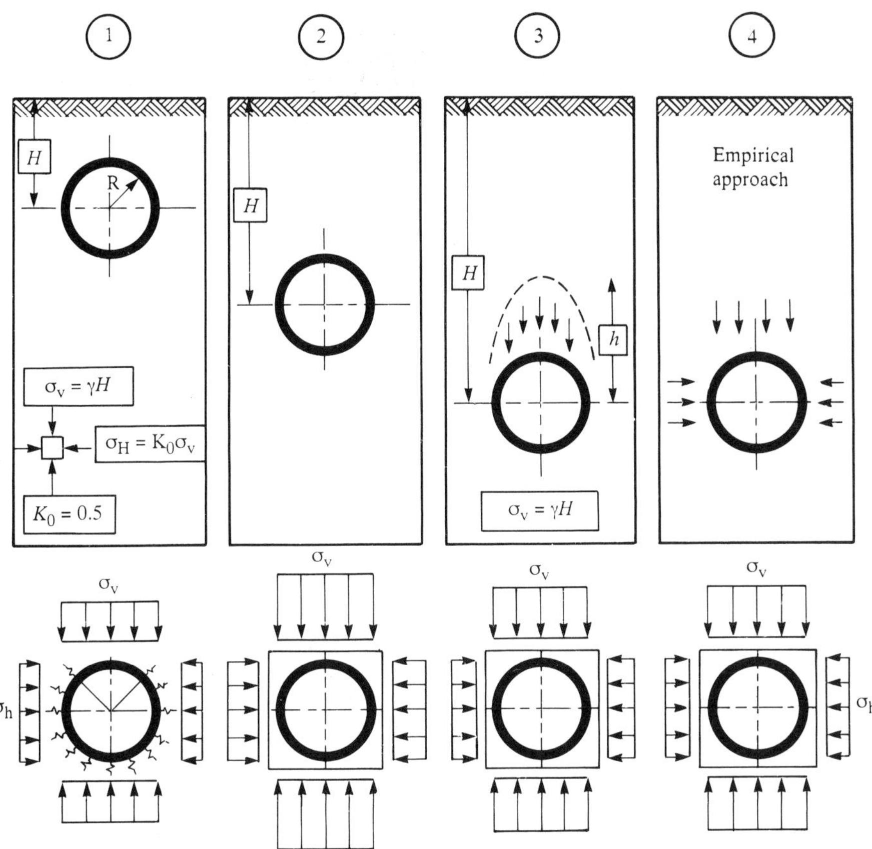

Figure 22.11 *Plane-strain design models for different depths and ground stiffnesses*

In case (2), for moderately stiff ground, the ground stiffness is employed by assuming a two-dimensional continuum model and complete bond between lining and ground. As in case (1), stress release due to predeformation of the ground is neglected. Inward displacements result in a reduction of the pressure on the lining.

Case (3) assumes that some stress release is caused by deformation that occurs before the lining participates. In medium-hard rock or in highly cohesive ground, the ground may be strong enough to allow a certain unsupported section at the tunnel face. Also, for tunnels having a high overburden, a reduction of the acting crown pressure (represented in Figure 22.11 by $h < H$) is taken into account. The ground pressure acting on the lining may be derived from a three-dimensional analysis (Kielbassa and Duddeck 1991), from *in situ* monitoring and experiences, or from ground models like those of Terzaghi (see the synopsis by Szechy 1966).

In case (4), the ground stresses acting on the lining are determined by an empirical approach, which may be based on previous experiences with the same ground and the same tunnelling method, on *in situ* observations and monitoring of initial tunnel sections, on interpretation of the observed data, and on continuous improvements of the design model.

If a plane model is not justified (as is the case for caverns, or for more complicated geometries of underground structures, or for an investigation directly at the tunnelling face) a three-dimensional model may be necessary. The three-dimensional model also may be conceived as consisting of discontinuous masses (block theory) or a continuum with discrete discontinuous fissures or faults. The ground may be modelled as homogeneous or heterogeneous, isotropic or anisotropic; as a two-dimensional, that is allowing some stress release before the lining is acting, or as a three-dimensional stiff medium (see Figure 22.12). The lining may be modelled either as a beam element with bending stiffness or as a continuum.

The capacity of the numerical approaches is large (Duddeck 1986), although it is still difficult to cover collapse modes by numerical analyses. Plasticity, viscosity, fracture of the rock, non-linear stress–strain and deformation behaviour etc. may be covered by special

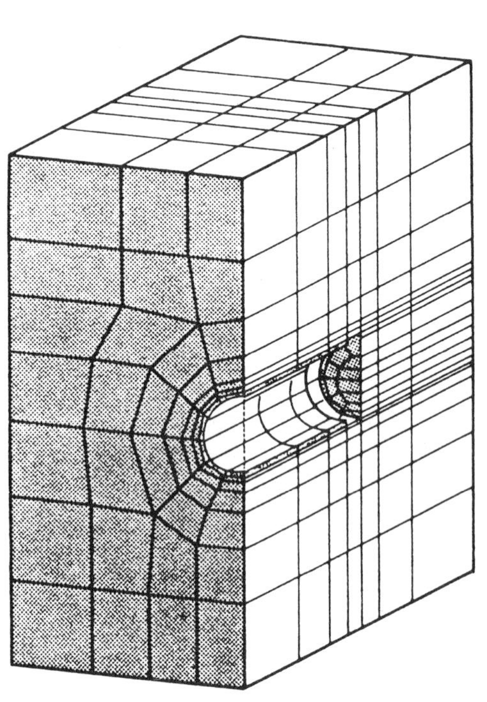

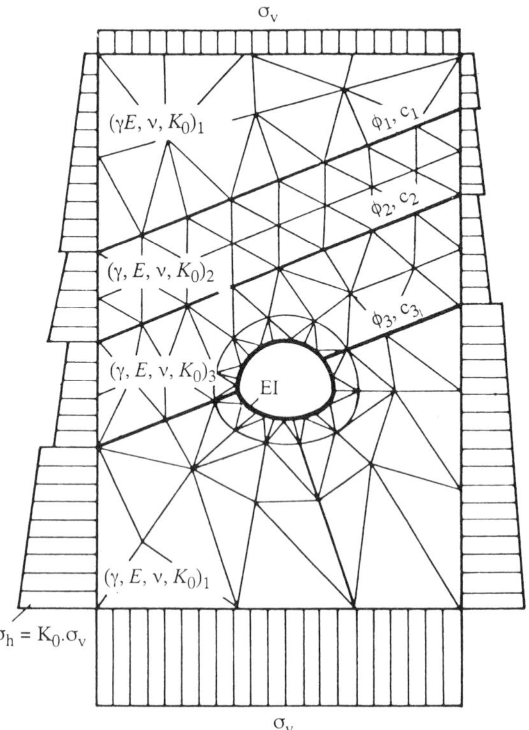

Figure 22.12 *Three-dimensional continuum model for applying the finite element method and an example for a two-dimensional FEM model*

assumptions for material laws and by applying numerical methods like those of the finite element method, the boundary element method, or a combination of both (Swoboda 1988). From their origins, the finite element method and the boundary element method are basically continuum methods. Thus, homogeneous media and stress–strain fields are evaluated best. In general, discontinua such as rock with fissures and faults, and failure modes, which are initiated by local rupture, shear failure, or full collapse, cannot be covered by continuum methods. For tunnel structures where the ground provides the principal stability of the opening (as in hard rock) or where the geometrical properties of the underground opening can be modelled only by numerical analysis — for example, in the case of closely spaced twin tunnels — a continuum or discontinuum model is necessary.

22.4.3 Empirical approach and observational design method

The structural elements and the excavation procedure, especially for the preliminary support of a tunnel, may be selected based mainly on experience and empirical considerations that rely more on direct observations than on numerical calculations. This procedure may be especially reasonable if experiences from a successful tunnelling project can be applied to a similar, new one, still to be designed. Such a transfer of information is justified only when:

(1) The ground conditions, including those of the groundwater, are comparable.
(2) The dimensions of the tunnel and its cross-sectional shape are similar.
(3) The depths of overburden are approximately the same.
(4) The tunnelling methods to be employed are the same.
(5) *In situ* monitoring yields results comparable to those for the preceding tunnelling project.

One disadvantage of the empirical approach is that, lacking an incentive to apply a more appropriate tunnelling design via a consistent safety assessment, the structure may be designed over-conservatively, resulting in higher construction costs. The simple empirical approach contributes little to the advancement of the state of the art in tunnelling.

The empirical approach to tunnel design also may be applied to larger projects in only slightly changing ground if provision is made (especially in the tender) for initial experiences to be extrapolated to the subsequent sections along the tunnel axis. Such situation justifies a measurement programme that is more intensive for the first sections, in order to gain experiences.

By combining numerical methods with the empirical approach and the immediate interpretation of *in situ* measurements, a tunnelling design procedure may be applied that is adjustable as the tunnel excavation proceeds. In this approach, the field measurements of ground movements, displacements and stresses are used on an ongoing basis to verify or modify the design of the tunnel. More intensively instrumented sections at the early stages of the tunnelling provide the data for these procedures. The interpretation of the measured data yields insight into the ground behaviour as a reaction to the tunnelling procedure.

In applying the observational method, the following conditions must be met:

(1) The chosen tunnelling process must be adjustable along the tunnel line.
(2) Owner and contractor must agree in advance on contractual arrangements that allow for modifications of the design on an on-going basis during the project.
(3) The field measurements should be interpreted on the basis of a suitable concept of a design model relating measurement data to design criteria. As a first approximation the convergence-confinement method (see Figure 22.10) may be applied, extended also for ground pressures that vary along the tunnel lining and for horseshoe-type cross-sections.
(4) The interpretation of a particular instrumented section must be used to draw conclusions about the other sections of the tunnel. Hence, the experiences are restricted to those tunnel sections that are comparable with respect to ground conditions.
(5) Field measurements should be provided throughout the entire length of the tunnel in order to check its assumed behaviour.

22.4.4 Special design features

Special considerations may be necessary if unusual ground behaviour is expected or if the quality of the rock is enhanced by employing ground improvement methods.

(a) Swelling ground

Stress release due to tunnelling and/or groundwater influx may cause swelling and a corresponding increase in pressure on the lining. In such cases, a circular cross-section or at least an invert arch is recommended. Swelling resulting from a chemical reaction, as when anhydrite hydrates to form gypsum, generally is much more pronounced than that due to the physical absorption of water, as in clay.

(b) Underground erosion, mining subsidence and sinkholes

Tunnelling in ground that is subject to settlements, as in the case of gypsum dissolution or mining subsidence, requires special design considerations. A flexible lining that follows the ground movements by utilizing its plastic deformation capacity is more suitable in these cases than is a too-rigid or brittle, failure-prone lining. If there is a possibility that sinkholes will form in the ground, a tunnel structure that can be repaired easily may be more economical than a structure designed to allow for bridging the sinkholes.

(c) Grouting and injections

Intensive grouting or injections of the rock may improve the ground characteristics considered in the design model. Although in most cases grouting is applied only for closing discontinuities in rock or for strengthening soft ground, in both cases the goal is to achieve better homogeneity.

(d) Drainage and compressed air

Usually the ground is stabilized by dewatering and by avoiding inflows of water. Ground failure may be avoided if the pore water pressure is minimized. The assumed ground characteristics may be valid only if successful drainage is possible or if water inflow is prevented, as in tunnelling under compressed air.

(e) Ground freezing

Improving the ground by freezing changes the ground properties. The time-dependent stress–strain behaviour of frozen ground can be significant. Freezing draws water toward the lining, causing an increase in water volume and heave at the surface. Concreting on frozen ground delays the strength development of the concrete.

22.5 *In situ* monitoring and its interpretation

In situ monitoring during excavation and at longer intervals after the tunnel is completed should be regarded as an integral part of the design, not only for checking the structural safety and the applied design model but also for verifying the basic conception of the response of the ground to tunnelling and the effectiveness of the structural support. The main objectives of *in situ* monitoring are:

(1) To control the deformations of the tunnel, including securing the open tunnel profile. The time-history development of displacements and convergences may be considered as a safety criterion, although field measurements do not yield the margins the structure can endure before failing.

(2) To verify that the appropriate tunnelling method was selected is appropriate.
(3) To control the settlements at the surface, for instance, in order to obtain information on the deformation pattern in the ground and on that part of settlement caused by lowering the groundwater level.
(4) To measure the development of stresses in the structural members, indicating sufficient strength or the possibility of strength failure.
(5) To indicate progressive deformations, which require immediate action for ground and support strengthening.
(6) To furnish evidence for insurance claims, for example by providing results of levelling the settlement at the surface in town areas.

A programme for monitoring the deformations and stresses during the excavation may comprise the following measurements (see Figure 22.13):

(1) Levelling the crown (at the least) inside the tunnel as soon as possible. With regard to interpretation of the data, Figure 22.9 reveals that often only a small fraction of the entire crown movement can be monitored, because a larger part occurs before the bolt can be set. For difficult tunnelling, the distance between two crown readings may be as close as 10–15 m. Levelling of the invert is recommended for rock having swelling potentials.
(2) Convergence readings (in triangular settings; *K* in Figure 22.13) should be the standard method for early information. They are easily applied and are accurate to within 1 mm.
(3) In a few cross-sections, the lining may be equipped with stress cells for reading the ground pressures and ring forces in the lining (*G* and *R* in Figure 22.13).
(4) Stress cells also should be installed in a few sections of the final second lining if long-term readings are desired after the tunnel has been completed.
(5) Surface levelling along the tunnel axis and perpendicular to it yields settlements and the correlation with measurements inside the tunnel.
(6) Extensometers, inclinometers and gliding micrometers may be installed from the surface well ahead of the tunnel face, yielding deformation measurements within the ground (see Figure 22.13). Monitoring ground deformation is especially appropriate for checking and interpreting the design model. Therefore, the installation should be combined with convergence readings and stress cells in the same cross-section.

The frequency of the readings depends on how far from the tunnel face the measurements are taken, and on the results. For example, readings may be performed initially twice a day; then be reduced to one reading per week four diameters behind the face; and end with one reading per month if the time-data curves justify this reduction in measurement readings.

The results of *in situ* monitoring should be interpreted with regard to the excavation steps, the structural support work, and the structural design model in conjunction with safety considerations. The actual readings normally show a broad scatter of values. Expectations of reliability may not be met, especially for pressure cells, because stresses and strains are very local characteristics. Deformations and convergence readings are more reliably obtainable because displacements register integrals along a larger section of the ground. The *in situ* measurements should be interpreted in consideration with the following aspects:

(1) The results should verify whether the tunnelling method is appropriate.
(2) Graphed time-history charts may reveal a decreasing rate of deformation, or uncover danger of collapse.
(3) Large discrepancies between the predicted and actually observed deformations may force revision of the design model. However, measurements are valid only

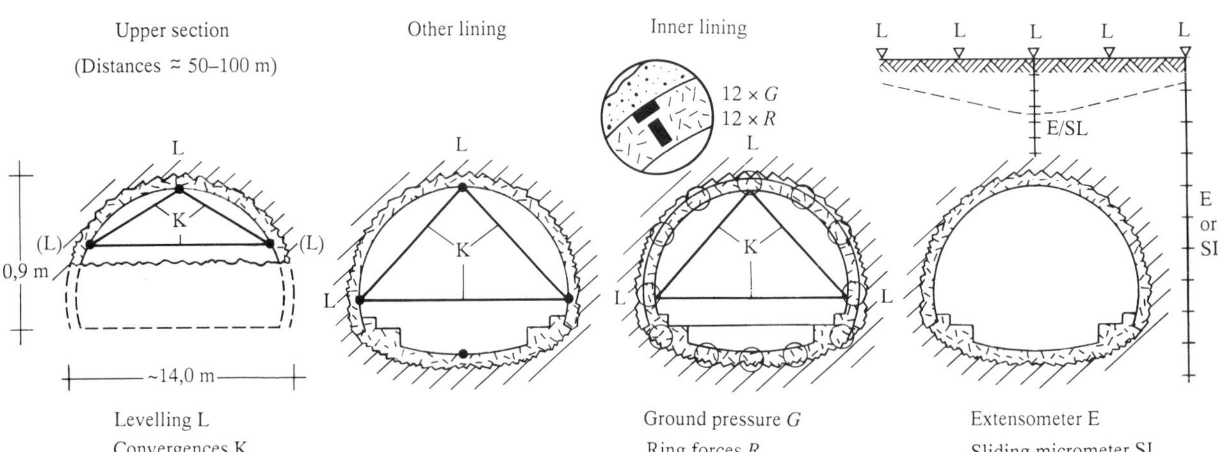

Figure 22.13 *Example of monitoring a tunnel excavation, the linings, and the surface settlements*

for the actual state at the time and the place where they are taken. Long-term influences such as rising water level, traffic vibrations, and long-term creep are not registered during excavation.

(4) The readings may promote visual understanding of the structural behaviour of ground and support interaction.

(5) The readings may cover only a fraction of the actual phenomena if bolts and stress cells are installed too late (see Figure 22.9).

(6) The tunnel may be considered stable when all the readings cease to increase. However, a safety margin against failure (especially against sudden collapse) cannot be deduced from measurements, except by extrapolation.

22.6 Structural detailing of the lining

For concrete linings, the following structural design specifications are suggested:

(1) The thickness of a second lining of cast-in-place concrete may have a lower limit of 250–300 mm to avoid concrete placing problems such as undercompaction or honey-combing of concrete. The following lower limits are recommended:

25 cm if the lining is unreinforced;
30 cm if the lining is reinforced;
35 cm for watertight concrete.

(2) Reinforcement may be desirable for crack control, even when it is not required for covering inner stresses. On the other hand, reinforcement may cause concrete-placing problems or long-term durability problems due to steel corrosion. If reinforcement in the second lining is provided for crack control, a closely spaced steel mesh reinforcement may have the following cross-sections in both directions:

At the outer surface, at least 150 mm^2/m of steel;
At the inner surface, at least 300 mm^2/m of steel.

(3) The recommended minimum cover of reinforcement is

3.0 cm	At the outer surface if a waterproof membrane is provided.
5.0–6.0 cm	At the outer surface if it is directly in contact with the ground and groundwater.
4.0–5.0 cm	At the inner tunnel surface.
5.0 cm	For the tunnel invert and where water is aggressive.

(4) For lining segments, specifications (1)–(3) above are not valid, especially if the segmented tunnel ring is the outer preliminary lining. For detailing the tunnel segments, special attention should be given to avoiding damage during transport and erection.

(5) Sealing against water (waterproofing sheets) may be necessary under the following conditions:

(a) When aggressive water action threatens to damage concrete and steel.
(b) When the water pressure level is more than 15 m above the crown.
(c) When there is a possibility of freezing of ingressing water along the tunnel section close to the portals.
(d) When the inner installations of the tunnel must be protected.

(6) In achieving watertightness of concrete, special specifications of the concrete mixture, avoidance of shrinkage stresses and temperature gradients during setting, and the final quality of the concrete are much more important than theoretical computations of crack widths.

(7) Temperature effects (tension stresses) may be somewhat controlled by working joints (as close as 5 m at the portals) and by additional surface reinforcement in concrete exposed to low temperatures.

(8) An outer lining of shotcrete may be considered to participate in providing stability of the tunnel only when the long-term durability of the shotcrete is preserved. Requirements for achieving long-term durability include the absence of aggressive water, the limitation of concrete additives for accelerating the setting (liquid accelerators), and avoiding shotcrete shadows behind steel arches and reinforcements.

22.7 Documents for tunnelling

The basic documents for tunnel design should include or cover:

(1) the geological report presenting the results of the geological and geophysical survey;
(2) the hydrogeological report;
(3) the geotechnical report on the site investigation, including the interpretation of the results of site and laboratory tests with respect to the tunnelling process, rock classification, etc.;
(4) information on line, cross-section, drainage, and structural elements affecting later use of the tunnel;
(5) a plan for and a description of the projected excavation or driving procedure, including the different cross-sections related to different ground conditions;
(6) design documents for the types of excavation methods and tunnel supports likely to be applied, considering, for instance, excavation advance and face support (types and number of rock bolts, anchors, amount of grouting, shotcrete stength, closure lengths, etc.); the corresponding items also for a tunnel boring machine;
(7) the programme for the *in situ* monitoring of the tunnel by field measurements;
(8) the analysis of stresses and deformation (for unlined tunnels as well as for single or double-lined tunnels),

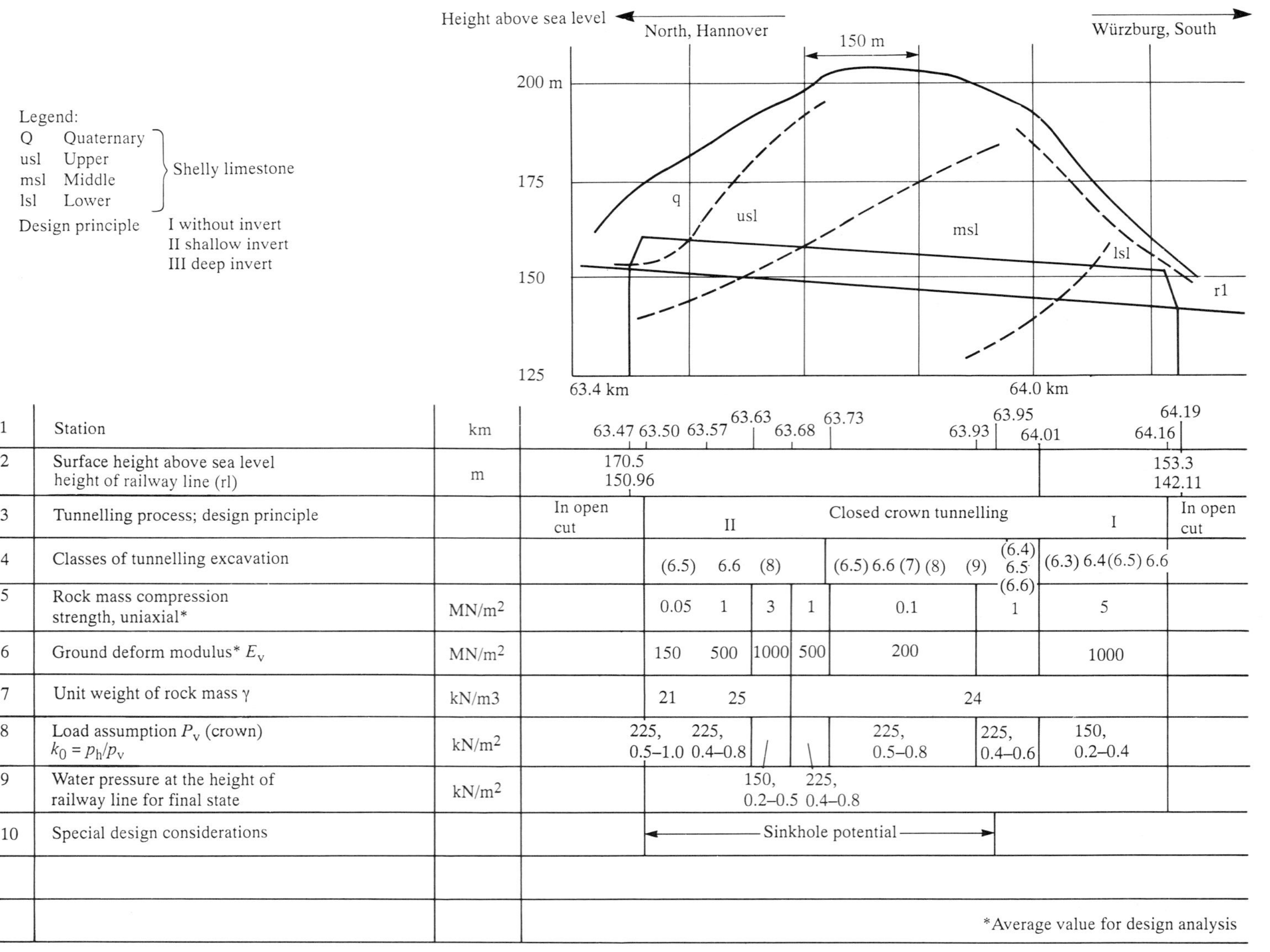

		Unit	Values
1	Station	km	63.47 63.50 63.57 63.63 63.68 63.73 63.93 63.95 64.01 64.16 64.19
2	Surface height above sea level; height of railway line (rl)	m	170.5, 150.96; 153.3, 142.11
3	Tunnelling process; design principle		In open cut; II; Closed crown tunnelling; I; In open cut
4	Classes of tunnelling excavation		(6.5) 6.6 (8); (6.5) 6.6 (7) (8) (9); (6.4) 6.5 (6.6); (6.3) 6.4 (6.5) 6.6
5	Rock mass compression strength, uniaxial*	MN/m^2	0.05 1; 3; 1; 0.1; 1; 5
6	Ground deform modulus* E_v	MN/m^2	150 500; 1000; 500; 200; 1000
7	Unit weight of rock mass γ	kN/m3	21 25; 24
8	Load assumption P_v (crown) $k_0 = p_h/p_v$	kN/m^2	225, 0.5–1.0; 225, 0.4–0.8; 225, 0.5–0.8; 225, 0.4–0.6; 150, 0.2–0.4
9	Water pressure at the height of railway line for final state	kN/m^2	150, 0.2–0.5; 225, 0.4–0.8
10	Special design considerations		← Sinkhole potential →

*Average value for design analysis

Figure 22.14 *Example of a condensed table of design conditions. Predicted ground conditions, tunnelling classes and design characteristics along a tunnel of the high-speed railway line in Germany.*

and the dimensioning of the tunnel support for intermediate phases and final linings;

(9) the design for waterproofing or drainage;

(10) structural documents for the final design of the tunnel project, including the detailing;

(11) during and after the excavation, reports on the field measurements and interpretation of their results with respect to the response of the ground and the structural safety of the tunnel;

(12) documentation of the problems encountered during the excavation and measures applied, for example, strengthening the ground or changing the projected type of support, based on monitoring results.

The above sequence of these basic documents also provides the general outline of the design procedure. In Figure 22.14 an example is given for a table of design conditions prepared for tendering.

References

Anon. (1980) 'Tunnel stability by convergence-confinement method', General Report, French Engineers, *Underground Space*, **4**, 225–232

ITA (1982) 'Views on structural design models for tunnelling', ITA Working Group on Design of Tunnels, *Advances in Tunnelling Technology and Subsurface Use*, **2**, 155–227

ITA (1984) Recommendations of the ITA Working Group 'contractual Sharing of Risks'. *Advances in Tunnelling Technology and Subsurface Use*, **3**, 195–203

Barton, N., Lien, R. and Lunde, J. (1974) 'Engineering classification of rock masses for the design of tunnel support', *Rock Mechanics*, **6**, 189–236

Bieniawski, Z. T. (1984) *Rock Mechanics Design in Mining and Tunnelling*, Balkema, Rotterdam

Duddeck, H. (1980) 'On the basic requirements of applying the convergence-confinement method', *Underground Space*, **4**, 241–247

Duddeck, H. (1986) 'Leistungsfähigkeit und Grenzen der Methode der finiten Elemente in der Geotechnik.' *Felsbau*, **4**, 126–133

Duddeck, H. (1987) 'General approaches to the design of underground openings'. In *Proc. Tunnel Australia 1987*, Melbourne, 159–172

Duddeck, H. (Ed.) (1988) 'Guide lines for the design of tunnels', ITA-Working Group in *Tunnelling and Underground Space Technology*, **3**, (3), Pergamon Press, Oxford, New York

Erdmann, J. (1983) *Comparison of Two-dimensional and Development of Three-dimensional Design Methods for Tunnels* (in German), Berichte Institut für Statik, Technical University Braunschweig, Germany

Hoek, E. and Brown, E. T. (1980) *Underground Excavations in Rock*, Institution of Mining Metallurgy, London

Kerisel, J. (1980) 'Commentary on the General Report', *Underground Space*, **4**, 233–239

Kielbassa, S. and Duddeck, H. (1991) *Stress–Strain Fields at the Tunnelling Face – Three-dimensional Analysis for Two-dimensional Technical Approach.* Rock Mechanics and Rock Engineering, **24**, 115–132

Leichnitz, W. (1987) *Tunnelbau im Nordabschnitt der Neubaustrecke Hannover-Würzburg* (tunnelling for the northern section of the new railway line from Hannover to Würzburg), Deutsche Bundebahn Hannover

Lombardi, G. (1980) 'Some comments on the convergence-confinement method', *Underground Space*, **4**, 249–258

Maidl, B. (1988) *Handbuch des Tunnel- und Stollenbaus*, I and II Verlag Glückauf, Essen,

Müller-Salzburg, L. (1978) *Der Felsbau Band III Tunnelbau*, Ferdinand Enke Verlag, Stuttgart

Swoboda, G. (1988) (Ed.), 'Numerical methods in geomechanics', *Proc. 6th Int. Conf.*, Innsbruck, Balkema, Rotterdam

Szechy, K. (1966) *The Art of Tunnelling*, Akademiai Kiado, Budapest

Wittke, W. (1984) *Felsmechanik*, Springer, Berlin

23 Underground chambers in hard rock masses

Professor O Stephansson
Royal Institute of Technology, Stockholm

23.1 Introduction

The rock mass in which an underground chamber is to be excavated can be regarded as a construction material, and the rock engineer should provide a design for the underground chamber which is capable of standing up with a minimum of support, if any. The task of the rock engineer is to ascertain that his assumptions are true and to modify the design and to support the rock mass where there is a need for it. However, the major problem in rock engineering compared to most other engineering branches is the unpredictable, complex, inhomogenous behaviour of the rock material. The extent to which the rock engineer can overcome the difficulties depends on his awareness of the geological conditions which exist on site and the extent to which he can take them into account. Therefore an accurate interpretation of the geology on site is an essential prerequisite to the design of underground chambers. Further, the geological data must focus on those properties which have a decisive influence on the location, size, shape, excavation sequence and reinforcement of the chamber.

Many places throughout the world have suitable geological conditions for underground construction. For this reason underground space should be regarded as a natural resource to be utilized wisely in the same way as mineral resources, hydropower and land for agriculture. The key to the functional and economic benefits of underground development is the availability of suitable ground conditions. This demand is often fulfilled when constructing chambers in hard rock. A second requirement is knowledge of the technology and its economic and technical characteristics.

23.2 Benefits of underground chambers

The growing demand for space in urban areas, new environmental demands and commercial possibilities provided by technical developments have all made the use of underground space competitive and sometimes necessary. Rock construction can enable the excavation of extensive rock spaces to be carried out near or even beneath the centres of densely populated areas. In addition the excavation of caverns in rock makes it possible to locate plants and activities which are potentially dangerous to the environment near population centres, for example, sewage treatment plant.

A unique characteristic of underground chambers is their large freedom of size and form. The size is in most cases limited by the quality of the rock mass. In good-quality rock masses chambers have been excavated to widths of more than 30 m with only minimum rock reinforcement. Spans are often restricted to 15–20 m in fair to poor-quality rocks and most sedimentary rock types. The geometrical form of an underground chamber is governed by the use to which it is put. The category, type and advantages of underground chambers in rock masses are presented in Table 23.1.

Subsurface construction belonging to categories art and culture, civil work and sport has been developing rapidly in many cities during the past two decades. As an example subsurface rock spaces in Helsinki, Finland, now account for nearly 20% of the space occupied by all public buildings (Anttikoski 1986). The constraints of the surface and related structures, which affect all surface buildings, do not apply below ground. Made-to-measure three-dimensional rock caverns offer incomparable advantages as demonstrated in a number of underground constructions by Winqvist and Mellgren (1988).

Both military and civil defence installations below ground have been in continuous development since the Second World War. An outline of existing installations in Sweden are listed in Table 23.1.

Many types of installations are constructed underground for economic reasons, for example, oil and gas storage and cold storage. The benefits of underground chambers compared with conventional structures at the ground surface are lower construction and operational costs, and less maintenance. A typical underground oil storage plant as built in Scandinavia is shown in Figure 23.1. Today there are over 200 such installations in use for storing crude oil and refined products from petrol to heavy fuel oil. Liquefied petroleum gas (LPG) is also stored in

Table 23.1 Utilization of underground chambers in rock masses

Category	*Type*	*Advantages of going underground*
Art and culture	Art centre Church Concert hall Museum	Steady temperature, constant humidity, low energy consumption, no dust, good acoustics and special atmosphere
Civil work	Electrical transfer Telephone exchange Sand silo Workshop Laboratory	Location in the city, minimize fire and short circuit, high level of security, constant temperature and humidity
Defence	Defence shelter Naval base Underground hangar Control centre	Shield against direct hits, shock waves, gas and radioactive fallout
Energy	Compressed air Heat storage in caverns Hydropower Oil and gas storage Oil and gas power station	Optimum tailoring for equipment and maintenance, low cost for large volumes and low operating cost
Food	Cold store Grain Fish Liquor and wine	High thermal inertia, good insulation, low energy consumption, minimum maintenance
Sport	Gymnastics Ice hockey rink Physical exercise Swimming hall	Good insulation, low energy consumption, little maintenance cost
Waste	(1) *Non-toxic* Refuse collection and sewage treatment plant Compacting plant	No discharge, minimum environmental conflict, limited land purchase, advantageous climate for the treatment process
	(2) *Toxic* Chemical waste plant	
	(3) *Radioactive* Low, medium and high level waste	Isolation from biosphere for very long time, high integrity and safety

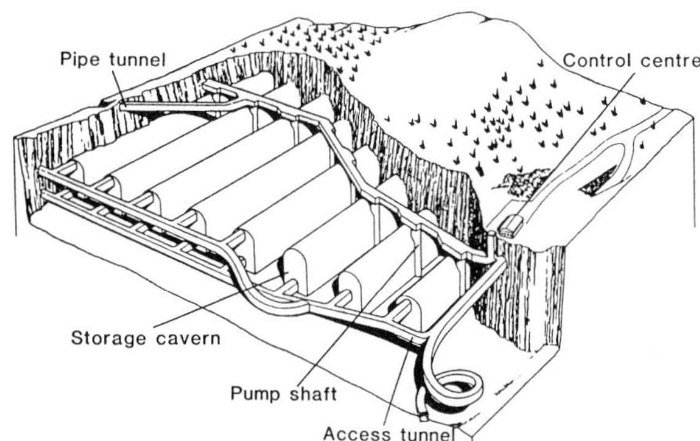

Figure 23.1 *Typical underground oil storage plant in Scandinavia. Over 200 installations are in use for storing crude oil and refined products.*

several of the installations. Rock chambers are often 16–20 m wide, 30 m high and many hundreds of metres long. Several have a volume of more than 1 million m^3 and they are typically located at the coast with access to modern harbours.

The main reasons why oil can be stored in unlined rock chambers are that oil and water do not mix and that oil is lighter than water. Therefore the hydrostatic pressure of the water in the rock mass is greater than that of the stored oil at all points. The large horizontal oil–water interface which develops from storage over a water surface, creates a favourable environment for microbial growth. This can make the petroleum product corrosive and affect the quality seriously. The specific problems which are faced when oil is stored in horizontal caverns (microbial attack, sludge deposits, polluted bedwater, etc.) can be resolved by using a vertical cavern design. For instance, vertical

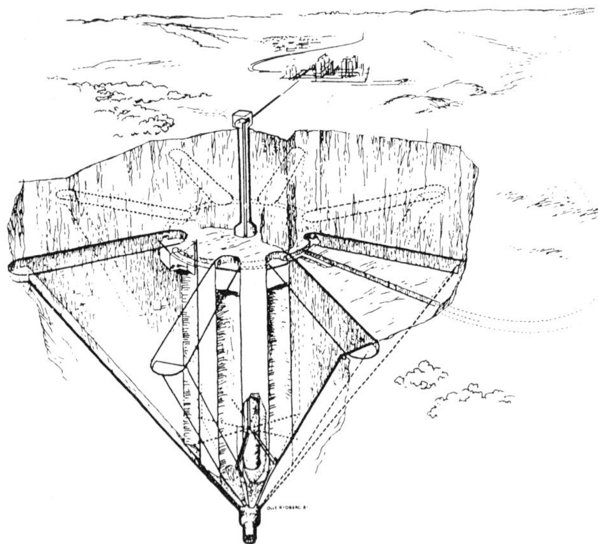

Figure 23.2 *Concept of large-scale underground storage plant. Several vertical caverns oriented around a central pump shaft minimize the oil–water contact and the microbial activity (after Daerga et al. 1986)*

cylindrical tanks with cone-shaped bases were suggested by Sagefors and Svemar (1986) and the 'funnel storage plant', (Daerga *et al.* 1986) consists of several vertical caverns oriented around a central pump shaft (Figure 23.2).

Construction costs for different oil storage alternatives are shown in Figure 23.3. This comparison is made for crystalline rocks and is representative of Scandinavia. The construction cost for a 'funnel store' with a volume of 4 million m^3 is found to be 3.5 USD/barrel calculated from prices which applied in 1981. The low cost and rational construction are due to the implementation of improved drilling and blasting techniques in a large-scale mining method.

23.3 Design and construction procedure for underground chambers

With such a vast range of underground usage as demonstrated in Table 23.1, many kinds of rock engineering and rock mechanics considerations need to be addressed. In most cases the rock mass at the site of the construction is hidden, covered with soil or located at depth far away from the exposed rock. Occasionally, when an existing underground installation is going to be enlarged, the rock engineer will have access to the site and is able to judge the quality of the rock mass. However, in most cases the rock engineer is restricted to surface exposures, geophysical information, drill hole logs, drill cores and sometimes exploration shafts and tunnels.

Another important aspect of underground working is that the construction of chambers in rock is initially affected by virgin stresses. This separates geomechanics engineering from most other engineering disciplines and calls for rock stress measurements to be done. When the chamber has been excavated it will cause a change in the initial stress in the vicinity of the opening. The new stress field will affect the overall stability of the chamber and will determine whether rock support is needed.

The majority of underground chambers are made below the water table. Therefore the amount of groundwater flow into the chamber, the weakening of the rock mass due to alteration and swelling, and the degradation of the rock reinforcement due to corrosion are matters to be considered.

The large temperature inertia of rock masses gives a fairly constant temperature for most underground openings. The absolute temperature is governed by the geological terrain and the temperature gradient. Finally, the humidity, which is controlled by the access of water and the temperature, sometimes has severe effects on the swelling or shrinkage of any clay materials associated with the rock mass.

Given the major characteristics of the rock mass for

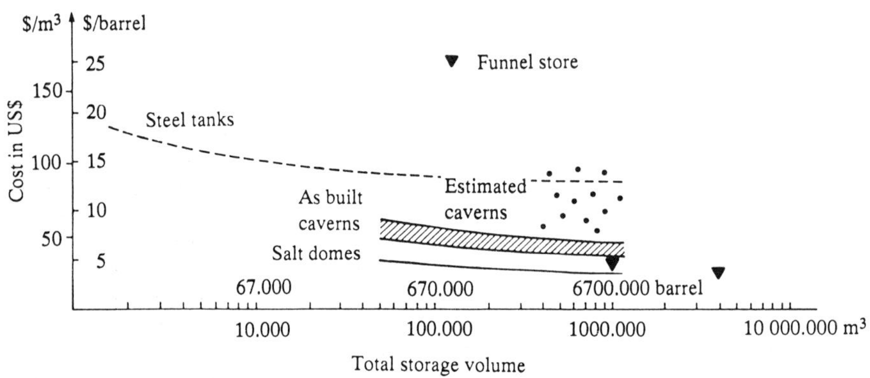

Figure 23.3 *Construction costs for different oil storage alternatives. Price and cost level for 1981. Funnel store refers to the concept presented in Figure 23.2 (after Daerga et al. 1986)*

underground construction one has to consider where to start and what are the basic steps involved in the design and construction procedure. In planning the design of an underground chamber one can follow different procedures. The one suggested by Hoek and Brown (1980) is to a large extent based on the design of underground support systems and four sources of possible instability:

(1) adverse structural geology;
(2) excessively high rock stress;
(3) weathering and/or swelling rock;
(4) excessive groundwater pressure or flow.

This design method calls for a good knowledge of the geology and rock-mass conditions at the site. Rock-mass classification schemes such as those due to Barton *et al.* (1974) and Bieniawski (1976) were developed as methods of estimating support requirements for underground excavations. They serve as examples of a slightly different design approach and will be discussed further in later sections.

The design and construction procedure presented in Table 23.2 will be followed in this chapter. This outlines a complete scheme, noting the various steps from the very first preliminary investigation or feasibility study to the final rock monitoring and control of the stability of the completed underground chamber.

Table 23.2 Design and construction procedure for underground chambers

Site characterization
Preliminary investigation
Geological survey
Applied geophysical survey
Drilling and excavation
Rock testing
Rock stress measurement
Hydraulic testing
Rock-mass classification
Rock-engineering consideration
Layout optimization
Excavation sequence
Stress and strength analysis
Chamber excavation
Rock support and reinforcement
Rock monitoring and control

23.4 Site characterization

While it is impossible to anticipate all the possible conditions that might cause problems in the construction of an underground chamber it is absolutely necessary that all possible efforts be made to obtain and present – in a readable form – the rock-mass characteristics at an early stage of an underground project. There are numerous examples where rock engineers have been faced with unexpected problems during excavation due to poor site investigations. To overcome or reduce the problems some countries provide codes of practice (e.g. the *British Standard Code of Practice for Site Investigations*, SI, BS 5930:1981). This helps the rock engineer to structure the information and defines the minimum that should be known about geology and site exploration. In a major rock-engineering project each of the stages listed under site characterization in Table 23.2 might be carried out and reported on.

23.4.1 Preliminary investigation

At an early stage of an underground chamber project it is important that any geological study recorded in scientific papers, technical reports, and so forth, are studied. This stage is sometimes called 'archive drilling' and means that all possible information is collected and reported on. The following maps can be studied and techniques applied:

(1) topographic maps;
(2) regional geological maps;
(3) tectonic and structural maps;
(4) geophysical and hydrogeology maps;
(5) air photographs;
(6) stereoscopic examination;
(7) contouring;
(8) sectioning;
(9) photogeology.

The objective of the preliminary investigation is to present the best possible three-dimensional characteristics of the rock volume of the area where the underground chamber can possibly be located. Notice that at this stage of the project several alternatives should be looked into and the optimum location should be determined based on the function of the chamber, the geological situation, rock-engineering considerations, excavation technique and economy.

23.4.2 Geological survey

Once the most likely site for the underground chamber has been located the more detailed investigations will start. Geological maps, typically to the scale of about 1:1000 or less, are produced from observations of outcrops, cuts and trenches and, if possible, from existing underground excavations. The maps should show the different rock types and their spatial distribution, the geological structures, any major weak zones and the degree to which the rocks have been weathered.

Graphical presentation of geological data is of utmost importance. From plan maps and sections the geologist and the rock engineer should produce three-dimensional block diagrams and/or transparent models of the rock mass and its major discontinuities. As the site characterization proceeds the data gathered from further additional investigations are added to the maps, sections and models.

23.4.3 Geophysical survey

There are several geophysical methods available that are capable of adding essential information to the overall geological situation at a potential site. Seismic refraction exploration along selected profiles is an old and well-known method in the study of rock and soil characteristics. The method gives as output the thickness of the overburden, the position of the water table, the quality of the rock mass and the location of major faults and fracture zones. Airborne electromagnetic methods and electrical resistivity measurements are other geophysical methods applicable.

Drill-hole seismic surveys in combination with tomography and drill-hole reflection radar are two new and promising geophysical methods for determination of the rock-mass quality and detection of major discontinuities. Both methods need drill holes but can scan fairly large rock volumes. The principle of the radar system is shown in Figure 23.4. The distance to a reflecting object is determined by measuring the difference in arrival time between the direct and the reflected pulse. As the radar is pushed into the drill hole the time difference varies in a typical manner determined by the reflector (Olson *et al.* 1985) drill-hole seismic surveys and drill-hole reflection radar have been applied in the underground research laboratories for radioactive waste disposal at Stripa in Sweden, Grimsel in Switzerland and Pinnowa in Canada.

23.4.4 Diamond drilling

Diamond drilling is one of the most important methods of subsurface exploration. The drill core from the rock mass at a potential site is used for verification of a proper site selection of the chamber. One or several vertical holes are usually drilled for this purpose. Inclined holes are drilled with the aim to penetrate weak zones (faults, shear zones) and to determine their orientation and dimensions at depth. The number of holes to be drilled for an underground project varies from a few to several tens depending upon the size of the chamber and length of the access tunnels, the amount of rock exposures, degree of weathering and frequency of weak zones.

Drill-bit sizes are chosen to suit the rock type and depth of hole being drilled. The common bit sizes for igneous and other hard rocks are EX (22 mm), AX (28 mm) and BX (41 mm), but for softer rocks larger core drills should be used, HX (76 mm) and larger, together with double-tube core barrel. It needs to be pointed out that a successful diamond drilling campaign needs highly skilled and experienced drill operators. The contract between the operator and the contractor should aim at encouraging a high percentage of core recovery and if possible the core should be orientated.

Logging cores for engineering purposes should be conducted with great care by an engineering geologist or a rock engineer. Each drill-hole log should contain the following information:

(1) drill-hole identification, grid reference, contract details;
(2) drilling method, equipment, orientation;

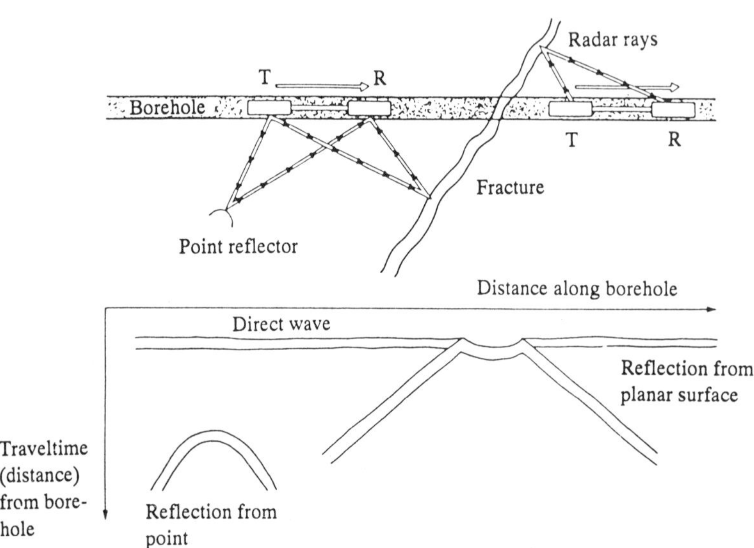

Figure 23.4 *Principle of drill-hole reflection radar and the pattern generated by a point and a plane reflector (after Olsson et al. 1985)*

(3) drilling progress, resistance, *in situ* testing;
(4) groundwater levels and changes in water level;
(5) geological description; rock type, discontinuities, core recovery, RQD, weathering, field tests.

A semi-automatic core-logging system called PETRO CORE was developed at the Swedish Research Mine at Kiruna (Ludvig *et al.* 1986). The latest version of the system, manufactured by Petro Bloc AB in Sweden, is shown in Figure 23.5. The core box is placed on a special table where an automatic tape measure is placed. True depth of geological features are defined and coded into the personal computer and the information can be applied in computer-aided design (CAD) of the chamber and its access tunnels or plotted as demonstrated in Figure 23.6.

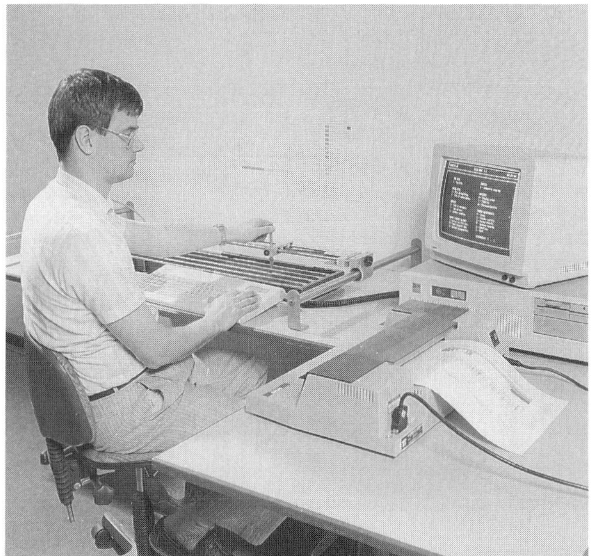

Figure 23.5 *PETRO-CORE: a semi-automatic core-logging system. Courtesy of Petroblock Co.*

At a late stage of the design of a large underground chamber, and often as a part of the final construction, exploratory adits and shafts are excavated. The additional information helps determine the final location and orientation of the chamber and these adits and shafts serve as test sites for optimization of drilling, blasting and rock reinforcement.

23.4.5 Rock testing

The rock-engineering-based approach to the design and construction of underground chambers used in this chapter requires prior definition of the stress–strain behaviour and strength of the rock mass. The most important parameters are the elasticity and the stress level at which yield, fracturing or slip occur. In designing rock chambers in hard rock it is the behaviour of the intact rock material that is of concern. Also the behaviour of large single discontinuities or a small number of discontinuities is of paramount importance.

Samples, especially core, from the field can be prepared and tested in the laboratory. Suggested techniques for determining the strength and deformability of rock material are given by the International Society for Rock Mechanics Commission on Standardization of Laboratory and Field Tests (Brown 1981) and an outline of rock testing is given in Chapter 8.

Instead of bringing the core boxes or samples to the laboratory the most essential tests can be conducted in the field by means of a portable rock tester (Figure 23.7). The system consists of a loading frame with a capacity of 10 tonnes and a number of test fixtures for three-point bending, four-point bending, uniaxial compression and tension testing (Röshoff 1986). The load actuator is driven by a small hydraulic pump. A bridge box for selection and balance of the test channels and strain gauges forms the interface between the sample and the read-out unit. A data logger monitors the load and signals from the strain gauges, and a microcomputer carries out the calculation, printing and plotting of elastic parameters and strength in accordance with international standards.

A complete testing procedure with the rock tester is performed as follows:

(1) A specimen of a core is selected from the core box.
(2) A core edge notch is prepared by a portable diamond saw.
(3) The specimen is loaded in three-point bending.
(4) Failure load is processed by the microcomputer and fracture toughness, K_{Ic}, is read.
(5) The resulting two halves are prepared by gluing axial and radial strain gauges to the core, the specimens are loaded in four-point bending, stress–strain curves are plotted and the microcomputer calculates Young's modulus, E, Poisson's ratio, ν, and modulus of rupture for both samples.
(6) A point load test can be carried out on the remaining four pieces of the core. The point-load index for appropriate core diameter is determined and the uniaxial compressive strength is calculated.

In rock-engineering problems other than those involving only fracture of intact rock, in particular those related to the shear behaviour along discontinuities will be important. The most commonly used method for shear testing of discontinuities in rock is the direct shear test. The test is commonly carried out in the laboratory, but it may also be carried out in the field using a portable shear box. Discontinuities contained in pieces of drill cores or small blocks are fixed inside the shear box using epoxy resin or plaster of Paris as encapsulating material. Normal stiffness, k_n, shear stiffness, k_s, peak strength τ_p, and residual strength, τ_r are determined from the shear test.

When a number of such tests are carried out at a range of normal stresses, a linear shear strength envelope is

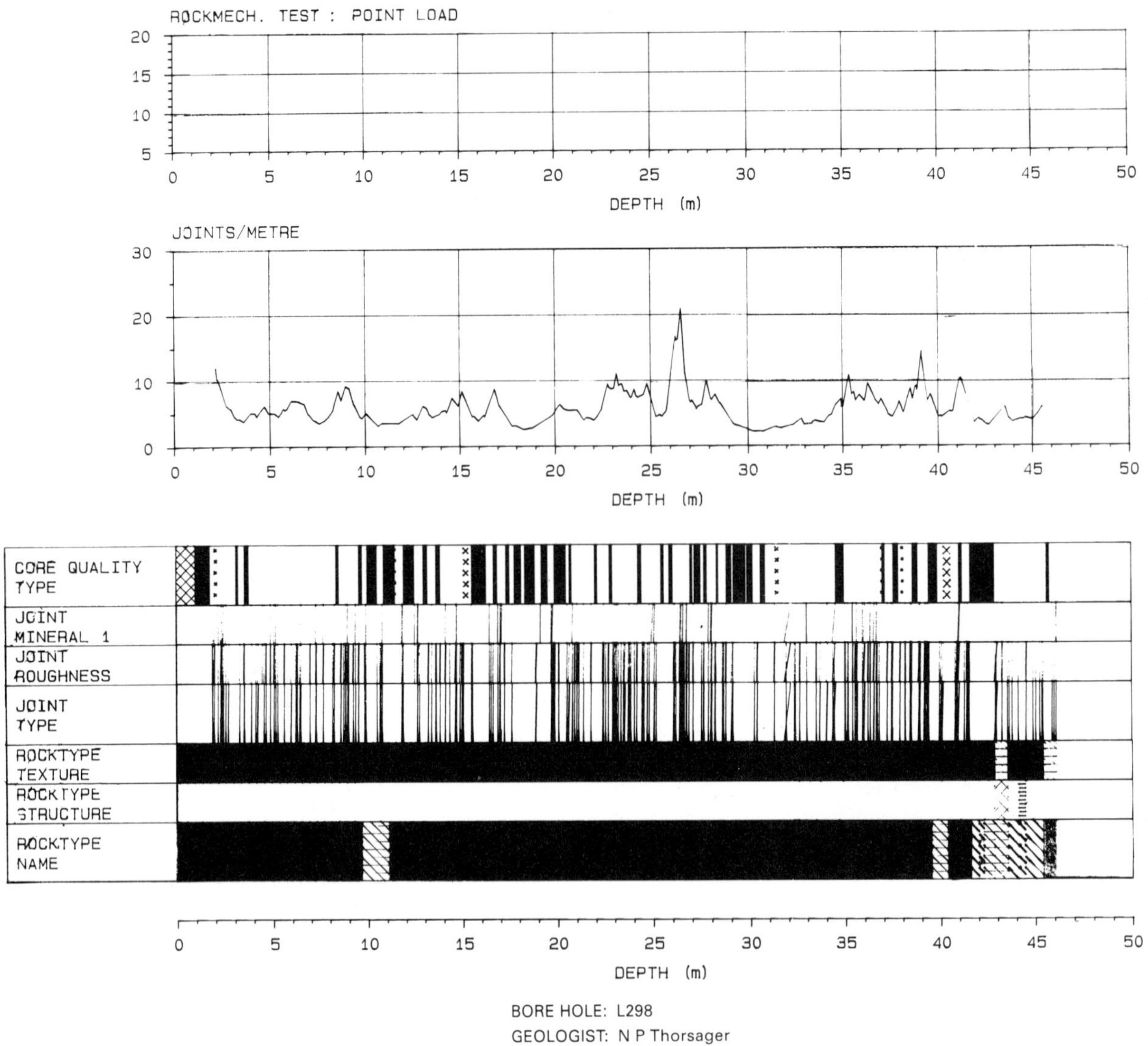

Figure 23.6 *Output from core-logging system PETRO- CORE*

obtained. The angle of friction ϕ, of the discontinuity surface is determined by the slope of the envelope in a diagram relating shear stress (peak or residual) versus normal stress.

23.4.6 Rock stress measurement

Critical to the design and construction of a major underground excavation is the stress regime of the rock mass in which it is to be built. The pre-existing state of stress at a potential site for an underground chamber can be estimated or more preferentially measured. Compilations of a large number of rock-stress measurements from many sites all over the world have shown that the measured vertical stress, S_v, is in fair agreement with the simple prediction given by calculating the vertical stress due to the overlying weight of rock at a particular depth from the equation

$$S_v = \gamma Z \qquad (23.1)$$

where γ is the unit weight of the rock, usually 27 kN/m^3, and Z is the true vertical depth at which the stress is required.

At shallow depths there is a considerable amount of

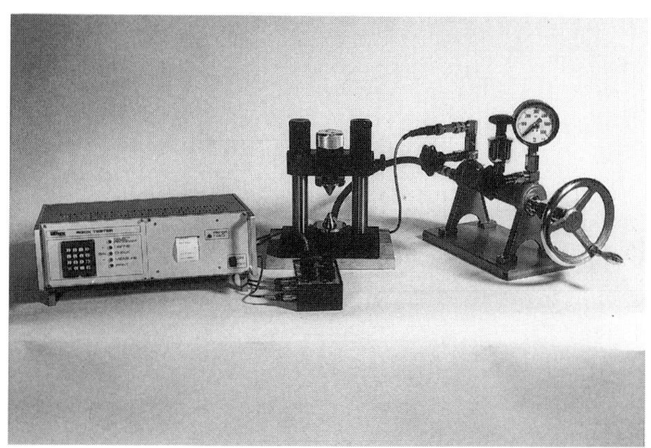

Figure 23.7 *Advanced portable ROCK TESTER for drill cores. The system consists of: (1) microcomputer, (2) bridge box, (3) loading frame, and a hydraulic pump. Elasticity, fracture toughness and strength of intact rock material are obtained in accordance with adopted standard test procedures. Courtesy of Bemek Co.*

scatter in the stress data and in particular for the horizontal stresses (cf. Brown and Hoek 1978; Stephansson *et al.* 1986). The scatter sometimes can be associated with the inaccuracy of the measuring techniques but most differences in stress state are associated with the global plate motions. Some areas are under compressional stresses from plate collision and push from the oceanic ridges while other areas are in tension due to diverging plate motions or graben formation. Therefore the rock engineer is referred to the world-wide compilation of rock stresses conducted by the World Stress Map Project (see Zoback *et al.* 1989) to obtain the first estimate of the stress regime and the direction of the major horizontal stress. Once the stress regime has been determined the ratio of the average horizontal to vertical stress against depth below surface is defined by

$$k = \frac{100}{Z} + 0.3 \tag{23.2}$$

for tension regimes, and

$$k = \frac{1500}{Z} + 0.5 \tag{23.3}$$

for compressional regimes, where Z is the depth in metres, and

$$k = \frac{0.5\,(S_H + S_h)}{S_v} \tag{23.4}$$

in which S_H and S_h are the major and minor horizontal stresses respectively.

The empirical Equations (23.2) and (23.3) are limiting equations and are based on a large set of data (Brown and Hoek 1978). However, it is important to bear in mind that factors like surface topography, erosion, major discontinuities and residual stresses will alter the state of stress.

It is clear from this discussion that the state of stress can vary a lot and for major underground chamber projects it is therefore recommended that stress measurements are carried out. In the design of the chamber it is the average state of stress in the zone of influence of the excavation that is of interest. Therefore, for most projects stress measurements will be conducted in a drill hole, and for the majority of projects in drill holes from the ground surface.

Techniques for measurement of rock stresses in water-filled, deep drill holes are available and have been used in several large underground projects (Hallbjörn 1986). Strains are determined in the wall of the drill hole by overcoring that part of the hole containing the measuring probe. From experimentally determined elastic properties of the rock material and the recorded strains from the overcoring procedure the field stress tensor can be obtained using procedures developed from elastic theory. This method, although by far the most accurate, is expensive to use and a large survey in a deep drill hole takes time due to the large-diameter drilling and the overcoring.

A slightly less complete but reliable and much cheaper method for stress determination is hydraulic fracturing. This method is recommended for stress measurement in vertical drill holes from the ground surface as it provides a vertical stress profile from a large number of tests and because any stress discontinuity from major faults or shear zones is easy to detect. The hydrofracturing stress measurement field unit at Luleå University of Technology (Bjarnason *et al.*, 1989) is shown in Figure 23.8. It consists of 1000 m long multihose containing high-pressure tubing, signal cables and a strength member, mounted on a truck. After drill core inspection, suitable test points are selected and a stradle-packer is lowered to the selected test point. The interval of the stradle-packer is pressurized until a fracture is formed at a pressure, p_b, when the pump is shut and the so-called shut-in pressure p_s, is recorded (cf. the pressure–time record in Figure 23.9). For the ideal case of axially oriented hydrofractures the following simple equations are applied:

$$S_h = p_s \tag{23.5}$$

$$S_H = 3S_h - p_r \tag{23.6}$$

where S_h and S_H are the minor and major horizontal stress at the test point γ and p_r is the re-opening pressure. The vertical stress is determined by the weight of the overburden according to Equation (23.1).

According to the theory of hydraulic fracturing the fracture will be oriented parallel with the azimuth of the major principal stress. The fracture orientation at the test point is achieved by inflating an impression packer and the orientation of the packer is recorded by a magnetic single-shot compass. Occasionally, horizontal fractures are

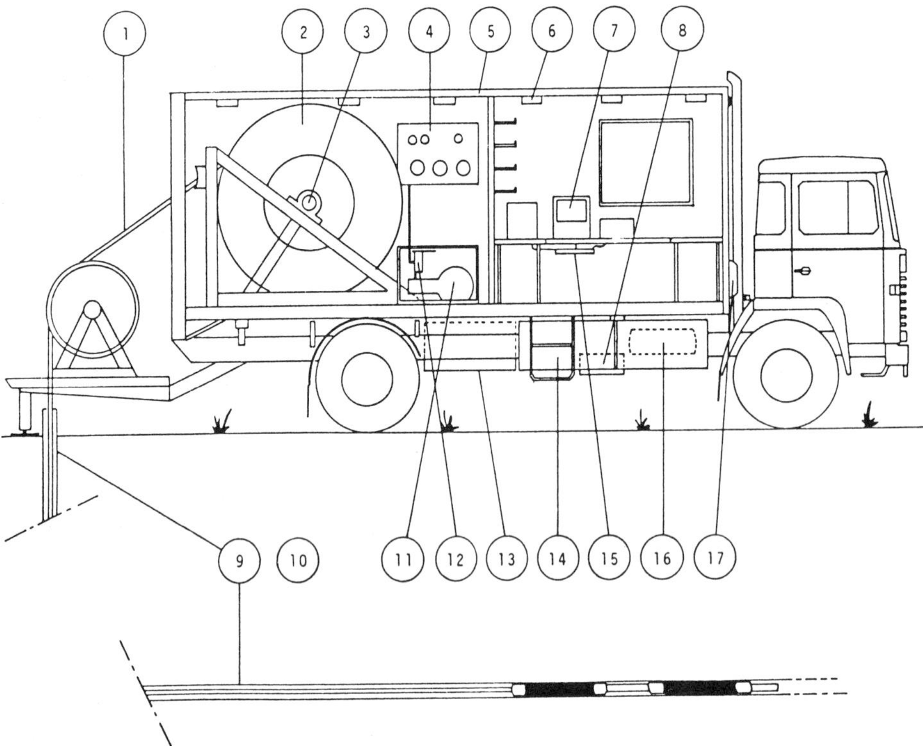

Figure 23.8 *The 1000 m hydrofracturing field unit at Luleå University of Technology: (1) multihose, 1000 m long; (2) drum, powered by the hydraulic system of the truck; (3) slip ring contact for signal cable, mounted inside the hub of the drum; (4) manifold and flow meters; (5) heat-insulated truck body; (6) lamps; (7) data-acquisition system; (8) main hydraulic pump; (9) drill hole; (10) roller, hydraulically driven to assist while running the hose uphole; (11) water pump, maximum pressure 80–100 MPa, maximum flow rate 15 l/min; (12) bypass valve; (13) water tank; (14) diesel tank; (15) electricity 220 V a.c.; (16) compressed air; (17) heat fan.*

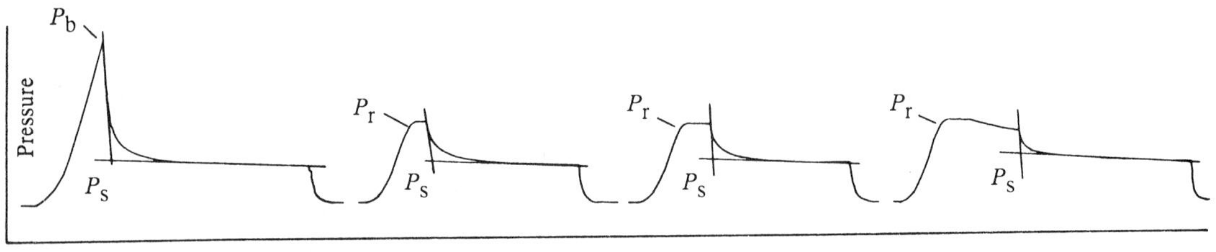

Figure 23.9 *Typical pressure versus time record from hydrofracturing measurement for stress determination. p_b is breakdown pressure at fracture initiation, p_r is the re-opening pressure and p_s is the shut-in pressure*

formed in drill holes where the majority of hydrofractures are vertical. The instantaneous shut-in pressures on these fractures have shown an excellent agreement with the vertical stress magnitude calculated from the weight of the overburden. Horizontal fractures are formed where the vertical stress is the minimum stress, being exceeded by the horizontal stresses which are large.

In conclusion, hydrofracturing stress measurements are highly recommended for the design and construction of underground chambers. Measurements can be conducted in deep drill holes from the ground surface or from shallow drill holes from drifts and tunnels. The measuring technique is robust and simple and the data evaluation straightforward. The relatively large number of data points recorded in a drill hole will give a reliable stress profile with depth.

23.4.7 Hydraulic testing

The important hydraulic properties of the rock mass are:

(1) the amount of water it can hold in its voids, expressed by its porosity;
(2) the amount that can be drained from it, expressed by its specific yield;
(3) the ease with which the water can flow through it, expressed by its permeability.

These parameters are of utmost importance for the success of an underground project and are dealt with in Chapter 4. The exploration drill holes at a site should be used as wells for pump tests with observation of the change in water tables in adjacent drill holes. The quantity of water yielded under steady-state flow conditions can be used to predict the extent of dewatering of the ground prior to excavation and the permeability of the rock mass around the pumped well can be determined. Water injection tests with single or double packers are other means of characterizing the hydraulics of single fractures, fracture zones and the rock mass. Data collected from pump tests, injection tests and piezometer tests of groundwater pressure are used for the rock-mass classification design and construction of the underground chamber.

23.5 Rock-mass classification

During and after collecting the information, listed under site characterization in Table 23.2, the data can be inserted into any of the existing rock-mass classification schemes. This does not mean that the rock-engineering problem should be solved by using analytical and/or numerical analysis techniques and approaches based on engineering mechanics. Instead the idea with the classification is to use previous experience gained in underground excavations elsewhere and to quantify this experience so that it may be extrapolated from one site to another. Some of the most significant steps in the classification systems for underground support are reviewed by Hoek and Brown (1980). The CSIR Geomechanics or Rock Mass Rating (RMR) scheme developed by Bieniawski (1976) and the NGI tunnelling quality index, Q, developed by Barton *et al.*, (1974) are used in this chapter. The reason is that these schemes are widely used and applicable to the design and the rock–support interaction of the chambers.

Bieniawski's Geomechanics classification scheme is based on data obtained mainly from civil engineering projects in sedimentary rocks in South Africa and uses five classification parameters (see Chapter 3). For each parameter or range of parameters a rating value is assigned and the overall rock mass rating (RMR) is obtained by adding all the ratings from the individual parameters (Table 3.8). The RMR value may be adjusted for the orientation of discontinuities by applying the corrections listed in Table 23.3. Interpretation of the ratings in terms of stand-up times of underground excavations is shown in the table and illustrated in Figure 23.10. Cohesion and friction angle of the rock mass are also shown in Table 23.3.

The Q system of Barton *et al.* (1974) combines six parameters in a multiplicative function (see Chapter 3).

Numerical values are assigned to each parameter of the Q system. For a detailed description of the parameters and the numerical values assigned to them the reader is referred to Barton *et al.*, 1974.

Because the Q system and the RMR system include slightly different parameters they cannot be strictly correlated. The following approximate relationship has been proposed by Bieniawski:

$$\mathrm{RMR} = 9 \log Q + 44 \tag{23.7}$$

In designing underground chambers it is recommended that both classifications are used. If there appears to be a large discrepancy in results after applying *Equation (3.2)* there is a need for a more thorough analysis of the rock mass conditions at the site. The two systems have been widely used all over the world and most rock engineers have found them to be simple to use and of assistance in making difficult practical decisions. How to use the rock-mass classification systems in the choice of support systems is discussed later in this chapter.

23.6 Rock-engineering considerations

An underground project requires cooperation between the clients, consultants, contractors and equipment manufacturers. The interaction between the different groups varies from country to country and also within a country. Often in Scandinavian countries the experience from an underground project is shared among the parties with the aim of improving both methods and equipment. The incentive

Table 23.3 Geomechanics classification of jointed rock masses (after Bieniawski 1976)

(a) The effects of joint strike and dip in tunnelling

Strike perpendicular to tunnel axis						
Drive with dip		*Drive against dip*		*Strike parallel to tunnel axis*		
Dip 45–90°	Dip 20–45°	Dip 45–90°	Dip 20–45°	Dip 45–90°	Dip 20–45°	Dip 0°–20° irrespective of strike
Very favourable	Favourable	Fair	Unfavourable	Very unfavourable	Fair	Unfavourable

(b) Rating adjustment for joint orientations

Strike and dip orientations of joints		Very favourable	Favourable	Fair	Unfavourable	Very unfavourable
Ratings	tunnels	0	−2	−5	−10	−12
	foundations	0	−2	−7	−15	−25
	slopes	0	−5	−25	−50	−60

(c) Rock mass classes determined from total ratings

Ratings	100 ← 81	80 ← 61	60 ← 41	40 ← 21	<20
Class no.	I	II	III	IV	V
Description	very good rock	good rock	fair rock	poor rock	very poor rock

(d) Meaning of rock classes

Class no.	I	II	III	IV	V
average stand-up time	10 years for 5 m span	6 months for 4 m span	1 week for 3 m span	5 hours for 1.5 m span	10 minutes for 0.5 m span
cohesion of the rock mass (kPa)	>300	200–300	150–200	<100	
friction angle of the rock mass	>45°	40–45°	35–40°	30–35°	<30°

to do this is contained in the stipulated risk-sharing in the contract between the owner and the contractor whereby both parties can gain from efficient and cost-reducing problem solving (Winqvist and Mellgren 1988).

The typical Scandinavian construction technique for underground chambers combines smooth blasting with bolting, shotcrete and cement grout injection where necessary. The very small amount of rock reinforcement used is the major difference compared to other design philosophies in other countries with very similar rock conditions.

23.6.1 Layout optimization

The rapid development of computer science and technology enables the rock engineer to apply computer-aided design (CAD) in the planning, design and construction of underground chambers. In Sweden the very first attempts to apply the new technique was for the design of the repository for low and intermediate radioactive waste at Forsmark, central Sweden (Hedman 1988). A diagrammatic representation of the underground excavations which consists of one silo, 70 m high and 31 m in diameter, four rock chambers and rooms for operational service and technical supply is shown in Figure 23.11. The repository is situated 65 m under the Baltic Sea and the quantity of rock excavated amounted to 430 000 m^3. Results from the site characterization (geology, structures, geophysical results, drilling and testing of drill cores and drill holes) should be stored in a computerized data base and the data base should be linked with the CAD system.

The design and orientation of an underground chamber in the rock space is governed by a number of criteria, of which the most important ones are function, available space, volume, accessibility, environment, excavation, maintenance and economics. The size and depth of location is to a large extent determined by the rock quality, adverse geological structures, the magnitude of rock stresses and the groundwater conditions.

The rock engineer has to find the optimum given on one hand by the design criteria and on the other hand the

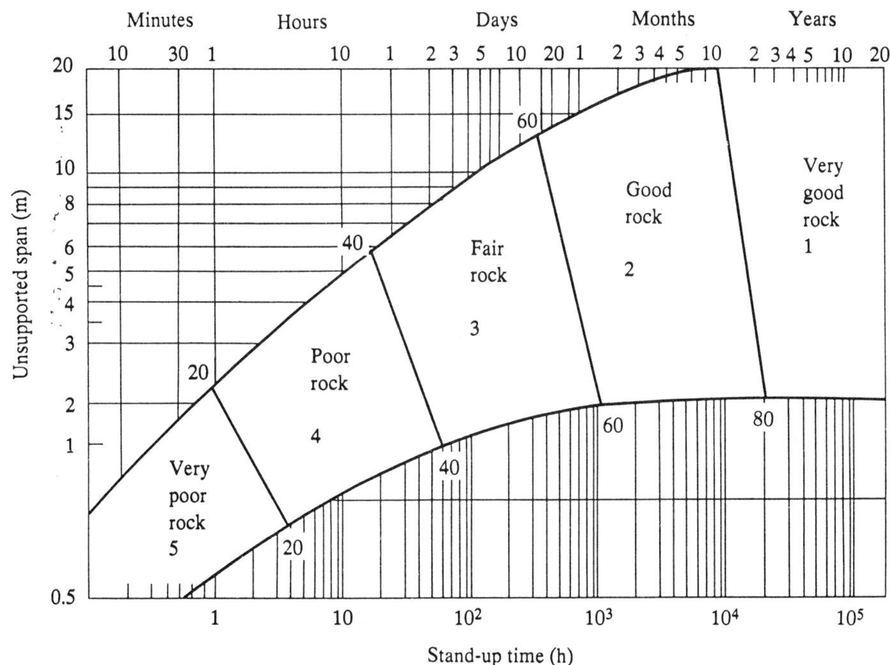

Figure 23.10 *Stand-up times of unsupported underground excavation spans for different rock mass rating (RMR) classes (after Bieniawski 1976)*

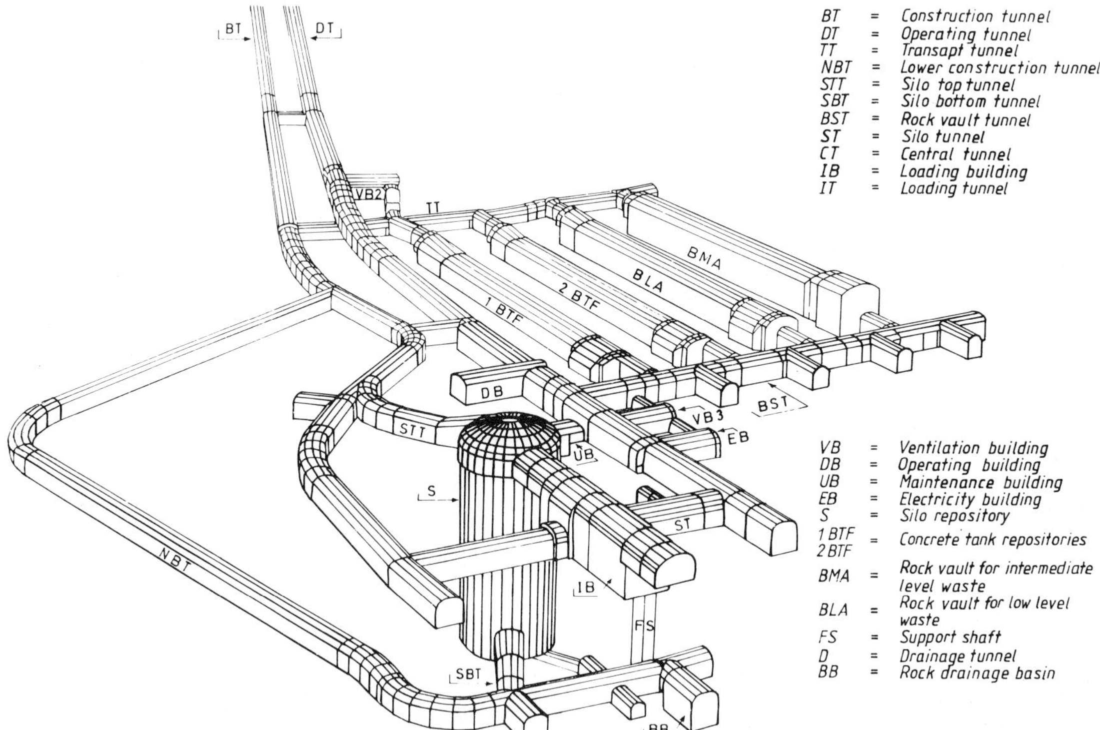

Figure 23.11 *View of underground construction for low and intermediate waste repository at Forsmark, central Sweden. CAD technique has been applied in the planning and design (after Hedman 1988)*

quality of the available rock mass. Very often the starting point in the design work is to define the space needed to fulfil the function and the access to the chamber. Then the structure is adjusted to the available rock volume which is often limited by major discontinuities like faults and shear zones.

The orientation of a rock chamber is governed by the rock quality, any adverse geological structures and the rock stresses. In blocky rock masses the excavation of the chamber may loosen and liberate blocks of various sizes from the roof or the sidewalls of the chamber. Structurally controlled failure of all kinds of planar structures in the rock mass can be analysed by means of stereographic projection techniques. Several examples of the application of this method are presented by Hoek and Brown (1980).

If potentially dangerous blocks are found prior to movement but subsequently their stability is assured, then no block movement will occur anywhere. This is the principle of block theory and the most dangerously located blocks are called key blocks. The block theory developed by Goodman and Shi (1988), establishes procedures for describing and locating key blocks and for establishing their support requirements. An introduction to the key block theory and its application to underground chambers has been presented by Goodman (1980).

The optimum orientation and shape of an underground chamber excavated in blocky rock masses are those which give the minimum volume of potentially unstable wedges. Hence, for the case of two sets of intersecting joints or two major discontinuities the optimum orientation of the long axis of the chamber is at the strike direction of the line of intersection of the two discontinuities. Furthermore access tunnels and the long axes of rock chambers preferentially should cross major discontinuities at right angles in order to minimize the required support.

The optimum depth of a chamber is governed by the rock quality, the state of stress, groundwater hydrology and cost of excavation. For fair to good rock mass quality, low stress magnitude and/or shallow location the long axis of the rock chamber should be oriented parallel to the direction of the minor horizontal stress. This guarantees the maximum compressive stress in the roof and floor of the chamber and therefore the best possible self-support of the rock mass in the roof (Figure 23.12). The relatively high permeability of the rock mass in low stress regimes at shallow depth enhances the groundwater flow and demands grouting prior to and during excavation. Therefore moving the chamber deeper down will reduce the cost for grouting but at the same time the lengthening of the access tunnels will increase the excavation costs. Rock chambers located at great depth and/or in high-stress regimes preferentially should be oriented with their long axes parallel to the azimuth of the major horizontal rock stress, (cf. Figure 23.12).

If several underground chambers are going to be located in one and the same area the minimum distance between them should not be less than half of the sum of the two adjacent openings. When small chambers are to be excavated next to large ones they should be placed in the low-stress zones. These are general rules of thumb to be applied in optimizing the layout of underground chambers.

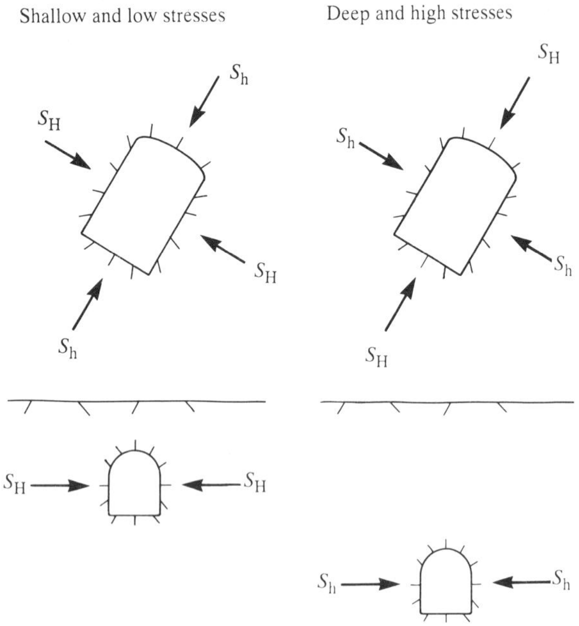

Figure 23.12 *Preferential orientation of rock chamber with respect to depth and stress regime in fair- to good-quality rock masses*

23.6.2 Excavation sequence

The excavation of an underground chamber normally is carried out in stages through a main access tunnel from the ground surface. A typical scheme for excavation of a dual oil and gas storage chamber is shown in Figure 23.13.

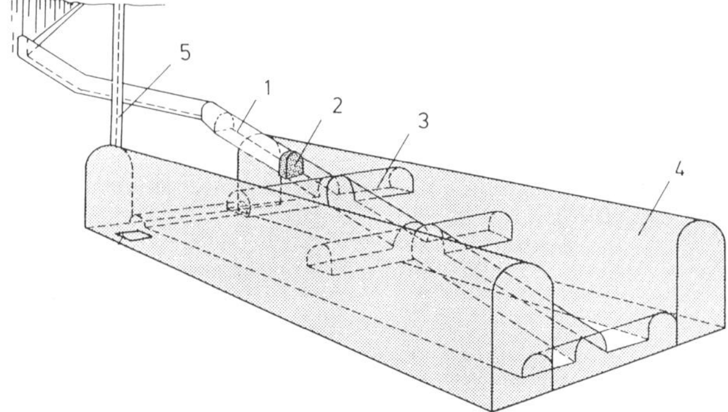

Figure 23.13 *Principal scheme for a dual cavern oil storage facility: (1) main access tunnel from ground surface; (2) concrete plug; (3) branch tunnel to excavation levels; (4) main storage cavern; (5) vertical shaft for pumps and instruments (after Johansson 1987)*

Typically a central head tunnel is excavated and advanced by two or more rounds before the side slashing. This is normally followed by benching with horizontal or vertical drilling and blasting. The bench height is typically 5–8 m and the blasted rock is mucked out through the main access tunnel. Rock vaults for low and intermediate level radioactive waste (BMA and BLA) and chambers for concrete tank repositories (1, 2 BTF) at the waste repository in Forsmark were excavated with a head tunnel and benches, (cf. Figure 23.11).

During the excavation of a rock chamber the stresses in the roof, walls and floor will change and the rock engineer has to take this into consideration in the design phase. Further, permanent rock reinforcements and instruments for rock monitoring have to be installed while the structure is still accessible. This can be illustrated by means of the excavation of the large silo for intermediate radioactive waste at Forsmark (cf. Figure 23.11). The excavation was made in four main steps (Larsson and Christiansson 1986).

(1) The dome of the silo was excavated with a rough contour, 2–3 m inside the final contour. Extensometers for monitoring the change in displacements and the finite displacement were installed in the roof. The smooth blasting to the finite contour was done with hand drilling and blasting.
(2) An access tunnel (NBT) and a silo bottom tunnel (SBT) were excavated to the bottom of the silo.
(3) From the dome a 50 m deep shaft was drilled in the centre of the silo. Blasting started from the bottom and proceeded upwards and only swelling rock masses from the blasting were loaded from the silo bottom tunnel (Figure 23.14).
(4) Benches, 6 m in height, were blasted from the top and loading was done from the silo bottom tunnel. Rock reinforcement and installation of extensometers in the silo wall took place as excavation proceeded downwards.

23.6.3 Stress and strength analysis

Rock engineering practice requires effective techniques for predicting rock mass response to excavation. The earliest attempts to develop a predictive tool for application in underground design involved studies of physical models of underground structures and later photoelastic models. The major disadvantages of physical and photoelastic modelling concern the expense and the time to design, construct and test the models.

Classical stress analysis is the basis for determining the stress and displacement distributions around openings. Today there exist a number of closed-form solutions for simple excavation shapes, primarily two-dimensional.

Many design problems in rock-engineering practice involve complex geometries, non-homogeneity of the rock mass and non-linear constitutive behaviour of the rock medium. Several computational schemes now are available which provide efficient methods for analysis of the state of stress and induced displacement around underground excavations. Boundary element methods provide the most efficient algorithms for analysis of elastic rock

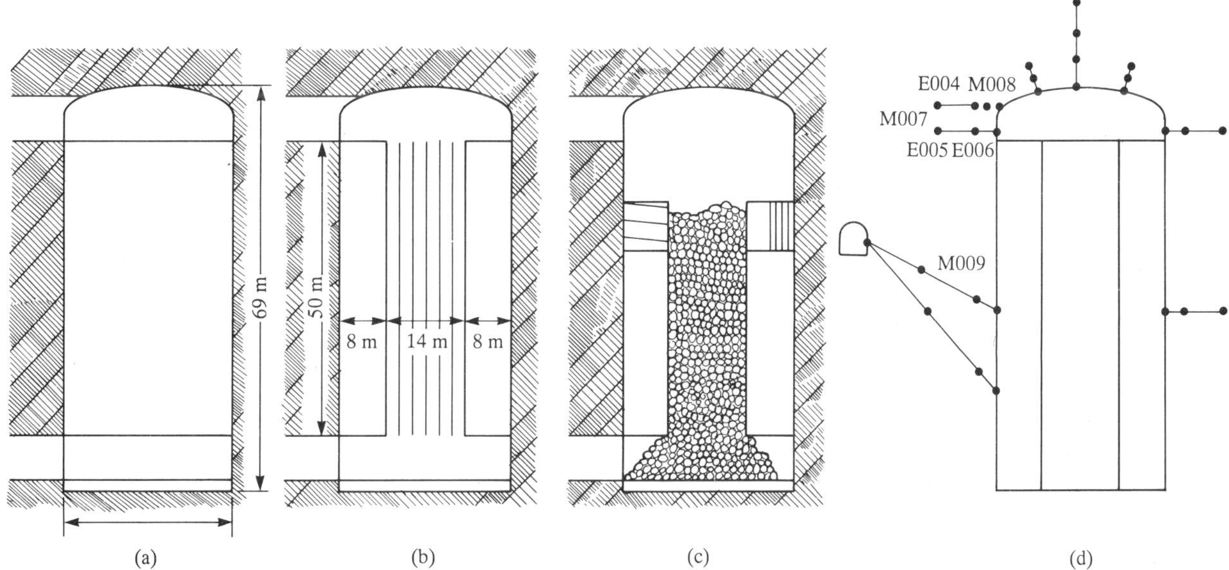

Figure 23.14 *Excavation phases of the silo for intermediate radioactive waste at Forsmark, central Sweden: (a) access tunnels to the top and bottom; drilling of the central shaft; (c) drilling and blasting of the complete silo; (d) installation of extensometers for rock monitoring*

masses where elastic domains are separated by continuous planes of weakness. The discrete element method provides a powerful method of analysis of jointed rock masses or rocks with highly non-linear properties. For other cases, linked boundary element – discrete element or linked boundary element – finite element schemes are appropriate (Brown 1987).

The two-dimensional computer codes FLAC and UDEC and the three-dimensional code 3DEC are the most widely used computational schemes for rock-engineering purposes. The codes are run on personal computers, and are very user-oriented and have powerful input–output routines.

Figure 23.15 has proved most useful for the practising rock engineer. The data are compiled from a large number of boundary element analyses of elastic rock material and shows the values of the maximum boundary stresses in the roof and sidewalls of excavations for different ratios of applied virgin stresses (Hoek and Brown 1980). In choosing the shape of a chamber the rock engineer should strive to achieve a uniform compressional stress distribution around the opening. The optimum shape – which sometimes is called the harmonic hole – is usually an ovaloid or an ellipse with the same axis ratio as the ratio of the virgin rock stresses (Barton *et al.* 1974).

Once the rock engineer has selected the most appropriate shape of the chamber the state of stress and its magnitude at all points in the vicinity of the opening has to be less than the failure limit for the particular rock type. If the stresses exceed the Mohr envelope, rock reinforcement is needed. Based on a large number of triaxial tests of intact rocks and deformability of rock masses Hoek and Brown (1980) presented an empirical failure criteria for intact rock specimens and jointed rock masses, of the form

$$\sigma_1 = \sigma_3 + \sqrt{(m\sigma_3 + s\sigma_c^2)} \tag{23.8}$$

Data about the constants m and s for different rock types and rock qualities are listed in Table 1.2, (Hoek and Brown 1988). The RMR and Q ratings for the different

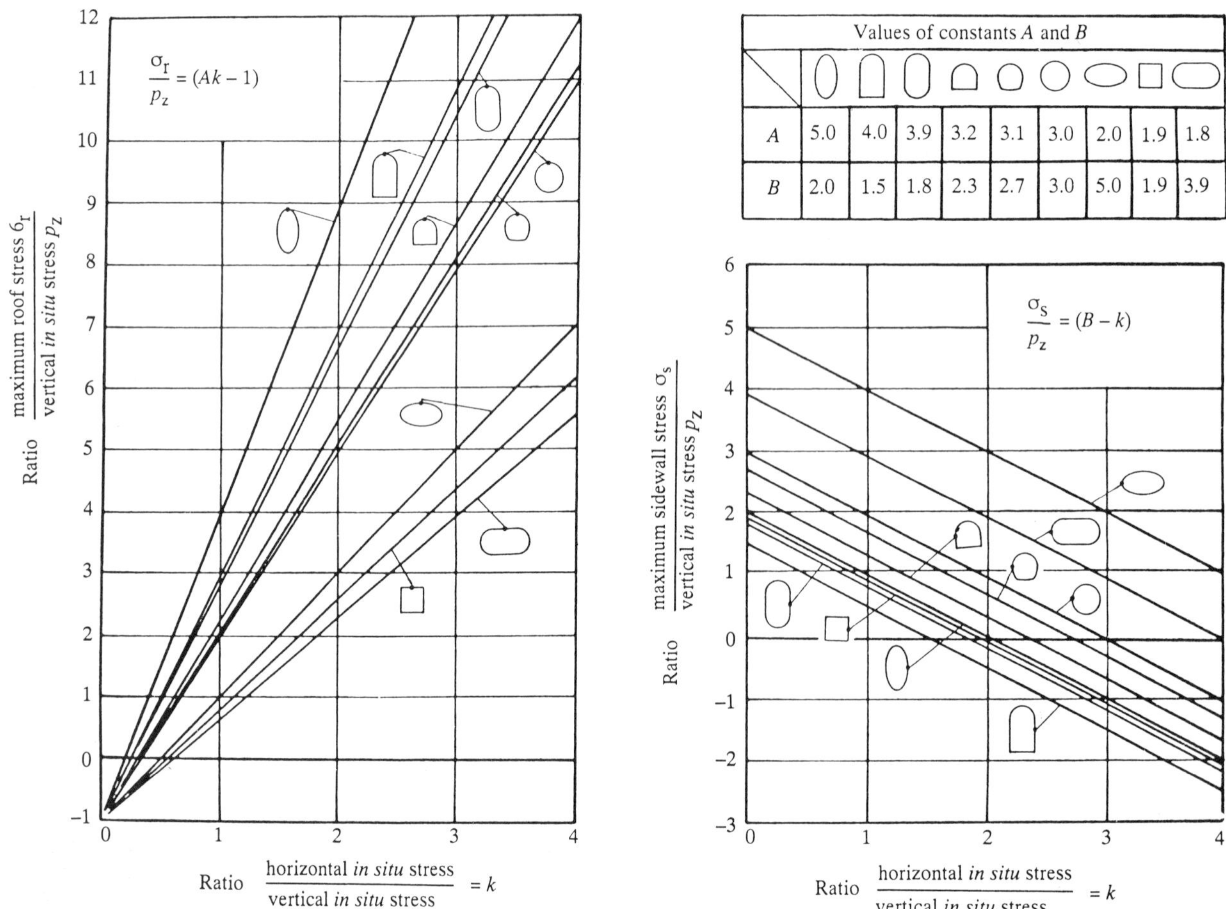

Values of constants A and B

A	5.0	4.0	3.9	3.2	3.1	3.0	2.0	1.9	1.8
B	2.0	1.5	1.8	2.3	2.7	3.0	5.0	1.9	3.9

Figure 23.15 *Influence of excavation shape and ratio of virgin rock stress upon maximum excavation boundary stress (after Hoek and Brown 1980)*

rock qualities are also given. This failure criteria is available in most computational schemes for stress analysis today and the rock engineer only has to present the appropriate values for the constants *m* and *s*. Outputs from the computational scheme will indicate areas where the Mohr failure envelope is exceeded and rock reinforcement needs to be installed. Where computational schemes are not available graphs of major and minor principal stresses σ_1 and σ_3 around single openings can be used (Hoek and Brown 1980).

23.7 Rock support and reinforcement

According to Hoek and Brown (1980) 'the principal objective in the design of underground excavation support is to help the rock mass to support itself.' Their design method for underground excavation is the best available and the main scheme will be followed in this presentation. The recommendation of support for rock chambers and tunnels based on the RMR and *Q* systems of rock-mass classification will be presented as well.

During the excavation phase temporary support or reinforcement are installed to ensure safe working conditions. Typically rock bolts and shotcrete – two of the most important reinforcements – are used, sometimes in combination with wire mesh (see Chapter 17). During or immediately after excavation permanent support installation is conducted. The temporary support will form part of the total support or reinforcement required and will be taken into consideration in any rock-support interaction analysis. Removal of temporary support prior to installation of permanent support should be avoided.

23.7.1 Rock-support interaction analysis

After emplacement, the reinforcement offers a radial support pressure to the excavated rock surface. The principle of rock-mass and -support interaction is illustrated in Figure 23.16. The support reaction by the reinforcement is activated after a certain amount of radial displacement of the rock mass. This is because the reinforcement is installed after the excavation. As radial deformation increases the support will carry more load and the stiffer the reinforcement the steeper the slope of the support reaction curve. The curve representing the behaviour of the rock mass under the given stress condition is known as the ground characteristic. As radial displacement increases the pressure required to limit the deformation will decrease. Equilibrium between the rock mass and the reinforcement is reached where the two curves intersect (Figure 23.16). The equations for the ground characteristics of different openings and the analysis of available support for various support systems are presented by Hoek and Brown (1980). A diagram of the maximum support pressure for various excavation

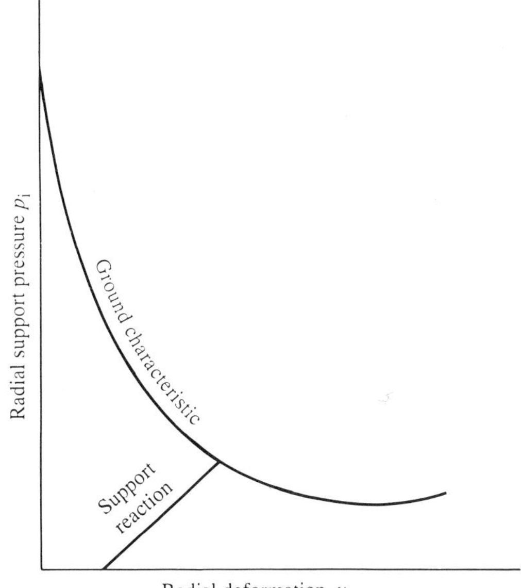

Figure 23.16 *Radial support pressure versus radial deformation for rock mass (ground characteristic) and reinforcement (support reaction)*

radii and support systems are shown in Figure 23.17. Notice the very low support pressure compared with the state of stress in the surrounding rock mass. An application of rock-support interaction analysis for a cavern 250 m below the surface in good-quality quartzitic rocks is presented in Figure 23.18 (Hoek and Brown 1980). The span of the excavation is 25 m and it is subjected to a uniform and equal horizontal and vertical stress of 10 MPa. The ground characteristic curve is calculated for a circular opening and the following rock-mass properties:

uniaxial compressive strength of intact rock	σ_c = 200 MPa
material constant for intact rock	m = 1.5
material constant for rock mass	s = 0.004
modulus of elasticity	E = 12 GPa
Poisson's ratio	ν = 0.2
material constant for broken rock	m_r = 0.1
material constant for broken rock mass	s_r = 0
unit weight of broken rock mass	γ_r = 0.02 MN/m^3
in situ stress	p_o = 10 MPa
radius of cavern	r_i = 12.5 m

Support reaction resulting from a 2 × 2 m pattern of 5 m long 25 mm diameter mechanically anchored rock bolts (curve 1) will not provide adequate support as shown in Figure 23.18. A reduction of the bolt spacing to 1.4 × 1.4 m pattern gives enough support pressure but at rather large roof deformation. Combination of rock bolts of 2 ×

Key

A SHOTCRETE – 5 cm thick shotcrete. $\sigma_{c.\ conc.}$ = 14 MPa after 1 day.

B SHOTCRETE – 5 cm thick shotcrete. $\sigma_{c.\ conc.}$ = 35 MPa after 28 days.

C CONCRETE – 30 cm thick concrete. $\sigma_{c.\ conc.}$ = 35 MPa after 28 days.

D CONCRETE – 50 cm thick concrete. $\sigma_{c.\ conc.}$ = 35 MPa after 28 days.

E STEEL SETS – (6 I 12) space 2 m. Blocked $2\theta = 22\ 1/2°$, σ_{ys} = 248 MPa.

F STEEL SETS – (8 I 23) space 1.5 m. Blocked $2\theta = 22\ 1/2°$, σ_{ys} = 248 MPa.

G STEEL SETS – (12 W 65) at 1 m. Blocked $2\theta = 22\ 1/2°$, σ_{ys} = 248 MPa.

H VERY LIGHT ROCKBOLTS – 16 mm dia. at 2.5 m centres. Mechanical anchor. T_{bf} = 0.11 MN.

I LIGHT ROCKBOLTS – 19 mm dia. at 2.0 m centres. Mechanical anchor. T_{bf} = 0.18 MN.

J MEDIUM ROCKBOLTS – 25 mm dia. at 1.5 m centres. Mechanical anchor. T_{bf} = 0.267 MN.

K HEAVY ROCKBOLTS – 34 mm dia. at 1 m centres. Resin anchored. T_{bf} = 345 MN.

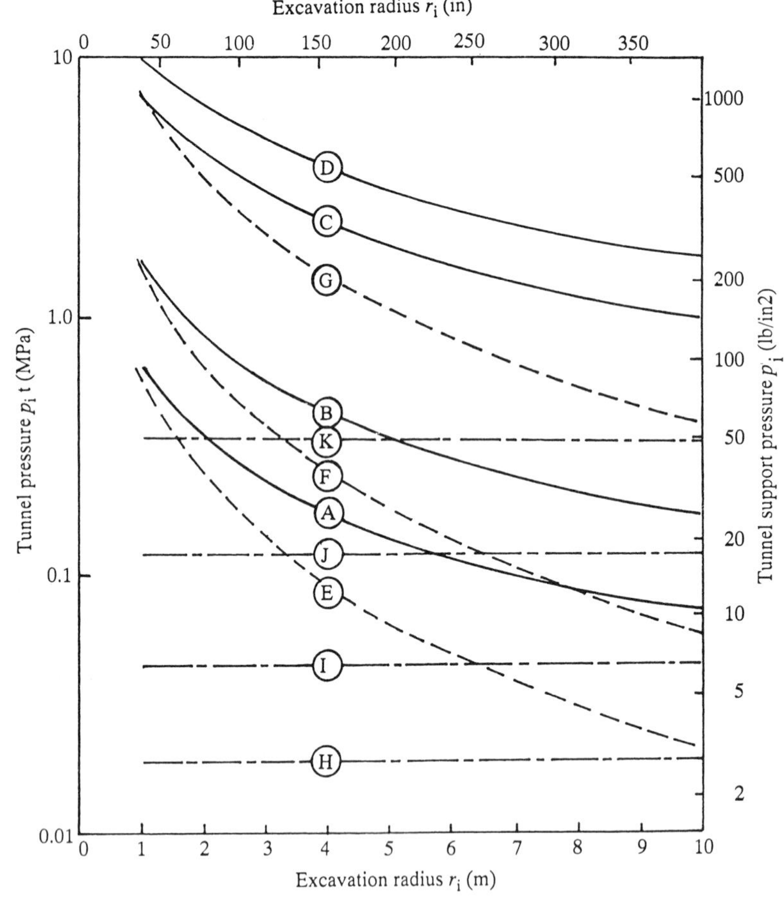

Figure 23.17 *Maximum support pressure versus excavation radius for various support systems (after Hoek and Brown 1980)*

2 m pattern and a 50 mm thick shotcrete layer gives the support reaction marked 4 in Figure 23.18 and this is also recommended as a temporary support for the cavern roof during excavation.

23.7.2 Guidelines for rock support

On the basis of the RMR classification scheme Bieniawski (1976) has presented guidelines for the choice of support systems for underground excavations. An example of his guide applicable to a horseshoe-shaped opening subjected to a vertical stress of less than 25 MPa and excavated with drilling and blasting is shown in Table 23.4. For good rock-quality emplacement of local bolts in the crown, 3 m long and spaced 2.5 × 2.5 m, together with 50 mm of shotcrete in the crown were required as a support system. This is slightly less than the calculated support by the rock-support interaction analysis in the previous example. The discrepancy is due to the smaller diameter of the excavation in the application of the Bieniawski scheme.

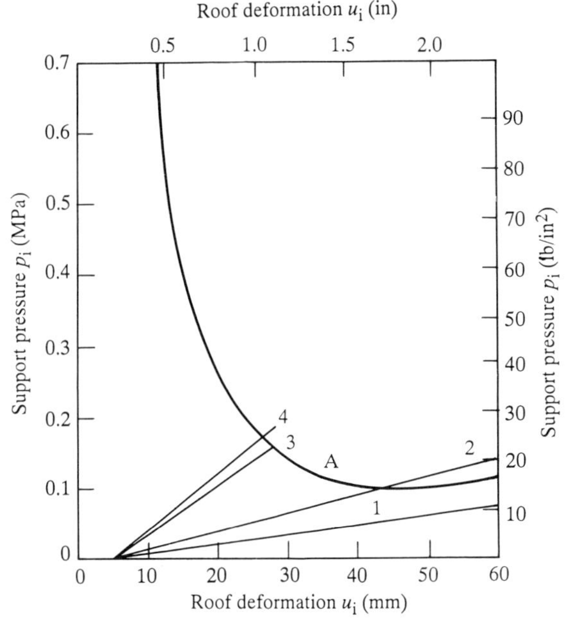

1 – 5 m × 25 mm bolts at 2 m × 2 m
2 – 5 m × 25 mm bolts at 1.4 m × 1.4 m
3 – 50 mm shotcrete
4 – shotcrete and bolts

Figure 23.18 *Ground characteristic and support reactions of four different support systems for a rock cavern with a span of 25 m located in good-quality quartzitic rocks and subjected to a uniform stress of 10 MPa (modified after Hoek and Brown 1980)*

Table 23.4 Guide for excavation and rock support for different rock-mass classes (after Bieniawski 1976)

Shape: horseshoe; width: 10 m; vertical stress: below 25 MPa; construction: drilling and blasting

Rock-mass class	*Excavation*	*Support* *Rockbolts (20 mm dia. fully bonded)*	*Shotcrete*	*Steel sets*
Very good rock I RMR: 81–100	Full face. 3 m advance	Generally no support required except for occasional spot bolting		
Good rock II RMR: 61–80	Full face. 1.0–1.5 m advance. Complete support 20 m from face	Locally bolts in crown, 3 m long, spaced 2.5 m with occasional mesh	50 mm in crown where required	None
Fair rock III RMR: 41–60	Top heading and bench, 1.5–3 m advance in heading. Commence support after each blast. Complete support 10 m from face	Systematic bolts 4 m long, spaced 1.5–2 m in crown and walls with mesh in crown	50–100 mm in crown, 30 mm in sidewalls	None
Poor rock IV RMR: 21–40	Top heading and bench, 1–1.5 m advance in heading. Install support concurrently with excavation 10 m from face	Systematic bolts 4–5 m long, spaced 1–1.5 m in crown and walls with wire mesh	100–150 mm in crown, and 100 mm in sides	Light ribs spaced 1.5 m where required
Very poor rock V RMR: <20	Multiple drifts. 0.5–1.5 m advance in top heading. Install support concurrently with excavation. Shotcrete as soon as possible after blasting	Systematic bolts 5–6 m long, spaced 1–1.5 m in crown and walls with wire mesh. Bolt invert	150–200 mm in crown, 150 mm on sides and 50 mm on face	Medium to heavy ribs spaced 0.75 m with steel lagging and forepoling if required. Close invert

By introducing a quantity called the equivalent dimension D_e, of an excavation the Q system for rock-mass characterization of Barton *et al.* (1974) can be applied for support. This dimension is obtained by dividing the span, diameter or height of the excavation by a quantity called the excavation support ratio, ESR, and the following relation holds:

$$D_e = \frac{\text{excavation span, diameter or height (m)}}{\text{Excavation Support Ratio (ESR)}} \tag{23.9}$$

Table 23.5

Excavation category	*ESR*
A. Temporary mine openings	3–5
B. Permanent mine openings, water tunnels for hydropower (excluding high-pressure penstocks) pilot tunnels, drifts and headings for large excavations)	1.6
C. Storage rooms, water treatment plants, minor road and railway tunnels, surge chambers, access tunnels	1.3
D. Power stations, major road and railway tunnels, civil defence chambers, portals, intersections	1.0
E. Underground nuclear power stations, railway stations, sports and public facilities, factories	0.8

The ESR is related to the use of the excavation as given in Table 23.5.

The ESR is roughly analogous to the inverse of the factor of safety used.

The relationship between the maximum equivalent dimension, D_e, of an unsupported chamber and the quality index, Q, is shown in Figure 23.19. In order to apply the system to support design the reader is referred to Barton *et al.* (1974) or a slightly modified presentation by Hoek and Brown (1980). The recommended support based on the Q system for the underground chamber in the previous example is found to be untensioned grouted dowels on a grid spacing 1.5–2.0 m and 20–30 mm shotcrete applied directly to the rock. The recommendation is based on the following data of the rock mass as given by the Q system.

$$\begin{aligned} Q &= 40\text{–}10 \\ D_e &= 5\text{–}14 \\ \mathrm{RQD}/J_n &< 10 \\ J_r/J_n &< 1.5 \\ p_i &\approx 0.05\ \mathrm{MPa} \end{aligned}$$

For increasing inter-block strength and/or increasing block size of the rock mass the shotcrete can be omitted. In conclusion, the three different approaches for support design give very similar results.

Brady and Brown (1985) have suggested the following set of guidelines relating to the performance of support and reinforcement practice.

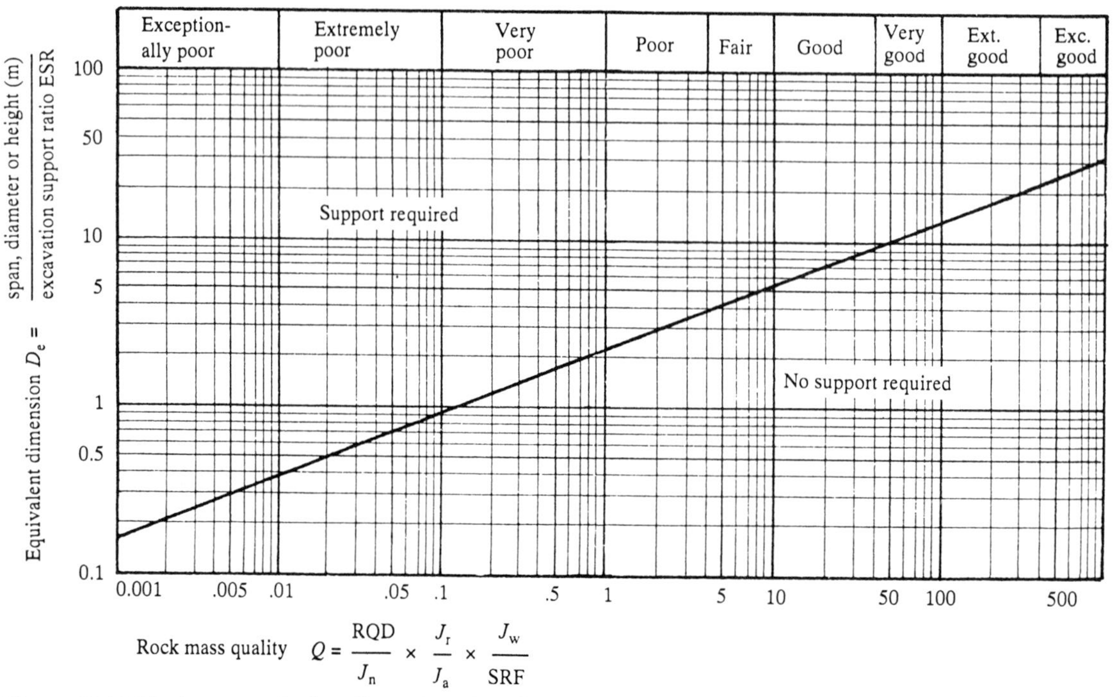

Figure 23.19 *Maximum equivalent dimension, D_e, of an unsupported underground excavation as a function of the rock-mass quality (after Barton et al. 1974)*

(1) The reinforcement should be installed close to the face soon after excavation.
(2) There should be good contact between the rock mass and the reinforcement.
(3) The deformability of the reinforcement should be such that it can conform to the displacements of the excavation surface.
(4) Ideally, the reinforcing system should help prevent deterioration in the mechanical properties of the rock mass with time due to weathering.
(5) Repeated removal and replacement of reinforcing elements should be avoided.
(6) The reinforcing system should be easily adaptable to changing rock-mass conditions and excavation cross-section.
(7) The reinforcing system should provide minimum obstruction of the working face.
(8) The rock mass surrounding the excavation should be disturbed as little as possible during the excavation process.

One always has to remember that the very best rock support is gentle and smooth blasting.

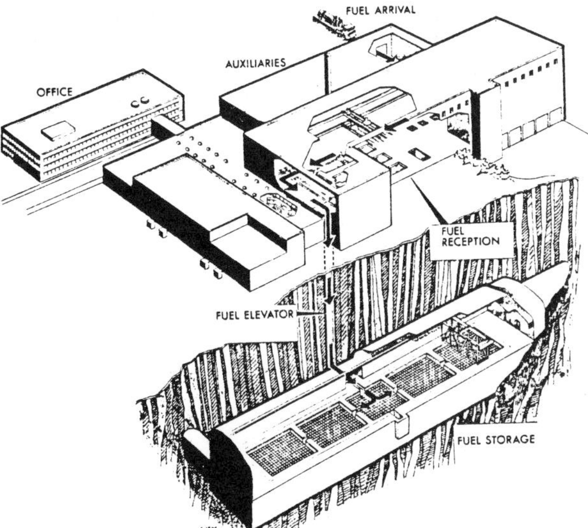

Figure 23.20 *Perspective view of the intermediate storage for spent nuclear fuel – CLAB – at Simpevarp, southern Sweden*

23.8 Application of design and construction procedure to an intermediate storage facility for spent nuclear fuel

The design and construction procedure presented and summarized in Table 23.2 will now be applied to the intermediate storage facility for spent nuclear fuel (CLAB) in Sweden (Röshoff *et al.* 1983). The storage is located in hard crystalline rocks adjacent to the Simpevarp nuclear power plant on the east coast of southern Sweden. Spent fuel is received from this plant and from the other three nuclear power stations in Sweden. The storage capacity is about 3000 tonnes of uranium at present and can be expanded to a capacity of 9000 tonnes. The storage has been in operation since 1987.

23.8.1 Lay-out, environment and security requirements of the storage

CLAB storage facilities comprise three main units on the ground: fuel reception, auxiliaries and office (Figure 23.20). The storage chamber is connected to the fuel reception by a fuel shaft and a service shaft.

The cavern is 117 m in length, 27 m in height and has a span of 21 m. It contains four water-filled pools for storage of spent fuel and one central pool. The pools are of concrete with an inside cover of stainless steel.

The cavern and the pool are designed and constructed to resist an earthquake which generates a ground acceleration of 0.1 g. CLAB is located in an area of Fennoscandia, which is characterized by very low seismicity. The rate of glacial uplift in the area is of the order of 2 mm/year.

The requirements for the stability of the roof and walls of the cavern is extremely high as no rock burst, rock falls or collapse are allowed. Therefore the site investigation for location and orientation of the cavern in relation to tectonic structures was more thorough than is normal for caverns in hard crystalline bedrock.

High demands are required for the environment inside the cavern, for example, humidity, temperature, dust and contamination. In order to fulfil these requirements the

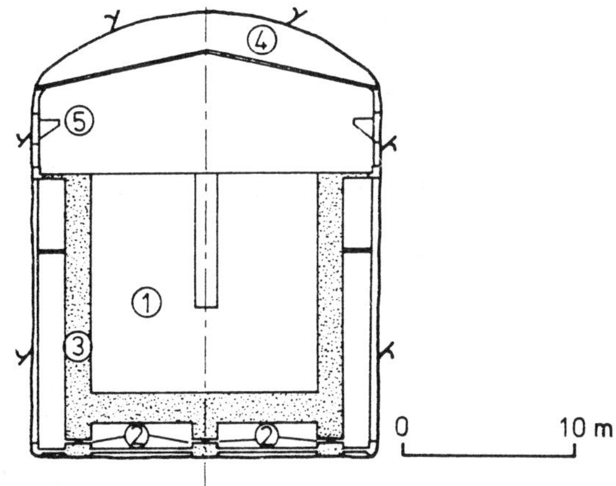

Figure 23.21 *Cross-section of the storage room at CLAB: (1) Storage pool; (2) sliding bearing; (3) concrete wall; (4) walls and ceiling; (5) cantilever for traverse*

facilities in the rock cavern are surrounded by free-standing walls and roof. The walls are made of concrete and the roof is a light-gauge steel construction (Figure 23.21). Roof and walls are air- and waterproof. The open space between the steel construction and the roof of the cavern are ventilated and drained separately.

The cavern must be safe against forces introduced from the outside, for instance, sabotage, bombs, etc. Therefore, the room has been placed at a depth of 30 m below the ground surface.

23.8.2 Site characterization

The site investigation started in 1978 with geological surface mapping, a few seismic profiles and one 80 m deep vertical drill hole (Figure 23.22). The drill hole was analysed by TV-inspection and hydraulic tests were conducted.

Later the same year this investigation was followed by a more detailed investigation and the sub-surface was investigated by eleven drill holes (D1–D11) with a core diameter of 32 mm and a total length of 750 m. Water pump tests were performed in each drill hole.

The main tectonic weak zones were detected from the seismic profiles and later confirmed by diamond drilling (Figure 23.22). The weak zone north of the cavern is 30 m wide and dips steeply towards the south. The mean joint frequency of the zone is 16 joints per metre. Soil was observed in the tectonic weak zone. The zone west of the cavern is 6 m wide and dips westwards. The zone to the south is 10 m wide, dips towards the south and cuts the cavern and the transport tunnel.

The orientation of the joints, which belonged to three sets, was N10°E/80°W, vertical; E-W to N45°E, vertical; and a third group with varying strike dipping at 20–30°. The joint frequency was rather high with a mean joint frequency of 7.2 joints per metre. Some parts were intensely fractured, 47% of the cores had a frequency of more than 6 joints per metre and only 15% were known to have a low frequency. Joint fillings were rare but a coating of chlorite on the joint surfaces was often observed.

Water pump tests were performed after drilling each hole. A double packer unit, with test section of 3 m length, was used. Some holes were analysed with a single packer system. The water pressure during a test was 5 bars except for the uppermost parts of the drill hole. The results of the testing showed that the bedrock was very tight as 94% of the drill holes tested had a leakage which was less than 0.1 l/min/m/bar.

The *in situ* stresses were measured in the vertical drill hole 01 (Figure 23.22). An overcoring technique for determining the three-dimensional stress field was used (Hallbjörn 1986). The measurements were taken at three different levels, all within the volume of the planned cavern. The residual stresses were measured and found to be low. The maximum principle stress, σ_1, was horizontal and oriented east–west with a magnitude of 6 MPa. The intermediate stress, σ_2, had a magnitude of 4 MPa and was oriented north–south in the horizontal plane. The minimum principal stress, σ_3, was vertical with a magnitude varying between zero and 2.5 MPa and was in accordance with the load of the overburden.

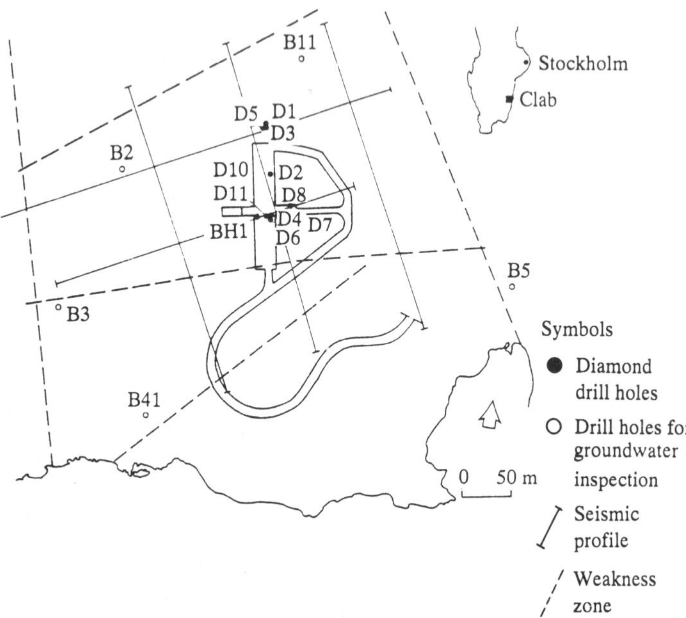

Figure 23.22 *Location of underground chamber for CLAB and access tunnels. Drill holes and seismically indicated weakness zones are shown*

The *uniaxial compressive strength* was determined on 30 core samples with a core diameter of 32 mm and 42 mm. The mean uniaxial compressive strength of the granitic volcanics was 200 MPa ± 40 MPa. Young's modulus (E_{lab}) and Poisson's ratio (ν) were determined from a compressive test and a bending test. The mean value for Young's modulus of the volcanics was 85 GPa ± 5 GPa and that of Poisson's ratio was 0.25.

23.8.3 Rock-mass classification

Geomechanics classification of the jointed rock masses according to Bieniawski (1976) gave RMR = 72 corresponding to class II, i.e. good rock quality. The *Q* system gave a fair to good rock quality with a *Q* value slightly below 10.

23.8.4 Rock-engineering considerations

The rock chamber for CLAB was oriented with its long axis parallel to the intermediate principal stress (σ_2 = 4 MPa) and in accordance with the principle presented in Figure 23.12. This means that the major principal stress is acting across the span of the chamber. The orientation was also governed by the two major sets of joints which intersected the chamber diagonally.

The chamber was located in the large block of jointed rocks surrounded by four major waekness zones (Figure 23.22). The 10 m wide weakness zone to the south intersects the chamber and the access tunnel.

A two-dimensional linear finite element analysis was performed to evaluate the stresses and displacements around the cavern. The model contained 1083 elements with 1150 nodal points.

The material parameters were chosen as E_{mass} = 30 GPa, which is one-third of Young's modulus for the intact rock, Poisson's ratio ν = 0.27 and the density of rock ρ = 2.6 t/m^3.

The excavation was simulated for three stages, namely, excavation of the gallery, bench 1 and the complete cavern. As the virgin horizontal stresses varied between 4 and 8 MPa two calculations were performed, where σ_{hi} was taken as 4 MPa and 8 MPa. The vertical stress was a function of the weight of the overburden.

Results of calculated displacements and stresses were presented by Röshoff *et al.* (1983). Tensile stresses of the order of 6 MPa were found in the wall of the chamber.

23.8.5 Chamber excavation

The chamber was blasted with a gallery and three benches. The gallery section is 125 m^2 and each bench is 6.3 m high. The safety requirements demanded smooth blasting and contour blasting. The contour holes were spaced at 0.6 m and loaded with Gurit 0.25 kg/m drill hole. The two next holes in the row demanded limited concentrations and were loaded with 0.52 kg/m and 0.91 kg/m drill-hole explosives to avoid damage of the rock mass. The floor of the cavern was smoothly blasted to limit the damage, as the water pools were founded directly on the intact rock mass.

The contour of the cavern was measured at 5 m intervals in the storage room at 10 m intervals in the transport tunnel. On average the contour was located 0.3–0.4 m outside the theoretical contour. The maximum deviation was 1.2 m and a block fall at the eastern corner of the transport placed the wall about 3 m outside the theoretical contour. This situation was caused by unfavourable intersection of rock structures.

The precision of the drilling for the blasting holes was good, as the drill-hole pattern has remained visible in full length at the contour. Sharp corners were rounded off to avoid stress concentrations.

There was 98 000 m^3 of rock excavated underground and 65 000 m^3 at the surface for CLAB. The excavation phase involved two shifts per day for drilling, blasting, loading and transportation. Rock support was placed during the night shift.

23.8.6 Rock support and reinforcement

The rock support for CLAB comprised grouting, rockbolting, anchoring and shotcreting. Concrete arches and columns were constructed in areas of poor rock quality.

The roof and the walls of the cavern were systematically grouted prior to blasting. The roof, gallery and bench 1 were grouted from seven drill holes in a fan-shape and the length of each drill hole was about 6.0 m. The grouted zone in the roof and walls was about 4.5 m thick.

The rock bolts were of steel (Ks 40) with a diameter of 25 mm and the length varied between 3 and 8 m. All bolts were fully grouted. The anchors were of Dywidag type with maximum lengths of 8 m and diameter 35 mm. The shotcrete was a mixture of cement, water and 8 mm sand.

The permanent support of the cavern consisted of 100 mm reinforced shotcrete in the roof and 50 mm of shotcrete covering both roof and walls. The thickness of the shotcrete was controlled at each 200 m^2 of emplaced shotcrete. Ten samples were taken from each test area. The strength of the shotcrete was tested for each 500 m^2 of shotcreting. More than 50% of the samples had a tensile strength greater than 1 MPa.

The roof was systematically anchored with two anchors per square metre. The length of the bolts varied between 4 and 6 m. Bench 1 was systematically bolted with slake bolts 6.4 m in length and a density of two bolts per square metre. The bolting of bench 2 and 3 was made selectively with bolts 6 m in length to prevent local rock fall (Figure 23.23).

The roof at the crosscut between the cavern and the transept tunnel was supported by six strongly reinforced concrete arches. Two of the arches are placed diagonally and four perpendicular to the cavern and the transept. The arches rest on reinforced concrete coloumns placed in the

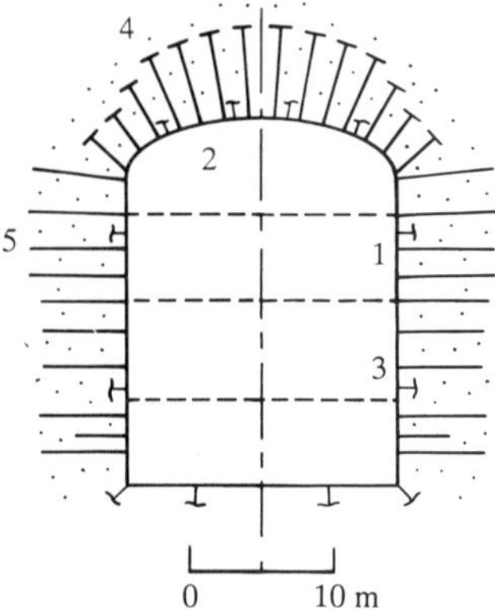

Figure 23.23 *Permanent support of the rock chamber at CLAB: (1) Grouted zone 4.5 m; (2) 100 mm reinforced shotcrete; (3) 50 mm shotcrete; (4) anchors, length 3.2–4.8 m; (5) rock bolts, length 3.0–6.4 m*

corners of the crosscut. Their thickness is 0.5–1.0 m and the length varies between 4 and 8 m. The rock mass at the corners was bolted selectively with slake fully grouted bolts 8 m in length.

The traverse rest on concrete cantilevers. Each cantilever was anchored to the rock wall by one rock anchor (type Dywiday ST110/125 GWS), 36 mm in diameter and stressed to 83 tonnes. Five 8 m long slake fully grouted bolts of diameter 25mm also supported the cantilever.

23.8.7 Rock monitoring and control

Movements of the rock mass around the cavern were measured in two sections of the walls and along two vertical drill holes in the roof. The convergency of the contour of the cavern was measured by the Distometer ISETH and the displacement of the rock mass in the walls was recorded by the Sliding Micrometer ISETH. The

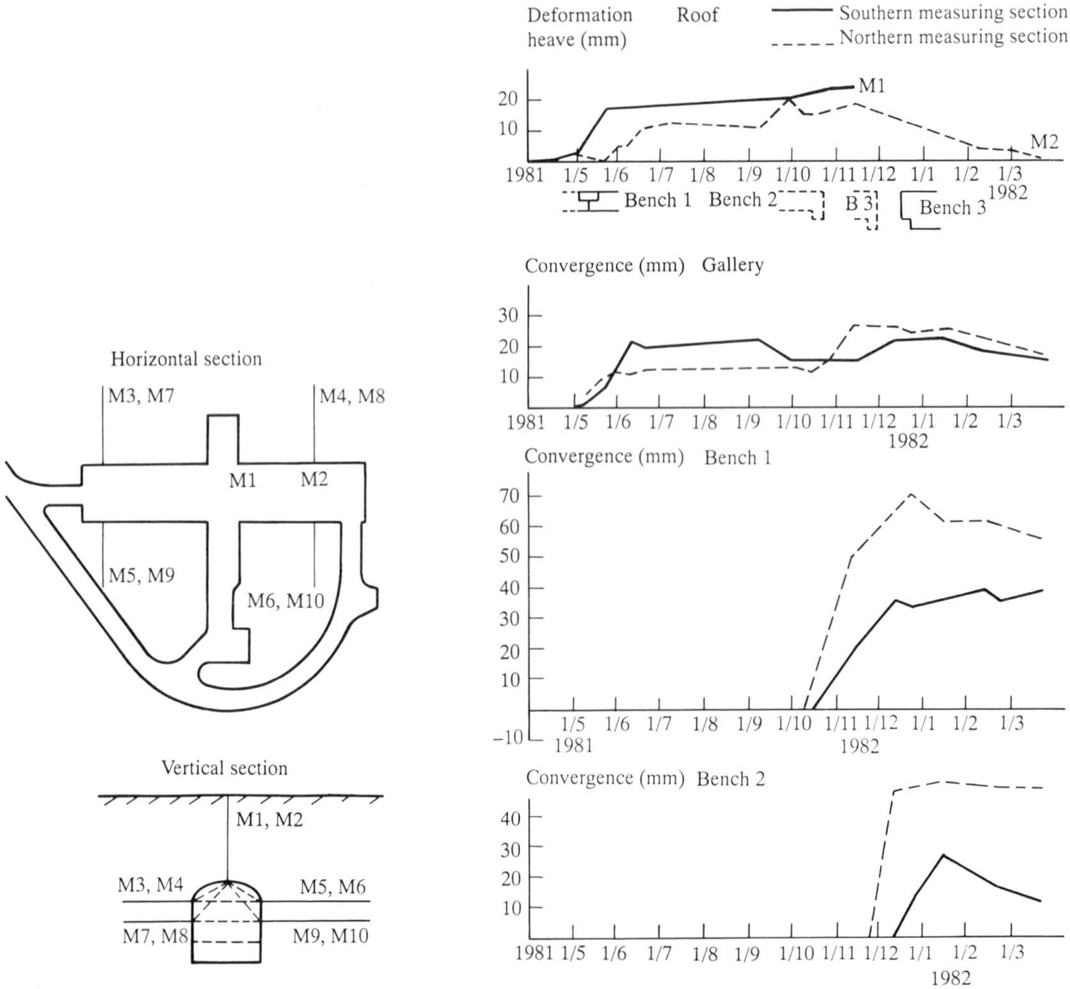

Figure 23.24 *Rock monitoring of the rock surface and rock mass for the rock chamber at CLAB*

accuracy of the Sliding Micrometer is ±0.003 mm within a distance of 1 m, and ±0.02 mm at a measuring length below 20 m for the Distometer.

The measuring programme involved 10 drill holes, M1-M10, of which M1 and M2 were vertically drilled from the surface (Figure 23.24). The holes M3–M6 were drilled in the gallery and M7–M10 in bench 1. These drill holes were 23 m in length and oriented horizontally. Convergency measuring points were located at each drill hole in the walls and in the roof below drill holes M1 and M2. Fourteen measuring lines were selected altogether. M1 and M5 were, however, both damaged during the excavation.

The total deformation of the cavern in relation to the excavation stages is presented in Figure 23.24. The result is based on convergence measurements in the roof and bench 2 and a mean value of data from the Sliding Mecrometer and the Distometer for the gallery and bench 1.

The roof heaved to a maximum value of 2 mm in the northern part of the cavern when benches 2 and 3 were passed. Later the roof sagged and almost recovered to 0.4–0.5 mm above original level after completion of the cavern.

The walls in the chamber showed the maximum deformation when benches 1 and 2 passed. In the southern section the deformation increased when benches 2 and 3 passed. The final convergence was small and was found to be 1.7 mm and 2.1 mm in the southern and the northern sections. The largest convergency occurred for bench 1 when excavating benches 2 and 3. Then the final deformation in the northern section was 3.5 mm and 6 mm in the southern section.

Since the completion of the excavation in 1982 until 1992 only minor deformations have been recorded and the groundwater table recorded in six of the drill holes around the chamber is stationary. The average water leakage into the chamber and access tunnels is 60 l/min and it has been constant for a long time with only minor fluctuation over the seasons.

References

Anttikoski, U. (1986) 'The city of Helsinki utilizes rock resources'. In *Rock Engineering in Finland*, J. Roinisto, J. (ed.), Finnish Tunnelling Association, Rakentajain Kustannus OY, Helsinki, 16–19.

Barton, N. R., Lien, R. and Lunde, J. (1974) 'Engineering classification of rock masses for the design of tunnel support'. *Rock Mechanics,* **6**, 189–239.

Bieniawski, Z. T. (1976) 'Rock mass classifications in rock engineering'. In *Exploration for Rock Engineering*, Z. T. Bieniawski, (ed.), Balkema, Rotterdam, pp. 97–106.

Bjarnason, B., Ljunggren, C. and Stephansson, O. (1989) 'New developments in hydrofracturing stress measurements at Luleå University of Technology'. *Int. J. Rock Mech. Min. Sci. & Geomech. Abstr.* **26**, 579–586

Brady, B. H. and Brown, E. T. (1985) *Rock Mechanics for Underground Mining*, Allen & Unwin, London, 527 pp.

Brown E. T. (ed). (1981) *Rock Characterization, Testing and Monitoring – ISRM Suggested Methods*, International Society for Rock Mechanics Commission on Standardization of Laboratory and Field Tests, Pergamon Press, Oxford, 211 pp.

Brown, E. T. (1987) *Analytical and Computational Methods in Engineering Rock Mechanics*, Allen & Unwin, London, 259 pp.

Brown, E. T. and Hoek, E. (1978) 'Trends in relationships between measured rock *in-situ* stresses and depth', *Int. J. Rock Mech. Min. Sci.,* **15**, 211–215.

Daerga, P-A., Stephansson, O. and Sagefors, I. (1986) 'Funnel Store – New concept for large rock caverns' In *Large Rock Caverns*, Saari (ed.), Vol. 1, Pergamon Press, Oxford, pp. 479–488.

Goodman, R. E. (1980) *Introduction to Rock Mechanics*, (2nd edn), Wiley, New York, 562 pp.

Goodman, R. E. and Shi, G. (1985) *Block Theory and its Application to Rock Engineering*, Prentice Hall, Englewood Cliffs, New Jersey.

Hallbjörn, L. (1986) 'Rock stress measurements performed by Swedish State Power Board'. In *Rock Stress and Rock Stress Measurements*, O. Stephansson (ed.), Centek Publishers, Luleå, pp. 197–205.

Hedman, T. 'An underground repository for radioactive waste in Sweden'. In *Waste Management '88'* R. G. Post (ed.), University of Arizona, Tucson, Arizona, pp. 853–859.

Hoek, E. and Brown, E. T. (1980) *Underground Excavations in Rock*, The Institution of Mining and Metallurgy, London, 527 pp.

Hoek, E. and Brown, E. T. (1988) 'The Hoek-Brown failure criterion – a 1988 update'. *Proc. 15th Canadian Rock Mechanics Symp.*, Toronto, Canada, October, 1–8.

Johansson, S. (1987) 'Excavation of large rock caverns for oil storage at Neste Oy Poruoo works in Finland, *Proc. 6th Int. Cong. on Rock Mechanics*, Balkema, Rotterdam, **1**, 147–153.

Larsson, H. and Christiansson, R. 'A silo in bedrock for nuclear waste'. In *Large Rock Caverns*, Vol. 2, K. Saari, (ed.) Pergamon Press, Oxford, pp. 817–828.

Ludvig, B., Stephansson, O., Tian, Y. S., Song, S. X. and Zhao, H. W. (1986) 'Rock mechanical investigation for the Sino-Swedish project in Jinchuan mine'. In *Engineering in Complex Rock Formations*, C. Li, and L. Yang (eds.), Science Press, Beijing, pp. 572–578.

Olsson, O., Falk, L. Forslund, O., Lundmark, L. and Sandberg, E. (1985) 'Radar investigations of fracture zones in crystalline rock'. In *Fundamentals of Rock Joints*, Stephansson, O. (ed.), Centek Publishers, Luleå, pp. 515–523.

Röshoff, K. (1986) 'BEMEK Rock Tester – a field and laboratory instrument for strength and elastic parameter testing of rock'. In *Engineering in Complex Rock Formations*, C. Li, and L. Yang, Science Press, Beijing, pp. 63–69.

Röshoff, K., Stephansson, O., Larsson, H., Stanfors, R. and Eriksson, K. (1983) 'CLAB – An intermediate storage of spent nuclear fuel in Sweden'. *Proc. 5th Int. Cong. on Rock Mechanics*, Balkema, Rotterdam, **2**, E151–E159.

Sagefors, I. and Svemar, C. (1986) 'Modern design for storing oil and liquified gas in underground rock caverns'. In *Large Rock Caverns*, Vol. 1, K. Saari (ed.), Pergamon Press, Oxford, pp. 597–608.

Stephansson, O., Särkkä, P. and Myrvang, A. (1986) 'State of stress in Fennoscandia'. In *Rock Stress and Rock Stress Measurements*, O. Stephansson (ed.), Centek Publishers, Luleå, pp. 21–32.

Winqvist, T. and Mellgren, K-E. (1988) *Going Underground*, Royal Swedish Academy of Engineering Sciences, Stockholm, 177 pp.

Zoback, M-L. *et al.* (1989) 'Global patterns of tectonic stress, *Nature*, 341, No. 6240, 291–298

24 Shafts and raises in rock masses

Dr A Auld
I W Farmer & Partners, Newcastle-upon-Tyne, UK

24.1 Introduction

A shaft is a vertical or steeply inclined excavation of limited width in relation to its depth, made to provide access to underground workings. This terminology is given in Anon (1974). From the same reference, a raise is a vertical or steeply inclined shaft which has been driven upwards. In this chapter, the design and construction of both shafts and raises are studied with regard to their installation through rock masses.

The importance of shaft and raise construction can be recognized from Table 24.1 by the number of different shaft purposes listed in relation to the various underground environments. Shafts primarily provide the means of access and egress for people, materials and equipment to most of the underground environments indicated, in addition to being the ventilation and services route.

The shafts referred to in this chapter are of the large diameter and deep type, typically up to 8 m internal diameter and reaching depths of over 1000 m for UK coalmine conditions (Auld 1987). Table 24.2 lists the deep shafts sunk from the surface in the UK over the last thirty years. Shafts with greater depths and larger internal diameters are to be found in the South African gold mines where the President Steyn Gold Mining Co. Ltd No. 4 shaft, located near Welkom in the Orange Free State, has the largest cross-sectional area in the world for a deep shaft. It was commissioned in 1972 and is 10.973 m × 10.210 m comprising two semi-circles of 5.105 m radius each with a 763 mm parallel section between the semi-circles (Moll 1973). The depth is 2365.2 m. Hartebeestfontein Gold Mining Co. Ltd No. 6 south shaft, sunk in 1967 in the Western Transvaal, has the longest single wind in the world at 2471 m. Its internal diameter is 8.53 m.

The deepest level of excavation in the world is also to be found in the Western Transvaal gold mines of South Africa at the Western Deep Levels Gold Mining Co. Ltd No. 2 shaft, where the tertiary vertical shaft has been sunk to 3580 m (2.22 miles) below the surface at an internal diameter of 7.93 m. It was commissioned in 1979. In addition, South Africa currently boasts the largest and longest raise bored shaft in the world at 6.02 m diameter excavated, with a length of 1033 m, which was constructed in 1986 for Rustenburg Platinum Mines in the Transvaal (Schmidt and Fletcher 1987).

Rock types vary immensely depending upon their location throughout the world. Typical of the sedimentary rocks encountered in UK coalmines are sandstones, siltstones, mudstones, shales, marls and occasional limestones (Auld 1987). These rocks are normally considered to be competent (self-standing when excavated) for the size and depths of shafts constructed in UK coalmines. However, the main problem is the presence of aquifers to depths of up to 650 m. The design and construction of shafts in these conditions must accommodate the means of controlling the water during sinking and, in the finished state, the lining must be capable of resisting hydrostatic pressure for the full head to surface. Ground freezing from the surface was used to seal off water inflow through the Bunter Sandstone for all ten, 7.315 m internal diameter, Selby coalmine shaft sinkings in North Yorkshire, UK during 1977–86 (Wild and Forrest 1981). The same method of water control was adopted at Asfordby coal mine in Leicestershire, UK, for the Keuper Waterstones when sinking the 8 m internal diameter pair of shafts during 1985–89 (Harvey 1988). However, in this case, although the freeze holes were drilled from the surface, freezing was carried out only locally to the aquifer using an underground freeze chamber excavated in the shaft wall above the water zone. Combinations of in-shaft pumping and strata grouting are the other methods of water control normally employed in UK coalmine shaft sinking.

South African conditions differ greatly from those found in the UK in so far as the rocks are much stronger and more massive, being predominantly of the igneous and metamorphic types. Henderson (1969) indicated that in the sinking of the 8.992 m (29 ft 6 in) internal diameter No. 5 shaft for Hartebeestfontein Gold Mining Co. Ltd in the Western Transvaal during 1966–67, sedimentary beds were first encountered consisting of 59 m of decomposed micaceous sandstone and shales, followed by 292 m of dolomite. Underlying the sedimentary beds were 1298 m of lava (igneous) and 372 m of quartzite (metamorphosed sandstone). In these conditions stability during excavation was also not a problem, although the lining was carried

Table 24.1 The importance of shafts and raises in relation to the underground environment

Shaft purpose	*Underground environment*							
	Deep mines	*Power stations*	*Storage*		*Oil platforms**	*Road and rail tunnels*	*Nuclear shelters*	*Testing of nuclear devices*
			Nuclear waste	*Liquid or gas*				
Mineral or oil egress	•				•			
People access	•	•	•		•		•	•
Materials access	•	•	•		•		•	•
Equipment access	•	•	•		•		•	•
Ventilation	•	•	•		•	•	•	•
Service (cables and pipelines)	•	•	•		•		•	•
Ore pass	•							
Underground bunker	•							
Rescue	•							
Nuclear waste transfer			•					
Water pressure and surge		•						
Penstock				•				
Nuclear device transfer								•

* Oil retrieval via an underground drilling site
† US Atomic Energy Commission test programme

Table 24.2 Deep shafts sunk from the surface in the UK over the last 30 years

Project	*Client*	*Contractor*	*Internal diameter (m)*	*Depth (m)*	*Date*
Kellingley colliery No. 1 and No. 2 shafts	British Coal	Thyssen (GB) Ltd	7.315	780	1958–62
Daw Mill colliery No. 2 shaft	British Coal	Cementation Mining Ltd	5.486	550	1968–70
Boulby mine No. 1 and No. 2 shafts	Cleveland Potash Ltd	Thyssen (GB) Ltd	5.5	1150	1969–74
Dinorwic pumped storage scheme ventilation shaft	Central Electricity Generating Board	Thyssen (GB) Ltd	5	255	1976–77
Dinorwic pumped storage scheme high-pressure shaft	Central Electricity Generating Board	Thyssen (GB) Ltd	10	440	1976–80
Selby Wistow No. 1 and No. 2 shafts	British Coal	Cementation Mining Ltd	7.315	411 (No. 1) 383 (No. 2)	1977–81
Selby Riccall No. 1 and No. 2 shafts	British Coal	Cementation Mining Ltd	7.315	823	1977–83
Selby Stillingfleet No. 1 and No. 2 shafts	British Coal	Thyssen (GB) Ltd	7.315	708	1978–82
North Selby No. 1 and No. 2 shafts	British Coal	Cementation Mining Ltd	7.315	1043	1978–86
Selby Whitemoor No. 1 and No. 2 shafts	British Coal	Thyssen (GB) Ltd	7.315	965	1979–85
Dearne Valley shaft	British Coal	Thyssen (GB) Ltd	3.658	300	1980–82
Castlebridge shaft	British Coal	Thyssen (GB) Ltd	6.1	407	1980–83
Dodworth-Redbrook shaft	British Coal	Amalgamated Construction Ltd	6.1	413	1981–83
Maltby No. 3 shaft	British Coal	Cementation Mining Ltd	8	1000	1981–87
Asfordby No. 1 and No. 2 shafts	British Coal	Cementation Mining Ltd	7.32	527	1985–89
Total = 23					

close to the shaft bottom in the weak quartzite. The presence of water was minimal, requiring only a nominal 380 mm thick concrete lining throughout the shaft length of 2021 m. Routine cover drilling and stage cement–sand injections were used for pore and fissure treatment through the dolomite and lava. The decomposed micaceous sandstone and shale zone at the surface was pre-cemented before sinking to improve its condition for supporting the headgear foundations.

Other much more difficult shaft sinking conditions can be encountered. Kelland and Black (1969) described the strata conditions encountered, the water and ground stability control measures needed and the linings installed during the construction of two potash shafts in Saskatchewan, Canada, during 1965–68 (Table 24.3). One shaft had an internal diameter of 4.877 m (16 ft) and the other 5.639 m (18 ft 6 in). The Saskatchewan potash shaft sinking problem, summed up in its simplest form by Kelland and Black, was 'to sink and secure a shaft through a quicksand type of formation down to a salt zone and to seal off sufficient groundwater en route to prevent uncontrolled leaching of the salt in the shaft bottom and the mining area'.

From the three sets of strata conditions described with their appropriate water and ground stability control measures and lining details, namely those relevant to the UK, some parts of South Africa and particularly the specific area of Canada, where the strata presented a great range of different sinking difficulties throughout the length of a single shaft, the design and construction of shafts and raises in rock masses can be placed properly into perspective. In designing shafts and their linings, the process must review fitness for purpose, strata and aquifer imposed loads on the lining once it is installed and any loading resulting from the construction process and the measures adopted for water and ground stability control. Similarly the sinking and lining method is related to and must be appropriate for the particular strata conditions to be encountered and the water and ground stability control measures required.

When considering the design of shaft linings through rock masses in this chapter, discussion is restricted to the design of the lining itself. The mechanisms of rock mass behaviour which induce stresses, strains and deformation in a lining are discussed in other chapters. In addition, the design parameters reviewed refer to competent rock masses.

All the shafts considered in this chapter are circular as this shape is the norm for inherent ground stability and affords the best lining structural strength.

Table 24.3 Strata conditions, water and ground stability control measures and lining details – sinking of two potash shafts in Saskatchewan by The Cementation Co. (Canada) Ltd for Cominco Ltd during 1965–68 (Kelland and Black 1969)[8]

Strata	*Depth* (m)	*Strata conditions*	*Water and ground stability control measures*	*Lining details*
Glacial till	0–62.5	Stiff boulder clay containing wet gravel zones and large boulders	Frozen from surface prior to sinking	610 mm (24 in) concrete
Cretaceous shales	62.5–517.6	Normally contains no free water. In this case 'contained' water content of 20–35% by volume, when frozen, produced inward expansion of up to 100 mm on radius during shaft excavation	Frozen from surface prior to sinking. Rock bolts, polyethylene sheets and wire mesh temporary support, with maximum open excavation length of 16.75 m	610 mm (24 in) concrete. Shale expansion pressure on freshly placed concrete lining alleviated by temporary support method adopted
Blairmore formation	517.6–632.8	Major geological hurdle. Mainly water-bearing unconsolidated and poorly cemented sands, silts and shales	Frozen from surface prior to sinking to consolidate fluid and mechanically weak areas	Combination of high strength, cast iron tubbing backed by high-strength concrete
Mississippian dolomites and limestones	632.8–1060.4	Competent rock formations containing a few high-pressure water-bearing zones in finely porous or finely fissured rock	Top section frozen from surface (bottom of freeze tubes 684.3 m deep). Highly penetrating chrome lignin chemical grout (T.D.M.) used to seal water-bearing zones below	760 mm (30 in), 915 mm (36 in) and 1070 mm (42 in) concrete up to 41 N/mm^2 (6000 lb/in^2) in strength
Prairie Evaporites	1060.4–1200.9	Salt bed containing potash layers in various degrees of concentration in its upper reaches. Shaft excavation exhibits time-dependent (creep) inward radial deformation under overburden pressure	Rockbolts and mesh temporary support	Unlined due to high rock creep pressure which would develop

24.2 Shaft and raise design

24.2.1 Design parameters

In reviewing the factors which need to be considered for the design of shafts and raises, it has been assumed in Section 24.2.1 that basic planning requirements have been fulfilled. Such items would include decisions on the number of shafts necessary and the selection of the most suitable sites (Unrug 1984).

For a mine, the number of shafts is dependent upon size of daily production and dimensions of the mining area. To obtain a minimum cost to produce the product, it is essential to find an optimum balance between capital expenditure and operating costs. The most suitable site for a mine shaft requires a compromise to accommodate the best position for mine planning, the best geological and hydrogeological conditions and the most favourable environmental situation.

The factors which need to be reviewed in shaft and raise design are contained in Table 24.4. Several of the parameters are inter-dependent and these relationships are indicated in the table together with the necessary steps in the design process. Each design parameter is examined briefly in Section 24.2.1.

(a) Shaft purpose

Based on the purpose for which the shaft is required (Table 24.2), shaft cross-section general arrangements for its permanent operating condition and the temporary sinking state can be prepared. These are also dependent upon the shape chosen and the size required to accommodate winding equipment, services and ventilation needs (upcast or downcast depending upon direction of flow). Cross-section, permanent operating condition general arrangements for the two Asfordby coal mine shafts in Leicestershire, UK, are shown in Figure 24.1. The temporary sinking general arrangements can be seen in Figure 24.8.

(b) Shaft shape, size and depth

For deep shafts with a large cross-sectional area, the best lining shape to resist hydrostatic pressure or rock deformation loading is circular, as the induced stresses in the lining are all in compression (see Section 24.2.2(a). The circular excavated shape is also superior in relation to inherent ground stability, as the ring shape minimizes the excavation inward radial closure effect exhibited by rocks which deform under high overburden stresses.

Shaft size is determined by what furnishings and services have to be installed in it, the area taken up by skips, cages, counterweights and ropes, and the additional space needed for ventilation purposes. Depth is controlled by the level to which access is required underground and also the limits on winding equipment. For the South African gold mining operations, it was shown in Section 24.1 that three separate stages of shaft sinking were required to reach the lowest extremity in the Western Deep Levels Gold Mining Co. Ltd No. 2 shaft. This resulted in primary, secondary and tertiary vertical shafts being sunk due to limitations on maximum single winding length.

(c) Geology and hydrogeology

It is essential to know as much about the geology and hydrogeology of the rock mass(es) to be sunk through as it

Table 24.4 Design parameters, their inter-relationship and steps in the design process

	Design parameter	*Design process*	*Relationship to other design parameters*
1	Shaft purpose	Purpose specified (see Table 24.1). Develop permanent and sinking shaft cross-section general arrangements	Inter-related with 2, 3 and 4
2	Shaft shape	Determine shape. Circular is normal for inherent ground stability and lining structural strength	Depends on 1
3	Shaft size	Determine size	Depends on 1
4	Shaft depth	Determine depth	Depends on 1
5	Strata geology	Define strata geology. Determine rock mechanics properties (strength, modulus of elasticity, Poisson's ratio, RQD)	
6	Strata hydrogeology	Define aquifer zones. Determine rock permeability and potential excavation water inflow	
7	Construction factors	Determine methods of excavation, lining installation, strata water control and maintaining ground stability	
8	Lining loads	Determine lining loads, either hydrostatic pressure or rock deformation load	
9	Lining type	Determine lining type	Depends on 2–8

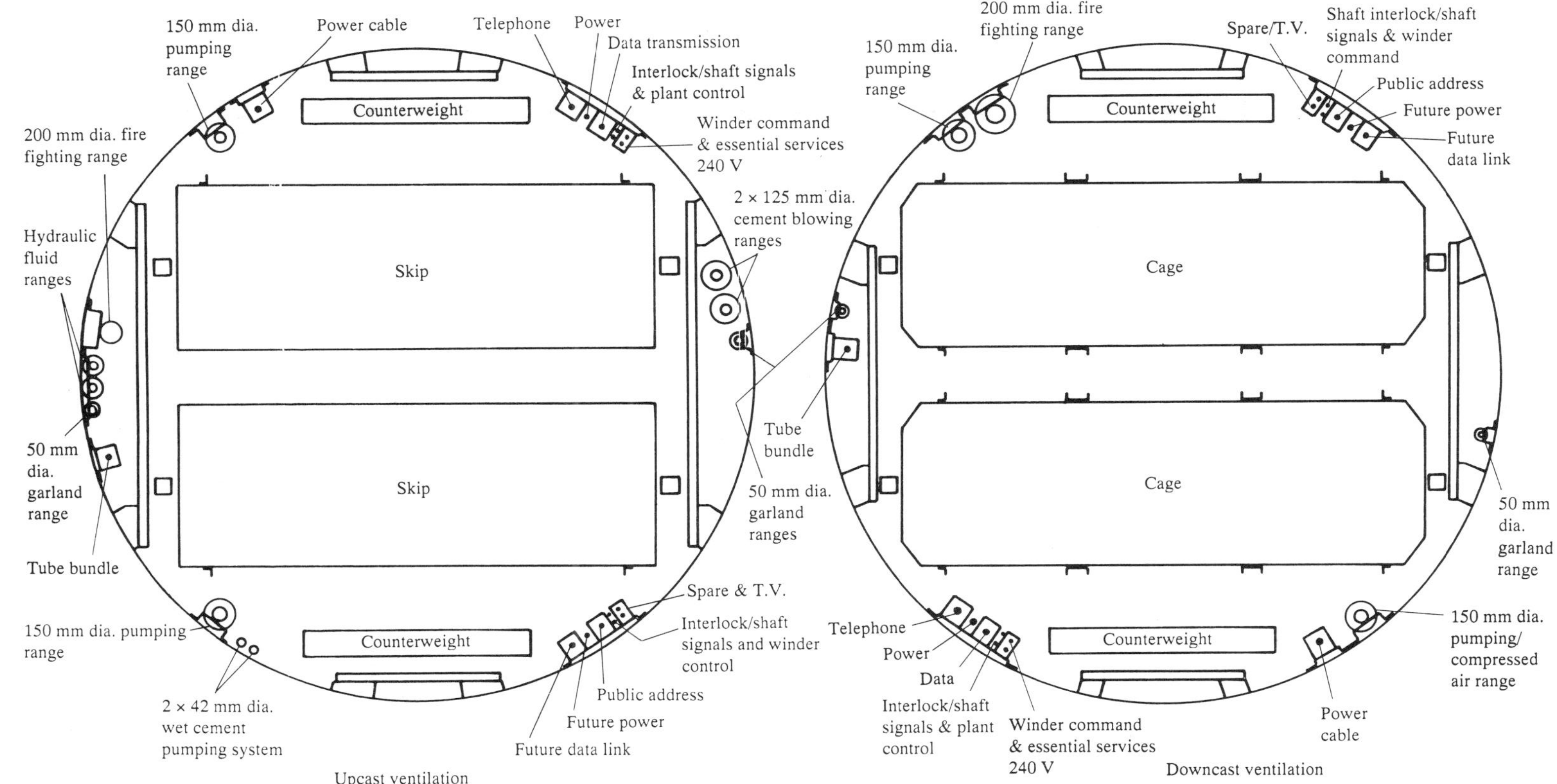

Figure 24.1 *Operating general arrangements for the two Asfordby coalmine shafts in Leicestershire, UK*

is possible to obtain within the practicability and economics of the specific project. This is very important from the pre-engineering aspect in terms of defining the best construction methods relating to excavation, lining installation, strata water control and maintenance of ground stability. Figure 24.2 illustrates the geology and hydrogeology for the North Selby Coal Mine No. 2 shaft in North Yorkshire, UK, constructed during 1978–86.

In order to obtain the necessary geological and hydrogeological information, it is essential to drill and retrieve the core from the full length of a shaft centre-line drill hole. A detailed geological log should be compiled. Observation of the core enables weak zones and potential water inflow areas to be pinpointed. Testing in the laboratory on selected samples from the core allows the rock mechanics and permeability properties of the various strata to be determined. Important rock mechanics properties are strength, modulus of elasticity, Poisson's ratio and rock quality designation (RQD). The driller's log gives important information at no additional cost, such as the location of lost circulation zones and drilling penetration rates.

Where groundwater is suspected it is vital to know its location, what the potential flow is likely to be into an untreated shaft excavation and also the chemical composition of the water. Estimates of potential flow can be prepared from data obtained by an oilfield drill stem test (DST) or alternatively from the more sensitive technique (Chalmers *et al.* 1979) which uses a well pump and pressure-recovery test over a longer time period. Pressure-recovery testing in probe holes during excavation is also very important to confirm predictions from the shaft centre-line drill hole and to pick up any other unexpected localized groundwater conditions.

The subject of geological and hydrogeological ground pre-assessment and in shaft testing is outside the scope of this chapter. Readers are referred to the authoritative papers by Adamson and Scott (1973); Black *et al.* (1982); Daw and Scott (1983); Daw (1984); Daw *et al.* (1988); together with other chapters in this book, for further guidance on the subject.

(d) Construction factors

Methods of excavation and lining installation are discussed in Section 24.3. Chapter 15 deals with the temporary treatment of rock masses. Such measures are in-shaft pumping, deep well dewatering from the surface, pre-grouting either from the surface or in shaft from the excavation face (Black *et al.* (1982); Harris and Pollard (1986), Daw and Pollard (1986)), ground freezing (Harris and Pollard (1986)), and strata water-pressure relief (Fotheringham and Black (1983), Scott and Daw (1983)).

With regard to maintaining ground stability during excavation, it has already been stated in the Introduction that shaft design and construction aspects in this chapter are restricted to installation in competent rock masses. Control of the distance of the installed lining from the sump during sinking and temporary support of the excavation walls using rockbolts and mesh would normally be sufficient for these conditions. Ground freezing is used here for stopping ingress of water and not for the additional purpose of strata strengthening which it is capable of where incompetent conditions, such as sands, silts and clays, are encountered.

(e) Lining loads

In strong, competent rocks no rock pressure is applied to the shaft lining. Being competent means that, after excavation has taken place, the redistribution and concentration of stress within the rock surrounding the opening is less than the inherent strength of the rock. Compression failure in the rock at the excavation face does not occur and inward deformation is minimal, being compatible with the elastic properties of the rock. The majority of the elastic movement takes place almost immediately during the excavation process, thus preventing subsequent rock pressure from imposing itself on the shaft lining. Altounyan (1982) confirmed the latter concept.

Design of shaft linings through sedimentary rock types in the UK does not normally allow for rock loading because of the competent nature of the strata, although exceptions do exist, such as those experienced when sinking the two 5.5 m internal diameter shafts at Boulby Mine in North Yorkshire to a depth of 1150 m during 1969–74 (Cleasby *et al.* 1975).

Effects of ground movement through mining close to the shaft are not applicable because protective shaft pillar zones are adhered to. Some of the rocks, such as marls anhydrite and halite, exhibit time-dependent deformation characteristics but *in situ* they are generally composed of thin beds sandwiched between stronger rock which prevents lateral squeeze through bed interface shearing resistance.

Where rocks contain aquifers, their hydrostatic pressure must be taken into account in the shaft lining design. The design pressure at any aquifer level is equivalent to the full head of water to the surface, plus any artesian head where appropriate, as this is the pressure which will develop on the lining after the aquifer is sealed in behind it. As mentioned above, in the UK the aquifers are generally situated within a zone which approaches a maximum depth of 650 m, below which the strata are normally dry. Figure 24.2 shows the 7.315 m finished diameter, hydrostatic pressure resisting concrete shaft lining down to a depth of 635 m in the North Selby Coal Mine No. 2 shaft. The lower end of the hydrostatic load-bearing section of the shaft is thoroughly cement grouted to prevent downward migration of groundwater behind the lining and the shaft below 635 m, in dry competent strata, is then lined with a nominal thickness of concrete.

It should be noted that, in competent rock, hydrostatic pressure and rock loading cannot be additive. Initially when the concrete is placed directly against the excavation face the ground is pressure relieved, and build up of groundwater pressure on the back of the shaft lining

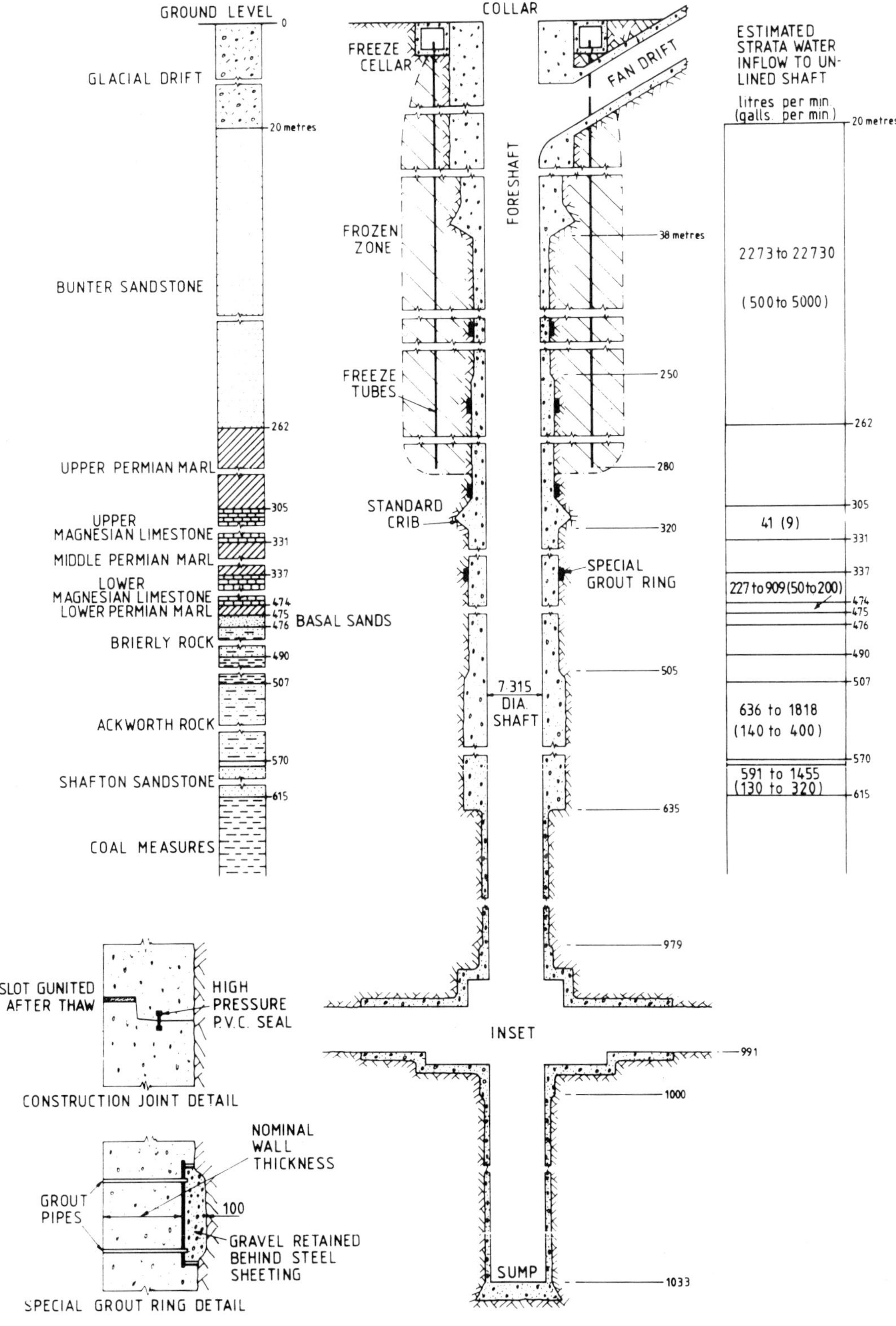

Figure 24.2 *Shaft section with geology and estimated water inflows for North Selby coalmine No. 2 (upcast) shaft, North Yorkshire, UK, constructed during 1978–86*

gradually takes place after the construction joints have been sealed. As pressure builds up during recharging of the surrounding strata, deformation of the lining allows water to penetrate the micro-annulus formed between the rock face and the lining until, for design purposes, it can be assumed that hydrostatic pressure completely surrounds the circumference. To seal the lining thoroughly and to establish tight contact between the back of the lining and the rock, thus forcing the water out of the micro-annulus and back into the rock, the process of backwall grouting is employed. It is during this backwall grouting stage that the shaft lining is at its most vulnerable, as grout pressures slightly above the design hydrostatic values have to be applied to chase the water back, and strict control must be exercised on the grout pump pressure. The residual interface pressure between the concrete lining and surrounding rock generally levels off at the hydrostatic pressure of the aquifer. Measurements of pressure on shaft linings taken *in situ* confirm the correctness of this statement (Altounyan 1982; Altounyan *et al.* 1982).

Only if the rock deformed inwards to such an extent, after the lining was installed, that an interface pressure greater than hydrostatic was developed would water be excluded from the interface. In this case, the higher rock pressure on its own becomes the design criterion.

(f) Lining types

Various lining systems can be used depending upon the ground conditions in which they are to be installed and what loadings they are to be designed for. Nine basic lining types are illustrated in Figure 24.3.

The most common type is illustrated in Figure 24.3(a). With this form of lining, concrete is placed directly against the competent ground as sinking proceeds and the weight of the lining is transferred to the rock by mechanical interlock. It is used as a watertight lining in competent ground containing aquifers and is designed to resist hydrostatic pressure only. Under this condition, the concrete lining does not need to include reinforcement as the resulting stresses are compressive. Except for the 3 m internal diameter drilled shaft constructed to a depth of 220 m from the surface for the British Coal Corporation at Betws, South Wales, during 1985–86, all 23 large-diameter shafts sunk from the surface in the UK since 1957 have used this form of lining (Auld 1987). The two 5.5 m internal diameter shafts at Boulby Mine in North Yorkshire, constructed to a depth of 1150 m during 1969–74 (Cleasby *et al.* 1975) had, in addition, through the Bunter Sandstone from 610 m to 914 m, short lengths of a combined concrete and steel lining (Figure 24.3(c)) in No. 1 shaft and cast iron tubbing (Figure 24.3(d)) in No. 2 shaft to withstand the high water pressure. A concrete lining, cast directly against the competent rock as sinking proceeds, is also typical of the lining type in general use for shafts in South Africa (Henderson 1969), Canada (Kelland and Black 1969) and the USA (Richards and Abel 1979).

Although a concrete lining in contact with the rock is regarded as a watertight lining, because of the necessity to have construction joints, only a fully welded steel membrane installed at the back provides a 100% watertight system (Figure 24.3(b)). With this type, the shaft must be sunk and lined first with a primary concrete lining before the steel membrane and inner concrete lining is constructed in an upwards direction supported from a foundation. (Figure 24.15). Concrete linings with rear steel membranes are also designed to resist hydrostatic pressure only.

The combined concrete and steel lining (Figure 24.3(c)) is designed to resist large hydrostatic pressures or ground loadings (competent or incompetent) and consists of inner and outer steel liners with concrete between. To avoid buckling of the inner steel liner, it must be anchored to the concrete using ties welded on the back. The construction method is similar to that for the concrete lining with a rear steel membrane.

Tubbing (Figure 24.3(d)) is another high-load-carrying capacity lining. The tubbing can be either spheroidal graphite cast iron or fabricated steel-bolted segments which may be installed quickly in difficult ground conditions as sinking proceeds downwards. This type of lining is designed to resist large hydrostatic pressures or ground loadings (competent or incompetent). Water sealing is achieved using lead gaskets, PVC gaskets or lead caulking.

The ring-stiffened steel lining (Figure 24.3(e)) is specially fabricated for installation in drilled shafts, where the excavation is kept stable by the drilling mud which remains to fill the hole on completion of the drilling. Short lengths of circular casing are manufactured, complete with ring stiffeners, for welding on top of each other at the shaft surface. The first section has a closed end and, as each section is welded on, the whole casing is floated in, displacing the drilling mud. After completion of the floating in process, the annulus between the lining and the rock is grouted. This type of lining is designed to resist hydrostatic pressure only.

The 'sliding' type of lining (Figure 24.3(f)) is standard German construction (Hegemann and Jessberger 1985) for use in alluvial ground, sands and clays. For this system, the shaft is first sunk through frozen ground, which is unstable when thawed, using a primary precast concrete block outer lining with infill concrete between it and the frozen ground. Frozen ground deforms considerably at depth under the overburden pressure, both immediately upon excavation and afterwards in relation to time, and therefore vertical and horizontal compressible plates of wood flax and glue are needed between the precast concrete blocks to accommodate the squeeze. Once the shaft is sunk through the unstable ground, a foundation is constructed in strong rock and the secondary reinforced concrete lining, with its rear thin steel membrane, is installed in an upwards direction. In Germany, the coal workings approach close to the shaft, without leaving a protective unworked zone (shaft pillar), and hence the asphalt ('sliding') layer is provided between the primary and secondary linings to accommodate ground movement

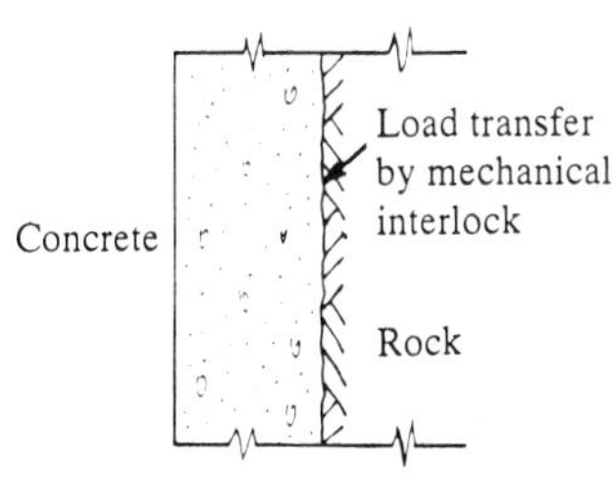

(a) Concrete

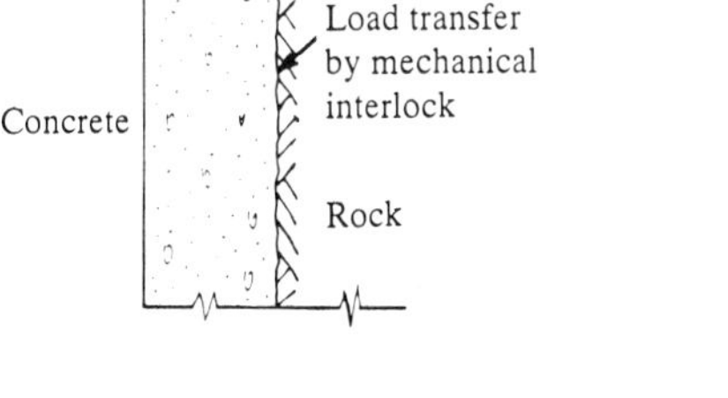

Bolted connection
Spheroidal graphite cast iron tubbing bolted segments
Load transfer by mechanical interlock
Rock
Infill concrete

(d) Tubbing

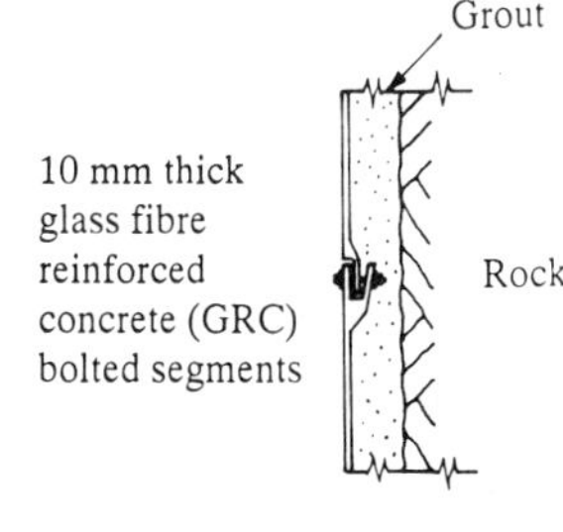

(g) GRC permanent formwork

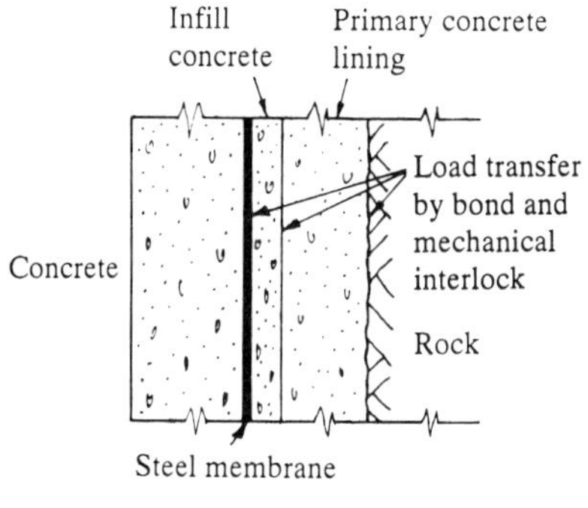

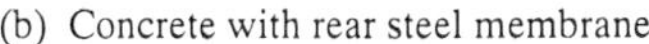

(b) Concrete with rear steel membrane

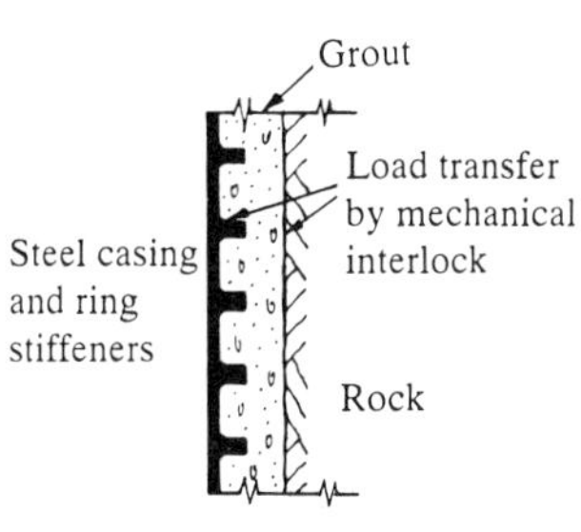

(e) Ring stiffened steel

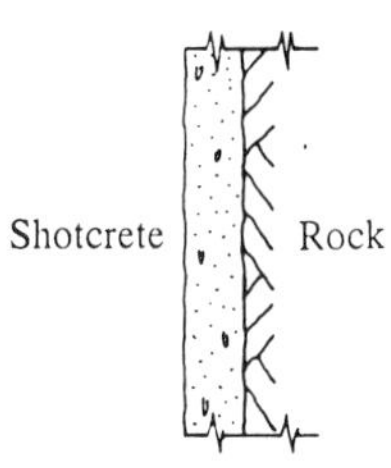

(h) Shotcrete

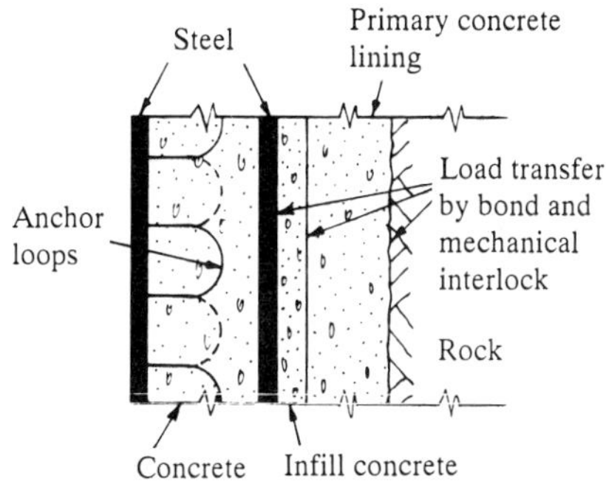

(c) Combined concrete and steel

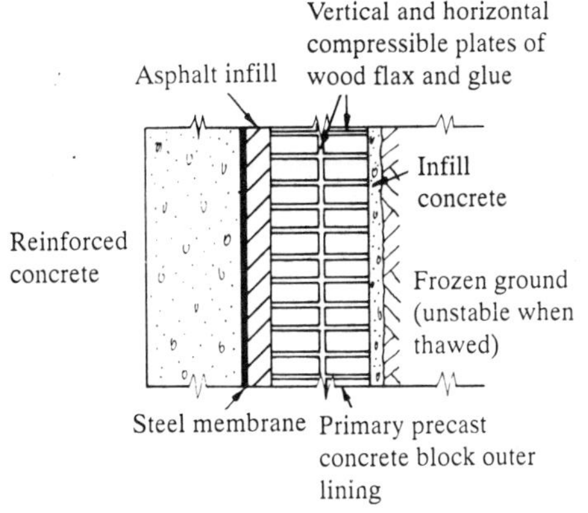

(f) Sliding

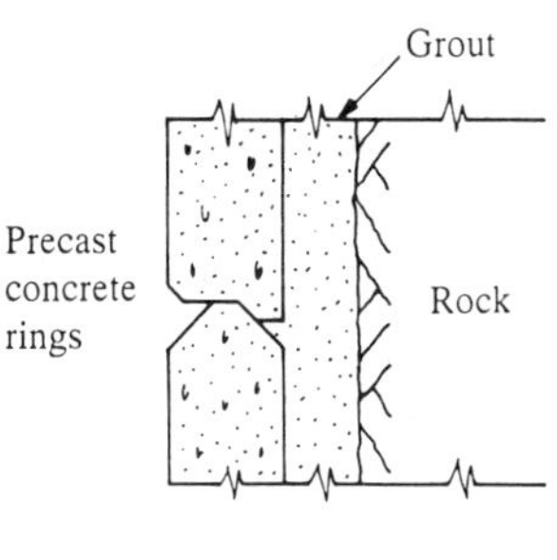

(i) Precast concrete

Figure 24.3 *Shaft lining types*

resulting from the coal workings. The inner secondary reinforced concrete lining is designed to resist asphalt pressure at 1.3 times hydrostatic pressure.

Figure 24.3(g) shows a lining constructed using glass-fibre-reinforced concrete (GRC) bolted segments. A cradle is used to support the lining as it is lowered into the excavation, and the segments are systematically bolted together at the shaft top to form the complete lining as the descent proceeds. Once the lining is installed throughout the shaft length, the outer annulus is grouted up in stages from the bottom upwards. Such linings provide nominal support for the excavation. They are generally adopted for small diameter (up to 2 m) shafts where they also give a smooth finish for ventilation purposes.

Shotcrete linings (Figure 24.3h) are used for temporary or nominal permanent support. In competent ground, rock bolts and mesh secure the rock face first, with the application of shotcrete on top providing additional security against falling fragments.

Precast concrete linings (Figure 24.3i), in the form of complete rings, can be lowered into a drilled shaft full of drilling mud as individual open-ended units (Skonberg 1980, Pliska 1984). They are located one on top of the other in the hole and after placing a concrete plug in the bottom of the shaft using a tremie pipe, the surrounding annulus is grouted up in stages by displacement of the drilling mud. The shaft becomes fit for service after baling out the drilling mud left inside.

24.2.2 Lining design

Nine different lining types have been discussed in the previous section. It is not possible to cover fully the design of each one in this chapter. Therefore attention is focused on concrete linings, which are the most widely used type, with guidelines and references provided for the design of combined concrete and steel, ring-stiffened steel and tubbing systems.

(a) *Concrete linings*

The design of concrete shaft linings is covered by Auld (1979), Auld (1982) and Black and Auld (1985). On the basis of an elastic analysis, if a uniform external pressure is applied to a thick cylinder, then the tangential stress developed is a maximum on the inside face (Figure 24.4). This is the governing criterion which is normally adopted for the design of concrete shaft linings. The value of the maximum tangential stress is given by

$$\sigma_{t\max} = 2pr_o^2/(r_o^2 - r_i^2) \qquad (24.1)$$

which can be converted to relate directly to the wall thickness, t:

$$\sigma_{t\max} = 2p(t + r_i)^2/[t(t + 2r_i)] \qquad (24.2)$$

Equation 24.2 is known as the Lamé formula in recognition of the original elastic theory development (Lamé and Clapeyron 1833). Transposing Equation (24.2) produces

$$t = r_i\left\{[\sigma_{t\max}/(\sigma_{t\max} - 2p)]^{1/2} - 1\right\} \qquad (24.3)$$

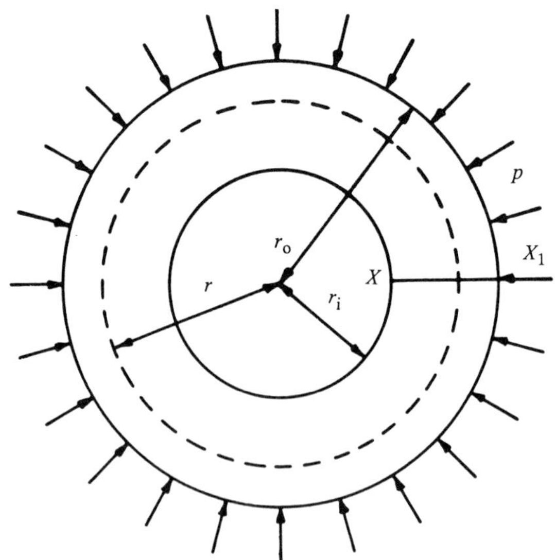

Notation

p = uniform external pressure
r_i = internal radius
r_o = external radius
r = radius to the point at which stresses, strains and deformation are determined
E = modulus of elasticity
υ = Poisson's ratio
σ_r = radial stress
σ_t = tangential stress
ε_r = radial strain
ε_t = tangential strain
δ = deformation

General expressions

Stresses

$$\sigma_r = \frac{pr_o^2}{r_o^2 - r_i^2}\left(1 - \frac{r_i^2}{r^2}\right) \qquad \text{(GE1)}$$

$$\sigma_t = \frac{pr_o^2}{r_o^2 - {}_i^2}\left(1 + \frac{r_i^2}{r^2}\right) \qquad \text{(GE2)}$$

Strains

$$\varepsilon_r = \frac{pr_o^2}{E(r_o^2 - r_i^2)}\left[(1-\upsilon) - \frac{r_i^2}{r^2}(1 + \upsilon)\right] \qquad \text{(GE3)}$$

$$\varepsilon_t = \frac{pr_o^2}{E(r_o^2 - r_i^2)}\left[(1 - \nu) + \frac{r_i^2}{r^2}(1 + \nu)\right] \qquad \text{(GE4)}$$

Deformation

$$\delta_r = \frac{pr_o^2}{rE(r_o^2 - r_i^2)}\left[r^2(1 - \upsilon) + r_i^2(1 + \upsilon)\right] \qquad \text{(GE5)}$$

Figure 24.4 *Elastic stresses, strains and deformation in a thick cylinder subjected to uniform external pressure*

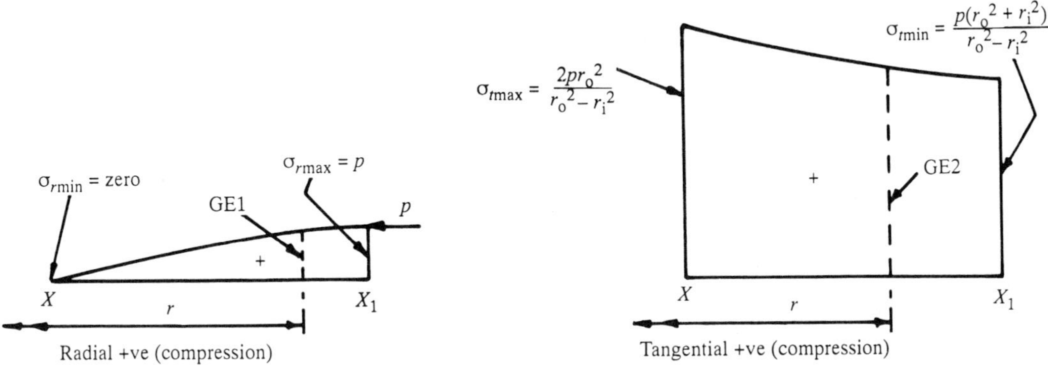

Elastic stress distribution at X–X_1

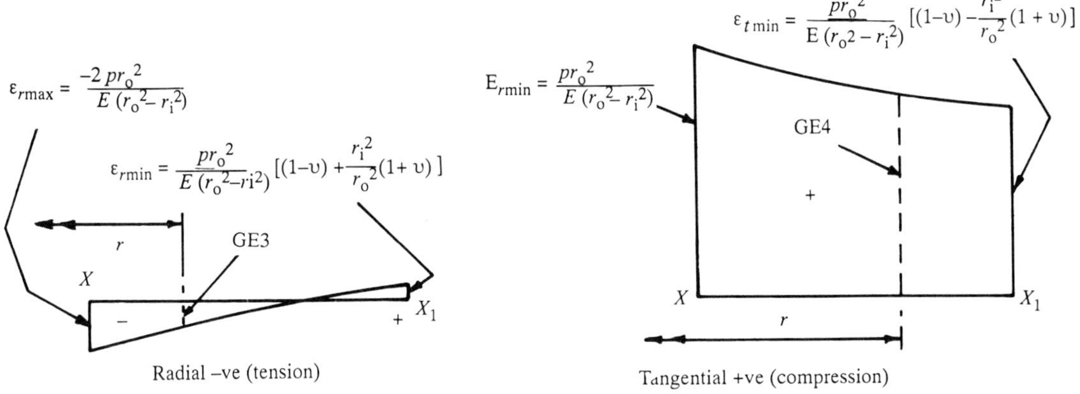

Elastic strain distribution at X–X_1

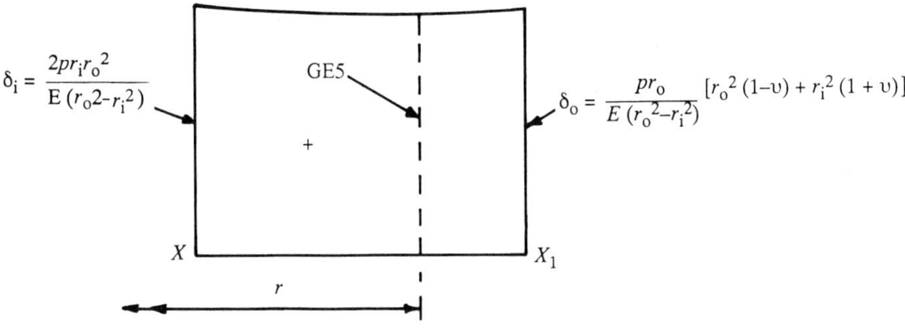

Inward elastic deformation across X–X_1

Figure 24.4 *continued*

The value $\sigma_{t\max}$ is the maximum induced tangential stress based on an elastic analysis. To put this fact into context with concrete design theory it is necessary to introduce parameters which relate to concrete strength, quality control and workmanship. In accordance with the British Standard for the structural use of concrete (Anon 1985) the maximum stress is determined from

$$\sigma_{t\max} = 0.67 f_{cu}/(1.4 \times 1.5) = 0.3190 f_{cu} \quad (24.4)$$

where f_{cu} is the concrete characteristic strength. This is the strength value obtained below which not more that 5% of all strength results fall for 150 mm, 28-day-old concrete cubes tested to failure in compression. The 0.67 figure relates the characteristic strength to the actual strength of the structure *in situ*, including taking account of the difference between instantaneous loads on the test specimens at the age of 28 days and loads applied for a longer duration on specimens of the same age. Anon (1985) adopts an ultimate limit state approach to the design of concrete structures and the values 1.4 and 1.5 are partial safety factors for load and the concrete material respectively. The latter is introduced to account for possible strength differences between test specimens and the actual structure caused by aspects such as insufficient compaction and differences in curing.

For design purposes, substituting the value of $\sigma_{t\max}$ from Equation (24.2) into Eq. (24.2) and reorganizing produces

$$d = 0.3190 f_{cu} t(t + 2r_i)/[2 \times 0.01\,(t + r_i)^2] \quad (24.5)$$

where the value of the pressure, p, has been replaced by $0.01d$, d being the depth in metres at which the pressure is applied and 0.01 N/mm^2 per metre of depth is the hydrostatic pressure. Equation (24.5) therefore indicates the depth to which any wall thickness under consideration can reach on the basis of resisting hydrostatic pressure.

Example 1 To what depth can a Grade 60 concrete, 1200 mm thick shaft lining extend in competent rock on the basis of resisting hydrostatic pressure if the shaft internal diameter is 7.3 m?

$$\begin{aligned} d &= 0.3190 \times 60 \times 1.2\,(1.2 + 7.3)/[2 \times 0.01\,(1.2 \\ &\quad + 3.65)^2] \\ &= 415\text{ m} \end{aligned}$$

Alternatively, if the concrete grade, the shaft internal diameter and the depth are specified, the required wall thickness can be found directly from Equation (24.3):

$$\begin{aligned} t &= 3650\,\{[0.3190 \times 60/(0.3190 \times 60 - 2 \times 415 \\ &\quad \times 0.01)]^{1/2} - 1\} \\ &= 1200\text{ mm} \end{aligned}$$

General expressions for the elastic stresses, strains and deformation in a thick cylinder subjected to uniform external pressure are contained in Figure 24.4. Their distributions throughout the wall thickness are also indicated. The deformation of the inside face of the lining for the example given, assuming a modulus of elasticity for the concrete of 32 kN/mm^2, would be

$$\begin{aligned} \delta_i &= 2pr_i r_o^2/[E(r_o^2 - r_i^2)] \\ &= 2 \times 415 \times 0.01 \times 3650 \times 4.85^2/[32\,000\,(4.85^2 \\ &\quad - 3.65^2)] \\ &= 2.2\text{ mm} \end{aligned}$$

It should be noted that low-pressure backwall grouting is used to fill any voids left behind the lining after the concrete is cast and high-pressure grouting subsequently is adopted for final tightening up and water sealing of the joints. Although the design pressure is hydrostatic, the sealing pressure needed is normally slightly over hydrostatic up to a maximum of 1.25 × hydrostatic.

The design of concrete shaft linings in this text has specifically been restricted to uniformly distributed hydrostatic pressure which is normal design for competent rock. In unstable conditions, such as sands, silts and clays at depth where ground stabilizing methods such as ground freezing need to be employed, non-uniform loads may occur on the lining and have to be catered for in design.

The subject of unstable ground conditions is not covered in this text, the attention being focused on competent rock. However, readers are referred to Link *et al.* (1976), which gives guidelines for the design of shaft linings in unstable ground conditions.

(b) *Combined concrete and steel linings*

There are no specific codes of practice for the design of shaft linings. The document referred to in Section 24.2.2(a) on concrete shaft linings, by Link *et al.* (1976), although giving guidance primarily for the design of shaft linings in unstable ground, also includes design criteria for uniform load conditions. It makes reference to combined concrete and steel linings and is a recognized document for the basis of their design.

Example 2 illustrates how the design of a combined concrete and steel lining should be approached when designing for uniform hydrostatic pressure in a competent rock situation. The first step is to equate the concrete area in the lining cross-section to an equivalent steel area based on the modular ratio, n, of the two materials. Two conditions are considered, a long-term situation where $n = 15$, which provides for concrete creep, and a short-term stress condition with $n = 10$. Section properties are evaluated and stresses calculated. The long-term condition governs the maximum depth to which the shaft lining can reach, while the short-term situation dictates the concrete grade required.

Buckling stability of the overall lining must be checked and tie bars need to be incorporated to prevent the inner steel shell from peeling away from the concrete. Additional factors for checking are the pull-out resistance of the ties from the concrete, the weld strength of the ties at the connection to the inner steel shell and the bending stresses in the steel shell between tie centres.

Example 2 To what depth can a 1200 mm thick combined concrete and steel lining extend in competent rock on a hydrostatic pressure-resisting basis if the shaft internal diameter is 7.3 m? Assume concrete Grade 55, steel Grade 50 and lining details as shown in Figure 24.5. Design in accordance with Link *et al*. (1976).

Lining section properties

	Area, A (mm^2)	x (mm)	Ax (mm^3)	Ax^2 (mm^4)	I_{NA} (mm^4)	$I_{NA} + Ax^2$ (mm^4)
1	60	30	1 800	0.054×10^6	0.018×10^6	0.072×10^6
2	40	1180	47 200	55.696×10^6	0.005×10^6	55.701×10^6
3 (LT)	73.3	610	44 733	27.287×10^6	7.394×10^6	34.681×10^6
Σ (LT)	173.3		93 733			90.454×10^6
3 (ST)	110	610	67 100	40.931×10^6	11.092×10^6	52.023×10^6
Σ (ST)	210		116 100			107.796×10^6

$$\bar{x} = \frac{\Sigma(Ax)}{\Sigma(A)} = 541\ (LT) \text{ or } 553\ (ST);\ I_{XX} = I_{NA} + Ax^2\ \therefore\ I_{NA} = I_{XX} - \Sigma(A)x^2;$$

$$I_{NA(LT)} = 90.454 \times 10^6 - 173.3 \times 541^2 = 39.732 \times 10^6\ \text{mm}^4;\ I_{NA(ST)} = 107.79 \times 10^6 - 210 \times 553^2 = 43.576 \times 10^6\ \text{mm}^4$$

Calculations Tangential stress, $\sigma_t = \dfrac{pr_o}{A}\left(1 + \dfrac{y}{r_s}\right)$

σ permissible for steel Grade 50 (N/mm^2)	Area, A (mm^2)	r_o (mm)	a $\left(1 + \frac{y}{r_s}\right)$	$p = \frac{\sigma A}{r_o a}$ (N/mm^2)	Depth $= p/0.01$ (m)	b $\left(1 + \frac{y'}{r_s}\right)$	Conc. stress $= \frac{pr_o b}{An}$ (N/mm^2)
210	173.3 (LT)	4850	1.1291	6.65	665	1.1148	13.83
210	210 (ST)	4850	1.1316	8.04	804	1.1173	17.16*

* For $p = 6.65$ N/mm². Therefore permissible depth = 665 m and required concrete Grade = 17.16/0.3190 from Equation (24.4) = 53.8, say 55.

Check stability

Slenderness ratio, $\lambda = \dfrac{1.8135}{i} r_s$

where

radius of inertia, $i = \sqrt{(I/A)}$

$$i_{(LT)} = \sqrt{\left(\frac{39.732 \times 10^6}{173.3}\right)} = 479\ \text{mm};$$

$$i_{(ST)} = \sqrt{\left(\frac{43.576 \times 10^6}{210}\right)} = 456\ \text{mm};$$

$$\lambda_{(LT)} = \frac{1.8135 \times 4191}{479} = 15.9 < 50;$$

$$\lambda_{(ST)} = \frac{1.8135 \times 4203}{456} = 16.7 < 50;$$

Buckling safety factor, $\gamma_k = \dfrac{p_k}{p} = \dfrac{\Sigma(A\sigma_k)}{pr_o}\ \dfrac{(r_i + r_o)}{2r_o} \geqslant \gamma_{kerf} \geqslant 1.5 + \dfrac{\lambda}{100}$

where

p_k = buckling pressure and
σ_k = buckling stress for each material

σ_k = 343 N/mm² for Grade 50 steel and 41 N/mm² for Grade 55 concrete

$$\therefore \gamma_k = \left[\frac{(60 + 40)\,343 + 1100 \times 41}{6.65 \times 4850}\right] \times \frac{(3650 + 4850)}{2 \times 4850}$$

$$= 2.2 > 1.5 + \frac{16.7}{100} = 1.67$$

Ties

Separating stress at $Y - Y$ = radial stress at $Y - Y$

$$\sigma_{rYY} = \frac{pr_o}{r} \frac{A_1}{\Sigma(A)} \left(1 + \frac{y_r}{r_s}\right)$$

$$= 6.65 \times \frac{4850}{3710} \times \frac{60}{173.3} \left(1 + \frac{511}{4191}\right)$$

$$= 3.38 \text{ N/mm}^2 \ (LT)$$

$$= 6.65 \times \frac{4850}{3710} \times \frac{60}{210} \left(1 + \frac{523}{4203}\right)$$

$$= 2.79 \text{ N/mm}^2 \ (ST)$$

Tie cross-section = 150 mm wide × 30 mm thick (Grade 43 steel)

$$\text{Spacing} = \pi \times \frac{7.420 \times 7.5°}{360°}$$

= 485 mm (48 around circumference)

$$\text{Tie force, } F = \frac{350 \times 485 \times 3.38}{1000} = 574 \text{ kN;}$$

$$\text{Tie stress} = \frac{574 \times 1000}{150 \times 30} = 128 \text{ N/mm}^2$$

<165 N/mm^2 permissible for Grade 43 steel

Additional factors to check

(1) pull-out resistance of tie from concrete;
(2) weld strength of tie to steel shell;
(3) bending stress in steel shell between tie centres.

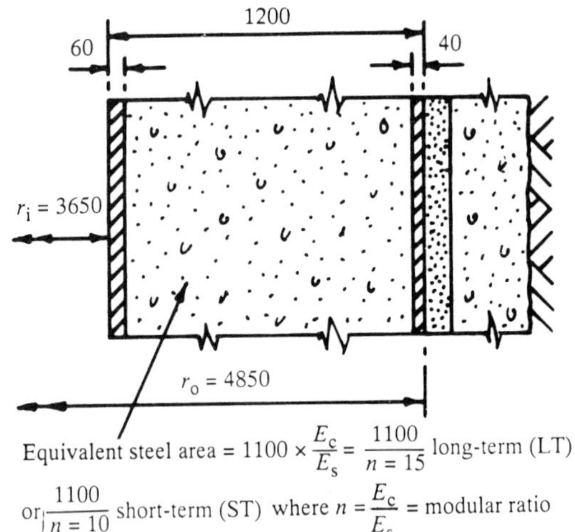

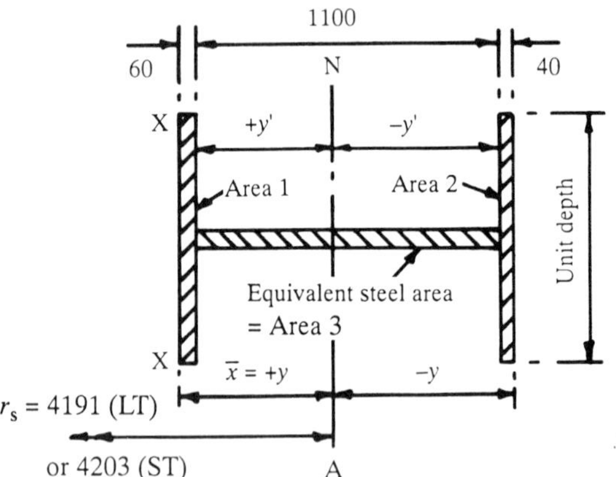

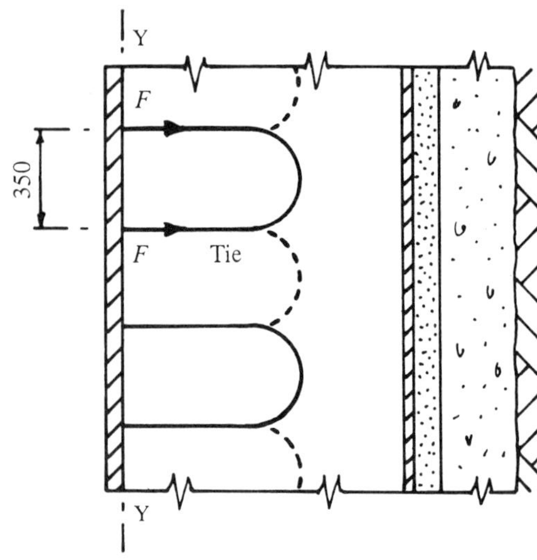

Figure 24.5 *Lining details for Example 2*

(c) Other types of lining

The design of ring-stiffened steel casings for drilled shafts is covered by Russell and DeHart (1967). These linings are also designed for a uniformly distributed pressure equal to hydrostatic, this being the imposed load in the final installed condition. The load imposed during the floating-in process does not exceed hydrostatic as the external drilling mud pressure is counterbalanced by internal water pressure which is regulated to control sinking. Grouting of the annulus is also carried out in stages to ensure the lining hydrostatic design pressure is not exceeded. Additional design aspects such as stresses due to casing weight during suspension, hole misalignment, point loads and shell out of roundness are also considered.

A stress and stability analysis for the casing is carried out on a threefold basis:

(1) The panel elastic collapse pressure of the shell between stiffeners is computed (Figure 24.3e).
(2) The overall elastic collapse pressure of the shell and the stiffeners is evaluated.
(3) Bending moments, shear forces and axial forces are determined for the interaction between the stiffeners and the shell through fixity at the welds only.

Factors of safety against collapse pressure and against yield in the stress calculations are applied. Additional relevant information is provided by Windenburg and Trilling (1934), Timoshenko and Woinowsky-Krieger (1959) and Link (1980).

The design of spheroidal graphite cast iron tubbing linings (Figure 25.3d) is carried out by first determining the position of the neutral axis for the cross-sectional area and then evaluating the moment of inertia for the same. A stress check follows at the inside face of the lining, based on the axial hoop stress induced by the imposed radial pressure plus the bending stress caused by the eccentricity between the line of action of the hoop force applied at the centre of bearing of the vertical side flanges and the neutral axis of the cross-sectional area. The value of this stress should not exceed the permissible stresses given by Gilbert (1974).

Table 24.5 Shaft and raise construction methods

	Conventional sinking and lining		*Alternative methods of shaft construction*			
Excavation method	Drill and blast	Mechanical excavators (USA impact breaker, Polish roadheader boom machines and German Schachtfräse	Large-diameter shaft drilling	Large-diameter shaft boring	Raise drilling	Box-hole drilling
Method of transferring muck from excavation face to removal system	Cactus grab, clam shell grab, Cryderman, backhoe, Eimco loader	As for drill-and-blast excavation	Mud flow over cutter head face	Mechanical or pneumatic lifting methods or pumped mud or free fall down pilot hole	Free fall	Free fall
Muck removal method	Skip wound to surface	Skip wound to surface	Reverse circulation mud flush, with air-lift assistance, up drill string to surface	Skip wound, pneumatic or pumped to surface or free fall down pilot hole for removal through underground system	Free fall for removal through underground system	As for raise drilling
Lining installation methods	1. Lining downwards with the permanent lining as sinking proceeds 1.1 *In situ* concrete using hanging rods 1.2 Precast concrete segments, fabricated steel tubbing or spheroidal graphite (SG) cast iron tubbing 2. Temporary lining downwards with inner permanent lining installed upwards from a foundation 2.1 Slipformed concrete permanent lining 2.2 Combined concrete and steel permanent lining with concrete or asphalt infill behind		1. Floated in and grouted up 1.1 Ring stiffened steel 1.2 Combined concrete and steel 1.3 Precast concrete 1.4 Continuous shaft lining (CSL) with concrete casting at surface 2. Open-ended precast concrete circular segments lowered in under drilling mud and grouted up 3. Slipformed from the bottom upwards under mud 4. Autoform method	1. As conventional lining installation method 1.1 2. Continuous shaft lining (CSL) within shaft concrete casting	1. Glass-reinforced concrete (GRC) permanent formwork grouted up behind 2. Cast *in situ* concrete with inflatable void former 3. Shotcrete using rotating spray machine	As for raise drilling

24.3 Shaft and raise construction

Shaft construction methods can be categorized under two distinct headings. One classification group covers the field of conventional sinking and lining whereas the second encompasses alternative mechanized systems. The differences between the two categories can be observed from Table 24.5.

Excavation in conventional sinking and lining is by the traditional method of drilling and blasting. In some cases, mechanical digger and cutter systems, such as impact breakers and roadheader booms, have been introduced to operate in the shaft sump as a means of speeding up the excavation rate. Removal of the muck (excavated material) is via skip winding to the surface. Lining takes place downwards as sinking proceeds with the installation of the permanent lining or by using temporary support prior to constructing the permanent lining back upwards from a foundation.

The need to reduce construction time (and hence cost) for major shaft sinking projects, in order to minimize interest payments and speed up return on investment, has led to the development of different kinds of specialist mechanized equipment. Examples are large-diameter shaft drilling rigs, large-diameter boring machines, raise borers and box-hole drills (see Section 24.3.2). Shaft construction utilizing these items of equipment falls into the alternative methods category of Table 24.5. Such methods necessitate their own specific muck removal and lining systems.

Longden (1967) gave an overall account of the techniques involved in shaft and raise construction. Douglas and Pfutzenreuter (1989) provided a detailed overview of current South African vertical circular shaft conventional sinking and lining practice.

24.3.1 Conventional sinking and lining

(a) Collars and foreshafts

The first part of the sinking process involves the construction of a shaft collar, foreshaft and any associated connecting surface structures such as ventilation drifts (Figure 24.2). Reinforced concrete is used for these structures with the collar and foreshaft serving a twofold purpose:

(1) to provide a rigid, load-carrying structure which passes through the soft surface soil deposits and transfers the headframe and other collar loads to the hard competent rock below;
(2) to give sufficient initial depth of shaft to enable the installation of the sinking scaffold to be carried out prior to the main excavation.

The depth of the foreshaft construction ranges from 20 to 50 m. Unrug (1984) gave guidance on the design of collars and foreshafts.

Figure 24.6 illustrates typical foreshaft construction. Excavation is generally carried out using a diesel back-actor and may be assisted by hydraulic impact breaker or limited blasting. The rock is hoisted to the surface in skips by mobile or derrick type cranes. Temporary support consists of rockbolts and mesh, liner plates or bolted precast concrete segments depending upon ground conditions. The permanent reinforced concrete lining is cast in short lengths (normally 3 m) from the bottom upwards.

(b) Headframes and winding facilities

Figure 24.7 shows a typical conventional shaft sinking and lining construction arrangement. Located on top of the shaft collar is the temporary sinking headframe with built-in muck disposal system. The headframe can have built-in adjustable legs to cope with heave in frozen ground and it may be erected to one side of the shaft to be moved into position over the shaft on rollers. It can also be moved into position below a permanent headframe. A separate winding house is indicated containing a ground-mounted four-drum scaffold hoist and a double drum sinking hoist. The former possesses low speed to raise and lower the sinking scaffold whereas the latter has a higher speed for muck removal and men and materials access via skip winding. Alternatively called a kibble or hoppit, the skip is a rugged general-purpose bucket, usually around 4 m^3 capacity in the UK, which is slung by a three-legged chain or steel hoop.

Suspended in the shaft on four ropes is the sinking scaffold which provides a working platform for men, storage space for equipment and materials and a safety cover for men standing in the sump (Figure 24.8). The scaffold is also used to suspend concrete shutters during lining construction and may have mucking equipment built in below the lowest deck. It moves at slow speeds independently from the main hoisting hoppits and provides a second means of egress from the shaft.

A twin muck hoppit system is normally employed, one being raised while simultaneously the other is lowered. The winding ropes that suspend the scaffold act as guides for the hoppits during their ascent and descent in the shaft, the control method being sliding steel crossheads loosely attached to each pair of stage ropes. During mucking cycles each hoppit of debris is raised to the top of the shaft and through a shaft top butterfly door which is then closed to prevent spillage from falling back into the shaft. Tipping takes place within the headframe into a holding chute with control door that delivers debris to trucks for disposal. The door and chute systems are interlocked with the winder to prevent collisions between moving hoppits and closed doors or extended chutes. For mechanical and electrical engineering aspects of shaft sinking, readers are referred to Blackwood (1981).

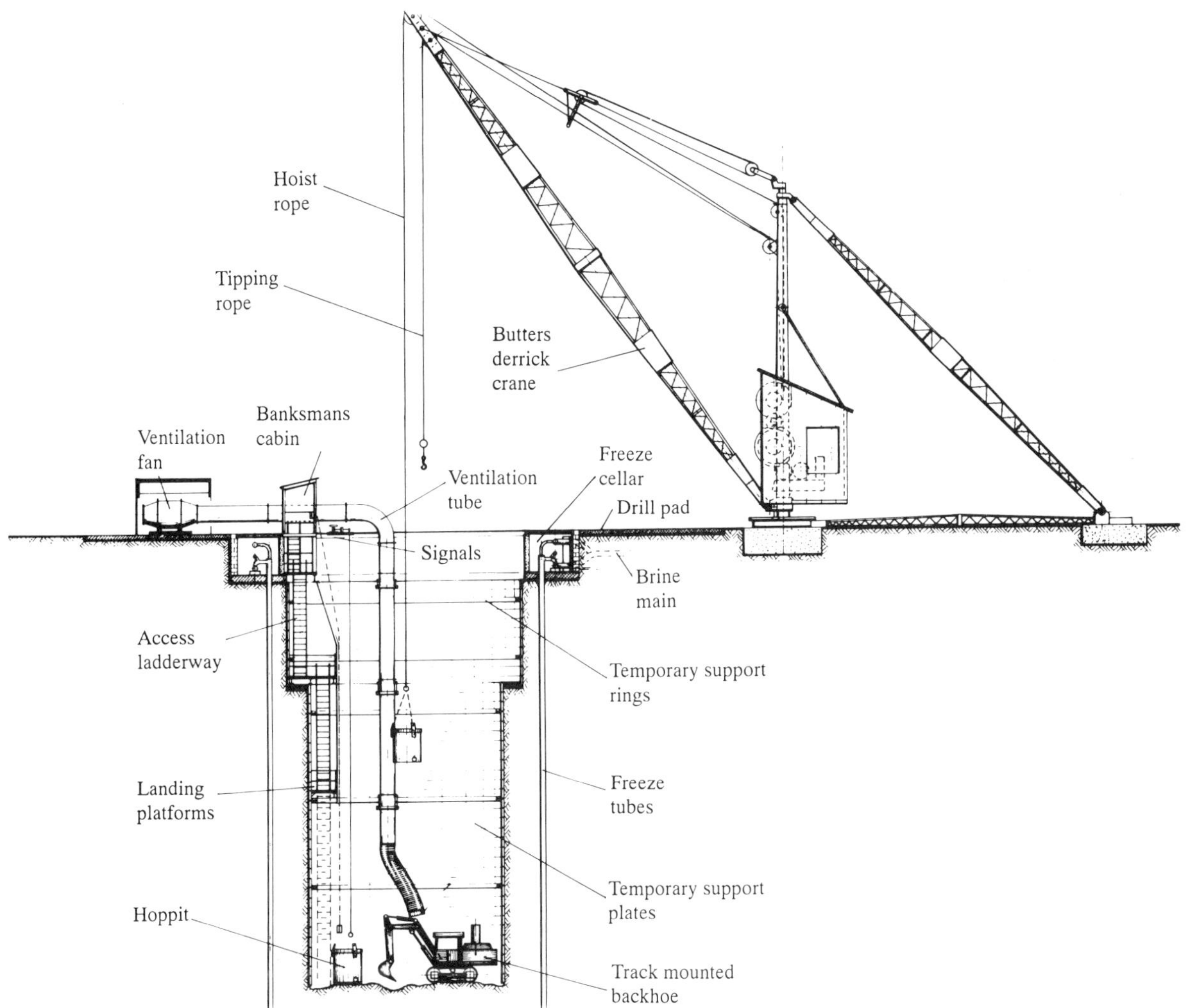

Figure 24.6 *Typical foreshaft construction*

(c) Drilling and firing

Drilling of the shaft sump can be carried out manually using pneumatic hand-held machines or with the use of multi-boom sinking jumbos. These are equipped with pneumatic drifters mounted on hydraulically operated booms which can drill 3.5 m long holes in a single pass. When equipped with extension rods, probe holes or grout injection holes can be drilled up to 40 m deep. Varying ground conditions require different types of drill bits. Full sump drilling and firing is employed for high-speed sinking in reasonably dry conditions. Benching is used where conditions are very wet. The latter technique provides a large sump for water to be collected below the level of the drilling bench and also obviates the need for complete cleaning in the sump which becomes increasingly difficult in direct proportion to the amount of water present.

Shaft sinkers tend to use the highest strength explosive cartridges permissible under the governing mining regulations. In the UK high-strength gelignite is used, except in Coal Measures where 'permitted' explosives have to be adopted (Tunnicliffe and Keeble 1981). Rounds are fired from the surface by the parallel mains system of initiation as the parallel circuit significantly reduces the possibility of misfires occurring through current leakage. This is particularly important because the shaft sump is very often wet and detonation wires tend to get trampled underfoot during charging. The use of an a.c. power source means that zero delay or instantaneous detonators cannot be used and, as the accuracy of delay times is adversely affected, half-second delay detonators only are used.

For further information on the subject of blasting, readers are referred tp explosive manufacturers' literature. Some examples are given in ICI Nobel's Explosives

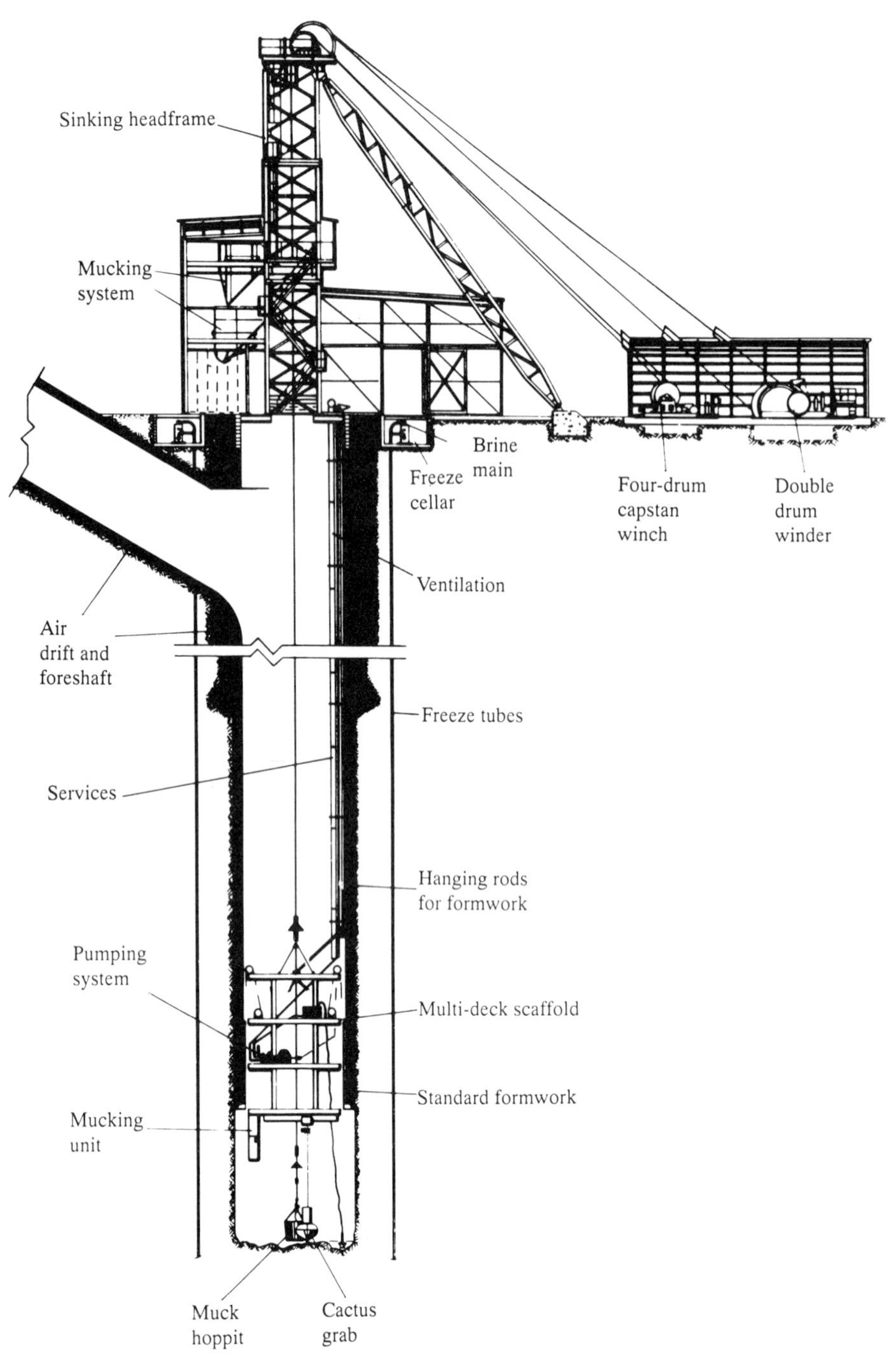

Figure 24.7 *Typical conventional shaft sinking and lining construction*

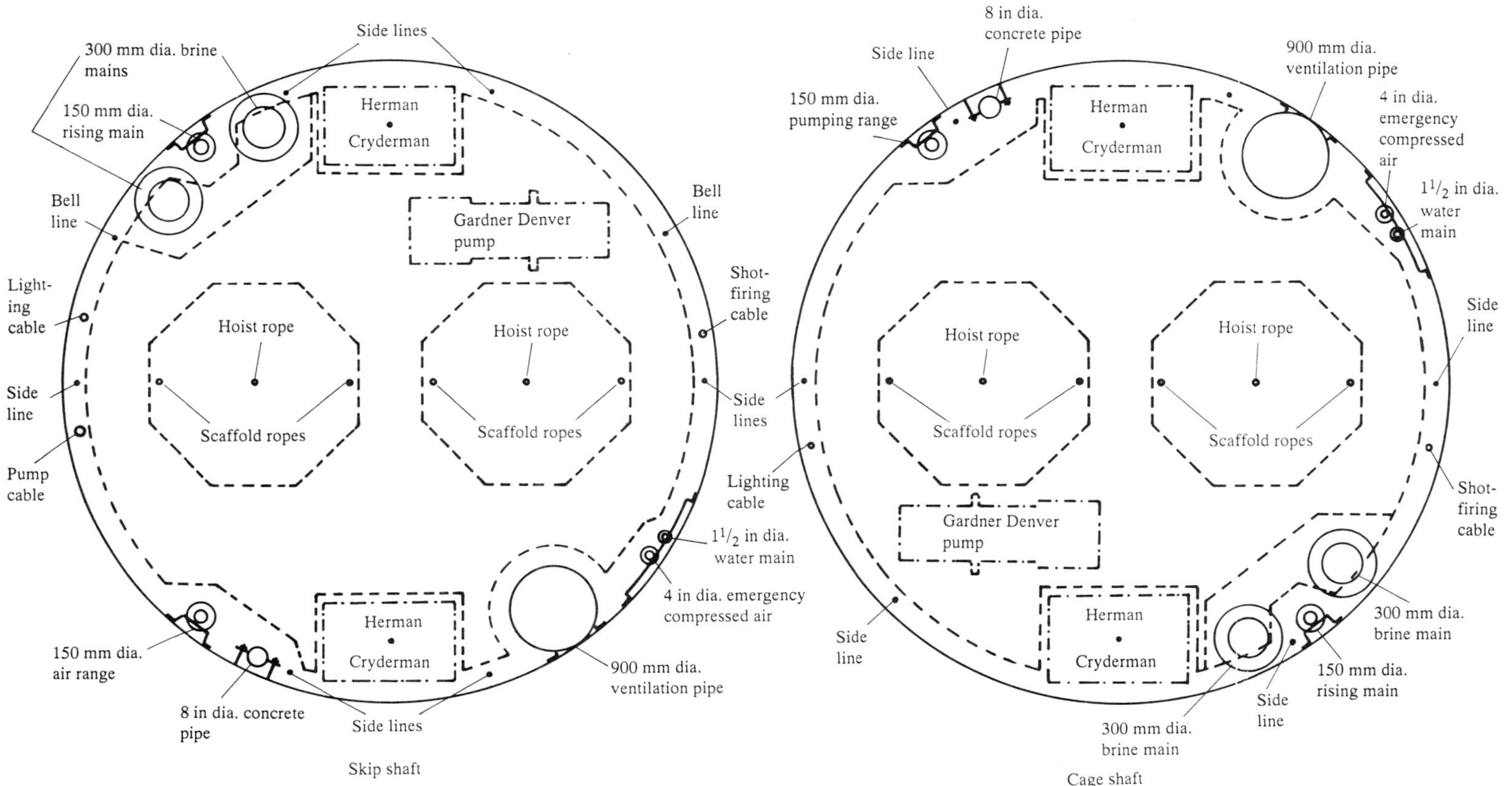

Figure 24.8 *Sinking general arrangements for the two Asfordby coalmine shafts in Leicestershire, UK*

Company Ltd (undated a,b,c) and Nitro Nobel Industries Sweden (undated).

Drilling and blasting suffers a number of problems when ground freezing is introduced:

(1) Drilling is slower as the ground is strengthened by freezing.
(2) Water flushing of drills is not possible as the holes freeze solid. Air flushing or scrolled drill rods must be used.
(3) Drilled holes close up and cannot be charged in some frozen rock such as Bunter Sandstone.
(4) Explosives need to be limited to protect freeze tubes from damage.
(5) Low-temperature lubricants need to be added to compressed air to prevent pneumatic equipment from icing up.
(6) The ventilation air supply to the shaft may require heating during cold ambient conditions to improve the working environment in the sump.

(d) Mucking systems

Various methods are in use today including equipment integral with the sinking scaffold, independently suspended equipment braced off the shaft wall and tracked loaders operating in the sump. The most common system, with the highest capacity, is the pneumatic cactus grab suspended from either a traversing or a rotating boom (Figure 24.9) built into the bottom deck of the scaffold. An operator controls the grab from a cab which is also slung below the bottom of the scaffold.

For smaller diameter shafts, wall-mounted Cryderman (Figure 24.10) or Alimak units offer alternatives. These are more positive in their action, the Cryderman employing a telescopic boom with grab or clam-shell, and the Alimak having a conventional back-hoe bucket arrangement.

Eimco style rocker shovel muckers (Figure 24.11) may be used in larger-diameter shafts, but they are poor performers in wet conditions and their tracks suffer badly from abrasive materials. However, they can be effectively used to clean up sumps that have been substantially mucked out by grabs, as grabs are inefficient at cleaning up when there is little muck left to remove.

(e) Mechanical excavators

Several attempts have been made to speed up the conventional sinking and lining process using mechanical digger and cutter systems. During the period 1975–1978, the Cementation Company of America, Inc., successfully developed and tested a mechanical impact breaker and back-hoe system (Figure 24.12) for the US Bureau of

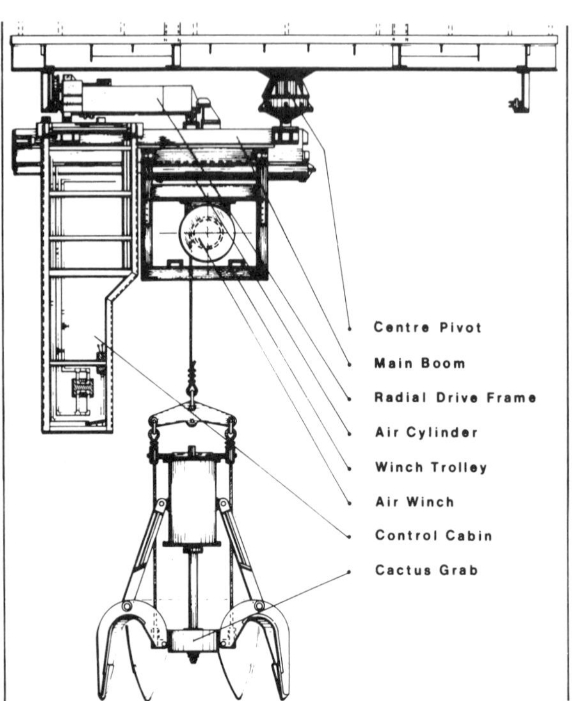

Figure 24.9 *Rotating boom pneumatic cactus grab-mucking system*

Figure 24.10 *Cryderman mucking unit*

Figure 24.11 *Eimco 630 rocker shovel mucker*

Mines (Hynd 1982). However, when installed for the construction of a shaft project in 1981 unexpected structural problems and a lower than expected breaking rate in the massive sandstone were encountered. The expectations of this shaft excavation method in other situations are still believed to be good.

A single header machine for shaft sinking was first applied in the Polish mines of the Legnica-Glogow copper basin, south western Poland, in 1974 and a two-header unit is now in use (White 1986). The headers are suspended on a ring connected to the shaft lining and lowered by hydraulic hacks to contact the shaft bottom. Muck is removed by a cactus grab. Shaft sizes in the basin range from 5.5 m to 9 m internal diameter with typically a 7.5 m internal diameter shaft being opened up to 9.2 m diameter excavated before lining. The strata consists of sedimentary rocks, mainly limestones, dolomites, anhydrites and sandstones as well as gravels.

A similar type of concept to the Polish header machine system has been developed in Germany (Anon 1987). Called a *Schachtfräse*, it consists of a swivelling cutting head mounted on a rotating frame suspended from a sinking scaffold. Hoisting and clamping facilities are provided for the cutting head frame independent of scaffold hoisting. A pneumatic system of muck removal from the face is incorporated. The first operation of a machine of this type was in rock salt at Asse Colliery which proved the validity of the method. A second generation machine (but still a prototype) was first used at Lohberg Colliery in Coal Measures to construct a 100 m deep staple shaft to an excavated diameter of 6.2 m with the spoil directed down a previously drilled pilot hole of 1.2 m diameter. Difficulties were experienced with the machine on alignment control, extreme pick wear and general handling and operating problems. However, when operating, average sinking rates of 3 m/day were achieved and initial experiences were encouraging enough to confirm that the original design objectives could be achieved.

Figure 24.12 *Cementation impact breaker and back-hoe*

(f) Excavation temporary support

During shaft sinking a major danger is the possibility of rock fragments falling on to men who are working below in a closely confined space. In the UK coal industry, where the rocks encountered are normally of the sedimentary type, temporary protection to the excavated face is required under the Mine Manager's support rules scheme. Support rules are detailed and written in the form of a method statement to describe precisely the individual steps to be taken for excavation and support during sinking. The following support rules were established for the sinking of British Coal's two 7.315 m internal diameter, Selby Wistow shafts which were constructed down to 411 m (No. 1) and 383 m (No. 2), during 1977–81 (Tunnicliffe and Keeble 1981):

(1) Supports to consist of rock bolts 610 mm long and Weldmesh, 75 mm × 75 mm × 10 gauge, in 2.44 m (8 ft) horizontal by 1.22 m (4 ft) vertical panels with 152 mm (6 in) overlaps.
(2) Rock bolting pattern to be 1.143 m (3 ft 9 in) centres horizontally and 1.067 m (3 ft 6 in) centres vertically, that is 6 per panel through the overlap sections.
(3) Maximum distance from last ring of bolts to floor to be 3 m.
(4) Maximum distance from finished shaft lining to floor excavation to be 20 m.

Depending upon ground conditions, varying degrees of temporary support may be required:

(1) *None*. Henderson (1969) indicated that no temporary support was employed during the sinking of the 8.992 m (29 ft 6 in) internal diameter No. 5 shaft for Hartebeestfontein Gold Mining Co. Ltd in the Western Transvaal of South Africa during 1966–67. The bottom of the concrete lining was carried between 10.973 m to 18.288 m (36 ft to 60 ft) from the shaft bottom in competent ground. However, in poor ground the lining was carried much closer to the face, and in bad ground to within 0.914 m (3 ft) of the shaft bottom. This condition applied when sinking through the weak quartzite from 1524 m to 2021 m (5000 ft to 6630 ft) in depth.
(2) *Rock bolts and mesh*. In some cases this type of temporary support may also be adopted as the permanent support which is provided as an additional safety measure in competent rock.
(3) *Rock bolts, mesh and shotcrete*. The latter is applied in weaker ground for additional strength and the prevention of face weathering.
(4) *Liner plates, precast concrete segments or fabricated steel tubbing*. These types of temporary support may be needed in weak ground. The latter type was required during sinking through the weakly cemented Basal Sands at a depth of 429.60–433.60 m in British Coal's two 7.215 m internal diameter, Selby Riccall shafts which were constructed during 1977–83 (Fotheringham and Black 1983).

(g) Permanent lining installation

Table 24.5 indicates the two basic methods of shaft lining for conventional sinking. These are lining downwards with the permanent lining as sinking proceeds or providing a temporary lining downwards with the inner permanent lining installed upwards from a foundation. Two different systems of installation are listed for the first method and two for the second.

As mentioned in Section 24.2.1(f) the most common type of shaft lining is that of *in situ* concrete placed directly against the rock as sinking proceeds. In wet conditions backsheets of PVC, fastened to the rock bolts and mesh for casting the concrete against, are used to channel the water behind, outflow being controlled via pipes to the grout ports in the formwork for pumping to the surface disposal points.

Auld (1983) described the method of lining downwards with concrete cast *in situ*. Figure 24.13 illustrates the principles. Shaft lining formwork is supported during concrete pouring by means of hanging rods suspended from the lift above. The kerb ring at the bottom, once it is lined up from the side lines and levelled in relation to the

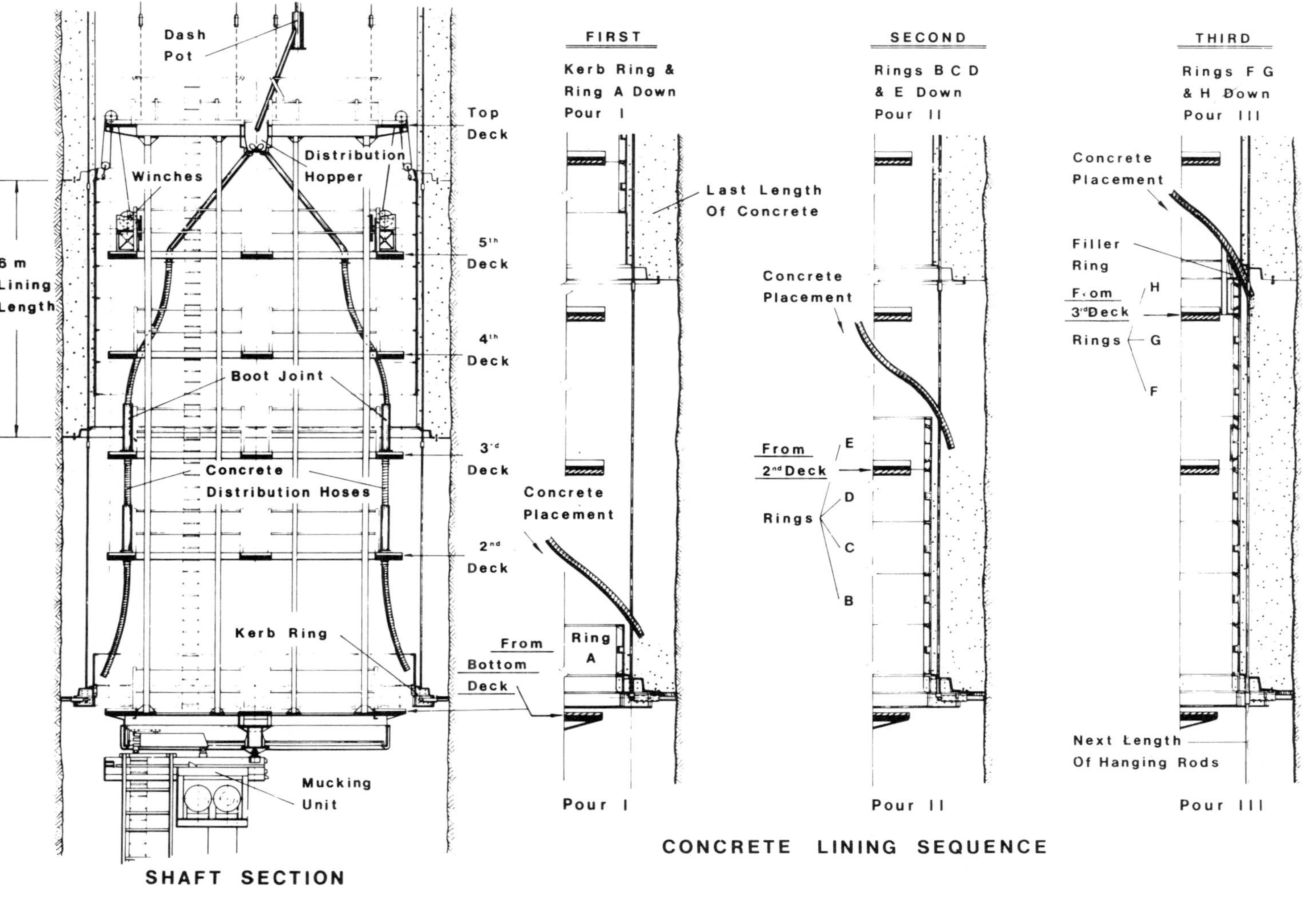

Figure 24.13 *Lining downwards with concrete cast in situ*

shaft depth markers, acts as a stop-end to the bottom of the lift being constructed and provides a seating upon which to lower the complete set of formwork rings from the previous pour. Lowering of the formwork is achieved by means of winches situated on the scaffold.

Figure 24.13 also shows the concrete lining sequence. Pour doors in the formwork assist with the placing of the concrete and the special shaped joint permits matching up to the previous pour. The concrete is transported from the surface via a 150 mm or 200 mm diameter vertical pipeline in the shaft, and is received first by a dash pot and then by a distribution box. Passage into the formwork is by means of flexible hoses attached to steel pipe outlets from the distribution box.

Further information on the transportation and placing of concrete for shaft linings has been given by Auld (1983). Details of concrete mix designs can also be found elsewhere (Auld 1983; Auld 1989). The latter references provide data on the use of admixtures such as plasticizers, retarding plasticizers and superplasticizers, the use of cement replacement materials such as pulverized fuel ash (PFA) and ground granulated blast-furnace slag (GGBFS) and the new high-strength concretes approaching 90 MPa in strength which incorporate microsilica.

Once the concrete has reached its design strength, the scaffold can be raised to carry out a backwall grouting operation throughout a particular length of lining. The grout is injected through steel pipes (normally 50 mm in diameter) which have been previously cast into the concrete lining. The purpose of this operation is twofold, the first being to introduce relatively thick cement grouts at low pressures for void filling at the rear of the lining, particularly behind the high-pressure PVC seals provided at the joints (Figure 24.13). Secondly, thin cement grouts can subsequently be injected up to pressures of 1.25 times hydrostatic pressure to chase water back into the strata

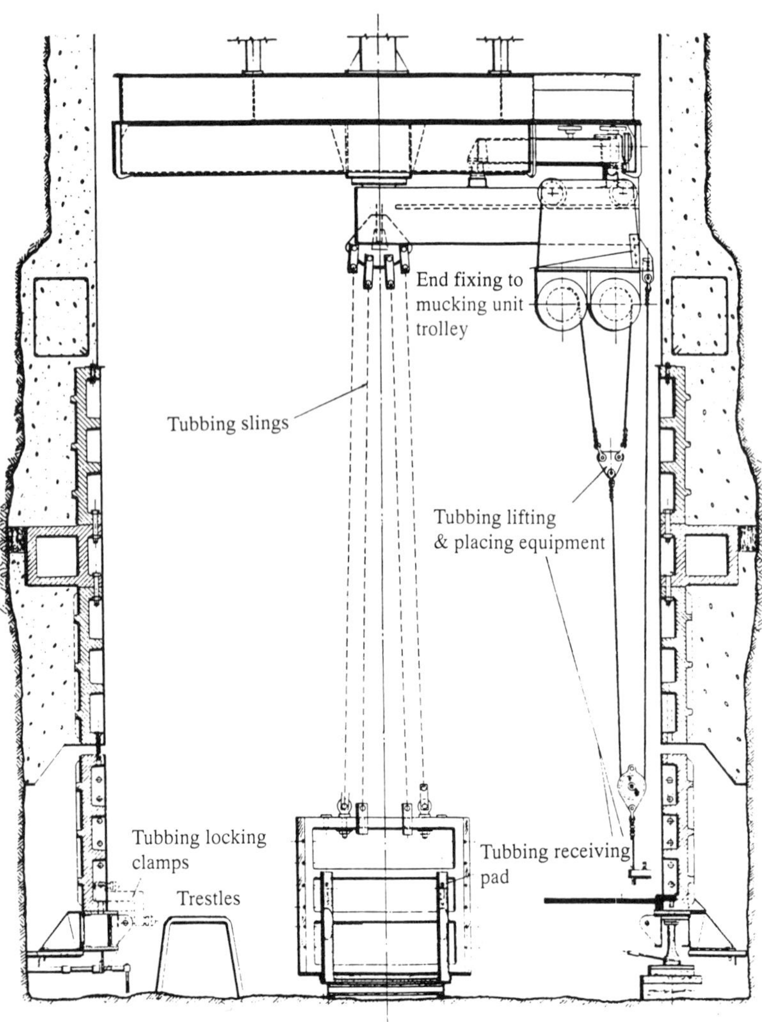

Figure 24.14 *Installation of cast iron tubbing*

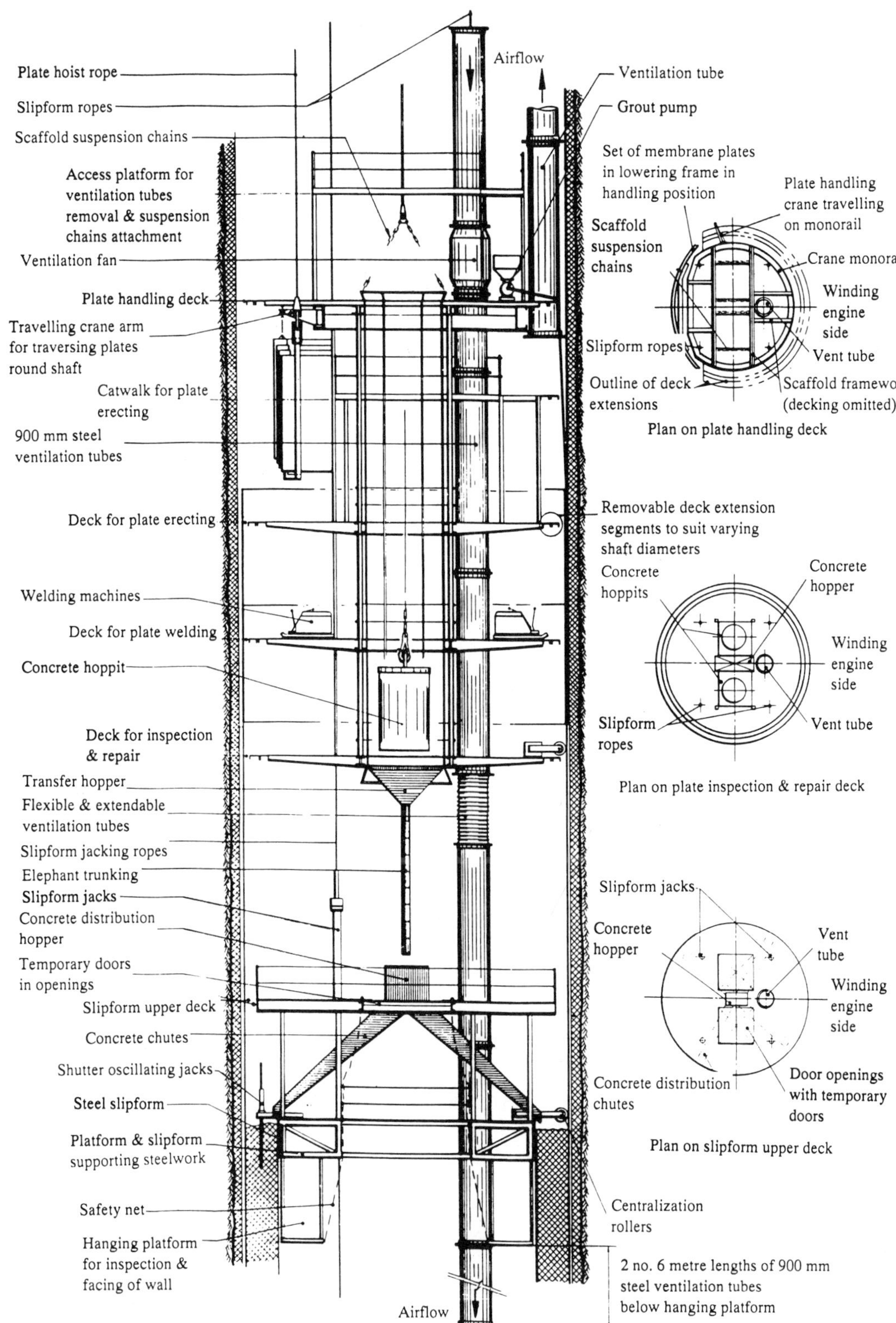

Figure 24.15 *Installation of combined concrete and steel lining*

from the interface with the lining, hence creating a permanent seal.

The second system indicated in Table 24.5 for lining downwards with the permanent lining as sinking proceeds is the installation of precast concrete segments, fabricated steel tubbing or spheroidal graphite (SG) cast iron tubbing. Figure 24.14 shows the method for installing cast iron tubbing. For the installation of any segmental system the same basic requirements prevail:

(1) transportation down the shaft;
(2) a means of lifting from the place of deposit and movement into position to connect up to the previously installed unit;
(3) backfilling with concrete or grout.

Where a temporary lining is provided first when sinking downwards, the method of installation follows that of Item 1 in Table 24.5. The two types of permanent lining installation from the bottom upwards, Item 2, both involve slipforming the inner concrete lining, but the second has the added complication of steel plate incorporation. Figure 24.15 indicates the principles of combined concrete and steel lining installation. The requirements in this case are:

(1) transportation of the steel plates and concrete down the shaft;
(2) a means of moving the steel plates into position onto those previously installed;
(3) scaffold decks correctly positioned for welding the plates and placing concrete;
(4) backfilling with concrete or grout.

(h) Services and ventilation

Shaft sinking service pipe ranges are fixed to the shaft wall on temporary brackets (Figure 24.8). These ranges are extended from the scaffold as sinking continues. They typically include 100 mm diameter compressed air, 150 mm diameter rising water main, 38 mm diameter water, 200 mm diameter concrete and 900 mm diameter ventilation pipes. Self-supporting power and parallel mains blasting cables are also provided. Other services are side (plumb) lines, bell (signal) lines, and signal, communications and lighting cables. Blackwood (1981) gave a full account of all these requirements.

(i) Insets

The construction of reinforced concrete insets requires particular mining expertise. Concrete pour sizes must be closely related to the excavation sequence. Figure 24.16 shows the stages in the construction of a typical inset with numbers in circles indicating the sequence.

Immediately before excavation commences for the inset, the shaft lining is secured down to a suitable level immediately above. Normally reinforcement would be provided within the shaft wall, in the section adjacent to the inset, to create a neck ring for the purpose of securement and to provide resistance to differential movement between the shaft and the inset. Once the ring has been constructed, excavation can proceed downwards in small stages with headings being driven transversely using steel arch temporary supports.

A special technique used in inset construction is the provision of a haunched arch roof (Figure 24.16). In addition to reducing the bending moment on the roof slab by means of the arch, the haunching allows it to be cast and 'locked' in to the ground, and subsequent excavation can proceed below without any other means of support.

(j) Shaft furnishing

Furnishing can be carried out simultaneously with sinking or alternatively completed as a separate operation when excavation has been completed. In the latter case, the shaft may be furnished in a downward or upward pass, using the sinking scaffold or a special furnishing scaffold depending upon the arrangement and complexity of the permanent equipment.

Buntons can be built into pockets left in the lining or bolted on to a smooth wall. It is very important to start off the installations accurately, with the assistance of specially manufactured jigs if necessary (Douglas and Pfutzenreuter 1989). Local alignment accuracy is vital irrespective of tightly specified overall top-to-bottom tolerances. Furnishing techniques are described by Thompson (1973) and the development, design and installation aspects have also been covered in detail by Glenday (1989).

24.3.2 Alternative methods of shaft construction

In addition to the conventional systems of sinking by drilling and blasting or using mechanical excavators, where for both the muck is transported up the shaft in a hoppit, there are four other major alternative methods for shaft excavation. These are shaft drilling, shaft boring, raise boring and box-hole drilling. The prime objective of using such machines is speed of construction. However, for large-diameter shafts, although the excavation can be carried out speedily, particularly in the case of a drilled shaft, lining the shaft remains a substantial exercise. In addition, the capital and mobilization costs for big rigs are large. The pros and cons for their use, when compared with conventional shaft sinking, must be carefully weighed up prior to any contract to ensure that cost, time and risk-reduction benefits can be achieved on an overall project basis.

(a) Shaft drilling

This is the process of excavating a shaft from the top downwards to its required diameter in a single pass using a shaft-drilling machine which is located at the shaft top. Construction of large-diameter shafts can also be carried

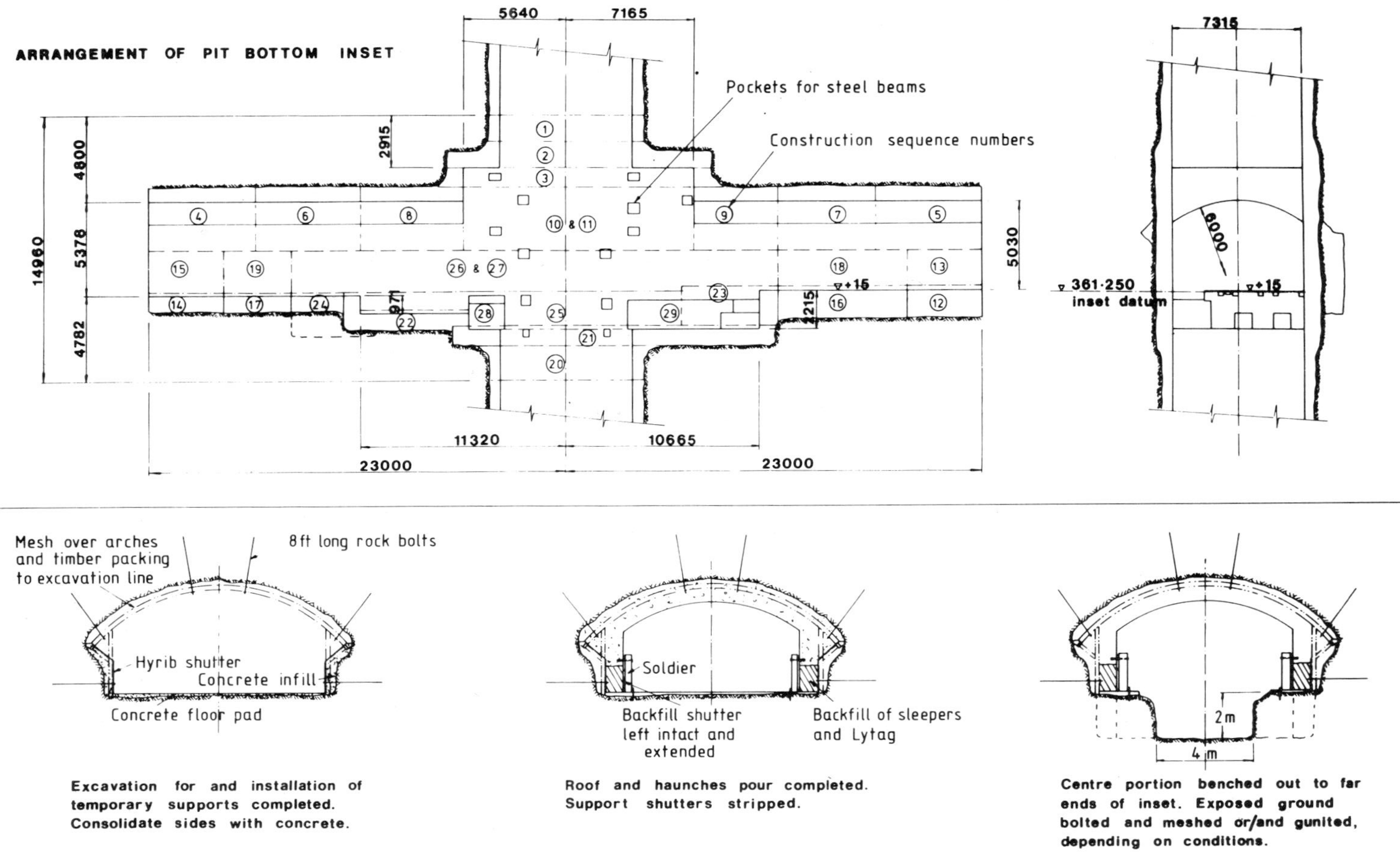

Figure 24.16 *Inset construction*

out with the machine by drilling a pilot hole or smaller-diameter shaft first. Using successively larger-diameter or under reaming bits, reaming from the shaft top downwards or from the bottom upwards respectively can be carried out in stages to increase the shaft diameter. Muck retrieval is normally reverse circulation mud flush, with air-lift assistance (Figure 24.17), the shaft excavation being kept full of drilling mud to maintain its stability prior to lining installation.

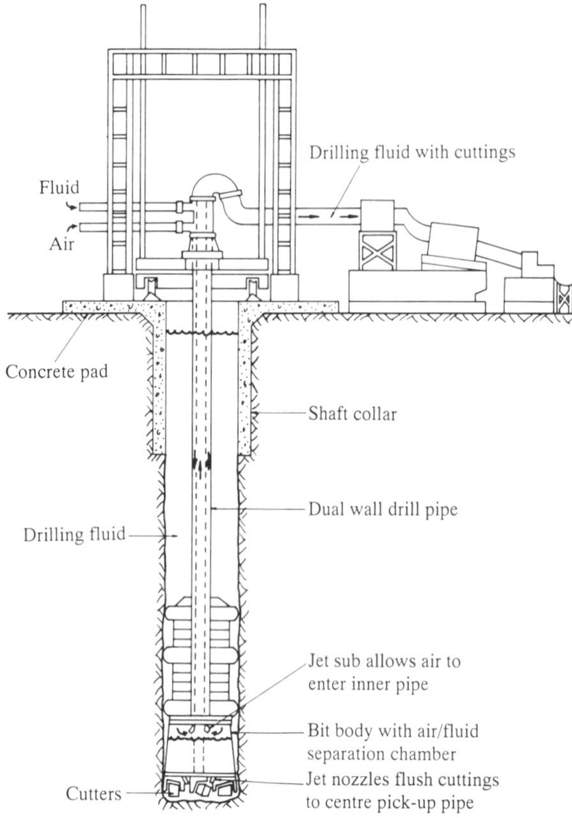

Figure 24.17 *Shaft drilling (Hughes Micon CSD 300 rig)*

There are three basic types of shaft drilling machine:

(1) modified oilfield drilling rig;
(2) purpose-built drilling rig;
(3) combination shaft drill and raise borer.

Figure 24.18 shows a modified oilfield drilling rig. It is the Rowan Drilling – US, Western Division Rig No. 14 in operation drilling the 723 m (2371 ft) deep, 3 m (10 ft) excavated diameter, shaft for the US Bureau of Mines oil shale research project in the Piceance Creek Basin, Colorado, during 1977 (Utter and Hawkins 1978). A ring-stiffened steel lining of 2.4 m (8 ft) internal diameter,

Figure 24.18 *Modified oilfield drilling rig (photograph courtesy Rowan Drilling – US)*

was installed in the hole by the floating-in method (Figure 24.19). Oilfield rigs have tall derrick structures with high hook-load capability to facilitate the joining together of long single lengths of drill pipe and also to be able to support the total weight of the drill stem and bit weight when drilling to deep levels. Rig 14 has a total derrick height of 55.474 m (182 ft) and a hook load of 703 tonnes (1 550 000 lbs). Rotation of the drill pipe is achieved using a rotary table, situated at the suspended floor level, to turn a kelly bar which moves up and down through the rotary table. Lengths of drill pipe are systematically attached between the last length in the hole and the kelly bar as drilling proceeds downwards. The raising and lowering operation of the sheave block is carried out using ground-mounted draw works to rope wind via the derrick top pulley wheel. A large-diameter bit assembly (Figure 24.23), which has an arrangement of cutters and stabilizers and is attached to the end of the drill stem, provides the means of excavation. The oilfield rig had to be modified to facilitate the attachment and removal of the large-diameter bit above ground level.

Figure 24.20 shows the typical operating layout of the rig. A mud pit with a storage capacity of 8.1 million litres

Figure 24.19 *Floating in a ring-stiffened steel drilled shaft lining (photograph courtesy Rowan Drilling – US)*

(51 000 bbl) was provided with a 1.5 million litres (9200 bbl) capacity cuttings pit. The mud is circulated down the shaft and up the centre of the annular drill stem to transport cuttings from the sump to surface, while at the same time being used to stabilize the excavation and prevent water ingress (Figure 24.17). In the case of an oilfield rig, this reverse circulation (down the shaft and up the drill stem) mud flush system operates at the top of the drill stem through a swivel located above the kelly bar and air-lift assistance is introduced into the drill stem via its outer annulus.

Tight drilling alignment tolerances were set for the project because running the heavy steel casing needed a very straight hole. Specifications required that no two points in the shaft with a vertical separation of 55 m (180 ft) could have a lateral displacement of more than 0.12 m (0.38 ft).

Another well-documented project using adapted oilfield type drilling equipment was that undertaken by the Wyoming Minerals–Conoco Crownpoint Joint Venture during 1980 (Presley 1981, Hunter 1981, 1983.). Three shafts, one 3.048 m (10 ft) in diameter and two 1.829 m (6 ft) in diameter, were drilled to depths of 683.7 m (2243 ft), 666.9 m (2188 ft) and 666.0 m (2188 ft) respectively and cased with hydrostatic designed steel casing. The project was for the development of an underground uranium mine.

In the late 1960s and through the 1970s Parker Drilling Company were involved in large-hole projects with the US Atomic Energy Commission in the State of Nevada and the Amchitka Island of Alaska. Using oilfield rigs, large boreholes 3.658 m (144 in) in diameter were drilled to approximately 1829 m (6000 ft) at Amchitka (Anon 1988).

Figures 24.21 and 24.22 show two purpose-built drilling rigs. The first is the Zeni Z 250 E rig and the second is the Wirth L 35 rig. Zeni's rig is rated to be capable of drilling holes with a diameter of 4.7 m down to 200 m depth, diameter 4 m down to 700 m depth and diameter 3.2 m down to 1000 m depth. Wirth quote their L 40 rig as being able to drill shafts up to 8 m diameter down to depths of 600–1000 m. These rigs operate in a similar fashion to the modified oilfield rigs, with rotary table, kelly swivel and ground level draw works, but the masts are purpose-built A-frames. They are much more compact and are designed for mobility and fast erection, being trailer mounted. Table 24.6 compares the Rowan, Zeni and Wirth rigs on a size, hook-load, machine torque and revolutions per minute (r.p.m.) basis.

The third type of drilling rig is also a purpose-built machine for large diameter shaft drilling but, in addition, it can also be used as a raise borer (see Section 24.3.2(c)). Two examples of combination shaft drill and raise borers are shown in Figures 24.23 and 24.24. The former is a Hughes Micon CSD 300 rig and the latter is a Robbins 121 BR machine. Hook load and machine torque values for the two combination rigs are contained in Table 24.6 where they can be compared against the purpose-built and modified oilfield rigs. A major basic difference between the combination rigs and the others is that the rotating action is developed through a power swivel, which moves up and down in the frame with the drill pipe, as opposed to being applied to a kelly bar via a static rotary table. A Hughes Micon CSD 300 rig blind-drilled a 4.267 m (14 ft) diameter shaft to a depth of 750 m (2460 ft) through hard rock with compressive strengths ranging from 207 to 345 N/mm^2 (30 000–50 000 lb/in^2) at the Agnew Mining Company nickel mine in Western Australia during 1982–83 (Kaesehagen 1983; Richardson 1984). The Robbins 121 BR is quoted by the manufacturers as being capable of excavating up to 5 m (16 ft) diameter shafts or raises to 600 m (2000 ft) depth.

Although the purpose-built and combination shaft drill and raise borers are more compact than the modified oilfield rigs, they require similar mud farm and cuttings disposal systems. One major advantage of using a shaft drilling machine with its mud flush system which also stabilizes the hole, is that no ground pre-treatment is necessary to prevent ingress of water into an open excavation.

A very important aspect to be considered when shaft drilling is the ability to maintain verticality. Using conventional drill string equipment, a verticality tolerance conventional drill string equipment. a verticality tolerance of 0.25% is achievable with normal good drilling practices (Anon 1989a). The recent development of steerable bit systems should provide the capability to control the

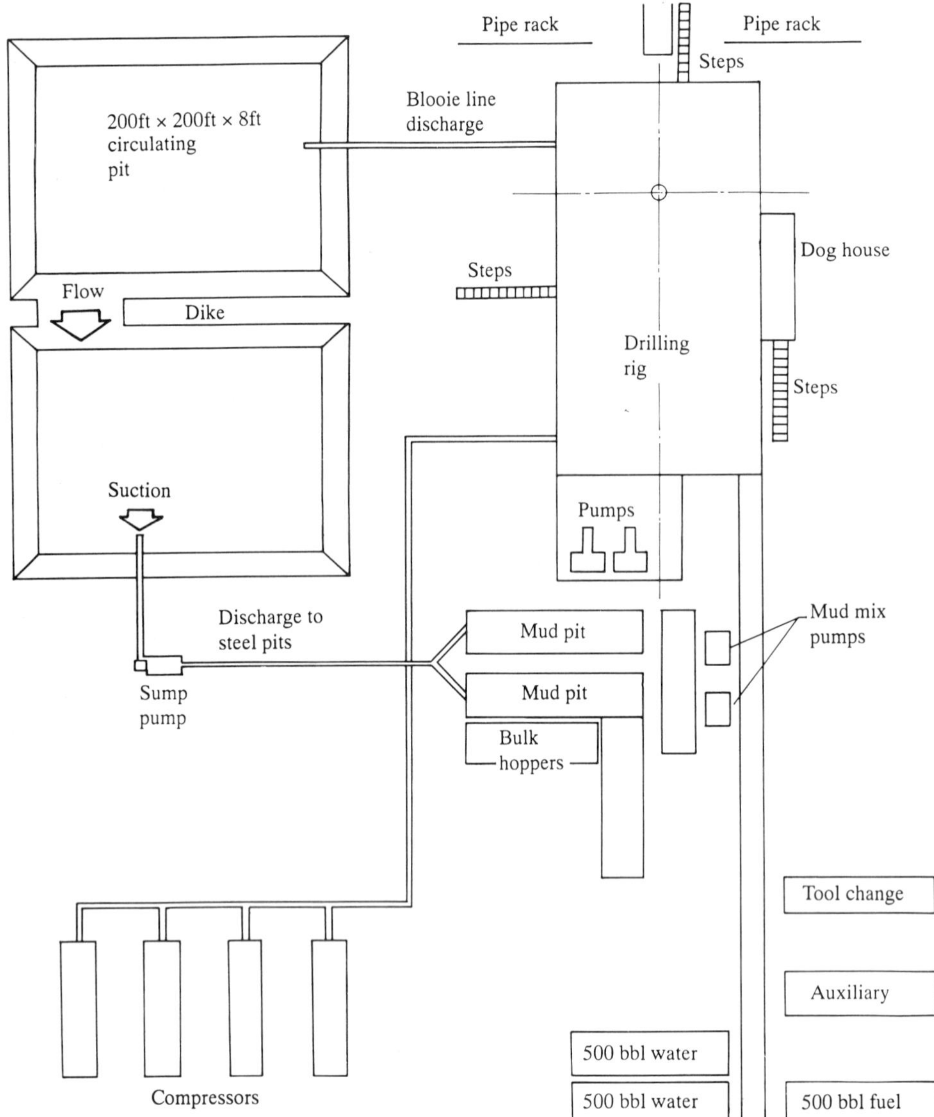

Figure 24.20 *Shaft-drilling site layout (Rowan No. 14 rig)*

verticality tolerance to within even smaller deviation limits. Two such systems are the Wirth steerable bit SVB 2100 (Figure 24.25) and the Zeni vertical drilling assembly (VDS) (Figure 24.26). The former is suspended down the hole from a non-rotating drill stem, rotation being developed from an electrically driven power swivel incorporated in the bit system with forward thrust created by weights sliding down inside a guide frame which clamps hydraulically to the drill-hole wall. With the Zeni system, the hydraulic sideways gripper system develops the forward thrust, but the roation is produced via a rotating drill stem driven by a rotary table located at the surface. Both of the steerable bit systems operate using the reverse circulation mud-flush method with air-lift assistance.

Mud technology is another critical factor for successful drilling. The drilling mud is required to resist groundwater inflow and to maintain stability of the excavated face, to support the cuttings whilst transporting them to surface and to have a gel strength sufficiently low to allow them to be removed from the mud using mechanical separation equipment. In addition, bits are prone to clogging ('bit-balling') in clay formations. It is therefore very important to design the drilling mud correctly to cater for all these characteristics.

The prediction of accurate drilling rates is a subject

Figure 24.21 *Zeni-250E purpose-built shaft-drilling rig (photograph courtesy Zeni Drilling Company)*

Figure 24.22 *Wirth L35 purpose-built shaft-drilling rig (photograph courtesy Wirth GmbH)*

Figure 24.23 *Hughes Micon CSD 300 combination shaft drill and raise borer (photograph courtesy Hughes Tool Company)*

which appears to present problems. Richardson (1984) quoted actual drilling rates versus predicted rates for the Agnew Mining Company shaft in Western Australia constructed in 1982–83. Large differences are apparent. TH Engineering Services Ltd (Anon 1989a) carried out an exercise whereby drilling rigs were classified in relation to their capability to construct shafts of 2, 3 and 4 m finished diameter down to depths of 300, 650 and 1000 m through water-bearing strata. Zeni and Hutchinson (1988) quoted that the total collar weight on the cutting head which is required to reach an economical drilling weight is determined by providing 44 tonnes per metre of drilled diameter. As a rough guide also, the rig-lifting capacity required is in the range of 100 tonnes per metre of diameter. Using the latter fact, and a method whereby required machine torque was evaluated from cutting performance and power needed based on rock strength, theoretical cutting energy, drilling rate and revolutions per minute, TH Engineering Services Ltd (Anon 1989a) were able to categorize the suitability of various drilling rigs for their particular project. However, a missing link in the approach was the question of bit thrust, and further work is required to confirm the validity of the method. Table

Table 24.6 Comparison of large-diameter shaft drilling rig types

Rig category	*Rig type*	*Derrick dimensions and rig overall length*	*Hook load* (tonnes)	*Method of applying rotation*	*Machine torque* (kN m)	*R.p.m.*
MO	Rowan No. 14	55.47 m (182 ft) high by 11.582 m (38 ft) base plus 9.144 m (30 ft) high floor, overall length = 26 m (85 ft 4 in)	703 (1 550 000 lb)	Rotary table		4–8 (low-speed range) 30–40 (high-speed range)
PB	Zeni Z 250 E	27.5 m high by 6 m wide base width A-frame, overall length = 24 m	250	Rotary table	476 max. at 7 r.p.m.	
PB	Wirth L 40	23 m high by 11 m wide base width A-frame, overall length = 21 m	408 (4000 kN)	Rotary table	0–420	0–17.5
C	Hughes Micon	19.279 m high by 12.573 m wide by 2.667 m long	907 (2 000 000 lb)	Power swivel	678 (500 000 ft lb) at 7 r.p.m.	0–17
C	Robbins 121 BR	10.543 m high by 2.324 m wide by 2.248 mlong	568 (568 000 kg)	Power swivel	496 at 4 r.p.m. (50 600 kg m) 124 at 16 r.p.m. 27 at 74 r.p.m. (2 800 kg m)	0–74, 0–25 and 0–6

MO Modified oilfield
PB Purpose built
C Combination shaft drill and raise border
r.p.m. Revolutions per minute

24.6 provides data on lifting capacity (hook load) and machine torque for the five different rigs mentioned previously.

(b) Shaft boring

This is the process of excavating a shaft from the top downwards to its required diameter in a single pass using a shaft-boring machine. In this case the machine is initially installed at the top of the proposed shaft but then travels downwards as the excavation proceeds. Being rodless, it has on board personnel for operation control and requires the provision of services and a muck retrieval system back to the shaft top. Construction of large-diameter shafts can also be carried out with the machine by drilling a pilot hole first. Using this method, free-fall muck retrieval via a pilot hole leading to an underground removal system can be adopted.

The two major manufacturers of shaft-boring machines are Wirth GmbH of Erkelenz in West Germany and The Robbins Company of Seattle, USA. Benjamin (1988) quoted Wirth's opinion, based on development since 1970 and the building of eight machines of this type, that all the engineering features which it is desirable to have cannot be accommodated in a machine built for shafts of less than 5 m diameter. The upper limit of diameter is set by what is called for and so far this has been 8.5 m.

During the period 1975–81, The Robbins Company was the prime contractor for the development, manufacture and demonstration of a blind shaft borer capable of vertical downward penetration through rock (Hynd 1982). The project had as its objective the production of a shaft-sinking machine and support system capable of boring and lining a coalmine shaft at a rate of 7.62 m (25 ft) per day. US Steel's Oakgrove No. 4 shaft was chosen for the demonstration site. Figure 24.27 shows the machine, called the Robbins 241SB-184 blind shaft borer, being tested in the workshops.

The boring and lining objective of 7.62 m (25 ft) per day was not achieved, the best being the equivalent of 4.877 m (16 ft). However, many things were achieved. Among these was the application of the hard rock technology of the standard horizontal tunnel borer to the vertical downward mode. A shaft of 178.9 m (587 ft) deep of 7.442 m (24 ft 5 in) excavation diameter (6.706 m or 22 ft finished diameter inside the concrete lining) was sunk at the demonstration site in Alabama.

Although the shaft was not sunk at an economic rate, it was a tremendous achievement. The chief problem with

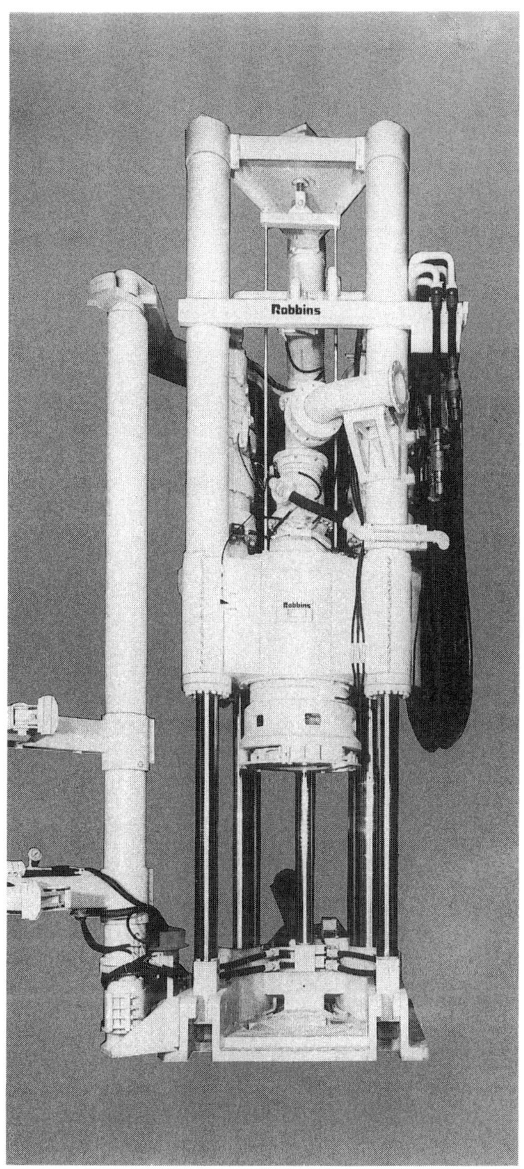

Figure 24.24 *Robbins 121 BR combination shaft drill and raise borer (photograph courtesy of The Robbins Company)*

Figure 24.25 *Wirth steerable bit SVB 2100 (photograph courtesy Wirth GmbH)*

the borer was the picking up and elevating of the broken muck to a height suitable for skip loading. All other problems were relatively minor and were or could be resolved. The project is well documented by Friant and Sands (1977), Sands and Little (1979) and Ruby and Sands (1979).

The Robbins 241SB-184 blind shaft borer featured a full-face cutter head. A system of scrapers, flight conveyors and bucket elevator removed cuttings from the bottom. Muck was discharged into a hopper whence it was hoisted from the shaft in skips. A pneumatic hoisting system up the shaft was also considered (Ruby and Sands 1979). To overcome the problem of picking up the broken muck off the face, and then elevating to a suitable height for transfer to the removal system up the shaft, a new machine was developed by The Robbins Company during 1983 to 1985. It is called the Redpath/Robbins Shaft Boring Machine (SBM) and incorporates a vertical cutter wheel which slews about the shaft vertical axis (Figure 24.28). The mucking system is a hydraulically operated clam bucket, attached to a telescopic boom, feeding a storage hopper which can be elevated to discharge into a bucket for hoisting out of the shaft. Details of the SBM, which can bore shafts in the range of 6.1–7.3 m (20–24 ft) diameter, are given in papers by Hendricks (1985) and Woods (1988). To date, as far as the author is aware, although a prototype machine has been built it has not been used for shaft construction.

Figure 24.26 *Zeni steerable vertical drilling assembly (VDS) (photograph courtesy Zeni Drilling Company)*

Figure 24.27 *Robbins 241SB-184 blind shaft borer (photograph courtesy The Robbins Company)*

In 1971 the first Wirth shaft boring machine was put into service in the German coalmining industry. The machine was capable of reaming shafts with a diameter of up to 5 m on a pilot hole. Nicknamed the V-mole, its technique of reaming shafts on a pilot hole has been very successful, culminating in the development of a third generation of machines designated SB V11 650/850s (Figure 24.29). Their capability is to construct shafts in the 6.5–8.5 m range. The number of shafts constructed by the Wirth V-mole method is 31 in Germany, 5 in the USA and 2 in Russia (Benjamin 1988). The average depth of shafts has ranged from 200 to 250 m, the deepest being 700 m. Daily rates of sinking, including the finished lining, lay between 8 and 10 m with peak performances of up to 18 m.

The world's deepest and South Africa's widest, vertical bored shaft was completed during the second half of 1989 (Anon 1989b). It is the No.1B ventilation shaft at the Oryx gold mine in the Orange Free State and has a depth of 972 m with a diameter of 6.5 m. A Wirth SB V11 650/850 was used and the whole operation took just six months including concrete lining through 410 m of shale. Total time for the completed shaft was twelve months because of the need to raise-bore the pilot hole prior to the actual shaft. Penetration rates in excess of 3 m/h were achieved in shale and quarzite although the advance was limited by the slower waste removal operation. Average advance was 206 m/month.

Adaption of the V-mole principle to blind shaft boring has also been undertaken. As with the Robbins machines, the main problem was how to collect the debris from the face and to convey it through the shaft. An hydraulic transport system was chosen with the debris collected using sludge pumps. The first opportunity for a full-face shaft boring arose at Gneisenau Colliery, Dortmund, in 1978 with the construction of a 60 m deep staple shaft to an excavated diameter of 5.1 m. It was sunk in 70 days with a 3 m best daily rate of boring. The knowledge gained

Figure 24.28 *Redpath/Robbins shaft-boring machine (SBM) (photograph courtesy of The Robbins Company)*

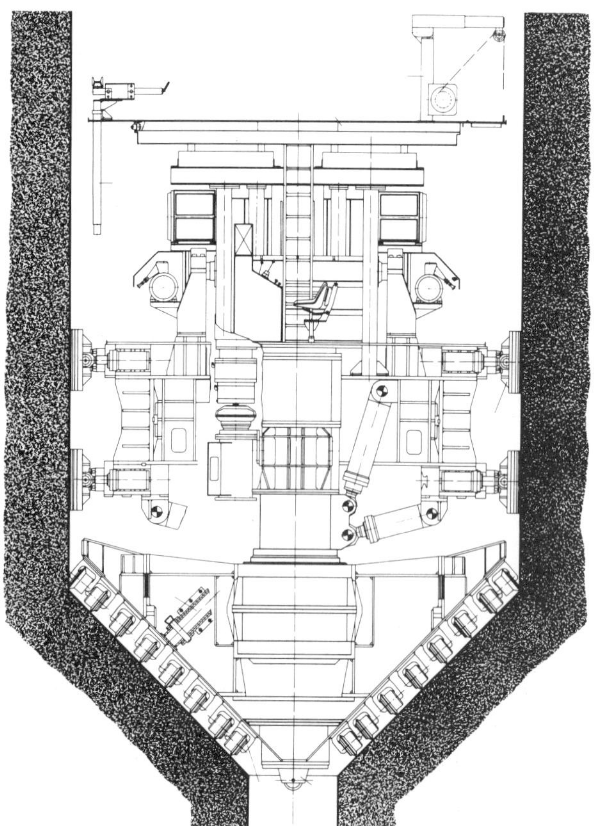

Figure 24.29 *Wirth SB VII 650/850 shaft-boring machine with pilot hole (diagram courtesy Wirth GmbH)*

enabled a second 5.8 m bored diameter, 180 m deep, underground shaft to be constructed by this method at Ruhrkohle AG's Heinrich Robert Colliery in 1983. A new machine, the Wirth VSB V1 580/750 (Figure 24.30) was developed for the project which achieved daily rates of boring of up to 7 m of completed shaft. The best monthly performance was 53.3 m of completed shaft at an average of 4.4 m per day of boring. However, further development of the system is required to achieve a technically reliable and efficient method for blind shaft boring in the future.

Numerous papers describe the development and application of the Wirth machines. The most recent ones are those by Brümmer (1981), Grieves (1981), Hanke (1982), Hanke and Gabitzsch (1983), Brümmer (1984, 1986).

One of the disadvantages of a shaft-boring machine, when considered against shaft drilling, is the requirement to operate in relatively dry, competent rock conditions since the machine descends the shaft with all its mechanisms and the operators on board. In the UK, although the strata are considered competent, their water-bearing nature almost certainly dictates the need for ground freezing to eliminate the possibility of water ingress into the shaft. Such ground treatment measures considerably add to the overall cost of such a project.

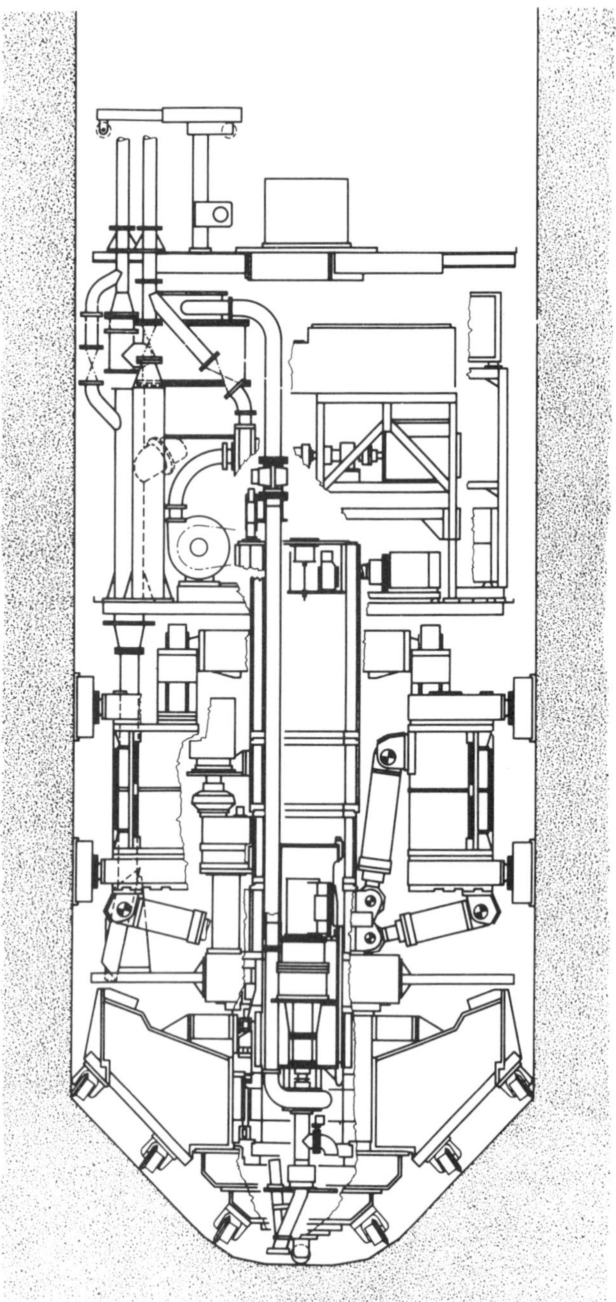

Figure 24.30 *Wirth VSB VI 580/750 blind shaft-boring machine (diagram courtesy Wirth GmbH)*

(c) Raise boring

By this method a shaft is excavated from the bottom upwards using a raise-boring machine. Location of the machine is at the shaft top to drill a pilot hole downwards first into a previously constructed access area situated at the shaft bottom. The raise boring head is then attached to the drill stem at the bottom and the pilot hole is reamed to the required shaft diameter by the machine pulling the head in an upwards direction back to the shaft top. Such a process can be carried out several times using different head sizes to ream the shaft in stages. Muck retrieval is free fall via the pilot hole or enlarged shaft for removal through the underground system.

The principle of raise boring is the most widely used of the alternative methods of shaft construction. Several manufacturers each produce a range of machines capable of raise-boring shafts of various diameter and with different depth capabilities. The smallest models will excavate shafts with diameters in the order of 0.75 m over a depth of 1150 m whereas the largest machines have a capability to raise bore 6 m diameter over depths of up to 1000 m. An example of the latter type is the Wirth HG 330-SP shown in Figure 24.31. Wirth have also now produced a sequential reaming head for diameters of 4.25–6.02 m (Figure 24.3). The two parts ream independently, thus reducing the cutting energy required as the shaft is excavated as a two-stage process and not as a single raise. Schmidt and Fletcher (1987) described their experiences with Wirth HG 330 machines and the sequential reaming head in South Africa where a 1033 m deep, 6.02 m diameter shaft was constructed for Rustenburg Platinum Mines in 1986.

In the UK coalmining industry the process of raise boring has been used to construct small-diameter (less than 2 m) ventilation shafts from one level to another underground. Maximum shaft depths are in the order of 200 m. Anything larger in diameter and length presents unsupported excavation stability problems in sedimentary and Coal Measures strata. The use of raise boring for large-diameter deep shafts is therefore generally restricted to very competent rock which does not need temporary support. Such conditions predominate in South Africa where the process is very widely used.

In addition to stability problems in UK sedimentary rocks, there are difficulties with aquifer zones. The strata must be either dry or subjected to ground pre-treatment

Figure 24.31 *Wirth HG 330-SP raise borer (photograph courtesy Wirth GmbH)*

Figure 24.32 *Wirth raise borer sequential reaming head (photograph courtesy Wirth GmbH)*

by grouting or freezing to prevent water ingress into the shaft during the pilot-hole drilling and raise-boring process.

(d) Box-hole drilling

For this method of construction a shaft is excavated from the bottom upwards using a box-hole drilling machine. Location of the machine is underground at the shaft bottom to drill the shaft upwards in a single pass or in successively larger reaming stages. However, this process is largely restricted to small-diameter shafts (2 m or less) and shallow depths (less than 400 m) (Anon 1989a). The difficulty of controlling the drill-stem direction when drilling upwards limits the use of this method.

(e) Lining installation methods for drilled, bored, raise bored and box-hole drilled shafts

Table 24.5 sets out various lining methods for the alternative methods of shaft construction. Most are described in Section 24.2.1(f). The continuous shaft lining (CSL) with concrete casting at the surface is a top-down slipform concept with stationary form and moving concrete liner which is floated in. It is believed to have good possibilities but requires extensive research and development work (Schalge 1981). The continuous shaft lining (CSL) with in-shaft concrete casting method, which is another top-down slipforming process but with a moving form, has also been the subject of study specifically to match the sinking rates expected from large-diameter shaft-boring machines (Torbin *et al.* 1983). In the case of the Autoform lining method, this involves the use of a special stage system which can be suspended on a drill pipe and has been developed by Zeni Drilling Company.

24.4 Scenario of shaft and raise construction

24.4.1 State of the art

Hynd (1982) discussed the changes which took place in conventional shaft sinking in the 30 years prior to 1982. The use of drill and blast methods to break up the rock saw no major changes, although there were improvements. It was in the size of underground crews where the main change took place. They dropped over the 30 years from 16–19 to 8–12 and then 5–7 per 8-hour shift. The first drop was due to improvements in the conventional operations such as:

(1) the introduction of mechanical mucking;
(2) the change to dropping the concrete form from one pour position to the next;
(3) the hanging of the form above the shaft bottom (as opposed to setting it up on the sump floor);
(4) the introduction of shaft jumbos (pieces of equipment with several drilling rigs mounted on them) permitting the drilling of deeper rounds with smaller crews.

The hanging of the form above the shaft bottom and the transportation of concrete via a vertical pipeline in the shaft made it possible to simultaneously pour concrete and muck out, although the latter did not reduce the crew size. In the UK coalmining industry, the practice of simultaneous sinking and lining is not normally permitted for safety reasons.

Hynd (1982) was of the opinion that the second drop in underground crew size was a result of increased competition in the industry and not from a change of method. Studying rates of advance is one way of reinforcing this fact. Although there are exceptions, conventional shaft-sinking rates reached a peak in the 1960s. At sinkings like Geco Mine's No. 5 shaft in Ontario, Canada, and in the USA at Magma Copper Company's San Manuel shafts 5, 3C and 3D above 610 m (2000 ft), rates between 3.35 m (11 ft) and 4.57 m (15 ft) per day were routinely achieved. In the Saskatchewan Potash sinkings in the 1960s, the advance rates reached better than 6.10 m (20 ft) per day in the first 305 m (1000 ft).

The change to a more economical approach to shaft sinking also took place in South Africa once the 305 m (1000 ft) per month target had been achieved. Grieves (1981) quoted record-breaking month's performances achieved at Hartebeestfontein No. 4 shaft in October 1960 of 337.1 m, Western Reefs No. 4 shaft in October 1961 of 340.7 m and Buffelsfontein eastern twin shaft in March 1962 of 381.2 m. The method was very labour intensive, requiring a crew of over 60 workers at the shaft bottom during the drill cycle. Safety precautions were strict, but in the drive to achieve rapid advance, cases of personal injury were still somewhat high because of the large number of people engaged in this potentially hostile environment.

In the UK, the South African method as it became known to the rest of the world, using a mechanical mucking unit and cactus grab, was adopted in the late 1950s in a modified form with greatly reduced manpower and non-simultaneous sinking and lining, which was insisted on by the British Mines Inspectorate. It was used successfully to sink the 7.3 m diameter concrete lined shafts at Kellingley to 770 m depth, with rates of advance of over 90 m per month achieved. This was a British record at that time.

The advent of the major shaft sinking work at British Coal's Selby New Mine Project in 1977 (Auld, 1987) saw shaft sinking records broken again during the early 1980s. At the North Selby site the upcast shaft sinking made consistent good progress during September–November 1980, with 166 metres completed through frozen ground (Tunnicliffe and Keeble, 1981). A European shaft sinking record for a 7.315 m (24 ft) diameter shaft was twice achieved at Stillingfleet in the Coal Measures, first in the No. 1 shaft at 111.1 m between 9 September and 10 October 1980, and secondly in the No. 2 shaft at 128.7 m

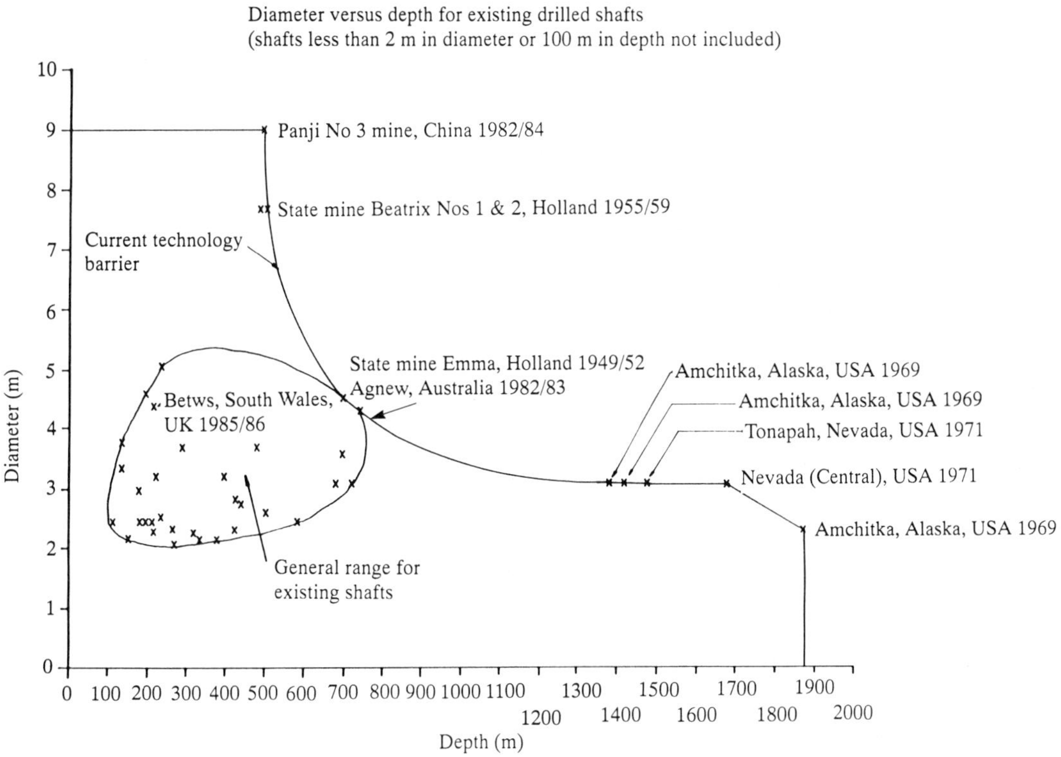

Figure 24.33 *Drilled shafts constructed to date (Anon 1989a)*

between 18 November and 19 December 1980. However, with the tremendous escalation in mining labour costs, the impact of health and safety legislation, and environmental regulations, together with a very real shortage of miners willing to work in this exposed situation, it is unlikely that these records will be broken in future.

Major improvements in concrete shaft-lining technology took place between 1977 and 1986, also associated with the Selby Project. Very good-quality concrete, with high strengths of 45–60 N/MM^2, was successfully used in UK shaft linings for the first time with the aid of plasticizing and superplasticizing admixtures to produce high workability for placing (Auld 1989). Cement replacement concrete mixes, incorporating ground granulated blast furnace slag, were also adopted. Subsequent to Selby, very high strength, superior durability microsilica concrete has been developed, with strengths up to 80 N/mm^2. This has been used in the deeper sections of British Coal's shafts at Maltby and Asfordby.

The previous comments relating to high mining labour costs, tight legislation and shortage of miners willing to work in exposed conditions point to the need to develop alternative methods of mechanized shaft sinking. Much effort has been concentrated on the development of shaft drilling, shaft boring and raise boring. Figures 24.33–24.35 put into perspective what has been achieved to date by way of constructed shafts. Current technology barriers (i.e. limits of diameter and depth not as yet exceeded) are shown for drilled shafts in Figure 24.33, for bored shafts in Figure 24.34 and for raise-bored shafts in Figure 24.35. It can be observed from these figures that the standard coalmine shaft in the UK at 7.3–8.0 m finished diameter, requiring an excavation diameter of up to 10–11 m, and constructed to a depth of 1000 m, remains in the domain of conventional shaft sinking and lining.

24.4.2 Costs

It is very difficult to generalize on the cost of shaft construction as there are so many variables to consider. Some of the major factors which affect the cost are:

(1) Shaft diameter.
(2) Shaft depth.
(3) Whether the shaft is constructed by the conventional drill and blast method or whether it is drilled, bored or raise bored.
(4) Whether ground pre-treatment is required such as freezing or grouting.
(5) Whether the lining is designed to withstand hydrostatic pressure, or some other load, and to what depth of the total shaft length this is required or whether the lining is nominal in thickness throughout.

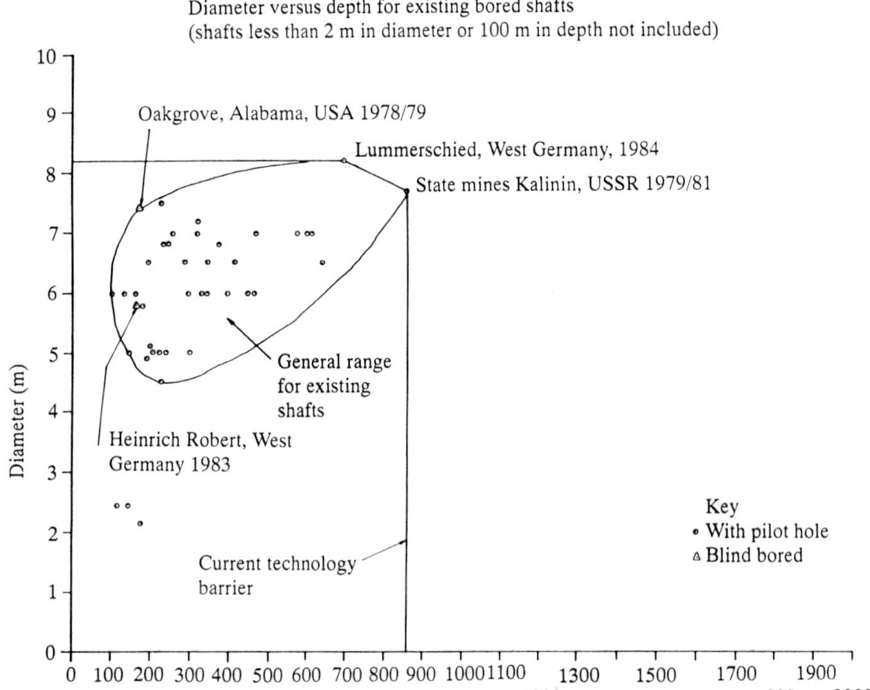

Figure 24.34 *Bored shafts constructed to date (Anon 1989a)*

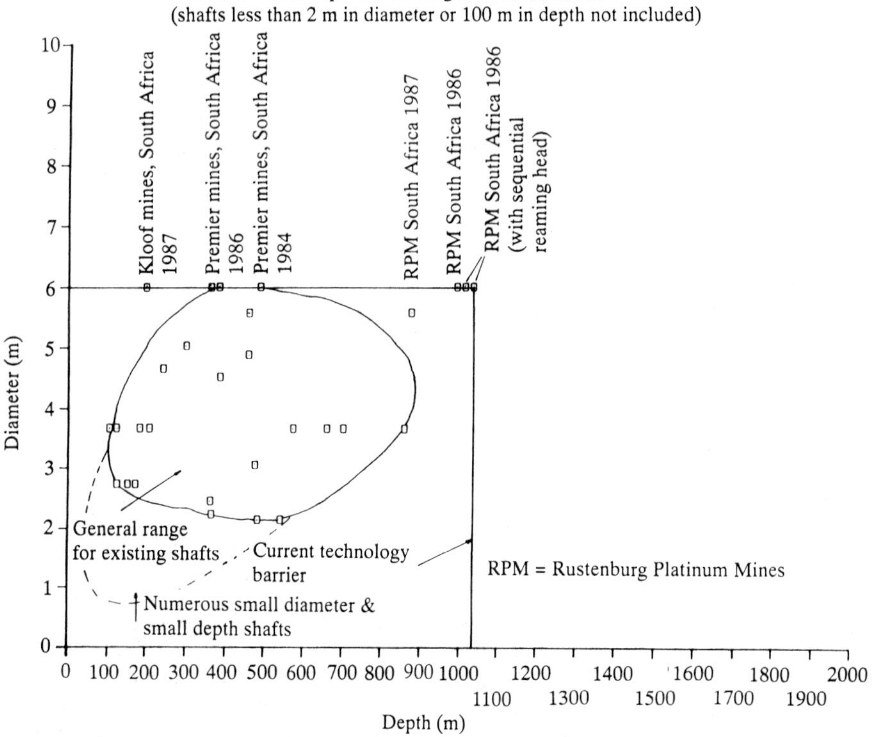

Figure 24.35 *Raise-bored shafts constructed to date (Anon 1989a)*

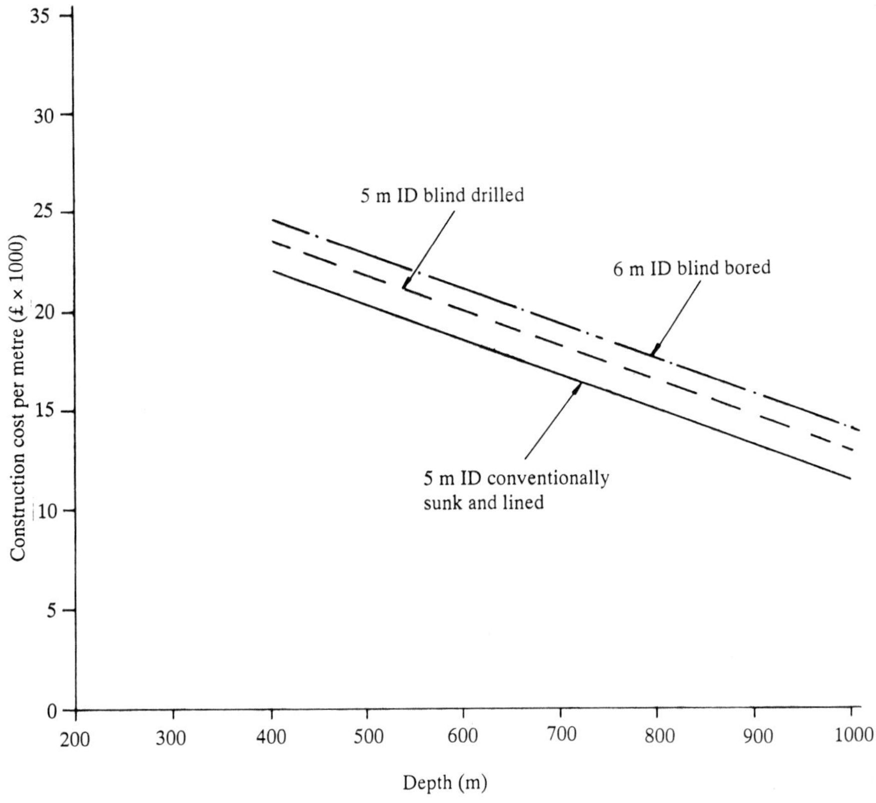

Figure 24.26 *Shaft construction (Anon 1989a)*

(6) What material is the lining constructed from. Is it concrete or does it require a combined concrete and steel lining? Alternatively a ring stiffened steel lining may be required which is floated in for a drilled shaft.
(7) Can the shaft be sunk and lined in one pass downwards or does it require to be sunk on temporary support going downwards with the permanent lining subsequently constructed from a bottom foundation upwards.

Each situation must be reviewed in its own right to arrive at the true cost. However, as a very general guide the information shown in Figure 24.36 may be referred to. Estimated construction costs per metre versus depth for a 5 m internal diameter (ID) conventionally sunk and lined shaft are compared with similarly estimated data for a 5 m ID blind drilled shaft and a 6 m ID blind bored shaft (Anon 1989a). It can be seen that the costs of the drilled and bored shafts are slightly higher than for the conventionally sunk and lined shaft. However, recourse to Figure 24.37 indicates that substantial time savings would be achieved with the drilled and bored shafts, compared with the conventional method, hence producing a return on investment much more quickly.

It should be noted that the cost information provided in Figure 24.36 has been produced by a series of plots from two different information sources and with shafts having different lengths of hydrostatic pressure resisting lining. When multiplying the total length by the cost per metre at the deeper levels, the total cost may be less than for similar calculations at shallower depths. This indicates the true relationship should not be a straight line but a curve of decreasing slope and hence slight increases in cost per metre may have to be made to arrive at the correct total cost for the deeper shafts.

Acknowledgements

The author wishes to thank Mr B. Myers, Managing Director of Trafalgar House Construction Holdings Limited, and Mr J. C. Black, Managing Director of Cementation Mining Limited, for permission to publish this chapter. Other credits are due to Hughes Tool Company, The Robbins Company, Rowan Drilling–US, Wirth GmbH and Zeni Drilling Company, to whom the author is indebted for the provision of illustrations. These are acknowledged with the appropriate figures. Many of the other figures are presented by courtesy of Cementation Mining Limited.

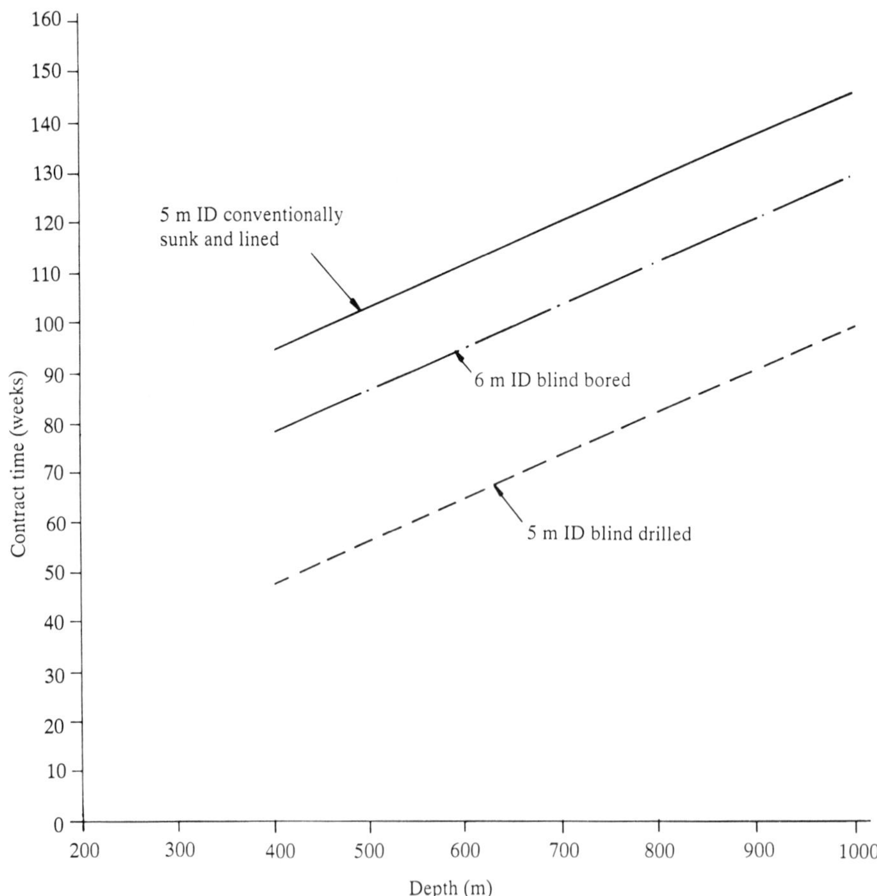

Figure 24.37 *Shaft construction time (Anon 1989a)*

References

Adamson, J. N. and Scott, R. A. (1973) 'Borehole investigations and logging methods in shaft sinking', *The Mining Engineer*, January, 181–189

Altounyan, P. F. R. (1982) 'Measurement of stresses, strain and temperature in concrete in shafts and insets', *Proc. Symp. on Strata Mechanics*, Newcastle upon Tyne, I. W. Farmer (ed.), Elsevier, Amsterdam, Developments in Geotechnical Engineering, Vol. 32

Altounyan, P. F. R., Bell, M. J., Farmer, I. W. and Happer, C. J. (1982) 'Temperature, stress and strain measurements during and after construction of concrete linings in frozen sandstone', *Proc. Int. Symp. on Ground Freezing*, Hanover, New Hampshire, 22–24 June, 343–348

Anon (1974) *Glossary of Mining Terms. Section 8. Winning and Working*, BS 3618: Section 8, British Standards Institution, London, 26 pp.

Anon (1985) *Structural Use of Concrete Part 1. Code of Practice for Design and Construction*, BS 8110: Part 1, British Standards Institution, London, 2/1–2/7

Anon (1987) 'Weiterentwickling der Schachtfräse und Ersteinsatz im Steinkohlengebirge auf der Schachtanlage Lohberg' (Further development of the shaft milling machine and first use in Coal Measures at Lohberg Colliery), Gewerkschaft Walter Aktiengesellschaft, *Bergbauspezialunternehmen*, Essen, December, 5–8

Anon (1988) *Parker Drilling Company Brochure*, Tulsa, Oklahoma, USA, pp. 1–35

Anon (1989a) *Report on Study into Alternative Methods of Shaft Sinking through Water Bearing Strata*, Vol. 4, TH Engineering Services Ltd, British Coal Headquarters Technical Department

Anon (1989b) 'Record bored shaft', *Mining J.*, **313**, (8041), London, 298

Auld, F. A. (1979) 'Design of concrete shaft linings', *Proc. Inst. Civil Engineers*, Part 2, **67**, 817–832

Auld, F. A. (1982) 'Ultimate strength of concrete shaft linings and its influence on design', *Proc. Symp. on Strata Mechanics*, Newcastle upon Tyne, , I. W. Farmer (ed.), Elsevier,

Amsterdam, Developments in Geotechnical Engineering, Vol. 32

Auld, F. A. (1983) *Concrete in Underground Works*, Concrete Society Technical Paper No. 105, The Concrete Society, London, 35 pp.

Auld, F. A. (1987) 'A decade of deep shaft concrete lining', *Concrete,* **21**(2), 4–8

Auld, F. A. (1989) 'High-strength, superior durability, concrete shaft linings', *Proc. Inst. Mining and Metall. Shaft Engineering Conference*, Harrogate, 179–193

Benjamin, J. R. (1988) '*Machines for Rotary Drilling of Shafts Part 1. Selection of Type*', Wirth GmbH, Erkelenz, West Germany, 1–17

Black, J. C. and Auld, F. A. (1985) 'Current and future U.K. practice for the permanent support of shaft excavations', *Int. J. Mining Eng.,* **3**(1), 35–48

Black, J. C., Pollard, C. A. and Daw G. P. (1982) 'Hydrogeological assessment and grouting at Selby', *Proc. Conf. on Grouting in Geotech. Eng.*, ASCE, New Orleans, 665–679

Blackwood, W. (1981) 'Shaft sinking: mechanical and electrical engineering aspects', *Mining Technology*, 227–239

Brümmer, K. H. (1981) 'Full face shaft boring without drill pipe', *Dielmann-Haniel GmbH unser Betrieb,* **29**, 1–8

Brümmer, K. H. (1984) 'Present state and prospects of development of shaft boring techniques using autonomous shaft boring machines', *Deilmann-Haniel GmbH unser Betrieb,* **38**, 1–7

Brümmer, K. H. (1986) 'Shaft sinking with autonomous shaft boring machine', *German Mining* **1**, 3–7

Chalmers, A., Daw, G. P. and Scott, R. A. (1979) 'A modified form of aquifer depletion/recovery test for assessing potential water makes into deep excavations', *Proc. 4th Int. Cong. on Rock Mechanics*, Montreux, **2**, 67–72 and Discussion Session **3**, 225

Cleasby, J. V., Pearse, G. E., Grieves, M. and Thorburn, G. (1975) 'Shaft-sinking at Boulby mine, Cleveland Potash Ltd', *Trans. Inst. Mining and Metall.* Section A, **84**, A7-A28

Daw, G. P. and Scott, R. A. (1983) 'Hydrological testing for deep shafts and tunnels', *Bull. Int. Assoc. Engg. Geol.* **26–27**, 391–395

Daw, G. P. (1984) 'Application of aquifer testing to deep shaft investigations', *Q. J. Engg. Geol.,* **17**, 367–379

Daw, G. P. and Pollard, C. A. (1986) 'Grouting for ground water control in underground mining', *Int. J. Mine Water,* **5**(4), 1–40

Daw, G. P., Fear, N. J., Jeffery, R. I. and Pollard, C. A. (1988) 'Hydrogeological investigations and ground treatment for shaft sinking at Asfordby New Mine', *Proc. 3rd Int. Mine Water Cong.* Melbourne, 683–692

Douglas, A. A. B. and Pfutzenreuter, F. R. B. (1989) 'Overview of current South African vertical circular shaft construction practice', *Proc. Shaft Engg. Conf.*, Harrogate, Institution of Mining and Metallurgy, 137–154

Fotheringham, J. B. and Black, J. C. (1983) 'Groundwater pressure relief in shaft sinking', *The Mining Engineer*, August, 85–91

Friant, J. E. and Sands, P. F. (1977) 'Development of blind shaft boring system', *3rd Symp. on Underground Mining*, Louisville

Gilbert, G. N. J. (1974) *Engineering Data on Nodular Cast Irons – SI Units*, Report 1160, British Cast Iron Research Association, Alvechurch, Birmingham, 27 pp.

Glenday, W. B. (1989) 'Development of shaft steelwork as applied to deep circular shafts: design and installation aspects' *Proc. Shaft Engg. Conf., Harrogate*, Institution of Mining and Metallurgy, 25–37

Grieves, M. (1981) 'Shaft sinking today – a boring business tomorrow', *Mining Engg*, December, 1705–1710

Hanke, K-P. M. (1982) 'Shaft sinking using the V-mole-description of the TMCI operation in Alabama', *Proc. First Mine Ventilation Symp.*, University of Alabama, Chapter 1, 10–18

Hanke, N. P. and Gabitzsch, B. (1983) 'Technical and economic aspects of rapid shaft construction with a V-mole', *Proc. Rapid Excavation and Tunnelling Conf.*, Chicago, **2**, 1047–1066

Harris, J. S. and Pollard, C. A. (1986) 'Some aspects of groundwater control by the ground freezing and grouting methods', In *Groundwater in Engineering Geology*, Engineering Geology Special Publication No. 3, J. C. Cripps, F. G. Bell and M. C. Culshaw (eds.), The Geological Society, London, pp. 455–467

Harvey, S. J. (1988) 'Construction of the Asfordby Mine shafts through the Bunter Sandstone by use of ground freezing', *The Mining Engineer*, August, 51–58

Hegemann, J. and Jessberger, H. L. (1985) 'Deep frozen shaft with gliding liner system', *Proc. 4th Int. Symp. on Ground Freezing*, Sapporo, Japan, 5–7 August, 357–373

Henderson, R. W. (1969) 'Sinking of No. 5 shaft, Hartebeestfontein Gold Mining Co. Ltd., with particular reference to the installation of the brattice wall', *Trans. Inst. Mining and Metall.,* Section A, **78**, A39-A58

Hendricks, R. S. (1985) 'Development of a mechanical shaft excavation system', *Proc. Rapid Excavation and Tunnelling Conf.,* **2**, New York, 1024–1045

Hunter, H. E. (1981) 'Shaft drilling-Crownpoint Project', *SME/AIME 5th Annual Uranium Seminar*, Albuquerque, 1–14

Hunter, H. E. (1983) 'Drilled shaft construction at Crownpoint, New Mexico', *Proc. the Rapid Excavation and Tunnelling Conf.,* Chicago, **1**, 544–565

Hynd, J. G. S. (1982) 'A review of recent shaft excavation methods', *Proc. 1st NMIMT Symp. on Mining Techniques, Shaft Sinking and Boring Technology,* Socorro, New Mexico, USA, 1–26

ICI Nobel's Explosives Company Ltd (a) Explosives in Tunnelling and Shaft Sinking, 12 pp.

ICI Nobel's Explosives Company Ltd (b) *The Initiation of Explosives*, 12 pp.

ICI Nobel's Explosives Company Ltd (c) *Magnadet Innovation to Initiation*, 2 pp.

Kaesehagen, F. E. (1983) 'Blind shaft drilling of the Agnew Mine No. 1 ventilation shaft', *The Australian Institution of Mining and Metallurgy, Sydney Branch, Project Development Symposium*, 391–401

Kelland, J. D. and Black, J. C. (1969) 'Cominco's Saskatchewan potash shafts', *Proc. 9th Commonwealth Mining and Metallurgical Congress*, Institution of Mining and Metallurgy, London, 1–20

Lamé and Clapeyron (1833) 'Mémoire sur l'équilibre intérieur des corps solides homogènes', *Mém. divers savans,* **4**,

Link, H. (1980) 'Berechnung ringversteifter Bohrschachtverrohrungen aus Stahl in den USA' (Calculations for ring-reinforced steel borehole casings in the USA), *Gluckauf-Forschungshefte* **41**(H3), 2–11

Link, H., Lütgendorf, H. O. and Stoss, K. (1976) *Richtlinien zur Berechnung von Schachtauskleidungen in nicht Standfestem Gebirge* (Instructions for the Design of Shaft Linings in Unstable Ground), Verlag Gluckauf GmbH, Essen, 43 pp.

Longden, H. A. (1967) 'Current techniques in deep shaft sinking and development', *The Mining Engineer,* **127**, Part 3,

Moll, C. (1973) 'Modern shaft systems in the Republic of South Africa', *Int. Conf. on Hoisting Men, Materials, Minerals,* L. R. Robinson, E. A. Bunt and K. Kraft (eds.) South African Institution of Mechanical Engineers, Johannesburg, 85–93

Nitro Nobel Industries Sweden (undated) NONEL Users' Manual, 17 pp.

Pliska, R. J. (1984) 'Large diameter blind shaft drilling', *Proc. Canad. Inst. of Mining & Metall. 86th Annual General Meeting*, Ottawa, Canada, 1–26

Presley, C. K. (1981) 'The drilled shaft approach to development of a new uranium mine'. *Proc. 38th Annual Meeting of the Canadian Association of Diamond Drilling Contractors*, Winnipeg, 1–13

Richards, D. P. and Abel, J. F. (1979) 'Shaft lining design in rock', *Proc. Mini-Symposium Shaft Design and Construction*, SME-AIME Annual Meeting, New Orleans, 1–7

Richardson, P. (1984) 'Australia's largest blind-drilled shaft', *Presentation to American Society of Civil Engineers*, Atlanta, 1–11

Ruby, K. and Sands, P. (1979) 'A pneumatic hoisting system for the blind shaft borer', *SME-AIME Mini Symp. on Shaft Design and Construction*, New Orleans, 13–19

Russell, J. E. and DeHart, R. C. (1967) *Design Considerations for Deep Hole Casings*, an interim report to US Atomic Energy Commission, Southwest Research Institute, San Antonio, Texas, 42 pp.

Sands, P. F. and Little, W. E. (1979) 'Blind shaft borer-status report', *SME-AIME Mini Symp. on Shaft Design and Construction*, New Orleans, 9–12

Schalge, R. (1981) *Development of large diameter drilling equipment (Phase II) Volume III: Shaft liner system. Final Report*, Prepared for U.S. Department of Energy by Fenix & Scisson, Inc., September

Schmidt, N. F. B. and Fletcher, A. E. (1987) 'Raiseboring experience with the Wirth two stage sequential reaming head and HG 330 raiseborer', *Proc. 6th Australian Tunnelling Conf.*, Melbourne, 307–317

Scott, R. A. and Daw, G. P. (1983) 'Ground water pressure relief wells in shaft sinking', *Int. J. Mining Engg,* **1**(3), 229–236

Skonberg, E. R. (1980) 'Precast concrete liners for blind drilled shafts', *Annual District Four Meeting 1980 Canadian Institute of Mining*, Flin Flon, Manitoba, 1–8

Thompson, M. H. (1973) 'Shaft sinking and equipping techniques', *Int. Conf. on Hoisting Men, Materials, Minerals*, L. R. Robinson, E. A. Bunt, and K. Kraft (eds.), The South African Institution of Mechanical Engineers, Johannesburg, 27–50

Timoshenko, S. P. and Woinowsky-Krieger, S. (1959) *Theory of Plates and Shells* (2nd ed.), Engineering Societies Monographs, International Students Edition, McGraw-Hill Kogakusha, Tokyo, 466–481

Torbin, R., Brunsing, T., Ounanian, D., Henderson, R., Kelly, J. and Kirby, G. (1983) *Development of a continuous shaft lining system Volumes I-III. Final Technical Report*, Prepared for US Department of Energy by Foster-Miller, Inc., May

Tunnicliffe, J. F. and Keeble, S. (1981) 'Shaft sinking at Selby', *The Mining Engineer*, August, 69–79

Unrug, K. F. (1984) 'Shaft design criteria', *Int. J. of Mining Engg,* **2**(2), 141–155

Utter, S. and Hawkins, J. E. (1978) 'Drilling and casing a large-diameter shaft in the Piceance Creek Basin', *Proc. 11th Oil Shale Symp.*, Colorado School of Mines, 292–310

White, L. (1986) 'Polish copper, Europe's biggest miner thrives on stratiform ores and integrated smelting and refining', *E & MJ* 187, No. 2 26–30

Wild, W. M. and Forrest, W. (1981) 'The application of the freezing process to ten shafts and two drifts at the Selby Project', *The Mining Engineer*, June, 895–904

Windenburg, D. F. and Trilling, C. (1934) Collapse by instability of thin cylindrical shells under external pressure', *Trans. Am. Soc. Mechanical Engineers*, 819–825

Woods, B. M. (1988) 'Future shaft sinking techniques', *The Mining Engineer,* August 97–99

Zeni, D. and Hutchinson, D. M. (1988) *Rapid Access of Gold Deposits Using Blind Drilled Shafts'*, Zeni Drilling Company, Morgantown, West Virginia, U.S.A., pp 1–16

25 Socketed foundations in rock masses

Professor F Kulhawy, Cornell University
Professor J P Carter, University of Sydney

25.1 Introduction

The variability of natural rock masses is the main reason that traditional methods for the design of foundations in rock have been largely empirical. Fortunately, there have been many recent studies that have improved vastly our understanding of the nature and engineering behaviour of rock masses. Other chapters of this book discuss at length some of these recent developments, and detailed descriptions are given on the topics of rock mass characterization and mechanical behaviour. These more recent advances in understanding have made possible more rational, analytical treatments of rock-foundation systems, which in some cases have substantiated the traditional empirical approaches, and in other cases replaced them by more soundly-based theory. Modern methods to predict the deformations of foundations in rock, such as the vertical and horizontal displacements, are examples that fit well into the latter category.

Foundations in rock must satisfy the same criteria as other types of foundations, including adequate stability and tolerable deformations. The problem of assessing the ultimate vertical load capacity of foundations bearing directly on rock has been covered in Chapter 12, and therefore will not be repeated here. In practice, many concrete foundation elements are actually cast into the rock (socketed foundations), either as a matter of construction convenience or to increase the resistance to vertical and lateral loading. Under such loadings, the ultimate resistance of the foundation also may be a function of the maximum shear and normal stresses that may be developed between the rock and the sides of the foundation element. The problems of assessing the maximum side shear stresses that may be mobilized on foundations in rock and the maximum resistance to lateral loading also have been addressed in Chapter 12.

Very often in practice, the design of a foundation in rock is governed by a displacement criterion, rather than by stability requirements. This is particularly relevant to structures that impose very large loads on the underlying rock, such as high-rise buildings and long-span bridges, or where unusually tight tolerances are set on the allowable foundation movements, for example, with particle accelerators, radar towers, and radio telescopes. In such cases, reliable methods for the accurate prediction of foundation movements are required. The primary aim of this chapter is to present some of the newer methods of analysis that may be applied to the prediction of the load–deformation response of foundations in rock masses under vertical and lateral loading. Simple closed-form expressions are presented and, for brevity, detailed derivations of these expressions have been omitted. However, the basic assumptions underlying each of the key equations have been discussed. The important input parameters are identified and, where possible, recommendations based on field evidence are made for the selection of their values. Finally, the use of the simple closed-form expressions in the design of a foundation in rock is illustrated by considering a typical example.

Cast-in-place concrete elements that are circular or square in plan are the most common type of foundations used in rock. Therefore, in the following sections, the problem of a single, solid, vertical, cylindrical shaft foundation embedded in the rock mass is treated in detail. Square or nearly square foundations can be analyzed approximately by these methods, assuming a cylindrical shaft with the same axial or flexural rigidity as the square or rectangle. It is assumed that the cylindrical foundation is wholly contained within the rock mass and that the foundation is constructed by casting concrete into a hole formed in the rock, so that initially there is perfect contact (or bonding) between the concrete and rock, along the entire embedded length of the shaft. Various contact conditions at the tip or base of the shaft will be presented. Foundations bearing directly on, but not embedded in, the rock mass have been considered in detail in Chapter 12 and elsewhere (Kulhawy and Goodman 1987), so the emphasis here is placed on the more general problem that includes embedment effects.

25.2 Axial loading

Most of the techniques proposed for the calculation of the vertical displacements of embedded shaft foundations are based on the theory of elasticity. It has been usual to assume that the foundation is essentially an elastic inclusion within a surrounding rock mass, as shown in Figures 25.1 and 25.2. In a recent study by Rowe and Armitage (1987a, b), the possibility of slip occurring at the interface between the shaft and the rock mass also was included. From these studies, design charts were prepared for the computation of shaft settlements under compression loading.

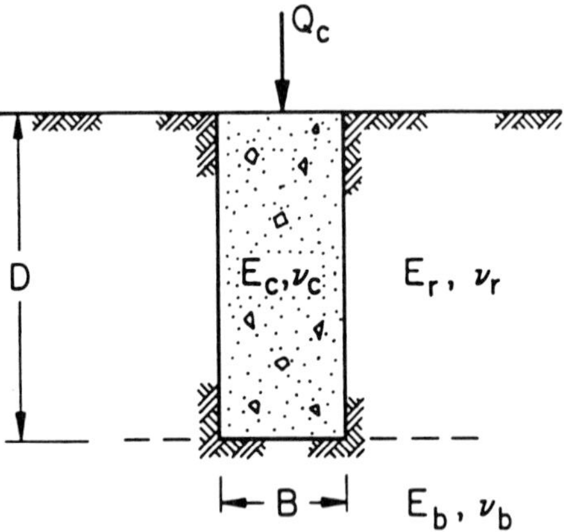

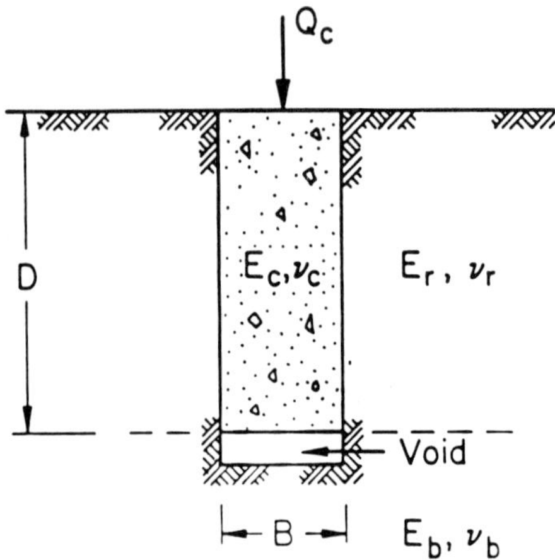

Figure 25.1 *Compression loading of a socketed shaft foundation in rock*

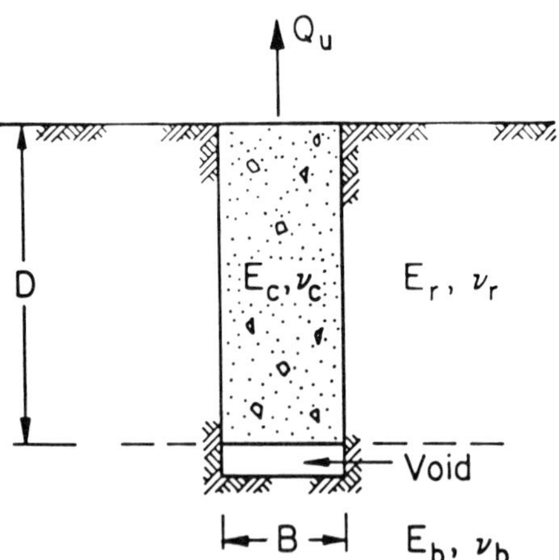

Figure 25.2 *Uplift loading of a socketed shaft foundation in rock*

In this section, a simple method for computing the load–displacement behaviour of a rock-socketed shaft is presented. Approximate equations are developed to describe the response of a shaft to axial compression or uplift loading. Separate equations are presented for perfect interface bonding (no slip) and for slip along the entire cylindrical interface. The simplicity of these approximate closed-form equations makes them attractive for design purposes.

25.2.1 Load transfer mechanisms

(a) Compression loading

When a 'complete' socketed foundation is loaded in axial compression (Figure 25.1), support is provided by shear transfer along the socket wall and vertical stress transfer at the tip of the shaft. The distribution of the load between side and tip resistance is a function of the socket geometry, relative stiffness of the shaft and rock mass, socket roughness and strength, and foundation settlement. When relatively small loads are applied, the rock socket behaves essentially in a linear manner, and the load transfer can be computed accurately using the theory of elasticity. This linear behaviour is illustrated in Figure 25.3 as the line OA. As the load is increased to point A on Figure 25.3, the shear stress at some point along the interface will reach the shear strength, and the socket 'bond' will begin to rupture and relative displacement (slip) will occur between the foundation and the surrounding rock. As the loading is increased further (beyond A), this process will continue along the shaft, more of the shaft will slip, and a greater proportion of the applied load will be transferred to the tip of the shaft. If loading is continued, eventually the entire shaft will slip (point B); beyond this point, a greater proportion of the total axial load will be transmitted directly to the tip.

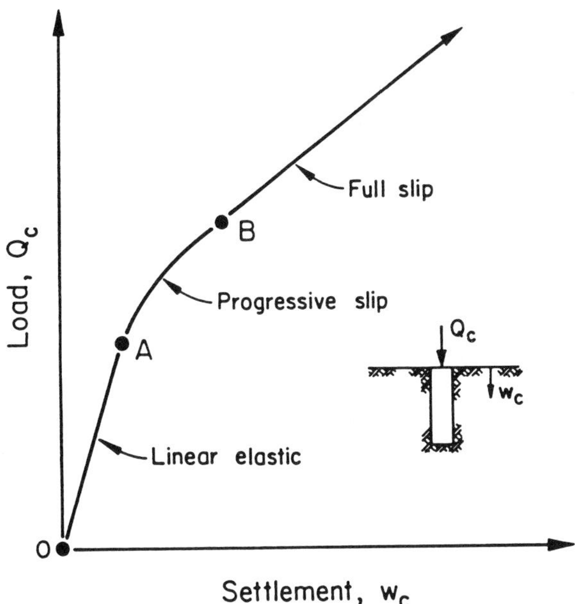

Figure 25.3 *Idealized axial load–displacement behaviour*

In field studies involving compression loading, Horvath (1980) and Williams (1980) showed that slip can occur within the working-load range of the shaft. As long as an adequate factor of safety on the tip resistance is maintained, it may be reasonable and economical to design these compression foundations for the partial or full slip conditions. In such cases, good construction practice will be required to ensure that soft compressible materials do not collect in significant quantities at the bottom of the hole, as they then may contribute to unacceptable settlements.

Clearly, it is desirable to be able to predict the entire load–displacement behaviour of a shaft and, in the case of compression loading, to be able to estimate the proportion of the applied load transmitted to the tip. Prediction of the full load–displacement curve, including progressive slip (region AB on Figure 25.3), requires the use of numerical techniques. However, for design purposes, the initial linear elastic response (OA) and the full slip condition (the region beyond point B) provide good bounds on the overall shaft behaviour. Approximate analytical solutions for the bounds are given in this chapter.

(b) Uplift loading

The manner in which axial uplift loads are transmitted to the rock mass depends upon the load magnitude and the way in which the load is applied to the socket. In practice, it is usual for the uplift load to be applied at the level of the rock surface, *that is*, at the butt of the shaft, to reinforcing members or stub anchors embedded in the concrete foundations. This is the loading case considered here; alternative uplift loading arrangements are given elsewhere by Carter and Kulhawy (1988). For loading at the butt, the axial stresses in the shaft will be tensile and, because of the Poisson's ratio effect, the radial stress changes at the interface also will be tensile. This loading will tend to reduce the frictional component of side shear resistance and any beneficial effects that may arise from dilation at the interface.

In principle, some resistance to uplift loading may also be provided in a 'complete' socket by the tensile strength of the interface between the tip of the shaft and the rock mass. However, in practice it is prudent to ignore this component of resistance because of the uncertainty about 'perfect clean-out' of the socket at the time that the concrete is placed. It is usual to design these foundations to resist uplift loading on the basis of a 'shear only' socket.

25.2.2 Analysis of load–displacement response

The problem of axial loading of a socketed shaft with either a 'complete' or 'shear' socket, as shown in Figures 25.1 and 25.2, has been investigated in detail by Carter and Kulhawy (1988). Some of the basic assumptions of the analysis and some of the results are presented here.

In this chapter, the presence of any overlying soil is ignored for brevity. However, the extension of the analysis to include an overlying soil layer is straightforward (e.g. Carter and Kulhawy 1988). The concrete shaft is modelled as an elastic cylindrical inclusion, with Young's modulus E_c and Poisson's ratio ν_c, embedded in an elastic rock mass. The shaft has depth D and diameter B. The rock mass surrounding the shaft side is homogeneous with Young's modulus E_r and Poisson's ratio ν_r. Beneath the shaft tip, the rock mass has a Young's modulus E_b and Poisson's ratio ν_b. A vertical load Q is applied to the shaft butt, which is considered to be distributed uniformly so that the average applied axial stress is $\sigma_b = 4Q/(\pi B^2)$.

An elastic model for the shaft is consistent with current structural engineering practice. If the shaft is reinforced concrete, then E_c should be assigned the value of the equivalent section modulus. An elastic model to represent the mechanical behaviour of the rock mass is a simplification of reality, and the selection of suitable properties to characterize the mass is a matter where considerable judgment and experience are required. Allowance must be made for the discontinuous nature of most natural rock masses. Modulus values determined from laboratory testing of intact core specimens are generally considered inappropriate, because they typically overestimate the stiffness of the discontinuous mass. Suggested methods for determining representative elastic properties of the rock mass are described in Chapter 12.

The analysis of two different stages in the loading history of the shaft are of interest. Initially, the case where perfect contact is maintained (no slip) along the interface will be considered. Then slip between the shaft and rock mass along the full length of the interface will be examined. Furthermore, solutions will be presented for two types of shaft. In the first case, the structural foundation will be considered to be compressible, so that its own axial shortening (or extension in the case of uplift loading) contributes to the overall displacement of the foundation–rock system. In addition, the limiting case of a rigid foundation element also will be considered, as it has considerable application to real foundations in rock.

25.2.3 Linear elastic behaviour

For linear elastic behaviour, perfect bonding is assumed along the entire shaft–rock mass interface. Under an applied axial load, the displacements in the rock mass are predominantly vertical, and the load is transferred from the shaft to the rock mass by vertical shear stresses acting on the cylindrical interface, with little change in vertical normal stress in the rock mass (except near the tip of a complete socket). The pattern of deformation around the shaft may be visualized as an infinite number of concentric cylinders sliding inside each other (Randolph 1977). Randolph and Wroth (1978) have shown that, for this type of behaviour, the displacement of the shaft w may be described adequately in terms of hyperbolic sine and cosine functions of depth z below the surface, as given below:

$$w(z) = A_1 \sinh[\mu z] + A_2 \cosh[\mu z] \tag{25.1}$$

in which the constant μ is given by

$$(\mu D)^2 = \left(\frac{2}{\zeta\lambda}\right)\left(\frac{2D}{B}\right)^2 \tag{25.2}$$

where

$$\zeta = \ln[5(1 - \nu_r)D/B] \tag{25.3}$$

$$\lambda = E_c/G_r \tag{25.4}$$

and

$$G_r = E_r/[2(1 + \nu_r)] \tag{25.5}$$

G_r is the elastic shear modulus of the rock mass.

The constants A_1 and A_2 can be determined from the boundary conditions of the problem. Some particular cases of practical interest are considered below.

(a) Complete socket under compression loading

When the shaft tip bears directly on the bottom of the socket hole, a contribution to the shaft settlement will arise from the displacement of the rock beneath the tip (Figure 25.1a). This case can be approximated as a rigid punch acting on the surface of an elastic half-space with Young's modulus E_b and Poisson's ratio ν_b. To retain generality, the elastic rock mass below the shaft tip may be different from that surrounding the shaft. If the standard solution for the displacement of a rigid punch resting on an elastic half-space (e.g. Poulos and Davis 1974) is used as one of the boundary conditions for the shaft, then the elastic settlement at the butt of the complete shaft w_c is given by (Randolph and Wroth 1978).

$$\frac{G_r B w_c}{2Q_c} = \frac{1 + \left(\frac{4}{1-\nu_b}\right)\left(\frac{1}{\pi\lambda\xi}\right)\left(\frac{2D}{B}\right)\left(\frac{\tanh[\mu D]}{\mu D}\right)}{\left(\frac{4}{1-\nu_b}\right)\left(\frac{1}{\xi}\right) + \left(\frac{2\pi}{\zeta}\right)\left(\frac{2D}{B}\right)\left(\frac{\tanh[\mu D]}{\mu D}\right)} \tag{25.6}$$

in which

$$\xi = G_r/G_b \tag{25.7}$$

and

$$G_b = E_b/[2(1 + \nu_b)] \tag{25.8}$$

The proportion of the applied load transmitted to the base (tip) is given by

$$\frac{Q_{tip}}{2Q_c} = \frac{\left(\frac{4}{1-\nu_b}\right)\left(\frac{1}{\xi}\right)\left(\frac{1}{\cosh[\mu D]}\right)}{\left(\frac{4}{1-\nu_b}\right)\left(\frac{1}{\xi}\right) + \left(\frac{2\pi}{\zeta}\right)\left(\frac{2D}{B}\right)\left(\frac{\tanh[\mu D]}{\mu D}\right)} \tag{25.9}$$

The solution given by Equation (25.6) is plotted in Figure 25.4 for cases where $\nu_r = \nu_b = 0.25$ and $E_r = E_b$. A number of values of the modulus ratio (E_c/E_r) and the slenderness ratio (D/B) are shown. Also plotted are finite element solutions by Pells and Turner (1979). The general agreement between the two solutions, though not perfect, is reasonable and could be considered satisfactory for design purposes.

(b) Shear socket under compression loading

For a shear socket under compression loading (Figure 25.1b), the boundary condition at the shaft tip is one of zero axial stress. For this case, the settlement at the shaft butt w_c is given by

$$\frac{E_r B w_c}{2Q_c} = \left(\frac{1}{\pi}\right)\left(\frac{E_r}{E_c}\right)\left(\frac{2}{\mu B}\right)\left(\frac{\cosh[\mu D]}{\sinh[\mu D]}\right) \tag{25.10}$$

This solution is plotted in Figure 25.5, where it also is compared with finite element solutions (Pells and Turner 1979). The accuracy of the approximate expression (Equation (25.10)) is quite clear, particularly at larger values of the modulus ratio (E_c/E_r).

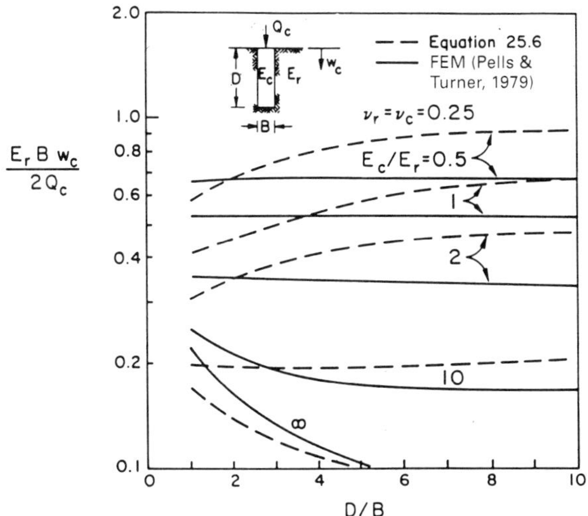

Figure 25.4 *Elastic settlement of a shaft in a complete rock socket*

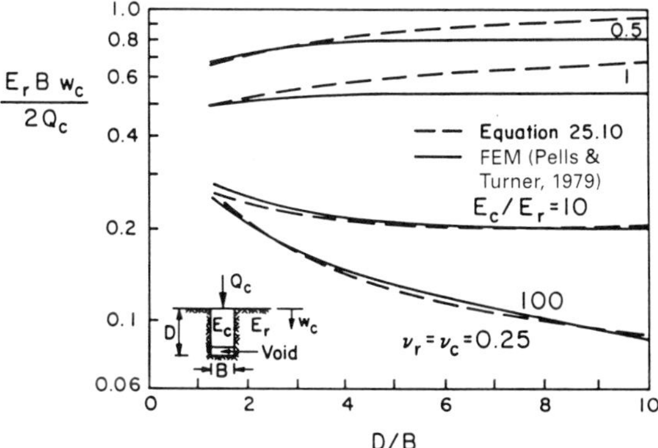

Figure 25.5 *Elastic settlement of a shaft in a shear-only socket*

(c) Shear socket under uplift loading

For uplift loading applied directly to the butt of the shaft (Figure 25.2), the uplift displacement is simply the reverse of the settlement of the same type of socket when subjected to a compressive load equal in magnitude to the uplift load. Therefore, the magnitude of the uplift displacement is given by Equation (25.10).

25.2.4 Full slip behaviour

The case of slip along the entire length of the shaft (beyond point B in Figure 25.3) also has been considered in detail by Carter and Kulhawy (1988). For this case, the shear strength of the interface is given by the Mohr–Coulomb criterion:

$$\tau = c + \sigma_r \tan \phi \tag{25.11}$$

in which c = interface cohesion, ϕ = interface friction angle, and σ_r = radial normal stress acting on the interface. As relative displacement (slip) occurs, the interface may dilate, and it is assumed that the displacement components follow the dilation law:

$$\frac{\Delta u}{\Delta v} = \tan \psi \tag{25.12}$$

In which Δu and Δv are the relative shear and normal displacements of the concrete-rock interface and ψ is the angle of dilation defined by Davis (1968). This type of behaviour is illustrated in Figure 25.6.

Cases where the full interface shear strength is mobilized, but where no dilation accompanies the slip displacements, have been considered previously by Kulhawy and Goodman (1987). It is possible to generalize their solutions to include interface dilation. Dilation is important because it will increase both the normal stress on the interface and the interface shear strength.

To determine the radial displacements at the interface, the procedure suggested by Goodman (1980) and Kulhawy and Goodman (1987) is followed, in which conditions of plane strain are assumed, as an approximation, independently in the rock mass and in the slipping shaft. The rock mass is considered to be linear elastic, even after full slip

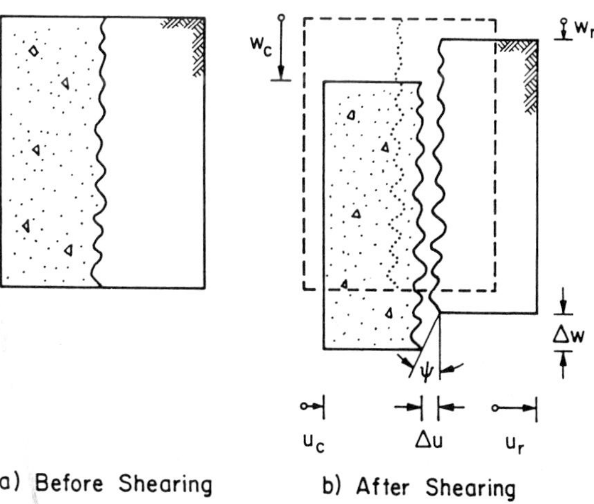

Figure 25.6 *Schematic illustration of dilatancy at the shaft–rock mass interface*

has taken place, and the shaft is considered to be an elastic column. These assumptions, together with the dilatancy law, permit the derivation of an expression for the variation of vertical stress in the compressible shaft (Carter and Kulhawy 1988). The distribution of shear stress acting on the shaft then can be calculated from equilibrium, and the vertical displacement can be determined as a function of depth z by treating the shaft as a simple elastic column. Solutions for some cases of practical interest are given below.

(a) Shear socket under compression loading

For a shear socket under compression loading, the 'full slip' displacement of the shaft butt w_c can be calculated from

$$w_c = F_1 \left(\frac{Q_c}{\pi E_r B}\right) - F_2 B \tag{25.13}$$

in which

$$F_1 = a_1 (\lambda_2 B C_2 - \lambda_1 B C_1) - 4a_3 \tag{25.14}$$

$$F_2 = a_2 \left(\frac{c}{E_r}\right) \tag{25.15}$$

$$C_{1,2} = \exp[\lambda_{2,1} D]/(\exp[\lambda_2 D] - \exp[\lambda_1 D]) \tag{25.16}$$

$$\lambda_{1,2} = \frac{-\beta \pm (\beta^2 + 4\alpha)^{1/2}}{2\alpha} \tag{25.17}$$

$$\alpha = a_1 \left(\frac{E_c}{E_r}\right) \left(\frac{B^2}{4}\right) \tag{25.18}$$

$$\beta = a_3 \left(\frac{E_c}{E_r}\right) B \tag{25.19}$$

$$a_1 = (1 + \nu_r)\zeta + a_2 \tag{25.20}$$

$$a_2 = \left[\left(1 - \nu_c\right)\left(\frac{E_r}{E_c}\right) + \left(1 + \nu_r\right)\right]\left(\frac{1}{2 \tan\phi \tan\psi}\right) \tag{25.21}$$

$$a_3 = \left(\frac{\nu_c}{2 \tan\psi}\right)\left(\frac{E_r}{E_c}\right) \tag{25.22}$$

(b) Complete socket under compression loading

For this case, the 'full slip' solution for the settlement of the shaft butt is found as

$$w_c = F_3 \left(\frac{Q_c}{\pi E_r B}\right) - F_4 B \tag{25.23}$$

in which

$$F_3 = a_1(\lambda_1 B C_3 - \lambda_2 B C_4) - 4a_3 \tag{25.24}$$

$$F_4 = \left[1 - a_1 \left(\frac{\lambda_1 - \lambda_2}{D_4 - D_3}\right) B\right] a_2 \left(\frac{c}{E_r}\right) \tag{25.25}$$

$$C_{3,4} = D_{3,4}/(D_4 - D_3) \tag{25.26}$$

$$D_{3,4} = \left[\pi(1 - \nu_b^2)\left(\frac{E_r}{E_b}\right) + 4a_3 + a_1 \lambda_{2,1} B\right] \exp[\lambda_{2,1} D] \tag{25.27}$$

The proportion of the applied load transmitted to the tip is given by

$$\frac{Q_{tip}}{Q_c} = P_3 + P_4 \left(\frac{\pi B^2 c}{Q_c}\right) \tag{25.28}$$

in which

$$P_3 = a_1(\lambda_1 - \lambda_2) B \exp[(\lambda_1 + \lambda_2)D]/(D_4 - D_3) \tag{25.29}$$

$$P_4 = a_2(\exp[\lambda_1 D] - \exp[\lambda_2 D])/(D_4 - D_3) \tag{25.30}$$

(c) Shear socket under uplift loading

Where the uplift force is applied to the butt of the shaft (Figure 25.2), the axial stresses will be tensile, and the Poisson's ratio effects induce tensile changes in the radial stress acting at the interface between the shaft and rock mass. The solution for 'full slip' uplift displacement of the shaft butt w_u can be obtained by an analysis similar to that for compression loading and is as follows:

$$w_u = F_5 \left(\frac{Q_u}{\pi E_r B}\right) + F_6 B \tag{25.31}$$

in which

$$F_5 = 4a_3 - a_1(\lambda_1 B C_5 - \lambda_2 B C_6) \tag{25.32}$$

$$F_6 = a_2 \left(\frac{c}{E_r}\right)$$

$$C_{5,6} = \exp[-\lambda_{2,1} D]/(\exp[-\lambda_1 D] - \exp[-\lambda_2 D]) \tag{25.33}$$

In this case, the uplift load Q_u is tensile and should be input to Equation (25.31) as a negative quantity, and a negative value of w_u indicates an upward displacement.

It also should be noted carefully that this analysis is strictly valid only as long as the radial stress acting at the concrete–rock interface remains compressive. For this condition to be satisfied, the tensile stress changes produced by the effect of Poisson's ratio of the shaft must be offset by the compressive changes induced by dilation and any initial, *in situ* compressive horizontal stress at the interface.

25.2.5 Rigid shafts

In practice, it is common for the stiffness of the shaft material to be much larger than that of the host rock ($E_c \gg E_r$). Shafts of this type may behave rigidly when subjected to axial loading and, for such cases, the governing equations are greatly simplified.

(a) Shear socket under compression loading

The initial elastic response of a compressible shaft is described by Equation (25.6), from which it is clear that the axial stiffness is a function of tanh $[\mu D]$. For small values of μD, the function tanh $[\mu D]$ is closely approximated by μD and, for such cases, the elastic settlement of the shear socket can be approximated as

$$w_c = \left(\frac{\zeta}{2\pi}\right)\left(\frac{1}{G_r D}\right) Q_c \tag{25.34}$$

Equation (25.34) will be accurate to within 10% whenever μD is less than at most about 0.5, that is, whenever

$$\left(\frac{E_c}{E_r}\right)\left(\frac{B}{2D}\right)^2 \geqslant 1 \tag{25.35}$$

For practical purposes, the inequality 25.35 can be considered as the definition of rigidity with respect to axial loading. A further implication of shaft rigidity is that the elastic stress transfer is linear down the shaft so that the interface shear stresses are constant with depth.

If it also is assumed that the shear stress distribution remains constant at full slip, and that $E_c \gg E_r$, then the load–displacement relationship for the butt of the shaft, for a dilatant, cohesive–frictional interface, can be written simply as

$$w_c = R_1 \left(\frac{Q_c}{\pi E_r B}\right) - R_2 B \tag{25.36}$$

in which

$$R_1 = \left(1 + \nu_r\right)\left(\zeta + \frac{1}{2 \tan \phi \tan \psi}\right)\left(\frac{B}{D}\right) \tag{25.37}$$

and

$$R_2 = \left(\frac{1 + \nu_r}{2 \tan \phi \tan \psi}\right)\left(\frac{c}{E_r}\right) \tag{25.38}$$

(b) Complete socket under compression loading

For a rigid shaft in contact with the rock at its tip, the elastic settlement is given by

$$w_c = \frac{Q_c}{\left(\frac{E_b B}{1 - \nu_b^2}\right) + \left(\frac{\pi}{\zeta}\right)\left(\frac{E_r D}{1 + \nu_r}\right)} \tag{25.39}$$

and the proportion of the applied load transmitted to the tip during the linear elastic response is

$$\frac{Q_{tip}}{Q_c} = \frac{\left(\frac{4}{1 - \nu_b}\right)\left(\frac{1}{\xi}\right)}{\left(\frac{4}{1 - \nu_b}\right)\left(\frac{1}{\xi}\right) + \left(\frac{2\pi}{\zeta}\right)\left(\frac{2D}{B}\right)} \tag{25.40}$$

Once full slip of the shaft has occurred, the load–displacement relationship becomes

$$w_c = R_3 \left(\frac{Q_c}{\pi E_r B}\right) - R_4 B \tag{25.41}$$

in which

$$R_3 = \frac{2R_1 R_5}{R_1 + 2R_5} \tag{25.42}$$

$$R_4 = \frac{2R_2 R_5}{R_1 + 2R_5} \tag{25.43}$$

and

$$R_5 = \left(\frac{\pi}{2}\right)\left(1 - \nu_b^2\right)\left(\frac{E_r}{E_b}\right) \tag{25.44}$$

and the load transmitted to the tip is given by

$$\frac{Q_{tip}}{Q_c} = \left(\frac{R_1}{R_1 + 2R_5}\right) - \left(\frac{R_6}{R_1 + 2R_5}\right)\left(\frac{\pi B^2 c}{Q_c}\right) \tag{25.45}$$

in which

$$R_6 = \left(\frac{1 + \nu_r}{2 \tan \phi \tan \psi}\right) \tag{25.46}$$

(c) Shear socket under uplift loading

For uplift loading applied to the butt, the linear elastic response of the butt of the shaft is given by Equation (25.34). Once full slip has occurred, the response is predicted by

$$w_u = R_1 \left(\frac{Q_u}{\pi E_r B} \right) + R_2 B \qquad (25.47)$$

For the uplift case, Q_u should be input into this equation as a negative quantity, and a negative value of w_u indicates an upward displacement.

(d) General remarks – rigid shafts

It is perhaps worth noting here that, for rigid shafts, the effects of the Poisson's ratio of the shaft material ν_c are negligible. This fact has several important implications. First, the strength mobilized at the initiation of shaft slip is almost entirely cohesive in nature (i.e. it arises mainly from interface bonding), because there is almost no normal stress acting to generate frictional resistance. Second, as the shaft slips, the changes in radial stress result almost entirely from dilation at the interface. For these reasons, the response of a rigid shaft, employing side resistance only, is essentially the same in uplift as it is under compression loading.

25.2.6 Comparisons with finite element solutions

The adequacy of the closed-form expressions can be demonstrated by comparing them with chart solutions, obtained using a nonlinear finite element analysis (Rowe and Armitage 1987 a, b). Special care must be taken when the closed-form solutions corresponding to $\psi = 0$ are evaluated; simplified closed-form expressions for this case have been presented elsewhere by Kulhawy and Goodman (1987). Comparisons are made for two different cases: in the first, parameters have been selected to represent a compressible shaft with low relative stiffness ($D/B = 10$, $E_c/E_r = 10$, $E_b/E_r = 1$); and in the second, a rigid shaft with large relative stiffness ($D/B = 2$, $E_c/E_r = 100$, $E_b/E_r = 1$), is investigated. In all cases, the Poisson's ratio of the rock mass was 0.3 and that of the shaft was 0.15. The interface is purely cohesive and non-dilatant ($\phi = \psi = 0$). The predictions of the load–displacement response are presented in Figures 25.7(a) and 25.8(a), and the predictions of the tip load are given in Figures 25.7(b) and 25.8(b). The overall agreement between the simple equations presented herein and the numerical solutions from finite element analysis is very good. However, a few points are worthy of further discussion.

The closed-form expressions cannot predict the load–displacement response between the occurrence of first slip and full slip of the shaft, but the finite element results indicate that the progression of slip along the socket takes place over a relatively small interval of displacement. (Finite element predictions of first slip and full slip have been indicated on Figures 25.7 and 25.8). Therefore it seems reasonable, at least for most practical cases, to ignore the small region of the curves corresponding to progressive slip and to assume that the load–displacement relationship is bilinear, with the slope of the initial portion given by Equation (25.6) and the slip portion by Equation (25.23) for complete sockets. Figures 25.7(a) and 25.8(a) indicate that this simplification is reasonably accurate for a range of shaft foundations in rock.

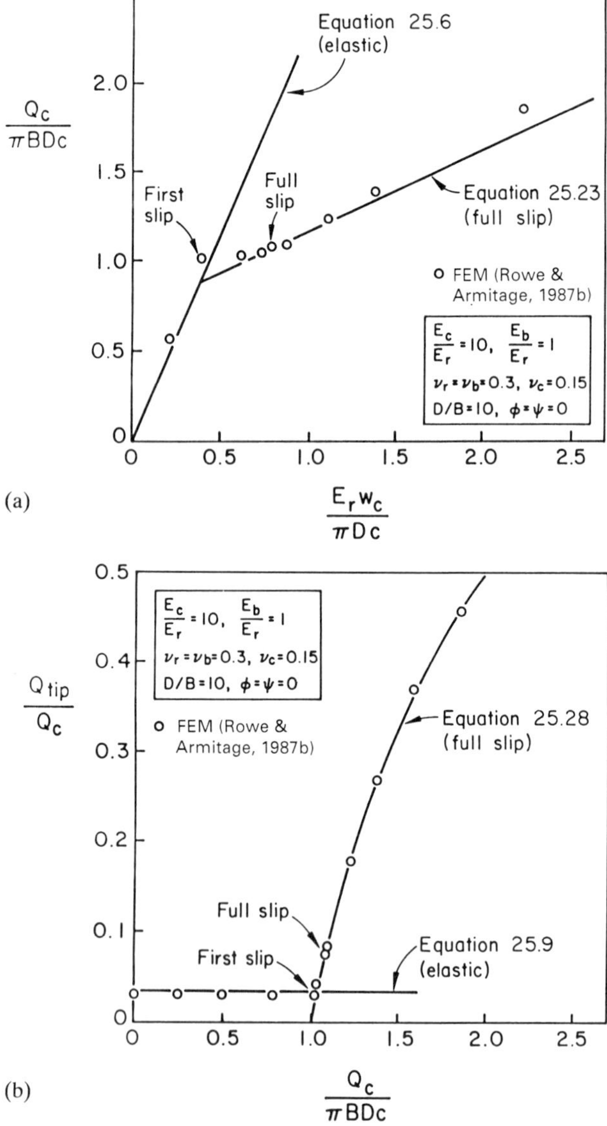

Figure 25.7 a and b *Comparison between finite element and closed-form predictions for an illustrative compressible shaft*

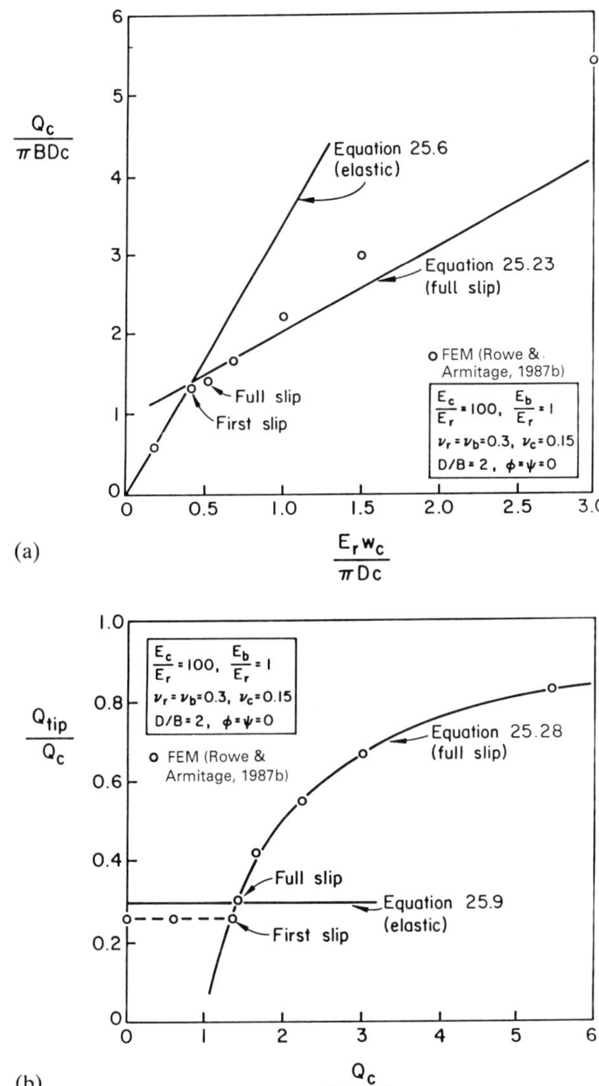

Figure 25.8 a and b *Comparison between finite element and closed-form predictions for an illustrative rigid shaft*

Figures 25.7(a) and 25.8(a) also indicate that the closed-form equations tend to overpredict settlements once full slip has occurred. However, the differences are small and, for design purposes, the simple equations should give a reasonable first approximation for many shaft-rock mass systems.

25.2.7 Trends in behaviour

The comparisons made in Figures 25.7 and 25.8 were for cases where the interface strength was purely cohesive and non-dilatant, and where the shear strength was unchanged by any interface slip. In reality, this is a somewhat unlikely possibility. The initial shear strength along the interface may have both cohesive and frictional components. Once rupture occurs, it is likely that the cohesive component of the strength (the 'bonding') may diminish, dilation will occur, and eventually (after sufficient slip has taken place) the interface shear resistance may become purely frictional and dilation may cease.

(a) *Purely frictional sockets*

Consider the case where the interface resistance is purely frictional, with an angle of friction (ϕ) that remains constant throughout the loading, and where the shaft tip rests directly on the bottom of the socket hole. This model will produce conservative predictions for compression loading, because it corresponds to a shaft that fits perfectly into the socket hole, but nowhere along the shaft is there any bonding between the rock mass and the shaft. It also will be conservative to assume that initially there is no transfer of normal stress across the interface between the shaft and the rock and, at least for shallow sockets, this will closely model reality. Under these circumstances, the entire shaft will be on the verge of slipping as soon as compression loading is applied to the shaft butt. A shaft in this kind of socket will exhibit no elastic side-shear behaviour, but will slip from the commencement of loading. Some typical solutions have been plotted in Figure 25.9 for cases where $E_c/E_r = 10$, $E_b/E_r = 1$, $\nu_r = \nu_b = 0.3$, $\nu_c = 0.15$, $\phi = 30°$, and $c = 0$. Solutions are plotted for a range of values of D/B and for a number of values of the dilation angle ψ. Dimensionless settlements are plotted in Figure 25.9a, and the proportion of the applied compression loading transmitted to the tip is plotted in Figure 25.9b. Also shown is the elastic solution for a perfectly bonded (no slip) interface. These results indicate that only a small rate of dilation is required to cause a marked increase in stiffness of the shaft under compression loading. A dilation angle of $\psi = 30°$ produces behaviour approaching the stiffness of a fully bonded (elastic) socket. If no dilatancy occurs as interface slip takes place ($\psi = 0$), the response of the shaft also is linear, but its stiffness is less than that for a fully bonded socket.

In Figure 25.10(a), predictions are plotted for sockets having the following properties: $E_c/E_r = 10$, $E_b/E_r = 1$, $\nu_b = \nu_r = 0.3$ and $\nu_c = 0.15$. Results are given for three different values of the interface friction angle ϕ and for a range of slenderness ratios D/B. It can be seen that the stiffness of most of these frictional but non-dilatant sockets is in the range from about 20 to 40% of the stiffness for an equivalent, perfectly bonded, elastic socket. It also is interesting (and perhaps surprising) to note that there is only a minor dependence of the stiffness of the non-dilatant, frictional shaft on the interface friction angle. This behaviour can be understood better when the predicted load distributions are plotted, as in Figure 25.10(b). For the typical case of $E_c/E_r = 10$ and $\nu_c = 0.15$, a large proportion of the shaft loading is actually

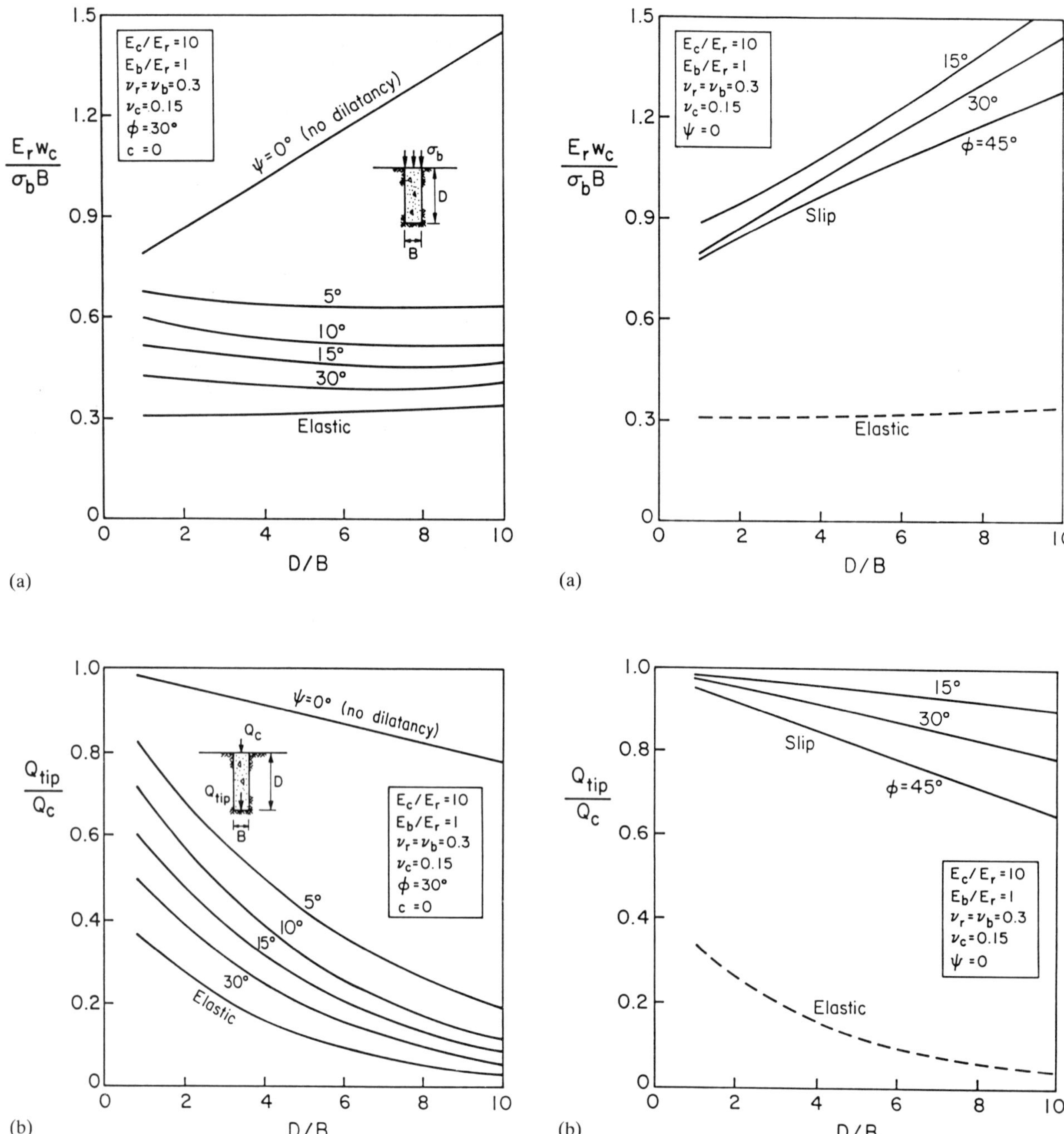

Figure 25.9 a and b *Predicted behaviour of a shaft in a purely frictional socket with dilatancy*

Figure 25.10 a and b *Predicted behaviour of a shaft in a purely frictional socket without dilatancy*

transmitted to the shaft tip, regardless of the friction angle, and therefore the predominant contribution to the overall settlement comes from displacement of the rock mass below the shaft tip. A comparison of Figures 25.9(b) and 25.10(b) demonstrates the significance of interface dilation on the load transfer in a socket; the lower the rate of dilatancy, the more compressive load is transmitted to the shaft tip.

(b) Cohesive–frictional sockets

Consider now those cases where cohesion is assumed at the interface (due mainly to actual bonding between the concrete and the rock, but also as apparent cohesion from mechanical interlocking of the asperities along the interface), and where the cohesion remains constant but no dilation occurs as slip takes place ($\psi = 0$).

Some predictions of the load–displacement response of complete sockets subjected to compression loading are given in Figure 25.11. The case where $E_c/E_r = 100$ and $D/B = 5$ is considered, and typical values of the parameters that might occur in practice have been used to add dimensions to the predictions. For the sockets where $c > 0$, the response is bilinear, following first the predictions for a linear elastic (fully bonded) socket, and eventually following the prediction based on the full slip analysis. A linear relationship between load and settlement is predicted for the full slip condition, with the slope of this line the same as that predicted for a purely frictional interface. For the cases considered, this slope is independent of the value of the interface cohesion and therefore the curves corresponding to the three values of c are all parallel.

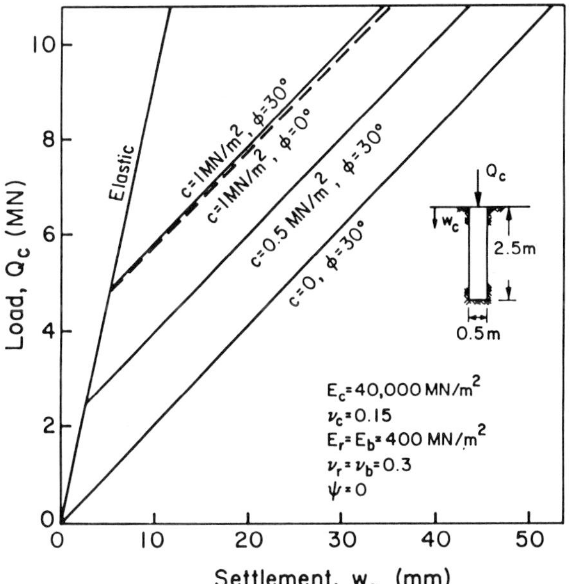

Figure 25.11 *Predicted behaviour of shafts in cohesive–frictional sockets*

Also plotted on Figure 25.11 is the curve predicted for the case where $c = 1$ MPa and $\phi = 0°$. This purely cohesive socket has almost the same load–settlement relationship as that of a socket with $c = 1$ MPa and $\phi = 30°$, indicating that the behaviour after slip is almost independent of the value of ϕ. It should be noted, however, that this is true only for very stiff sockets with non-dilating interfaces. As the modulus ratio E_c/E_r becomes large and the slenderness ratio D/B becomes small, the curves for cohesive-frictional and purely cohesive interfaces are practically the same.

The analytical solutions for the load–settlement response of socketed shaft foundations under compression loading assume a cohesive–frictional shear strength at the socket walls, with the possibility of dilatancy contributing to the shear resistance. The analysis can only be applied where the strength parameters and the dilatancy are constant and independent of the magnitude of the relative shear displacement at the interface. These assumptions are reasonable for small slip displacements, but they tend to be unrealistic as larger slip displacements occur. In particular, it is obvious that the true cohesive component of the strength is diminished after the bond between the concrete and rock is first ruptured. Similarly, it is likely that the tendency for the interface to dilate with shearing will abate as the shearing progresses and the interface asperities are worn down. With the present analysis, it is not possible to model this reduction of cohesive strength and dilatancy exactly. However, it is possible to use the analysis to produce bounds on the likely load–settlement response, as illustrated schematically in Figure 25.12. The analysis assuming $c = 0$ and $\psi = 0°$ (see Kulhawy and Goodman, 1987 for precise details) should provide a lower bound on the mobilized socket strength and an upper bound on the shaft displacements. Conversely, with the selection of appropriate (non-zero) values of c, ϕ, and ψ, the corresponding load–displacement curve should provide an upper bound on mobilized strength and a lower bound on the settlements. The actual load–settlement

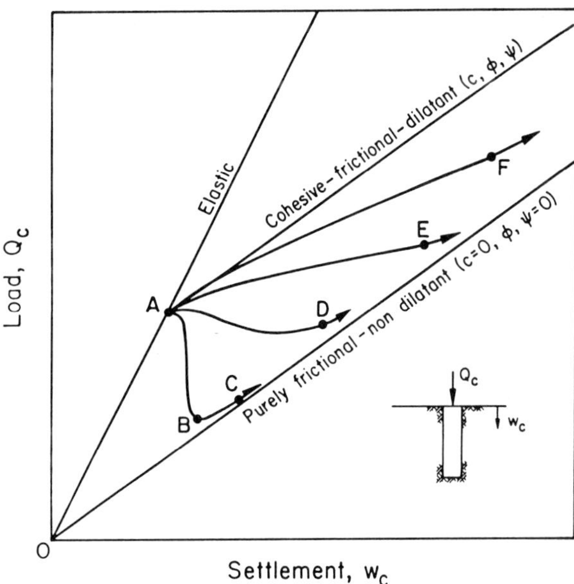

Figure 25.12 *Schematic representation of the behaviour of socketed shafts in compression*

curve is likely to fall between these limits; just where it falls depends on the brittleness and the dilatancy of the interface. In an extreme case, if the loss of cohesive strength is rapid, then a curve like OABC could be observed. If the interface exhibits a plastic or work-hardening shear response (which is more likely in well-constructed, rough sockets that ensure some dilation), then the shaft behaviour may be more like that represented by the curve OAF. At larger displacements, the sockets should tend toward a purely frictional, non-dilatant behaviour.

When the shaft is subjected to uplift loads, the tensile tip resistance normally is ignored. The behaviour in this case is illustrated in Figure 25.13 for a relatively rigid shaft with $D/B = 5$, $\nu_r = 0.25$, and interface friction angle $\phi = 30°$. It can be seen that the load-displacement response is initially linear elastic. Full slip of the shaft occurs when the shear stresses along the interface reach a magnitude of c and hence $Q_u = -\pi BDc$. A negative value of Q_u denotes tensile force and a negative value of w_u denotes an upward displacement. Under uplift loading, tensile normal stresses are induced at the shaft–rock interface from Poisson's ratio effects. The initial normal stresses on the interface are likely to be negligible for shallow sockets. Therefore the load to commence full slip will effectively be independent of the interface friction angle.

Once slip has commenced, the stiffness of the shaft is a function of the interface dilation angle ψ, as illustrated for typical values in Figure 25.13. According to this simple model, the stiffness corresponding to full slip is constant and this follows from the assumption of constant c, ϕ, and ψ. The slip response is stiffer for larger values of ψ. In reality, the dilation must cease after sufficient slip displacement has occurred, and ultimately a maximum uplift load will be reached. The value of this ultimate load is given by $Q_u = -\pi BD\tau_{max}$, where τ_{max} is the peak unit side shear resistance described in Chapter 12. In Figure 25.13, the case where $\tau_{max} = 2c$ has been considered.

It is clear from the above discussion that guidelines for the selection of preliminary design values of the strength and deformation parameters would be useful, particularly if full use is to be made of the simple and convenient expressions for the load–displacement behaviour. Guidelines such as these are presented below.

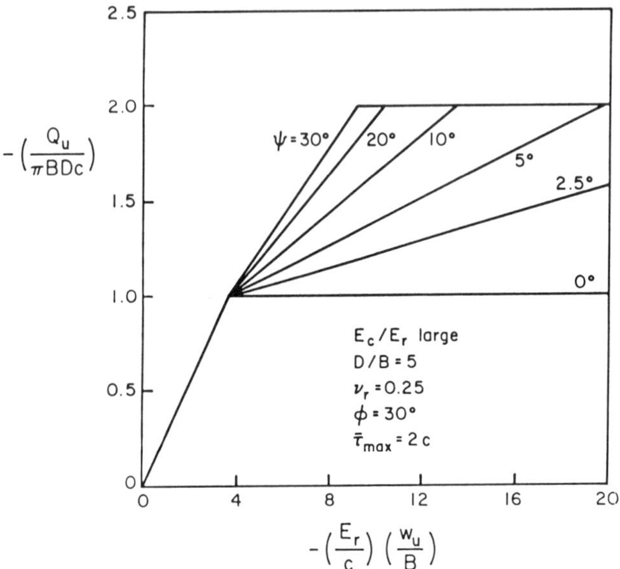

Figure 25.13 *Predicted uplift behaviour of a relatively rigid shaft in a cohesive–frictional socket*

25.2.8 Interpretation of field tests

Carter and Kulhawy (1988) examined a number of reported field tests on rock-socketed shafts and interpreted the data in terms of the simple analytical model presented in previous sections. The field tests included both compression and uplift tests on 'complete' and 'shear' sockets. The interpretation is illustrated in Figures 25.14 and 25.15. In each case, at least a bilinear fit was made to

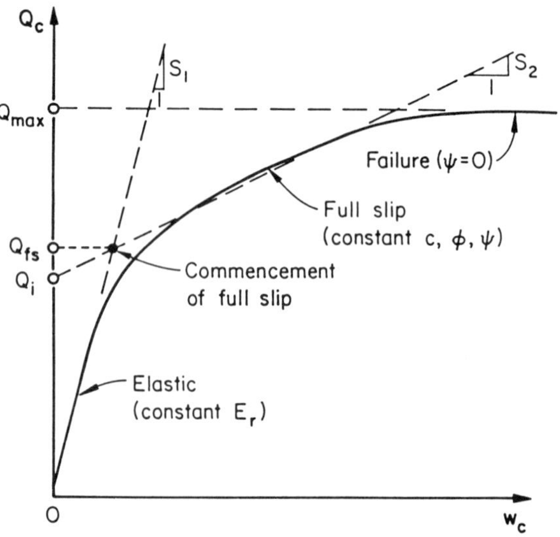

Figure 25.14 *Interpretation of a field test in a shear-only rock socket*

the observed behaviour. In some cases where 'shear' sockets were tested, a trilinear fit was possible, with the third horizontal portion of the curve corresponding to the ultimate condition of full slip at constant shear stress τ_{max} and zero dilation ($\psi = 0$). The slopes of the elastic and 'full slip' portions of each measured load–deflection curve (S_1, S_2, S_3) were determined, as shown in Figures 25.14 and 25.15, and these then were used to calculate the

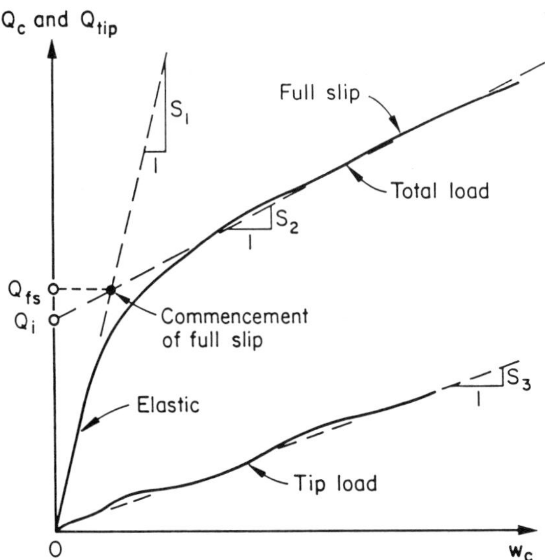

Figure 25.15 *Interpretation of a field test in complete rock socket*

appropriate model parameters. In each case examined, it was assumed that the shaft was effectively rigid. The theoretical expressions relating the model parameters to the slopes of the various portions of the load–deflection curve for rigid shafts are summarized in Table 25.1.

In many of the test loadings, values of the rock uniaxial compressive strength (q_u) were reported. These data allowed correlations between c, ($\tan \phi \tan \psi$), τ_{max}, and q_u to be made, as shown in Figures 25.16–25.18. In all figures, the stresses have been normalized by the atmospheric pressure, p_a. Figure 25.16 shows a positive correlation between c and q_u, with all data points above or on the curve given by

$$\frac{c}{p_a} = 0.1 \left(\frac{q_u}{p_a}\right)^{2/3} \tag{25.48}$$

Figure 25.17 also shows a positive correlation between ($\tan \phi \tan \psi$) and q_u, although intuitively this may have been less obvious. A positive correlation between ($\tan \phi \tan \psi$) and the interface roughness would be expected. However, a reasonable lower bound to almost all points in Figure 25.17 is given by

$$\tan \phi \tan \psi = 0.001 \left(\frac{q_u}{p_a}\right)^{2/3} \tag{25.49}$$

The fact that the same exponent appears in both Equations (25.48) and (25.49) may only be fortuitous, and the confident application of these equations in anything but preliminary design must await further corroborating evidence.

Table 25.1 Interpretation of axial load tests on rigid shafts in rock

(a) Shear-only socket in compression or uplift

$$E_r = \left(\frac{(1 + \nu_r)\zeta}{\pi D}\right) S_1$$

$$\tan \phi \tan \psi = \left(\frac{1}{2\zeta}\right) \left(\frac{S_2}{S_1 - S_2}\right)$$

$$c = (2\zeta \tan \phi \tan \psi + 1) \left(\frac{Q_i}{\pi BD}\right)$$

(b) Complete socket in compression

$$E_r = \left(\frac{(1 + \nu_r)\zeta}{\pi D}\right) (S_1 - S_3)$$

$$E_b = \frac{(1 - \nu_b^2)}{B} S_3$$

$$\tan \phi \tan \psi = \left(\frac{1}{2\zeta}\right) \left(\frac{S_2 - S_3}{S_1 - S_2}\right)$$

$$c = (2\zeta \tan \phi \tan \psi + 1) \left(\frac{Q_i}{\pi BD}\right)$$

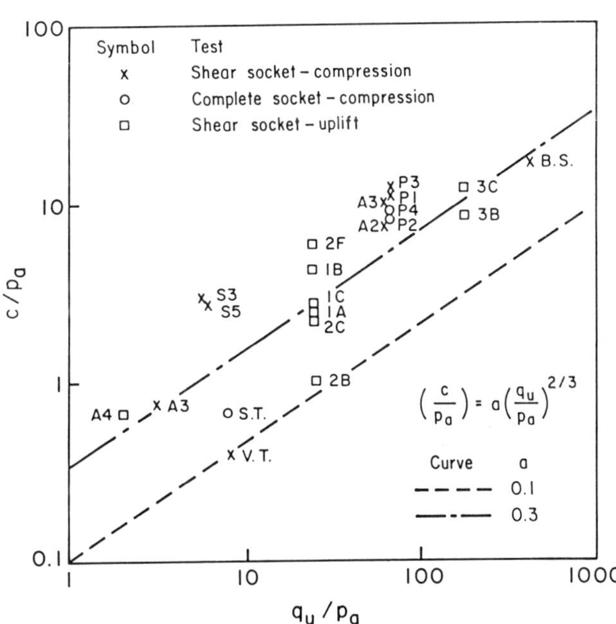

Figure 25.16 *Interface cohesion versus uniaxial compressive strength of intact rock*

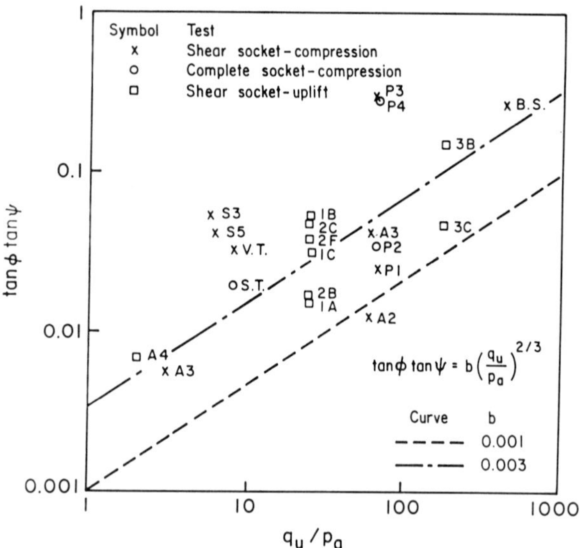

Figure 25.17 *Interface friction–dilation parameters versus uniaxial compressive strength of intact rock*

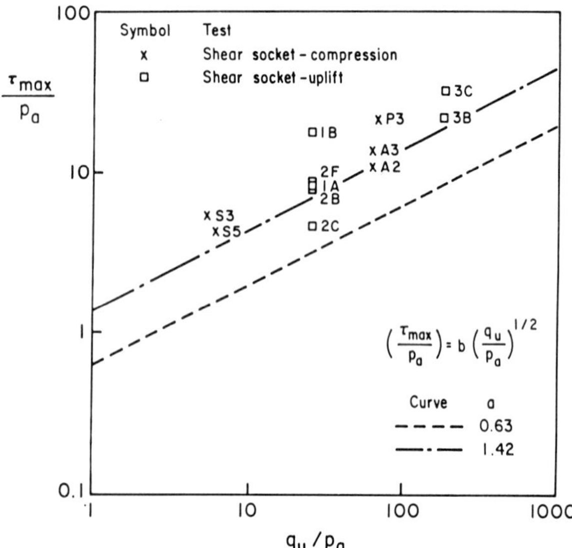

Figure 25.18 *Peak value of average side shear stress versus uniaxial compressive strength of intact rock*

From the cases studied in the literature only a limited set of values are available for τ_{max}. These have been plotted in Figure 25.18. The curve corresponding to

$$\frac{\tau_{max}}{p_a} = 0.63 \left(\frac{q_u}{p_a}\right)^{2/3} \tag{25.50}$$

has been plotted on Figure 25.18, and it is found to be a lower bound to all the data points. The equation is consistent with prior data presented in Chapter 12.

For most of the examined cases of axial loading, it was found that the assumption of rigid shaft behaviour was reasonable. This is no doubt because of the stubby geometry of the test shafts, but also may result from the fact that the data were obtained for shafts in relatively soft sedimentary rock masses. The assumption of rigidity may be less acceptable for deeper shafts in harder rocks where the modulus values for the rock mass and the shaft material are closer.

25.3 Lateral loading

Drilled shaft foundations are often employed to transmit lateral (horizontal) forces and overturning moments to the ground. As in most design, an adequate margin of safety against collapse must be ensured, and the displacements resulting from this form of loading must be tolerable. In this section, some of the existing methods for predicting the lateral displacements of embedded foundations will be reviewed briefly, and some new and simple closed-form solutions for the response of rock-socketed drilled shafts will be presented.

23.3.1 Recent methods

There appear to be no published methods of analysis specifically for use in predicting the response of rock-socketed shafts subjected to lateral loading. In recent years, theoretical approaches for predicting the lateral displacements of slender piles in soil have been developed extensively. The most recent and significant development in the analysis of laterally loaded piles was made by modelling the soil as an elastic continuum and the pile as an elastic beam. Numerical solutions were developed, first with the use of the integral equation (or boundary element) method (Poulos 1971a, b, 1972; Banerjee and Davies 1978) and second with the use of the finite element method (Randolph 1977, 1981; Kuhlemeyer 1979a, b). Most of these elastic solutions were presented in the form of charts. Recently, however, convenient, closed-form expressions for the response of flexible piles to lateral loading were given by Randolph (1987). The description 'flexible' was applied to piles in which the induced displacements and bending moments are confined to the upper part of the pile and in which the overall length of the pile does not significantly affect its response. The expressions are only approximate, but they provide a close fit to the more rigorous solutions obtained by the finite element method.

The designer of laterally loaded piles in soil now has available solutions for the pile response that are very simple to use. Unfortunately for the designer of laterally loaded, rock-socketed shafts, these solutions do not cover all cases occurring in practice. In particular, many of the theoretical solutions are not applicable to the case of a relatively short, stubby, rigid foundation embedded in rock. In the following, new solutions are presented to cover these important cases.

25.3.2 Problem idealization

Figure 25.19 represents cases where either rock is at the ground surface or the lateral loading on the shaft at the level of the rock surface can be specified completely. The

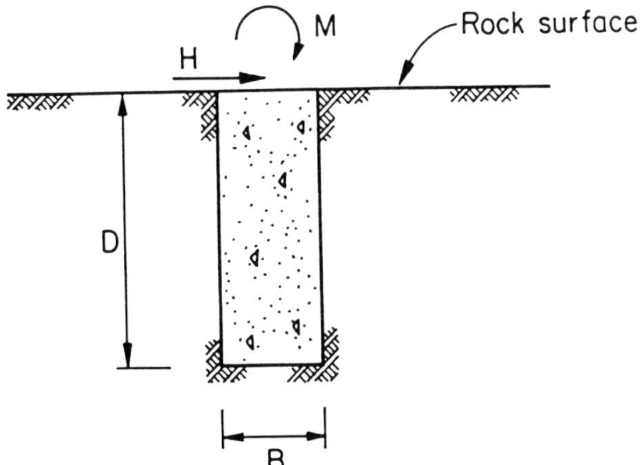

Figure 25.19 *Lateral loading of a rock-socketed shaft*

shaft is idealized as a cylindrical elastic inclusion, with an effective Young's modulus E_e and Poisson's ratio ν_c, depth D, and diameter B. For a solid shaft, having an actual bending rigidity $(EI)_c$, the effective Young's modulus is given by

$$E_e = (EI)_c/(\pi B^4/64) \tag{25.51}$$

The elastic shaft is embedded in a homogeneous, isotropic elastic rock mass, with properties E_r and ν_r and, at the surface of the rock mass, it is subjected to a known lateral (horizontal) force H and moment M. The more general case of lateral loading of a shaft socketed into a rock mass, with an overlying soil layer, is relatively straightforward (Carter and Kulhawy 1988), but it is not considered here for brevity.

23.3.3 Analysis of lateral loading

In the following, some simple closed-form expressions are presented for both relatively flexible and relatively rigid shafts subjected to lateral loading. These equations have been developed from a finite element study of the behaviour of axisymmetric bodies subjected to non-symmetric loading. The technique has been described by

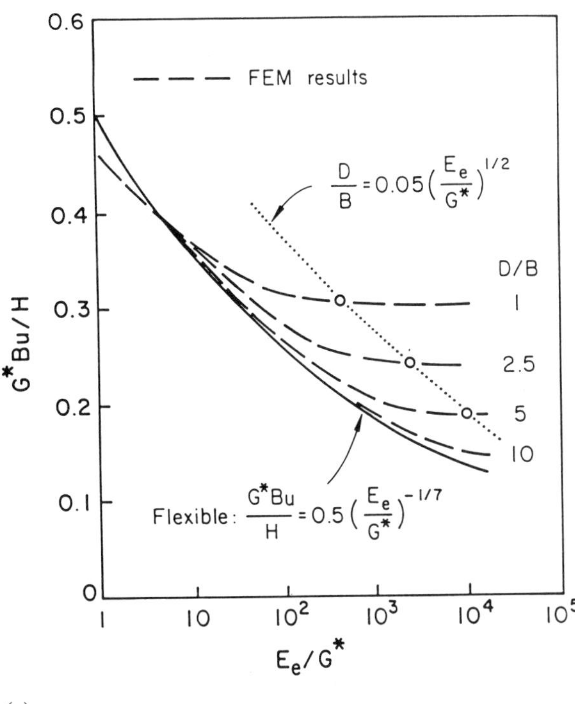

(a)

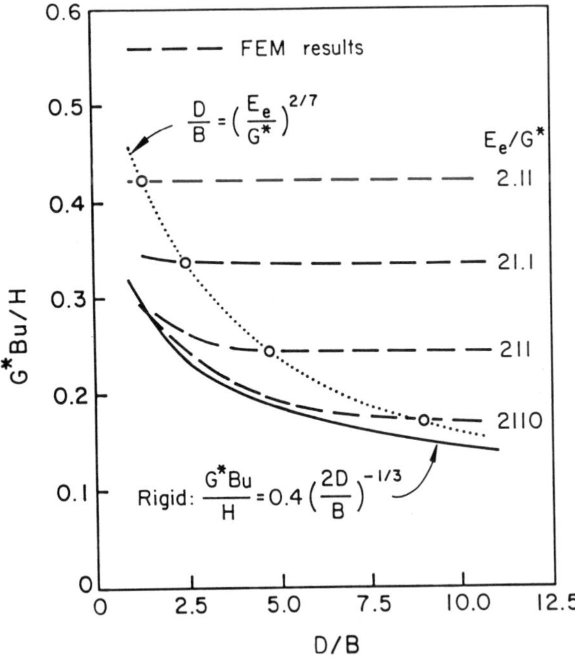

(b)

Figure 25.20 a and b *Lateral load–displacement relations*

Wilson (1965) and is similar to that used for the study of the lateral loading of 'flexible' piles (Randolph 1977) and for the study of the effects of surface loadings on pile behaviour (Carter 1982).

An extensive parametric study was performed for socketed shafts covering a large range of relative stiffnesses, and it was found that the effects of variations in Poisson's ratio of the rock mass ν_r could be represented approximately by considering an equivalent shear modulus of the rock mass G^*, defined by Randolph (1981) as

$$G^* = G_r(1 + 3\nu_r/4) \qquad (25.52)$$

in which G_r is the shear modulus of the elastic rock mass (Equation (25.5)).

For a homogeneous rock mass, it was found that the horizontal displacement u and rotation θ of the shaft at the rock surface depend on the relative moduli of the shaft and rock mass E_e/G^* and on the geometry of the shaft D/B. The results are presented in Figures 25.20–25.22. Figure 25.20 shows the lateral load–displacement relationships; Figure 25.21 shows the relationship between moment and rotation; and Figure 25.22 indicates the load–rotation and moment–displacement relations. Two types of plot of the same data are provided in each figure, showing the dimensionless displacements plotted against the modulus ratio E_e/G^* and slenderness ratio D/B. The finite element results have been plotted as the dashed curves on all figures.

Randolph (1981) suggested that a shaft would behave as if it were infinitely long when

$$(D/B) \geqslant (E_e/G^*)^{2/7} \qquad (25.53)$$

For these cases, the shaft response is dependent only on the modulus ratio E_e/G^* and Poisson's ratio of the rock mass ν_r. Dotted curves corresponding to the equality condition in expression (25.53) have been plotted on Figures 25.20(b), 25.21(b) and 25.22(b), from which it can be seen that the finite element predictions are independent of D/B whenever condition (25.53) holds. Such a shaft is 'flexible', and the following closed-form expressions provide accurate approximations for the deformations (Randolph 1981):

$$u = 0.5\frac{H}{G^*B}\left(\frac{E_e}{G^*}\right)^{-1/7} + 1.08\frac{M}{G^*B^2}\left(\frac{E_e}{G^*}\right)^{-3/7} \qquad (25.54)$$

$$\theta = 1.08\frac{H}{G^*B^2}\left(\frac{E_e}{G^*}\right)^{-3/7} + 6.4\frac{M}{G^*B^3}\left(\frac{E_e}{G^*}\right)^{-5/7} \qquad (25.55)$$

The appropriate forms of these equations have been plotted on Figures 25.20(a), 25.21(a) and 25.22(a). Equations (25.54) and (25.55) provide adequate predictions of the behaviour of 'flexible' shafts socketed into

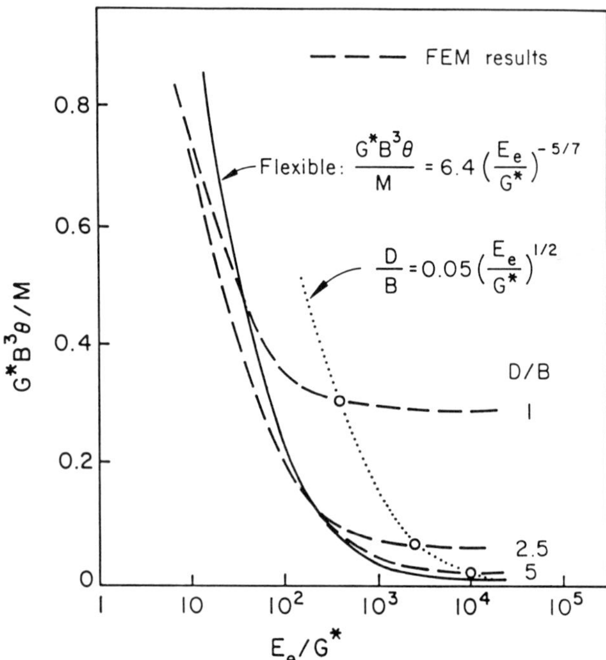

(a)

(b)

Figure 25.21 a and b *Moment-rotation relations*

elastic rock masses. Randolph (1981) verified their accuracy for the following ranges of parameters: $10^2 \leq E_e/E_r \leq 10^6$ and $D/B \geq 10$. Carter and Kulhawy (1988) verified that the range of applicability can be extended to $1 \leq E_e/E_r \leq 10^6$ and $D/B \geq 1$.

There are cases encountered in practice, particularly when short stubby shafts are socketed into weaker rock, where the shafts behave as rigid structural members. In these cases, the displacements of the shaft are independent of the modulus ratio E_e/E_r and depend only on the slenderness ratio D/B and Poisson's ratio of the rock mass ν_r.

The dotted curves in Figures 25.20(a), 25.21(a) and 25.22(a) indicate that a shaft will behave as a rigid member when

$$(D/B) \leq 0.05(E_e/G^*)^{1/2} \tag{25.56}$$

It has been shown (Carter and Kulhawy 1988) that the displacements of these rigid shafts can be expressed, to sufficient accuracy, by the simple equations below:

$$u = 0.4\frac{H}{G^*B}\left(\frac{2D}{B}\right)^{-1/3} + 0.03\frac{M}{G^*B^2}\left(\frac{2D}{B}\right)^{-7/8} \tag{25.57}$$

$$\theta = 0.3\frac{H}{G^*B^2}\left(\frac{2D}{B}\right)^{-7/8} + 0.8\frac{M}{G^*B^3}\left(\frac{2D}{B}\right)^{-5/3} \tag{25.58}$$

Appropriate forms of these equations have been plotted as solid curves on Figures 25.20(b), 25.21(b), and 25.22(b), where satisfactory agreement with the finite element solutions can be seen. Because the shaft displaces as a rigid body in the elastic rock mass, the depth beneath the surface to its centre of rotation can be computed as

$$z_c = \frac{0.4\left(\frac{2D}{B}\right)^{-1/3} + 0.3\left(\frac{e}{B}\right)\left(\frac{2D}{B}\right)^{-7/8}}{0.3\left(\frac{2D}{B}\right)^{-7/8} + 0.8\left(\frac{e}{B}\right)\left(\frac{2D}{B}\right)^{-5/3}} \tag{25.59}$$

in which $e = M/H$ is the vertical eccentricity of the applied horizontal force H. When applying Equations (25.56)–(25.59), it should be noted that their accuracy has been verified only for the following ranges of parameters: $1 \leq D/B \leq 10$ and $E_e/E_r \geq 1$.

Traditionally, the influence factors for laterally loaded piles and shafts have been presented in numerous charts. The approximate equations presented above are more attractive for design because of their succinctness.

Shafts can be described as having 'intermediate' stiffness whenever the slenderness ratio is bounded approximately as follows:

$$0.05(E_e/G^*)^{1/2} < (D/B) < (E_e/G^*)^{2/7} \tag{25.60}$$

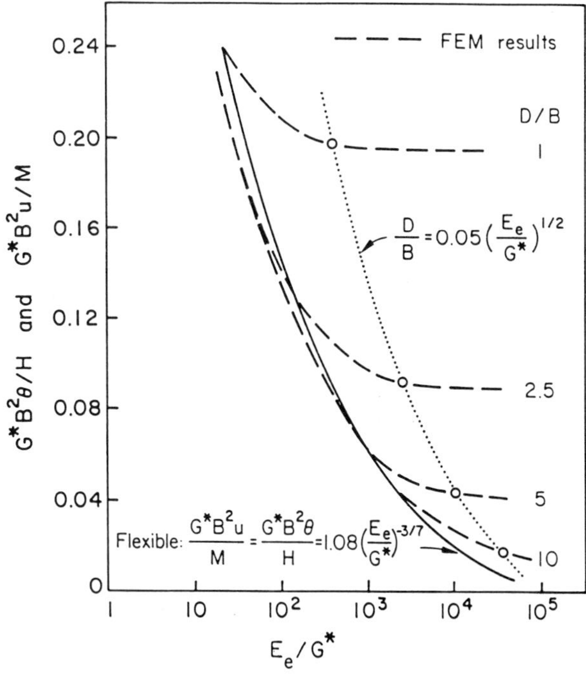

(a)

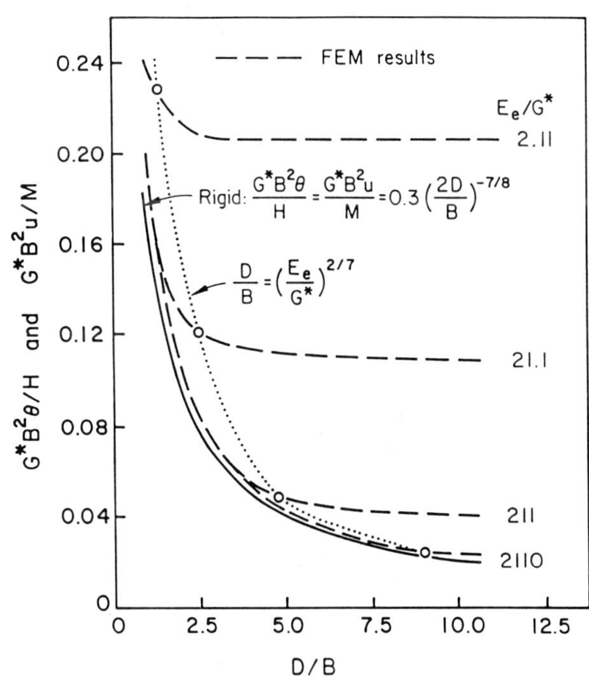

(b)

Figure 25.22 a and b *Lateral load–rotation and moment-displacement relations*

Figures 25.20–25.22 show that, in these cases, the finite element predictions are almost always larger than the predictions from Equations (25.54) and (25.55) for flexible shafts and Equations (25.54) and (25.58) for rigid shafts. Typically, displacements for an intermediate case exceed the maximum of the predictions for corresponding rigid and flexible shafts by no more than about 25%, and often by much less. For the sake of simplicity, without the sacrifice of much accuracy, it is suggested that the displacements in the intermediate case be taken as 1.25 times the maximum of either (a) the predicted displacement of a rigid shaft with the same slenderness ratio D/B as the actual shaft, or (b) the predicted displacement of a flexible shaft with the same modulus ratio E_e/G^* as the actual shaft. Values calculated in this manner should, in most cases, be slightly larger than those given by the more rigorous finite element analysis for a shaft of intermediate stiffness.

25.3.4 Range of application

In practice, shafts that are socketed into rock normally have a slenderness ratio falling typically within the range 1–10. Furthermore, the modulus ratio E_e/E_r generally lies in the range of 1–10 000. The combinations of modulus and slenderness ratios that produce flexible, intermediate, and rigid shafts are shown in Figure 25.23 for $\nu_r = 0.25$. It can be seen that each case covers a significant portion of those likely to occur in practice.

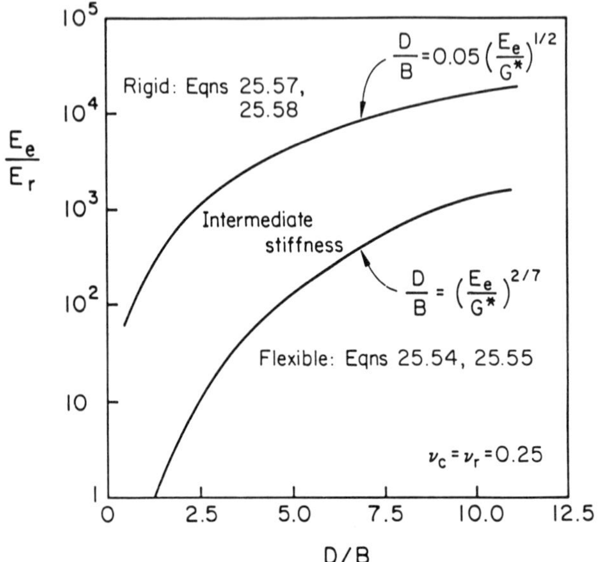

Figure 25.23 *Ranges of application of closed-form expressions for lateral deformations*

25.4 Design example

In this section, some of the analytical techniques and the empirical data presented earlier will be used to illustrate how the preliminary design of a drilled shaft foundation for one leg of a lattice tower transmission structure may be conducted. It is important to realize that the calculations presented below illustrate only the preliminary stages of the design of a shaft in rock, and in no way do they replace the need for load testing or other means of validating the design. In general, it is good practice to test load some foundation elements at any new site until sufficient experience with the natural materials and confidence in the design have been gained.

25.4.1 Geotechnical data and design criteria

The design loadings for this case are shown in Figure 25.24. The foundation to be used is a drilled shaft in rock, and it is assumed that the rock outcrops at the surface. The rock mass is described as being of good quality, with tightly interlocking, undisturbed rock blocks and rough unweathered joints spaced at about 1–3 m. It is to be assumed that Young's modulus of the rock mass is $E_r = 100\ \mathrm{MN/m^2}$, and Poisson's ratio, $\nu_r = 0.25$, and that the uniaxial compressive strength of the intact rock is $q_u = 5\ \mathrm{MN/m^2}$, indicating that it is quite a soft rock material.

The foundation is to be designed according to the following specifications. It must have an adequate margin of safety against failure under each mode of loading: axial compression, axial uplift, and lateral loading. An overall factor of safety of at least 2.5 is required for all modes of loading. Furthermore, both the axial and lateral displace-

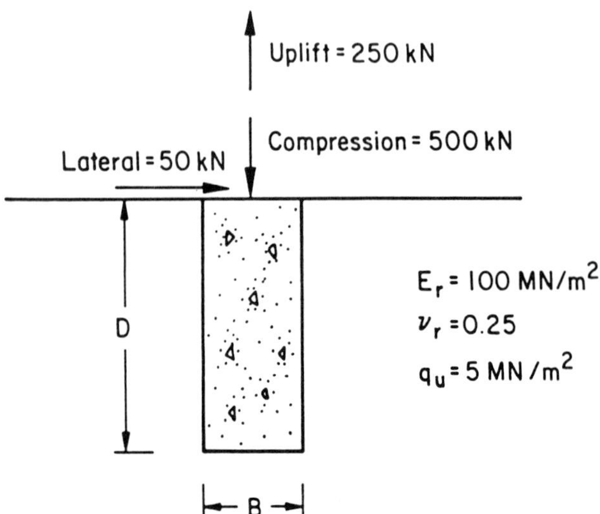

Figure 25.24 *Loading and rock mass properties for a socketed shaft foundation in rock*

ments under the respective working loads must not exceed 20 mm and the groundline rotation under the lateral force must not exceed 1°.

25.4.2 Design procedure for axial loading

The aim of the design exercise is to determine suitable shaft dimensions to meet all the design criteria. For the given loadings, it is likely that axial compression will govern the design, and a suitable starting point for the design process involves an examination of the ultimate condition. In the following, it will be assumed that resistance to axial loading is provided by only the side shear. Tip resistance is specifically ignored, even for compression loading.

Knowledge of the rock strength, together with the preliminary design Equation 25.50, allows a conservative estimate to be made of the peak unit side shear resistance

$$\tau_{max} \approx 0.63 \times \left(\frac{5000}{100}\right)^{1/2} \times 100 = 445\ \text{kN/m}^2 \quad (25.61)$$

in which atmospheric pressure is assumed to be $p_a = 100\ \text{kN/m}^2$.

A factor of safety of 2.5 implies that the maximum axial load to be considered is $Q_{max} = 1250\ \text{kN}$. The maximum axial load is related to the unit side shear resistance by

$$Q_{max} = \pi B D \tau_{max} \quad (25.62)$$

The selection of the shaft diameter can be made on the basis of construction considerations. For the present case, it is assumed that $B = 0.5$ m, and therefore, from Equation (25.62),

$$D = 1250/(\pi \times 0.5 \times 445) = 1.79\ \text{m} \quad (25.63)$$

The design value of D is selected to be 1.8 m, in which case the factor of safety for axial compression is slightly larger than 2.5.

Before predicting the load–displacement response of a shaft with $B = 0.5$ m and $D = 1.8$ m, the relative rigidity of the shaft is considered. For the present case,

$$\left(\frac{E_c}{E_r}\right)\left(\frac{B}{2D}\right)^2 = \left(\frac{35\,000}{100}\right) \times \left(\frac{0.5}{2 \times 1.8}\right)^2$$
$$= 6.75 > 1 \quad (25.64)$$

in which the modulus of the concrete shaft has been assumed as $E_c = 35\ \text{GN/m}^2$. Because the relative rigidity is greater than 1, it is reasonable to assume the shaft behaves rigidly under axial loading.

Consider first the linear elastic response of this rigid shaft in a 'shear only' socket. The prediction of its elastic axial stiffness is given below.

From Equation (25.3), the parameter ζ is given by

$$\zeta = \ln[5(1 - \nu_r)D/B] = \ln[5 \times 0.75 \times 1.8/0.5]$$
$$= 2.603 \quad (25.65)$$

From Equation (25.34), the elastic stiffness is

$$\frac{Q_c}{w_c} = \left(\frac{\pi}{(1 + \nu_r)\zeta}\right) E_r D$$
$$= \left(\frac{\pi}{1.25 \times 2.603}\right) \times 100 \times 1.8$$
$$= 174\ \text{MN/m} \quad (25.66)$$

To determine the response of the shaft once it slips, values of the parameters c and $(\tan\phi \tan\psi)$ are required. From Equation (25.48), a preliminary design value of c is given by

$$c \approx 0.1 \times \left(\frac{5000}{100}\right)^{2/3} \times 100 = 135\ \text{kN/m}^2 \quad (25.67)$$

Equation 25.49 gives a value of $(\tan\phi \tan\psi)$ suitable for use in preliminary design,

$$\tan\phi \tan\psi \approx 0.001 \left(\frac{5000}{100}\right)^{2/3} = 0.0136 \quad (25.68)$$

The load–displacement relationship for full slip is given in Equation (25.36). For the present case,

$$R_1 = 1.25 \times \left(2.603 + \frac{1}{2 \times 0.0136}\right) \times \left(\frac{0.5}{1.8}\right)$$
$$= 13.67 \quad (25.69)$$

$$R_2 = \left(\frac{1.25}{2 \times 0.0136}\right) \times \left(\frac{135}{100\,000}\right)$$
$$= 0.062 \quad (25.70)$$

so that

$$w_c = \left(\frac{13.67}{\pi \times 100\,000 \times 0.5}\right) Q_c - 0.062 \times 0.5$$

that is,

$$w_c = 0.000087 Q_c - 0.031\ (m \text{ with } Q_c \text{ in kN}) \quad (25.71)$$

The predicted overall axial load–displacement behaviour therefore can be represented as in Figure 25.25, which shows an initial linear elastic portion, a linear slip portion, and ultimate failure. Under the design compressive load of 500 kN, the predicted displacement is 12.5 mm, which lies within the slip region and also meets the design specification.

It will be appreciated that, in the design calculations presented above, the predicted slip displacements were based on what are likely to be lower-bound estimates of c and ($\tan \phi \tan \psi$). Because this is a conservative approach, it is likely to overpredict the actual displacements at any given load.

The same general axial response can be expected under uplift loading, provided the same values of E_r, τ_{max}, c, and ($\tan \phi \tan \psi$) can be relied upon. If this is reasonable, then the uplift displacement under an uplift load of 250 kN can be expected to be about 1.4 mm, which lies within the linear elastic region (Figure 15.25). It will not always be reasonable to adopt the same values for the model parameters when examining both compression and uplift behaviour. For example, adverse joint orientations may necessitate the reduction of the design values of any or all of E_r, τ_{max}, c, and ($\tan \phi \tan \psi$) for the uplift case. In the present case, it also would be advisable to examine the possibility of wedge failure during uplift (Pease and Kulhawy 1984) as the aspect ratio $D/B = 3.6$ is relatively low, but this is not pursued here.

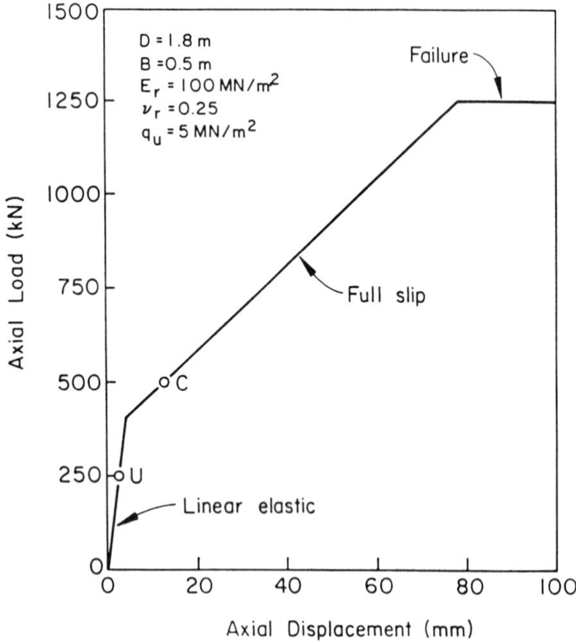

Figure 25.25 *Predicted axial behaviour of a socketed shaft foundation in rock*

25.4.3 Design procedure for lateral loading

The behaviour under lateral loading is now considered. Initially proceeding under the assumption that the shaft is effectively rigid with respect to bending, Equations (25.57) and (25.58) may be used to compute the displacements. The effective shear modulus of the rock is $G^* = G_r(1 + 3\nu_r/4)$ and, since $G_r = E_r/[2(1 + \nu_r)]$, their values are calculated as

$$G_r = \frac{100}{2 \times 1.25} = 40 \text{ MN/m}^2 \tag{25.72}$$

and

$$G^* = 40 \times (1 + 3 \times 0.25/4) = 47.5 \text{ MN/m}^2 \tag{25.73}$$

Therefore the displacement and rotation at the groundline caused by a horizontal force of 50 kN are

$$u = 0.4 \times \left(\frac{50}{47\,500 \times 0.5}\right) \times \left(\frac{3.6}{0.5}\right)^{-1/3} \times 1000 = 4 \text{ mm} \tag{25.74}$$

and

$$\theta = 0.3 \times \left(\frac{50}{47\,500 \times 0.5 \times 0.5}\right) \times \left(\frac{3.6}{0.5}\right)^{-7/8} \times \frac{180}{\pi} = 0.013° \tag{25.75}$$

For the shaft to be effectively rigid in bending, the quantity $(E_c/G^*)(B/2D)^2$ should be larger than 100 (see Equation 25.56). In the present case it is only 14.25, and so the shaft is of intermediate flexibility. As suggested above, the displacement and rotation for shafts of this type should be increased by about 25%. Even the increased values still meet the design specification.

In the calculations presented above, the same value of rock mass modulus E_r was used to calculate both the axial and lateral displacements. This may not always be justified, however, and the designer needs to determine that this assumption is reasonable for each particular site. Axial displacement of the shaft essentially involves shearing of the rock mass on vertical, cylindrical surfaces, while under lateral loading, the deformation pattern in the rock mass is more complex. Therefore, in the application of the linear theories to predict axial and lateral displacements of a shaft, it may be more reasonable on occasion to adopt different values of E_r for the axial and lateral modes.

It is also necessary to check that adequate capacity can be provided by the rock mass to resist the lateral loading, as suggested earlier. The shaft considered here has a ratio of $D/B = 3.6$, and it is likely that plane strain conditions will be approximated at a depth of about 3 diameters. This could be included in the determination of the lateral capacity; however, it will be simpler and more conservative to assume that the limiting normal stress on the projected area of the face of the shaft is equal to the uniaxial compressive strength of the rock mass, or

$$p_{max} = s^{1/2} q_u \tag{25.76}$$

in which s is the parameter used in the Hoek and Brown (1980) failure criterion for jointed rock masses. Based on the description of the rock mass given previously, the recommendations made by Hoek (1983) for the strength of jointed rock masses indicate a representative value of $s = 0.1$, and therefore

$$p_{max} = (0.1)^{1/2} \times 5 = 1.58\ \text{MN/m}^2 \tag{25.77}$$

The ultimate lateral force H_{max} can be estimated conservatively (given that values of s and q_u are representative) by

$$H_{max} = p_{max} BD = 1580 \times 0.5 \times 1.8$$
$$= 1420\ \text{kN} \tag{25.78}$$

The factor of safety against failure under lateral loading then can be calculated as

$$F = H_{max}/H = 1420/50 = 28.4 \tag{25.79}$$

which more than adequately meets the design specification.

It should be noted that, for lower-quality rock masses, the values of s suggested by Hoek (1983) decrease rapidly and, for such cases, it could be necessary to increase the shaft dimensions to achieve an adequate margin of safety for the lateral loading.

Acknowledgments

This chapter draws liberally on previous work by the authors (Carter and Kulhawy, 1988). Appreciation is expressed to the Electric Power Research Institute (USA) which supported this previous work. The text was typed by L. Crouse and R. Venkatachalam, and A. Avcisoy drafted the figures.

References

Banerjee, P. K. and Davies, T. G. (1978) 'The behaviour of axially and laterally loaded single piles embedded in non-homogeneous soils', *Geotechnique,* **28**, 309–326

Carter, J. P. (1982) 'A numerical method for pile deformations due to nearby surface loadings', *Proc. 4th Int. Conf. on Numerical Methods in Geomech.,* Edmonton, Alberta, **2**, 811–817

Carter, J. P. and Kulhawy, F. H. (1988) 'Analysis and design of drilled shaft foundations socketed into rock', *Report EL-5918*, Electric Power Research Institute, Palo Alto, California, 188 pp.

Davis, E. H. (1968) 'Theories of plasticity and the failure of soil masses'. Chapter 6 in *Soil Mechanics: Selected Topics*, I. K. Lee (ed.), Butterworths, London, pp. 341–380

Goodman, R. E. (1980) *Introduction to Rock Mechanics*, Wiley, New York

Hoek, E. (1983) 'Strength of jointed rock masses', *Geotechnique,* **33**, 187–223

Hoek, E. and Brown, E. T. (1980) *Underground Excavations in Rock*, Institution of Mining and Metallurgy, London

Horvath, R. G. (1980) 'Load transfer system for rock-socketed drilled pier foundations', *Research Project Report for the National Research Council of Canada*, DSS File No. 10X5.31155-9-4420, Contract Serial No. 1SX79.000531

Kuhlemeyer, R. L. (1979a) 'Static and dynamic laterally loaded floating piles', *J. Geotech. Eng. Div.,* American Society of Civil Engineers, **105**, 289–304

Kuhlemeyer, R. L. (1979b) 'Bending element for circular beams and piles', *J. Geotech. Eng. Div.*, American Society of Civil Engineers, **105**, 325–330

Kulhawy, F. H. and Goodman, R. E., (1987) 'Foundations in rock'. Chapter 55 in *Ground Engineers Reference Book*, F. G. Bell (ed.), Butterworths, London

Pease, K. A. and Kulhawy, F. H. (1984) 'Load transfer mechanisms in rock sockets and anchors', *Report EL-3777*, Electric Power Research Institute, Palo Alto, California

Pells, P. J. N. and Turner, R. M. (1979) 'Elastic solutions for the design and analysis of rock-socketed piles', *Candia. Geotech. J.*, **16**, 481–487

Poulos, H. G. (1971a) 'Behavior of laterally loaded piles: I – single piles'. *J. Soil Mechanics and Foundations Div.*, American Society of Civil Engineers, **97**, 711–731

Poulos, H. G. (1971b) 'Behavior of laterally loaded piles: II – pile groups'. *J. Soil Mechanics and Foundations Div.*, American Society of Civil Engineers, **97**, 733–751

Poulos, H. G. (1972) 'Behaviour of laterally loaded piles: III – socketed piles'. *J. Soil Mechanics and Foundations Div.*, American Society of Civil Engineers, **98**, 341–360

Poulos, H. G. and Davis, E. H. (1974) *Elastic Solutions for Soil and Rock Mechanics*, Wiley, New York

Randolph, M. F. (1977) *A Theoretical Study of the Performance of Piles*, Ph.D. Thesis, University of Cambridge, UK

Randolph, M. F. (1981) 'The response of flexible piles to lateral loading', *Geotechnique,* **31**, 247–259

Randolph, M. F. and Wroth, C. P. (1978) 'Analysis of deformation of vertically loaded piles', *J. Geotech. Eng. Div.*, American Society of Civil Engineers, **104**, 1465–1488

Rowe, R. K. and Armitage, H. H. (1987a) 'Theoretical solutions for axial deformation of drilled shafts in rock', *Can. Geotech. J.*, **24**, 114–125

Rowe, R. K. and Armitage, H. H. (1987b) 'A design method for drilled piers in soft rock', *Can. Geotech. J.*, **24**, 126–142

Williams, A. F. (1980) *The Design and Performance of Piles Socketed into Weak Rock*, Ph.D. Thesis, Monash University, Melbourne, Australia

Wilson, E. L. (1965) 'Structural analysis of axisymmetric solids', *J. Amer. Inst. Aeronautics and Astronautics,* **3**, 2269–2274

26 Retaining structures for rock masses

Professor H Brandl
Technische Universität Wien, Austria

26.1 Introduction: general aspects

This contribution focuses on various types of retaining structures and accompanying stabilizing measures for rock and its weathered products. Only basic hints on calculation methods and design aspects are given; for further details see Chapters 5, 6 and 11.

A retaining structure is designed to sustain the lateral (and vertical) pressure of rock, and owes its stability primarily to:

(1) its own weight and to the weight of any rock or soil located directly above its base (gravity walls or blocks);
(2) its passive 'earth' pressure (e.g. pile walls, pier walls);
(3) the bond strength of the structure and the rock (e.g. rock reinforcements, systematic rock nailing);
(4) the forces of non-prestressed anchors (e.g. anchored walls, ribs, beams, slabs).

In a wider sense, revetments, facings, shotcrete covers, wire meshes and rock fall fences should be mentioned.

Accordingly, due to this variety of retaining measures their design methods differ and are also influenced by the purpose the structure has to fulfil. Influence factors of significant importance are:

(1) the local situation and the purpose of the retaining structure;
(2) the allowable risks;
(3) the scatter of the rock parameters;
(4) the possible effects of seepage water;
(5) the possible and allowable deformation of rock bodies and retaining structures.

For example, supporting a slope which is likely to slide immediately below an existing (critical) building or a densely populated area, certainly needs a different consideration from a mere retaining structure for a small road in an uninhabited area. So the degree of 'calculated risk' taken differs widely, and is influenced by the extent to which rock investigations are taken, the assumptions of 'plausible' rock (or soil) parameters, etc. Retaining structures (including backfill) and surrounding rock masses always have to be regarded as a complex whole.

Therefore rigid rules or guidelines for safety factors are doubtful and should not be considered generally valid. Especially for large-scale sliding slopes, safety factors have to be accepted which in many cases are below usual values. Furthermore, experience frequently has shown that critical deformation of rock masses can be reduced effectively by increasing the safety factor to $F \geqslant 1.1$. As the actual ('absolute') stability of many rock slopes is rather difficult to evaluate, the relative improvement by means of retaining structures (and accompanying measures, e.g. drainage etc.) is of special importance.

Designing retaining structures for rock masses requires not only an advanced knowledge of geology but also an experienced engineering judgement. In rock engineering theoretical assumptions and sophisticated calculations are more frequently only an auxiliary tool for a final decision than in soil mechanics.

26.2 Failure modes in rock masses

In order to design a proper retaining structure the possible failure modes of the specific rock masses should be known. These depend primarily on:

(1) the type and structure of the rock;
(2) the seepage water conditions and other characteristics;
(3) the geometric conditions of the slope;
(4) possible external loads (including seismic forces).

Mechanically, rock masses are multiple-body systems being neither homogeneous nor isotropic (Jumikis 1983). They are characterized by structural features such as joints, bedding planes, cleavage, schistosity and shear zones. 'Voids' may be present in the form of rock pores, cavities, fissures and cracks, being either isolated or interconnected, fully or partially filled with air, gas and/or water. Therefore rock is not a continuum but a regulated discontinuum and most rock-slope failures occur along discontinuities or zones of weakness. Accordingly, in rock-slope stability problems, the actual rupture or sliding surface depends upon the spatial orientation, frequency and distribution of the discontinuities, and the involved shear and interlock resistance to shear along them (see Jumikis 1983). Therefore statistical discontinuity surveys are of great importance.

The stress–strain properties of rock bodies largely depend on the load direction related to the orientation of the discontinuities.

If rock is extensively weathered, highly disintegrated, densely jointed, loosely fragmented or decomposed, it behaves more or less like 'soil': in such cases the theories and calculation methods of soil mechanics are suitable tools for evaluating the slope stability of rock masses and designing retaining structures. Anyway, the engineer is less concerned with identifying a particular rock type than with estimating its strength and elastic properties.

The theoretical assumptions for designing retaining structures have to be adapted to the local conditions. Pre-existing planes of weakness in rock represent potential slip lines requiring modification of those conventional analyses commonly used in soil mechanics.

26.3 Strength parameters

Shear strength is the most important rock parameter used in stability analyses, which are necessary for the design of restraining structures. The investigation of small rock samples in the laboratory provides only approximate information; usually small specimens represent stronger rock material than the rock mass *in situ*. On the other hand, the exploration costs increase with the size, that is, the diameter of the rock sample. *In situ* tests are economically justifiable only in the case of large and important structures, and the application of locally gained results to the whole complex rock mass may include certain risks. Therefore a sufficient number of laboratory specimens should be investigated, focusing on the effect of planes or zones of weakness in the rock. In the case of heterogeneously layered texture, the shear strength of rock depends essentially on the shear orientation. Rock material, gouge or mylonitic zones show a typical post-failure behaviour which should be taken into account when designing retaining structures.

Weathering and decomposition products on the joint surfaces represent materials which frequently develop slickensides with very low angles of internal friction, forming potential failure lines.

For such cases extensive laboratory shear tests on specimens from the weak zones should be performed. In slopes prone to slippage, or with a high clay mineral content (especially montmorillonite) in bedding, joints or faults etc. small movements may cause a gradual decrease of the angle of internal friction. Limit values of residual shear strength have been measured down to $\phi_r = 5°$. Though the knowledge of mineral composition provides some hints on the slippage-proneness of slopes, direct testing in the triaxial apparatus or shear box is still the most reliable method to determine shear parameters. These tests allow the investigation of a great number of samples, thus enabling a statistical analysis to form the basis of the final design parameter assumption.

The shear strength of filled joints is not only influenced by the properties of the fillings but also by their thickness. Clay seams of even a fraction of 1 mm reduce the strength significantly. Therefore shear parameters of the fill material should commonly be considered as decisive in designing practice – independently of the actual thickness of the fillings (for safety reasons).

When determining the residual shear strength in laboratory or field tests, attention should be paid to the following facts:

(1) The shear strain should be increased up to the limit value of ϕ_r. This can be performed more easily by the circular ring-shear apparatus or in shear boxes (e.g. to and fro shear directions) than by the triaxial apparatus.
(2) If the normal stress at the beginning of the shear test is too small, the measured value of ϕ_r is not the theoretical limit value. As ϕ_r mostly decreases with increasing normal stress, the overburden should be taken into account when assessing the possible residual shear strength in the field. Deep-lying slide planes are more critical than those near to the surface.
(3) A decreasing degree of saturation reduces the tendency towards slickensides and decreasing ϕ_r.

In most cases of engineering practice it would be too expensive or even impossible to design retaining structures on the basis of the residual shear strength of the rock. But on the other hand it cannot be justifiable to use the peak value, ϕ, if ϕ_r is extremely small. Thus each structure must be considered unique, and especially in mountainous regions generally valid rules for design can hardly be given.

Laboratory tests and practical experience show that cohesion, c, and angle of internal friction, ϕ, are not proper constants but may alter (with time, caused by external influences, etc.). Frequently cohesion must be considered an unreliable parameter and therefore be taken into account only cautiously when evaluating the forces acting on retaining walls.

The knowledge of the residual shear strength is also of importance when evaluating the urgency of support measures. If a slope starts moving critically (and also the retaining structure), stabilizing works might become necessary to avoid a progressive decrease of shear strength and consequent overall failure (Brandl 1979).

The *wall friction* of retaining structures, δ, depends on the following factors:

(1) roughness of the rear side of the structure;

(2) internal friction, ϕ, of the backfill or rock (in case of a direct structural contact by casting concrete or spraying shotcrete);
(3) relative deformation between retaining structure and rock/backfill;
(4) inclination of the ground surface behind the wall.

The angle of wall friction is further influenced by the method of construction on site, by settlements of the backfill of the structure, and dynamic loads. Hence the fundamentals of soil mechanics can be applied to rock mechanics in this case.

26.4 Lateral pressures on retaining structures

The lateral pressure on retaining structures depends on various factors, mainly:

(1) movement and deformation of the complex system;
(2) wall friction and structural details;
(3) rock properties;
(4) properties of the backfill;
(5) water pressures.

Moreover the lateral pressure on retaining structures is not constant but changes more or less seasonally due to *temperature changes*. Usually the maximum value occurs during summer, when structure and rock dilate; minimum pressures occur during cold winter months, when contraction takes place. This effect is more evident for rigid walls, but is commonly not considered in design practice, unless the retaining structure is located in a climatically extreme place.

The above listed factors influence each other more or less mutually. Therefore they are treated partly together in the following sections.

26.4.1 Effect of movements or deformation

The theoretical relations between lateral pressure and movement/deformation of the structure are similar to those of soil mechanics only in case of heavily fissured, weathered or decomposed rock. The larger the structure is, the larger may be the single rock bodies within such a retained mass. Measurements taken from a specific retaining structure up to 65 m height and 600 m length confirmed the common limit values of active 'earth' pressure and pressure at rest for such cases. In this case single rock bodies had a length of several metres but behaved as giant 'grains' within the weathered, decomposed mass.

In the case of sound massive rock retaining structures, these fulfil more or less only a facing function, mainly as a protection against long-term weathering. The lateral pressure may be neglected.

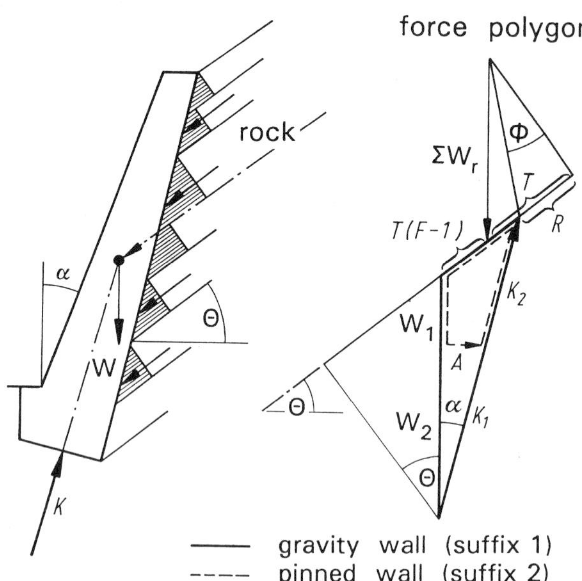

Figure 26.1 *Stability analysis of a rock retaining wall; comparison of different structures: W, weight of the retaining wall (general); W_1, weight of a non-tied back gravity wall; W_2, weight of a 'pinned' wall (rock nailing); ΣW_r, weight of the hatched rock wedges which might move; T.F, weight of the downward (tangential) component of ΣW_r, multiplied by the safety factor F; A, anchor forces; K, reaction force in the wall base; R, friction force $R = N.tg\phi$*

Between these limits all intermediate values are possible according to the specific conditions. Figure 26.1 indicates the stability analysis for a structure sustaining medium fissured or jointed rock. The overall stability of the whole rock mass (slope) should be given, but single rock bodies might slide. The difficulties in calculation, viz. uncertainties, lie in recognizing slippage-prone rock bodies and their possible failure mechanisms, and in the mutual influence of structural stiffness, anchor forces and reaction forces. According to Figure 26.1 it is supposed that the rock mass behind the hatched wedges does not slide, if those are retained. This leads to a safety factor of

$$F = \frac{R + K \sin(\theta + \alpha) - W \sin\theta}{T}$$

If anchors or nails are installed at relatively dense centres, this multiple tied-back structure is called a *pinned wall*. From the dotted lines of the force polygon in Figure 26.1 it can be seen how only small anchor forces (A) are sufficient in this case to provide equilibrium in spite of the greatly reduced weight of the structure. Accordingly, the nails have to be anchored over only a short length in the stable rock mass. Contrary to classical 'rock nailing' a certain prestressing is recommended in case of 'pinned' retaining structures in order to avoid a possible slide.

26.4.2 Excessive lateral pressures

Depending on the discontinuities of rock masses, a stress concentration or release may occur locally. Arching or wedge effects cannot be calculated precisely but should be roughly evaluated.

In heavily weathered or decomposed rock masses the earth pressure or the pressure at rest may be exceeded significantly under certain conditions, such as

(1) hard rock bodies in a weak matrix;
(2) long-term creep;
(3) freezing pressures, swelling pressures;
(4) significant stiffness differences between the elements of a retaining structure (see Section 26.4.3).

Hard rock bodies within a decomposed rock mass lead to a stress concentration on the stiffer material as indicated in Figure 26.2. The massive blocks act statically like abutments for arches – and they may transfer concentrated loads to the retaining structure.

In order to avoid such an irregular stressing of the retaining wall it is recommended that the heterogeneous rock is removed by excavation or blasting for at least 1 m width and replaced by backfill of sandy gravel etc.

The effects of creep on the lateral pressure of retaining structures are described in Section 26.6.

Freezing pressures occur if frost-susceptible rock or its weathered products are in close contact with the rear side of thin retaining walls, revetment structures, or behind shotcrete facings (see Section 26.14). Comprehensive freeze–thaw tests with various rock specimens show that ice lenses segregate not only in weathered, decomposed rock masses, or in fine-grained gouge and mylonite, but also in relatively undisturbed material. Critical materials are, for instance,

(1) Sedimentary rock such as greywacke with marked graded bedding, associated with shale or flysch;
(2) Metamorphic rock such as slate, schist (especially if sericitic or graphitic), gneiss or sericitic quartzite.

Freeze–thaw action associated with water transport is possible in firm-looking rock if it contains traces of clay minerals. Contrary to soil, the ice lenses in rock may also grow in a plane inclined normal to the temperature gradient due to the presence of discontinuities. Water transport to the ice lenses occurs not only via the absorbed water of clay minerals but also by capillarity within the joints. The pressure of the freezing water in cracks, crevices, pores, joints or bedding planes causes not only a mechanical disintegration, splitting or break-up of rock, but also a local overstressing of retaining structures.

The frost susceptibility of rock masses can only be

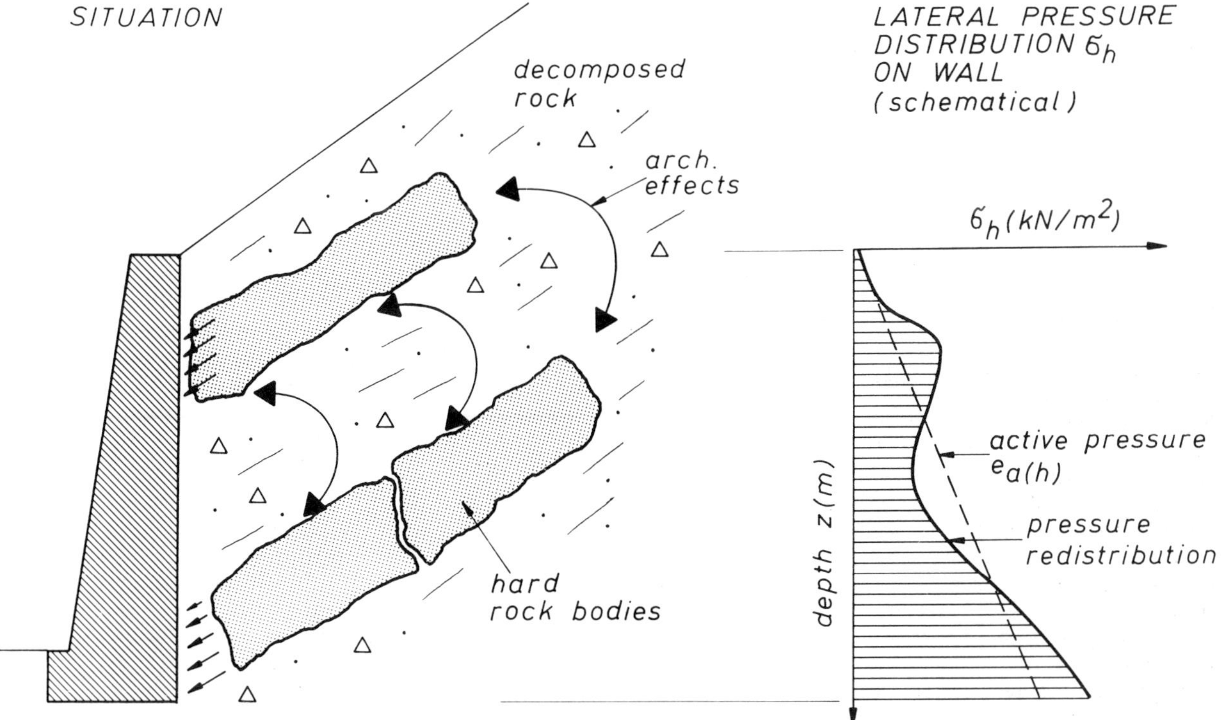

Figure 26.2 *Locally increased lateral earth pressure on retaining walls due to stress concentrations on hard rock bodies within coarsely crushed, decomposed rock masses*

evaluated by experience and/or freeze–thaw tests in the laboratory. As ice lenses commonly develop in a rather irregular manner, the lateral pressure on retaining structures or rock facings varies locally to a significant extent. Practically, such pressures are not taken into account in the design calculations for a retaining structure. Therefore proper construction is necessary to avoid cracks or even failure such as

(1) excavating frost-susceptible rock zones and refilling with shotcrete or sandy gravel;
(2) dewatering rock slope, drainage of backfill;
(3) total width of retaining structure plus granular backfill above frost penetration depth;
(4) stengthening of the reinforcement (e.g. two or three layers of wire mesh for shotcrete which is sprayed directly on the rock face as a revetment);
(5) rock nailing of the critical zone.

The lateral pressure due to growing ice lenses may amount to nearly 1 MPa under extreme conditions.

Swelling pressures may occur if an excavated slope liberates swelling-prone rock zones. This problem should be treated similarly to freezing pressures, checking the risk by laboratory tests and experience, and performing proper design and construction.

26.4.3 Effect of the type of retaining structure

In the case of a conventional retaining wall with constant stiffness the lateral 'earth' pressure may be calculated similar to soil mechanics or with a rock-body analysis. But a pressure rearrangement occurs if the retaining structure consists of elements with differing stiffness, or if it is tied back with prestressed anchors. Figure 26.3 shows such a retaining system which consists of crib elements (prefabri-

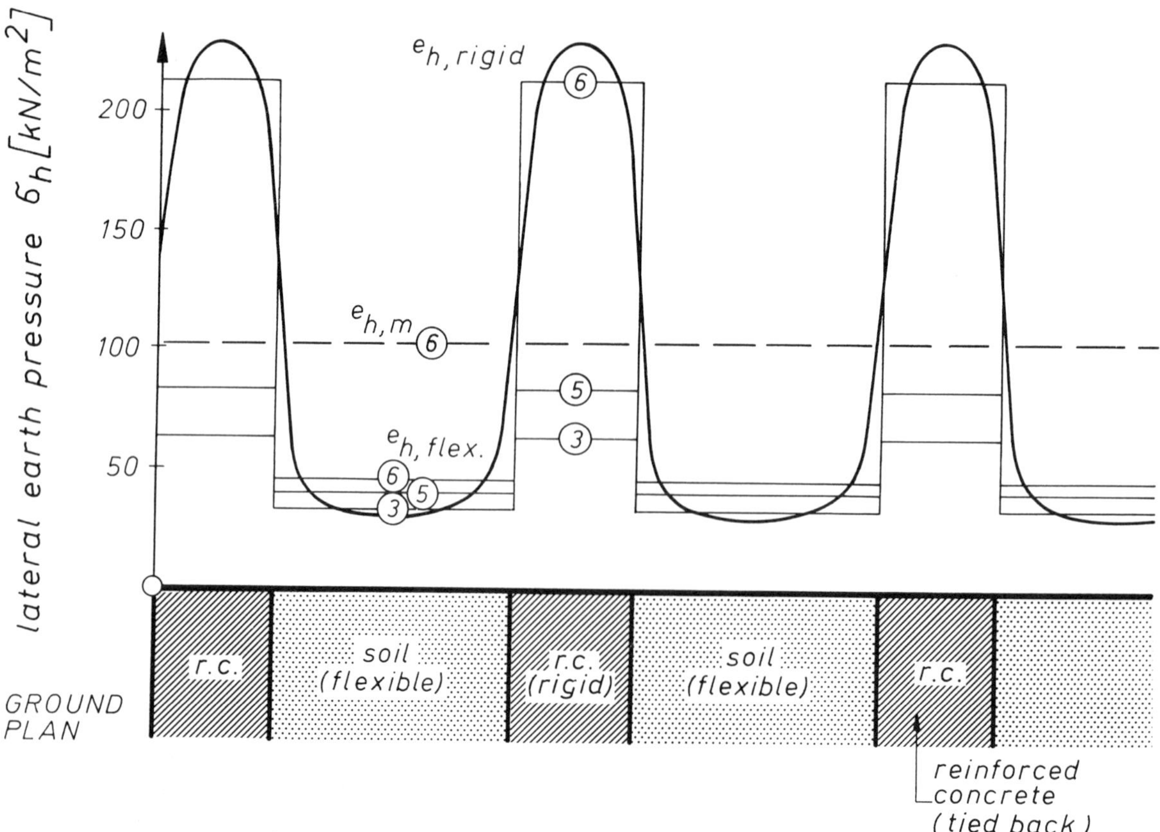

Figure 26.3 *Stress distribution on the rear side of a crib wall consisting of elements with strongly differing stiffness (Ground plan similar to Figure 26.26, No. 4 (a)). In situ measurement results: f, flexible cells filled with soil; r, rigid cells filled with reinforced concrete and tied back with prestressed anchors. Lateral earth pressures e_h during the construction stages No. 3, 5, 6 (corresponding to wall heights of 13.5 m, 16.5 m, 20.8 m; mean values over the wall height according to a rectangular earth pressure redistribution: $e_{h,f}$, horizontal pressure on the flexible crib elements; $e_{h,r}$, horizontal pressure on the rigid cells; $e_{h,m}$, mean value of the horizontal pressure along the wall (back-calculated ficticious value for comparison); corresponds well with the earth pressure at rest (= e_o in half height of the wall)*

cated r.c.) between anchored cast-in-place ribs. Such structures have proved very successful for several reasons:

(1) adaptable to local conditions (e.g. rock properties, surface geometry);
(2) rapid and easy construction, even in difficult terrain;
(3) excellent drainage of slope and/or backfill;
(4) aesthetic as the crib elements provide sufficient spaces for plant growth;
(5) subsequent strengthening is easily possible, if increasing slope forces make this necessary, by installing additional anchors.

Comprehensive *in situ* measurements on numerous retaining structures of this type have disclosed that the lateral pressure distribution differs widely from the classical assumptions (Brandl 1984, 1985). Figure 26.3 shows that the stiffer elements attract the pressure, and the flexible ones are unloaded. The lateral pressure behind the tied-back ribs is strongly influenced by the anchor forces. If these are very high (exceeding the pressure at rest), in order to obtain sufficient large-scale slope stability, the lateral pressure is practically a reaction force. Consequently, single elements of the retaining structure must be designed to take over the respective lateral pressure. Moreover the resultant earth pressure is larger than for conventional, non-anchored retaining walls.

The design of such complex structures should be based on limit value assumptions to cover the influence of possible rock inhomogeneities and different construction stages. Site measurements have shown that the lateral pressure distribution may change significantly during the construction work (not only the magnitude of horizontal stress but also the 'earth' pressure). This does not affect the overall stability of the rock slope–retaining structure system, but does affect the internal stability of the structure.

26.4.4 Effect of the backfill

Conventional retaining walls are backfilled with proper granular material. If the backfill is wide enough for full earth pressure to develop, the classical earth pressure theories are valid unless the rock masses transfer an excessive sliding pressure onto this backfill.

In rock engineering there is frequently only a narrow space between the rock face and the retaining structure. Consequently, the theoretically assumed wedge of sliding soil cannot develop. This causes a reduction of the earth pressure, similar to silo cells (Figure 26.4). According to the silo theory the vertical pressure in a certain depth, z, becomes an exponential function:

$$\sigma_z = p_{vz} = \gamma z_0 (1 - e^{-z/z_0}) \qquad (26.1)$$

$$z_0 = \frac{a\,b}{2(a+b)K \tan \delta_s} \qquad (26.2)$$

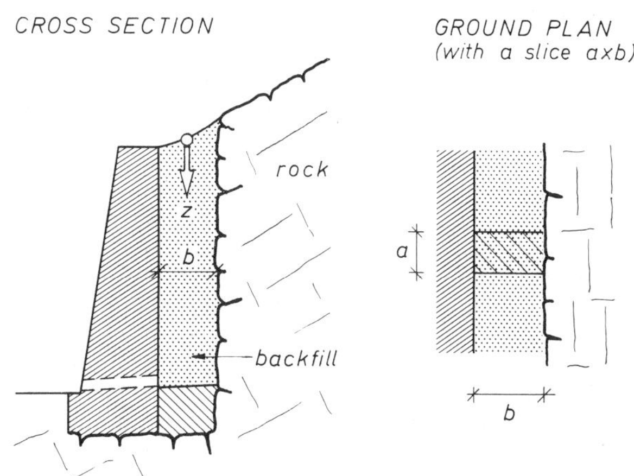

Figure 26.4 *Silo pressure within the narrow backfill between the rock face and the retaining structure*

where z_0 = the fictive depth where the geostatic pressure is equal to the maximum value of silo pressure, that is:

$$P_{vz,\,max} = \gamma\, z_0$$

If $a \rightarrow \infty$ (long wall, so-called 'trench conditions'):

$$p_{vz} = \frac{\gamma\, b}{2K \tan \delta_s} \cdot (1 - e^{-2K \tan \delta_s \cdot z/b}) \qquad (26.3)$$

The horizontal silo pressure will then be

$$p_{hz} = K\, p_{vz} \qquad (26.4)$$

where

γ = density of the backfill,
K = coefficient of earth pressure (frequently at rest: K_0),
δ_s = angle of wall friction within the 'silo' (commonly $\delta_s = 2/3\phi$),
ϕ = friction angle of the fill material,
b = width of narrow backfill

The coefficient K depends on the possible wall movements and varies between the active limit K_a and the pressure at rest, K_o. Usually $K_o = 1 - \sin \phi$ is preferred according to the rock-face behaviour. For calculating the inner stability of some retaining types (e.g. crib walls), a limit value investigation with K_a may be useful.

If a uniform surcharge load p_o is acting, the additional silo pressure will be (Brandl 1980a, 1982, 1984)

$$\Delta p_{vz} = p_0 \cdot e^{[2(a+b)/ab]K \tan \delta_s z} \qquad (26.5)$$

where the lateral earth pressure $E_h = P_{hz} + \Delta p_{hz}$; and for a long narrow backfill

$$\Delta p_{vz} = p_0 e^{(2K \tan \delta_s \cdot z)/b} \qquad (26.5a)$$

For further details about silo pressure, see (Brandl 1980a, 1982, 1984, 1985, 1987b).

These formulae are inevitably approximations, because the angles of wall friction, δ, need not be the same for the retaining wall and the rock face; and the earth pressure coefficient, *K*, varies with depth (viz. the ratio *z/b*, as can be verified by model tests and on-site measurements; see Brandl 1980a, 1982, 1984). Furthermore the silo theory for active conditions provides relatively high pressures on top of the wall which contradict to the Rankine theory. Actually increased values have been measured in these top zones on many construction sites. These may be explained by constraint effects due to backfill compaction. To sum up, in spite of theoretically weak points, the silo analogy with convenient assumptions has proved suitable in practice. As an alternative, statically admissible stress fields in the fill may be considered for the active case.

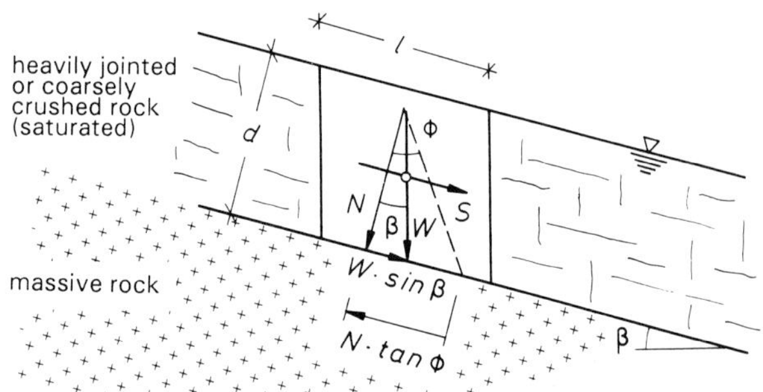

Figure 26.5 *Stability analysis for slope-parallel slide planes (cohesionless) and groundwater flow. Hydraulic gradient i = sin β; flow pressure S = l.d.* $\gamma_w.i$; γ_{sat} *= unit weight of saturated rock mass;*

$$\textit{safety factor } F = \frac{\gamma_{sat}\cdot\tan\phi}{(\gamma_{sat.}+\gamma_w)\cdot\tan\beta}$$

26.4.5 Effect of water pressures

In rock masses three forms of water effects on stability dominate, namely, seepage pressure; lateral water pressures in tension cracks; and shear strength deterioration of gouge, weathered products etc. with increasing water content. Practically in all cases the slope stability decreases. This reduction in safety must be compensated for by retaining measures (besides proper drainage!). Some hints are given below.

The static effect of *seepage pressures* can be calculated conventionally as indicated in Figure 26.5. This simple example illustrates the effect of seepage parallel to the surface as it frequently occurs in rock masses. Consequently, proper drainage (e.g. drainage drill holes – see Section 26.17) may increase the stability more effectively than costly retaining structures.

Due to the pattern and structure of their discontinuities rock masses show an anisotropic permeability behaviour. If the hydraulic gradient is not parallel to the discontinuity, its normal component has no influence on the seepage forces. In case of a multiple joint pattern the seepage may be calculated by the tensor method (Wittke *et al.* 1987).

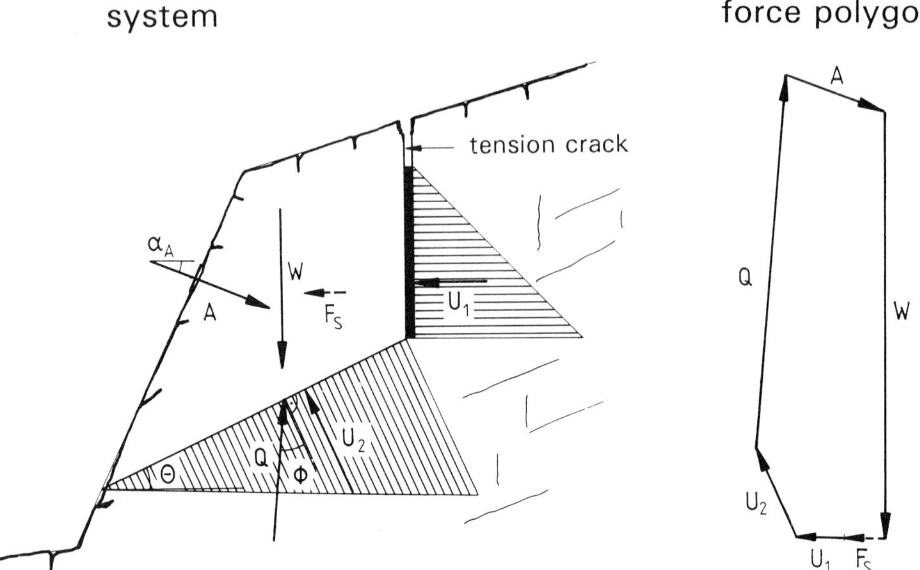

Figure 26.6 *Stability analysis of a planar rock wedge on an inclined planar discontinuity filled with a permeable gouge. Tension crack filled with water. Stability analysis with and without anchor forces A;* F_s*, seismic force as earthquake effect in case of equivalent static forces (see section 4.7). (a) Cross section; (b) Force polygon (schematical)*

The influence of lateral water pressures in *tension cracks* or joints is illustrated in Figure 26.6.

In the case of a translatory movement the safety factor against sliding, F, for a non-anchored block is given by

$$F = \frac{(W \cos \theta - U_1 \sin \theta - U_2) \tan \theta + c\,l}{W \sin \theta + U_1 \cos \theta} \qquad (26.6)$$

where c and ϕ are the shear parameters in the slide plane. If the rock block is tied back by anchor forces, A, the system's overall factor of safety F may be calculated as

$$F = \frac{[W \cos \theta - U_1 \sin \theta - U_2 + A \sin(\alpha + \theta)] \tan \theta + c\,l}{W \sin \theta + U_1 \cos \theta - A \cos(\alpha + \theta)} \qquad (26.7)$$

If in Equation 26.7 the values c, U and A are zero, the safety factor becomes

$$F = \frac{\tan \phi}{\tan \theta}$$

The factor of safety F against sliding, and the required anchor forces may also be determined graphically from the force polygons in Figure 26.10.

Instead of an overall factor of safety, F, partial factors F_i are sometimes used. This theoretical aspect takes into account the different degrees of confidence which the designer has in the particular parameters, and the different degree of mobilization of the various forces within the system (anchor tension, friction resistance, cohesion, water pressures cannot be assumed to be all fully mobilized simultaneously).

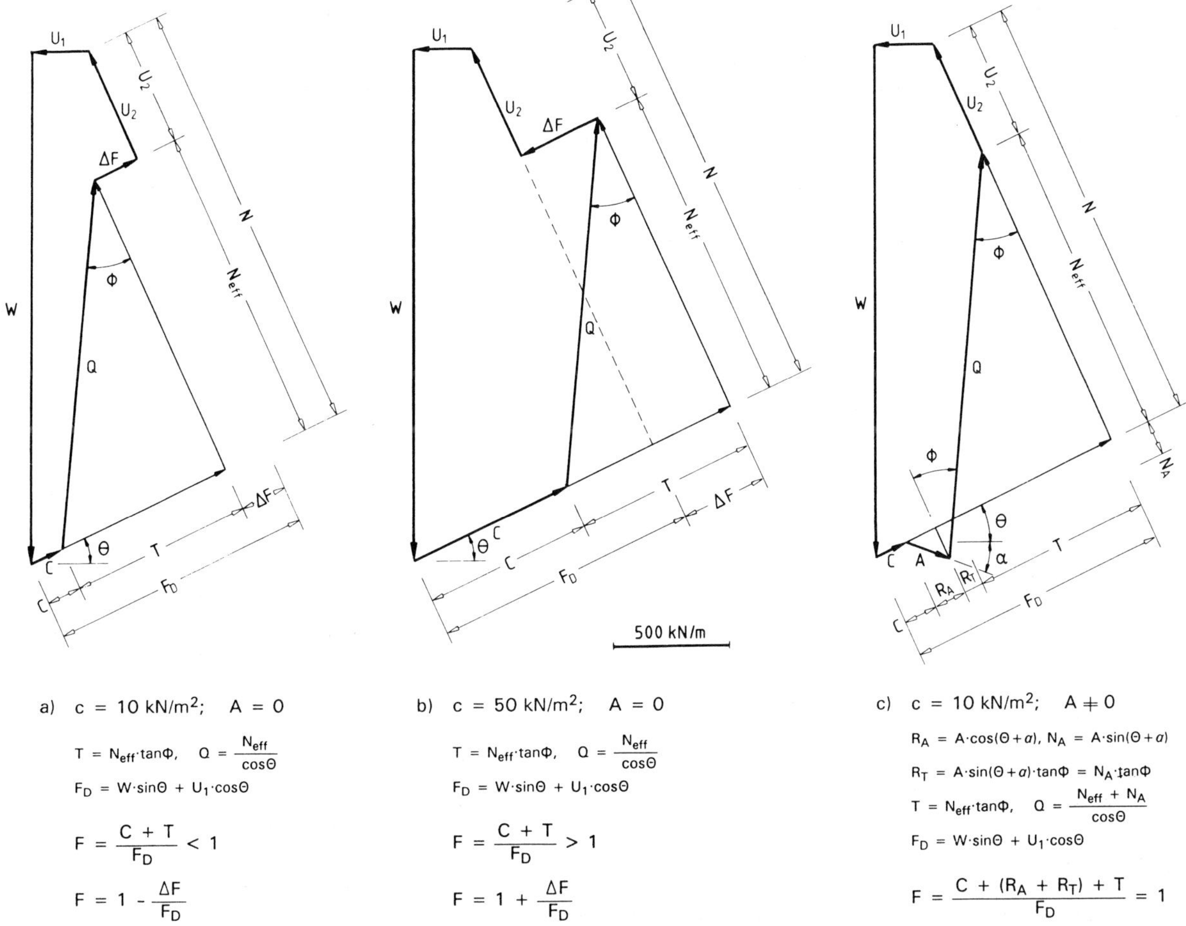

Figure 26.7 *Deducing the safety factor, F, from force polygons of Figure 26.6. (a) Low cohesion; no anchors. Safety factor F<1; (b) High cohesion; no anchors. Safety factor F>1; (c) Low cohesion; prestressed anchors. Safety factor F = 1*

Failure can occur along an inclined stratified/jointed rock mass due to breakdown as a result of the water pressure in an open fissure. The resistant forces in the slide plane are insufficient, unless the water pressure is relieved by drainage holes.

Tension cracks filled with water significantly reduce the safety factor of a rock–block system. Consequently, retaining structures should be designed in connection with proper accompanying measures which may obviate costly anchors or structural reinforcements such as

(1) drainage to reduce the water pressure;
(2) covering, plugging and sealing the tension cracks (with cement, bentonite etc.);
(3) grouting of joints.

Using unslaked lime to fill tension cracks is a simple method which may be very effective if the cracks contain a fine-grained gouge with clay minerals. The chemical–physical reactions are similar to soil stabilization with lime, thus increasing the friction angle along the slide plane. Shatter zones forming highly permeable aquifers adjacent to faults facilitate the transport of improving additives.

Placing of impermeable geomembranes over tension cracks keeps rainfall from penetrating into the ground. This emergency measure has proved a very successful expedient until the definitive stabilizing measures are finished.

26.4.6 Dynamic loading, earthquakes

Dynamic loading of retaining structures or anchors may occur as a result of blasting (for rock slope excavation) dynamic surcharge loads, or earthquakes.

Blasting near to anchors may be dangerous if they are highly stressed and not designed to withstand temporary load increases. For short prestressed anchors the load changes to be expected would be greater according to Hooke's law for the free length of the anchor tendon. Furthermore, the dilatancy occurring in a particular rock due to blasting works depends on the rock properties. A generalization is therefore hardly possible (Hanna 1982). A comprehensive report on anchor prestress load fluctuations during blasting is given by Littlejohn *et al.* (1977) for a reinforced slope of an opencast coal mine. The largest increases were 7% of the service load.

Dynamic surcharge loads and *earthquakes* may increase the lateral pressure on retaining structures and reduce the earth/rock resistance, as they may reduce the shear strength of joint filling such as uniform sands. In the laboratory, dynamic properties of jointed rock masses have been determined on models consisting of cubes by means of an impact device. However, the results obtained show a great deal of scatter, and an extrapolation from small-scale laboratory tests to the site behaviour of rock masses is questionable.

In practical design dynamic effects are taken into account by assuming reduced shear parameters in the rock discontinuities (ϕ), or shock coefficients (similar to statics), or by raising the required safety factor, derived from conventional static computations.

In situ measurements on retaining walls along streets, highways and railways have shown that the influence of the moving traffic is less than theoretically expected, especially if the road surface lies higher than the top of the wall.

Coulomb's theory is also valid for dynamic stresses. The internal friction of a cohesionless fragmented rock body decreases continuously with increasing stress or strain amplitude from a critical frequency almost linearly with increasing frequency.

For cohesionless backfills with $\phi = 35°$ the additional active earth pressure force can be evaluated according to the simplified formula (Kezdi 1972)

$$E_{a,\mathrm{dyn}} = \frac{1}{2}\gamma H^2(0.75)\frac{b_h}{g} \tag{26.8}$$

acting in $0.4H$ below the head of the wall, where H = height of the wall, b_h = horizontal acceleration, and g = acceleration due to gravity.

The sudden shock of an earthquake may 'liquefy' extensively cracked, loosely fragmented rock masses which behave like cohesionless uniform soil with megagrains. In many cases large-scale failures are induced by the local breakdown of loosened zones causing the loss of support for larger rock masses.

There are very few reports in the literature of the effect of earthquakes on prestressed anchors. The author's own site experience comprises about 500 000 m run of prestressed permanent anchors in seismic zones up to intensity 10 on the Mercalli–Sieberg scale. The actual earthquake intensity since construction (2 to 20 years) has not yet exceeded 4–6°. Up until now, all structures have operated reliably.

In regions of seismic activity the stability calculation of important retaining structures should include seismic forces. Under certain conditions, both horizontal and vertical seismic forces must be considered. This is commonly performed by assuming equivalent static forces, but dynamic FE-calculations (based on amplitude, frequency, viz. period of oscillation) are increasingly used for critical projects. The finite element method provides a displacement evaluation and enables valuable parametric studies to be made.

Figure 26.8 illustrates the equivalent static effect of an earthquake on a retaining wall which supports a weak rock. When considering a densely jointed or extensively weathered mass approximating to a 'soil', Coulomb's theory may be used. The seismic force, F_s, is applied at the centre of gravity of the slide-impending rock body (rupture wedge). From the force polygons in Figure 26.8 it is obvious that seismic forces increase the active earth pressure significantly.

Stabilizing measures to raise the safety factors of dynamically loaded retaining structures and/or adjacent rock masses include flattening of the rock slope, retaining

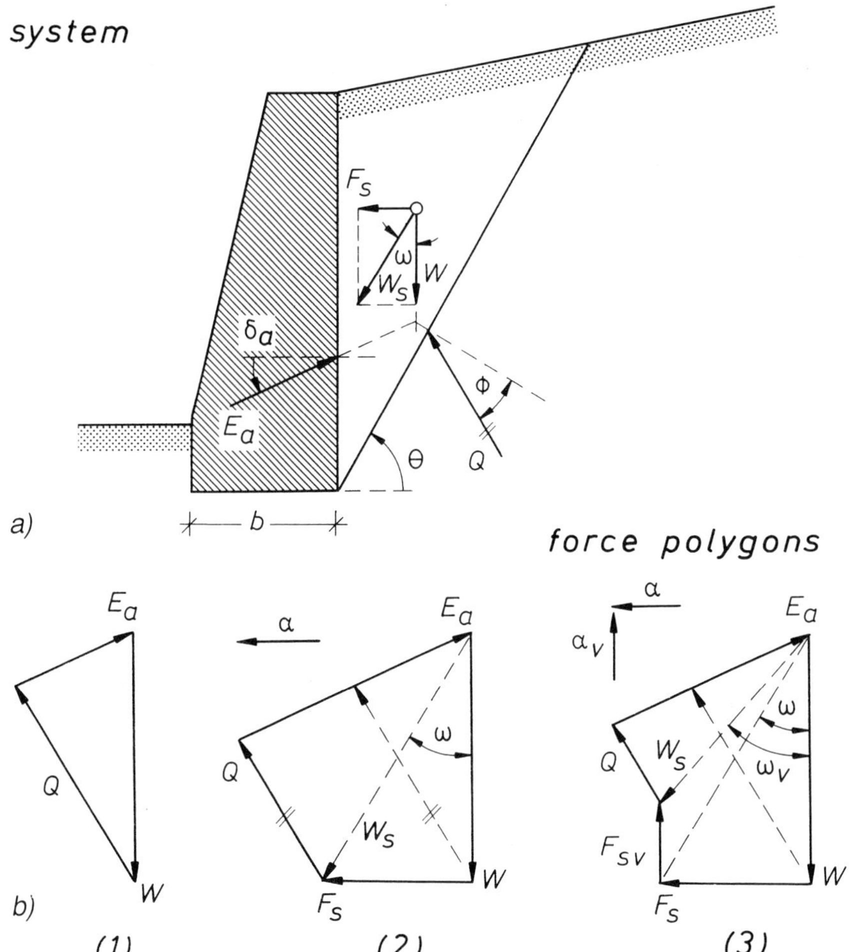

Figure 26.8 *Effect of seismic forces on the magnitude of active earth pressure: (a) Geometry and notations; (b) Force polygons: (1) without seismic forces: (2) with horizontal earthquake acceleration (force F_s); (3) with horizontal and vertical earthquake acceleration (Forces F_s and F_{sv})*

structures of increased load-carrying capacity, anchoring (with prestressed anchors), rock reinforcement and rock grouting.

Figure 26.6 indicates how the stability, F, of a slide-prone rock body can be increased by means of prestressed anchors:

$$F = \frac{(W\cos\theta - F_s\sin\theta - U_1\sin\theta - U_2)\tan\phi + A\sin(\theta+\alpha)\tan\phi + cl}{W\sin\theta + F_s\cos\theta + U_1\cos\theta - A\cos(\theta+\alpha)} \quad (26.9)$$

where

F_s = seismic force = $\alpha_e W$
W = weight of the rock body
α_e = seismic (earthquake) factor, comprising the seismic acceleration.

Details on the computation of the seismic stability of rock wedges are given by Ghosh and Haupt (1989). Aydan *et al.* (1989) have given an account of toppling and sliding of columnar rock slopes using dynamic equilibrium equations (see also Section 26.7).

Earthquakes are sudden loads of rare occurrence in relation to short-term service loads of structures. Accordingly the required safety factor for stability analyses (slopes, retaining structures) can be lower than for continuous loads or repeated short-term loading. In design practice values of $F \geqslant 1.0$–1.2 are frequently considered sufficient if all negative effects are superimposed in a plausible manner. An over-pessimistic superposition of all driving forces at their particular maximum (e.g. seepage pressures) and the worst rock parameters (e.g. minimum value of residual shear strength) would provide calculated safety factors which were too low in most practical cases. Therefore realistic assumptions for a proper calculation must be based on engineering judgement, which includes unavoidably a calculated risk.

26.5 Effects of surcharge loading

26.5.1 General aspects

The effect of surcharge on the rock below it is to increase the intensity of the lateral earth pressure. Comparative calculations have shown that the results gained by the various methods have an extremely wide scatter. Therefore it is not admissible to choose any method at will. Furthermore it has to be taken into account that nearly each structure has its individuality, meaning that certain design methods have to be eliminated.

If the rock is heavily weathered or closely jointed, theories which have proved successful in soil mechanics may be used. The following factors afford sufficient accuracy in practice (Brandl 1987b):

(1) uniformly distributed surcharge;
(2) line loads and single loads;
(3) irregular surcharge loads, internal forces and uneven ground surface;
(4) dynamic surcharge loads and earthquakes.

In addition, the classical fundamentals of soil mechanics provide a suitable theoretical aid if a massive and fairly homogeneous rock mass is loaded.

In all the other cases, effects of surcharge loading should be taken into account according to the possible failure modes in connection with discontinuities such as bedding, schistosity, cleavage, joints and faults. The relevant failure lines are usually dominated by these discontinuities or even by pre-existing rupture planes. This can be taken into account by basic (rock) statics as indicated in Figure 26.9. Commonly the calculation is based on the failure criteria of Coulomb and Mohr as in soil mechanics. The significant difference lies in the possible rupture bodies.

If the rock mass is closely jointed or heavily weathered, the influence of surcharge loads (single loads, strip loads) may be calculated in similar fashion to soil mechanics. Discontinuities, however, lead to an alternation of the stress distribution, rather different from the elastic–isotropic half-space. Steep joint sets cause a stress concentration beneath the load (Figure 26.10) and horizontal joints or stratification favour a widened stress distribution. This effect can roughly be assessed by assuming a modified half- or quarter-space with a 'concentration factor' ν. Boussinesq's equation for the horizontal stress, σ_h changes then to

$$\sigma_h = \frac{\nu P}{2\pi r^2} \cos^{\nu-2}\theta \sin^2\theta \qquad \text{for single loads } (P) \qquad (26.10)$$

$$\sigma_h = \frac{\bar{p}}{r} \cos^{\nu-2}\theta \sin^2\theta \qquad \text{for strip loads } (\bar{p}) \qquad (26.11)$$

where f is a coefficient dependent on ν, according to the following table (linear interpolation is permissible):

ν	3	4	5	6
f	$\frac{2}{\pi}$	$\frac{3}{4}$	$\frac{8}{3\pi}$	$\frac{15}{16}$

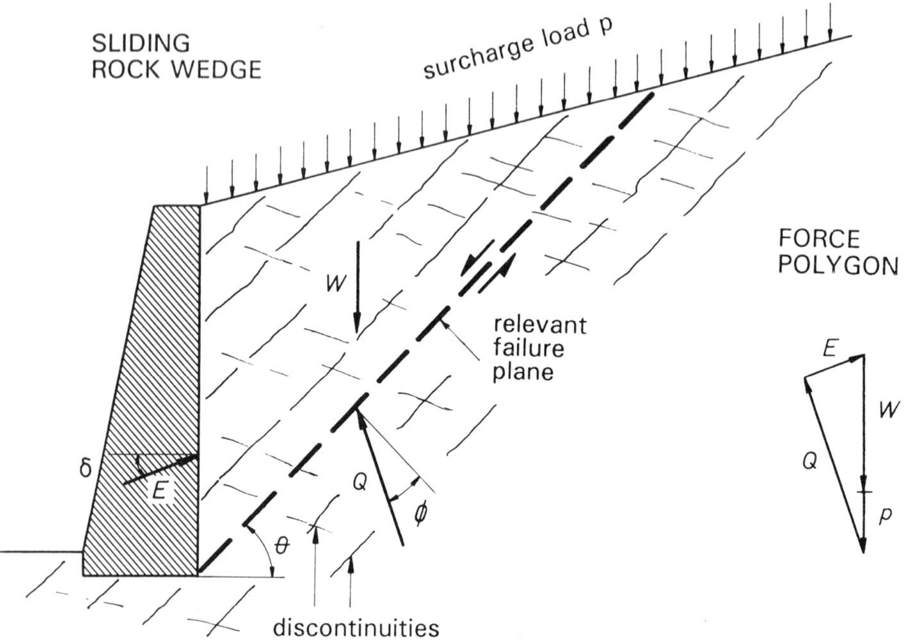

Figure 26.9 *Determination of 'earth' pressure of jointed rock and uniformly distributed surcharge; schematical. Inclination of θ geologically pre-determined. The angle of failure plane $\theta_{geolog.}$ may be larger or smaller than the theoretically active limit value θ_a*

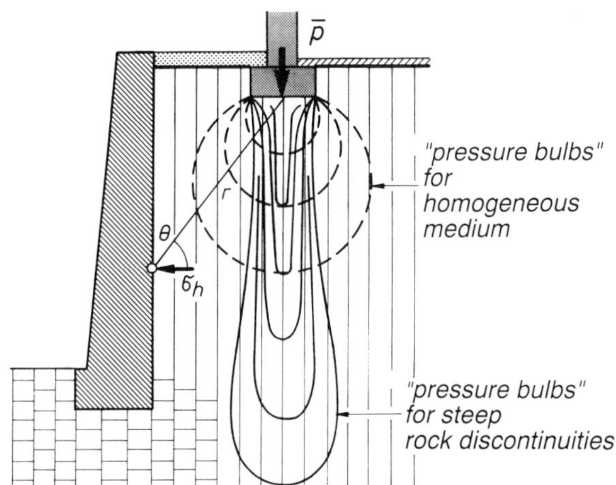

Figure 26.10 *Influence of rock discontinuities on the stress distribution (pressure bulbs) caused by surcharge loads. Modified half space with a concentration factor* ν

The factor ν describes the stress concentration, hence ν = 1–2 for more or less horizontal discontinuities, ν = 3 for the ideal isotropic half space, and ν = 4–6 for steeply inclined discontinuities. If the joint set dips steeper than the connection line between load P and the toe of the retaining wall, the calculation for ν = 3 provides safe results. But if the joints daylight, a stress concentration on the structure should be taken into account (inclined stretched pressure bulb).

26.5.2 Conventional retaining walls

Figure 26.11 shows an example of a kinematic calculation method if rigid rock bodies undergo translation as (quasi) monolithic blocks. Additionally, the rock bodies themselves may consist of parted blocks which move against each other along inner sliding lines.

The first step is to assume a kinematically possible failure mechanism, and to determine the relative and absolute displacements between the single rock bodies by means of a displacement plan. This plan starts with drawing the displacement vector, u_1 (given, allowable or assumed) from the zero point, acting parallel to the outer slide planes of rock body 1. Then the direction of the outer slide planes of the other rock bodies are drawn. At the end point of the u_1-line a straight line is drawn parallel to the contact surface (inner failure plane) of this rock body, the displacement of which is given (in this example No. 1, but may be also No. 3 with regard to an allowable wall movement). The displacement vectors parallel to the inner failure planes represent the relative movements $u_{i,j}$ between adjacent sliding rock bodies i and j. The vectors u_i (from the zero point of the displacement plan) represent the movements of the rock bodies in relation to the stable ground.

The next step is the determination of the direction of the friction forces, Q (resultant reaction forces against failure plane), due to the derived relative displacements. Their inclination is given by $\pm\,\phi$ (viz. $\phi' = \arctan(\tan\phi/F)$), to the perpendicular line on the failure planes. Practically, one force polygon of the rock bodies is drawn after the other. Commonly, the overall force polygon will not close because no limit equilibrium exists. This can only be provided by installing an auxiliary force (e.g. vertical, P'). In order to establish limit equilibrium, the shear parameters are modified by varying the safety factor, F, till P' becomes zero:

$$\tan\phi' = \frac{\tan\phi}{F} \qquad \text{and} \qquad c' = \frac{c}{F}$$

The solution may be obtained analytically or graphically.

A variation of the friction angle, ϕ, along the several failure planes can be considered by applying an additional driving force ΔT_i in the particular plane i. This equivalent force corresponds to a reduction of ϕ by $\Delta\phi_i$ (Figure 26.12). Hence, ϕ is the maximum friction angle of all failure planes as a basic value. Another method of calculation is the use of partial safety factors, by which an overall safety factor can be deduced. In the example of Figure 26.12, four partial safety factors are obtained, i.e. F_a, F_b, F_c, F_d. For deducing an overall safety factor F from these partial values, a certain probability of the specific failures must be assumed, leading to a specific valuation of the various failure causes, e.g.:

$$F = x_A \cdot F_a + x_b \cdot F_b + x_c \cdot F_c \qquad (26.12)$$
$$\text{with } \Sigma x_i = 1{\cdot}0$$

A relevant unforeseen increase of the extrenal load P is unlikely, and the friction in the wall base can be assessed relatively accurately. Hence, an exceeding of the shear strength in the pane AB is the most probable failure cause, and the factors x_i are choosen in this case

$$x_A = 0{\cdot}10;\ x_b = 0{\cdot}75;\ x_c = 0{\cdot}15$$

The lateral pressure on the retaining wall is conventionally derived from the adjacent sliding body (3 in Figure 26.11). The resistant body before the wall toe (4 in Figure 26.11) should be neglected in practice for safety reasons.

26.5.3 Rock nailing

Figure 26.13 shows the possible failure mechanisms for rock nailing, depending on the rock properties, geometry of the complex retaining structure (e.g. length and inclination of the 'nails'), and on the surcharge loads, V. In such systems failures along rupture lines dominate; zoned fractures hardly occur. In the case of high loads near the

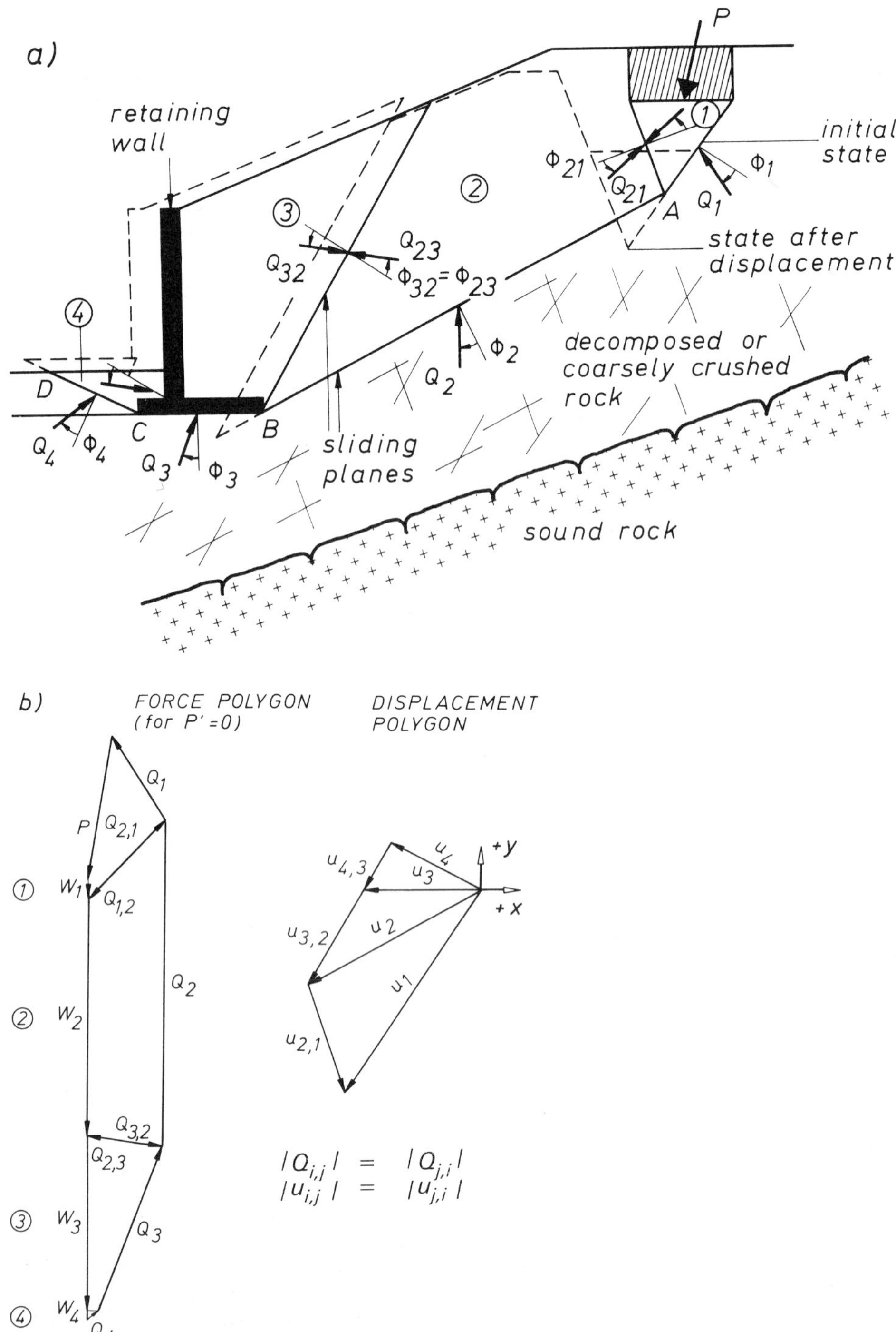

Figure 26.11 *Kinematic calculation method of conventional retaining walls sustaining rigid rock bodies which move as (quasi) monolithic blocks:*
(a) System and notations: P, external load; Q_i, resultant reaction force against failure planes; ϕ_i, angle of internal friction along the failure planes; W_i, weight of sliding blocks
(b) Force polygon (for $P' = 0$) and displacement polygon: u_i 'absolute' displacement of the block i; u_{ij} relative displacement between adjacent sliding blocks i and j

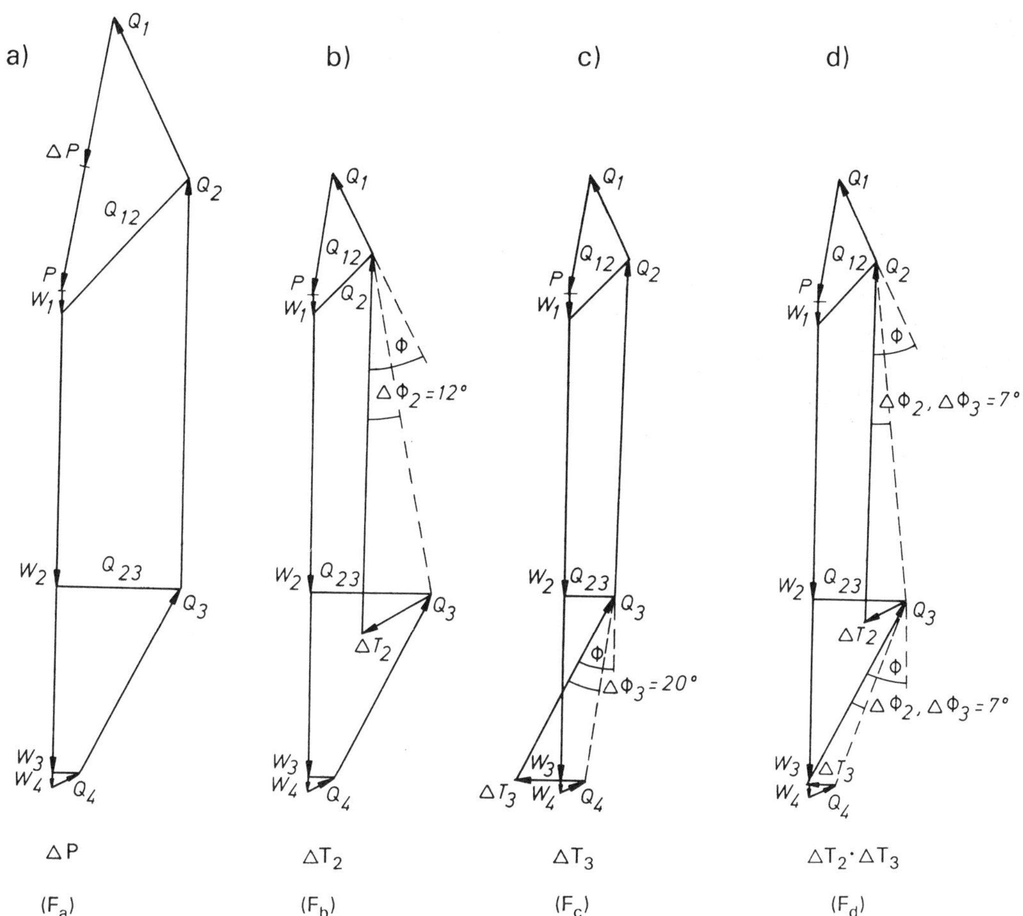

Figure 26.12 *Force polygons according to different failure causes in Figure 26.11: (a) Increase of the load P by ΔP; (b) Reduction of ϕ by $\Delta\phi_1$ in the slide plane A-B by installing an additional driving shear force ΔT_1; (c) Reduction of ϕ by $\Delta\phi_2$ in the slide plane B-C by installing an additional driving shear force ΔT_2; (d) Reduction of ϕ by $\Delta\phi_3$ in the slide planes A-B and B-C by installing additional driving shear forces ΔT_1 and ΔT_2*

wall face, slip circles are the critical failure mode. For densely fissured, heavily jointed or decomposed rock without cohesion and for high loads at the rear end of the nails, translation with two sliding bodies provides the lowest safety factor. Figure 26.13 shows an example for evaluating the stability of such a retaining structure (see also Gässler 1987 for soil nailing). The safety factors F may be defined as Gudehus 1981:

$$F_\phi = \frac{\phi_{\text{existing}}}{\phi_{\text{limit}}} \quad \text{(Fellenius rule)} \qquad (26.13a)$$

$$F_N = \frac{A_{\text{existing}}}{A_{\text{limit}}} \qquad (26.13b)$$

where A_{existing} = sum of the axial nail forces which may be mobilized at the maximum.

$$F_p = \frac{P_{\text{limit}}}{P_{\text{existing}}} \qquad (26.13c)$$

where P_{existing} = actual existing surcharge load.

The suffix limit means the specific values in the state of limit equilibrium.

According to the hypothesis of minimum stability, sliding geometry is relevant for the state of limit equilibrium, which leads to the safety factor of $F = 1$ as a minimum.

The inclination of the main failure planes, θ_1 and θ_2, in Figure 26.13 are either variable in the case of heavily decomposed rock or they are given by the joint pattern or by pre-existing rock defects. Accordingly, the geometry of sliding planes has to be varied more extensively, if a decomposed rock behaves similarly to soil masses.

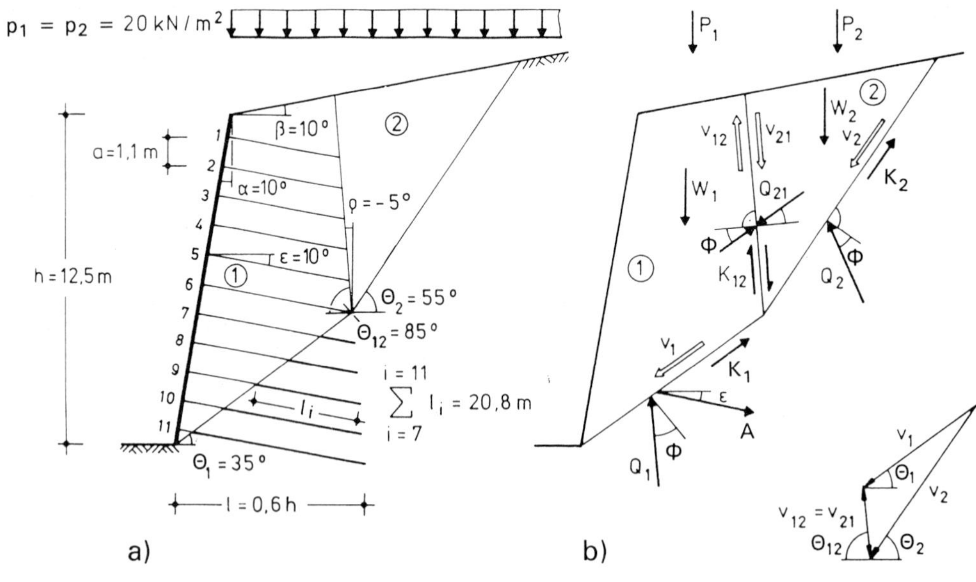

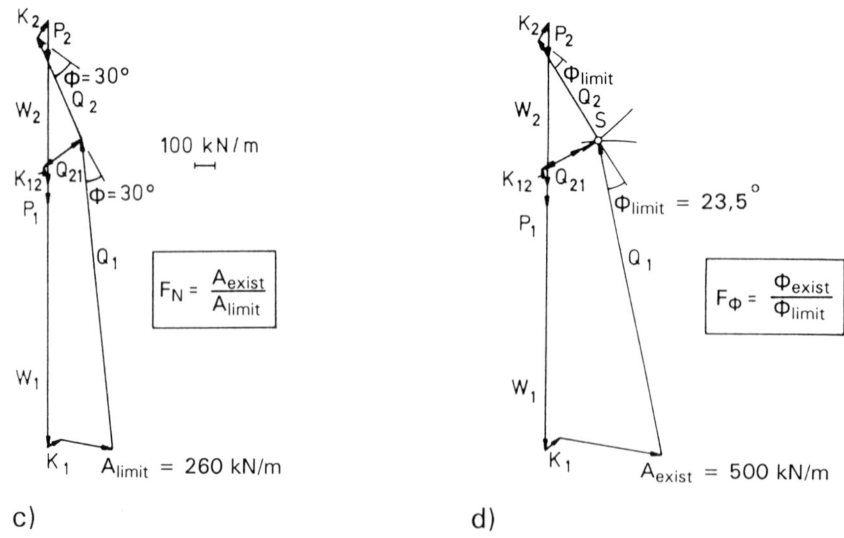

Figure 26.13 *Stability analysis of a rock nailing structure (densely fissured or heavily decomposed or coarsely crushed rock without cohesion). Example of a translatory failure mechanism consisting of two rupture bodies: (a) Cross section; (b) Forces acting on the system and displacement polygon; (c) Force polygon, closed with A_{limit} leads to F_N; (d) Force polygon, closed with ϕ_{limit} leads to $F\phi$; Cohesion forces $K_i = C_i$ and $K_{ij} = C_{ij}$; v_i, v_{ij} displacements in the failure plane i, ij (also u_i, u_{ij} in use for kinematic analyses, e.g. Figure 26.11*

26.5.4 Pinned or anchored retaining walls

Surcharge loads have to be superimposed to the dead loads (weight) of sliding rock bodies. Starting from the unloaded slopes as indicated in Sections 26.4 and 26.11, the calculation or graphical determination is more or less fundamental rock statics.

26.6 Effect of slope creep

Creep occurs relatively frequently in rock slopes, due to steep inclination, gradual deterioration, progressive overloading, water pressures etc. Creep may occur in weathered, decomposed rock with fine-grained zones, along faults, but also in firm rock strata over silt–clay seams, and in discontinuities filled with gouge.

In a slope undergoing creep, retaining structures may be stressed by a lateral pressure, E_{cr}, that significantly exceeds the theoretical earth pressure at rest, E_0. This creep pressure, E_{cr}, may also be considered as 'sliding pressure' or 'stagnation pressure' on a retaining structure. Its magnitude can be evaluated according to Figure 26.14. A sliding surface AB is assumed, which enables, even for a nearly immovable wall, further creep of the slope. Corresponding with this assumption, *in situ* measurements have shown greater deformation of the decomposed rock than of the retaining structure (Brandl 1980b). For theoretical simplification the curved sliding surface is replaced by the plane A′B, which is surcharge loaded by the mass above. Within the wedge A′BC, the secondary failure surface has to be determined, which provides the maximum earth pressure force, E_{cr}. This can be calculated like the classical theory (body NMB as surcharge). Because of this surcharge the weathered rock in close vicinity to the retaining wall is compressed. Depending on the inclination, δ_1, of the pressure force E, the creep pressure, E_{cr}, varies, according to the limit values $0 \leq \delta_1 \geq \phi$. If assuming a (quasi) cohesionless mass and limit equilibrium $\beta = \phi$ (Figure 24.14), the creep pressure becomes a special case of an increased Rankine earth pressure

$$E_{cr} = m(\phi)\gamma \frac{h^2}{2} \cos^2 \phi \qquad (26.14)$$

The multiplication factor $m(\phi)$ also depends on the stiffness of the retaining structure. The limit values of Table 26.1 were obtained from numerous *in situ* measurements. They have proved very useful for practical design in the Alpine sliding areas of Austria for about 20 years.

In the special case of short plane or curved retaining structures the friction forces on both ends of the wall cause

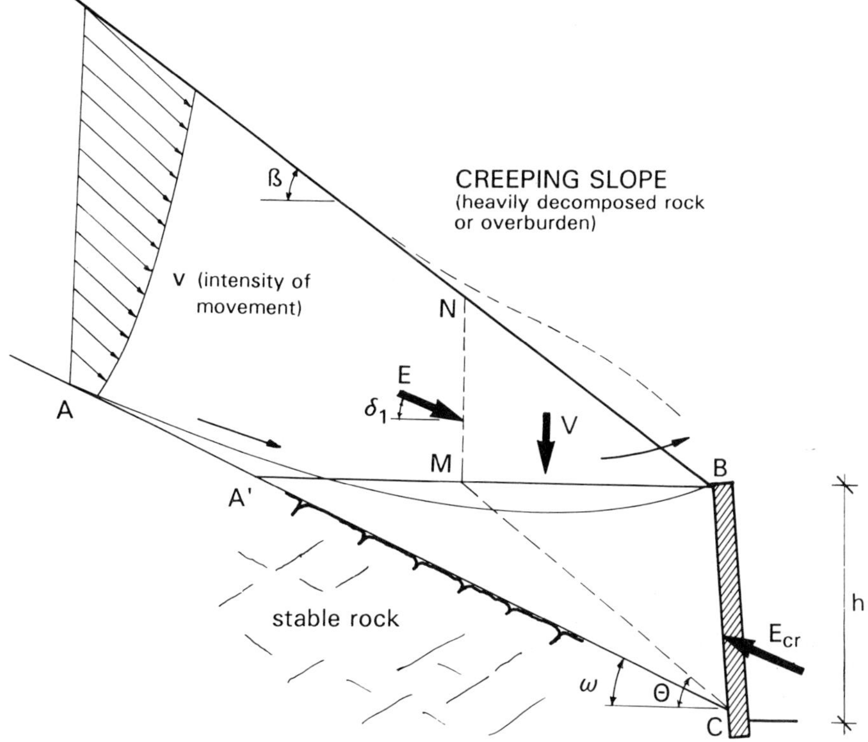

Figure 26.14 *Calculation of the 'creep pressure', E_{cr}, in a creeping slope (see Brandl, 1980b). Displacement profile in point A indicates a discontinuous creep with slip on a solid base*

Table 26.1 Factor m (φ) for evaluating the creep pressure in weathered, decomposed, or closely jointed rock masses according to Figure 26.14. Slope inclination β = angle of internal friction φ. Cohesion negligible

Stiffness of retaining structure	*Friction φ or inclination β (degrees)* 20	22.5	25	27.5	30	32.5	35	37.5	40	42.5	45
Flexible	1.14	1.18	1.24	1.35	1.46	1.59	1.73	1.89	2.07	2.26	2.46
Rigid	1.28	1.41	1.55	1.71	1.89	2.10	2.34	2.62	2.94	3.30	3.70

an increasing pressure. This can be taken into account by assuming an increased breadth of the retaining structure, depending on the geometry of the wall, the rock parameters and inclination of the slope: mainly $B' = 1.2B$ to $1.5B$.

Creep increases with the ratio of acting shear stress to residual shear strength. The time until a creep ruptures occurs, is directly proportional to stress under the following underground conditions:

(1) joints filled with weathered and decomposed products;
(2) silty-clayey seams within rock strata
(3) fault zones occupied by mylonite;
(4) heavily weathered, decomposed rock masses which physically-mechanically behave like soils.

Therefore creep may be checked in laboratory tests at a relatively high stress level – in order to shorten the test period. The failure behaviour for lower stress levels then may be determined by extrapolation.

The influence of the water content within weak zones increases with the content of clay minerals and may be nearly negligible in case of sandy gouges. In materials of medium to high plasticity the creep velocity may increase exponentially with increasing water content, especially if the stress level exceeds about two-thirds of the residual strength. Furthermore a linear correlation between the logarithm of creep velocity and the logarithm of time exists in most cases.

With regard to the long-term stability of rock slopes and retaining structures creep-tests should be performed under varying boundary conditions (e.g. water content, stress level). For this purpose direct shear tests on layered samples (rock and gouge) have proved suitable. Besides these laboratory investigations field measurements are unavoidable, if a slope is likely to slide. Geodetic surveys, inclinometers and extensometers provide the most valuable results for checking creep behaviour.

There are three stages of creep behaviour namely, primary, secondary and tertiary creep, starting with an initial deformation, Δl_0. In most cases a secondary phase with a constant creep velocity does not clearly occur at higher stress levels in gouge or decomposed rock. The primary transitional stage changes more or less directly into the creep-acceleration phase which finally causes failure.

For a rough first evaluation of the beginning of the tertiary creep phase, $t_{tert.}$, the following empirical formula can be used (Höwing 1984):

$$t_{tert} = 7.6 \times 10^{89} \times d^{-23.5} e^{-186R} e^{-12.5T} \text{ (min)} \tag{26.15}$$

where

d (mm) is the thickness of the gouge, filled joints or fine-grained seams where sliding or creeping occurs
R is the stress level
T is the clay content (<0.002 mm) related to the respective soil weight.

This approximate formula should be limited to rock masses with filled discontinuities and is not applicable to heavily weathered or decomposed rock with shear zones.

A knowledge of creep behaviour of rock masses is of significant importance for a proper design of retaining structures. They should be fairly flexible, and the possible need at any time for strengthening measures (e.g. additional anchors) should be provided for. Figure 26.15 illustrates the structural details of the head of an anchored pile wall consisting of twin piles with 2.8 m spacing and a continuous capping beam in the longitudinal direction. Upon this beam another retaining wall is placed, being connected with the capping beam by a hinged bearing and making sufficient allowance for the possible need for additional anchors. The flexible top wall avoids the development of full creep pressure. This measure has proved successful for more than 15 years as confirmed by long-term monitoring.

If site monitoring indicates that creep is accelerating to a critical value (i.e. towards the tertiary phase), urgent stabilizing measures are needed to avoid failure: e.g. rapid placing of a counterweight (earth or rock fill) in front of the retaining structure, installation of prestressed anchors, even with resin etc. in the bonding length to enable a quick prestressing.

26.7 Stability of retaining structures

26.7.1 External and internal stability

The stability of retaining walls can be checked according to the methods in soil mechanics (see Brandl 1980a, 1982, 1984, 1985, 1987a,b; Brandl and Dalmatiner 1986).

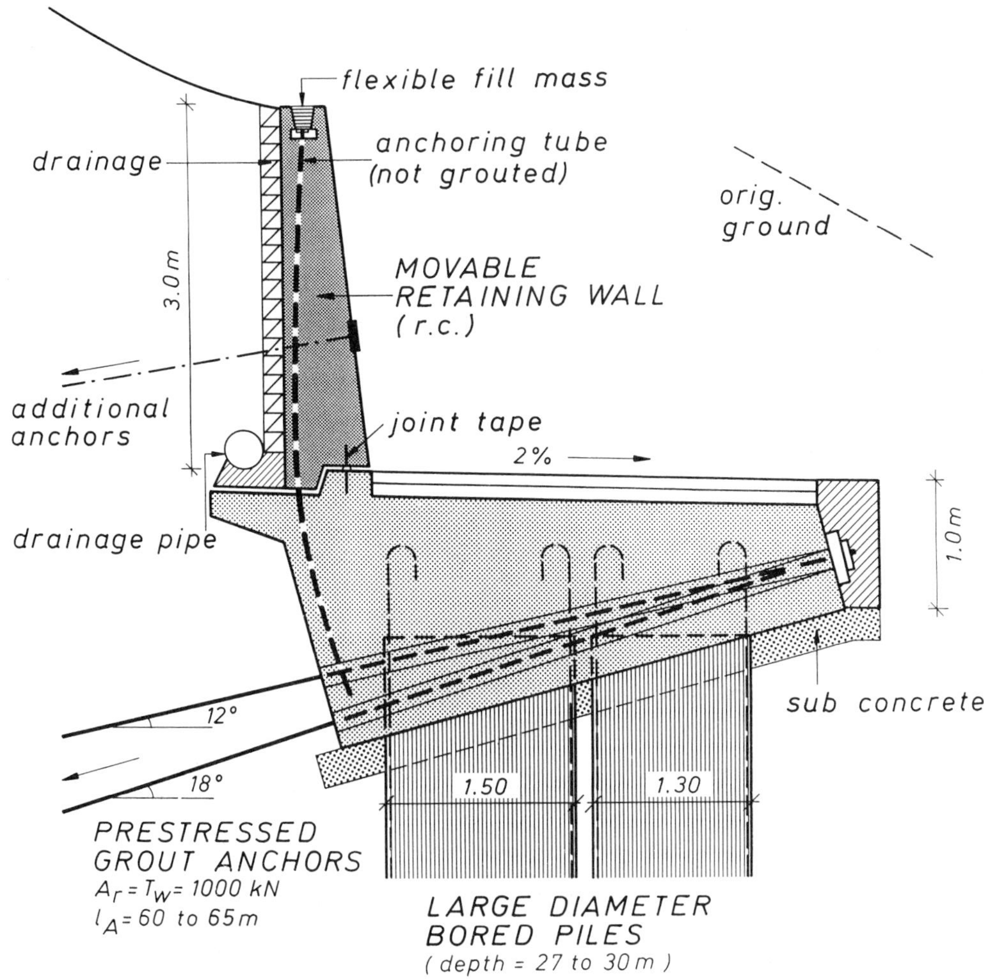

Figure 26.15 *Top of a twin-pile wall in an unstable slope (deep seated failure planes and creep zone near the surface); heavily weathered and decomposed schists. Capping beam of the pile wall with movable reinforced concrete wall for lateral pressure reduction; precautionary features include the installation of additional prestressed anchors. Capping beam used as permanent access way for slope monitoring*

Distinction should be made between *external* and *internal instability*. Cases of failure attributable to external instability include:

(1) sliding in connection with a more or less horizontal displacement of the wall;
(2) too large settlements and/or overturning,
(3) failure of wall base (allowable soil pressures exceeded);
(4) curved failure surface behind and under the wall (soil/rock shear failure);

Cases of internal instability include:

(5) failure caused by exceeding the strength or deformability of the wall material or of wall members (wall section failure);
(6) failure of an individual anchor of tied-back walls.

Both the external and internal stability can be proved conventionally, as for 'monolithic' retaining walls, pile walls and pier walls. Anchored walls, crib walls, soil nailings, reinforced earth and 'dead men' construction require special theoretical treatment. Overall stability comprises a special case of internal plus external stability, e.g. sliding along a deep-seated failure plane and rupture (of anchors). Furthermore for retaining structures in steeply inclined slopes or landslide areas, slope stability analyses must be performed.

For spaced restraining structures or local rock supports, and for rock revetment walls specific analyses are required.

26.7.2 Toppling of rock

Toppling is a failure mode which differs significantly from conventional sliding along an existing or an induced slip plane. It involves rotation of columns or blocks of rock about some fixed base. Consequently the design of a retaining structure must be based on an analysis providing the required forces to withstand toppling (and sliding) of the sustained rock mass.

Block toppling may occur in a hard rock mass with widely spaced orthogonal joints but it is also possible in columnar rock underlain by weak material. Goodman and Bray (1976) described a number of different types of toppling failures, and Hoek and Bray (1981) developed a theoretical model for limiting equilibrium analysis of toppling on a stepped base. According to Figures 26.16 and 26.17, the slope is unstable if the toe block exerts a force $P_o > 0$. If an anchor A is installed through block 1 at a distance l_1 above its base, the tension to prevent toppling of block 1 is (Hoek and Bray 1981):

$$A_t = \frac{(W_1/2)(y_1 \sin\alpha - \Delta x \cos\alpha) + P_1(y_1 - \Delta x \tan\theta)}{l_1 \cos(\alpha + \alpha_A)} \tag{26.16}$$

while the tension in the anchor to prevent sliding is

$$A_s = \frac{P_1(1 - \tan^2\phi) - W_1(\tan\phi\cos\alpha - \sin\alpha)}{\tan\phi\sin(\alpha + \alpha_A) + \cos(\alpha + \alpha_A)} \tag{26.17}$$

In case of an anchored retaining structure, the normal and shear forces on the base of the concrete wall or block increase the restraining forces as indicated in Figure 26.16.

The calculation procedure starts with the uppermost block of the toppling set (or sliding set), determining the lateral forces $P_{n-1,t}$ required to prevent toppling and $P_{n-1,s}$ to prevent sliding. The lower blocks are treated step by step till the toe block is reached.

Aydan *et al.* (1989) have developed a more sophisticated stability analysis for toppling (and sliding) by assuming dynamic-equilibrium conditions. According to Figure 26.18 the equations for each column may be written as follows:

$$W_i \sin\alpha + P_{i+1} - P_{i-1} - S_i = \frac{W_i}{g} a_x^i, \tag{26.18}$$

$$W_i \cos\alpha + T_{i+1} - T_{i-1} - N_i = \frac{W_i}{g} a_y^i, \tag{26.19}$$

$$W_i \sin\alpha \frac{h_i}{2} - W_i \cos\alpha \frac{t_i}{2} + P_{i+1} h_i - T_{i+1} t_i - P_{i-1}(h_{i-1} - x_{i-1}) - N_i e_i = \frac{W_i}{g} \frac{t_i^2 + h_i^2}{3} a_\theta^i, \tag{26.20}$$

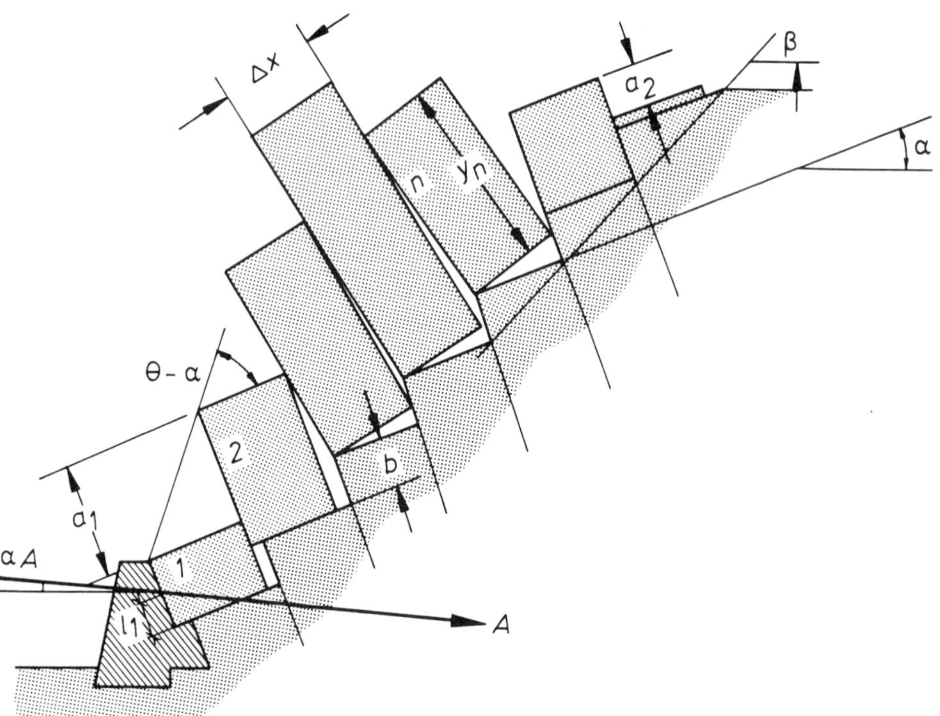

Figure 26.16 *Model for limit equlibrium analysis of toppling on a stepped base (see Hoek and Bray, 1981). Anchored retaining structure on toe of the unstable rock assembly*

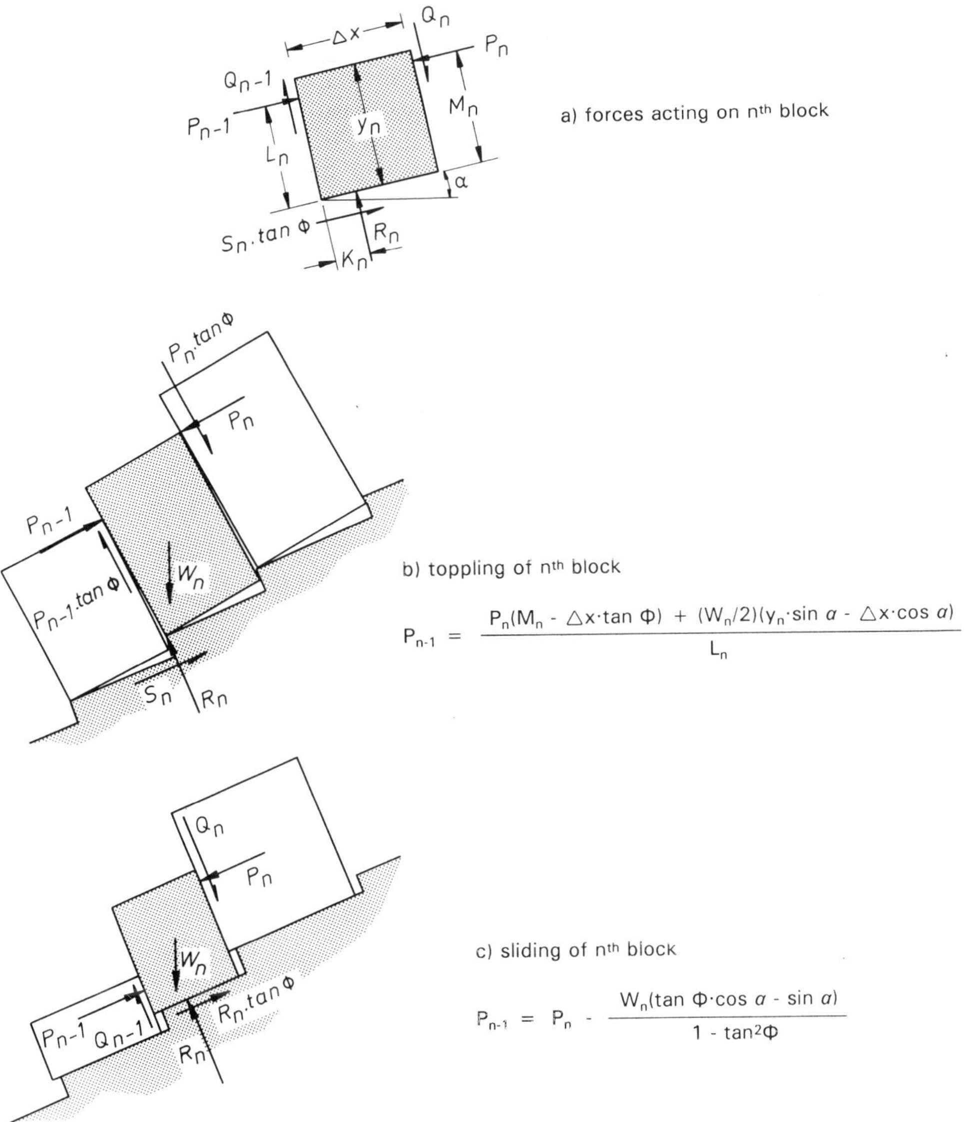

Figure 26.17 *Limit equilibrium conditions for toppling and for sliding of the nth block (Hoek and Bray, 1981)*

where

α = inclination of discontinuity set
g = acceleration of gravity
a_x^i = x component of acceleration acting on column i
a_y^i = y component of acceleration acting on column i
a_θ^i = angular acceleration acting on column i
t_i = width of column i
W_i = weight of column i
h_{i-1} = height of column $i - 1$
x_{i-1} = relative height of between bases of columns $i - 1$ and i
h_i = height of column i
e_i = distance of the point of reaction at the base of column i
N_i = the normal reaction force at the base of column i
S_i = the shear reaction force at the base of column i
P_{i-1} = the normal force exerted by column $i - 1$
T_{i-1} = the shear force exerted by column $i - 1$
P_{i+1} = the normal force exerted by column $i + 1$
T_{i+1} = the shear force exerted by column $i + 1$

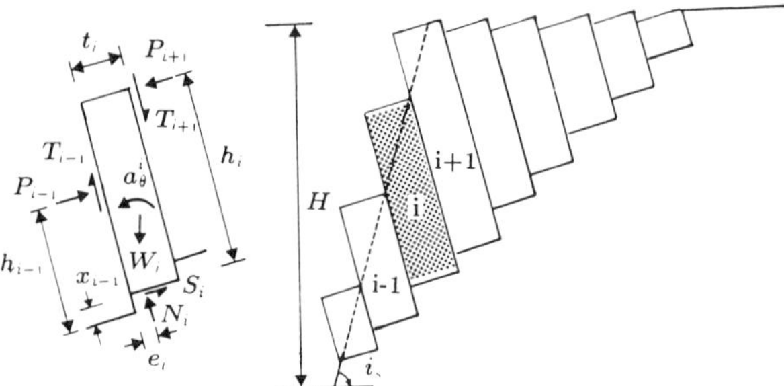

Figure 26.18 *Stability analysis of a blocky column and discontinuous rock slope on the basis of dynamic equilibrium equations (combined sliding and toppling failure of slope – see Aydan, 1989)*

The static condition for the initiation of toppling failure occurring on a critically orientated basal plane daylighting near the toe of the slope is:

$$P_{i-1} = \frac{P_{i+1}\,(h_i - t_i\,\mu_2) + W_i\,(\sin\alpha h_i - \cos\alpha\, t_i)/2}{h_{i-1}} \qquad (26.21)$$

where

μ_1 = friction coefficient between columns and their base

μ_2 = friction coefficient between columns

respectively

$$\left.\begin{aligned} &\tan\alpha \leqslant \mu_1 = \tan\phi_1 \\ &\text{and} \\ &\frac{S_1}{N_1} \leqslant \mu_1 \end{aligned}\right\} \text{relevant for sliding}$$

Starting from the topmost column the equation is solved step by step. The criteria for the slope stability F are:

$P_o < 0 \quad F > 1$

$P_o = 0 \quad F = 1$

$P_o > 0 \quad F < 1$

Consequently, in case of an unstable slope, a retaining structure has to take over normal forces exceeding P_o. This may be achieved by monolyithic walls and/or prestressed anchors.

The static stability against sliding may be checked for a columnar slope by the following expression for column 1:

$$P_o = \sum_{j=i}^{n} W_i \left\{ \frac{[\sin\alpha - \cos\alpha\,\mu_1]}{1 - \mu_2\mu_1} \right\} \qquad (26.22)$$

Further details on the stability of columnar slopes are given in Aydan *et al.* (1989).

26.7.3 Buckling of rock

Another special failure mode, which requires a specific analysis for designing restraining structures, is buckling of rock plates (Figure 26.19). Such stability problems cannot

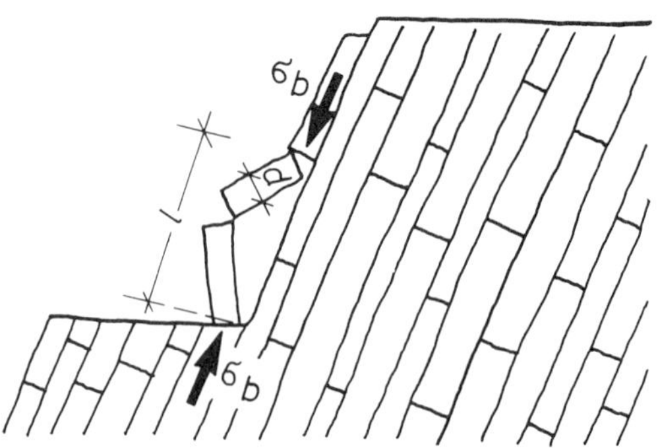

Figure 26.19 *Buckling of rock slabs stressed in plane by* σ_b

be solved by means of conventional theory. According to Figure 26.19 rock slabs of a length l and width d are stressed uniaxially in their longitudinal direction. The critical zone is near the toe of the slope, regarding the maximum dead weight of the overlying rock slabs.

Assuming a hinged support on both sides of the buckling rock member (2nd Euler case), the buckling stress is

$$\sigma_b = \frac{\pi^2 E}{12(ld)^2} \qquad (26.23)$$

where

E = modulus of elasticity (Young's modulus of the rock mass).

When exceeding the buckling load, slices of rock mass burst off from the rock surface. Buckling may be avoided by tying back the critical rock slabs. Unlike the case in Figure 26.21, relatively short anchors or nails are sufficient, and the tendons should be installed perpendicular to the slabs. The buckling length l is reduced effectively.

26.8 Monolithic retaining walls

Monolithic retaining walls are predominantly used for supporting heavily weathered rock, or rock masses with very close joint spacings. In these cases the principles of soil mechanics are practically fully valid with regard to design, calculation and construction (Brandl 1987b).

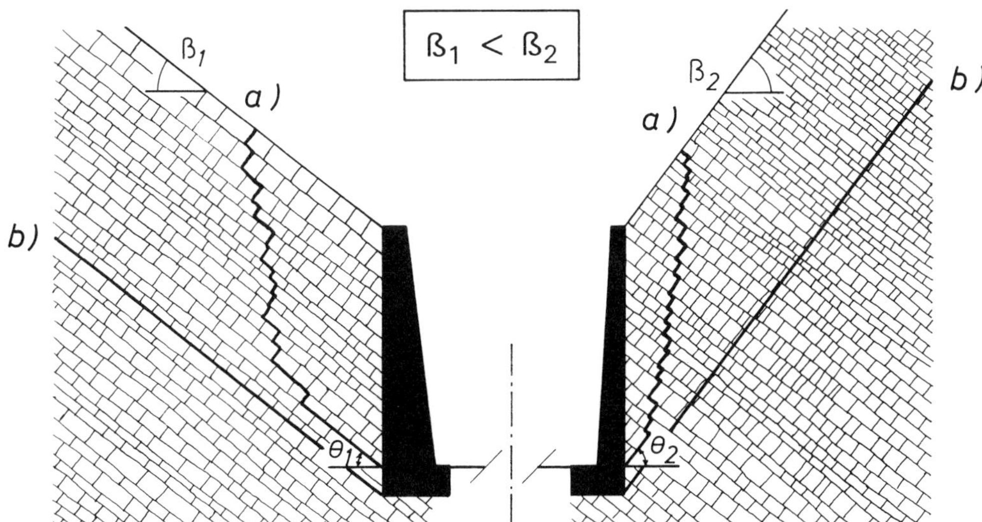

Figure 26.20 *Possible failure curves in a stratified jointed rock mass supported by retaining walls.* $\beta_1 < \beta_2$ *according to the foliation planes, viz. joint pattern*
(a) Failure curve in steps with θ_1 *and* θ_2 *as inclination of the rock discontinuities*
(b) Continuous failure plane with geologically given angle θ_1 *or* θ_2*;* ϕ*, angle of internal friction;* β*, slope angle, where* $\beta \leq \phi$*;* θ_1*,* θ_2*, angles of joints, viz. failure planes*

If the rock mass shows a relevant stratification or discontinuities, the conventional theories of soil mechanics should be modified as indicated in Section 26.4. Generally, the failure planes are more or less controlled by rock bedding or defects and their inclination cannot be calculated by Coulomb's theory or similar methods. According to Figure 26.20 the sliding angle, θ, is not only influenced by the rock properties but also by the situation of the retaining wall in relation to the discontinuities. The slip may occur along bedding planes and/or joints, thus forming either straight or stepped rupture lines. Consequently, boundary investigations with different rock wedges should be performed to check the possible lateral pressures on a retaining wall. Figure 26.20 also demonstrates schematically the different wall strength required on the right and left side of a cut slope.

If shearing and crushing of blocks is possible, the conventional earth theory (e.g. Coulomb) becomes relevant; and even slip circles may occur.

The common types of reinforced concrete walls are gravity walls, semi-gravity walls (with prestressed anchors), cantilever walls, counterfort walls, buttressed walls, and shelf-retaining walls. Depending on the subsoil conditions, retaining walls may be founded on shallow footings or on piles, diaphragm walls, etc. Thus numerous combinations of structural elements are possible.

In rock engineering semi-gravity walls and buttressed walls are preferred. Counterfort walls and shelf-retaining walls require wider excavations to construct the reinforced concrete members on the backfill side. This may increase the costs significantly and inititiate rock movements.

Very economical solutions may be achieved by tying back semi-gravity walls with one or two rows of prestressed grouted anchors. In critical cases pipes should be placed within the reinforced concrete of semi-gravity walls to enable a quick subsequent installation of anchors at any time. The reinforcement must be designed for such a contingency.

Semi-gravity walls with advance slope anchoring have proved successful if the rock mass is fairly slide-prone even on a small scale. According to Figure 26.21 this may be achieved by installing an anchored continuous reinforced concrete beam at first. Under that protection the cut for casting the semi-gravity wall is excavated cautiously in steps.

26.9 Special types of retaining walls

26.9.1 Pile walls; diaphragm walls

Pile walls are predominantly used as retaining structures in steeply inclined slopes and landslide areas, to reduce the cut slope and the danger of inducing slides during the construction period. Furthermore they have proved successful if failure surfaces are running deep below the ground surface on the front face of a retaining wall. According to their construction they are preferably installed in heavily weathered rock, commonly aiming at a bed of hard rock at their toe.

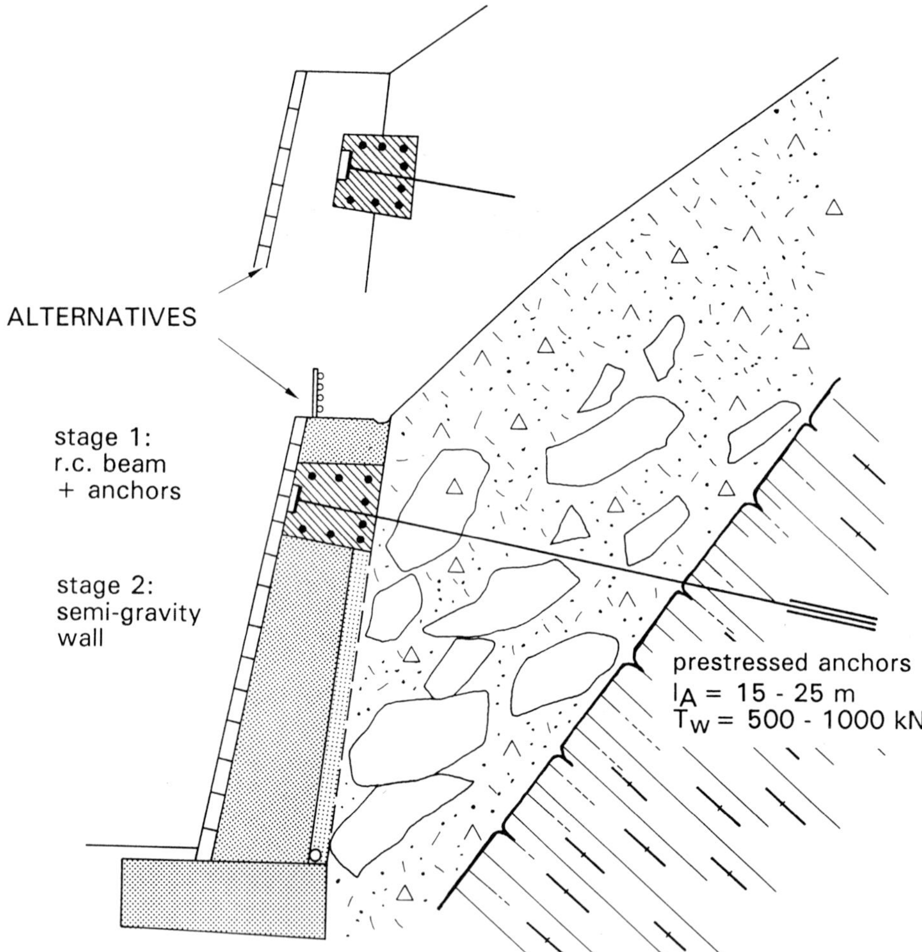

Figure 26.21 *Semi-gravity wall with preliminary anchoring to avoid slides during full-face slope excavation. First construction stage: small slope cut with immediate casting the anchored reinforced concrete beam. Second construction stage: overall slope cut (in longitudinal sections) and casting the monolithic wall.* l_A, *anchor length;* T_w, *working load*

Pile walls usually consist of bored piles with a diameter of at least 60 cm. Root pile walls with micropiles of 15–25 cm are special types (see Section 26.12.2). The geometric placement depends on the pressure, which has to be taken over, and on the 'earth' resistance available, (Figure 26.22). Single piles with shotcrete arches between them are loaded by the full arch pressure. Within structures consisting of secant piles, reinforced piles and non-reinforced ones alternate.

Counterforts of secant piles installed parallel to the slope inclination may be considered as a monolithic (vertical 'plate') unit, thus providing a large common resisting moment in the line of the slope.

After completing the piles, the soil and/or rock along the front face is excavated. During this construction period prestressed anchors may be installed step by step. A tied-back structure is frequently more economical than more piles of larger diameters and depths. The front face of pile walls may be left untreated or beautified by various revetments.

The calculation for a pile wall may be based on the principles of soil mechanics but considering possible rock discontinuities. Generally, three failure modes should be investigated:

(1) Failure of the weak rock in the sliding zone. The resistance corresponds to the passive 'earth' pressure.
(2) Failure of the piles by excessive bending moments. This may be calculated according to the dowel theory or with a subgrade reaction model.
(3) Shear failure of the piles, which may occur along a sharp discontinuity in case of a fixed bedding in hard underlying rock. This failure mode is relevant for short piles and large-diameter piers.

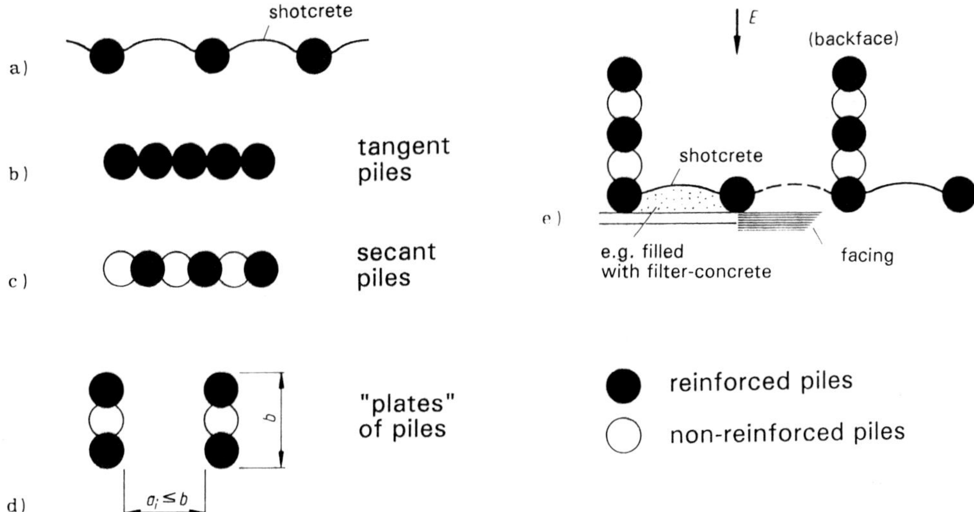

Figure 26.22 *Several patterns of pile walls, according to the required resistant moment. Bored piles in weak rock, diameter d ⩾ 60 cm (usually 90 to 120 cm)*

The following theories and calculation methods have proved successful in practice (Brandl 1987a); earth pressure theory, slope failure theory, dowel theory and palisade theory (for pile patterns according to Figure 26.22d and e).

As an example, the safety factor against slope failure of an anchored retaining wall on piles (Figure 26.23) may be evaluated as follows:

$$F = \frac{\Sigma T + A \sin(\theta + \delta) \tan \phi + S \cos \theta_s}{\Sigma W \sin \theta - A \cos(\theta + \delta)} \qquad (26.24)$$

where

ΣT is the resultant of the resisting forces of the rock and
$\Sigma W \sin \theta$ is the resultant of the driving forces

Each calculation method is based on theoretical simplifications and assumptions. Therefore at least two different theories should be investigated to check possible boundaries and facilitate engineering judgement.

Diaphragm walls are not often used for retaining structures in steep, sliding slopes. Design of the structure is performed similarly to pile walls or sheet pile walls, according to its flexibility and pattern. Special care must be given to the stability of open slurry trenches.

Diaphragm walls of 0.6–1.5 m width may be constructed even in hard rock (with a crushing strength of 500–1000 bars; 50–100 MPa), if a hydrofraise is used as cutting equipment for trench excavation. Depths of 50–100 m are available.

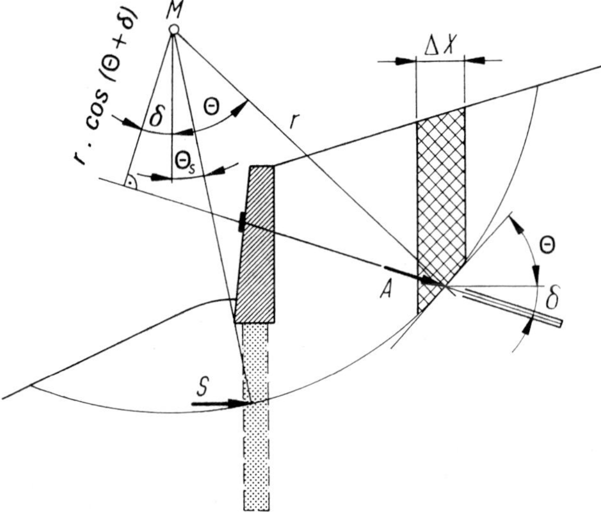

Figure 26.23 *Slope failure calculation of an anchored retaining wall on piles; heavily weathered, extensively jointed or decomposed rock: A, anchor force; S, shear resistance of the 'dowel' (pile or pier)*

26.9.2 Pier walls

Pier walls may be considered as pile walls with extremely large pile diameters. According to the lateral earth or rock pressure, elliptical cross-sections of the piers are generally preferable. A prerequisite for arbitrary cross-sectional

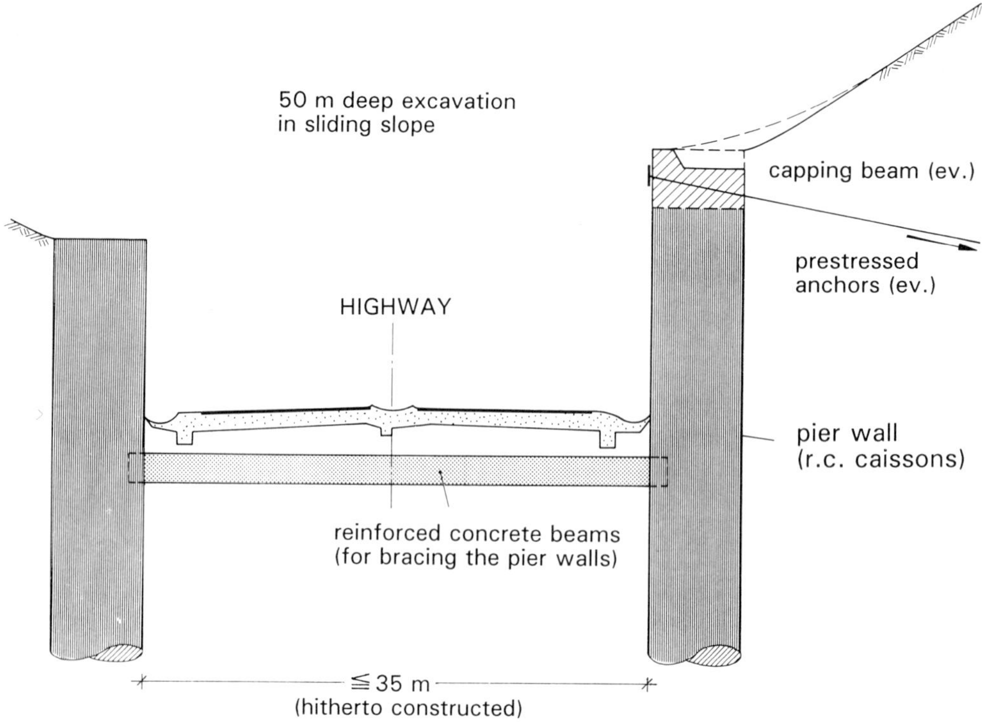

Figure 26.24 *Pier walls strutted against each other by reinforced concrete beams on the bottom of an unstable cut in heavily weathered decomposed schists with clayey mylonites; residual shear strength* $\phi_r \geqslant 7°$

areas is an open-shaft construction, using shotcrete for the lining, of thickness about 5–30 cm (usually 10–20 cm), frequently reinforced (1–3 vertical mats of wire mesh – weldmesh), depending on the pier diameter, rock parameters, seepage water, depth, etc. The distance between the piers should usually not exceed the length of the main axis of the pier ellipse. The free-standing rock should be covered with shotcrete, its thickness and reinforcement (wire mesh) depending on the properties and arch forces which have to be transmitted to the piers (by experience: at least 15 cm). Furthermore a shotcrete lining must be provided with weepholes (pipes) for draining away water. The pier wall may be left as it is or be given a facing.

Pier walls may be tied-back or are connected to other restraining measures. Sometimes they are even anchored in several levels along the shaft (Brandl 1985; Brandl and Brandecker 1982). In such cases rearrangements of the lateral pressure have to be taken into account (differing from the classical earth-pressure theory).

If pier walls in a cut slope are stressed by extremely high lateral forces, a structural strengthening is possible not only by installing prestressed anchors but also by horizontal props or other elements which increase the forces of resistance. Figure 26.24 shows a retaining structure on either side of a cut slope in sliding schists. The pier walls are strutted with reinforced concrete beams beneath the excavation line for the highway.

For evaluating the shear forces and bending moments in pier walls, the 'earth'-pressure theory and the theory of subgrade reaction (k_s) have proved useful, but both methods must be compared and used critically as they are based on theoretical simplifications. In engineering practice these approximations are usually allowable, but when designing and calculating complicated structures, the following hints should be considered:

(1) limit analyses with varying parameters;
(2) comparison of the results gained by both methods.

Usually the 'earth' pressure theory can be interpreted more clearly from the soil and rock mechanics point of view – even if the lateral pressure distribution is rearranged. The modulus of subgrade reaction is not a constant, but depends on several factors. The change of the modulus with depth frequently influences the results more than its absolute value.

Pier walls may be constructed to great depths by sinking shafts with shotcrete lining. From the static point of view 100 m and more are commonly available, unless seepage or groundwater inflow becomes too extensive. Another advantage of pier walls is the possibility of *in situ* ground investigation during excavating the shafts.

26.9.3 Anchored element walls

Anchor walls (multiple-anchored element walls) are predominantly used as retaining structures in steep, sliding slopes, frequently for highway and railroad construction work (Brandl 1979, 1980b, 1982, 1985). Further examples of application are deep and large excavations in urban areas (e.g. subways).

Tied-back walls are the most suitable structures for semi-empirical design, as the anchor forces can be raised or relieved according to the results of control measurements. An essential prerequisite is the use of prestressed anchors (grouted). One of the greatest advantages of such walls is the step-by-step construction from the original ground. Thus the slope is excavated very cautiously in stages, and the danger of inducing slide momements decreases significantly. Furthermore a later strengthening of the retaining structure by installing supplementary anchors is easily possible, if monitoring makes this necessary.

Anchored walls usually consist of reinforced concrete panels of about 2–3 m height, depending on the stability of the rock and slope. The suitable length of panels varies between 4 and 8 m according to the mutual distance of the anchors. The panels may be of cast-in-place concrete or of precast elements. If the rock is heavily weathered the panels are preferably placed close together, thus covering the whole excavation surface. In this case, for drainage purposes a 20–50 cm thick filter concrete should be placed before installing the reinforced panels.

Anchored walls have proved reliable in mountainous regions for more than 20 years, reaching heights of about 50 m and some hundred metres in length (Brandl 1979, 1980b, 1982, 1985).

Overall stability for several anchor levels must be proved mainly for structures with relatively short anchors, but should be investigated anyway (similar to Section 26.7.1). Additional slope stability analyses are especially important to determine the theoretically required anchor lengths. But definitely the lengths must be optimized on the construction site, depending on the local rock properties, which should be investigated continuously by taking at least random rock samples from representative anchor drillings. In design practice an approximately rectangular 'earth' pressure distribution on the back face of a multi-anchor wall is frequently assumed. The prestressed anchors then are positioned in a regular pattern, which is sufficient in most cases to ensure slope stability. Exceptions to this occur if major discontinuities, particularly faults or other local rock defects, complicate the situation.

The optimum position and inclination of the anchors depends on the slope geometry and on the rock discontinuities. Long prestressed grout anchors should be predominantly stressed only by tension, whereas 'nails' may involve shear forces. Consequently a horizontally stratified rock mass should be tied back. If the rock strata are thick enough to enable a 'homogeneous' bedding of the fixed anchor length within one rock layer, the anchors should be inclined. This facilitates drilling and leads to an increase in bearing capacity.

If the lateral pressures or the sliding forces of a slope are not too great, the anchored wall may be divided into single element panels with shotcrete between them. The general stability analyses have to be extended, proving the safety against rock failure in the horizontal direction, as the shotcrete is rather weak. The main advantages claimed for such element walls are their flexibility, speed of construction and low cost. Furthermore changes in anchor forces make possible a good adaptation to non-homogeneous rock, local discontinuities, sliding rock masses, various loads, etc.

An intermediate solution between closed and spaced anchor walls is achieved by placing rows of panels with a vertical spacing between. Such retaining structures may be considered as horizontal 'beams'.

In cases of extremely difficult terrain and sliding rock masses, various retaining structures must be combined (Figure 26.25).

26.9.4 Walls of reinforced concrete ribs and beams

Walls of ribs and/or beams consist of horizontal, vertical or inclined reinforced concrete members which are commonly cast in place and tied back. The elements are placed in a more or less regular pattern (hence 'wall'), but single structures for a local support of unstable rock bodies are also in use (see Section 26.14). The spaces between the reinforced concrete elements may remain untreated if the rock is not sensitive to weathering or if limited rockfalls are allowed (e.g. in connection with a protective wire mesh). But in extensively jointed or weathered rock the spacing must be stabilized and covered, thus leading to a more or less continuous (but not monolithic) 'wall'. The following measures and structural elements have proved successful in practice, depending on the lateral slope pressures and the weathering susceptibility of the rock:

(1) Covering of the rock face with (reinforced) shotcrete.
(2) Arches of reinforced shotcrete; alternatives are sketched in Figure 26.26(2a). A thrust line arch is preferred.
(3) Arch of reinforced shotcrete and rock bolting (Figure 26.26(2b).
(4) Anchored ribs and conventional rock nailing walls with spacings (Figure 26.26(3).
(5) Anchored ribs (cast-*in-situ*) combined with prefabricated crib wall elements. The cells between the stretchers may be filled with soil (Figure 26.24d) or reinforced concrete (Figure 26.26(4a).

In case of Figure 26.26(1–2a) anchored ribs have to take over the whole lateral pressure and provide sufficient

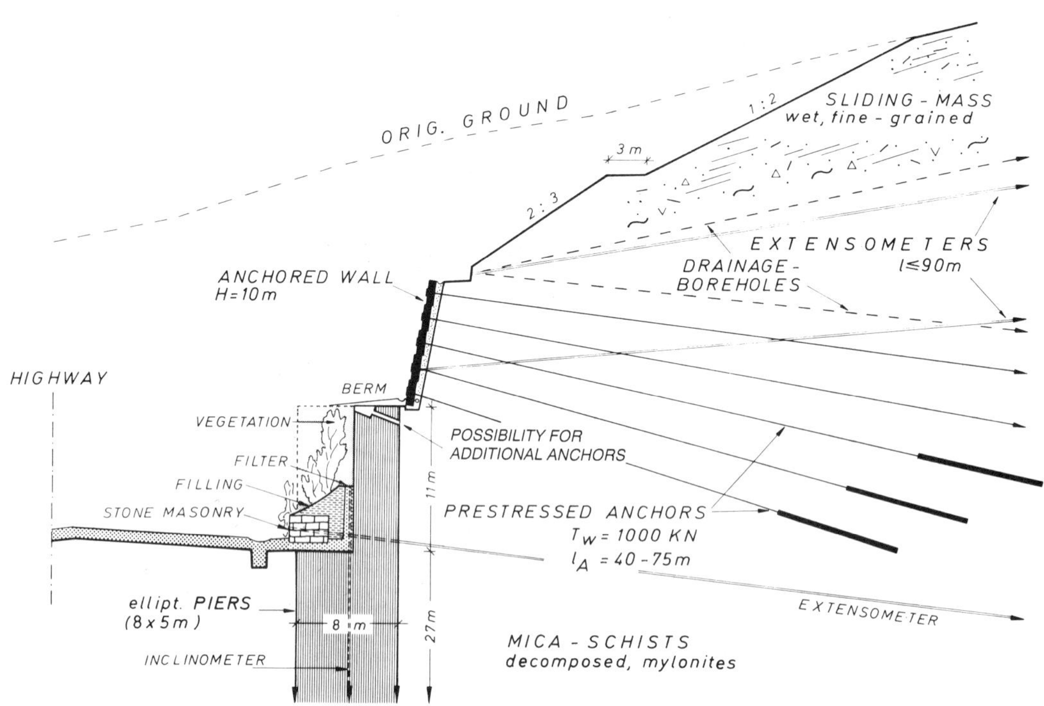

Figure 26.25 *Combined pier wall (reinforced concrete caissons) and anchored element wall to support an unstable slope cut: weathered mica schists with mylonites; residual shear strength $\phi_r \geq 5°$. Height of retaining structure 22 m above highway, length 265 m; total depth of piers 33 to 40 m, cross section of piers 8 × 5 m. Face of piers treated with silica for preservation. l_A, total anchor length; T_w, working load*

safety against slope failure. Nailed rock bodies, however, may be considered a retaining element, so that the anchored ribs are stressed only by reduced differential *H*-forces (Figure 26.26(3)).

As already indicated in Section 26.4.3 the distribution of lateral ('earth') pressures on such retaining structures differ widely from the classical one. The rearrangement is concentrated on the stiff, anchored ribs.

The thickness and reinforcement of the shotcrete depends on geotechnical and static requirements. At the minimum it is 5 cm for a rock face, and it should be 7.5–10 cm for reinforced shotcrete facings. For arches with a static function the required thickness increases to about 20–30 cm and two layers of wire mesh (weld mesh, rather than chainlink mesh) should be placed as a reinforcement. As an alternative to vertical anchor ribs, continuous anchor beams may be placed horizontally. This scheme is fairly similar to anchored element walls with vertical spacings between the horizontal panel rows.

Commonly, the reinforcement of shotcrete consists of wire mesh or welded fabrics, but steel fibre reinforced shotcrete is used more and more, especially for supporting rough rock faces and for rock slopes where access is difficult (difficult installation of common steel reinforcement). Another advantage is the increase in fracture toughness, thus avoiding brittle failure. Accordingly the gradual onset of rupture is indicated by large deformations, which allow stabilizing, strengthening measures to be applied in time.

26.9.5 Rock nailing walls

Nailed retaining structures are not to be confused with tied-back walls consisting of long grouted anchors (usually prestressed), though they may be combined with them. The essential difference from reinforced earth is the use of the *in situ* rock, which is covered with shotcrete or cast-in-place concrete. Moreover, the bars are relatively rigid and can therefore withstand tensile forces, shear forces and bending moments.

The bars ('nails') are installed by drilling, but can be driven into decomposed weak rock. To gain a sufficient bond between nail and surrounding rock, the drill holes are grouted with cement mortar. Depending on loads and geometric conditions, the length of the nails varies from 0.5*H* up to 0.7*H* (*H* being the height of 'wall'), their spacing is about 0.5–1.5 m^2 of wall-face as a rule for

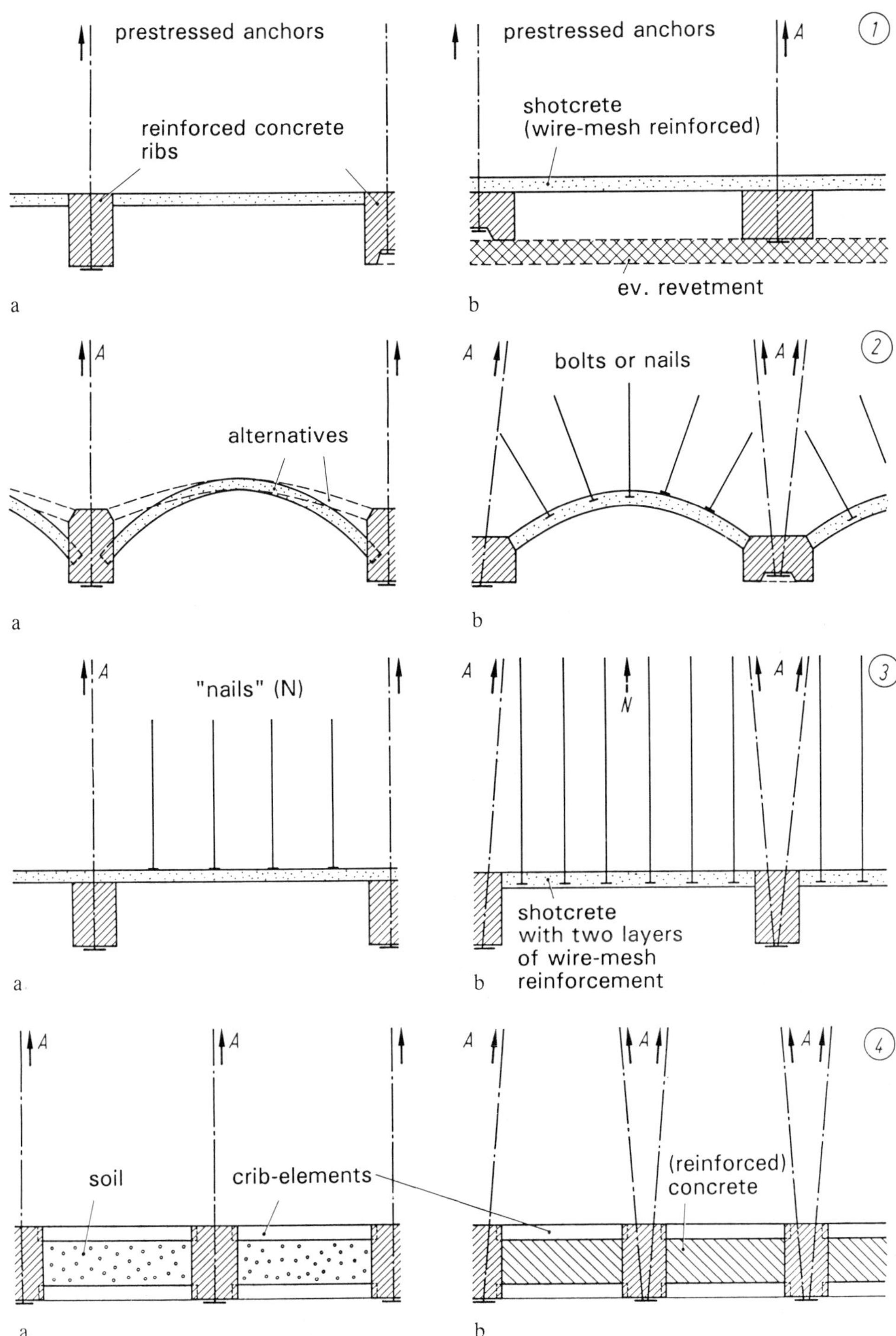

Figure 26.26 *Alternatives of 'rib walls', schematical; several combinations among 1 to 4 and a to b possible, depending on the local conditions:*
A, prestressed grout anchors (with free length of tendon); N, rock 'nails' (steel bars or fibre glass rock bolts);
1, weak structure for low lateral pressures; 3b, strongest alternative; 4, anchored r.c. ribs in connection with crib wall elements

narrow-jointed, heavily weathered rock. In rock masses with distinct discontinuities, the length and spacing of the nails depend on the joint pattern and/or stratification.

In spite of the complex mechanism of the rock–bar interaction nailed retaining structures behave like a monolith. Thus the external stability can be calculated conventionally (sliding, tilting, base failure). The internal safety factor, F, is evaluated by assuming possible sliding bodies (see Section 26.5.3). For supervising and site control the tensile forces should be measured on about 2–5% of the nails (pull-out tests, etc.).

Rock nailing is more common for restraining single rock bodies than for constructing continuous walls (see Section 26.14).

26.9.6 Revetment walls

Revetment walls are rather slender structures which predominantly protect the rock surface from erosion and weathering, and prevent single rock bodies from sliding. The overall stability of the slope is supposed to be sufficient, hence only local instabilities should be avoided. Contrary to conventional rock nailing, the nails or rockbolts should be prestressed to avoid slip of single rock bodies behind a 'pinned' wall. A pinned or multiple tied-back revetment wall is superior to a single anchored wall or a gravity-like structure, because it may be flexibly adapted to local rock discontinuities and reduces the internal forces in the reinforced concrete.

Very economical retaining structures are available if revetment walls and anchored reinforced concrete ribs are combined.

If the rock mass is covered with decomposed slope wash, heavily weathered products, talus or soil, structures such as illustrated in Figure 26.27 have proved suitable. In such cases the retaining or revetment walls are commonly anchored near their crest.

The stability of revetment walls depends to a great extent on an effective system of drainage. Several measures are possible to prevent the development of water pressure behind the structure:

(1) drainage pipes or slits, mole drains in the concrete or shotcrete;
(2) drainage trenches behind the wall;
(3) drainage-filter concrete at the back face of the wall.

If the rock is not too frost-susceptible *shotcrete facings* are a very economical alternative to conventional revetment walls. The thickness of shotcrete should be about 10–25 cm. It should be reinforced with one or two layers of wire mesh. Rock bolts (or nails) should be used for local rock strengthening and pinning the wire mesh. New developments in shotcrete technology include synthetics for the reinforcement: warp-knitted fabrics as chain-link meshes and fibre shotcrete instead of mesh reinforcing.

Geotextile structures represent special types of revetment walls or rock facings (Figure 26.28). They provide a proper rock protection against weathering and erosion, and excellent drainage. Furthermore they enable planting of grass, flowers and bushes. Such structures are very flexible and may be easily adapted to local conditions. Due to their composite effect (fill material + geotextile) they can be designed to take over relevant lateral forces (earth or rock pressure). The calculation is based on theoretical assumptions similar to 'reinforced earth' (Brandl 1987a).

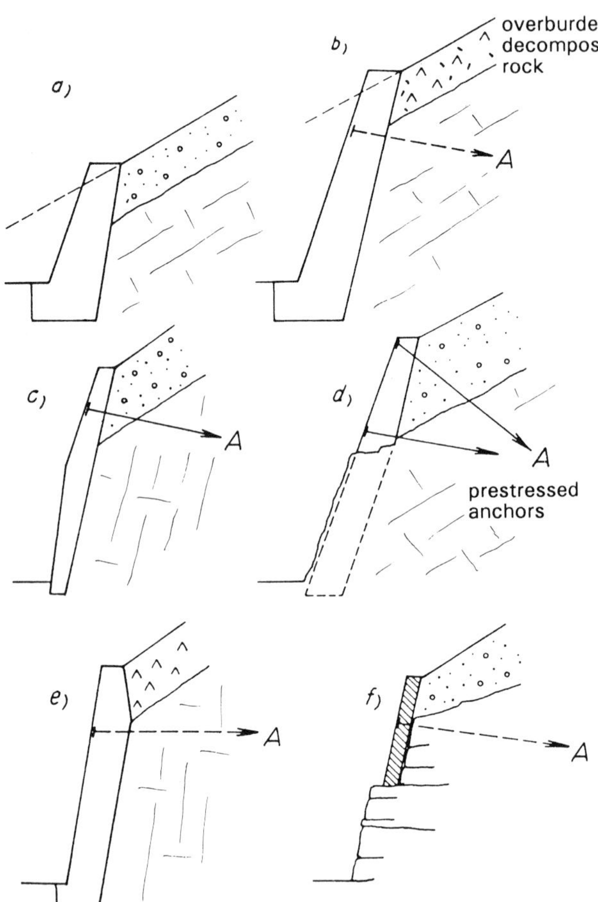

Figure 26.27 *Retaining (and revetment) walls for slope cuts of different quality in rock with an overburden of heavily weathered or decomposed rock, or soil.*
A, prestressed anchors

26.10 Composite retaining structures

Retaining walls of the composite body-principle are structures which utilize the bond effect between rock and reinforcing or enclosing materials. Besides rock nailing or rockbolting and geotextile walls, the following retaining structures belong to this group: crib walls, reinforced earth

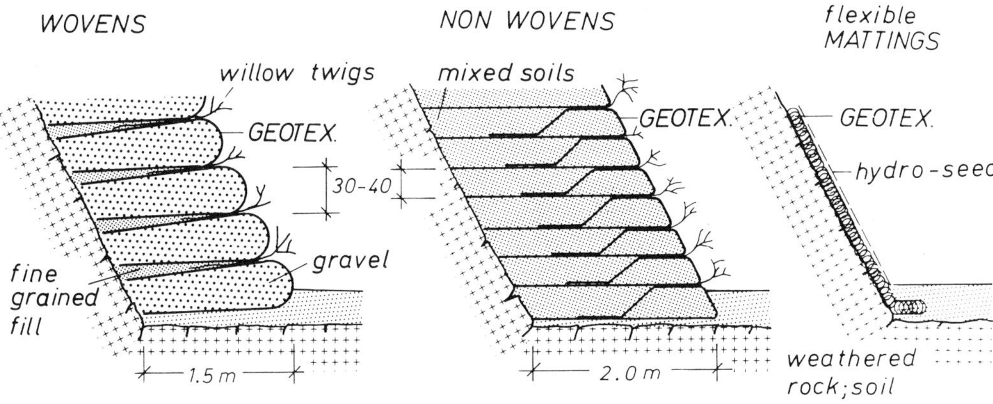

Figure 26.28 *Geotextile structures as revetment and/or retaining walls for rock facings; width depends on rock properties and geotextile characteristics*

walls, gabion walls, and retaining structures of strengthened earth (e.g. with synthetic filaments) or stabilized soil (e.g. with cement, lime or chemicals).

Grouting or dowelling of rock also provides a composite body which is capable of slope support.

Crib walls vary from thin revetments to strong retaining structures. The maximum wall height hitherto constructed in Austria is about 45 m. Crib structures have proved especially suitable in mountainous areas to stabilize steeply inclined and/or creeping slopes. The advantages of crib walls, their design and calculation are described by Brandl (1980a, 1982, 1984, 1985, 1987b).

Gabions are long strong wire baskets or large cages of wickerwork or strap iron filled with stones. The flexible and highly permeable structures act like gravity walls.

A *deadman wall* is illustrated in Figures 26.29 and 26.30. As in reinforced earth, the anchor forces due to the lateral earth or rock pressure are taken over by the backfill. The fundamental difference is the use of loop-shaped anchor strips, which connect the L-shaped reinforced concrete elements on the front face with semi-circular concrete anchor elements on the backface of the structure. Thus the effect of the straps is not a soil reinforcement, and friction is of merely secondary importance. To avoid long-term corrosion problems and provide a flexible behaviour of the retaining wall, geotextiles are preferred as tie elements. The precast concrete elements are loosely stacked, with some flexible material such as rubber or impregnated soft board between the joints. The assembly of such walls requires only a light crane. The fill material has to be placed and compacted in layers: near to the concrete elements lighter compaction machinery should be used.

Such composite walls (Brandl and Dalmatiner 1986) have proved successful especially for high retaining structures, being economical, well adapted to locally differing conditions (geometry, soil and rock) and fairly insensitive to differential settlements. Moreover a footing is not necessary in most cases, as the vertical pressure is transferred predominantly to the ground via the fill material. Just as with crib walls quick constuction is possible. Similarly, the structure provides natural drainage and sufficient root space for plant growth, greenery frequently covering virtually all the concrete.

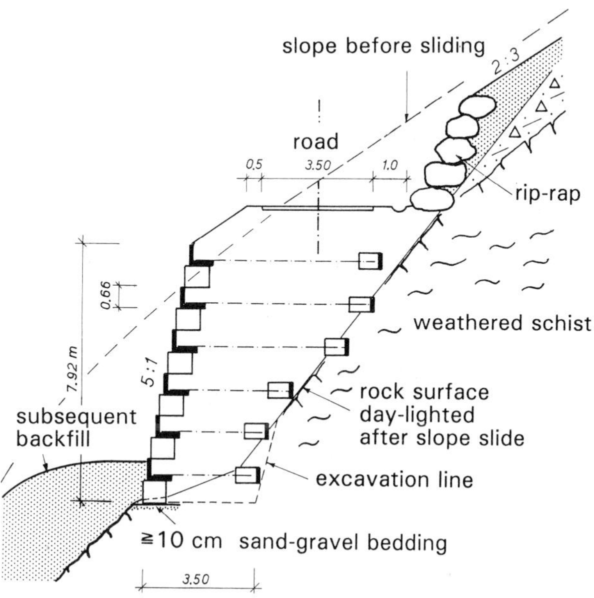

Figure 26.29 *Composite retaining wall, tied back according to 'deadman's' principle. Stabilization of a steeply inclined slope for road construction*

For design, the external and internal stability must be investigated. The external stability is calculated similarly to conventional retaining walls, the structure being considered a 'quasi-monolith', but with a statically

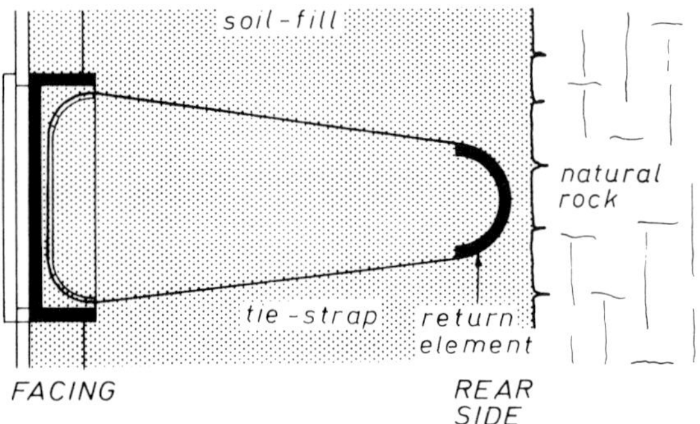

Figure 26.30 *Ground plan of a 'deadman' retaining wall with anchoring straps connecting prefabricated elements on the facing and rear end of the composite structure*

reduced, assumed width of the wall. The resultant of all acting forces should always run within the core of the cross-section. The internal stability refers to the tie-straps and the concrete elements. The calculation is similar to that of cellular cofferdams, assuming a lateral earth pressure between an increased active value and the earth pressure at rest (depending on the extension of the straps).

26.11 Rock reinforcement, rock anchoring

Rock anchoring and rock reinforcement use structural members which transmit tensile forces into the rock mass. Commonly *rock reinforcement* means the installation of rock bolts or nails for the following purposes:

(1) Local support of a rock zone which is not in equilibrium by itself or exhibits an insufficient factor of safety: tying back and/or dowelling causes tension and/or shear stresses in the reinforcing members.
(2) Rock improvement by a dense 'nailing' in a rather regular pattern: hence a new 'composite' material is created behaving like a quasi-monolithic block (e.g. nailing walls, root pile walls).

Rock reinforcement comprises predominantly passive reinforcing members. Rockbolts and rock nails are commonly made from special steels, but fibreglass is used more and more. Glassfibre materials are corrosion-proof and flexible, have a high strength, and can be cut; their low weight enables easy handling.

In *rock anchoring* prestressed tensile elements of greater length and load carrying capacity are employed. Consequently they are installed in a smaller number – as compared to rockbolts and nails – and they may be considered a substructural member which transmits a tensile force from the main structure to the surrounding rock. Whilst rock bolts and nails are used in all directions, long anchors should possibly be installed with a downward inclination of at least 5–10°. Exceptions occur in rock cavern construction.

Further distinctive features include the following:

(1) *Rockbolts*: length 1–3(5) m
Most commonly comprising a fully grouted bar (steel or synthetic) with cement or resin grouting agents. Friction-anchored rockbolts and mechanically anchored rockbolts are also in use, but should be limited to temporary measures.
(2) *Rock nails*: length 5(3)-15 m
May be fully grouted or contain a free anchor length. Full grouting is preferred for corrosion protection, causing nail behaviour like a long rockbolt. Nails with a free length may be active ('dead anchors') or passive (prestressed).
(4) *Rock anchors*: length 15(10)-150 m
Most commonly comprising a fixed length for load transfer into firm ground, and a free length to enable deformation without excessive constraints. Re-stressing, relieving and long-term monitoring are possible; an increased corrosion protection for permanent anchors is required. Usually rock anchors are prestressed to reduce movements, so reducing the likelihood of decrease of shear strength in the rock mass.

For on-site checking of the tension bearing capacity of rockbolts or nails, pull-out tests have proved suitable. Rock anchors are controlled by precontract tests, site suitability tests and routine acceptance tests to varying degrees. Instead of pull-out tests the load–displacement relationship and creep coefficients are determined. Multi-strand stressing has found favour, especially for high-capacity anchors.

Details on rock reinforcement by nailing are given in Sections 21.9.5 and 21.14.

The fundamentals of *anchoring* are illustrated in Figure 26.31, where a rock wedge is supposed to have too low a safety factor. The weight of the rock wedge is

$$W = \tfrac{1}{2}\gamma \left(\frac{h}{\sin \beta_1}\right)^2 \frac{\sin(\beta_1 - \beta_2)\sin(\beta_1 - \theta_i)}{\sin(\theta_i - \beta_2)} \tag{26.25}$$

Assuming translation, the safety factor, F, against sliding along the discontinuity (joint plane) D_i is obtained by comparing the resisting and active force:

$$F = \frac{(W\cos\theta_i + A_n)\tan\phi_i + A_t}{W\sin\theta_i} \tag{26.26}$$

No cohesion is effective along the failure plane in this case. This conventional analysis also may be performed graphically as indicated in Figure 26.31(b).

The original stability (without anchors) may be determined conventionally or graphically with the reference

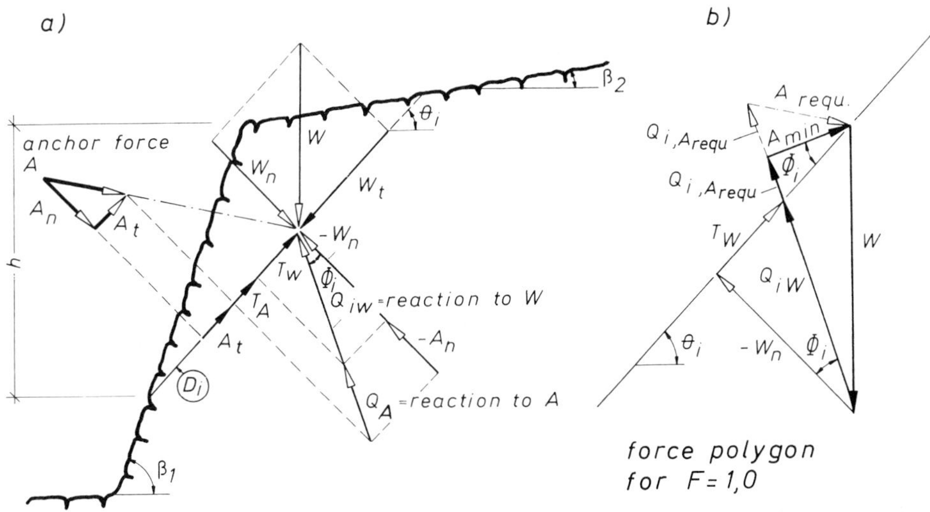

Figure 26.31 *Stability analysis of a tied back rock wedge (modified after Wittke, 1987); (a) Situation and forces; (b) Force polygon; minimum anchor force A_{min}; weight W; friction angle ϕ_i; angle of slip plane θ_i*

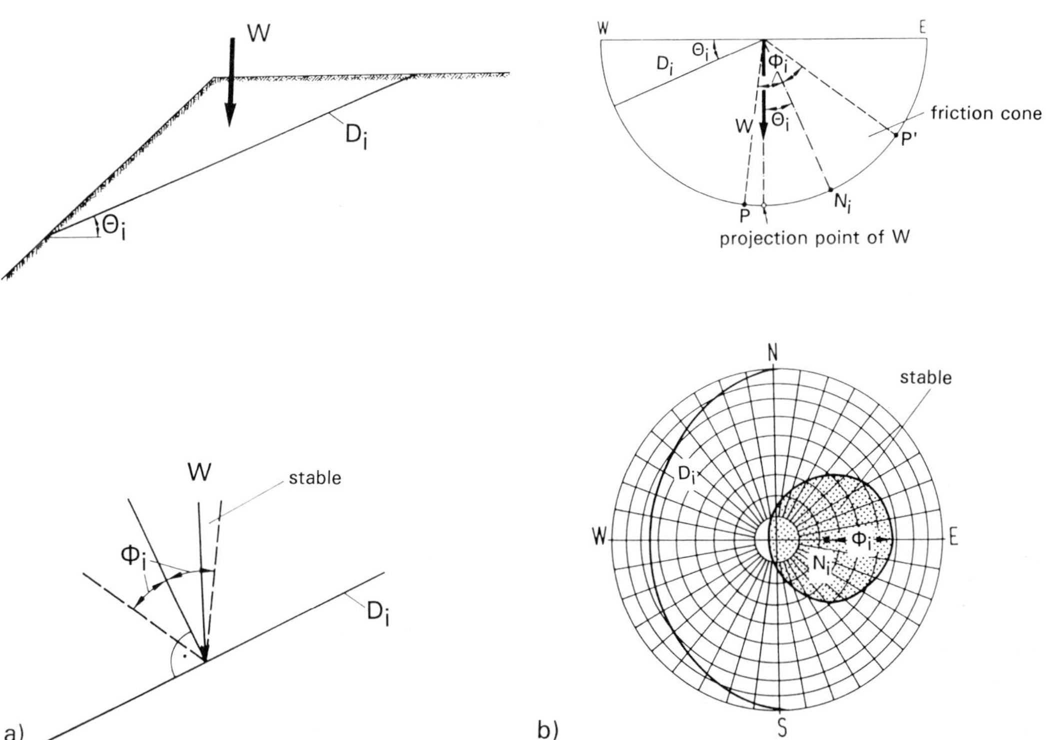

Figure 26.32 *Stability analysis by means of geotechnical friction cone (Wittke, 1987). (a) Scheme; (b) Presentation in the reference hemisphere and in the polar equal-area net*

hemisphere and friction cone (Figure 26.32). Due to its dead weight, the rock wedge shows limit equilibrium if the dip angle θ_i and the friction angle ϕ_i of the joint have the same magnitude. In the case of additional surcharge loads, limit equilibrium is possible as long as the resultant, R, of all forces acting on the failure wedge does not run steeper than ϕ_i to the normal line on the joint plane D_i (see Figure 26.32b). In case of limit equilibrium the inclination of the resultant, R, is ϕ_i to the normal line on the plane D_i. Accordingly it cuts the lower reference hemisphere in either point P or P′ for two-dimensional conditions. If the resultant, R, comprises a component normal to the intersection plane in the reference hemisphere of Figure 26.32(b), the inclinations of R must lie within the 'friction cone', to provide equilibrium of the rock wedge. The penetrating points of these inclinations are situated within a circle on the surface on the reference hemisphere; its centre point is obtained by the projection point N_i of the joint plane D_i. This circle is drawn with the aid of the Schmidts net (John and Deutsch 1974).

By means of this graphical method a rapid evaluation of the safety factor and of necessary anchor forces (and their most efficient inclination) is possible. Similar methods exist for three-dimensional rock wedges based on two failure planes (Wittke *et al.* 1987). In Figure 26.33 the failure planes (discontinuities) are indicated by D_i and D_j, and the static splitting of an anchor force into several components is illustrated.

An essential static advantage of prestressed anchors is the increase of friction resistance in the failure line as indicated in Figure 26.34, considering the conventional slice method. Assuming a circular slip surface (in heterogeneous massive or closely jointed, heavily weathered rock) the safety factor, F, as a ratio of driving to resisting forces is, according to the Fellenius method,

$$F = \frac{\Sigma(\Delta W_i \cos\theta_i \tan\phi + c\,\Delta l_i)}{\Sigma \Delta W_i \sin\theta_i} \quad \text{before anchoring} \tag{26.27}$$

By installing prestressed anchors, the safety factor, F_A, increases to

$$F_A = \frac{\Sigma(\Delta W_i \cos\theta_i \tan\phi + c\,\Delta l_i) + A \sin\alpha_A \tan\phi}{\Sigma \Delta W_i \sin\theta_i - A\cos\alpha_A} \tag{26.28}$$

where

A = anchor force

α_A = anchor inclination towards the chord of the slip circle (Figure 26.34).

$A \sin\alpha_A$ is the increase in normal forces on the slip surface (mean value), which causes an additional resisting friction force of $A \sin\alpha_A \tan\phi$. Strictly speaking it is the integral over the compression stresses coming from the anchor head and spreading within the rock mass. Equation (26.28) assumes a similar degree of mobilization of the rock shear forces and the shear forces induced by the anchor tension; the differential lateral pressures uphill and downhill of the slices are neglected. Equation 26.28 contains another approximation by assuming α_A between anchor and chord of the sliding surface. This provides a required anchor force which is theoretically independent of its location.

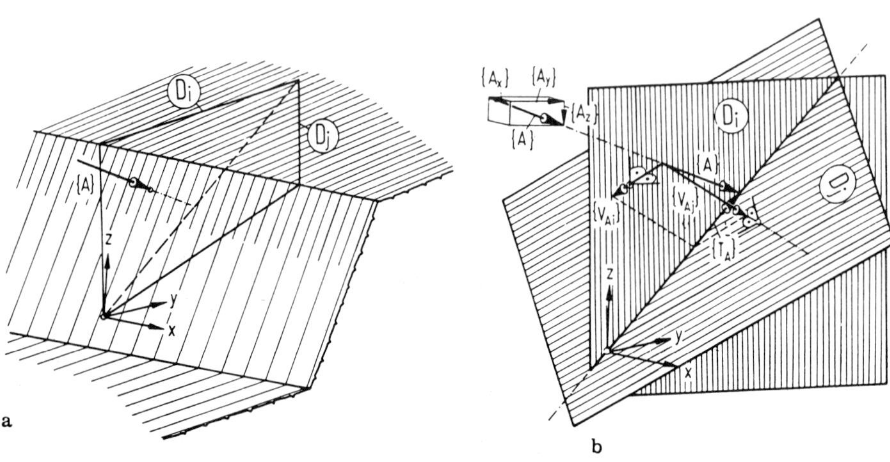

Figure 26.33 *Statical consideration of a prestressed anchor A acting on a rock wedge sliding parallel to the intersection line x (see Wittke, 1987) (a) Three dimensional rock wedge; (b) Force polygon.*
D_i, D_j, discontinuities; V_{Ai} V_{Aj}, components perpendicular to the discontinuities; T_A, component parallel to the direction of sliding

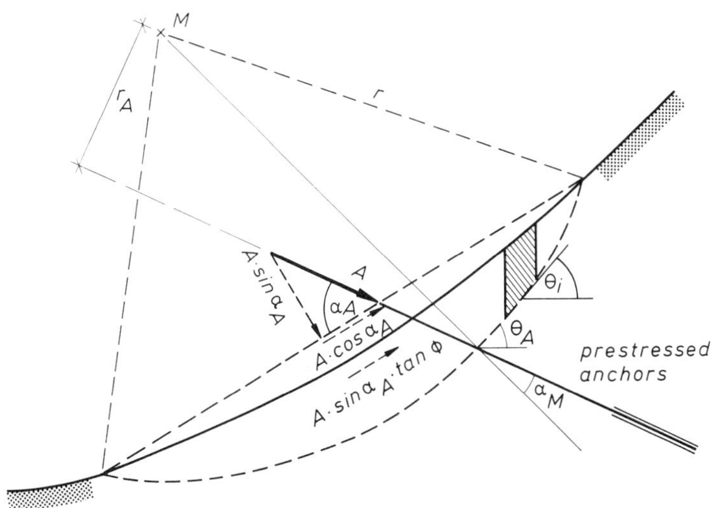

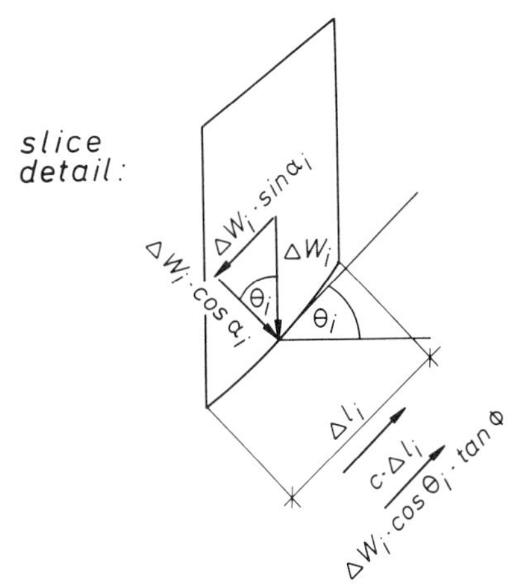

Figure 26.34 *Stabilizing effect of prestressed anchors on the slope stability of closely jointed, heavily weathered rock. (a) System (b) Detail of considered slice i*

When comparing theresisting and driving moment, Equation 26.28 changes to

$$F_A = \frac{\Sigma(\Delta W_i \cos\theta_i \tan\phi + c\,\Delta l_i) + A\cos\alpha_M \tan\phi}{\Sigma\Delta W_i \sin\theta_i - A\, r_A/r} \tag{26.28a}$$

Hence both formula provide the needed anchor force for a required safety factor, F_A.

Two definitions are commonly used to describe the safety factor:

(1) the Fellenius rule

$$F_1 = \frac{\tan\phi_{\mathrm{existing}}}{\tan\phi_{\mathrm{eq}}}$$

(2) comparing the actual total anchor force $A_{,\mathrm{existing}}$ with the equilibrium anchor force, $A_{,\mathrm{eq}}$, by

$$F_2 = \frac{A_{,\ \mathrm{existing}}}{A_{,\mathrm{eq}}}$$

The first definition is preferred in rock mechanics, the second one in statics. Consequently, both alternatives should be considered simultaneously in practice. The final decision requires an experienced engineering judgement.

The *internal stability* of a structure-anchor(s)-ground system is of special interest, if the anchors are relatively short in relation to the height of the retaining structure. There are several methods in use; one of the most common ones is the sliding-block analysis originally postulated by Kranz (1940) for sheet pile walls. According to Figure 26.35 failure planes and a 'deadman' are assumed. By use of a force polygon the anchor force $A_{\mathrm{posssible}}$ is obtained which provides equilibrium with the driving force. The required anchor force, A_{required}, is the minimum anchor force necessary to be in equilibrium with the earth pressure or a magnitude derived from an overall stability analysis with curved slip surfaces. This leads to the safety factor

$$F = \frac{A_{\mathrm{possible}}}{A_{\mathrm{required}}} \geqslant 1.5$$

where the resisting force of the passive wedge is commonly neglected for flat founded retaining structures. In creeping slopes an increased lateral pressure on the retaining structure should be assumed (see Section 26.6). The method is also applicable to several anchor levels and has proved suitable in engineering practice for about 50 years, in spite of various theoretical shortcomings. A comprehensive collection of data, long-term experience and simplicity of use are the justification of the Kranz method, which has been adapted by German and Austrian Standards, and by the French Bureau Securitas, (Anon 1972) and Hanna (1982). It is therefore acceptable to continue using it, though the safety factor cannot in any way be compared with the value obtained by other stability calculations (e.g. according to Bishop (1955) and Janbu (1959)). Therefore a second analysis should be performed simultaneously for comparison (e.g. kinematic method – Figure 26.36).

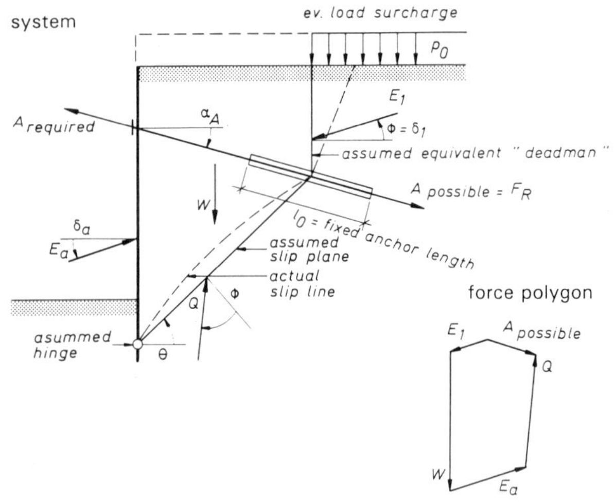

Figure 26.35 *Kranz method of overall stability analysis for an anchored retaining structure.*

- E_1 *earth pressure against imaginary continuous 'deadman' or anchor wall (substitute wall)*
- W *total weight of assumed sliding body*
- E_a *total active earth poressure on wall*
- ϕ *angle of internal friction*
- δ *angle of wall friction (commonly $\delta_1 = \phi$)*
- Q *resultant reaction force against lower failure plane*

The kinematic method is based on a failure mechanism which consists of several rigid rupture bodies (see Goldscheider and Gudehus, 1974). In a graph all forces acting on the rupture bodies are drawn (see Figure 26.36 for example). A virtual displacement is imposed then on the system, and the work of the external forces (W_f) is compared with the energy which is dissipated due to the displacement (E_d). Stability is given, if the displacement requires more energy by friction ad cohesion than the external forces can raise. Hence the safety factor, F is defined as

$$F = \frac{E_d}{W_f} = \frac{\Sigma(Q_i \sin \phi_i\, u_i) + C_i\, u_i}{\Sigma_{K^i} \cdot \bar{u}_i} \qquad (26.29)$$

where

- Q_i = resultant reaction force against failure plane i
- C_i = resistant cohesion force along failure plane i
- K_i = external forces (self weight W_i, load P_i, anchor forces A_i, etc.)
- $\bar{u}_i$ = component of the displacement u_i in the direction of K_i

According to the example of Figure 26.36 the following steps have proved suitable for a practicable stability analysis:

(1) Drawing the displacement polygon by assuming a unit-displacement for u_i = '1'. This leads to all the

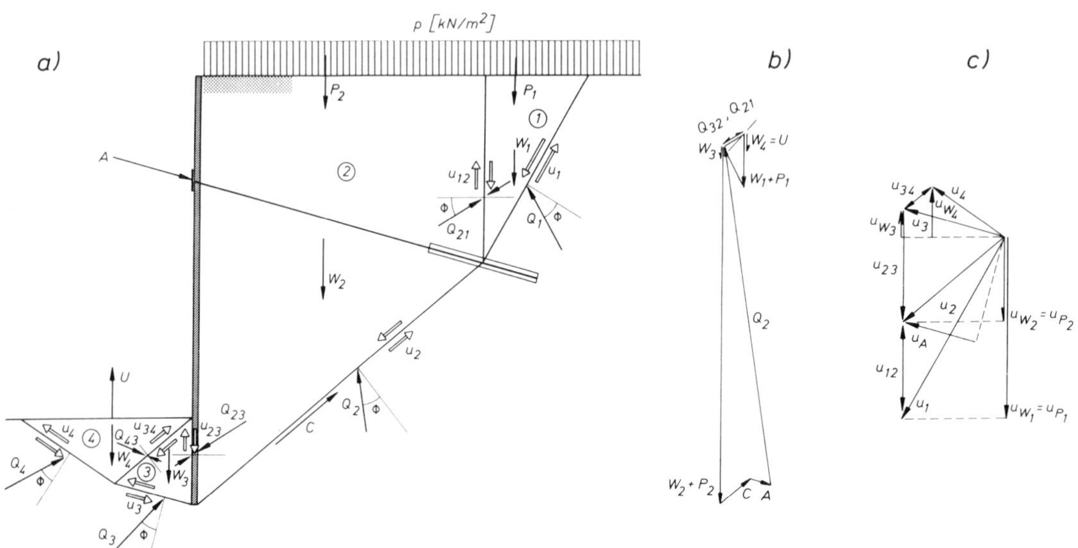

Figure 26.36 *Kinematic method of overall stability analysis:*
(a) System with multi-body failure mechanism; (b) Force polygon; (c) Displacement polygon.
A, working load of prestressed anchor; P_i, p_i = surcharge loads; W_i, self weight of rock body i; Q_i, resultant reaction force against failure plane i; ϕ_i friction angle; C_i, cohesion resistant force; U, fictive force to obtain limit equilibrium (e.g. trench excavation as failure cause);

other displacements u_i, which are realtive data unless one value is pregiven or limited by an allowable maximum.
Accordingly, the displacement of the anchor in Figure 26.36 is obtained from the displacement polygon by $u_A = 0{\cdot}38$ (relative to $U_i = 1$), and $u_{W1} = u_{P1} = 0{\cdot}87$, etc.)

(2) Assuming a relevant failure cause (or more causes). In Figure 26.36 the self weight of W_4 is reduced (e.g. by excavating a trench); this can be statically simulated by imposing an additional force U acting upwards.
(3) Drawing the force polygon. The failure cause(s) must lead to a closing of the force polygon. Consequently, another failure cause leads to another safety factor.

26.12 Dowelling of rock bodies

Dowels increase the shear resistance in rock discontinuities or along possible failure planes. Consequently filling of open cracks with mortar may also be considered as dowelling. This is not, however, the commonly accepted concept of dowels as used in Britain (see Chapter 18).

26.12.1 Vertical dowelling with piles or piers

'Dowels' may be micropiles or remaining grout pipes on the one hand to large-diameter bored piles or (elliptic) piers with diameters of several metres on the other. For rock slopes micropiles and piers have proved especially suitable. The first may be installed even under extremely steep surface conditions, and piers provide a very great shear resistance. The dowels are placed in regular or irregular patterns, as single elements or as continuous walls – depending on the local requirements. Bored piles of a standard diameter (0.6–1.5 m) are commonly used to stabilize creeping slopes or intensively weathered rock.

The bearing-deformation characteristics of dowels depend on their slenderness; largely microdowels withstand shear movements only along a short section (about 1 m or less), and the upper parts of the dowel do not take over substantial forces. Large-diameter dowels, however, tilt. On creeping slopes, the economically optimal pile diameter is about one-tenth of the slide plane depth beneath the surface. This rough estimation assumes a failure plane fairly parallel to the surface and facilitates a first design. If only small deformations are allowable, the diameter and the embedding length of the dowels should be increased.

For piles with a diameter of $d = 0.6$–1.5 m placed in a regular pattern, it has proved successful to connect them by capping beams. Compression beams arranged like trusses provide a fairly uniform load transfer from the unstable slope to the retaining structure.

The design loads on dowels depend on various factors which partially influence each other and include

(1) rock characteristics;
(2) slope stability (including eventual seepage pressures);
(3) geometry and rate of movement of the sliding rock mass;
(4) relative movements between dowel and rock;
(5) diameter and length of the dowels;
(6) ratio of stiffness between dowel and rock;
(7) allowable risks and 'residual movements' of a creeping slope after dowelling.

A survey of some calculation methods is given in Figures 26.37 and 26.38. Due to the different theoretical assumptions, the results may vary over a wide range. Consequently the design should comprise at least two hypotheses to check possible boundaries. Parametric studies with different rock parameters and static systems are also recommended, especially with regard to the group effect of the dowels. This group effect is influenced by several factors (Brandl 1987a):

(1) dimensions of the dowel group;
(2) pile spacing;
(3) support of pile head and toe (free, hinged, fixed);
(4) stiffness and skin friction of the piles;
(5) driving and resisting rock forces;
(6) rock parameters and discontinuities.

Local failures in the dowelling group may initiate a gradual instability of the whole retaining system similar to a long-term zip effect. Such progressive failures are known only from micropiles or grouted pipes in connection with rock gouge of small residual strength. In such cases the safety factor preferably should be increased by using larger pile diameters rather than increasing the dowel number.

The stabilizing effect of dowelling can be checked by conventional slope failure analyses (e.g. Figure 26.39). As an example, for slip circles, Bishop's method provides a safety factor of

$$F = \frac{r\Sigma(T_{f,i} + T_{s,i})}{\Sigma M_i} \tag{26.30}$$

where

$$T_{f,i} = \frac{V_i - u_i\, b_i + c_i\, b_i \cot \phi_i}{\cos \theta_i \cot \phi_i + \dfrac{\sin \theta_i}{F}}$$

= the resisting shear force of the ground in the slip surface within the slice i

$T_{s,i}$ = an additional resisting shear force from dowels in the slip surface within the slice i

M_i = driving moment of the vertical and horizontal loads of the slice i, related to the centre point of the slip circle with radius r

V_i = sum of vertical components of all loads acting on slice i (Figure 26.39b)

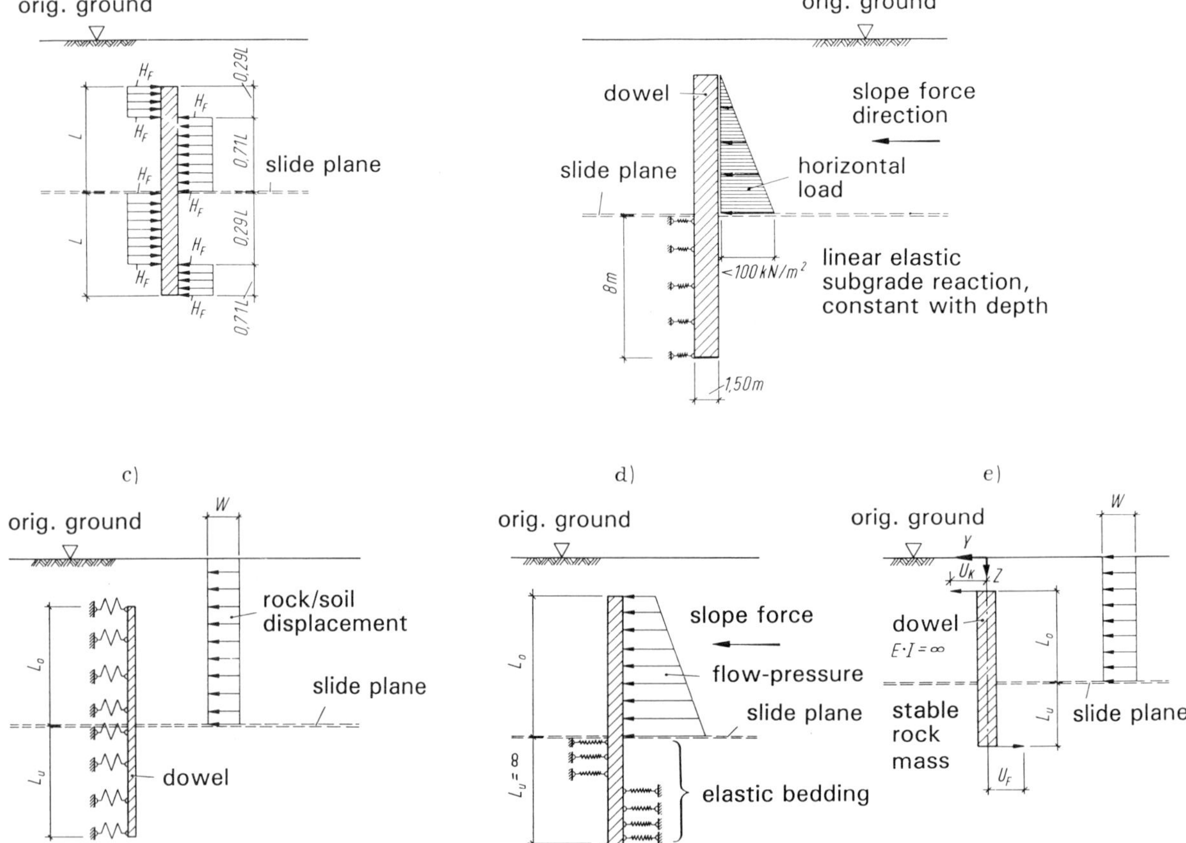

Figure 26.37 *Several hypotheses for dowelling a slide plane (Brandl, 1987a): (a) Rigid dowels, loaded by a creeping or stagnation pressure; (b) Conventional method with lateral 'earth' pressure and resistance according to the subgrade reaction mode; (c) Head deformation and maximum internal forces of the dowel derived from the differential equation for the flexural (bending) member; (d) Differential equation of the flexural (bending) member for dowels of a limited length, block sliding and variable moduli of subgrade reaction; (e) After Sommer and Buczek (1987): Rigid dowels, block sliding, hyperbolic load distribution (see Figure 26.38)*

$U_i = u_i . b_i =$ the resultant pore water pressure in the slip surface within the slice i
$F_{s,i}$ = seismic force on the slice i
W_i = self weight of the slice i
P_i = additional loads acting on the slice i (e.g. foundations, anchors etc.)
θ_i = angle between slip surface in slice i and horizontal line
ϕ_i, c_i = relevant shear parameters on the slip surface along slice i

Arbitrarily curved slip surfaces lead to a safety factor of

$$F = \frac{\Sigma(T_{f,i} + T_{s,i})}{\Sigma(V_i \sin \theta_i + H_i \cos \theta_i)} \qquad (26.31)$$

where

H_i = sum of horizontal components of all loads acting on slice i.
$V_i = W_i + F_{s,i,v} + P_{i,v}$
$H_i = F_{s,i.h} + P_{i,h}$
$M_i = W_i . r . \sin \theta_i$ (if only the self weight is to be considered)

Slope dowelling is certainly not as flexible as anchoring. Calculation and design include more uncertainties and a supplementary strengthening is rather difficult. Therefore long-term monitoring should be performed in critical cases, especially if the calculated safety factor against slope failure is $F \leqslant 1.2$; for large scale projects; for projects where excessive deformation (or even a failure)

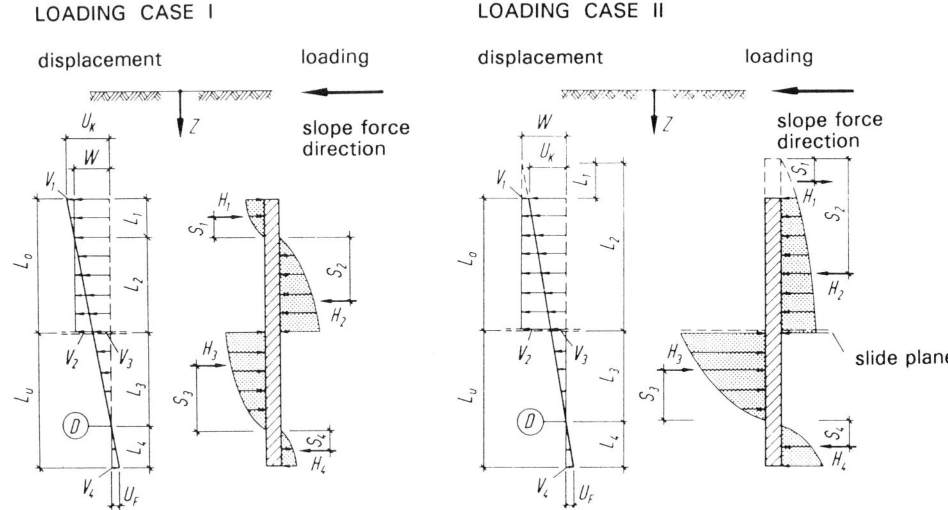

Figure 26.38 *Calculation of dowels – detail to Figure 26.37. Dowel movements (tilting), rock/soil movement (block sliding), forces on dowel and resultants; D, rotary point. Loading case I: dowel with fixed head and toe; Loading case II: dowel only with fixed toe (embedded in stable rock mass)*

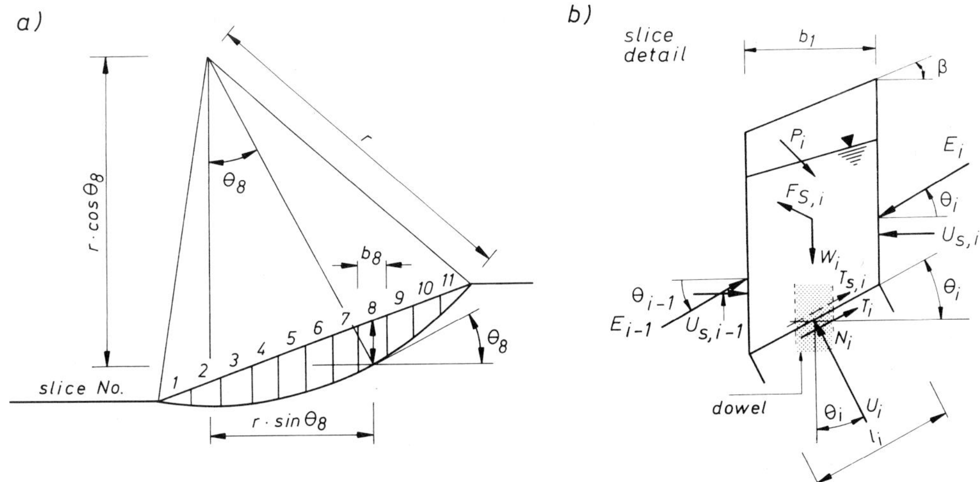

Figure 26.39 *Circularly curved failure surface in a highly weathered, disintegrated, loosely fragmented, coarsely crushed or extensively jointed rock mass.*
(a) Situation; (b) Loads and forces of the slice i (forces on vertical slice planes commonly neglected). Dowelling indicated by resisting force $T_{s,i}$.
E_i; E_{i-1}, earth pressures on the side planes of slice i; $U_{s,i}$; $U_{s,i-1}$, resultant pore water pressures on the slide planes of slice i

would have serious consequences; and if the rock parameters and seepage conditions vary over a wide range.

26.12.2 Vertical and inclined dowelling by micropiles or grouting

Rock-slope stabilization against failure also may be achieved by means of reticulated micropiles or grouts forming a three-dimensional lattice structure. Exact knowledge of the potential slide plane is of secondary importance as far as spacing and diameter of the dowels are concerned, but it determines their length. The more or less vertical installation of slender dowels throughout the slope height is not optimal for static reasons. Flexible vertical dowels are most effective near to the toe of a slope but rather ineffective as reinforcement in the upper part. Therefore horizontally inclined members are recommended especially near the crest of the slope.

Reticulated dowelling or reinforcement of rock goes back to the 'root pile' system (Lizzi 1977). Root piles ('*pali radice*') are bored pressure piles with a standard diameter of 15–25 cm with proper reinforcement. Mini-piles with a diameter of only 8–12 cm have no helical reinforcement, and high-quality concrete is difficult to cast. Therefore a minimum diameter of 15–20 cm is recommended for rock strengthening and dowelling.

In very steep rock faces the root piles are installed with a flat inclination achieving more a rock reinforcement than a dowelling. No pile-head capping structures are necessary, except for some wire mesh to prevent the fall of small rock bodies. Here it is preferable to use a large number of small-diameter piles than a small number of large-diameter piles. In the case of Figure 26.40 the dowelling effect dominates, hence achieving a safety factor of

$$F = \frac{(T_R + S)}{T_D}$$

where

T_R = resultant of resistant forces in the critical slide surface
S = additional shear forces provided by the piles
T_D = resultant of driving forces in the critical slide surface

To increase the common bearing effect of the pile group, capping beams have proved successful.

Reticulated root piles also can be arranged in a scheme forming a quasi-retaining wall. The piles act as reinforcing members, thus providing a composite body according to the 'knot' effects among the nearby piles. The capping beam increases the quasi-monolithic behaviour of the retaining structure which may be designed in a similar way to gravity walls (e.g. Figure 26.40(b)).

Bar walls consist of root piles which are installed vertically in a regular pattern and tied back with prestressed grout anchors. If such pile groups are laterally loaded by the slope pressure, the effect of bond is taken into account by using an analogy with a reinforced concrete beam with compression and tensile steel. Root piles simulate steel bars; rock simulates concrete, but without tensile stresses (Brandl 1987a, 1988). Assuming a common neutral axis for the whole 'bar wall', the maximum bending moment can be reduced by a semi-empirical factor which indicates the intensity of bond within the structure. The rest of the bending moment will be transformed into inner axial loads (equal compressive and tensile forces).

An essential advantage of root piles or various micropiles is the rather small equipment necessary, so that the piles can be installed in confined areas. Their construction does not cause any significant destabilization, and they provide a progressive stabilization of a slope as the work is completed.

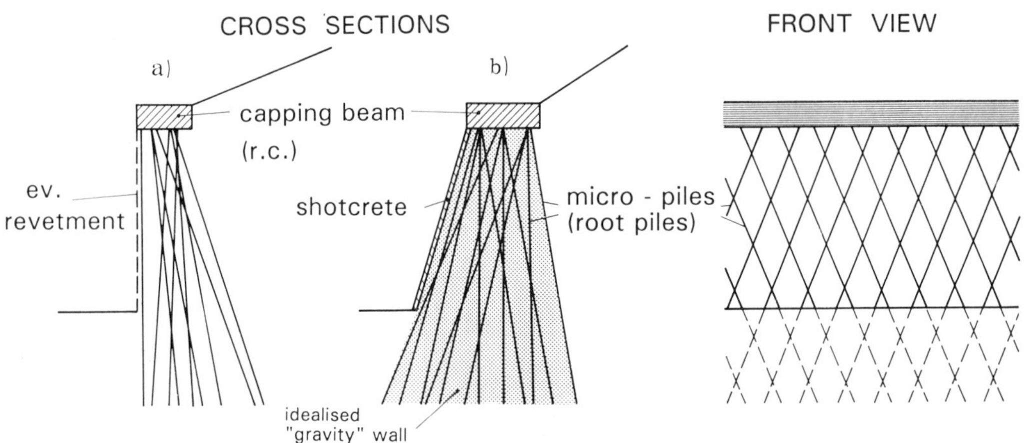

Figure 26.40 *Reticulated micro pile structure (commonly root piles) as retaining wall for jointed weak rock*

26.12.3 Horizontal dowelling

Highly inclined discontinuities in rock bodies may be locally strengthened by horizontal dowelling. The method is similar to compound timber girders but only seldom used in rock engineering.

26.13 Rock grouting

Massive bodies of rock can be produced by grouting and act as retaining structures like gravity walls or abutments for unstable rock zones. Furthermore incompetent near-surface rock can be reinforced or supported, which otherwise would be subjected to excessive weathering or failure. The injected grout solidifies the rock giving it a more or less monolithic structure. The applied grout pressure should avoid lifting the rock or widening discontinuities. Commonly used grout materials are cement, bentonite, sand, clay pozzolan or certain chemicals. When appropriate, sand, clay, rock flour and other inert material can be applied as fillers. Chemical grouts such as silica gels or synthetic resins are suitable in finely fissured rock but require comprehensive suitability tests regarding the long-term behaviour and environmental aspects.

Rock improvement is predominantly achieved by consolidation grouting; grouting of rock joints, faults, fissures and cavities; and by flushing out clay film and gouge from rock defects and subsequently concreting with grout under an appropriate pressure. The success of rock improvement by grouting may be checked by the following investigations before and after the grout operation: rock sampling; water tests (e.g. Lugeon's test); geophysical investigations (e.g. cross-hole seismic); and compressive tests, shear tests and triaxial tests on sample material. The degree of rock improvement allows an assessment of the increase in safety factor and whether supplementary post-grouting is needed.

Rock improvement by grouting is commonly used in connection with other stabilizing or retaining measures. So the overall and internal stability of the multiple-body system must be investigated.

26.14 Spaced and single restraining structures

The commonly used restraining structures and remedial methods against local rock weathering, erosion, fall and slide are:

(1) Surface protection with wire mesh (chainlink mesh, weld mesh), steel straps, flexible rope fences; fixed with rockbolts (or nails) or loosely hanging. Instead of steel wire mesh, geosynthetic products have been increasingly used for about five years. Their main advantages are no corrosion problems, light weight, ease of handling and great flexibility. In the case of fabrics, chainlink meshes (warp-knitted fabrics) are preferred. The interlacing points are not specially bonded. For special applications extra threads can be added in both directions. If the fabrics are not covered by shotcrete, they must be stable against long term *ultra violet* radiation.

(2) Local rock bolting, nailing, anchoring, frequently in combination with the following measures (3 to 8).

(3) Surface covering with shotcrete, usually reinforced with weld mesh (fixed with rockbolts). Chainlink steel mesh has the advantage of being adaptable to an uneven rock surface, but the reinforcing effect is not as good as in the case of commonly used weld meshes.

New developments are fibre shotcrete and polymer modified shotcrete of increased tensile strength, thus avoiding conventional steel reinforcement.

(4) Local rock fillings (of voids, cracks, etc.), rock supports or brackets, struts and stiffening elements of reinforced concrete; frequently tied back.

(5) Anchored blocks or panels in a more or less regular pattern; the reinforced concrete elements are designed to prevent local rock break-outs and to distribute the load at the anchor head uniformly to the surrounding rock.

(6) Anchored ribs of reinforced concrete as local support of weak or overhanging rock bodies and/or to increase the overall slope stability.

(7) Anchored beams of reinforced concrete, preferably along berms.

(8) Tied-back beams of reinforced concrete, placed as girder grille. Such structures are statically indeterminate and therefore somewhat sensitive to differential deformation. The girder grille consists of shotcrete beams (reinforced with weld mesh), and the spacings are covered with chainlink mesh to protect against rock erosion and rock fall. Rock nails (length 5–15 m) are placed in the knots of the girder grille.

(9) Rock-supporting arches of masonry, mass concrete or reinforced concrete.

(10) Grouting of rock bodies. These restraining structures are designed semi-empirically in most cases. Usually they are placed from top to bottom consecutively during excavation, thus maintaining the slope stability in an efficient manner. Special care must be given to the optimum direction of the prestressed anchors, depending on the discontinuities, faults, the required forces, the mutual distance between the anchors, etc., and the method of construction (drilling method). Therefore great practical experience, as well as *in situ* measurements and control systems, are necessary.

Figure 26.41 shows a combination of several restraining structures for a highway construction in a steeply inclined rock slope:

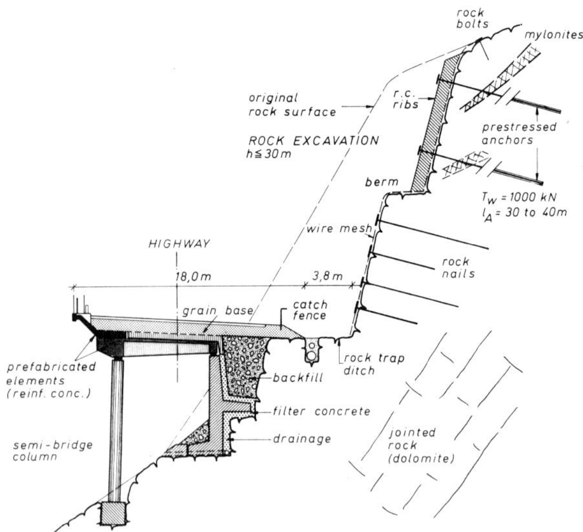

Figure 26.41 *Several restraining measures for highway construction on a steeply inclined rock slope. T_W = working loads; l_A = anchor length*

- Long prestressed anchors in the mylonitic rock zone near the crest of the slope cut;
- Rock nails (non-prestressed) in the lower part of the slope cut;
- Shotcrete and wire-mesh locally covering the weak zones of the new rock surface;
- A shelf retaining wall to carry the semi-bridge and the highway lane;
- A free space (rock trap) between slope cut and highway (with a catch fence) as safety against rockfall.

26.15 Protective structures for bridges and masts

In order to protect sensitive structures such as bridges and masts from lateral pressures on steep slopes or those on which creep occurs, tied back restraining structures are used (Brandl 1979, 1980b, 1982). These include reinforced concrete beams, placed vertically, horizontally or inclined, with shotcrete between the anchored elements; and anchored walls (on filter concrete).

The technique makes use of flexible walls which cause heavily weathered rock to 'flow' around the structures. Such retaining structures may be plane-shaped or curved. In many cases drainage holes, drilled uphill from the wall, increase the stability. Usually the anchors are prestressed, but rock nails (length 3–10 m) may be used (mainly along the shotcreted areas).

Bridge abutments in steeply inclined, slide-prone slopes may require extensive anchoring and deep foundations. Piers have proved especially successful. The lateral pressure along their circumference is not only influenced by the surface inclination but also by the rock discontinuities. Hence it might be necessary to tie back such structures at several levels to achieve a sufficient safety factor. Further details can be found in Brandl and Dalmatiner (1988).

26.16 Accompanying measures

Retaining structures should always be designed in relation to an appropriate slope surfacing (earthwork) and an effective drainage system.

High slope cuts should generally be constructed by sections separated by *berms*, the spacing depending on overall slope stability; local rock surface stability (weathering, erosion); rock properties and seepage conditions; inclination and total height of the slope; geometry of the cut slope (e.g. longitudinal inclination of the cut bottom); accessibility; and aesthetic aspects. Hence the spacing of berms commonly varies between 8 and 20 m, with an optimum of about 12–15 m.

From a theoretical point of view berms are unfavourable, as they cause local stress concentrations (notch stresses). But in engineering practice they have proved very successful for several reasons:

(1) They provide permanent access to the slope.
(2) They facilitate maintenance and remedial works.
(3) They facilitate long-term monitoring.
(4) They allow the construction of supplementary retaining structures at several levels (e.g. anchored beams, ribs) and are therefore an essential prerequisite for a semi-empirical design.
(5) They permit proper surface drainage and subsurface drainage.
(6) They provide a braking effect on local slides or rockfalls and may act as traps with catch fences).

Berms used on high rock slopes should have a minimum width of about 3 m, and their surface should be sloped for drainage. In jointed, weathered rock the downhill edge should be protected (by shotcrete, wire mesh, rockbolts etc.) to avoid a gradual disintegration.

26.17 Final remarks

In mountainous regions the rock parameters frequently exhibit wide variation, even within a small area, to such an extent that geotechnical design procedures seem to provide not more than limit values and serve for reference only. Due to the steeply inclined slopes, there is also the problem of the seepage flow. Therefore the safety of such slopes cannot be proved in the usual manner by theoretical methods. The results of evaluating lateral pressure and slope stability are less influenced by the method of calculation than by the assumption of relevant rock properties and seepage flow conditions. Therefore sophisticated design methods are not warranted, but parametric studies with certain boundaries are very important.

The optimal solution for retaining structures can frequently be achieved only step by step in connection with *in situ* measurements (Brandl 1979). It would be economically unjustifiable to construct most expensive protective structures, whilst throughout assuming the most unfavourable parameters.

'Calculated risks' are to be accepted in the design of roads and expressways through valleys in mountainous areas where hillsides with a slide potential extend over a distance of several kilometres. Risk assessment has to distinguish between the possibility of local slides and the stability against general failure. In order to reduce construction costs as well as to save time, the use of supplementary construction methods (mainly anchors) should be considered. These are less costly than an 'absolutely safe' structure which seeks to avoid the possibility of additional measures taken at a later time. Finally one should bear in mind that 'absolute safety' cannot be provided under such extreme topographical and geotechnical conditions.

In such cases flexible retaining structures are preferred, adaptable step by step, both technologically as well as economically, to the locally prevailing rock pressures, slope movements and ground conditions. This practical approach is based on continuous measurements and observations of the retaining structure, the ground surface and the rock mass (e.g. by extensometers and inclinometers) during the entire construction period. Subsequent random monitoring is recommended. Calculations and theoretical considerations are only the basis for the first design and for interpreting the results obtained. This 'semi-empirical' design method has proved itself under most difficult conditions for more than 20 years (Brandl and Brandecker 1982).

Experience has shown that subsequent strengthening of retaining structures may be necessary due to critical construction stages; hazardous natural events (heavy rainfalls, earthquake); long-term creep of slopes; long-term deterioration of rock masses or the retaining structure itself, and alteration of external loads. Hence the possibility of strengthening retaining structures at any time after their construction should be taken into account in the design and calculation stages.

The semi-empirical design of retaining structures requires the following prerequisites:

(1) comprehensive *rock* (and *soil*) investigations;
(2) calculations with parametric studies;
(3) design of possible supplementary measures;
(4) reliable monitoring;
(5) practical experience and proper engineering judgement and intuitive feel for the subject;
(6) joint willingness of all involved persons, clients and contractors to take a calculated risk.

In many cases of rock engineering under difficult conditions this philosophy provides the only technical solution – not to mention the cost savings.

References and further reading

Anon (1972) 'Ground anchors', *French Code of Practice*, Recommendation TA.72, Editions Eyrolles.

Aydan, Ö., Shimizu, Y. and Ichikawa, Y. (1989) 'The effective failure modes and stability of slopes in rock masses with two discontinuity sets', *Rock Mechanics and Rock Engg,* **22**, 163–188

Bishop, A. W. (1955) The use of the slip circle in the stability analysis of slopes, *Geotechnique,* **5**, 7–17

Brandl, H. (1973) *'Die Verwendung von anstehendem Felsgestein im Straßenbau*, Straßen- und Tiefbau, No. 8, Switzerland

Brandl, H. (1979) 'Design of high, flexible retaining structures in steeply inclined, unstable slopes', *Proc. 7th Eur. Conf. Soil Mech. Found. Engg*, Brighton

Brandl, H. (1980a) *Raumgitter-Stutzmauern (Crib Walls and Other Retaining Structures)*, Series Strassenforschung (Road Research) No. 141, Bundesministerium für Bauten und Technik (Federal Ministry for Buildings and Techniques), Vienna

Brandl, H. (1980a) 'Erd- und kriechdrucktheorie für rutschhänge mit praktischen anwendungen' ('Theory of earth pressure and creep pressure of sliding slopes with practical examples'), *Proc. 6th Danube-European Conf. Soil Mech. Found. Engg*, Varna, 95–104

Brandl, H. (1982) *Raumgitter-Stutzmauern (Crib Walls and Other Retaining Structures)*, Series Strassenforschung (Road Research) No. 208, Bundesministerium für Bauten und Technik (Federal Ministry for Buildings and Techniques), Vienna

Brandl, H. (1984) *Raumgitter-Stützmauern (Crib Walls and Other Retaining Structures)*, Series Strassenforschung (Road Research) No. 251, Bundesministerium für Bauten und Technik (Federal Ministry for Buildings and Techniques), Vienna

Brandl, H. (1985) 'Slope stabilization and support by crib walls and prestressed anchors', *Proc. 3rd Int. Geotech.* Seminar, Nanyang Technological Institute, Singapore, 179–198

Brandl, H. (1987a) 'Konstruktive Hangsicherungen' ('Structural slope stabilization methods'). In *Grundbau-Taschenbuch, Dritte Auflage*, Teil 3, Smoltczyk, U. (ed.), Verlag Ernst & Sohn, Berlin, pp. 317–426

Brandl, H. (1987b) 'Retaining walls and other restraining structures'. In *Ground Engineer's Reference Book*, F. G. Bell (ed.), Butterworths, London, 47/1–47/34

Brandl, H. (1988) 'The interaction between soil and groups of small diameter bored piles', *Proc 1st Int. Geotech. Seminar*, Ghent, 3–16

Brandl, H. and Brandecker, H. (1982) 'Autobahnbau unter extremen geotechnischen Bedingungen', *Mitteilungen für Grundbau, Bodenmechanik und Felsbau*, Technische Universität Wien; Heft 1, Vienna

Brandl, H. and Dalmatiner, J. (1986) *'Stützmauer-System' und andere Konstruktionen nach dem Boden-Anker-Verbundprinzip (Earth-Retaining Structures)*. No. 280 Series Strassenforschung des Bundesministeriums für Bauten und Technik (Federal Ministry for Buildings and Techniques), Vienna

Brandl, H. and Dalmatiner, J. (1980) 'Brunnenfundierungen von Bauwerken in Hängen' ('Pier foundation of structures in slopes') (No. 352) and 'Geotechnische Baustellenmessungen und Langzeit-Überwachungen von Hangbrücken und Talübergängen' ('Geotechnical site control and long term monitoring of slope bridges') (No. 353) Series Strassenforschung Bundesministerium für wirtschaftliche Angelegenheiten, Vienna

Gässler, G. (1987) 'Vernagelte Geländesprünge – Tragverhalten und Standsicherheit', *Reports of Institut für Bodenmechanik und Felsmechanik*, Universität Fridericana, Karlsruhe, 108

Gässler, G. and Gudehus, G. (1983) 'Soil nailing – statistical design', *Proc. 8th Eur. Conf. Soil Mech. Found. Engg.*, Helsinki, 491–494

Ghosh, A. and Haupt, W. (1989) 'Computation of the seismic stability of rock wedges', *Rock Mechanics and Rock Engineering,* **22**, 109–125

Goldscheider, M., Gudehus, G. (1974) 'Verbesserte Standsicherheits-nachweise'. Baugrundtagung, Frankfurt. Deutsche Gesellschaft für Erd- und Grundbau, Essen

Goodman, R. E. and Bray, J. W. (1976) 'Toppling of rock slopes', *Proc. Speciality Conf. Rock Engg. for Foundations and Slopes*, Boulder, Colorado, American Society of Civil Engineers **2**, 201–234

Gudehus, G. (1981) *Bodenmechanik*, Enke Verlag, Stuttgart

Hanna, T. H. (1982) *Foundations in Tension*, Trans Tech Publications, Clausthal-Zellerfeld, BRD

Hausmann, M. R. (1989) *Engineering Principles of Ground Modification*, Vols. 1 and 2, University of Technology, Sydney, Australia

Hoek, E. and Bray, J. W. (1980) *Rock Slope Engineering*, The Institution of Mining and Metallurgy (3rd Edn.), London

Hoek, E. and Londe, P. (1974) 'Surface workings in rock', *Proc 3rd Int. Conf. Rock Mechanics* , Denver **1**, 613–654

Howing, K. D. (1984) 'Das Kriechverhalten gefüllter Gesteinstrennflächen und dessen Auswirking auf die Langzeitstabilität von Feslböschungen', *Reports of Institut für Geologie*, Ruhr-Universität, Bochum

John, K. W. and Deutsch, R. (1974) 'Die Anwendung der Lagenkugel in der Geotechnik', *Festschrift L. Müller-Salzburg* zum 65. Geburtstag. Universität Fridericana, Karlsruhe

Jaubu, N. (1954) Application of composite slip circles for stability analysis. Proceedings European Conference on Stability of Earth Slopes, Stockholm, **4**, 43–49

Jumikis, A. R. (1983) *Rock Mechanics*, Trans Tech Publications, Clausthal-Zellerfeld, BRD

Leventhal, A. R. and Mostyn, G. R. (1987) 'Slope stabilization techniques and their application' (in Walker, 1987) 183–230

Kezdi, A. (1972) 'Stability of rigid structures', *General Report. Proc. 5th Eur. Conf. Soil. Mech. Found. Engg.*, Madrid

Kovari, K. and Fritz, P. (1976) 'Stabilitätsberechnung ebener und räumlicher Felsböschungen', *Rock Mechanics*, Springer-Verlag, Vienna/New York, **8**, 73–113

Kranz, E. (1940) Uber die Veraukerung von Spundvauden Mittlelungen aus den Gebiete des Wasserbanse und der Baugrundforschung, Heft II, Verlage Wilhelm Ernst, Berlin

Littlejohn, G. S., Norton, P. J. and Turner, M. J. (1977) 'A study of rock slope reinforcement at Westfield (Scotland) open pit and the effect of blasting on prestressed anchors', *Proc. Conf. Rock Engg*, University of Newcastle upon Tyne, **1**, 293–310

Lizzi, F. (1977) 'Practical engineering in structurally complex formations (The *in situ* reinforced earth)'. *Proc. Int. Symp. Geotech. Struct. Complex Formations*, Capri, 327–333

Müller, L. (1963) *Der Felsbau*, Vol. 1, F. Enke Verlag, Stuttgart

Pacher, F. (1957) 'Über die Berechnung von Felssicherungen, verankerter Stüzmauern und Futtermauern', *Geologie und Bauwesen*, Vienna, 1

Siller, TH. I. (1988) 'Seismic response of tiedback retaining walls,' *Doctoral Dissertation*, Carnegie-Mellon University

Sommer, H. and Buczek, H. (1987) 'Zur Stabilisierung von Rutschungen in Tonhängen mit biegesteifen Elementen', *Heft 1 der Mitteilungen des Fachgebietes Grundbau, Boden- und Felsmechanik*, Universität-Gesamthochschule Kassel

Stillborg, B. (1986) *Rock Bolting*, Trans Tech Publications, Clausthal–Zellerfeld, BRD

Sturm, U. (1987) 'Ein allgemeines Berechnungsverfahren für Grenzlastzustände im Grundbau, dargestellt am Beispiel des Grund- und Böschungsbruches'. Heft 17 der Veröffentlichungen des Grund-bauinstitutes der Technischen Universität Berlin

Walker, B. F. and Fell, R. (ed.) (1987), *Soil Slope Instability and Stabilisation*, A. A. Balkema, Rotterdam and Brookfield

Wichter, L. and Meininger, W. (1985) 'Stabilization of a cutting by reinforced concrete piles and prestressed anchors near Stuttgart', *Proc. European Sub-Committee on Stabilization of Landslides in Europe*, ISSMFE. Boğaziçi University, Istanbul, 93–109

Wittke, W., Breder, R. and Erichsen, C. (1987) 'Böschungsgleichgewicht im Fels'. In *Grundbau-Taschenbuch*, Dritte Auflage/Teil3, U. Smoltczyk, (ed.), Ernst & Sohn, Berlin, 71–153

Index